Essentials of Medical Language

David M. Allan, MA, MD

Karen D. Lockyer, BS, RHIT, CPC

Higher Education

Boston Burr Ridge, IL Dubuque, IA New York San Francisco St. Louis
Bangkok Bogotá Caracas Kuala Lumpur Lisbon London Madrid Mexico City
Milan Montreal New Delhi Santiago Seoul Singapore Sydney Taipei Toronto

ESSENTIALS OF MEDICAL LANGUAGE

Published by McGraw-Hill, a business unit of The McGraw-Hill Companies, Inc., 1221 Avenue of the Americas, New York, NY, 10020.

Some ancillaries, including electronic and print components, may not be available to customers outside the United States.

This book is printed on acid-free paper.

1 2 3 4 5 6 7 8 9 0 DOW/DOW 0 9

ISBN 978-0-07-337414-7
MHID 0-07-337414-8

Vice president/Editor in chief: *Elizabeth Haefele*
Vice president/Director of marketing: *John E. Biernat*
Senior sponsoring editor: *Debbie Fitzgerald*
Director of development, Allied Health: *Patricia Hesse*
Executive marketing manager: *Roxan Kinsey*
Lead media producer: *Damian Moshak*
Media producer: *Marc Mattson*
Director, Editing/Design/Production: *Jess Ann Kosic*
Project manager: *Marlena Pechan*
Senior production supervisor: *Janean A. Utley*
Senior designer: *Srdjan Savanovic*
Senior photo research coordinator: *Carrie K. Burger*
Photo researcher: *Pam Carley*
Media project manager: *Mark A. S. Dierker*
Outside development house: *Patricia Gillivan, Triple SSS Press Media Development, Inc.*
Outside consultant: *Adrianne Rippinger*
Typeface: *10.5/12 ITC Giovanni*
Compositor: *Laserwords Private Limited*
Printer: *R. R. Donnelley*
Cover credit: © *Richard Smith/Alamy*
Credits: The credits section for this book begins on page C1 and is considered an extension of the copyright page.

Library of Congress Cataloging-in-Publication Data

Allan, David, 1942-
Essentials of medical language / David M. Allan, Karen D. Lockyer.
p. ; cm.
Based on: Medical language for modern health care / David Allan, Karen Lockyer, Michelle Buchman. 1st ed. c2008.
Includes index.
ISBN-13: 978-0-07-337414-7 (alk. paper)
ISBN-10: 0-07-337414-8 (alk. paper)
1. Medicine—Terminology—Programmed instruction. 2. Communication in medicine—Programmed instruction. I. Lockyer, Karen. II. Allan, David, 1942- Medical language for modern health care. III. Title.
[DNLM: 1. Clinical Medicine. 2. Communication. 3. Terminology as Topic. WB 102 A417e 2010]
R123.A44 2010
610.1'4—dc22

2008040571

The Internet addresses listed in the text were accurate at the time of publication. The inclusion of a Web site does not indicate an endorsement by the authors or McGraw-Hill, and McGraw-Hill does not guarantee the accuracy of the information presented at these sites.

David Allan

David Allan received his medical training at Cambridge University and Guy's Hospital in England. He was Chief Resident in Pediatrics at Bellevue Hospital in New York City before moving to San Diego, California.

Dr. Allan has worked as a family physician in England, a pediatrician in San Diego, and Associate Dean at the University of California, San Diego School of Medicine. He has designed, written, and produced more than 100 award-winning multimedia programs with virtual reality as their conceptual base. Dr. Allan resides happily in San Diego and walks the beach most days.

Karen Lockyer

Karen Lockyer holds a degree in Health Information (RHIT), a national coding certification (CPC), and a BS from Rutgers University. She is also a credentialed member of AHIMA (American Health Information Management Association) and AAPC (American Academy of Professional Coders).

Mrs. Lockyer has worked in medical practice administration and the Health Information Management fields for many years. She has taught medical terminology for high school, community college, and workforce development areas at the National Institutes of Health and the federal government's Office of Personnel Management. She has also taught coding and billing for undergraduate and certificate programs at the community college level.

Residing in Southlake, Texas, Karen enjoys the sights and flavors of the Southwest.

BRIEF CONTENTS

APPENDIXES

DETAILED CONTENTS

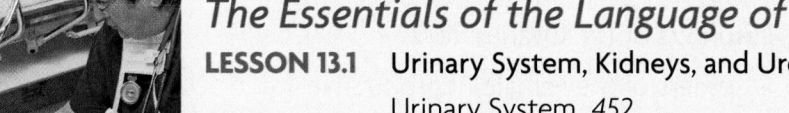

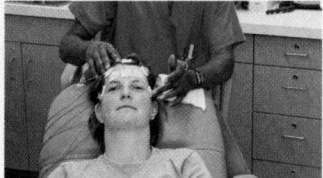

PREFACE

Essentials of Medical Language is designed for you, the student. The development of many medical terminology textbooks and learning programs begins with the question, "What topics should this book or program cover?" This question has been the basis of a host of textbooks available today. There is only one problem: where do *you*, the student, fit into this question?

To put the focus back on the student, a new question guided the design and writing of *Essentials of Medical Language:*

What medical terminology knowledge and skills do students preparing for careers in modern health care need to be successful?

Time and time again, instructors and students alike have indicated their belief that students learn medical language best when it is connected to real life: real health professionals interacting with real patients in a real medical setting. Just as one of the best ways to learn a foreign language is to be immersed in the language and culture of the country where it is spoken, one of the best ways to learn medical language is to be immersed within a vibrant, authentic, modern health care community.

Essentials of Medical Language helps students learn the terminology and language of modern health care in a way that bridges the gap between the classroom and the clinical setting.

HOW STUDENT NEEDS ARE MET

This book was designed with your needs in mind. You are a student preparing for a career as an allied health professional. You may have already had a few health care–related courses, or you may just be beginning your studies in the field. While your background and interests may differ, you share the need to understand medical terminology.

To make sure your needs were addressed in this book, we asked both students and experienced medical terminology instructors, "What helps students learn medical terminology?" Overwhelmingly, the responses pointed to three common factors:

- motivation to learn
- retention of the material
- opportunities for application and practice
- readily available information

THIS TEXTBOOK INCORPORATES FEATURES DESIGNED TO ADDRESS THESE FOUR FACTORS.

Motivation to learn	→	In order for students to be motivated to learn, what they are learning must be meaningful and relevant. To ensure the chapters in *Essentials of Medical Language* fit these criteria, the student is asked to step into the role of an allied health professional in each chapter. Authentic patient cases are used to illustrate how medical language is used on the job.
Retention of the material	→	When students encounter new medical terms within the context of a patient case, they are able to remember it more effectively. In addition, each chapter presents medical terms from one body system or medical specialty, which further serves to "tie it all together" to help students retain the knowledge and skills.
Opportunities for application and practice	→	Practice makes perfect. This is especially true for learning medical terminology. This textbook provides many opportunities for students to apply what they are learning. Exercises are included in the lessons, as well as at the end of each chapter. Additional exercises are available on the student Online Learning Center (www.mhhe.com/AllanEssMedLanguage).
Readily available information	→	In this book, all the information needed for a specific topic is presented in self-contained two-page spreads. On the left-hand page, new medical terms are introduced. On the right-hand page, for each new term, the pronunciation, color-coded word elements, and definition are provided in a **Word Analysis and Definition (WAD) Table.**

HOW INSTRUCTORS' NEEDS ARE MET

When you use *Essentials of Medical Language,* you will be supported at every point in the program. Each chapter in the book is broken down into lessons, and the Instructor's Manual provides lesson plans and additional materials for each lesson. Following are features of the textbook designed to address student needs:

Lesson-Based Approach

Each chapter of *Essentials of Medical Language* is divided into lessons covering different aspects of the overall chapter subject. Lessons within a chapter break down into topics. Each topic is designed so your students will not have to flip back and forth when completing exercises or looking at figures, tables, and boxes. All main concepts and ideas presented in topics begin and end within a two-page "spread." These spreads help learning flow smoothly by ensuring that valuable class and reading time is not wasted on flipping pages.

You Are . . . Your Patient Is . . . Case Scenarios

Each chapter and most lessons begin by immediately placing your students in the role of an allied health professional faced with a situation in which medical communication is necessary. Many different professional allied health and LPN-level nursing roles are utilized so your students can "experience" various specialties and positions. The patient cases introduced at the beginning of the chapters and lessons are referenced throughout the lessons to further unify the students' experience.

Chapter Outcomes and Lesson Objectives

The major learning outcomes for each chapter are previewed in the beginning so you and your students can focus on what they need to know and be able to do by the end of the chapter. Each lesson has outcome-based learning objectives. Accomplishing each lesson's objectives helps ensure students will be able to achieve the chapter outcomes and, ultimately, the goal of the textbook: to help them learn the essential terminology and language of modern health care.

Word Analysis and Definition Tables (WAD)

Each lesson contains tables listing important medical terms and their pronunciation, elements, and definition. Prefixes, suffixes, and combining forms are color-coded. These tables provide your students with an at-a-glance view of the terms covered. The tables are excellent for reference as well as for studying and reviewing.

End-of-Lesson and End-of-Chapter Exercises

At the end of each lesson is a series of exercises. The end-of-lesson exercises provide your students with immediate practice using the terms in the lesson. These exercises focus on basic understanding and ability to apply the terms. They are an excellent foundation for the end-of-chapter exercises, which are often based on authentic situations, such as interactions with patients, physicians, or medical documentation. The end-of-chapter exercises will require your students to understand, accurately apply, and think critically about the medical language they use. Throughout the text, frequent opportunities for application and reinforcement of medical language skills and concepts are provided to help your students build confidence and knowledge. A wide variety of exercises and activities are included to address different medical settings and levels of learning (including knowledge, comprehension, application, analysis, synthesis, and evaluation).

TEACHING AND LEARNING SUPPLEMENTS

FOR THE INSTRUCTOR:

INSTRUCTOR'S MANUAL (007-335228-4)

The Instructor's Manual (available in print and in the online Instructor Resources) is an invaluable resource for new and experienced medical terminology instructors. All of the components of the *Essentials of Medical Language* textbook program are designed to be coherent and connected in order to create a consistent environment in which students can learn. The Instructor's Manual shows how each component of the textbook program works together to support and reinforce the content and strengths of the other components, from art to exercises to content to test bank questions.

The Instructor's Manual contains the following sections:

- **Your Medical Terminology Course—An Introduction to Teaching Medical Terminology**

  The Instructor's Manual contains a helpful introduction to teaching medical terminology, as well as other helpful resources such as:

  - information about student learning styles and corresponding instructor strategies
  - innovative learning activities
  - assessment techniques and strategies
  - classroom management tips
  - techniques for teaching limited-English-proficiency students

- **Lesson Planning Guide**

  In addition, the Instructor's Manual contains a Lesson Planning Guide with a complete and customizable lesson plan for each of the 53 lessons in the book. Each lesson plan contains a step-by-step 50-minute teaching plan and master copies of handouts. These lessons may be used alone or combined to accommodate different class schedules. The lessons can easily be revised to reflect your preferred topic or sequence or to add or delete topics entirely. Each of the lesson plans is designed to be used with a corresponding PowerPoint® presentation that is available in the online Instructor Resources.

- **Internet-Based Research Activities**

  The Instructor's Manual also includes Internet-based research activities for each chapter in the book.

INSTRUCTOR RESOURCES, www.mhhe.com/AllanEssMedLanguage

The instructor resources contain:

- **Instructor's Manual**—written by Teleologic Learning Company.
- **McGraw-Hill's EZ-Test Test Generator**—The flexible electronic testing program allows instructors to create tests from book-specific items. It accommodates a wide range of question types, and instructors may add their own questions. Multiple versions of the test can be created and any test can be exported for use with course management systems such as WebCT, BlackBoard, or PageOut. EZ-Test Online is a new service and gives you a place to easily administer your EZ-Test–created exams and quizzes online. The program is available for Windows and Macintosh environments.
- **PowerPoint® Lecture Outlines**—PowerPoint® lectures with speaking notes are available for the chapters in the textbook. Each 50-minute lesson plan in the Instructor's Manual Lesson Planning Guide dedicates approximately 20 to 25 minutes to the use of the corresponding

ready-made PowerPoint presentations. The PowerPoint presentations, which combine art and lecture notes, are designed to help instructors discuss with students the important points of the lessons. The slides are customizable, allowing instructors to modify lectures to ensure that the needs of their unique students and curricula are met.

- **Image Bank**—features selected textbook images.

COURSE DELIVERY SYSTEMS

With help from our partners, WebCT, Blackboard, TopClass, eCollege, and other course management systems, instructors can take complete control over their course content. These course cartridges also provide online testing and powerful student tracking features.

HOW TO TEACH MEDICAL TERMINOLOGY

Online Course for Instructors to Support *Essentials of Medical Language* is found on the instructor resources of the Online Learning Center, www.mhhe.com/AllanEssMedLanguage.

The **How to Teach Medical Terminology online course** provides instructors with the introductory knowledge and resources they need to begin effectively using the *Essentials of Medical Language* textbook and related materials. This course is designed to cover the "basics" of how to effectively teach medical terminology.

How to Teach Medical Terminology allows instructors to choose for themselves which module they wish to take, or they may opt to take a self-assessment survey that will recommend one of the three modules.

- **Module 1** is designed for the inexperienced instructor.
- **Module 2** is designed for the instructor who has previous classroom experience but who has never taught Medical Terminology.
- **Module 3** is designed for the experienced Medical Terminology instructor who has not used a contextualized approach to teach before.

Upon completion of a given module, instructors will take a final assessment designed to demonstrate their understanding and achievement of the learning objectives for that module. Those who score 70% or higher on the final assessment will receive a certificate that can be printed for professional development purposes.

FOR THE STUDENT:

ONLINE LEARNING CENTER (OLC)
www.mhhe.com/AllanEssMedLanguage

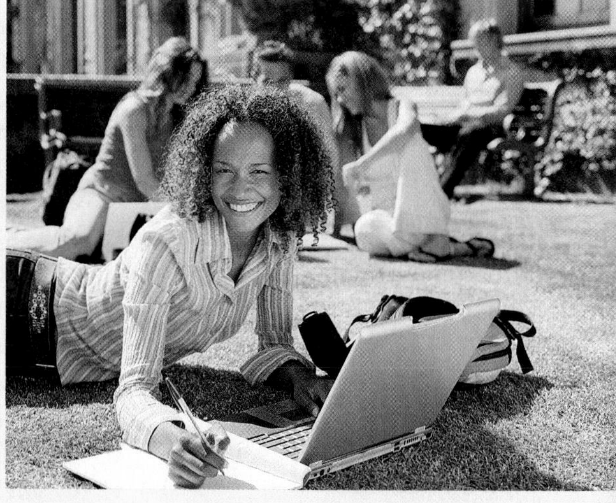

The OLC offers an extensive array of learning and teaching tools. The site includes quizzes for each chapter, links to websites, and interactive activities. Students also will be able to access chapter-specific interactive exercises via McGraw-Hill CONNECT, where they can apply medical language to realistic patient scenarios. These exercises provide multiple opportunities for practice and the mastery of core concepts. The exercises are designed to

- Help students learn medical terms, including specifically their definitions, roots, prefixes, and suffixes, plus accurate spelling.
- Help students understand the meaning and use of medical terms.
- Help students learn how and when to correctly apply medical terms in written and verbal communication.

ACKNOWLEDGMENTS

The uniqueness, beauty, and high standards of this book are due to the skills and devotion of a team of people who worked closely and happily together. The team includes: Adrianne Rippinger, William Thomas, and Kari Sandhass of the Teleologic Learning Company; Patricia Gillivan of Triple SSS Press Media Development; and many talented people from McGraw-Hill Higher Education editorial and production. Our deepest thanks to all of them.

We would also like to thank the dedicated staff of Greater Annapolis Medical Group, Annapolis, Maryland, for opening their practice to our photography team.

David Allan
Karen Lockyer

For insightful reviews, criticisms, helpful suggestions, and information, we would like to acknowledge the following:

Vanessa J. Austin, RMA, CAHI
Clarian Health

Cynthia Boles, MBA
Bradford School

William J. Burke, BA
Madison Area Technical College
Blackhawk Technical College

Jennifer Campbell, M.Ed., OT/L
Tulsa Community College

Carmen Carpenter, BSN, MS
South University

Marie Cissell, MN, RN
South Dakota State University, College of Nursing

Christina Rauberts Conklin, AA, RMA
Keiser University

Brian E. Conroy, MD
Dean, Allied Health Dept,
Lehigh Valley College

Kimberly Corsi, LRCP, CCS
Davenport University

Patricia A. Dudek, Diploma in Nursing
McCann School of Business and Technology

Jane W, Dumas, MSN
Remington College

Rhonda K. Epps, AS
National College of Business and Technology

Jean Fennema, BA
Pima Medical Institute

Walter E. Flowers
Lamson Institute, San Antonio

Anna E. Fritz, MPH, MT (ASCP)
Medical Assisting Program Chair
South College, Knoxville, Tennesee

Tammy R. Gockman, CBCS, MA
Assistant Director of Education
American Professional Institute

Darlene S, Grayson, BS
Remington College

JoAnne E. Habenicht, MPA, RT, (R)(T)(M)
Manhattan College

Elizabeth Hoffman, MA Ed., BS
Baker College Of Clinton Township

Diana Hollwedel, LPN
Career Institute of Florida

Janet Hunter, MS, MBA
Northland Pioneer College

Judith B. Johnson, RN
Nashville State Community College

Timothy J. Jones, BA, MA
Oklahoma City Community College

Judith Karls, RN, BSN, M.Ed.
Madison Area Technical College

Heather Lane, BS
Missouri College

Sandra A. Lehrke, RN, MS, CMA
Anoka Technical College

Leigh Ann Long, RN
Brookstone College of Business

Nelly Mangarova, MD
Ohlone College

Wilsetta McClain, MBA, ABD
Baker College of Auburn Hills

Pam McConnell, MA, AS
High Tech Institute

Sue B. Meeks, CPC-A
Milan Institute

Cathleen A. Murphy, DC
Katharine Gibbs College

Fred R. Pearson, Ph.D.
Brigham Young University Idaho

Adrienne L. Reaves, BS, M.Ed.
Westwood College Institute of Healthcare Program Director

Shawn Marie Russell, BA, CPC
University of Alaska

Becky Schonberger, RN, CMA
Ivy Tech Community College

Rebecca L. Schultz, PhD
University of Sioux Falls

Gene Simon, RHIA, RMD
Florida Career College

Lynn G. Slack, BS CMA
Kaplan Career Institute—ICM Campus

Donna J. Slovensky, PhD, RHIA, FAHIMA
University of Alabama at Birmingham

Gregory V. Smith, MSW
Brown Mackie College, Tucson

Catherine A. Teel, AST, Health Care Technology, RMA
McCann School of Business and Technology

Lynne A. Thomas, BA
Clarita Career College

Kathryn Whitley, RN, MSN, NP-C
Patrick Henry Community College

Kathy Wishon, RN
North Metro Technical College

Mindy Wray, CS, CMA, RMA
ECPI College of Technology

Daphne Zito, MS
Katharine Gibbs School

Susan K. Zolvinski, BS, MBA
Brown Mackie College

Contextual Approach Promotes Active Learning

Chapters in the textbook are organized by body system in accordance with an overall anatomy and physiology (A & P) approach. Lessons introduce and define terminology through the context of A & P, pathology, and clinical and diagnostic procedures/tests. The organization of the body systems into chapters is based on an "outside to inside" sequence that reflects a physician's differential diagnosis method used during an examination.

To provide students with an authentic context, the medical specialty associated with each body area or system is introduced along with relevant anatomy and physiology. Students actually step into the role of an allied health professional associated with each specialty. Patient cases and documentation are used to illustrate the real-life application of medical terminology in modern health care: to care for and communicate with patients and to interact with other members of the health care team.

The A & P organizational approach, used in conjunction with an authentic medical setting and patient cases, encourages student motivation and facilitates active, engaged learning.

Innovative Pedagogical Aids Provide a Coherent Learning Program

Each chapter is structured around a consistent and unique framework of pedagogic devices. No matter what the subject matter of a chapter, the structure enables students to develop a consistent learning strategy, making *Essentials of Medical Language* a superior learning tool.

YOU ARE COMMUNICATING WITH

Each chapter opens by placing the student in the role of an allied health professional related to the specialty and associated body systems/areas covered by the chapter. The student is also introduced to a patient and given information about the patient's case.

LEARNING OUTCOMES

At the same time, **Learning Outcomes** are presented to let students know what they will learn in the chapter. This technique immediately engages students, motivating them to read on to learn how this patient's case (and their role in the patient's care) relates to the medical terminology being introduced in the chapter.

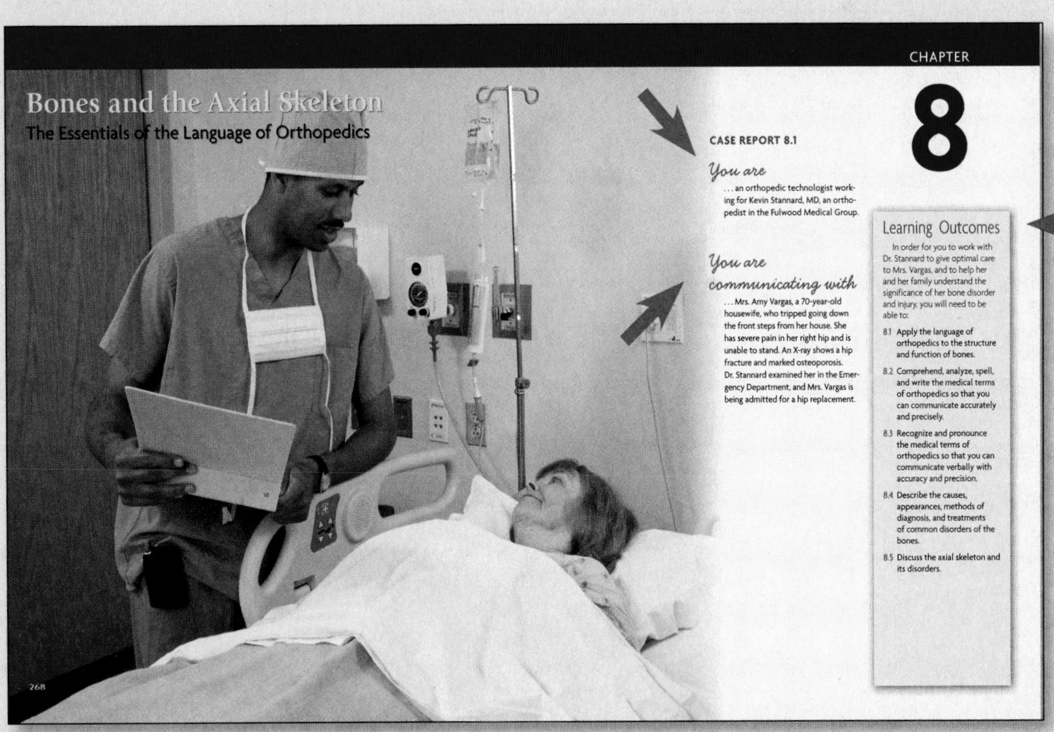

LESSON-BASED ORGANIZATION

The chapter content is broken down into chunks, or lessons, to help students digest new information and relate it to previously learned information. Rather than containing many various topics within a chapter, these lessons group the chapter material into logical, streamlined learning units designed to help students achieve the chapter outcomes. Lessons within a chapter build on one another to form a cohesive, coherent experience for the learner.

Each lesson is based on specific **Lesson Objectives** designed to support the students' achievement of the overall chapter outcomes.

Each lesson in a chapter contains an Introduction, Lesson Objectives, Lesson Topics, Word Analysis and Definition Tables, and Lesson Exercises. Within each lesson, all topics and information are presented in **self-contained two-page spreads.** This means students will no longer have to flip back and forth to see figures on one page that are described on another.

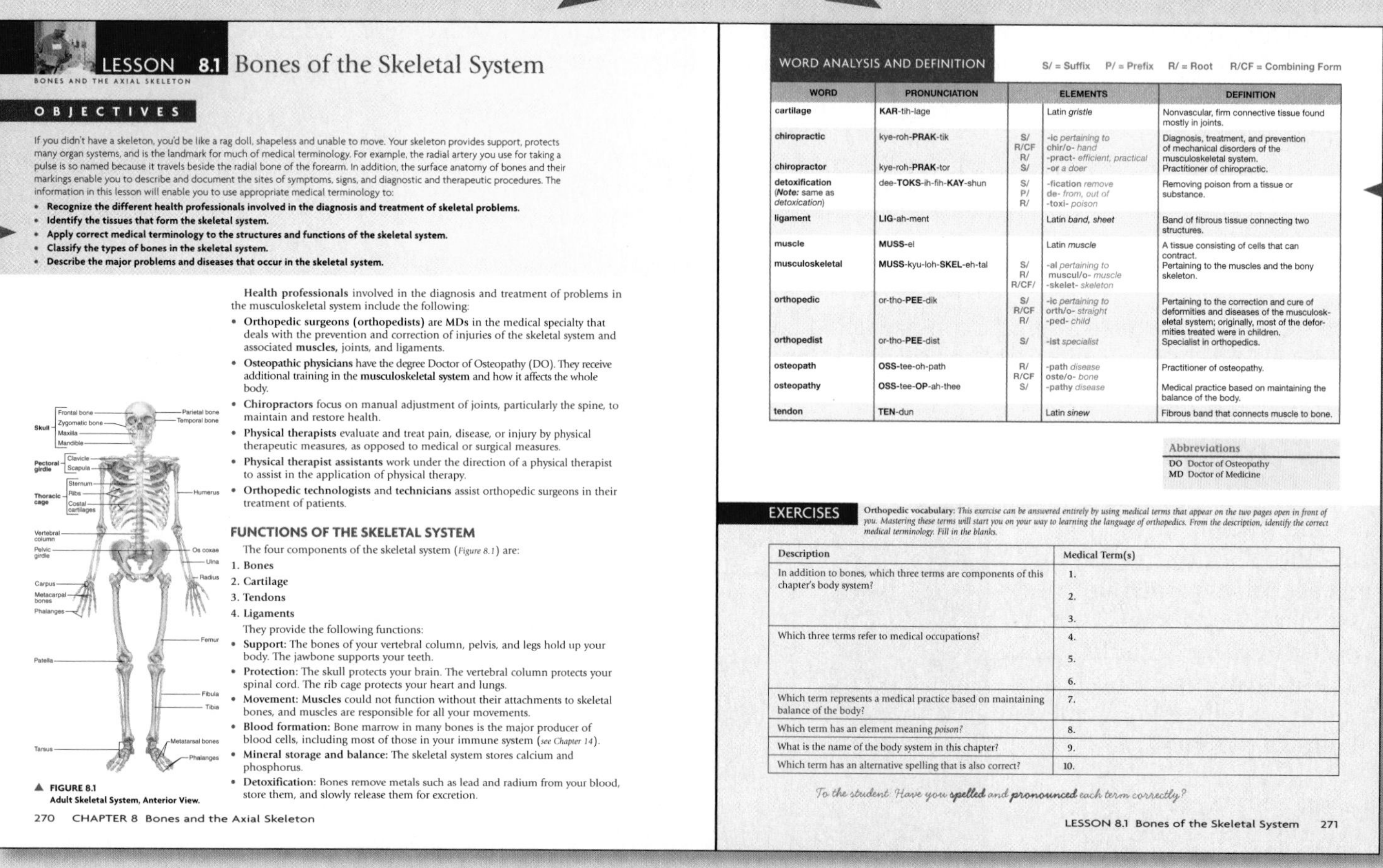

LESSON 8.1 Bones of the Skeletal System

BONES AND THE AXIAL SKELETON

OBJECTIVES

If you didn't have a skeleton, you'd be like a rag doll, shapeless and unable to move. Your skeleton provides support, protects many organ systems, and is the landmark for much of medical terminology. For example, the radial artery you use for taking a pulse is so named because it travels beside the radial bone of the forearm. In addition, the surface anatomy of bones and their markings enable you to describe and document the sites of symptoms, signs, and diagnostic and therapeutic procedures. The information in this lesson will enable you to use appropriate medical terminology to:

- **Recognize the different health professionals involved in the diagnosis and treatment of skeletal problems.**
- **Identify the tissues that form the skeletal system.**
- **Apply correct medical terminology to the structures and functions of the skeletal system.**
- **Classify the types of bones in the skeletal system.**
- **Describe the major problems and diseases that occur in the skeletal system.**

Health professionals involved in the diagnosis and treatment of problems in the musculoskeletal system include the following:

- **Orthopedic surgeons (orthopedists)** are **MDs** in the medical specialty that deals with the prevention and correction of injuries of the skeletal system and associated **muscles,** joints, and ligaments.
- **Osteopathic physicians** have the degree Doctor of Osteopathy (DO). They receive additional training in the **musculoskeletal system** and how it affects the whole body.
- **Chiropractors** focus on manual adjustment of joints, particularly the spine, to maintain and restore health.
- **Physical therapists** evaluate and treat pain, disease, or injury by physical therapeutic measures, as opposed to medical or surgical measures.
- **Physical therapist assistants** work under the direction of a physical therapist to assist in the application of physical therapy.
- **Orthopedic technologists** and **technicians** assist orthopedic surgeons in their treatment of patients.

▲ **FIGURE 8.1**
Adult Skeletal System, Anterior View.

FUNCTIONS OF THE SKELETAL SYSTEM

The four components of the skeletal system (*Figure 8.1*) are:

1. **Bones**
2. **Cartilage**
3. **Tendons**
4. **Ligaments**

They provide the following functions:

- **Support:** The bones of your vertebral column, pelvis, and legs hold up your body. The jawbone supports your teeth.
- **Protection:** The skull protects your brain. The vertebral column protects your spinal cord. The rib cage protects your heart and lungs.
- **Movement: Muscles** could not function without their attachments to skeletal bones, and muscles are responsible for all your movements.
- **Blood formation:** Bone marrow in many bones is the major producer of blood cells, including most of those in your immune system (*see Chapter 14*).
- **Mineral storage and balance:** The skeletal system stores calcium and phosphorus.
- **Detoxification:** Bones remove metals such as lead and radium from your blood, store them, and slowly release them for excretion.

270 CHAPTER 8 Bones and the Axial Skeleton

WORD ANALYSIS AND DEFINITION

S/ = Suffix P/ = Prefix R/ = Root R/CF = Combining Form

WORD	PRONUNCIATION	ELEMENTS		DEFINITION
cartilage	**KAR**-tih-lage		Latin *gristle*	Nonvascular, firm connective tissue found mostly in joints.
chiropractic	kye-roh-**PRAK**-tik	S/ R/CF R/	-ic *pertaining to* chir/o- *hand* -pract- *efficient, practical*	Diagnosis, treatment, and prevention of mechanical disorders of the musculoskeletal system.
chiropractor	kye-roh-**PRAK**-tor	S/	-or *a doer*	Practitioner of chiropractic.
detoxification (***Note:*** same as *detoxication*)	dee-**TOKS**-ih-fih-**KAY**-shun	S/ P/ R/	-fication *remove* de- *from, out of* -toxi- *poison*	Removing poison from a tissue or substance.
ligament	**LIG**-ah-ment		Latin *band, sheet*	Band of fibrous tissue connecting two structures.
muscle	**MUSS**-el		Latin *muscle*	A tissue consisting of cells that can contract.
musculoskeletal	**MUSS**-kyu-loh-**SKEL**-eh-tal	S/ R/ R/CF/	-al *pertaining to* muscul/o- *muscle* -skelet- *skeleton*	Pertaining to the muscles and the bony skeleton.
orthopedic	or-tho-**PEE**-dik	S/ R/CF R/	-ic *pertaining to* orth/o- *straight* -ped- *child*	Pertaining to the correction and cure of deformities and diseases of the musculoskeletal system; originally, most of the deformities treated were in children.
orthopedist	or-tho-**PEE**-dist	S/	-ist *specialist*	Specialist in orthopedics.
osteopath	**OSS**-tee-oh-path	R/ R/CF	-path *disease* oste/o- *bone*	Practitioner of osteopathy.
osteopathy	**OSS**-tee-**OP**-ah-thee	S/	-pathy *disease*	Medical practice based on maintaining the balance of the body.
tendon	**TEN**-dun		Latin *sinew*	Fibrous band that connects muscle to bone.

Abbreviations

DO Doctor of Osteopathy
MD Doctor of Medicine

EXERCISES

Orthopedic vocabulary: *This exercise can be answered entirely by using medical terms that appear on the two pages open in front of you. Mastering these terms will start you on your way to learning the language of orthopedics. From the description, identify the correct medical terminology. Fill in the blanks.*

Description	Medical Term(s)
In addition to bones, which three terms are components of this chapter's body system?	1. 2. 3.
Which three terms refer to medical occupations?	4. 5. 6.
Which term represents a medical practice based on maintaining balance of the body?	7.
Which term has an element meaning *poison?*	8.
What is the name of the body system in this chapter?	9.
Which term has an alternative spelling that is also correct?	10.

*To the student: Have you **spelled** and **pronounced** each term correctly?*

LESSON 8.1 Bones of the Skeletal System 271

WORD ANALYSIS AND DEFINITION TABLES

The medical terms covered in each lesson are introduced in context, either within a patient case or in the lesson topics. To facilitate easy reference and review, the terms are also listed in tables as a group. The **Word Analysis and Definition (WAD) Tables** list the term and its pronunciation, elements, and definition in a concise, color-coded, at-a-glance format.

LESSON AND CHAPTER-END EXERCISES

Each lesson within a chapter ends with exercises designed to allow students to check their basic understanding of the terms they just learned. These "checkpoints" can be used by instructors as assignments or for self-evaluation by students.

At the end of each chapter you will find 10–15 pages of exercises that ask students to apply what they learned in all lessons of a chapter. These chapter-end exercises reinforce learning and help students go beyond mere memorization to think critically about the medical language they use. In addition to reviewing and recalling the definitions of terms learned in the chapter, students are asked to use medical terms in new and different ways to ensure a thorough understanding.

EXERCISES **Elements** *remain your best clue for understanding a medical term. In this exercise, the meaning of each element is given below the line—this is your clue to constructing the term. Write the correct element on the line above its meaning. After you have constructed the term, give its definition in the space provided.*

Study Hint More than one element can have the same meaning.

1. ________ / ________
cortex pertaining to
The term is ________ and means ________.

2. ________ / ________ / ________
around bone structure
The term is ________ and means ________.

3. ________ / ________ / ________
upon, above growth pertaining to
The term is ________ and means ________.

4. ________ / ________ / ________
middle pertaining to
The term is ________ and means ________.

LESSON 8.1 Bones of the Skeletal System 273

STUDY HINT BOXES

Study Hint boxes are found throughout the review exercises. They reinforce and remind students to use basic study skills.

CHAPER 8 REVIEW

BONES AND THE AXIAL SKELETON

E. **Terminology challenge:** *suture.* Medical terms can have more than one meaning/usage. *Use the Glossary, your library, or an online medical dictionary if you need help answering these questions.*

1. Define *suture* as it is used in this chapter.
2. Now use this meaning of *suture* in a sentence that is not a definition or taken directly out of the text.

Suture can also be a noun and a verb with another meaning. Can you identify them?

3. *Suture* as a noun (person, place, or thing) can also mean (definition) ________.
4. Write a sentence with *suture* having this meaning.
5. *Suture* as a verb (action) can also mean (definition)
6. Write a sentence with *suture* having this meaning.

Study Hint First, read the sentences and underline or highlight any medical terms (or abbreviations) you will need to explain. Then, rewrite the sentence in non-medical language.

See your medical vocabulary increase as you now know one term with three different meanings!

F. **Translate** the following sentences into layperson's language a patient can understand.

1. A patient with osteopenia is at risk for osteoporosis.

284 CHAPTER 8 REVIEW Bones and the Axial Skeleton

HAPER 8 REVIEW

D THE AXIAL SKELETON

ollowing terms from this chapter are particularly difficult to spell and pronounce. Correct pronun- edical terms is the mark of an educated professional. Circle the correct spelling, and then check (✓) d the pronunciation. *Remember:* Pronunciations are on the Student Online Learning Center EssMedLanguage).

			Pronunciation ✓
cocyx	coccyx	coccyz	
cartilage	carrtilage	cartilege	
skoliosis	scolliosis	skolioses	
osteomielitis	osteomyelitis	osteomyelites	
khyphosis	kyphosis	kyiphosis	
achondroplasia	acondroplasia	achodroplasia	
occipitel	ocippital	occipital	
spenoid	sphenoid	phenoid	
epiphysial	epiphyseal	epifiseal	
chirropractic	chiropracctic	chiropractice	

le the correct answer.

r *low bone density* is:
d. osteomalacia
e. osteopenia

Study Hint Immediately cross off any answer you know is not correct. In your remaining choices, there is only *one best answer.*

ones are determined by their:
d. weight
e. number

ffix meaning *disease?*
d. osteopath
e. osteogenic

tured, blood vessels bleed into the fracture site and form a(n):
d. osteoblast
e. condyle

ns that are *both* diagnoses:
medullary d. periosteum osteoporosis
b. cortex osteopathy e. osteomalacia rickets
c. orthopedic osteomyelitis

290 CHAPTER 8 REVIEW Bones and the Axial Skeleton

VIVID ILLUSTRATIONS AND PHOTOS

Colorful, precise anatomical illustrations and photos lend a realistic view of body structures and correlate to the clinical context of the lessons.

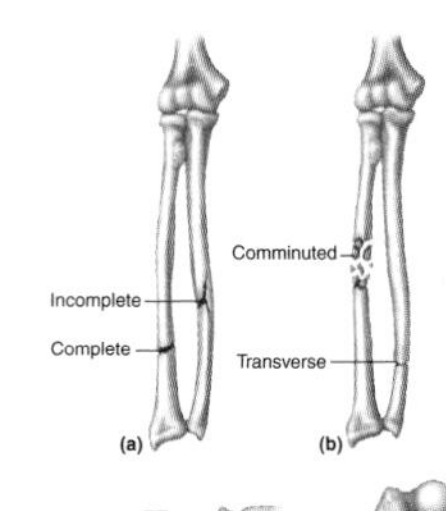

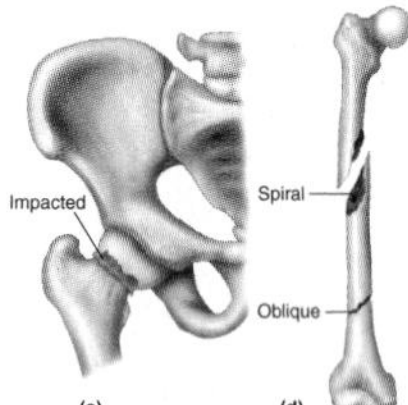

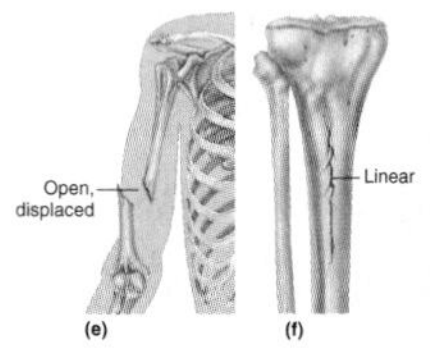

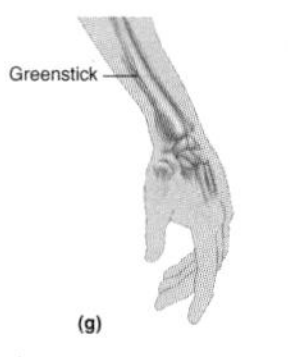

▲ **FIGURE 8.7** **Bone Fractures.**

BONE FRACTURES (FXS)

TABLE 8.1 Classification and Definition of Bone Fractures

Name	Description	Reference
Closed (also called **simple** fracture)	A bone is broken, but the skin is not broken.	Figure 8.7g
Open (also called **compound** fracture)	A fragment of the fractured bone breaks the skin, or a wound extends to the site of the fracture.	Figure 8.7e
Displaced	The fractured bone parts are out of line.	Figure 8.7e
Complete	A bone is broken into at least two fragments.	Figure 8.7a
Incomplete	The fracture does not extend completely across the bone. It can be **hairline,** as in a **stress** fracture in the foot, when there is no separation of the two fragments.	Figure 8.7a
Comminuted	The bone breaks into several pieces, usually two major pieces and several smaller fragments.	Figure 8.7b
Transverse	The fracture is at right angles to the long axis of the bone.	Figure 8.7b
Impacted	The fracture consists of one bone fragment driven into another, resulting in shortening -of a limb.	Figure 8.7c
Spiral	The fracture spirals around the long axis of the bone.	Figure 8.7d
Oblique	The fracture runs diagonally across the long axis of the bone.	Figure 8.7d
Linear	The fracture runs parallel to the long axis of the bone.	Figure 8.7f
Greenstick	This is a partial fracture. One side breaks, and the other bends.	Figure 8.7g
Pathologic	The fracture occurs in an area of bone weakened by disease, such as cancer.	—
Compression	The fracture occurs in a vertebra from trauma or pathology, leading to the vertebra being crushed.	—
Stress	This is a fatigue fracture caused by repetitive, local stress on a bone, as occurs in marching or running.	—

Healing of Fractures

When a bone is fractured, blood vessels bleed into the fracture site, forming a hematoma. After a few days, bone-forming cells called **osteoblasts** move in and start to produce new bone matrix, which develops into **osteocytes** (bone cells). Eventually the new bone fuses together the segments of the fracture.

Surgical Procedures for Fractures

The initial goal of fracture treatment is to bring the ends of the bone at the break back opposite each other so that they fit together as they did in the original bone. This is called **alignment.**

External manipulation is used frequently. The bone is pulled from the distal end back into alignment. This process is called **reduction.** Anesthesia may be used.

In **external fixation,** the alignment is maintained by immobilizing the bone

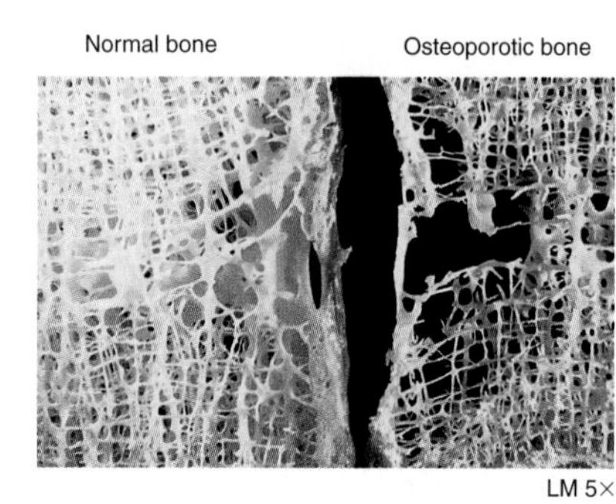

▲ **FIGURE 8.4**
Normal Bone and Osteoporotic Bone.

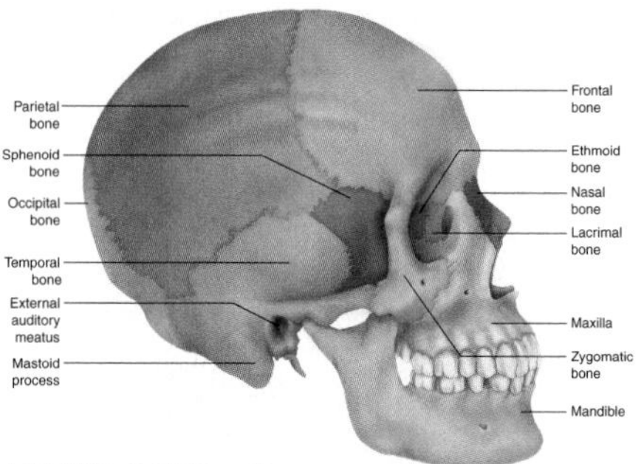

▲ **FIGURE 8.10** **Skull, Right Lateral View.**

SKULL AND FACE

The Skull

The human skull (*Figure 8.10*) has 22 bones, 8 of which make up the **cranium,** the upper part of the skull that encloses the **cranial cavity** and protects the brain. The bones of the cranium are the following:

1. The **frontal** bone (1) forms the forehead, the roofs of the orbits, and part of the floor of the cranium and contains a pair of right and left frontal sinuses above the orbits.
2. **Parietal** bones (2) form the bulging sides and roof of the cranium.
3. The **occipital** bone (1) forms the back of and part of the base of the cranium.
4. **Temporal** bones (2) form the sides of and part of the base of the cranium.
5. The **sphenoid** bone (1) forms part of the base of the cranium and the orbits.
6. The **ethmoid** bone (1) forms part of the nose and the orbits and is hollow, forming the ethmoid sinuses.

The bones of the cranium are joined together by **sutures,** joints that appear as seams, covered on the inside and outside by a thin layer of connective tissue.

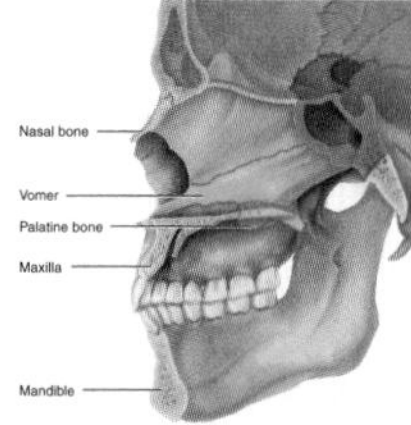

▲ **FIGURE 8.11** **Facial Bones.**

The lower part of the skull comprises the 14 bones of the facial skeleton (*Figure 8.11*):

1. **Maxillary** bones (2) form the upper jaw **(maxilla),** hold the upper teeth, and are hollow, forming the maxillary sinuses.
2. **Palatine** bones (2) are located behind the maxilla and cannot be seen on a lateral view of the skull.
3. **Zygomatic** bones (2) are the prominences of the cheeks below the eyes.
4. **Lacrimal** bones (2) form the medial wall of each orbit.
5. **Nasal** bones (2) form the sides and bridge of the nose.
6. The **vomer** bone (1) separates the two nasal cavities.
7. Inferior nasal **conchae** (2) are fragile bones in the lower nasal cavity.
8. The **mandible** (1) is the lower jawbone, which holds the lower teeth. The mandible articulates (joins) with the temporal bone to form the **temporomandibular joint (TMJ).**

The third component of the axial skeleton, the rib cage, is discussed in *Chapter 7,* "Respiratory System."

TABLES

Meaningful tables aid in summarizing concepts and lesson topics.

Incomplete, Complete (a); Comminuted, Transverse (b); Impacted (c); Spiral, Oblique (d); Open, displaced (e); Linear (f); Greenstick (g)

▲ FIGURE 8.7 Bone Fractures.

BONE FRACTURES (FXS)

TABLE 8.1 Classification and Definition of Bone Fractures

Name	Description	Reference
Closed (also called **simple** fracture)	A bone is broken, but the skin is not broken.	Figure 8.7g
Open (also called **compound** fracture)	A fragment of the fractured bone breaks the skin, or a wound extends to the site of the fracture.	Figure 8.7e
Displaced	The fractured bone parts are out of line.	Figure 8.7e
Complete	A bone is broken into at least two fragments.	Figure 8.7a
Incomplete	The fracture does not extend completely across the bone. It can be **hairline**, as in a **stress** fracture in the foot, when there is no separation of the two fragments.	Figure 8.7a
Comminuted	The bone breaks into several pieces, usually two major pieces and several smaller fragments.	Figure 8.7b
Transverse	The fracture is at right angles to the long axis of the bone.	Figure 8.7b
Impacted	The fracture consists of one bone fragment driven into another, resulting in shortening of a limb.	Figure 8.7c
Spiral	The fracture spirals around the long axis of the bone.	Figure 8.7d
Oblique	The fracture runs diagonally across the long axis of the bone.	Figure 8.7d
Linear	The fracture runs parallel to the long axis of the bone.	Figure 8.7f
Greenstick	This is a partial fracture. One side breaks, and the other bends.	Figure 8.7g
Pathologic	The fracture occurs in an area of bone weakened by disease, such as cancer.	—
Compression	The fracture occurs in a vertebra from trauma or pathology, leading to the vertebra being crushed.	—
Stress	This is a fatigue fracture caused by repetitive, local stress on a bone, as occurs in marching or running.	—

Healing of Fractures

When a bone is fractured, blood vessels bleed into the fracture site, forming a hematoma. After a few days, bone-forming cells called **osteoblasts** move in and start to produce new bone matrix, which develops into **osteocytes** (bone cells). Eventually the new bone fuses together the segments of the fracture.

Surgical Procedures for Fractures

The initial goal of fracture treatment is to bring the ends of the bone at the break back opposite each other so that they fit together as they did in the original bone. This is called **alignment**.

External manipulation is used frequently. The bone is pulled from the distal end back into alignment. This process is called **reduction.** Anesthesia may be used.

In **external fixation,** the alignment is maintained by immobilizing the bone through the use of:

- **Plaster casts.**
- **Splints.**
- **Traction,** which is the gentle but continuous application of a pulling force that can align a fracture, reduce muscle spasm, and relieve pain.
- **External fixators,** by which the bone fragments are secured to a strong external steel rod by means of steel pins.

276 CHAPTER 8 Bones and the Axial Skeleton

KEYNOTES AND ABBREVIATIONS

Keynotes and Abbreviations offer students additional information correlating to the lesson.

▲ FIGURE 8.4 Normal Bone and Osteoporotic Bone.

Keynote

Osteomalacia occurs in some developing nations and occasionally in this country when children drink soft drinks instead of milk fortified with vitamin D.

Case Report 8.1 (continued)

On questioning, Amy Vargas demonstrated many of the risk factors for osteoporosis including family history, lack of exercise, cigarette smoking, inadequate diet, postmenopause, and increasing age.

Diseases of Bone

Osteoporosis results from a loss of bone density (*Figure 8.4*). It is more common in women than in men, 10 million people alread density **(osteopenia)** an

In women, productio its protection against bo reduction in testosteron

Women at risk for o screening using a **DEXA** 1,200 milligrams **(mg)** o vitamin D and to expose

There are several **FD** osteoporosis.

Osteomyelitis is an i usually with a staphyloc

Osteomalacia, know deficiency. When bones

Abbreviations

BMD bone mineral density
DEXA dual energy x-ray absorptiometry
FDA Food and Drug Administration
IU international unit(s)
mg milligram

EXERCISES

Suffixes: *The combining form oste/o means bone, and it is the main element in each of the following terms. You choose the correct suffix to complete the term. Fill in the blanks.*

genesis genic penia malacia porosis myelitis

1. Disease caused by vitamin D deficiency osteo/________
2. Low bone density osteo/________
3. Porous, brittle, fragile bones osteo/________
4. Most common malignant bone tumor osteo/________
5. Rare genetic disorder producing easily fractured bones, often in utero osteo/________
6. Inflammation of bone tissue osteo/________

Note: *The meaning of the combining form never changes. The addition of six different suffixes has helped you learn six new terms in orthopedic vocabulary!*

ONLINE LEARNING CENTER (OLC)

www.mhhe.com/AllanEssMedLanguage

This online resource offers an extensive array of quizzing and learning tools that will help students master the topics covered in their textbook.

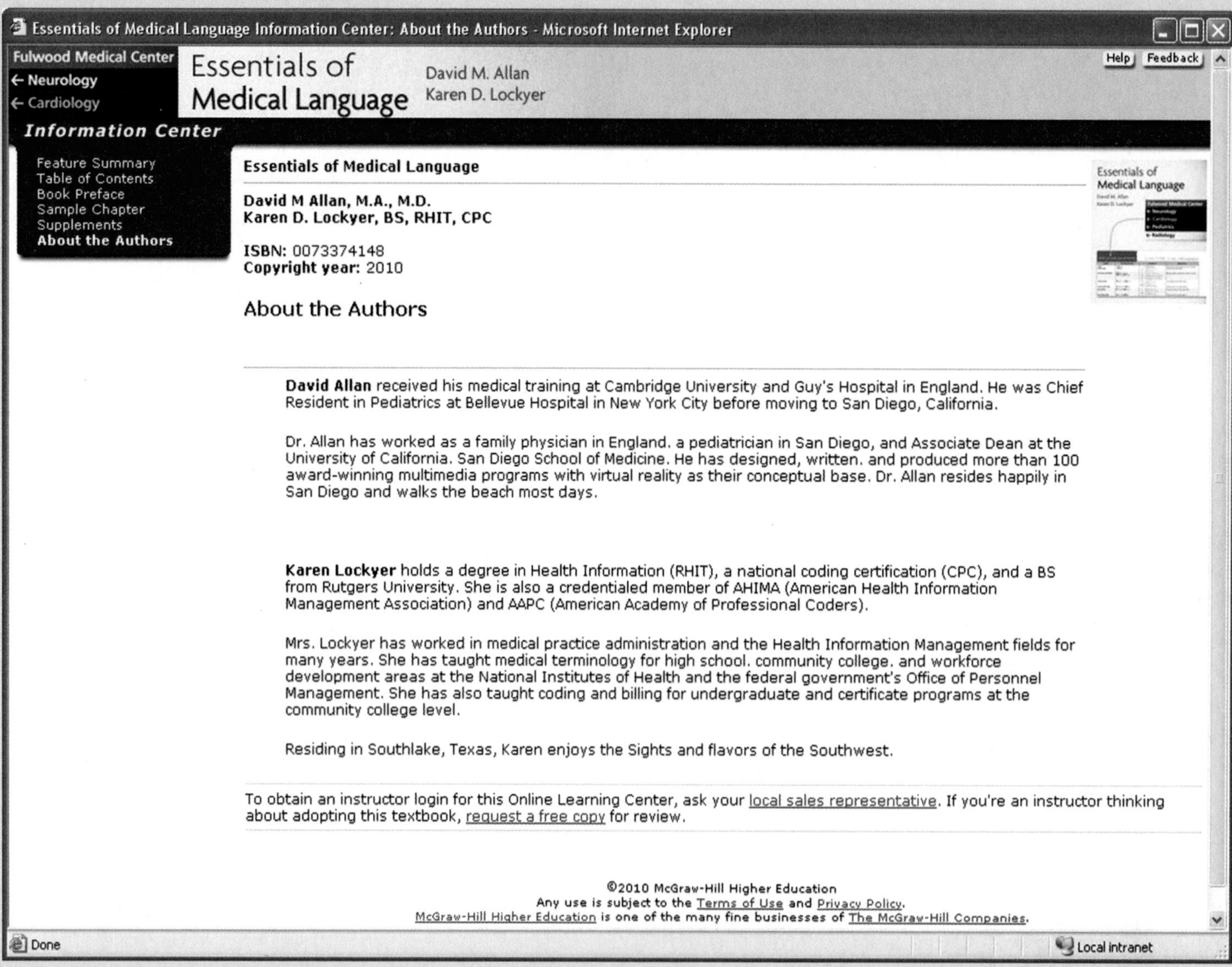

Mc Graw Hill connect™

McGraw-Hill *Connect*™ *ALLIED HEALTH*

McGraw-Hill *Connect ALLIED HEALTH* is a web-based assignment and assessment platform that gives students the means to better connect with their coursework, with their instructors, and with the important concepts that they will need to know for success now and in the future. With *Connect ALLIED HEALTH* instructors can deliver assignments, quizzes and tests easily online. Students can practice important skills at their own pace and on their own schedule. With *Connect ALLIED HEALTH Plus,* students also get 24/7 online access to an eBook – an online edition of the text – to aid them in successfully completing their work, wherever and whenever they choose.

McGraw-Hill LearnSmart™: Medical Terminology

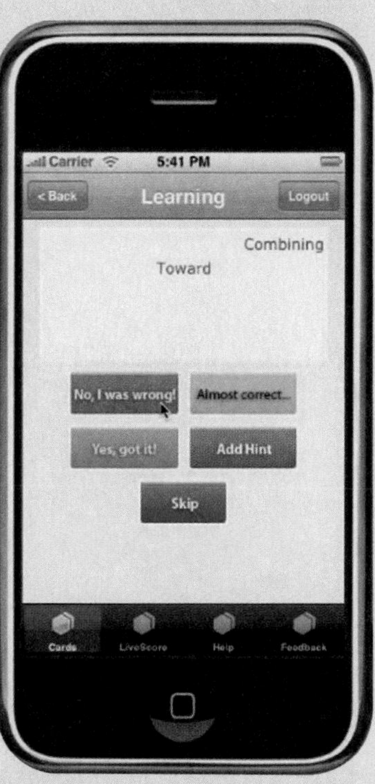

McGraw-Hill LearnSmart is a diagnostic learning system that determines the level of student knowledge, then feeds the student appropriate content. Students learn faster and study more efficiently.

As a student works within the system, LearnSmart develops a personal learning path adapted to what the student has learned and retained. LearnSmart is also able to recommend additional study resources to help the student master topics.

In addition to being an innovative, outstanding study tool, LearnSmart has features for instructors. There is a Course Gauge where the instructor can see exactly what students have accomplished as well as a built-in assessment tool for graded assignments.

Students and instructors will be able to access LearnSmart anywhere via a web browser. And for students on the go, it will also be available through any iPhone or iPod Touch.

Learning the Essentials of Medical Language

Welcome

CASE REPORT W.1

You are

. . . a student preparing for a career as a health professional and allied health care worker.

You are communicating with

. . . many different health professionals in health care teams as you go through an externship at Fulwood Medical Center. The center comprises a medical office building with physicians in a wide range of primary care, medical specialties, and complementary medicine therapies; a 300-bed hospital with a busy Emergency Room and operating rooms; a laboratory, pharmacy, X-Ray Department, Physiotherapy Department, and Patient Education Unit that serve both the hospital and the medical offices.

Between attending classes, doing your externship, working part-time, and bringing up two children, you have a full schedule. The knowledge and skills you are learning in the classroom and at Fulwood Medical Center will prepare you for a successful future.

Learning Outcomes

In order to get the most out of your learning experiences and this textbook, you need to:

- W-1 Recognize the need to learn medical terminology.
- W-2 Comprehend the value of learning medical terminology through the use of the realistic health care settings presented in this book.
- W-3 Develop effective study habits and organizational strategies for learning and work.
- W-4 Identify with the need for lifelong active learning as a health professional.

(*Note:* The pronunciations and meanings of the medical terms used in this Case Report are on the last page of this chapter.)

You are

... Luis Guitterez, a certified medical assistant (CMA) working with Susan Lee, MD, a primary care physician at Fulwood Medical Center.

You are communicating with

... Dr. Lee and Mrs. Martha Jones, a patient.

CASE REPORT W.2

Luiz Guitterez, CMA: Dr. Lee, this is Mrs. Martha Jones, who is a type 2 **diabetic** with **retinopathy** and **neuropathy.** She had a routine appointment with us today. Her temperature is 97.8, pulse 120, respirations 24, blood pressure 100/50.

Mrs. Martha Jones: Dr. Lee, I've had a cough and cold for the past few days, and today I'm feeling drowsy and nauseous and my chest hurts.

Dr. Lee: Did you give yourself your morning insulin?

Mrs. Jones: I can't remember.

Dr. Lee: Luis, she's confused, has **tachycardia** and **tachypnea,** and is **hypotensive.** I'm concerned she is going into diabetic **ketoacidosis.** Get the glucometer and test her blood glucose, while I examine her. She may have **pneumonia.**

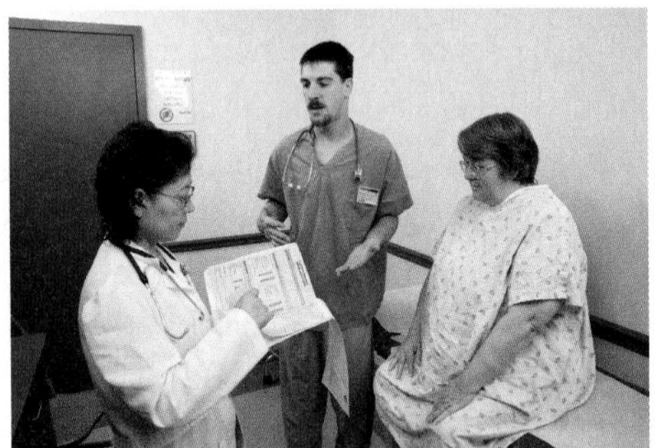

▲ **FIGURE W.1 Direct Communication with Doctor and Patient.**

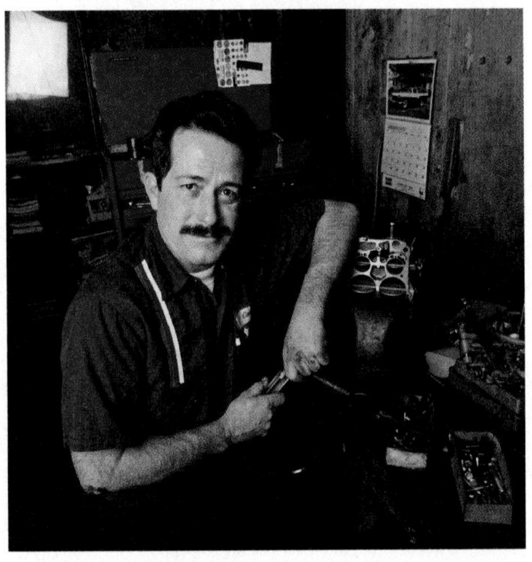

▲ **FIGURE W.2 Every Profession Has Its Own Language**
You may have difficulty understanding your auto mechanic when he tells you that the expansion valve, evaporator core, and orifice tubes in your air-conditioning system need replacing.

"WHY DO I NEED TO LEARN MEDICAL TERMINOLOGY?"

Communication Needs

Throughout your career as a health professional, you will need to communicate with other health professionals. This need is present whether you are providing direct patient care—for example, as a CMA like Luis Guitterez—or whether you are providing indirect patient care—for example, as a medical transcriptionist, biller, or coder. In this book, you will find all the medical terms necessary to equip yourself with the essential medical vocabulary needed for work and further study in any of the allied health professional careers.

As you can see in Case Report W.2, health professionals use specific terms and a different language to describe to each other situations they encounter each day. You need to be able to understand, spell, and pronounce the terms they use.

Modern medical terminology is an artificial language constructed over centuries using words and elements from Greek and Latin origins (where healing professions began). Some 15,000 or more words are formed from 1200 Greek and Latin roots. New words are being added continually as new medical discoveries are made. Medical terminology enables health professionals from different fields, different specialties, and different countries to communicate clearly and precisely with each other. Every profession has its own language (*Figure W.2*).

Listening, Speaking, Reading, Writing, and Critical Thinking

Daily in your practice as a health professional you will:

Listen to information from physicians about patient care, and carry out their instructions.

Listen to patients describing their symptoms, and translate their descriptions into medical terms.

Speak to physicians and other health professionals to report information and ask questions.

Speak to patients to translate and clarify information given to them by physicians and other health professionals.

Read physicians' comments and treatment plans in patient medical records and insurance reports.

Read the results of physical examinations, procedures, and laboratory and diagnostic tests.

Write to document actions taken by yourself and other members of the health care team *(Figure W.3)*.

Write to precisely record verbal orders, test results given over the phone, and other phone messages.

Think critically to evaluate medical documentation for accuracy.

Think critically to analyze and discover the meaning of unfamiliar medical terms using the strategies outlined in *Chapter 1* of this book.

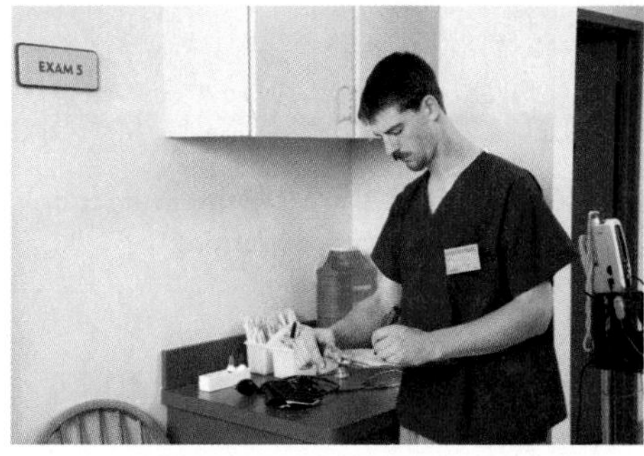

▲ **FIGURE W.3 Accurate Documentation of Care Is Critical.**

If you cannot speak the language, you cannot join the club.

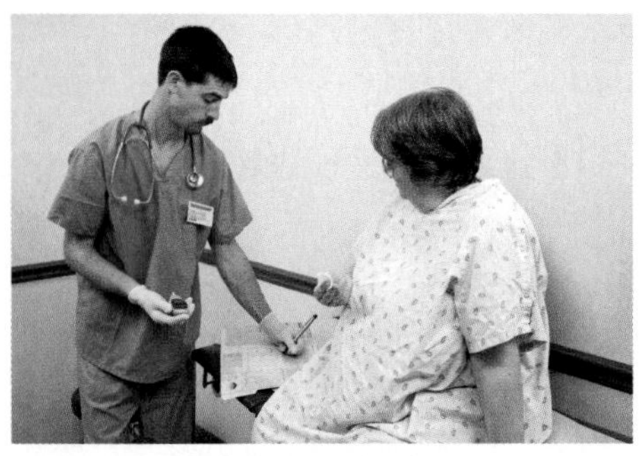

▲ **FIGURE W.4 Every Patient Interaction Is an Opportunity for Learning.**

"WHAT IS LIFELONG, ACTIVE LEARNING?"

Lifelong Learning

Your current training in medical terminology is necessary for you to be able to continue your education in your health care profession. But it is important to recognize that school is only one of the many places where you acquire knowledge. You also acquire knowledge:

- Each time you ask a question about a patient or a report and receive an answer.
- Each time you analyze an unfamiliar medical term and discover its meaning.
- Each time you interact with a patient and see how that patient is coping with his or her problems (*Figure W.4*).

All these are opportunities for learning to discover ***your own*** answers to ***your own*** problems or lack of knowledge.

This type of knowledge—discovered through your own experience and driven by your own needs and goals—is genuine, real, and trustworthy for you. It is not like what you learn in school, which is determined by some distant authority.

The authentic knowledge you gain from solving your own problems, whether by yourself or with the help of other people or resources, motivates you to acquire still more knowledge and helps you grow as a person and as a professional.

Throughout your working life, additional classroom training will be needed to keep your skills and professional knowledge up to date with new developments in medicine. You will also continue to learn through your own experience. Everything you do in life can result in learning.

Your own experience and judgment become your most valuable resources for making your life vibrant, strong, creative, and what ***you*** want it to be.

Your own experience and judgment maximize your professional and personal success.

Keynote

Novelist Lillian Smith said, "When you stop learning, stop listening, stop looking, and stop asking new questions, then it is time to die."

ACTIVE LEARNING

It's no good sitting back and expecting someone else to pour knowledge into your head. You have to **actively work at learning.**

Get the Most Out of Lectures

- *Prepare* for your classroom experiences. Preview the book chapter before class, and the material will be much easier to understand.
- *Listen actively.* You cannot do this if you are looking at your cell phone, daydreaming, or worrying about what you have to get for dinner.
- *Ask* a question if you do not comprehend something the instructor is saying.
- *Write* good notes. Focus on the main points, and capture key ideas; review and edit your notes within 24 hours of the class.

Study Hint

The SQ3R model for reading is a successful equation for studying:

Survey what you are going to read.

Question what you are going to learn after the preview.

Read the assignment.

Recite. Stop every once in a while, look up from the book, and put what you've just read into your own words.

Review. After you've finished, review the main points.

Get the Most Out of Reading

- *Concentrate* on what you are reading. Review the titles, objectives, headings, and visuals for each lesson to identify what the lesson is all about.
- *Read actively* using the SQ3R method (see the Study Hint) to help you.
- *Write* down any questions you have.

Study with a Partner or Group

- *Find* a study partner. Schedule study dates, compare notes, talk through concepts and questions, and quiz each other.
- *Establish* a small study group, including your study partner. Again, compare notes and quiz each other.

Perform Well on Tests

- *Read* the directions carefully, and scan the entire test so that you know how long it is and what types of activities it contains.
- *Answer* the easy questions or sections first so that you finish as much as possible before doing the difficult questions, which might slow you down.
- *Use* any extra time, after you have finished the test, to check that you have answered all the questions and then to confirm your answers.

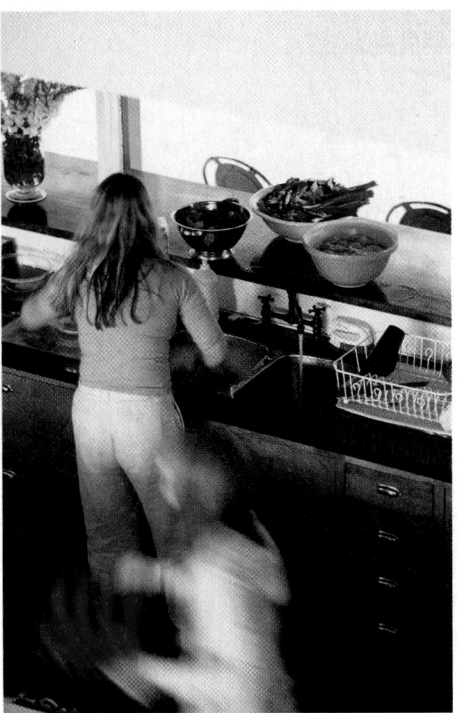

▲ **FIGURE W.5** **An Evening at Home.**

Keynote

You are what you put into your studies.

Case Report W.3

Your first day of externship at Fulwood Medical Center went well. You enjoyed being in the Primary Care Clinic with Dr. Lee and Luis Guitterez, CMA. You wonder if this could be a career choice for you. Now it's 6:15 p.m. at home, and you have yet to feed the kids, get them to bed, pay some bills, pick up around the house, and review a whole chapter in your medical terminology text to prepare for a test in class tomorrow. How are you going to do all this?

"HOW CAN I HELP MYSELF LEARN BETTER?"

You have a lot of time and money invested in your education. To succeed, you need to be able to focus and manage your time and your studies. To manage the difficulties described in Case Report W.3 (*Figure W.5*), you need to:

- ***Recognize*** the stresses in your life at different times.
- ***Prioritize*** mentally, and handle each task in the order of importance.

In this case, eat a healthy meal with your kids, enjoy putting them to bed, pay the bills, and then relax (or meditate) for 10 minutes. When you are relaxed, settle down to review the text, and go to bed at a reasonable hour. Picking up around the house will have to wait, since study and sleep are a higher priority. Sounds too easy? What other choices do you have to be able to study in an effective way?

- ***Actively develop a support group.*** Enlist the support of your spouse, mother, siblings, friends—any people you can trust and rely on. If you have a test every Thursday, get one of them to come over Wednesday night and put the kids to bed while you go over to his or her house or the library to study.
- ***Find your own space.*** Create a place where you keep everything for your courses at your fingertips, clutter-free.
- ***Study when you are most productive.*** Are you a night owl or an early bird? Set a daily study time for yourself.
- ***Balance your life.*** While studying should be a main focus, plan time for family, friends, leisure, exercise, and sleep.
- ***Resist distractions.*** Avoid the temptation to surf the Web, send instant messages, and make phone calls. Stick to your schedule.
- ***Be realistic*** when planning—know your limits and priorities.
- ***Be prepared*** for the unexpected (child's illness, your illness, inclement weather) that can turn your schedule into shambles.
- ***Reprioritize*** daily on the basis of schedule disruptions and other conflicts.
- ***Identify*** clear goals for what you need to get done today, this week, this month, before the end of the semester, and so on.

EXERCISES **Write out** all of your activities for a typical week. On average, how many hours each week do you spend sleeping, grooming, eating, working, running errands, studying, attending your children's activities, and watching TV? Add all the hours up. There are 168 hours in the week. How many hours do you have left for studying? A sample time budget is shown below.

Activity	Number of Hours per Day	Number of Days per Week	Number of Hours per Week
Sleeping	8	7	56
Grooming	1	7	7
Meals: preparation, eating, cleanup	1	7	7
Cleaning, laundry	1	3	3
Commuting to and from school	1	5	5
In class	4	5	20
Doing errands	1	3	3
Family time	3	7	21
Church, workout, hobbies			5
Job			30
Friends, going out, TV, entertainment			6
TOTAL			163
TOTAL HOURS IN A WEEK			168
Hours remaining for study			5

- *Are 5 hours enough for study?*
- *When are they available?*
- *What can you do to increase them?*

Study hours should be spent in a setting that allows you to concentrate on your work and not be distracted. Turn off your cell phone and TV. The biggest question to ask yourself is, "Am I investing my time wisely?" If not, how can you budget your time differently so that more time is spent on higher-priority activities?

"WHY SHOULD I CHOOSE THIS BOOK?"

Although the chapters in this book are organized by body system, as in many other textbooks on medical terminology, this book has many unique features that enhance learning, create interest, and provide a consistent learning strategy for you.

Each chapter is broken down into lessons; each lesson is broken down into self-contained topic areas so that there are smaller "chunks" of information to master.

You Are . . . You Are Communicating With . . .

CHAPTER

6

CASE REPORT 6.1

You are

. . . a **cardiovascular** technologist (CVT) employed by the **Cardiology** Department at Fulwood Medical Center. You have been called to the Emergency Department (ED) to take an **electrocardiogram (ECG** or **EKG)**, STAT.

You are communicating with

. . . Mr. Hank Johnson, the 64-year-old owner of a printing company. Eight months ago he had a left total hip replacement. In the past 3 months, Mr. Johnson has returned to his daily workouts. This morning, while riding his exercise bike, he felt tightness in his chest. He kept on cycling. He developed pain in the center of his chest, radiating down his left arm and up into his jaw, and became **diaphoretic.** His personal trainer called 911.

You perform the ECG and the automatic report describes abnormalities in the chest leads. As you remove the **electrodes,** Mr. Johnson complains that he is feeling faint and short of breath **(SOB)**. You are the only person in the room.

Learning Outcomes

No matter which discipline or setting you work in as a health professional, the condition of the patient's cardiovascular system will always be a factor during diagnosis, treatment, and communication. In order to be able to understand, communicate, and document conditions affecting the heart, blood vessels, and blood, you need to be able to:

6.1 Apply the language of cardiology to the structure and functions of the cardiovascular system.

6.2 Comprehend, analyze, spell, and write the terms of cardiology so that you communicate and document accurately.

6.3 Recognize and pronounce the medical terms of cardiology so that you communicate verbally with accuracy and precision.

6.4 Specify the correct medical terminology for common disorders of the cardiovascular system.

At the beginning of each chapter and lesson, you are placed in the role of a health professional in a field related to the body system and medical specialty covered in the material. At the same time, learning objectives are presented for each chapter and lesson. These techniques immediately engage your attention, motivate you to read on to discover how this patient's diagnosis and care progress, and illustrate the medical terminology being introduced in the lessons.

Word Analysis And Definition

All the information needed for a topic area is presented in self-contained two-page spreads. On the left-hand page, the new medical terms are introduced. On the right-hand page, for each new medical term the pronunciation, color-coded word elements, and definition are provided in a **Word Analysis and Definition (WAD)** box. For example, in Case Report W.2 earlier in this chapter, the medical terms *diabetic, retinopathy, neuropathy, tachycardia, tachypnea, hypotensive, ketoacidosis, glucometer,* and *pneumonia* were used. On the right-hand page here, you can see an example of how these terms are analyzed. All these terms will appear again in the appropriate body-system chapter.

Also, below each WAD are exercises that test your understanding of key components of the terminology analyzed in the WAD.

Exercises

In addition to the exercises at the end of each topic area, there are six pages of exercises at the end of each chapter and twenty pages of exercises in Appendix A at the end of the book. After every four chapters, four pages of exercises review the key material in those four chapters. More exercises are found on the Student Online Learning Center (www.mhhe.com/AllanEssMedLanguage).

The exercises are designed to help different styles of learners understand the logic of medical terminology. Exercises take comprehension beyond the basic level, and show you the practical application of the terminology you are learning in the classroom, and its use in the workplace. Constructing and deconstructing medical terms into their elements is the basic foundation on which this knowledge is built. Attention is given to developing skills in pronunciation, spelling, forming plurals, using abbreviations, and writing medical language. The exercises take you beyond memorization and teach you to think critically about the realistic application of the medical language you are learning.

Abbreviations

AMA	American Medical Association
CMA	certified medical assistant

Additional Unique Features

Keynotes emphasize key points in the text or offer additional important information.

Study hints show you how to use basic study skills, and they provide ways by which to retain information.

Abbreviation boxes show common abbreviations accepted by The Joint Commission and AMA.

Illustrations and **photos** are vivid and colorful and are correlated precisely to the medical terminology of the lessons.

Keynote

Key points and additional information are offered here.

Study Hint

Anything that is referred to as the "most powerful," " largest," "smallest," "most common," and the like, is probably going to be a test question.

WORD ANALYSIS AND DEFINITION

S/ = Suffix P/ = Prefix R/ = Root R/CF = Combining Form

WORD	PRONUNCIATION	ELEMENTS		DEFINITION
diabetes mellitus	dye-ah-**BEE**-teez **MEL**-ih-tus		diabetes, Greek *a siphon* mellitus, Latin *sweetened with honey*	Metabolic syndrome caused by absolute or relative insulin deficiency and/or insulin ineffectiveness.
diabetic (adj)	dye-ah-**BET**-ik	S/ R/	**-ic** *pertaining to* **diabet-** *diabetes*	Pertaining to or suffering from diabetes.
hypotension	**HIGH**-poh-**TEN**-shun	S/ P/ R/	**-ion** *action, condition* **hypo-** *below* **-tens-** *pressure*	Persistent low arterial blood pressure.
hypotensive (adj)	**HIGH**-poh-**TEN**-siv	S/	**-ive** *pertaining to, quality of*	Pertaining to or suffering from hypotension.
ketoacidosis	**KEY**-toe-ass-ih-**DOE**-sis	S/ R/CF R/CF	**-sis** *abnormal condition* **ket/o-** *ketone* **-acid/o-** *acid*	Excessive production of ketones, making the blood acid.
neuropathy	nyu-**ROP**-ah-thee	S/ R/CF	**-pathy** *disease* **neur/o-** *nerve*	Any disorder affecting the nervous system.
pneumonia (***Note:*** The initial "p" is silent.)	new-**MOH**-nee-ah	S/ R/	**-ia** *condition* **pneumon-** *air, lung*	Inflammation of the lung parenchyma.
retinopathy	ret-ih-**NOP**-ah-thee	S/ R/CF	**-pathy** *disease* **retin/o-** *retina*	Any disease of the retina.
tachycardia	tak-ih-**KAR**-dee-ah	S/ P/ R/	**-ia** *condition* **tachy-** *rapid* **-card-** *heart*	Rapid heart rate, above 100 beats per minute.
tachypnea	tak-ip-**NEE**-ah	P/ R/	**tachy-** *rapid* **-pnea** *breathe*	Rapid breathing.

The elements of a term are discussed in Chapter 1.

EXERCISES

Elements *are your best tool for understanding medical terms. In the chart below, the elements are listed in column 1. Identify the meaning of each element in column 2, and give an example of a term containing that element in column 3. Some terms will apply to more than one element. The first one is done for you.*

Element	Meaning of Element	Medical Term Containing This Element
hypo	below	hypotension
tens		
ion		
neuro		
retino		
pathy		
ia		
pneumon		
pnea		
tachy		

1. Choose any term from column 3, and use it in a sentence of your choice:

___

___

___

The Anatomy of Word Construction

The Essential Elements of the Language of Medicine

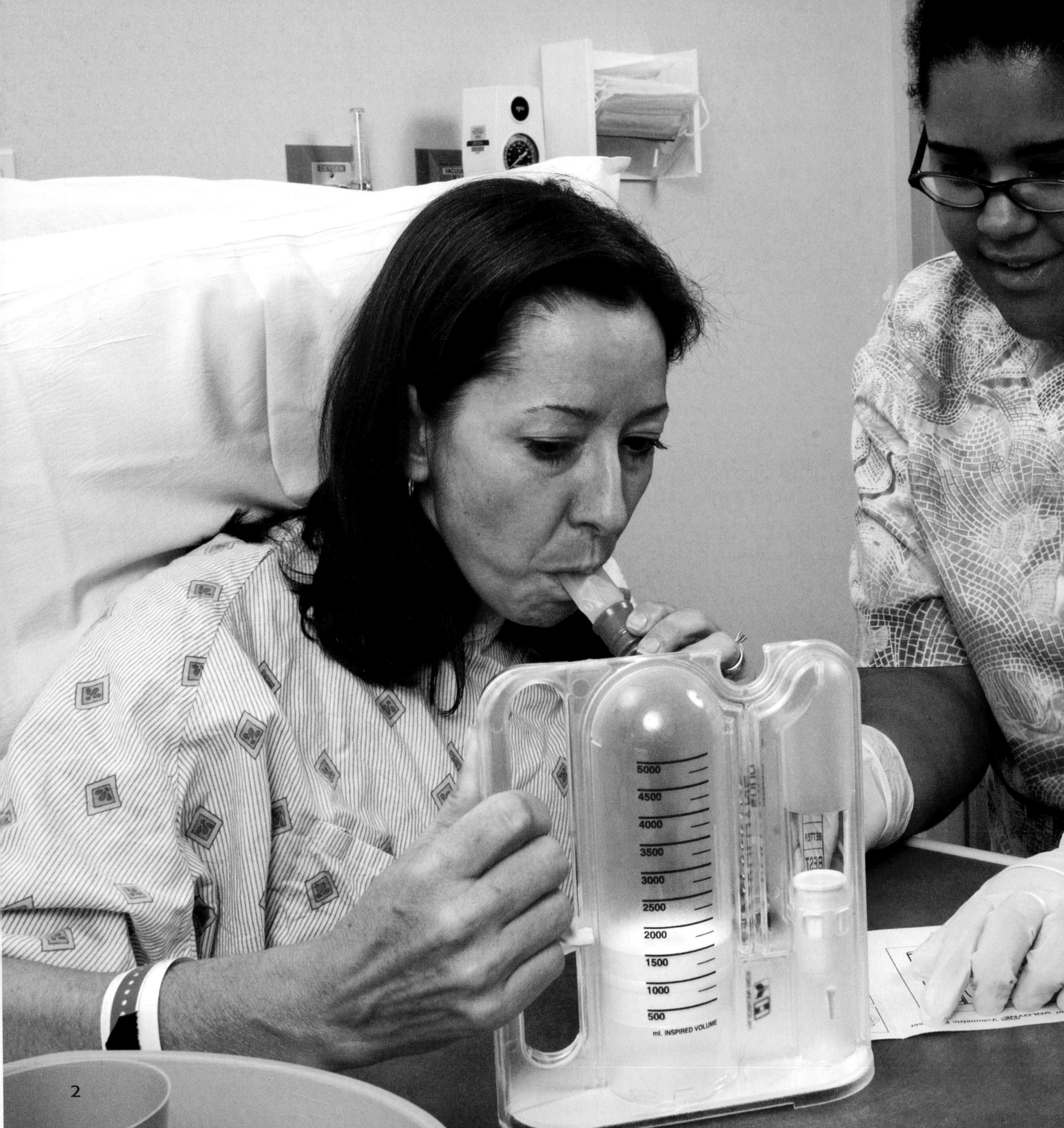

CHAPTER

CASE REPORT 1.1

You are

. . . a respiratory therapist working with Tavis Senko, MD, a pulmonologist at Fulwood Medical Center.

You are communicating with

. . . Mrs. Sandra Schwartz, a 43-year-old woman referred to Dr. Senko by her primary care physician, Dr. Andrew McDonald, an internist. She has a persistent abnormality on her chest x-ray. You have been asked to determine her pulmonary function prior to a scheduled bronchoscopy.

This summary of a Case Report illustrates for you the use of some simple medical terms. Modern health care and medicine has its own language. The medical terms all have precise meanings, which enable you, as a health professional, to communicate clearly and accurately with other health professionals involved in the care of a patient. This communication is critical for patient safety and the delivery of high-quality patient care.

Learning Outcomes

The technical language of medicine has been developed logically from Latin and Greek roots, for it was in Latin and Greek cultures that the concept of treating patients began. Medical terms are built from their individual parts, or **elements,** which form the **anatomy** of the word. The information in this chapter will enable you to:

1.1 Identify the root, combining vowels, and combining forms of medical terms.

1.2 Understand the importance of suffixes and prefixes in forming medical terms.

1.3 Link word elements together to form medical terms.

1.4 Break down or deconstruct a medical term into its elements.

1.5 Connect the singular and plural forms of medical terms.

1.6 Verbalize the pronunciation of medical terms by employing the system used in the textbook and the Student Online Learning Center (www.mhhe.com/AllanEssMedLanguage) for reference.

LESSON 1.1 The Construction of Medical Words

THE ANATOMY OF WORD CONSTRUCTION

OBJECTIVES

Your confidence in using and understanding medical terms will increase as you understand the logic of how these terms are built from their individual parts or elements. The information in this lesson will enable you to:

- **Select and identify the meaning of the *roots* of essential medical terms.**
- **Define the elements *combining vowel* and *combining form*.**
- **Identify the combining vowel and combining form of essential medical terms.**
- **Define the elements *suffix* and *prefix*.**
- **Select and identify the meaning of the suffixes and prefixes of essential medical terms.**

ROOTS AND COMBINING VOWELS

Case Report 1.1 (continued)

From her medical records, you see that 2 months ago Mrs. Schwartz developed a right upper lobe (RUL) *pneumonia*. After treatment with an antibiotic, a follow-up chest x-ray (CXR) showed some residual collapse in the right upper lobe and a small right *pneumothorax*. Mrs. Schwartz has smoked a pack a day since she was a teenager. Dr. Senko is concerned that she has lung cancer, and he has scheduled her for bronchoscopy.

ROOTS:

- A root is the constant foundation and core of a medical term.
- Roots are usually of Greek or Latin origin.
- All medical terms have *one* or *more* roots.
- A root can appear anywhere in the term.
- More than one root can have the same meaning.
- A root plus a combining vowel creates a combining form.

Keynote

Different roots can have the same meaning. Pulmon- and pneumon- both mean *lung*.

Every medical term has a root, the element that provides the core meaning of the word. For example, in Case Report 1.1:

- The word *pneumonia* has the root *pneumon-*, taken from the Greek word meaning *lung* or *air. Pneumonia* is an infection of the lung tissue. The Greek root *pneum-* is also used to mean *lung* or *air.*
- Dr. Tavis Senko is a *pulmonologist;* the root *pulmon-* is taken from the Latin word meaning *lung.* A *pulmonologist* is a specialist in treating diseases of the lung.

Combining Forms

Roots are often joined to other elements in the medical term by placing a combining vowel on the end of the root.

For example, if the vowel "o" is added to the root pneum-, the combining form pneum/o- is made. Throughout this textbook, the combining vowel will be separated from the root by a slash (/) whenever the term is being analyzed.

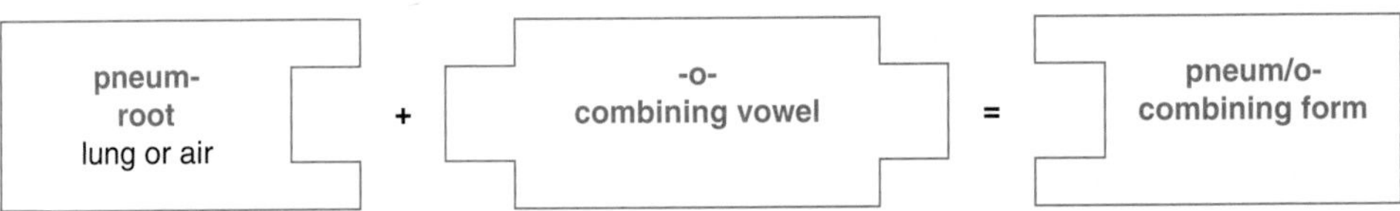

- The Latin root pulmon- can also have the combining vowel "o" added to make the combining form pulmon/o-:

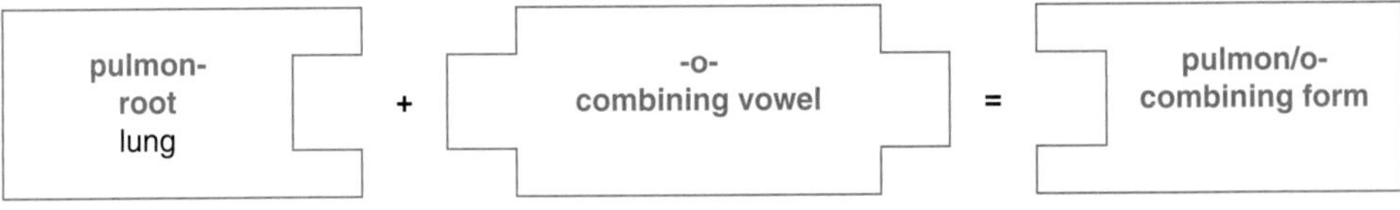

Any vowel, "a," "e," "i," "o," or "u," can be used as a combining vowel.

- The root respir- means *to breathe.* Adding the combining vowel "a" makes the combining form respir/a-:

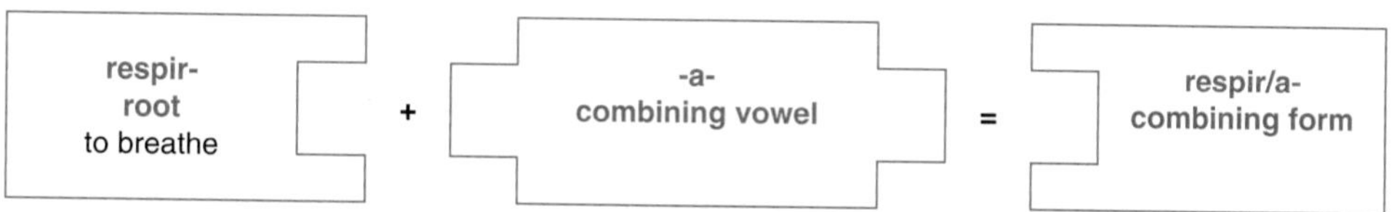

COMBINING FORMS:

- Combine a root and a combining vowel.
- Can be attached to another root or combining vowel.
- Can precede another word element called a suffix.
- Can follow a prefix.

Many medical terms contain more than one root; when two roots occur together, they are always joined by a combining vowel:

- The word **pneumothorax** has the root *pneum-*, from the Greek word meaning *air* or *lung,* and the root *-thorax*, from the Greek word meaning *chest.* A **pneumothorax** is the presence of air in the space that surrounds the lungs in the chest. The combining vowel "o" is used to join the two roots together. The root and the combining vowel together make the combining form, pneum/o-.

pneum- root lung or air + -o- combining vowel + -thorax root chest = **pneumothorax** air in the chest

EXERCISES

Review: *what you have just learned about the roots and combining forms on the two pages spread open in front of you. Fill in the blanks.*

1. Which element is the core or foundation of every medical term? ______
2. Give two examples of the element named in question #1 above: ______
3. If a combining vowel is added to the element in question #1, what is the name of the new element? ______
4. Give an example of a root that has become a combining form:

   ______ + ______ = ______

   Root | Combining vowel | Combining form (Don't forget the slash!)

5. More than one element can have the same meaning in medical terminology. Give an example from the elements on these two pages:

   ______ and ______ both mean ______

6. Give an example of a term using each element in the above question:

   ______ means ______

   ______ means ______

7. Practice using the terms in question #6 above. Write one sentence (that is not directly out of the text) for each of the terms.

   (a) ______

   (b) ______

8. The following terms have not been introduced yet, but the principle remains the same. The root/combining form will always carry the same meaning. Circle the roots/combining forms that you recognize, and provide the meaning on the blank line.

8a. *Pneumonectomy, pneumonitis,* and *pneumococcal* all pertain to ______

8b. *Respirator* and *respiratory* both pertain to ______

Study Hint

Even though both these elements mean the same thing, they are not interchangeable. Only certain elements belong with certain terms, and you must know them.

SUFFIXES

A suffix is an element added to the end of a root or combining form to give it a new meaning. Different suffixes can be added to the same root to build new words, all with different meanings. For example:

- Add the suffix *-ary* to the root *pulmon-* to form the term **pulmonary.** The suffix -ary means *pertaining to* or *relating to.* The adjective **pulmonary** means *pertaining to the lung,* as in the term *pulmonary circulation,* the passage of blood through the lungs.

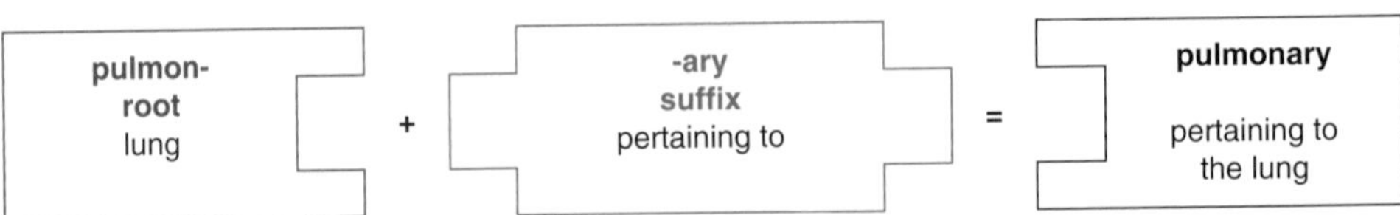

- Add the suffix -logy to the combining form pulmon/o- to form the term **pulmonology.** The suffix -logy means *study of,* and **pulmonology** is the study of the structure, functions, and diseases of the lungs.

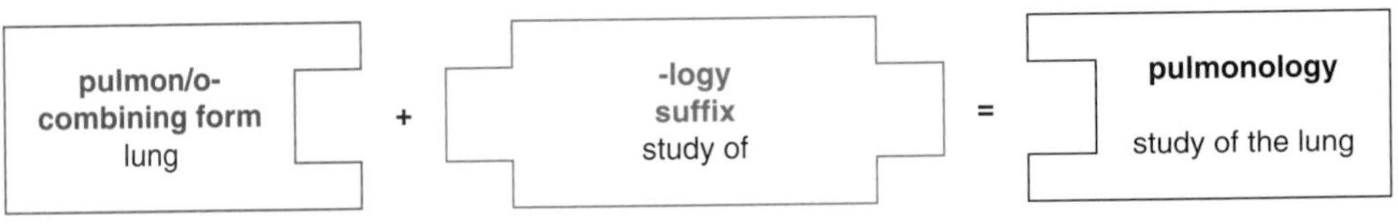

- Add the suffix -ia to the root pneumon- to form the term **pneumonia.** The suffix -ia means *a condition of,* and **pneumonia** is a condition of the lungs. The term is used to describe an infection of the lung tissue.

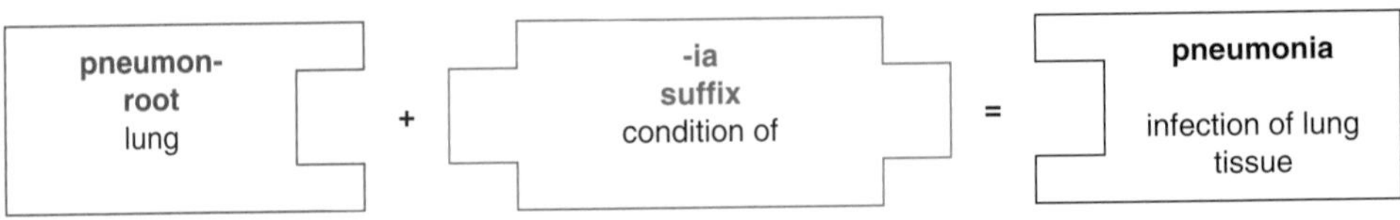

- Add the suffix -ation to the root respir- to make the term **respiration.** The suffix -ation means a process, and the term **respiration** is the process of breathing in and out.

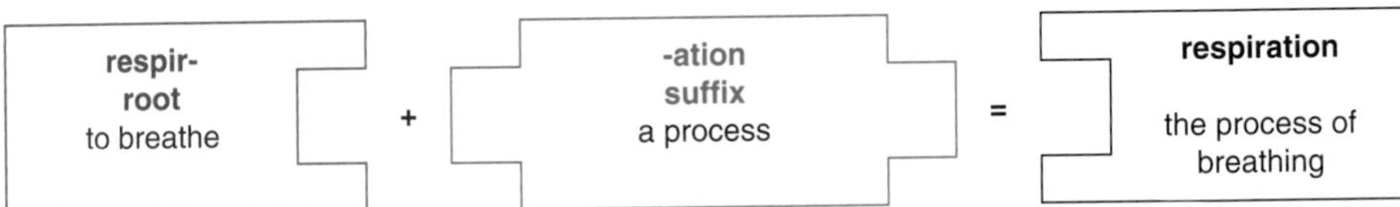

Whereas most roots are specific to body systems and medical specialties, suffixes are universal and are used across all body systems and specialties.

SUFFIX:

- A suffix is a group of letters attached to the end of a root or combining form.
- A suffix changes the meaning of the word.
- If the suffix begins with a consonant, it must follow a combining vowel.
- If the suffix begins with a vowel, no combining vowel is needed.
- A few medical terms can have two suffixes.
- A suffix always appears at the end of a term.
- Suffixes that are different can have the same meaning.

EXERCISES

Elements: *It is important for you to recognize the identity of an element. Is it a root, combining form, or suffix? This will help you to determine its place in the term when you are building terms.*

1. Build the appropriate medical term to match the definitions given. The placement of the elements is noted for you under the line; each different element is separated on the line. Write the correct elements on the line.

   a. Study of the lungs:

   ________________________ / ________________________

   Root or combining form — Suffix

   b. Pertaining to the lung:

   ________________________ / ________________________

   Root or combining form — Suffix

   c. The process of breathing:

   ________________________ / ________________________

   Root or combining form — Suffix

   d. Condition of the lung:

   ________________________ / ________________________

   Root or combining form — Suffix

   e. Use any one of the preceding terms in a sentence of your choice—one that is *not* a definition from above.

   ____

   ____

   ____

   ____

   ____

   ____

   f. Choose another term from above and use it in patient documentation that you write below.

   ____

   ____

   ____

   ____

   ____

   ____

PREFIXES:

- A prefix precedes a root to change its meaning.
- Prefixes can have more than one meaning.
- Prefixes never require a combining vowel.
- An occasional medical term can have two prefixes.
- A prefix always appears at the beginning of a term.
- Not every term has a prefix.

PREFIXES

A prefix is an element added to the beginning of a root or combining form to continue to expand the meaning of medical terms. Prefixes usually indicate time, number, color, or location. Examples of prefixes defining time are described below:

- The term ***mature*** can refer to an infant born after a normal length of pregnancy of between 37 and 42 weeks.
  - An infant born before 37 weeks is called ***premature***; the prefix pre- means *before,* and **premature** means that the infant was *born before 37 weeks.*
  - An infant born after 42 weeks is called ***postmature***; the prefix post- means *after,* and **postmature** means that the infant was *born after 42 weeks.*

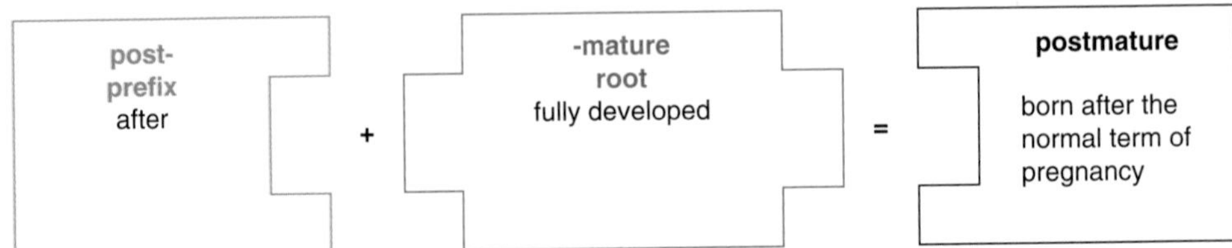

- The term ***natal*** contains the root nat-, which means *birth* or *born,* and the suffix -al, which means *pertaining to.* The term **natal** means *pertaining to birth.*
  - Add the prefix pre-, which means *before,* to form the term **prenatal,** which means *the time before the birth.*
  - Add the prefix post-, which means *after,* to form the term **postnatal,** which means *the time after birth.*
  - Add the prefix peri-, which means *around,* to form the term **perinatal,** which means *around the time of birth* and includes *the time immediately before, during, and immediately after birth.*

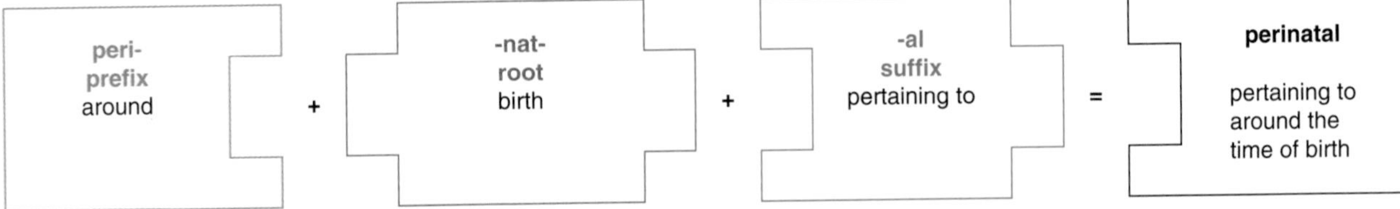

Examples of prefixes indicating number are described below:

- The term ***lateral*** contains the root later-, which means *side,* and the suffix -al, which means *pertaining to.* The term **lateral** means *pertaining to a side of the body.*
  - Add the prefix uni-, which means *one,* to form the term **unilateral,** which means *pertaining to one side of the body only.*
  - Add the prefix bi-, which means *two,* to form the term **bilateral,** which means *pertaining to both sides of the body.*

PRACTICAL POINTS:

- A root can start a term and does not become a prefix.
- A root can end a term and does not become a suffix.
- An example of both of these is *pneumothorax.*

Examples of prefixes indicating location are described below:

- The term ***gastric*** contains the root gastr-, which means *stomach,* and the suffix -ic, which means *pertaining to.* The term **gastric** means *pertaining to the stomach.*
  - Add the prefix epi-, which means *above,* to form the term **epigastric,** which means *pertaining to above the stomach.*
  - Add the prefix hypo-, which means *below,* to form the term **hypogastric,** which means *pertaining to below the stomach.*

Examples of prefixes indicating size are described below:

- The root -cyte means *cell.*
  - Add the prefix macro-, which means *large,* to form the term **macrocyte,** which means *a large red blood cell.*
  - Add the prefix micro-, which means *small,* to form the term **microcyte,** which means *a small red blood cell.*

WORD ANALYSIS AND DEFINITION

S/ = Suffix   P/ = Prefix   R/ = Root   R/CF = Combining Form

WORD	PRONUNCIATION	ELEMENTS		DEFINITION
gastric	**GAS**-trik	S/ R/	**-ic** *pertaining to* **gastr-** *stomach*	Pertaining to the stomach.
epigastric **hypogastric**	ep-ih-**GAS**-trik high-poh-**GAS**-trik	P/ P/	**epi-** *above* **hypo-** *below*	Abdominal region above the stomach. Abdominal region below the stomach.
lateral	**LAT**-er-al	S/ R/	**-al** *pertaining to* **later-** *side*	Pertaining to one side of the body.
bilateral **unilateral**	by-**LAT**-er-al you-nih-**LAT**-er-al	P/ P/	**bi-** *two* **uni-** *one*	Pertaining to both sides of the body. Pertaining to one side of the body only.
macrocyte	**MACK**-roh-site	P/ R/	**macro-** *large* **-cyte** *cell*	Large red blood cell.
macrocytic (adj) (***Note:*** The "e" in *cyte* is deleted to allow the word to flow.)	mack-roh-**SIT**-ik	S/	**-ic** *pertaining to*	Pertaining to a macrocyte.
mature	mah-**TYUR**		Latin *ready*	Fully developed.
postmature	post-mah-**TYUR**	P/ R/	**post-** *after* **-mature** *fully developed*	Infant born after 42 weeks of gestation.
premature	pree-mah-**TYUR**	P/	**pre-** *before*	Occurring before the expected time; e.g., an infant born before 37 weeks of gestation.
microcyte	**MY**-kroh-site	P/ R/	**micro-** *small* **-cyte** *cell*	Small red blood cell.
microcytic (adj) (***Note:*** The "e" in *cyte* is deleted to allow the word to flow.)	my-kroh-**SIT**-ik	S/	**-ic** *pertaining to*	Pertaining to a small red blood cell.
natal	**NAY**-tal	S/ R/	**-al** *pertaining to* **nat-** *birth, born*	Pertaining to birth.
perinatal **postnatal** **prenatal**	per-ih-**NAY**-tal post-**NAY**-tal pree-**NAY**-tal	P/ P/ P/	**peri-** *around* **post-** *after* **pre-** *before*	Around the time of birth. After the birth. Before the birth.

One of the key design concepts of this book is that all the textual and visual information that you need for any given topic will be on the two pages spread open in front of you. As part of this, in the top right-hand quarter of each two-page spread will be a Word Analysis and Definition (WAD) box designed to provide the elements, definition, and pronunciation of every new and repeated significant medical term that is given in the two pages you are reviewing. This box is always shown on the right-hand page.

Review all the terms in the WAD before you start any exercise.

EXERCISES

Prefixes: *Solid knowledge of prefixes will quickly help increase your medical vocabulary. A good example is the Word Analysis and Definition (WAD) entry for* natal. *The addition of three different prefixes builds three new medical terms. To begin this exercise, underline every prefix in the bolded terms below. Answer the first question, and then write the correct term on the line next to the definitions in 2 through 4. Follow the instructions for question 5.*

natal: **prenatal** **postnatal** **perinatal**

1. The term *natal* means ____________________
2. Pertaining to around the time of birth: ____________________
3. Pertaining to after the birth: ____________________
4. Pertaining to before the birth: ____________________
5. In 2 through 4 above, underline the word in the definition that is the clue to the correct prefix to use in the term.

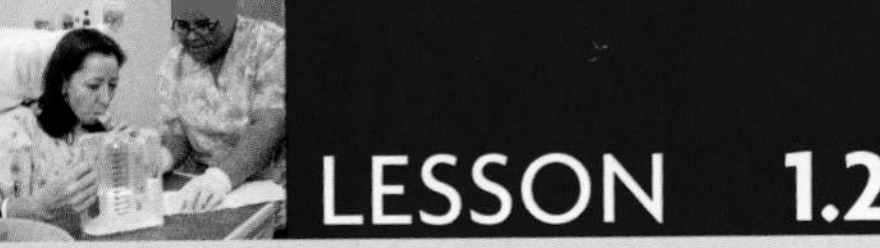

LESSON 1.2 Word Analysis and Deconstruction

OBJECTIVES

When you see an unfamiliar medical term, the process of analyzing or deconstructing the term involves breaking it down into its component elements to give you the meaning of the term. In this lesson you will learn to:

- **Break down or deconstruct a medical term into its elements.**
- **Use the word elements to analyze and determine the meaning of the term.**
- **Recognize the importance of precision in both verbal and written communication.**
- **Utilize word analysis to help ensure the precise use of medical terms.**
- **Apply the correct pronunciation to medical terms.**

You are

... a medical assistant working in the office of Lokesh Bannerjee, MD, a cardiologist in Fulwood Medical Center.

You are communicating with

... the 70-year-old wife and the 45-year-old son of James Donovan, a 75-year-old man who is to be admitted to the hospital's acute care *cardiology* unit.

CASE REPORT 1.2

For Mr. Donovan, Dr. Bannerjee has diagnosed an acute *myocardial infarction* (AMI), confirmed by changes in the patient's *electrocardiogram* (ECG/EKG). One of your tasks is to explain the diagnosis and reasons for admission to the hospital to Mrs. Donovan and her son.

As part of the prehospital care given to Mr. Donovan, the advanced cardiac life support (ACLS) protocol treatment of morphine, oxygen, nitroglycerin, and aspirin (MONA) was administered. This protocol is used for all patients with chest pain caused by myocardial *ischemia.*

The *italicized* terms in the Case Report are used as examples in the following text and are deconstructed in the Word Analysis and Definition box on the opposite page.

WORD DECONSTRUCTION

Keynote

Always begin deconstructing a medical term by identifying the **suffix.**

When you see an unfamiliar medical term, first identify the suffix. For example, in the term **cardiologist** the suffix at the end of the word is -logist, which means *one who studies and is a specialist in.* That leaves the element **cardi/o-**, which is the combining form for *heart.* The term **cardiologist** means *a specialist in the heart and its diseases.* It has a combining form and a suffix.

In the term **myocardial,** the suffix at the end of the word is -al, which you learned earlier in this chapter means *pertaining to.* At the beginning of the word is the combining form my/o-, which means *muscle.* In the middle of the word is the root -cardi-, which means *heart.* The term **myocardial** means *pertaining to the heart muscle.* It has a combining form, a root, and a suffix.

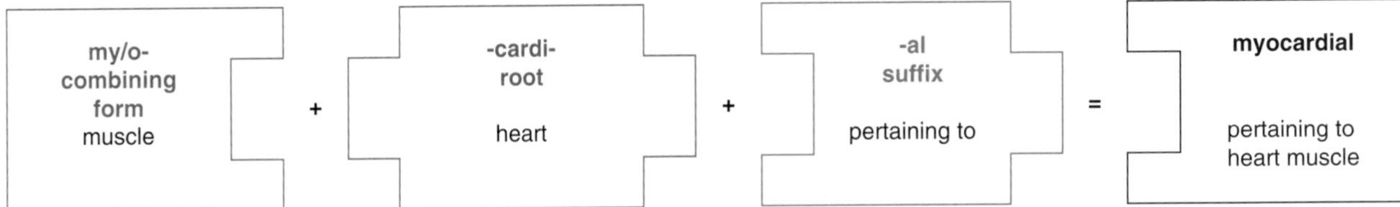

If the suffix is changed to -um, meaning *a structure,* the new term **myocardium** means *the structure called the heart muscle.*

The term **cardiomyopathy** contains the suffix -pathy, meaning *a disease,* the combining form cardi/o-, meaning the *heart,* and the combining form my/o-, meaning *muscle.* The term **cardiomyopathy** means a *disease of the heart muscle.*

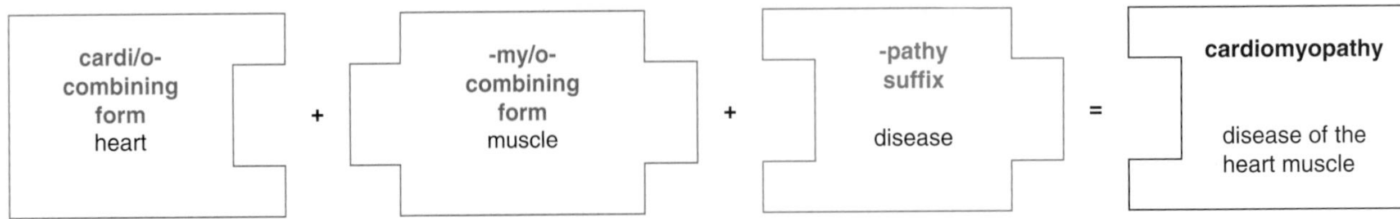

WORD ANALYSIS AND DEFINITION

S/ = Suffix P/ = Prefix R/ = Root R/CF = Combining Form

WORD	PRONUNCIATION	ELEMENTS		DEFINITION
cardiologist	kar-dee-**OL**-oh-jist	S/ R/CF	**-logist** *one who studies and is a specialist in* **cardi/o-** *heart*	A medical specialist in the diagnosis and treatment of disorders of the heart.
cardiology	kar-dee-**OL**-oh-jee	S/	**-logy** *study of*	Medical specialty of diseases of the heart.
cardiomyopathy	**KAR**-dee-oh-my-**OP**-ah-thee	S/ R/CF R/CF	**-pathy** *disease* **cardi/o-** *heart* **-my/o-** *muscle*	Disease of the heart muscle, the myocardium.
diagnosis (noun)	die-ag-**NO**-sis	P/ R/	**dia-** *complete* **-gnosis** *knowledge of an abnormal condition*	The determination of the cause of a disease.
diagnoses (pl)	die-ag-**NO**-sees			
diagnostic (adj) (The "is" in *gnosis* is deleted to allow the word to flow.)	die-ag-**NOS**-tik	S/	**-tic** *pertaining to*	Pertaining to or establishing a diagnosis.
diagnose (verb)	die-ag-**NOSE**	R/	**-gnose** *recognize an abnormal condition*	To make a diagnosis.
prognosis (noun)	prog-**NO**-sis	P/ R/	**pro-** *before, project forward* **-gnosis** *knowledge of an abnormal condition*	A forecast of the probable course and outcome of a disease.
electrocardiogram	ee-lek-troh-**KAR**-dee-oh-gram	S/ R/CF R/CF	**-gram** *record* **electr/o-** *electricity* **-cardi/o-** *heart*	Record of the heart's electrical signals.
infarct	in-**FARKT**	P/ R/	**in-** *in* **-farct** *area of dead tissue*	An area of cell death resulting from blockage of its blood supply.
infarction	in-**FARK**-shun	S/	**-ion** *action, condition*	Sudden blockage of an artery.
ischemia	is-**KEY**-me-ah	S/ R/	**-emia** *a blood condition* **isch-** *to block*	Lack of blood supply to tissue.
ischemic (adj)	is-**KEY**-mik	S/	**-emic** *pertaining to a condition of the blood*	Pertaining to the lack of blood supply to tissue.
myocardial (adj)	MY-oh-**KAR**-dee-al	S/ R/CF R/	**-al** *pertaining to* **my/o-** *muscle* **-cardi-** *heart*	Pertaining to heart muscle.
myocardium	MY-oh-**KAR**-dee-um	S/	**-um** *structure*	All the heart muscle.

The term **ischemia** has the suffix -emia, which means *a blood condition*. The root isch- means *to block*. **Ischemia** means a *blockage of the blood flow*. The term **myocardial ischemia** means a *blockage of the blood flow to the heart muscle*. In lay terms this is called a "heart attack."

If the suffix -emia is changed to -emic, which means *pertaining to a condition of the blood*, the new term is an adjective, **ischemic**, meaning *pertaining to a blockage of blood flow*. It has a root and a suffix.

Throughout the book, abbreviations that are used in the text will be listed and defined in an **Abbreviations box**.

Abbreviations

ACLS	advanced cardiac life support
AMI	acute myocardial infarction
CXR	chest x-ray
ECG/EKG	electrocardiogram
MONA	morphine, oxygen, nitroglycerine, aspirin
RUL	right upper lobe

EXERCISES

Precision in communication: *In addition to using the precise medical terms and speaking and spelling them correctly, you must use the appropriate form of the term as well. Reread the WAD entry for* diagnosis. *Note that there are singular and plural forms of the term, as well as the noun, adjective, and verb forms. Insert the correct form of the term in the documentation below.*

Note: A noun is a person, place, or thing.
A verb denotes action.
An adjective usually describes something.

Singular: One
Plural: More than one

1. The primary ______________________ for this patient is myocardial ischemia.
2. Dr. Bannerjee is unable to ______________________ this patient until he gets the lab results confirmed.
3. The ______________________ tests have been ordered for this patient first thing in the morning.
4. It is possible for this patient to have multiple ______________________ if there is more than one condition present.

- Some words that are pronounced the same are spelled differently. For example:
  - Both *ilium* and *ileum* are pronounced **ILL**-ee-um. *Ilium* is a bone in the pelvis; *ileum* is a segment of the small intestine.
  - Both *mucus* and *mucous* are pronounced **MYU**-kus. *Mucus* is a noun and is the name of a fluid secreted by *mucous* (adjective) membranes that line body cavities.
- Some words if incorrectly pronounced sound the same. For example:
  - The term *prostate*, pronounced **PROS**-tate, refers to the gland at the base of the male bladder. The term *prostrate* means to be physically weak or exhausted or to lie flat on the ground.
- Train your ear to hear the differences—*reflex* is not *reflux*.
- Many medical terms form a verb, a noun, a plural, and an adjective, and you have to know them all—for example, *diagnose, diagnosis, diagnoses,* and *diagnostic* (*see the WAD on the previous spread*).

PRONUNCIATION

Correct pronunciation of medical terms is essential so that other health professionals can understand what you are saying. It is a most important component in ensuring patient safety and providing high-quality patient care.

Throughout this textbook the pronunciation of medical terms is spelled out phonetically using modern English forms. The part of the word to be emphasized is shown in bold, uppercase letters.

For example, the term *pulmonary* is phonetically written **PUL**-moh-nar-ee. The term *pulmonology* is written **PUL**-moh-**NOL**-oh-jee. This illustrates that words derived from the same root can have their emphasis placed on different parts of the word and that the part being emphasized can be from different elements. The syllable of emphasis ***NOL*** is derived partly from the combining form pulmon/o- and partly from the suffix -logy.

Refer to the Student Online Learning Center (www.mhhe.com/AllanEssMedLanguage) to hear the glossary terms pronounced correctly. WWW

PLURALS

For many words in the English language, you can change the word from singular to plural by adding an "s." For medical terms, this happens only occasionally. The plurals are formed in ways that were logical to Greeks and Romans but have to be learned by memory in English. Examples of medical terms with Greek and Latin plurals are shown in *Table 1.1*.

Throughout this book, the Greek and Latin plurals of medical terms are given in the Word Analysis and Definition box with the singular medical term, as with the term *diagnosis* in the previous spread.

TABLE 1.1 Singular and Plural Forms

Singular Ending	Plural Ending	Examples
-a	-ae	axilla axillae
-is	-es	diagnosis diagnoses
-on	-a	ganglion ganglia
-um	-a	septum septa

Adapted from Kenneth S. Saladin, *Anatomy and Physiology,* 3rd ed., fig. 1.2, p. 21. Copyright © 2004 The McGraw-Hill Companies, Inc. Reprinted with permission.

WORD ANALYSIS AND DEFINITION

S/ = Suffix P/ = Prefix R/ = Root R/CF = Combining Form

WORD	PRONUNCIATION	ELEMENTS		DEFINITION
axilla **axillae (pl)** **axillary (adj)**	**AK**-sill-ah **AK**-sill-ee **AK**-sill-air-ee	 S/ R/	Latin *armpit* **-ary** *pertaining to* **axill-** *armpit*	Medical term for the armpit. Pertaining to the armpit.
dementia	dee-**MEN**-she-ah	S/ P/ R/	**-ia** *condition* **de-** *without* **-ment-** *mind*	Chronic, progressive, irreversible loss of intellectual and mental functions.
ganglion **ganglia (pl)**	**GANG**-lee-on **GANG**-lee-ah		Greek *a swelling or knot*	A fluid-filled cyst or a collection of nerve cells outside the brain and spinal cord.
ileum **ilium** **ilia (pl)**	**ILL**-ee-um **ILL**-ee-um **ILL**-ee-ah		Latin *to twist or roll up* Latin *groin*	Third portion of the small intestine. Large wing-shaped bone at the upper and posterior part of the pelvis.
mucus (noun) **mucous (adj)** **mucosa**	**MYU**-kus **MYU**-kus myu-**KOH**-sah	 S/ R/ S/	Greek *slime* **-ous** *pertaining to* **muc-** *mucus* **-osa** *full of; like*	Sticky secretion of cells in mucous membranes. Pertaining to mucus or the mucosa. Lining of a tubular structure that secretes mucus.
prostate **prostrate** **prostration (noun)**	**PROS**-tate pros-**TRAYT** pros-**TRAY**-shun		Greek *one who stands before* Latin *to stretch out*	Organ surrounding the urethra at the base of the male urinary bladder. To lay flat or to be overcome by physical weakness and exhaustion.
reflex **reflux**	**REE**-fleks **REE**-fluks		Latin *bend back* Latin *backward flow*	An involuntary response to a stimulus. Backward flow.
septum **septa (pl)**	**SEP**-tum **SEP**-tah		Latin *a partition*	A thin wall separating two cavities or two tissue masses.

EXERCISES

Medical language: *Many terms in medicine sound and/or look very similar. The difference of only one letter can make a new term. Train your eye and ear to know the difference. Circle the correct choice of terms in the following documentation.*

1. The patient's nasal (mucus/mucous) membrane is severely infected.
2. Schedule this patient for a (prostrate/prostate) exam at his next annual physical.
3. The doctor checked the (reflex/reflux) in the patient's knee.
4. The patient's (ilium/ileum) was severely fractured in the motor vehicle accident.

Plurals: *Insert the correct form of the plural in the following sentences.*

5. Because of additional medical problems needing treatment, this patient's insurance claim form will have multiple (diagnoses/diagnosis) ______________.
6. Check both (axilla/axillae) ______________ for any evidence of enlarged lymph nodes.
7. Several (septa/septum) ______________ exist in the body—e.g., in the heart and in the nose.
8. A cluster of (ganglia/ganglion) ______________ has formed on her left wrist.

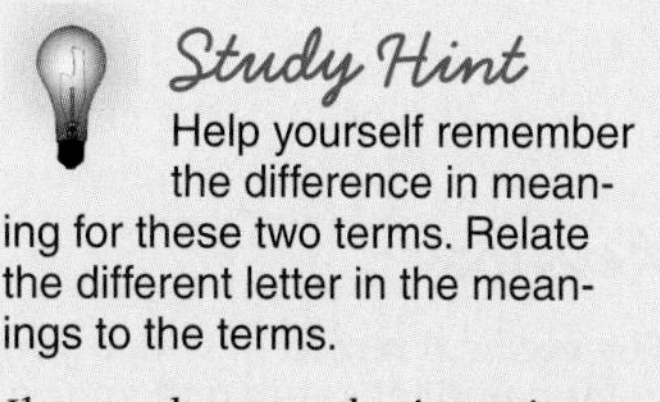

Study Hint

Help yourself remember the difference in meaning for these two terms. Relate the different letter in the meanings to the terms.

Ileum relates to the intestine.

Ilium relates to the pelvis.

PRECISION IN COMMUNICATION

Case Report 1.2 (continued)

While Mr. Donovan is waiting for *admission* to the hospital, oxygen is being administered through *nasal* prongs. He is *hypotensive*, and an *intravenous* (IV) *infusion* of normal saline has been started. From his medical record, it is noted that he is being seen in the *neurology* clinic for early *dementia*.

Keynote

Many words, when they are written or pronounced, have an element that if misspelled or mispronounced gives the intended word an entirely different meaning. A treatment response to the different meaning could cause a medical error and perhaps the death of a patient.

Each year in the United States, more than 400,000 people die because of drug reactions and medical errors. Many of these errors are due to inaccurate or imprecise written or verbal communication between different members of the health care team.

In the above Case Report, if **hypotension** (low blood pressure) were confused with **hypertension** (high blood pressure) incorrect and dangerous treatment could be prescribed.

- In the word **hypotension,** the **suffix -ion** means *a condition*. The **prefix hypo-** means *below or less than normal*. The **root -tens-** is from the Latin word for *pressure*. So **hypotension** is a condition of below-normal pressure, or low blood pressure.
- In the word **hypertension,** the **prefix hyper-** means *above or more than normal*. So **hypertension** is a condition of above-normal pressure, or high blood pressure.

Your ability to identify, spell, and pronounce the different **prefixes hypo-** and **hyper-** is essential to ensure accuracy and precision in your written and verbal communication.

Keynote

Precision in written and verbal communication is essential to prevent errors in patient care.

Also in the above Case Report, the term **neurology,** the specialty of the nervous system (*see Chapter 12*), can sound very similar to **urology,** the study of the urinary system (*see Chapter 13*). In the urinary system, if a patient's **ureter,** the tube from the kidney to the bladder, were confused with the **urethra,** the tube from the bladder to the outside, the consequences could be serious.

A medical term may also relate to more than one anatomical structure. The term **cervical** means *relating to a neck* in any sense. It can pertain to the neck that joins the head to the trunk with the **cervical vertebrae.** It can also relate to the **cervix** of the **uterus** with its **cervical canal.**

Keynote

The medical record in which you document a patient's care and your actions is a legal document. It can be used in court as evidence in professional medical liability cases.

Being a health professional requires the utmost attention to detail and precision in both written documentation and verbal communication. A patient's life can be in your hands. Any incorrect spelling can reflect badly on the whole health team. Any incorrect pronunciation and spelling can reflect badly on you as a health professional.

WORD ANALYSIS AND DEFINITION

S/ = Suffix P/ = Prefix R/ = Root R/CF = Combining Form

WORD	PRONUNCIATION	ELEMENTS		DEFINITION
cervical (adj)	**SER**-vih-kal	S/ R/	**-al** *pertaining to* **cervic-** *neck*	Pertaining to the cervix or to the neck region.
cervix	**SER**-viks		Latin *neck*	Lower part of the uterus.
hypertension	**HIGH**-per-**TEN**-shun	S/ P/ R/	**-ion** *condition, action* **hyper-** *above normal* **-tens-** *pressure*	Persistent high arterial blood pressure.
hypertensive (adj)	**HIGH**-per-**TEN**-siv	S/	**-ive** *pertaining to*	Pertaining to or suffering from high blood pressure.
hypotension	**HIGH**-poh-**TEN**-shun	P/	**hypo-** *below normal*	Persistent low arterial blood pressure.
hypotensive (adj)	**HIGH**-poh-**TEN**-siv			Pertaining to or suffering from low blood pressure.
infusion	in-**FYU**-zhun	P/ R/	in- *in* **-fusion** *to pour*	Introduction of a substance other than blood intravenously.
transfusion	trans-**FYU**-zhun	P/	**trans-** *across, through*	Transfer of blood or a blood component from a donor to a recipient.
intravenous	**IN**-trah-**VEE**-nus	S/ P/ R/	**-ous** *pertaining to* **intra-** *within, inside* **-ven-** *vein*	Inside a vein.
neurology	nyu-**ROL**-oh-jee	S/ R/CF	**-logy** *study of* **neur/o-** *nerve*	Medical specialty of disorders of the nervous system.
neurologist	nyu-**ROL**-oh-jist	S/	**-logist** *one who studies and is a specialist in*	Medical specialist in disorders of the nervous system.
ureter	you-**REE**-ter		Greek *urinary canal*	Tube that connects a kidney to the urinary bladder.
urethra	you-**REE**-thrah		Greek *passage for urine*	Canal leading from the bladder to the outside.
urology	you-**ROL**-oh-jee	S/ R/CF	**-logy** *study of* **ur/o-** *urine*	Medical specialty of disorders of the urinary system.
uterus	**YOU**-ter-us		Latin *womb*	Organ in which an egg develops into a fetus.
vertebra **vertebrae (pl)**	**VER**-teh-brah **VER**-teh-bree		Latin *bone in the spine*	One of the bones of the spinal column.

Abbreviation	
IV	intravenous

EXERCISES

Patient documentation: *Read the following excerpts from patient charts, and insert the correct medical term from the above Word Analysis and Definition (WAD) box.* ***Always review the WAD before you start the exercise.***

1. This patient has several badly fractured ________________________ in his spinal column.
2. This patient has nerve damage. Refer him to the department of ________________________.
3. Schedule this patient for an ________________________ of chemotherapy drugs today.
4. This patient has low blood pressure—he is ________________________ and anemic.
5. I am ordering an immediate ________________________ of 2 units of whole blood for this patient.
6. Send this patient for ________________________ x-rays of his neck immediately.

Brain teaser: If a medical specialist in the study of disorders of the *nervous* system is a *neurologist,* what is a medical specialist in the study of disorders of the *urinary* system called? ________________________

(***Hint:*** *Use your knowledge of suffixes and roots to help you.*)

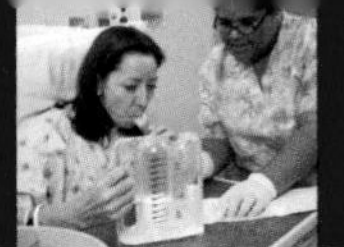

CHAPTER 1 REVIEW

THE ANATOMY OF WORD CONSTRUCTION

CHALLENGE YOUR KNOWLEDGE

A. **Keynotes** contain useful information and will often be the source of test questions. All the answers for this exercise on medical term elements can be found in Keynotes earlier in this chapter. For each statement, circle T (true) or (F) false. *If the statement is false, rewrite the statement so that it is true on the lines below.*

1.	A suffix changes the meaning of a term.	T	F
2.	Different suffixes can have the same meaning.	T	F
3.	The core foundation of every term is a root or combining form.	T	F
4.	Every term must have a prefix.	T	F
5.	If the suffix begins with a consonant, it must follow a root.	T	F
6.	A root plus a combining vowel equals a combining form.	T	F
7.	A medical term will never have more than one suffix.	T	F
8.	A prefix always comes at the beginning of the term.	T	F
9.	Combining forms can precede a suffix.	T	F
10.	Always begin deconstructing a term by identifying the root.	T	F

Corrected statements:

___

___

___

___

___

___

___

B. Build medical terms in this exercise. The first column in the chart below presents statements relating to the terms you will build. Look for clues in the statement words that will help you select the elements from the list. Build the term by inserting the correct elements in the appropriate columns (some elements you will use more than once, and some you will not use at all). The first term has been built, and the elements highlighted, to help you understand.

To complete the exercise, use any one of the terms in a brief sentence that is not a definition.

pneum/o-	-cyte	-logy	trans-	-ic	-thorax
-um	macro-	gastr-	-tension	-emia	isch-
-logy	cardio-	-mature	bi-	peri-	-ation
-al	respir-	-my/o-	-pathy	intra-	-later-
-nat-	post-	hypo-	pulmon/o-	epi-	card/i-
neur/o-	cervic-				

Statement	Prefix	Root/CF	Suffix
air in the chest		pneum/o-, -thorax	
the heart muscle			
pertaining to around the time of birth			
large red blood cell			
pertaining to both sides of the body			
study of the lung			
disease of the heart muscle			
low blood pressure			
process of breathing			
study of the nervous system			
pertaining to above the stomach			
lack of blood supply to tissue			
infant born after more than 42 weeks of gestation			

Sentence:

____

____

____

____

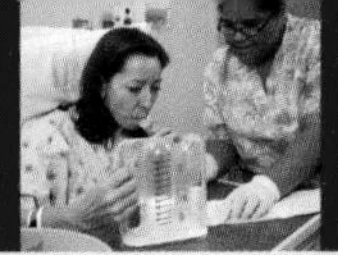

THE ANATOMY OF WORD CONSTRUCTION

C. **Partner exercise:** Ask your study partner to close his or her text. Dictate the following sentences to your partner, and then ask him or her to write the sentences and show them to you. Check your partner's sentences against the text below. The sentence is not correct unless every word is present and everything is spelled correctly. When you have finished checking your partner's answers, close your book and ask your partner to dictate the sentences to you and you write them down.

1. Mr. Donovan's chest pain is caused by myocardial ischemia, and an intravenous infusion has been started.
2. In addition to his hypotension, the record notes that Mr. Donovan is also being seen in the neurology clinic for dementia.
3. Mr. Donovan's physician has diagnosed an acute myocardial infarction based on a diagnostic electrocardiogram.

D. **Layperson's language:** Patients may request that you "translate" medical language into language they can more easily understand. Practice communicating the correct information with this exercise. If there are abbreviations in the sentence, "translate" them as well.

1. The pulmonologist read the patient's CXR and diagnosed a cancer in her RUL.

____

____

____

2. The patient's past medical history includes pneumonia and pneumothorax, as well as problems with her ileum.

____

____

____

3. The ACLS protocol treatment is used for all patients with chest pain caused by myocardial ischemia.

____

____

____

E. **Latin and Greek terms** do not deconstruct into the elements of prefix, root, combining form, and suffix the way most medical terms do. You just have to know them for what they are. Test yourself by matching the literal or defined meaning in column l with the correct Latin or Greek term in column 2.

	Column 1	Column 2
_____	1. armpit	A. ganglion
_____	2. one who stands before	B. mature
_____	3. bend back	C. prostate
_____	4. neck	D. reflex
_____	5. backward flow	E. mucus
_____	6. slime	F. ileum
_____	7. wing-shaped bone in pelvis	G. axilla
_____	8. fluid-filled cyst	H. reflux
_____	9. ripe	I. cervix
_____	10. 3rd portion of small intestine	J. ilium

F. **Identify and define** the elements in this table. Then give an example of these elements in a medical term from this chapter. Fill in the chart.

Element	Identify as P, R, CF, or S	Define Element	Medical Term Containing This Element
uro			
tens			
later			
nat			
cervic			
gram			
gastr			
logist			
gnos			
de			
micro			
ven			

G. **Spelling demons:** The following terms from this chapter are particularly difficult to spell and pronounce. Correct pronunciation and spelling of medical terms is the mark of an educated professional. Circle the correct spelling, and then check (✓) that you have practiced the pronunciation. Remember that pronunciations are on the Student Online Learning Center (www.mhhe.com/AllanEssMedLanguage).

				Pronunciation ✓
1.	diagnosis	diagnossis	diagnosiss	________
2.	infart	infarct	infarrct	________
3.	isscemia	iskchemia	ischemia	________
4.	miocardium	myocardeum	myocardium	________
5.	axila	axilla	axeila	________
6.	septtum	siptum	septum	________
7.	dementia	dimentia	dementea	________
8.	intraveinous	intravenous	intravinous	________
9.	vertebrae	verteebrae	vertebray	________
10.	pnumothorax	pneumothorax	pneumonthorax	________

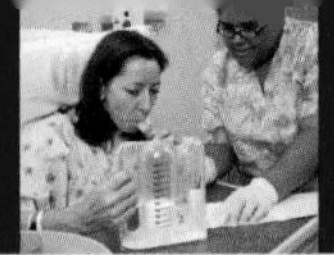

THE ANATOMY OF WORD CONSTRUCTION

H. **Grouping opposites:** Fill in the chart, and then answer the questions that follow. After you have completed this exercise, use it for study review. Grouping opposite elements or terms into pairs will make them easier to remember.

Element	Meaning of Element	Medical Term Containing This Element	Meaning of the Medical Term
1. pre			
2. post			
3. epi			
4. hypo			
5. macro			
6. micro			
7. hyper			
8. hypo			

9. These elements are all (P, R, CF, S) ____________

The meaning of the above elements will help you determine the correct answers to the following questions, even though you have never seen these terms before!

10. If a complication occurs after surgery, is it *pre* or *post* operative?

    ____________ operative

11. Which would be the topmost layer of skin (the one above everything else)—the epidermis or the hypodermis?

    ____________ dermis

12. Organisms that are too small to be seen with the naked eye are called ____________ scopic.

13. Would too much sugar in the blood be hyperglycemia or hypoglycemia?

    ____________ glycemia.

Finish this exercise by circling the word or words in questions 10 through 13 that led you to choose the correct element.

I. **Difference between:** If you really understand a term, you can explain it to someone else. Briefly explain these terms to a patient, in language he or she can understand.

1. transfusion:

   ____

2. infusion:

   ____

3. diagnosis:

   ____

4. prognosis:

   ____

J. **Elements** will always remain your best clue to understanding a medical term. The following terms have one element underlined and bolded—define that element and define the term. Fill in the blanks.

1. cardio**<u>myo</u>**pathy element defined: ____________________

   term defined: ____________________

2. **<u>isch</u>**emia element defined: ____________________

   term defined: ____________________

3. myocardi**<u>um</u>** element defined: ____________________

   term defined: ____________________

4. **<u>cervic</u>**al element defined: ____________________

   term defined: ____________________

5. hypertens**<u>ion</u>** element defined: ____________________

   term defined: ____________________

6. **<u>trans</u>**fusion element defined: ____________________

   term defined: ____________________

7. **<u>neuro</u>**logist element defined: ____________________

   term defined: ____________________

8. **<u>bi</u>**lateral element defined: ____________________

   term defined: ____________________

9. **<u>gastr</u>**ic element defined: ____________________

   term defined: ____________________

10. **<u>intra</u>**venous element defined: ____________________

    term defined: ____________________

K. **Terminology challenge.**

1. Write the precise medical language for what, in layperson's terms, is called a "heart attack": ____________________

   ____

   ____

   ____

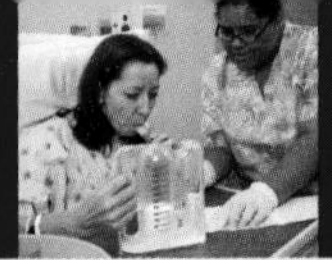

THE ANATOMY OF WORD CONSTRUCTION

L. **Proofread** the following sentences for errors in fact *or* spelling. Underline any misspelled terms or errors in fact in a sentence; then rewrite the incorrect sentences correctly on the lines below. There is only one sentence that is entirely correct.

1. An acute myocardial infraction can be confirmed by an EGK.
2. Patients with chest pain could possibly have myocardeal ischemic.
3. An electrocardiogram is a diagnostic tool to check for possible heart attack.
4. The pulmonalogist ordered an IV transfusion of saline and nasal oxygen.
5. Mr. Donovan's dimentia may be a complicating factor in the treatment for his heart attack.

Rewrites:

____

____

____

____

____

____

____

____

____

____

____

M. **Plurals:** Because many medical terms are directly from Greek or Latin, they do not form their plurals just by adding "s" as happens in English. In this exercise, you are given the singular form of five terms. Choose one of the endings to form the correct plural of each term (some endings you will use more than once, and some you will not use at all).

ae   es   ides   ges   ies   era   a

Singular	Plural
1. diagnosis	________________
2. axilla	________________
3. ganglion	________________
4. septum	________________
5. vertebra	________________

N. Roots/Combining Forms: The meaning of the R/CF is given in column 1. Match the meaning to the correct term containing that R/CF in column two.

_____	1. chest	A. dementia
_____	2. armpit	B. microcyte
_____	3. heart	C. hypertension
_____	4. nerve	D. prognosis
_____	5. birth	E. neurologist
_____	6. pressure	F. pneumothorax
_____	7. knowledge	G. infarct
_____	8. mind	H. myocardial
_____	9. area of dead tissue	I. axillary
_____	10. cell	J. perinatal

Use any two terms from column 2 in patient documentation of your choice.

1. ____

____

____

2. ____

____

____

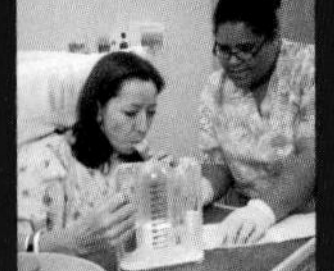

CHAPTER 1 REVIEW

THE ANATOMY OF WORD CONSTRUCTION

O. Chapter challenge: Read all the possible choices before you circle the correct answer.

> *Study Hint*
> Immediately cross off any answer you know is *not* correct. In your remaining choices, there is only ***one best answer.***

1. Which answer best describes *pulmono* and *pneumo?*
   a. They are both roots and mean *chest.*
   b. They are both combining forms but have different meanings.
   c. One is a root, and the other is a combining form.
   d. They are both combining forms and mean *lung.*
   e. One is a suffix, and the other is a prefix.

2. Based on their *elements,* pick the pair of terms that logically belong together:
   a. reflex and urology
   b. pulmonology and pulmonologist
   c. prostate and prostrate
   d. ganglia and ganglion
   e. multilateral and epigastric

3. Which two terms have prefixes denoting numbers?
   a. hypotension and hypertension
   b. premature and postmature
   c. bilateral and unilateral
   d. pericardial and perinatal
   e. epigastric and hypogastric

4. The body system concerned with air or breathing is the ______________________, and an organ in that system is the ______________________.

   a. urinary — ureter
   b. musculoskeletal — ilium
   c. respiratory — lung
   d. cardiovascular — heart
   e. digestive — stomach

5. The term *cardiomyopathy* has a suffix meaning:
   a. condition
   b. disease
   c. action
   d. structure
   e. pertaining to

6. Which pair contains terms that are both diagnoses?
   a. pneumothorax and pulmonologist
   b. cervix and ureter
   c. pulmonology and pneumonia
   d. reflux and reflex
   e. pneumonia and pneumothorax

7. Which word element appears at the beginning of the term?
   a. root
   b. combining vowel
   c. combining form
   d. suffix
   e. prefix

8. In the term *dementia,* the prefix means:
   a. below normal
   b. without
   c. above normal
   d. through
   e. within

9. In the abbreviation *RUL,* the "R" stands for:
   a. rigid
   b. regular
   c. right
   d. removed
   e. radical

10. Circle the choice in which *both* the terms are spelled correctly:
    a. pullmonologist pneumonia
    b. perinatil bilateral
    c. cardiomyopathy dimentia
    d. urreter urethra
    e. cervical respiration

11. Circle the term that has a suffix meaning *study of:*
    a. myocardial
    b. macrocyte
    c. cardiologist
    d. postmature
    e. neurology

12. Find the only term that is an adjective:
    a. mucus
    b. ganglion
    c. prostration
    d. ischemic
    e. prognosis

13. Which set contains elements that are opposites?
    a. hyper and hypo
    b. micro and macro
    c. epi and hypo
    d. None of these are opposites.
    e. They are all opposites.

14. "Graphic record of the heart's electrical currents" is a definition for:
    a. cardiovascular
    b. electrocardiogram
    c. cardiologist
    d. cardiomyopathy
    e. myocardium

15. The term containing a suffix meaning *structure* is:
    a. ischemia
    b. axillary
    c. myocardium
    d. cervical
    e. infusion

16. Which term does *not* contain a prefix?
    a. diagnosis
    b. bilateral
    c. infarct
    d. transfusion
    e. mucous

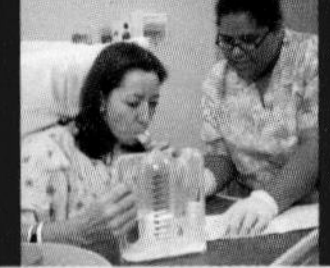

P. **Case Report challenge:** Now that you are more comfortable with the terms in this chapter, you can apply that knowledge and briefly answer the questions about the case report.

Study Hint

If you read the case report through first, and then go back and underline or highlight all the medical terminology, you will find it easier to answer the questions.

CASE REPORT 1.1

You are

. . . a respiratory therapist working with Tavis Senko, MD, a pulmonologist at Fulwood Medical Center.

You are communicating with

. . . Mrs. Sandra Schwartz, a 43-year-old woman referred to Dr. Senko by her primary care physician, Dr. Andrew McDonald, an internist. She has a persistent abnormality on her chest x-ray. You have been asked to determine her pulmonary function prior to a scheduled bronchoscopy.

From her medical records, you see that 2 months ago Mrs. Schwartz developed a right upper lobe (RUL) *pneumonia.* After treatment with an antibiotic, a follow-up chest x-ray (CXR) showed some residual collapse in the right upper lobe and a small right *pneumothorax*. Mrs. Schwartz has smoked a pack a day since she was a teenager. Dr. Senko is concerned that she has lung cancer, and he has scheduled her for bronchoscopy.

1. What type of specialist is Dr. Senko? ____________________
2. What symptom did Mrs. Schwartz have that meant she needed to see a specialist?
____
3. What disease or condition is in Mrs. Schwartz's past medical history?
____
4. Give a brief definition of the condition in question 3 above: ____________________
5. What diagnostic test did Mrs. Schwartz have done? ____________________
6. What part of Mrs. Schwartz's lung shows residual collapse? ____________________
7. Approximately how many years has Mrs. Schwartz been a smoker? ____________________
8. What procedure has Dr. Senko scheduled for Mrs. Schwartz? ____________________
9. Based on her past medical history, history of smoking, and current diagnostic findings, what is a probable diagnosis for Mrs. Schwartz? ____________________

Q. **Chapter challenge:** Read all the possible choices before you circle the correct answer.

1. The terms *myocardial* and *myocardium* both refer to the:
   - a. lung
   - b. hip
   - c. pelvis
   - d. heart
   - e. blood
2. The prefix in *intravenous* means:
   - a. across
   - b. around
   - c. within
   - d. two
   - e. before
3. Which terms do not deconstruct into word elements?
   - a. axillary and axilla
   - b. pulmonologist and pulmonology
   - c. ileum and ilium
   - d. respiration and pulmonary
   - e. cervix and cervical
4. **Brain teaser:** Circle the terms used to describe newborn babies' *development.*
   - a. unilateral and bilateral
   - b. epigastric and hypogastric
   - c. premature and postmature
   - d. perigastric and perinatal
   - e. prenatal and postnatal
5. Circle the only choice that does not contain a combining form:
   - a. pneumothorax
   - b. pulmonology
   - c. pneumonia
   - d. myocardial
   - e. cardiomyopathy
6. The medical term for "armpit" is:
   - a. septum
   - b. mucosa
   - c. axilla
   - d. ganglion
   - e. ilium

Congratulations! You are on your way to learning medical terminology.

The Body as a Whole

The Essentials of the Language of Anatomy

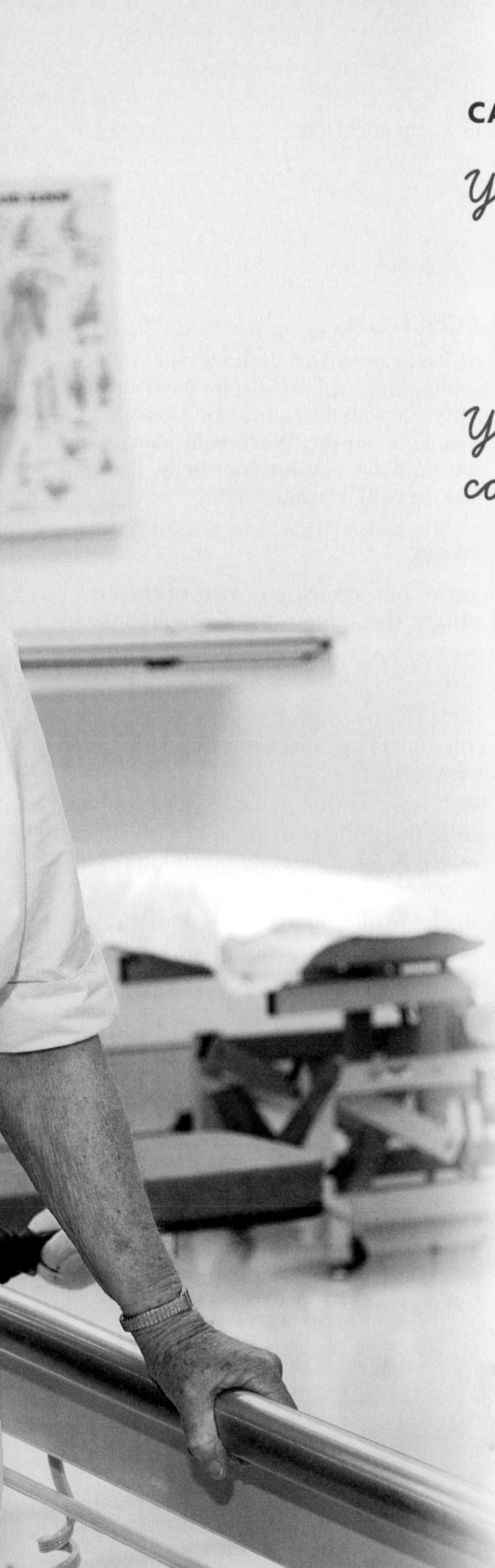

CASE REPORT 2.1

You are

. . . a physical therapy assistant (*PTA*) employed in the Rehabilitation Unit at Fulwood Medical Center.

You are communicating with

. . . Mrs. Amy Vargas, a 70-year-old housewife, who is 2 weeks postop following an emergency right hip replacement for a hip fracture. Your task is to help her increase her walking ability and increase strength and mobility in her hip joint and upper arms.

You know that, in order to prevent her hip replacement from dislocating, she has been instructed to:

- Not bring the right leg or knee ***medially*** across the ***sagittal*** plane, for example, not to cross the right leg over the left leg.
- Not lift the right knee so that it is superior to the right hip.
- Not bend the trunk ***anteriorly*** so that it is at more than a 90-degree angle to the thigh.

Learning Outcomes

Effective medical treatment recognizes that each organ, tissue, and cell in your body, no matter where it is situated, functions in harmony with and affects every other organ, tissue, and cell. To understand these concepts, you need to be able to:

2.1 Describe the medical terms of the different anatomical planes, directions, and body regions.

2.2 Integrate individual body systems into the organization and function of the body as a whole.

2.3 Comprehend, spell, and write medical terms pertaining to the body as a whole so that you communicate and document accurately and precisely.

2.4 Recognize and pronounce medical terms pertaining to the body as a whole so that you communicate verbally with accuracy and precision.

LESSON 2.1 THE BODY AS A WHOLE

Anatomical Positions, Planes, and Directions

OBJECTIVES

Terms have been developed over the past several thousand years to enable you to describe clearly where different anatomical structures and lesions are in relation to each other. To do this, you need to be able to:

- **Define the fundamental anatomical position on which all descriptions of anatomical locations are based.**
- **Describe the medical terminology of the different anatomical planes and directions.**
- **Relate these terms to physical sites on the body.**
- **Locate the body cavities.**
- **Identify the medical terminology of the four abdominal quadrants and nine regions.**

Keynote

The transverse plane is the only horizontal body plane.

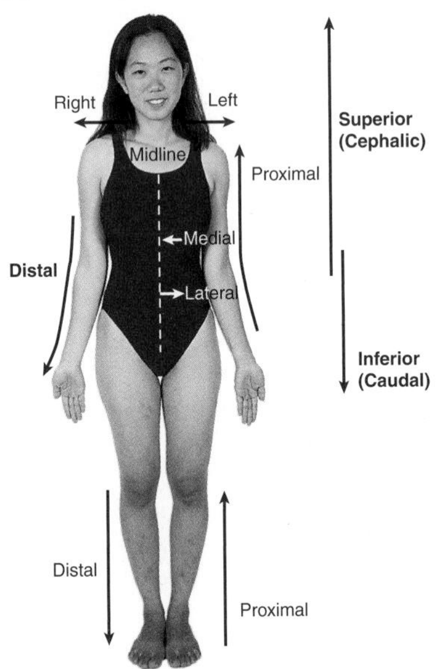

▲ **FIGURE 2.1 Anatomical Position, with Directional Terms.**

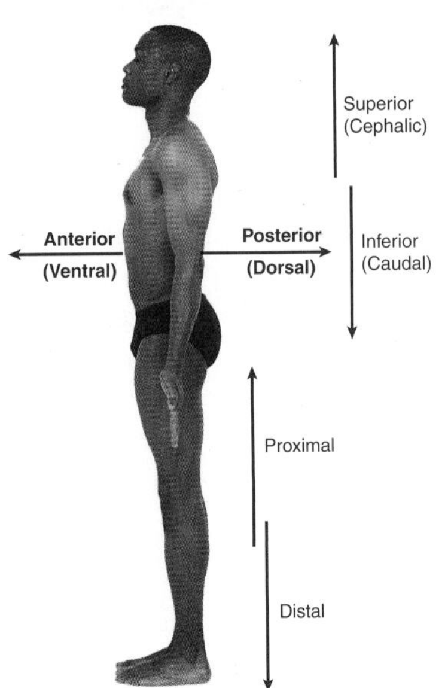

▲ **FIGURE 2.2 Other Directional Terms.**

FUNDAMENTAL ANATOMICAL POSITION

When all anatomical descriptions are used, it is assumed that the body is in the "anatomical position" (*Figure 2.1*). The body is standing erect with feet flat on the floor, face and eyes are facing forward, and arms are at the side with the palms facing forward.

When you lie down flat on your back, you are **supine.** When your palms face forward, the forearm is supine. When you lie down flat on your belly, you are **prone.** When your palms face backward, the forearm is prone.

ANATOMICAL DIRECTIONAL TERMS

Directional terms describe the position of one structure or part of the body relative to another structure or part of the body. These directional terms are shown in *Figures 2.1 and 2.2.*

ANATOMICAL PLANES

Different views of the body are based on imaginary "slices" producing flat surfaces (planes) that pass through the body (*Figure 2.3*).

The **three major anatomical planes** are:

- **Transverse or horizontal:** A plane passing across the body parallel to the floor and perpendicular to the body's long axis. It divides the body into an upper or superior portion and a lower or inferior portion.
- **Saggital:** A vertical plane that divides the body into right and left portions.
- **Frontal (coronal):** A vertical plane that divides the body into front (**anterior**) and back (**posterior**) portions.

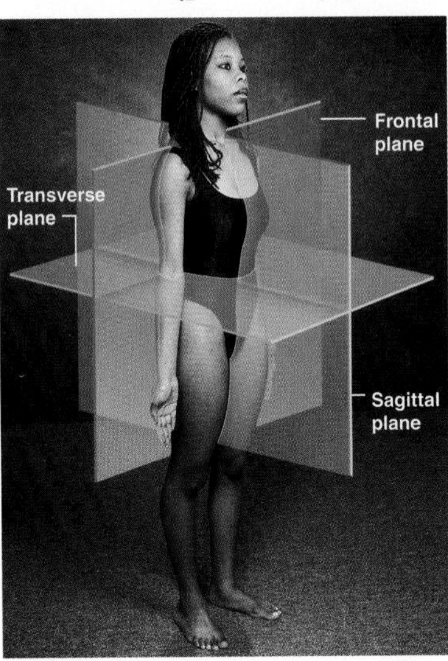

▲ **FIGURE 2.3 Anatomical Planes.**

Study Hint

Help your memory with little tricks of association for medical terms. *Example:* The medical term ***supine*** has the word *up* in it. The meaning of supine is *lying with the face and the anterior part of the body UP.* Associate *UP* with *sUPine,* and you will have no trouble remembering its definition.

Then associate the opposite term, and you will know the meaning of *prone* as well.

Abbreviation	
PTA	physical therapy assistant

WORD ANALYSIS AND DEFINITION

S/ = Suffix P/ = Prefix R/ = Root R/CF = Combining Form

WORD	PRONUNCIATION	ELEMENTS		DEFINITION
abdomen **abdominal (adj)**	**AB**-doh-men ab-**DOM**-in-al	 S/ R/	Latin *abdomen* **-al** *pertaining to* **abdomin-** *abdomen*	Part of the trunk between thorax and pelvis. Pertaining to the abdomen.
anterior (opposite of *posterior*)	an-**TEER**-ee-or	S/ R/	**-ior** *pertaining to* **anter-** *before, front part*	The front surface of the body; situated in front.
caudal (opposite of *cephalic,* same as *inferior*)	**KAW**-dal	S/ R/	**-al** *pertaining to* **caud-** *tail*	Pertaining to or nearer to the tailbone.
cephalic (opposite of *caudal,* same as *superior*)	seh-**FAL**-ik	S/ R/	**-ic** *pertaining to* **cephal-** *head*	Pertaining to or nearer to the head.
coronal (same as *frontal*)	**KOR**-oh-nal	S/ R/	**-al** *pertaining to* **coron-** *crown*	Pertaining to the vertical plane dividing the body into anterior and posterior portions.
distal (opposite of *proximal*)	**DISS**-tal	S/ R/	**-al** *pertaining to* **dist-** *away from the center*	Situated away from the center of the body.
dorsal (same as *posterior*)	**DOOR**-sal	S/ R/	**-al** *pertaining to* **dors-** *back*	Pertaining to the back or situated behind.
lateral (opposite of *medial*)	**LAT**-er-al	S/ R/	**-al** *pertaining to* **later-** *side*	Situated at the side of a structure.
medial (opposite of *lateral*)	**ME**-dee-al	S/ R/	**-al** *pertaining to* **medi-** *middle*	Nearer to the middle of the body.
posterior (opposite of *anterior*)	pohs-**TEER**-ee-or	S/ R/	**-ior** *pertaining to* **poster-** *back part*	Pertaining to the back surface of the body; situated behind.
prone (opposite of *supine*)	PROHN		Latin *bending forward*	Lying face down, flat on your belly.
proximal (opposite of *distal*)	**PROK**-sih-mal	S/ R/	**-al** *pertaining to* **proxim-** *nearest to the center*	Situated nearest to the center of the body.
sagittal	**SAJ**-ih-tal	S/ R/	**-al** *pertaining to* **sagitt-** *arrow*	Vertical plane through the body dividing it into right and left portions.
supine (opposite of *prone*)	soo-**PINE**		Latin *lying on the back*	Lying face up, flat on your spine.
transverse	trans-**VERS**		Latin *crosswise*	Horizontal plane dividing the body into upper and lower portions.
ventral (same as *anterior*)	**VEN**-tral	S/ R/	**-al** *pertaining to* **ventr-** *belly*	Pertaining to the belly or situated nearer the surface of the belly.

EXERCISES

Group *the opposites in the WAD above for ease of study. Fill in the chart.*

Term	Meaning of Term	Opposite Term	Meaning of Opposite Term
Anterior			
Caudal			
Distal			
Prone			

BODY CAVITIES

FIGURE 2.4 ▶ **Major Body Cavities.**

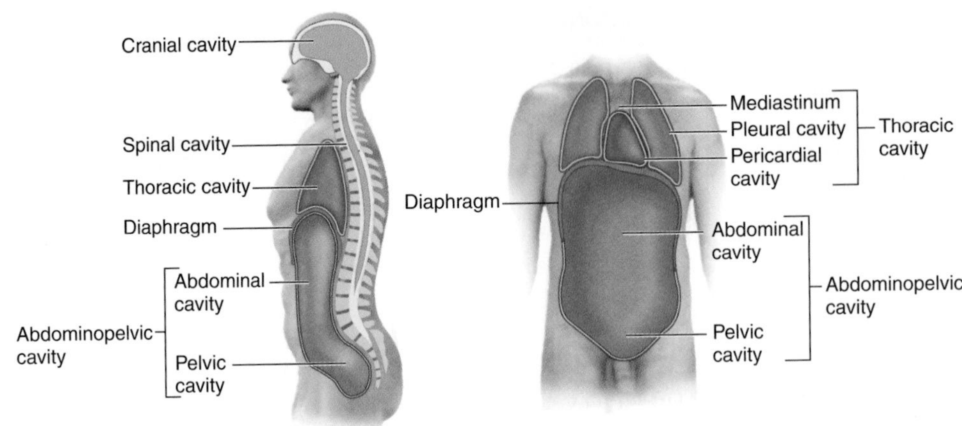

The body contains many **cavities.** Some, like the nasal cavity, open to the outside. Five **cavities** do not open to the outside; they are shown in *Figure 2.4* and listed below.

1. **Cranial cavity:** Contains the brain within the skull.
2. **Thoracic cavity:** Contains the heart, lungs, thymus gland, trachea and esophagus, and numerous blood vessels and nerves.
3. **Abdominal cavity:** Is separated from the thoracic cavity by the **diaphragm** and contains the stomach, intestines, liver, spleen, pancreas, and kidneys.
4. **Pelvic cavity:** Is surrounded by the pelvic bones and contains the urinary bladder, part of the large intestine, the rectum and anus, and the internal reproductive organs.
5. **Spinal cavity:** Contains the spinal cord.

The abdominal cavity and pelvic cavity can be combined as the **abdominopelvic cavity.**

ABDOMINAL QUADRANTS

One way of referring to the locations of abdominal structures and to the site of abdominal pain and other abnormalities is to divide the abdomen into **quadrants,** as shown in *Figure 2.5a*. The locations are right upper quadrant (**RUQ**), left upper quadrant (**LUQ**), right lower quadrant (**RLQ**), and left lower quadrant (**LLQ**).

In addition, there are nine regions in the abdomen, as shown in *Figure 2.5b*.

Abbreviations

LLQ	left lower quadrant
LUQ	left upper quadrant
RLQ	right lower quadrant
RUQ	right upper quadrant

FIGURE 2.5 ▶ **Regional Anatomy.**

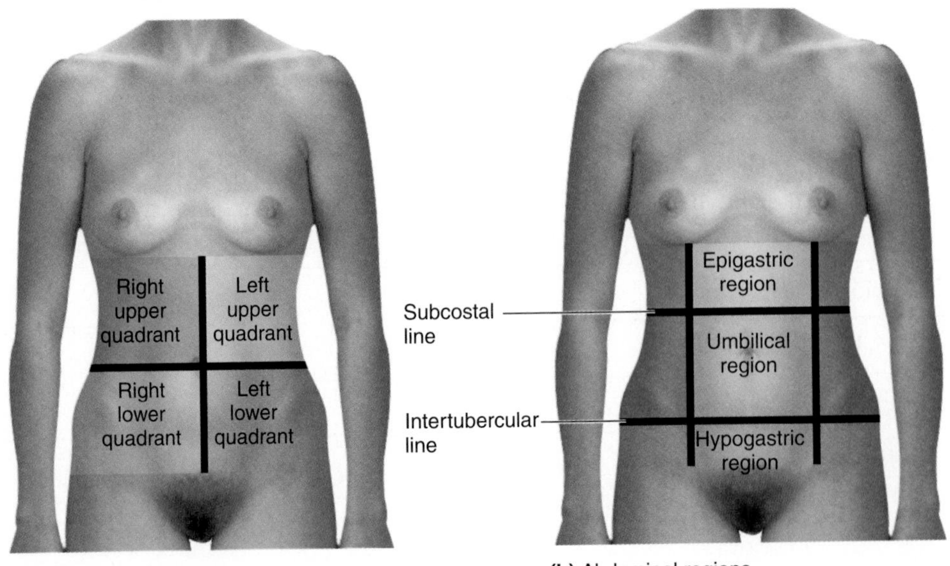

(a) Abdominal quadrants

(b) Abdominal regions

WORD ANALYSIS AND DEFINITION

S/ = Suffix P/ = Prefix R/ = Root R/CF = Combining Form

WORD	PRONUNCIATION		ELEMENTS	DEFINITION
abdominopelvic	ab-**DOM**-ih-no-**PEL**-vik	S/ R/CF R/	**-ic** *pertaining to* **abdomin/o-** *abdomen* **-pelv-** *pelvis*	Pertaining to the abdomen and pelvis.
cavity **cavities (pl)**	**KAV**-ih-tee **KAV**-ih-tees	S/ R/	**-ity** *state, condition* **cav-** *hollow space*	A hollow space or body compartment.
cranial (adj)	**KRAY**-nee-al	S/ R/	**-al** *pertaining to* **crani-** *skull*	Pertaining to the cranium.
cranium	**KRAY**-nee-um	S/	**-um** *structure*	The skull.
diaphragm	**DIE**-ah-fram		Greek *diaphragm, fence*	Muscular sheet separating the abdominal and thoracic cavities.
diaphragmatic (adj)	**DIE**-ah-frag-**MAT**-ik	S/ R/	**-ic** *pertaining to* **diaphragmat-** *diaphragm*	Pertaining to the diaphragm.
quadrant	**KWAD**-rant		Latin *one quarter*	One quarter of a circle; one of four regions of the surface of the abdomen.
spine	SPYN		Latin *spine*	The vertebral column *or* a short bony projection.
spinal (adj)	**SPY**-nal	S/ R/	**-al** *pertaining to* **spin-** *spine*	Pertaining to the spine.
thoracic (adj)	**THOR**-ass-ik	S/ R/	**-ic** *pertaining to* **thorac-** *chest*	Pertaining to the chest (thorax).
thorax	**THOR**-acks		Greek *chest*	The part of the trunk between the abdomen and neck.
umbilical (adj)	um-**BILL**-ih-kal	S/ R/	**-al** *pertaining to* **umbilic-** *navel (belly button)*	Pertaining to the umbilicus or the center of the abdomen.
umbilicus	um-**BILL**-ih-kuss	R/	**umbilicus** *navel (belly button)*	Pit in the abdomen where the umbilical cord entered the fetus.

EXERCISES

Deconstruct the following terms into their basic elements. *Note that not every type of element will appear in every term. The only element every term needs is a root or a combining form. Fill in the blanks.*

1. diaphragmatic ________/________/________
   P R/CF S

2. abdominopelvic ________/________/________
   P R/CF S

3. umbilical ________/________/________
   P R/CF S

4. cranial ________/________/________
   P R/CF S

5. thoracic ________/________/________
   P R/CF S

6. cavity ________/________/________
   P R/CF S

7. spinal ________/________/________
   P R/CF S

LESSON 2.2 Organization of the Body

OBJECTIVES

All the different elements of your body interact with each other to enable your body to be in constant change as it reacts to your environment and to the nourishment you give it. To understand the structure and function of the elements of your body, you need to be able to:

- **Name the medical terms associated with cells, tissues, and organs.**
- **Discuss the medical terminology for the major structures and functions of a cell.**
- **Describe the structures and functions of the nucleus and mitochondria.**

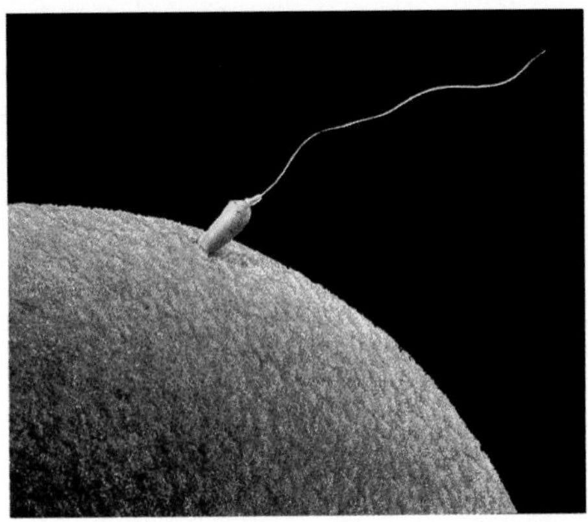

▲ **FIGURE 2.6 Fertilization of Egg by Single Sperm.**

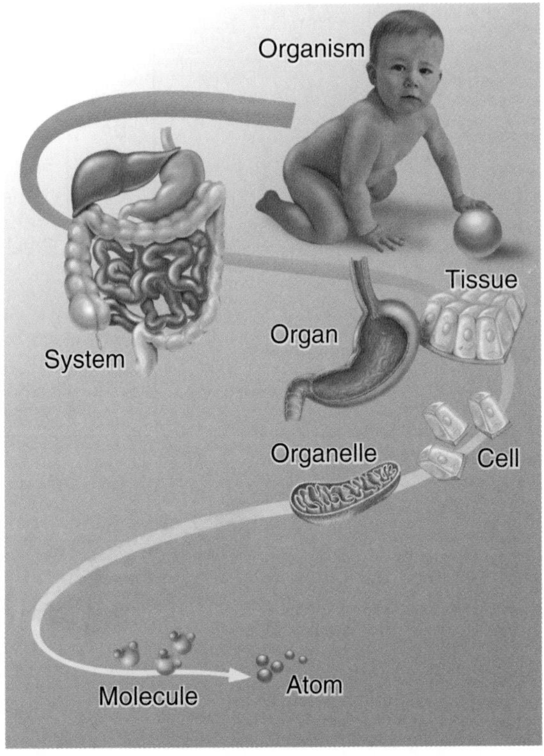

COMPOSITION OF THE BODY

- The whole body or organism is composed of *organ* systems.
  - Organ systems are composed of *organs.*
    - Organs are composed of *tissues.*
      - Tissues are composed of *cells.*
        - Cells are composed in part of *organelles.*
          - Organelles are composed of *molecules.*
            - Molecules are composed of *atoms.*

THE CELL

The result of the **fertilization** of an egg by a sperm is a single fertilized cell, the **zygote** (*Figure 2.6*). This cell is the origin of every cell in your body. It divides and multiplies into millions of cells that are the basic unit of every tissue and organ. The structure and all the functions of your tissues and organs are due to their cells. **Cytology** is the study of cell structure and function.

WORD ANALYSIS AND DEFINITION

S/ = Suffix P/ = Prefix R/ = Root R/CF = Combining Form

WORD	PRONUNCIATION	ELEMENTS		DEFINITION
cell	SELL		Latin *a storeroom*	The smallest unit of the body capable of independent existence.
cytology	**SIGH-**tol-oh-jee	S/ R/CF	**-logy** *study of* **cyt/o-** *cell*	Study of the cell.
cytologist	**SIGH-**tol-oh-jist	S/	**-logist** *one who studies, a specialist*	Specialist in the structure, chemistry, and pathology of the cell.
fertilization (noun)	**FER**-til-eye-**ZAY**-shun	S/ R/	**-ation** *process* **fertiliz-** *to make fruitful* Greek *to bear*	Union of a male sperm and a female egg.
fertilize (verb)	**FER**-til-ize			Penetration of the egg by sperm.
organ	**OR**-gan		Latin *instrument, tool*	Structure with specific functions in a body system.
organelle	**OR**-gah-nell	S/ R/	**-elle** *small* **organ-** *organ*	Part of a cell having specialized function(s).
organism	**OR**-gan-izm	S/	**-ism** *condition, process*	Any whole living, individual plant or animal.
tissue	**TISH**-you		Latin *to weave*	Collection of similar cells.
zygote	**ZYE**-goat		Greek *yolk*	Cell resulting from the union of sperm and egg.

EXERCISES

Review *the terms in the WAD box above and the text on the opposite page before answering the questions. Pay careful attention to word elements and meanings. Fill in the blanks.*

1. Put the following terms in the ascending order of their size:

organism **cells** **molecules** **organs**
organ systems **organelles** **atoms** **tissues**

a. ____________________
b. ____________________
c. ____________________
d. ____________________
e. ____________________
f. ____________________
g. ____________________
h. ____________________

2. The suffix ____________ means *study of.* The suffix that means *specialist (in the study of)* is ____________.
3. What part of *cyt/o* makes it a combining form rather than a root? ____________ ____________
4. Write two definitions of any two terms in question 1.

a. ____________________
____

b. ____________________
____

STRUCTURE AND FUNCTION OF CELLS

As the zygote divides, every cell derived from it becomes a small, complex factory that carries out these basic functions of life:

- *Manufacture* of proteins and **lipids.**
- *Production* and use of energy.
- *Communication* with other cells.
- *Replication* of **deoxyribonucleic acid (DNA)**.
- *Reproduction* of itself.

Abbreviation

DNA deoxyribonucleic acid

All your cells contain a fluid called **cytoplasm** (intracellular fluid) surrounded by a cell **membrane** (*Figure 2.7*).

The cell membrane is made of **proteins** and **lipids** and allows water, oxygen, glucose, **electrolytes, steroids,** and alcohol to pass through it. On the outside of the cell membrane are receptors that bind to chemical messengers such as **hormones** sent by other cells. These are the chemical signals by which your cells communicate with each other.

Organelles are small structures in the cytoplasm of the cell that carry out special **metabolic** tasks, the chemical processes that occur in the cell. They include the nucleus, nucleolus, and **mitochondria.** These organelles are defined, and their functions detailed, in the succeeding pages.

Keynote

The cytoplasm is a clear, gelatinous substance crowded with different organelles.

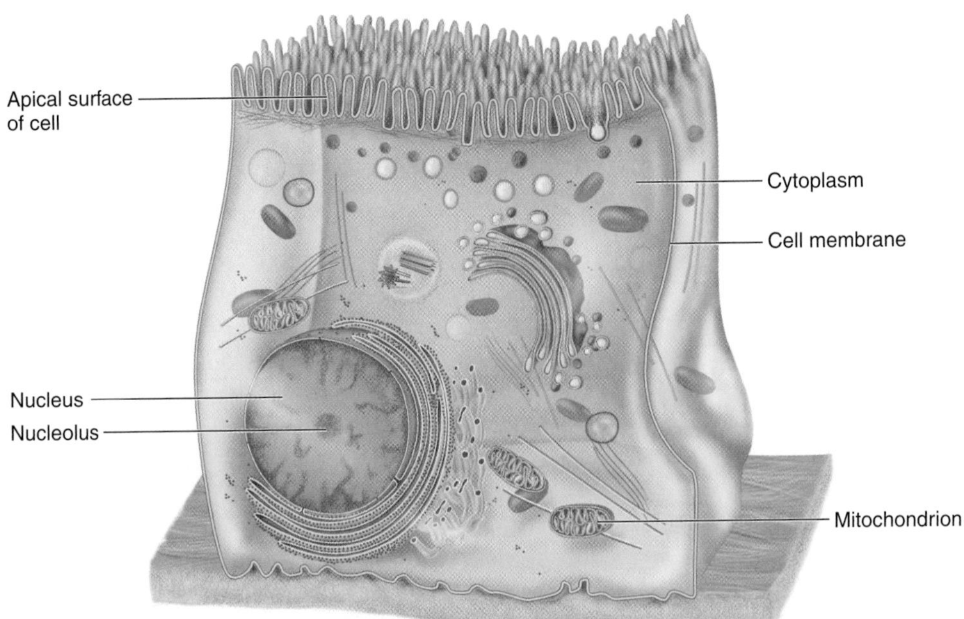

▲ **FIGURE 2.7 Structure of a Representative Cell.**

WORD ANALYSIS AND DEFINITION

S/ = Suffix P/ = Prefix R/ = Root R/CF = Combining Form

WORD	PRONUNCIATION	ELEMENTS		DEFINITION
cytoplasm	**SIGH**-toh-plazm	S/ R/CF	**-plasm** *something formed* **cyt/o-** *cell*	Clear, gelatinous substance that forms the substance of a cell, except for the nucleus.
deoxyribonucleic acid (DNA)	dee-**OCK**-see-rye boh-noo-**KLEE**-ik **ASS**-id		*deoxyribose* (a sugar) *nucleic acid* (a protein)	Source of hereditary characteristics found in chromosomes.
electrolyte	ee-**LEK**-troh-lite	S/ R/CF	**-lyte** *soluble* **electr/o-** *electricity*	Substance that, when dissolved in a suitable medium, forms electrically charged particles.
hormone	**HOR**-mohn		Greek *set in motion*	Chemical formed in one tissue or organ and carried by the blood to stimulate or inhibit a function of another tissue or organ.
hormonal (adj)	hor-**MOHN**-al	S/ R/	**-al** *pertaining to* **hormon-** *hormone*	Pertaining to a hormone.
lipid	**LIP**-id		Greek *fat*	General term for all types of fatty compounds; for example, cholesterol, triglycerides and fatty acids.
membrane	**MEM**-brain		Latin *parchment*	Thin layer of tissue covering a structure or cavity.
membranous (adj)	**MEM**-brah-nus	S/ R/	**-ous** *pertaining to* **membran-** *cover, skin*	Pertaining to a membrane.
metabolism	meh-**TAB**-oh-lizm	S/ R/	**-ism** *condition, process* **metabol-** *change*	The constantly changing physical and chemical processes occurring in the cell that are the sum of anabolism and catabolism.
metabolic (adj)	met-ah-**BOL**-ik	S/	**-ic** *pertaining to*	Pertaining to metabolism.
mitochondria (pl) **mitochondrion (singular)**	my-toe-**KON**-dree-ah my-toe-**KON**-dree-on	S/ R/CF R/ S/	**-ia** *condition* **mit/o-** *thread* **-chondr-** *granule* **-ion** *condition*	Organelles that generate, store, and release energy for cell activities.
protein	**PRO**-teen		Greek *protein*	Class of food substances based on amino acids.
steroid	**STER**-oyd	S/ R/	**-oid** *resembling* **ster-** *solid*	Large family of chemical substances found in many drugs, hormones, and body components.

EXERCISES

Elements: *Knowledge of elements is your best clue to determining the meaning of medical terminology. Analyze the elements in these questions to find your answers. Fill in the blanks.*

1. What do the terms *metabolism* and *mitochondria* have in common?

   (Hint: They both lack the same thing.)

   ____

   Use either term in a sentence of your choice that is not a definition.

   ____

2. Which term relates to electrically charged particles? *(Circle the best answer.)*

   protein membrane electrolyte

3. Which term relates to change?

   steroid metabolic lipid

4. Which term is a condition?

   metabolism cytoplasm hormone

STRUCTURE AND FUNCTION OF CELLS (continued)

Organelles

The **nucleus** is the largest organelle (*Figure 2.8*). It directs all the activities of the cell. The nucleus is surrounded by its own membrane. The 46 molecules of DNA in the nucleus form 46 **chromosomes** (*Figure 2.9*).

Each nucleus contains a **nucleolus,** a small dense body composed of **ribonucleic acid (RNA)** and protein. It is involved in the manufacture of proteins from simple materials—a process called **anabolism.**

Abbreviation	
RNA	ribonucleic acid

Mitochondria are the powerhouses of the cell. They produce energy by breaking down compounds such as glucose and fat—a process called **catabolism.**

Metabolism is the sum of the constructive processes of anabolism and the destructive processes of catabolism within a cell (**intracellular**).

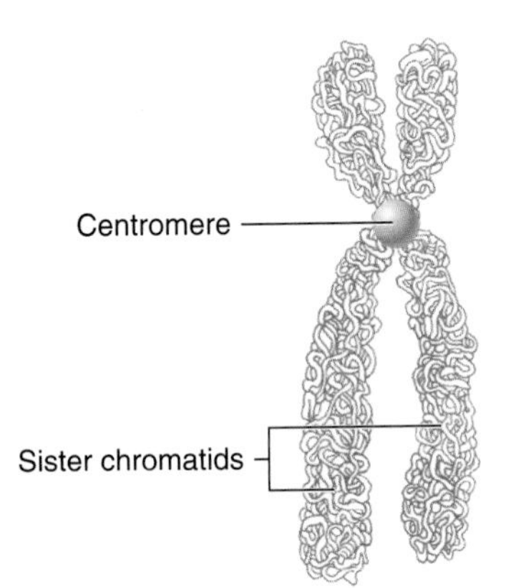

▲ **FIGURE 2.9**
Chromosome Structure.

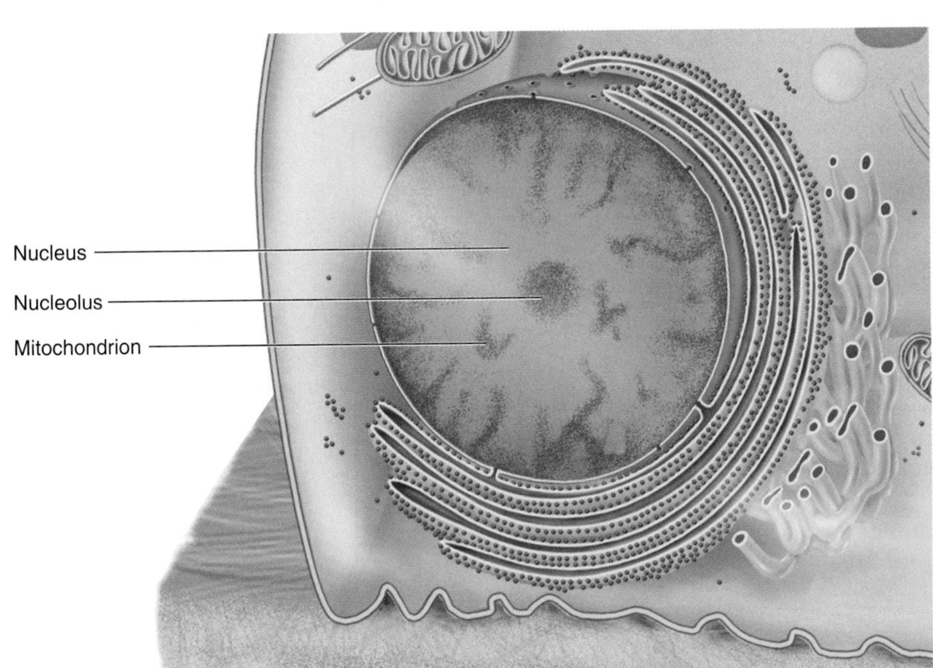

▲ **FIGURE 2.8** **The Nucleus.**

WORD ANALYSIS AND DEFINITION

S/ = Suffix   P/ = Prefix   R/ = Root   R/CF = Combining Form

WORD	PRONUNCIATION	ELEMENTS		DEFINITION
anabolism	an-**AB**-oh-lizm	S/ R/	**-ism** *process, condition* **anabol-** *build up*	The buildup of complex substances in the cell from simpler ones as a part of metabolism.
catabolism	kah-**TAB**-oh-lizm	S/ R/	**-ism** *process, condition* **catabol-** *break down*	The breakdown of complex substances into simpler ones as a part of metabolism.
chromosome	**KROH**-moh-sohm	S/ R/CF	**-some** *body* **chrom/o-** *color*	Body in the nucleus that contains DNA and genes.
intracellular	in-trah-**SELL**-you-lar	S/ P/ R/	**-ar** *pertaining to* **intra-** *within* **-cellul-** *small cell*	Within the cell.
nucleolus	nyu-**KLEE**-oh-lus	S/ R/CF	**-lus** small **nucle/o-** *nucleus*	Small mass within the nucleus.
nucleus	**NYU**-klee-us		Latin *command center*	Functional center of a cell or structure.
nuclear (adj)	**NYU**-klee-ar	R/ S/	**nucle-** *nucleus* **-ar** *pertaining to*	Pertaining to a nucleus.
ribonucleic acid (RNA)	**RYE**-boh-nyu-**KLEE**-ik **ASS-id**	S/ P/ R/	**-ic** *pertaining to* **ribo-** *from ribose, a sugar* **-nucle-** *nucleus*	The information carrier from DNA in the nucleus to an organelle to produce protein molecules.

EXERCISES

Match *the definition in column 1 with the correct medical term in column 2.* (Note: *Several terms are very similar in appearance, but their elements make them different.)*

	Definition	Medical Term
_____	1. part of nucleus with DNA and genes	A. nuclear
_____	2. destructive process in the cell	B. intracellular
_____	3. functional center of the cell	C. chromosome
_____	4. pertaining to the nucleus	D. catabolism
_____	5. constructive process in the cell	E. nucleus
_____	6. small mass within the nucleus	F. nucleolus
_____	7. within the cell	G. RNA
_____	8. information carrier for DNA	H. anabolism

LESSON 2.3 Tissues, Organs, and Organ Systems

THE BODY AS A WHOLE

OBJECTIVES

The information in this lesson will enable you to:

- **Define the four primary tissue groups.**
- **Discuss the medical terminology for the structure and functions of each tissue group.**
- **Name the organ systems.**
- **Describe the medical terminology for the functions of each organ system.**

CASE REPORT 2.2

You are

. . . a physical therapy assistant employed in the Rehabilitation Unit in Fulwood Medical Center.

You are communicating with

. . . Mr. Richard Josen, a 22-year-old man who injured tissues in his left knee while playing football (*Figure 2.10*). Using *arthroscopy,* the orthopedic surgeon removed his torn anterior *cruciate ligament* (ACL) and replaced it with a *graft* from his *patellar* tendon. The torn medial collateral ligament was sutured together. The tear in his medial *meniscus* was repaired. Rehabilitation focused on strengthening the *muscles* around his knee joint and regaining joint mobility and stability.

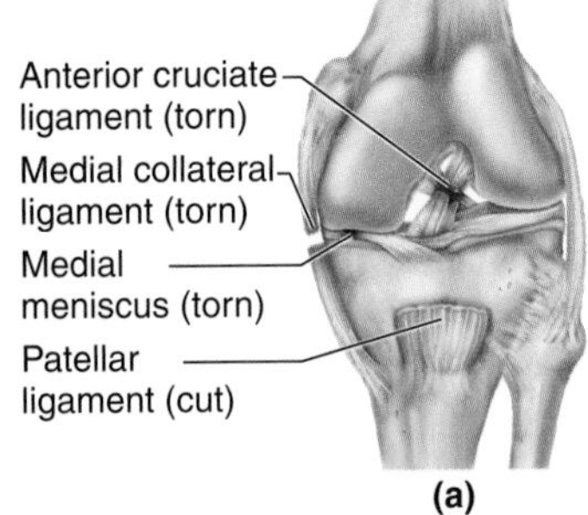

FIGURE 2.10 Knee Anatomy.
(a) Injury to left knee. (b) Normal knee.

TISSUES

Tissues hold your body together. The many tissues of your body have different structures for specialized functions. Each different tissue is made of similar cells with unique materials around them that are manufactured by the cells. **Histology** is the study of the structure and function of tissues. The four primary tissue groups are outlined in *Table 2.1.*

TABLE 2.1 The Four Primary Tissue Groups

Type	Function	Location
Connective	Bind, support, protect, fill spaces, store fat	Widely distributed throughout the body, e.g., in blood, bone cartilage, and fat
Epithelial	Protect, secrete, absorb, excrete	Cover body surface, cover and line internal organs, compose glands
Muscle	Movement	Attached to bones, in the walls of hollow internal organs, in the heart
Nervous	Transmit impulses for coordination, sensory reception, motor actions	Brain, spinal cord, nerves

Adapted from Shier, Butler, and Lewis, *Hole's Human Anatomy and Physiology,* 10th ed. Copyright © 2004 The McGraw-Hill Companies, Inc. Adapted with permission.

WORD ANALYSIS AND DEFINITION

S/ = Suffix P/ = Prefix R/ = Root R/CF = Combining Form

WORD	PRONUNCIATION	ELEMENTS		DEFINITION
arthroscopy	ar-**THROS-**koh-pee	S/ R/CF	**-scopy** *to examine, to view* **arthr/o-** *joint*	Visual examination of the interior of a joint.
connective tissue	koh-**NECK**-tiv **TISH**-you	S/ R/	**-ive** *pertaining to* **connect-** *join together* **tissue** Latin *to weave*	The supporting tissue of the body.
cruciate	**KRU**-she-ate		Latin *cross*	Shaped like a cross.
graft	GRAFT		French *transplant*	Transplantation of living tissue.
histology **histologist**	his-**TOL**-oh-jee his-**TOL**-oh-jist	S/ R/CF S/	**-logy** *study of* **hist/o-** *tissue* **-logist** *one who studies, specialist*	Study of the structure and function of cells, tissues, and organs. Specialist in histology.
ligament	**LIG**-ah-ment		Latin *band*	Band of fibrous tissue connecting two structures.
meniscus	meh-**NISS**-kuss		Greek *crescent*	Disc of cartilage between the bones of a joint.
muscle	**MUSS**-el		Latin *muscle*	A tissue consisting of contractile cells.
patella (singular) **patellae (pl)**	pah-**TELL**-ah pah-**TELL**-ee		Latin *small plate*	Thin, circular bone embedded in the patellar tendon in front of the knee joint; also called the *kneecap.*
patellar (adj)	pah-**TELL**-ar	S/ R/	**-ar** pertaining to **patell-** *patella*	Pertaining to the patella.

Abbreviation

ACL anterior cruciate ligament

EXERCISES

Dictionary exercise: *When you are working in the medical field, you will be exposed to medical terms you may not recognize. Learn to use a good medical dictionary, or practice going online to find the definitions you need. The Case Report on the opposite page contains some terms that are not defined in the WAD above.*

Use a dictionary (or go online) to define each of the following terms and identify it as noun, verb, or adjective.

1. *orthopedic* (noun, verb, adjective) ____________

   Definition:

   ____

   ____

2. *rehabilitation* (noun, verb, adjective) ____________

   Definition:

   ____

   ____

3. *collateral* (noun, verb, adjective) ____________

   Definition:

   ____

   ____

4. *sutured* (noun, verb, adjective) ____________

   Definition:

   ____

   ____

Learn to help yourself for lifelong learning!

CONNECTIVE TISSUES

To understand the relation of structure to function in the different tissues, the knee joint is used in this lesson to illustrate the structures and functions of the different tissues found in the joint.

Connective Tissues in the Knee Joint

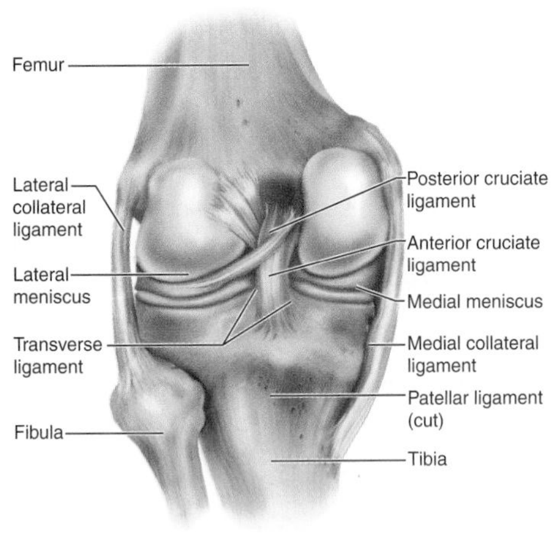

▲ **FIGURE 2.11 Ligaments of Knee Joint.**

- The **bones** of the knee joint are the **femur, tibia,** and **patella** (*see Chapter 9*). Bone is the hardest connective tissue due to the presence of calcium mineral salts, mostly calcium phosphate. Bones have a good blood supply that enables them to heal after a fracture. Bones as a whole are covered with a thick fibrous tissue called the **periosteum.**

- **Cartilage** has a flexible, rubbery **matrix** that allows it to function as a shock absorber (in the knee, as a **meniscus**) and a gliding surface where two bones meet to form a joint. Cartilage has very few blood vessels and heals poorly or not at all. When it is injured or torn, surgical repair is usually necessary. Cartilage also forms the shape of your ear, the tip of your nose, and your larynx.

- **Ligaments** are strips or bands of fibrous connective tissue made of **collagen** fibers. The knee joint has a complex array of 11 ligaments that hold it together. The blood supply to ligaments is poor, so they do not heal well without surgery (*Figure 2.11*).

- **Tendons** are thick, strong ligaments that attach muscles to bone.

- The **joint capsule** of the knee joint encloses the joint cavity and is made of thin, fibrous connective tissue. It is strengthened by fibers that extend over it from the ligaments and muscles surrounding the knee joint. These features are common to most joints.

- **The synovial membrane** lines many joint capsules and secretes **synovial fluid.** This fluid is a slippery lubricant retained in the joint cavity by the capsule. It makes joint movement almost friction-free and distributes **nutrients** to the cartilage on the joint surfaces of bone.

- **Muscle tissue** stabilizes the joint. Extensions of the tendons of the large muscles in front of, and in the rear of, the thigh are major stabilizers of the knee joint. The muscles themselves respectively extend and flex the joint (*see Chapter 9*).

- **Nervous tissue** carries messages between the brain and the knee structures. All the knee structures are well supplied with nerves, which is why a knee injury is excruciatingly painful.

WORD ANALYSIS AND DEFINITION

S/ = Suffix   P/ = Prefix   R/ = Root   R/CF = Combining Form

WORD	PRONUNCIATION		ELEMENTS	DEFINITION
capsule	**KAP**-syul		Latin *little box*	Fibrous tissue layer surrounding a joint or other structure.
capsular (adj)	**KAP**-syu-lar	S/ R/	**-ar** *pertaining to* **capsul-** *box*	Pertaining to a capsule.
cartilage	**KAR**-tih-lage		Latin *gristle*	Nonvascular, firm connective tissue found mostly in joints.
collagen	**KOLL**-ah-jen	S/ R/CF	**-gen** *produce, form* **coll/a-** *glue*	Major protein of connective tissue, cartilage, and bone.
matrix	**MAY**-triks		Latin mater *mother*	Substance that surrounds and protects cells, is manufactured by the cells, and holds them together.
nutrient	**NYU**-tree-ent	S/ R/	**-ent** *end result* **nutri-** *nourish*	A substance in food required for normal physiologic function.
periosteum	**PER**-ee-**OSS**-tee-um	S/ P/ R/	**-um** *tissue* **peri-** *around* **-oste-** *bone*	Fibrous membrane covering a bone.
synovial	si-**NOH**-vee-al	S/ P/ R/CF	**-al** *pertaining to* **syn-** *together* **-ov/i-** *egg*	Pertaining to the synovial membrane or fluid.
tendon	**TEN**-dun		Latin *sinew*	Fibrous band that connects muscle to bone.

EXERCISES

Match *the element in column 1 with the meaning in column 2. Give an example of a medical term containing that element in column 3. Some terms in the third column will appear more than once.*

	Element	Meaning	Medical Term
______	1. colla	A. box	____________
______	2. peri	B. together	____________
______	3. al	C. around	____________
______	4. oste	D. nourish	____________
______	5. nutri	E. tissue	____________
______	6. um	F. form	____________
______	7. capsul	G. pertaining to	____________
______	8. syn	H. egg	____________
______	9. ovi	I. bone	____________
______	10. gen	J. glue	____________
______	11. ent	K. pertaining to	____________
______	12. ar	L. end result	____________

ORGANS AND ORGAN SYSTEMS

An **organ** is a structure composed of several tissues that work together to carry out specific functions. For example, the skin is an organ that has different tissues in it such as epithelial cells, hair, nails, and glands (*see Chapter 3*).

An **organ system** is a group of organs with a specific collective function such as digestion, circulation, or respiration. For example, the nose, pharynx, larynx, trachea, bronchi, and lungs work together to achieve the total function of respiration (*see Chapter 7*).

The different organs in an organ system are usually interconnected. For example, in the **urinary** organ system (*Figure 2.12*), the organs are the kidneys, ureters, bladder, and urethra, and they are all connected (*see Chapter 9*).

All your **organ systems** work together to ensure that your body's internal environment remains relatively constant. This process is called **homeostasis.** It ensures that cells receive adequate nutrients and oxygen and that their waste products are removed. Your cells can then function normally. Disease affecting an organ or organ system disrupts this game plan of homeostasis.

The body has 11 organ systems, shown in *Table 2.2*. Muscular and skeletal are considered one organ system, the musculoskeletal system (*see Chapter 9*). Each body system has a chapter in this book where the terms associated with it are defined.

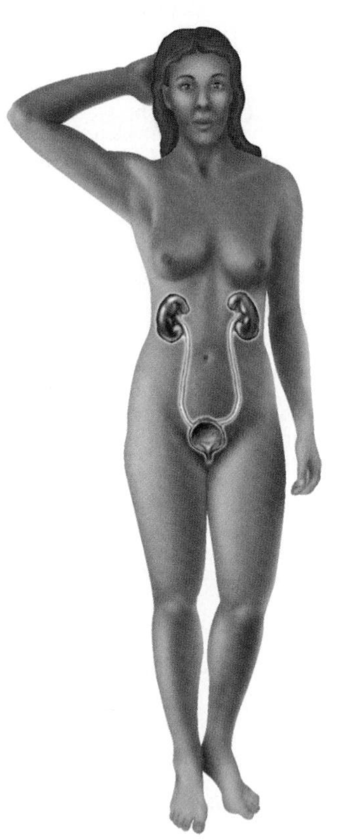

▲ **FIGURE 2.12 The Urinary System.**

Keynote

Homeostasis is the coordinated response of all the organs to maintain the internal physiologic stability of an organism.

TABLE 2.2 Organ Systems

Organ System	Major Organs	Major Functions
Integumentary	Skin, hair, nails, sweat glands, sebaceous glands	Protect tissues, regulate body temperature, support sensory receptors
Skeletal	Bones, ligaments, cartilages	Provide framework, protect soft tissues, provide attachments for muscles, produce blood cells, store inorganic salts
Muscular	Muscles	Cause movements, maintain posture, produce body heat
Nervous	Brain, spinal cord, nerves, sense organs	Receive and interpret sensory information, stimulate muscles and glands
Endocrine	Glands that secrete hormones: pituitary, thyroid, parathyroid, adrenal, pancreas, ovaries, testes, pineal, thymus	Control metabolic activities of organs
Cardiovascular	Heart, blood vessels	Move blood and transport substances throughout body
Lymphatic	Lymph vessels and nodes, thymus, spleen	Defend body against infection, return tissue fluid to blood, carry certain absorbed food molecules
Digestive	Mouth, tongue, teeth, salivary glands, pharynx, esophagus, stomach, liver, gallbladder, pancreas, small and large intestines	Receive, break down, and absorb food; eliminate unabsorbed material
Respiratory	Nasal cavity, pharynx, larynx, trachea, bronchi, lungs	Intake and output air, exchange gases between air and blood
Urinary	Kidneys, ureters, urinary bladder, and urethra	Remove wastes from blood, maintain water and electrolyte balance, store and transport urine
Reproductive	*Male:* scrotum, testes, epididymides, vasa deferentia, seminal vesicles, prostate, bulbourethral glands, urethra, penis	Produce and maintain sperm cells, transfer sperm cells into female reproductive tract
	Female: ovaries, fallopian tubes, uterus, vagina, vulva	Produce and maintain egg cells, receive sperm cells, support development of an embryo, function in birth process

WORD ANALYSIS AND DEFINITION

S/ = Suffix P/ = Prefix R/ = Root R/CF = Combining Form

WORD	PRONUNCIATION	ELEMENTS		DEFINITION
cardiovascular	**KAR**-dee-oh-**VAS**-kyu-lar	S/ R/CF R/	**-ar** *pertaining to* **cardi/o-** *heart* **-vascul-** *blood vessel*	Pertaining to the heart and blood vessels.
digestion	die-**JEST**-shun	S/ R/	**-ion** *action* **digest-** *break down food*	Breakdown of food into elements suitable for cell metabolism.
digestive (adj)	die-**JEST**-iv	S/	**-ive** *pertaining to*	Pertaining to digestion.
endocrine	**EN**-doh-krin	P/ R/	**endo-** *within* **-crine** *to secrete*	A gland that produces an internal or hormonal substance.
homeostasis (***Note:*** *Hemo*stasis is the arrest of bleeding.)	hoh-mee-oh-**STAY**-sis	S/ R/CF	**-stasis** *standstill, control* **home/o-** *the same*	Maintaining the stability of a system or the body's internal environment.
integument	in-**TEG**-you-ment		Latin *a covering*	Organ system that covers the body, the skin being the main organ within the system.
integumentary (adj)	in-**TEG**-you-**MENT**-ah-ree	S/ R/	**-ary** *pertaining to* **integument-** *-covering of the body*	Pertaining to the covering of the body.
lymph	LIMF		Latin *clear spring water*	Clear fluid collected from body tissues and transported by lymph vessels to the venous circulation.
lymphatic (adj)	lim-**FAT**-ic	S/ R/	**-atic** *pertaining to* **lymph-** *lymph, lymphatic system*	Pertaining to lymph or the lymphatic system.
nervous	**NER**-vus	S/ R/	**-ous** *pertaining to* **nerv-** *nerve*	Pertaining to a nerve or the nervous system; *or* easily excited or agitated.
nervous system	**NER**-vus **SIS**-tem		**system** Greek *an organized whole*	The whole, integrated nerve apparatus.
respiration	**RES**-pih-**RAY**-shun	S/ R/	**-ation** *process* **respir-** *to breathe*	Process of breathing; fundamental process of life used to exchange oxygen and carbon dioxide.
respiratory (adj)	**RES**-pih-rah-tor-ee	S/	**-atory** *pertaining to*	Pertaining to respiration.
skeleton	**SKEL**-eh-ton		Greek *skeleton or mummy*	The bony framework of the body.
skeletal (adj)	**SKEL**-eh-tal	S/ R/	**-al** *pertaining to* **skelet-** *skeleton*	Pertaining to the skeleton.
urinary (adj)	**YUR**-in-ary	S/ R/	**-ary** *pertaining to* **urin-** *urine*	Pertaining to urine.

EXERCISES

Build *the correct medical terms by working with the literal meanings of the elements in the WAD above. Write the correct elements on the line to complete the term.*

1. ______________ / ______________ / ______________
   lymph — pertaining to

2. ______________ / ______________ / ______________
   heart and blood vessels — pertaining to

3. ______________ / ______________ / ______________
   skeleton — pertaining to

4. ______________ / ______________ / ______________
   skin — pertaining to

5. ______________ / ______________ / ______________
   digestion — pertaining to

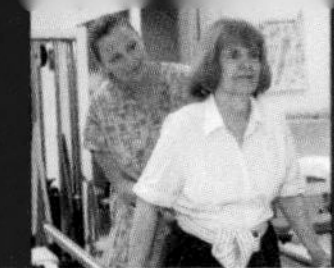

CHAPTER 2 REVIEW

THE BODY AS A WHOLE

CHALLENGE YOUR KNOWLEDGE

A. **Construct** the following medical terminology, using the elements as your guide. The word in capitals will be your clue to the missing element. Fill in the blanks.

1. RESEMBLING a solid ______________________________ /oid
2. SMALL organ ______________________________ /elle
3. pertaining to CHANGE ______________________________/ic
4. process of BREAKING DOWN ______________________________ /ism
5. use of an instrument to examine a JOINT ______________________________ /scopy
6. one who studies TISSUE ______________________________/logist
7. tissue AROUND bone ______________________________/oste/ um
8. standing THE SAME ______________________________ / stasis
9. pertaining to THE BACK ______________________________/al
10. pertaining to a NERVE ______________________________ /ous

B. **Word attack exercise:** This exercise will help you develop a method of analyzing medical terminology questions to determine the correct answer. Work the exercise step-by-step.

Question:

Which of the following terms means fibrous membrane covering a bone?

a. collagen
b. synovial
c. integumentary
d. endocrine
e. periosteum

1. The question concerns a "fibrous membrane covering a bone." Do you recognize any of these words in the question as having an element that matches an answer? Write them here:

   ____

2. Read the answer choices again, and cross off any that you know are not correct.

   The incorrect ones are ____.

3. Of the remaining answer choices, look for one that contains any element that matches any of the words in the question.

   (*See answer 1 above.*) ____.

4. The correct answer to this question is ____.

   because it contains the element(s) ____.

C. **Spelling correctly** is *always important* and is the mark of an educated professional. Choose the correct spelling to complete the patient's documentation. Circle the best choice.

1. This patient's (electrolite/electrolyte) balance should be checked once a day.
2. Mr. Josen has torn his anterior (cruxiate/cruciate) ligament and will require a surgical repair.
3. The medial (meniscus/menniscus) will be repaired at the same time.
4. The patient has worn away the (cartiledge/cartilage) in his kneecap.
5. This injectable drug should help replace the loss of (sinovial/synovial) fluid in the patient's knee.
6. Mr. Rose's (umbilical/umbellical) hernia surgery should be scheduled as soon as possible.
7. The patient's broken ribs have punctured his (thoracic/thorascic) cavity.
8. The doctor has ordered an x-ray of the patient's left (patella/patela).
9. Because she is bedridden, the patient's (integementary/integumentary) system is severely compromised.
10. Her (resperation/respiration) is shallow and labored.

D. **Roots/combining forms** are the foundation of every medical term. List here the 10 roots/combining forms in this chapter that you have the most difficulty remembering. Be sure to include their meanings, and provide an example of each in a medical term.

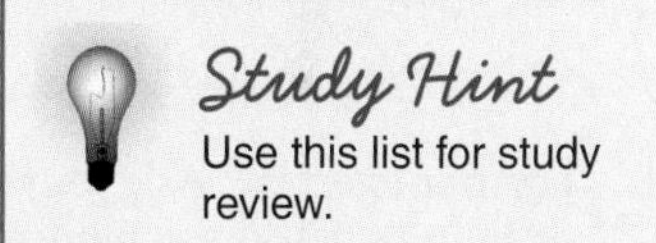
Study Hint
Use this list for study review.

Root/Combining Form	Meaning of R/CF	Example of a Medical Term

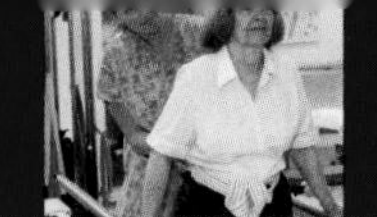

THE BODY AS A WHOLE

E. **Difference between:** If you know the meaning of the term, you can briefly explain it to someone else. Write your explanations below.

1. organ

____

____

____

2. organelle

____

____

____

3. The difference in these two terms is the element ______________________________,

   which means ______________________________.

F. **Partner exercise:** Ask your study partner to close his or her text. Dictate the following sentences to your partner, and then ask him or her to spell the sentences back to you. Check your partner's sentences against the text below. The sentence is not correct unless every word is present and *everything* is spelled correctly. When you have finished checking your partner's answers, close your book, ask your partner to dictate the sentences to you, and you write them down and then check them.

1. In order to prevent her hip replacement from dislocating, she has been instructed to not bring the right leg or knee medially across the sagittal plane.
2. The transverse, or horizontal, plane divides the body into an upper or superior portion and a lower or inferior portion.
3. The abdominal cavity is separated from the thoracic cavity by the diaphragm and contains the stomach, intestines, liver, spleen, pancreas, and kidneys.

G. **Layperson's language:** Translate the following sentences into language your patient can understand. Be brief, but be sure you are conveying the complete information. Rewrite the sentences on the lines below.

1. The doctor has told the patient her patellar joint lacks synovial fluid and this is the cause of her severe pain.

____

____

____

____

2. The orthopedic surgeon has recommended an arthroscopy to repair the torn ACL ligament and the torn medial meniscus.

____

____

____

H. Group the elements to make studying easier. Fill in the chart below with the meaning of the given element, and list a medical term containing that element. Remember to answer the question at the end of the exercise.

Element	Meaning of Element	Medical Term with This Element
al		
ar		
ation		
elle		
ia		
ic		
ior		
ity		
logist		
logy		
lus		
oid		
um		

1. These elements are all ______________________________.

I. What am I? A brief clue is given as to the identity of the term. Can you name them all? Fill in the blanks.

1. largest organelle ______________________________
2. part of the trunk between abdomen and neck ______________________________
3. powerhouse of the cell ______________________________
4. stable internal environment of the body ______________________________
5. the skin ______________________________
6. belly button (navel) ______________________________
7. fertilized egg ______________________________
8. smallest unit of the body capable of independent existence ______________________________
9. clear fluid collected from body tissues ______________________________
10. chemical messenger ______________________________

THE BODY AS A WHOLE

J. **Terminology challenge:** More than one element can have the same meaning. For example, there are many elements that all mean *pertaining to*. Find more pairs of elements in this chapter that both mean the same thing. Do *not* use any examples of *pertaining to*.

1. ______________ and ______________ both mean ______________
2. ______________ and ______________ both mean ______________
3. ______________ and ______________ both mean ______________
4. ______________ and ______________ both mean ______________

K. **Recall and review:** There is a large volume of medical terminology to learn. This exercise reviews terms that appear in this chapter and the previous one. S-t-r-e-t-c-h your memory—try to answer the questions without turning back in your book. Your knowledge of roots and two different suffixes will help you fill in the blanks and the chart.

Suffix for *study of* = ______________

Suffix for *specialist* = ______________

Base of Term	Medical term for the *Study of* This Field	Medical Term for a *Specialist* in This Field
the lungs		
the heart		
the nervous system		
cells		
the urinary system		
tissues		

L. **Identification** of elements, and knowledge of their meaning, will aid your understanding of the meaning of a term. Know your elements to increase your medical vocabulary! Fill in the chart.

Element	Meaning of Element	Medical Term Example	Meaning of Medical Term
caud			
cephal			
colla			
electro			
endo			
metabol			
oste			
poster			
scopy			
vascul			

The Integumentary System

The Essentials of the Language of Dermatology

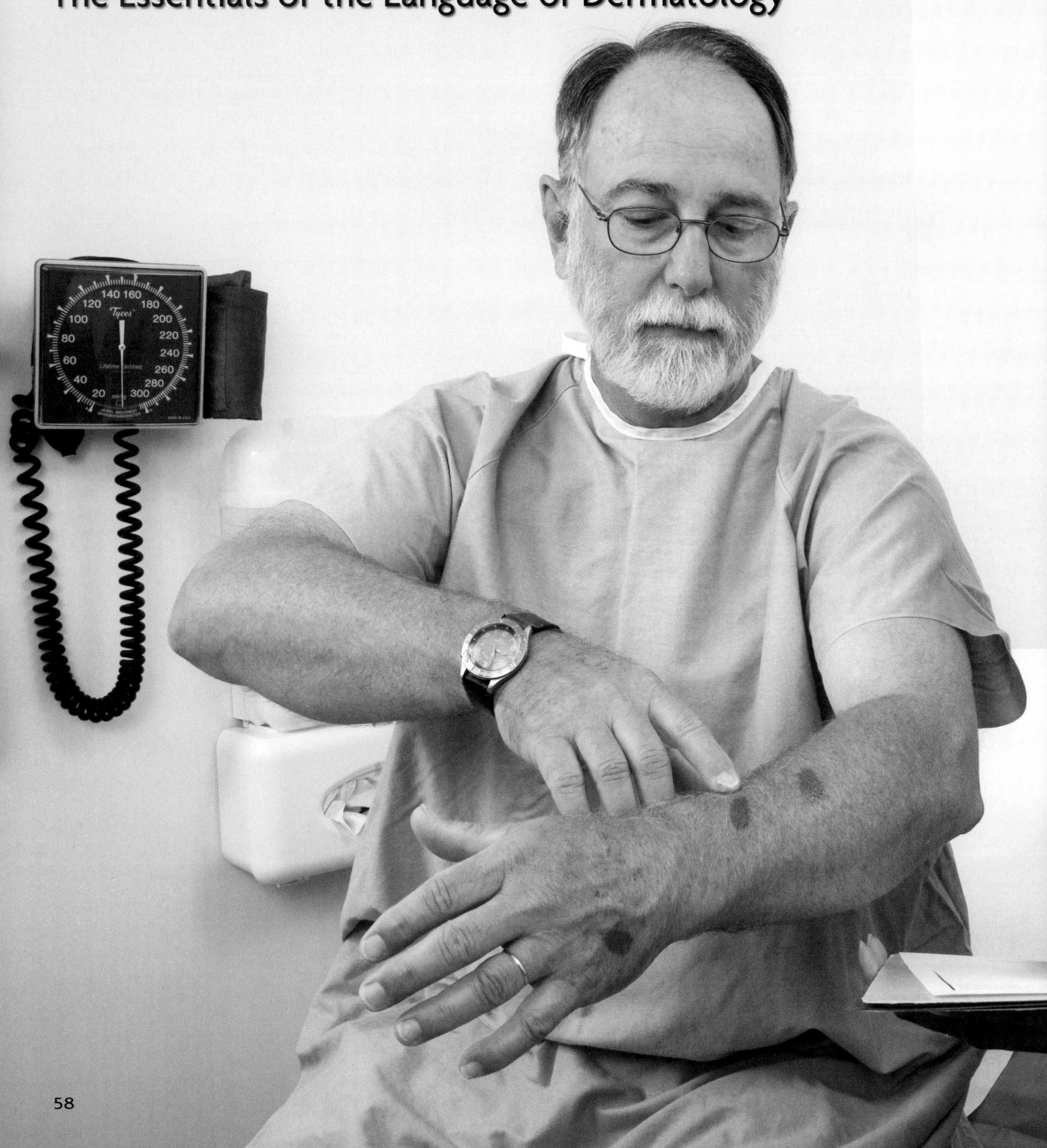

7. What two parts of this patient's knee were repaired?

____

____

8. What is the function of a *meniscus?*

____

____

____

9. What is the meaning of the term *"rehabilitation"*?

____

____

10. Describe "joint mobility and stability":

____

____

____

S. Chapter challenge: Circle the correct answer.

1. Circle the pair of terms that relate to the knee joint:

a.	skeleton	lymphatic
b.	integumentary	ligament
c.	coronal	quadrant
d.	diaphragm	zygote
e.	cruciate	meniscus

2. The suffix *stasis* means:

a. breaking down
b. stand still, control
c. to secrete
d. something formed
e. producing

3. Choose the term that has a root meaning *blood vessel:*

a. integument
b. cardiovascular
c. respiratory
d. lymph
e. dorsal

4. **Brain teaser:** Which term functions as a "shock absorber"?

a. ligament
b. membrane
c. tendon
d. meniscus
e. muscle

THE BODY AS A WHOLE

R. **Case Report challenge:** Now that you are more comfortable with the terms in this chapter, you can apply that knowledge and briefly answer the questions about the case report. If you read the report through once and then go back and underline all the medical terminology, this will make it easier to answer the questions. Fill in the blanks.

CASE REPORT 2.2

You are

. . . a physical therapy assistant employed in the Rehabilitation Unit in Fulwood Medical Center.

You are communicating with

. . . Mr. Richard Josen, a 22-year-old man who injured tissues in his left knee while playing football. Using ***arthroscopy,*** the orthopedic surgeon removed his torn anterior ***cruciate ligament*** (ACL) and replaced it with a ***graft*** from his ***patellar*** tendon. The torn medial collateral ligament was sutured together. The tear in his medial ***meniscus*** was repaired. Rehabilitation focused on strengthening the ***muscles*** around his knee joint and regaining joint mobility and stability.

1. What type of procedure is an *arthroscopy?*

____

____

____

____

2. What is the function of a *ligament?*

____

____

3. A *cruciate* ligament forms what shape?

____

4. What was removed from this patient? ____________________

5. What is *transplanted,* and where did it come from?

____

6. Describe the location of *medial:*

____

____

9. *Epigastric, hypogastric,* and *umbilical* describe:

a. body planes
b. directional terms
c. body cavities
d. anatomical positions
e. body regions

10. Which term is the opposite of *caudal* and the same as *superior:*

a. anterior
b. cephalic
c. coronal
d. distal
e. sagittal

11. Circle the term that has a root that means *chest:*

a. umbilicus
b. cranial
c. thoracic
d. nucleolus
e. collagen

12. What does a histologist specialize in studying?

a. blood
b. cells
c. tissues
d. heart
e. skin

13. "A chemical formed in one tissue or organ and carried by the blood to stimulate or inhibit a function of another tissue or organ" is the definition of:

a. lipid
b. electrolyte
c. protein
d. hormone
e. steroid

14. Which term has an element that means *color?*

a. metabolism
b. chromosome
c. intracellular
d. collagen
e. endocrine

15. Which of the following is the only horizontal plane?

a. posterior
b. anterior
c. superior
d. transverse
e. coronal

16. Find the only pair of incorrectly spelled medical terms:

a. mitochondria organelle
b. membraneous chromosone
c. muscle cruciate
d. patellar cartilage
e. synovial cytology

CHAPTER 2 REVIEW

THE BODY AS A WHOLE

Q. Chapter challenge: Circle the correct answer.

1. Which statement is most correct?
   a. Another name for cytoplasm is *intracellular fluid.*
   b. Organelles have no specific functions.
   c. The cell membrane is made of cartilage and lipids.
   d. The nucleus, nucleolus, and electrolytes are organelles.
   e. The cell membrane does not allow alcohol to pass through.

2. *Patella* is the correct medical term for:
   a. the thigh
   b. the muscle between two body cavities
   c. the kneecap
   d. the membrane covering a bone
   e. the disc of connective tissue in the knee

3. In the abbreviation *LLQ,* the first "L" stands for:
   a. lower
   b. left
   c. limp
   d. ligament
   e. lateral

> **Study Hint**
> Immediately cross off any answer you know is not correct. In your remaining choices, there is only *one best answer.*

4. Circle the pair of correct opposites:
   a. superior and lateral
   b. transverse and horizontal
   c. coronal and frontal
   d. distal and proximal
   e. cephalic and caudal

5. The diaphragm separates:
   a. the pelvic cavity and the spinal cavity
   b. the lungs and the heart
   c. the cranial cavity and the thoracic cavity
   d. the stomach and the intestines
   e. the thoracic cavity and the abdominopelvic cavity

6. Choose the correct body system and the organ it contains:
   a. integumentary — eyelash
   b. endocrine — pineal
   c. urinary — pancreas
   d. digestive — uterus
   e. reproductive — urethra

7. Circle the term with a combining form meaning *egg:*
   a. synovial
   b. metabolism
   c. homeostasis
   d. urinary
   e. cranial

8. Maintaining the body's internal environment is:
   a. hemostasis
   b. metabolism
   c. anabolism
   d. homeostasis
   e. catabolism

P. **Precision in communication:** The terminology for positions, planes, and directions includes very important terms you will hear often in medical practice, and you need to use them precisely. Some of the following statements are incorrect; rewrite them correctly on the lines below.

1. *Inferior* pertains to being situated above something.
2. Directional terms describe the position of one structure or part of the body relative to another structure or part of the body.
3. The heart is superior to the abdominopelvic cavity.
4. Saggital is a horizontal plane that divides the body into right and left portions.
5. Another name for a coronal plane is *frontal plane.*
6. Distal is opposite to dorsal.
7. Posterior is the opposite of cephalic.
8. When you lie flat on your back, you are prone.
9. Ventral is the same as posterior.
10. The transverse plane is the only one that is not vertical.
11. The spinal cavity is a posterior body cavity.
12. A flat surface that passes through the body is a plane.

Rewrite the incorrect statements:

THE BODY AS A WHOLE

O. Build the correct medical term using the appropriate combination of elements. There are more elements than you need. Some elements you may use twice. Fill in the blanks.

caud	cav	homeo	cyto
ism	cyt	cellul	arthro
ity	stasis	logist	syn
lyte	endo	intra	ar
catabol	logy	anabol	al
proxim	scopy	histo	ovi
ic	hemo	cephal	ia

1. pertaining to, or near, the head ______ / ______ / ______
   P R/CF S
2. nearest to the center of the body ______ / ______ / ______
   P R/CF S
3. hollow space ______ / ______ / ______
   P R/CF S
4. study of the cell ______ / ______ / ______
   P R/CF S
5. within the cell ______ / ______ / ______
   P R/CF S
6. buildup process in metabolism ______ / ______ / ______
   P R/CF S
7. visual examination of a joint ______ / ______ / ______
   P R/CF S
8. one who studies tissue ______ / ______ / ______
   P R/CF S
9. fluid that lubricates a joint ______ / ______ / ______
   P R/CF S
10. maintaining body's internal environment ______ / ______ / ______
    P R/CF S

M. **Latin and Greek terms** cannot be further deconstructed into prefix, root, or suffix. You must know them for what they are. Test your knowledge of these terms with this exercise. Match the meaning in column 1 with the correct medical term in column 2.

	Meaning	Medical Term
_____	1. yolk	A. lipid
_____	2. sinew	B. supine
_____	3. band	C. patella
_____	4. fence	D. membrane
_____	5. fat	E. prone
_____	6. bending forward	F. zygote
_____	7. small plate	G. tendon
_____	8. parchment	H. medial
_____	9. lying on the back	I. diaphragm
_____	10. middle	J. ligament

N. **Brain teaser:** From the description given, can you determine what the medical term is?

1. "The body is standing erect with feet flat on the floor, face and eyes are facing forward, and the arms are at the side with the palms facing forward."

   Medical term: ______________________________

2. "Opposite of 'caudal,' same as 'superior':

   Medical term: ______________________________

3. "Contains the heart, lungs, thymus gland, trachea, esophagus, and numerous blood vessels and nerves."

   Medical term: ______________________________

THE BODY AS A WHOLE

J. **Terminology challenge:** More than one element can have the same meaning. For example, there are many elements that all mean *pertaining to.* Find more pairs of elements in this chapter that both mean the same thing. Do *not* use any examples of *pertaining to.*

1. ______________ and ______________ both mean ______________
2. ______________ and ______________ both mean ______________
3. ______________ and ______________ both mean ______________
4. ______________ and ______________ both mean ______________

K. **Recall and review:** There is a large volume of medical terminology to learn. This exercise reviews terms that appear in this chapter and the previous one. S-t-r-e-t-c-h your memory—try to answer the questions without turning back in your book. Your knowledge of roots and two different suffixes will help you fill in the blanks and the chart.

Suffix for *study of* = ______________

Suffix for *specialist* = ______________

Base of Term	Medical term for the *Study of* This Field	Medical Term for a *Specialist* in This Field
the lungs		
the heart		
the nervous system		
cells		
the urinary system		
tissues		

L. **Identification** of elements, and knowledge of their meaning, will aid your understanding of the meaning of a term. Know your elements to increase your medical vocabulary! Fill in the chart.

Element	Meaning of Element	Medical Term Example	Meaning of Medical Term
caud			
cephal			
colla			
electro			
endo			
metabol			
oste			
poster			
scopy			
vascul			

H. Group the elements to make studying easier. Fill in the chart below with the meaning of the given element, and list a medical term containing that element. Remember to answer the question at the end of the exercise.

Element	Meaning of Element	Medical Term with This Element
al		
ar		
ation		
elle		
ia		
ic		
ior		
ity		
logist		
logy		
lus		
oid		
um		

1. These elements are all ______________________________.

I. What am I? A brief clue is given as to the identity of the term. Can you name them all? Fill in the blanks.

1. largest organelle ______________________________
2. part of the trunk between abdomen and neck ______________________________
3. powerhouse of the cell ______________________________
4. stable internal environment of the body ______________________________
5. the skin ______________________________
6. belly button (navel) ______________________________
7. fertilized egg ______________________________
8. smallest unit of the body capable of independent existence ______________________________
9. clear fluid collected from body tissues ______________________________
10. chemical messenger ______________________________

CHAPTER

CASE REPORT 3.1

You are

. . . a dermatology technician working with dermatologist Laura Echols, MD, a member of the Fulwood Medical Group.

You are communicating with

. . . Mr. Rod Andrews, a 60-year-old man, who shows you three skin lesions, two on his left forearm and one on the back of his left hand. You learn that he has been living for the past 10 years in Arizona and has come back home to be near his daughter and young grandchildren. You find no other skin lesions on his body.

Learning Outcomes

In addition to anticipating Dr. Echols' needs for equipment to biopsy, diagnose, and treat the lesions, you also have to be able to communicate clearly with her in medical terms and understand her language as she communicates with you and the patient about the etiology (cause) and structure of the lesions. You will then need to document the medical history and treatment and communicate clearly with Mr. Andrews about the treatment of his lesions and their prognosis.

To perform these tasks, you must be able to:

3.1 Apply the language of dermatology to the skin and its associated organs.

3.2 Comprehend, analyze, spell, and write the medical terms of dermatology.

3.3 Recognize and pronounce the medical terms of dermatology.

3.4 Describe the etiology, treatment, and prognosis of common dermatologic conditions.

LESSON 3.1 Structure and Function of the Skin

THE INTEGUMENTARY SYSTEM

OBJECTIVES

The three lesions on Mr. Andrews' arm and hand developed in the outer layer of the skin called the **epidermis.** This lesson looks at the structure and function of the skin so that you will be able to:

- **Identify the medical terminology for describing the functions of the skin.**
- **Name the tissues in the different layers of the skin.**
- **Specify the medical terminology for the functions of the different layers of the skin.**

Case Report 3.1 *(continued)*

When Dr. Echols examined Mr. Andrews, she determined clinically that two of his lesions were basal cell **carcinomas,** and she treated them with **cryosurgery.** She believed that the third lesion was a **squamous cell carcinoma,** and she performed a **biopsy removal** of the **cutaneous** lesion. You sent this to the laboratory with a request for pathologic diagnosis and determination of whether the lesion had been completely removed.

Keynote

Skin is the largest and most vulnerable organ in the body and accounts for 7% to 8% of body weight.

The **integumentary system** consists of the skin and its associated organs (*Figure 3.1*). The study and treatment of the integumentary system is called **dermatology.** The medical specialist in disorders of the skin is a **dermatologist.**

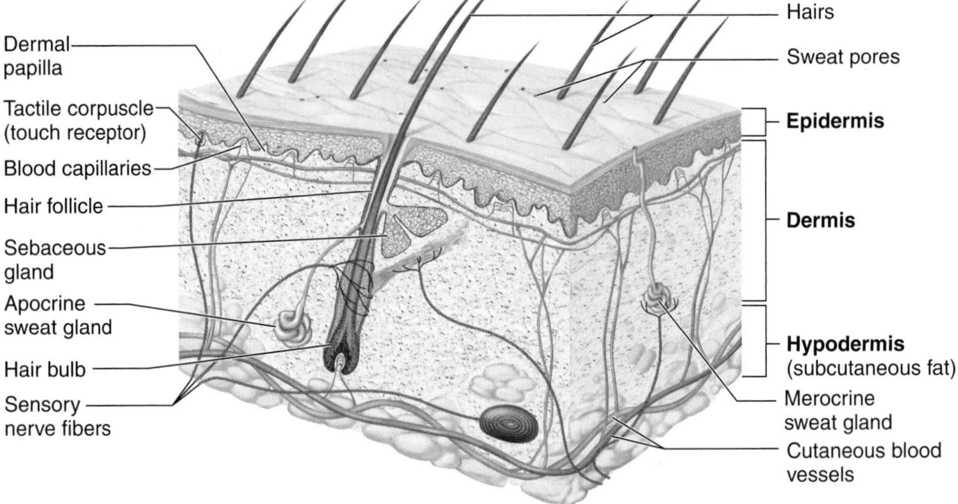

FIGURE 3.1
Structure of the Skin and Subcutaneous Tissue.

WORD ANALYSIS AND DEFINITION

S/ = Suffix P/ = Prefix R/ = Root R/CF = Combining Form

WORD	PRONUNCIATION	ELEMENTS		DEFINITION
biopsy	**BI**-op-see	S/ R/	**-opsy** *to view* **bi-** *life*	Removing tissue from a living person for laboratory examination.
carcinoma	kar-sih **-NOH-** *mah*	S/ R/	**-oma** *tumor, mass* **carcin-** *cancer*	A malignant and invasive epithelial tumor.
cryosurgery	cry-oh-**SUR**-jer-ee	S/ R/ R/	**-ery** *process of* **cryo-** *icy cold* **-surg-** *operate*	Use of liquid nitrogen or argon gas in a probe to freeze and kill abnormal tissue.
cutaneous	kyu-**TAY**-nee-us	S/ R/CF	**-ous** *pertaining to* **cutan/e-** *skin*	Pertaining to the skin.
dermatology **dermatologist**	der-mah-**TOL**-oh-jee der-mah-**TOL**-oh-jist	S/ R/CF S/	**-logy** *study of* **dermat/o-** *skin* **-logist** *one who studies, specialist*	Medical specialty concerned with disorders of the skin. Medical specialist in diseases of the skin.
dermatologic (adj) **dermis**	der-mah-toh-**LOJ**-ik **DER-miss**	S/ R/	**-ic** *pertaining to* **-log-** *to study* Latin *skin*	Pertaining to the skin and dermatology. Connective tissue layer of the skin beneath the epidermis.
epidermis **epidermal (adj)**	ep-ih-**DER**-miss ep-ih-**DER**-mal	P/ R/ S/ R/	**epi-** *above, upon* **-dermis** *skin* **-al** *pertaining to* **-derm-** *skin*	Top layer of the skin. Pertaining to the epidermis.
etiology	ee-tee-**OL**-oh-jee	S/ R/CF	**-logy** *study of* **eti/o-** *cause*	The study of the causes of a disease.
integument **integumentary (adj)**	in-**TEG**-you-ment in-**TEG**-you-**MENT**-ah-ree	 S/ R/	Latin *a covering* **-ary** *pertaining to* **integument-** *covering of the body*	Organ system that covers the body, the skin being the main organ within the system. Pertaining to the covering of the body.
prognosis	prog-**NO**-sis	P/ R/	**pro-** *projecting forward* **-gnosis** *knowledge*	Forecast of the probable future course and outcome of a disease.
squamous cell	**SKWAY**-mus SELL		Latin *scaly*	Flat, scale-like epithelial cell.

EXERCISES

***Identify** the following elements, and define their meaning. These elements will appear in more terminology throughout this book. Learn them now to increase your vocabulary. Fill in the chart, and answer the question below it.*

Element	Identity of Element (P, R, CF, or S)	Meaning of Element
bio		
carcin		
cryo		
cutane		
dermato		
dermis		
epi		
etio		
gnosis		
oma		

Note: More than one element can have the same meaning, but the elements are not interchangeable. For example, *dermatology* always has the element *dermat/o*—no other element meaning *skin* will be used for *study of skin.*

1. List all the elements in the WAD above that can mean *skin:*

____

____

____

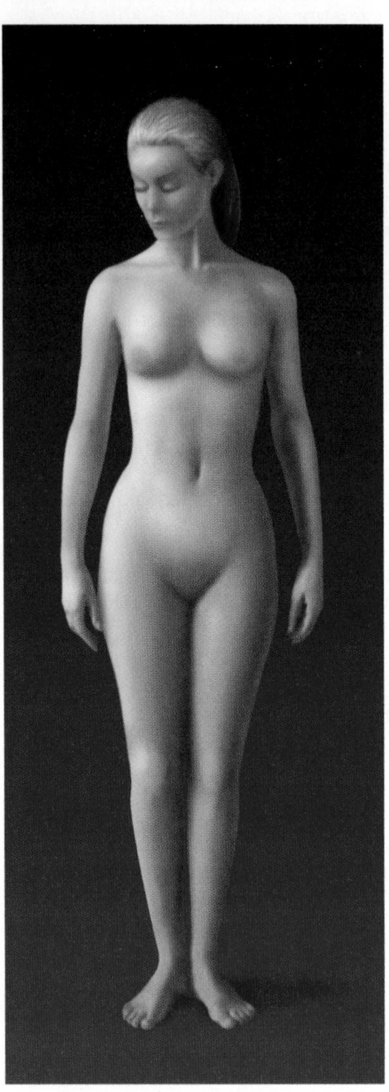

▲ **FIGURE 3.2**
Integumentary System.
The skin provides protection, contains sensory organs, and helps control body temperature.

FUNCTIONS OF THE SKIN

- **Protection:** The skin is a physical barrier against injury, chemicals, ultraviolet rays, microbes, and toxins that is not easily breached (*Figure 3.2*). The surface of the skin is populated by bacteria and other pathogens, called normal **flora.**
- **Water resistance:** You don't swell up every time you take a bath because your skin is water resistant. It also prevents water from leaking out of the body tissues.
- **Temperature regulation:** A network of capillaries in the skin opens up or dilates **(vasodilation)** when your body is too hot. When your body is cold, the capillary network narrows **(vasoconstriction),** blood flow decreases, and heat is retained in your body (*see Chapter 6*).
- **Vitamin D synthesis:** As little as 15 to 30 minutes of sunlight daily allows your skin cells to initiate the metabolism of vitamin D, which is essential for bone growth and maintenance.
- **Sensation:** Nerve endings that detect touch, pressure, heat, cold, pain, vibration, and tissue injury are particularly numerous in the skin of your face, fingers, palms, soles, nipples, and genitals.
- **Excretion and secretion:** Water and small amounts of waste products from cell metabolism are lost through the skin by **excretion** and by **secretion** from your sweat glands.
- **Social functions:** The skin reflects your emotions, blushing when you are self-conscious, going pale when you are frightened, wrinkling when you dislike something.

WORD ANALYSIS AND DEFINITION

S/ = Suffix P/ = Prefix R/ = Root R/CF = Combining Form

WORD	PRONUNCIATION	ELEMENTS		DEFINITION
excrete (verb) **excretion (noun)**	eks-**KREET** eks-**KREE**-shun	P/ R S/	**ex-** *out of, away from* **-crete** *separate* **-ion** *action*	To pass waste products of metabolism out of the body. Removal of waste products of metabolism out of the body.
flora	**FLO**-rah		Latin *flower*	The population of microorganisms covering the exterior and interior surfaces of healthy animals.
regulation (noun) **regulate (verb)**	reg-you-**LAY**-shun reg-you-LATE	S/ R/ S/	**-ation** *process* **regul-** *to rule* **-ate** *pertaining to*	Control of the way in which a process progresses. To control the way in which a process progresses.
resistance **resistant**	ree-**ZIS**-tants ree-**ZIS**-tant	S/ R/ S/	**-ance** *state of* **resist-** *to withstand* **-ant** *pertaining to, forming*	The ability of an organism to withstand the effects of an antagonistic agent. Able to resist.
secrete (verb) **secretion (noun)**	seh-**KREET** seh-**KREE**-shun	R/ S/	**secret-** *produce* **-ion** *action*	To produce a chemical substance in a cell and release it from the cell.
sensation (noun) **sense (verb)**	sen-**SAY**-shun	S/ R/	**-ation** *process* **sens-** *to feel*	The conscious feeling of the effects of a stimulation.
synthesis **synthetic (adj)**	**SIN**-the-sis sin-**THET**-ik	P/ R/ S/ R/	**syn-** *together* **-thesis** *to organize, arrange* **-ic** *pertaining to* **-thet-** *arrange, organize*	The process of building a compound from different elements. Built up or put together from simpler compounds.
vaso*constriction* **vaso*dilation***	**VAY**-soh-con-**STRIK**-shun **VAY**-soh-dih-**LAY**-shun	S/ R/CF R/ R/	**-ion** *action* **vas/o-** *blood vessel* **-constrict-** *narrow* **-dilat-** *widen, open up*	*Reduction* in diameter of a blood vessel. *Increase* in diameter of a blood vessel.

EXERCISES

***Construct** the correct medical term from the elements given. Every term needs a root or combining form; every term may not have a prefix. Fill in the blanks, and write a sentence. Some elements you may need to use twice.*

thesis	constrict	cret/e	resist	ance	syn
vas/o	ex	sens	ation	ion	

1. ability to withstand the effects of another substance ________ / ________ / ________
   P R/CF S
2. narrowing of a blood vessel ________ / ________ / ________
   P R/CF S
3. to pass waste products out of the body ________ / ________ / ________
   P R/CF S
4. building a compound from different elements ________ / ________ / ________
   P R/CF S
5. conscious feeling of the effects of stimulation ________ / ________ / ________
   P R/CF S
6. Use any one of these terms in a sentence of your own choice that is *not* a definition.

____

____

____

STRUCTURE OF THE SKIN

Epidermis

The three lesions that Mr. Andrews had were present in the epidermis, the most **superficial** layer of his skin.

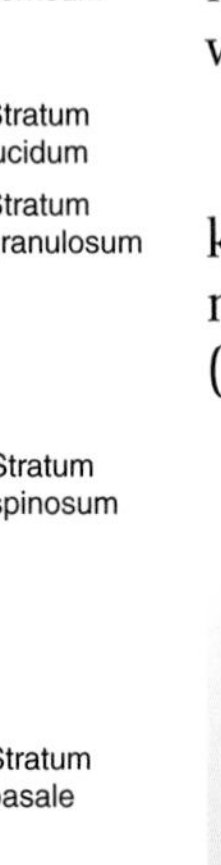

▲ **FIGURE 3.3**
Epidermis.

Abbreviation	
UV	ultraviolet

The outer layer of the epidermis (*Figure 3.3*) is a layer of compact, dead cells packed with **keratin** that are continually shed. Keratin is a tough, scaly protein that is also the basis for hair and nails. **Dandruff** is clumps of these cells stuck together with **sebum,** oil from **sebaceous glands.**

In the lower layers of the epidermis, cells are filled with a protein that becomes keratin. Other cells produce a brown/black pigment called **melanin,** which determines the color of the skin and also protects the skin from damage by **ultraviolet** (UV) light.

Case Report 3.1 *(continued)*

Dr. Echols learned that Mr. Andrews had driven extensively in Arizona wearing a short-sleeved shirt. His left forearm and hand were exposed to sunlight through the untinted car window to his left. This was an important factor in causing his skin cancers, all of which responded to treatment.

WORD ANALYSIS AND DEFINITION

S/ = Suffix P/ = Prefix R/ = Root R/CF = Combining Form

WORD	PRONUNCIATION	ELEMENTS		DEFINITION
dandruff	**DAN**-druff		Source unknown	Scales in hair from shedding of the epidermis.
keratin	**KER**-ah-tin	S/ R/	**-in** *substance* **kerat-** *hard protein*	Protein present in skin, hair, and nails.
melanin	**MEL**-ah-nin	S/ R/	**-in** *substance* **melan-** *black pigment*	Black pigment found in skin, hair, and the retina.
sebaceous glands **sebum**	seh-**BAY**-shus GLANZ **SEE**-bum	S/ R/CF	**-ous** *pertaining to* **sebac/e-** *wax* Latin *tallow*	Glands in the dermis that open into hair follicles and secrete a waxy fluid called *sebum.* Waxy secretion of the sebaceous glands.
superficial	soo-per-**FISH**-al		Latin *surface*	Situated near the surface.
ultraviolet	ul-trah-**VIE**-oh-let	P/ R/	**ultra-** *beyond* **-violet** *violet, bluish purple*	Light rays at a higher frequency than the violet end of the spectrum.

EXERCISES

Practice *using your medical terminology in the following exercise. When possible, be sure to deconstruct the term, using the slashes provided. Fill in the blanks.*

1. This pigment is responsible for skin color:

_______________ / _______________ / _______________

P R/CF S

2. Skin needs protection from this type of light:

_______________ / _______________ / _______________

P R/CF S

3. These glands secrete an oily substance:

_______________ / _______________ / _______________

P R/CF S

4. This hard protein is present in skin and nails:

_______________ / _______________ / _______________

P R/CF S

Complete the following statements.

5. The opposite of a deep wound is a _______________ wound.
6. Scales in the hair from the shedding of epidermis are called _______________.
7. The waxy secretion of sebaceous glands is called _______________.

STRUCTURE OF THE SKIN (continued)

Dermis

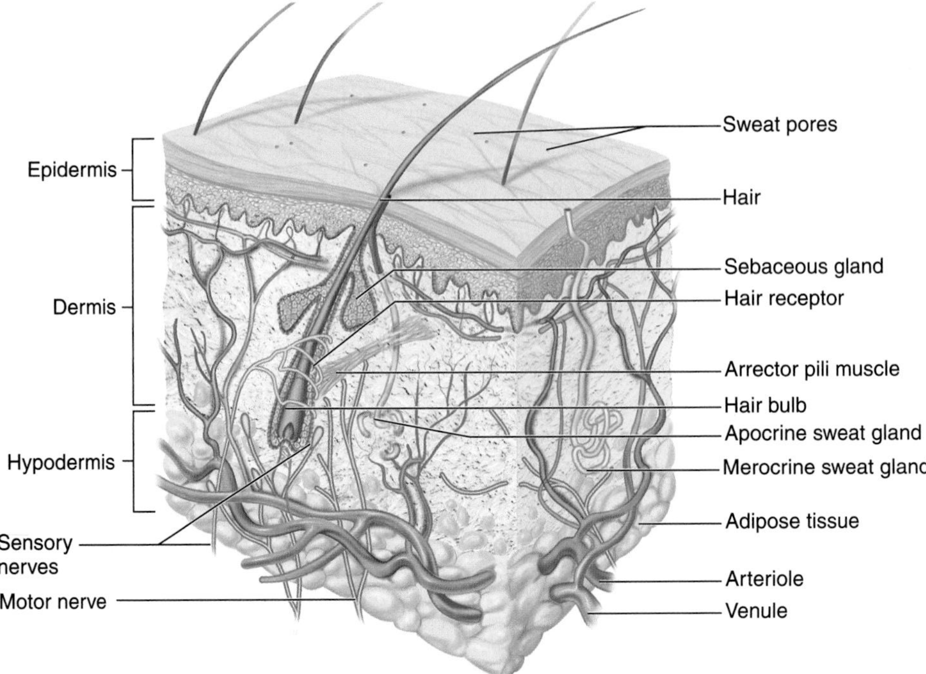

FIGURE 3.4
Dermis and Its Organs.

Figure 3.4 shows that the **dermis** is a much thicker connective tissue layer than the epidermis. It consists mostly of **collagen** fibers. It is well supplied with blood vessels and nerves and contains the other skin organs: sweat glands, sebaceous glands, hair **follicles,** and nail roots.

Hypodermis or Subcutaneous Tissue Layer

This layer beneath the dermis is the site of **subcutaneous** fat (**adipose** tissue). It is also called the *subcutaneous tissue layer.*

Clinical Applications

Injections are given into three areas of the skin:

- **Intradermal,** in which a short, thin needle is introduced into the epidermis, raising a small **wheal.** This site is used for allergy testing or a tuberculosis (**TB**) test.
- **Subcutaneous (SC),** in which a longer needle pierces the epidermis and dermis to reach the hypodermis (subcutaneous) layer. This site is used for insulin injections and for some immunizations.
- **Intramuscular (IM),** in which a long needle penetrates the epidermis, dermis, and hypodermis to reach into the muscles underneath. Some antibiotics and some immunizations are given by this route.

In adition there are **transdermal** applications, in which some medications are administered through the skin by an adhesive transdermal patch that is applied to the skin. The medication diffuses across the epidermis and enters the blood vessels in the dermis. Contraceptive hormones, **analgesics,** and antinausea/antiseasickness medications are examples.

Abbreviations

IM	intramuscular
SC	subcutaneous
TB	tuberculosis

WORD ANALYSIS AND DEFINITION

S/ = Suffix P/ = Prefix R/ = Root R/CF = Combining Form

WORD	PRONUNCIATION	ELEMENTS		DEFINITION
adipose	**ADD**-ih-pose	S/ R/	**-ose** *full of* **adip-** *fat*	Containing fat.
analgesic (adj) **analgesia**	an-al-**JEE**-sic an-al-**JEE**-zee-ah	S/ P/ R/ S/	**-ic** *pertaining to* **an-** *without* **-alges-** *sensation of pain* **-ia** *condition*	Substance that reduces or relieves the response to pain without producing loss of consciousness. State in which pain is reduced or relieved.
collagen	**KOLL**-ah-jen	S/ R/CF	**-gen** *source, producer* **coll/a-** *glue*	The major protein of connective tissue, cartilage and bone.
dermis	**DER**-miss		Greek *skin*	Connective tissue layer of the skin beneath the epidermis.
dermal	**DER**-mal	S/ R/	**-al** *pertaining to* **derm-** *skin*	Pertaining to the skin.
follicle	**FOLL**-ih-kull		Latin *small sac*	Mass of cells containing a cavity or a small cul-de-sac, such as a hair follicle.
hypodermis **hypodermic (adj)** (same as *subcutaneous*)	high-poh-**DER**-miss high-poh-**DER**-mik	P/ R/ S/ R/	**hypo-** *below* **-dermis** *skin* **-ic** *pertaining to* **-derm-** *skin*	Tissue layer of skin below the dermis. Pertaining to the hypodermis.
intradermal	in-trah-**DER**-mal	S/ P/ R/	**-al** *pertaining to* **intra-** *within* **-derm-** *skin*	Within the epidermis.
intramuscular	in-trah-**MUSS**-kew-lar	S/ P/ R/	**-ar** *pertaining to* **intra-** *within* **-muscul-** *muscle*	Within the muscle.
subcutaneous (same as *hypodermic*)	sub-kew-**TAY**-nee-us	S/ P/ R/CF	**-ous** *pertaining to* **sub-** *below* **-cutan/e-** *skin*	Below the skin.
transdermal	trans-**DER**-mal	S/ P/ R/	**-al** *pertaining to* **trans-** *across, through* **-derm-** *skin*	Going across or through the skin.
wheal (same as *hives*)	WHEEL		Old English *wheal*	Small, itchy swelling of the skin. (Wheals raised by an injection do not itch.)

EXERCISES

Prefixes are often used to denote numbers, location, and colors. Concentrate on prefixes from this and previous chapters, because they will appear in many medical terms. Fill in the chart, and answer the questions below it.

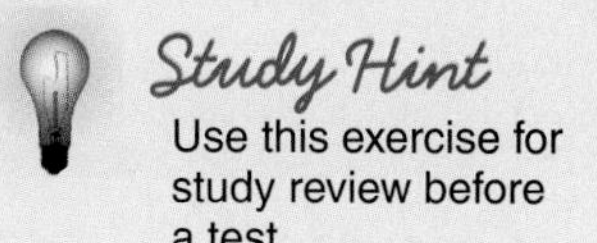

Use this exercise for study review before a test.

Prefix	Meaning of Prefix	Medical Term with This Prefix
epi		
hypo		
intra		
sub		
trans		

1. As a group, these prefixes all describe ____________________

2. Write a sentence about a patient, using any one term from the above table.

____

____

____

LESSON 3.2 Disorders of the Skin

THE INTEGUMENTARY SYSTEM

OBJECTIVES

Your skin provides the first line of defense against injury, disease, **allergens,** and pollutants. Because your skin is constantly exposed to the elements, it is susceptible to many different problems and disorders. The skin shows the same types of disease as most organs—infections, tumors, cancers—but, in addition, because of its protective covering, it is the first responder to many irritant and **allergenic** agents. The information in this lesson will enable you to:

- **Describe common disorders of the skin.**
- **Apply correct medical terminology in describing these disorders.**

You are

. . . a medical assistant working with dermatologist Dr. Lenore Echols in Fulwood Medical Center.

You are communicating with

. . . Ms. Cheryl Fox, a 37-year-old nursing assistant working in a surgical unit in Fulwood Medical Center.

CASE REPORT 3.2

Recently her fingers have become red and itchy, with occasional **vesicles.** She has also noticed irritation and swelling of her earlobes and a generalized **pruritus.** Over the weekends both the itching and the **rash** on her hands worsen. A patch test by Dr. Echols showed that Ms. Fox is allergic to nickel in rings that she wears on both hands and in her earrings. She had been wearing these on weekends and not during her workdays.

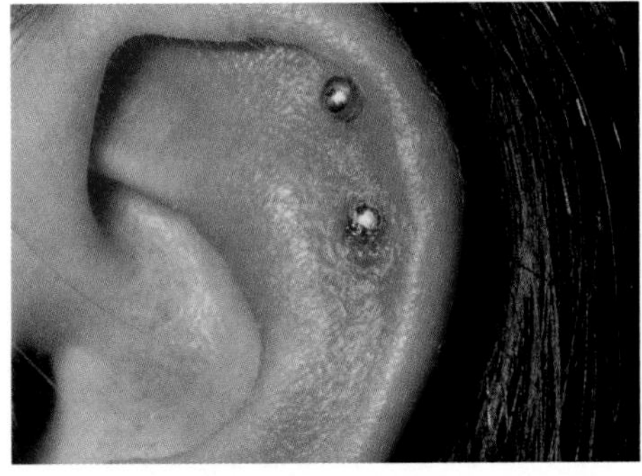

▲ FIGURE 3.5
Dermatitis of the Ear.

DERMATITIS

Dermatitis, also called **eczema,** is an inflammation that produces swollen, red, itchy skin. Ms. Fox has a dermatitis (*Figure 3.5*), resulting from direct exposure to an irritating agent.

The types of dermatitis include:

- **Contact dermatitis,** which results from direct contact with irritants or allergens. Irritants include soaps, detergents, cleaning products, and solvents.
- **Atopic** or **allergic dermatitis,** which is due to allergens that include nickel in jewelry (as with Ms. Fox), perfume, cosmetics, poison ivy, and latex.

Case Report 3.2 (continued)

For Ms. Fox, the allergy was not just a local reaction to an irritant. Her form of atopic or allergic dermatitis develops when the whole body becomes sensitive to an allergen. This whole-body involvement is shown by her systemic symptoms of pruritus distant from the local irritant site. Ms. Fox has stopped wearing the rings and earrings.

- **Seborrheic dermatitis,** which produces a red **rash** overlaid with a yellow, oily scale. This type of dermatitis is common in people with oily skin or hair.
- **Stasis dermatitis,** which occurs in the lower leg when varicose veins slow the return of blood and the accumulation of fluid interferes with the nourishment of the skin.

Eczema is a general term used for inflamed, itchy skin conditions. When the itchy skin is scratched, it becomes **excoriated** and produces the dry, red, scaly patches characteristic of eczema. The atopic dermatitis that Ms. Fox developed is a common form of eczema.

WORD ANALYSIS AND DEFINITION

S/ = Suffix P/ = Prefix R/ = Root R/CF = Combining Form

WORD	PRONUNCIATION		ELEMENTS	DEFINITION
allergen (The duplicate "g" is deleted.) **allergenic (adj)**	**AL**-er-jen al-er-**JEN**-ik	S/ R/ R/ S/	**-gen-** *produce* **all-** *strange, other* **-erg-** *work, activity* **-ic** *pertaining to*	Substance producing a hypersensitivity (allergic) reaction. Pertaining to the capacity to produce an allergic reaction.
allergy **allergic (adj)**	**AL**-er-jee ah-**LER**-jik			Hypersensitivity to an allergen. Pertaining to being hypersensitive.
atopy **atopic (adj)**	**AT**-oh-pee ay-**TOP**-ik	 S/	Greek *strangeness* **-ic** *pertaining to*	State of hypersensitivity to an allergen—allergic. Pertaining to an allergy.
dermatitis	der-mah-**TYE**-tis	S/ R/	**-itis** *inflammation* **dermat-** *skin*	Inflammation of the skin.
eczema **eczematous (adj)**	**EK**-zeh-mah **EK**-zem-ah-tus	 S/ R/CF	Greek *to boil or ferment* **-tous** *pertaining to* **eczem/a-** *eczema*	Inflammatory skin disease, often with a serous discharge. Pertaining to or marked by eczema.
excoriate (verb) **excoriation (noun)**	eks-**KOR**-ee-ate eks-**KOR**-ee-**AY**-shun	S/ P/ R/ S/	**-ate** *pertaining to* **ex-** *away from* **-cori-** *skin* **-ation** *process*	To scratch. Scratch mark.
pruritus **pruritic (adj)** **antipruritic**	proo-**RYE**-tus proo-**RIT**-ik **AN**-tee-proo-**RIT**-ik	 S/ R/ P/	Latin *to itch* **-ic** *pertaining to* **prurit-** *itch* **anti-** *against*	Itching. Itchy. Medication against itching.
rash	RASH		French *skin eruption*	Skin eruption.
seborrhea **seborrheic (adj)** (The "a" is deleted to enable the word to flow.)	seb-oh-**REE**-ah seb-oh-**REE**-ik	S/ R/CF S/	**-rrhea** *flow* **seb/o-** *sebum* **-ic** *pertaining to*	Excessive amount of sebum. Pertaining to seborrhea.
stasis	**STAY**-sis		Greek *staying in one place*	Stagnation in the flow of any body fluid.
vesicle	**VES**-ih-kull		Latin *blister*	Small sac containing liquid; e.g., a blister.

EXERCISES

Medical documentation *should always be neat, legible, and spelled correctly for patient safety. Also remember that the patient record is a legal document. The WAD above contains some medical terms that are difficult to spell and pronounce. After you finish this exercise listen to the pronunciation of these terms on the Student Online Learning Center (www.mhhe.com/AllanEssMedLanguage) and practice them yourself. Below, insert the correct spelling of the term on the line.*

1. Medication has been prescribed for this patient's case of ______________________ dermatitis.
   seborheic   seborrheic

2. The ______________________ rash on the patient's face is slowly clearing after he tried the new medication.
   eczematous   exzematous

3. Ms. Fox's ______________________ is an allergic reaction to her jewelry, which contains nickel.
   purritis   pruritus

4. Severe itching and scratching has produced ______________________ of this patient's skin.
   excoreiation   excoriation

5. Ms. Fox's ______________________ has cleared up nicely after 2 weeks of treatment.
   dermattitis   dermatitis

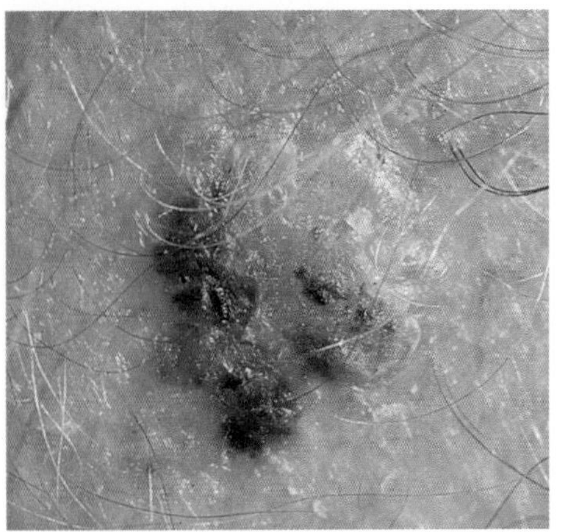

▲ **FIGURE 3.6**
Basal Cell Carcinoma.

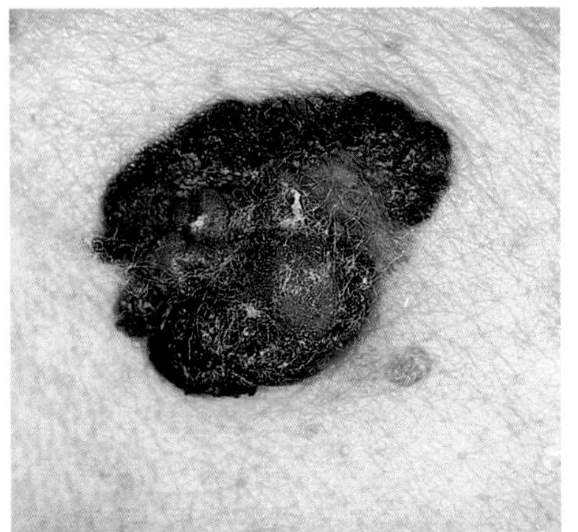

▲ **FIGURE 3.7**
Malignant Melanoma.

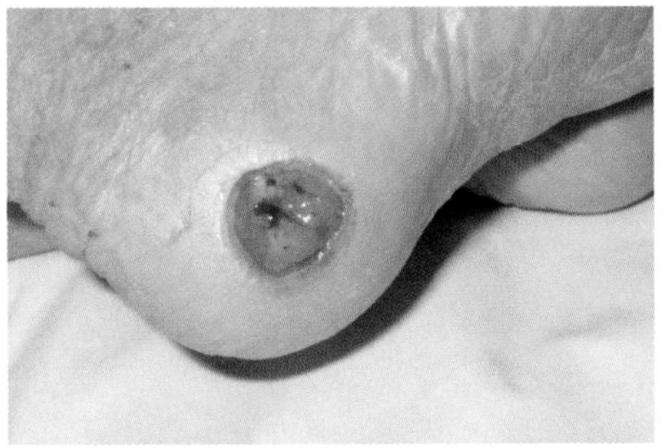

▲ **FIGURE 3.8**
Decubitus (Pressure) Ulcer on the Heel.

DISORDERS OF THE SKIN (continued)

Skin Cancers

The two basal cell carcinomas (*Figure 3.6*) on Mr. Andrews' arm arose from the **basal** (bottom) layer of the epidermis. This is the most common skin cancer and the least dangerous because it does not **metastasize.**

The squamous cell carcinoma on his hand arose from cells in the middle layers of the epidermis. This skin cancer responds well to surgical removal but can metastasize to lymph glands if neglected.

Malignant melanoma (*Figure 3.7*) is the least common skin cancer but is the most dangerous. It arises in **melanin**-producing cells in the basal layer of the epidermis. It metastasizes quickly and is fatal if neglected.

Sunlight in excess can also be an irritant to the skin. Too much sunlight not only can burn the skin but also can lead to cancer, as it did for Mr. Rod Andrews.

Any congenital (present at birth) lesion of the skin, including various types of birthmarks and all moles, is referred to as a **nevus.**

Pressure Ulcers

When a patient lies in one position for a long period, the pressure between the bed and bony projections, such as the lower spine or heel, cuts off the blood supply to the skin and **pressure** or **decubitus ulcers** can appear (*Figure 3.8*). The protective function of the skin is broken, and germs can enter the body. The elderly are at risk for pressure ulcers because their skin is often thin and dry. In addition, poor nutritional status can deplete the fatty protective layer in the hypodermis under the skin.

WORD ANALYSIS AND DEFINITION

S/ = Suffix P/ = Prefix R/ = Root R/CF = Combining Form

WORD	PRONUNCIATION	ELEMENTS		DEFINITION
decubitus **ulcer** (pressure ulcer)	deh-**KYU**-bit-us **UL**-ser	P/ R/ R/	**de-** *from* **-cubitus** *lying down* **ulcer** *sore*	Sore caused by lying down for long periods of time.
malignant **malignancy**	mah-**LIG**-nant mah-**LIG**-nan-see	S/ R/ S/	**-ant** *forming, pertaining to* **malign-** *harmful, bad* **-ancy** *state of*	Tumor that invades surrounding tissues and metastasizes to distant organs. State of being malignant.
melanin **melanoma**	**MEL**-ah-nin mel-ah-**NO**-mah	 S/- R/	Greek *black* **oma** *tumor, mass* **melan-** *black pigment*	Black pigment found in skin, hair, and retina. Malignant neoplasm formed from cells that produce melanin.
metastasis (noun) **metastasize (verb)** **metastatic (adj)**	meh-**TAS**-tah-sis meh-**TAS**-tah-size meh-tah-**STAT**-ik	R/ P/ S/ R/ S/	**-stasis** *stagnate, stay in one place* **meta-** *beyond, subsequent to* **-ize** *affect in a specific way* **-stat-** *stationary* **-ic** *pertaining to*	Spread of a disease from one part of the body to another. To spread to distant parts. Pertaining to the character of cells that can metastasize.
nevus **nevi (pl)**	**NEE**-vus **NEE**-veye		Latin *mole, birthmark*	Congenital lesion of the skin.

EXERCISES

Apply your knowledge of elements. Identify the italicized element in the first column, define it, and use it to determine the meaning of the medical term. Then use any two terms from the chart in sentences of your choice that are not definitions.

Medical Term	Identity of Element (P, S, R, or CF)	Meaning of Element	Meaning of Medical Term
meta*stasis*			
*malign*ant			
melan*oma*			
de*cubitus*			

Sentence:

1. ______________________________

2. ______________________________

DISORDERS OF THE SKIN

Infections of the Skin

The skin can be susceptible to many different types of infections. The following descriptions are examples.

Viral Infections

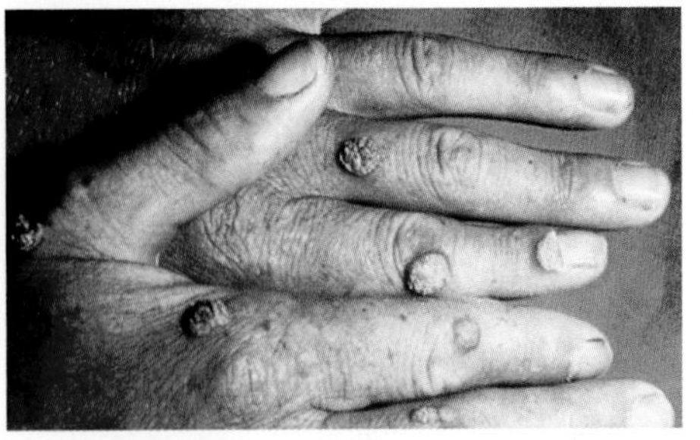

▲ **FIGURE 3.9**
Warts of Hands.

Warts (verrucas) are caused by the human **papillomavirus** invading the epidermis (*Figure 3.9*).

Varicella-zoster virus causes chickenpox in unvaccinated people, forming **macules** (small, flat spots different in color from the surrounding skin), **papules** (small, solid elevations), and **vesicles** (small sacs containing fluid). The virus can then remain dormant in the peripheral nerves for decades before erupting as the painful vesicles of **herpes zoster,** also called **shingles** (*Figure 3.10*).

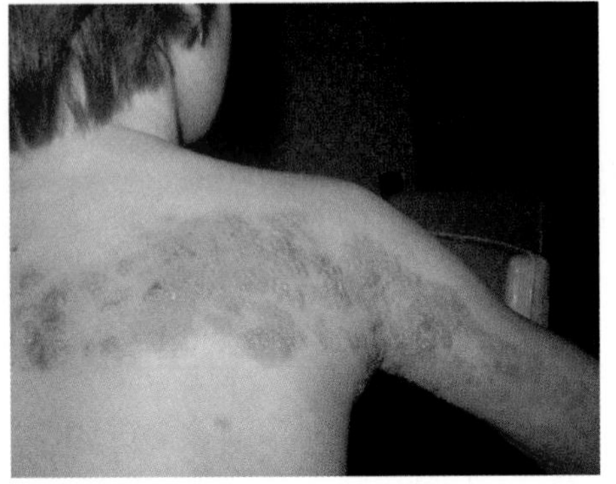

▲ **FIGURE 3.10**
Shingles.

Fungal Infections

Tinea is a general term for a group of related skin infections caused by different species of **fungi.**

Tinea pedis (athlete's foot) causes itching, redness, and peeling of the skin of the foot, particularly between the toes (*Figure 3.11*). The term **tinea capitis** describes infection of the scalp (ringworm); **tinea corporis** is the name for infections of the body. **Tinea cruris,** "jock itch," is the name for infections of the groin. The fungus spreads from animals, from the soil, and by direct contact with infected individuals.

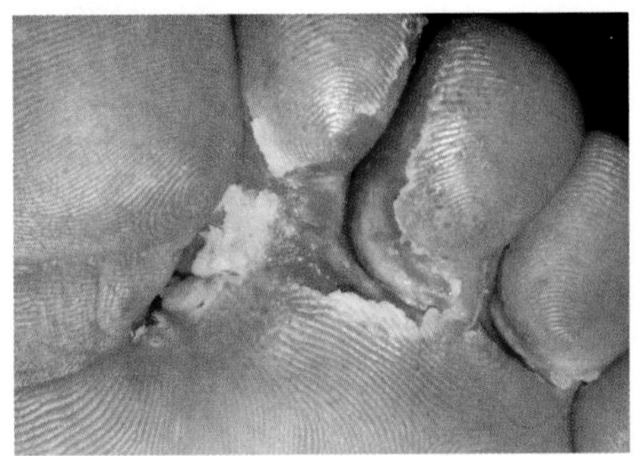

▲ **FIGURE 3.11**
Tinea Pedis Between the Toes.

A yeastlike fungus, ***Candida albicans,*** can produce recurrent infections of the skin, nails, and mucous membranes. The first sign can be a recurrent diaper rash or oral **thrush** in infants. Older children can show recurrent or persistent lesions on the scalp. In adults, chronic **mucocutaneous candidiasis** can affect the mouth (thrush) (*Figure 3.12*) and vagina as well as the skin. It can also occur with diseases of the immune system (*see Chapter 11*).

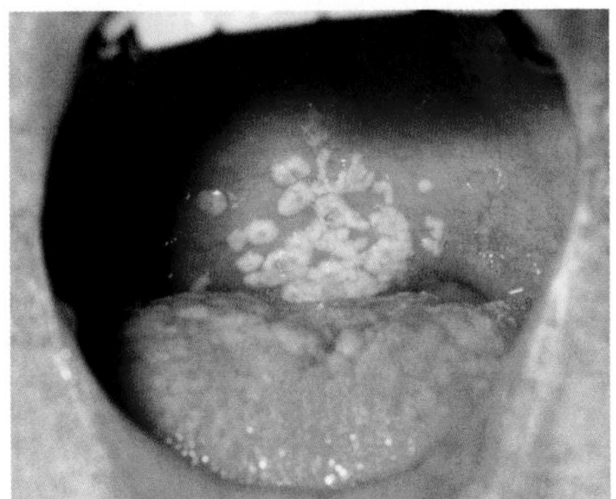

▲ **FIGURE 3.12**
Thrush.

WORD ANALYSIS AND DEFINITION

S/ = Suffix   P/ = Prefix   R/ = Root   R/CF = Combining Form

WORD	PRONUNCIATION	ELEMENTS		DEFINITION
Candida candidiasis *Candida albicans* thrush	**KAN**-did-ah kan-dih-**DIE**-ah-sis **KAN**-did-ah **AL**-bih-kanz THRUSH	 S/ R/	Latin *dazzling white* **-iasis** *state of, condition* **candid-** *Candida* albicans *white*	A yeastlike fungus. Infection with the yeastlike fungus. The most common form of *Candida.* Another name for infection with *Candida.*
fungus **fungi (pl)**	**FUN**-gus **FUN**-jee *or* **FUN**-gee		Latin *mushroom*	General term used to describe yeasts and molds.
herpes zoster (shingles)	**HER**-pees **ZOS**-ter		herpes Greek *to creep or spread* zoster Greek *belt, girdle*	Painful eruption of vesicles that follows a nerve root on one side of the body.
macule	**MACK**-yul		Latin *spot*	Small, flat spot or patch on the skin.
mucocutaneous	**MYU**-koh-kyu-**TAY**-nee-us	S/ R/CF R/CF	**-ous** *pertaining to* **muc/o-** *mucous membrane* **-cutan/e-** *skin*	Junction of skin and mucous membrane; e.g., the lips.
papillomavirus	pap-ih-**LOH**-mah-vi-rus	S/ R/CF	**-oma mass, tumor** **papill/o-** *papilla, pimple* virus Latin *poison*	Virus that causes warts and is associated with cancer.
papule	**PAP**-yul		Latin *pimple*	Small, circumscribed elevation on the skin.
tinea	**TIN**-ee-ah		Latin *worm*	General term for a group of related skin infections caused by different species of fungi.
verruca (wart)	ver-**ROO**-cah		Latin *wart*	Wart caused by a virus.

EXERCISES

***Latin and Greek terms** do not deconstruct into prefix, root, and suffix that give clues to their meaning. You must know them for what they are. Test your knowledge of these terms by matching the definition in column 1 with the correct medical term in column 2.*

	Definition	Medical Term
____	1. to creep or spread	A. tinea
____	2. mushroom	B. verruca
____	3. worm	C. papule
____	4. belt	D. macule
____	5. dazzling white	E. fungus
____	6. wart	F. herpes
____	7. pimple	G. virus
____	8. spot	H. candida
____	9. poison	I. zoster

DISORDERS OF THE SKIN (continued)

▲ **FIGURE 3.13**
Body LIce.

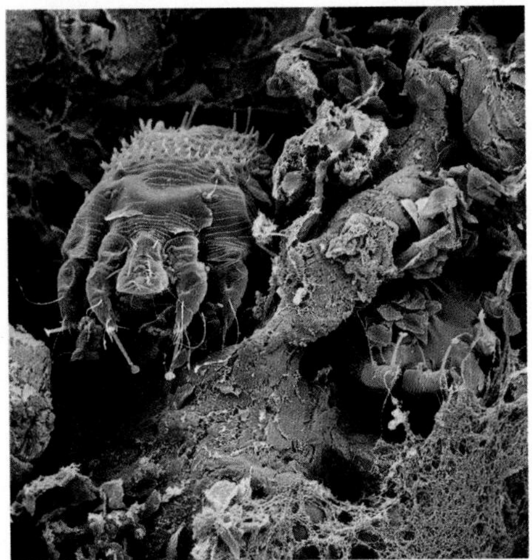

▲ **FIGURE 3.14**
The Itch Mite of Scabies.

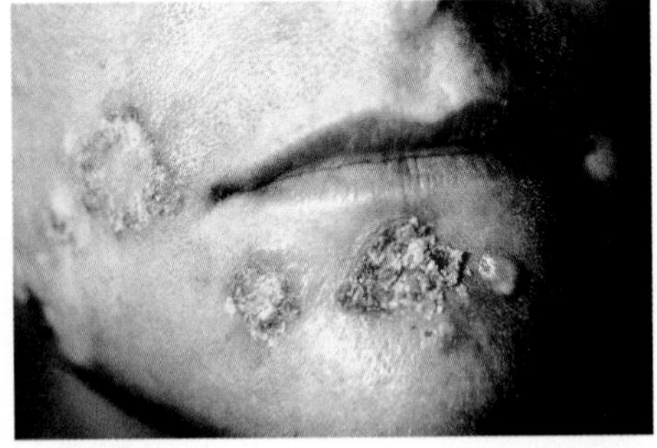

▲ **FIGURE 3.15**
Impetigo.

Parasitic Infestations

A **parasite** is an organism that lives in contact with and feeds off another organism, the host. This process is caused an **infestation.** It is different from an **infection.**

Lice (*Figure 3.13*) are small, wingless, blood-sucking parasites that produce the disease **pediculosis** by attaching their **nits** (eggs) to hair and clothing.

Itch mites produce an intense, itching rash, called **scabies** (*Figure 3.14*), often in the genital area, waist, breast, and armpits. The mites live and lay eggs under the skin.

Bacterial Infections

Staphylococcus aureus (commonly called "staph") is the most common bacterium to invade the skin and is the cause of pimples, boils, **carbuncles,** and **impetigo** (*Figure 3.15*). Staph can cause a **cellulitis** of the epidermis and dermis. *Group A Streptococcus* (strep) can also cause cellulitis.

Very occasionally, some strains of both staph and strep can be very **toxic,** and their enzymes digest the connective tissues and spread into muscle layers. This condition is called **necrotizing fasciitis.** Very aggressive treatment with surgery and antibiotics is needed.

WORD ANALYSIS AND DEFINITION

S/ = Suffix P/ = Prefix R/ = Root R/CF = Combining Form

WORD	PRONUNCIATION	ELEMENTS		DEFINITION
carbuncle	**KAR**-bunk-ul		Latin *carbuncle*	Infection of many hair follicles in a small area, often on the back of the neck.
cellulitis	sell-you-**LIE**-tis	S/ R/	**-itis** *inflammation* **cellul-** *cell*	Infection of subcutaneous connective tissue.
impetigo	im-peh-**TIE**-go		Latin *scabby eruption*	Infection of the skin producing thick, yellow crusts.
infection	in-**FECK**-shun	S/ R/	**-ion** *action* **infect-** *internal invasion, infection*	Invasion of the body by disease-producing microorganisms.
infectious (adj)	in-**FECK**-shus	S/	**-ious** *pertaining to*	Capable of being transmitted, or a disease caused by the action of a microorganism.
infestation	in-fes-**TAY**-shun	S/ R/	**-ation** *process* **infest-** *invade*	Act of being invaded on the skin by a troublesome other species, such as a parasite.
louse **lice (pl)**	LOWSE LISE		Old English *louse*	Parasitic insect.
necrotizing fasciitis	neh-kroh-**TIZE**-ing fash-eh -**EYE**-tis	S/ R/CF S/ S/ R/CF	**-ing** *quality of* **necr/o-** *death* **-tiz-** *pertaining to* **-itis** *inflammation* **fasc/i** – *fascia*	Inflammation of fascia producing death of the tissue.
parasite	**PAR**-ah-site		Greek *guest*	An organism that attaches itself to, lives on or in, and derives its nutrition from another species.
parasitic (adj)	par-ah-**SIT**-ik	S/ R/	**-ic** *pertaining to* **parasit-** *parasite*	Pertaining to a parasite.
pediculosis	peh-dick-you-**LOH**-sis	S/ R/	**-osis** *condition* **pedicul-** *louse*	An infestation with lice.
scabies	**SKAY**-bees		Latin *to scratch*	Skin disease produced by mites.
toxin	**TOK**-*sin*		Greek *poison*	Poisonous substance formed by a cell or organism.
toxic (adj)	**TOK**-sick	S/ R/	**-ic** *pertaining to* **tox-** *poison*	Pertaining to a toxin.
toxicity (contains two suffixes)	toks-**ISS**-ih-tee	S/	**-ity** *state, condition*	The state of being poisonous.

EXERCISES

***Proofread** your documentation. Each of the following sentences has either an error in fact or an error in spelling. Correct each sentence on the line below.*

1. Necrotising fascitis is an inflammation of the fascia that produces death of tissue.

2. An infection of lice can be troublesome to eliminate.

3. Cellulitis is an infection of the epidermis.

4. Impetigo is produced by the itch mite.

5. Pidiculosis is associated with lice.

DISORDERS OF THE SKIN (continued)

Collagen Diseases

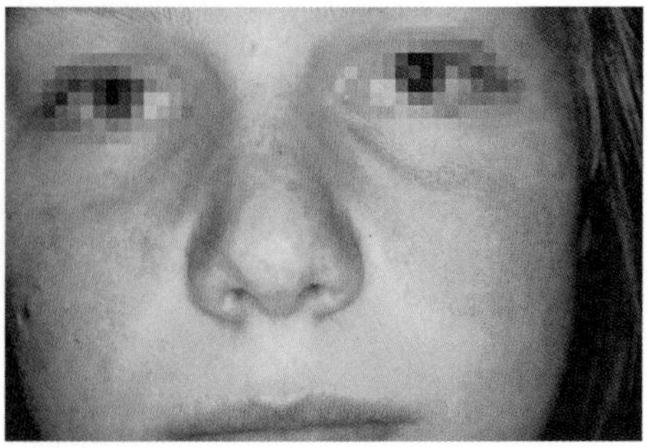

▲ **FIGURE 3.16**
Systemic Lupus Erythematosus

Collagen, a fibrous protein, comprises 30% of total body protein. Therefore, collagen diseases can have a dramatic effect all over the body, as well as in the skin.

Systemic lupus erythematosus (SLE), an **autoimmune** disease, occurs most commonly in women and produces characteristic skin lesions. A butterfly-shaped, red rash on both cheeks and joined across the bridge of the nose is commonly seen (*Figure 3.16*). The disease also affects multiple organs, including the kidneys, brain, heart, and joints.

Rosacea produces a similar facial rash to that of SLE, but there are no systemic complications.

Scleroderma is a chronic, persistent autoimmune disease, occurring more often in women, characterized by hardening and shrinking of the skin that makes it feel leathery (*Figure 3.17*). Joints show swelling, pain, and stiffness. Internal organs such as the heart, lungs, kidneys, and digestive tract are involved in a similar process. The etiology is unknown, and there is no effective treatment.

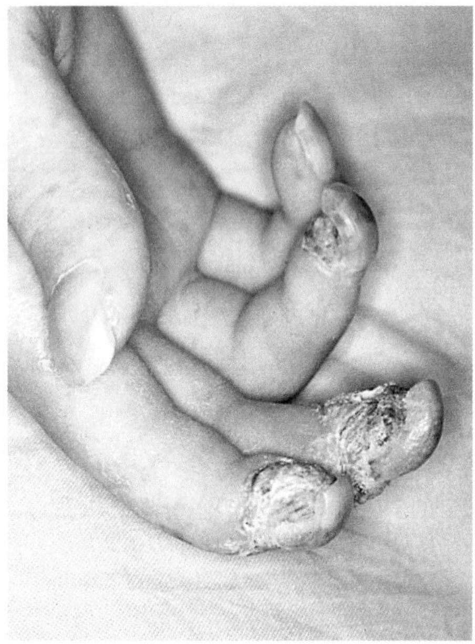

▲ **FIGURE 3.17**
Scleroderma.

Other Skin Diseases

Psoriasis (*Figure 3.18*) is marked by itchy, flaky, red patches of skin of various sizes covered with white or silvery scales. It appears most commonly on the scalp, elbows, and knees. Its cause is unknown.

Skin Manifestations of Internal Disease

The presence of cancer inside the body is often shown by skin lesions, even before the cancer or other disease has produced **symptoms** or been diagnosed.

Dermatomyositis (*Figure 3.19*) is often associated with ovarian cancer, which can appear within 4 to 5 years after the skin disease is diagnosed. The skin disease presents with a reddish-purple rash around the eyes that often precedes muscle weakness by weeks or months.

Kaposi sarcoma mostly develops in association with human immunodeficiency virus (HIV) infection. Raised red or brown blotches or bumps in tissues occur below the skin surface and eventually spread throughout the body.

Herpes zoster, referred to earlier in this chapter, is common in immunocompromised patients, including the elderly, individuals with HIV, and patients on chemotherapy (*see Chapter 11*).

Abbreviations

HIV human immunodeficiency virus
SLE systemic lupus erythematosus

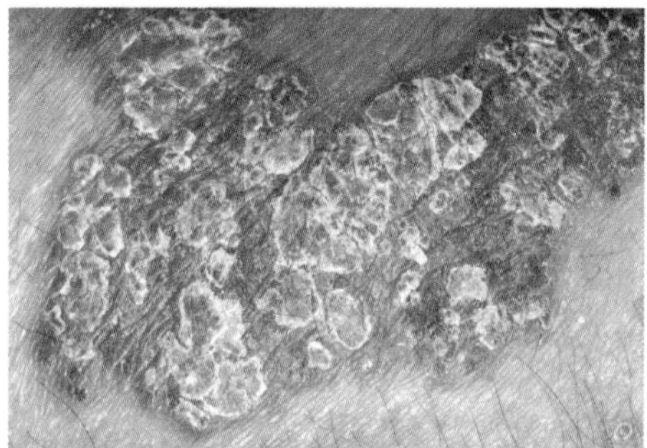

▲ **FIGURE 3.18**
Psoriasis.

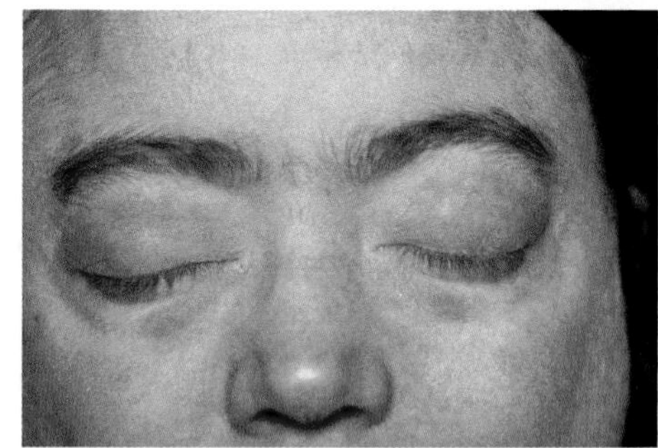

▲ **FIGURE 3.19**
Periorbital Rash of Dermatomyositis.

WORD ANALYSIS AND DEFINITION

S/ = Suffix P/ = Prefix R/ = Root R/CF = Combining Form

WORD	PRONUNCIATION	ELEMENTS		DEFINITION
autoimmune	awe-toe-im-**YUNE**	P/ R/	**auto-** *self* **-immune** *protected from*	Diseases in which the body makes antibodies directed against its own tissues.
dermatomyositis	**DER**-mah-toe-**MY**-oh-site-is	S/ R/CF R/	**-itis** *inflammation* **dermat/o-** *skin* **-myos-** *muscle*	Inflammation of the skin and muscles.
Kaposi sarcoma	kah-**POH**-see sar-**KOH**-mah		Moritz Kaposi, Hungarian dermatologist, 1837–1902	A form of skin cancer seen in AIDS patients.
psoriasis	so-**RYE**-ah-sis		Greek *the itch*	Rash characterized by reddish, silver-scaled patches.
rosacea	roh-**ZAY**-she-ah		Latin *rosy*	Persistent erythematous rash of the central face.
scleroderma	sklair-oh-**DERM**-ah	S/ R/CF	**-derma** *skin* **scler/o-** *hard*	Thickening and hardening of the skin due to new collagen formation.
symptom (subjective)	**SIMP**-tum		Greek *event or feeling that has happened to someone*	Departure from normal health experienced by the patient.
symptomatic (adj)	simp-toe-**MAT**-ik	S/ R/	**-ic** *pertaining to* **symptomat-** *symptoms*	Pertaining to the symptoms of a disease.
sign (objective)	SINE		Latin *mark*	Physical evidence of a disease process.
systemic lupus erythematosus	sis-**TEM**-ik **LOO**-pus er-ih-**THEE**-mah-toe-sus	S/ R/ S/ R/	**-ic** *pertaining to* **system-** *the body as a whole* lupus Latin *wolf* **-osus** *condition* **erythemat-** *redness*	Inflammatory connective tissue disease affecting the whole body.

EXERCISES

***Document** using the language of dermatology by inserting the correct term in the space provided. Always review the WAD before you start any exercise.*

1. According to the patient's (departure from normal health felt by the patient) ______________________, the physician diagnosed an (body attacks its own tissues) ______________________ disease.
2. Several medications were prescribed for the patient's (inflammation of the skin and muscles) ______________________.
3. What we thought was originally a localized problem has now spread, and the patient's final diagnosis is (inflammatory connective tissue disease affecting the whole body) ______________________.
4. Because of this patient's past history of HIV, his new tumors have been diagnosed by the pathologist as (a form of skin cancer seen in AIDS patients) ______________________.
5. Please schedule this patient to be seen in the (thickening and hardening of the skin due to collagen formation) ______________________ Clinic as soon as possible.
6. Patients with (rash with reddish, silver-scaled patches) ______________________ often wear long sleeves to hide their elbows.

LESSON 3.3 Accessory Skin Organs

THE INTEGUMENTARY SYSTEM

OBJECTIVES

Hair follicles, sebaceous glands, sweat glands, and **nails** are ***accessory organs*** located in your skin. Each has specific anatomical and physiological characteristics. This lesson will enable you to:

- **Identify the accessory skin organs.**
- **Describe certain disorders affecting the accessory skin organs.**
- **Apply correct medical terminology to the accessory skin organs and their disorders.**

You are

... a pharmacist working in the pharmacy at Fulwood Medical Center.

You are communicating with

... Mr. Wayne Winter, an 18-year-old man, who will be starting college in a few months' time.

CASE REPORT 3.3

Mr. Winter has had *acne* since the age of 15 and has tried several over-the-counter products. Retinoic acid has also been unsuccessful. He has numerous *comedones*, papules, *pustules*, and *scars* on his face and forehead and has severe *cystic* lesions and scars on his back. His social life is nonexistent and he is teased by his peers. He wishes to change all this before he gets to college.

Your role is to explain to him how to use the medications Dr. Echols, a dermatologist, has prescribed, what their effects will be, and what possible complications can occur.

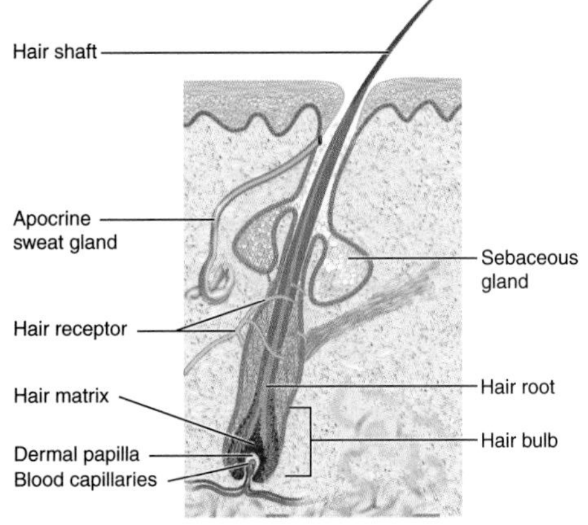

▲ **FIGURE 3.20**
Hair Follicle and Sebaceous Gland.

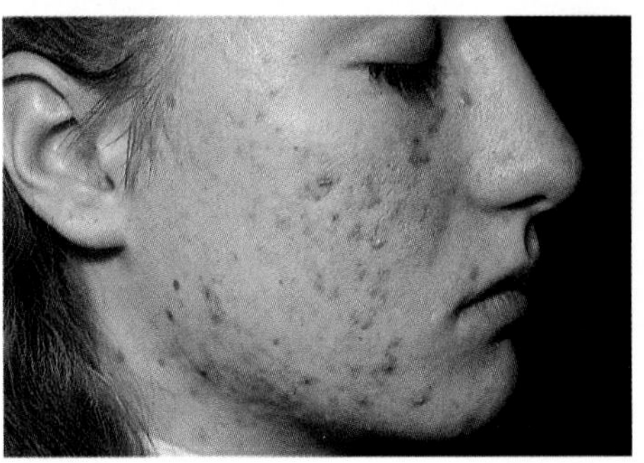

▲ **FIGURE 3.21** **Acne.**

HAIR FOLLICLES AND SEBACEOUS GLANDS

Each hair follicle has a sebaceous gland opening into it (*Figure 3.20*). The gland secretes into the follicle a mixture of oily, acidic sebum and broken-down cells from the base of the gland.

Around puberty, **androgens** are thought to trigger excessive production of sebum from the sebaceous glands, and sebum brings with it excessive numbers of broken-down cells. This blocks the follicle, forming a **comedo** (whitehead or blackhead). Comedones can stay closed, leading to papules, or can rupture, allowing bacteria to get in and produce **pustules.** These are the classic signs of **acne,** which is said to affect, in different degrees, about 85% of people between 12 and 25 years (*Figure 3.21*).

A different skin problem involving the sebaceous glands is seborrheic dermatitis. The glands are thought to be inflamed and to produce a different sebum. The skin around the face and scalp is reddened and covered with yellow, greasy scales. In infants, this condition is called "cradle cap." Seborrheic dermatitis of the scalp produces dandruff.

WORD ANALYSIS AND DEFINITION

S/ = Suffix P/ = Prefix R/ = Root R/CF = Combining Form

WORD	PRONUNCIATION	ELEMENTS		DEFINITION
inflammation	in-flah-**MAY**-shun	S/ P/ R/	**-ion** *action, condition* **in-** *in* **-flammat-** *flame*	A complex of cell and chemical reactions occurring in response to an injury, or chemical, or biologic agent.
inflammatory (adj)	in-**FLAM**-ah-tor-ee	S/	**-ory** *having the function of*	Causing or affected by inflammation.
scald	SKAWLD		Latin *wash in hot water*	Burn from contact with hot liquid or steam.
shock	SHOCK		German *to clash*	Sudden physical or mental collapse or circulatory collapse.

EXERCISES

Lesson objectives: Meet the lesson objectives with this exercise on burns. Employ your knowledge of medical terms from the integumentary system. Fill in the blanks.

1. **Hit by lightning:**

   Degree of burn: ____________________

   Symptoms: ____

   ____

   Layers of skin involved: ____

   What other specific type of injury/accident could produce this burn? ____________________

2. **Scald:**

   Degree of burn: ____________________

   Symptoms: ____

   ____

   Layers of skin involved: ____

   How quickly does this heal? ____

3. **Prolonged flame contact in house fire:**

   Degree of burn: ____________________

   Symptoms: ____

   ____

   Layers of skin involved: ____

   What specific surgical procedure is used to promote healing? ____________________

4. **Sunburn:**

   Degree of burn: ____________________

   Symptoms: ____

   ____

   Layers of skin involved: ____

   Why is this burn only superficial? ____

Case Report 3.4 (continued)

Mr. Hapgood's burns were mostly third-degree. The burned, dead tissue formed an **eschar** that can have toxic effects on the digestive, respiratory, and cardiovascular systems. The eschar was surgically removed by **debridement**.

Keynote

In third- and fourth-degree burns there is no dermal tissue left for regeneration, and skin grafts are necessary.

In full-thickness burns there is no dermal tissue left for **regeneration,** and skin grafts are needed. The ideal graft is an **autograft,** taken from another location on the patient, because it is not rejected by the immune system. Mr. Hapgood had autografts taken from his unburned legs and back.

If the patient's burns are too extensive, **allografts,** which are grafts from another person, are needed. These are provided by skin banks, which acquire them from deceased people (**cadavers**). **Homograft** is another name for allograft. A **xenograft** or **heterograft** is a graft from another species, such as a pig for heart valves.

Artificial skin is being developed commercially and can stimulate the growth of new connective tissues from the patient's underlying tissue.

The treatment and prognosis for a burn patient also depend on the extent of the body surface that is affected. This is estimated by subdividing the skin's surface into regions, each one of which is a fraction of or multiple of 9% of the total surface area (*Figure 3.28*).

- Head and neck are assigned 9% (4½% anterior and posterior).
- Each arm is 9% (4½% anterior and 4½% posterior).
- Each leg is 18% (9% anterior and 9% posterior).
- The anterior trunk is 18%.
- The posterior trunk is 18%.
- Genitalia are 1%.

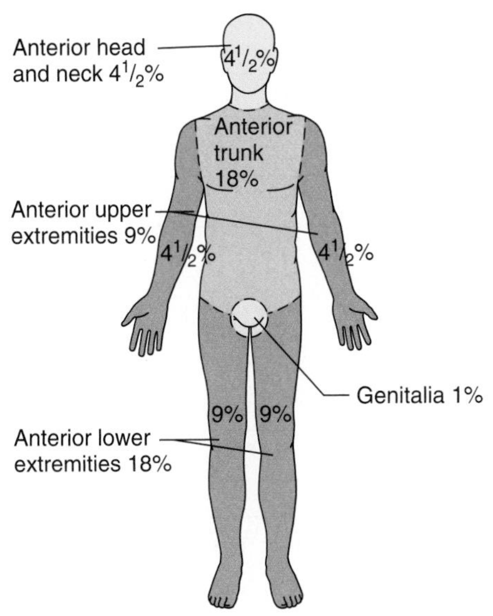

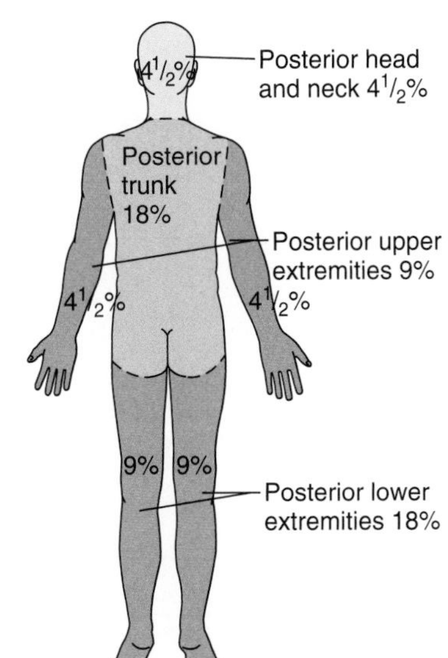

▲ **FIGURE 3.28**
Rule of Nines.

WORD ANALYSIS AND DEFINITION

S/ = Suffix P/ = Prefix R/ = Root R/CF = Combining Form

WORD	PRONUNCIATION	ELEMENTS		DEFINITION
autograft	**AWE**-toe-graft	P/ R/	**auto-** *self, same* **-graft** *transplant*	A graft removed from the patient's own skin.
cadaver	kah-**DAV**-er		Latin *dead body*	A dead body or corpse.
debridement	day-**BREED**-mon (French pronunciation of -ment)	S/ P/ R/	**-ment** *resulting state* **de-** *take away* **-bride-** *rubbish*	The removal of injured or necrotic tissue.
eschar	**ESS**-kar		Greek *scab of a burn*	The burnt, dead tissue lying on top of third-degree burns.
homograft (same as *allograft*) **allograft**	**HOH**-moh-graft **AL**-oh-graft	P/ R/ P/	**homo-** *same, alike* **-graft** *transplant* **allo-** *other*	Skin graft from another person or a cadaver.
regenerate (verb) **regeneration (noun)**	ree-**JEN**-eh-rate ree-**JEN**-eh-**RAY**-shun	S/ P/ R/ S/	**-ate** *composed of* **re-** *again* **-gener-** *produce* **-ation** *process*	Reconstitution of a lost part. The process of reconstitution.
xenograft **heterograft** (same as xenograft)	**ZEN**-oh-graft **HET**-er-oh-graft	P/ R/ P/	**xeno-** *foreign* **-graft** *transplant* **hetero-** *different*	A graft from another species (not human).

EXERCISES

***Employ** medical language in written communication. Insert the correct term from the spread on the appropriate blank; then proofread your paragraph for spelling errors.*

1. Recovery from severe burns is a painful and difficult process. The burnt, dead tissue lying on top of third-degree burns is known as ________________________. Before the burn can heal, this must be surgically removed in a procedure known as ________________. If the burn is deep enough, the tissue cannot ________________________ itself, and a skin graft is needed. This graft can come from the patient, ________________________, or from another person, ________________________, or even from another species ________________________. Grafts may also come from skin banks or ________________________.

2. An ________________________ is taken from the patient; an ________________________ can come from another person. Donor skin comes from skin banks, which acquire skin from ________________________. A ________________________ is from a different species, not human. An example of this would be a ________________________ for the heart.

3. Two of the terms in question 2 are also known by another name. Write them below.

____ is the same as ____

____ is the same as ____

WOUNDS AND TISSUE REPAIR

FIGURE 3.29 Wound Healing. ▶
(a) Bleeding into the wound.
(b) Scab formation and macrophage activity.

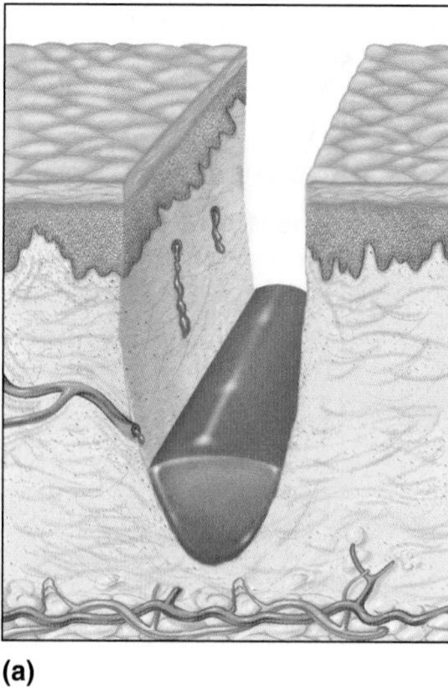

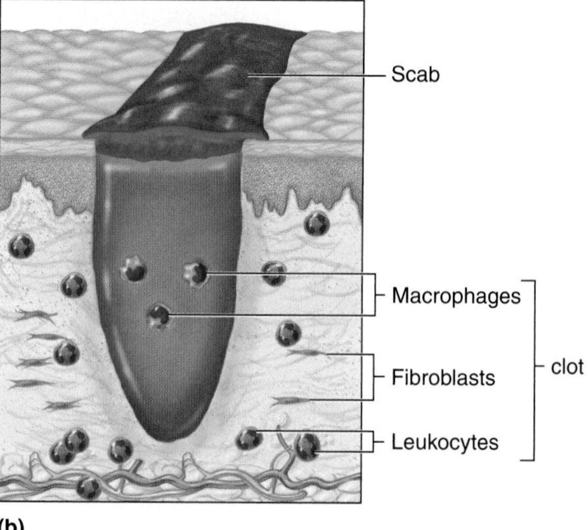

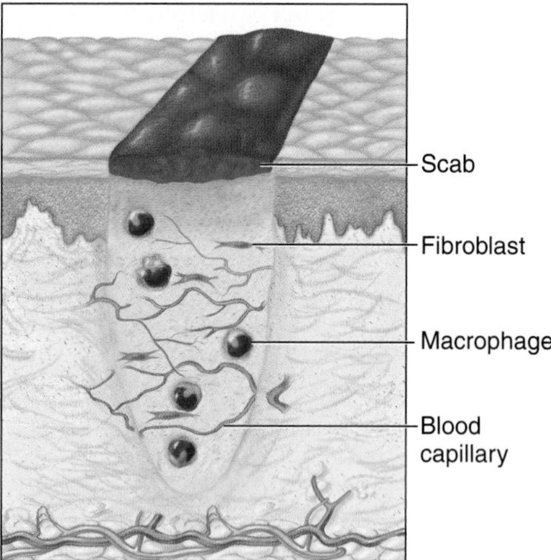

▲ **FIGURE 3.30**
Formation of Granulation Tissue.

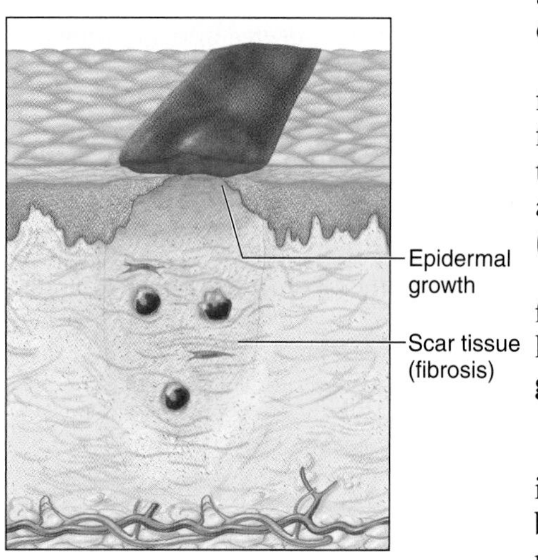

▲ **FIGURE 3.31 Scar Formation.**

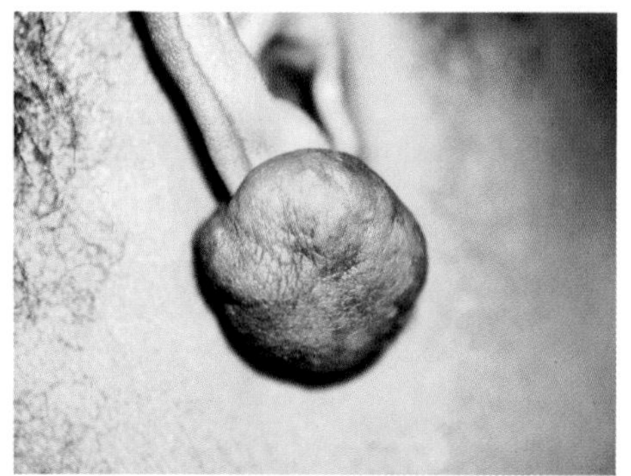

▲ **FIGURE 3.32**
A Keloid of the Earlobe.
This scar resulted from piercing the ear for earrings.

If you cut yourself when shaving and produce a superficial **laceration** in the epidermis, the epithelial cells along the edges of the laceration will divide rapidly and fill in the gap to heal it.

If you cut yourself more deeply, extending the **wound** into the dermis or hypodermis, or if a surgeon makes an **incision**, then blood vessels in the dermis break and blood escapes into the wound (*Figure 3.29a*). The surgeon would perform an **excision** if he or she removed a lesion from the skin or any other tissue.

The escaped blood forms a **clot** in the wound. The clot consists of the protein fibrin together with platelets, blood cells, and dried tissue fluids trapped in the fibers. Cells that digest and clean up the tissue debris come into the wound with the escaped blood (*see Chapter 11*). The surface of the clot dries and hardens in the air to form a **scab.** The scab seals and protects the wound from becoming infected (*Figure 3.29b*).

The clot is invaded by new capillaries from the surrounding dermis. Three or four days after the injury, other cells migrate into the wound and form new collagen fibers that pull the wound together. This soft tissue in the wound is called **granulation** tissue (*Figure 3.30*), which is later replaced by a scar (*Figure 3.31*).

Suturing brings together the edges of the wound to enhance tissue healing.

In some people, there is excessive fibrosis and scar tissue formation, producing raised, irregular, lumpy, shiny scars called **keloids** (*Figure 3.32*). They can extend beyond the edges of the original wound and often return if they are surgically removed. They are most common on the upper body and earlobes.

A superficial scraping of the skin, a mucous membrane, or the cornea (*Chapter 4*) is called an **abrasion.**

Surgery on the skin is now being performed using focused light beams called lasers. They remove lesions such as birthmarks and tattoos and create a fresh surface over which new skin can grow.

Cosmetic Procedures

Abdominoplasty: "Tummy tuck."

Blepharoplasty: Correction of defects in the eyelids.

Dermabrasion: Removal of upper layers of skin using a high-powered rotating brush.

Lipectomy: Surgical removal of fatty tissue.

Liposuction: Surgical removal of fatty tissue using suction.

Mammoplasty: Surgical procedure to alter the size or shape of breasts.

Rhinoplasty: Surgical procedure to alter the size or shape of the nose.

WORD ANALYSIS AND DEFINITION

S/ = Suffix P/ = Prefix R/ = Root R/CF = Combining Form

WORD	PRONUNCIATION		ELEMENTS	DEFINITION
abdominoplasty (tummy tuck)	ab-**DOM**-ih-noh-plas-tee	S/ R/CF	**-plasty** *surgical repair* **abdomin/o-** *abdomen*	Surgical removal of excess subcutaneous fat from abdominal wall.
abrasion	ah-**BRAY**-shun		Latin *to scrape*	Area of skin or mucous membrane that has been scraped off.
blepharoplasty	**BLEF**-ah-roh-plas-tee	S/ R/CF	**-plasty** *surgical repair* **blephar/o-** *eyelid*	Surgical repair of an eyelid.
clot	KLOT		German *to block*	The mass of fibrin and cells that is produced in a wound.
dermabrasion	der-mah-**BRAY**-shun	S/ R/ R/	**-ion** *action* **derm-** *skin* **-abras-** *scrape off*	Removal of upper layers of skin by rotary brush.
granulation	gran-you-**LAY**-shun	S/ R/	**-ation** *process* **granul-** *small grain*	New fibrous tissue formed during wound healing.
incision **excision**	in-**SIZH**-un ek-**SIZH**-un	S/ R/ R/	**-ion** *action, condition* **incis-** *cut into* **excis-** *cut out*	A cut or surgical wound. Surgical removal of part or all of a structure.
keloid	**KEY**-loyd		Greek *stain*	Raised, irregular, lumpy scar due to excess collagen fiber production during healing of a wound.
laceration	lass-eh-**RAY**-shun	S/ R/	**-ation** *process* **lacer-** *to tear*	A tear of the skin.
lipectomy **liposuction**	lip-**ECK**-toe-me **LIP**-oh-suck-shun	S/ R/ S/ R/CF R/	**-ectomy** *surgical excision* **lip-** *lipid, fat* **-ion** *action* **lip/o-** *fat* **-suct-** *suck*	Surgical removal of adipose tissue. Surgical removal of adipose tissue using suction.
mammoplasty	**MAM**-oh-plas-tee	S/ R/CF	**-plasty** *surgical repair* **mamm/o-** breast	Surgical reshaping of the breast.
rhinoplasty	**RYE**-no-plas-tee	S/ R/CF	**-plasty** *surgical repair* **rhin/o-** *nose*	Surgical procedure to change size or shape of nose.
scab	SKAB		Old English *crust*	Crust that forms over a wound or sore during healing.
suture (noun) **suture (verb)**	**SOO**-chur		Latin *seam*	Stitch to hold the edges of a wound together. To stitch the edges of a wound together.
wound	WOOND		Old English *wound*	Any injury that interrupts the continuity of skin or a mucous membrane.

EXERCISES

Wounds: Use your knowledge of the meaning of the following medical terms to put them in the correct order of their occurrence and give a brief description of each term. Fill in the chart.

keloid **scab** **clot** **laceration** **wound** **granulation**

Medical Term	Brief Description
1.	
2.	
3.	
4.	
5.	
6.	

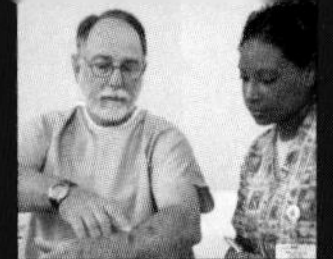

CHAPTER 3 REVIEW

THE INTEGUMENTARY SYSTEM

CHALLENGE YOUR KNOWLEDGE

A. **Prefixes** can work across all body systems and still have the same meaning. *Hypo**dermic*** means *pertaining to below the **skin*** while *hypo-gastric* means *pertaining to below the **stomach.*** Prefix and suffix remain the same; only the body system root (*derm and gastr*) changes. Increase your medical vocabulary with a working knowledge of the following prefixes. Fill in the blanks.

Prefix	Meaning of Prefix	Medical Term with This Prefix	Meaning of Medical Term
allo			
pro			
auto			
cryo			
de			
epi			
ex			
hetero			
homo			
hypo			
an			
syn			

B. **Train your eye and ear** to hear and see the slight differences in medical terms. *System* is not *symptom,* and a *macule* is not a *papule.* Briefly explain to your patient the differences in the following terms.

1. system: ____

   symptom: ____

2. macule: ____

   papule: ____

C. **Latin and Greek terms** cannot be further deconstructed into prefix, root, or suffix. You must know them for what they are. Test your knowledge of these terms with this exercise. There are more answer choices than you need.

parasite	rosacea	tinea	impetigo
armpit	vesicle	cyst	fungus
psoriasis	toxin	verruca	macule

1. small sac containing liquid ______________________
2. organism living off another species ______________________
3. scabby eruption ______________________
4. wart ______________________
5. worm ______________________
6. poison ______________________
7. axilla ______________________
8. spot ______________________
9. sac, bladder ______________________
10. rosy ______________________

D. **Spelling demons:** Every chapter contains some very difficult spelling demons. Challenge your ability to spell all these terms correctly. Circle the best choice below; then listen to the pronunciations on the Student Online Learning Center (www.mhhe.com/AllanEssMedLanguage) to be sure you can pronounce them correctly. 

1.	allopecia	alopecia	allopechia
2.	onchomycosis	onychomycosis	onycomycosis
3.	zenograft	xenograft	zennograft
4.	mamoplasty	manoplasty	mammoplasty
5.	syntethis	synthetis	synthesis
6.	sebaceous	sebeceous	sebacious
7.	puritis	pruritis	pruritus
8.	subcutaneous	subcutenious	subcutanous
9.	sebborheic	seborrheic	seborheic
10.	blefaroplasty	belpharoplasty	blepharoplasty

E. Create your own study hint for the medical term *rhinoplasty*.

Study Hint
Equate it to an animal in English.

____

____

____

____

F. Roots/Combining Forms are the core foundations of every medical term. You are given a brief definition of a term. Construct the complete term by adding the correct R/CF. Fill in the blanks.

1. inflammation of the skin ______________________________/itis
2. to scratch ex/ ______________________________/ate
3. medication against itching anti/______________________________/ic
4. infestation with lice ______________________________/osis
5. pertaining to poison ______________________________/ic
6. thickening of the skin due to excess collagen ______________________________/derma
7. infection alongside a nail para/ ______________________________/ia
8. surgical removal of injured or necrotic tissue de/______________________________/ment
9. skin graft from a cadaver homo/______________________________
10. a tear of the skin ______________________________/ation

Remember: A root or combining form can begin or end a medical term, and it does not become a prefix or suffix. Its identity always remains a root, and every term needs at least one root.

G. Difference between: If you know the meaning of a term, you will be able to explain it to someone else. Briefly write how you would explain to a fellow student the difference between:

1. vasoconstriction:

____

____

2. vasodilation:

____

____

3. incision:

____

____

4. excision:

____

____

H. Review and recall, from Chapters 1 through 3, all the suffixes that mean *pertaining to.* List them here.

Study Hint
Frequent review of elements from previous chapters will aid your recall on a test and make learning new terms easier.

Pertaining to	Pertaining to	Pertaining to	Pertaining to

I. Seek and find: Some terms in this book are defined in context and do not appear in a WAD. These terms could also appear on a test, so pay attention to them! Here is a sample of terms that have appeared in this chapter. Challenge yourself to answer without turning back to check in the book. Fill in the blanks, and then check your spelling!

What is another name for:

subcutaneous fat	
hives	
pressure ulcer	
allengic dermatitis	
shingles	
lice eggs	
itch mites	
cradle cap	
blackhead/whitehead	
thrush	
subcutaneous	
dead body	
allergic reaction	

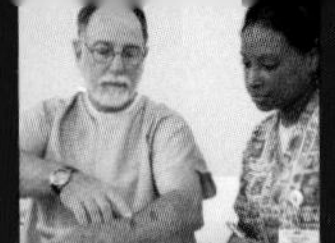

THE INTEGUMENTARY SYSTEM

J. Prefixes: Continue working with the important prefixes in this chapter. These prefixes will appear again in succeeding chapters and terminology for other body systems. Knowing them will greatly increase your medical vocabulary. Fill in the blanks.

Prefix	Meaning of Prefix	Medical Term with This Prefix	Meaning of Medical Term
in			
intra			
para			
re			
sub			
trans			
ultra			
xeno			

K. Terminology challenge: Express yourself in medical vocabulary. Briefly explain how the terms *diagnosis, prognosis,* and *etiology* are related in the disease process.

____

____

____

____

____

____

L. Partner exercise: Ask your study partner to close his or her text. Dictate the following sentences to your partner, so that he or she can write them down and then hand them back to you. Check your partner's sentences against the text below. The sentence is not correct unless every word is present and every word is spelled correctly. When you have finished checking your partner's answers, review them with him or her. Then close your book and ask your partner to dictate the sentences to you. Write them down and have him or her check them and review them with you.

1. Suturing brings together the edges of the wound to enhance tissue healing. In some people there is excessive fibrosis and scar tissue formation, which may produce a keloid.
2. A subcutaneous injection consists of using a longer needle to pierce the epidermis and dermis to reach the hypodermis or subcutaneous layer of skin.
3. The varicella-zoster virus causes chickenpox in unvaccinated people and can remain dormant in the peripheral nerves for decades before erupting as the painful vesicles of shingles.

Brain teaser: In the preceding sentence, what is the meaning of the word ***dormant?*** (Use your dictionary or go online to find out.) Define it on the lines below.

Dormant: ____

____

____

M. Make the connection: In this exercise, each set of terms has something in common. Use the elements to help you determine what that is. Fill in the blanks.

Medical Terms	All Relate To:
1. keratin, paronychia, onychomycosis	______________________
2. cadavers, skin banks, homograft	______________________
3. mammoplasty, abdominoplasty, blepharoplasty	______________________
4. integumentary, dermatitis, subcutaneous	______________________
5. comedo, papule, vesicle	______________________

N. Demonstrate your knowledge of the language of dermatology by indicating whether the following medical terms are a diagnosis or a procedure. Place a check mark (✔) in the appropriate column.

Medical Term	Diagnosis	Procedure
pruritus		
pediculosis		
biopsy		
debridement		
laceration		
abdominoplasty		
carcinoma		
autograft		
verruca		
stasis dermatitis		
rhinoplasty		
SLE		
lipectomy		

CHAPTER 3 REVIEW

THE INTEGUMENTARY SYSTEM

O. Chapter challenge: Circle the correct answer.

> **Study Hint**
> Immediately cross off any answer you know is not correct. In your remaining choices, there is only *one best answer.*

1. A *transdermal* patch can be used to administer:
   - **a.** contraceptive hormones
   - **b.** analgesics
   - **c.** antinausea medications
   - **d.** a, b and c
   - **e.** only a and c
2. The abbreviations **SC** and **IM** relate to:
   - **a.** lobes of the lung
   - **b.** body regions
   - **c.** injection sites
   - **d.** layers of a wound
   - **e.** accessory skin organs
3. Tummy tuck, breast alteration, and nose alteration are:
   - **a.** abdominoplasty, mammoplasty, and rhinoplasty
   - **b.** liposuction, lipectomy, and blepharoplasty
   - **c.** abdominoplasty, liposuction, and blepharoplasty
   - **d.** dermabrasion, mammoplasty, and lipectomy
   - **e.** abdominoplasty, dermabrasion, and blepharoplasty
4. Burnt, dead tissue lying on top of third-degree burns is called:
   - **a.** granulation
   - **b.** keloid
   - **c.** necrotizing
   - **d.** eschar
   - **e.** keratin
5. *Tinea pedis, capitis, corporis,* and *cruris* are all:
   - **a.** bacterial infections
   - **b.** viral infections
   - **c.** fungal infections
   - **d.** systemic infections
   - **e.** parasitic infestations
6. Which three terms represent a process?
   - **a.** abdominoplasty, biopsy, liposuction
   - **b.** host, parasite, infestation
   - **c.** cellulitis, pediculosis, dermatitis
   - **d.** sebaceous glands, sweat glands, hair follicles
   - **e.** intradermal, hypodermic, subcutaneous
7. *Comedones, papules, pustules*—if these three signs are present, what is the condition?
   - **a.** necrotizing fasciitis
   - **b.** acne
   - **c.** onychomycosis
   - **d.** scleroderma
   - **e.** squamous cell carcinoma
8. A synonym for *subcutaneous* is:
   - **a.** epidermal
   - **b.** dermal
   - **c.** intradermal
   - **d.** hypodermic
   - **e.** transdermal

9. Which medical term contains an element meaning *self?*

   a. allograft
   b. autograft
   c. homograft
   d. xenograft
   e. heterograft

10. The most common skin cancer and the least dangerous is:

   a. Kaposi sarcoma
   b. basal cell carcinoma
   c. melanoma
   d. squamous cell carcinoma
   e. herpes zoster

11. "Can extend beyond the edges of the original wound and often return if they are surgically removed." This describes:

   a. papules
   b. macules
   c. pustules
   d. scars
   e. keloids

12. Which two terms refer to dead tissue?

   a. cryosurgery carcinoma
   b. necrotizing eschar
   c. keratin sebum
   d. analgesic subcutaneous
   e. atopy excoriate

13. Circle the pair of terms that have roots or combining forms meaning *skin:*

   a. seborrheic allergenic
   b. dermatitis subcutaneous
   c. metastasize decubitus
   d. nevus hypodermic
   e. integumentary shingles

14. Which pair of terms is spelled correctly?

   a. fascitis pediculosis
   b. staphylococus aureus
   c. scleroderma soriasis
   d. dermatomyositis alopecia
   e. onycomycosis paronchia

15. *Pimples, boils, carbuncles,* and *impetigo* can be caused by:

   a. papillomavirus
   b. herpes zoster
   c. staph
   d. varicella
   e. Candida

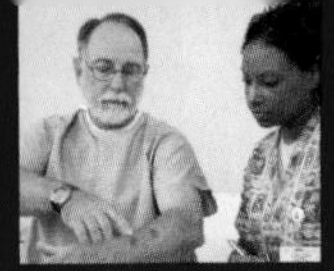

THE INTEGUMENTARY SYSTEM

P. Case Report challenge: Now that you are more comfortable with the terms in this chapter, you can apply that knowledge and briefly answer the questions about the Case Report. If you read the report through and underline all the medical terminology, this will make it easier to answer the questions.

you are

... a burn technologist employed in the Burn Unit at Fulwood Medical Center.

You are communicating with

... the son and daughter of Mr. Steven Hapgood, a 52-year-old man admitted to the Fulwood Burn Unit with severe burns over his face, chest, and abdomen. After an evening of drinking, he had been smoking in bed and fell asleep. His next-door neighbors in the apartment building smelled smoke and called 911. In the Burn Unit, his initial treatment included large volumes of intravenous fluids to prevent *shock*.

Mr. Hapgood's burns were mostly third-degree. The burned, dead tissue formed an *eschar* that can have toxic effects on the digestive, respiratory, and cardiovascular systems. The eschar was surgically removed by **debridement**.

1. Explain to Mr. Hapgood's son and daughter exactly which layers of skin are affected by his third-degree burn. Use nonmedical language they can understand.

____

____

____

____

____

____.

2. Explain to them also the surgical procedure their father is about to undergo.

____

____

____

____

3. Why does this procedure need to be performed?

____

____

____

____

4. What other body systems could be affected by the deep tissue burn Mr. Hapgood has suffered?

5. What is meant by "*toxic* effects"?

6. Define *shock:*

7. Describe *eschar:*

8. Why do you think burn patients are kept in a separate unit and not mixed in with other medical/surgical patients?

Special Senses of the Eye and Ear

The Essentials of the Languages of Ophthalmology and Otology

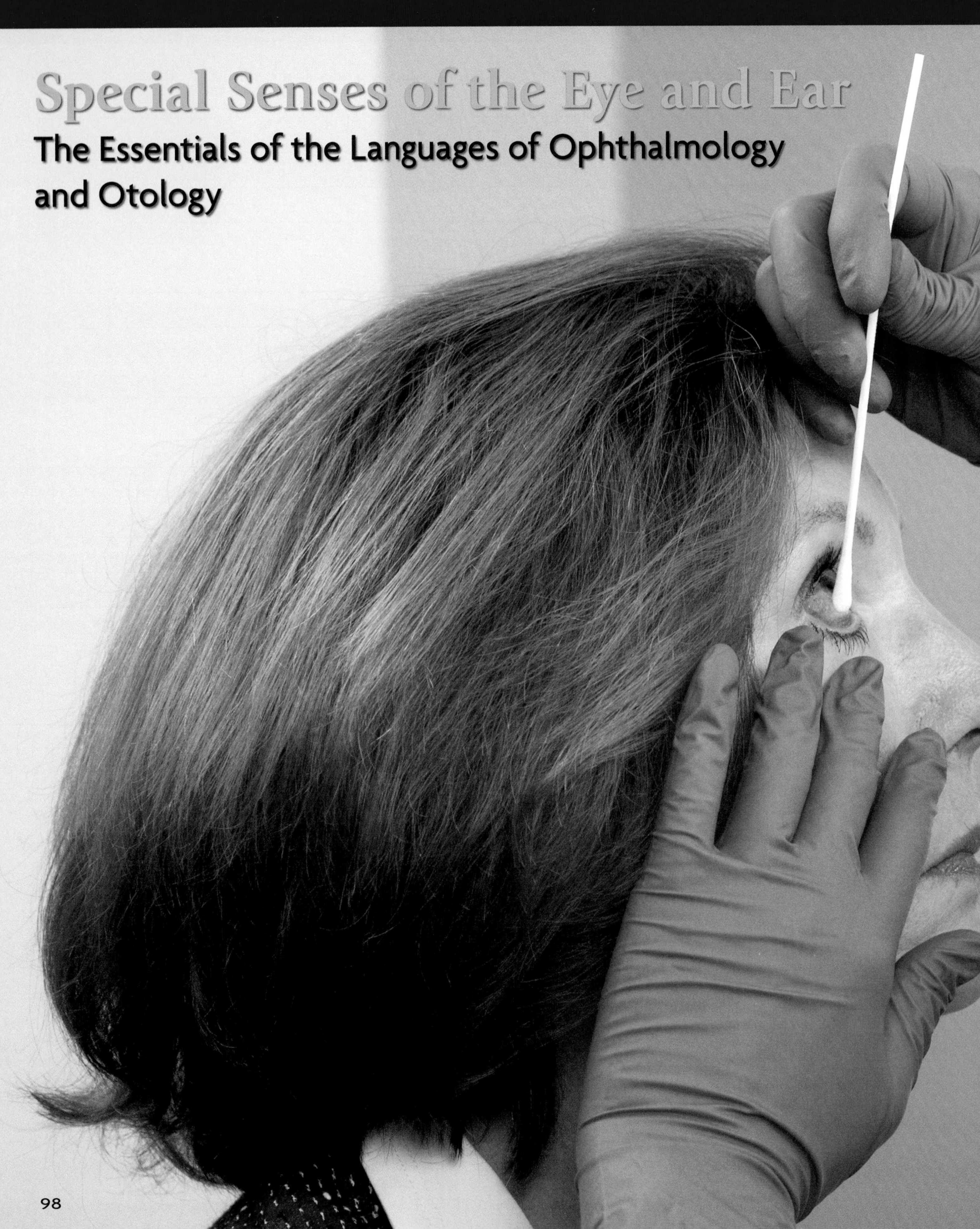

CHAPTER

THE EYE AND SEEING
CASE REPORT 4.1

You are

... an **ophthalmic** technician (OT) working in the office of **ophthalmologist** Angela Chun, MD, a member of the Fulwood Medical Group.

You are communicating with

... Mrs. Jenny Hughes, a 30-year-old computer software consultant, who walked into the office with painful, red, swollen eyelids and a sticky, purulent discharge from both eyes. The administrative medical assistant did not hold her in the reception area but brought her directly to you.

Mrs. Hughes complains of headache and **photophobia,** and her eyelids were stuck together when she woke up this morning. She tells you that a couple of days earlier she had gone into a small business office to install software at the firm's 10 workstations. One of the employees was absent with pink eye. Mrs. Hughes wants to know if she could have contracted the disease from that employee's keyboard and how to prevent her husband and two children from getting it.

You have in your hand a clipboard, pen attached, with the office Notice of Privacy Practices and sign-in sheet for her to sign. How do you proceed?

Learning Outcomes

In order to make correct decisions in situations like this, to communicate with Dr. Chun about the patient, to participate in patient education, and to document the patient's care, you need to be able to:

4.1 Apply the language of **ophthalmology** in everyday work situations.

4.2 Comprehend, analyze, spell, and write the medical terms of ophthalmology.

4.3 Recognize and pronounce the medical terms of ophthalmology.

NOTE: The sense of smell is discussed as an integral part of the respiratory system; the sense of taste, as an integral part of the digestive system; and the sense of touch, as an integral part of the nervous system.

LESSON 4.1 Accessory Structures of the Eye

OBJECTIVES

Mrs. Jenny Hughes's **"pink eye"** involved her **conjunctiva** and eyelids, two of the **periorbital** accessory structures of the eye, located around the **orbit** and in front of the eyeball. The other accessory structures are the eyebrows, eyelashes, and the **lacrimal** (tear) apparatus. All these structures support and protect the exposed front surface of the eye.

The information in this lesson will enable you to:

- **Link the composition of the accessory structures of the eye to their functions.**
- **Explain the roles of the accessory structures in protecting the eye.**
- **Apply the correct medical terminology to the accessory structures and their disorders.**

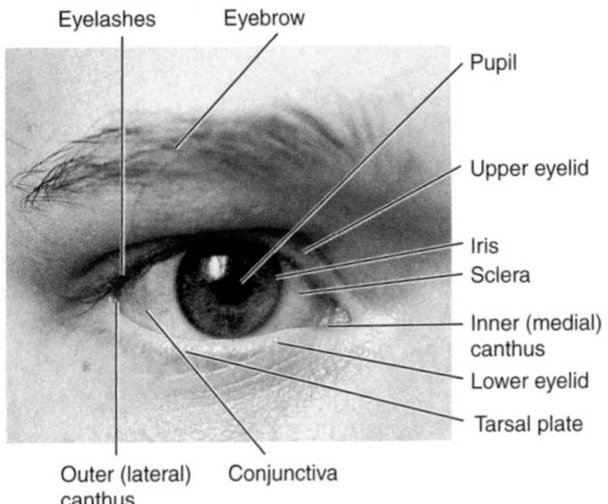

▲ **FIGURE 4.1 External Anatomy of the Eye.**

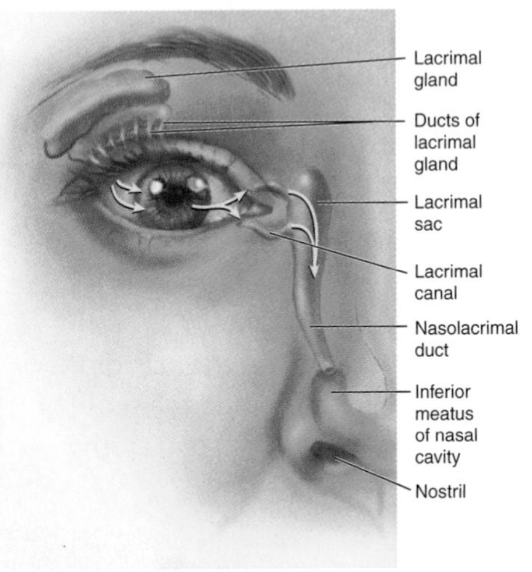

▲ **FIGURE 4.2 Lacrimal Apparatus.**

The accessory structures of the eye are the following:

Eyebrows keep sweat from running into the eye and function in nonverbal communication (*Figure 4.1*).

Eyelids protect the eye from foreign objects. They blink to move tears across the surface of the eye and sweep debris away. They close in sleep to keep out visual stimuli. They are covered in the body's thinnest layer of skin.

Eyelashes are strong hairs that help keep debris out of the eyes. They arise from hair follicles with their sebaceous glands on the edge of the lids.

The **conjunctiva** is a transparent mucous membrane that lines the inside of both eyelids and covers the front of the eye except the central portion, the **cornea.** In the conjunctiva, numerous goblet cells secrete a thin film of mucin that prevents the eyeball from drying. The conjunctiva is freely movable over the eyeball. It has numerous small blood vessels and is richly supplied with nerve endings that make it very sensitive to pain.

The **lacrimal apparatus** (*Figure 4.2*) consists of the **lacrimal (tear) gland** located in the superolateral corner of the orbit. The gland secretes tears, and short ducts carry the tears to the surface of the conjunctiva. After washing across the conjunctiva, the tears leave the eye at the medial corner of the eye by draining into the **lacrimal sac.** They then flow into the **nasolacrimal duct,** which carries the tears into the nose, from where they are swallowed.

The functions of tears are to:

- **Clean and lubricate** the surface of the eye.
- **Deliver** nutrients and oxygen to the conjunctiva.
- **Prevent infection** through bactericidal enzymes.

WORD ANALYSIS AND DEFINITION

S/ = Suffix P/ = Prefix R/ = Root R/CF = Combining Form

WORD	PRONUNCIATION	ELEMENTS		DEFINITION
conjunctiva **conjunctival (adj)** **conjunctivitis** **"pink eye"**	kon-junk-**TIE**-vah kon-junk-**TIE**-val kon-junk-tih-**VI**-tis PINK EYE	 S/ R/ S/	Latin *inner lining of eyelids* **-al** *pertaining to* **conjunctiv-** *conjunctiva* **-itis** *inflammation* Lay term for *conjunctivitis*	Inner lining of eyelids. Pertaining to the conjunctiva. Inflammation of the conjunctiva. Conjunctivitis.
cornea **corneal (adj)**	**KOR**-nee-ah **KOR**-nee-al		Latin *web, tunic*	The central, transparent part of the outer coat of the eye covering the iris and pupil.
lacrimal **nasolacrimal duct**	**LAK**-rim-al **NAY**-zoh-**LAK**-rim-al DUKT	S/ R/ R/CF R/	**-al** *pertaining to* **lacrim-** *tear* **nas/o-** *nose* **duct** *to lead*	Pertaining to tears and the tear apparatus. Passage from the lacrimal sac to the nose.
ophthalmology **ophthalmologist** **ophthalmic (adj)**	off-thal-**MALL**-oh-jee off-thal-**MALL**-oh-jist off-**THAL**-mick	S/ R/CF S/ R/ S/	**-logy** *study of* **ophthalm/o-** *eye* **-logist** *one who studies, specialist* **ophthalm-** *eye* **-ic** *pertaining to*	Diagnosis and treatment of diseases of the eye. Medical specialist in ophthalmology. Pertaining to the eye.
orbit **orbital (adj)** **periorbital**	**OR**-bit **OR**-bit-al per-ee-**OR**-bit-al	 S/ R/ P/	Latin *circle* **-al** *pertaining to* **orbit-** orbit **peri-** *around*	The bony socket that holds the eyeball. Pertaining to the orbit. Pertaining to tissues around the orbit.
photophobia **photophobic (adj)**	foh-toe-**FOH**-bee-ah foh-toe-**FOH**-bik	S/ R/CF R/ S/	**-ia** *condition* **phot/o-** *light* **-phob-** *fear* **-ic** *pertaining to*	Fear of the light because it hurts the eyes. Pertaining to or suffering from photophobia.

EXERCISES

Build *your medical vocabulary. Many of the prefixes and suffixes you see in this chapter you will meet in later chapters and use to build your knowledge of additional medical terms. Fill in the chart, and then answer the question below.*

Medical Term	Prefix	Meaning of Prefix	Root(s)/CF(s)	Meaning of R/CF	Suffix	Meaning of Suffix
nasolacrimal						
conjunctivitis						
periorbital						
ophthalmologist						
photophobia						

1. Take the term *ophthalmologist,* add two different suffixes to the same root/combining form, and make two new medical terms:

_______ / _______ / _______ means _______
P R/CF S

_______ / _______ / _______ means _______
P R/CF S

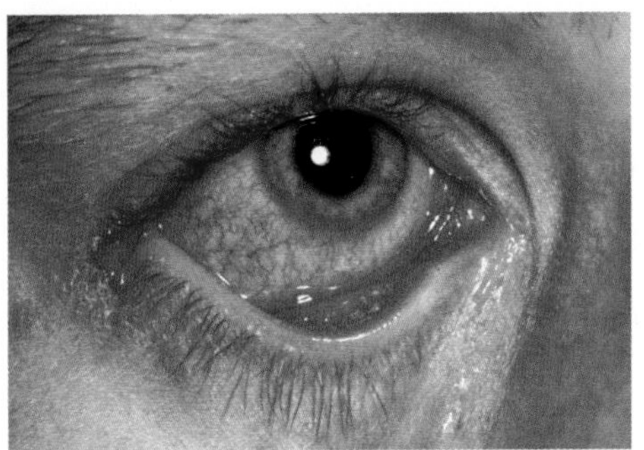

▲ **FIGURE 4.3** **Conjunctivitis.**

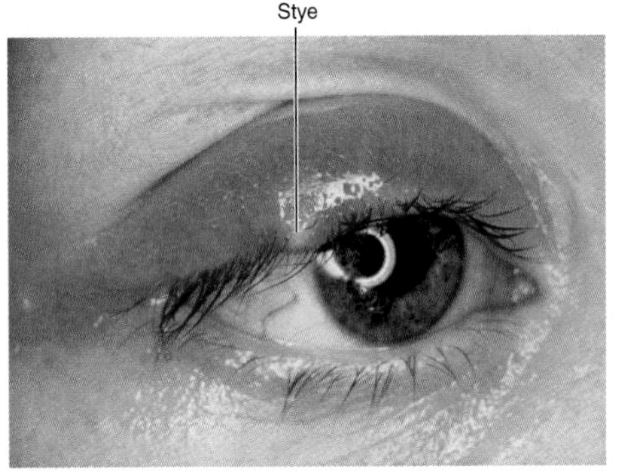

▲ **FIGURE 4.4** **Stye Showing Pus-Filled Cyst.**

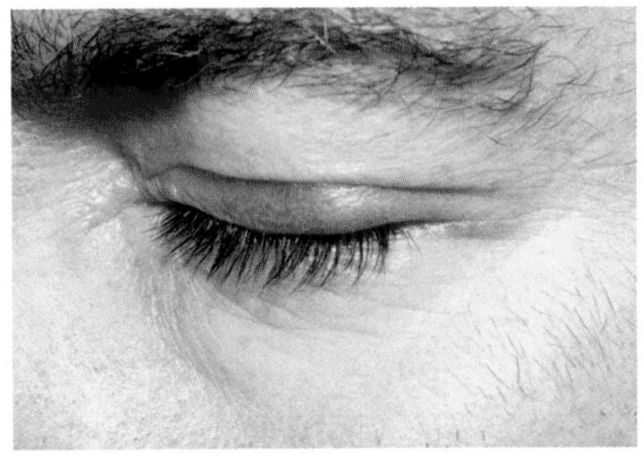

▲ **FIGURE 4.5** **Blepharitis.**

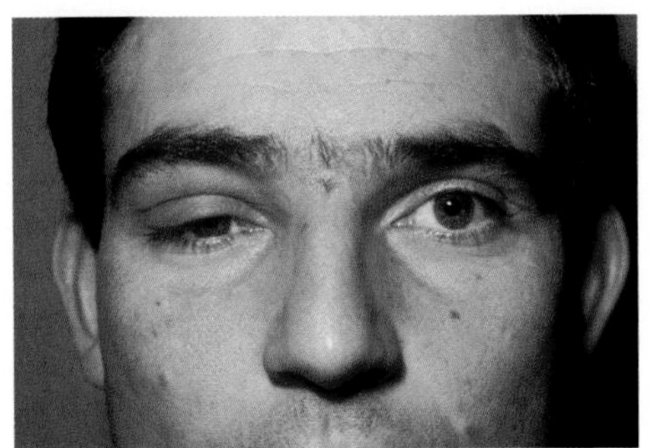

▲ **FIGURE 4.6** **Ptosis of Right Eyelid.**

Case Report 4.1 *(continued)*

Mrs. Hughes's pink eye is called acute **contagious** conjunctivitis (*Figure 4.3*). It responds well to **antibiotic** eyedrops. Her hands were **contaminated** from the keyboard of the employee who had left work and gone home with pink eye. She transmitted the infection to her eye by touching or rubbing.

Your documentation of Mrs. Hughes's office visit could read:

Progress Note: 04/10/08
Mrs. Jenny Hughes was brought directly into the clinical area at 1030 hrs with what appeared to be conjunctivitis, "pink eye." Both eyelids were red and swollen with a **purulent discharge.** She complained of headache and photophobia. Dr. Chun prescribed Neosporin eyedrops, three drops q4h. A swab of the discharge was sent to the laboratory. I instructed and watched Mrs. Hughes wash her hands and use an alcohol-based hand gel. I then had her sign in and sign our Notice of Privacy Practices. I instructed her in the use of the eyedrops and emphasized home care and hand care measures to prevent the infection from spreading to her family. She was given a return appointment in 1 week and told to call the office if the eyedrops do not help. **Daphne** Butras, OT. 1055 hrs.

DISORDERS OF THE ACCESSORY GLANDS

Conjunctivitis (*Figure 4.3*), inflammation of the conjunctiva, is more commonly viral than bacterial; it can also be caused by irritants such as chlorine, soaps, fumes, and smoke.

Eyelid edema, generalized swelling of the eyelids, is often produced by an allergic reaction (*see Chapter 11*) due to cosmetics, pollen in the air, or stings and bites from insects.

A **stye** or **hordeolum** is an infection of an eyelash follicle producing an abscess (*Figure 4.4*), with localized pain, swelling, redness, and pus formation at the edge of the eyelid.

Blepharitis occurs when multiple eyelash follicles become infected. The margin of the eyelid shows persistent redness and crusting and may become ulcerated (*Figure 4.5*).

Ptosis, in which the upper eyelid is constantly drooped over the eye, is due to **paresis** of the muscle that raises the upper lid (*Figure 4.6*). The term **blepharoptosis** is used specifically for sagging of the eyelids due to excess skin. The plastic surgery procedure of **blepharoplasty** is used for the repair of the eyelid.

WORD ANALYSIS AND DEFINITION

S/ = Suffix P/ = Prefix R/ = Root R/CF = Combining Form

WORD	PRONUNCIATION		ELEMENTS	DEFINITION
antibiotic	**AN**-tih-bye-**OT**-ik	S/ P/ R/	**-tic** *pertaining to* **anti-** *against* **-bio-** *life*	A substance that has the capacity to inhibit the growth of or destroy bacteria and other microorganisms.
blepharitis	blef-ah-**RYE**-tis	S/ R/	**-itis** *inflammation* **blephar-** *eyelid*	Inflammation of the eyelid.
blepharoptosis	**BLEF**-ah-**ROP**-toe-sis	S/ R/CF	**-ptosis** *drooping* **blephar/o-** *eyelid*	Drooping of the upper eyelid.
blepharoplasty	**BLEF**-ah-roh-plas-tee	S/	**-plasty** *surgical repair*	Surgical repair of the eyelid.
contagious	kon-**TAY**-jus		Latin *touch closely*	Infection can be transmitted from person to person or from a person to a surface to a person.
contaminate (verb)	kon-**TAM**-in-ate	S/ P/ R/	**-ate** *composed of, pertaining to* **con-** *together* **-tamin-** *touch*	To cause the presence of an infectious agent to be on any surface.
contamination (noun)	**KON**-tam-ih-**NAY**-shun	S/	**-ation** *process*	The presence of an infectious agent on any surface.
hordeolum (also called *stye*)	hor-**DEE**-oh-lum		Latin *stye in the eye*	Abscess in an eyelash follicle.
paresis	par-**EE**-sis		Greek *paralysis*	Partial paralysis.
ptosis (***Note:*** When a word begins with two consonants, the first is silent) (***Note:*** *Ptosis* can also be used as a suffix.)	**TOE**-sis		Greek *drooping*	Sinking down of the upper eyelid or an organ.
purulent	**PURE**-you-lent	S/ R/	**-ulent** *abounding in* **pur-** *pus*	Showing or containing a lot of pus.

Study Hint
Note that the term *ptosis* is a stand-alone term and has a meaning in its own right. It can also function as a suffix specifically relating to the eyelids when it is added to the combining form *blephar/o.*

EXERCISES

*The **chapter objectives** can be met by inserting the correct medical terminology in the following patient documentation. Use only the terms contained in the WAD box above.* Remember to proofread your work for correct spelling! *Fill in the blanks.*

1. The patient's drooping left eyelid has been diagnosed as ______________________, and the surgical procedure ______________________ will correct this condition.
2. ______________________ of the patient's left eye muscle has resulted in constant drooping of her eyelid over her eye.
3. The patient has been advised to wipe down all household surfaces with a disinfectant to reduce the chance of further ______________________ in the household.
4. The ______________________ on the patient's right eye has a ______________________ drainage.
5. Mrs. Hughes's eye infection will be helped by the prescribed ______________________.

You are

. . . an ophthalmic technician employed by Angela Chun, MD, an ophthalmologist at Fulwood Medical Center.

You are communicating with

. . . Mrs. Jenny Hughes, the mother of Sam Hughes, a 2½–year-old boy, who has been referred by his pediatrician to Dr. Chun.

CASE REPORT 4.2

Mrs. Hughes states that for the past couple of months Sam's right eye has turned in. The only visual difficulty she has noticed is that he sometimes misses a Cheerio when he tries to grab it.

You are responsible for documenting Sam's diagnostic and therapeutic procedures and explaining the significance of these to his mother.

EXTRINSIC MUSCLES OF THE EYE

Humans, with their two eyes working closely together, have developed very good three-dimensional perception **(stereopsis)** and hand-eye coordination. Stereopsis depends on an accurate alignment of the two eyes.

This alignment is held in place by the coordination of six **extrinsic** eye muscles in each eye that are attached to the inner wall of the orbit and to the outer surface of the eyeball (*Figure 4.7*). These muscles move the eye in all directions.

When there is muscle imbalance in one eye, as in Sam's case, the alignment breaks down and the resulting condition is called **strabismus** (*Figure 4.8*), known in lay terms as "cross-eyed."

Esotropia is the eye turned in toward the nose. In congenital or infantile esotropia, both eyes look in toward the nose—the right eye looks to the left and the left eye looks to the right (*Figure 4.9*). Children with this condition require surgical intervention.

Accomodative esotropia is an inward eye turn, usually noticed around 2 years of age in 1% to 2% of children. In Sam's case, he had an accommodative esotropia in one eye. He will probably respond to glasses and perhaps a patch over the stronger eye.

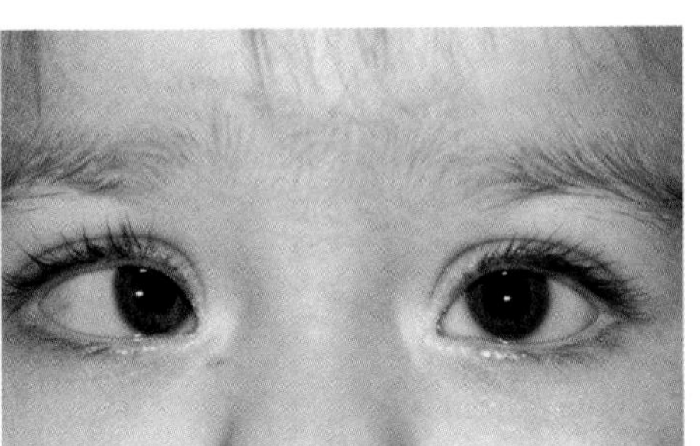

▲ **FIGURE 4.8 Strabismus.**

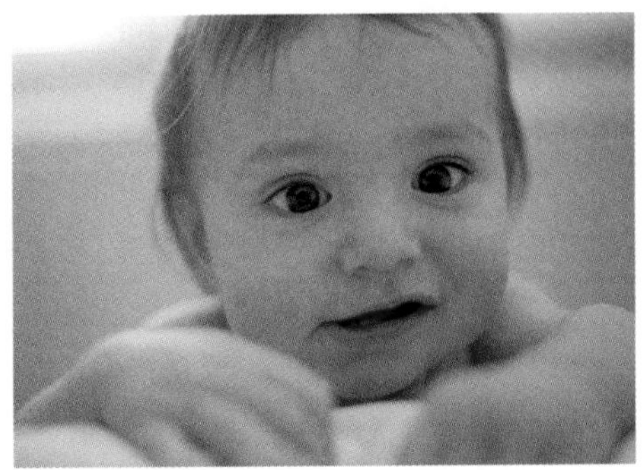

▲ **FIGURE 4.9 Congenital Esotropia.**

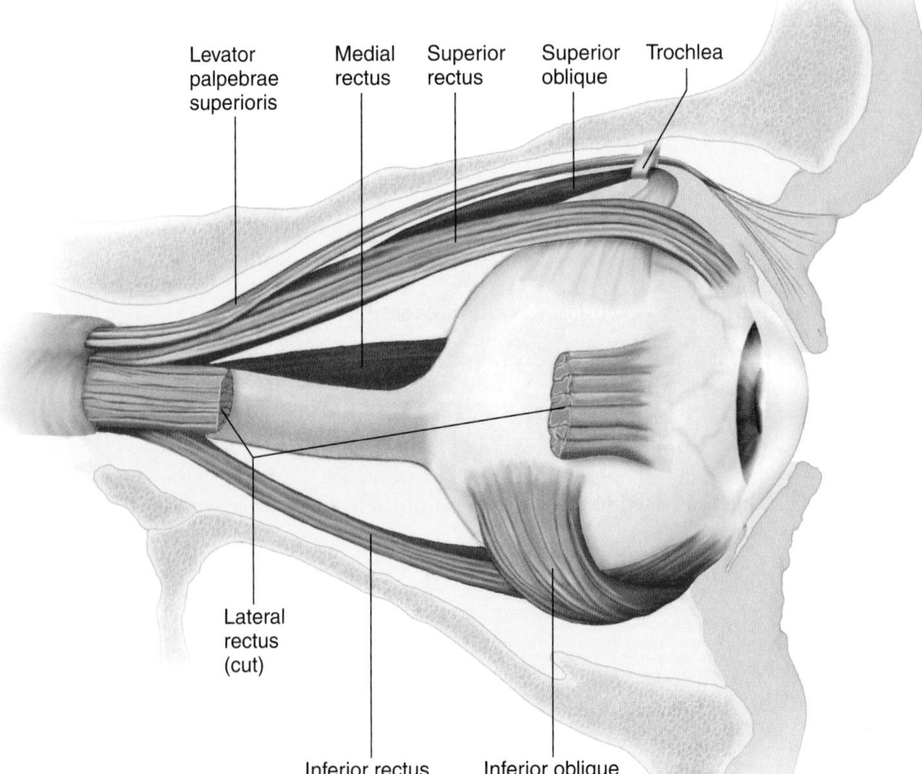

FIGURE 4.7 Extrinsic Muscles of the Right Eye: Lateral View. ▶

WORD ANALYSIS AND DEFINITION

S/ = Suffix P/ = Prefix R/ = Root R/CF = Combining Form

WORD	PRONUNCIATION	ELEMENTS		DEFINITION
accommodation (noun)	ah-kom-oh-**DAY**-shun	S/ P/ R/	**-ion** *action* **ac-** *toward* **-commodat-** *adjust*	The act of adjusting something to make it fit the needs; in this case the lens of the eye adjusts itself.
accommodate (verb)	ah-**KOM**-oh-date	S/	**-ate** *pertaining to, composed of*	To adapt to meet a need.
accommodative (adj)	ah-**kom**-oh-**DAY**-tiv	S/	**-ive** *pertaining to*	Pertaining to accommodation.
amblyopia	am-blee-**OH**-pee-ah	P/ R/	**ambly-** *dull* **-opia** *sight*	Failure or incomplete development of the pathways of vision to the brain.
esotropia	es-oh-**TROH**-pee-ah	S/ P/ R/	**-ia** *condition* **eso-** *inward* **-trop-** *turn*	Turning the eye inward toward the nose.
exotropia	ek-soh-**TROH**-pee-ah	S/ P/	**-ia** *condition* **exo-** *outward*	Turning the eye outward away from the nose.
extrinsic	eks-**TRIN**-sik		Latin *on the outer side*	Any muscle located entirely on the outside of the structure under consideration; e.g., the eye.
intrinsic	in-**TRIN**-sik		Latin *on the inner side*	Any muscle located entirely within (inside) the structure under consideration; e.g., the eye.
ocular	**OCK**-you-lar	S/ R/	**-ar** *pertaining to* **ocul-** *eye*	Pertaining to the eye.
optometrist	op-**TOM**-eh-trist	R/CF S/	**opt/o-** *vision* **-metrist** *skilled in measurement*	Someone skilled in the measurement of vision but who cannot treat eye diseases or prescribe medication.
optometry	op-**TOM**-eh-tree	S/	**-metry** *process of measuring*	The profession of the measurement of vision.
stereopsis	ster-ee-**OP**-sis	S/ R/	**-opsis** *vision* **stere-** *three-dimensional*	Three-dimensional vision.
strabismus	strah-**BIZ**-mus	S/ R/	**-ismus** *take action* **strab-** *squint*	Turning of an eye away from its normal position.

Exotropia, an outward turning of one eye, is noticed around 2 to 4 years of age. It will often respond to vision therapy, which includes eye exercises and glasses, from an **optometrist.** Eye muscle surgery can establish good **ocular** alignment.

Strabismus is not the same as **amblyopia,** or "lazy eye," which occurs in children when vision in one eye has not developed as well as vision in the other. It occurs because the eye and the brain are not cooperating for the affected eye.

Keynote

Amblyopia is treated with a patch over the stronger eye to develop the vision in the weaker eye.

EXERCISES

Deconstruct *the following medical terms into their elements; then answer the questions. Complete the chart, and fill in the blanks. Notice in this exercise that not every medical term needs a prefix and/or a suffix* but *every medical term does contain one or more roots and/or combining forms.*

Medical Term	Prefix	Root(s)/Combining Form(s)	Suffix
exotropia			
optometrist			
esotropia			
ambylopia			
stereopsis			
ocular			

1. Which term is a medical occupation? ______________________________

2. Which terms in the WAD are opposites? ____________________ and ____________________

LESSON 4.2 The Eyeball and Seeing

OBJECTIVES

The eyeball is a hollow sphere about 1 inch in diameter. Knowledge of its terminology, structure, and function enables you to understand how we see and what major problems and disorders can arise with the eye.

In this lesson, the information will enable you to:

- **Identify the principal components of the eyeball and their functions.**
- **Explain the role of the cornea and the problems that can occur in that structure.**
- **Use correct medical terminology to describe the structures and functions of the lens and its associated structures.**
- **Link the different components of the retina to their functions.**
- **Apply correct medical terminology to the structures and disorders of the eyeball.**

THE EYEBALL (GLOBE)

The functions of the eyeball are to:

1. **Adjust** continuously the amount of light it lets in to reach the retina.
2. **Focus** continuously on near and distant objects.
3. **Produce** images continuously of those objects and instantly transmit them to the brain.

The front of the eyeball is covered by the conjunctiva, a thin layer of tissue that covers the inside of the eyelids and curves over the eyeball to meet the **sclera** (*Figure 4.10*), the tough, white outer layer of the eye.

Keynote

The cornea protects the eye and, by changing shape, provides about 60% of the eye's focusing power.

The center of the front of the eye is a transparent, dome-shaped membrane called the cornea. The cornea has no blood supply and obtains its nutrients from tears and from fluid in the anterior chamber behind it.

When light rays strike the eye, they pass through the cornea. Because of its dome curvature, those rays striking the edge of the cornea are bent toward its center. The light rays then go through the **pupil**, the black opening in the center of the colored area in the front of the eye.

Keynote

The iris controls the amount of light entering the eye.

This colored area is called the **iris.** It controls the amount of light entering the eye. When you are in a dark place, the iris opens **(dilates)** to allow more light to enter. When you are in bright light, the iris closes **(constricts)** to admit less light.

After passing through the pupil, the light rays pass through the transparent **lens.** The lens can become thicker and thinner, enabling it to bend the light rays and focus them on the **retina** at the back of the eye. This process of changing focus is called accommodation. The process of bending the light rays by the cornea and lens is called **refraction.**

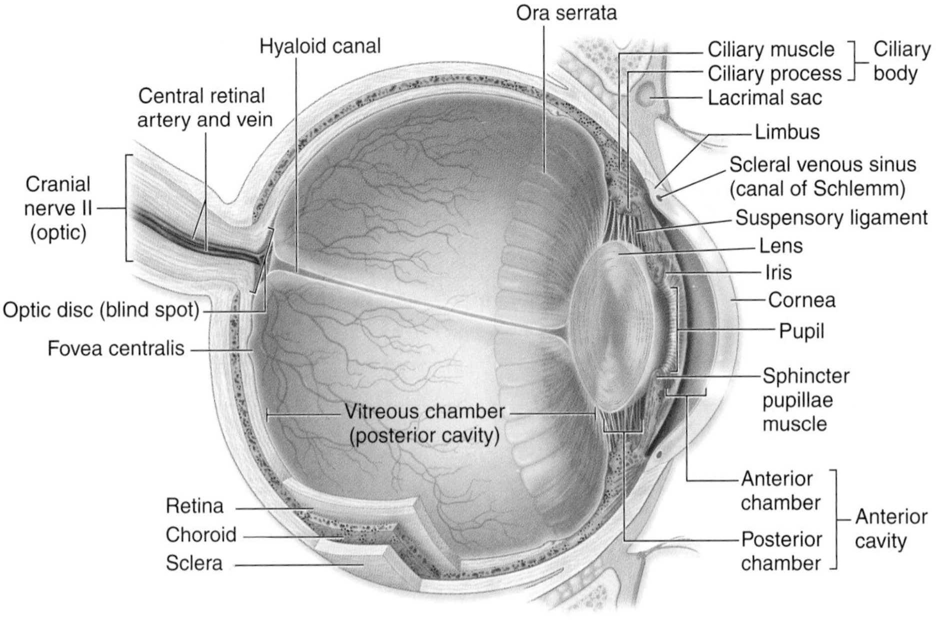

▲ **FIGURE 4.10 Anatomy of the Eyeball.**

WORD ANALYSIS AND DEFINITION

S/ = Suffix P/ = Prefix R/ = Root R/CF = Combining Form

WORD	PRONUNCIATION	ELEMENTS		DEFINITION
avascular	a-**VAS**-cue-lar	S/ P/ R/	**-ar** *pertaining to* **a-** *without* **-vascul-** *blood vessel*	Without a blood supply.
constrict (verb) **constriction (noun)**	kon-**STRIKT** kon-**STRIK**-shun	P/ R/ S/	**con-** *with, together* **-strict** *narrow* **-ion** *action, condition*	Become or make narrow. A narrowed portion of a structure.
dilate (verb) **dilation (noun)**	**DIE**-late die-**LAY**-shun	 S/ R/	Latin *dilate* **-ion** *action, condition* **dilat-** *dilate*	To perform or undergo dilation. Stretching or enlarging an opening or a structure.
iris	**EYE**-ris		Greek *diaphragm of the eye*	Colored portion of the eye with the pupil in its center.
lens	LENZ		Latin *lentil shape*	Transparent refractive structure behind the iris.
presbyopia	prez-bee-**OH**-pee-ah	S/ R/	**-opia** sight **presby-** *old man*	Difficulty in nearsighted vision occurring in middle and old age.
pupil **pupillary (adj)** (***Note:*** Change to *ll.)*	**PYU**-pill **PYU**-pill-ah-ree	 S/ R/	Latin *pupil* **-ary** *pertaining to* **pupill-** *pupil*	The opening in the center of the iris that allows light to reach the lens. Pertaining to the pupil.
refract (verb) **refraction (noun)**	ree-**FRACT** ree-**FRAK**-shun	 S/ R/	Latin *break up* **-ion** *condition* **refract-** *bend*	Make a change in direction or bend a ray of light. The bending of light.
retina **retinal (adj)**	**RET**-ih-nah **RET**-ih-nal	 S/ R/	Latin *net* **-al** *pertaining to* **retin-** *retina*	Light-sensitive innermost layer of the eyeball. Pertaining to the retina.
sclera **scleral (adj)** **scleritis**	**SKLAIR**-ah **SKLAIR**-al sklair-**RI**-tis	 S/ R/ S/	Greek *hard* **-al** *pertaining to* **scler-** *hardness, white of eye* **-itis** *inflammation*	Fibrous outer covering of the eyeball and the white of the eye. Pertaining to the sclera. Inflammation of the sclera.

The lens has no supply of blood vessels **(avascular)** or nerves. With increasing age, the lens loses its elasticity. When you reach your forties, your eyes may have difficulty focusing on near objects, a condition called **presbyopia.**

Medical shorthand for a quick, normal eye examination can be **PERRLA: p**upils **e**qual, **r**ound, **r**eactive to **l**ight and **a**ccommodation.

Keynote

The lens changes its shape to focus rays of light on the retina.

EXERCISES

Define *the statements in column 1, using the* language of ophthalmology *in column 2. These medical terms all relate to the vision process. Fill in the blanks.*

_____	1. colored portion of the eye	A. lens
_____	2. bending a ray of light	B. retina
_____	3. opening in the iris	C. avascular
_____	4. narrowed portion of a structure	D. pupil
_____	5. transparent, refractive structure	E. constriction
_____	6. difficulty in nearsighted vision	F. refract
_____	7. innermost layer of eyeball	G. iris
_____	8. fibrous, outer covering of the eye	H. presbyopia
_____	9. without a blood supply	I. sclera

THE EYEBALL (GLOBE) (continued)

The Retina

Keynote

Rods of the retina perceive only dim light and not color. Cones of the retina perceive bright light and color.

The final destination of the light rays is the retina, at the back of your eye (*Figure 4.11a*). It is the size of a postage stamp and has ten layers of cells. The retina has 130 million **rods** (*Figure 4.11b*), which perceive only light, not color, and function mostly when the light is dim. The retina has 6.5 million **cones** (*Figure 4.11b*), which are activated by light and color and have precise **visual acuity** (sharpness). Different cones respond to red, blue, and green light. Your perception of color is based on the intensity of different mixtures of colors from the three types of cones.

Some people have a hereditary lack of response by one or more of the three types of cones and show color blindness. The most common form is red-green color blindness, in which these colors and related shades cannot be distinguished from each other.

Keynote

Rods and cones are called **photoreceptor cells.**

The rods and cones convert the energy of the light rays into electrical impulses, and the **optic nerve,** a bundle of more than a million nerve fibers, transmits these impulses to the visual cortex at the back of the brain. The area where the optic nerve leaves the retina is called the **optic disc.** Because it has no rods and cones, the optic disc cannot form images and thus is called the "blind spot."

Just lateral to the optic disc at the back of the retina is a circular, yellowish region called the **macula lutea** (*Figure 4.11a*). In the center of the macula is a small pit called the **fovea centralis,** which has 4000 tiny cones and no rods. Each cone has its own nerve fiber, and this makes the fovea the area of sharpest vision. As you read this text, the words are precisely focused on your fovea centralis.

Behind the **photoreceptor** layer of the retina is a very vascular layer called the **choroid.** This layer, together with the iris and its muscle, is called the **uvea.**

Segments of Eye

The eyeball is divided into two fluid-filled segments separated by the lens. The fluids maintain the shape of the eyeball. In the back, the posterior segment extends from the back of the lens to the retina and contains a transparent jelly called the **vitreous humor,** which helps maintain the shape of the eyeball.

In front, the anterior segment extends from the cornea to the lens. It contains the **aqueous humor,** which is continually produced and then moves from the eye into the venous bloodstream. The aqueous humor also removes waste products and helps maintain the internal chemical environment of the eye.

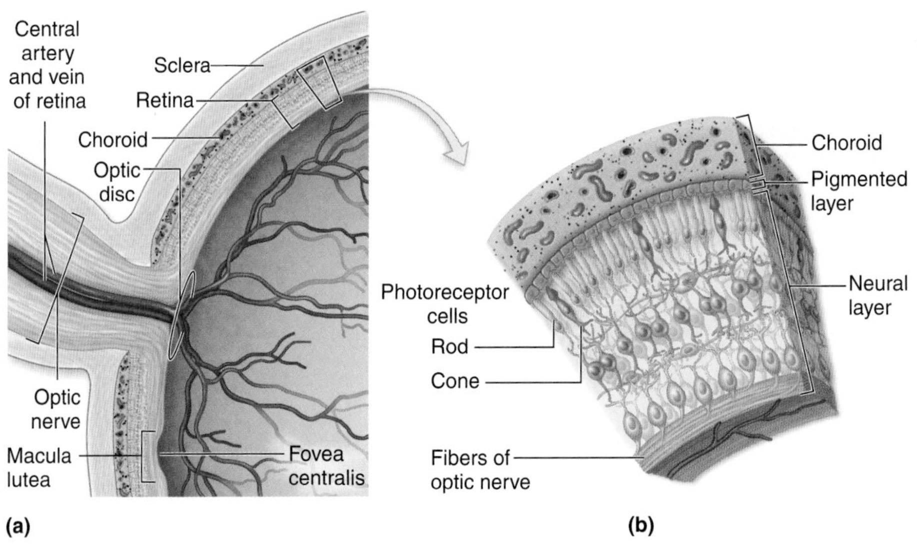

▲ **FIGURE 4.11 Structure of the Retina.**
(a) retina. (b) rods and cones.

WORD ANALYSIS AND DEFINITION

S/ = Suffix P/ = Prefix R/ = Root R/CF = Combining Form

WORD	PRONUNCIATION	ELEMENTS		DEFINITION
aqueous humor	**ACHE**-we-us **HEW**-mor		aqueous Latin *watery* humor Latin *liquid*	Watery liquid in the anterior and posterior chambers of the eye.
choroid	**KOR**-oid		Greek *membrane*	Region of the retina and uvea.
fovea centralis	**FOH**-vee-ah sen-**TRAH**-lis	 R/ S/	fovea Latin *a pit* central- *center* **-is** *pertaining to*	Small pit in the center of the macula that has the highest visual acuity.
macula lutea	**MACK**-you-lah **LOO**-tee-ah		macula Latin *small spot* lutea Latin *yellow*	Yellowish spot on the back of the retina; contains the fovea.
optic **optical (adj)**	**OP**-tick **OP**-tih-kal	 S/ R/	Greek *eye* **-al** *pertaining to* **optic-** *eye*	Pertaining to the eye or vision. Pertaining to the eye or vision.
photoreceptor	foh-toe-ree-**SEP**-tor	S/ R/CF R/	**-or** *that which does something* **phot/o-** *light* **-recept-** *receive*	A photoreceptor cell receives light and converts it into electrical impulses.
uvea **uveitis**	**YOU**-vee-ah you-vee-**EYE**-tis	 S/ R/	Latin *vascular layer* **-itis** *inflammation* **uve-** uvea	Middle coat of the eyeball; includes the iris, ciliary body, and choroid. Inflammation of the uvea.
visual acuity	**VIH**-zhoo-wal ah-**KYU**-ih-tee		acuity Latin *sharpen*	Sharpness and clearness of vision.
vitreous humor	**VIT**-ree-us **HEW**-mor		vitreous Latin *glass* humor Latin *liquid*	Vitreous humor is a gelatinous liquid in the posterior segment of the eyeball with the appearance of glass.

EXERCISES

The **language of ophthalmology** *will be your answers in this exercise. Review the WAD before you begin the exercise. Circle the best answer.*

1. Sharpness and clearness of vision is called:
   a. optical
   b. acuity
   c. vascular
   d. aqueous
   e. choroid
2. The term meaning *watery liquid* is:
   a. fovea centralis
   b. macula lutea
   c. aqueous humor
   d. conjunctival
   e. pupillary
3. Because it has no rods and cones, the optic disc cannot form images and thus is called the:
   a. fovea
   b. blind spot
   c. uvea
   d. visual cortex
   e. optic nerve
4. The gel contained in the *posterior segment* is called:
   a. vitreous humor
   b. fovea centralis
   c. visual cortex
   d. aqueous humor
   e. macula lutea
5. The term *photoreceptor* has:
   a. two combining forms and a suffix
   b. a prefix, a root, and a suffix
   c. a root, a combining form, and a suffix
   d. two roots and a suffix
   e. a suffix and a root
6. The area of sharpest vision is the:
   a. cornea
   b. rods
   c. macula lutea
   d. fovea
   e. cones

FIGURE 4.12 **Visual Pathway.**

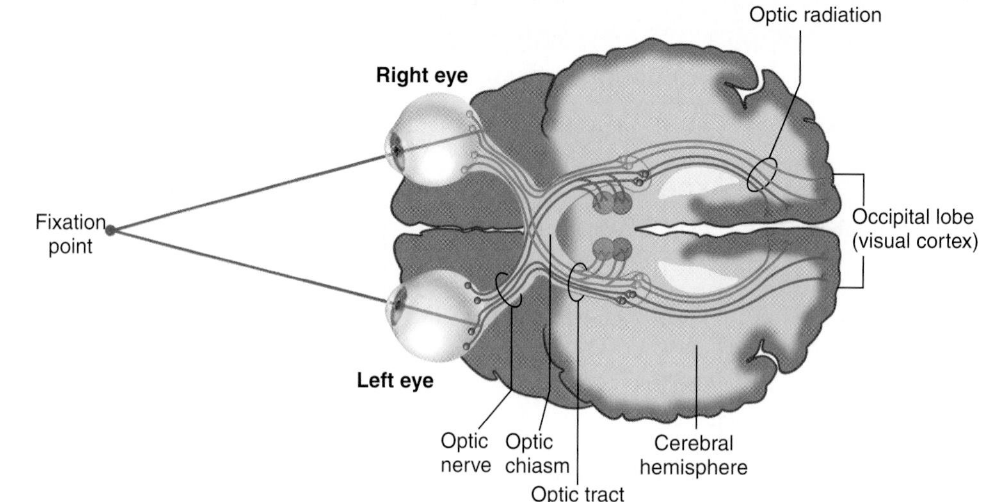

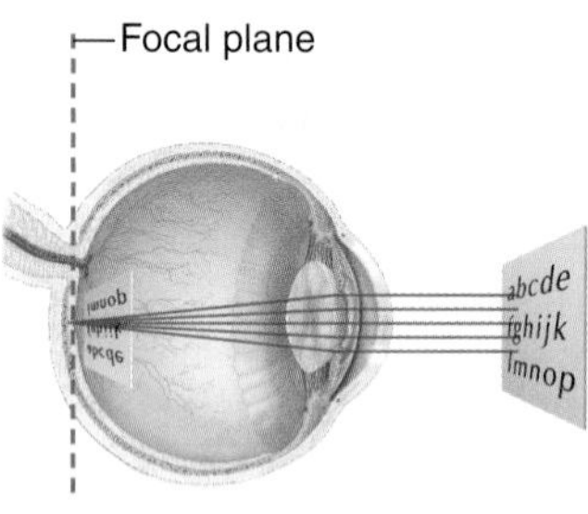

FIGURE 4.13 Normal Vision.

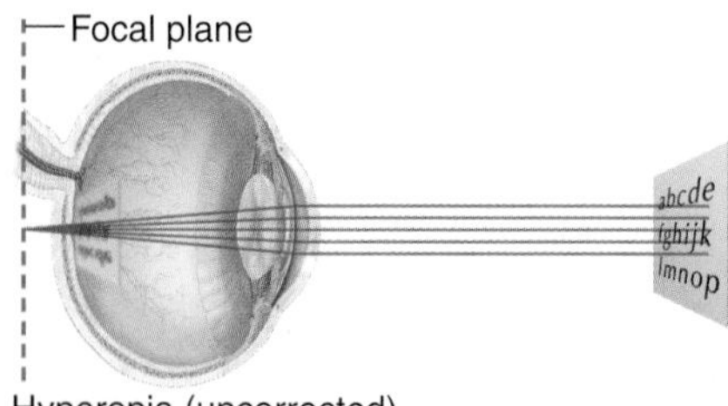

FIGURE 4.14 Hyperopia (Farsightedness).

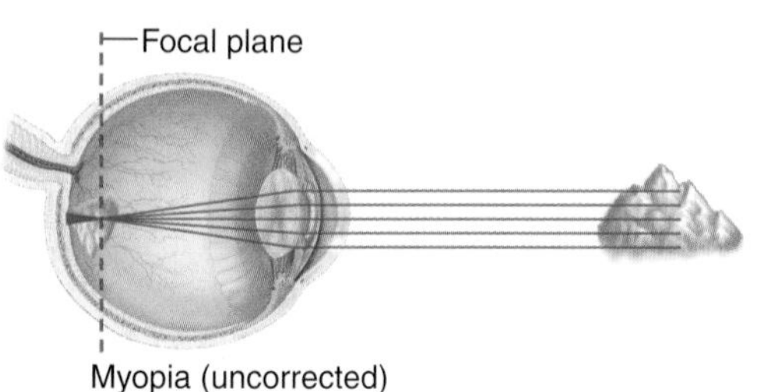

FIGURE 4.15 Myopia (Nearsightedness).

VISUAL PATHWAY

The optic nerves leave each orbit through the optic **foramen.** They converge into an "X" called the optic **chiasm.** Here, the fibers from the medial half of each retina cross the opposite side of the brain (*Figure 4.12*). After leaving the chiasm, the fibers form the optic **tract** and then the optic **radiation** to take the nerve impulses to the **visual cortex** in the occipital lobes at the back of your brain. Here the incoming visual stimuli are interpreted.

Refraction

Light is traveling at a speed of 186,000 miles per second when it hits your eye. Light rays that hit the center of your cornea pass straight through. Rays that hit away from the center are bent toward the center. The light rays then hit your lens and are bent again, and in normal vision, the image is focused sharply on your retina (*Figure 4.13*).

Farsighted people are said to have **hyperopia** (*Figure 4.14*). Because the eyeball is shortened, objects close to the eye are focused behind the retina and vision is blurred. Convex lenses are needed to correct the problem.

Nearsighted people are said to have **myopia** (*Figure 4.15*). Because the eyeball is elongated, far-away objects are focused in front of the retina. Vision is blurred. Concave lenses are needed to correct the problem.

In **presbyopia,** when you reach your forties, the lens loses its flexibility and there is difficulty focusing for near vision. Convex bifocal or transitional lenses are needed for this problem.

In **astigmatism,** unequal curvatures of the cornea cause unequal focusing and blurred images. Cylindrical lenses, which refract light more in one plane than another, are needed to correct this problem.

A surgical procedure, **radial keratotomy,** is used to treat myopia. Radial cuts, like the spokes of a wheel, flatten the cornea and enable it to refract the light rays to focus on the retina.

Laser surgery can also change the shape of the cornea. It can flatten it to correct myopia or can alter the outer edges of the cornea to correct hyperopia.

Laser-assisted in situ keratomileusis (LASIK) is being used to treat myopia, hyperopia, and astigmatism. A computer-controlled laser, using a cold beam of ultraviolet light, alters the shape of the cornea by vaporizing the tissue.

WORD ANALYSIS AND DEFINITION

S/ = Suffix P/ = Prefix R/ = Root R/CF = Combining Form

WORD	PRONUNCIATION		ELEMENTS	DEFINITION
cataract	**KAT**-ah-ract		Latin *to break down*	Complete or partial opacity of the lens.
fluorescein	flor-**ESS**-ee-in	P/ R/	**fluo-** *fluorine* **-rescein** *resin*	Dye that produces a vivid green color under a blue light to diagnose corneal abrasions and foreign bodies.
glaucoma	glau-**KOH**-mah	S/ R/	**-oma** *mass, tumor* **glauc-** *lens opacity*	Increased intraocular pressure.
intraocular	in-trah-**OCK**-you-lar	S/ P/ R/	**-ar** *pertaining to* **intra-** *inside* **-ocul-** *eye*	Pertaining to the inside of the eye.
ophthalmia neonatorum	off-**THAL**-me-ah ne-oh-nay-**TOR**-um	S/ R/ S/ P/ R/	**-ia** *condition* **ophthalm-** *eye* **-orum** *function of* **neo-** *new* **-nat-** *born*	Conjunctivitis of the newborn.
photosensitivity **photosensitive (adj)**	foh-toe-**SEN**-sih-tiv-ih-tee **FOH**-toe-**SEN**-sih-tiv	S/ R/CF R/	**-ity** *condition* **phot/o-** *light* **-sensitiv-** *feeling*	When light produces pain in the eye.
pollution	poh-**LOO**-shun		Latin *to defile*	Condition that is unclean, impure, and a danger to health.
pollutant	poh-**LOO**-tant	S/ R/	**-ant** *pertaining to* **pollut-** *unclean*	Substance that makes an environment unclean or impure.

EXERCISES

Review *the medical terms contained on the two pages open in front of you. Use these terms to complete the following sentences. You may use a term only one time. Fill in the blanks.*

1. ____________________ can be the cause of bloodshot eyes.
2. Dust mites and pollen are ____________________.
3. Scratching your eye with a tree branch produces a(n) ____________________.
4. ____________________ produces increased intraocular pressure.
5. When light causes pain in the eye, this is termed ____________________.
6. ____________________ is a dye used to diagnose corneal abrasions.
7. Another name for *conjunctivitis of the newborn* is ____________________.
8. ____________________ is a complete or partial opacity of the lens.
9. ____________________ refers to anything pertaining to the inside of the eye.
10. ____________________ cataract is present at birth.
11. ____________________ can be part of seasonal hay fever.
12. A corneal abrasion can grow into a(n) ____________________.

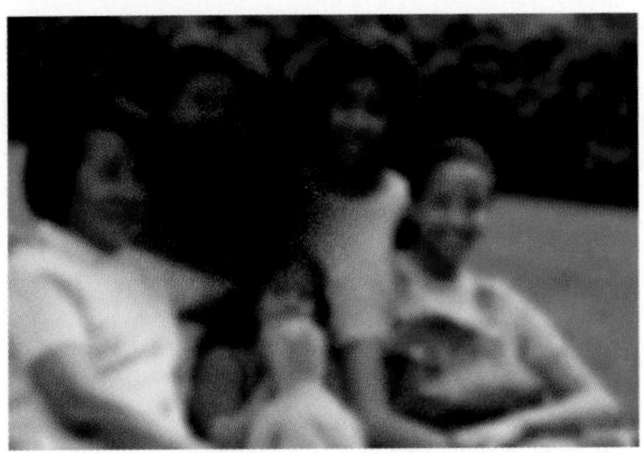

▲ **FIGURE 4.20 Vision with Macular Degeneration.**

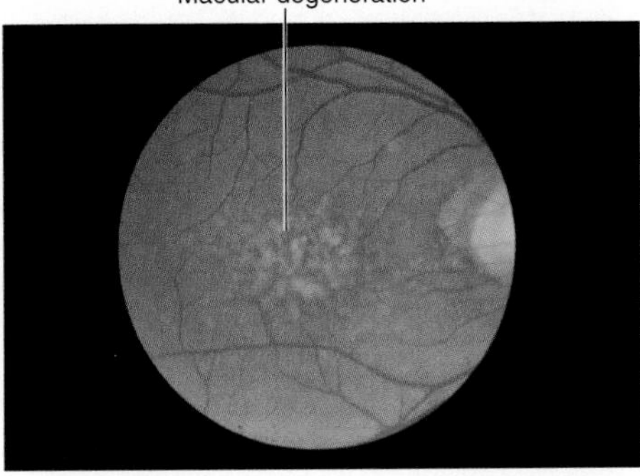

▲ **FIGURE 4.21 Ophthalmoscopic View of Macular Degeneration.**

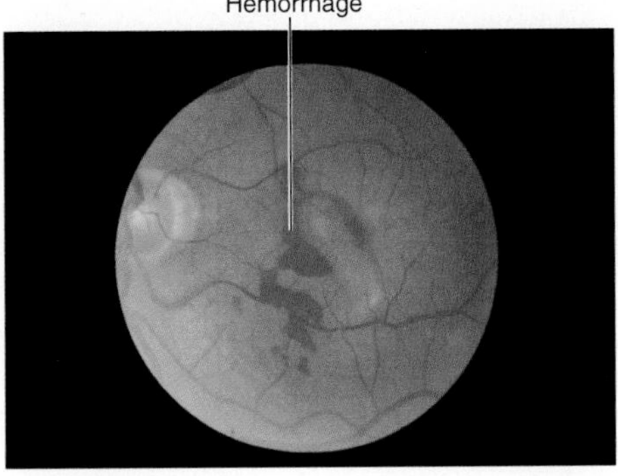

▲ **FIGURE 4.22 Ophthalmoscopic View of Diabetic Retinopathy.**

▲ **FIGURE 4.23 Vision with Diabetic Retinopathy.**

DISORDERS OF THE RETINA

Macular Degeneration

Degeneration of the central macula results in loss of visual acuity, with a dark blurry area of vision loss in the center of the visual field (*Figure 4.20*).

There is photoreceptor cell loss and bleeding with capillary proliferation and scar formation (*Figure 4.21*). It can progress to blindness. Most cases occur in people over 55. Currently, there is no known cure, but laser **photocoagulation** destroys the abnormal capillaries and this slows the pace of the visual loss.

Retinal Detachment

Separation of the retina from its underlying choroid layer may be partial or complete and produces a retinal tear or hole. The detachment can happen suddenly, without pain. The patient sees a dark shadow invading his or her peripheral vision. The detachment can be seen on **ophthalmoscopic** examination.

This condition is a surgical emergency. Treatment of small lesions is by **laser surgery** to "weld" the retina back into place.

Diabetic Retinopathy

Some 50% of diabetics have **retinopathy.** Hemorrhages (bleeding) can occur, leading to destruction of the photoreceptor cells (rods and cones) and visual difficulties.

Ophthalmoscopic examination shows the disease (*Figure 4.22*), and fluorescein **angiography** with pictures taken as the dye passes through the retina reveals more details.

Laser photocoagulation is usually effective in controlling the lesions, but once vision is lost from an area of the retina, it usually does not return (*Figure 4.23*).

Cancer of the Eye

Tumors of the skin of the eyelids include the **squamous cell** and basal cell carcinomas and melanoma described in *Chapter 3.*

Retinoblastoma is the most common cancer in children and is diagnosed most commonly around 18 months of age. Of those affected, 20% have the cancer in both eyes. The condition can be hereditary. With early detection and aggressive treatment based on chemotherapy and laser surgery, 90% are cured.

In adults, the most common cancers in the eye are metastases to the eye from cancer of the lung in men and cancer of the breast in women.

WORD ANALYSIS AND DEFINITION

S/ = Suffix P/ = Prefix R/ = Root R/CF = Combining Form

WORD	PRONUNCIATION	ELEMENTS		DEFINITION
angiography	an-jee-**OG**-rah-fee	S/ R/CF	**-graphy** *process of recording* **angi/o-** *blood vessel*	Radiography of vessels after injection of contrast material.
angiogram	**AN**-jee-oh-gram	S/	**-gram** *a record*	Radiogram obtained after injection of radi-opaque material into blood vessels.
laser surgery	**LAY**-zer **SUR**-jer-ee	 S/ R/	**L**ight **A**mplification by **S**imulated **E**mission of **R**adiation **-ery** *process of* **surg-** *operate*	Use of a concentrated, intense narrow beam of electromagnetic radiation for surgery.
ophthalmoscope	off-**THAL**-moh-skope	S/ R/CF	**-scope** *instrument for viewing* **ophthalm/o-** *eye*	Instrument for viewing the retina.
ophthalmoscopy **ophthalmoscopic (adj)**	**OFF**-thal-**MOS**-koh-pee **OFF**-thal-**MOS**-koh-pik	S/ S/	**-scopy** *to view, examine* **-ic** *pertaining to*	The process of viewing the retina. Pertaining to the use of an ophthalmoscope.
photocoagulation	foh-toe-koh-ag-you-**LAY**-shun	S/ R/CF R/	**-ation** *process* **phot/o** *light* **-coagul-** *clotting*	Using light (laser beam) to form a clot.
retinoblastoma	**RET**-in-oh-blas-**TOE**-mah	S/ R/CF R/	**-oma** *tumor, mass* **retin/o-** *retina* **-blast-** *immature cell*	Malignant neoplasm of primitive retinal cells.
retinopathy	ret-ih-**NOP**-ah-thee	S/	**-pathy** *disease*	Degenerative disease of the retina.

EXERCISES

Elements *help build your knowledge of the* language of ophthalmology. *One element in each of the following medical terms is bolded. Identify the type of element (P, R/CF, S) in column 2, and then write the meaning of the element in column 3. Fill in the chart, and answer the questions below it.*

Medical Term	Type of Element P, R, CF, S	Meaning of Element
retino**blast**oma		
ophthalmo**scope**		
angio**graphy**		
ophthalmoscopy		
photo**coagul**ation		
retino**pathy**		
angiogram		
ophthalmoscop**ic**		

1. List all the terms from the above table that are medical procedures:

___

___

2. List all the terms from the above table that are diagnoses:

___

___

OPHTHALMIC PROCEDURES

Procedures used to examine the eye include the following.

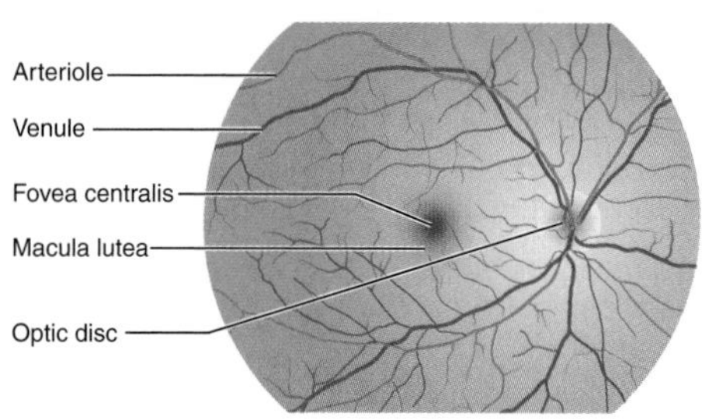

▲ **FIGURE 4.24 Anatomy of the Fundus.**

Examination of the Retina

Fundoscopy examines the retina with an ophthalmoscope. It identifies the optic disc (*Figure 4.24*), where the optic nerve leaves the back of the eye. The optic disc has no receptor cells and therefore produces a blind spot in the visual field of each eye. In the middle of the optic disc, a retinal artery enters to supply the intraocular structures, and a retinal vein leaves the eye.

Distance Vision

The **Snellen letter chart** is used to test distance vision (*Figure 4.25a*). The results are recorded as a fraction. For example, when the chart is viewed from 20 feet, line 8 is the smallest line a person with standard vision can read. This is recorded as 20/20. If the patient using her or his left eye misses two letters on line 8, document it as **OS** 20/20 –2. The right eye is documented as **OD,** and use of both eyes as **OU.**

Near Vision

Use handheld charts or **Jaeger reading cards** with printed paragraphs of different sizes of print to test near vision (*Figure 4.25b*).

Color Vision

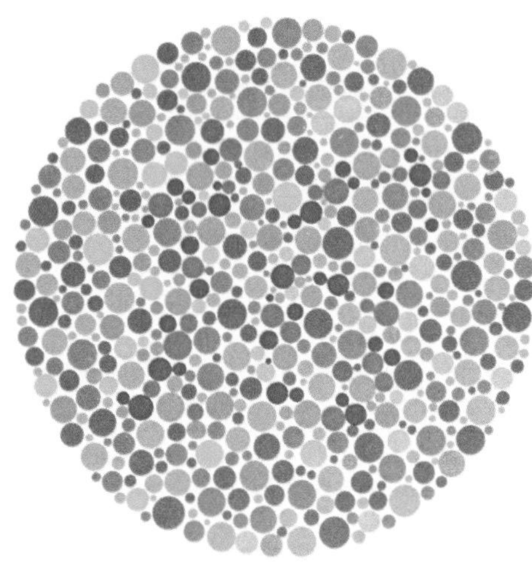

▲ **FIGURE 4.26**
Test for Color Blindness.
Reproduced with permission from *Ishihara's Tests for Color Deficiency,* published by Kanehara Trading Inc., Tokyo, Japan. Tests for color deficiency cannot be conducted with this figure. For accurate testing, the original plates should be used.

The **Ishihara color system** is used to test color vision. In the example in *Figure 4.26*, people with red-green color blindness would not be able to see the number 16 among the colored dots.

Visual Fields

Sit 2 feet in front of your patient, who covers one eye. Cover your own opposite eye and bring a pencil into the horizontal and vertical fields to compare the patient's **peripheral vision** with your own.

Intraocular Pressure

Intraocular pressure is measured with a **tonometer,** which determines the eyeball's resistance to indentation or tension. Increased intraocular pressure indicates glaucoma.

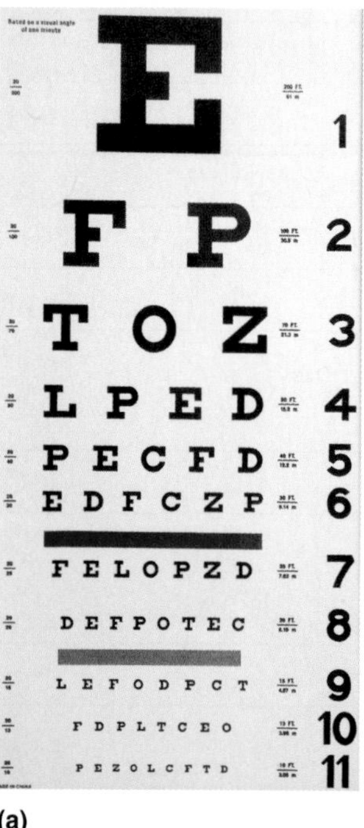

(a)

V = .50 D.

The fourteenth of August was the day fixed upon for the sailing of the brig Pilgrim, on her voyage from Boston round Cape Horn, to the western coast of North America. As she was to get under way early in the afternoon, I made my appearance on board at twelve o'clock in full sea-rig, and with my chest, containing an outfit for a two or three years voyage, which I had undertaken from a determination to cure, if possible, by an entire change of life, and by a long absence from books and study, a weakness of the eyes which had obliged me to give up my pursuits, and which no medical aid seemed likely to cure. The change from the tight dress coat, silk cap and kid gloves of an undergraduate at Cambridge, to the

V = .75 D.

loose duck trousers, checked shirt and tarpaulin hat of a sailor, though somewhat of a transformation, was soon made, and I supposed that I should pass very well for a Jack tar. But it is impossible to deceive the practiced eye in these matters; and while I supposed myself to be looking as salt as Neptune himself, I was, no doubt, known for a landsman by every one on board, as soon as I hove in sight. A sailor has a peculiar cut to his clothes, and a way of wear-

V = 1. D.

ing them which a green hand can never get. The trousers, tight around the hips, and thence hanging long and loose around the feet, a superabundance of checked shirt, a low-crowned, well-varnished black hat, worn on the back of the head, with half a fathom of black ribbon hanging over the left eye, and a peculiar tie to the black silk neckerchief, with sundry other *details*, are signs the want of which betray the beginner at once.

V = 1.25 D.

Beside the points in my dress which were out of the way, doubtless my complexion and hands would distinguish me from the regular *salt*, who, with a sun-browned cheek, wide step and rolling gait, swings his bronzed and toughened hands athwartships half open, as though just to ready to grasp a rope. "With all my imperfections

V = 1.50 D.

on my head," I joined the crew, and we hauled out into the stream and came to anchor for the night. The next day we were employed in preparation for sea, reeving and studding-sail gear, crossing royal yards, putting on chafing gear, and taking on board our powder. On the

(b)

▲ **FIGURE 4.25 Visual Acuity Tests.**

WORD ANALYSIS AND DEFINITION

S/ = Suffix P/ = Prefix R/ = Root R/CF = Combining Form

WORD	PRONUNCIATION	ELEMENTS		DEFINITION
fundus	**FUN**-dus		Latin *bottom*	Part farthest from the opening of a hollow organ.
fundoscopy	fun-**DOS**-koh-pee	S/ R/CF	**-scopy** *to examine* **fund/o-** *fundus*	Examination of the fundus (retina) of the eye.
fundoscopic (adj)	fun-doh-**SKOP**-ik	S/	**-ic** *pertaining to*	Pertaining to fundoscopy.
Ishihara color system	ish-ee-**HAR**-ah		Shinobu Ishihara, Japanese ophthalmologist, 1879–1963	Test for color vision defects.
Jaeger reading cards	**YA**-ger		Eduard Jaeger, Austrian ophthalmologist, 1818–1884	Type of different sizes for testing near vision.
peripheral	peh-**RIF**-er-al	S/ R/	**-al** *pertaining to* **peripher-** *external boundary*	Pertaining to the periphery or external boundary.
peripheral vision	peh-**RIF**-er-al **VIZH**-un	S/ R/	**-ion** *action, condition* **vis-** *sight*	Ability to see objects as they come into the outer edges of the visual field.
Snellen letter chart	**SNEL**-en		Hermann Snellen, Dutch ophthalmologist, 1834–1908	Test for acuity of distant vision.
tonometer	toe-**NOM**-eh-ter	S/ R/CF	**-meter** *measure* **ton/o** *pressure, tension*	Instrument for determining intraocular pressure.
tonometry	toe-**NOM**-eh-tree	S/	**-metry** *process of measuring*	The measurement of intraocular pressure.

Abbreviations

OD	right eye
OS	left eye
OU	both eyes

EXERCISES

Communication *is key. The OT in Dr. Chun's office needs to be familiar with all the WAD terms in order to communicate with Dr. Chun and her patients. Show your understanding of the* language of ophthalmology *by circling the correct answer.*

1. Which of these instrument(s) can be found in an ophthalmologist's office?
- **a.** tonometer
- **b.** cystoscope
- **c.** ophthalmoscope
- **d.** a and b
- **e.** a and c

2. Which is used to test color vision?
- **a.** Snellen
- **b.** Jaeger
- **c.** Ishihara system
- **d.** visual fields
- **e.** otoscope

3. *Peripheral vision* measures the outer edge of the:
- **a.** anterior segment
- **b.** vitreous humor
- **c.** aqueous humor
- **d.** posterior segment
- **e.** visual field

4. The abbreviation *OD* means:
- **a.** right eye
- **b.** both eyes
- **c.** left eye
- **d.** does not pertain to the eye
- **e.** normal eye

5. A test for near vision is:
- **a.** Snellen chart
- **b.** ophthalmoscope
- **c.** Jaeger cards
- **d.** Ishihara
- **e.** visual fields

6. *Tonometry* measures: *(Be precise!)*
- **a.** interocular pressure
- **b.** arterial pressure
- **c.** venous pressure
- **d.** intraocular pressure
- **e.** capillary pressure

Consolable
Consuelo
Pain
Sin dolor
WelchAllyn

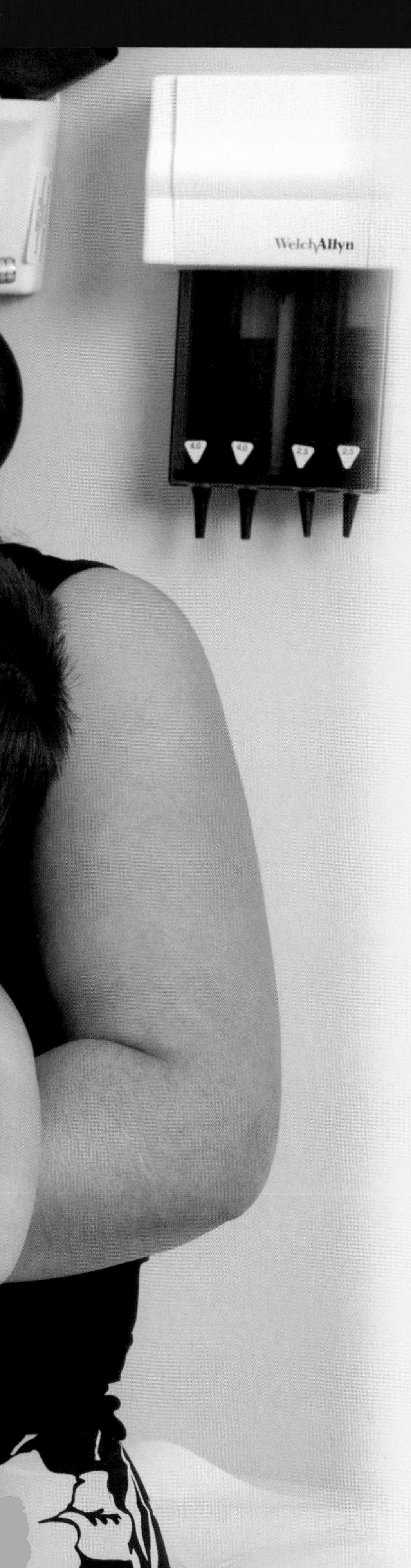

THE EAR AND HEARING
CASE REPORT 4.3

You are

. . . a medical assistant working for primary care physician Susan Lee, MD, of the Fulwood Medical Group.

You are communicating with

. . . Mrs. Carmen Cardenas, who has brought in her 3-year-old son, Eddie. She tells you that Eddie has had a cold for a couple of days. Early this morning he woke up screaming, felt hot, and was tugging his ears. She gave him **acetaminophen** with some orange juice, and he threw up. She also tells you this is the third similar episode in the past year, and, since the last time, she is concerned that he is not hearing normally. You see a worried mother and a restless toddler with a green, nasal discharge. His oral temperature taken with an electronic digital thermometer is 102.4°F, pulse 100. You tell her that Dr. Lee will be in to see Eddie as soon as possible.

Learning Outcomes

In order to understand what is going on with Eddie, to communicate with Dr. Lee about him, to respond to the mother's concerns, and to document the office visit, you need to be able to:

4.4 Comprehend, analyze, spell, and write the medical terms of **otology.**

4.5 Recognize and pronounce the medical terms of otology.

4.6 Discuss the cause, appearance, diagnosis, and treatment of common disorders of the ear.

LESSON 4.4 The Ear and Hearing

SPECIAL SENSES OF THE EYE AND EAR

OBJECTIVES

To understand the specific problem that Eddie has, you must be able to:

- **Recognize the structures and functions of the three regions of the ear.**
- **Explain how sound waves progress through the ear, are transferred to the brain, and are recognized as sounds.**
- **Identify common diseases of the ear that interfere with the process of hearing.**

Case Report 4.3 *(continued)*

Documentation. 05/10/08

Examination by Dr. Lee showed that Eddie has a **bilateral acute otitis media** (BOM) with an upper respiratory infection (URI). Dr. Lee is also concerned that Eddie has a **chronic** otitis media with **effusion** (OME) that is causing a hearing loss. She prescribed Amoxicillin 250 mg q.i.d. with acetaminophen 160 mg p.r.n. for 10 days, when she will see Eddie again. If, after the acute infection subsides, there remains an effusion with hearing loss, Dr. Lee may need to refer Eddie to an **otologist.** I explained this to Mrs. Cardenas.

Signed: Luis Guittierez, CMA. 1115 hrs.

The ear has three major sections (*Figure 4.27*):

- External ear
- Middle ear
- Inner ear

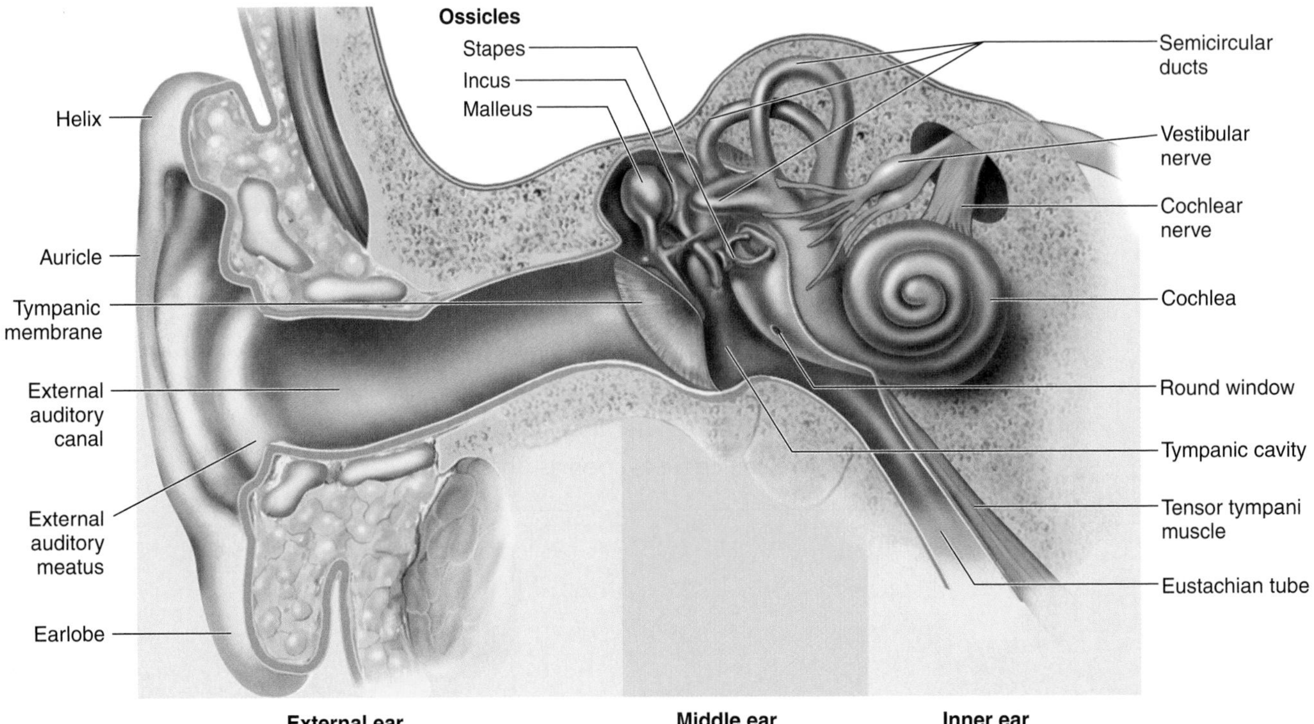

▲ **FIGURE 4.27 Anatomical Regions of the Ear.**

WORD ANALYSIS AND DEFINITION

S/ = Suffix P/ = Prefix R/ = Root R/CF = Combining Form

WORD	PRONUNCIATION	ELEMENTS		DEFINITION
acetaminophen	ah-seat-ah-**MIN**-oh-fen		Generic drug name	Medication that is an analgesic and antipyretic.
acute	ah-**KYUT**		Latin *sharp*	Disease of sudden onset.
bilateral	by-**LAT**-er-al	S/ P/ R/	**-al** *pertaining to* **bi-** *two* **-later-** *side*	On two sides; e.g., in both ears.
chronic	**KRON**-ik		Greek *time*	A persistent, long-term disease.
effusion	eh-**FYU**-shun		Latin *pouring out*	Collection of fluid that has escaped from blood vessels into a cavity or tissues.
otitis media	oh-**TIE**-tis **ME**-dee-ah	S/ R/ R/	**-itis** *inflammation* **ot-** *ear* **media** *middle*	Inflammation of the middle ear.
otologist	oh-**TOL**-oh-jist	S/ R/CF	**-logist** *specialist* **ot/o-** *ear*	Medical specialist in diseases of the ear.
otology	oh-**TOL**-oh-jee	S/	**-logy** *study of*	Diagnosis and treatment of disorders of the ear.
otorhinolaryngologist	oh-toe-**rhino**-lah-rin-**GOL**-oh-jist	R/CF R/CF	**-rhin/o-** *nose* **-laryng/o-** *larynx*	Ear, nose, and throat medical specialist.

Abbreviations

BOM	bilateral otitis media
mg	milligram
OME	otitis media with effusion
p.r.n.	when necessary
q.i.d.	four times each day
URI	upper respiratory infection

EXERCISES

Deconstruct *these medical terms into their basic elements. Analyzing each term will help you answer the following questions. Complete the chart, and fill in the blanks.*

Medical Term	Meaning of Prefix	Meaning of Root(s)/CF(s)	Meaning of Suffix
otology			
otitis media			
otologist			
otorhinolaryngologist			

1. Which term contains three combining forms? ____________________

2. What is the difference between an *otologist* and an *otorhinolaryngologist?*

____

____

3. Where in the ear does *otitis media* occur? ____________________

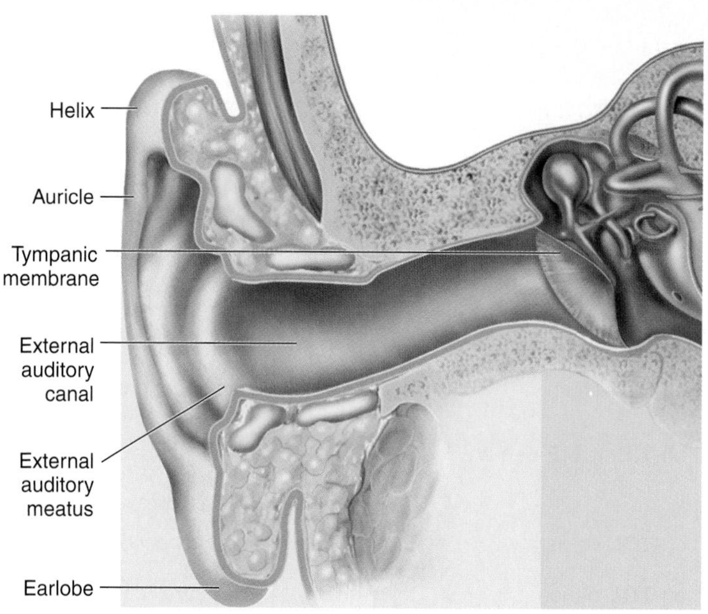

FIGURE 4.28 External Ear.

EXTERNAL EAR

The external ear comprises several structures.

The **auricle** or **pinna** is a wing-shaped structure that directs sound waves into the external **auditory meatus** and canal. The external auditory canal ends at the **tympanic** membrane (*Figure 4.28*).

Keynote

The **external auditory canal** is the only skin-lined cul-de-sac in the body.

The meatus and canal are lined with skin that contains many modified sweat glands called **ceruminous** glands, which secrete **cerumen.** Cerumen combines with dead skin cells to form earwax.

If a foreign body, such as a small bead, gets into the auditory canal, or if cerumen becomes **impacted** in the canal, hearing loss can occur.

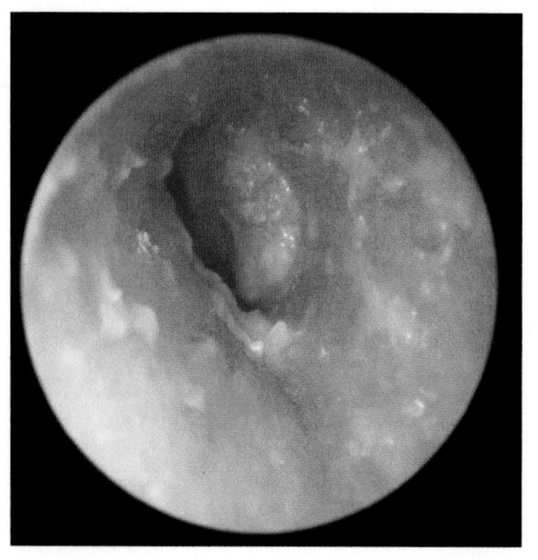

▲ **FIGURE 4.29 Otoscopic View of Otitis Externa.**

Disorders of the External Ear

Otitis externa (*Figure 4.29*) is an infection of the lining of the external auditory canal. **Otoscopic** examination shows a painful, red, swollen ear canal, sometimes with purulent drainage. The infection can be bacterial or fungal. Treatment entails thorough cleansing of the canal, acidification with a **topical** solution of 2% acetic acid in a hydrocortisone solution, and the use of antibiotic drops. Occasionally a wick is needed to enable the **topical** medications to penetrate down the canal.

Swimmer's ear is a form of otitis externa resulting from swimming, particularly if the water is polluted.

Excessive earwax can be removed in your physician's office by ear **irrigation** or with a **curette,** a small metal ring at the end of a handle.

WORD ANALYSIS AND DEFINITION

S/ = Suffix P/ = Prefix R/ = Root R/CF = Combining Form

WORD	PRONUNCIATION		ELEMENTS	DEFINITION
auditory (adj) **audiology** **audiologist**	**AW**-dih-tor-ee aw-dee-**OL**-oh-jee aw-dee-**OL**-oh-jist	 S/ R/CF S/	Latin *hearing* **-logy** *study of* **audi/o-** *hearing* **-logist** *one who studies, specialist*	Relating to hearing or the organs of hearing. Study of hearing disorders. Specialist in evaluation of hearing function.
auricle	**AW**-ri-kul		Latin *ear*	The shell-like external ear.
cerumen **ceruminous (adj)**	seh-**ROO**-men seh-**ROO**-mih-nus	 S/ R/	Latin *wax* **-ous** *pertaining to* **cerumin-** *cerumen*	Waxy secretion of glands of the external ear. Pertaining to cerumen.
curette **curettage** (The final "e" of *curette* is dropped because the suffix *-age* begins with a vowel.)	kyu-**RET** kyu-reh-**TAHZH**	S/ R/ S/	**-ette** *little* **cur-** *cleanse, cure* **-age** *pertaining to*	Scoop-shaped instrument for scraping or removal of new growths (or earwax). The use of a curette.
impacted	im-**PAK**-ted		Latin *driven in*	Immovably wedged, as with earwax blocking the external canal.
irrigation	ih-rih-**GAY**-shun	S/ R/	**-ation** *process* **-irrig-** *to water*	Use of water; e.g., to remove wax from the external ear canal.
meatus **meatal (adj)**	me-**AY**-tus me-**AY**-tal		Latin *go through*	Passage or channel; also the external opening of a passage.
otoscope **otoscopy** **otoscopic (adj)**	**OH**-toe-skope oh-**TOS**-koh-pee oh-toe-**SKOP**-ik	S/ R/CF S/ S/	**-scope** *instrument for viewing* **ot/o-** *ear* **-scopy** *to examine* **-ic** *pertaining to*	Instrument for examining the ear. Examination of the ear. Pertaining to examination with an otoscope.
pinna **pinnae (pl)**	**PIN**-ah **PIN**-ee		Latin *wing*	Another name for *auricle.*
topical	**TOP**-ih-kal	S/ R/	**-al** *pertaining to* **topic-** *local*	Medication applied to the skin to obtain a local numbing effect.
tympanic	tim-**PAN**-ik	S/ R/	**–ic** *pertaining to* **tympan-** *eardrum*	Pertaining to the tympanic membrane (eardrum) or tympanic cavity.

EXERCISES

Test *your knowledge of the* language of otology *by matching correct answers. Every part of the body has its own specialized vocabulary. Fill in the blanks.*

_____ 1. pertaining to the eardrum — A. meatus

_____ 2. external opening of a passage — B. curette

_____ 3. instrument for examining the ear — C. curettage

_____ 4. procedure for scraping or removing growths — D. auricle

_____ 5. scoop-shaped instrument — E. tympanic

_____ 6. shell-like external ear — F. otoscope

FIGURE 4.30
Middle Ear. (*m.* = muscle)

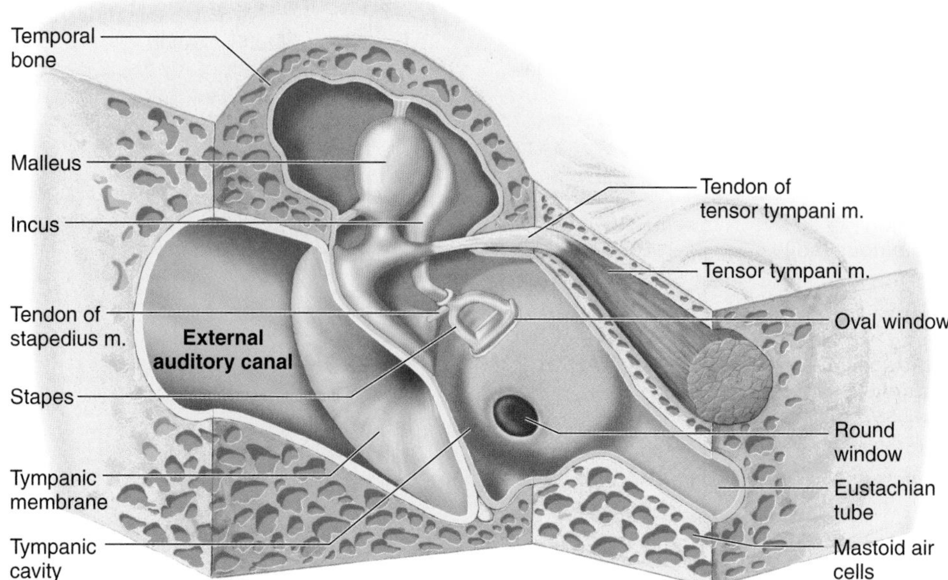

MIDDLE EAR

The middle ear has four components (*Figure 4.30*):

1. The **tympanic membrane** (eardrum) is located at the inner end of the external auditory canal. It vibrates freely as sound waves hit it. It has a good nerve supply and is very sensitive to pain. When examined through the otoscope, it is transparent and reflects light (*Figure 4.31*).
2. The **tympanic cavity** is immediately behind the tympanic membrane. It is filled with air that enters through the eustachian tube, and the cavity is continuous with the **mastoid** air cells in the bone behind it. The cavity contains the **ossicles.**
3. The three **ossicles,** the **malleus, incus,** and **stapes,** are attached to the wall of the tympanic cavity by tiny ligaments. The malleus is attached to the tympanic membrane and vibrates with the membrane when sound waves hit it. The malleus is attached to the incus, which also vibrates and passes the vibrations onto the stapes. The stapes is attached to the oval window, an opening that transmits the vibrations to the inner ear.
4. The **eustachian (auditory) tube** connects the middle ear with the **nasopharynx** (throat), into which it opens close to the pharyngeal **tonsils (adenoids)** (*Figure 4.32*). In children under 5 years, the tube is not fully developed. It is short and horizontal, and valvelike flaps in the throat that protect it are not developed.

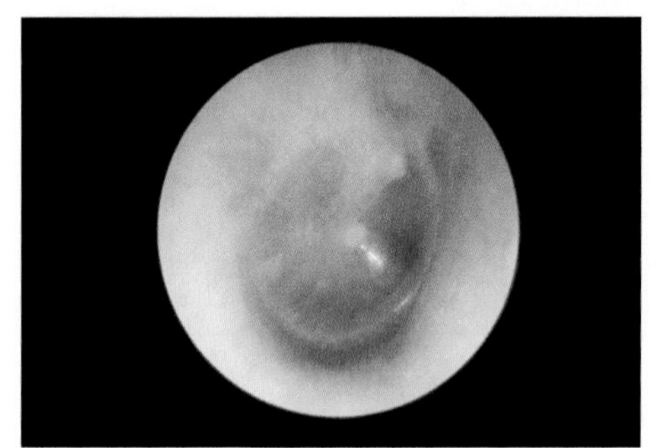

▲ **FIGURE 4.31 Otoscopic View of Normal Tympanic Membrane.**

The three ossicles amplify sound so that soft sounds can be heard. The stapes is the smallest bone in the body.

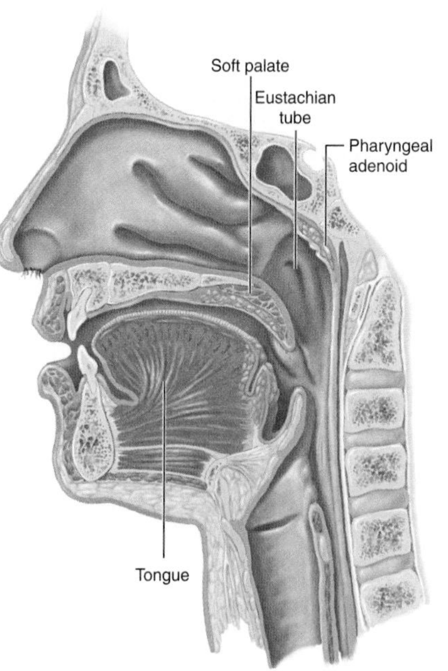

▲ **FIGURE 4.32 Nasopharynx (Throat).**

WORD ANALYSIS AND DEFINITION

S/ = Suffix P/ = Prefix R/ = Root R/CF = Combining Form

WORD	PRONUNCIATION	ELEMENTS		DEFINITION
adenoid	**ADD**-eh-noyd	S/ R/	**-oid** *resembling* **aden-** *gland*	Single mass of lymphoid tissue in the midline at the back of the throat.
eustachian tube (syn: *auditory tube*)	you-**STAY**-shun TYUB		Bartolomeo Eustachio, Italian anatomist, 1524–1574	Tube that connects the middle ear to the nasopharynx.
incus	**IN**-cuss		Latin *anvil*	Middle one of the three ossicles in the middle ear; shaped like an anvil.
malleus	**MAL**-ee-us		Latin *hammer*	Outer (lateral) one of the three ossicles in the middle ear; shaped like a hammer.
mastoid	**MASS**-toyd	S/ R/	**-oid** *resembling* **mast-** *breast*	Small bony protrusion immediately behind the ear.
nasopharynx **nasopharyngeal (adj)**	**NAY**-zoh-fair-inks **NAY**-zoh-fair-**RIN**-jee-al	R/CF R/ S/ R/	**nas/o-** *nose* **-pharynx** *throat* **-eal** *pertaining to* **-pharyng-** *pharynx*	Region of the pharynx at the back of the nose and above the soft palate. Pertaining to the nasopharynx.
ossicle	**OSS**-ih-kel	S/ R/CF	**-cle** *small* **oss/i-** *bone*	A small bone, particularly relating to the three bones in the middle ear.
stapes	**STAY**-peas		Latin *stirrup*	Inner (medial) one of the three ossicles of the middle ear; shaped like a stirrup.

EXERCISES

Review *the terms in the WAD box before you start this exercise. Pay special attention to the spelling. Circle the best choice.*

1. In the term *mastoid,* the suffix means:
   condition   resembling   inflammation
2. The element *mast* means:
   throat   breast   ear
3. The element *pharynx* means:
   throat   nose   tongue
4. The term *nasopharynx* is composed of:
   prefix + suffix   root + root   combining form + root
5. Shaped like a hammer:
   malleus   malleolus   maleus

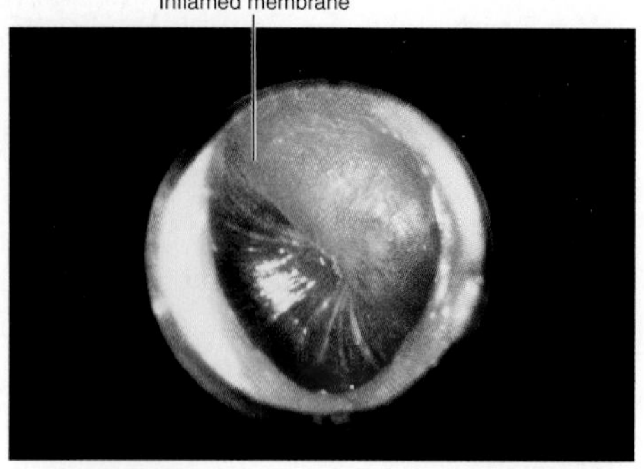

▲ **FIGURE 4.33 Otoscopic View of Acute Otitis Media.**

Case Report 4.3 (continued)

Eddie Cardenas's ear problems began with his eustachian tube. His cold (upper respiratory infection, **URI,** and **coryza**) inflamed the mucous membranes of his throat and eustachian tube. Because he is a young child, his eustachian tube is short and horizontal and thus the inflammation spread easily from his throat into the middle ear, causing his acute otitis media **(AOM).** The inflammatory process produced fluid (effusion) in the middle ear. His tympanic membrane became inflamed and painful, which you could see through an otoscope (*Figure 4.33*).

DISORDERS OF THE MIDDLE EAR

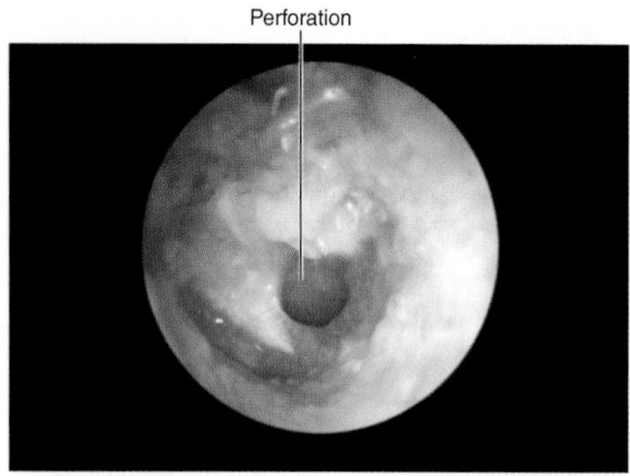

▲ **FIGURE 4.34 Otoscopic View of Chronic Otitis Media with Perforation.**

- **Acute otitis media** (AOM) is the presence of pus in the middle ear with pain in the ear, fever, and redness of the tympanic membrane. AOM occurs most often in the first 2 to 4 years of age. If the infection is viral, it will go away on its own. If bacterial, oral antibiotics may be necessary.
- **Chronic otitis media** occurs when the acute infection subsides and the eustachian tube is blocked. The effusion (fluid) in the middle ear cannot drain out, and it gradually becomes stickier. This is called **chronic otitis media with effusion (OME)** and produces hearing loss because the sticky fluid prevents the ossicles from vibrating. You can see the fluid through the otoscope (*Figure 4.34*). Dr Lee was concerned that this had happened to Eddie in his previous ear infection.
- If the sticky fluid persists, a **myringotomy** can be performed and a small, hollow plastic tube may have to be inserted through the tympanic membrane to allow the effusion to drain. These ear tubes have several names: **tympanostomy** tubes, **pressure-equalization tubes,** and, most commonly, **PE tubes** (*Figure 4.35*). The tubes are inserted under general anesthesia as outpatient surgery. They remain in the ear for 6 to 18 months before they drop out on their own.
- A **perforated** tympanic membrane (*Figure 4.35*) can occur in acute and chronic otitis media (AOM) when pus in the middle ear cannot escape down the eustachian tube. It builds up pressure and punctures the eardrum. Most perforations will heal spontaneously in a month, leaving a small scar. Other causes of perforation include a puncture by a Q-tip, an open-handed slap to the ear, or induced pressure as in scuba diving.
- **Cholesteotoma** is a complication of chronic otitis media with effusion (OME). Chronically inflamed cells in the middle ear multiply and collect into a tumor. They damage the ossicles and can spread to the inner ear. Surgical removal is required.
- **Otosclerosis** is a middle-ear disease that usually affects people between 18 and 35 years. It can affect one ear or both and produces a gradual hearing loss for low and soft sounds. Its etiology is unknown. Spongy bone forms around the junction of the oval window and stapes. This makes the stapes unable to conduct the sound vibrations to the inner ear. The only treatment is to replace the stapes with a metal or plastic **prosthesis.**

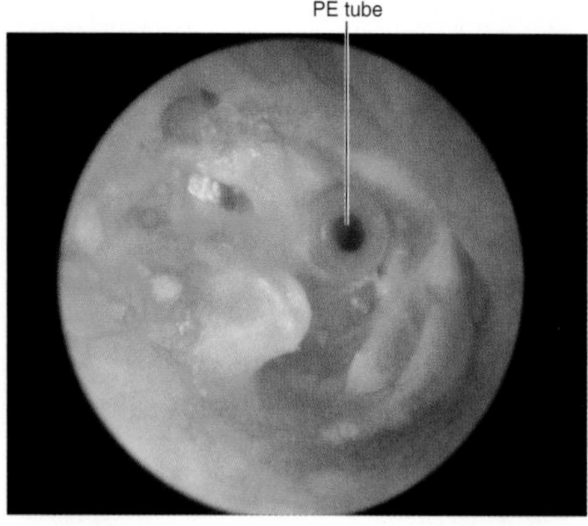

▲ **FIGURE 4.35 Otoscopic View of a PE Tube in the Tympanic Membrane.**

WORD ANALYSIS AND DEFINITION

S/ = Suffix P/ = Prefix R/ = Root R/CF = Combining Form

WORD	PRONUNCIATION	ELEMENTS		DEFINITION
cholesteatoma	koh-less-tee-ah-**TOE**-mah	S/ R/CF R/	**-oma** *tumor, mass* **chol/e-** *bile* **-steat-** *fat*	Yellow, waxy tumor arising in the middle ear.
coryza (syn: *acute rhinitis*)	koh-**RYE**-zah		Greek *catarrh*	Viral inflammation of the mucous membrane of the nose.
myringotomy	mir-in-**GOT**-oh-me	S/ R/CF	**-tomy** *surgical incision* **myring/o-** *tympanic membrane*	Incision in the tympanic membrane.
otosclerosis	oh-toe-sklair-**OH**-sis	S/ R/CF R/CF	**-sis** *abnormal condition* **-scler/o-** *hard* **ot/o-** *ear*	Hardening at the junction of the stapes and oval window that causes loss of hearing.
perforated **perforation**	**PER**-foh-ray-ted per-foh-**RAY**-shun	 S/ R/	Latin *to bore through* **-ion** *action* **perforat-** *bore through*	Punctured with one or more holes. A hole through the wall of a structure.
prosthesis	**PROS**-thee-sis		Greek *addition*	Manufactured substitute for a missing or diseased part of the body.
tympanostomy	tim-pan-**OS**-toe-me	S/ R/CF	**-stomy** *new opening* **tympan/o-** *eardrum*	Surgically created new opening in the tympanic membrane to allow fluid to drain from the middle ear.

Abbreviations

AOM	acute otitis media
PE tube	pressure-equalization tube
URI	upper respiratory infection

EXERCISES

Build medical terms. *This is a two-step exercise. Fill in the blanks with the correct element to complete the term, and under the line write the type of element (prefix, root, combining form, suffix) you have used.*

1. hardening at the junction of the stapes and oval window: oto/ ______________________ /sis

2. incision into the tympanic membrane: myringo/ ______________________________

3. yellow, waxy tumor in the middle ear: ______________________ / ______________________ /oma

Match the following terms to their meanings:

_____	4. manufactured body part	A. coryza
_____	5. fluid in a cavity	B. perforated
_____	6. acute rhinitis	C. prosthesis
_____	7. punctured	D. effusion

INNER EAR FOR HEARING

The inner ear is a **labyrinth** of complex, intricate systems of passages. The passages in the **cochlea,** a part of the labyrinth (*Figure 4.36*), contain receptors to translate vibrations into electrical nerve impulses so that the brain can interpret them as different sounds.

Sound waves cause the tympanic membrane and the ossicles to vibrate (*Figure 4.37*). The stapes moves the membrane of the oval window to generate pressure waves in the fluid inside the cochlea of the inner ear. The pressure waves cause **basilar** membranes inside the cochlea to vibrate. Hair cells attached to the membrane convert this motion into nerve impulses, which travel via the cochlear nerve to the brain. The excess pressure waves in the cochlea escape to the middle ear via the round window (*Figure 4.37*).

Today, the most common cause of hearing loss is damage to the fine hairs in the cochlea by repeated loud noise related either to work, such as the use of jackhammers or leaf blowers, or to leisure, such as amplified music at concerts, personal listening devices, and motorcycles. This is a **sensorineural hearing loss.**

Hearing aids are becoming more sophisticated and smaller, but they do not help people with cochlear damage. **Cochlear implants** are used to bypass the damaged hair cells and directly stimulate cochlear nerve endings.

A **conductive hearing loss** occurs when sound is not conducted efficiently through the external auditory canal to the tympanic membrane and the ossicles. Causes include:

- Middle-ear pathology such as acute otitis media, otitis media with effusion, perforated eardrum
- Infected external auditory canal
- Foreign body in the external canal

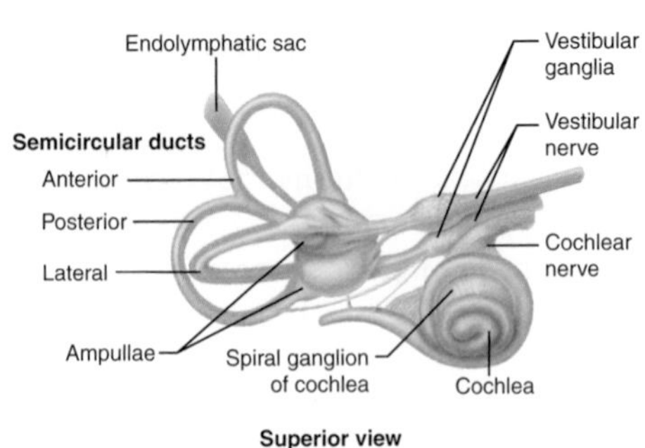

▲ **FIGURE 4.36**
Inner Ear.

Keynote

Repeated loud noise causes hearing loss in young people.

Hearing Test Procedures

- **Whispered speech testing:** Ask the patient to cover one ear. Stand 2 feet away from the uncovered ear, whisper words, and ask the patient to repeat them. If the patient cannot repeat them, say the words more loudly. This is a simple, but unmeasured, screening method.
- **Weber test:** Place a vibrating tuning fork in the middle of the patient's forehead, and ask whether the tone is louder in one ear or equal on both sides. This determines on which side a hearing loss is located.
- **Rinne test:** Place the vibrating tuning fork on the mastoid process. Then hold it opposite the ear canal. Normally, sound is heard longer by air conduction at the ear canal than by bone conduction at the mastoid process. The reverse indicates a conductive hearing loss.
- **Audiometer:** After proper training, use an audiometer to test for hearing loss. The audiometer is an electronic device that generates sounds in different frequencies and intensities and can print out the patient's responses.

In recording the results of hearing testing, **AD** is shorthand for the right ear, **AS** for the left ear, and **AU** for both ears.

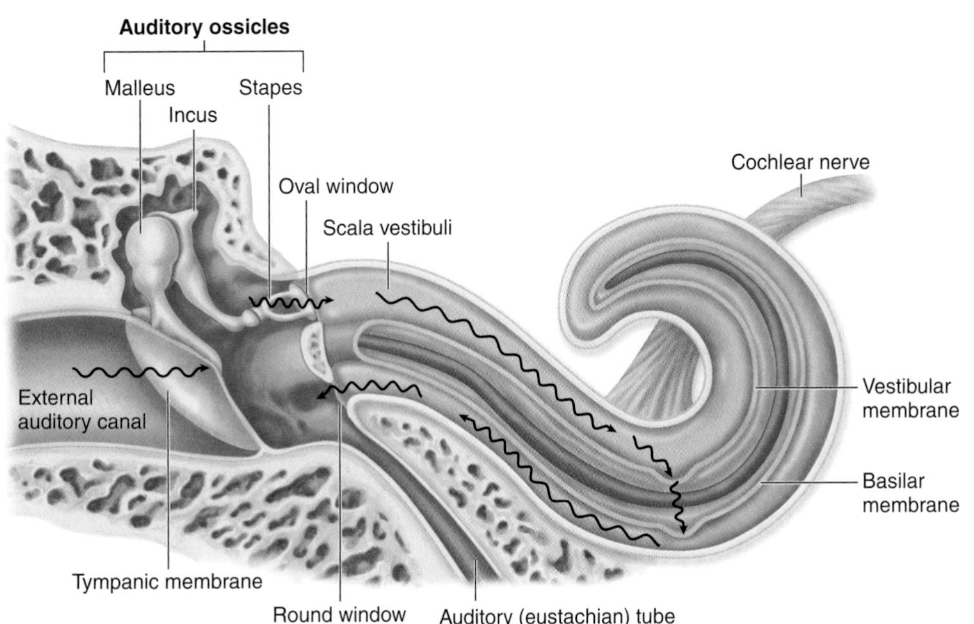

▲ **FIGURE 4.37 Model of the Hearing Process.**

WORD ANALYSIS AND DEFINITION

S/ = Suffix P/ = Prefix R/ = Root R/CF = Combining Form

WORD	PRONUNCIATION	ELEMENTS		DEFINITION
audiometer	aw-dee-**OM**-ee-ter	S/ R/CF	**-meter** *measure* **audi/o-** *hearing*	Instrument to measure hearing.
audiometric (adj)	**AW**-dee-oh-**MET**-rik	S/	**-metric** *pertaining to measurement*	Pertaining to the measurement of hearing.
basilar	**BAS**-ih-lar	S/ R/	**-ar** *pertaining to* **basil-** *base, support*	Pertaining to the base of a structure.
cochlea **cochlear (adj)**	**KOK**-lee-ah **KOK**-lee-ar		Latin *snail shell*	An intricate combination of passages; used to describe the inner ear.
conductive hearing loss	kon-**DUK**-tiv		Latin *to lead*	Hearing loss caused by lesions in the outer ear or middle ear.
implant	im-**PLANT**		Latin *to plant*	To insert material into tissues; or the material inserted into tissues.
labyrinth **labyrinthitis**	**LAB**-ih-rinth **LAB**-ih-rin-**THI**-tis	 S/ R/	Greek *labyrinth* **-itis** *inflammation* **labyrinth** *-inner ear*	The inner ear. Inflammation of the inner ear.
Rinne test	**RIN**-eh TEST		Friedrich Rinne, German otologist, 1819–1868	Test for conductive hearing loss.
sensorineural hearing loss	**SEN**-sor-ih-**NYUR**-al	S/ R/CF R/	**-al** *pertaining to* **sensor/i-** *sensory* **-neur-** *nerve*	Hearing loss caused by lesions of the inner ear or the auditory nerve.
Weber test	**VA**-ber TEST		Ernst Weber, German physiologist, 1794–1878	Test for sensorineural hearing loss.

Abbreviations

AD	right ear
AS	left ear
AU	both ears

EXERCISES

Match *the meaning of the element in column 1 with the correct element in column 2. One meaning appears twice, and has two different answers. Fill in the blanks.*

_____	1. hearing	A. meter
_____	2. nerve	B. al
_____	3. pertaining to	C. metric
_____	4. measure	D. audi/o
_____	5. inner ear	E. itis
_____	6. sensory	F. basil
_____	7. base, support	G. neur
_____	8. pertaining to measurement	H. sensor/i
_____	9. inflammation	I. labyrinth
_____	10. pertaining to	J. ar

You are

... Sonia Ramos, a medical assistant working with Sylvia Thompson, MD, an otolaryngologist at Fulwood Medical Center.

You are communicating with

... Mr. Ernesto Santiago, a 44-year-old man who was referred to Dr. Thompson.

CASE REPORT 4.4

Mr. Santiago complains of recurrent attacks of nausea, vomiting, a sense of spinning or whirling, and ringing in his ears. The attacks last about 24 hours and are getting more frequent. He has been having trouble hearing quiet speech on his left side.

Your role is to document his examination, diagnosis, and care and to act as translator between Mr. Santiago and Dr. Thompson.

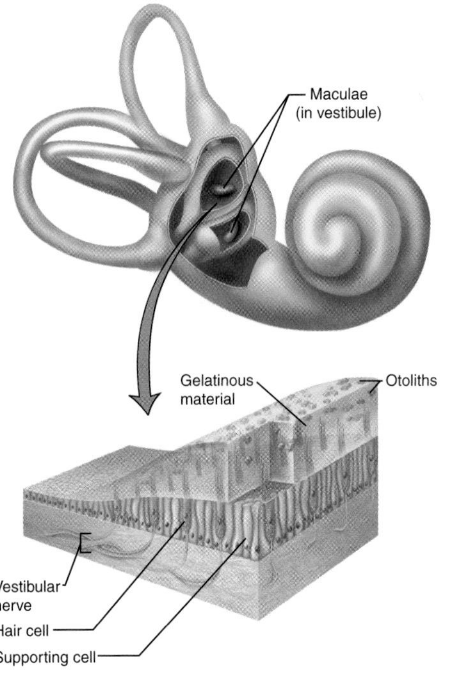

▲ **FIGURE 4.38 Vestibule of the Inner Ear.**

INNER EAR FOR EQUILIBRIUM AND BALANCE

The **vestibule** and the three semicircular canals (*Figure 4.38*) are the organs of balance. Inside the fluid-filled vestibule are two raised, flat areas covered with hair cells and a jelly-like material. This gelatinous material contains crystals of calcium and protein called **otoliths.** The position of the head alters the pressure applied to the hair cells by the gelatinous mass. The hair cells respond to horizontal and vertical changes and send impulses to the brain indicating the position to which the head has tilted.

Each of the three fluid-filled semicircular canals has a dilated end called an **ampulla** that contains a mound of hair cells embedded in a gelatinous material that together are called the **crista ampullaris** (*Figure 4.39*). It detects rotational movements of the head that distort the hair cells and lead to stimulation of connected nerve cells. The nerve impulses travel via the vestibular nerve and go to the brain. From the brain, nerve impulses travel to the muscles to maintain **equilibrium** and balance.

The sensation of spinning or whirling that Mr. Santiago experiences is called **vertigo,** often described by patients as dizziness. The ringing in his ears is called **tinnitus.** Both sensations arise in the inner ear.

Case Report 4.4 (continued)

The recurrent attacks that Mr. Santiago suffered are called **Ménière disease.** The disease involves the destruction of inner-ear hair cells, but the etiology is unknown and there is no cure. Dr. Thompson prescribed medication to control Mr. Santiago's nausea and vomiting.

Benign paroxysmal positional vertigo (BPPV) is another type of intermittent vertigo caused by fragments of the otoliths in the vestibule migrating into the semicircular canals. There they brush against the hair cells, sending conflicting signals to the brain and thereby producing vertigo.

Acute labyrinthitis is an acute viral infection of the labyrinth producing extreme vertigo, nausea, and vomiting. It usually lasts 1 to 2 weeks.

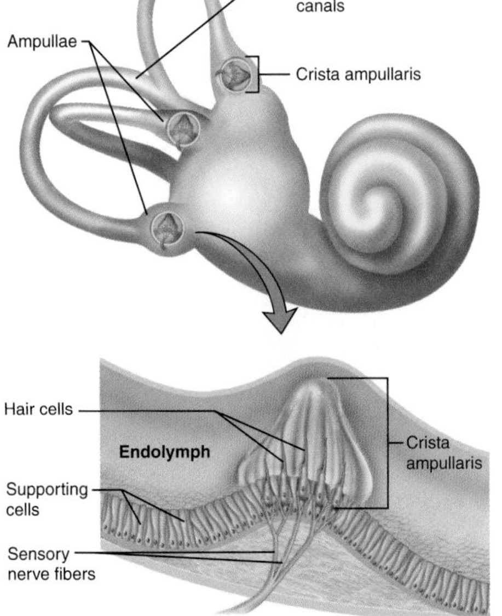

◀ **FIGURE 4.39 Semicircular Canals.**

WORD ANALYSIS AND DEFINITION

S/ = Suffix **P/** = Prefix **R/** = Root **R/CF** = Combining Form

WORD	PRONUNCIATION		ELEMENTS	DEFINITION
ampulla	am-**PULL**-ah		Latin *two-handled bottle*	Dilated portion of a canal or duct.
crista ampullaris	**KRIS**-tah am-**PULL**-air-is	R/ S/ R/	**crista** *crest* **-aris** *pertaining to* **ampull-** *bottle-shaped*	Mound of hair cells and gelatinous material in the ampulla of a semicircular canal.
equilibrium	ee-kwi-**LIB**-ree-um	P/ R/	**equi-** *equal* **-librium** *balance*	Being evenly balanced.
Ménière disease	men-**YEAR** diz-**EEZ**		Prosper Ménière, French physician, 1799–1862	Disorder of the inner ear with acute attacks of tinnitus, vertigo, and hearing loss.
otolith	**OH**-toe-lith	R/ R/CF	**-lith** *stone* **ot/o-** *ear*	A calcium particle in the vestibule of the inner ear.
paroxysmal	par-ock-**SIZ**-mal	S/ R/	**-al** *pertaining to* **paroxysm-** *sudden, sharp attack*	Occurring in sharp, spasmodic episodes.
tinnitus	**TIN**-ih-tus		Latin *jingle*	Persistent ringing, whistling, clicking, or booming noise in the ears.
vertigo	**VER**-tih-go		Latin *dizziness*	Sensation of spinning or whirling.
vestibule **vestibular (adj)**	**VES**-tih-byul ves-**TIB**-you-lar	 S/ R/	Latin *entrance chamber* **-ar** *pertaining to* **vestibul-** *vestibule*	Space at the entrance to a canal. Pertaining to the vestibule.

Abbreviation

BPPV benign paroxysmal positional vertigo

EXERCISES

Definitions *are provided in this exercise. The medical vocabulary answers are contained on the two-page spread open in front of you. Insert the correct answers in the blanks.*

1. crystals of calcium and protein ______
2. sense of balance ______
3. ringing in the ears ______
4. dizziness ______
5. occurring in spasmodic episodes ______
6. dilated portion of a canal or duct ______
7. space at the entrance to a canal ______
8. organs of balance ______
9. disorder of the inner ear with vertigo ______

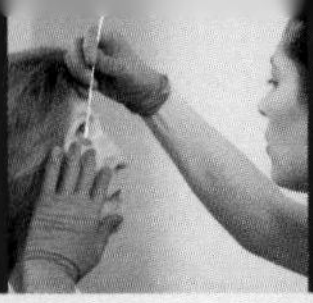

CHAPTER 4 REVIEW

SPECIAL SENSES OF THE EYE AND EAR

CHALLENGE YOUR KNOWLEDGE

A. **Suffixes:** The following terms all have a suffix with a common meaning. Circle the suffix, define each term on the line below, and then answer the question.

1. periorbital: ______________________

2. nasopharyngeal: ______________________

3. accommodate: ______________________

4. intraocular: ______________________

5. antibiotic: ______________________

6. optic: ______________________

7. pupillary: ______________________

8. pollutant: ______________________

9. ceruminous: ______________________

10. ampullaris: ______________________

11. List each of the individual suffixes here:

____

____

____

These suffixes all mean ______________________, and the terms come from the languages of ______________________ and ______________________.

B. **Word elements** are your most valuable tool for increasing your medical vocabulary. Use your knowledge of word elements to answer the following questions. Circle the correct answer.

1. Which of the following terms refers to *tears?*

   a. otolith
   b. lacrimal
   c. purulent
   d. uvea
   e. vestibule

2. This term is used to indicate partial *paralysis:*

   a. parietal
   b. periorbital
   c. paresis
   d. ptosis
   e. presbyopia

3. Based on its suffix, you can tell that a *keratotomy* is:

   a. a body part
   b. a procedure
   c. a diagnosis
   d. a medication
   e. an infection

4. The location of *periorbital* is:

   a. outside the eye
   b. around the eye
   c. beside the eye
   d. within the eye
   e. behind the eye

5. The term *amblyopia* signifies:

   a. sound
   b. light
   c. sight
   d. movement
   e. pain

6. *In situ* is a Latin phrase that means: *(Be precise!)*

   a. in this place
   b. in another place
   c. in the correct place
   d. in the place
   e. in place of

7. The lens of the eye is *avascular* because it has no:

   a. connective tissue
   b. aqueous humor
   c. blood supply
   d. fibrous outer covering
   e. mucous membrane

8. *Angiography* is an x-ray visualization of:

   a. organs
   b. bones
   c. blood vessels
   d. muscles
   e. glands

CHAPTER 4 REVIEW

SPECIAL SENSES OF THE EYE AND EAR

C. Patient education: Your patient is confused by some medical terms the doctor has used. Explain to the patient in simple language the difference between:

1. *esotropia* and *exotropia:*

____

____

____

____

2. *amblyopia* and *presbyopia:*

____

____

____

____

3. *refraction* and *accommodation:*

____

____

____

____

D. Spelling correctly is the mark of an educated professional. Proofread the following documentation to find the errors. Circle the misspelled terms, and rewrite the correctly spelled terms on the lines below.

1. The optalmoscopic examination revealed scleriitis and retinopathy.

____

2. Perulent discharge is coming from both eyes, and the patient also complains of photofobia.

____

3. The patient's blepharotosis can be remedied surgically. Schedule a blepharopasty as soon as possible.

____

4. Irigation and curetage have been ordered for the patient with impackted ceruman.

____

5. The patient suffers from chronick otitis medial with efusion.

____

____

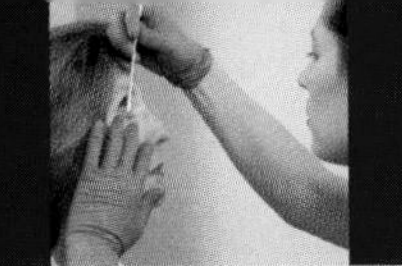

CHAPTER 4 REVIEW

SPECIAL SENSES OF THE EYE AND EAR

J. Recall and review: How well do you remember these word elements from the previous chapter? Try to answer without looking back to check. Fill in the chart.

Element	Type of Element (P, R, CF, or S)	Meaning of Element
dermato		
logist		
hypo		
cutaneo		
um		

K. Build terms from the language of otology. Identify the following elements by placing a check mark in the appropriate column. Give the meaning of the element, and then give an example of a medical term containing that element. The first one is done for you. Fill in the chart.

Element	Prefix	Root	CF	Suffix	Meaning of Element	Medical Term
nas		✓			nose	nasal
anti						
steat						
sis						
lith						
rhino						
tympan						
stomy						
sclero						
audio						

L. Terminology challenge: List below all the various medical procedures detailed in this text for otology and ophthalmology.

a. Otology ________________________________

b. Ophthalmology ________________________________

2. Caroline Mason has had many ear problems since she was a child. Frequent infections necessitated a *myringotomy* with PE tubes at a young age. Even after tube removal she continued to have frequent URIs, *tonsillitis, impacted cerumen, labyrinthitis,* and *vertigo* later in life.

   a. Briefly explain what a *myringotomy* is:

   ____

   ____

   ____

   ____

   Define the following terms and abbreviations that appear above:

   b. PE tubes: ______________________________

   c. URI ______________________________

   d. tonsillitis ______________________________

   e. impacted ______________________________

   f. cerumen ______________________________

   g. labyrinthitis ______________________________

   h. vertigo ______________________________

I. **Deconstruct** the following medical terms. A portion of the term is italicized and bolded—identify that element and give its meaning. The first one is done for you. Fill in the blanks.

Medical Term	Element	Meaning of Element
1. ***ot***itis	root	ear
2. ***audio***meter	________	________
3. mast***oid***	________	________
4. ***chole***steatoma	________	________
5. ***myring***otomy	________	________
6. oto***scler***osis	________	________
7. ***neur***al	________	________
8. ***equi***librium	________	________
9. oto***lith***	________	________
10. ***paroxysm***al	________	________

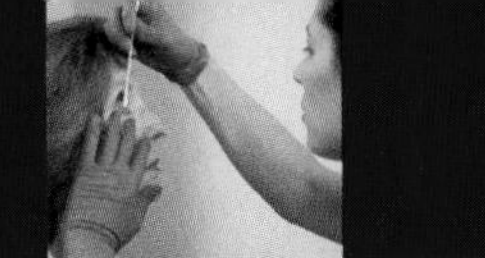

SPECIAL SENSES OF THE EYE AND EAR

G. **Identify** the following medical terms or abbreviations by specialty; then identify them as either a diagnosis or a procedure. Fill in the chart with a check mark (✓) in the appropriate columns.

Medical Term	Ophthalmology	Otology	Diagnosis	Procedure
1. uveitis				
2. BOM				
3. myringotomy				
4. tinnitus				
5. fundoscopy				
6. cholesteatoma				
7. strabismus				
8. otoscopy				
9. photocoagulation				

H. **Short answers for patient education:** Patients will ask you for clarification of certain terms they do not understand or for more explanation of body processes. Be prepared to answer the following questions for your patients.

1. Andrew Baker has severe otosclerosis in his left ear. Dr. Lee has recommended replacement of his stapes with a plastic *prosthesis.* Can you explain to Mr. Baker what a *prosthesis* is, and compare it to other body part replacements he may already have?

   Look up *prosthesis* in the glossary, a medical dictionary, or online. Define *prosthesis:*

   ____

   Name 3 other types of *prostheses* that can be inserted into the body.

   Prostheses: ______________ ______________ ______________

   How will this particular prosthesis help Mr. Baker?

   ____

   ____

   ____

   ____

E. Master your documentation—IT IS A LEGAL RECORD. Circle the most appropriate choice and insert the correct abbreviation where indicated.

1. Patient complains of sticky eyelids with (purulent/perulent) discharge, both eyes (abbrev. ______________).
   Diagnosis: (scleritis/conjunctivitis)
2. (Refraction/Accommodation) reveals patient's vision now 20-40 in the right eye, (abbrev. ______________) with correction.
3. Mr. Baker has continued decreasing vision in his left eye (abbrev. ______________). If his diabetes remains uncontrolled, his (retinopathy/retinoblastoma) will worsen.
4. (Opthalmoscopic/Ophthalmoscopic) examination of the left eye reveals (catarracks/cataracts) forming.
5. (Vertigo/Tinnitus) is often described by patients as dizziness.
6. (BPPV/AOM) ______________ is another type of intermittent vertigo.

F. Match the Latin and Greek terms in column 1 to their meanings in column 2 to increase your knowledge of the *language of ophthalmology.* Fill in the blanks.

_____	1. conjunctiva	A. paralysis
_____	2. lutea	B. on the outer side
_____	3. chiasm	C. circle
_____	4. hordeolum	D. hard
_____	5. foramen	E. yellow
_____	6. extrinsic	F. inner lining of eyelids
_____	7. ptosis	G. hole
_____	8. sclera	H. stye
_____	9. orbit	I. cross
_____	10. paresis	J. falling or drooping

Use any one term from column 1 in a sentence of patient documentation.

____

____

____

____

CHAPTER 4 REVIEW

SPECIAL SENSES OF THE EYE AND EAR

C. Patient education: Your patient is confused by some medical terms the doctor has used. Explain to the patient in simple language the difference between:

1. *esotropia* and *exotropia:*

____

2. *amblyopia* and *presbyopia:*

____

3. *refraction* and *accommodation:*

____

D. Spelling correctly is the mark of an educated professional. Proofread the following documentation to find the errors. Circle the misspelled terms, and rewrite the correctly spelled terms on the lines below.

1. The optalmoscopic examination revealed scleriitis and retinopathy.

2. Perulent discharge is coming from both eyes, and the patient also complains of photofobia.

3. The patient's blepharotosis can be remedied surgically. Schedule a blepharopasty as soon as possible.

4. Irigation and curetage have been ordered for the patient with impackted ceruman.

5. The patient suffers from chronick otitis medial with efusion.

B. **Word elements** are your most valuable tool for increasing your medical vocabulary. Use your knowledge of word elements to answer the following questions. Circle the correct answer.

1. Which of the following terms refers to *tears?*

   a. otolith
   b. lacrimal
   c. purulent
   d. uvea
   e. vestibule

2. This term is used to indicate partial *paralysis:*

   a. parietal
   b. periorbital
   c. paresis
   d. ptosis
   e. presbyopia

3. Based on its suffix, you can tell that a *keratotomy* is:

   a. a body part
   b. a procedure
   c. a diagnosis
   d. a medication
   e. an infection

4. The location of *periorbital* is:

   a. outside the eye
   b. around the eye
   c. beside the eye
   d. within the eye
   e. behind the eye

5. The term *amblyopia* signifies:

   a. sound
   b. light
   c. sight
   d. movement
   e. pain

6. *In situ* is a Latin phrase that means: *(Be precise!)*

   a. in this place
   b. in another place
   c. in the correct place
   d. in the place
   e. in place of

7. The lens of the eye is *avascular* because it has no:

   a. connective tissue
   b. aqueous humor
   c. blood supply
   d. fibrous outer covering
   e. mucous membrane

8. *Angiography* is an x-ray visualization of:

   a. organs
   b. bones
   c. blood vessels
   d. muscles
   e. glands

M. **Partner exercise:** Ask your study partner to close his or her text. Dictate the following sentences to your partner, and have him or her write them down on a blank sheet of paper. Check your partner's written sentences and the spelling of each word to ensure all are written exactly as shown in the sentences below. The sentence is not correct unless every word is present and *everything* is spelled correctly. When you have finished checking your partner's answers, close your book, ask your partner to dictate the sentences to you, and then write them down yourself.

1. Stye or hordeolum is an infection of an eyelash follicle producing an abscess with localized pain, swelling, redness, and pus formation at the edge of the eyelid.
2. Laser-assisted in situ keratomileusis (LASIK) is being used to treat myopia, hyperopia, and astigmatism.
3. The eustachian tube connects the middle ear with the nasopharynx, into which it opens close to the pharyngeal tonsils.
4. A perforated tympanic membrane can occur in acute otitis media when pus in the middle ear cannot escape down the auditory tube.

N. **Precision in communication:** Using the correct form (noun, verb, adjective) of the term is as important as using the correct term itself. Choose the correct form of the term from the word bank and insert it *on the line* in the appropriate sentence. After you fill in the term, write *under the line* what form of the term you have used (noun, verb, adjective).

Remember:

noun	person, place, or thing
verb	action
adjective	describes detail

refract	**accommodation**	**refractive**
accommodate	**accommodative**	**refraction**

1. The process of bending the light rays by the cornea and lens is called ______________________________.
2. ______________________________ esotropia is an inward eye turn, usually noticed around 2 years of age in 1% to 2% of children.
3. This process of the eye changing focus is called ______________________________.
4. The lens can ______________________________ itself to light rays by becoming thicker or thinner.
5. The cornea and lens work together to ______________________________ light in the vision process.

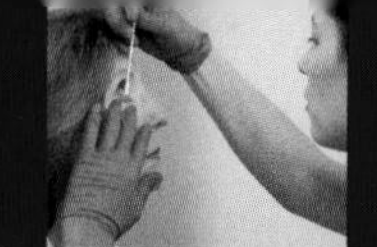

CHAPTER 4 REVIEW

SPECIAL SENSES OF THE EYE AND EAR

O. Roots/combining forms remain the core foundation of every term. Of the three possible answers, circle the correct term related to the question about the root.

1. The R/CF meaning *hearing* appears in the term:

   vestibule audiometric auricle

2. The R/CF meaning *fear* appears in the term:

   periorbital photophobia nasolacrimal

3. The R/CF meaning *eyelid* appears in the term:

   blepharitis paresis ptosis

4. The R/CF meaning *pus* appears in the term:

   extrinsic stereopsis purulent

5. The R/CF meaning *cornea* appears in the term:

   myopia keratotomy glaucoma

6. The R/CF meaning *blood vessel* appears in the term:

   angiography fundoscopy retinopathy

7. The R/CF meaning *side* appears in the term:

   otitis media periorbital bilateral

8. The R/CF meaning *nose* appears in the term:

   otorhinolaryngologist ceruminous curettage

9. The R/CF meaning *gland* appears in the term:

   adenoid cornea tonsil

10. The R/CF meaning *throat* appears in the term:

    tonsillectomy nasopharynx myringotomy

> **Study Hint**
> The term ***pur**ulent* means *discharging* ***pus.*** An easy way to remember how to spell it is that the term starts with the same two letters as does the word *pus.* This term is frequently misspelled "perulent." Remember *pus* and you will start the term correctly.

P. Chapter challenge: Circle the best answer.

1. *Ophthalmia neonatorum* is a type of:
   a. uveitis
   b. retinitis
   c. conjunctivitis
   d. corneal abrasion
   e. scleritis

2. The three terms that are *all* spelled correctly are:
   a. occular, stereopsis, cornia
   b. blepharoptosis, sty, contagious
   c. pupilary, cornial, avascular
   d. optalmologist, ptosis, paresis
   e. presbyopia, pupil, chiasm

3. A *tonometer* is an instrument used to measure:
   a. peripheral vision
   b. aqueous humor
   c. intraocular pressure
   d. sound waves in the eardrum
   e. fluid in the middle ear

> **Study Hint**
> Immediately cross off any answer you know is not correct. In your remaining choices, there is only *one best answer.*

4. The external opening of a passage is:

a. a pinna
b. an auricle
c. a meatus
d. an adenoid
e. a labyrinth

5. Which set of three terms is most likely to appear in an otorhinolaryngologist's dictation?

a. meatus, ceruminous, periorbital
b. curettage, retina, impacted
c. photosensitivity, Ishihara, otitis
d. foramen, presbyopia, hyperopia
e. nasopharynx, labyrinth, tonsillectomy

6. The term *tympanic* is associated with the:

a. auricle
b. eardrum
c. pinna
d. ossicles
e. stapes

7. Choose the three *ossicles:*

a. otolith, crista ampullaris, vestibule
b. malleus, incus, stapes
c. coryza, cochlea, labyrinth
d. tonsils, adenoids, mastoids
e. iris, uvea, sclera

8. The abbreviations *PERRLA, OD, OU,* and *OS* all relate to:

a. the nose
b. therapeutic procedures
c. the eye
d. radiology procedures
e. the ear

9. Name a symptom of *BPPV:*

a. ringing in the ears
b. dizziness
c. fever
d. rash
e. effusion

10. The term that means the same as "pink eye" is:

a. blepharitis
b. conjunctivitis
c. blepharoptosis
d. esotropia
e. scleritis

11. Which term relates to *pus?*

a. ambylopia
b. avascular
c. purulent
d. aqueous
e. strabismus

12. What is the medical term for three-dimensional perception?

a. exotropia
b. accommodation
c. presbyopia
d. stereopsis
e. refraction

SPECIAL SENSES OF THE EYE AND EAR

Q. Case Report challenge: Now that you are more comfortable with the terms in this chapter, you can apply that knowledge and answer the questions about the Case Report. If you read the report through and underline all the medical terminology, this will make it easier to answer the questions.

You are

. . . a medical assistant working for primary care physician Susan Lee, MD, of the Fulwood Medical Group.

You are communicating with

. . . Mrs. Carmen Cardenas, who has brought in her 3-year-old son, Eddie. She tells you that Eddie has had a cold for a couple of days. Early this morning he woke up screaming, felt hot, and was tugging his ears. She gave him acetaminophen with some orange juice, and he threw up. She also tells you this is the third similar episode in the past year, and, since the last time, she is concerned that he is not hearing normally. You see a worried mother and a restless toddler with a green, nasal discharge. His oral temperature taken with an electronic digital thermometer is 102.4°F, pulse 100. You tell her that Dr. Lee will be in to see Eddie as soon as possible.

Documentation. 05/10/08

Examination by Dr. Lee showed that Eddie has a bilateral acute otitis media (BOM) with an upper respiratory infection (URI). Dr. Lee is also concerned that Eddie has a chronic otitis media with effusion (OME) that is causing a hearing loss. She prescribed amoxicillin 250 mg q.i.d. with *acetaminophen* 160 mg p.r.n. for 10 days, when she will see Eddie again. If, after the acute infection subsides, there remains an effusion with hearing loss, Dr. Lee may need to refer Eddie to an otologist. I explained this to Mrs. Cardenas.

Signed, Luis Guittierez, CMA. 1115 hrs.

1. What are Eddie's symptoms?

____

____

2. What does the term *bilateral* mean for Eddie?

____

____

3. What is an *effusion?*

____

____

4. Which of Eddie's symptoms will be reduced by the *acetaminophen?*

____

____

5. One medication is to be given q.i.d., while the other is given p.r.n. What is the difference?

____

____

6. Find the two terms in the case report that are opposites, and define them.

Term: ______________________

Means: ______________________________________

Term: ______________________

Means: ______________________________________

7. What type of specialist will Eddie be referred to if his condition does not improve after medication?

____

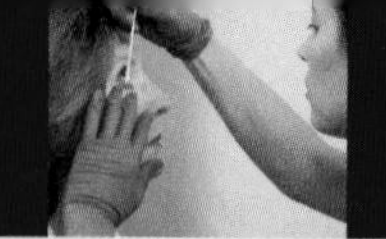

CUMULATIVE EXERCISES

These exercises are built around terms and elements from this and the preceding three chapters. They are an excellent source of review for tests—be sure to do every exercise!

A. **Suffixes:** List here all the suffixes from the first four chapters that mean *pertaining to,* and give an example of a term containing each suffix.

Suffix Meaning *Pertaining To*	Medical Term Containing This Suffix

B. **Abbreviations:** For each abbreviation listed below, write out in words the meaning of the abbreviation. *An answer is not correct unless all the words are spelled correctly!* Fill in the blanks.

1. LLQ ______________________________
2. OU ______________________________
3. TB ______________________________
4. ACLS ______________________________
5. URI ______________________________
6. IM ______________________________
7. AOM ______________________________
8. SLE ______________________________
9. HIV ______________________________
10. RUQ ______________________________

C. **Diagnosis/procedure:** The following medical terms are either a diagnosis or a procedure. Place a check mark (✓) in the appropriate column for diagnosis or procedure. In the last column, write the name of the specialist who would be using that medical term in dictation.

dermatologist **otologist** **ophthalmologist** **pulmonologist** **cardiologist**

Medical term	**Diagnosis**	**Procedure**	**Specialist**
scleritis			
LASIK			
myocardial infarction			
otolith			
retinopathy			
eczema			
tonsillectomy			
retinoblastoma			
pneumonia			
photophobia			
tinea pedis			
refraction			
labyrinthitis			
otoscopy			
carbuncle			
fundoscopy			
cataract			
verucca			
cryosurgery			
shingles			

D. **Roots/Combining Forms** are always your best clue for understanding the meaning of a term. For each R/CF listed below, define its meaning in column 2, and in column 3 give an example of a term in which it appears. Write the meaning of the medical term in column 4.

Root/CF	Meaning of R/CF	Medical Term Containing This R/CF	Meaning of Medical Term
photo			
cutane			
lacrim			
cardio			
tympan			
dermat			
thorax			
audio			
naso			
pneumon			
blephar			
crani			
dermis			
ophthalmo			
pulmon			

Choose terms from the chart above and use each in a sentence of patient documentation.

1. ______________________________

2. ______________________________

E. **Prefixes** generally add description to a medical term—color, location, size, and number. Complete the terms defined below by writing the appropriate prefix *on the line.* Then write the meaning of the prefix *under the line.* The first one is done for you.

exo	bi	hypo	trans
eso	pre	sub	an
hyper	peri	a	
intra	neo		

1. around the orbit (of the eye) ___peri/ orbital___
   around

2. same as subcutaneous ____________________/dermic

3. turning eye inward toward the nose ____________________/tropia

4. inability to focus light rays ____________________/stigmatism

5. ability to see distant objects but not close ones ____________________/opia

6. on both sides ____________________/lateral

7. newborn ____________________/natal

8. across or through the skin ____________________/dermal

9. before birth ____________________/natal

10. below the skin ____________________/cutaneous

11. turning the eye outward ____________________/tropia

12. without pain ____________________/algesia

13. within the cell ____________________/cellular

F. **Spelling demons:** Some medical terms are particularly difficult to pronounce and spell. Develop your ability to recognize a misspelling when you see it, and know how to correct it. Only one term in this exercise is spelled correctly—write the correct spelling on the line next to each incorrectly spelled term.

1. curetage ______
2. integamentary ______
3. tosis ______
4. umbillicus ______
5. ceruminus ______
6. iskemia ______
7. perulent ______
8. subcutaneous ______
9. eustacian ______
10. crusciate ______
11. sebaseous ______
12. ophalmoscope ______
13. thoraxic ______
14. infarktion ______
15. carttilage ______
16. diafram ______

G. **Chapter challenge:** Circle the best answer.

1. Which term refers to something that is "watery"?
   - a. vitreous
   - b. centralis
   - c. neonatal
   - d. ophthalmoscopic
   - e. aqueous
2. Who fits eyeglasses but cannot prescribe medication?
   - a. otologist
   - b. otorhinolaryngologist
   - c. optometrist
   - d. ophthalmologist
   - e. none of these
3. *Astigmatism* is the inability to:
   - a. see objects close at hand
   - b. see objects on the outer edge of the visual field
   - c. see distant objects
   - d. focus light rays from different planes
   - e. look at bright light

4. The correct plural for the term *foramen* is:
   - a. foramens
   - b. foramena
   - c. foramina
   - d. foraminae
   - e. none of these is correct
5. Muscle imbalance in one eye causes alignment to break down and results in:
   - a. "lazy eye"
   - b. strabismus
   - c. "cross-eyes"
   - d. a and b
   - e. b and c
6. Which abbreviation would appear with medication directions?
   - a. BOM
   - b. p.r.n.
   - c. URI
   - d. OME
   - e. OU
7. The term with a root meaning *fear* is:
   - a. conjunctivitis
   - b. photocoagulation
   - c. sensorineural
   - d. photophobia
   - e. otolith
8. **Brain teaser:** Fluid produced by an inflammatory process is called:
   - a. coryza
   - b. aqueous
   - c. vitreous
   - d. ptosis
   - e. effusion

Monitor My Progress An honest self-assessment of your progress with this material will make it easier for you to learn from your mistakes. Jot down below any wrong answers you have corrected; terms, elements, or spellings you find difficult to remember; and so on. The more times you see, speak, or write the terms, the better you will come to know them.

Based on these review exercises, I need to spend more time on:

The Digestive System

The Essentials of the Language of Gastroenterology

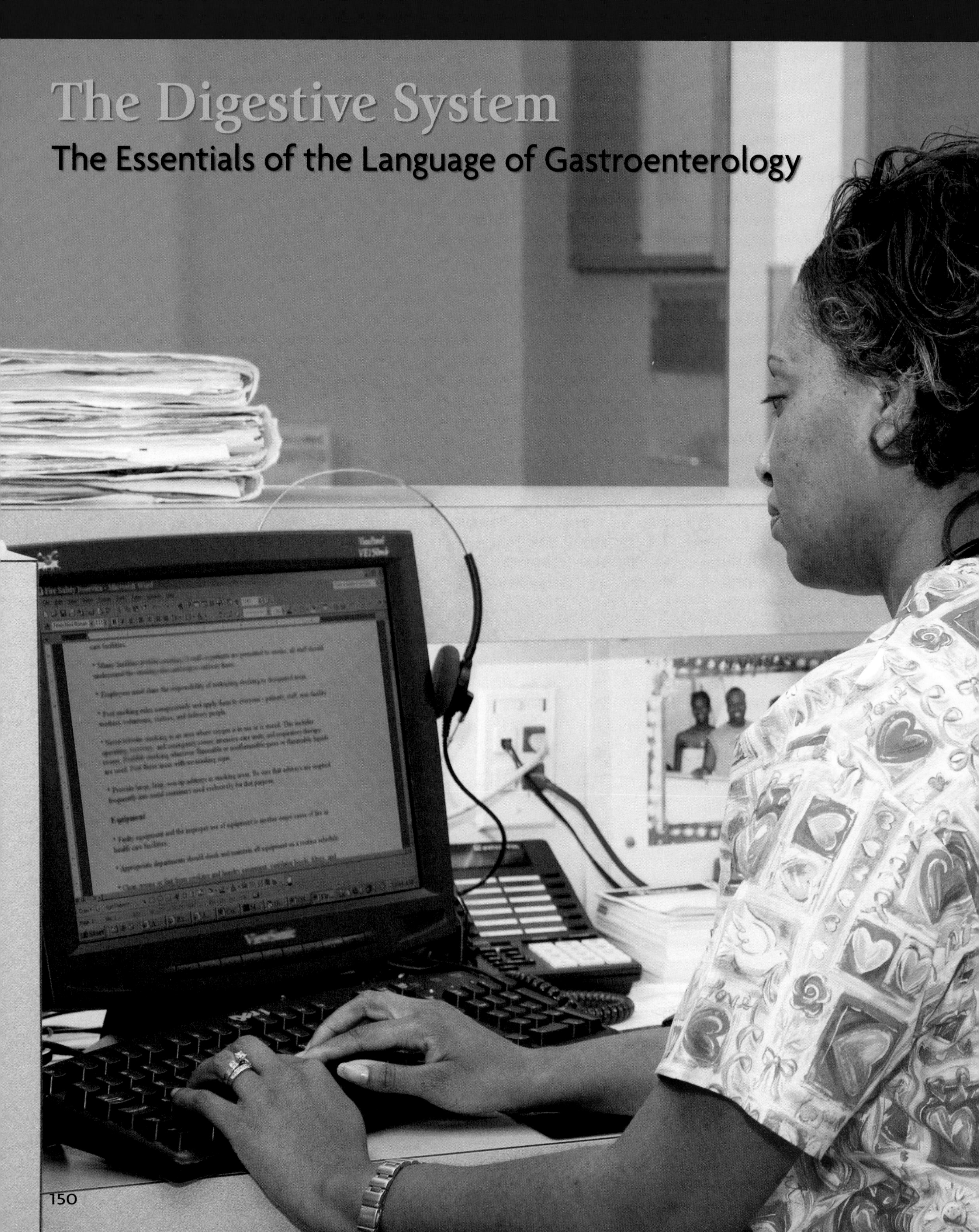

CHAPTER

CASE REPORT 5.1

You are

. . . a medical transcriptionist at Fulwood Medical Center.

You are communicating with

. . . one of your physicians, Dr. Stewart Walsh, who has dictated this letter to request authorization for a surgical procedure:

Fulwood Medical Center
3333 Medical Parkway, Fulwood, MI 01234
555-247-6100

Department of Bariatric Surgery

To: Charles Leavenworth, MD
Medical Director
Lombard Insurance Company

From: Stewart Walsh, MD, FACS
Chief of Surgery
Center for Bariatric Surgery
Fulwood Medical Center

10/06/08

Dear Doctor Leavenworth,

Request for authorization of Surgery

Re: Mrs. Martha Jones
Subscriber ID 056437

Mrs. Jones is a 52-year-old former waitress, recently divorced. She is 5 feet 4 inches tall and weighs 275 pounds. She has type 2 diabetes with frequent episodes of hypoglycemia and also ketoacidosis, requiring three different hospitalizations. She now has diabetic retinopathy and peripheral vasculitis. Complicating this are hypertension (185/110), coronary artery disease, and pulmonary edema. In spite of monthly meetings with our nutritionist, she has gained 25 pounds in the past six months.

To reduce and control her weight, I am proposing to perform a gastric bypass using a laparoscopic approach. We will need to admit her two days prior to surgery to control her blood sugar and cardiovascular problems, and we anticipate that she will remain in the hospital for two days after surgery, barring any complications. She is also very aware of the necessary follow-up to the procedure and the counseling required for a new lifestyle.

We believe not only that this is an essential procedure medically but that it will reduce in the long term the financial burden of her multiple therapies and improve the quality of the patient's life. Enclosed is supportive documentation of her current history and medical problems.

Your company has designated our hospital as a Center of Excellence for weight-loss surgery, and I look forward to your prompt agreement with this approach for this patient.

Sincerely,

Stewart Walsh, MD, FACS
Chief of Surgery
Center for Bariatric Surgery
Fulwood Medical Center

Learning Outcomes

As a medical transcriptionist, you and all health care professionals directly and indirectly involved with Mrs. Jones's care need to be able to:

5.1 Apply the language of gastroenterology to the gastrointestinal tract, liver, gallbladder, and pancreas.

5.2 Comprehend, analyze, spell, and write the medical terms of gastroenterology.

5.3 Recognize and pronounce the medical terms of gastroenterology.

In Mrs. Jones's case, the gastric bypass procedure reduced the size of her available stomach from 2 quarts to 2 ounces. The bypass was taken to the mid-ileum. This resulted in her being able to eat less and to absorb less. She had no complications from the **laparoscopic** procedure. In the succeeding 2 months, she lost 15 pounds in weight.

LESSON 5.1 The Digestive System

OBJECTIVES

There are basic terms and elements of medical terminology that apply throughout the different parts of the digestive system. These are discussed in this lesson and will enable you to:

- **List the organs and accessory organs of the digestive system.**
- **Identify the components of the alimentary canal.**
- **Select the correct medical terminology to describe the structure and functions of the digestive system.**

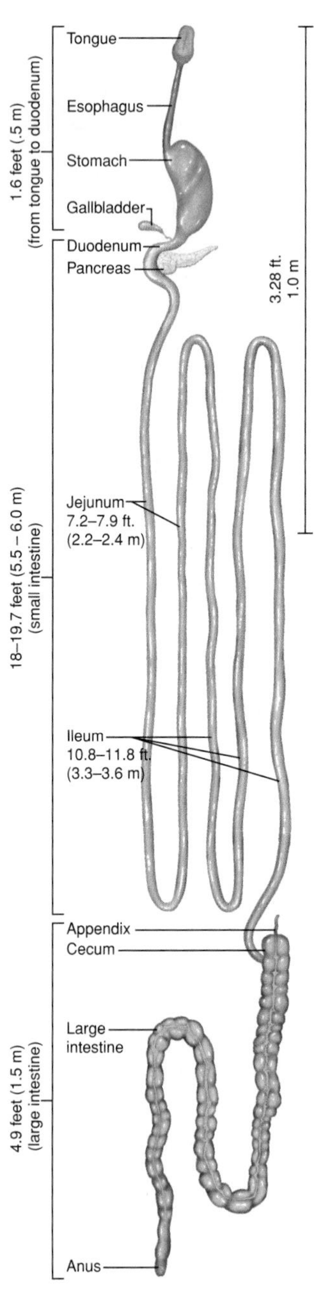

▲ **FIGURE 5.1 Alimentary Canal.**

ALIMENTARY CANAL AND ACCESSORY ORGANS

Every cell in your body requires a constant supply of nourishment in a form that can be absorbed across its cell membrane. The **digestive system** breaks down the nutrients in food into elements that can be transported to the cells via the blood and lymphatics.

The digestive system consists of the **alimentary canal,** or digestive tract, which extends from the mouth to the **anus,** and **accessory organs** connected to the canal to assist in digestion.

The term **gastrointestinal (GI)** technically refers to the stomach and intestines but is often used to mean the whole digestive system.

Gastroenterology is the study of the digestive system.

A **gastroenterologist** is a physician who specializes in diseases of the digestive system.

The **alimentary canal** (*Figure 5.1*) includes the:

- **Mouth**
- **Pharynx**
- **Esophagus**
- **Stomach**
- **Small intestine**
- **Large intestine**

The **accessory organs** of digestion include the:

- **Teeth**
- **Tongue**
- **Salivary glands**
- **Liver**
- **Gallbladder**
- **Pancreas**

WORD ANALYSIS AND DEFINITION

S/ = Suffix P/ = Prefix R/ = Root R/CF = Combining Form

WORD	PRONUNCIATION	ELEMENTS		DEFINITION
alimentary	al-ih-**MEN**-tar-ee	S/ R/	**-ary** *pertaining to* **aliment-** *nourishment, food*	Pertaining to the digestive tract.
alimentary canal	kah-**NAL**		canal, Latin *a duct or channel*	Digestive tract.
anus	**A**-nus		Latin *a ring*	Terminal opening of the digestive tract through which feces are discharged.
anal (adj)	**A**-nal	S/ R/	**-al** *pertaining to* **an-** *anus*	Pertaining to the anus.
bariatric	bar-ee-**AT**-rik	S/ R/	**-atric** *treatment* **bari-** *weight*	Treatment of obesity.
digestion	die-**JEST**-shun	S/ R/	**-ion** *action* **digest-** *to break down food*	Breakdown of food into elements suitable for cell metabolism.
digestive (adj)	die-**JEST**-iv	S/	**-ive** *nature of, quality of*	Pertaining to digestion.
esophagus	ee-**SOF**-ah-gus		Greek *gullet*	Tube linking the pharynx and the stomach.
gastric	**GAS**-trik	S/ R/	**-ic** *pertaining to* **gastr-** *stomach*	Pertaining to the stomach.
gastroenterology	**GAS**-troh-en-ter-**OL**-oh-gee	S/ R/CF R/CF	**-logy** *study of* **gastr/o-** *stomach* **-enter/o-** *intestine*	Medical specialty of the stomach and intestines.
gastroenterologist	**GAS**-troh-en-ter-**OL**-oh-jist	S/	**-logist** *one who studies, specialist*	Medical specialist in gastroenterology.
gastrointestinal	**GAS**-troh-in-**TESS**-tin-al	S/ R/CF R/	**-al** *pertaining to* **gastr/o-** *stomach* **intestin-** *gut, intestine*	Pertaining to the stomach and intestines.
intestine	in-**TESS**-tin		Latin *intestine, gut*	The digestive tube from stomach to anus.
intestinal (adj)	in-**TESS**-tin-al	S/ R/	**-al** *pertaining to* **intestin-** *gut*	Pertaining to the intestines.
laparoscopy	lap-ah-**ROS**-koh-pee	S/ R/CF	**-scopy** *to view, to examine* **lapar/o-** *abdomen in general*	Examination of contents of abdomen using an endoscope.
laparoscope	**LAP**-ah-roh-skope	S/	**-scope** *instrument for viewing*	Instrument (endoscope) used for viewing abdominal contents.
laparoscopic (adj)	**LAP**-ah-roh-**SKOP**-ik	S/	**-ic** *pertaining to*	Pertaining to laparoscopy.

Abbreviation

GI gastrointestinal

EXERCISES

Analyzing elements can tell you a lot about a medical term. Look closely at the following medical terms, and let the elements be your guide. Review the WAD before you start the exercise. Fill in the blanks.

gastric **gastroenterology** **gastrointestinal** **gastroenterologist** **gastroscope**

1. Based on the root, the term *gastric* pertains to the ________________.
2. Analyzing the two combining forms, the term *gastroenterology* pertains to the ________________ and the ________________.
3. In the term *gastrointestinal,* the root *intestin* has the same meaning as the combining form ________________ in *gastroenterology.*
4. A *gastroscope* would be inserted into the ________________ for examination or biopsy.

Good work! You have mastered these terms.

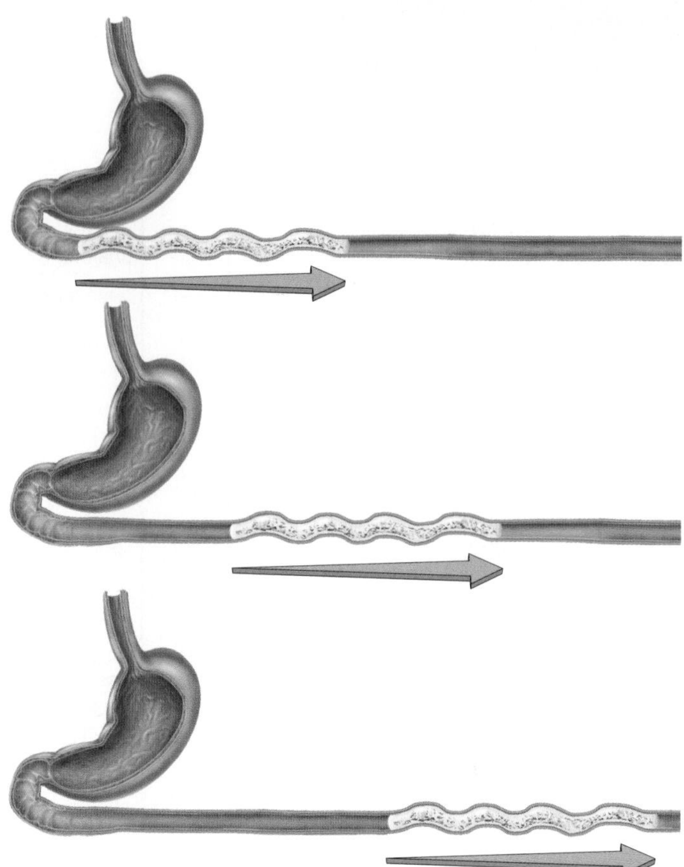

FIGURE 5.2 ▶
Peristalsis of the Small Intestine.
Successive waves of peristalsis overlap each other and move the contents along.

ACTIONS AND FUNCTIONS OF THE DIGESTIVE SYSTEM

The actions and functions have five components:

1. **Propulsion:** The mechanical movement of food from the mouth to the **anus.** Normally, this takes 24 to 36 hours.
2. **Digestion:** The breakdown of foods into forms that can be transported to and absorbed into cells. This process has two components:
   a. **Mechanical** digestion breaks larger pieces of food into smaller ones without altering their chemical composition. **Mastication** (chewing) breaks down the food into smaller particles so that digestive enzymes have a larger surface area with which to interact. **Deglutition** (swallowing) moves the **bolus** of food from the mouth into the esophagus. **Peristalsis,** or waves of contraction and relaxation, moves material through most of the alimentary canal (*Figure 5.2*).
   b. **Chemical** digestion breaks down large molecules of food into smaller and simpler chemicals. This process is carried out by digestive enzymes produced by the salivary glands, stomach, small intestine, and pancreas.

   The digestive enzymes break down the three main groups of foods: **carbohydrates, proteins,** and fats.
3. **Secretion:** The addition throughout the digestive tract of secretions that lubricate, liquefy, and digest the food. Mucus lubricates the food and the lining of the tract. Water liquefies the food to make it easier to digest and absorb. Enzymes break down the food.
4. **Absorption:** The movement of nutrient molecules out of the digestive tract, through the epithelial cells lining the tract, and into the blood or lymph for transportation to body cells.
5. **Elimination:** The process by which the undigested residue of food is removed from the body.

Keynote

Digestion is both mechanical and chemical.

WORD ANALYSIS AND DEFINITION

S/ = Suffix P/ = Prefix R/ = Root R/CF = Combining Form

WORD	PRONUNCIATION	ELEMENTS		DEFINITION
bolus	**BOH**-lus		Greek *lump*	A single mass of a substance.
carbohydrate	kar-boh-**HIGH**-drate	S/ R/CF R/	**-ate** *pertaining to, composed of* **carb/o-** *carbon* **-hydr-** *water*	Group of organic food compounds that includes sugars, starch, glycogen, and cellulose.
deglutition	dee-glue-**TISH**-un	S/ R/	**-ion** *action, condition* **deglutit-** *to swallow*	The act of swallowing.
masticate (verb)	**MAS**-tih-kate	S/ R/	**-ate** *pertaining to, composed of* **mastic-** *chew*	To chew.
mastication (noun)	mas-tih-**KAY**-shun	S/	**-ation** *process*	The process of chewing.
peristalsis	per-ih-**STAL**-sis	P/ R/	**peri-** *around* **-stalsis** *constrict*	Waves of alternate contraction and relaxation of the intestinal wall to move food along the digestive tract.
protein	**PRO**-teen		Greek *first, primary*	Class of food substances based on amino acids.
secrete (verb)	seh-**KREET**		Latin *to separate*	To produce a chemical substance in a cell and release it from the cell.
secretion (noun)	seh-**KREE**-shun	S/ R/	**-ion** *action, condition* **secret-** *secrete, produce*	The production of a chemical substance in a cell and its release from the cell.

EXERCISES

Match *the correct medical term in column 2 to the definition in column 1. The body process, or action, is described for you below. Fill in the blanks.*

_____ 1. swallowing A. mastication

_____ 2. releasing products of metabolism B. elimination

_____ 3. chewing C. deglutition

_____ 4. removal of waste material D. secretion

Review the text on page 154 to obtain your answers to the following questions.

5. What is the mechanical movement of food from mouth to anus called? ____________________

6. Which organs produce digestive enzymes? ____________________

7. What lubricates the food in the digestive tract? ____________________

8. What liquefies the food in the digestive tract? ____________________

9. What breaks down food in the digestive tract? ____________________

10. What are the three main groups of food broken down by digestive enzymes? ____________________

LESSON 5.2 Mouth, Pharynx, and Esophagus

THE DIGESTIVE SYSTEM

OBJECTIVES

When you pop a piece of chicken and some vegetable into your mouth, you start a cascade of digestive tract events that occur during the following 24 to 36 hours. In this chapter, you will follow the food as it goes through the digestive tract. In this lesson, you will review the first stages in the cascade, when the food is in the mouth and then is swallowed. The information in this lesson will enable you to:

- **Identify the medical terminology for the structures and functions of the teeth, tongue, and salivary glands.**
- **Document the processes and outcomes of mastication and swallowing of food.**
- **Select the correct medical terminology to describe the structures, functions, and disorders of the mouth, pharynx and esophagus.**

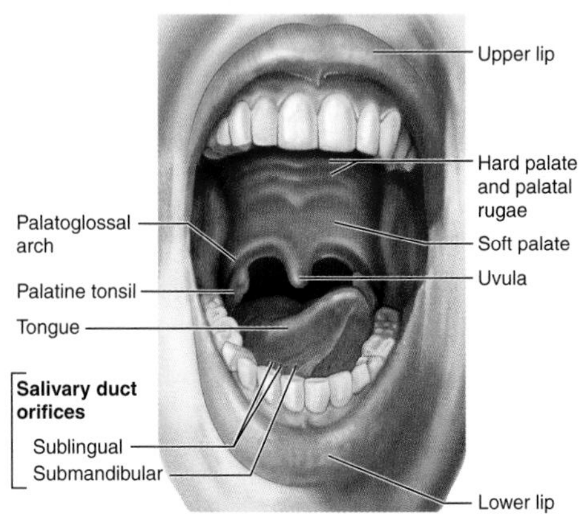

▲ **FIGURE 5.3 Mouth.**

▲ **FIGURE 5.4 Tongue.**

THE MOUTH AND MASTICATION

The mouth, or **oral** cavity (*Figure 5.3*), is the entrance to your digestive tract and is the first site of mechanical digestion, through mastication **(chewing),** and of chemical digestion, through an **enzyme** in your saliva.

The roof of the mouth is called the **palate.** The anterior two-thirds is the bony **hard palate.** The posterior one-third is the muscular **soft palate.** The skeletal muscle of the soft palate has a projection called the **uvula,** which closes off the **nasopharynx** during swallowing.

The **tongue** moves food around your mouth and helps the cheeks, lips, and gums hold the food in place while you chew it. Small, rough, raised areas on the tongue, called **papillae,** contain some 4,000 taste buds that react to the chemical nature of the food to give you the different sensations of taste (*Figure 5.4*). A tastebud cell lives for 7 to 10 days and is then replaced.

Adult Teeth

The normal adult has 32 teeth, 16 rooted in the upper jaw (maxilla) and 16 in the lower jaw (mandible) (*Figure 5.3*). The bulk of a tooth is composed of **dentine** (also spelled *dentin*), a substance like bone but harder, that is covered in **enamel.** The dentine surrounds a central **pulp** cavity, containing blood vessels, nerves, and connective tissue. The blood vessels and nerves reach this cavity from the jaw through tubular root canals.

Salivary Glands

Salivary glands secrete **saliva.** The two **parotid** glands, the two **submandibular** glands, the two **sublingual** glands (*Figure 5.5*), and numerous minor salivary glands scattered in the mucosa of the tongue and cheeks secrete more than a quart of saliva each day.

Saliva is 95% water, and its major functions are to begin the digestion of starch and fat and lubricate food to make it easier to swallow.

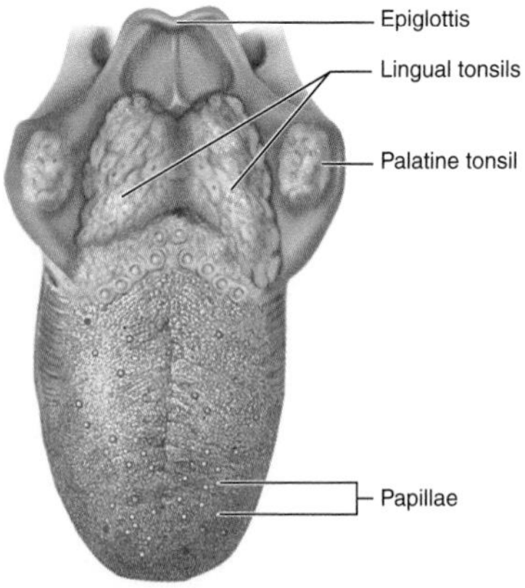

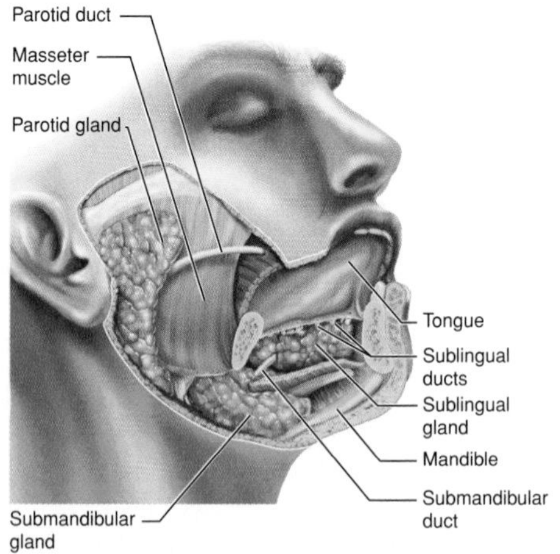

▲ **FIGURE 5.5 Salivary Glands.**

WORD ANALYSIS AND DEFINITION

S/ = Suffix   P/ = Prefix   R/ = Root   R/CF = Combining Form

WORD	PRONUNCIATION	ELEMENTS		DEFINITION
dentine (also spelled *dentin*)	**DEN**-tin	S/ R/	**-ine** *pertaining to, substance* **dent-** tooth	Dense, ivory-like substance located under the enamel in a tooth.
enamel	ee-**NAM-**el		French *enamel*	Hard substance covering a tooth.
enzyme	**EN**-zime	P/ R/	**en-** *in* **-zyme** *enzyme, fermenting*	Protein that induces changes in other substances.
mouth	MOWTH		Old English *mouth*	External opening of a cavity or canal.
nasopharynx	**NAY**-zoh-**FAIR**-inks	R/CF R/	**nas/o-** *nose* **-pharynx** *throat*	Region of the pharynx at the back of the nose and above the soft palate.
oral	**OR**-al	S/ R/	**-al** *pertaining to* **or- (os)** *mouth*	Pertaining to the mouth.
palate	**PAL**-ate		Latin *palate*	Roof of the mouth.
papilla **papillae (pl)**	pah-**PILL**-ah pah-**PILL**-ee		Latin *small pimple*	Any small projection.
parotid	pah-**ROT**-id	S/ P/ R/	**-id** *having a particular quality* **par-** *beside* **-ot-** *ear*	Parotid gland is the salivary gland beside the ear.
pulp	PULP		Latin *flesh*	Dental pulp is the connective tissue in the cavity in the center of the tooth.
saliva (noun) **salivary (adj)**	sa-**LIE**-vah **SAL**-ih-var-ee	 S/ R/	Latin *spit* **-ary** *pertaining to* **saliv-** *saliva*	Secretion in mouth from salivary glands. Pertaining to saliva.
sublingual	sub-**LING**-wal	S/ P/ R/	**-al** *pertaining to* **sub-** *underneath* **-lingu-** *tongue*	Underneath the tongue.
submandibular	sub-man-**DIB**-you-lar	S/ P/ R/	**-ar** *pertaining to* **sub-** *underneath* **-mandibul-** *mandible*	Underneath the mandible.
tongue	TUNG		Latin *tongue*	Mobile muscle mass in the mouth; bears the taste buds.
uvula	**YOU**-vyu-lah		Latin *grape*	Fleshy projection of the soft palate.

EXERCISES

Apply your knowledge *of the same elements to various terms, and increase your medical vocabulary. Focus on what is the same and what is different about the following terms. First, slash each term into its elements, and then fill in the chart.*

Medical term	Meaning of Prefix	Meaning of Root	Meaning of Suffix
submandibular			
sublingual			
subcutaneous (from previous chapter)			

Answer the following questions based on the above chart.

1. What is similar about every term? ______________________________

2. Which element makes every term different? ______________________________

3. Do all these terms describe size, number, location, or color? ______________________________

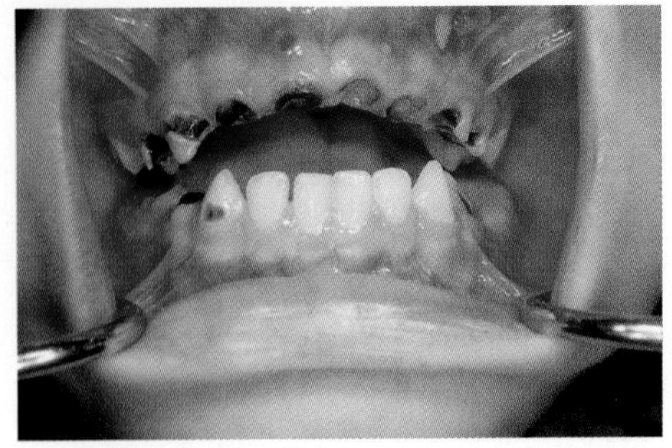

▲ **FIGURE 5.6** **Dental Caries.**

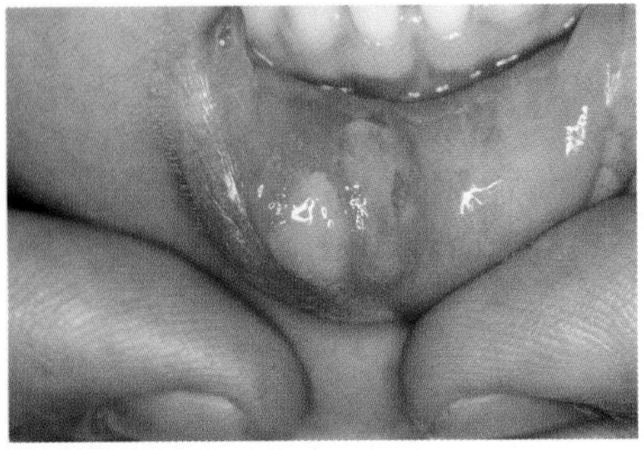

▲ **FIGURE 5.7** **Cold Sores.**
Ulcer inside lower lip.

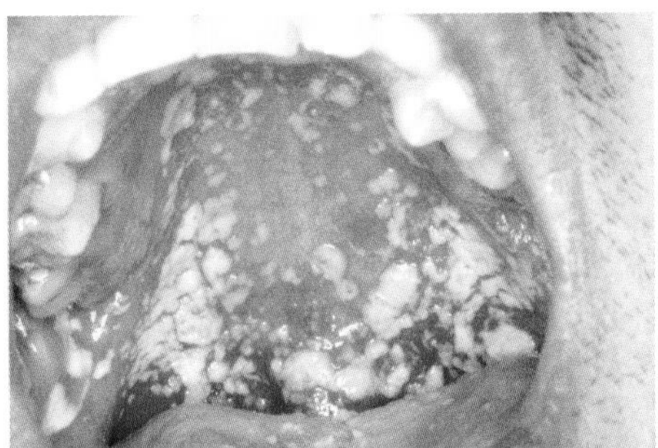

▲ **FIGURE 5.8** **Oral Thrush.**

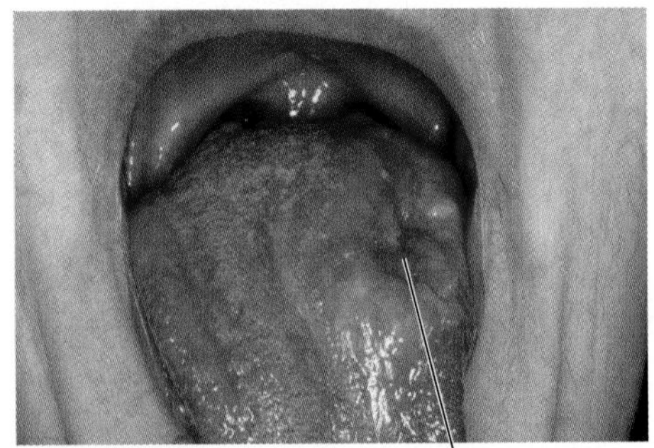

▲ **FIGURE 5.9** **Oral Cancer.**
Cancer of the tongue.

DISORDERS OF THE MOUTH

An accumulation of **dental plaque,** a collection of oral microorganisms and their products, or **tartar,** calcified deposits at the gingival margin of the teeth, is a precursor to invasion by bacteria that cause dental disease.

Dental caries, tooth decay and cavity formation, is an erosion of the tooth surface caused by bacteria (*Figure 5.6*). If untreated, it can lead to an abscess at the root of the tooth. **Gingivitis** is an infection of the gums. **Periodontal disease** occurs when the gums and the jawbone are involved in a disease process. In **periodontitis,** infection causes the gums to pull away from the teeth, forming pockets that become sources of infection. The infection can spread to the underlying bone. Infection of the gums with a purulent discharge is called **pyorrhea.**

The term **stomatitis** is used for any infection of the mouth:

- **Mouth ulcers,** also called **canker** sores, are erosions of the mucous membrane lining the mouth. The most common type are **aphthous** ulcers, which occur in clusters of small ulcers and last for 3 or 4 days. They are usually related to stress or illness but can also be caused by trauma.
- **Cold sores,** or fever blisters (*Figure 5.7*), are recurrent ulcers of the lips, lining of the mouth, and gums due to infection with the virus **herpes simplex type 1 (HSV-1).** The ulcers usually clear spontaneously.
- **Thrush** (*Figure 5.8*) is an infection occurring anywhere in the mouth that is caused by the fungus *Candida albicans.* This fungus is found normally in the mouth, but it can multiply out of control as a result of prolonged antibiotic or steroid treatment, cancer chemotherapy, or diabetes. Newborn babies can acquire oral thrush from the mother's vaginal yeast infection during the birth process. Treatment with antifungal agents is usually successful.
- **Oral cancer** (*Figure 5.9*) occurs most often on the lip, but can also occur on the tongue. Eighty percent of oral cancers are associated with smoking or chewing tobacco. Metastasis occurs to lymph nodes, bone, lung, and liver.

Halitosis is the medical term for "bad breath," which occurs in association with any of the above mouth disorders.

WORD ANALYSIS AND DEFINITION

S/ = Suffix P/ = Prefix R/ = Root R/CF = Combining Form

WORD	PRONUNCIATION	ELEMENTS		DEFINITION
aphthous	**AF**-thus		Greek *ulcer*	Painful small oral ulcers (canker sores).
canker	**KANG**-ker		Latin *crab*	Nonmedical term for aphthous ulcer.
caries	**KARE**-eez		Latin *dry rot*	Bacterial destruction of teeth.
gingiva	**JIN**-jih-vah		Latin *gum*	Tissue surrounding the teeth and covering the jaw.
gingival (adj)	**JIN**-jih-val	S/ R/	**-al** *pertaining to* **gingiv-** *gum*	Pertaining to the gums.
gingivitis	jin-jih-**VI**-tis	S/	**-itis** *inflammation*	Inflammation of the gums.
gingivectomy	jin-jih-**VEC**-toe-me	S/	**-ectomy** *surgical excision*	Surgical removal of diseased gum tissue.
halitosis	hal-ih-**TOE**-sis	S/ R/	**-osis** *condition* **halit-** *breath*	Bad odor of the breath.
periodontal	**PER**-ee-oh-**DON**-tal	S/ P/ R/	**-al** *pertaining to* **peri-** *around* **-odont-** *tooth*	Around a tooth.
periodontics	**PER**-ee-oh-**DON**-tiks	S/	**-ics** *knowledge*	Branch of dentistry specializing in disorders of tissues around the teeth.
periodontist	**PER**-ee-oh-**DON**-tist	S/	**-ist** *specialist*	Specialist in periodontics.
periodontitis	**PER**-ee-oh-don-**TIE**-tis	S/	**-itis** *inflammation*	Inflammation of tissues around a tooth.
plaque	PLAK		French *plate*	Patch of abnormal tissue.
pyorrhea	pie-oh-**REE**-ah	R/ R/CF	**-rrhea** *flow* **py/o-** *pus*	Purulent discharge.
tartar	**TAR**-tar		Latin *crust on wine casks*	Calcified deposit at the gingival margin of the teeth.
thrush	THRUSH		Root unknown	Infection with *Candida albicans.*

Abbreviation

HSV-1 Herpes simplex virus, type 1

EXERCISES

Suffixes: *Find the correct suffix to complete the medical terms, which all appear on this two-page spread. Fill in the blanks.*

ist itis ectomy rrhea osis al ics

1. inflammation of the gums gingiv/ __________
2. specialized branch of dentistry periodont/ __________
3. around a tooth periodont/ __________
4. bad breath halit/ __________
5. surgical removal of diseased gum tissue gingiv/ __________
6. specialist in periodontics periodont/ __________
7. purulent discharge pyo/ __________

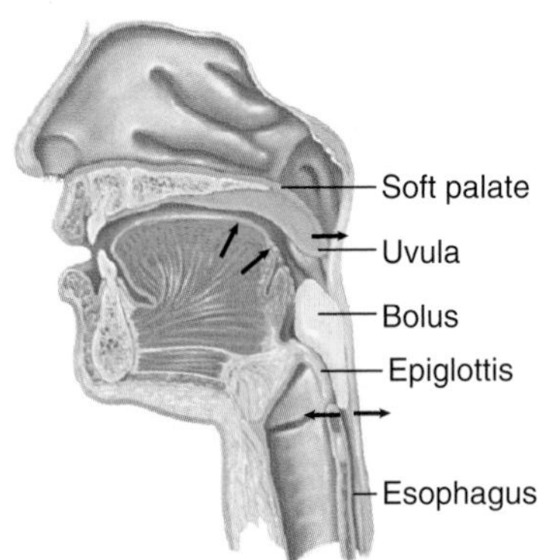

▲ **FIGURE 5.10 Swallowing.** The bolus of food is in the oropharynx.

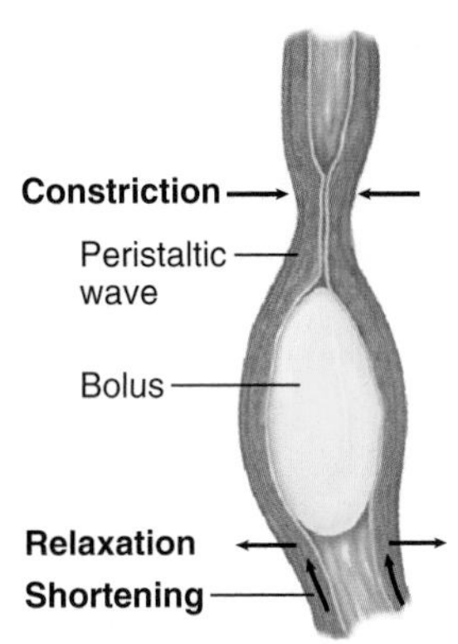

▲ **FIGURE 5.11 Swallowing.** The bolus of food is in the esophagus.

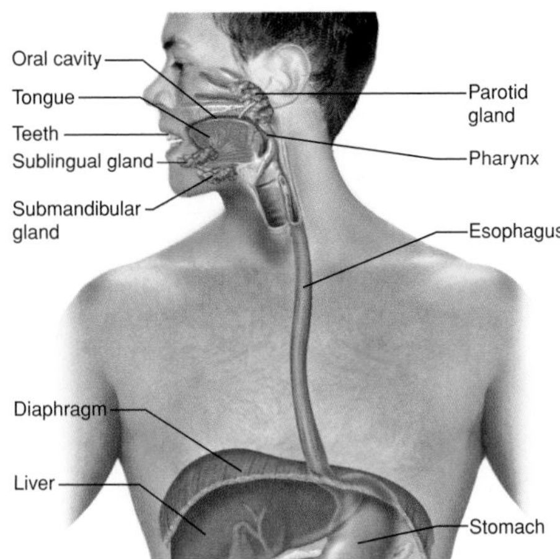

▲ **FIGURE 5.12 Esophagus.**

Abbreviation
GERD gastroesophageal reflux disease

ESOPHAGUS

The pieces of chicken and vegetable that you ingested have now been sliced and ground into small particles by the teeth, partly digested and lubricated by saliva, and rolled into a bolus between the tongue and the hard palate, the bony roof of the mouth. The bolus is now ready to be swallowed down the oropharynx (*Figure 5.10*) into the **esophagus.**

From the lower end of the pharynx, the bolus enters the esophagus where peristaltic contractions of the muscles in the wall of the esophagus move the bolus down the esophagus (*Figure 5.11*). At the lower end of the esophagus the **cardiac sphincter** relaxes to allow the bolus to enter the stomach.

The **esophagus** (*Figure 5.12*) is a tube 9 to 10 inches long, and it pierces the diaphragm at the esophageal **hiatus** to go from the thoracic cavity to the abdominal cavity (*Figure 5.12*).

Disorders of the Esophagus

Esophagitis is inflammation of the lining of the esophagus, producing a **postprandial** burning chest pain **(heartburn),** pain on swallowing, and occasional vomiting of blood **(hematemesis).** The most common cause is **reflux** of the stomach's acid contents into the esophagus, **gastroesophageal reflux disease (GERD).**

Hiatal hernia occurs when a portion of the stomach protrudes through the diaphragm alongside the esophagus at the esophageal hiatus. Surgical repair is sometimes necessary, and it is called a hiatal **herniorrhaphy.**

Esophageal varices are varicose veins of the esophagus. They are **asymptomatic** until they rupture, causing massive bleeding and hematemesis. They are a complication of cirrhosis of the liver (*Lesson 5.4*).

Cancer of the esophagus arises from the lining of the tube. Symptoms are **dysphagia** (difficulty swallowing), a burning sensation in the chest, and weight loss. Risk factors include cigarettes, alcohol, betel-nut chewing, and esophageal reflux. The cancer metastasizes to liver, bone, and lung.

WORD ANALYSIS AND DEFINITION

S/ = Suffix P/ = Prefix R/ = Root R/CF = Combining Form

WORD	PRONUNCIATION		ELEMENTS	DEFINITION
asymptomatic **symptomatic** *(with symptoms)*	**A**-simp-toe-**MAT**-ik simp-toe-**MAT**-ik	S/ P/ R/	**-ic** *pertaining to* **a-** *without* **symptomat-** *symptom*	Without any symptoms or abnormalities experienced by the patient. Pertaining to the symptoms of a disease.
dysphagia	dis-**FAY**-jee-ah	P/ R/	**dys-** *difficulty* **-phagia** *swallowing*	Difficulty in swallowing.
emesis **hematemesis**	**EM**-eh-sis he-mah-**TEM**-eh-sis	 R/ R/	emesis Greek *to vomit* **hemat-** *blood* **-emesis** *to vomit*	Vomit. Vomiting of red blood.
esophagus **esophageal (adj)** **esophagitis**	ee-**SOF**-ah-gus ee-**SOF**-ah-**JEE**-al ee-**SOF**-ah-**JI**-tis	 S/ R/ S/	Greek *gullet* **-eal** *pertaining to* **esophag-** *esophagus* **-itis** *inflammation*	Tube linking pharynx to stomach. Pertaining to the esophagus. Inflammation of the lining of the esophagus.
hernia **herniorrhaphy**	**HER**-nee-ah **HER**-nee-**OR**-ah-fee	 S/ R/CF	Latin *rupture* **-rrhaphy** *suture* **herni/o-** *hernia*	Protrusion of a structure through the tissue that normally contains it. Repair of a hernia.
hiatus (noun) **hiatal (adj)**	high-**AY**-tus high-**AY**-tal	 S/ R/	Latin *an aperture* **-al** *pertaining to* **hiat-** *opening*	An opening through a structure. Pertaining to a hiatus.
postprandial	post-**PRAN**-dee-al	S/ P/ R/	**-ial** *pertaining to* **post-** *after* **-prand-** *breakfast*	Following a meal.
reflux	**REE**-fluks	P/ R/	**re-** *back* **-flux** *flow*	Backward flow.
sphincter	**SFINK**-ter		Greek *a band*	A band of muscle that encircles an opening; when it contracts, the opening squeezes closed.
varix **varices (pl)** **varicose (adj)**	**VAIR**-iks **VAIR**-ih-seez **VAIR**-ih-kose	 S/ R/	Latin *dilated vein* **-ose** *full of* **varic-** *varicosity; dilated, tortuous vein*	Dilated, tortuous vein. Characterized by or affected with varices.

EXERCISES

Knowledge of elements *is your best tool for increasing your medical vocabulary. Each of the following terms has an element in bold. Identify that element and give its meaning. Then use any one term in a sentence that is not a definition. Fill in the chart.*

Medical Term	Identity of Element (P, R, CF, or S)	Meaning of Element
1. **re**flux		
2. dys**phagia**		
3. hernio**rrhaphy**		
4. hemat**emesis**		
5. a**symptomat**ic		
6. **hiat**al		
7. post**prand**ial		
8. varic**ose**		
9. esophag**itis**		

Sentence:

____

LESSON 5.3

THE DIGESTIVE SYSTEM

Digestion—Stomach and Small Intestine

OBJECTIVES

The bolus of food that you swallowed has passed down the esophagus. It enters the stomach, and the process of digestion begins in earnest. The stomach continues the mechanical breakdown of the food particles and begins the chemical digestion of protein and fats. But it is in the small intestine that the greatest amount of digestion and absorption of food occurs. The information in this lesson will enable you to:

- **Apply the correct medical terminology for the functions of the stomach and small intestine in digestion.**
- **Explain how food is propelled through the stomach and small intestine.**
- **Use the correct medical terminology to describe the process of digestion.**

You are

. . . a medical interpreter working in Fulwood Medical Center.

You are communicating with

. . . Mr. Xavier Ramirez, a 45-year-old farm worker, who has come to Dr. Susan Lee's primary care clinic.

CASE REPORT 5.2

Mr. Ramirez complains of persistent burning epigastric pain for several months. He has been a chain-smoker since the age of 14. His pain is eased by antacids but quickly returns. He has also been taking aspirin because of joint pain in his fingers while he works.

Dr. Lee has decided to refer him to a gastroenterologist for gastroscopy. Your role is to explain the procedure to Mr. Ramirez and ensure that he keeps his appointments.

DIGESTION: THE STOMACH

The stomach's peristaltic contractions mix different boluses of food together and also push the contents toward the **pylorus** (*Figure 5.13*) to produce a mixture of semidigested food called **chyme.**

The cells of the lining of the stomach secrete (*Figure 5.14*):

1. **Mucus:** Continues to lubricate food and protect the stomach lining.
2. **Hydrochloric acid (HCl):** Breaks up the connective tissue of the chicken and the cell walls of the vegetable that you ingested.
3. **Pepsin** (an active enzyme): Digests the protein in the chicken and vegetable.
4. **Intrinsic factor:** Is essential for the absorption of vitamin B_{12} in the small intestine.
5. **Gastrin** (a chemical): Stimulates the production of HCl and **pepsinogen** and stimulates the peristaltic contractions of the stomach.

A typical meal like your chicken and vegetable takes 3 to 4 hours to exit the stomach as a mixture called chyme. Peristaltic waves squirt 2 to 3 milliliters **(ml)** of the chyme at a time through the **pyloric sphincter** into the **duodenum,** the first part of the small intestine (*Figure 5.13*).

▲ **FIGURE 5.13 Stomach.**

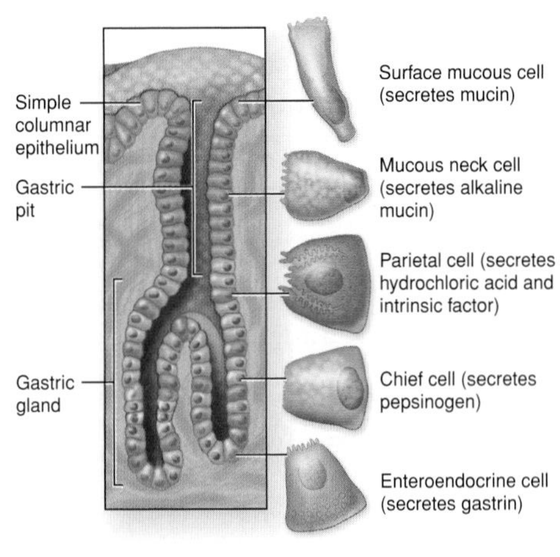

▲ **FIGURE 5.14 Gastric Cells and Their Secretions.**

WORD ANALYSIS AND DEFINITION

S/ = Suffix P/ = Prefix R/ = Root R/CF = Combining Form

WORD	PRONUNCIATION	ELEMENTS		DEFINITION
chyme	KYME		Greek *juice*	Semifluid, partially digested food passed from the stomach into the duodenum.
duodenum	du-oh-**DEE**-num	S/ R/	**-um** *structure* **duoden-** *twelve*	The first part of the small intestine; approximately twelve finger-breadths (9 to 10 inches) in length.
duodenal (adj)	du-oh-**DEE**-nal	S/	**-al** *pertaining to*	Pertaining to the duodenum.
gastrin	**GAS**-trin	S/ R/	**-in** *substance, chemical compound* **gastr-** *stomach*	Hormone secreted in the stomach that stimulates secretion of HCl and increases gastric motility.
hydrochloric acid (HCl)	high-droh-**KLOR**-ic **ASS**-id	S/ R/CF R/	**-ic** *pertaining to* **hydr/o-** *water* **-chlor-** *green*	The acid of gastric juice.
intrinsic factor	in-**TRIN**-sik **FAK**-tor	S/ R/ R/	**-ic** *pertaining to* **intrins-** *on the inside* **factor** *maker*	Makes the absorption of vitamin B_{12} happen.
mucus (noun)	**MYU**-kus		Latin *slime*	Sticky secretion of cells in mucous membranes.
mucous (adj)	**MYU**-kus			Relating to mucus or the mucosa.
mucin	**MYU**-sin			Protein element of mucus.
pepsin	**PEP**-sin		Greek *to digest*	Enzyme produced by the stomach that breaks down protein.
pepsinogen	pep-**SIN**-oh-jen	S/ R/CF	**-gen** *produce* **pepsin/o-** *pepsin*	Converted by HCl in the stomach to pepsin.
pylorus	pie-**LOR**-us	S/ R/	**-us** *pertaining to* **pylor-** *gate, pylorus*	Exit area of the stomach.
pyloric (adj)	pie-**LOR**-ik	S/	**-ic** *pertaining to*	Pertaining to the pylorus.

Abbreviations

HCl hydrochloric acid
ml milliliter

EXERCISES

Grammatical forms *can change with the addition of just one letter. Be aware of the correct spelling of these forms and also how they are used in a sentence. For example,* mucus *is the* noun *(person, place, or thing) form of the term, and* mucous *is the* adjective *(descriptive) form of the term. Insert the correct grammatical form (noun or adjective) in each of the following sentences. Check your spelling when you are finished.*

1. The sticky secretion of the ______________________ membranes is ______________________.
2. A sinus infection will produce a lot of ______________________ in the nasal passages.
3. Some internal organs have a ______________________ lining or covering.

Demonstrate *that you know the difference in the forms of these terms. Create a sentence of patient documentation using any form of these terms.*

Sentence:

____

____

____

DISORDERS OF THE STOMACH

Gastroesophageal reflux disease (GERD) refers to the regurgitation of stomach contents back into the esophagus. The acidity can irritate and ulcerate the lining of the esophagus with bleeding. Scar tissue can cause an esophageal **stricture** and dysphagia.

Keynote

Difficulty in swallowing is called *dysphagia*.

Vomiting can result from overdistension or irritation of any part of the digestive tract. The muscles of the diaphragm and abdominal wall forcefully contract and expel the stomach contents upward into the esophagus and mouth.

Gastritis can be acute or chronic. It is an inflammation of the lining of the stomach, producing symptoms of epigastric pain, feeling of fullness, nausea, and occasional bleeding. Gastritis can be caused by common medications such as aspirin and **NSAIDs,** by radiotherapy and chemotherapy, and by alcohol and smoking. Treatment is directed at removal of the factors causing the gastritis, acid neutralization, and suppression of gastric acid (*see below*).

Peptic ulcers occur in the stomach and duodenum when the mucosal lining breaks down (*Figure 5.15*). Most peptic ulcers are caused by the bacterium *Helicobacter pylori (H. pylori),* which produces enzymes that weaken the protective mucus. These ulcers respond to antibiotics. **Dyspepsia,** epigastric pain with bloating and nausea, is the most common symptom.

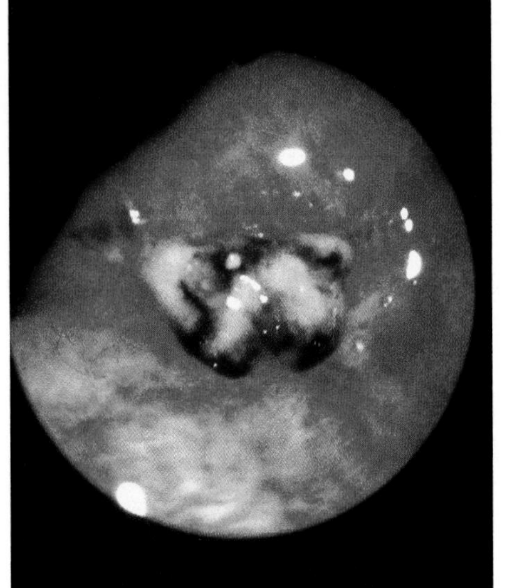

▲ **FIGURE 5.15 Bleeding Peptic Ulcer.**

Gastric ulcers are peptic ulcers occurring in the stomach. Bleeding can occur from erosion of a blood vessel. If untreated, the ulcer can erode through the entire wall, causing a **perforation.**

Gastric cancer can be asymptomatic for a long period and then cause **indigestion, anorexia,** abdominal pain, and weight loss. It affects men twice as often as women. It metastasizes to lymph nodes, liver, peritoneum, chest, and brain. It is usually treated with surgical removal and chemotherapy.

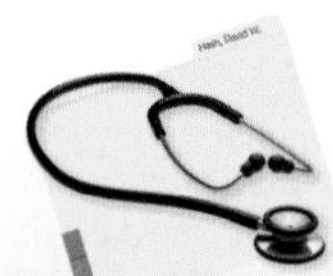

Case Report 5.2 *(continued)*

Gastroscopy on Mr. Ramirez reveals a gastric ulcer.

WORD ANALYSIS AND DEFINITION

S/ = Suffix P/ = Prefix R/ = Root R/CF = Combining Form

WORD	PRONUNCIATION	ELEMENTS		DEFINITION
anorexia	an-oh-**RECK**-see-ah	S/ P/ R/	**-ia** *condition* **an-** *without* **-orex-** *appetite*	Without an appetite; or an aversion to food.
dyspepsia	dis-**PEP**-see-ah	S/ P/ R/	**-ia** *condition* **dys-** *difficult, bad* **-peps-** *digestion*	"Upset stomach," epigastric pain, nausea, and gas.
gastritis	gas-**TRY**-tis	S/ R/	**-itis** *inflammation* **gastr-** *stomach*	Inflammation of the lining of the stomach.
gastroesophageal	**GAS**-troh-ee-sof-ah-**JEE**-al	S/ R/CF R/CF	**-al** *pertaining to* **gastr/o-** *stomach* **-esophag/e-** *esophagus*	Pertaining to the stomach and esophagus.
gastroscope	**GAS**-troh-skope	S/ R/CF	**-scope** *instrument for viewing* **gastr/o-** *stomach*	Endoscope for examining the inside of the stomach.
gastroscopy	gas-**TROS**-koh-pee	S/	**-scopy** *to examine, to view*	Endoscopic examination of the stomach.
indigestion	in-dih-**JESS**-chun	S/ P/ R/	**-ion** *action, condition* **in-** *in, not* **-digest-** *to break down*	Symptoms resulting from difficulty in digesting food.
peptic	**PEP**-tik	S/ R/	**-ic** *pertaining to* **pept-** *digest*	Relating to the stomach and duodenum.
perforation	per-foh-**RAY**-shun	S/ R/	**-ion** *action, condition* **perforat-** *bore through*	A hole through the wall of a structure.
stricture	**STRICK**-shur		Latin *draw tight*	Narrowing of a tube.

EXERCISES

Deconstruct *the following medical terms into basic elements. The definitions of the elements will help you understand the meaning of the term. Not every type of element will appear in every term. First slash the terms in column 1 into elements; then complete the chart.*

Medical Term	Meaning of Prefix	Meaning of Root/CF	Meaning of Suffix	Definition of Medical Term
gastroesophageal				
peptic				
anorexia				
gastroscopy				
indigestion				
dyspepsia				
gastritis				
perforation				
gastroscope				

SMALL INTESTINE

The small intestine, called *small* because of its diameter, finishes the process of chemical digestion and is responsible for the **absorption** of most of the nutrients. The small intestine extends from the pylorus of the stomach to the beginning of the large intestine, and it has three segments (*Figure 5.16*):

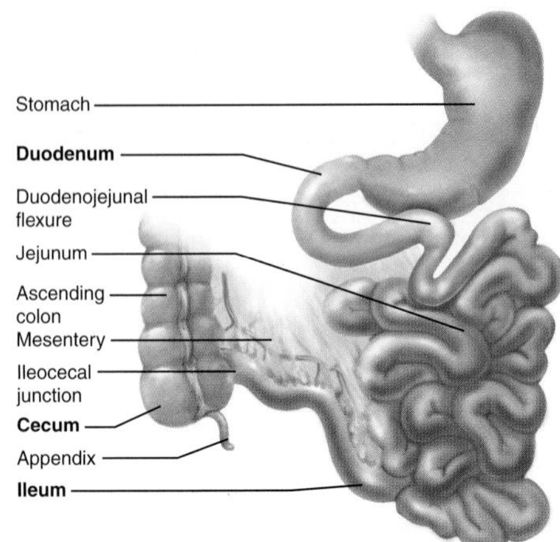

▲ **FIGURE 5.16 Small Intestine.**

1. The **duodenum** is the first 9 to 10 inches of the small intestine. It receives chyme from the stomach, together with pancreatic juices and bile.

2. The **jejunum** makes up about 40% of the small intestine's length. It is the primary region for chemical digestion and nutrient absorption.

3. The **ileum** makes up about 55% of the small intestine's length. It ends at the **ileocecal** valve, a **sphincter** that controls entry into the large intestine.

Digestion in the Small Intestine

After leaving the stomach as chyme, the food spends 3 to 5 hours in the small intestine, where most of the nutrients are absorbed.

Peristaltic movements of the small intestine have three functions:

1. **Mix** chyme with intestinal and pancreatic juices and with bile to facilitate absorption and digestion.

2. **Churn** chyme to make contact with the mucosa for digestion and absorption.

3. **Move** the residue toward the large intestine for eventual elimination.

From the duodenum to the middle of the ileum, the lining of the small intestine is thrown into circular folds, called **plicae.** Along their surface, the plicae have tiny finger-like **villi.** The plicae and villi increase the surface area over which secretions can act on the food and through which nutrients can be absorbed. The folds also act as "speed bumps" to slow down the movement of chyme through the small intestine. The cells at the tip of the villi are shed and renewed every three or four days.

WORD ANALYSIS AND DEFINITION

S/ = Suffix P/ = Prefix R/ = Root R/CF = Combining Form

WORD	PRONUNCIATION	ELEMENTS		DEFINITION
absorption	ab-**SORP**-shun	S/ R/	**-ion** *action, condition* **absorpt-** *to swallow*	Uptake of nutrients and water by cells in the GI tract.
cecum (noun) **cecal (adj)**	**SEE**-kum **SEE**-kal	S/ R/ S/	**-um** *structure* **cec-** *cecum* **-al** *pertaining to*	Blind pouch that is the first part of the large intestine. Pertaining to the cecum.
ileum **ileocecal**	**ILL**-ee-um **ILL**-ee-oh-**SEE**-cal	S/ R/ S/ R/CF R/	**-um** *structure* **ile-** *ileum* **-al** *pertaining to* **ile/o-** *ileum* **cec-** *cecum*	Third portion of small intestine. Pertaining to the junction of the ileum and cecum.
jejunum (noun) **jejunal (adj)**	je-**JEW**-num je-**JEW**-nal	S/ R/ S/	**-um** *structure* **jejun-** *jejunum* **-al** *pertaining to*	Segment of small intestine between the duodenum and the ileum. Pertaining to the jejunum.
plica **plicae (pl)**	**PLEE**-cah **PLEE**-key		Latin *fold*	Fold in a mucous membrane.
villus **villi (pl)**	**VILL**-us **VILL**-eye		Latin *shaggy hair*	Thin, hairlike projection, particularly of a mucous membrane lining a cavity.

EXERCISES

Practice the language of gastroenterology. *Review this spread and the WAD before you start the exercise. Use the terms found there to answer the following questions.*

1. Write the three segments of the small intestine in their correct order of location in the intestine.

_______________, _______________, _______________

2. A term can often designate more than one organ (example: *gastrointestinal*).

Find a similar kind of term in the WAD above.

____

This term means:

____

____

The two organs referred to in this term are:

_______________ and _______________

3. **Patient education:**

Explain to your patient the function of a *sphincter* in the intestine.

____

____

____

____

LESSON 5.4 Digestion—Liver, Gallbladder, and Pancreas

THE DIGESTIVE SYSTEM

OBJECTIVES

The liver and pancreas secrete enzymes that are responsible for most of the digestion that occurs in the small intestine. You will need to be able to:

- **Apply correct medical terminology to describe the functions and common disorders of the liver, gallbladder, bile ducts, and pancreas.**

You are

. . . a coder in the Health Information Management department at Fulwood Medical Center.

You are communicating with

. . . Dr. Susan Lee, and you are questioning the documentation of the care given to Mrs. Sandra Jacobs.

CASE REPORT 5.3

Mrs. Jacobs, a 46-year-old mother of four, presented in Dr. Lee's primary care clinic with episodes of cramping pain in her upper abdomen associated with nausea and vomiting. Physical examination reveals an obese white woman with tenderness over her gallbladder. Her BP is 170/90 and she has slight pedal edema. A **provisional diagnosis** of gallstones has been made. She has been referred for an ultrasound examination and an appointment has been made to see Dr. Walsh in the surgery department.

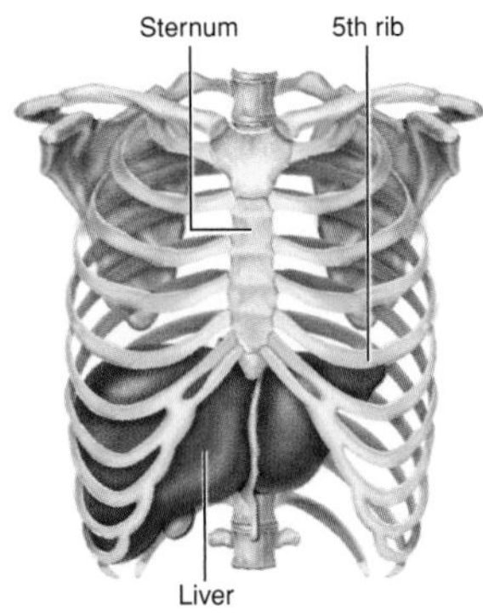

▲ **FIGURE 5.17 Location of Liver.**

Keynote

Only the production of bile relates the liver to digestion.

Keynote

Vaccines are available to prevent hepatitis A and B.

THE LIVER

The **liver,** the body's largest internal organ, is a complex structure located under the right ribs below the diaphragm (*Figure 5.17*).

The liver has multiple functions, including to:

1. **Manufacture** and **excrete bile.** Although it contains no digestive enzymes, bile plays key roles in digestion.
2. **Remove** the pigment **bilirubin** from the bloodstream, and excrete it in bile. Bilirubin is produced as a breakdown product of the hemoglobin in red blood cells (*Chapter 6*).
3. **Remove** excess glucose (sugar) from the blood and store it as **glycogen.**
4. **Convert** proteins and fats into glucose.
5. **Store** fat and the fat-soluble vitamins A, D, E, and K.
6. **Manufacture** blood proteins, including those necessary for clotting.
7. **Remove** toxins from the blood.

A major reason that the liver can perform all these functions is that venous blood is returned from all the intestines directly to the liver.

Disorders of the Liver

Hepatitis is an inflammation of the liver causing **jaundice.** Viral hepatitis is the most common cause of hepatitis and is related to three major types of virus:

1. **Hepatitis A virus (HAV)** is highly contagious and causes a mild to severe infection. It is transmitted by contaminated food.
2. **Hepatitis B virus (HBV),** or serum hepatitis, is transmitted through contact with blood, semen, vaginal secretions, or saliva, as well as by a needle prick and the sharing of contaminated needles.
3. **Hepatitis C virus (HCV)** is the most common blood-borne infection in the United States. Like HBV, the disease is transmitted by blood and body fluids.

WORD ANALYSIS AND DEFINITION

S/ = Suffix P/ = Prefix R/ = Root R/CF = Combining Form

WORD	PRONUNCIATION	ELEMENTS		DEFINITION
bile	BILE		Latin *bile*	Fluid secreted by the liver into the duodenum.
bile acids	BILE **AH**-sids			Steroids synthesized from cholesterol.
biliary (adj)	**BILL**-ee-air-ee	S/ R/CF	**-ary** *pertaining to* **bil/i-** *bile*	Pertaining to bile or the biliary tract.
bilirubin	bill-ee-**RU**-bin	S/ R/CF	**-rubin** *rust colored* **bil/i-** *bile*	Bile pigment formed in the liver from hemoglobin.
cirrhosis	sir-**ROE**-sis	S/ R/	**-osis** *condition* **cirrh-** *yellow*	Extensive fibrotic liver disease.
glycogen	**GLYE**-koh-gen	S/ R/CF	**-gen** *produce, create* **glyc/o-** *sugar, glycogen*	The body's principal carbohydrate reserve, stored in the liver and skeletal muscle.
hemangioma	he-**MAN**-jee-oh-mah	S/ R/ R/	**-oma** *tumor* **hem-** *blood* **-angi-** *blood vessel*	Abnormal mass of proliferating blood vessels.
hepatic	hep-**AT**-ik	S/ R/	**-ic** *pertaining to* **hepat-** *liver*	Pertaining to the liver.
hepatitis	hep-ah-**TIE**-tis	S/	**-itis** *inflammation*	Inflammation of the liver.
jaundice	**JAWN**-dis		French *yellow*	Yellow staining of tissues with bile pigments, including bilirubin.
liver	**LIV**-er		Old English *liver*	Body's largest organ, located in the right upper quadrant of the abdomen.
provisional diagnosis (also called preliminary diagnosis)	pro-**VISH**-un-al die-ag-**NO**-sis	S/ R/	**-al** *pertaining to* **provision-** *provide*	A temporary diagnosis pending further examination or testing.

In addition, **hepatitis D** can occur in association with hepatitis B, making the infection worse. **Hepatitis E** is similar to hepatitis A and occurs mostly in underdeveloped countries.

Chronic hepatitis occurs when the acute hepatitis is not healed after 6 months. It progresses slowly, can last for years, and is difficult to treat.

Cirrhosis of the liver is a chronic irreversible disease, replacing normal liver cells with hard, fibrous scar tissue. Its most common cause is alcoholism. There is no known cure.

Cancer of the liver as a primary cancer usually arises in patients with chronic liver disease, usually from HBV infection.

Abbreviations

HAV hepatitis A virus
HBV hepatitis B virus
HCV hepatitis C virus

EXERCISES

Elements: *Use this exercise as a quick review of the elements in the WAD. Circle the correct choices, and then answer the question.*

1. In the term *hepatic,* the root means:
   pancreas stomach liver
2. The suffix in *cirrhosis* indicates this is:
   a condition a surgical excision a structure
3. *Bilirubin* is the color of:
   blood milk rust
4. The root in *cirrhosis* indicates a:
   color size location
5. Which three terms in the WAD refer to color?
   ____________, ____________, ____________

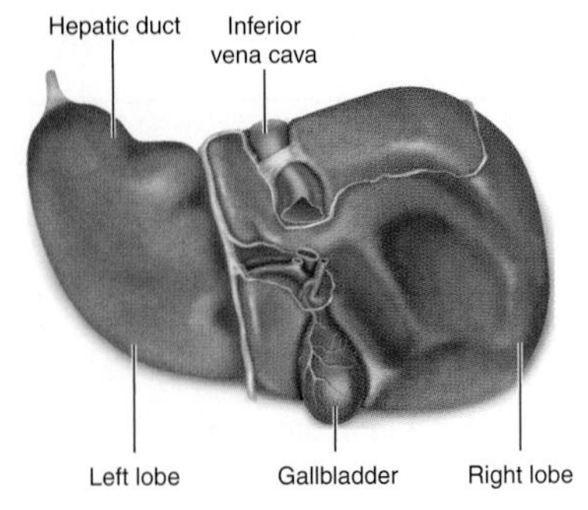

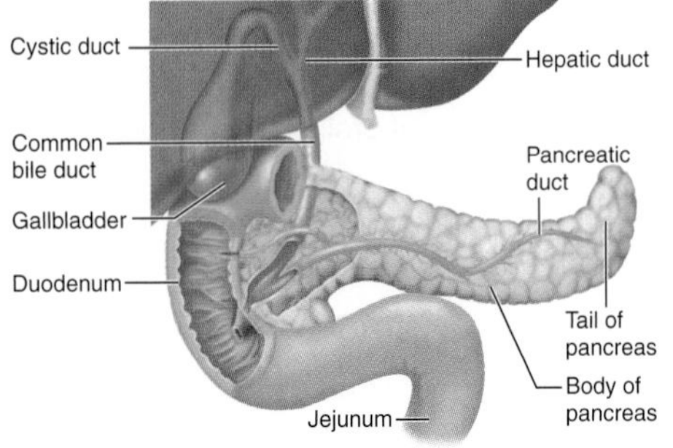

▲ **FIGURE 5.18** (a) Underside of liver. (b) Anatomy of gallbladder, pancreas, and biliary tract.

Keynote

Between meals, the gallbladder absorbs water and electrolytes from the bile and concentrates it 10 to 20 times.

Keynote

Jaundice is caused by deposits of bilirubin in the tissues.

▲ **FIGURE 5.19 Gallstones.**

GALLBLADDER, BILIARY TRACT, AND PANCREAS

Gallbladder and Biliary Tract

On the underside of the liver is the **gallbladder,** which stores and concentrates the bile that the liver produces. The **cystic duct** from the gallbladder joins with the **hepatic duct** to form the **common bile duct.** The system of ducts to get the bile from the liver to the duodenum is called the **biliary tract** (*Figure 5.18*).

Disorders of the Gallbladder

Gallstones (cholelithiasis) can form in the gallbladder from excess cholesterol, bile salts, and bile pigment (*Figure 5.19*). The stones can vary in size and number. Risk factors are obesity, high-cholesterol diets, multiple pregnancies, and rapid weight loss.

Case Report 5.3 (continued)

Mrs. Jacobs presented in the emergency department with the classic gallstone symptoms of severe waves of right upper quadrant pain (biliary colic), nausea, and vomiting.

If small stones become impacted in the common bile duct, this is called **choledocholithiasis.** This can cause biliary colic and jaundice (*see below*).

Cholecystitis is an acute or chronic inflammation of the gallbladder, usually associated with cholelithiasis and obstruction of the cystic duct with a stone.

Jaundice (icterus) is a symptom of many different diseases in the biliary tract and liver. It is a yellow discoloration of the skin (*Chapter 3*) and sclera of the eyes (*Chapter 4*) caused by deposits of bilirubin just below the outer layers of the skin.

THE PANCREAS

The pancreas is a spongy gland, the head of which is encircled by the duodenum (*Figure 5.18b*). Most of the pancreas secretes digestive juices, but smaller areas of **pancreatic islet cells** secrete the hormones **insulin** and **glucagon** (*Chapter 13*).

The pancreas is called an **exocrine** gland, and the pancreatic digestive juices it forms are excreted through the pancreatic duct. The pancreatic duct joins the common bile duct shortly before it opens into the duodenum (*Figure 5.18b*). Pancreatic and bile juices then enter the duodenum. Other pancreatic cells secrete the hormones insulin and glucagon, which go directly into the bloodstream. This part of the pancreas is an **endocrine** gland (*Figure 5.20*).

Pancreatic juices contain alkaline electrolytes, which help neutralize the acid chyme as it comes from the stomach, and enzymes, which break down starches, fats, and proteins into their simpler elements.

Disorders of the Pancreas

Pancreatitis is inflammation of the pancreas. The acute disease ranges from a mild, self-limiting episode to an acute life-threatening emergency. In the chronic form there is a progressive destruction of pancreatic tissue leading to malabsorption and diabetes.

Pancreatic cancer is the fourth leading cause of cancer-related death. Treatment is surgical resection of the cancer. The prognosis is poor.

Cystic fibrosis (CF) is an inherited disease that becomes apparent in infancy or childhood. The pancreas, liver, intestines, sweat glands, and lungs all produce abnormally thick mucous secretions. There is malabsorption of fat and protein, leading to large, bulky, foul-smelling stools. Problems with thick mucous secretions in the lungs lead to chronic lung disease.

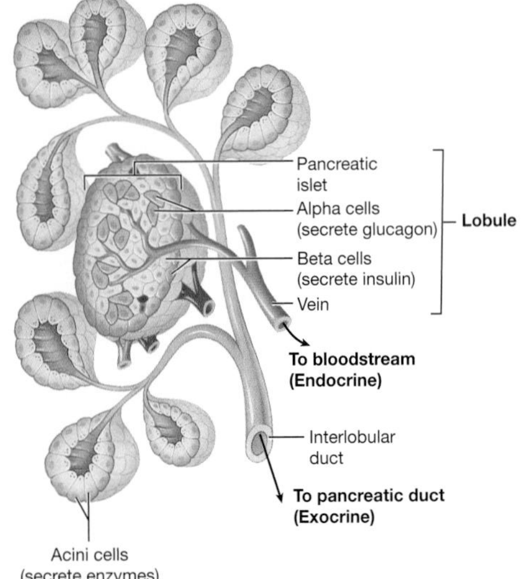

◀ **FIGURE 5.20 Exocrine and Endocrine Aspects of Pancreas.**

WORD ANALYSIS AND DEFINITION

S/ = Suffix P/ = Prefix R/ = Root R/CF = Combining Form

WORD	PRONUNCIATION	ELEMENTS		DEFINITION
cholecystitis	**KOH**-leh-sis-**TIE**-tis	S/ R/CF R/	**-itis** *inflammation* **chol/e** - *bile* **-cyst-** *bladder*	Inflammation of the gallbladder.
cholecystectomy	**KOH**-leh-sis-**TECK**-toe-me	S/	**-ectomy** *surgical excision*	Surgical removal of the gallbladder.
choledocholithiasis	koh-leh-**DOH**-koh-lih-**THIGH**-ah-sis	S/ R/CF R/	**-iasis** *condition* **choledoch/o-** *common bile duct* **-lith-** *stone*	Presence of a gallstone in the common bile duct.
cholelithiasis	**KOH**-leh-lih-**THIGH**-ah-sis	S/ R/CF R/	**-iasis** *condition* **chol/e-** *bile* **-lith-** *stone*	Condition of having bile stones (gallstones).
cholelithotomy	**KOH**-leh-lih-**THOT**-oh-me	S/	**-otomy** *surgical incision*	Surgical removal of a gallstone(s).
endocrine	**EN**-doh-krin	P/ R/	**endo-** *within, inside* **-crine** *secrete*	A gland that produces an internal or hormonal substance and secretes it into the bloodstream.
exocrine	**EK**-soh-krin	P/	**exo-** *outward, outside*	A gland that secretes substances outwardly through excretory ducts.
gallstone	**GAWL**-stone	R/ R/	**gall-** *bitter* **-stone** *pebble*	Hard mass of cholesterol, calcium, and bilirubin that can be formed in the gallbladder and bile duct.
gallbladder	**GAWL**-blad-er		bladder, Old English *receptacle*	Receptacle on the inferior surface of the liver for storing bile.
glucagon	**GLU**-kah-gon	R/ R/	**gluc-** *glucose, sugar* **-agon** *to fight*	Hormone that mobilizes glucose from body storage.
insulin	**IN**-syu-lin	S/ R/	**-in** *chemical compound* **insul-** *island*	Pancreatic hormone that suppresses blood glucose levels and transports glucose into cells.
pancreas	**PAN**-kree-as		Greek *sweetbread*	Lobulated gland, the head of which is tucked into the curve of the duodenum.
pancreatic (adj)	pan-kree-**AT**-ik	S/ R/	**-ic** *pertaining to* **pancreat-** *pancreas*	Pertaining to the pancreas.
pancreatitis	**PAN**-kree-ah-**TIE**-tis	S/	**-itis** *inflammation*	Inflammation of the pancreas.

Keynote

The pancreas is the only organ that is both an endocrine and an exocrine gland.

Abbreviation

CF	cystic fibrosis

EXERCISES

Roots or combining forms *can remain the same, but the addition of other elements will construct entirely different medical terms. Use this group of elements, in addition to chol/e, to form the medical terms that are defined. Some elements you will use more than once; some elements you will not use at all. Fill in the blanks.*

The element chol/e *means* ______________________.

Add these elements to chol/e *to form the medical terms defined below:*

lith choledoch/o iasis chol ectomy osis cyst otomy cyst/o itis

1. condition of bile stones (gallstones):

   ______________/ ______________/ ______________

2. surgical removal of gallbladder:

   ______________/ ______________/ ______________

3. gallstone in the common bile duct:

   ______________/ ______________/ ______________

4. surgical incision into gallbladder to remove gallstones:

   ______________/ ______________/ ______________

LESSON 5.5 Absorption and Malabsorption

OBJECTIVES

In the previous lessons in this chapter, you learned about the digestive secretions of the different segments of the digestive tract, liver, and pancreas. This information can now be brought together to review the overall process of digestion so that you will be able to:

- **Use correct medical terminology to explain the chemical digestion and absorption of proteins, carbohydrates, and fats.**
- **Select the correct medical terminology to describe disorders of chemical digestion, absorption, and malabsorption.**

You are

. . . a dietician working at Fulwood Medical Center.

You are communicating with

. . . Mrs. Jan Stark, a 36-year-old pottery maker, and her husband, Mr. Tom Stark.

CASE REPORT 5.4

From her medical record, you see that Mrs. Stark has been referred to you by Dr. Cameron Grabowski, a gastroenterologist.

For the past 10 years she has had spasmodic episodes of **diarrhea** and **flatulence** associated with severe headaches and fatigue. During those episodes her stools were greasy and pale. She has seen several physicians who have recommended low-fat, high-carbohydrate diets without relief.

Dr. Grabowski has performed an intestinal biopsy through oral **endoscopy** and diagnosed **celiac disease**. This condition is a sensitivity to the protein **gluten** that is found in wheat, rye, barley, and oats. He has asked you to ensure that Mrs. Stark accepts a diet free of gluten-containing foods such as breads, cereals, cookies, and beer.

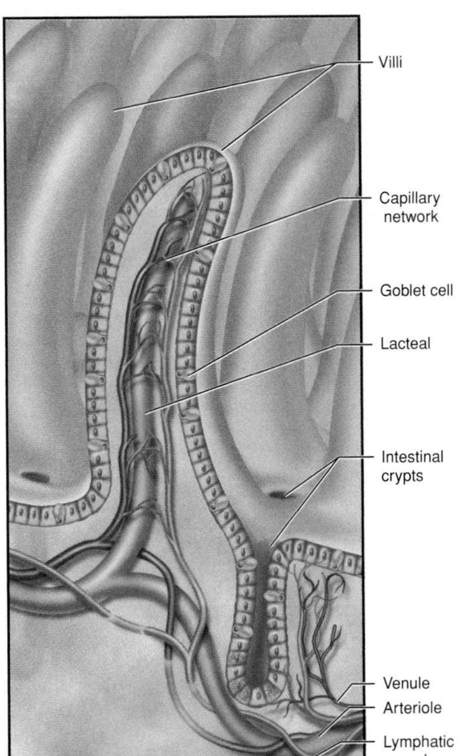

▲ **FIGURE 5.21 Intestinal Villi.**

CHEMICAL DIGESTION, ABSORPTION, AND TRANSPORT

Carbohydrates

In the small intestine, carbohydrates, such as **starches,** are broken down into simple sugars, such as glucose and fructose, which are taken up by the lining cells and transferred to the capillaries of the villi (*Figure 5.21*). They are then carried by the **portal** vein to the liver, where the nonglucose sugars are converted to glucose.

Glucose is the major source of energy for all cells.

Glycogen, the storage form of carbohydrate, is found in the liver and skeletal muscle. Glycogen in muscles supplies glucose during high-intensity and endurance exercise.

The leading sources of carbohydrate intake are bread, soft drinks, cookies, cakes, donuts, syrups, jams, potatoes, and rice.

WORD ANALYSIS AND DEFINITION

S/ = Suffix P/ = Prefix R/ = Root R/CF = Combining Form

WORD	PRONUNCIATION		ELEMENTS	DEFINITION
celiac disease	**SEE**-lee-ack diz-**EEZ**	S/ R/ P/ R/	**-ac** *pertaining to* **celi-** *abdomen* **dis-** *apart* **-ease** *normal function*	Disease caused by sensitivity to gluten.
diarrhea	die-ah-**REE**-ah	P/ R/	**dia-** *complete* **-rrhea** *flow, discharge*	Abnormally frequent and loose stools.
endoscopy **endoscope**	en-**DOS**-koh-pee **EN**-doh-skope	P/ R/ R/	**endo-** *inside, within* **-scopy** *to examine, view* **-scope** *instrument for viewing*	The use of an endoscope. Instrument used to examine the interior of a tubular or hollow organ.
flatulence	**FLAT**-you-lents	S/ R/	**-ence** *forming* **flatul-** *excessive gas, flatus*	Excessive amount of gas in the stomach and intestines.
flatus	**FLAY**-tus		Latin *blowing*	Gas or air expelled through the anus.
gluten	**GLU**-ten		Latin *glue*	Insoluble protein found in wheat, barley, and oats
glycogen	**GLYE**-koh-jen	S/ R/CF	**-gen** *to produce* **glyc/o** *sugar/glucose*	The body's principal carbohydrate reserve, stored in the liver and skeletal muscle.
malabsorption	mal-ab-**SORP**-shun	S/ P/ R/	**-ion** *action, condition* **mal-** *bad, difficult* **-absorpt-** *swallow, take in*	Inadequate gastrointestinal absorption of nutrients.
portal vein	**POR**-tal VANE		portal, Latin *gate* vein, Latin *vein*	The vein that carries blood from the intestines to the liver.
starch	STARCH		Anglo-Saxon *stiffen*	Complex carbohydrate made of multiple units of glucose attached together.

EXERCISES

Construct *the correct medical term to match the definitions. Review the WAD before you fill in the blanks.*

1. inadequate gastrointestinal absorption of nutrients

   ______________ / ______________ / ______________
   P R/CF S

2. abnormally frequent and loose stools

   ______________ / ______________ / ______________
   P R/CF S

3. examination of a hollow structure with a special instrument

   ______________ / ______________ / ______________
   P R/CF S

4. excessive amount of gas in stomach and intestines

   ______________ / ______________ / ______________
   P R/CF S

5. the instrument used for endoscopy

   ______________ / ______________ / ______________
   P R/CF S

Study Hint

Endoscope is a generic (general) term that means any instrument *(scope)* used to examine the inside *(endo)* of a tubular or hollow organ. The instrument obtains its specific name from the organ it is used to examine. Thus, an instrument used to view a stomach is a gastroscope specifically, but it is also an endoscope in general.

CHEMICAL DIGESTION, ABSORPTION, AND TRANSPORT (continued)

Proteins

Keynote

Proteins are broken down into amino acids.

Proteins arrive in the duodenum and small intestine only 10% to 20% digested. Enzymes produced by cells in the small intestine, together with the pancreatic enzyme trypsin, break down the remaining proteins into **amino acids.**

The amino acids are carried away in the blood and are transported to cells all over the body to be used as building blocks for new tissue formation.

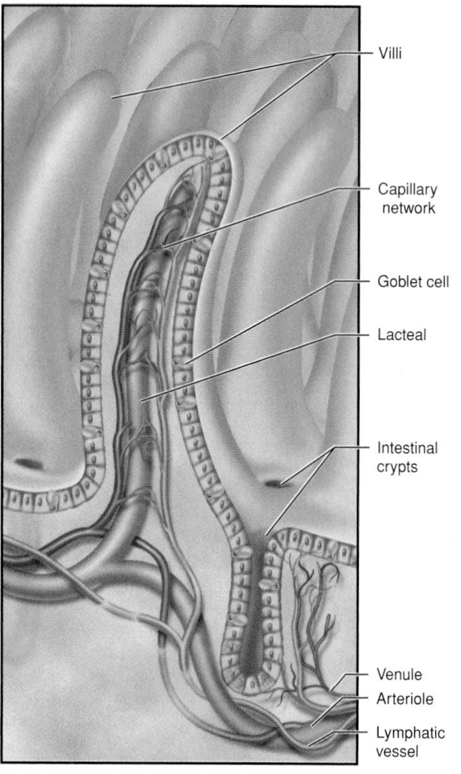

▲ **FIGURE 5.22 Intestinal Villi.**

Lipids

Lipids (fats) enter the duodenum and small intestine as large globules. These have to be **emulsified** by the bile salts into smaller droplets so that pancreatic lipase can digest the fats into very small droplets of free **fatty acids** and **monoglycerides.**

The very small droplets are taken up by the **lacteals** inside the villi and then into the lymphatic system. The white, fatty lymphatic **chyle** eventually reaches the thoracic duct and is transferred into the bloodstream (*Chapter 7*). It is stored in adipose tissue. The fat-soluble vitamins A, D, E, and K are absorbed with the lipids.

Water

Water is 92% absorbed by the small intestine and taken into the bloodstream through the capillaries in the villi (*Figure 5.22*). Water-soluble vitamins, C and the B complex, are absorbed with water except for B_{12}. This is a large molecule that has to bind with intrinsic factor in the stomach so that cells in the distant ileum can receive it and absorb it into the bloodstream.

Water has no caloric value. It constitutes approximately 60% of body weight (about 10 gallons). You can survive six to eight weeks without food, but only a few days without water. This is because water is an integral part of all tissues, but is not stored in a particular place from which it can be released.

Minerals

Keynote

Electrolytes are minerals.

Minerals are absorbed along the whole length of the small intestine. Iron and calcium are absorbed according to the body's needs. The other minerals are absorbed regardless of need, and the kidneys excrete the surplus.

The major minerals are sodium (**Na**), potassium (**K**), calcium (**Ca**), and magnesium (**Mg**).

Abbreviations

Ca	calcium
K	potassium
Mg	magnesium
Na	sodium

WORD ANALYSIS AND DEFINITION

S/ = Suffix P/ = Prefix R/ = Root R/CF = Combining Form

WORD	PRONUNCIATION	ELEMENTS		DEFINITION
amino acid	ah-**ME**-no **ASS**-id	R/CF	**amin/o-** *nitrogen containing* acid, Latin *sour*	The basic building block for protein.
chyle	KYLE		Greek *juice*	A milky fluid that results from the digestion and absorption of fats in the small intestine.
emulsify	ee-**MUL**-sih-fye	S/ R/	**-ify** *to become* **emuls-** *suspend in a liquid*	Break up into very small droplets to suspend in a solution (emulsion).
emulsion (noun)	ee-**MUL**-shun	S/	**-ion** *condition, action*	The system that contains small droplets suspended in a liquid.
lacteal	**LAK**-tee-al	S/ R/CF	**-al** *pertaining to* **lact/e-** *milk*	A lymphatic vessel carrying chyle away from the intestine.
lipid	**LIP**-id		Greek *fat*	General term for all types of fatty compounds; e.g., cholesterol, triglycerides, and fatty acids.
mineral	**MIN**-er-al	S/ R/	**-al** *pertaining to* **miner-** *mines*	Inorganic compound usually found in the earth's crust.

EXERCISES

Seek and find *the correct medical terms to fill in the blanks. The answers to the following questions can all be found on the two-page spread open in front of you.*

1. What proteins break down into: ______________________
2. Another name for minerals: ______________________
3. Milky fluid that results from the digestion and absorption of fats: ______________________
4. Term used to describe all types of fatty compounds: ______________________
5. Break up into small droplets to suspend in a liquid: ______________________
6. Lymphatic vessel that carries chyle away from the intestine: ______________________
7. Inorganic compounds found in the earth's crust: ______________________
8. Number of calories in an ounce of water: ______________________
9. Percentage of the body's weight that is water: ______________________
10. List the four major minerals: ______________________

DISORDERS OF ABSORPTION

The term **malabsorption syndromes** refers to a group of diseases in which intestinal absorption of nutrients is impaired.

Case Report 5.4 *(continued)*

Mrs. Jan Stark, who presented in the emergency department (ED) with **dehydration**, was found to have celiac disease, a sensitivity to the protein gluten. This disease involves the destruction of epithelial cells of the digestive tract lining, so intestinal enzymes are not being produced and absorption is not taking place.

The diagnosis was made through an intestinal biopsy by oral endoscopy. A diet free of gluten-containing foods, such as breads, cereals, cookies, and beer, relieved her symptoms.

Abbreviation	
ED	emergency department

Malnutrition can arise from malabsorption but can also result from insufficient food in famine and poverty areas of the world. Malnutrition can also result from loss of appetite in people with cancer or with terminal illness.

Keynote

In malnutrition, the body breaks down its own tissues to meet its nutritional and metabolic needs.

Lactose intolerance occurs when the small intestine is not producing enough of the enzyme **lactase** to break down the milk sugar lactose. The result is diarrhea and cramps. Lactase can be taken in pill form before eating dairy products, and milk products can be avoided.

Keynote

Milk sugar is lactose. The enzyme lactase breaks down lactose into glucose.

Crohn disease (or **regional enteritis**) is an inflammation of the small intestine, frequently in the ileum and occasionally also in the large intestine. The symptoms are abdominal pain, diarrhea, fatigue, and weight loss. There is no cure.

Constipation occurs when fecal movement through the large intestine is slow and thus too much water is reabsorbed by the large intestine. The feces become hardened. Factors causing constipation are lack of dietary fiber, lack of exercise, and emotional upset.

Keynote

Diarrhea is caused by irritation of the intestinal lining that causes feces to pass through the intestine too quickly for adequate amounts of water to be reabsorbed.

Gastroenteritis (stomach "flu") is an infection of the stomach and intestine that can be caused by a large number of bacteria and viruses. It causes vomiting, diarrhea, and fever. An outbreak of gastroenteritis can sometimes be traced to contaminated food or water.

Dysentery is a severe form of bacterial gastroenteritis with blood and mucus in frequent, watery stools.

Malnutrition, malabsorption, and severe forms of diarrhea and vomiting can cause dehydration and electrolyte imbalance, possibly leading to coma and death.

WORD ANALYSIS AND DEFINITION

S/ = Suffix P/ = Prefix R/ = Root R/CF = Combining Form

WORD	PRONUNCIATION	ELEMENTS		DEFINITION
constipation	kon-stih-**PAY**-shun	S/ R/	**-ation** *process* **constip-** *press together*	Hard, infrequent bowel movements.
Crohn disease	KRONE diz-**EEZ**		Burrill Crohn, New York gastroenterologist, 1884–1983	Narrowing and thickening of terminal small bowel.
(also known as **regional enteritis**)	**REE**-jun-al en-ter-**I**-tis	S/ R/	**-itis** *inflammation* **enter-** *intestine*	
dehydration	dee-high-**DRAY**-shun	S/ P/ R/	**-ation** *a process* **de-** *without* **-hydr-** *water*	Process of losing body water.
dysentery	**DIS**-en-tare-ee	P/ R/	**dys-** *bad, difficult* **-entery** *condition of the intestine*	Disease with diarrhea, bowel spasms, fever, and dehydration.
gastroenteritis	**GAS**-troh-en-ter-**I**-tis	S/ R/ R/CF	**-itis** *inflammation* **-enter-** *intestine* **gastr/o-** *stomach*	Inflammation of the stomach and intestines.
lactose	**LAK**-toes		Latin *milk sugar*	The disaccharide found in cow's milk.
lactase	**LAK**-tase	S/ R/	**-ase** *enzyme* **lact-** *milk*	Enzyme that breaks down lactose (milk sugar) to glucose and galactose.
malnutrition	mal-nyu-**TRISH**-un	S/ P/ R/	**-ion** *process* **mal-** *bad, difficult, inadequate* **-nutrit-** *nourishment*	Inadequate nutrition from poor diet or inadequate absorption of nutrients.

EXERCISES

Elements build terms. *Notice that some terms can be formed with only a prefix and a root. Others are formed with only a root and a suffix. The only element that needs to be present in every term is a root or combining form. The definitions for the terms are given below. Write the correct element on the line to build the term, and label the element (P, R, CF, or S) under the line.*

1. inflammation of the stomach and intestines

_______________/_______________/_______________

2. process of losing body water

_______________/_______________/_______________

3. severe form of bacterial gastroenteritis

_______________/_______________/_______________

4. inadequate nutrition from poor diet

_______________/_______________/_______________

5. enzyme that breaks down lactose to glucose

_______________/_______________/_______________

LESSON 5.6 The Large Intestine and Elimination

THE DIGESTIVE SYSTEM

OBJECTIVES

After the nutrients have been digested and absorbed in the small intestine, the residual materials have to be prepared in the large intestine for elimination from the body. The information in this lesson will enable you to:

- **Apply correct medical terminology to describe the structure and functions of the large intestine.**
- **Select the correct medical terminology to describe disorders of the large intestine.**

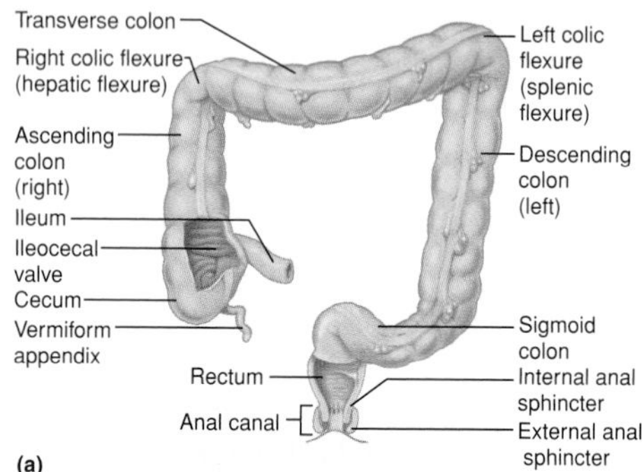

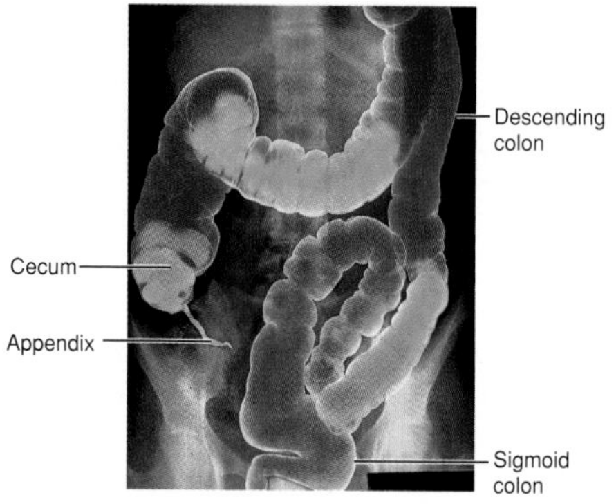

▲ **FIGURE 5.23 Large Intestine.** (a) Surface anatomy. (b) X-ray of large intestine following barium enema.

Keynote

A sphincter is a ring of smooth muscle that forms a one-way valve.

Keynote

Colon: Ascending to transverse to descending to sigmoid to rectum.

STRUCTURE AND FUNCTIONS OF THE LARGE INTESTINE

Structure of the Large Intestine

The **large intestine** is so named because its diameter is much greater than that of the small intestine. Its location forms a perimeter in the abdominal cavity around the central mass of the small intestine.

At the junction between the small and large intestines, a ring of smooth muscle called the **ileocecal sphincter** forms a one-way valve. This allows chyme to pass into the large intestine and prevents the contents of the large intestine from backing into the ileum.

At the beginning of the large intestine, the **cecum** is a pouch in the right lower quadrant of the abdomen. A narrow tube with a closed end, the **vermiform appendix,** projects downward from the cecum (*Figure 5.23a and 5.23b*). The function of the appendix is not known.

The ascending **colon** begins at the cecum and extends upward until, just underneath the liver, it makes a sharp turn at the hepatic **flexure** and becomes the transverse colon. At the left side of the abdomen, near the spleen at the splenic flexure, the transverse colon turns downward to form the descending colon. At the pelvic brim, the descending colon makes an S-shaped curve, the **sigmoid** colon, which descends in the pelvis to become the **rectum** and then the **anal canal.**

The **rectum** has three transverse folds, rectal valves that enable it to retain **feces** while passing gas (flatus).

The anal canal (*Figure 5.24*) is the last 1 to 2 inches of the large intestine, opening to the outside as the **anus.** An internal anal sphincter, composed of smooth muscle from the intestinal wall, and an external anal sphincter, composed of skeletal muscle that you can control voluntarily, guard the exit of the anus.

Functions of the Large Intestine

- **Absorption** of water and electrolytes. The large intestine receives more than 1 liter (1,000 ml) of chyme each day from the small intestine and reabsorbs water and electrolytes to reduce the volume to 100 to 150 ml of feces to be eliminated by **defecation.**
- **Secretion** of mucus that protects the intestinal wall and holds particles of fecal matter together.
- **Digestion,** by the bacteria that inhabit the large intestine, of any food remnants that have escaped the digestive enzymes of the small intestine.
- **Peristalsis** that in the large intestine happens only a few times a day to produce mass movements toward the rectum.
- **Elimination** of materials that were not digested or absorbed.

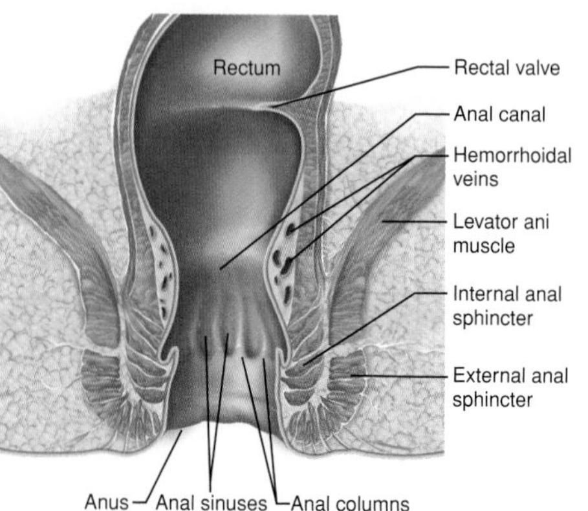

◀ **FIGURE 5.24 Anal Canal.**

WORD ANALYSIS AND DEFINITION

S/ = Suffix P/ = Prefix R/ = Root R/CF = Combining Form

WORD	PRONUNCIATION	ELEMENTS		DEFINITION
anus	**A**-nus		Latin *ring*	Terminal opening of the digestive tract through which feces are discharged.
anal (adj)	**A**-nal	S/	**-al** *pertaining to*	Pertaining to the anus
		R/CF	**an/o-** *anus*	
anorectal junction	**A**-no-**RECK**-tal **JUNK**-shun	R/	**-rect-** *rectum*	The junction between the anus and rectum.
		S/	**-ion** *condition, action*	
		R/	**junct-** *joining together*	
appendix	ah-**PEN**-dicks		Latin *appendage*	Small blind projection from the pouch of the cecum.
appendicitis	ah-pen-dih-**SIGH**-tis	S/	**-itis** *inflammation*	Inflammation of the appendix.
		R/	**appendic-** *appendix*	
appendectomy	ah-pen-**DEK**-toe-me	S/	**-ectomy** *surgical excision*	Surgical removal of the appendix.
colon	**KOH**-lon		Greek *colon*	The large intestine, extending from the cecum to the rectum.
colic (adj)	**KOL**-ik	S/	**-ic** *pertaining to*	Spasmodic, crampy pains in the abdomen.
		R/	**col-** *colon*	
colitis	koh-**LIE**-tis	S/	**-itis** *inflammation*	Inflammation of the colon.
feces	**FEE**-sees		Latin *dregs*	Undigested, waste material discharged from the bowel.
fecal (adj)	**FEE**-kal	S/	**-al** *pertaining to*	Pertaining to feces.
		R/	**fec-** *feces*	
defecation	def-eh-**KAY**-shun	S/	**-ation** *process*	Evacuation of feces from rectum and anus.
		P/	**de-** *removal*	
flexure	**FLEK**-shur		Latin *bend*	A bend in a structure.
ileum	**ILL**-ee-um		Latin *roll up, twist*	Third portion of the small intestine.
ileocecal sphincter	**ILL**-ee-oh-**SEE**-cal **SFINK**-ter	S/	**-al** *pertaining to*	Band of muscle that encircles the junction of ileum and cecum.
		R/CF	**ile/o-** *ileum*	
		R/	**-cec-** *cecum*	
			sphincter, Greek *band*	
rectum	**RECK**-tum		Latin *straight*	Terminal part of the colon from the sigmoid to the anal canal.
rectal (adj)	**RECK**-tal	S/	**-al** *pertaining to*	Pertaining to the rectum.
sigmoid	**SIG**-moyd		Greek *letter "S"*	Sigmoid colon is shaped like an "S."

EXERCISES

Deconstruct: *Attack a medical term with your analytical skills. Break down these terms into their elements to define the word. Know the meaning of the elements = know the meaning of the term. Fill in the chart; then fill in the blanks.*

Medical Term	Meaning of Prefix	Meaning of Root/CF	Meaning of Suffix
colitis			
defecation			
anorectal			
colic			
ileocecal			
appendicitis			

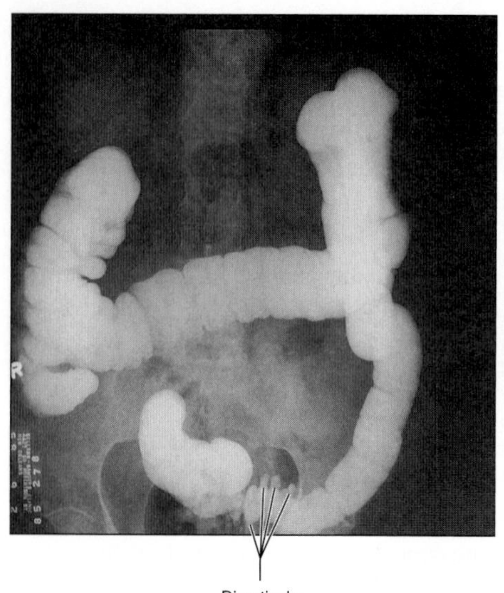

▲ **FIGURE 5.25 Barium Enema Showing Diverticulosis.**

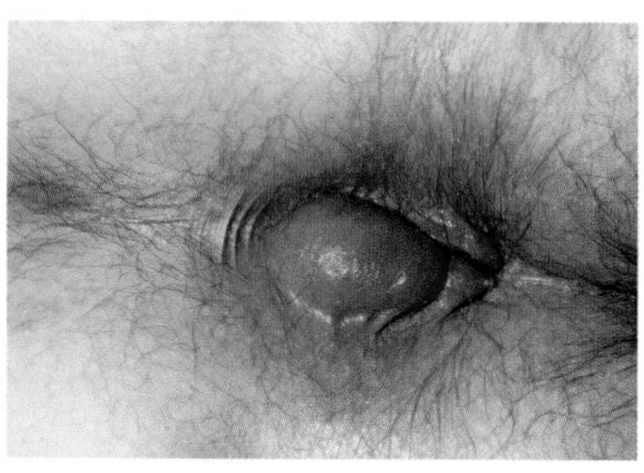

▲ **FIGURE 5.26 External Hemorrhoid.**

Keynote

Fissures are tears; fistulas are abnormal passages.

Keynote

Consuming black licorice, Pepto-Bismol, or blueberries can produce black stools.

DISORDERS OF THE LARGE INTESTINE AND ANAL CANAL

Disorders of the Large Intestine

Appendicitis is the most common cause of acute abdominal pain in the right lower quadrant. If neglected, the inflamed appendix can rupture, leading to **peritonitis.** A surgical appendectomy, usually performed through laparoscopy, is the treatment for appendicitis.

Diverticulosis is the presence of small pouches bulging outward through weak spots in the lining of the large intestine (*Figure 5.25*). They are asymptomatic until the pouches become infected and inflamed, a condition called **diverticulitis.** The most likely cause of **diverticular disease** (diverticulosis and diverticulitis) is a low-fiber diet.

Ulcerative colitis is an extensive inflammation and ulceration of the lining of the large intestine. It produces bouts of bloody diarrhea, crampy pain, and, often, weight loss and electrolyte imbalance.

Irritable bowel syndrome (IBS) is an increasingly common large-**bowel** disorder presenting with crampy pains, gas, and changes in bowel habits to either constipation or diarrhea. There are no anatomical changes seen in the bowel and the cause is unknown.

Polyps are masses of tissue arising from the wall of the large intestine and protruding into the bowel lumen. They vary in size and shape. Most are benign. Endoscopic biopsy can determine if they are precancerous or cancerous.

Colon and rectal cancers are the second cause of cancer deaths after lung cancer. The majority occur in the rectum and sigmoid colon. These cancers spread by:

1. Direct extension through the bowel wall.
2. Metastasis to regional lymph nodes.
3. Movement down the lumen of the bowel.
4. Blood-borne metastases to liver, lung, bone, and brain.

Obstruction of the large bowel can be caused by cancers, large polyps, or diverticulitis.

Intussusception is a form of obstruction that occurs when one segment of bowel slips inside another segment, causing an obstruction.

Proctitis is inflammation of the lining of the rectum, often associated with ulcerative colitis, Crohn disease, or radiation therapy.

Disorders of the Anal Canal

Hemorrhoids are dilated veins in the submucosa of the anal canal, often associated with pregnancy, chronic constipation, diarrhea, or aging. They protrude into the anal canal **(internal)** or bulge out along the edge of the anus **(external)** (*Figure 5.26*), producing pain and bright red blood from the anus. A **thrombosed** hemorrhoid, in which blood has clotted, is very painful.

Anal fissures are tears in the lining of the anal canal, perhaps due to difficult bowel movements (**BMs**). **Anal fistulas** occur following abscesses in the anal glands and are an abnormal passage (fistula) between the anal canal and the skin outside the anus. The surgical procedures of **fistulectomy** and **fistulotomy** are used in the treatment of anal fistulas.

Gastrointestinal (GI) Bleeding

Bleeding can occur anywhere in the gastrointestinal tract from a variety of causes. The bleeding can be internal and painless. It can present in different ways to provide a clue as to the site of bleeding:

- **Hematemesis** is the vomiting of bright red blood, which indicates an upper GI source of bleeding (esophagus, stomach, duodenum) that is brisk.
- **Vomiting of "coffee grounds"** occurs when bleeding from an upper GI source has slowed or stopped.

WORD ANALYSIS AND DEFINITION

S/ = Suffix P/ = Prefix R/ = Root R/CF = Combining Form

WORD	PRONUNCIATION		ELEMENTS	DEFINITION
bowel	**BOUGH**-el		Latin *sausage*	Another name for *intestine*.
diverticulum	die-ver-**TICK**-you-lum	S/ R/	**-um** *tissue, structure* **diverticul-** *byroad*	A pouchlike opening or sac from a tubular structure (e.g., intestine).
diverticula (pl) **diverticulosis**	di-ver-**TICK**-you-lah **DIE**-ver-tick-you-**LOW**-sis	S/	**-osis** *condition*	Presence of a number of small pouches in the wall of the large intestine.
diverticulitis	**DIE**-ver-tick-you-**LIE**-tis	S/	**-itis** *inflammation*	Inflammation of the diverticula.
fissure	**FISH**-ur		Latin *slit*	Deep furrow or cleft.
fistula	**FIS**-tyu-lah		Latin *pipe, tube*	Abnormal passage.
hemorrhoid	**HEM**-oh-royd	S/ R/CF	**-rrhoid** *flow* **hem/o-** *blood*	Dilated rectal vein producing painful anal swelling.
hemorrhoids (pl) **hemorrhoidectomy** (***Note:*** This term has two suffixes.)	**HEM**-oh-roy-**DEK**-toh-me	S/	**-ectomy** *surgical excision*	Surgical removal of hemorrhoids.
intussusception	**IN**-tuss-sus-**SEP**-shun	S/ P/ R/	**-ion** *action, condition* **intus-** *within* **-suscept-** *to take up*	The slipping of one part of bowel inside another to cause obstruction.
melena	mel-**EN**-ah		Greek *black*	The passage of black, tarry stools.
occult blood	oh-**KULT** BLUD		occult, Latin *to hide*	Blood that cannot be seen in the stool but is positive on a fecal occult blood test.
Hemoccult test	**HEEM**-o-kult TEST			Trade name for a fecal occult blood test.
peritoneum	per-ih-toe-**NEE**-um	S/ R/CF	**-um** *tissue* **periton/e-** *stretch over*	Membrane that lines the abdominal cavity.
peritoneal (adj) **peritonitis**	**PER**-ih-toe-**NEE**-al **PER**-ih-toe-**NIE**-tis	S/ S/	**-al** *pertaining to* **-itis** *inflammation*	Pertaining to the peritoneum. Inflammation of the peritoneum.
polyp	**POL**-ip		Latin *foot*	Mass of tissue that projects into the lumen of bowel.
polyposis	pol-ih-**POH**-sis	S/ R/	**-osis** *condition* **polyp-** *polyp*	Presence of several polyps.
polypectomy	pol-ip-**ECK**-toh-mee	S/	**-ectomy** *surgical excision*	Excision or removal of a polyp.
proctitis	prok-**TIE**-tis	S/ R/	**-itis** *inflammation* **proct-** *rectum*	Inflammation of the lining of the rectum.

- **Melena,** the passage of black, tarry stools, usually indicates upper GI bleeding.
- **Occult blood** cannot be seen in the stool, but when a chemical fecal occult blood test (Hemoccult) is positive, the source of the bleeding can be anywhere in the GI tract.

Abbreviations

BM bowel movement
IBS irritable bowel syndrome

EXERCISES

Suffixes: *This is a two-part exercise. First, insert the correct suffix to complete the term. Then circle the word in the definition that has the same meaning as the suffix. (There are more suffixes than you need.)*

osis **ectomy** **um** **itis** **al** **ion**

1. excision or removal of a polyp polyp/ ____________
2. inflammation of the diverticula diverticul/ ____________
3. membranous structure lining the abdominal cavity peritone/ ____________
4. condition of having multiple polyps polyp/ ____________
5. pertaining to the peritoneum peritone/ ____________

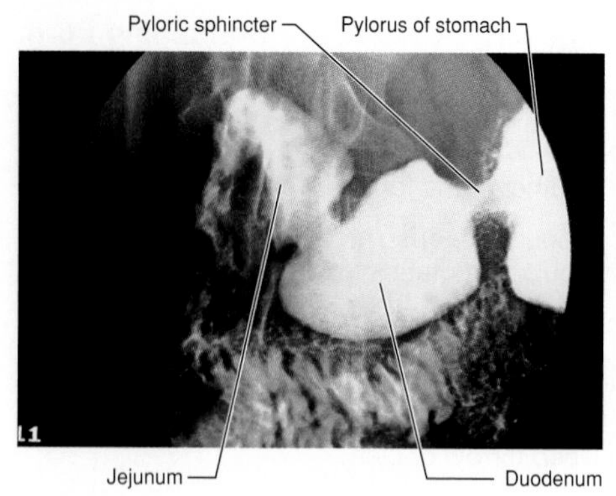

▲ **FIGURE 5.27 Barium Meal.**

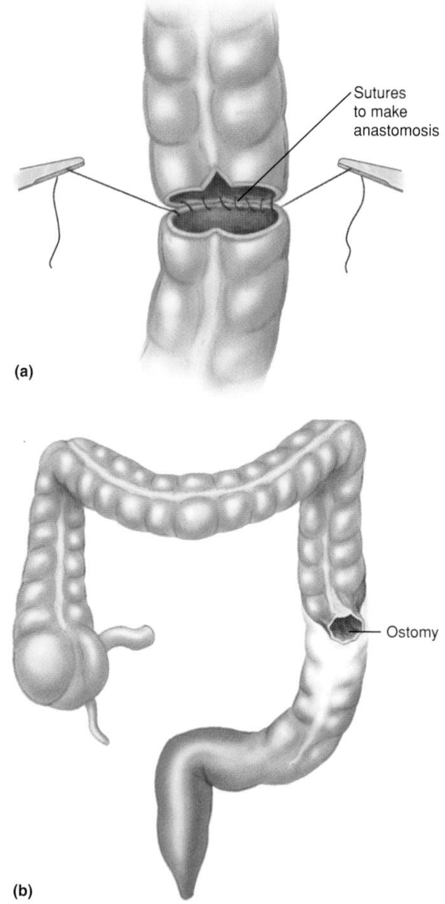

▲ **FIGURE 5.28 Intestinal Resections.** (a) Anastomosis. (b) Ostomy.

DIAGNOSTIC AND THERAPEUTIC PROCEDURES

Procedures for the Biliary System

Cholangiography requires injection of a dye intravenously, followed by x-ray of the biliary system so that the biliary system can be visualized.

Cholecystectomy is surgical removal of the gallbladder through an open incision or by laparoscopic cholecystectomy.

Cholelithotomy is a surgical incision into the gallbladder to remove gallstones.

Diagnostic Procedures for the Digestive Tract

Barium swallow involves ingestion of barium sulfate, a contrast material, to show details of the pharynx and esophagus on x-ray.

Barium meal (*Figure 5.27*) uses barium sulfate to study the distal esophagus, stomach, and duodenum on x-ray.

Enteroscopy through an oral endoscope is used to visualize and biopsy tumors and ulcers and to control bleeding from the esophagus, stomach, and duodenum. This procedure is also called **esophagogastroduodenoscopy.**

Angiography uses dye to highlight blood vessels. It can be used to define the site of a bleed.

Fecal occult blood test (Hemoccult) is used to detect the presence of blood not visible to the naked eye.

Nasogastric aspiration and **lavage** that shows bright red blood indicates active upper GI bleeding. Emesis of coffee-ground material indicates the bleeding has slowed or stopped.

Upper GI barium x-rays are less accurate than panendoscopy at identifying the bleeding lesion.

Barium enema involves the injection of a radiographic contrast material into the large intestine as an enema, and x-ray films are taken.

Endoscopy enables direct visual examination of the intestine with a flexible tube containing light-transmitting glass fibers or a video transmitter. **Panendoscopy** examines the esophagus, stomach, and duodenum.

Anoscopy examines the anus and lower rectum with a rigid instrument. **Flexible sigmoidoscopy** examines the rectum and sigmoid colon. **Colonoscopy** examines the whole length of the colon. **Gastroscopy** examines the stomach. **Proctoscopy** examines the anus. Endoscopy can also be used to perform a biopsy, remove polyps (polypectomy), and coagulate bleeding lesions.

Digital rectal exam is performed by the physician, who palpates the rectum and prostate gland with an index finger.

Surgical Procedures for the Digestive Tract

Intestinal resections are used to surgically remove diseased portions of the intestine. The remaining portions of the intestine can be joined together through an **anastomosis** (*Figure 5.28a*). If there is insufficient bowel remaining, an **ostomy** (*Figure 5.28b*) can be performed, in which the end of the bowel opens onto the skin at a **stoma. Ileostomy** and **colostomy** are two common procedures.

WORD ANALYSIS AND DEFINITION

S/ = Suffix   P/ = Prefix   R/ = Root   R/CF = Combining Form

WORD	PRONUNCIATION		ELEMENTS	DEFINITION
anastomosis	ah-**NAS**-toh-**MOH**-sis	S/ R/	**-osis** *condition* **anastom-** *provide a mouth*	A surgically made union between two tubular structures.
anastomoses (pl)	ah-**NAS**-toh-**MOH**-seez			
endoscope	**EN**-doh-skope	P/ R/	**endo-** *within, inside* **-scope** *instrument for viewing*	Instrument to examine the inside of a hollow or tubular organ.
endoscopy	en-**DOS**-koh-pee	P/ R/	**endo-** *within, inside* **-scopy** *to examine, to view*	The use of an endoscope.
colonoscopy	koh-lon-**OSS**-koh-pee	R/CF	**colon/o** *colon*	Examination of the inside of the colon by endoscopy.
panendoscopy (*Note:* two prefixes)	pan-en-**DOS**-koh-pee	P/ P/	**pan-** *all* **endo-** *within, inside*	Examination of the inside of the esophagus, stomach, and upper duodenum using a flexible fiber-optic endoscope.
proctoscopy (also called *anoscopy*)	prok-**TOSS**-koh-pee	R/CF	**proct/o-** *anus*	Examination of the inside of the anus by endoscopy.
sigmoidoscopy	sig-moi-**DOS**-koh-pee	S/ R/ R/CF	**-oid** *resembling* **sigm-** *Greek"S"* **sigmoid/o-** *sigmoid colon*	Examination of the sigmoid colon by endoscopy.
gastroscopy	gas-**TROSS**-koh-pee	R/CF	**gastr/o-** stomach	Examination of the inside of the stomach by endoscopy.
enema	**EN**-eh-mah		Greek *injection*	An injection of fluid into the rectum.
ostomy	**OSS**-toe-me	S/- R/	**-stomy** *new opening* **os-** *mouth*	Artificial opening into a tubular structure.
colostomy	koh-**LOSS**-toe-me	R/CF	**col/o-** *colon*	Artificial opening from the colon to the outside of the body.
ileostomy	ill-ee-**OS**-toe-me	R/CF	**ile/o-** *ileum*	Artificial opening from the ileum to the outside of the body.
stoma	**STOW**-mah		Greek *mouth*	Artificial opening.

EXERCISES

Test yourself *on the elements contained in the WAD. The prefixes and suffixes you will see again in terms in other chapters. Match the element in column 1 with its definition in column 2.*

_____	1. os	A. colon
_____	2. gastro	B. instrument
_____	3. endo	C. all
_____	4. colo	D. mouth
_____	5. scope	E. condition
_____	6. procto	F. ileum
_____	7. stomy	G. stomach
_____	8. ileo	H. within
_____	9. pan	I. new opening
_____	10. osis	J. anus

CHAPTER 5 REVIEW

THE DIGESTIVE SYSTEM

CHALLENGE YOUR KNOWLEDGE

A. **Roots** are the core meaning of every term. These terms all end in *itis,* which means *inflammation.* The root will tell you exactly what is inflamed. If you slash the term first, this will help isolate the roots. Fill in the chart.

Medical Term	Root	Meaning of Root	Meaning of Medical Term
appendicitis			
cholecystitis			
colitis			
diverticulitis			
esophagitis			
gastritis			
gastroenteritis			
gingivitis			
hepatitis			
pancreatitis			
periodontitis			
peritonitis			
proctitis			

B. **Spelling demons:** Have a fellow student dictate the following terms to you. Cover column 1 with a piece of paper, and write your terms in column 2. *Try your best to spell them correctly on the first attempt!*

1. intussusception ______
2. hemorrhoidectomy ______
3. dyspepsia ______
4. gastroenterologist ______
5. peristalsis ______
6. cirrhosis ______
7. aphthous ______
8. pyorrhea ______
9. herniorrhaphy ______
10. hematemesis ______

C. **Identification:** The following word bank contains one- and two-word definitions for the terms in the chart. Put the correct definition next to each term in the chart. There are more words in the bank than medical terms.

Word Bank:

procedure	symptom	disease
adjective	plural term	noun
singular term	structure	instrument
process	inflammation	organ

Medical Term	Definition
1. diarrhea	
2. duodenum	
3. dysentery	
4. endoscope	
5. gingivectomy	
6. mucus	
7. tongue	
8. varicose	
9. villi	

D. **Precision in documentation:** Get the stone in the right place: *cholelithiasis* or *choledocholithiasis?* Make the correct choice of medical terminology. Fill in the blanks.

1. Patient's films revealed a stone in the common bile duct.

   Diagnosis: ______________________________

2. The presence of a stone in the patient's gallbladder was confirmed by the radiologist.

   Diagnosis: ______________________________

3. Acute or chronic inflammation of the gallbladder.

   Diagnosis: ______________________________

CHAPTER 5 REVIEW

THE DIGESTIVE SYSTEM

E. Some of the terms in this chapter are particularly hard to spell and pronounce. Refer to the Student Online Learning Center (www.mhhe.com/AllanEssMedLanguage) for pronunciation practice, and remember that spelling is important! Read these sentences aloud after you have circled your choices for the correct spelling.

1. The (nasopharynx/nasopharyix) is behind the nose and above the soft palate.
2. (Hyatus/Hiatus) is rooted in the Latin word for *opening* and is the opening through the diaphragm for the (esophagis/esophagus).
3. (Sphincter/Sphinchter) is a root meaning *a band* and is a band of muscle that encircles a tube.
4. (Uvula/Vuvula) comes from the Latin meaning *small grape.*
5. (Postprandial/Postperandial) medication, meaning *medication taken after meals,* may be ordered.
6. (Hemotemasis/Hematemesis) is the (vomiting/vomitting) of blood.
7. Throat cancer would produce (dysphagia/dyspagia).
8. (Varricces/Varices) is the plural of (varix/varex) and means *dilated vein.* The adjective is (varicose/varricose).
9. (Hernioraphy/Herniorrhaphy) is the surgical fixation of a hernia.

F. **Roots:** Master all forms of the same term, including the plurals. Always analyze a term beginning with the suffix. Fill in the blanks.

1. diverticulitis **diverticulum** **diverticulosis** **diverticula**

1. What starts out as a single ____________________, if left untreated, can lead to many ____________________ ____________. The condition of having a number of these small pouches in the wall of the large intestine is known as ____________________. Should these pouches become inflamed, ____________________ will result.

2. polypectomy **polyposis** **polyp (singular)**

2. The first polyp was found on sigmoidoscopy. A follow-up colonoscopy 6 months later found several more (plural) ____________________ in the large intestine. Diagnosis is ____________________. Proposed treatment is ____________________.

3. peritoneum **peritonitis** **peritoneal**

3. The ____________________ laceration sliced completely through the ____________________. Because of an infection in the wound, the patient developed ____________________.

4. **gingivectomy** **gingival** **gingivitis** **gingiva**

4. The patient's ____________________ tissue is severely infected. Diagnosis is ____________________. A ____________________ has been scheduled; I will try to preserve as much of the healthy ____________________ as possible.

G. **Translate** these sentences, which come directly from this chapter. Express the same thought in language your patient can understand.

"Choledocholithiasis can cause biliary colic and jaundice."

"Endoscopy can be used to perform a biopsy of a polyp, polypectomy, and to coagulate bleeding lesions."

H. **The two suffixes *scope* and *scopy* are** attached to many medical terms you will meet in later chapters. *Scope* is the actual instrument used in the procedure. The procedure itself is the *scopy.* Start by slashing the terms in the first column, and then utilize the *language of gastroenterology* to fill in the chart for this exercise.

Name of Instrument	Name of Procedure	Definition of Procedure
anoscope		
colonoscope		
endoscope		
gastroscope		
laparoscope		
panendoscope		
proctoscope		
sigmoidscope		

CHAPTER 5 REVIEW

THE DIGESTIVE SYSTEM

I. **Greek and Latin** are the origins for many medical terms. Give a meaning for each term; then use any three terms in sentences of your choice. Fill in the chart.

Medical Term	Meaning
bolus	
canker	
caries	
esophagus	
intestine	
pulp	
saliva	
ulcer	
uvula	

1. ______________________________

2. ______________________________

3. ______________________________

J. **Prefixes** make the difference in precision. Analyze the two medical terms in each pair below. Write a brief description of how they differ, based on their prefixes.

1. *emesis* and *hematemesis*

____

____

2. *symptomatic* and *asymptomatic*

____

____

K. **Recall and review:** How well do you remember these word elements from the previous chapter? Try to answer without looking back to check. Fill in the chart.

Element	Type of Element (P, R, CF, or S)	Meaning of Element
ophthalmo		
peri		
audio		
exo		
oto		

L. **Terminology challenge:** Train your eye and ear to know the difference. Words can be very similar looking or sounding but mean entirely different things. If you really understand the following medical terms, you can explain them to a fellow student. Explain to your classmate the difference between them.

1. *chyle* and *chyme*

____

____

2. *fissure* and *fistula*

____

____

3. *stricture* and *sphincter*

____

____

THE DIGESTIVE SYSTEM

M. Word attack: Work through this exercise step-by-step for tips on how to apply your knowledge of elements to answer multiple-choice questions.

Circle the term with the element meaning *difficult/bad:*

a. anorexia

b. perforation

c. indigestion

d. dyspepsia

e. stricture

> *Study Hint*
> As soon as you discard a choice (for whatever reason) in multiple-choice questions, cross off that choice. Your eyes will then focus only on the possible correct choices when you choose your answer.

1. First, discard any answer choice(s) that do *not* break down into elements, since you are looking for a term with an element.

   Discard: ______________________________

2. Since prefixes are the most usual element for description (size, color, location, etc.), start by analyzing the prefixes. The prefixes are:

   ____________________ means ____________________

   ____________________ means ____________________

   ____________________ means ____________________

   ____________________ means ____________________

3. The correct answer is ______________________________.

Nice work! Remember to apply this approach when you answer the multiple-choice questions on a test.

N. **Construct** medical terms to match the definitions given. You are given more elements than you need for possible answers. Fill in the blanks.

bari	abdomino	heme	deglutit
par	ary	ar	ot
mastic	lingu	dys	halit
laparo	odont	emesis	peri
mandibul	gingiv	phagia	rrhea
sub	rrhaphy	hemat	hiat

1. pertaining to food or nutrition aliment/ ______________________
2. removal of diseased gum tissue ______________________/ectomy
3. salivary gland beside the ear ______________________/id
4. underneath the tongue sub/______________________/al
5. examination of the contents of the abdomen ______________________/scopy
6. around a tooth ______________________/ondontal
7. purulent discharge pyo ______________________
8. process of swallowing ______________________/ion
9. bad breath ______________________/osis
10. management of obesity ______________________/atrics
11. pertaining to beneath the lower jawbone ______________________
12. difficulty swallowing ______________________
13. vomiting of red blood ______________________

CHAPTER 5 REVIEW

THE DIGESTIVE SYSTEM

O. Chapter challenge: Circle the best answer.

Study Hint
Immediately cross off any answer you know is not correct. In your remaining choices, there is only *one best answer.*

1. What is the adjective used to describe a *dilated vein?*
   a. adipose
   b. varicose
   c. edematous
   d. alimentary
   e. sublingual

2. What substance in the body is harder than bone?
   a. dentine
   b. papilla
   c. pulp
   d. plaque
   e. tartar

3. The fungus *Candida albicans* causes:
   a. caries
   b. aphthous ulcers
   c. thrush
   d. canker sores
   e. fever blisters

4. Icterus is another term for:
   a. peristalsis
   b. cirrhosis
   c. jaundice
   d. diverticulosis
   e. GERD

5. The set of terms that are all diagnoses is:
   a. cholangiography, jaundice, cholelithiasis
   b. endoscopy, icterus, cholecystectomy
   c. hepatitis, diverticulosis, choledocholithiasis
   d. pancreatitis, polypectomy, anoscopy
   e. diabetes, cystic fibrosis, cholelithotomy

6. If you have *postprandial* burning chest pain, it is coming:
   a. after you go to bed
   b. before a meal
   c. when you wake up
   d. after a meal
   e. at none of these times

7. Which is the only term that signifies the "bile duct."
   a. cholelithiasis
   b. cholecystitis
   c. cholelithotomy
   d. choledocholithiasis
   e. cholecystectomy

8. The "R" in the abbreviation *GERD* stands for:
   a. reflex
   b. removal
   c. reflux
   d. resection
   e. resorption

9. What do all these terms have in common: *diarrhea, flatulence, pyorrhea?*

a. They are all symptoms.

b. None of them has a prefix.

c. They are all misspelled.

d. They all relate to the large intestine.

e. They have all of the above in common.

10. The one term that is *not* a body process is:

a. peristalsis

b. secretion

c. deglutition

d. mastication

e. halitosis

11. *Regional enteritis* is another name for:

a. diverticulosis

b. IBS

c. celiac disease

d. peritonitis

e. Crohn disease

12. Hiatal hernia occurs when:

a. a portion of the intestine telescopes in on itself

b. a portion of the stomach protrudes through the diaphragm

c. a portion of the liver protrudes through the stomach

d. a portion of the stomach protrudes through the intestine

e. a portion of the bowel protrudes through the stomach

13. The term that contains two prefixes is:

a. panendoscopy

b. gastroenterology

c. nasopharynx

d. submandibular

e. periodontal

14. *Cholecystectomy* and *cholelithotomy* are both procedures on the:

a. small intestine

b. pancreas

c. liver

d. large intestine

e. gallbladder

15. Proofread the following sentences. Which is the only one that is correct?

a. A gingivectomy creates a new opening outside the body.

b. Melena is the passage of black, tarry stools.

c. Dysphagia is difficulty in chewing.

d. The roof of the mouth is called the uvula.

e. The maxilla is the lower jaw.

CHAPTER 5 REVIEW

THE DIGESTIVE SYSTEM

P. **Case Report Challenge:** Now that you are more comfortable with the terms in this chapter, you can apply that knowledge and briefly answer the questions about the Case Report.

You may someday find yourself doing medical transcription in a hospital or physician's office. **Proofread** the beginning of the letter Dr. Walsh has sent to the patient's insurance company, asking for preauthorization for her surgery. Underline or highlight *all the errors;* then rewrite the correct terms on the lines below the letter. Then add a *brief* definition of the term, using the Glossary or a medical dictionary.

Dear Doctor Leavenworth,

Request for Authorization of Surgery

Re: Mrs. Martha Jones.

Subscriber ID # FMC 056437

Mrs. Jones is a 52-year-old former waitress, recently divorced. She is 5 feet 4 inches tall and weighs 275 pounds. She has Type III diabeetes with frequent episodes of hyperglycemia and also ketoacidoses requiring three different hospitalizations. She now has diabetick retinnopathy and peripheral vascullitis. Complicating this is hypertension (185/110), coronery artery disease, and pulmonary edema.

1. Write the correct form of the misspelled words and a brief definition here:

____

____

____

____

____

____

____

____

____

____

____

Reread the following Case Report on Mrs. Jan Stark. Remember to highlight or underline the medical terms before you answer the questions.

> For the past 10 years she has had spasmodic episodes of *diarrhea* and *flatulence* associated with severe headaches and fatigue. During those episodes her stools were greasy and pale. She has seen several physicians who have recommended low-fat, high-carbohydrate diets without relief.
>
> Dr. Grabowski has performed an intestinal biopsy through oral endoscopy and diagnosed celiac disease.

2. List all of Mrs. Stark's symptoms:

____

____

____

____

____

3. Define *flatulence:*

____

4. If this was an *intestinal biopsy,* what *specific* type of endoscope was used?

____

5. *Oral endoscopy* means the scope was inserted through the patient's _______________.

6. What does the term *spasmodic* mean? Check the Glossary or a medical dictionary, and write the definition on the lines below.

____

____

____

____

7. What does Mrs. Stark need to eliminate from her diet to improve her condition?

____

____

____

Cardiovascular and Circulatory Systems

The Essentials of the Language of Cardiology

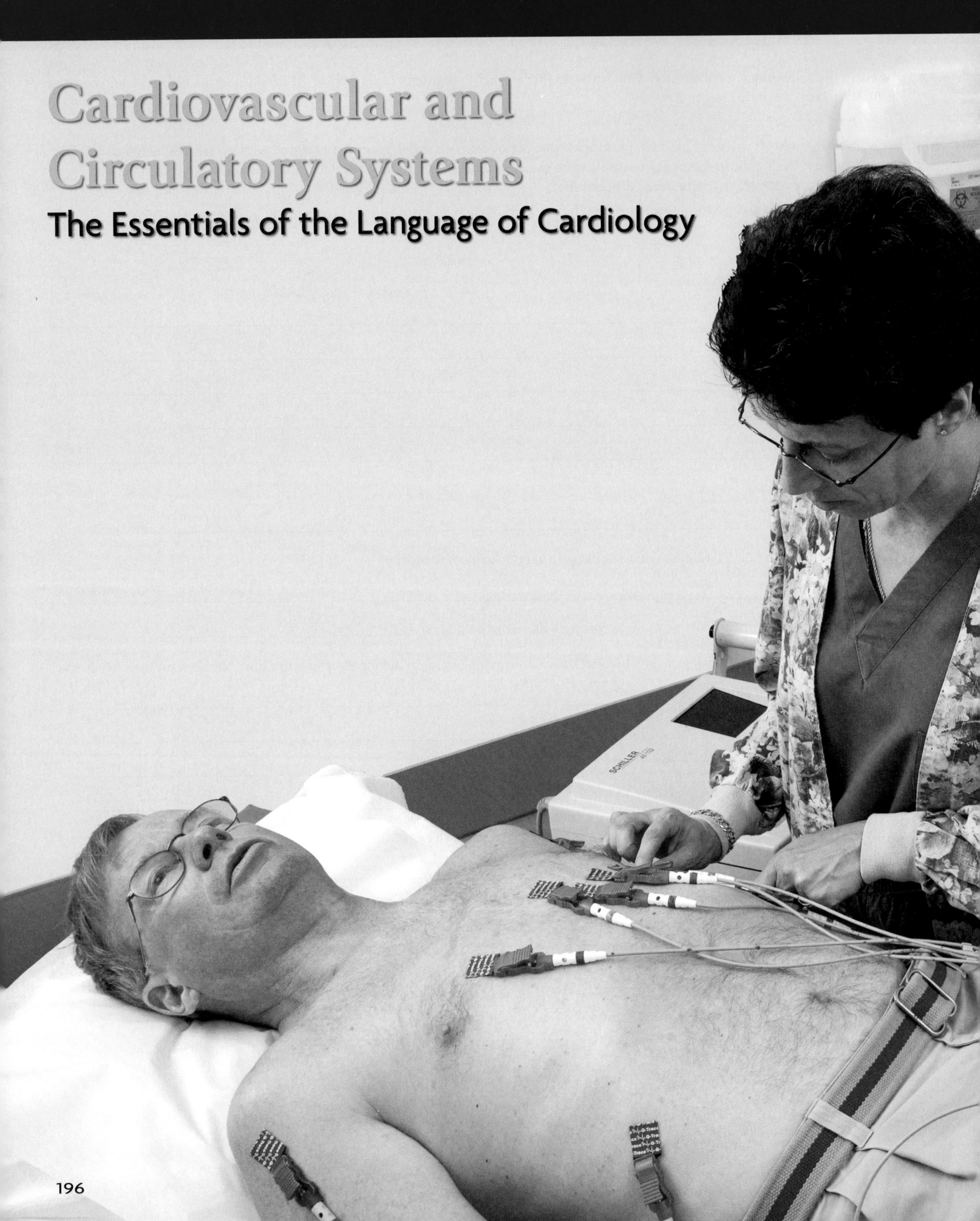

CHAPTER 6

CASE REPORT 6.1

You are

. . . a **cardiovascular** technologist (CVT) employed by the **Cardiology** Department at Fulwood Medical Center. You have been called to the Emergency Department (ED) to take an **electrocardiogram (ECG** or **EKG)**, STAT.

You are communicating with

. . . Mr. Hank Johnson, the 64-year-old owner of a printing company. Eight months ago he had a left total hip replacement. In the past 3 months, Mr. Johnson has returned to his daily workouts. This morning, while riding his exercise bike, he felt tightness in his chest. He kept on cycling. He developed pain in the center of his chest, radiating down his left arm and up into his jaw, and became **diaphoretic.** His personal trainer called 911.

You perform the ECG and the automatic report describes abnormalities in the chest leads. As you remove the **electrodes,** Mr. Johnson complains that he is feeling faint and short of breath **(SOB)**. You are the only person in the room.

Learning Outcomes

No matter which discipline or setting you work in as a health professional, the condition of the patient's cardiovascular system will always be a factor during diagnosis, treatment, and communication. In order to be able to understand, communicate, and document conditions affecting the heart, blood vessels, and blood, you need to be able to:

6.1 Apply the language of cardiology to the structure and functions of the cardiovascular system.

6.2 Comprehend, analyze, spell, and write the terms of cardiology so that you communicate and document accurately.

6.3 Recognize and pronounce the medical terms of cardiology so that you communicate verbally with accuracy and precision.

6.4 Specify the correct medical terminology for common disorders of the cardiovascular system.

LESSON 6.1 The Heart

CARDIOVASCULAR AND CIRCULATORY SYSTEMS

OBJECTIVES

If you have a healthy heart rate of 60 beats per minute and you live to be 80 years old, your heart will beat (contract and relax) 2,522,880,000 times. Your heart pumps approximately 2,000 gallons of blood each day. In 80 years of life, it will have pumped a total of 58,400,000 gallons of blood. When your heart fails, there is no circulation of blood, tissues are deprived of oxygen and nutrients, and metabolic wastes accumulate. Your cells die. The information in this lesson will enable you to:

- **Describe the medical terminology for the location, structure, and functions of the heart.**
- **Explain the heart cycle.**
- **Identify the medical terminology for the blood supply to the heart.**
- **Detail the medical terminology for the electrical properties of the heart.**

Abbreviations	
CPR	cardiopulmonary resuscitation
CVT	cardiovasular technician
ECG	electrocardiogram
EKG	electrocardiogram
SOB	short(ness) of breath
STAT	immediately

LOCATION OF THE HEART

It is important to know the position of the heart so that you can perform effective **cardiopulmonary resuscitation (CPR)**.

The heart is a blunt cone, pointing down and to the left. It lies obliquely with one-third of its mass behind the **sternum** and two-thirds to the left of the sternum (*Figure 6.1a*). The heart is roughly the size of your fist and weighs approximately 10 ounces. It is located in the **thoracic cavity** between the two lungs, in an area called the **mediastinum** (*Figure 6.1b*).

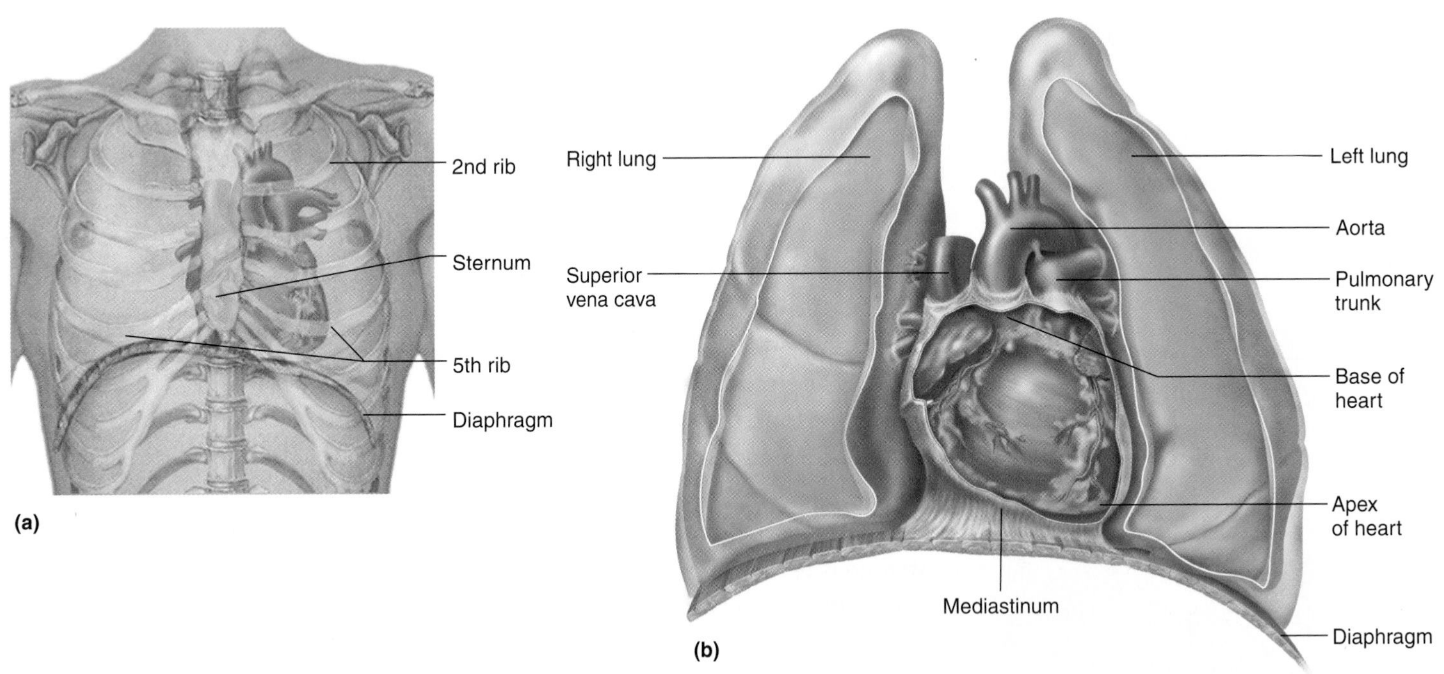

▲ **FIGURE 6.1 Position of Heart in Thoracic Cavity.**
(a) Relationship of heart to sternum. (b) Position of heart in mediastinum.

WORD ANALYSIS AND DEFINITION

S/ = Suffix P/ = Prefix R/ = Root R/CF = Combining Form

WORD	PRONUNCIATION	ELEMENTS		DEFINITION
cardiologist	kar-dee-**OL**-oh-jist	S/ R/CF	**-logist** *one who studies, specialist* **cardi/o-** *heart*	A medical specialist in the diagnosis and treatment of the heart (cardiology).
cardiology	kar-dee-**OL**-oh-jee	S/	**-logy** *study of*	Medical specialty of diseases of the heart.
cardiopulmonary resuscitation (CPR)	**KAR**-dee-oh-**PUL**-mo-nar-ee ree-sus-ih-**TAY**-shun	S/ R/ R/CF S/ R/-	**-ary** *pertaining to* **-pulmon-** *lung* **cardi/o-** *heart* **-ation** *a process* **resuscit-** *revive from apparent death*	The attempt to restore cardiac and pulmonary function.
cardiovascular	**KAR**-dee-oh-**VAS**-kyu-lar	S/ R/CF R/	**-ar** *pertaining to* **cardi/o-** *heart* **-vascul-** *blood vessel*	Pertaining to the heart and blood vessels.
diaphoresis (noun)	**DIE**-ah-foh-**REE**-sis	R/ S/	**diaphor-** *sweat* **-esis** *condition*	Sweat or perspiration.
diaphoretic (adj)	**DIE**-ah-foh-**RET**-ic	S/	**-etic** *pertaining to*	Pertaining to sweat or perspiration.
electrocardiogram	ee-lek-troh-**KAR**-dee-oh-gram	S/ R/CF R/CF	**-gram** *a record* **electr/o-** *electricity* **-cardi/o-** *heart*	Record of the electrical signals of the heart.
electrocardiograph **electrocardiography**	ee-lek-troh-**KAR**-dee-oh-graf ee-**LEK**-troh-kar-dee-**OG**-rah-fee	S/ S/	**-graph** *to record* **-graphy** *process of recording*	Machine that makes the electrocardiogram. Interpretation of electrocardiograms.
electrode	ee-**LEK**-trode	S/ R/	**-ode** *way, road* **electr-** *electricity*	A device for conducting electricity.
mediastinum	**ME**-dee-ass-**TIE**-num	S/ P/ R/	**-um** *structure* **media-** *middle* **-stin-** *partition*	Area between the lungs containing the heart, aorta, venae cavae, esophagus, and trachea.
sternum	**STIR**-num		Latin *the chest*	Long, flat bone forming the center of the anterior wall of the chest.
thoracic cavity	**THOR**-ass-ik **KAV**-ih-tee	S/ R/	**-ic** *pertaining to* **thorac-** *chest* cavity, Latin *hollow*	Space within the chest containing the lungs, heart, esophagus, trachea, aorta, venae cavae, and pulmonary vessels.

EXERCISES

Apply *the language of cardiology in the following documentation. Be sure to use the correct form (noun, adjective) of the term. There is only one best answer for each blank.*

cardiologist **cardiovascular** **cardiology** **cardiopulmonary**

1. The root/combining form in all these terms is ______________ and means ______________________________.
2. The ________________ Department sent a specialist to examine the patient in the Emergency Room because of his symptoms. The ________________ ordered an angioplasty, which found that the patient had three obstructed arteries in his heart, so the ________________ surgeon prepared to operate immediately. Before surgery could begin, the patient suffered a heart attack and ________________ resuscitation was needed.

electrodes **electrocardiogram** **electrocardiograph** **electrocardiography**

3. The root/combining form in all these terms is ________________ and means ________________.
4. The patient was scheduled for an ______________________________ today. The cardiovascular technician attached the ______________________________ to the patient's chest and proceeded to turn on the ______________________________. Unfortunately, he was unable to obtain the ______________________________ due to a malfunction of the machine. This study will have to be rescheduled for the patient.

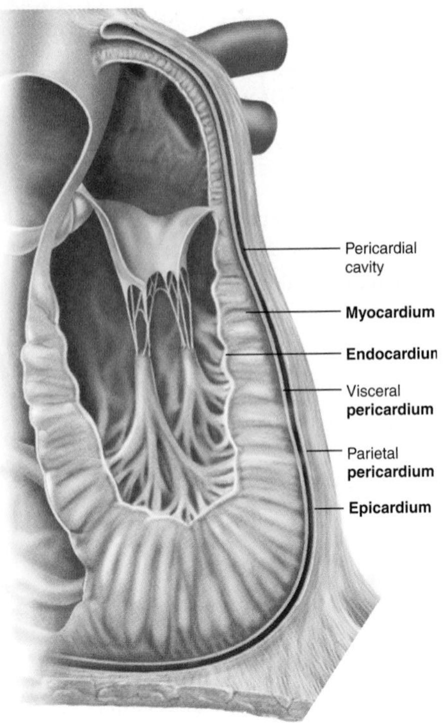

▲ **FIGURE 6.2**
Heart Wall.

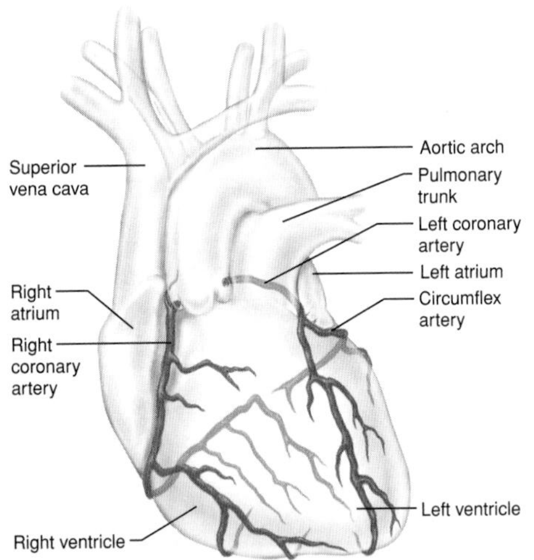

▲ **FIGURE 6.3**
Coronary Arterial Circulation.

Abbreviation	
MI	myocardial infarction

FUNCTIONS AND STRUCTURE OF THE HEART

Functions of the Heart

1. **Pump blood.** Contractions of the heart generate the pressure to produce movement of blood through the blood vessels.
2. **Route blood.** The heart can be described as two pumps: a pump on the right side of the heart that sends blood through the **pulmonary** circulation of the lungs and back to the pump on the left side, which sends blood through the **systemic** circulation of the body. The valves of the heart ensure this one-way flow of blood.
3. **Regulate blood supply.** The changing metabolic needs of tissues and organs (for example, when you exercise) are met by changes in the rate and force of the heart's contraction.

Structure of the Heart

The heart wall consists of three layers (*Figure 6.2*):

1. **Endocardium:** Connective tissue lining the inside of the heart.
2. **Myocardium:** Cardiac muscle cells that enable the heart to contract.
3. **Epicardium:** An outer single layer of cells overlying a thin layer of connective tissue.

The **pericardium** is a double-layered connective tissue sac that surrounds and protects the heart.

Blood Supply to Heart Muscle

Because the heart beats continually and forcefully, it requires an abundant supply of oxygen and nutrients. To meet this need, the cardiac muscle has its own blood circulation, the **coronary circulation** (*Figure 6.3*), the **arteries** of which arise directly from the **aorta.**

If any of the coronary arteries become blocked, the blood supply to a part of the cardiac muscle is cut off **(ischemia)** and the cells supplied by that artery die **(necrosis)** within minutes. This is a **myocardial infarction (MI),** what many call a "heart attack."

Case Report 6.1 *(continued)*

This is what was happening to Mr. Johnson at the beginning of this chapter. It was shown by the changes on the ECG. Cardiac muscle cells are very vulnerable to ischemia, and the repair of muscle cell death is mainly by **fibrosis.** This limits cardiac function by causing a diminished ability of Mr. Johnson's heart muscle to contract normally.

WORD ANALYSIS AND DEFINITION

S/ = Suffix P/ = Prefix R/ = Root R/CF = Combining Form

WORD	PRONUNCIATION	ELEMENTS		DEFINITION
aorta **aortic (adj)**	a-**OR**-tuh a-**OR**-tik	 S/ R/	Greek *lift up* **-ic** *pertaining to* **aort-** *aorta*	Main trunk of the systemic arterial system. Pertaining to the aorta.
coronary circulation	**KOR**-oh-nair-ee **SER**-kyu-**LAY**-shun	S/ R/ S/ R/	**-ary** *pertaining to* **coron-** *crown, coronary* **-ion** *action, condition* **circulat-** *circular route*	Blood vessels supplying the heart muscle.
endocardium **endocardial (adj)**	**EN**-doh-kar **DEE**-um **EN**-doh-kar-**DEE**-al	S/ P/ R/CF S/	**-um** *structure* **endo-** *inside* **-card/i-** *heart* **-al** *pertaining to*	The inside lining of the heart. Pertaining to the endocardium.
epicardium **epicardial (adj)**	**EP**-ih-kar-**DEE**-um **EP**-ih-kar-**DEE**-al	S/ P/ R/CF S/	**-um** *structure* **epi-** *upon, above* **-card/i-** *heart* **-al** *pertaining to*	The outer layer of the heart wall. Pertaining to the epicardium.
fibrosis (noun) **fibrotic (adj)**	fie-**BROH**-sis fie-**BROT**-ik	S/ R/CF S/	**-sis** *abnormal condition* **fibr/o-** *fiber* **-tic** *pertaining to*	Repair of dead tissue cells by formation of fibrous tissue. Pertaining to or affected by fibrosis.
infarct **infarction**	in-**FARKT** in-**FARK**-shun	P/ R/ S/	**in-** *in* **-farct-** *area of dead tissue* **-ion** *action, condition*	Area of cell death resulting from an infarction. Sudden blockage of an artery.
ischemia **ischemic (adj)**	is-**KEY**-me-ah is-**KEY**-mik	R/ R/ S/	**-emia** *a blood condition* **isch-** *to keep back* **-emic** *a condition of the blood*	Lack of blood supply to a tissue. Pertaining to or affected by the lack of blood supply to a tissue.
myocardium **myocardial (adj)**	**MY**-oh-**KAR**-dee-um my-oh-**KAR**-dee-al	S/ R/CF R/CF S/	**-um** *structure* **my/o-** *muscle* **-card/i-** *heart* **-al** *pertaining to*	All the heart muscle. Pertaining to heart muscle.
necrosis **necrotic (adj)**	neh-**KROH**-sis neh-**KROT**-ik	S/ R/ R/CF S/	**-osis** *condition* **necr-** *death* **necr/o-** *death* **-tic** *pertaining to*	Pathologic death of cells or tissue. Pertaining to or affected by necrosis (death).
pericardium (noun) **pericardial (adj)**	per-ih-**KAR**-dee-um per-ih-**KAR**-dee-al	S/ P/ R/CF S/	**-um** *structure* **peri-** *around* **-card/i-** *heart* **-al** *pertaining to*	A double layer of membranes surrounding the heart. Pertaining to the pericardium.
pulmonary	**PULL**-moh-**NAR**-ee	S/ R/	**-ary** *pertaining to* **pulmon-** *lung*	Pertaining to the lungs and their blood supply.

EXERCISES

Elements: Solid knowledge of elements will help increase your medical vocabulary. Many of the elements in the WAD above will appear throughout later chapters. Learn these elements now, and you will recognize them later. Put a ✓ in the appropriate column to identify the type of element; then fill in the last two columns.

Element	Prefix	Root	CF	Suffix	Meaning of Element	Medical Term with This Element
peri						
um						
emia						
isch						
myo						
pulmon						

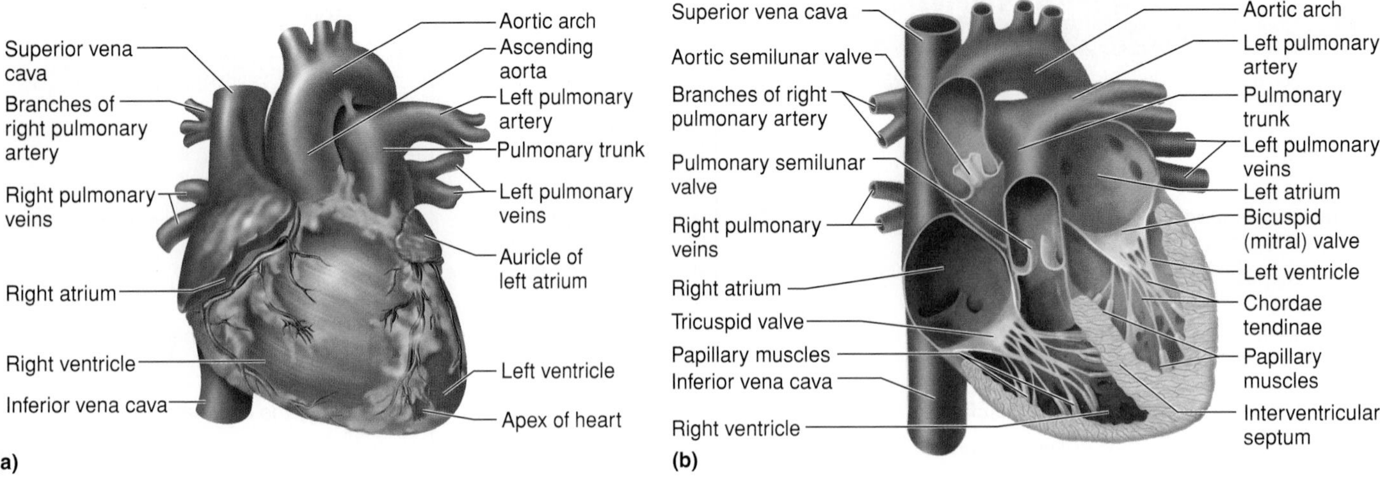

▲ **FIGURE 6.4 Anatomy of the Heart.**
(a) External anatomy of heart. (b) Chambers of the heart.

Keynote

The pulmonary circulation is the only place in the body where deoxygenated blood is carried in arteries and oxygenated blood is carried in veins.

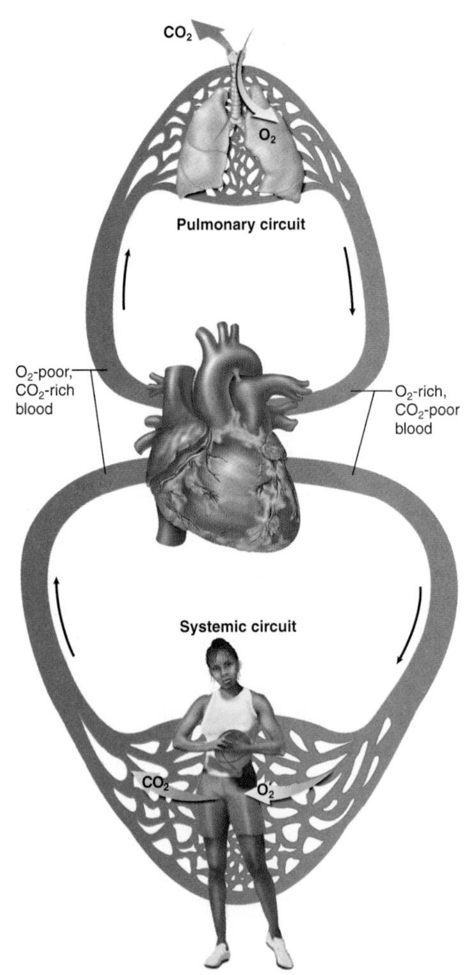

▲ **FIGURE 6.5**
Schematic of Cardiovascular System.

BLOOD FLOW THROUGH THE HEART

The heart has four chambers (*Figure 6.4*):

1. Right **atrium**
2. Right **ventricle**
3. Left atrium
4. Left ventricle

The right and left sides of the heart are separated by the cardiac **septum,** which can be described in two sections:

- The right and left atria are separated by a thin muscle wall called the **interatrial** septum.
- The right and left ventricles are separated by a thicker muscle wall called the **interventricular** septum.

In addition, the **atrioventricular (AV) septum** is located between and behind the right atrium and left ventricle.

Blood circulates to all parts of the body except the lungs through the **systemic circulation** to deliver **oxygen** (O_2) and nutrients to the body cells and pick up carbon dioxide (CO_2) and metabolic waste products. This **deoxygenated** blood returns to the heart and is then sent out into the **pulmonary circulation** to the lungs, where the waste material **carbon dioxide** is exchanged for oxygen from inhaled air (*Figure 6.5*).

The left ventricle pumps blood out through the **aorta** into the systemic circulation, and the right ventricle pumps blood out to the pulmonary circulation through the pulmonary arteries.

Four valves ensure the correct flow of blood through the heart (*see Figure 6.4b*):

1. The **tricuspid** valve controls the opening between the right atrium and the right ventricle.
2. The **pulmonary semilunar** valve controls the opening between the right ventricle and the pulmonary artery trunk.
3. The **mitral** valve, also known as the **bicuspid valve,** controls the opening between the left atrium and the left ventricle.
4. The **aortic semilunar** valve controls the opening between the left ventricle and the aorta.

WORD ANALYSIS AND DEFINITION

S/ = Suffix   P/ = Prefix   R/ = Root   R/CF = Combining Form

WORD	PRONUNCIATION	ELEMENTS		DEFINITION
atrioventricular (AV)	**A**-tree-oh-ven-**TRICK**-you-lar	S/ R/CF R/	**-ar** *pertaining to* **atri/o-** *entrance, atrium* **-ventricul-** *ventricle*	Pertaining to both the atrium and the ventricle.
atrium **atria (pl)** **atrial (adj)**	**A**-tree-um **A**-tree-ah **A**-tree-al	S/ R/ S/	**-um** *structure* **atri-** *entrance, atrium* **-al** *pertaining to*	Chamber where blood enters the heart on both the right and left sides. Pertaining to the atrium.
bicuspid	by-**KUSS**-pid	S/ P/ R/	**-id** *having a particular quality* **bi-** *two* **-cusp-** *point*	Having two points; a bicuspid heart valve has two flaps.
interatrial	**IN**-ter-**AY**-tree-al	S/ P/ R/	**-al** *pertaining to* **inter-** *between* **-atri-** *atrium*	Between the atria of the heart.
interventricular (IV)	**IN**-ter-ven-**TRIK**-you-lar	S/ P/ R/	**-ar** *pertaining to* **inter-** *between* **-ventricul-** *ventricle*	Between the ventricles of the heart.
mitral	**MY**-tral		Latin *turban*	Shaped like the headdress of a Catholic bishop.
semilunar	sem-ee-**LOO**-nar	S/ P/ R/	**-ar** *pertaining to* **semi-** *half* **-lun-** *moon*	Appears like a half moon.
septum **septa (pl)**	**SEP**- tum **SEP**-tah		Latin *partition*	A thin wall dividing two cavities.
tricuspid	try-**KUSS**-pid	S/ P/ R/	**-id** *having a particular quality* **tri-** *three* **-cusp-** *point*	Having three points; a tricuspid heart valve has three flaps.
ventricle	**VEN**-trih-kel		Latin *small belly*	Chamber of the heart (pumps blood) or a cavity in the brain (produces cerebrospinal fluid).

Abbreviations

AV	atrioventricular
CO_2	carbon dioxide
IV	interventricular
O_2	oxygen

EXERCISES

Prefixes *can generally answer a question: Where is it? What color is it? How big is it? How many are there? This exercise groups some prefixes from the WAD and asks you to identify what makes them similar. Studying elements in similar groups will make them easier to remember. Fill in the chart.*

Prefix	Meaning of Prefix	Medical Term with This Prefix	Meaning of Medical Term
bi			
semi			
tri			

These prefixes are all similar because:

____

Prefix	Meaning of Prefix	English Term with This Prefix	Meaning of English Term
bi			
semi			
tri			

Study Hint

Try to relate these same prefixes to common English words. It will help you remember the meaning of the prefix in the medical terms.

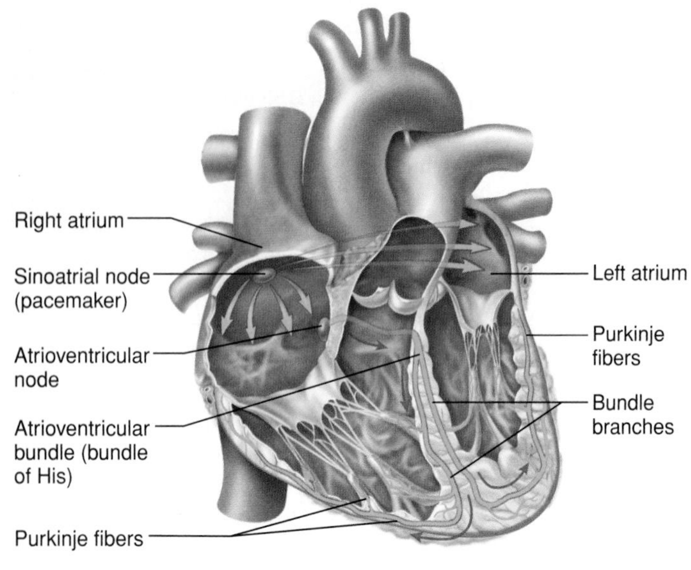

▲ **FIGURE 6.6**
Cardiac Conduction System.

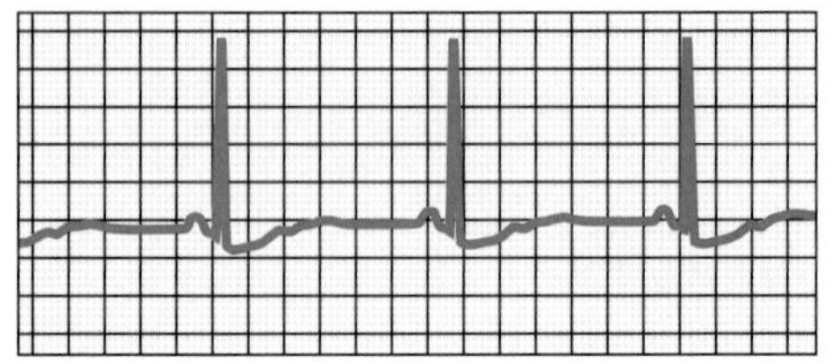

▲ **FIGURE 6.7**
Bradycardia.

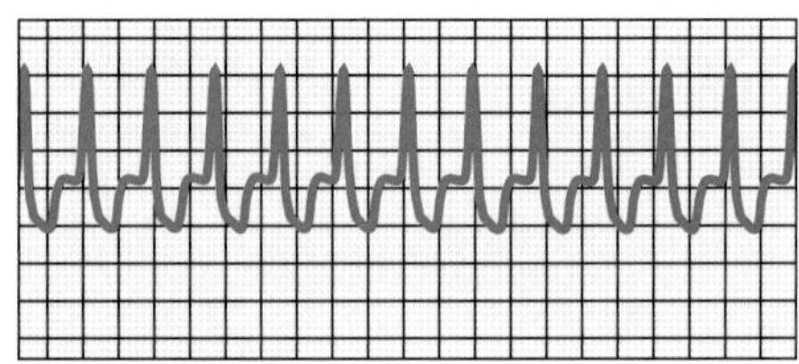

▲ **FIGURE 6.8**
Tachycardia.

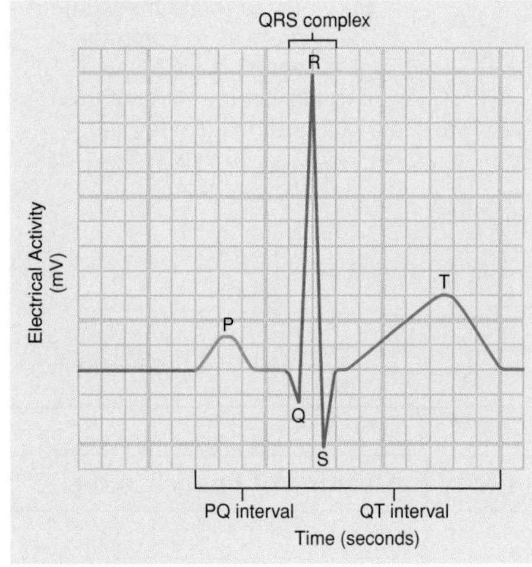

▲ **FIGURE 6.9**
Normal Electrocardiogram.

THE HEARTBEAT

The actions of the four heart chambers are coordinated. When the atria contract (atrial **systole),** the ventricles relax (ventricular **diastole**). When the atria relax (atrial diastole), the ventricles contract (ventricular systole). Then the atria and ventricles all relax briefly. This series of events is a complete cardiac cycle, or heartbeat.

The "lub-dub, lub-dub" sounds heard through the stethoscope are made by the snap of the heart valves as they close. If there is an abnormality in valve closure, it will produce an added-on, abnormal sound called a **murmur.** (You can listen to both sounds on the Student Online Learning Center (www.mhhe.com/AllanEssMedLanguage).

Electrical Properties of the Heart

As the cardiac muscles contract, they generate a small electrical current. To keep the heart beating in a rhythm, a **conduction system** is in place. It consists of five components (*Figure 6.6*):

1. A small region of specialized muscle cells in the right atrium initiates the heartbeat. This area is called the **sinoatrial (SA) node.** It is the **pacemaker** of heart rhythm.
2. Electrical signals from the SA node spread out through the atria and come back together at the **atrioventricular (AV) node.** This is the electrical gateway to the ventricles.
3. Electrical signals leave the AV node to reach the ventricles through the **atrioventricular bundle,** called the **bundle of His.**
4. The AV bundle divides into the right and left **bundle branches,** which supply the two ventricles.
5. From the two bundle branches, **Purkinje fibers** spread through the ventricular myocardium and distribute the electrical stimuli to cause contraction of the ventricular myocardium.

The normal rate of heartbeat is 60 to 80 beats per minute. A heart rate slower than 60 is called **bradycardia** (*Figure 6.7*). A heart rate faster than 100 is called **tachycardia** (*Figure 6.8*).

The normal heartbeat with normal electrical conduction through the heart leading to a ventricular rate of around 60 to 80 beats per minute is called **sinus rhythm.** Any abnormal cardiac rhythm is called an **arrhythmia** or a **dysrhythmia.**

An instrument called an electrocardiograph picks up the electrical changes in the heart muscle and amplifies them to record an electrocardiogram in the form of waves (*Figure 6.9*).

WORD ANALYSIS AND DEFINITION

S/ = Suffix P/ = Prefix R/ = Root R/CF = Combining Form

WORD	PRONUNCIATION	ELEMENTS		DEFINITION
arrhythmia (***Note:*** double "rr")	a-**RITH**-me-ah	S/ P/ R/	**-ia** *condition* **a-** *without* **-rrhythm-** *rhythm*	Condition when the heart rhythm is abnormal.
bradycardia	brad-ee-**KAR**-dee-ah	S/ P/ R/	**-ia** *condition* **brady-** slow **-card-** *heart*	Slow heart rate, below 60 beats per minute.
bundle of His	**BUN**-del of HISS		Wilhelm His, Jr., German physician, 1863–1934	Pathway for electrical signals to be transmitted to the ventricles.
diastole (noun)	die-**AS**-toe-lee		Greek *dilation*	Dilation of heart cavities, during which they fill with blood.
diastolic (adj)	die-as-**TOL**-ik	S/ R/	**-ic** *pertaining to* **diastol-** *diastole*	Pertaining to diastole.
dysrhythmia (***Note:*** single "r")	dis-**RITH**-me-ah	S/ P/ R/	**-ia** *condition* **dys-** *bad, difficult* **-rhythm-** *rhythm*	An abnormal heart rhythm.
murmur	**MUR**-mur		Latin *low voice*	Abnormal heart sound heard with a stetho-scope when a valve closes or opens abnormally.
pacemaker	**PACE**-may-ker	S/ R/	**-maker** *one who makes* **pace-** *step*	Device that regulates cardiac electrical activity.
Purkinje fibers	per-**KIN**-jee **FIE**-bers		Johannes von Purkinje, Bohemian (German) anat-omist and physiologist, 1787–1869	Network of nerve fibers in the myocardium.
sinoatrial (SA) node	sigh-noh-**AY**-tree-al NODE	S/ R/CF R/	**-al** *pertaining to* **sin/o-** *sinus* **-atri-** *atrium*	The center of modified cardiac muscle fibers in the wall of the right atrium that acts as the pacemaker for the heart rhythm.
sinus rhythm	**SIGH**-nus **RITH**-um		sinus *channel, cavity* rhythm, Greek *to flow*	The normal (optimal) heart rhythm arising from the sinoatrial node.
systole (noun) **systolic (adj)**	**SIS**-toe-lee sis-**TOL**-ik	 S/ R/	Greek *contraction* **-ic** *pertaining to* **systol-** *systole, contraction*	Contraction of the heart muscle. Pertaining to systole.
tachycardia	tak-ih-**KAR**-dee-ah	S/ P/ R/	**-ia** *condition* **tachy-** *rapid* **-card-** *heart*	Rapid heart rate (above 100 beats per minute).

Abbreviation

SA sinoatrial

EXERCISES

Spelling: *The terms in the WAD can be particularly difficult to spell, yet they occur all the time in a cardiologist's dictation. Circle the correct choice for the documentation.*

1. This patient's symptoms relate to her (tachycardia/tachicardia).
2. An (arhythmia/arrhythmia) has been confirmed with the EKG.
3. Her (sysstolic/systolic) blood pressure is dangerously high.
4. The (murrmur/murmur) was detected during ventricular (dyastole/diastole).
5. A pacemaker will be inserted to counteract his (bradycardia/brachycardia).
6. Cardiac (dysrrhythmia/dysrhythmia) can be a symptom of a serious underlying condition.

LESSON 6.2 Disorders of the Heart

CARDIOVASCULAR AND CIRCULATORY SYSTEMS

OBJECTIVES

Any loss of consciousness precipitated by exertion can be due to a cardiac arrhythmia or a **cardiomyopathy.** In this lesson, the information will enable you to:

- **Name common cardiac arrhythmias.**
- **Discuss the medical terminology of disorders of the heart valves.**
- **Apply medical terminology to describe coronary heart disease.**
- **Use correct medical terminology to explain hypertensive heart disease.**

You are

. . . an EMT-P called to the gymnasium of Fulwood University.

You are communicating with

. . . Danny Gitlin, a 21-year-old guard on the university basketball team.

CASE REPORT 6.2

Danny lost consciousness during a strenuous practice. He had no pulse but was revived by the coach, who used an automatic external **defibrillator (AED).** Danny never lost consciousness before, but he has noticed episodes of rapid **palpitations** after games. On examination, he is fully conscious and looks fit. His pulse is 70 but irregular in rate and force, and his blood pressure is 110/65 mmHg. He has no known family history of heart disease.

DISORDERS OF THE HEART

Abnormal Heart Rhythms

Arrhythmias are abnormal or irregular heartbeats, and six types are commonly seen:

1. **Premature** beats occur most often in elderly people and are often associated with caffeine and stress.
2. **Atrial fibrillation (A-fib)** occurs when the two atria quiver rather than contract correctly to pump blood. This causes blood to pool in the atria and sometimes clot.
3. **Ventricular tachycardia (V-tach)** is a rapid heartbeat arising in the ventricles.
4. **Ventricular arrhythmias:**
   a. **Premature ventricular contractions (PVCs)** occur when extra impulses arise from a ventricle.
   b. **Ventricular fibrillation (V-fib)** occurs when the ventricles go out of control and beat ineffectively instead of pumping.
5. **Heart block** occurs when interference in cardiac electrical conduction causes the contractions of the atria to fail to coordinate with the contractions of the ventricles.
6. **Palpitations** are unpleasant sensations of a rapid or irregular heartbeat that last a few seconds or minutes. They can be brought on by exercise, anxiety, and stimulants such as caffeine.

Arrhythmias can be treated with medications, but some patients require mechanical **pacemakers.** Pacemakers consist of a battery, electronic circuits, and computer memory to generate electronic signals. The signals are carried along thin, insulated wires to the heart muscle. The most common need for a pacemaker is a very slow heart rate (bradycardia).

Abbreviations

AED	automatic external defibrillator
A-fib	atrial fibrillation
EMT-P	Emergency Medical Technician–Paramedic
ICD	implantable cardioverter/defibrillator
PVC	premature ventricular contraction
V-fib	ventricular fibrillation
V-tach	ventricular tachycardia

WORD ANALYSIS AND DEFINITION

S/ = Suffix   P/ = Prefix   R/ = Root   R/CF = Combining Form

WORD	PRONUNCIATION	ELEMENTS		DEFINITION
arrhythmia (***Note:*** double "rr")	a-**RITH**-me-ah	S/ P/ R/	**-ia** *condition* **a-** *without* **-rrhythm-** *rhythm*	Condition when the heart rhythm is abnormal.
bradycardia	brad-ee-**KAR**-dee-ah	S/ P/ R/	**-ia** *condition* **brady-** slow **-card-** *heart*	Slow heart rate, below 60 beats per minute.
bundle of His	**BUN**-del of HISS		Wilhelm His, Jr., German physician, 1863–1934	Pathway for electrical signals to be transmitted to the ventricles.
diastole (noun)	die-**AS**-toe-lee		Greek *dilation*	Dilation of heart cavities, during which they fill with blood.
diastolic (adj)	die-as-**TOL**-ik	S/ R/	**-ic** *pertaining to* **diastol-** *diastole*	Pertaining to diastole.
dysrhythmia (***Note:*** single "r")	dis-**RITH**-me-ah	S/ P/ R/	**-ia** *condition* **dys-** *bad, difficult* **-rhythm-** *rhythm*	An abnormal heart rhythm.
murmur	**MUR**-mur		Latin *low voice*	Abnormal heart sound heard with a stetho-scope when a valve closes or opens abnormally.
pacemaker	**PACE**-may-ker	S/ R/	**-maker** *one who makes* **pace-** *step*	Device that regulates cardiac electrical activity.
Purkinje fibers	per-**KIN**-jee **FIE**-bers		Johannes von Purkinje, Bohemian (German) anat-omist and physiologist, 1787–1869	Network of nerve fibers in the myocardium.
sinoatrial (SA) node	sigh-noh-**AY**-tree-al NODE	S/ R/CF R/	**-al** *pertaining to* **sin/o-** *sinus* **-atri-** *atrium*	The center of modified cardiac muscle fibers in the wall of the right atrium that acts as the pacemaker for the heart rhythm.
sinus rhythm	**SIGH**-nus **RITH**-um		sinus *channel, cavity* rhythm, Greek *to flow*	The normal (optimal) heart rhythm arising from the sinoatrial node.
systole (noun) **systolic (adj)**	**SIS**-toe-lee sis-**TOL**-ik	 S/ R/	Greek *contraction* **-ic** *pertaining to* **systol-** *systole, contraction*	Contraction of the heart muscle. Pertaining to systole.
tachycardia	tak-ih-**KAR**-dee-ah	S/ P/ R/	**-ia** *condition* **tachy-** *rapid* **-card-** *heart*	Rapid heart rate (above 100 beats per minute).

Abbreviation

SA   sinoatrial

EXERCISES

Spelling: *The terms in the WAD can be particularly difficult to spell, yet they occur all the time in a cardiologist's dictation. Circle the correct choice for the documentation.*

1. This patient's symptoms relate to her (tachycardia/tachicardia).
2. An (arhythmia/arrhythmia) has been confirmed with the EKG.
3. Her (sysstolic/systolic) blood pressure is dangerously high.
4. The (murrmur/murmur) was detected during ventricular (dyastole/diastole).
5. A pacemaker will be inserted to counteract his (bradycardia/brachycardia).
6. Cardiac (dysrrhythmia/dysrhythmia) can be a symptom of a serious underlying condition.

LESSON 6.2 Disorders of the Heart

CARDIOVASCULAR AND CIRCULATORY SYSTEMS

OBJECTIVES

Any loss of consciousness precipitated by exertion can be due to a cardiac arrhythmia or a **cardiomyopathy.** In this lesson, the information will enable you to:

- **Name common cardiac arrhythmias.**
- **Discuss the medical terminology of disorders of the heart valves.**
- **Apply medical terminology to describe coronary heart disease.**
- **Use correct medical terminology to explain hypertensive heart disease.**

You are

... an EMT-P called to the gymnasium of Fulwood University.

You are communicating with

... Danny Gitlin, a 21-year-old guard on the university basketball team.

CASE REPORT 6.2

Danny lost consciousness during a strenuous practice. He had no pulse but was revived by the coach, who used an automatic external **defibrillator (AED).** Danny never lost consciousness before, but he has noticed episodes of rapid **palpitations** after games. On examination, he is fully conscious and looks fit. His pulse is 70 but irregular in rate and force, and his blood pressure is 110/65 mmHg. He has no known family history of heart disease.

DISORDERS OF THE HEART

Abnormal Heart Rhythms

Arrhythmias are abnormal or irregular heartbeats, and six types are commonly seen:

1. **Premature** beats occur most often in elderly people and are often associated with caffeine and stress.
2. **Atrial fibrillation (A-fib)** occurs when the two atria quiver rather than contract correctly to pump blood. This causes blood to pool in the atria and sometimes clot.
3. **Ventricular tachycardia (V-tach)** is a rapid heartbeat arising in the ventricles.
4. **Ventricular arrhythmias:**
   a. **Premature ventricular contractions (PVCs)** occur when extra impulses arise from a ventricle.
   b. **Ventricular fibrillation (V-fib)** occurs when the ventricles go out of control and beat ineffectively instead of pumping.
5. **Heart block** occurs when interference in cardiac electrical conduction causes the contractions of the atria to fail to coordinate with the contractions of the ventricles.
6. **Palpitations** are unpleasant sensations of a rapid or irregular heartbeat that last a few seconds or minutes. They can be brought on by exercise, anxiety, and stimulants such as caffeine.

Arrhythmias can be treated with medications, but some patients require mechanical **pacemakers.** Pacemakers consist of a battery, electronic circuits, and computer memory to generate electronic signals. The signals are carried along thin, insulated wires to the heart muscle. The most common need for a pacemaker is a very slow heart rate (bradycardia).

Abbreviations

AED	automatic external defibrillator
A-fib	atrial fibrillation
EMT-P	Emergency Medical Technician–Paramedic
ICD	implantable cardioverter/defibrillator
PVC	premature ventricular contraction
V-fib	ventricular fibrillation
V-tach	ventricular tachycardia

WORD ANALYSIS AND DEFINITION

S/ = Suffix P/ = Prefix R/ = Root R/CF = Combining Form

WORD	PRONUNCIATION	ELEMENTS		DEFINITION
cardiomyopathy	**KAR**-dee-oh-my-**OP**-ah-thee	S/ R/CF R/CF	**-pathy** *disease* **cardi/o** *-heart* **-my/o-** *muscle*	Disease of the heart muscle, the myocardium.
cardioversion (also called *defibrillation*)	**KAR**-dee-oh-**VER**-shun	R/CF S/	**cardi/o-** *heart* **-version** *change*	Restoration of a normal heart rhythm by electric shock.
defibrillation	dee-fib-rih-**LAY**-shun	S/ P/ R/	**-ation** *process* **de-** *from, out of* **-fibrill-** *small fiber*	Restoration of uncontrolled twitching of cardiac muscle fibers to normal rhythm.
defibrillator	dee-fib-rih-**LAY**-tor	S/	**-ator** *instrument*	Instrument for defibrillation.
fibrillation	fi-brih-**LAY**-shun	S/ R/	**-ation** *a process* **fibrill-** *small fiber*	Uncontrolled quivering or twitching of the heart muscle.
implantable	im-**PLAN**-tah-bul	S/ P/ R/	**-able** *capable* **im-** *in* **-plant-** *insert*	A device that can be inserted into tissues.
pacemaker	**PACE**-may-ker	S/ R/	**-maker** *one who makes* **pace-** *step*	Device that regulates cardiac electrical activity.
palpitation	pal-pih-**TAY**-shun	S/ R/	**-ation** *a process* **palpit-** *throb*	Forcible, rapid beat of the heart felt by the patient.

In emergency situations, external **cardioversion** is performed through **automatic external defibrillators,** or **AEDs** (*Figure 6.10*), that send an electric shock to the heart to stop the heart and allow a normal contraction rhythm to resume. This was used for Danny Gitlin.

People with life-threatening arrhythmias may need an **implantable cardioverter/defibrillator (ICD),** which senses abnormal rhythms and gives the heart a small electric shock to return the rhythm to normal.

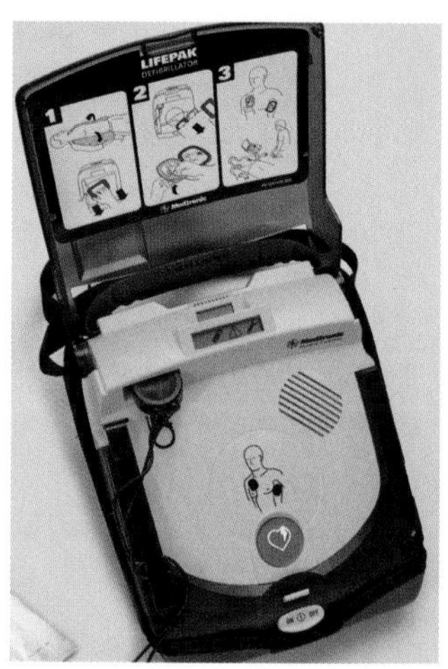

▲ **FIGURE 6.10 Automatic External Defibrillator.**

EXERCISES

Construct *the correct medical terms to match the definitions given. Notice, in particular, that some terms do not have prefixes or suffixes but every term must have a root and/or combining form. Fill in the blanks.*

1. forceful, rapid beat of the heart

_______________ / _______________ / _______________
P R/CF S

2. uncontrolled heart muscle twitching

_______________ / _______________ / _______________
P R/CF S

3. device that restores uncontrolled twitching of cardiac muscle to normal rhythm

_______________ / _______________ / _______________
P R/CF S

4. implanted device that regulates cardiac electrical activity

_______________ / _______________ / _______________
P R/CF S

DISORDERS OF THE HEART (continued)

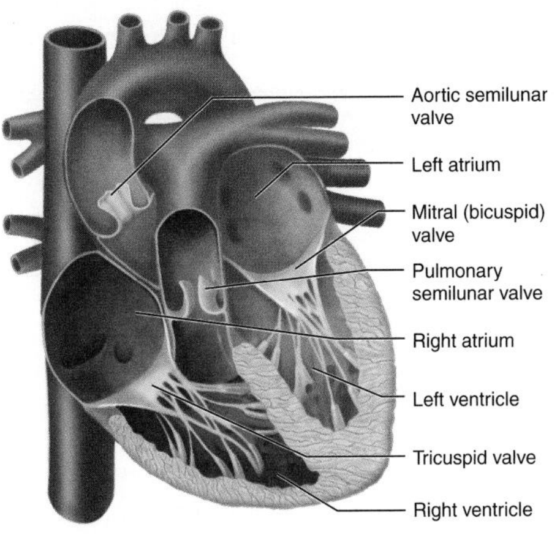

▲ **FIGURE 6.11**
Heart Valves.

Disorders of Heart Valves

Heart valves can **malfunction** in two basic ways:

1. **Stenosis:** The valve cannot open fully, and its opening is narrowed (constricted). Blood cannot flow freely through the valve and accumulates behind the valve.
2. **Incompetence** or **insufficiency:** The valve cannot close fully, and blood can **regurgitate** (flow back) through the valve to the chamber from where it started.

Mitral valve stenosis can occur following rheumatic fever. Because the blood cannot flow freely through the valve, the left atrium becomes dilated (enlarged). Eventually, chronic **heart failure** results. (*Figure 6.11* enables you to review the locations of the valves and chambers.)

Mitral valve incompetence occurs when there is leakage back through the valve (regurgitation) as the left ventricle contracts. The left atrium becomes dilated. Again, chronic heart failure results.

Mitral valve prolapse occurs when the cusps of the valve bulge back into the left atrium when the left ventricle contracts. This allows blood to flow back into the atrium.

Aortic valve stenosis is common in the elderly when the valves become calcified due to **atherosclerosis.** Blood flow into the systemic circulation is diminished, leading to dizziness and fainting. The left ventricle dilates, **hypertrophies,** ceases to beat strongly, and ultimately fails.

Aortic valve incompetence initially produces few symptoms other than a murmur. Eventually the left ventricle is unable to cope with the excess volume of blood and fails.

Malfunctions of the valves on the right side of the heart are much less common.

A prolapsed or incompetent mitral valve can often be repaired. If a valve replacement is necessary, there are two types of artificial valve to choose from:

1. Mechanical or **prosthetic** valves, which are made from different metal alloys and plastics.
2. **Tissue** valves, which can come from a pig or cow, or occasionally from a human cadaver (dead person), or can be constructed of tissue from the patient's own pericardium.

Disorders of the Heart Wall

Endocarditis is inflammation of the lining of the heart. It is usually secondary to an infection elsewhere. Intravenous drug users and people with damaged heart valves are at high risk for endocarditis.

Myocarditis is inflammation of the heart muscle. It can be bacterial, viral, or fungal in origin or can arise as a complication of other diseases such as influenza.

Pericarditis is inflammation of the covering of the heart. The inflammation causes an exudate (pericardial effusion) to be released into the pericardial space between the two layers of the pericardium. This interferes with the heart's ability to contract and expand normally, and cardiac output **(CO)** falls—a condition called **cardiac tamponade.**

Cardiomyopathy is a weakening of the heart muscle that causes it to pump inadequately. It causes **cardiomegaly** and heart failure.

Abbreviation	
CO	cardiac output

Case Report 6.2 (continued)

The physical findings on Danny Gitlin suggested a cardiomyopathy, and this was confirmed by echocardiography. Exercise testing (a cardiac stress test), with strong medical supervision, produced a ventricular arrhythmia that returned to normal on cessation of the test. Danny was treated with beta blockers and restricted to nonstrenuous sports.

Cor pulmonale is failure of the right ventricle to pump properly. Almost any chronic lung disease causing low blood oxygen **(hypoxia)** can cause this disorder.

WORD ANALYSIS AND DEFINITION

S/ = Suffix P/ = Prefix R/ = Root R/CF = Combining Form

WORD	PRONUNCIATION	ELEMENTS		DEFINITION
cardiomegaly	**KAR**-dee-oh-**MEG**-ah-lee	S/ R/CF	**-megaly** *enlargement* **cardi/o-** *heart*	Enlargement of the heart.
cor pulmonale	**KOR** pul-moh-**NAH**-lee	S/ R/ R/	**-ale** *pertaining to* **cor** *heart* **pulmon-** *lung*	Right-sided heart failure arising from chronic lung disease.
endocarditis	**EN**-doh-kar-**DIE**-tis	S/ P/ R/	**-itis** *inflammation* **endo-** *within* **-card-** *heart*	Inflammation of the lining of the heart.
hypertrophy (can be a noun or a verb)	high-**PER**-troh-fee	P/ R/	**hyper-** *above, excessive* **-trophy** *development*	Increase in size, but not in number, of an individual tissue element.
incompetence	in-**KOM**-peh-tense	S/ P/ R/	**-ence** *quality of* **in-** *not* **-compet-** *strive together*	Failure of a valve to close completely.
insufficiency	in-suh-**FISH**-en-see	S/ P/ R/CF	**-ency** *quality of* **in-** *not* **–suffic/i-** *enough*	Lack of completeness of function; e.g., for a heart valve to fail to close properly.
myocarditis	**MY**-oh-kar-**DIE**-tis	S/ R/CF R/	**-itis** *inflammation* **my/o-** *muscle* **-card-** *heart*	Inflammation of the heart muscle.
pericarditis	**PER**-ih-kar-**DIE**-tis	S/ P/ R/	**-itis** *inflammation* **peri-** *around* **-card-** *heart*	Inflammation of the pericardium, the covering of the heart.
prolapse	pro-**LAPS**		Latin *a falling*	An organ slips out of its normal position.
prosthesis (noun)	**PROS**-thee-sis		Greek *an addition*	A manufactured substitute for a missing or diseased part of the body.
prosthetic (adj)	pros-**THET**-ik	S/ R/	**-ic** *pertaining to* **prosthet-** *artificial part*	Pertaining to a prosthesis.
regurgitate	ree-**GUR**-jih-tate	S/ P/ R/	**-ate** *pertaining to* **re-** *back* **-gurgit-** *flood*	To flow backward; e.g., blood through a heart valve.
stenosis	ste-**NOH**-sis	S/ R/CF	**-sis** *abnormal condition* **sten/o-** *narrow*	Narrowing of a canal or passage, e.g., of a heart valve.
tamponade	tam-poh-**NAID**	S/ R/	**-ade** *a process* **tampon-** *plug*	Pathologic compression of an organ, such as the heart.

EXERCISES

Demonstrate *that you can use the two terms below correctly. Write a sentence of patient documentation for each of the terms.*

1. *Hyper**trophy*** means an increase in *size,* not number.
   (*trophy* = development; *hyper* = excessive) A *trophy* for a sports event comes in all *sizes.*
2. *Regurgitation* is a term that appeared in Chapter 5, on the digestive system, relative to GERD (gastroesophageal reflux disease). It is a body process that can be applicable to different body systems, but in all cases it means essentially the same thing: Something is flowing backward, and it shouldn't be. In one case, it is stomach contents; in another case, it is blood flowing backward through a heart valve.

   a. ______________________________

   ____

   b. ______________________________

   ____

Study Hint

Create easy tricks for yourself to help you remember medical terms.

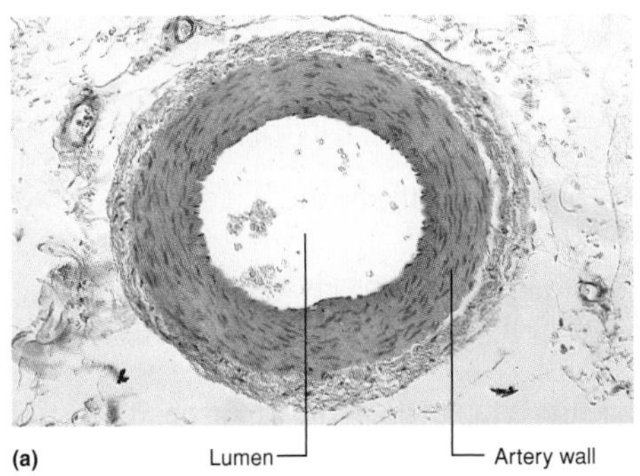

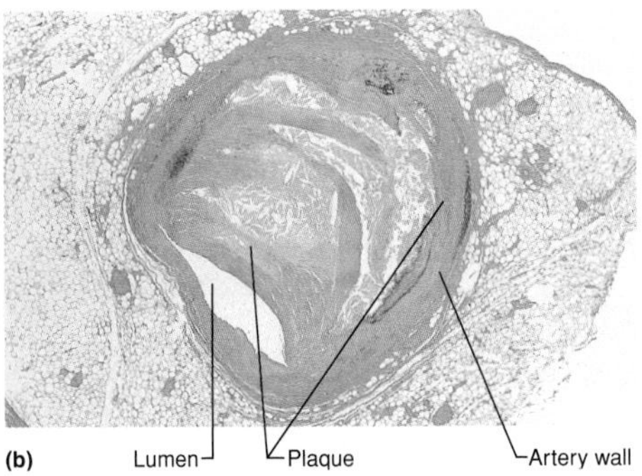

▲ **FIGURE 6.12**
Arterial Structure.
(a) Normal coronary artery.
(b) Advanced atherosclerosis.

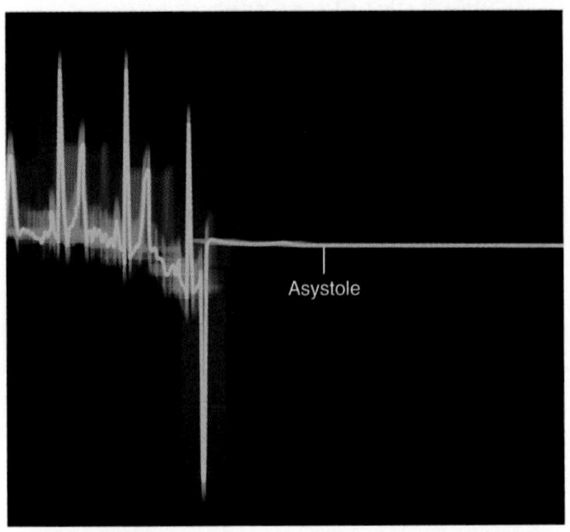

▲ **FIGURE 6.13**
Electrocardiogram (ECG) Showing Asystole.

Keynote

All these risk factors, except heredity and age, can be reduced by changes in lifestyle.

DISORDERS OF THE HEART (continued)

Coronary Artery Disease (CAD)

The coronary arteries supplying the myocardium become narrowed by **atherosclerotic plaques,** called **atheroma.** As the atheroma increases, the **lumen** of the artery becomes more and more narrow, or constricted (*compare Figure 6.12a to 6.12b*). The blood supplied to cardiac muscle by the artery is reduced. Platelet clumping can occur on the plaque and form a blood clot **(coronary thrombosis). Atherosclerosis** is the most common form of **arteriosclerosis** (hardening of the arteries), and it can lead to **arteriosclerotic** heart disease **(ASHD).**

Angina pectoris (pain in the chest on exertion) is often the first symptom of reduced oxygen supply to the myocardium. The pain goes away if the exertion is stopped or if a nitroglycerin **(NTG)** tablet is placed under the tongue **(sublingual).**

Myocardial infarction (MI) is the death of myocardial cells and is caused by the lack of blood supply (ischemia) when an artery eventually becomes blocked **(occluded).** Sudden, severe, crushing **substernal** or left chest pain is experienced. If the ischemia is not reversed within 4 to 6 hours, the myocardial cells die (necrosis).

Case Report 6.1 *(continued)*

For Mr. Johnson in the Emergency Department, the ECG (EKG) indicated that he was having an MI affecting the anterior wall of his left ventricle. The cardiovascular technician did not want to leave the patient and so used the call system to obtain nursing and medical help.

Circulatory shock occurs when cardiac output is insufficient to meet the body's metabolic needs. There are two main categories:

1. **Cardiogenic shock** occurs when the heart fails to pump effectively and organs and tissues are **perfused** inadequately. The pulse is weak and rapid, and blood pressure drops. The patient becomes pale, cold, sweaty, and anxious.
2. **Hypovolemic shock** occurs when there is a loss of blood volume, often from excessive bleeding (hemorrhage) or dehydration.

Cardiac arrest is the sudden cessation of cardiac activity that results from **anoxia.** Most patients show **asystole** on the cardiac monitor (*Figure 6.13*). A person in cardiac arrest has no pulse and is not breathing and can be referred to as a **pulseless nonbreather (PNB).**

Risk factors for CAD include:

- Heredity
- Age
- Obesity
- Lack of exercise
- Tobacco
- Diabetes mellitus
- Stress
- High blood pressure
- Elevated serum cholesterol

WORD ANALYSIS AND DEFINITION

S/ = Suffix P/ = Prefix R/ = Root R/CF = Combining Form

WORD	PRONUNCIATION	ELEMENTS		DEFINITION
anoxia (noun) **anoxic (adj)**	an-**OCK**-see-ah an-**OCK**-sik	S/ P/ R/ S/	**-ia** *condition* **an-** *without* **-ox-** *oxygen* **-ic** *pertaining to*	Without oxygen. Pertaining to or suffering from lack of oxygen.
arteriosclerosis **arteriosclerotic (adj)**	ar-**TIER**-ee-oh-skler-**OH**-sis ar-**TIER**-ee-oh-skler-**OT**-ik	S/ R/CF R/CF S/	**-sis** *abnormal condition* **arteri/o-** *artery* **-scler/o-** *hardness* **-tic** *pertaining to*	Hardening of the arteries. Pertaining to or affected by arteriosclerosis.
asystole	a-**SIS**-toe-lee	P/ R/CF	**a-** *without* **-systol/e** *contraction*	Absence of contractions of the heart.
atheroma (plaque) **atherectomy** **atherosclerosis**	ath-er-**ROE**-mah ath-er-**EK**-toe-me **ATH**-er-oh-skler-**OH**-sis	S/ R/ S/ S/ R/CF R/CF	**-oma** *tumor, mass* **ather-** *porridge, gruel* **-ectomy** *surgical excision* **-sis** *abnormal condition* **ather/o-** *porridge, gruel* **-scler/o-** *hardness*	Fatty deposit in the lining of an artery. Surgical removal of the atheroma. Hardening of the arteries due to atheroma (plaque).
cardiogenic	**KAR**-dee-oh-**JEN**-ik	S/ R/ R/CF	**-ic** *pertaining to* **-gen-** *produce* **cardi/o-** *heart*	Of cardiac origin.
hypovolemic	**HIGH**-poh-vo-**LEE**-mick	S/ P/ R/	**-emic** *a condition in the blood* **hypo-** *below* **-vol- volume**	Decreased blood volume in the body.
lumen	**LOO**-men		Latin *window*	The interior space of a tube-like structure.
occlude (verb) **occlusion (noun)**	oh-**KLUDE** oh-**KLU**-zhun		Latin *to close*	To close, plug, or completely obstruct. A complete obstruction.
perfuse (verb) **perfusion (noun)**	per-**FYUSE** per-**FYU**-shun	 S/ R/	Latin *to pour* **-ion** *action, condition* **perfus-** *to pour*	To force blood to flow through a lumen or vascular bed. The act of perfusing.
substernal	sub-**STER**-nal	S/ P/ R/	**-al** *pertaining to* **sub-** *under* **-stern-** *chest*	Under (behind) the sternum or breastbone.

Abbreviations

ASHD	arteriosclerotic heart disease
CAD	coronary artery disease
MI	myocardial infarction
NTG	nitroglycerin
PNB	pulseless nonbreather

EXERCISES

Recall *the use of the prefix in the following terms. You have seen this prefix before, and you will see it again—same prefix, different elements, more terms. Fill in the blanks.*

1. The prefix *sub* means ______________________________.
2. *Substernal* means ______________________________.
3. *Sublingual* means ______________________________.
4. *Subcutaneous* means ______________________________.
5. *Submandibular* means ______________________________.
6. *Submucosa* means ______________________________.

Keynote

Hypertension is the major cause of heart failure, stroke, and kidney failure.

Keynote

All these risk factors can be reduced by changes in lifestyle.

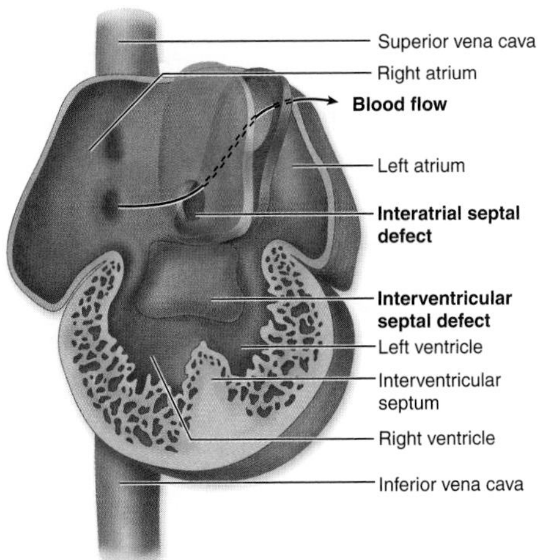

▲ **FIGURE 6.14**
Atrial and Ventricular Septal Defects.

Abbreviations

ASD	atrial septal defect
CHD	congenital heart disease
CHF	congestive heart failure
HTN	hypertension
PDA	patent ductus arteriosus
SOB	shortness of breath
TOF	tetralogy of Fallot
VSD	ventricular septal defect

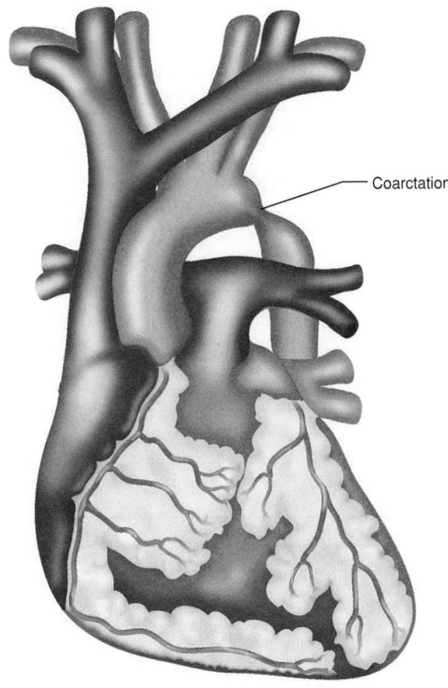

▲ **FIGURE 6.15 Coarctation of Aorta**

DISORDERS OF THE HEART (continued)

Hypertensive Heart Disease

Hypertension (HTN) is the most common cardiovascular disorder in this country, affecting more than 20% of the adult population. It results from a prolonged elevated blood pressure throughout the vascular system. The high pressure forces the ventricles to work harder to pump blood.

High blood pressure is currently defined as a blood pressure reading at or above 140/90 mmHg (millimeters of mercury). A normal blood pressure is below 120/80 mmHg. The first number, or **systolic** reading, reflects the blood pressure when the heart is contracting. The second number, or **diastolic** reading, reflects the blood pressure when the heart is relaxed between contractions.

Primary or **essential hypertension** is the most common type of hypertension. Its etiology is **idiopathic** (unknown). Its risk factors are:

- Overweight
- Lack of exercise
- Tobacco
- Alcohol
- Stress

Secondary hypertension results from other diseases such as kidney disease, atherosclerosis, and hyperthyroidism.

Malignant hypertension is a rare, severe, life-threatening form of hypertension in which the blood pressure reading can be greater than 200/120 mmHg. Aggressive intervention is indicated to reduce the blood pressure.

Prehypertension, with a systolic pressure between 120 and 139 mmHg and a diastolic pressure between 80 and 90 mmHg, may indicate an increased risk for cardiovascular disease.

Congestive Heart Failure (CHF)

CHF occurs with the inability of the heart to supply enough cardiac output to meet the body's metabolic needs. The patient shows shortness of breath **(SOB),** particularly when lying down **(orthopnea),** and **rales** can be heard at the bases of the lungs.

The most common conditions leading to CHF are:

- Cardiac ischemia
- Severe hypertension
- Valvular regurgitation
- Aortic stenosis
- Cardiomyopathy

Congenital Heart Disease (CHD)

CHD is the result of abnormal development of the heart in the fetus. Common **congenital defects** include the following.

Atrial septal defect (ASD), a hole in the interatrial septum, allows blood to **shunt** from the higher-pressure left atrium to the lower-pressure right atrium (*Figure 6.14*).

Ventricular septal defect (VSD), a gap in the interventricular septum, allows blood to shunt from the higher-pressure left ventricle to the lower-pressure right ventricle (*Figure 6.14*).

Patent ductus arteriosus (PDA) arises from a defect in the ductus arteriosus, a normal blood vessel in the fetus that usually closes within 24 hours of birth. When the artery remains open (patent), blood can shunt from the aorta to the pulmonary artery and the higher pressure causes damage to the lungs.

Coarctation of the aorta is a narrowing of the aorta shortly after the artery to the left arm branches from the aorta (*Figure 6.15*). This causes **hypertension** in the arms behind the narrowing and **hypotension** in the lower limbs and organs such as the kidney below the narrowing.

Congenital abnormalities can usually be surgically repaired.

Tetralogy of Fallot (TOF) is a **syndrome** in which four congenital heart defects all shunt blood away from the lungs. Children with TOF do not grow normally, are cyanotic (blue discoloration of the skin, lips, and nail beds), and are dyspneic (having difficulty breathing), and they require open heart surgery to correct the defects.

WORD ANALYSIS AND DEFINITION

S/ = Suffix   P/ = Prefix   R/ = Root   R/CF = Combining Form

WORD	PRONUNCIATION	ELEMENTS		DEFINITION
coarctation	koh-ark-**TAY**-shun	S/ R/	**-ation** *process* **coarct-** *press together, narrow*	Constriction, stenosis, particularly of the aorta.
congenital	kon-**JEN**-ih-tal	S/ P/ R/	**-al** *pertaining to* **con-** *together, with* **-genit-** *bring forth*	Present at birth, either inherited or due to an event during gestation up to the moment of birth
defect	**DEE**-fect		Latin *to lack*	An absence, malformation, or imperfection.
hypertension **hypotension** **prehypertension**	**HIGH**-per-**TEN**-shun **HIGH**-poh-**TEN**-shun pree-**HIGH**-per-**TEN**-shun	S/ P/ R/ P/ P/	**-ion** *condition, action* **hyper-** *excessive* **-tens-** *pressure* **hypo-** *low* **pre-** *before*	Persistent high arterial blood pressure. Persistent low arterial blood pressure. Precursor to hypertension.
idiopathic	**ID**-ih-oh-**PATH**-ik	S/ R/CF R/	**-ic** *pertaining to* **idi/o-** *unknown* **-path-** *disease*	Pertaining to a disease of unknown etiology.
orthopnea **orthopneic (adj)**	or-**THOP**-nee-ah or-**THOP**-nee-ik	S/ R/CF S	**-pnea** *breathe* **orth/o-** *straight* **-ic** *pertaining to*	Difficulty in breathing when lying flat. Pertaining to or affected by orthopnea.
patent ductus arteriosus (PDA) (***Note:*** This term is composed only of roots.)	**PAY**-tent **DUK**-tus ar-**TER**-ee-oh-sus	R/ R/ R/	**patent** *lie open* **ductus** *leading* **arteriosus** *artery*	An open, direct channel between the aorta and the pulmonary artery.
rale **rales (pl)**	RAHL RAHLS		French *rattle*	Crackle heard through a stethoscope when air bubbles through liquid in the lungs.
shunt	SHUNT		Middle English *divert*	A bypass or diversion of fluid, in this case, blood.
syndrome	**SIN**-drohm	P/ R/	**syn-** *together* **-drome** *running*	Combination of signs and symptoms associated with a particular disease process.
tetralogy of Fallot (TOF)	te-**TRAL**-oh-jee of fah-**LOW**	P/ R/	**tetra-** *four* **-logy** *study of* Etienne-Louis Fallot, French physician, 1850–1911	Set of four congenital heart defects occurring together.

EXERCISES

Abbreviations *need to be used carefully so that you communicate exactly what is necessary. The following sentences contain abbreviations—translate the abbreviations into their correct medical terms. Rewrite the sentences without the abbreviations and convey the same message. Fill in the blanks.*

1. ASD, TOF, VSD, and PDA are all examples of CHD.

____

____

____

2. The patient's symptoms were HTN and SOB.

____

____

____

✓ *your spelling!*

CARDIOLOGIC INVESTIGATIONS AND PROCEDURES

Blood Tests

A **lipid profile** helps determine the risk of CAD and comprises:

- Total cholesterol.
- High-density **lipoprotein (HDL)** ("good cholesterol").
- Low-density lipoprotein **(LDL)** ("bad cholesterol").
- **Triglycerides.**

Homocysteine is an amino acid in the blood. Elevated levels are related to a higher risk of CAD, stroke, and peripheral vascular disease.

Troponin I and T are part of a protein complex in muscle that is released into the blood during muscle injury. Troponin I is found in heart muscle but not in skeletal muscle. It is therefore a highly sensitive indicator of a recent MI.

Diagnostic Tests

An **electrocardiogram (ECG** or **EKG)** is a paper record of the electrical signals of your heart.

Cardiac stress testing is an exercise tolerance test to raise your heart rate and monitor its effect on cardiac function. **Nuclear imaging** of the heart, using an injection of a radioactive substance, can be used with the stress test.

Echocardiography uses ultrasound waves to study cardiac function.

A **Holter monitor** is a continuous ECG recorded on a tape-recorder cassette as you work, play, and rest for at least 24 hours.

An **ambulatory blood pressure monitor** provides a record of your blood pressure over a 24-hour period as you go about your daily activities.

Magnetic resonance imaging (MRI) can produce detailed images of the heart and identify sections of cardiac muscle that are not receiving an adequate blood supply.

Cardiac catheterization detects patterns of pressures and blood flows in the heart. A thin tube is threaded into the heart under x-ray guidance after being inserted into a vein or an artery.

A **coronary angiogram** uses a contrast dye injected during cardiac catheterization to identify coronary artery blockages.

Treatment Procedures

The most immediate need in the treatment of MI is to get blood and oxygen to the affected myocardium. This can be attempted in several ways:

1. **Injection of clot-busting (thrombolytic) drugs:** These drugs are injected within a few hours of the MI to dissolve the **thrombus.**
2. **Artery-cleaning angioplasty (percutaneous transluminal coronary angioplasty, or PTCA):** A balloon-tipped catheter is guided to the site of the blockage and inflated to expand the artery from the inside by compressing the plaque against the walls of the artery.
3. **Stent placement:** To reduce the likelihood that the artery will close up again (occlude), a wire-mesh tube, or **stent,** is placed inside the vessel. Some stents (**drug-eluting** stents) are covered with a special medication to help keep the artery open.
4. **Coronary artery bypass surgery (CABG):** Healthy blood vessels harvested from the leg, chest, or arm are used to bypass (detour) the blood around blocked coronary arteries.

Other procedures used in cardiology include:

- **Cardioversion:** An arrhythmia is converted back to normal sinus rhythm with an electric shock from a defibrillator.
- **Defibrillation:** V-fib is terminated and normal rhythm restored by delivery of an electric shock to the heart.
- **Heart transplant:** The heart of a recently deceased person (donor) is transplanted to the recipient after the recipient's diseased heart has been removed.

Abbreviations

CABG	coronary artery bypass graft
HDL	high-density lipoprotein
LDL	low-density lipoprotein
MRI	magnetic resonance imaging
PTCA	percutaneous transluminal coronary angioplasty

WORD ANALYSIS AND DEFINITION

S/ = Suffix P/ = Prefix R/ = Root R/CF = Combining Form

WORD	PRONUNCIATION	ELEMENTS		DEFINITION
angiogram	**AN-**jee-oh-gram	S/ R/CF	**-gram** *a record* **angi/o-** *blood vessel*	Radiograph obtained after injection of radiopaque contrast material into blood vessels.
angiography	**AN**-jee-**OG**-rah-fee	S/	**-graphy** *process of recording*	Radiography of blood vessels after injection of contrast material.
angioplasty	**AN**-jee-oh-**PLAS**-tee	S/ R/CF	**-plasty** *surgical repair* **angi/o-** *blood vessel*	Recanalization of a blood vessel by surgery.
catheter	**KATH-**eh-ter		Greek *to send down*	Hollow tube to allow passage of fluid into or out of a body cavity, organ, or vessel.
catheterize (verb)	**KATH**-eh-teh-**RIZE**	S/ R/	**-ize** *action* **catheter-** *catheter*	To introduce a catheter.
catheterization (noun)	**KATH**-eh-ter-ih-**ZAY**-shun	S/	**-ization** *process of inserting*	Introduction of a catheter.
echocardiography	**EK**-oh-kar-dee-**OG**-rah-fee	S/ R/CF R/CF	**-graphy** *process of recording* **ech/o-** *sound wave* **-cardi/o-** *heart*	Ultrasound recording of heart function.
lipoprotein	**LIE**-poh-pro-teen	R/CF R/	**lip/o-** *fat* **-protein** *protein*	Bonding of molecules of fat and protein.
percutaneous	**PER**-kyu-**TAY**-nee-us	S/ P/ R/CF	**-ous** *pertaining to* **per-** *through* **-cutan/e-** *skin*	Passage through the skin, in this case, by needle puncture.
stent	STENT		Charles Stent, English dentist, 19th century	Wire-mesh tube used to keep arteries open.
thrombus **thrombi (pl)**	**THROM**-bus **THROM**-bee		Latin *clot*	A clot attached to a diseased blood vessel or heart lining.
thrombolytic (adj)	throm-boh-**LIT**-ik	S/ R/CF R/	**-tic** *pertaining to* **thromb/o-** *blood clot* **-ly-** *break down*	Able to dissolve or break up a blood clot.
thrombolysis	throm-**BOL**-ih-sis	S/	**-lysis** *dissolve*	Dissolving of a thrombus (clot).
triglyceride	try-**GLISS**-eh-ride	S/ P/ R/	**-ide** *having a particular quality* **tri-** *three* **-glycer-** *sweet, glycerol*	Lipid containing three fatty acids.

EXERCISES

Elements *are your best clue to understanding a medical term. Test yourself on the elements in column 1: Write the meaning of the element in its appropriate column. The first one is done for you.*

Element	Meaning of Prefix	Meaning of Root/CF	Meaning of Suffix
angio		blood vessel	
cardio			
cutane			
echo			
gram			
lipo			
lysis			
ous			
per			
plasty			
thrombo			
tri			

LESSON 6.3 Circulatory Systems

CARDIOVASCULAR AND CIRCULATORY SYSTEMS

OBJECTIVES

In order to understand the **etiologies** and effects of Mrs. Jones's problems (*see Case Report 6.3*) and communicate with her and Dr. Bannerjee about them, you need to have the medical terminology and knowledge to be able to:

- **Specify the medical terminology for the functions of the systemic and pulmonary circulations.**
- **Identify the major arteries and veins in the body.**
- **Use correct medical terminology to explain the *hemodynamics* and control of blood flow.**
- **Apply correct medical terminology to describe common disorders of the circulatory system.**

You are

. . . a certified medical assistant working for Dr. Lokesh Bannerjee, a cardiologist in Fulwood Medical Center.

You are communicating with

. . . Mrs. Martha Jones. You are documenting her medical record after Dr. Bannerjee interviewed and examined her.

CASE REPORT 6.3

Fulwood Medical Center
Consultation Request and Report Form

Patient's Name: Jones, MARTHA Age: 52
To: Dr. LOKESH BANNERJEE Department: Cardiology
From: Dr. Susan Lee Department: Primary Care
Patient's Location: FULWOOD MEDICAL CENTER
Type of Consultation Desired:
☐ Consultation Only
☐ Consulation and follow Jointly
☒ Accept in Transfer
Referring Diagnoses: CLAUDICATION, POSSIBLE DVT
Reason for Consultation: Severe pain in both legs on walking
Signature: [signature] Date: 2/21/09 Time 1105 hrs
Consultation Report: by Lokesh Bannerjee
Chief complaint: Pt. c/o severe pain in both legs on walking about 100 yards or climbing a flight of stairs. Pain is so severe she must stop and wait 5 mins. before she can go on. For the past two weeks she has noticed soreness and hardness along a vein in her left calf.
Past medical history: Known type 2 diabetic with hypertension, CAD, diabetic retinopathy, and OA of her hips and knees. Several episodes of ketoacidosis and one of pulmonary edema. Bariatric surgery performed 8 months prior at 275 lbs.
Medications: metformin, verapamil, propanolol, Mevacor
Allergies: NKA
Physical examination: Ht: 5'2" Wt: 190 lbs. BP: 170/100 sitting, P: 80, regular. Both feet show slight pitting edema and skin is pale, cold, and dry. Small ulcer on lateral margin of each big toe. Varicosities, both legs. Tender cord in superficial vein of left calf. Flexion of left foot produces pain in left calf. Chest clear. Heart sounds unremarkable. No loss of sensation in legs or feet.
Impression:
1. varicose veins, both legs.
2. severe claudication, both legs.
3. probable deep vein thrombosis, left leg
4. possible peripheral neuropathy
5. H/O diabetes type 2. CAD. hypertension, retinopathy, OA
Plan: Admit patient to cardiology unit stat for IV heparin and conversion to oral anticoagulant therapy with Coumadin. Doppler studies, venogram, and angiogram have been ordered.
Signature: Lokesh Bannerjee MD Date 2/21/09 Time 1250 hrs

ORDER # **267116** ANDRUS CLINI-REC ® PRIMARY CARE CHARTING SYSTEM • © BIBBERO SYSTEMS, INC. • PETALUMA, CA
TO REORDER CALL TOLL FREE: (800) BIBBERO (800-242-9330) MFG IN U.S.A.

WORD ANALYSIS AND DEFINITION

S/ = Suffix P/ = Prefix R/ = Root R/CF = Combining Form

WORD	PRONUNCIATION	ELEMENTS		DEFINITION
artery	**AR**-ter-ee		Greek *artery*	Thick-walled blood vessel carrying oxygenated blood away from the heart.
claudication	klaw-dih-**KAY**-shun	S/ R/	**-ation** a process **claudic-** *limping*	Intermittent leg pain and limping.
Doppler	**DOP**-ler		Johann Doppler, Austrian mathematician and physician, 1803–1853	Diagnostic instrument that sends an ultrasonic beam into the body.
hemodynamics	**HE**-mo-die-**NAM**-iks	S/ R/CF R/	**-ics** *knowledge* **hem/o-** *blood* **-dynam-** *power*	The science of the blood flow through the circulation.
vein	VANE		Latin *vein*	Blood vessel carrying blood toward the heart.
venous (adj)	**VEE**-nuss	R/ S/	**ven-** *vein* **-ous** *pertaining to*	Pertaining to a vein.
venogram	**VEE**-noh-gram	S/ R/CF	**-gram** *recording* **ven/o-** *vein*	Radiograph of veins after injection of radiopaque contrast material.

Abbreviations

BP	blood pressure
CMA	certified medical assistant
HPI	history of present illness
NKA	no known allergies
OA	osteoarthritis
P	pulse rate

EXERCISES

Singular and plurals, nouns and adjectives: *Terms from the WAD can all be correctly inserted into the following paragraph of radiology documentation. Fill in the blanks.*

veins varicosities varix vein varicose varices venogram venous

1. The ____________ performed on this patient's left saphenous and popliteal ____________ shows the following: One slightly engorged and dilated ____________ at the midpoint of the saphenous ____________ and several ____________ at the terminal end of the popliteal vein where ____________ blood is pooling.

   Diagnosis: ____________ veins. These ____________ need immediate attention by a vascular surgeon.

2. This can be described as a **hemo**____________ study of the patient's ____________ **tion.**

Match the following elements from this WAD with their correct meanings.

_____	3. ven/o	A. recording
_____	4. hem/o	B. power
_____	5. ous	C. vein
_____	6. claudic	D. blood
_____	7. gram	E. limping
_____	8. dynam	F. pertaining to

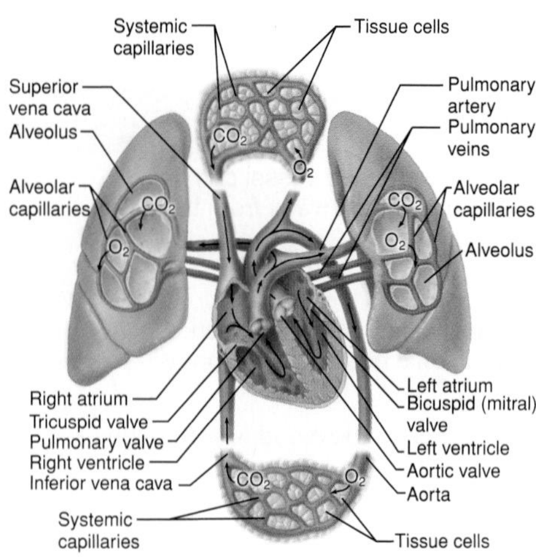

▲ **FIGURE 6.16**
Systemic and Pulmonary Circulations.

Keynote

All arteries of the systemic circulation arise either directly or indirectly from the aorta.

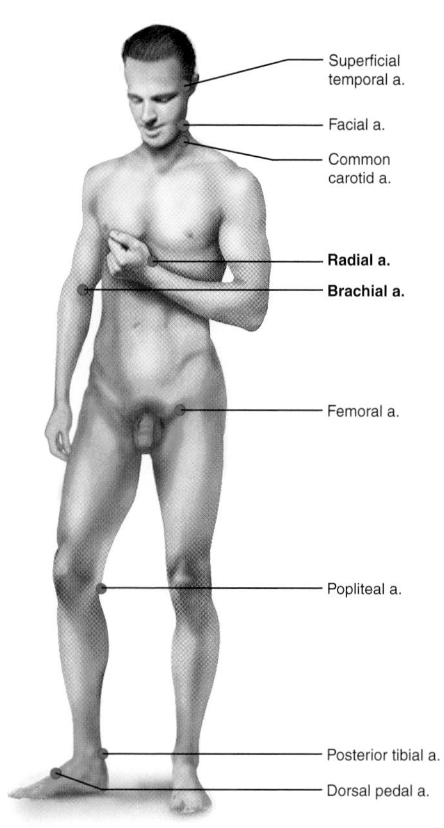

▲ **FIGURE 6.17**
Arterial Pulses—9 locations.
(a. = artery)

Keynote

Blood flows from arteries to veins through capillary beds.

CIRCULATORY SYSTEM

The term **circulatory system** refers to the heart and blood vessels and has two major divisions (*Figure 6.16*):

1. The **pulmonary circulation,** which carries deoxygenated blood from the heart to the lungs and returns oxygenated blood to the heart.
2. The **systemic circulation,** which supplies oxygenated blood to every organ except the lungs and then returns deoxygenated blood to the heart, which pumps it into the pulmonary circulation.

Functions of the Circulatory System

- **Transport.** The circulatory system carries oxygen, nutrients, hormones, and enzymes that **diffuse** from the blood into the cells. Waste products and carbon dioxide diffuse back from the cells into the system and are carried to the lungs, liver, and kidney for excretion.
- **Maintain homeostasis.** The systemic circulation directs blood flow to tissues to enable them to meet their metabolic needs.
- **Regulate blood pressure.** The ability of the arteries in the systemic circulation to expand and contract in coordination with the systole and diastole of the heartbeat maintains a steady flow of blood and blood pressure to the tissues.

Pulmonary Circulation

Deoxygenated blood from the body flows into the right side of the heart, from where it is pumped into the lungs. In the lungs, carbon dioxide is removed from the blood and excreted into the air. Oxygen is taken into the blood from the air in the lungs. The oxygenated blood returns to the heart via the pulmonary veins.

Systemic Circulation

Oxygenated blood enters the left side of the heart from the lungs. The blood is pumped out into the **aorta** that takes the blood to **arteries** in all areas of the body. This is the systemic circulation.

Arterial Pulses

The pulse is always part of a clinical examination because it can show heart rate, heart rhythm, and the state of the arterial wall by **palpation** (*Figure 6.17*).

The most easily accessible is the **radial artery** at the wrist, where the pulse is usually taken.

Blood Pressure (BP)

Blood pressure is the force the blood exerts on arterial walls as it is pumped around the circulatory system by the left ventricle. The pressure is measured using a **sphygmomanometer** and a **stethoscope,** usually at the brachial antery (*Figure 6.17*). This measurement is depicted on the Student Online Learning Center (www.mhhe.com/AllanEssMedLanguage).

Arterioles, Capillaries, and Venules

As the arteries branch farther away from the heart and distribute blood to specific organs, they become smaller, muscular vessels called **arterioles.** By contracting and relaxing, these arterioles are the primary controllers by which the body directs the relative amounts of blood that organs and structures receive.

From the arterioles the blood flows into **capillaries** and **capillary beds** (*Figure 6.18*). Red blood cells flow through the small capillaries in single file.

From the capillaries, tiny **venules** accept the blood and merge to form **veins.** The veins form reservoirs for blood, and at any moment 60% to 70% of the total blood volume is contained in the venules and veins.

WORD ANALYSIS AND DEFINITION

S/ = Suffix P/ = Prefix R/ = Root R/CF = Combining Form

WORD	PRONUNCIATION	ELEMENTS		DEFINITION
arteriole	ar-**TER**-ee-ole	S/ R/	**-ole** *small* **arteri-** *artery*	Small terminal artery leading into the capillary network.
brachial	**BRAY**-kee-al	S/ R/	**-al** *pertaining to* **brachi-** *arm*	Pertaining to the arm.
capillary	**KAP**-ih-lair-ee	S/ R/	**-ary** *pertaining to* **capill-** *hairlike structure*	Minute blood vessel between the arterial and venous systems.
diffuse	di-**FUSE**		Latin *to pull in different directions*	To disseminate or spread out.
homeostasis	hoh-mee-oh-**STAY**-sis	S/ R/CF	**-stasis** *stand still* **home/o-** *the same*	Maintaining the stability, or equilibrium, of a system or the body's internal environment.
palpate (verb) **palpation (noun)**	**PAL**-pate pal-**PAY**-shun	 S/ R/	Latin *touch, stroke* **-ion** *action, condition* **palpat-** *touch, stroke*	To examine with the fingers and hands. Examination with the fingers and hands.
sphygmomanometer	**SFIG**-moh-mah-**NOM**-ih-ter	S/ R/CF R/CF	**-meter** *instrument to measure* **sphygm/o-** *pulse* **-man/o-** *pressure*	Instrument for measuring arterial blood pressure.
stethoscope	**STETH**-oh-skope	S/ R/CF	**-scope** *instrument to examine* **steth/o-** *chest*	Instrument for listening to respiratory and cardiac sounds.
vena cava **venae cavae (pl)**	**VEE**-nah **KAY**-vah **VEE**-nee **KAY**-vee	R/CF R/	**ven/a** *vein* **cava** *cave*	One of the two largest veins in the body. The two largest veins in the body.
venule	**VEN**-yule or **VEEN**-yule	S/ R/	**-ule** *small* **ven-** *vein*	Small vein leading from the capillary network.

Systemic Venous Circulation

There are three major types of veins:

1. **Superficial,** such as those you can see under the skin of your arms and hands.
2. **Deep,** which run parallel to arteries and drain the same tissues that the arteries supply.
3. **Venous sinuses,** which are in the head and heart and have specific functions.

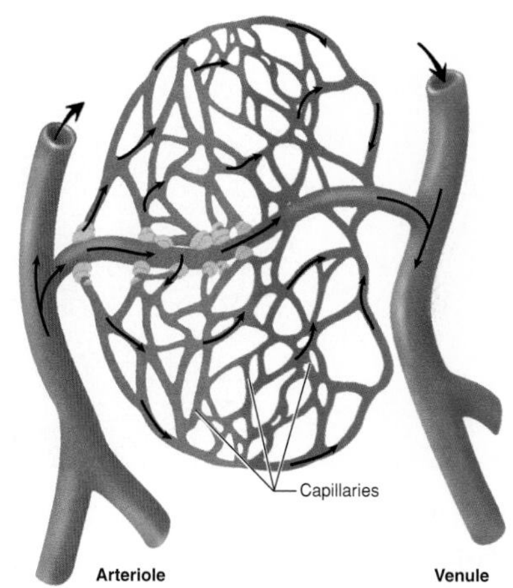

FIGURE 6.18 ▶
Capillary Bed.

EXERCISES

Build your knowledge *of the elements and terms that make up the language of the cardiovascular system. All your answers can be found in the WAD. Fill in the blanks.*

1. The two suffixes in the WAD that both mean *small* are ________________ and ________________.
2. List the four general terms in the WAD that are *blood vessels:* ________________, ________________, ________________ and ________________
3. Name two instruments found in an internal medicine practice: ________________ and ________________

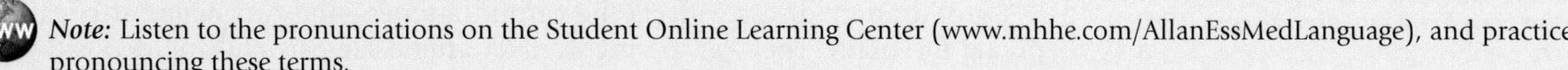
Note: Listen to the pronunciations on the Student Online Learning Center (www.mhhe.com/AllanEssMedLanguage), and practice pronouncing these terms.

4. State of equilibrium in the body is called ________________.

CIRCULATORY DISORDERS

Abbreviations

DVT	deep vein thrombosis
PVD	peripherial vascular disease

Case Report 6.3 (continued)

Mrs. Martha Jones, who had been referred to Dr. Bannerjee's cardiovascular clinic, has several circulatory problems related to her diabetes and obesity. She was diagnosed previously with hypertension, CAD, and diabetic retinopathy. She now has severe pain in her legs on walking.

Doppler studies and angiograms showed significant blockage of blood flow due to arteriosclerosis of the large arteries in her legs. This blockage produces the pain on walking (intermittent claudication), and is part of her peripheral vascular disease.

The ulcers on the edges of her big toes result from thickening of the walls of her capillaries and arterioles and the resulting poor circulation to her feet. Again, this is due to her diabetes.

In the venous system of her legs, the tender cordlike lesion is due to **thrombophlebitis** of a superficial vein in her left leg. A venogram showed a deep vein thrombosis (DVT).

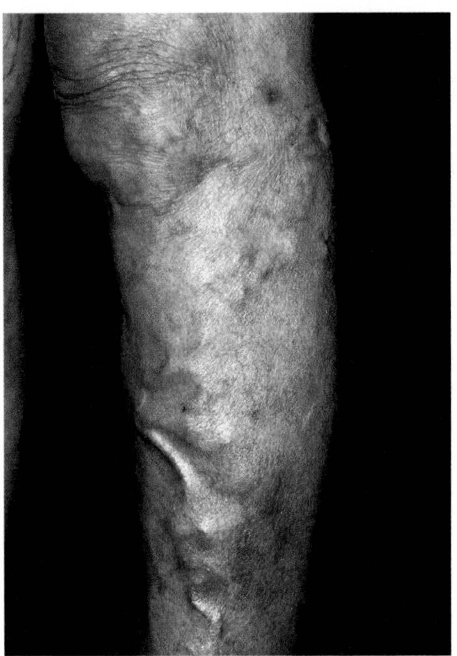

▲ **FIGURE 6.19**
Varicose Veins of the Leg.

Disorders of Veins

Thrombophlebitis is an inflammation of the lining of a vein, allowing clots (thrombi) to form.

Deep vein thrombosis (DVT) is thrombus formation in a deep vein. Risk factors include immobility, surgery, prolonged travel, and contraception (estrogen). The increased pressure in the capillaries due to back pressure from the blocked blood flow in the veins causes a collection of fluid in the tissues called **edema.**

A major complication of **thrombus** (clot) formation is that a piece of the clot can break off **(embolus)** and be carried in the bloodstream to another organ and block blood flow. It often lodges in the lungs, causing a pulmonary embolus (*see Chapter 7*).

Varicose veins are superficial veins that have lost their elasticity and appear swollen and tortuous (*Figure 6.19*). Their valves become incompetent, and blood flows backward and pools. Smaller, more superficial varicose veins are called **spider** veins. Varicose veins are associated with a family history, obesity, and prolonged standing. Treatments offered include laser technology and **sclerotherapy,** in which solutions that scar the veins are injected into them. **Collateral** circulations develop to take the blood through alternative routes.

A **phlebotomist** is a technician who draws blood **(phlebotomy).**

Disorders of Arteries

An **aneurysm** is a localized **dilation** of an artery. A common site is the aorta, mostly the abdominal aorta. Aneurysms can **rupture,** leading to severe bleeding and hypovolemic shock. Surgical repair consists of excision of the aneurysm and replacement with a **synthetic** graft.

Intracranial aneurysms are an important cause of bleeds into the cranial cavity and brain tissue.

Thromboangiitis obliterans (Buerger disease) is an inflammatory disease of the arteries with clot formation, usually in the legs. The occlusion of arteries and impaired circulation leads to intermittent claudication.

Raynaud disease is episodes of spasm of the small arteries supplying the fingers, hands, and feet following exposure to cold. It can be associated with connective tissue disorders such as scleroderma and lupus.

Carotid artery disease affects the carotid arteries—the two major arteries supplying the brain. They can be involved in arteriosclerosis and the deposition of plaque. This puts the patient at risk for a stroke. A carotid **endarterectomy** can be performed to surgically remove the plaque.

Keynote

All the disorders of the systemic arterial and venous systems are grouped under the term ***peripheral vascular disease*** **(PVD).**

WORD ANALYSIS AND DEFINITION

S/ = Suffix P/ = Prefix R/ = Root R/CF = Combining Form

WORD	PRONUNCIATION	ELEMENTS		DEFINITION
aneurysm	**AN**-yur-izm		Greek *dilation*	Circumscribed dilation of an artery or cardiac chamber.
collateral	koh-**LAT**-er-al	P/ R/	**col-** *with, together* **-lateral** *at the side*	Situated at the side, often to bypass an obstruction.
dilation	die-**LAY**-shun	S/ R/	**-ion** *action, condition* **dilat-** *widen, dilate, open up*	Stretching or enlarging an opening.
edema (noun)	ee-**DEE**-mah		Greek *swelling*	Excessive accumulation of fluid in cells and tissues.
edematous (adj)	ee-**DEM**-ah-tus	S/ R/	**-tous** *pertaining to* **edema-** *swelling*	Pertaining to or affected by edema.
endarterectomy	**END**-ar-ter-**EK**-toe-me	S/ P/ R/	**-ectomy** *surgical excision* **end-** *within* **-arter-** *artery*	Surgical removal of plaque from an artery.
phlebitis	fleh-**BIE**-tis	S/ R/	**-itis** *inflammation* **phleb-** *vein*	Inflammation of a vein.
phlebotomist	fleh-**BOT**-oh-mist	S/ R/CF R/	**-ist** *specialist in* **phleb/o-** *vein* **-tom-** *incise, cut*	Person skilled in taking blood from veins.
phlebotomy	fleh-**BOT**-oh-me	S/	**-tomy** *surgical incision*	Taking blood from a vein.
rupture	**RUP**-tyur		Latin *break*	Break or tear of any organ or body part.
sclerotherapy	**SKLEH-r**oh-**THAIR**-ah-pee	S/ R/CF	**-therapy** *treatment* **scler/o-** *hardness*	Injection of a solution into a vein to thrombose it.
sclerose (verb)	skleh-**ROSE**	S/ R/	**-ose** *full of* **scler-** *hardness*	To harden or thicken.
sclerosis (noun)	skleh-R**OH**-sis	S/	**-osis** *condition*	Thickening or hardening of a tissue.
synthetic	sin-**THET**-ik	S/ P/ R/	**-ic** *pertaining to* **syn-** *together* **-thet-***arrange*	Built up or put together from simpler compounds.
thromboembolism	**THROM**-boh-**EM**-boh-lizm	S/ R/CF R/	**-ism** *condition* **thromb/o-** *clot* **-embol-** *plug*	A piece of detached blood clot (embolus) blocking a distant blood vessel.
thrombophlebitis	**THROM**-boh-fleh-**BY**-tis	S/ R/CF R/	**-itis** *inflammation* **thromb/o-** *clot* **-phleb-** *vein*	Inflammation of a vein with clot formation.

EXERCISES

Analyze *the italicized terms and their elements to answer the following questions. Review the WAD, and circle the correct term.*

1. The suffix tells you *thrombophlebitis* is:
   a clot an inflammation an excision
2. *Endarterectomy* means the plaque has been:
   incised removed repaired
3. The combining form *phleb/o* means:
   artery vein capillary
4. The symptom *edema* is:
   a rash a swelling a lesion
5. *Aneurysm* describes a blood vessel that is:
   dilated constricted collapsed
6. The combining form *thromb/o* means:
   clot lump plug

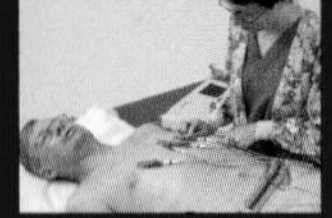

CARDIOVASCULAR SYSTEM

CHALLENGE YOUR KNOWLEDGE

A. **Deconstruct** the following medical terms into their basic elements by slashing the terms in the first column. Then fill in the rest of the chart. Knowledge of elements will help you understand the meaning of the term.

Medical Term	Meaning of Prefix	Meaning of Root/CF	Meaning of Suffix	Meaning of Medical Term
1. bradycardia				
2. endocardium				
3. cardiomyopathy				
4. bicuspid				
5. pericarditis				
6. diaphoresis				
7. pulmonary				
8. semilunar				
9. arteriosclerosis				
10. ischemia				

11. Use any one term from the chart in a sentence that is not a definition:

____

____

____

B. **Determine** the correct medical vocabulary to complete the following sentences in patient documentation. An explanation of the term is given in parentheses as a clue. Fill in the blanks.

1. The (under the tongue) ________________ medication did not dissolve completely, so liquid medication was given to the patient instead.
2. The patient's (hardening of the arteries) ________________ puts him at risk for a stroke. I have ordered a blood thinner.
3. Schedule the patient for a(n) (surgical removal of lipid deposit in artery lining) ________________ as soon as possible.
4. (Absence of heart contractions) ________________ occurred at 10:51 p.m., and the patient was pronounced dead.
5. Angioplasty showed a large clot (completely obstructing) ________________ his left coronary artery.
6. The bullet entered his left (under the breastbone) ________________ area and exited his back.
7. (Being without oxygen) ________________ has caused permanent brain damage to this infant.
8. The patient's diagnostic studies show clear evidence of (death of tissue) ________________ of the heart muscle.
9. Cardiogenic shock occurred, and the patient's tissues were not (forcing blood through a vascular bed or lumen) ________________ adequately.
10. Treatment options offered to the patient for her varicose veins include laser therapy and (injection of a solution into a vein to sclerose it) ________________.

C. **Translate:** Rewrite the following sentence using medical terms instead of the abbreviations. Make sure you are communicating the same information either way. Check your spelling after you have filled in the blanks.

1. The CVT started CPR after checking the patient's EKG. He paged the doctor STAT. Apparently, the patient had suffered an MI.

____

____

____

____

D. **Suffixes:** The following procedures could all be performed by a cardiovascular surgeon. First, slash the term into its elements, and fill in the blanks. Then use the suffix to help you analyze the meaning of the term. Write a brief description of the procedure on the lines below.

1. *angioplasty* ____________ / ____________ / ____________
   P R/CF S

   Description:

   ____

2. *endarterectomy* ____________ / ____________ / ____________
   P R/CF S

   Description:

   ____

3. *(cardiac)*
   catheterization ____________ / ____________ / ____________
   P R/CF S

   Description:

   ____

E. **Dictionary or online research:** Look up the term *perfusionist* in a dictionary or online. Write a brief job description for a *perfusionist.*

____

____

____

____

____

____

____

____

____

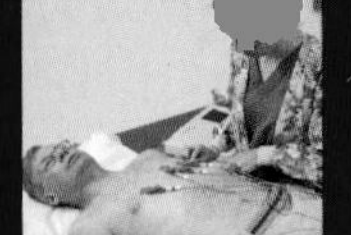

CARDIOVASCULAR SYSTEM

F. **Word attack:** Process of elimination can help with multiple-choice questions. Work through the following exercise step-by-step, and practice analyzing multiple-choice questions *and* answers.

If a patient is *hypovolemic,* he or she has:

a. decreased blood pressure

b. increased blood pressure

c. decreased blood volume

d. increased blood volume

e. low red blood cells

1. Read the question and *all* the answer choices. Then read the question again.
2. Look for subtle differences in the answers. Note that some answers relate to "increased" or "decreased." These answers *also* relate to "blood *pressure*" or "blood *volume*" (two different things).
3. Analyze the elements to know whether you are looking for "increased" *or* "decreased" and for "blood pressure" *or* "blood volume."
4. The prefix in the question is __________________ and means __________________, so you can immediately discard answer choices __________________ because they do not relate to the prefix.
5. Next, analyze the root. The root is __________________ and means ____________________________________.
6. Therefore, based on the meanings of the prefix and the root, the answer has to be __________________ and the meaning of *hypovolemic* is ____________________________________.

G. **Elements** remain the single best tool for increasing your medical vocabulary. Identify what the element in column 1 is, and define its meaning in the appropriate column. Then give an example of a medical term with that element. The first one is done for you. Fill in the chart.

Element	Meaning of Prefix	Meaning of Root/CF	Meaning of Suffix	Medical Term
cardio		*heart*		*cardiovascular*
a				
media				
ism				
lysis				
emia				
thorac				
myo				
phleb				
sclero				

H. **Interpret** the following paragraph from a patient's chest x-ray report, and answer the questions. Read it once, and then go back and read it again, underlining or highlighting the medical terms.

> The left ventricle is slightly enlarged. Right atrium and right ventricle appear to be dilated. There is tortuosity of the thoracic aorta, with arteriosclerosis. The hilar and interstitial structures are somewhat accentuated. There are large pericardial fat pads. Tricuspid and pulmonic valves appear normal but are not well seen. Mitral valve appears grossly normal for age. Aortic valve appears grossly normal for age.

Using your knowledge of the terminology in the above report, circle the correct answers for the following questions.

1. Where are the *pericardial fat pads?*
   a. in the heart
   b. around the heart
   c. beneath the heart
   d. beside the heart
   e. above the heart
2. What is a likely diagnosis for this patient?
   a. CABG
   b. ASHD
   c. PTCA
   d. DVT
   e. MI
3. Which two terms refer to chambers of the heart?
   a. tricuspid pulmonic
   b. atrium ventricle
   c. hilar interstitial
   d. mitral pulmonic
   e. aortic pericardial
4. Which heart valve has three flaps?
   a. pulmonic
   b. mitral
   c. tricuspid
   d. hilar
   e. bicuspid
5. A synonym for *dilated* is:
   a. expanded
   b. twisted
   c. thrombosed
   d. occluded
   e. constricted

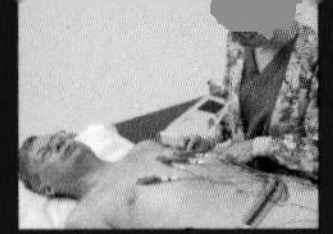

CHAPTER 6 REVIEW

CARDIOVASCULAR SYSTEM

I. Prefixes: Each of the following terms has similar roots and suffixes, but the prefix makes the difference. Deconstruct the term with slashes. Analyze the terms, and explain how they are different.

1. *arrhythmia*

_______________ / _______________ / _______________
prefix root suffix

2. *dysrhythmia*

_______________ / _______________ / _______________
prefix root suffix

3. Difference between *arrhythmia* and *dysrhythmia:*

____

____

____

J. Spelling demons: This chapter contains some particularly difficult terms to pronounce and spell. *You can master these terms with practice.* Write below the five most difficult terms for you to pronounce and spell correctly. Listen to the Glossary on the Student Online Learning Center (www.mhhe.com/AllanEssMedLanguage) for correct pronunciations, and check your spelling. Compare your list with your study partner's list to see if you have similar difficult terms.

_____ 1. Pronunciation correct _______________

_____ 2. Pronunciation correct _______________

_____ 3. Pronunciation correct _______________

_____ 4. Pronunciation correct _______________

_____ 5. Pronunciation correct _______________

K. You are mentoring a new CVT who has just been hired in the Cardiology Department at Fulwood Medical Center. Because you have been on the job for a while, you should be able to explain to him *the difference between:*

1. heart valve *insufficiency* and *stenosis*

____

____

____

____

____

2. a *thrombus* and an *embolus*

____

____

____

____

L. **Abbreviations** are present throughout medical documentation, and you must be absolutely certain you are interpreting them correctly. Fill in the correct abbreviations in the following patient documentation. All of the abbreviations contain some combination of the following letters. You will have to use some letters more than once:

A C D F H I M P S V

1. Studies show the patient has a hole in the interventricular septum, allowing blood to shunt from the higher-pressure left ventricle to the lower-pressure right ventricle. Diagnosis: ___________________
2. The pediatric cardiologist was called to the Neonatal Unit because the baby's fetal blood vessel had not closed normally. The baby was diagnosed with ___________________.
3. The patient's arterial vessels have become dangerously narrow due to his ___________________, and an angioplasty will be scheduled.
4. Due to her sedentary lifestyle, obesity, hypertension, and smoking history, the patient is at great risk for ___________________.
5. This patient's left ventricle is failing because it cannot pump out the blood it receives. He is going into ___________________.
6. Infant male was born with tetralogy of Fallot (___________________) syndrome. This is a form of ___________________.
7. This patient's ischemic attack resulted in occlusion of her coronary artery, and a(n) ___________________ followed.

M. **Layperson's language:** Mrs. Jones has been given the following diagnoses. Translate them into plain English for her.

Impression:

1. Varicose veins, both legs.
2. Severe claudication, both legs.
3. Probable deep vein thrombosis, left leg.
4. Possible peripheral neuropathy.
5. H/O CAD, hypertension, retinopathy.

Rewrite the medical terms into layperson's language for your patient.

1. ____
2. ____
3. ____
4. ____
5. ____.

____

____

N. **Recall and review:** How well do you remember these word elements from the previous chapter? Try to answer without looking back to check. Fill in the chart.

Element	Type of Element (P, R, CF, or S)	Meaning of Element
ase		
lacte		
celi		
enter		
hydr		

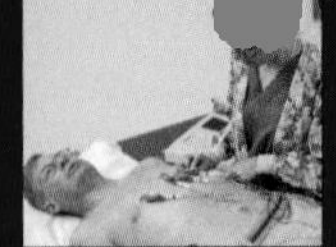

CARDIOVASCULAR SYSTEM

O. Seek and find: The following terms were defined in context in the text but did not necessarily appear in a WAD. Match the appropriate meaning in column 1 with the correct medical term in column 2.

_____ 1. blocked	A. patent
_____ 2. unknown	B. hemorrhage
_____ 3. lack of blood supply	C. coronary thrombosis
_____ 4. under the tongue	D. occluded
_____ 5. detour	E. thrombus
_____ 6. excessive bleeding	F. necrosis
_____ 7. pain in the chest on exertion	G. idiopathic
_____ 8. dead person	H. sublingual
_____ 9. open	I. ischemia
_____ 10. blood clot in the heart	J. angina pectoris
_____ 11. clot	K. bypass
_____ 12. death of cells	L. cadaver

P. Chapter challenge: Circle the best answer.

> *Study Hint*
> Immediately cross off any answer you know is not correct. In your remaining choices, there is only *one best answer.*

1. Find the abbreviations that are *cardiac diagnoses:*
   a. SOB, ECG, CVT
   b. STAT, IV, ASHD
   c. CPR, MI, AV
   d. SA, SC, CT
   e. CAD, CHF, VSD

2. The medical term meaning *an instrument for measuring arterial blood pressure* is:
   a. stethoscope
   b. sphygmomanometer
   c. defibrillator
   d. ventilator
   e. tonometer

3. Choose the correct abbreviation for "heart attack":
   a. CO
   b. AV
   c. ASHD
   d. MI
   e. CAD

4. Mr. Hank Johnson became *diaphoretic* in the ED. This means he:
   a. began to vomit
   b. started sweating
   c. hemorrhaged
   d. had a fever
   e. became itchy all over

5. Identify the pair of terms that are *incorrectly* spelled:

a. thrombolytic — arteriosclerosis
b. substernal — sublingual
c. oclude — oclusion
d. plaque — patent
e. varix — varices

6. The medical term *STAT* means something is to be performed:

a. after the patient consents
b. tomorrow
c. immediately
d. at the doctor's convenience
e. when the schedule permits

7. Which phrase best describes the terms *infusion, transfusion,* and *perfusion?*

a. The suffix is the same.
b. The root is the same.
c. There is no combining form.
d. Only the prefix makes them different.
e. All these are true.

8. Pick the set of terms that are all blood vessels:

a. vena cava, arteriole, varix
b. artery, edema, rupture
c. venule, arteriole, capillary
d. aneurysm, phlebitis, sclerose
e. platelet, thrombocyte, colloid

9. The medical term for a wire-mesh tube inserted to keep an artery open is:

a. shunt
b. pacemaker
c. stent
d. prosthesis
e. electrode

10. Which diagnosis correctly describes failure of the right ventricle to pump properly?

a. cardiomegaly
b. palpitation
c. hypertrophy
d. myocarditis
e. cor pulmonale

11. The "A" in the abbreviation *AED* stands for:

a. arteriosclerotic
b. automatic
c. asystole
d. aortic
e. atrial

12. Identify the set of incorrectly paired terms:

a. plaque — atheroma
b. thrombocyte — platelet
c. stenosis — constricted
d. mitral valve — bicuspid valve
e. bleeding — dehydration

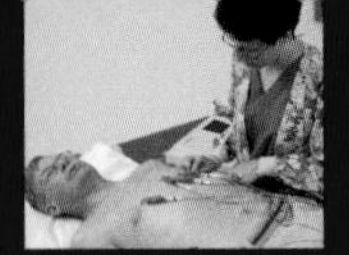

CHAPTER 6 REVIEW

CARDIOVASCULAR SYSTEM

Q. Case Report challenge: Now that you are more comfortable with the terms in this chapter, you can apply that knowledge and briefly answer the questions about the Case Report. If you read the report through and underline or highlight all the medical terminology, this will make it easier to answer the questions.

Case Report 6.3 *(continued)*

Mrs. Martha Jones, who had been referred to Dr. Bannerjee's cardiovascular clinic, has several circulatory problems related to her diabetes and obesity. She was diagnosed previously with hypertension, CAD, and diabetic retinopathy. She now has severe pain in her legs on walking.

Doppler studies and angiograms showed significant blockage of blood flow due to arteriosclerosis of the large arteries in her legs. This blockage produces the pain on walking *(intermittent claudication).*

The ulcers on the edges of her big toes result from thickening of the walls of her capillaries and arterioles and the resulting poor circulation to her feet. Again, this is due to her diabetes.

In the venous system of her legs, the tender cordlike lesion is due to thrombophlebitis of a superficial vein in her left leg. A venogram showed a deep vein thrombosis (DVT).

1. What does the term *hypertension* mean?

____

2. What part of the patient's body is affected by *retinopathy?*

____

3. List here all the terms used in the Case Report that relate to different types of blood vessels:

____

____

4. What is the opposite of a vein that is *superficial?*

____

5. In layperson's language, what is a *thrombosis?*

____

6. If the patient's pain is *intermittent,* what does that mean?

____

7. What specific diagnostic studies were ordered for Mrs. Jones?

____

8. What symptoms did Mrs. Jones present with when she saw Dr. Bannerjee?

____

____

R. **Chapter challenge:** Circle the best answer.

1. What do *erythrocyte, leukocyte,* and *bilirubin* all have in common?

   a. They are all diagnoses.
   b. They all refer to a color.
   c. They all have the same suffix.
   d. They are all blood cells.
   e. They are all cardiac terms.

2. A normal heartbeat of around 60 to 80 beats per minute is called:

   a. tachycardia
   b. bradycardia
   c. sinus rhythm
   d. dysrhythmia
   e. arrhythmia

3. "Good cholesterol" is:

   a. HDL
   b. MRI
   c. TOF
   d. CHF
   e. LDL

4. A *phlebotomist* is a technician who:

   a. takes x-rays
   b. circulates blood during surgery
   c. assists with dialysis
   d. administers breathing tests
   e. draws blood

5. The medical terms *tachycardia* and *bradycardia* basically describe:

   a. open closed
   b. fast slow
   c. top bottom
   d. back front
   e. right left

6. **Brain teaser:** Which statement has no mistakes?

   a. A localized dilation of an artery is called a *graft.*
   b. Carotid endarterectomy removes plague.
   c. Intercranial aneurysms can cause bleeding into the brain.
   d. Raynaud disease is associated with scleroderma.

Respiratory System

The Essentials of the Language of Pulmonology

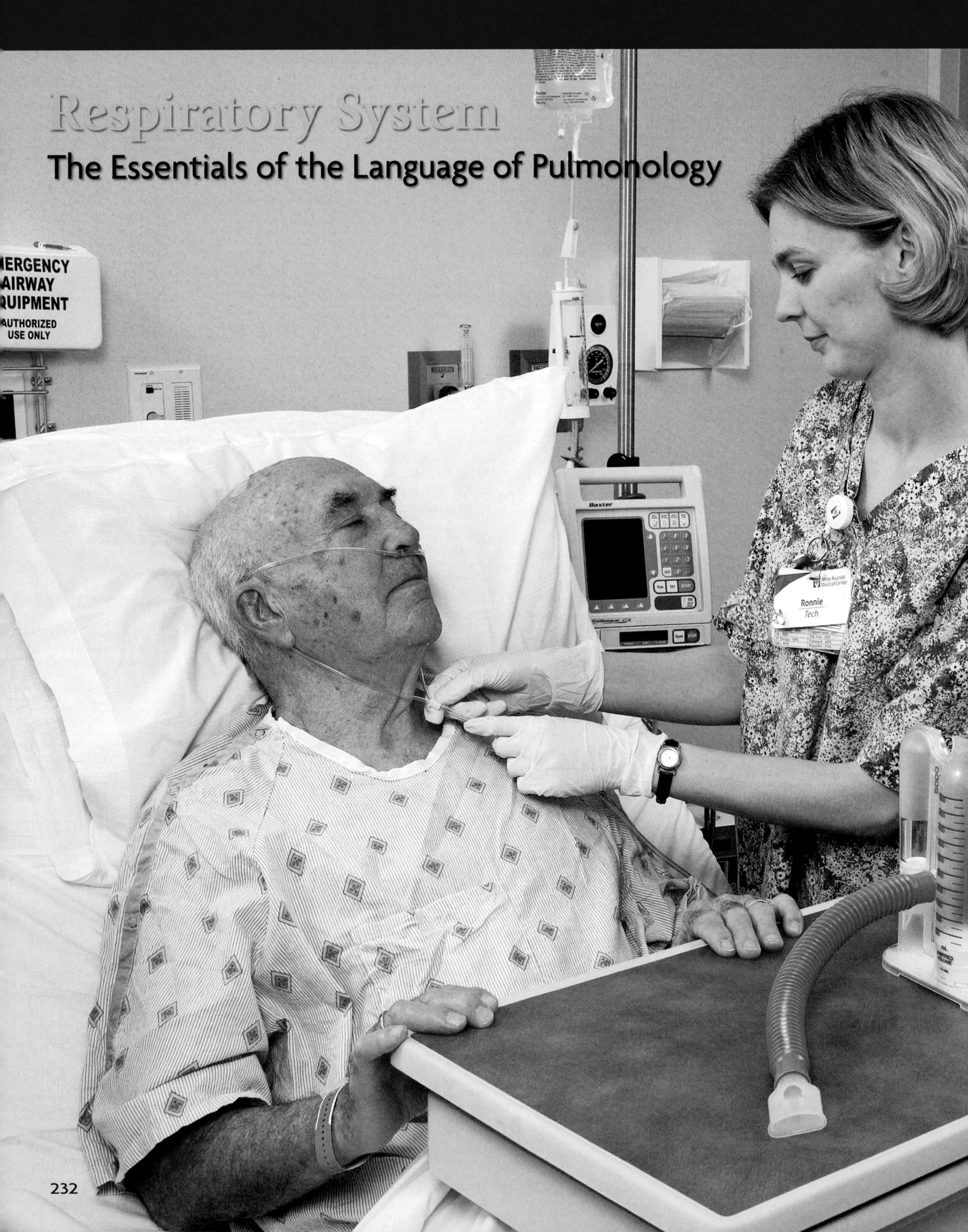

CASE REPORT 7.1

You are

. . . an advanced-level respiratory therapist working with pulmonologist Tavis Senko, MD, in the Acute Respiratory Care unit of Fulwood Medical Center.

You are communicating with

. . . Mr. Jude Jacobs, a 68-year-old white retired mail carrier, who has chronic obstructive pulmonary disease **(COPD)** and is on continual **oxygen** by nasal prongs. He has smoked two packs per day during all his adult life.

Last night, he was unable to sleep because of increased shortness of breath **(SOB)** and cough. His cough is productive of yellow **sputum.** He had to sit upright in bed to be able to breathe.

Vital signs are temperature **(T)** 101.6, pulse **(P)** 98, respirations *(R)* 36, blood pressure **(BP),** 150/90. On examination, he is **cyanotic** and frightened and is on oxygen by nasal prongs. Air entry is diminished in both lungs, and there are **rales** (crackles) at both bases.

You have been ordered to draw blood for arterial blood gases **(ABGs)** and to measure the amount of air entering and leaving his lungs by using **spirometry.**

Learning Outcomes

In order to provide optimal care to Mr. Jacobs, to determine what is causing his symptoms and signs, and to communicate with the other health professionals involved in his care, you need to be able to:

7.1 Apply the language of pulmonology to the functions of the respiratory system.

7.2 Comprehend, analyze, spell, and write the medical terms of pulmonology to communicate and document accurately and precisely.

7.3 Recognize and pronounce the medical terms of pulmonology so that you can communicate verbally with accuracy and precision.

7.4 Explain the effects of common respiratory disorders on health.

7.5 Translate medical terms of pulmonology into lay language in order to communicate with patients and their families.

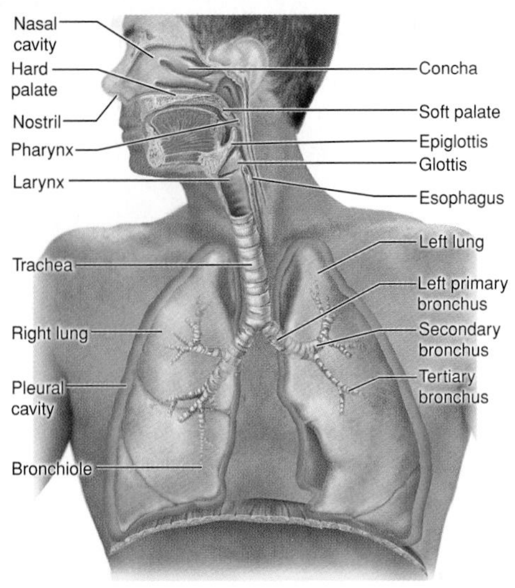

▲ **FIGURE 7.1 The Respiratory System.**

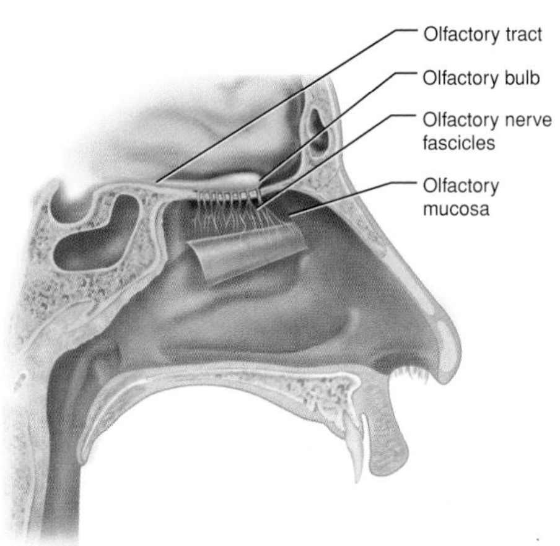

▲ **FIGURE 7.2 Olfactory Region of the Nose.**

Abbreviations

ABG arterial blood gas
COPD chronic obstructive pulmonary disease
CO_2 carbon dioxide
O_2 oxygen
RT respiratory therapist

INTRODUCTION TO THE RESPIRATORY SYSTEM

It sounds like a good scheme. Humans and animals breathe in **oxygen** and breathe out **carbon dioxide;** plants and trees breathe in carbon dioxide and breathe out oxygen; and nature would stay in balance.

Unfortunately, we humans have generated increasing amounts of carbon dioxide in the air by burning coal, oil, and natural gas and cutting down forests. We have disturbed the balance. In addition, we have created organic and inorganic chemicals and small particles of solid matter in the air that can damage the respiratory tract and be taken into our bodies and cause cancer, brain damage, and birth defects. These materials are called **pollutants.**

The Respiratory Tract

The respiratory tract (*Figure 7.1*) has six connected elements:

1. **Nose**
2. **Pharynx**
3. **Larynx**
4. **Trachea**
5. **Bronchi and bronchioles**
6. **Alveoli**

Respiration has two components:

- **Ventilation,** that is, the movement of air and its gases into **(inspiration)** and out of the lungs **(expiration).**
- The **exchange of gases** between air and blood and between blood and interstitial fluids.

Functions of the Respiratory System

1. **Exchange of gases:** All the body cells need oxygen and produce carbon dioxide. The respiratory system allows oxygen from the air to enter the blood and carbon dioxide to leave the blood and enter the air.
2. **Regulation of blood pH:** Regulation occurs by changing blood carbon dioxide levels.
3. **Protection:** The respiratory system protects against foreign bodies and against some microorganisms.
4. **Voice production:** Movement of air across the vocal cords makes voice and sound possible.
5. **Olfaction:** The 12 million receptor cells for smell are in a patch of epithelium, the size of a quarter, in the extreme superior region of the nasal cavity, the **olfactory region** (*Figure 7.2*). Each cell has 10 to 20 hairlike structures called **cilia** that project into the nasal cavity in a thin mucous film.

   Because the olfactory region is right at the top of the nose, you often have to sniff the air right up there to stimulate the sense of smell.

   A dog has 4 billion receptor cells, which is why dogs can be trained to sniff for drugs and explosives.

WORD ANALYSIS AND DEFINITION

S/ = Suffix P/ = Prefix R/ = Root R/CF = Combining Form

WORD	PRONUNCIATION	ELEMENTS		DEFINITION
alveolus **alveoli (pl)** **alveolar (adj)**	al-**VEE**-oh-lus al-**VEE**-oh-lee al-**VEE**-oh-lar	 S/ R/	Latin *hollow sac* **-ar** *pertaining to* **alveol-** *alveolus (air sac)*	Terminal element of respiratory tract where gas exchange occurs. Pertaining to the alveoli.
bronchus **bronchi (pl)** **bronchiole**	**BRONG**-kuss **BRONG**-key **BRONG**-key-ole	 S/ R/CF	Greek *windpipe* **-ole** *small* **bronch/i-** *bronchus*	One of two subdivisions of the trachea. Increasingly smaller subdivisions of bronchi.
cilium **cilia (pl)**	**SILL**-ee-um **SILL**-ee-ah		Latin *eyelash*	Hairlike motile projection from the surface of a cell.
expiration (***Note:*** The "s" is deleted from the root *spirat* because the prefix *ex* already has the "s" sound.)	**EKS**-pih-**RAY**-shun	S/ P/ R/	**-ion** *action, condition* **ex-** *out* **-spirat-** *breathe*	Breathe out.
inspiration (***Note:*** the opposite of *expiration*)	in-spih-**RAY**-shun	S/ P/ R/	**-ion** *action, condition* **in-** *into* **-spirat-** *breathe*	Breathe in.
olfaction **olfactory (adj)**	ol-**FAK**-shun ol-**FAK**-toh-ree	 S/ R/	Latin *to smell* **-ory** *having the function of* **olfact-** *smell*	Sense of smell. Relating to the sense of smell.
oxygen	**OCK**-see-jen	S/ R/	**-gen** *create* **oxy-** *oxygen*	The gas essential for life.
pharynx **pharyngeal (adj)**	**FAH**-rinks fah-**RIN**-jee-al	 S/ R/	Greek *throat* **-eal** *pertaining to* **pharyng-** *pharynx*	Tube from the back of the nose to the larynx. Pertaining to the pharynx.
pulmonary **pulmonology** **pulmonologist**	**PULL**-moh-**NAR**-ee **PULL**-moh-**NOL**-oh-jee **PULL**-moh-**NOL**-oh-jist	S/ R/ S/ R/CF S/	**-ary** *pertaining to* **pulmon-** *lung* **-logy** *study of* **pulmon/o-** *lung* **-logist** *one who studies, specialist*	Pertaining to the lungs. Study of the lungs, or the medical specialty of disorders of the lungs. Specialist in treating disorders of the lungs.
rale **rales (pl)**	RAHL RAHLS		French *rattle*	Crackle heard through a stethoscope when air bubbles through liquid in the lungs.
respiration **respirator** **respiratory (adj)**	**RES**-pih-**RAY**-shun **RES**-pih-**RAY**-tor **RES**-pih-rah-tor-ee	S/ P/ R/ S/ S/	**-ation** *a process* **re-** *again* **-spir-** *to breathe* **-ator** *person or thing that does something* **-atory** *pertaining to*	Process of breathing; fundamental process of life used to exchange oxygen and carbon dioxide. Another name for ventilator. Pertaining to respiration.
sputum	**SPYU**-tum		Latin *to spit*	Matter coughed up and spat out by individuals with respiratory disorders.
trachea **trachealis (adj)**	**TRAY**-kee-ah tray-kee-**AY**-lis	 S/ R/	Greek *windpipe* **-alis** *pertaining to* **trache-** *trachea*	Air tube from the larynx to the bronchi. Pertaining to the trachea.

EXERCISES

Language of pulmonology: *Medical terms taken directly from Latin, Greek, or other languages do not deconstruct into prefix, root, or suffix elements the way most medical terms do. You simply have to know them for what they are. Match the medical term in column 1 with its definition in column 2. Fill in the blanks.*

_____ 1. pharynx	A. matter coughed up and spat out
_____ 2. alveolus	B. rattle
_____ 3. olfaction	C. gas exchange occurs here
_____ 4. sputum	D. throat
_____ 5. rale	E. to smell

LESSON 7.1 RESPIRATORY SYSTEM

Upper Respiratory Tract

OBJECTIVES

The **upper respiratory tract** consists of the nose, pharynx and trachea. It is the first site that brings air and its pollutants inside our bodies. The information in this lesson will enable you to:

- **Trace the flow of air from the nose through the pharynx and larynx.**
- **Define the medical terminology for the protective mechanisms of the upper respiratory tract.**
- **Describe how sound is produced.**
- **Use correct medical terminology to identify common disorders of the upper respiratory tract.**

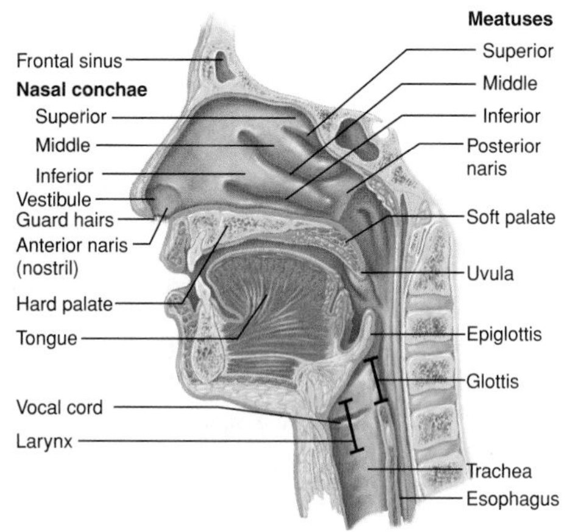

▲ **FIGURE 7.3 Upper Respiratory Tract.**

Abbreviation	
URI	upper respiratory infection

THE NOSE

When you breathe in air through your nose, the air goes through the nostrils **(nares)** into the **nasal cavity.** The nares are guarded by internal hairs to prevent the entry of large particles.

The nasal cavity is divided by the nasal **septum** into right and left compartments. The **palate** forms the floor of the nose (*Figure 7.3*). The **paranasal frontal** and **maxillary sinuses** (*see Chapter 8*) open into the nose.

Functions of the Nose

1. **Passageway for air.**
2. **Air cleanser:** The nasal hairs and the mucus secreted by the nasal mucous membrane trap particles of dust and solid pollutants.
3. **Air moisturizer:** Moisture secreted by the nasal mucosa, and from tears that drain into the cavity through the nasolacrimal duct (*see Chapter 4*), is added to the air.
4. **Air warmer:** The blood flowing through the nasal cavity beneath the mucous membrane lining also warms the air. This prevents damage from cold to the more fragile lower respiratory passages.
5. **Sense of smell (olfaction):** The olfactory region recognizes some 4,000 separate smells (*see previous pages*).

Disorders of the Nose

A **common cold** is a viral upper respiratory infection **(URI).** It is contagious, being transmitted from person to person in airborne droplets from coughing and sneezing. There is no proven effective treatment.

Rhinitis is an inflammation of the nasal mucosa, usually viral. It is also called **coryza.**

Allergic rhinitis affects 15% to 20% of the population. There is swelling of the mucous membranes of the nose, pharynx, and sinuses, with a clear, watery discharge. Treatment entails defining and removing the agent causing the allergy.

Sinusitis is an infection of the paranasal sinuses, often following a cold. Treatment with **antibiotics** and **decongestants** is indicated.

A **deviated nasal septum** occurs when the partition between the two nostrils is pushed to one side, leading to a partially obstructed airway in one nostril. Treatment is by surgery.

Nasal polyps are benign growths arising from the mucosa of the nasal cavity or a sinus. Surgical removal is indicated.

Epistaxis is bleeding from the septum of the nose, usually as a result of trauma.

WORD ANALYSIS AND DEFINITION

S/ = Suffix P/ = Prefix R/ = Root R/CF = Combining Form

WORD	PRONUNCIATION		ELEMENTS	DEFINITION
coryza (also called *acute rhinitis*)	koh-**RYE**-zah		Greek *catarrh*	Viral inflammation of the mucous membrane of the nose.
decongestant	dee-con-**JESS**-tant	S/ P/ R/	**-ant** *pertaining to* **de-** *take away, remove* **-congest-** *accumulation of fluid*	Agent that reduces the swelling and fluid in the nose and sinuses.
epistaxis	ep-ih-**STAK**-sis	S/ P/ R/	**-is** *pertaining to* **epi-** *above, upon* **-stax-** *fall in drops*	Nosebleed.
naris **nares (pl)**	**NAH**-ris **NAH**-rez		Latin *nostril*	Nostril.
nasal	**NAY**-zal	S/ R/	**-al** *pertaining to* **nas-** *nose*	Pertaining to the nose.
palate	**PAL**-ate		Latin *palate*	Roof of the mouth, floor of the nose.
paranasal	**PAR**-ah **NAY**-zal	S/ P/ R/	**-al** *pertaining to* **para-** *adjacent to* **-nas-** *nose*	Adjacent to the nose.
polyp	**POL**-ip		Latin *many feet*	Any mass of tissue that projects outward.
rhinitis (also called *coryza*)	rye-**NIE**-tis	S/ R/	**-itis** *inflammation* **rhin-** *nose*	Inflammation of the nasal mucosa.
septum **septa (pl)**	**SEP**-tum **SEP**-tah		Latin *partition*	Thin wall separating two cavities or tissue masses.
sinus	**SIGH**-nus		Latin *cavity*	Cavity or hollow space in a bone or other tissue.
sinusitis	sigh-nyu-**SIGH**-tis	S/ R/	**-itis** *inflammation* **sinus-** *sinus*	Inflammation of the lining of a sinus.

EXERCISES

Elements: *Work with elements to build your knowledge of the language of pulmonology. One element in each of the following medical terms is boldfaced. Identify the type of element (P, R, CF, or S) in column 2; then write the meaning of that element in column 3.*

Medical Term	Type of Element	Meaning of Element
1. nasal	________	________
2. decongestant	________	________
3. rhinitis	________	________
4. epistaxis	________	________
5. paranasal	________	________
6. sinusitis	________	________

7. There are two elements in this list that have the same meaning. Which are they? ________________ and ________________ both mean ________________.

You are

... a sleep technologist in the Sleep Disorders Clinic at Fulwood Medical Center. You are about to position the electrodes on your patient for an overnight **polysomnography** (sleep study).

You are communicating with

... Mr. Tye Gawlinski, a 29-year-old professional football player, and his wife.

CASE REPORT 7.2

Mrs. Gawlinski states that her husband snores loudly and has 40 or 50 periods in the night when he stops breathing. The snoring is so loud that she cannot sleep even in the adjoining bedroom. Mr. Gawlinski complains of being tired all day and not having the energy he needs for his job.

The sleep study is being performed to confirm a diagnosis of obstructive sleep **apnea**.

THE PHARYNX

The pharynx is a muscular funnel that receives air from the nasal cavity and food and drink from the oral cavity. It is divided into three regions (*Figure 7.4*):

1. **Nasopharynx:** Located at the back of the nose and above the soft palate and uvula. The posterior surface contains the pharyngeal **tonsil (adenoid).** Only air moves through this region.
2. **Oropharynx:** Located below the soft palate and above the epiglottis. It contains two sets of tonsils called the *palatine* and *lingual* tonsils. Air, food, and drink all pass through this region.
3. **Laryngopharynx:** Located below the tip of the epiglottis. It is the pathway to the esophagus. Only food and drink pass through the laryngopharynx.

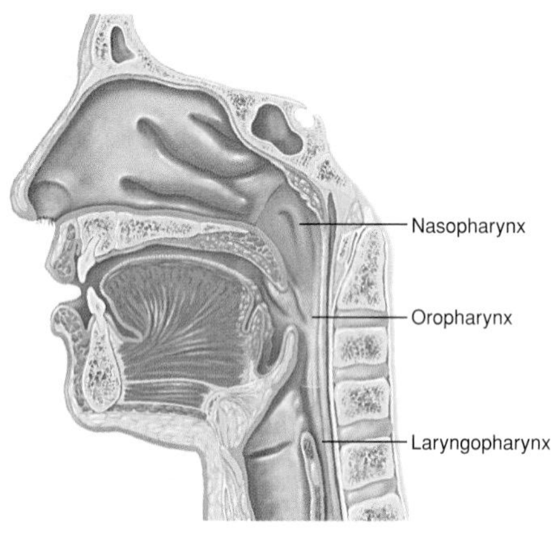

▲ **FIGURE 7.4 Regions of the Pharynx.**

Disorders of the Pharynx

Snoring occurs habitually in 25% of normal adults. The condition is most frequent in overweight males, and it becomes worse with age. The noises of snoring are made at the back of the mouth and nose where the tongue and upper pharynx meet the soft palate and uvula (*Figure 7.5*).

Obstructive sleep apnea is the condition Tye Gawlinski has. Bulky neck tissue from his football training causes obstruction by the soft tissues at the back of the nose and mouth. This leads to frequent episodes of gasping for breath, followed by complete cessation of breathing **(apnea).** The episodes reduce the level of oxygen in the blood **(hypoxia),** causing the heart to pump harder. After several years with this problem, hypertension and cardiac enlargement can occur.

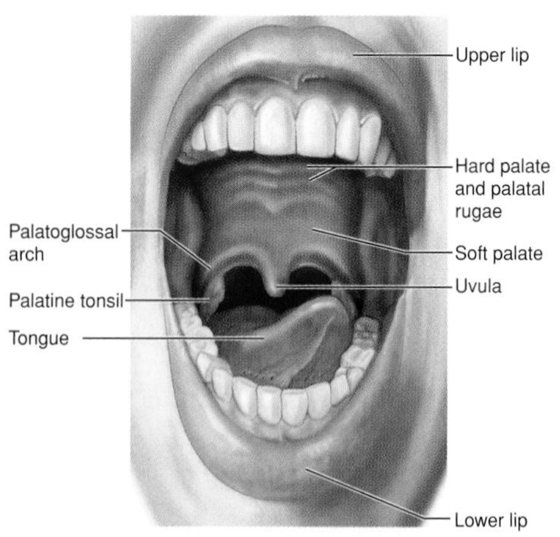

▲ **FIGURE 7.5 Soft Tissues at the Back of the Mouth.**

Case Report 7.2 (continued)

Initially, Mr. Gawlinski was instructed to sleep on his side and to use a device that, through a mask over his nose and mouth, produces a "continuous positive airway pressure" (CPAP) in his airways. He found this very uncomfortable and was kept awake by it. He chose to have the Pillar procedure, in which three tiny pillars were placed in the back of his throat to support the tissues. Mr. Gawlinski has stopped snoring.

Pharyngitis is an acute or chronic infection involving the pharynx, tonsils, and uvula. It is usually viral in children. Increasing the humidity of the air and getting rest are effective treatments.

Tonsillitis is an infection of the tonsils in the oropharynx by a virus or, in less than 20% of cases, a streptococcus. A rapid strep test and throat culture are used to identify the strep sore throat or pharyngitis.

Nasopharyngeal carcinoma is a rare form of cancer that occurs mostly in males between the ages of 50 and 60. Radiation and chemotherapy are used in treatment.

Abbreviation

CPAP continuous positive airway pressure

WORD ANALYSIS AND DEFINITION

S/ = Suffix P/ = Prefix R/ = Root R/CF = Combining Form

WORD	PRONUNCIATION		ELEMENTS	DEFINITION
adenoid	**ADD**-eh-noyd	S/ R/	**-oid** *resembling* **aden-** *gland*	Single mass of lymphoid tissue in the midline at the back of the throat.
apnea	**AP**-nee-ah	P/ R/	**a-** *without* **-pnea** *breathe*	Absence of spontaneous respiration.
hypoxia (noun) **hypoxic (adj)**	high-**POCK**-see-ah high-**POCK**-sik	S/ P/ R/ S/	**-ia** *condition* **hyp-** *below* **-ox-** *oxygen* **-ic** *pertaining to*	Decrease below normal levels of oxygen in tissues, gases, or blood. Deficient in oxygen.
laryngopharynx	lah-**RIN**-go-**FAH**-rinks	R/ R/CF	**-pharynx** *pharynx, throat* **laryng/o-** *larynx*	Region of the pharynx below the epiglottis that includes the larynx.
nasopharynx **nasopharyngeal (adj)**	**NAY**-zoh-**FAH**-rinks **NAY**-zoh-fah-**RIN**-jee-al	R/ R/CF S/ R/CF	**-pharynx** *pharynx, throat,* **nas/o-** *nose* **-eal** *pertaining to* **-pharyng-** *pharynx, throat*	Region of the pharynx at the back of the nose and above the soft palate. Pertaining to the nasopharynx.
oropharynx **oropharyngeal (adj)**	**OR**-oh-fah-rinks **OR**-oh-fah-**RIN**-jee-al	R/ R/CF S/ R	**-pharynx** *pharynx, throat* **or/o-** *mouth* **-eal** *pertaining to* **-pharyng-** *pharynx, throat*	Region at the back of the mouth between the soft palate and the tip of the epiglottis. Pertaining to the oropharynx.
pharyngitis	fair-in-**JIE**-tis	S/ R/	**-itis** *inflammation* **pharyng-** *pharynx, throat*	Inflammation of the pharynx.
polysomnography	**POLL**-ee-som-**NOG**-rah-fee	S/ P/ R/CF	**-graphy** *process of recording* **poly-** *many* **-somn/o-** *sleep*	Test to monitor brain waves, muscle tension, eye movement, and oxygen levels in the blood as the patient sleeps.
tonsil **tonsillitis** (***Note:*** double "ll") **tonsillectomy**	**TON**-sill ton-sih-**LIE**-tis ton-sih-**LEK**-toh-me	 S/ R/ S/	Latin *tonsil* **-itis** *inflammation* **tonsill-** *tonsil* **-ectomy** *surgical excision*	Mass of lymphoid tissue on either side of the throat at the back of the tongue. Inflammation of the tonsils. Surgical removal of the tonsils.

EXERCISES

Deconstruct the medical terms into their basic elements. Fill in the chart.

Medical Term	Meaning of Prefix	Meaning of Root(s), Combining Form(s)	Meaning of Suffix
hypoxia			
pharyngitis			
polysomnography			
tonsillectomy			
tonsillitis			

Use the medical terms from this chart to fill in the blanks.

1. Two terms that mean "inflammation of" are ______________________ and ______________________.
2. If left untreated, chronic obstructive sleep apnea could cause ______________________.
3. The medical term for a common sore throat is ______________________.
4. Surgical excision of the tonsils is ______________________.
5. Mr. Gawlinski underwent the diagnostic procedure of ______________________.

Study Hint

While the term *tonsil* has only one "l," notice that in the terms *tonsillitis* and *tonsillectomy,* you must double the "l."

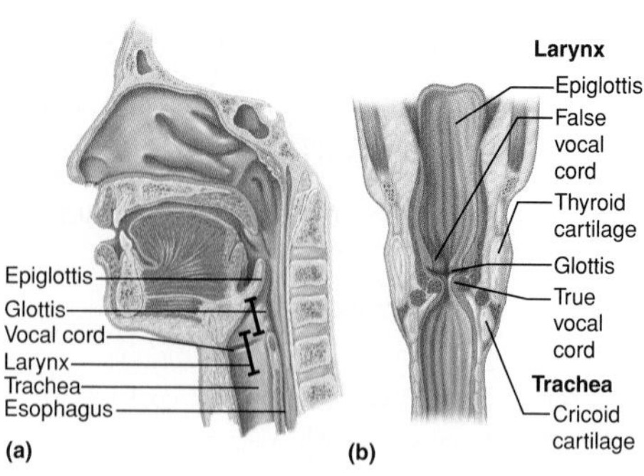

▲ **FIGURE 7.6 Larynx.**
(a) Location. (b) Structure.

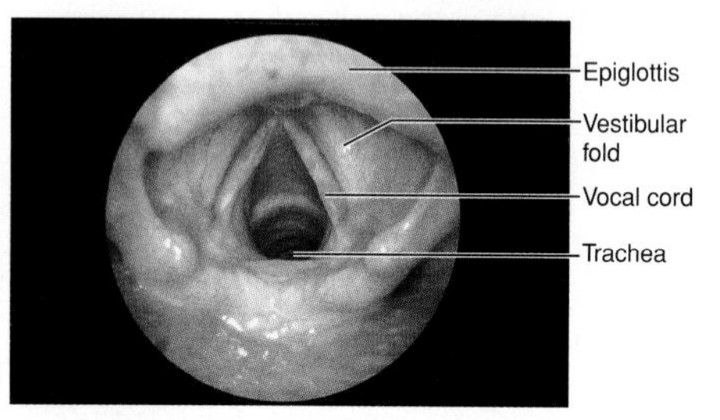

▲ **FIGURE 7.7 View of Larynx Using Laryngoscope.**

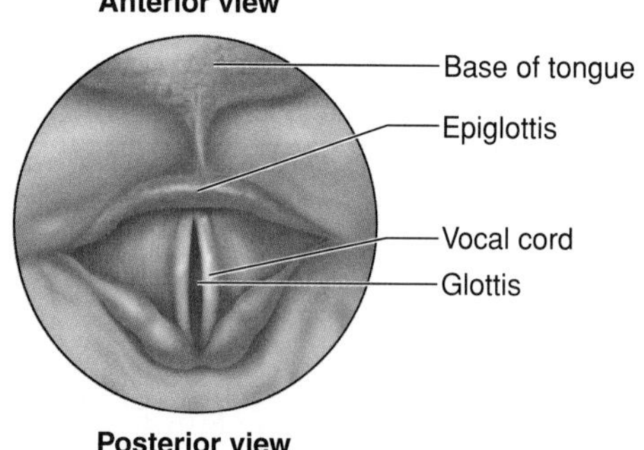

▲ **FIGURE 7.8 Vocal Cords Pulled Close and Taut.**

Keynote

On December 14, 1799, President George Washington died of an obstructed airway due to acute epiglottitis.

Keynote

Croup occurs in children aged 3 months to 5 years.

Keynote

Heavy smokers and drinkers have a 200-times greater risk of developing cancer of the larynx.

THE LARYNX

The flow of inhaled air moves on from the pharynx to the **larynx.** The upper opening into the larynx from the oropharynx is called the **glottis.** The spoon-shaped **epiglottis** guards the glottis. During the swallowing of food, the epiglottis is pushed down by the tongue to close the glottis and direct the food into the esophagus that lies behind it (*Figure 7.6a*). The **thyroid** cartilage, your Adam's apple, forms the anterior and lateral walls of the larynx.

Inside the larynx, two pairs of horizontal ligaments stretch across the lateral walls. These are the **vocal** cords (*Figures 7.7 and 7.8*), which enable sounds to be made as air passes between them.

Functions of the Larynx

1. Maintain an open passage for the movement of air to and from the trachea.
2. Prevent food and drink from entering the larynx (*Figure 7.7*).
3. Produce sounds through the vocal cords.

Sound Production Air moving past the vocal cords makes them vibrate to produce sound. The force of the air moving past the vocal cords determines the loudness of the sound. Muscles in the cords pull them closer together with varying degrees of tautness (*Figure 7.8*). A high-pitched sound is produced by taut cords and a lower pitch by more relaxed cords. Males' vocal cords are longer and thicker than those of females, vibrate more slowly, and produce lower-pitched sounds.

The crude sounds produced by the larynx are transformed into words by the actions of the pharynx, tongue, teeth, and lips.

Disorders of the Larynx

Laryngitis is inflammation of the mucosal lining of the larynx, producing hoarseness and sometimes progressing to loss of voice.

Epiglottitis is inflammation of the epiglottis. **Acute epiglottitis** is seen most commonly in children between ages 2 and 7 years and can cause acute airway obstruction and the need for **intubation.** It is preventable, as there is a vaccine available.

Croup (laryngotracheobronchitis) is a group of viral diseases causing inflammation and obstruction of the upper airway. In severe cases, a child makes a high-pitched, squeaky, inspiratory noise called **stridor.** Humidity is the initial treatment.

Papillomas or polyps are benign tumors of the larynx due to overuse or irritation and are treated by surgical excision using a **laryngoscope.**

Carcinoma of the larynx produces a persistent hoarseness. Its incidence peaks among people in their fifties and sixties. Treatment can be radiation and/or chemotherapy.

WORD ANALYSIS AND DEFINITION

S/ = Suffix P/ = Prefix R/ = Root R/CF = Combining Form

WORD	PRONUNCIATION	ELEMENTS		DEFINITION
croup (also called *laryngotracheobronchitis*)	KROOP		Old English *to cry out loud*	Infection of the upper airways in children characterized by a barking cough.
epiglottis	ep-ih-**GLOT**-is	P/ R/	**epi-** *above* **-glottis** *mouth of windpipe*	Leaf-shaped plate of cartilage that shuts off the larynx during swallowing.
epiglottitis	ep-ih-**GLOT**-eye-tis	S/ R/	**-itis** *inflammation* **-glott-** *mouth of windpipe*	Inflammation of the epiglottis.
intubation	**IN**-tyu-**BAY**-shun	S/ P/ R/	**-ation** *a process* **in-** *in* **-tub-** *tube*	Insertion of a tube into the trachea.
laryngotracheobronchitis (also called *croup*)	lah-**RING**-oh-**TRAY**-kee-oh-brong-**KIE**-tis	S/ R/CF R/CF R/	**-itis** *inflammation* **laryng/o-** *larynx* **-trache/o-***trachea* **-bronch-** *bronchus*	Inflammation of the larynx, trachea, and bronchi.
larynx **laryngeal (adj)**	**LAH-**rinks lah-**RIN**-jee-al	 S/ R/	Greek *larynx* **-eal** *pertaining to* **laryng-** *larynx*	Organ of sound production. Pertaining to the larynx.
laryngitis	lah-rin-**JEYE**-tis	S/ R/CF	**-itis** *inflammation* **laryng/o-** *larynx*	Inflammation of the larynx.
laryngoscope	lah-**RING**-oh-skope	S/	**-scope** *instrument for viewing*	Hollow tube with a light and camera used to visualize or operate on the larynx.
stridor	**STRY-**door		Latin *a harsh, creaking sound*	High-pitched noise made when there is a respiratory obstruction in the larynx or trachea.
vocal	**VOH**-kal	S/ R/	**-al** *pertaining to* **voc-** voice	Pertaining to the voice.

EXERCISES

Deconstruction: *For long or short medical terms, deconstruction into word elements is your key to solving the meaning of the term. Follow the directions and fill in the blanks.*

1. laryngotracheobronchitis

Slash the term into all its elements:

____

2. Write the meaning of each element under the line in question 1.

3. Combine the meanings of the elements, and define the term.

Laryngotracheobronchitis means:

____

____

4. vocal

Slash the term into all its elements:

____

5. Write the meaning of each element under the line in question 4.

6. Combine the meanings of the elements, and define the term.

Vocal means:

____

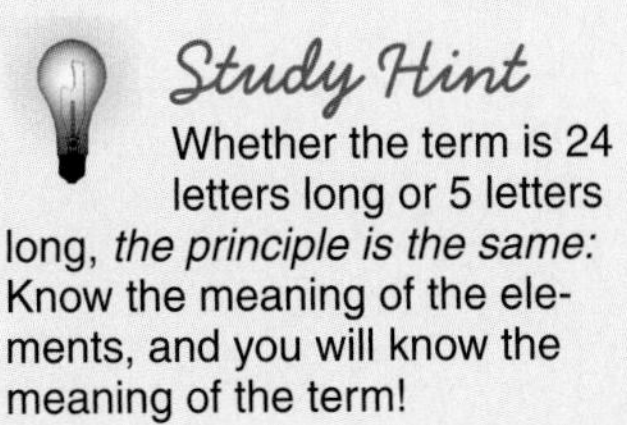

Study Hint

Whether the term is 24 letters long or 5 letters long, *the principle is the same:* Know the meaning of the elements, and you will know the meaning of the term!

LESSON 7.2 Lower Respiratory Tract

RESPIRATORY SYSTEM

OBJECTIVES

Once the air you inhale has passed through the upper airway and many of the pollutants and impurities have been filtered out, the major needs still remain: to get oxygen into the blood and remove carbon dioxide from the blood. To do these, the inhaled air has to get down into the **alveoli** of the lungs, where these exchanges can occur. In this lesson, you will learn to:

- **Trace the passage of air from the larynx into the alveoli and back.**
- **Describe the mechanics of ventilation.**
- **Integrate the medical terminology for the functions of the different elements of the lower airway with their structures.**
- **Use correct medical terminology to discuss the effects of common disorders of the lungs on health.**

TRACHEA

The flow of inhaled air now moves into the trachea (windpipe). This is a rigid tube that descends from the larynx and divides into the two main **bronchi** (*Figure 7.9*) going to the right and left lungs.

THE LUNGS

The two lungs are the main organs of respiration and are located in the thoracic cavity. Each lung is a soft, spongy, conical organ with its **base** resting on the **diaphragm** and its **apex** above and behind the clavicle. Its outer convex, **costal** surface presses against the **rib cage.** Its inner concave surface presses against the **mediastinum.**

The right lung has three **lobes,** superior, middle, and inferior. The left lung has two lobes, superior and inferior (*Figure 7.9a*). Each lobe is separated from the other by **fissures.**

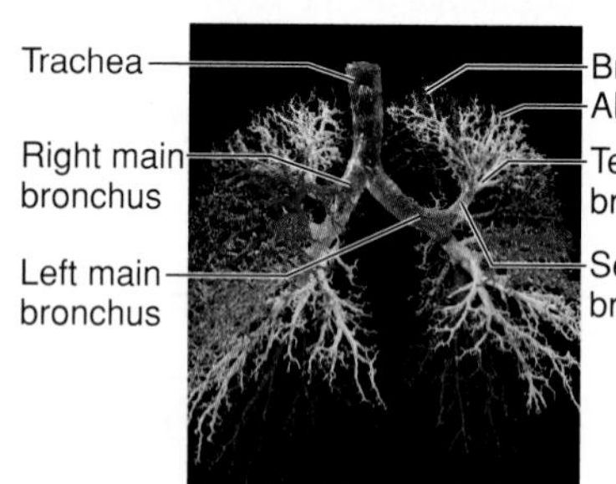

▲ **FIGURE 7.10 Latex Cast of Tracheobronchial Tree.**

TRACHEOBRONCHIAL TREE

As the inhaled air continues down the respiratory tract, the main bronchi divide into a **secondary (lobar)** bronchus for each lobe. Each secondary bronchus divides into **tertiary** bronchi that supply **segments** of each lobe (*Figure 7.10*).

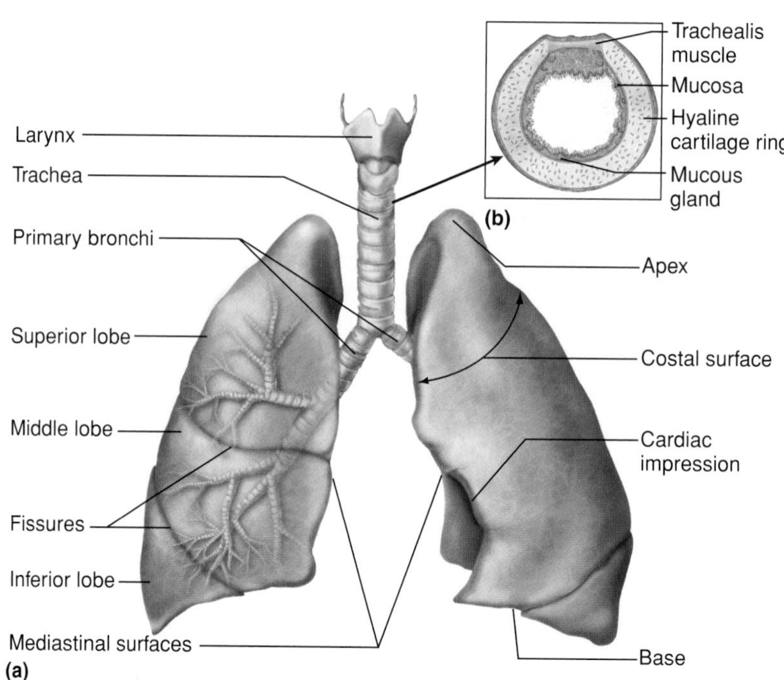

▶ **FIGURE 7.9 Lower Respiratory Tract.** (a) Gross anatomy. (b) C-shaped tracheal cartilage.

WORD ANALYSIS AND DEFINITION

S/ = Suffix P/ = Prefix R/ = Root R/CF = Combining Form

WORD	PRONUNCIATION	ELEMENTS		DEFINITION
diaphragm	**DIE**-ah-fram		Greek *diaphragm*	The muscular sheet separating the abdominal and thoracic cavities.
diaphragmatic (adj)	**DIE**-ah-frag-**MAT**-ik	S/ R/CF	**-tic** *pertaining to* **diaphragm/a-** *diaphragm*	Pertaining to the diaphragm.
fissure **fissures (pl)**	**FISH**-ur	S/ R/	**-ure** *process, result of* **fiss-** *split*	Deep furrow or cleft.
lobe **lobar (adj)**	LOBE **LOW**-bar	 S/ R/	Greek *lobe* **-ar** *pertaining to* **lob-** *lobe*	Subdivision of an organ or other part. Pertaining to a lobe.
lobectomy	low-**BECK**-toe-me	S/	**-ectomy** *surgical excision*	Surgical removal of a lobe.
mediastinum	**ME**-dee-ass-**TIE**-num	S/ P/ R/	**-um** *tissue* **media-** *middle* **-stin-** *partition*	Area between the lungs containing the heart, aorta, venae cavae, esophagus, and trachea.
mediastinal (adj)	**ME**-dee-ass-**TIE**-nal	S/	**-al** *pertaining to*	Pertaining to the mediastinum.
pleura **pleurae (pl)** **pleural (adj)**	**PLUR**-ah **PLUR**-ee **PLUR**-al	 S/ R/	Greek *rib* **-al** *pertaining to* **pleur-** *pleura*	Membrane covering the lungs and lining the ribs in the thoracic cavity. Pertaining to the pleura.
pleurisy	**PLUR**-ih-see	S/	**-isy** *inflammation*	Inflammation of the pleura.

Bronchioles and Alveoli

The tertiary bronchi divide into **bronchioles,** which in turn divide into several thin-walled **alveoli** (*Figure 7.11*).

Each **alveolus** is a thin-walled sac supported by a thin **respiratory membrane** that allows the exchange of gases with the surrounding pulmonary capillary network (*Figure 7.11*).

The Pleurae

The surface of each lung is covered with a double-layered serous membrane called the **pleura.** The space between the two layers is called the **pleural cavity,** which contains a thin film of lubricant fluid. This lubricant enables the lungs to expand (inspiration) and deflate (expiration) with minimal friction.

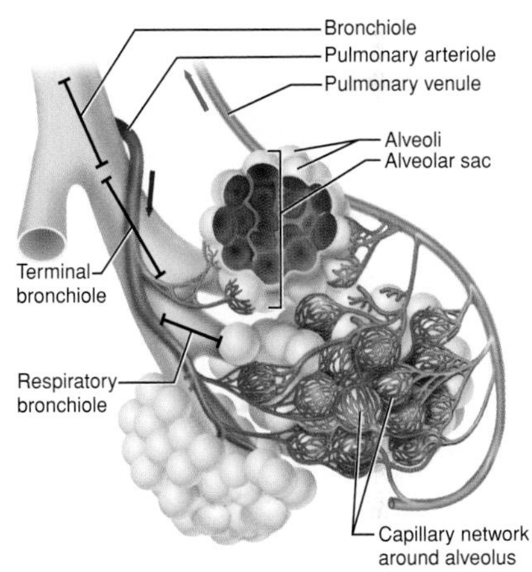

▲ **FIGURE 7.11 Bronchioles and Alveoli.**

Be precise!

EXERCISES

***Precision in communication** means using the correct form of the medical term, as well as the correct spelling. Test your knowledge of plurals, adjectives, and spelling with this exercise. Circle the correct choice.*

1. The median partition of the thoracic cavity is the:
   mediastenum mediasternum mediastinum midiasternum
2. The patient was diagnosed with ______________________ pneumonia.
   lobe lobular lobar lumbar
3. The ______________________ tissue was sent to pathology for a diagnosis.
   midiastenal mediastenal mediastinal mediasternal
4. The muscle separating the abdominal and thoracic cavities is the:
   diaphram diaphragm diaphracum diaphragmm
5. Removal of a portion of a lung is a:
   lobotomy lobectomy lobarectomy lobarotomy
6. Each lobe is separated from the other by:
   fishures fissures fisures fishurrs

Study Hint

Frequent errors are made on test questions regarding the term *pleural, meaning pertaining to the pleura.* It is not spelled "plural," which means *more than one.* Be especially careful when dealing with this term, and always check your spelling. *The right answer (pleural) is wrong if it is spelled "plural."* Don't lose points on a test because of carelessness.

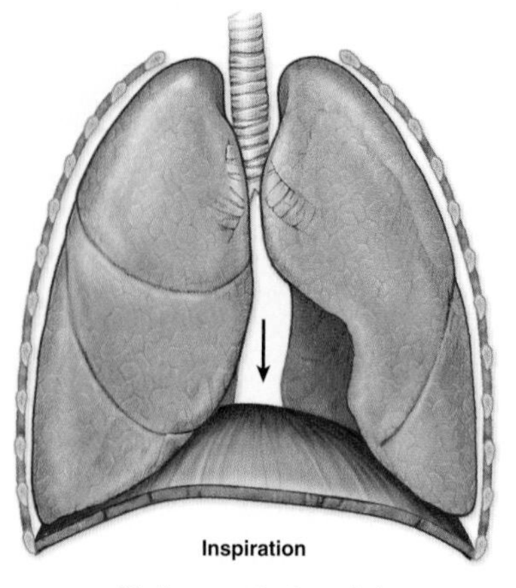
Inspiration
Diaphragm contracts; vertical dimensions of thoracic cavity increase.
(a)

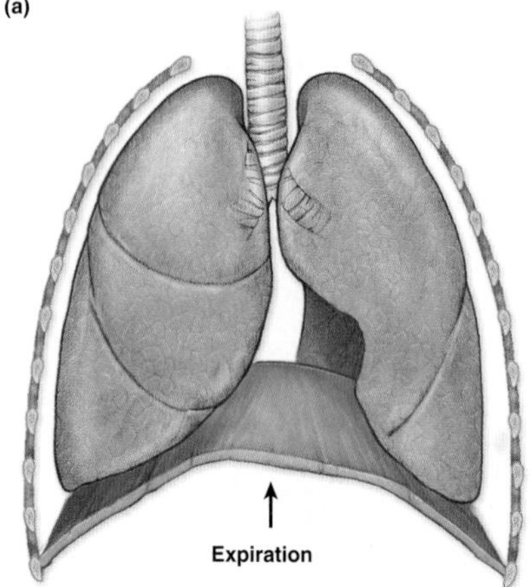
Expiration
Diaphragm relaxes; vertical dimensions of thoracic cavity decrease.
(b)

▲ **FIGURE 7.12 Inspiration and Expiration.**

Abbreviation	
SOB	short of breath

MECHANICS OF RESPIRATION

A resting adult breathes 10 to 15 times per minute and **inhales** about 500 mL of air during inspiration and **exhales** it during expiration. The mission is to get air into and out of the alveoli so that oxygen can get into the bloodstream and carbon dioxide can get out.

The diaphragm does most of the work. In inspiration it drops down and flattens to expand the thoracic cavity and reduce the pressure in the airways. In addition, the external intercostal muscles lift the chest wall up and out to further expand the thoracic cavity (*Figure 7.12a*).

Expiration is a process of letting go. The diaphragm and the intercostal muscles relax, and the thoracic cavity springs back to its original size (*Figure 7.12b*).

COMMON SIGNS AND SYMPTOMS OF RESPIRATORY DISORDERS

1. **Cough** is triggered by irritants in the respiratory tract. The irritants can be cigarette smoke, as with Mr. Jacobs, or infection, or tumors as in lung cancer. A **productive** cough produces **sputum,** which can be swallowed or **expectorated.** Bloody sputum is called **hemoptysis.** Thick, yellow **(purulent)** sputum indicates infection. A **nonproductive** cough is dry and hacking.

   Abnormal amounts of mucus arising from the upper respiratory tract and expectorated are called **phlegm.**

2. **Dyspnea,** or shortness of breath **(SOB),** can occur on exertion or, in severe disorders, during rest when all the respiratory muscles are used to exchange only a small volume of air.

3. **Cyanosis** is seen when the blood has increased levels of **unoxygenated hemoglobin** and has a characteristic dark red-blue color.

4. **Changes in the rate of breathing** may occur. **Eupnea** is the normal, easy respiration, around 15 breaths per minute, in a resting adult. Both **tachypnea** (rapid rate of breathing) and **hyperpnea** (breathing deeper and more rapidly than normal) are signs of respiratory difficulty, as is **bradypnea** (slow breathing).

5. Sneezing is caused by irritants in the nasal cavity.

6. Hiccups are reflex spasms of the diaphragm. The etiology is unknown, and there is no specific medical cure.

7. **Yawning** is a reflex that originates in the brainstem in response to hypoxia, boredom, or sleepiness. The exact mechanisms are not known.

WORD ANALYSIS AND DEFINITION

S/ = Suffix P/ = Prefix R/ = Root R/CF = Combining Form

WORD	PRONUNCIATION	ELEMENTS		DEFINITION
bradypnea (opposite of *tachypnea*)	brad-ip-**NEE**-ah	P/ R/	**brady-** *slow* **-pnea** *breathe*	Slow breathing.
cyanosis	sigh-ah-**NO**-sis	S/ R/	**-osis** *condition* **cyan-** *dark blue*	Blue discoloration of the skin, lips, and nail beds due to low levels of oxygen in the blood.
cyanotic (adj)	sigh-ah-**NOT**-ik	S/ R/CF	**-tic** *pertaining to* **cyan/o** *dark blue*	Pertaining to or marked by cyanosis.
dyspnea	disp-**NEE**-ah	P/ R/	**dys-** *bad, difficult* **-pnea** *breathe*	Difficult breathing.
eupnea	yoop-**NEE**-ah	P/ R/	**eu-** *normal, good* **-pnea** *breathe*	Normal breathing.
exhale	**EKS**-hail	P/ R/	**ex-** *out* **-hale** *breathe*	Breathe out.
expectorate	ek-**SPEK**-toh-rate	S/ P/ R/	**-ate** *pertaining to* **ex-** *out* **-pector-** *chest*	Cough up and spit out mucus from the respiratory tract.
hemoptysis	he-**MOP**-tih-sis	R/CF R/	**hem/o-** *blood* **-ptysis** *spit*	Bloody sputum.
hyperpnea	high-perp-**NEE**-ah	P/ R/	**hyper-** *excessive* **-pnea** *breathe*	Deeper and more rapid breathing than normal.
inhale	**IN**-hail	P/ R/	**in-** *in* **-hale** *breathe*	Breathe in.
phlegm	FLEM		Greek *flame*	Abnormal amounts of mucus expectorated from the respiratory tract.
tachypnea (opposite of *bradypnea*)	tak-ip-**NEE**-ah	P/ R/	**tachy-** *rapid* **-pnea** *breathe*	Rapid breathing.

EXERCISES

Construct terms: *Knowing just one element will enable you to build more terms with the addition of other elements. Practice building your pulmonology terms with the following root and various prefixes. Fill in the blanks.*

1. The root *pnea* means ______________________________.

Add the following prefixes to the root pnea to form the new terms.

tachy **brady** **dys** **eu** **hyper**

2. difficult breathing ______________________________ pnea
3. deeper breathing than normal ______________________________ pnea
4. slow breathing ______________________________ pnea
5. normal breathing ______________________________ pnea
6. rapid breathing ______________________________ pnea

Demonstrate that you can use breath *and* breathe *correctly. Write a brief sentence for each.*

7. ____

____

8. ____

____

Study Hint

The English word *breath* is a noun and is the air you inhale and exhale. ("Take a deep breath.") The English word ***breathe*** has an "e" on the end of it and is the verb meaning *to inhale and exhale.* ("She was unable to breath*e*.")

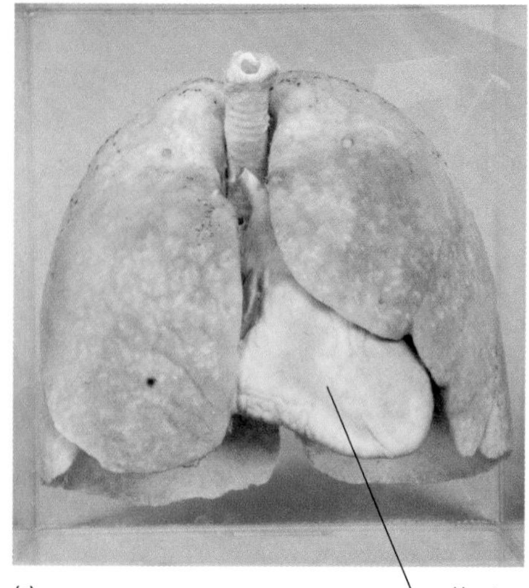

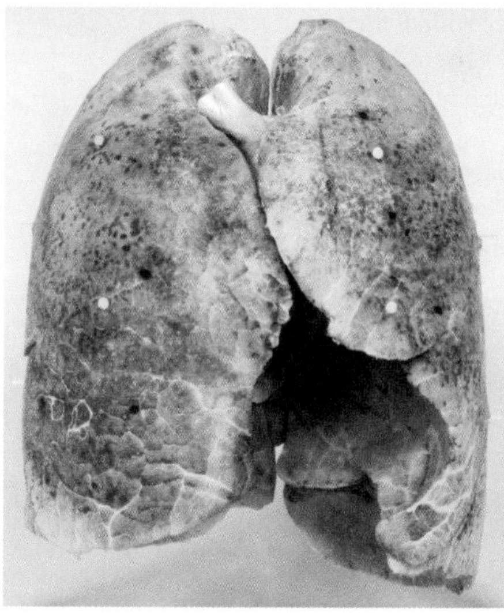

▲ **FIGURE 7.13 Whole Lungs.**
(a) Nonsmoker's lungs. (b) Smoker's lungs.

DISORDERS OF THE LOWER RESPIRATORY TRACT

Acute bronchitis can be viral or bacterial, leading to the production of excess mucus with some obstruction of airflow. It resolves without significant residual damage to the airway.

Chronic bronchitis is the most common obstructive disease and is due to cigarette smoking or repeated episodes of acute bronchitis. In addition to excess mucus production, cilia are destroyed. A pattern develops, involving chronic cough, dyspnea, and recurrent acute infections.

In advanced chronic bronchitis, hypoxia and **hypercapnia** (excess carbon dioxide) are produced and heart failure follows.

Bronchiolitis, inflammation of the small airway bronchioles, occurs in the adult as the early and often unrecognized beginning of airway changes in cigarette smokers or in those exposed to "secondhand smoke," who inhale the smoke produced by other peoples' cigarettes. Bronchiolitis also affects children under the age of 2, because their small airways become blocked very easily. The disease is viral and in severe cases can cause marked respiratory distress.

Pulmonary emphysema is a disease of the respiratory bronchioles and alveoli. These airways become enlarged, and the septa between the alveoli are destroyed, forming large sacs **(bullae).** There is a loss of surface area for gas exchange.

Chronic airway obstruction (CAO) is also called **chronic obstructive pulmonary disease (COPD).** It is a progressive disease, as Mr. Jacobs's history shows. It involves both chronic bronchitis and emphysema. A history of heavy cigarette smoking, with chronic cough and sputum production, is followed by increasing dyspnea and need for oxygen (*Figure 7.13*). Right-sided heart failure (cor pulmonale) is the end result, due to pulmonary hypertension and backup of blood into the right ventricle.

Bronchiectasis is the abnormal dilatation of the small bronchioles due to repeated infections. The damaged, dilated bronchi are unable to clear secretions, so additional infections and more damage can occur.

Bronchial asthma is a disorder with recurrent acute episodes of bronchial obstruction due to constriction of bronchioles **(bronchoconstriction), hypersecretion** of mucus, and inflammatory swelling of the bronchiolar lining (*Figure 7.14*). Between attacks, breathing can be normal. The etiology of asthma is an allergic response to substances such as pollen, animal dander, or the feces of house dust mites.

Cystic fibrosis (CF) is caused by an increased **viscosity** of secretions from the pancreas, salivary glands, liver, intestine, and lungs. In the lungs a particularly thick mucus obstructs the airways and causes repeated infections. Respiratory failure is the cause of death, often before the age of 30. The disorder is genetic.

Pulmonary edema is the collection of fluid in the lung tissues and alveoli. It is most frequently the result of left ventricular failure or mitral valve disease with congestive heart failure **(CHF).**

During **auscultation** of the chest, the air bubbling through abnormal fluid in the alveoli and small bronchioles, as in pulmonary edema, produces a noise called **rales.** When the bronchi are partly obstructed and air is being forced past the obstruction, a high-pitched noise called a **rhoncus** is heard.

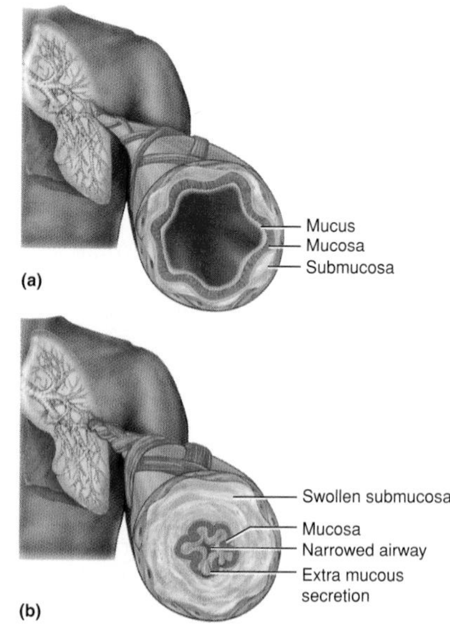

▲ **FIGURE 7.14 Bronchiole Diameter.**
(a) Normal airway. (b) Airway during an asthma attack.

Abbreviations

CAO	chronic airway obstruction
CF	cystic fibrosis
COPD	chronic obstructive pulmonary disease
CHF	congestive heart failure

WORD ANALYSIS AND DEFINITION

S/ = Suffix P/ = Prefix R/ = Root R/CF = Combining Form

WORD	PRONUNCIATION		ELEMENTS	DEFINITION
asthma	**AZ**-mah		Greek *asthma*	Episodes of breathing difficulty due to narrowed or obstructed airways.
asthmatic (adj)	az-**MAT**-ic	S/ R/	**-atic** *pertaining to* **asthm-** *asthma*	Pertaining to or suffering from asthma.
auscultation	aws-kul-**TAY**-shun	S/ R/	**-ation** *a process* **auscult-** *listen to*	Diagnostic method of listening to body sounds with a stethoscope.
bronchiectasis	brong-key-**ECK**-tah-sis	S/ R/CF	**-ectasis** *dilation* **bronch/i-** *bronchus*	Chronic dilation of the bronchi following inflammatory disease and obstruction.
bronchiolitis (***Note:*** This term has two suffixes—the "e" is dropped before the "i" in *itis*.)	brong-key-oh-**LYE**-tis	S/ S/ R/CF	**-itis** *inflammation* **-ole** *small* **bronch/i-** *bronchus*	Inflammation of the small bronchioles.
bronchitis	brong-**KI**-tis	S/ R/	**-itis** *inflammation* **bronch-** *bronchus*	Inflammation of the bronchi.
bronchoconstriction	**BRONG**-koh-kon-**STRIK**-shun	S/ R/CF R/	**-ion** *action, condition* **bronch/o-** *bronchus* **-constrict-** *to narrow*	Reduction in diameter of a bronchus.
bulla **bullae (pl)**	**BULL**-ah **BULL-ee**		Latin *bubble*	Bubble-like dilated structure.
cystic fibrosis	**SIS**-tik fie-**BRO**-sis	S/ R/ S/ R/	**-ic** *pertaining to* **cyst-** *cyst* **-osis** *condition* **fibr-** *fiber*	Genetic disease in which excessive viscid mucus obstructs passages, including bronchi.
emphysema	em-fih-**SEE**-mah	P/ R/	**em-** *in, into* **-physema** *blowing*	Dilation of respiratory bronchioles and alveoli.
hypercapnia	**HIGH**-per-**KAP**-nee-ah	S/ P/ R/	**-ia** *condition* **hyper-** *excessive* **-capn-** *carbon dioxide*	Abnormal increase of carbon dioxide in the arterial bloodstream.
hypersecretion	**HIGH**-per-seh-**KREE**-shun	S/ P/ R/	**-ion** *action, condition* **hyper-** *excessive* **-secret-** *secrete*	Excessive secretion of mucus (or enzymes or waste products).
rhoncus **rhonci (pl)**	**RONG**-kuss **RONG**-key		Greek *snoring*	Wheezing sound heard on auscultation of the lungs; made by air passing through a constricted lumen.
viscosity	viss-**KOS**-ih-tee	S/ R/	**-ity** *condition* **viscos-** *viscous, sticky*	The resistance of a fluid to flow.
viscous (adj) (cf. *viscus,* any internal organ)	**VISS**-kus			Sticky fluid that is resistant to flow.

EXERCISES

Elements. *The elements below all appear in the* **language of pulmonology.** *Challenge your knowledge of these elements, and make the correct match. Fill in the blanks.*

_____ 1. ectasis A. listen to

_____ 2. capn B. sticky

_____ 3. physema C. carbon dioxide

_____ 4. viscos D. blowing

_____ 5. auscult E. process

_____ 6. ole F. dilation

DISORDERS OF THE LOWER RESPIRATORY TRACT (continued)

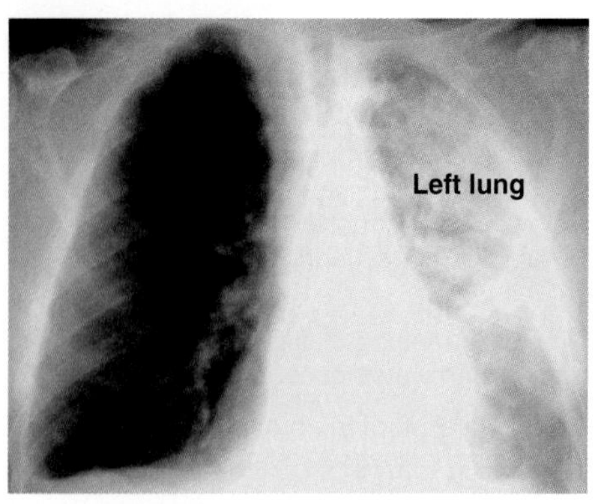

▲ **FIGURE 7.15 Chest X-ray of Patient with Pneumonia in the Left Lung.**
A normal lung appears as a black space on an x-ray because its spongy structure is filled with air. In contrast, a pneumonia lung appears white or opaque on an x-ray due to accumulation of fluid and cells in the alveoli.

Pneumonia (*Figure 7.15*) is an acute infection affecting the alveoli and lung parenchyma (functional cells of the lung). Bacterial infections focus on the alveoli; viral infections, on the parenchyma. **Lobar pneumonia** is an infection limited to one lung lobe. **Bronchopneumonia** is used to describe an infection in the bronchioles that spreads to the alveoli.

When an area of the lung **(segment)** or a lobe becomes airless as a result of the infection, the lung is **consolidated.** When an area of the lung collapses as a result of bronchial obstruction, this is called **atelectasis.**

Pleurisy, an inflammation of the pleurae, can be a complication of pneumonia. This condition is very painful on breathing because the parietal pleura is very pain-sensitive. The inflammation often leads to fluid accumulating in the pleural cavity. This is a **pleural effusion.** If the pleural effusion contains pus, the condition is called **empyema.** If it contains blood, the condition is called **hemothorax.** When pleural fluid is drawn off for therapeutic reasons or for laboratory analysis, the procedure is **aspiration** or **thoracentesis.**

Lung abscess can be a complication of bacterial pneumonia or cancer. Long-term antibiotics are used, and surgical resection of the abscess may be required.

Pneumothorax is the entry of air into the pleural cavity (*Figure 7.16*). The cause can be unknown, called a **spontaneous pneumothorax,** but it often results from trauma when a fractured rib, knife blade, or bullet lacerates the pleura.

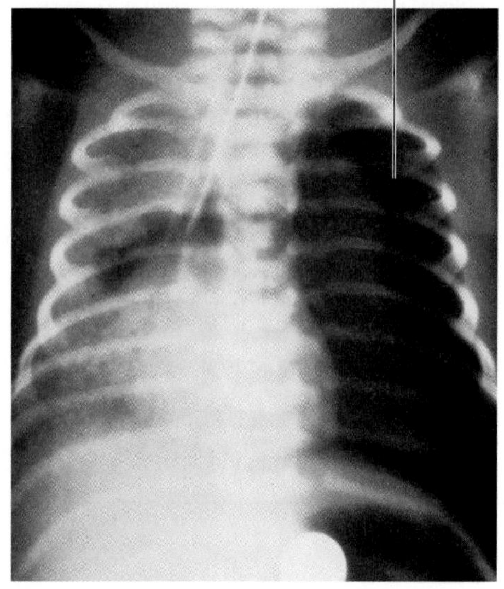

▲ **FIGURE 7.16 Left Pneumothorax.**
There are no lung markings seen in the area of the pneumothorax.

Adult respiratory distress syndrome (ARDS) is sudden, life-threatening lung failure caused by a variety of underlying conditions from major trauma to sepsis. The alveoli fill with fluid and collapse, and gas exchange is shut down. Hypoxia results. **Mechanical ventilation** has to be provided. The mortality is from 35% to 50%.

Neonatal respiratory distress syndrome (NRDS) is seen in premature babies whose lungs have not matured enough to produce surfactant (*see Chapter 16*). The alveoli collapse, and mechanical ventilation is needed to keep them open.

Chronic infections of the lung parenchyma are the result of prolonged exposure to infection or to occupational irritant dusts or droplets. These disorders are called **pneumoconioses.** Levels of dust inhalation overwhelm the airways' particle-clearing abilities, and the dust particles accumulate in the alveoli and parenchyma, leading to fibrosis. **Asbestosis** results from inhaling asbestos particles and can lead to a cancer in the pleura called **mesothelioma. Silicosis** from silica particles is called "stonecutters' disease." **Anthracosis** from coal dust particles is called "coal miners' disease." (**Anthrax** is a different disease, caused by toxins produced by the anthrax bacillus.) **Sarcoidosis** is a fibrotic disorder of the lung parenchyma.

Pulmonary tuberculosis is a chronic, infectious disease of the lungs.

Lung cancer, related to tobacco use, used to be a male disease, but now fatalities in women from lung cancer exceed those from breast cancer. Ninety percent of lung cancers arise in the mucous membranes of the larger bronchi and are called **bronchogenic carcinomas.** A subgroup of bronchogenic carcinomas called **adenocarcinoma** accounts for 30% to 50% of all lung cancers and is the most common in women. The lung cancer obstructs the bronchus, spreads into the surrounding lung tissues, and metastasizes to lymph nodes, liver, brain, and bone. This disease is associated with cigarette smoking.

Abbreviations

ARDS adult respiratory distress syndrome
NRDS neonatal respiratory distress syndrome

WORD ANALYSIS AND DEFINITION

S/ = Suffix P/ = Prefix R/ = Root R/CF = Combining Form

WORD	PRONUNCIATION	ELEMENTS		DEFINITION
adenocarcinoma	**ADD**-eh-noh-kar-sih-**NOH**-mah	S/ R/CF R/	**-oma** *tumor* **aden/o-** *gland* **-carcin-** *cancer*	A cancer arising from glandular epithelial cells.
anthracosis	an-thra-**KOH**-sis	S/ R/	**-osis** condition **anthrac-** coal	Lung disease caused by the inhalation of coal dust.
anthrax	**AN**-thraks		Greek *carbuncle*	A severe, malignant infectious disease.
asbestosis	as-bes-**TOE**-sis	S/ R/	**-osis** *condition* **asbest-** *asbestos*	Lung disease caused by the inhalation of asbestos particles.
aspiration	as-pih-**RAY**-shun	S/ R/	**-ion** *process* **aspirat-** *to breathe on*	Removal by suction of fluid or gas from a body cavity.
atelectasis	at-el-**ECK**-tah-sis	S/ R/	**-ectasis** *dilation* **atel-** *incomplete*	Collapse of part of a lung.
bronchogenic	**BRONG**-koh-**JEN**-ik	S/ R/CF	**-genic** *creation* **bronch/o-** *bronchus*	Arising from a bronchus.
bronchopneumonia	**BRONG**-koh-new-**MOH**-nee-ah	S/ R/ R/CF	**-ia** *condition* **-pneumon-** *air, lung* **bronch/o-** *bronchus*	Acute inflammation of the walls of smaller bronchioles with spread to lung parenchyma.
empyema	**EM**-pie-**EE**-mah	S/ P/ R/	**-ema** *result* **em-** *in, into* **-py-** *pus*	Pus in a body cavity, particularly in the pleural cavity.
hemothorax	he-moh-**THOR**-ax	R/CF R/	**hem/o-** *blood* **-thorax** *chest*	Blood in the pleural cavity.
pneumoconiosis **pneumoconioses (pl)**	new-moh-koh-nee-**OH**-sis	S/ R/ R/CF	**-osis** *condition* **-coni-** *dust* **pneum/o-** *lung, air*	Fibrotic lung disease caused by the inhalation of different dusts.
pneumonia **pneumonitis** (same as *pneumonia*)	new-**MOH**-nee-ah new-moh-**NI**-tis	S/ R/ S/	**-ia** *condition* **pneumon-** *lung, air* **-itis** *inflammation*	Inflammation of the lung parenchyma (tissue).
pneumothorax	new-moh-**THOR**-ax	R/CF R/	**pneum/o-** *air, lung* **-thorax** *chest*	Air in the pleural cavity.
sarcoidosis (***Note:*** two suffixes)	sar-koy-**DOH**-sis	S/ S/ R/	**-osis** *condition* **-oid-** *resembling* **sarc-** *sarcoma, flesh*	Granulomatous lesions of the lungs and other organs; cause is unknown.
silicosis	sil-ih-**KOH**-sis	S/ R/	**-osis** *condition* **silic-** *silicon, glass*	Fibrotic lung disease from inhaling silica particles.
thoracentesis (same as *pleural tap*)	**THOR**-ah-sen-**TEE**-sis	S/ R/	**-centesis** *to puncture* **thora-** *chest*	Insertion of a needle into the pleural cavity to withdraw fluid or air.
tuberculosis	too-**BER**-kyu-**LOW**-sis	S/ R/	**-osis** *condition* **tubercul-** *nodule, swelling, tuberculosis*	Infectious disease that can infect any organ or tissue.

EXERCISES

Suffixes: *The WAD above contains a lot of suffixes you will see again in other medical terms you will meet in later chapters. Confirm your knowledge of suffixes by filling in the chart with the correct meaning for the element; then answer the questions.*

Suffix	Meaning of Suffix	Medical Term with This Suffix
centesis		
ectasis		
ema		
genic		
ia		
ion		
oid		
osis		

LESSON 7.3 RESPIRATORY SYSTEM

Diagnostic and Therapeutic Procedures

OBJECTIVES

There are specific diagnostic and therapeutic procedures and a wide range of pharmacologic agents available to help patients with lung diseases. In this lesson, the information will enable you to:

- **Identify the terminology of specific pulmonary function tests (PFTs) and other diagnostic procedures.**
- **Use correct medical terminology to describe common therapeutic procedures.**
- **List different classes of pharmacologic agents and their effects on the lungs.**

Case Report 7.1 (continued)

Mr. Jacobs's forced expiratory volume was only 40% of normal because the fibrotic effects of repeated lung infections had reduced the volume of his airways. When he was off oxygen, Mr. Jacobs' arterial oxygen levels were below 50% of normal. Even with nasal prongs and oxygen, his arterial oxygen levels were only 75% of normal.

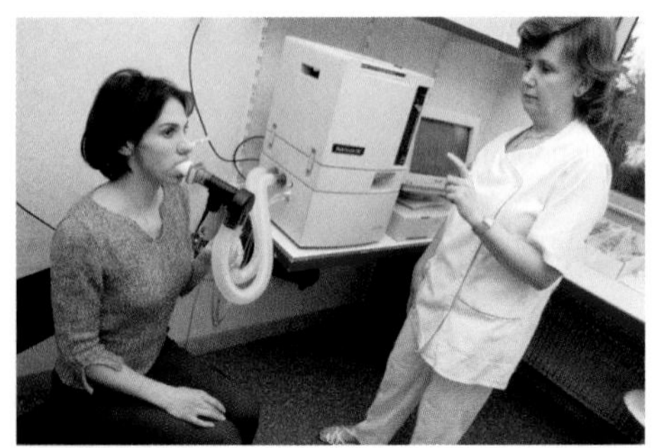

▲ **FIGURE 7.17 Spirometer.**

Keynote

Measure pulmonary function with a spirometer, a peak flow meter, and arterial blood gases.

DIAGNOSTIC PROCEDURES

Pulmonary Function Tests (PFTs)

A **spirometer** is a device for measuring the **volume** of air that patients move in and out of their respiratory systems (*Figure 7.17*). The volume of air expired at the end of the test is the patient's **forced expiratory vital capacity (FVC).** The spirometer also measures **flow rates,** and the **forced expiratory volume in 1 second (FEV1)** is the amount of air expired in the first second of the test.

In **obstructive** lung disorders such as asthma or COPD, the airways are constricted and resistant to airflow. This will cause a reduction in the FEV1. In **restrictive** lung disorders, in which the lung tissue is fibrotic or scarred and resists expansion, there will be a reduction in the FVC. Mr. Jacobs has both obstructive and restrictive disease.

A **peak flow meter** records the greatest flow of air that can be sustained for 10 milliseconds on forced expiration, the **peak expiratory flow rate (PEFR).** It is of value in following the course of asthma and in postoperative care to monitor the return of lung function after anesthesia.

Arterial blood gases, the measure of the levels of oxygen and carbon dioxide in the blood, are good indicators of respiratory function.

A pulse oximeter is a sensor placed on the finger to measure the oxygen saturation of the blood.

Other Diagnostic Procedures

Chest x-ray (CXR) is a radiographic image of the chest taken in **anteroposterior (AP), posteroanterior (PA),** lateral, and sometimes oblique and lateral **decubitus** positions.

Computed tomography (CT), angiography of the pulmonary circulation using contrast materials, **magnetic resonance angiography (MRA)** to define emboli in the pulmonary arteries, and **ultrasonography** of the pleural space are chest-imaging techniques in modern use. **Positron emission tomography (PET)** can sometimes distinguish benign from malignant lesions.

Bronchoscopy is a procedure for inserting a fiber optic into the bronchial tubes to visually examine the tubes, take a tissue biopsy, or do a washing for secretions.

Mediastinoscopy is used to stage lung cancer and diagnose mediastinal masses. The **mediastinoscope** is inserted through an incision in the suprasternal notch.

WORD ANALYSIS AND DEFINITION

S/ = Suffix P/ = Prefix R/ = Root R/CF = Combining Form

WORD	PRONUNCIATION	ELEMENTS		DEFINITION
bronchoscopy	brong-**KOS**-koh-pee	S/ R/CF	**-scopy** *to examine, view* **bronch/o** *bronchus*	Examination of the interior of the tracheobronchial tree with an endoscope.
bronchoscope	**BRONG**-koh-skope	S/	**-scope** *instrument for viewing*	Endoscope used for bronchoscopy.
endotracheal	en-doh-**TRAY**-kee-al	S/ P/ R/	**-al** *pertaining to* **endo-** *inside* **-trache-** *trachea*	Pertaining to being inside the trachea.
mediastinoscopy	**ME**-dee-ass-tih-**NOS-** koh-pee	S/ R/CF	**-scopy** *to examine, view* **mediastin/o** *mediastinum*	Examination of the mediastinum using an endoscope.
spirometer	spy-**ROM**-eh-ter	S/ R/CF	**-meter** *measure* **spir/o-** *to breathe*	An instrument used to measure respiratory volumes.
spirometry	spy-**ROM**-eh-tree	S/	**-metry** *process of measuring*	Use of a spirometer.
thoracotomy	thor-ah-**KOT-**oh-me	S/ R/CF	**-tomy** *surgical incision* **thorac/o** *chest*	Incision through the chest wall.
tomography	toe-**MOG**-rah-fee	S/ R/CF	**-graphy** *process of recording* **tom/o-** *cut, slice, layer*	Radiographic image of a selected slice of tissue.
transthoracic	tranz-thor-**ASS-**ik	S/ P/ R/	**-ic** *pertaining to* **trans-** *across* **-thorac-** *chest*	Going through the chest wall.
ultrasonography	**UL**-trah-soh-**NOG**-rah-fee	S/ P/ R/CF	**-graphy** *process of recording* **ultra-** *beyond* **-son/o-** sound	Delineation of deep structures using sound waves.

Tracheal aspiration uses a soft catheter that allows brushings and washings to be performed to remove cells and secretions from the trachea and main bronchi. The catheter can be passed through a tracheostomy or **endotracheal** tube but can also be passed through the mouth and nose.

Thoracentesis is the insertion of a needle through an intercostal space to remove fluid from a pleural effusion for laboratory study or to relieve pressure. The procedure is also called a **pleural tap.**

Percutaneous transthoracic needle aspiration is the insertion of a needle with a cutting chamber through an intercostal space to hook a specimen of parietal pleura for laboratory examination.

Thoracotomy is used to obtain an open biopsy of tissue from the lung, hilum, pleura, or mediastinum. It is performed through an intercostal incision under general anesthesia.

Abbreviations

AP anteroposterior
CT computerized tomography
CXR chest x-ray
FEV1 forced expiratory volume in 1 second
FVC forced vital capacity
MRA magnetic resonance angiography
PA posteroanterior
PEFR peak expiratory flow rate
PET positron emission tomography
PFTs pulmonary function tests

EXERCISES

Abbreviations *are helpful only if you know how to interpret them. This exercise asks questions about the abbreviations in the Abbreviations box on this page. Your knowledge of the language of pulmonology will assist you. Fill in the blanks.*

1. Insert the abbreviations that are directional terms:

____

2. List the abbreviations that stand for diagnostic radiological procedures:

____

3. Name the abbreviations that are measurements of PFTs:

____

4. Write a sentence (that is not directly out of the text) using any of these abbreviations:

____

____

____

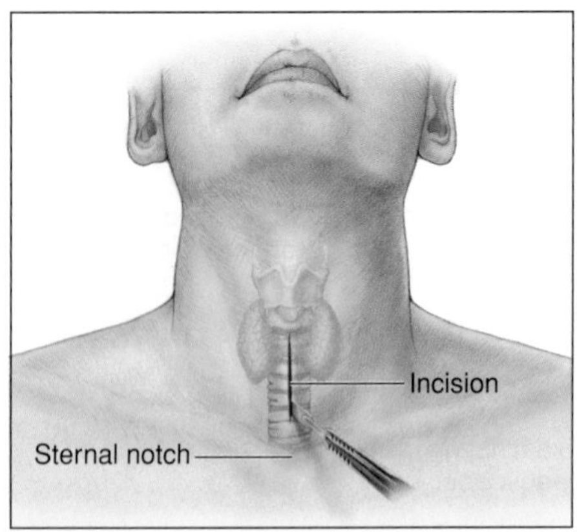

① Tracheotomy incision is made superior to sternal notch.

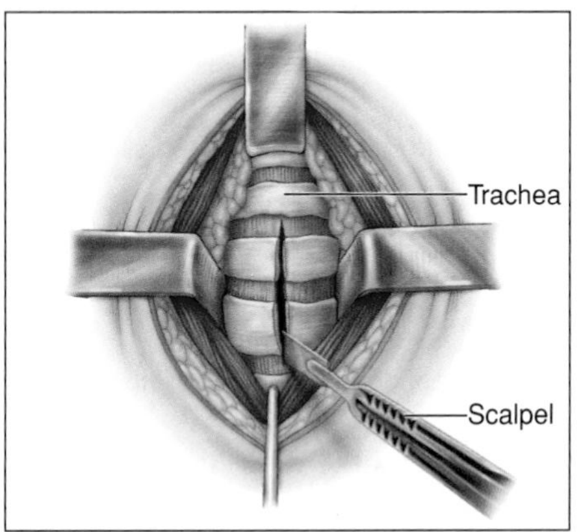

② Retractors separate the tissue, and an incision is made through the third and fourth tracheal rings.

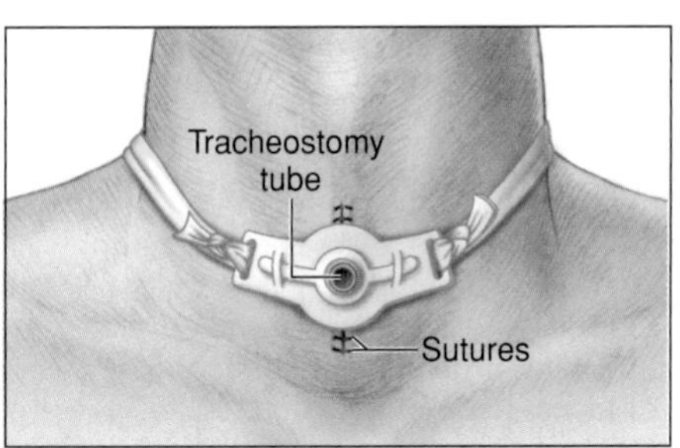

③ A tracheostomy tube is inserted, and the remaining incision is sutured closed.

▲ **FIGURE 7.18 Tracheostomy Procedure.**

THERAPEUTIC PROCEDURES

Pulmonary rehabilitation includes education, breathing exercises and retraining, exercises for the upper and lower extremities, and psychosocial support.

Nutritional support is critical for patients who have difficulty breathing or who lose a lot of weight.

Immunizations are available against influenza and the pneumococcus bacterium that is the most common cause of bacterial pneumonia.

Postural drainage therapy (PDT), by positioning and tilting the patient, uses gravity to promote drainage of secretions from lung segments. Chest **percussion** (tapping) on the chest wall can help loosen, mobilize, and drain the retained secretions. The two procedures are part of **chest physiotherapy.**

Constant positive airway pressure (CPAP) is an attempt to keep alveoli open by maintaining a positive pressure in the airways. A mask is fitted over the nose and mouth and attached to a **ventilator.** This can be used at night when sleeping or used in acute situations in COPD.

Positive end expiratory pressure (PEEP) is a technique in ventilation to keep the alveoli from collapsing in ARDS and neonatal respiratory distress syndrome.

In **intubation,** an **oropharyngeal airway** is used in the unconscious patient during bag and mask ventilation to maintain an open airway. A tube is inserted to prevent the tongue from falling back and obstructing the airway and to facilitate suctioning the airway. An **endotracheal intubation** involves the placement of a tube into the trachea. This allows patients to be placed on a ventilator and their breathing controlled.

Pulmonary resection is the surgical removal of lung tissue.

- **Wedge resection** is the removal of a small, localized area of diseased lung.
- **Segmental resection** is the removal of lung tissue attached to a bronchiole.
- **Lobectomy** is the removal of a lobe.
- **Pneumonectomy** is the removal of an entire lung.

Tracheotomy is an incision made into the trachea (windpipe) so that a temporary or permanent opening into the windpipe, called a **tracheostomy,** is created (*Figure 7.18*). A tube is placed into the opening to provide an airway. A tracheostomy is used to maintain an airway when there is obstruction or paralysis in the respiratory structures above it.

Mechanical ventilation is a process by which gases are moved into and out of the lungs via a device that is set to meet the respiratory requirements of the patient. It requires that a tracheostomy or endotracheal tube be attached to the mechanical device **(ventilator).** It can augment or replace the patient's own **ventilatory** efforts.

Pulmonary Pharmacology

- **Bronchodilators** relax the smooth muscles of the bronchioles. Examples are theophylline, beta2-agonists such as salbutamol and terbutaline, and anticholinergics such as ipratropium bromide.
- **Anti-inflammatory** drugs, such as corticosteroids, are best given by inhalation, but can be used orally or intravenously in acute episodes of asthma or COPD.
- **Mucolytics** are agents that attempt to break up mucus to allow it to be cleared more effectively from the airways. Examples are guaifenesin (common in over-the-counter cough medications), potassium iodide, and *N*-acetylcysteine taken through a **nebulizer.**
- **Antibiotics** are used when a bacterial infection is present. Penicillin, erythromycin, cefotaxime, and flucloxacillin are frequently used
- **Oxygen** is used in hypoxia and can be given by nasal **cannula** or by mask and intubation. Patients with severe, chronic COPD can be attached to a portable cylinder of oxygen.

WORD ANALYSIS AND DEFINITION

S/ = Suffix P/ = Prefix R/ = Root R/CF = Combining Form

WORD	PRONUNCIATION	ELEMENTS		DEFINITION
bronchodilator	**BRONG**-koh-die-**LAY**-tor	S/ R/CF R/	**-or** *one who does, that which does something* **bronch/o-** *bronchus* **-dilat-** *expand, open up*	Agent that increases the diameter of a bronchus.
cannula	**KAN**-you-lah		Latin *reed*	Tube inserted into a blood vessel or cavity as a channel for fluid or gas.
immunization	im-you-nih-**ZAY**-shun	S/ R/	**-ation** *process* **immuniz-** *make immune*	Administration of an agent to provide immunity.
mucolytic	**MYU**-koh-**LIT**-ik	S/ R/CF R/	**-ic** *pertaining to* **muc/o-** *mucus* **-lyt-** *dissolve*	Agent capable of dissolving or liquefying mucus.
nebulizer	**NEB**-you-liz-er	S/ R/	**-izer** *line of action, affects in a particular way* **nebul-** *cloud*	Device used to deliver liquid medicine in a fine mist.
pneumonectomy	**NEW**-moh-**NECK**-toe-me	S/ R/	**-ectomy** *surgical excision* **pneumon-** *lung, air*	Surgical removal of a lung.
resection **resect (verb)**	ree-**SEK**-shun ree-**SEKT**	S/ P/ R/	**-ion** *action, condition* **re-** *back* **-sect-** *cut off*	Removal of a specific part of an organ or structure.
tracheostomy **tracheotomy**	tray-kee-**OST**-oh-me tray-kee-**OT**-oh-me	S/ R/CF S/	**-stomy** *new opening* **trache/o-** *trachea* **-tomy** *surgical incision*	Insertion of a tube into the windpipe to assist breathing. Incision made into the trachea to create a tracheostomy.
ventilation **ventilator**	ven-tih-**LAY**-shun **VEN**-tih-lay-tor	S/ R/ S/	**-ation** *a process* **ventil-** *wind* **-ator** *person or thing that does something*	Movement of gases into and out of the lungs. Device that breathes for the patient.

Abbreviations

CPAP continuous positive airway pressure
PDT postural drainage therapy
PEEP positive end expiratory pressure

EXERCISES

Deconstruct: *Slash each of these medical terms into basic elements. Define the elements, and then use any two terms in one sentence each (that is not directly out of the text) to demonstrate that you understand the term's meaning. Fill in the blanks.*

Medical Term	Meaning of Prefix	Meaning of Root/CF	Meaning of Suffix
immunization			
nebulizer			
mucolytic			
bronchodilator			

1. ______________________________

____

2. ______________________________

____

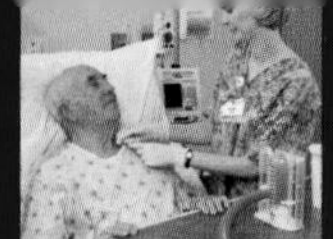

CHAPTER 7 REVIEW

RESPIRATORY SYSTEM

CHALLENGE YOUR KNOWLEDGE

A. **Word elements** will always remain the best tool for deconstructing medical terminology. Test your knowledge of respiratory word elements by breaking down the following medical terms. Fill in the chart.

Medical Term	Prefix	Root/Combining Form	Suffix	Meaning of Term
hemothorax				
aspiration				
pneumonia				
bronchitis				
empyema				
pneumoconiosis				
tachypnea				
bronchogenic				

Take five of the above terms and apply them appropriately in the following sentences:

1. Rapid breathing is ____________________.
2. Another term for *thoracentesis* is ____________________.
3. *Pleural effusion* containing pus is also termed ____________________.
4. *Anthracosis* is a form of ____________________.
5. Bloody *pleural effusion* is also termed ____________________.

B. **Terminology challenge:** For this exercise, remember terminology from a body system studied previously. Fill in the blanks.

1. What is the difference between the following terms?

   hemoptysis

   ____

   hematemesis

   ____

2. What do both these terms have in common? ____________________

   ____

3. Which two different body systems do they represent? ____________________

   ____

LESSON 8.1 Bones of the Skeletal System

BONES AND THE AXIAL SKELETON

OBJECTIVES

If you didn't have a skeleton, you'd be like a rag doll, shapeless and unable to move. Your skeleton provides support, protects many organ systems, and is the landmark for much of medical terminology. For example, the radial artery you use for taking a pulse is so named because it travels beside the radial bone of the forearm. In addition, the surface anatomy of bones and their markings enable you to describe and document the sites of symptoms, signs, and diagnostic and therapeutic procedures. The information in this lesson will enable you to use appropriate medical terminology to:

- **Recognize the different health professionals involved in the diagnosis and treatment of skeletal problems.**
- **Identify the tissues that form the skeletal system.**
- **Apply correct medical terminology to the structures and functions of the skeletal system.**
- **Classify the types of bones in the skeletal system.**
- **Describe the major problems and diseases that occur in the skeletal system.**

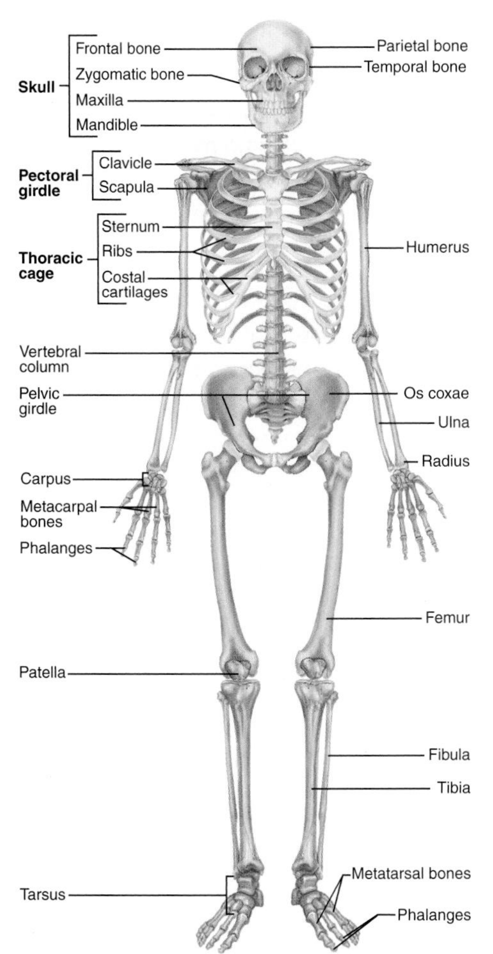

▲ **FIGURE 8.1**
Adult Skeletal System, Anterior View.

Health professionals involved in the diagnosis and treatment of problems in the musculoskeletal system include the following:

- **Orthopedic surgeons (orthopedists)** are **MDs** in the medical specialty that deals with the prevention and correction of injuries of the skeletal system and associated **muscles,** joints, and ligaments.
- **Osteopathic physicians** have the degree Doctor of Osteopathy (DO). They receive additional training in the **musculoskeletal system** and how it affects the whole body.
- **Chiropractors** focus on manual adjustment of joints, particularly the spine, to maintain and restore health.
- **Physical therapists** evaluate and treat pain, disease, or injury by physical therapeutic measures, as opposed to medical or surgical measures.
- **Physical therapist assistants** work under the direction of a physical therapist to assist in the application of physical therapy.
- **Orthopedic technologists** and **technicians** assist orthopedic surgeons in their treatment of patients.

FUNCTIONS OF THE SKELETAL SYSTEM

The four components of the skeletal system (*Figure 8.1*) are:

1. **Bones**
2. **Cartilage**
3. **Tendons**
4. **Ligaments**

They provide the following functions:

- **Support:** The bones of your vertebral column, pelvis, and legs hold up your body. The jawbone supports your teeth.
- **Protection:** The skull protects your brain. The vertebral column protects your spinal cord. The rib cage protects your heart and lungs.
- **Movement: Muscles** could not function without their attachments to skeletal bones, and muscles are responsible for all your movements.
- **Blood formation:** Bone marrow in many bones is the major producer of blood cells, including most of those in your immune system (*see Chapter 14*).
- **Mineral storage and balance:** The skeletal system stores calcium and phosphorus.
- **Detoxification:** Bones remove metals such as lead and radium from your blood, store them, and slowly release them for excretion.

CHAPTER 8

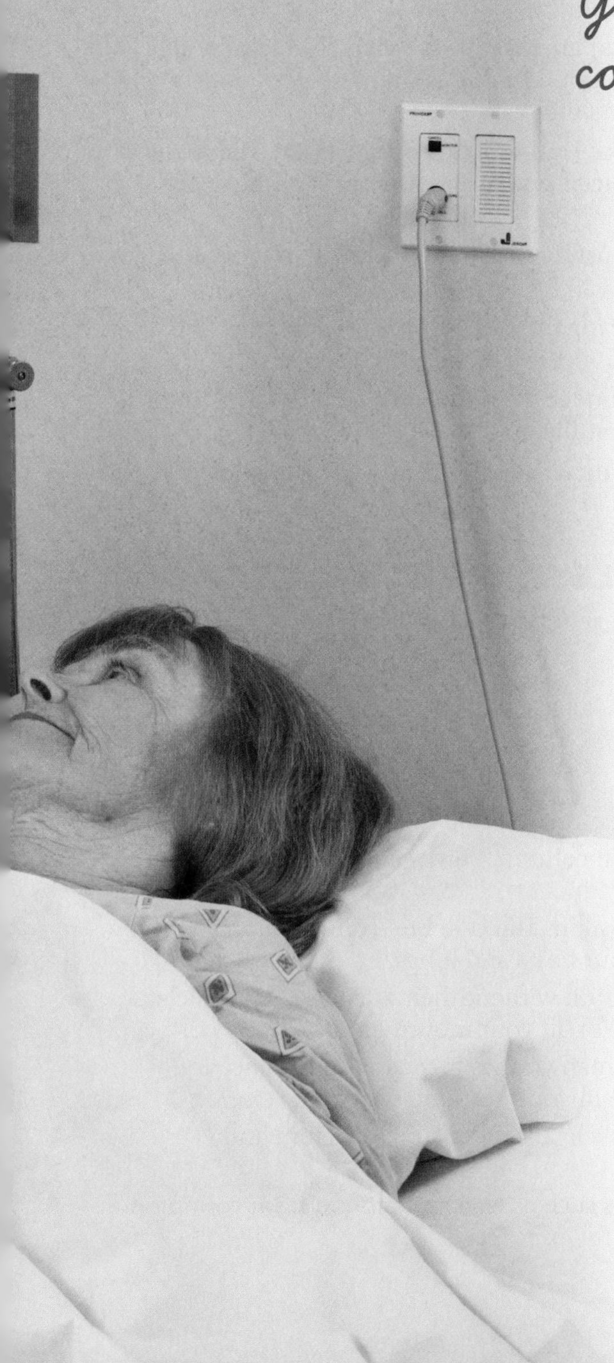

CASE REPORT 8.1

You are

. . . an orthopedic technologist working for Kevin Stannard, MD, an orthopedist in the Fulwood Medical Group.

You are communicating with

. . . Mrs. Amy Vargas, a 70-year-old housewife, who tripped going down the front steps from her house. She has severe pain in her right hip and is unable to stand. An X-ray shows a hip fracture and marked osteoporosis. Dr. Stannard examined her in the Emergency Department, and Mrs. Vargas is being admitted for a hip replacement.

Learning Outcomes

In order for you to work with Dr. Stannard to give optimal care to Mrs. Vargas, and to help her and her family understand the significance of her bone disorder and injury, you will need to be able to:

8.1 Apply the language of orthopedics to the structure and function of bones.

8.2 Comprehend, analyze, spell, and write the medical terms of orthopedics so that you can communicate accurately and precisely.

8.3 Recognize and pronounce the medical terms of orthopedics so that you can communicate verbally with accuracy and precision.

8.4 Describe the causes, appearances, methods of diagnosis, and treatments of common disorders of the bones.

8.5 Discuss the axial skeleton and its disorders.

Bones and the Axial Skeleton

The Essentials of the Language of Orthopedics

7. Define what makes a condition "chronic":

____

____

8. What is the opposite of a chronic condition?

____

____

R. Chapter challenge: Circle the correct answer.

1. Identify the pair of terms that can be diagnoses:
   a. epiglottis — pleura
   b. intubation — fissure
   c. alveoli — bronchioles
   d. pleurisy — epiglottitis
   e. diaphragm — eupnea
2. What can be said about the terms *endotracheal, transthoracic,* and *pulmonary:*
   a. All their suffixes mean the same thing.
   b. None of them has a prefix.
   c. All of them are adjectives.
   d. Answers a, b, and c are all correct.
   e. Answers a and c are correct.
3. *Bradypnea, dyspnea, eupnea, tachypnea,* and *hyperpnea* all describe:
   a. auscultation
   b. bronchoconstriction
   c. cor pulmonale
   d. respiration
   e. viscosity
4. Only one pair of terms has the correct spelling *and* plural form. Which is it?
   a. rhoncus — rhonci
   b. ronkus — ronki
   c. rhonchus — rhonchi
   d. ronchus — ronchi
   e. runchus — runchi
5. **Brain teaser:** When an area of the lung collapses as a result of bronchial obstruction, this is called:
   a. pneumoconiosis
   b. asbestosis
   c. atelectasis
   d. hemothorax
   e. mesothelioma

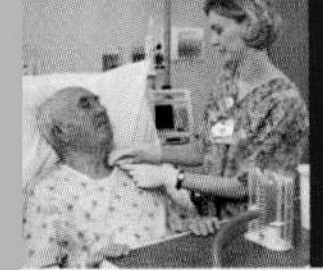

CHAPTER 7 REVIEW

RESPIRATORY SYSTEM

Q. Case Report challenge: Now that you are more comfortable with the terms in this chapter, you can apply that knowledge and briefly answer the questions about the Case Report. If you read the report through and underline all the medical terminology, this will make it easier to answer the questions.

You are communicating with

. . . Mr. Jude Jacobs, a 68-year-old white retired mail carrier, who has chronic obstructive pulmonary disease (COPD) and is on continual oxygen by nasal prongs. He has smoked two packs per day during all his adult life.

Last night, he was unable to sleep because of increased shortness of breath and cough. His cough is productive of yellow *sputum*. He had to sit upright in bed to be able to breathe.

Vital signs are temperature *(T)* 101.6, pulse *(P)* 98, respirations *(R)* 36, blood pressure *(BP)* 150/90. On examination, he is cyanotic and frightened and is on oxygen by nasal prongs. Air entry is diminished in both lungs, and there are rales at both bases.

You have been ordered to draw blood for arterial blood gases and to measure the amount of air entering and leaving his lungs by using *spirometry*.

1. What disease appears in Mr. Jacobs's past history?

   ____

2. Which particular symptom indicates Mr. Jacobs has an infection?

   ____

3. *Cyanotic* indicates an outward sign the physician can detect. What is it?

   ____

4. *Rales* heard through a stethoscope indicate the presence of what in the lungs?

   ____

5. Mr. Jacobs "is on oxygen by nasal prongs." What is the medical term for that device?

   ____

6. Where can abbreviations be substituted in this documentation? Write them here:

   ____

   ____

   ____

   ____

   ____

8. Which sentence is correct?

   a. Lung abcess can be a complication of bacterial pnemonia.

   b. Pneumonconiosis is caused by the inhalation of certain pollen.

   c. The cause of sarcoidosis is unknown.

   d. The synonym for *thoracentesis* is *plural tap.*

   e. Tuberculosis can affect any organ or tissue.

9. When a lung segment becomes airless as a result of infection, the lung is said to be:

   a. constricted
   b. consolidated
   c. compromised
   d. congested
   e. choked up

10. What is the commonality of the abbreviations *CXR, PFT,* and *PET?*

    a. They are all surgical procedures.
    b. They are all diagnostic tests.
    c. They are all diseases.
    d. They are all symptoms.
    e. They are all drugs.

11. An element in *hemothorax* can tell you that the pleural cavity contains:

    a. pus
    b. mucus
    c. phlegm
    d. blood
    e. water

12. Which set of terms is spelled correctly?

    a. oropharingeal bronchoscopy
    b. canula ventilator
    c. sinusitis decongestant
    d. spitum polyp
    e. tonsil tonsilectomy

13. *Polysomnography* is:

    a. a diagnostic test
    b. a sleep study
    c. a surgical procedure
    d. both a and b
    e. none of these

14. Find the medical term with an element that means *puncture:*

    a. thoracotomy
    b. pneumonectomy
    c. tomography
    d. thoracentesis
    e. rhinoplasty

15. Benign tumors of the larynx can be:

    a. polyps
    b. papillomas
    c. both a and b
    d. neither a nor b
    e. malignant

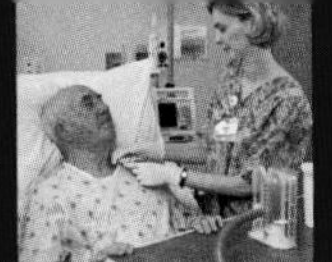

RESPIRATORY SYSTEM

P. **Chapter challenge:** Circle the correct answer.

1. *Epistaxis* is the medical term for:
   a. fainting
   b. vomiting
   c. nosebleed
   d. difficulty breathing
   e. productive cough

2. Another term for *heart failure* is:
   a. cyanosis
   b. conchae
   c. choana
   d. cor pulmonale
   e. chordae tendineae

Study Hint
Immediately cross off any answer you know is not correct. In your remaining choices, there is only *one best answer.*

3. What is the total number of *lobes* in *both* lungs?
   a. 4
   b. 5
   c. 6
   d. 2
   e. 3

4. What term is used for a reflex spasm of the diaphragm?
   a. coughing
   b. sneezing
   c. hiccups
   d. yawning
   e. blinking

5. The medical term for *croup* is:
   a. bronchiectasis
   b. bronchitis
   c. laryngotracheobronchitis
   d. laryngitis
   e. pharyngitis

6. Identify the pair of terms that both have elements meaning *lung:*
   a. pneumonectomy pulmonologist
   b. bronchitis laryngitis
   c. pulmonology pharynx
   d. trachea pneumonia
   e. pneumonic mediastinal

7. *Stridor, rales,* and *rhonchi* can all be considered:
   a. fissures
   b. signs
   c. diseases
   d. conditions
   e. infections

N. Word association can help you remember the definition and form a mental picture of what the term means. Use all the mental tricks you can to help you remember vocabulary.

Example:

The medical term *viscous* means ____________________.

Equate this to something seen in everyday life that has the same consistency or property.

Something from the supermarket: ____________________

Something from the gas station: ____________________

O. Train your eye and ear to see and hear the difference in medical terms. Remember that patient safety depends on you! Describe the difference in the following terms:

emphysema:

____

empyema:

____

rhoncus:

____

bronchus:

____

mucous:

____

mucus:

____

aspiration:

____

inspiration:

____

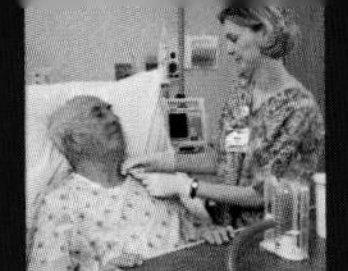

RESPIRATORY SYSTEM

M. **Translate** the following statements into layperson's language. Reconstruct each sentence into language a nonmedical person could understand. Fill in the blanks, and then check your spelling!

1. Sputum from a productive cough should be expectorated.

____

____

____

____

____

____

2. Tachypnea, hyperpnea, and bradypnea are all signs of respiratory difficulty.

____

____

____

____

____

____

3. Acute respiratory failure results in inadequate tissue oxygenation or carbon dioxide elimination.

____

____

____

____

____

____

L. **Procedures:** You are the coder in the general surgery department at Fulwood Medical Center. Using your medical terminology skills, fill in the correct codes for the following procedures.

Procedure	CPT Code
Laryngectomy	31360
Lobectomy	32480
Pneumonectomy	32440
Polypectomy	31237
Thoracentesis	32000
Thoracotomy	32100
Tonsillectomy	42825
Tracheostomy	31600

Explanation of Procedure	CPT Code
1. Patient was admitted for removal of lymphatic tissue at the back of his throat.	__________
2. A new airway opening was created for the patient.	__________
3. Benign growths from the mucosa were removed.	__________
4. Pleural fluid was drawn off for lab analysis.	__________
5. Entire lung was removed due to metastasis.	__________
6. Left lower lobe was removed from lung.	__________
7. Patient's chest was entered to explore for cause of bleeding.	__________
8. Tumor was removed; vocal cords were sacrificed.	__________

9. Use any of the above terms in a patient documentation sentence that is not a definition.

____

____

____

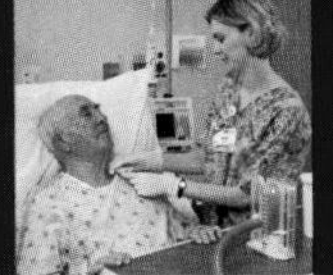

RESPIRATORY SYSTEM

K. **Suffixes:** Knowledge of elements aids you in choosing the appropriate medical term for the correct meaning you want to convey either verbally or in documentation. Suffixes will help you choose the correct term. Use the following medical terms to fill in the statements relating to the bronchus.

bronchogenic	bronchi
bronchopneumonia	bronchioles
bronchiolitis	bronchus
bronchoconstriction	bronchial
bronchitis	bronchiectasis

NOTE: In the term bronchopneumonia, the combining form is used, and it is one word. In the term bronchial asthma, the suffix al makes it an adjective, and the two words are separated.

1. One of 2 subdivisions of the trachea ____________________
2. Plural of the word above ____________________
3. Pertaining to the bronchus ____________________
4. Increasingly smaller subdivisions of bronchi ____________________
5. Inflammation of the bronchus ____________________
6. Chronic dilation of bronchi following inflammatory disease ____________________
7. Inflammation of the small bronchioles ____________________
8. Reduction in diameter of a bronchus ____________________
9. Arising from a bronchus ____________________
10. Acute inflammation of bronchioles with spread to lung parenchyma ____________________

Now, **build new terms,** also relating to the bronchus, using the word elements below. Your root or combining form will be bronch/o. You have more elements than you need. Fill in the blanks.

stenosis	plasty	gram
scope	pathy	pulmonary
dilation	dilator	scopy

1. An instrument used to view into the bronchus ____________________
2. Drug meant to open bronchial passages ____________________
3. Surgical procedure doing plastic repair on a bronchus ____________________
4. Examination of the bronchus ____________________
5. Pertaining to the bronchus and lung ____________________

I. Translate these physician's orders into medical terms, and test how well you understand what the physician has written. Fill in the blanks.

1. **Physician order:** This patient is to have AP and PA CXRs, followed by a CT, MRI, and PET scan STAT.

____

____

____

2. **Physician order:** Schedule the patient with COPD for ABGs and PFTs as soon as possible.

____

____

____

J. Recall and review: How well do you remember these word elements from the previous chapter? Try to answer without looking back to check. Fill in the blanks.

Element	Type of Element (P, R, CF, or S)	Meaning of Element
1. diaphor	________	________
2. media	________	________
3. emia	________	________
4. myo	________	________
5. peri	________	________
6. ic	________	________
7. hypo	________	________
8. intra	________	________

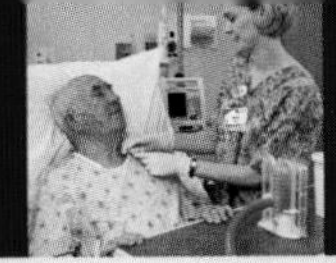

RESPIRATORY SYSTEM

H. Noun, adjective, verb: Meet a chapter objective by using the correct form of a term in precise communication. Insert the appropriate terms ON the line. UNDER the line, write the form (noun, adjective, verb) that you have used.

1. **resect** **resection**

   The patient is scheduled for a lung ______________________________ tomorrow.

   The surgeon is not sure he can ______________________________ the tumor.

2. **cyanosis** **cyanotic**

   This patient is showing signs of ______________________________ .

   Mr. Jacobs appeared ______________________________ to the respiratory therapist.

3. **respirator** **respiratory**

   Illnesses can spread quickly.

   The comatose patient will be taken off the ______________________________ .

4. Take the terms (a) *pharynx* and (b) *pharyngeal* and compose a sentence for each term. At the end of each sentence, write which form of the term you have used (noun, adjective, verb).

Sentence:

a. ______________________________

____

____

This form of the term is a(n) ____________________.

Sentence:

b. ______________________________

____

____

This form of the term is a(n) ____________________.

6. Therefore, an "abnormal increase of carbon dioxide in the arterial bloodstream" is

____.

Good job!

You are training yourself to think logically about medical terms.

G. **Construct** medical terms with the correct combination of elements. Write each term on the line provided after the definition. Some elements you will use more than once; some elements you will not use at all.

endo	ectomy	tomy	ation
bronch/o	trans	scopy	trache/o
lob	al	graphy	thorac/o
re	pneum/on	ventil	tom/o
son/o	osis	stomy	ion
ic	ator	ia	pharyng/e

1. surgical removal of a lung ____________/ ____________/ ____________
P R/CF S

2. examination of a bronchus ____________/ ____________/ ____________
P R/CF S

3. image of a selected slice of tissue ____________/ ____________/ ____________
P R/CF S

4. pertaining to being inside the trachea ____________/ ____________/ ____________
P R/CF S

5. examination of the mediastinum ____________/ ____________/ ____________
P R/CF S

6. incision through the chest wall ____________/ ____________/ ____________
P R/CF S

7. mechanical device that breathes for the patient ____________/ ____________/ ____________
P R/CF S

8. pertaining to across the chest ____________/ ____________/ ____________
P R/CF S

9. removal of part of a lung ____________/ ____________/ ____________
P R/CF S

10. new opening in the neck to the trachea ____________/ ____________/ ____________
P R/CF S

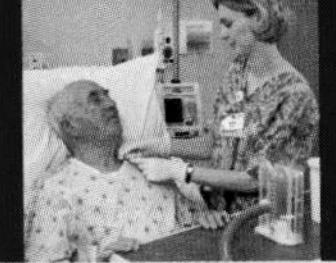

RESPIRATORY SYSTEM

E. **Suffix:** The suffix *itis* is one that you will meet over and over again in this text. Build the correct medical term to match its definition. Fill in the blanks.

1. inflammation of the bronchus ______________________ itis
2. inflammation of the organ of voice production ______________________ itis
3. inflammation of the tonsils ______________________ itis
4. inflammation of the throat ______________________ itis
5. inflammation of the nose ______________________ itis
6. inflammation of the epiglottis ______________________ itis
7. croup ______________________ itis
8. inflammation of the small bronchioles ______________________ itis
9. synonym for *pneumonia* ______________________ itis
10. inflammation of the sinuses ______________________ itis

F. **Word attack:** This is an exercise in the thought process required to arrive at the correct answer to a multiple-choice question. Look at the roots of the following terms to find the clues. Fill in the blanks.

Abnormal increase of carbon dioxide in the arterial bloodstream is:

a. hypertension

b. hyperglycemia

c. hypercapnia

d. hypersecretion

e. hyperpnea

1. Read the question and all the possible answers.
2. Read the question and answers again. Look for key words in the question. In this particular case, "abnormal" and "increase" should catch your eye. An increase of anything will require ______________________ as an element.
3. All of these terms begin with the same element, so you need to look at other elements for your clues.
4. Ask yourself: What is there an increase of? ______________________

   That is an important part of the question. Since a root is the foundation of every term, check your roots.

5. tens = ______________

   glyc = ______________

   capn = ______________

   secret = ______________

   pnea = ______________

C. **Roots:** Sometimes in medical terminology there can be two roots or combining forms with the same meaning. *Pneum/o* and *pulmon/o* are examples. These elements are not interchangeable—the medical term takes either one combining form or the other. Demonstrate your knowledge of the difference between the two elements by choosing the correct form for the terms listed below.

1. PFT is the abbreviation for what diagnostic test? ______________________________

   ____

2. Acute infection affecting the alveoli and lung parenchyma is______________________________.

3. Entry of air into the pleural cavity is called ______________________________.

4. A specialist in the study of the lung is called a ______________________________.

5. Surgical removal of a lung is called a ______________________________.

6. COPD is the abbreviation for what lung disease? ______________________________

   ____

7. Sarcoidosis is a form of the disease called ______________________________.

8. Study of the lungs and lung diseases is ______________________________.

D. **Explain the difference:** Do you understand the following terms well enough to explain the difference to patients if they should ask? Write a brief explanation of each term.

What is the difference between:

1. inspiration

   ____

   expiration

   ____

   aspiration

   ____

2. pneumothorax

   ____

   hemothorax

   ____

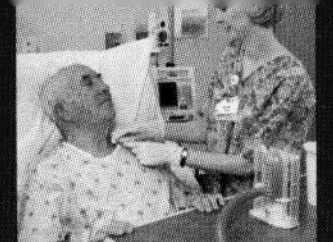

CHAPTER 7 REVIEW

RESPIRATORY SYSTEM

CHALLENGE YOUR KNOWLEDGE

A. **Word elements** will always remain the best tool for deconstructing medical terminology. Test your knowledge of respiratory word elements by breaking down the following medical terms. Fill in the chart.

Medical Term	Prefix	Root/Combining Form	Suffix	Meaning of Term
hemothorax				
aspiration				
pneumonia				
bronchitis				
empyema				
pneumoconiosis				
tachypnea				
bronchogenic				

Take five of the above terms and apply them appropriately in the following sentences:

1. Rapid breathing is ______________________.
2. Another term for *thoracentesis* is ______________________.
3. *Pleural effusion* containing pus is also termed ______________________.
4. *Anthracosis* is a form of ______________________.
5. Bloody *pleural effusion* is also termed ______________________.

B. **Terminology challenge:** For this exercise, remember terminology from a body system studied previously. Fill in the blanks.

1. What is the difference between the following terms?

   hemoptysis

   ____

   hematemesis

   ____

2. What do both these terms have in common? ______________________

   ____

3. Which two different body systems do they represent? ______________________

   ____

WORD ANALYSIS AND DEFINITION

S/ = Suffix P/ = Prefix R/ = Root R/CF = Combining Form

WORD	PRONUNCIATION		ELEMENTS	DEFINITION
bronchodilator	**BRONG**-koh-die-**LAY**-tor	S/ R/CF R/	**-or** *one who does, that which does something* **bronch/o-** *bronchus* **-dilat-** *expand, open up*	Agent that increases the diameter of a bronchus.
cannula	**KAN**-you-lah		Latin *reed*	Tube inserted into a blood vessel or cavity as a channel for fluid or gas.
immunization	im-you-nih-**ZAY**-shun	S/ R/	**-ation** *process* **immuniz-** *make immune*	Administration of an agent to provide immunity.
mucolytic	**MYU**-koh-**LIT**-ik	S/ R/CF R/	**-ic** *pertaining to* **muc/o-** *mucus* **-lyt-** *dissolve*	Agent capable of dissolving or liquefying mucus.
nebulizer	**NEB**-you-liz-er	S/ R/	**-izer** *line of action, affects in a particular way* **nebul-** *cloud*	Device used to deliver liquid medicine in a fine mist.
pneumonectomy	**NEW**-moh-**NECK**-toe-me	S/ R/	**-ectomy** *surgical excision* **pneumon-** *lung, air*	Surgical removal of a lung.
resection **resect (verb)**	ree-**SEK**-shun ree-**SEKT**	S/ P/ R/	**-ion** *action, condition* **re-** *back* **-sect-** *cut off*	Removal of a specific part of an organ or structure.
tracheostomy **tracheotomy**	tray-kee-**OST**-oh-me tray-kee-**OT**-oh-me	S/ R/CF S/	**-stomy** *new opening* **trache/o-** *trachea* **-tomy** *surgical incision*	Insertion of a tube into the windpipe to assist breathing. Incision made into the trachea to create a tracheostomy.
ventilation **ventilator**	ven-tih-**LAY**-shun **VEN**-tih-lay-tor	S/ R/ S/	**-ation** *a process* **ventil-** *wind* **-ator** *person or thing that does something*	Movement of gases into and out of the lungs. Device that breathes for the patient.

Abbreviations

CPAP continuous positive airway pressure
PDT postural drainage therapy
PEEP positive end expiratory pressure

EXERCISES

Deconstruct: *Slash each of these medical terms into basic elements. Define the elements, and then use any two terms in one sentence each (that is not directly out of the text) to demonstrate that you understand the term's meaning. Fill in the blanks.*

Medical Term	Meaning of Prefix	Meaning of Root/CF	Meaning of Suffix
immunization			
nebulizer			
mucolytic			
bronchodilator			

1. ______________________________

2. ______________________________

WORD ANALYSIS AND DEFINITION

S/ = Suffix P/ = Prefix R/ = Root R/CF = Combining Form

WORD	PRONUNCIATION	ELEMENTS		DEFINITION
cartilage	**KAR**-tih-lage		Latin *gristle*	Nonvascular, firm connective tissue found mostly in joints.
chiropractic	kye-roh-**PRAK**-tik	S/ R/CF R/	**-ic** *pertaining to* **chir/o-** *hand* **-pract-** *efficient, practical*	Diagnosis, treatment, and prevention of mechanical disorders of the musculoskeletal system.
chiropractor	kye-roh-**PRAK**-tor	S/	**-or** *a doer*	Practitioner of chiropractic.
detoxification (***Note:*** same as *detoxication*)	dee-**TOKS**-ih-fih-**KAY**-shun	S/ P/ R/	**-fication** *remove* **de-** *from, out of* **-toxi-** *poison*	Removing poison from a tissue or substance.
ligament	**LIG**-ah-ment		Latin *band, sheet*	Band of fibrous tissue connecting two structures.
muscle	**MUSS**-el		Latin *muscle*	A tissue consisting of cells that can contract.
musculoskeletal	**MUSS**-kyu-loh-**SKEL**-eh-tal	S/ R/ R/CF	**-al** *pertaining to* **muscul/o-** *muscle* **-skelet-** *skeleton*	Pertaining to the muscles and the bony skeleton.
orthopedic	or-tho-**PEE**-dik	S/ R/CF R/	**-ic** *pertaining to* **orth/o-** *straight* **-ped-** *child*	Pertaining to the correction and cure of deformities and diseases of the musculoskeletal system; originally, most of the deformities treated were in children.
orthopedist	or-tho-**PEE**-dist	S/	**-ist** *specialist*	Specialist in orthopedics.
osteopath	**OSS**-tee-oh-path	R/ R/CF	**-path** *disease* **oste/o-** *bone*	Practitioner of osteopathy.
osteopathy	**OSS**-tee-**OP**-ah-thee	S/	**-pathy** *disease*	Medical practice based on maintaining the balance of the body.
tendon	**TEN**-dun		Latin *sinew*	Fibrous band that connects muscle to bone.

Abbreviations

DO Doctor of Osteopathy
MD Doctor of Medicine

EXERCISES

Orthopedic vocabulary: *This exercise can be answered entirely by using medical terms that appear on the two pages open in front of you. Mastering these terms will start you on your way to learning the language of orthopedics. From the description, identify the correct medical terminology. Fill in the blanks.*

Description	Medical Term(s)
In addition to bones, which three terms are components of this chapter's body system?	1. 2. 3.
Which three terms refer to medical occupations?	4. 5. 6.
Which term represents a medical practice based on maintaining balance of the body?	7.
Which term has an element meaning *poison?*	8.
What is the name of the body system in this chapter?	9.

*To the student: Have you **spelled** and **pronounced** each term correctly?*

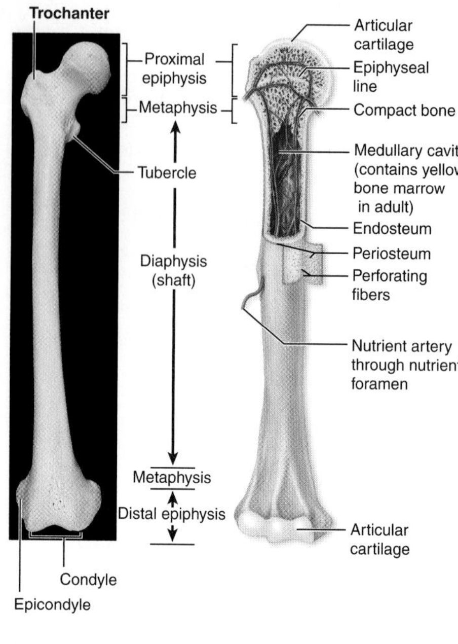

FIGURE 8.2
Femur: Long Bone of the Thigh.
(a) Anterior view. (b) Interior view.

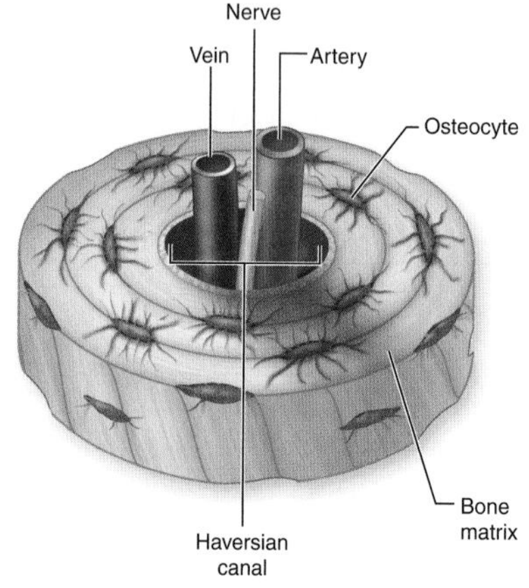

FIGURE 8.3
Blood Supply to Bone.

STRUCTURE OF BONES

Classification of Bones

The four classes of bones, as determined by their shape, are:

- **Long,** like the main bones of the limbs, palms, soles, fingers, and toes. They have a greater length than their width.
- **Short,** like the bones of the wrists and ankles, and the patella (kneecap). They have a length nearly equal to their width.
- **Flat,** like the bones of the skull and the ribs.
- **Irregular,** like the vertebrae. They do not fit any of the previous categories.

Structure of Bones

Long bones are the most common type of bone in the body (*Figure 8.2*). The shaft of a long bone is called the **diaphysis;** it contains a type of bone called compact bone. Each end of the bone is called the **epiphysis;** it contains spongy bone. Sandwiched between the diaphysis and epiphysis, thin layers of cartilage cells in the **epiphysial plate** or **line** enable the bones to grow in length.

A tough connective tissue sheath called **periosteum** covers the outer surface of all bones and is attached to the **cortical** bone by tough collagen fibers. The periosteum protects the bone and anchors blood vessels and nerves to the surface of the bone.

The hollow cylinder inside the diaphysis is called the **medulla** (*Figure 8.2b*). It contains bone **marrow,** a fatty tissue in adults. The epiphyseal ends contain red bone marrow with blood cells in different stages of development, as do the flat bones of the skull, the sternum, and the hip bones. The latter two places are sources for **bone marrow aspiration.**

Most bones are well supplied with blood (*Figure 8.3*). The blood vessels travel through the bone in a system of small **Haversian canals.** Because of this good blood supply, bone heals well.

WORD ANALYSIS AND DEFINITION

S/ = Suffix P/ = Prefix R/ = Root R/CF = Combining Form

WORD	PRONUNCIATION	ELEMENTS		DEFINITION
cortex **cortical (adj)**	**KOR**-teks **KOR**-tih-cal	 S/ R/	Latin *bark* **-al** *pertaining to* **cortic-** *cortex*	Outer portion of an organ, such as bone. Pertaining to a cortex.
diaphysis	die-**AF**-ih-sis		Greek *growing between*	The shaft of a long bone.
epiphysis	eh-**PIF**-ih-sis	P/ R/	**epi-** *upon, above* **-physis** *growth*	Expanded area at the proximal and distal ends of a long bone to provide increased surface area for attachment of ligaments and tendons.
epiphysial (adj) (***Note:*** The part "is" is deleted to enable the word to flow.)	eh-**PIF**-ih-see-al	S/	**-ial** *pertaining to*	Pertaining to an epiphysis.
Haversian canals	hah-**VER**-shan ka-**NALS**		Clopton Havers, English physician, 1655–1702	Vascular canals in bone.
marrow	**MAH**-roe		Old English *marrow*	Fatty, blood-forming tissue in the cavities of long bones.
medulla **medullary (adj)**	meh-**DULL**-ah meh-**DULL**-ah-ree	 S/ R/	Latin *marrow* **-ary** *pertaining to* **medulla-** *medulla*	Central portion of a structure surrounded by cortex. Pertaining to a medulla.
periosteum **periosteal (adj)**	**PER**-ee-**OSS**-tee-um **PER**-ee-**OSS**-tee-al	S/ P/ R/ S/	**-um** *structure* **peri-** *around* **-oste-** *bone* **-al** *pertaining to*	Strong membrane surrounding a bone. Pertaining to the periosteum.

EXERCISES

Elements *remain your best clue for understanding a medical term. In this exercise, the meaning of each element is given below the line—this is your clue to constructing the term. Write the correct element on the line above its meaning. After you have constructed the term, give its definition in the space provided.*

Study Hint
More than one element can have the same meaning.

1. ______________ / ______________
cortex pertaining to
The term is ______________ and means ______________.

2. ______________ / ______________ / ______________
around bone structure
The term is ______________ and means ______________.

3. ______________ / ______________ / ______________
upon, above growth pertaining to
The term is ______________ and means ______________.

4. ______________ / ______________ / ______________
middle pertaining to
The term is ______________ and means ______________.

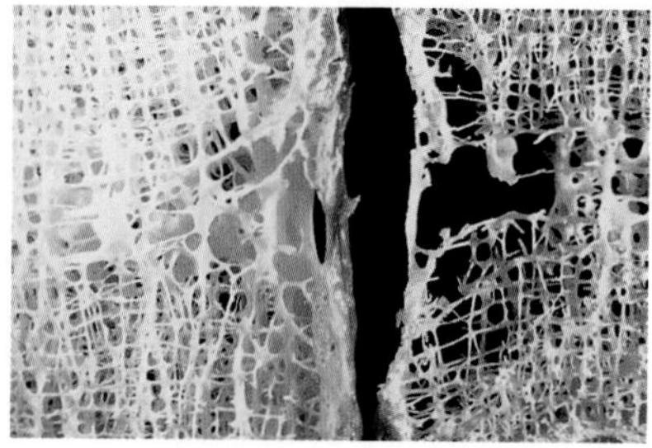

▲ **FIGURE 8.4**
Normal Bone and Osteoporotic Bone.

Keynote

Osteomalacia occurs in some developing nations and occasionally in this country when children drink soft drinks instead of milk fortified with vitamin D.

▲ **FIGURE 8.5**
Achondroplastic Dwarf with College Roommate.

Case Report 8.1 *(continued)*

On questioning, Amy Vargas demonstrated many of the risk factors for osteoporosis including family history, lack of exercise, cigarette smoking, inadequate diet, postmenopause, and increasing age.

DISEASES OF BONE

Osteoporosis results from a loss of bone density (*Figure 8.4*). It is more common in women than in men, and its incidence increases with age. In the United States, 10 million people already have osteoporosis, and 18 million more have low bone density **(osteopenia)** and are at risk for developing osteoporosis.

In women, production of the hormone estrogen decreases after menopause, and its protection against bone loss is lost. This leads to fragile, brittle bones. In men, reduction in testosterone has a similar but less marked effect.

Women at risk for osteoporosis should have bone mineral density **(BMD)** screening using a **DEXA** scan. Men and women over 50 are often advised to take 1,200 milligrams **(mg)** of calcium daily and 400 to 600 international units **(IU)** of vitamin D and to expose their bodies to the sun for 15 minutes daily.

There are several **FDA**-approved medications available for the treatment of osteoporosis.

Osteomyelitis is an inflammation of an area of bone due to bacterial infection, usually with a staphylococcus.

Osteomalacia, known as **rickets** in children, is a disease caused by vitamin D deficiency. When bones lack calcium, they become soft and flexible. They are not strong enough to bear weight, and they become bowed.

Achondroplasia occurs when the long bones stop growing in childhood, but the bones of the axial skeleton are not affected (*Figure 8.5*). This leads to short-stature individuals, about 4 feet tall. Intelligence and life span are normal. The disease is caused by a spontaneous gene mutation that then becomes a dominant gene for succeeding generations.

Osteogenic sarcoma is the most common malignant bone tumor. Peak incidence is between 10 and 15 years of age, and the tumor often occurs around the knee joint.

Osteogenesis imperfecta is a rare genetic disorder producing very brittle bones that are easily fractured, often **in utero** (while inside the uterus) (*Figure 8.6*).

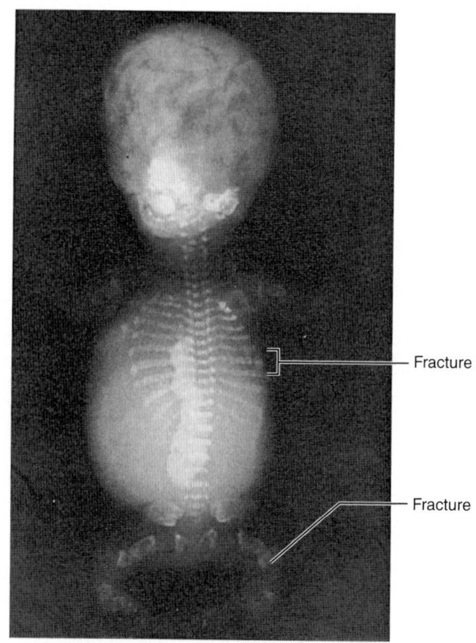

▲ **FIGURE 8.6** **Osteogenesis Imperfecta in Utero, Showing Fractures.**

WORD ANALYSIS AND DEFINITION

S/ = Suffix P/ = Prefix R/ = Root R/CF = Combining Form

WORD	PRONUNCIATION	ELEMENTS		DEFINITION
achondroplasia	a-kon-droh-**PLAY**-zee-ah	S/ P/ R/CF	**-plasia** *formation* **a-** *without* **-chondr/o-** *cartilage*	Condition with abnormal, early conversion of cartilage into bone, leading to dwarfism.
osteogenesis imperfecta	**OSS**-tee-oh-**JEN**-eh sis im-per-**FEK**-tah	S/ R/CF	**-genesis** *creation, formation* **oste/o-** *bone* **imperfecta,** Latin *unfinished*	Inherited condition when bone formation is incomplete, leading to fragile, easily broken bones.
osteomalacia	**OSS**-tee-oh-mah-**LAY**-she-ah	S/ R/CF	**-malacia** *abnormal softness* **oste/o-** *bone*	Soft, flexible bones lacking in calcium (rickets).
osteomyelitis	**OSS**-tee-oh-my-eh-**LIE**-tis	S/ R/CF R/	**-itis** *inflammation* **oste/o-** *bone* **-myel-** *bone marrow*	Inflammation of bone tissue.
osteopenia	**OSS**-tee-oh-**PEE**-nee-ah	S/ R/CF	**-penia** *deficient* **oste/o-** *bone*	Decreased calcification of bone.
osteoporosis	**OSS**-tee-oh-poh-**ROE**-sis	S/ R/CF R/CF	**-sis** *condition* **oste/o-** *bone* **-por/o-** *opening*	Condition in which the bones become more porous, brittle, and fragile and more likely to fracture.
rickets	**RICK**-ets		Old English *to twist*	Disease due to vitamin D deficiency, producing soft, flexible bones.
sarcoma	sar-**KOH**-mah	S/ R/	**-oma** *tumor, mass* **sarc-** *flesh*	Malignant tumor originating in connective tissue.
osteogenic sarcoma	**OSS**-tee-oh-**JEN**-ik sar-**KOH**-mah	S/ R/ R/CF	**-ic** *pertaining to* **-gen-** *creation* **oste/o-** *bone*	Malignant tumor originating in bone-producing cells.

Abbreviations

BMD bone mineral density
DEXA dual energy x-ray absorptiometry
FDA Food and Drug Administration
IU international unit(s)
mg milligram

EXERCISES

Suffixes: *The combining form* oste/o *means* bone, *and it is the main element in each of the following terms. You choose the correct suffix to complete the term. Fill in the blanks.*

genesis **genic** **penia** **malacia** **porosis** **myelitis**

1. Disease caused by vitamin D deficiency osteo/_______________
2. Low bone density osteo/_______________
3. Porous, brittle, fragile bones osteo/_______________
4. Most common malignant bone tumor osteo/_______________
5. Rare genetic disorder producing easily fractured bones, often in utero osteo/_______________
6. Inflammation of bone tissue osteo/_______________

Note: *The meaning of the combining form never changes. The addition of six different suffixes has helped you learn six new terms in orthopedic vocabulary!*

BONE FRACTURES (FXS)

TABLE 8.1 Classification and Definition of Bone Fractures

Name	Description	Reference
Closed (also called **simple** fracture)	A bone is broken, but the skin is not broken.	Figure 8.7g
Open (also called **compound** fracture)	A fragment of the fractured bone breaks the skin, or a wound extends to the site of the fracture.	Figure 8.7e
Displaced	The fractured bone parts are out of line.	Figure 8.7e
Complete	A bone is broken into at least two fragments.	Figure 8.7a
Incomplete	The fracture does not extend completely across the bone. It can be **hairline,** as in a **stress** fracture in the foot, when there is no separation of the two fragments.	Figure 8.7a
Comminuted	The bone breaks into several pieces, usually two major pieces and several smaller fragments.	Figure 8.7b
Transverse	The fracture is at right angles to the long axis of the bone.	Figure 8.7b
Impacted	The fracture consists of one bone fragment driven into another, resulting in shortening of a limb.	Figure 8.7c
Spiral	The fracture spirals around the long axis of the bone.	Figure 8.7d
Oblique	The fracture runs diagonally across the long axis of the bone.	Figure 8.7d
Linear	The fracture runs parallel to the long axis of the bone.	Figure 8.7f
Greenstick	This is a partial fracture. One side breaks, and the other bends.	Figure 8.7g
Pathologic	The fracture occurs in an area of bone weakened by disease, such as cancer.	—
Compression	The fracture occurs in a vertebra from trauma or pathology, leading to the vertebra being crushed.	—
Stress	This is a fatigue fracture caused by repetitive, local stress on a bone, as occurs in marching or running.	—

Incomplete
Complete
(a)
Comminuted
Transverse
(b)

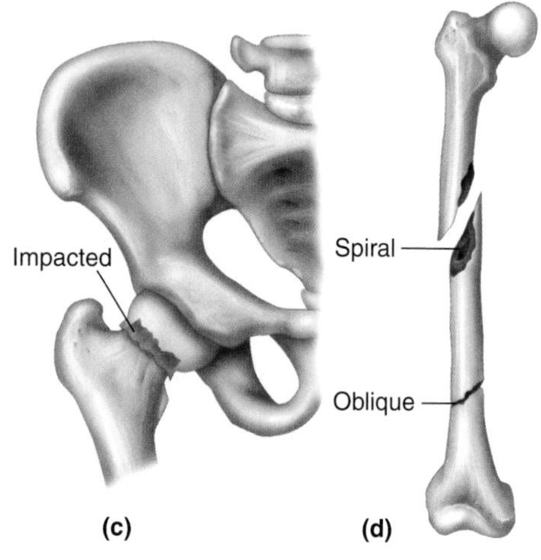

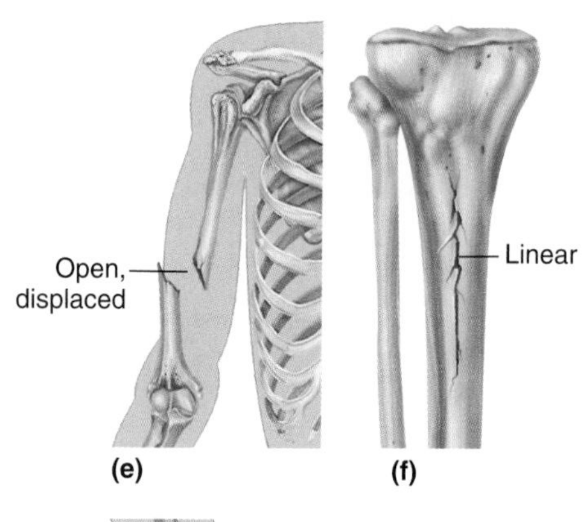

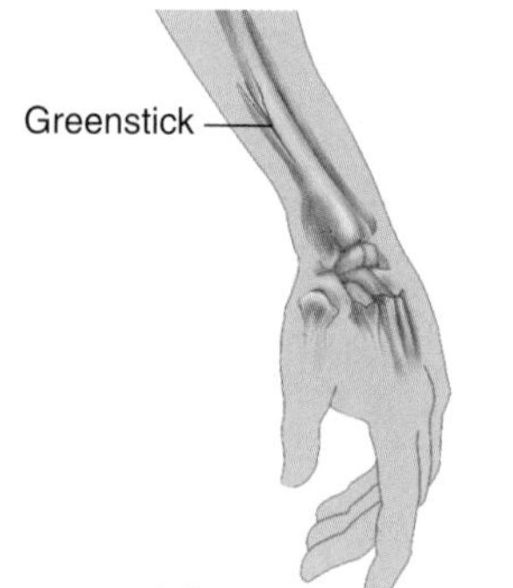

▲ **FIGURE 8.7** **Bone Fractures.**

Healing of Fractures

When a bone is fractured, blood vessels bleed into the fracture site, forming a hematoma. After a few days, bone-forming cells called **osteoblasts** move in and start to produce new bone matrix, which develops into **osteocytes** (bone cells). Eventually the new bone fuses together the segments of the fracture.

Surgical Procedures for Fractures

The initial goal of fracture treatment is to bring the ends of the bone at the break back opposite each other so that they fit together as they did in the original bone. This is called **alignment.**

External manipulation is used frequently. The bone is pulled from the distal end back into alignment. This process is called **reduction.** Anesthesia may be used.

In **external fixation,** the alignment is maintained by immobilizing the bone through the use of:

- **Plaster casts.**
- **Splints.**
- **Traction,** which is the gentle but continuous application of a pulling force that can align a fracture, reduce muscle spasm, and relieve pain.
- **External fixators,** by which the bone fragments are secured to a strong external steel rod by means of steel pins.

WORD ANALYSIS AND DEFINITION

S/ = Suffix P/ = Prefix R/ = Root R/CF = Combining Form

WORD	PRONUNCIATION	ELEMENTS		DEFINITION
alignment	a-**LINE**-ment	S/ P/ R/	**-ment** *resulting state* **a-** variant of **ad,** *into* **-lign-** *line*	Having a structure in its correct position relative to others.
comminuted	**KOM**-ih-nyu-ted	S/ R/	**-ed** *pertaining to* **comminut-** *break into pieces*	A fracture in which the bone is broken into pieces.
malunion	mal-**YOU**-nee-un	S/ P/ R/	**-ion** *condition, action* **mal-** *bad, difficult* **-un-** *one*	When the two bony ends of the fracture fail to heal together correctly.
nonunion	non-**YOU**-nee-un	P/	**non-** *not*	Total failure of healing of a fracture.
osteoblast	**OSS**-tee-oh-blast	S/ R/CF	**-blast** *immature cell* **oste/o-** *bone*	A bone-forming cell.
osteocyte	**OSS**-tee-oh-site	S/	**-cyte** *cell*	A bone-maintaining cell.
pathologic fracture	path-oh-**LOJ**-ik **FRAK**-chur	S/ R/CF R/ S/ R/	**-ic** *pertaining to* **path/o-** *disease* **-log-** *to study* **-ure** *result of* **fract-** *to break*	Fracture occurring at a site already weakened by a disease process, such as cancer.
reduction	ree-**DUCK**-shun	S/ P/ R/	**-ion** *action, condition* **re-** *backward* **-duct-** *lead*	Restore a structure to its normal position.
traction	**TRAK**-shun		Latin *to pull*	Pulling or dragging force.

Uncomplicated fractures take 8–12 weeks to heal.

Abbreviation

Fx	fracture

EXERCISES

Deconstruct *the following medical terms into their basic elements with slashes. Then provide a brief definition for each term. Fill in the chart, and fill in the blanks at the end of the exercise.*

Medical Term	Prefix	Root/CF	Suffix	Definition of Medical Term
reduction				
alignment				
malunion				
pathologic				
nonunion				

Demonstrate your understanding of the terms by finishing this exercise.

1. Use both *reduction* and *alignment* in one sentence.

____

2. Briefly explain the difference between a *malunion* and a *nonunion* of a fracture.

____

3. What type of preexisting condition might likely cause a pathologic fracture to occur? _______________

LESSON 8.2 Axial Skeleton

BONES AND THE AXIAL SKELETON

OBJECTIVES

In order to treat patients and educate them about their problems, you must understand the medical terminology for the structures and functions of the vertebral column and its joints, ligaments, and muscles The vertebral column is part of the axial skeleton. In this lesson, information about the axial skeleton will enable you to:

- **Name the regions and bones of the vertebral column.**
- **Discuss the joints, ligaments, and muscles of the vertebral column.**
- **Describe the bones of the skull.**
- **Explain the major problems and diseases that affect the vertebral column and skull.**

You are

. . . a physical therapist assistant working in the physical therapy department of Fulwood Medical Center.

You are communicating with

. . . Ms. Nancy Cardenas, a 27-year-old jeweler, who was waiting in her car at a traffic light 3 days ago when her car was rear-ended.

CASE REPORT 8.2

Ms. Cardenas now has severe neck pain radiating down her left arm, with dizziness and headaches. She is unable to go to work. Dr. Stannard has examined her and diagnosed her condition as a **whiplash** injury. An MRI shows herniation (rupture) of intervertebral discs between C5-C6 and C6-C7. Your role is to assist with a regimem of physiotherapy to relieve her symptoms.

STRUCTURE OF THE AXIAL SKELETON

The axial skeleton is composed of:

- **Vertebral column**
- **Skull**
- **Rib cage**

The axial skeleton is the upright axis of the body and protects the brain, **spinal** cord, heart, and lungs, which are most of the major centers of our physiology.

The **vertebral column** has 26 bones divided into five regions (*Figure 8.8*):

- **Cervical** region, with 7 vertebrae, labeled C1 to C7 and curved anteriorly.
- **Thoracic** region, with 12 vertebrae, labeled T1 to T12 and curved posteriorly.
- **Lumbar** region, with 5 vertebrae, labeled L1 to L5 and curved anteriorly.
- **Sacral** region, with 5 bones that in early childhood fuse into 1 bone curved posteriorly.
- **Coccyx** (tailbone), with 4 small bones fused together into 1 bone curved posteriorly.

The **spinal cord** lies protected in the vertebral canal of the vertebral column. Spinal nerves leave the spinal cord through the intervertebral foramina to travel to other parts of the body.

Intervertebral discs consist of fibrocartilage and inhabit the intervertebral space between the bodies of adjacent vertebrae. They provide additional support and cushioning (acting as a shock absorber) for the vertebral column.

Muscles that support and move the vertebral column consist of a superficial group that extends from the vertebrae to the ribs and maintains, bends, and straightens the **spine.** A deeper group of muscles connects the vertebrae to each other and extends and rotates the vertebral column.

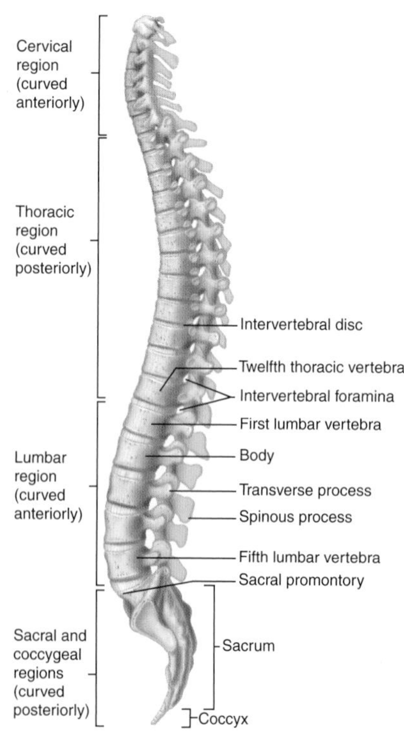

▲ **FIGURE 8.8**
Vertebral Column, Lateral View.

Case Report 8.2 (continued)

In Ms. Cardena's car accident, the whiplash caused a sprain of the spinal neck muscles together with protrusion of two cervical intervertebral discs. The centers of the discs bulge into the vertebral canal and pinch the nerves so that the pain radiates to her left arm.

WORD ANALYSIS AND DEFINITION

S/ = Suffix P/ = Prefix R/ = Root R/CF = Combining Form

WORD	PRONUNCIATION		ELEMENTS	DEFINITION
cervical	**SER**-vih-kal	S/ R/	**-al** *pertaining to* **cervic-** *neck*	Pertaining to the neck region.
coccyx	**KOK**-sicks		Greek *coccyx*	Small tailbone at the lowest end of the vertebral column.
kyphosis **kyphotic (adj)**	ki-**FOH**-sis ki-**FOT**-ik	S/ R/ S/ R/CF	**-osis** *condition* **kyph-** *bent, humpback* **-tic** *pertaining to* **kyph/o-** *bent, humpback*	A normal posterior curve of the spine that can be exaggerated in disease. Pertaining to or suffering from kyphosis.
lumbar	**LUM**-bar		Latin *loin*	The region of the back and sides between the ribs and pelvis.
sacrum **sacral (adj)**	**SAY**-crum **SAY**-kral	 S/ R/	Latin *sacred* **-al** *pertaining to* **sacr-** *sacrum*	Segment of the vertebral column that forms part of the pelvis. Pertaining to or in the neighborhood of the sacrum.
scoliosis **scoliotic (adj)**	skoh-lee-**OH**-sis **SKOH**-lee-**OT**-ik	S/ R/ S/ R/CF	**-osis** *condition* **scoli-** *crooked* **-tic** *pertaining to* **scoli/o-** *crooked*	An abnormal lateral curvature of the vertebral column. Pertaining to or suffering from scoliosis.
spine **spinal (adj)**	SPINE **SPY**-nal	 S/ R/	Latin *spine* **-al** *pertaining to* **spin-** *spine*	Vertebral column *or* a short projection from a bone. Pertaining to the spine.
vertebra **vertebrae (pl)** **vertebral (adj)**	**VER**-teh-brah **VER**-teh-bree **VER**-teh-bral	 S/ R/	Latin *spinal joint* **-al** *pertaining to* **vertebr-** *vertebra*	One of the bones of the spinal column. Pertaining to a vertebra.
whiplash	**HWIP**-lash	R/ R/	**whip-** *to swing* **-lash** *end of whip*	Symptoms caused by sudden, uncontrolled extension and flexion of the neck, often in an automobile accident.

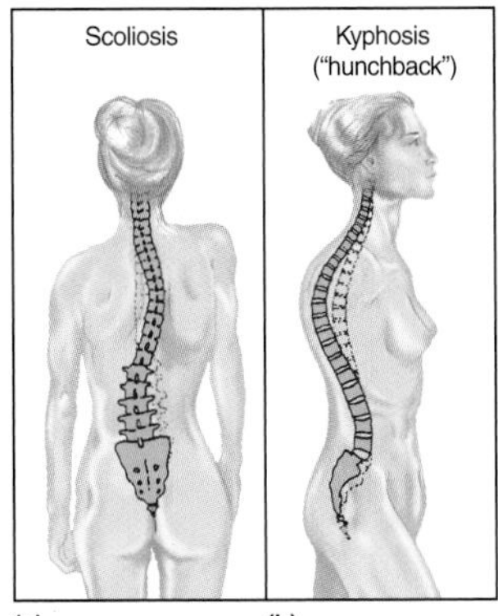

FIGURE 8.9
Abnormal Spine Curvatures.
(a) Scoliosis. (b) Kyphosis.

Scoliosis is an abnormal lateral curvature of the spine (*Figure 8.9a*). In older people, particularly those with osteoporosis, the normal anteriorly *concave* curvature in the thoracic region **(kyphosis)** is exaggerated (*Figure 8.9b*).

Abbreviations

C5	the fifth cervical vertebra
C5-C6	the intervertebral space between the fifth and sixth cervical vertebrae
C6	the sixth cervical vertebra
MRI	magnetic resonance imaging (diagnostic technique that produces focused slices of images of structures)

EXERCISES

Apply *the correct form of the similar terms appropriately in the following documentation, and you will meet a chapter objective! Fill in the blanks.*

vertebral **vertebra** **vertebrae**

1. The patient's C5 ____________________ was fractured in the accident.
2. A part of the axial skeleton is the ____________________ column.
3. The patient's C5 and C6 ____________________ were fractured in the accident.

Now supply the missing terms to complete the following sentence.

4. The designations C5-C6 and C6-C7 are for locations of ____________________.

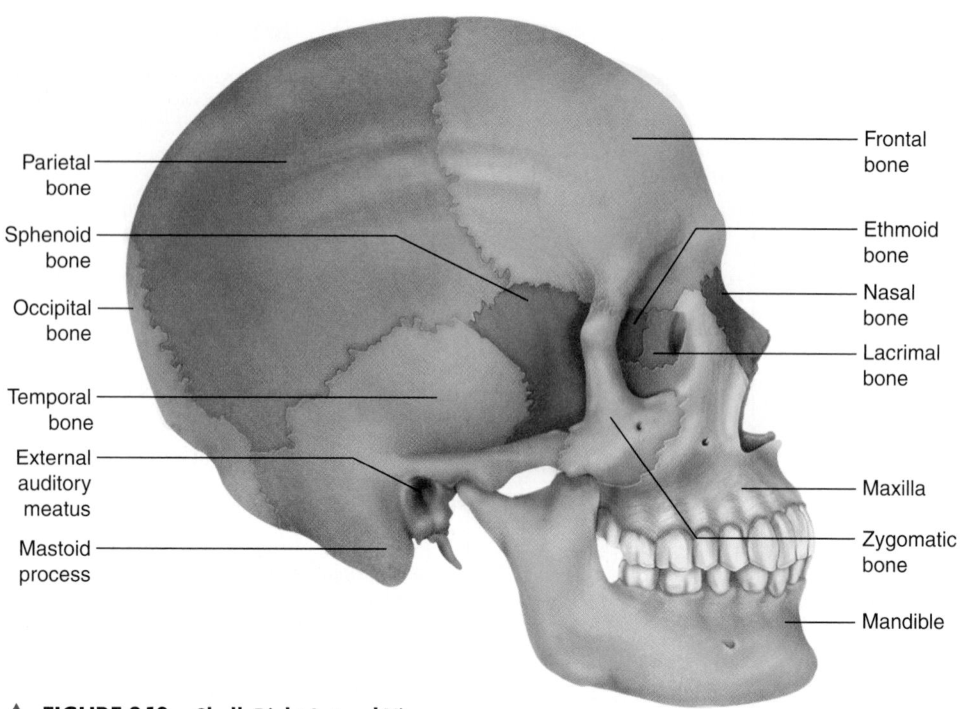

▲ **FIGURE 8.10 Skull, Right Lateral View.**

SKULL AND FACE

The Skull

The human skull (*Figure 8.10*) has 22 bones, 8 of which make up the **cranium,** the upper part of the skull that encloses the **cranial cavity** and protects the brain. The bones of the cranium are the following:

1. The **frontal** bone (1) forms the forehead, the roofs of the orbits, and part of the floor of the cranium and contains a pair of right and left frontal sinuses above the orbits.
2. **Parietal** bones (2) form the bulging sides and roof of the cranium.
3. The **occipital** bone (1) forms the back of and part of the base of the cranium.
4. **Temporal** bones (2) form the sides of and part of the base of the cranium.
5. The **sphenoid** bone (1) forms part of the base of the cranium and the orbits.
6. The **ethmoid** bone (1) forms part of the nose and the orbits and is hollow, forming the ethmoid sinuses.

The bones of the cranium are joined together by **sutures,** joints that appear as seams, covered on the inside and outside by a thin layer of connective tissue.

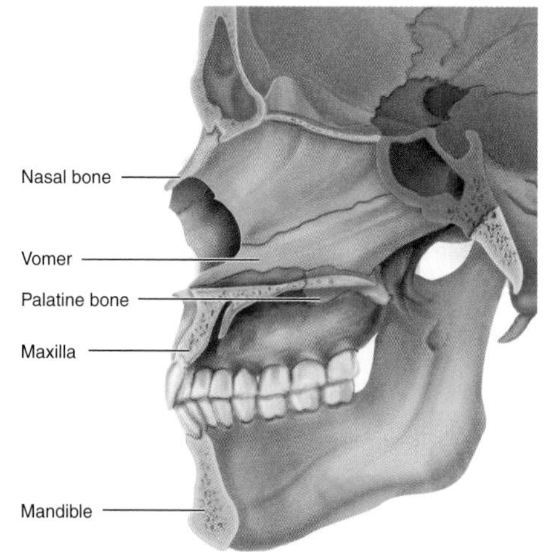

▲ **FIGURE 8.11 Facial Bones.**

The lower part of the skull comprises the 14 bones of the facial skeleton (*Figure 8.11*):

1. **Maxillary** bones (2) form the upper jaw **(maxilla),** hold the upper teeth, and are hollow, forming the maxillary sinuses.
2. **Palatine** bones (2) are located behind the maxilla and cannot be seen on a lateral view of the skull.
3. **Zygomatic** bones (2) are the prominences of the cheeks below the eyes.
4. **Lacrimal** bones (2) form the medial wall of each orbit.
5. **Nasal** bones (2) form the sides and bridge of the nose.
6. The **vomer** bone (1) separates the two nasal cavities.
7. Inferior nasal **conchae** (2) are fragile bones in the lower nasal cavity.
8. The **mandible** (1) is the lower jawbone, which holds the lower teeth. The mandible articulates (joins) with the temporal bone to form the **temporomandibular joint (TMJ).**

The third component of the axial skeleton, the rib cage, is discussed in *Chapter 7,* "Respiratory System."

WORD ANALYSIS AND DEFINITION

S/ = Suffix P/ = Prefix R/ = Root R/CF = Combining Form

WORD	PRONUNCIATION		ELEMENTS	DEFINITION
concha **conchae (pl)**	**KON**-kah **KON**-key		Latin *spiral seashell*	Shell-shaped bone on the medial wall of the nasal cavity.
cranium	**KRAY**-nee-um		Greek *skull*	The upper part of the skull that encloses and protects the brain.
cranial (adj)	**KRAY**-nee-al	S/ R/	**-al** *pertaining to* **crani-** *skull*	Pertaining to the skull.
ethmoid	**ETH**-moyd	S/ R/	**-oid** *resembling* **ethm-** *sieve*	Bone that forms the back of the nose and encloses numerous air cells.
lacrimal	**LAK**-rim-al	S/ R/	**-al** *pertaining to* **lacrim-** *tears*	Lacrimal bone forms part of the medial wall of the orbit, *or* pertaining to tears.
mandible **mandibular (adj)**	**MAN**-di-bel man-**DIB**-you-lar	 S/ R/	Latin *jaw* **-ar pertaining to** **mandibul-** *mandible*	Lower jaw bone. Pertaining to the mandible.
maxilla	mak-**SILL**-ah		Latin *jawbone*	Upper jawbone, containing right and left maxillary sinuses.
maxillary (adj)	mak-**SILL**-ah-ree	S/ R/	**-ary** *pertaining to* **maxilla-** *maxilla*	Pertaining to the maxilla.
occipital	ock-**SIP**-it-al	S/ R/	**-al** *pertaining to* **occipit-** *back of the head*	The back of the skull.
palatine	**PAL**-ah-tine	S/ R/	**-ine** *pertaining to* **palat-** *palate*	Bone that forms the hard palate and parts of the nose and orbits.
parietal	pah-**RYE**-eh-tal	S/ R/	**-al** *pertaining to* **pariet-** *wall*	The two bones forming the sidewalls and roof of the cranium.
sphenoid	**SFEE**-noyd	S/ R/	**-oid** *resemble* **sphen-** *wedge*	Wedge-shaped bone at the base of the skull.
temporal	**TEM**-pore-al	S/ R/	**-al** *pertaining to* **tempor-** *time*	Bone that forms part of the base and sides of the skull.
temporomandibular joint (TMJ)	**TEM**-pore-oh-man-**DIB**-you-lar JOYNT	S/ R/CF R/	**-ar** *pertaining to* **tempor/o-** *temple* **-mandibul-** *mandible*	The joint between the temporal bone and the mandible.
vomer	**VOH**-mer		Latin *ploughshare*	Lower nasal septum.
zygoma	zye-**GO**-mah		French *yoke*	Bone that forms the prominence of the cheek.
zygomatic (adj)	zye-go-**MAT**-ik	S/ R/	**-ic** *pertaining to* **zygomat-** *cheekbone*	Pertaining to the cheekbone.

Abbreviation

TMJ temporomandibular joint

EXERCISES

Roots: *The following medical terms from the WAD box are alike in that they have similar suffixes, but their roots make them different terms. Define each term after you have defined the suffix. Fill in the blanks.*

1. *Oid* means ______________________.

   ethmoid: ______________________

   sphenoid: ______________________

2. *Al* means ______________________.

   lacrimal: ______________________

   cranial: ______________________

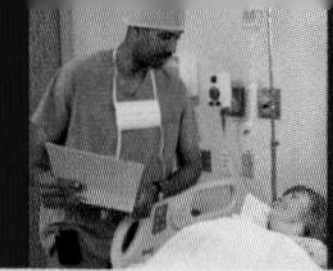

CHAPER 8 REVIEW

BONES AND THE AXIAL SKELETON

CHALLENGE YOUR KNOWLEDGE

A. **Identify** the type of element by placing a ✓ in the appropriate column. Define the element, and then give an example of a medical term using that element. Fill in the chart. The first one is done for you.

	Type of Element				
Element	**Prefix**	**Root/CF**	**Suffix**	**Meaning of Element**	**Medical Term Using This Element**
toxi		✓		poison	detoxification
ortho					
osteo					
peri					
malacia					
chiro					
chondro					
sarc					
pathy					
genesis					
penia					
plasia					

B. **Word attack:** First, read the question completely through, including all the possible answer choices. Be careful of words like *none, every,* and *all* because they restrict the answer. *Remember to cross off what you have determined to be an incorrect answer before you make your final choice.*

What do the terms *epiphysial, osteogenic, mandibular,* and *palatine* have in common?

1. *None* of them has a prefix.
2. Their suffixes *all* mean the same thing.
3. They are *all* diagnostic terms.
4. *All* three elements (P, R/CF, S) are present in *every* term.

Questions to ask yourself about the answer choices:

a. If any one of these terms has a prefix, this answer is not the correct choice because it states that *none* of them have a prefix.
   Do any of these terms have a prefix? (yes/no) ___________________
   If yes, which one(s)? ____

b. Determine whether there is a suffix in each term. Do *all* the suffixes mean the same thing? (yes/no) ___________________
   If yes, what do they *all* mean? ____

c. Are *all* these terms diagnoses? (yes/no) ___________________
   If no, which ones are not? ____

d. Does *every* term contain a prefix, root/CF, *and* suffix? (yes/no) ___________________
   Therefore, the answer is (1, 2, 3, or 4) ___________________ because

   ____

C. Create a test question for a fellow student. Write below a multiple-choice or fill-in-the-blank question about anything in this chapter. *Make the question a challenge, and try to stump a fellow student!* (If the question is multiple choice, it must have five possible answers.)

Question:

____

____

____

____

____

____

____

____

____

D. Prefixes: The following prefixes all appear in this chapter's terms. Use the correct prefix to build the terms; then define the terms. There are more prefixes than questions.

a re de inter trans non peri epi mal intra

1. ________________ /condyle
   (upon, above)
   Definition: ________________

2. ________________ /duct/ion
   (backward)
   Definition: ________________

3. ________________ /chondro/plasia
   (without)
   Definition: ________________

4. ________________ /union
   (bad)
   Definition: ________________

5. ________________ /oste/al
   (around)
   Definition: ________________

6. ________________ /toxi/fication
   (from, out of)
   Definition: ________________

7. ________________ /union
   (not)
   Definition: ________________

8. ________________ /vertebr/al
   (between)
   Definition: ________________

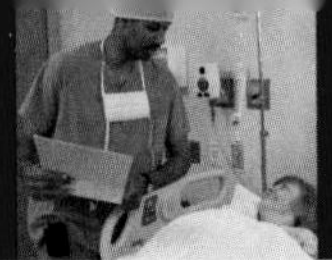

BONES AND THE AXIAL SKELETON

E. **Terminology challenge:** *suture.* Medical terms can have more than one meaning/usage. *Use the Glossary, your library, or an online medical dictionary if you need help answering these questions.*

1. Define *suture* as it is used in this chapter.

____

____

2. Now use this meaning of *suture* in a sentence that is not a definition or taken directly out of the text.

____

____

____

Suture can also be a noun and a verb with another meaning. Can you identify them?

3. *Suture* as a noun (person, place, or thing) can also mean (definition) ________________________.
4. Write a sentence with *suture* having this meaning.

____

____

5. *Suture* as a verb (action) can also mean (definition)

____.

6. Write a sentence with *suture* having this meaning.

____

____

____

See your medical vocabulary increase as you now know one term with three different meanings!

F. **Translate** the following sentences into layperson's language a patient can understand.

1. A patient with osteopenia is at risk for osteoporosis.

____

____

____

> *Study Hint*
> First, read the sentences and underline or highlight any medical terms (or abbreviations) you will need to explain. Then, rewrite the sentence in nonmedical language.

2. Intervertebral discs consist of fibrocartilage and inhabit the intervertebral space between the bodies of adjacent vertebrae.

____

____

____

3. Osteogenesis imperfecta is a rare genetic disorder producing very brittle bones that are easily fractured, often in utero.

____

____

____

G. **Definitions** for various medical terms are given below. Write the name of the correct medical term on the line following the definition. "C" if you can identify all the terms!

1. Nonvascular, firm connective tissue found mostly in joints:

   c______________________________

2. Outer portion of an organ, such as bone:

   c______________________________

3. Another name for a simple fracture:

   c______________________________

4. Pertaining to the neck:

   c______________________________

5. Fracture in which the bone is broken into several pieces:

   c______________________________

6. Upper part of the skull that encloses and protects the brain:

   c______________________________

7. Fracture occurs in a vertebra from trauma or pathology, leading to the vertebra being crushed:

   c______________________________

8. Shell-shaped bone on the medial wall of the nasal cavity:

   c______________________________

9. Another name for the tailbone:

   c______________________________

10. Another name for an open fracture:

   c______________________________

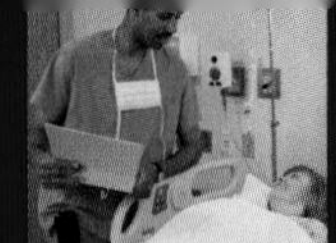

BONES AND THE AXIAL SKELETON

H. Roots and combining forms are the foundation of every medical term. Deconstruct the following medical terms to discover their basic foundations. Slash the term into its appropriate elements. *Notice that not every term needs a prefix.*

1. orthopedic ______/______/______ (P, R/CF, S)
2. osteopathy ______/______/______ (P, R/CF, S)
3. epicondyle ______/______/______ (P, R/CF, S)
4. cortical ______/______/______ (P, R/CF, S)
5. periosteum ______/______/______ (P, R/CF, S)
6. chiropractor ______/______/______ (P, R/CF, S)
7. osteomyelitis ______/______/______ (P, R/CF, S)
8. osteoporosis ______/______/______ (P, R/CF, S)
9. sarcoma ______/______/______ (P, R/CF, S)
10. detoxification ______/______/______ (P, R/CF, S)
11. epiphysial ______/______/______ (P, R/CF, S)
12. achondroplasia ______/______/______ (P, R/CF, S)
13. mandibular ______/______/______ (P, R/CF, S)

I. Patient education: Explain the difference among these abnormal spinal curvatures to patients. Be sure to use language a patient can understand.

1. scoliosis:

____

____

____

____

2. kyphosis:

____

____

____

____

3. Which one is the *most common* defect? (*See the Study Hint.*)

____

4. Which defect is seen in patients with osteoporosis? _______________

5. What is the *most common* type of bone in the body? _______________

6. Where is cartilage found *most often* in the body? _______________

7. What is the *most common* malignant bone tumor? _______________

Study Hint

"Most." Anything that is the "most powerful," "largest," "smallest," "most common," and so on, is probably going to be a test question! *Make sure you know them, and review them before a test.*

J. Suffixes: The following suffixes have all appeared in this chapter. The meaning of the suffix is given to you—identify the suffix, and give an example of a medical term that contains that suffix. Fill in the chart.

Suffix	Meaning of Suffix	Medical Term
	condition	
	resemble	
	a doer (one who does)	
	remove	
	process	
	specialist	
	disease	
	structure	
	formation	
	softness	
	deficient	
	inflammation	
	tumor, mass	

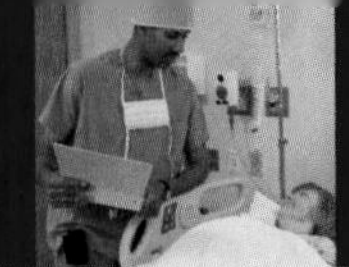

BONES AND THE AXIAL SKELETON

K. **Abbreviations** are frequently used in patient documentation. For patient safety, you must know exactly what they mean. Rewrite each sentence, translating the abbreviations into the correct medical terms. *Watch your spelling—the answer is not correct unless all the spelling is correct!*

1. Women at risk for osteoporosis should have BMD screening using a DEXA scan.

____

____

2. An MRI shows herniated discs at C5-C6 and C6-C7, with a fx at C2.

____

____

L. **Build** your orthopedic terminology by completing the medical terms defined. After you fill in the element on the line, write the type of element (prefix, root, combining form, suffix) you have used below the line. Fill in the blanks.

1. removing poison from tissue	de/ __________ / __________
2. soft bones lacking in calcium	osteo/ __________
3. projection above the condyle	__________ /condyle
4. membrane surrounding a bone	peri/ __________ / __________
5. bone broken in several pieces	__________ /ed
6. having a structure in its correct position	__________ / __________ /ment
7. bone-forming cell	osteo/ __________
8. pertaining to the neck	__________ /al
9. space between two vertebrae	__________ /vertebr/ __________

M. **Latin and Greek terms** cannot be further deconstructed into prefix, root, or suffix. You must know them for what they are. Test your knowledge of these terms with the following exercise. Match the meaning in column 1 with the correct medical term in column 2.

_____ 1. upper jawbone	A. cartilage
_____ 2. holds bones together or organs in place	B. vomer
_____ 3. outer portion of an organ	C. mandible
_____ 4. bone that forms prominence of the cheek	D. diaphysis
_____ 5. shaft of a long bone	E. coccyx
_____ 6. connective tissue found in joints	F. ligament
_____ 7. lower jawbone	G. maxilla
_____ 8. small tailbone at the end of the vertebral column	H. cortex
_____ 9. attaches muscle to bone	I. zygoma
_____ 10. lower nasal septum	J. tendon

N. Recall and review: This exercise on word elements contains elements from the previous chapter. Try to recall the previous elements without turning back in your book. Identify the type of element by placing a ✓ in the appropriate column; then write its meaning. Fill in the blanks.

	Type of Element			
Element	**Prefix**	**Root/CF**	**Suffix**	**Meaning of Element**
pulmono				
nas				
cyan				
pnea				
bio				
graphy				
rhin				
poly				
olfact				
pneumon				

O. Explanation please: If you really understand a term, you can explain it to someone else. Briefly explain the difference between the following terms.

1. External manipulation:

____

____

____

2. External fixation:

____

____

____

3. Reduction:

____

____

____

BONES AND THE AXIAL SKELETON

P. Spelling demons: The following terms from this chapter are particularly difficult to spell and pronounce. Correct pronunciation and spelling of medical terms is the mark of an educated professional. Circle the correct spelling, and then check (✓) that you have practiced the pronunciation. *Remember:* Pronunciations are on the Student Online Learning Center (www.mhhe.com/AllanEssMedLanguage).

					Pronunciation ✓
1.	cockyx	cocyx	coccyx	coccyz	______
2.	cartiledge	cartilage	carrtilage	cartilege	______
3.	scoliosis	skoliosis	scolliosis	skolioses	______
4.	osteomyilitis	osteomielitis	osteomyelitis	osteomyelites	______
5.	kiphosis	khyphosis	kyphosis	kyiphosis	______
6.	acondroplasea	achondroplasia	acondroplasia	achodroplasia	______
7.	ocipital	occipitel	ocippital	occipital	______
8.	sfenoid	spenoid	sphenoid	phenoid	______
9.	epipysial	epiphysial	epiphyseal	epifiseal	______
10.	chiropractic	chirropractic	chiropracctic	chiropractice	______

Q. Chapter challenge: Circle the correct answer.

Study Hint
Immediately cross off any answer you know is not correct. In your remaining choices, there is only *one best answer.*

1. The medical term for *low bone density* is:
   - a. osteocyte
   - b. osteomyelitis
   - c. periosteum
   - d. osteomalacia
   - e. osteopenia
2. The four classes of bones are determined by their:
   - a. length
   - b. shape
   - c. size
   - d. weight
   - e. number
3. Which term has a suffix meaning *disease?*
   - a. orthopedist
   - b. periosteum
   - c. chiropractor
   - d. osteopath
   - e. osteogenic
4. When a bone is fractured, blood vessels bleed into the fracture site and form a(n):
   - a. sarcoma
   - b. osteosarcoma
   - c. hematoma
   - d. osteoblast
   - e. condyle
5. Find the pair of terms that are *both* diagnoses:
   - a. achondroplasia medullary
   - b. cortex osteopathy
   - c. orthopedic osteomyelitis
   - d. periosteum osteoporosis
   - e. osteomalacia rickets

6. Which term has neither a suffix nor a prefix?
   a. fracture
   b. sacral
   c. periosteum
   d. osteopath
   e. lacrimal
7. What is the *medulla?*
   a. the shaft of a long bone
   b. the outer surface covering of bone
   c. the hollow cylinder inside the diaphysis
   d. the end of a bone
   e. the membrane surrounding a bone
8. What does the "O" in the abbreviation *DO* stand for?
   a. osteopath
   b. orthopedic
   c. orthopedist
   d. osteopathy
   e. occipital
9. Having a structure in its correct position relative to others is called:
   a. nonunion
   b. alignment
   c. malunion
   d. detoxification
   e. resorption
10. Which of these is *not* a bone in the cranium?
    a. frontal
    b. sphenoid
    c. parietal
    d. vomer
    e. occipital
11. Proofread the following statements for spelling and factual errors. Find the only correct statement.
    a. Ciropractors focus on manual adjustment of joints.
    b. Scoliosis is an abnormal lateral curvature of the vertebral column.
    c. Alignment is removing poison from tissues.
    d. An epicondyle is a medial projection at the ends of flat bones.
    e. A DEXA scan screens for bone marrow.
12. Which disease is a genetic disorder?
    a. osteogenic sarcoma
    b. osteoporosis
    c. osteogenesis imperfecta
    d. osteopenia
    e. osteomalacia
13. Which suffix signifies a deficiency of something?
    a. malacia
    b. penia
    c. osis
    d. oma
    e. itis

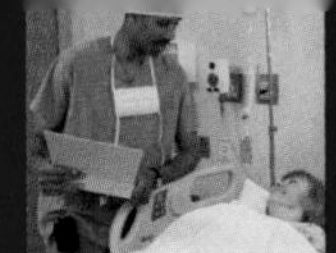

BONES AND THE AXIAL SKELETON

R. **Case Report challenge:** Now that you are more comfortable with the terms in this chapter, you can apply that knowledge and briefly answer the questions about the Case Report. If you read the report through and underline all the medical terminology, this will make it easier to answer the questions.

You are

. . .a physical therapist assistant working in the physical therapy department of Fulwood Medical Center.

You are communicating with

. . .Ms. Nancy Cardenas, a 27-year-old jeweler, who was waiting in her car at a traffic light 3 days ago when her car was rear-ended. Ms. Cardenas now has severe neck pain radiating down her left arm, with dizziness and headaches. She is unable to go to work. Dr. Stannard has examined her and diagnosed her condition as a whiplash injury. An MRI shows herniation of intervertebral discs between C5-C6 and C6-C7. Your role is to assist with a regimen of physiotherapy to relieve her symptoms.

1. Define *radiating pain:*

___

___

2. The term *whiplash* injury forms a mental picture. Describe that picture:

___

___

___

3. The abbreviation *MRI* stands for

___.

4. The prefix *inter* in the term *intervertebral* means that the discs are ___ certain vertebrae in the vertebral column.

5. List the *symptoms* Ms. Cardenas presented with:

___

___

6. What happens when a disc herniates?

___

___

___

7. What diagnosis has the doctor assigned to this patient?

___

S. **Chapter challenge:** Circle the correct answer.

1. Rickets is a disease caused by:
   a. poor bone marrow
   b. vitamin A deficiency
   c. excessive potassium
   d. infection
   e. vitamin D deficiency

2. Which term is *not* a diagnosis?
   a. osteomalacia
   b. osteoporosis
   c. osteopenia
   d. osteoblast
   e. osteomyelitis

3. The terms *comminuted, transverse,* and *impacted* all apply to:
   a. dental diagnoses
   b. radiologic procedures
   c. bone fractures
   d. surgical procedures
   e. intervertebral discs

4. Choose the incorrect pair of terms:

a.	curvature	scoliosis
b.	fixation	screws
c.	vertebra	disc
d.	coccyx	tailbone
e.	closed	compound

5. Using the elements as your guide, which term relates to bone marrow?
   a. osteomalacia
   b. osteoblast
   c. osteomyelitis
   d. osteopenia
   e. osteoporosis

6. Which term has a prefix meaning *without?*
   a. imperfecta
   b. achondroplasia
   c. alignment
   d. reduction
   e. intervertebral

7. **Brain teaser:** *Reduction* can be accomplished by:
   a. osteoblasts
   b. external fixation
   c. external manipulation
   d. pathologic fracture
   e. herniation

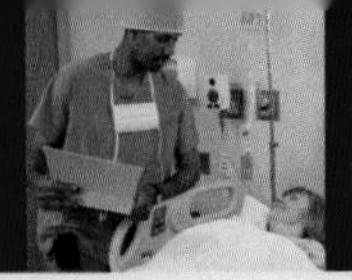

CUMULATIVE REVIEW

These exercises are built around terms and elements from this and the preceding three chapters. They are an excellent source of review for tests—be sure to do every exercise!

A. **Suffixes:** Medical terms are given in the first column—begin by slashing each term into its elements. Write the suffix in the second column and its meaning in the third column. Challenge yourself to think of another medical term from any previous chapter with this suffix—write that term in the fourth column, and write its meaning in the last column. *Your goal is to complete the entire chart!*

Medical Term	Suffix	Meaning of Suffix	Another Medical Term with This Suffix	Meaning of This Medical Term
pulmonary				
cardiomyopathy				
periosteum				
submandibular				
gastric				
cyanosis				
atheroma				
gingivectomy				
osteomyelitis				
intubation				
laparoscopy				
sacral				

B. **Abbreviations:** Write out in words the meaning of each abbreviation below. *Remember, an answer is not correct unless all the words are spelled correctly!* Fill in the blanks.

1. SOB ______________________
2. BMD ______________________
3. ABG ______________________
4. IU ______________________
5. STAT ______________________
6. COPD ______________________
7. DO ______________________
8. HSV-1 ______________________
9. AED ______________________
10. GI ______________________

C. **Diagnosis/procedure:** Each of the medical terms in the table is either a diagnosis or a procedure. Place a check mark (√) in the appropriate column for diagnosis or procedure. In the last column, write the name of the specialist (from the terms below) who would be using that medical term in his or her dictation.

gastroenterologist **periodontist** **cardiologist** **pulmonologist** **orthopedist**

Medical Term	Diagnosis	Procedure	Specialist
gingivectomy			
CHD			
laparoscopy			
PFTs			
pacemaker insertion			
lordosis			
thrush			
osteomyelitis			
COPD			
ischemia			
external fixation			
polysomnography			
periodontitis			
manipulation			
dyspnea			
MI			
pyorrhea			
URI			
osteogenic sarcoma			
cardiomyopathy			

Take any one term from the above chart and use it in a sentence of patient documentation.

Sentence: ______________________________

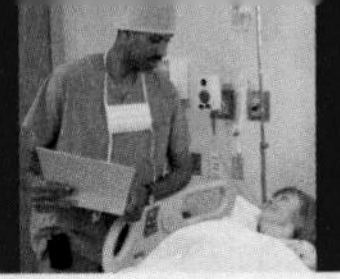

CUMULATIVE REVIEW

D. **Roots/combining forms** are always your best clue for understanding the meaning of a term. In the chart, the R/CF is given to you. Define its meaning in column 2, and in column 3 give an example of a term in which it appears. In the final column, write the meaning of the medical term.

Root/CF	Meaning of R/CF	Medical Term Containing This R/CF	Meaning of Medical Term
gastr			
my/o			
pulmon			
nas/o			
toxi			
isch			
oste			
bari			
pneum/o			
olfact			
diaphor			
chir/o			
necr/o			
enter/o			
chondr/o			
cyan			

E. **Prefixes** generally add description to a medical term—color, location, size, and the like. Complete the following terms by writing the appropriate prefix *on the line.* Then write the meaning of the prefix.

peri	sub	mal	a	endo	media
hypo	post	con	non	epi	
inter	bi	para	semi	re	

	Prefix		Meaning of Prefix
1. nosebleed	______________	/stax/is	______________
2. backward flow	______________	/flux	______________
3. under the tongue	______________	/lingu/al	______________
4. space between two vertebrae	______________	/vertebr/al	______________
5. inside lining of the heart	______________	/cardi/um	______________

	Prefix	Meaning of Prefix
6. without breath	______________/pnea	______________
7. present at birth	______________/genit/al	______________
8. around a tooth	______________/odont/al	______________
9. appear like a half moon	______________/lun/ar	______________
10. adjacent to the nose	______________/nas/al	______________
11. heart valve with two flaps	______________/cusp/id	______________
12. low blood pressure	______________/tens/ion	______________
13. area between the lungs containing the heart	______________/stin/um	______________

F. **Spelling demons:** Some medical terms are particularly difficult to pronounce and spell. Develop your ability to recognize a misspelling when you see it, and know how to correct it. Only one term in this exercise is spelled correctly—for all the incorrectly spelled terms, write the correct spelling on the line after the term.

1. esopagus ______________________________

2. arhythmia ______________________________

3. polip ______________________________

4. cocyx ______________________________

5. paristalsis ______________________________

6. tachicardia ______________________________

7. cominuted ______________________________

8. pharnygeal ______________________________

9. epifyseal ______________________________

10. ventrical ______________________________

Monitor My Progress: An honest self-assessment of your progress with this material will make it easier for you to learn from your mistakes. Jot down below any wrong answers you have corrected, any terms, elements, or spellings you find difficult to remember, and so on. The more times you see, speak, or write these terms, the better you will come to know them.

Based on these review exercises, I need to spend more time on:

____

____

____

The Appendicular Skeleton, Joints, and Muscles

The Essentials of the Languages of Orthopedics and Rehabilitation

CHAPTER 9

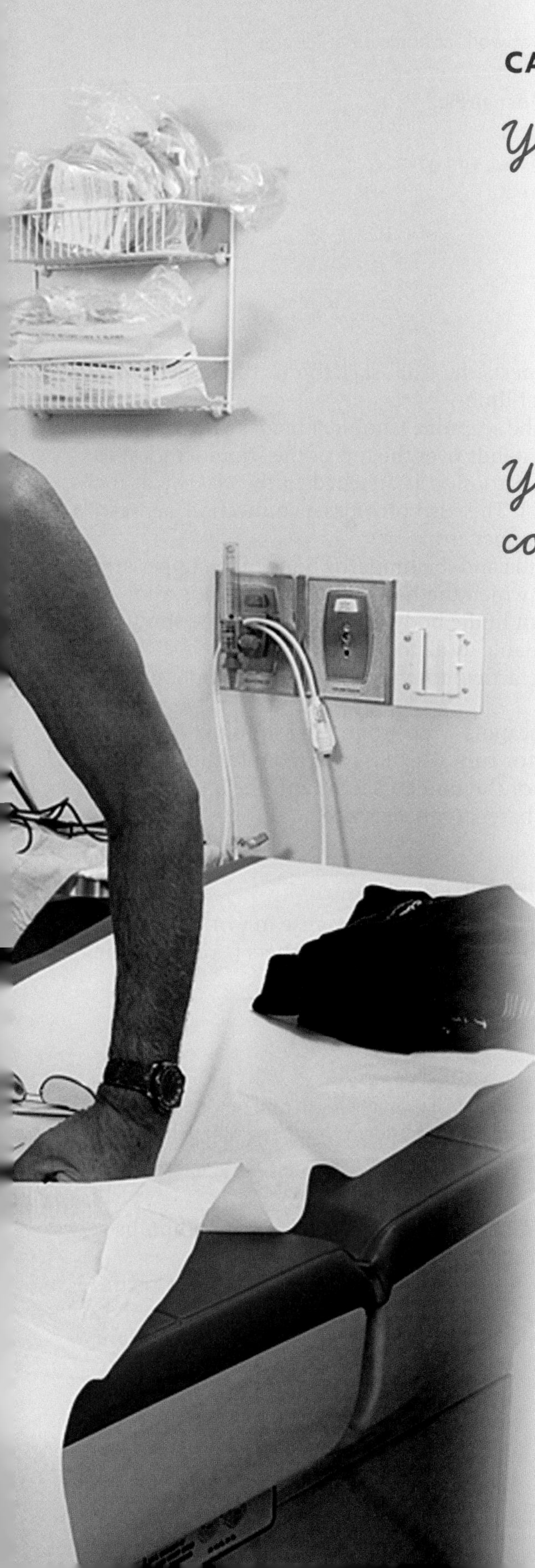

CASE REPORT 9.1

You are

. . . an orthopedic technologist working with orthopedist Kenneth Stannard, MD, in Fulwood Medical Center.

You are communicating with

. . . Mr. Bruce Adams, a 55-year-old construction worker who presents with severe pain in his right shoulder that has made him leave work and seek relief. The pain began 3 or 4 months ago; it is worse at the end of the workday and when he has to work with his arm above the shoulder. In the past week the pain has woken him from sleep.

Mr. Adams' primary care physician has given him pain medication, advised him to stop working, and referred him to Dr. Stannard for diagnosis and treatment. Physical examination shows marked limitation by pain of all passive and active movements of his right shoulder and weakness in all lifting movements.

Learning Outcomes

Attached to the axial skeleton through joints and muscles is the **appendicular skeleton,** the bones of the upper and lower limbs. An understanding of the terminology of the bones, joints, and muscles **(musculoskeletal system)** of the limbs and their disorders is a vital part of your overall knowledge of the human body for your work as a health professional. Information in this chapter will enable you to:

9.1 Describe the appendicular skeleton.

9.2 Select correct medical terminology to describe the structures and functions of the bones, joints, and muscles of the shoulder girdle and upper limbs.

9.3 Learn correct medical terminology to describe the structures and functions of the bones, joints, and muscles of the pelvic girdle and lower limbs.

9.4 Name the major problems and diseases that affect mobility and other functions of the limbs.

LESSON 9.1 THE APPENDICULAR SKELETON, JOINTS, AND MUSCLES

Shoulder Girdle, Arm, and Hand

OBJECTIVES

Your arms and hands are being used constantly throughout every day of your life, whether at work, at home, or at leisure. The information in this lesson will enable you to **use correct medical terminology to:**

- **Describe the structures and functions of the bones, joints, and muscles of the shoulder girdle.**
- **Identify common disorders of the bones, joints, and muscles of the shoulder girdle.**
- **Explain the structures and functions of the bones, joints, and muscles of the arm, elbow, and wrist.**
- **Name common disorders of the bones, joints, and muscles of the arm, elbow, and wrist.**
- **Specify the structures and functions of the bones, joints, and muscles of the hand.**
- **Define common disorders of the bones, joints, and muscles of the hand.**

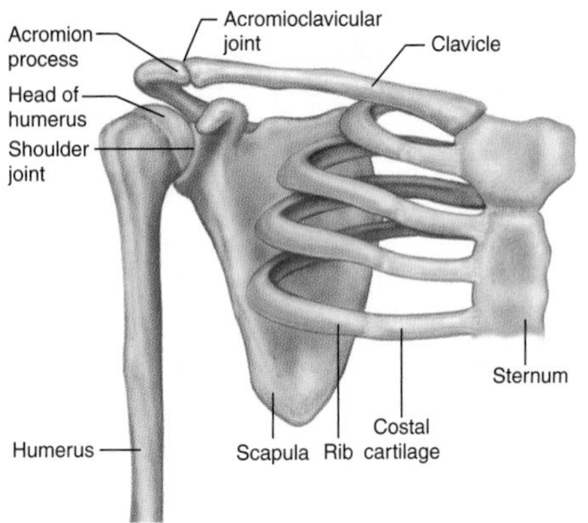

▲ **FIGURE 9.1 Pectoral Girdle and Humerus.**

Abbreviation	
AC	acromioclavicular

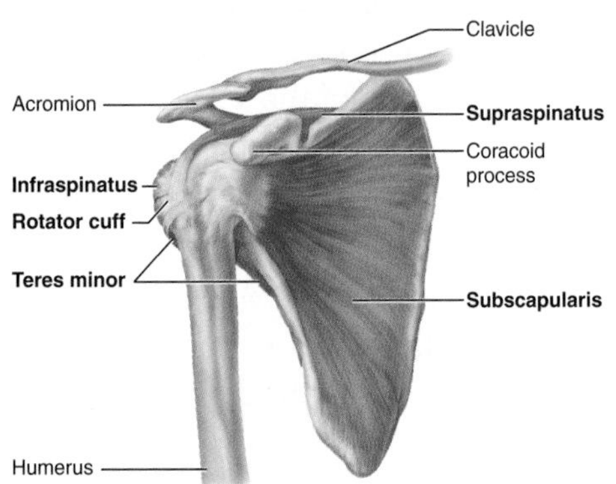

▲ **FIGURE 9.2 Rotator Cuff Muscles (labeled in bold).**

SHOULDER GIRDLE

The **pectoral** (shoulder) **girdle** connects the axial skeleton to the upper limbs and helps with movements of the upper limbs.

The bones of the pectoral girdle are the **scapulae** (shoulder blades) and **clavicles** (collarbones) (*Figure 9.1*). The scapula extends over the top of the shoulder joint to form a roof called the **acromion.** The acromion is attached to the clavicle at the **acromioclavicular (AC)** joint (*Figure 9.1*). This also provides a connection between the axial skeleton, pectoral girdle, and upper arm.

The shoulder joint between the scapula and the **humerus** bone of the upper arm (*Figure 9.1*) is a ball-and-socket joint in which the head of the humerus is allowed the greatest range of motion of any joint in the body. Because of this, the shoulder joint is the most unstable joint and is liable to **dislocation.**

Several ligaments hold together the **articulating** surfaces of the humerus and scapula.

Four muscles that **originate** on the scapula wrap around the shoulder joint and fuse to form one large tendon, the **rotator cuff,** which is **inserted** into the humerus (*Figure 9.2*). This tendon keeps the ball of the humerus tightly in the socket of the scapula and provides the strength that baseball pitchers need.

Common Disorders of the Shoulder Girdle

Rotator cuff tears are the result of wear and tear from overuse in work situations or in certain sports, such as the throwing done by baseball pitchers (*Figure 9.2*). The tears can be partial or complete.

Case Report 9.1 (continued)

When Mr. Adams was evaluated by Dr. Stannard, an MRI revealed a full-thickness tear of the rotator cuff. Mr. Adams has been scheduled for **ambulatory** surgery to repair the tear.

Shoulder separation is a dislocation of the acromioclavicular (AC) joint, usually due to a fall on the point of the shoulder.

Shoulder dislocation occurs when the ball of the humerus slips out of the socket of the scapula, usually anteriorly.

Shoulder subluxation occurs when the ball of the humerus slips partially out of position and then moves back in.

WORD ANALYSIS AND DEFINITION

S/ = Suffix P/ = Prefix R/ = Root R/CF = Combining Form

WORD	PRONUNCIATION	ELEMENTS		DEFINITION
acromion	ah-**CROW**-mee-on		Greek *tip of the shoulder*	Lateral end of the scapula, extending over the shoulder joint.
acromioclavicular	ah-**CROW**-mee-oh-klah-**VICK**-you-lar	S/ R/CF R/	**-ar** *pertaining to* **acromi/o-** *acromion* **-clavicul-** *clavicle*	The joint between the acromion and the clavicle.
ambulatory	am-byu-**LAY**-tor-ee	S/ R/	**-ory** *having the function of* **ambulat-***walking*	Surgery or any other care provided without an overnight stay in a medical facility.
articulate	ar-**TIK**-you-late	S/ R/	**-ate** *composed of* **articul-***joint*	Two separate bones have formed a joint.
articulation	ar-tik-you-**LAY**-shun	S/	**-ation** *process*	A joint.
clavicle	**KLAV**-ih- kul		Latin *collarbone*	Curved bone that forms the anterior part of the pectoral girdle.
clavicular (adj)	klah-**VICK**-you-lah	S/ R/	**-ar** *pertaining to* **clavicul-** *clavicle*	Pertaining to the clavicle.
dislocation	dis-low-**KAY**-shun	S/ P/ R/	**-ion** *action, condition* **dis-** *apart, away from* **-locat-** *place*	Completely out of joint.
humerus	**HYU-**mer-us		Latin *shoulder*	Single bone of the upper arm.
insertion **insert (verb)**	in-**SIR**-shun in-**SIRT**	S/ R/	**-ion** *action, condition* **insert-** *put together*	The insertion of a muscle is the attachment of a muscle to a more movable part of the skeleton, as distinct from the origin.
musculoskeletal system	**MUSS**-kyu-loh-**SKEL**-eh-tal **SIS**-tem	S/ R/CF R/	**-al** *pertaining to* **muscul/o-** *muscle* **-skelet-** *skeleton* system, Greek *an organized whole*	Pertaining to the muscles and the bony skeleton, functioning as an organized whole.
origin	**OR**-ih-gin		Latin *source of*	Fixed source of a muscle at its attachment to bone.
pectoral	**PEK**-tor-al	S/ R/	**-al** *pertaining to* **pector-** *chest*	Pertaining to the chest.
pectoral girdle	**PEK**-tor-al **GIR**-del		girdle, Old English *encircle*	Incomplete bony ring that attaches the upper limb to the axial skeleton.
rotator cuff	roh-**TAY**-tor CUFF	S/ R/	**-or** *one who does* **rotat-** *rotate* cuff, Old English *band*	Part of the capsule of the shoulder joint.
scapula **scapulae (pl)**	**SKAP**-you-lah **SKAP**-you-lee		Latin *shoulder blade*	Shoulder blade.
scapular (adj)	**SKAP**-you-lar	S/ R/	**-ar** *pertaining to* **scapul-** *scapula*	Pertaining to the shoulder blade.
subluxation	sub-luck-**SAY**-shun	S/ P/ R/	**-ion** *action, condition* **sub-** *under, below, slightly* **-luxat-** *dislocate*	An incomplete dislocation when some contact between the joint surfaces remains.

EXERCISES

Build *medical terms using the language of orthopedics to complete this exercise. Each term is defined and partially complete. Add the rest of the elements to complete the term, and write under the line the element(s) you have used. The first one is done for you. Fill in the blanks.*

1. incomplete dislocation sub / luxat / ion
   P R S

2. joint between acromion and clavicle ________ / ________ / ar

3. a joint ________ / ________ / ation

4. pertaining to the shoulder blade ________ / ________ / ar

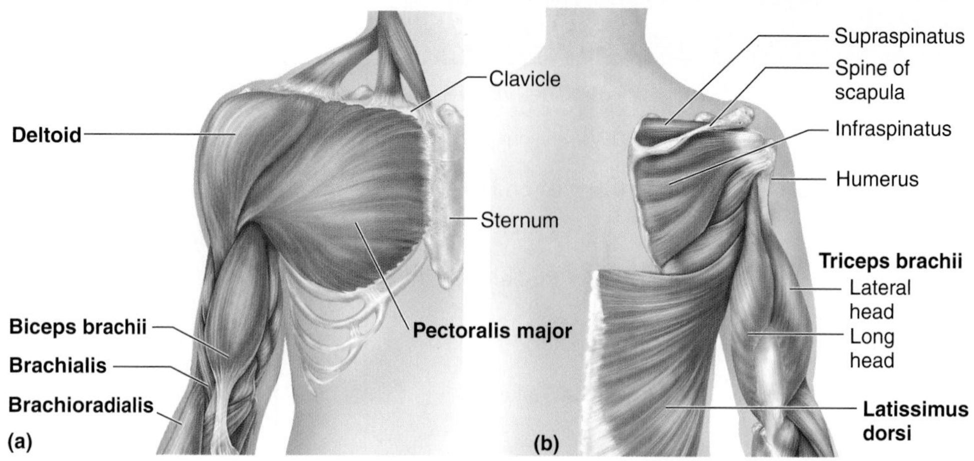

▲ **FIGURE 9.3** **Muscles Joining Arm to Body.** (a) Anterior view. (b) Posterior view.

UPPER ARM AND ELBOW JOINT

Muscles connect the humerus to the pectoral girdle, vertebral column, and ribs. These muscles enable the arm to be freely movable at the shoulder joint. The major anterior muscles are the **deltoid** and **pectoralis major** (*Figure 9.3a*), and among the major posterior muscles is the **latissimus dorsi** (*Figure 9.3b*).

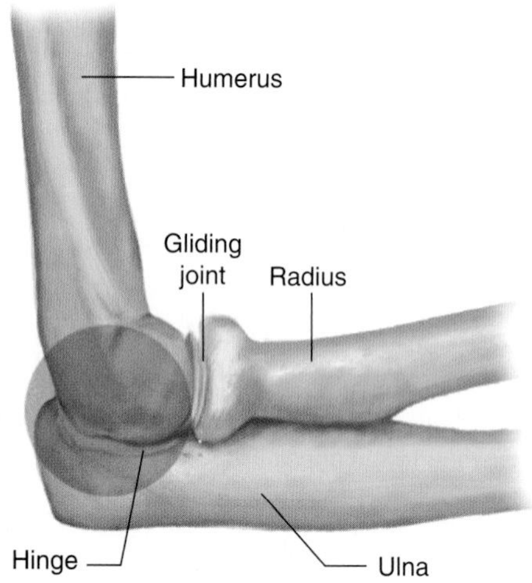

▲ **FIGURE 9.4** **Elbow Joint.**

Elbow Joint

The elbow joint (*Figure 9.4*) has two articulations:

1. A hinge joint between the humerus and **ulna** bone of the forearm, which allows flexion and extension of the elbow.
2. A gliding joint between the humerus and **radius** bone of the forearm, which allows pronation and supination of the forearm and hand.

A joint capsule and ligaments hold the two articulations together.

Muscles that move the elbow joint and forearm have their origins on the humerus or pectoral girdle and are inserted into the bones of the forearm. On the front of the arm, a group of three muscles (*see Figure 9.3a*)—**biceps brachii, brachialis,** and **brachioradialis**—flexes the forearm at the elbow joint and rotates the forearm and hand laterally (supination). On the back of the arm, a single muscle, the **triceps brachii,** extends the elbow joint and forearm (*see Figure 9.3b*).

Common Disorders of the Elbow Joint

Tennis elbow is a common joint injury. Tendons of upper-arm and forearm muscles are inserted into the humerus just above the elbow joint. Small tears in the tendons at their attachments are caused by trauma; or by overuse of the elbow joint, such as in weight lifting; or by poor techniques playing tennis or golf. Pain occurs when straightening the elbow or opening and closing the fingers. Treatment is rest, ice, pain medication, massage, and stretch exercises.

Ligament strains and bone fractures due to a heavy fall or a blow to the elbow are also common injuries.

LESSON 9.2 Pelvic Girdle, Leg, and Foot

OBJECTIVES

Your legs and feet are being used constantly throughout every day of your life, whether at work, at home, or at leisure, whether standing, walking, or changing positions. The information in this lesson will enable you to:

- **Use correct medical terminology to describe the structures and functions of the bones, joints, and muscles of the pelvic girdle.**
- **Name common disorders of the bones, joints, and muscles of the pelvic girdle.**
- **Choose the correct medical terminology for the structures and functions of the bones, joints, and muscles of the leg, knee, and ankle.**
- **Recognize common disorders of the bones, joints, and muscles of the knee and ankle.**
- **Specify the medical terminology for the structures and functions of the bones, joints, and muscles of the foot.**
- **Name common disorders of the bones, joints, and muscles of the foot.**

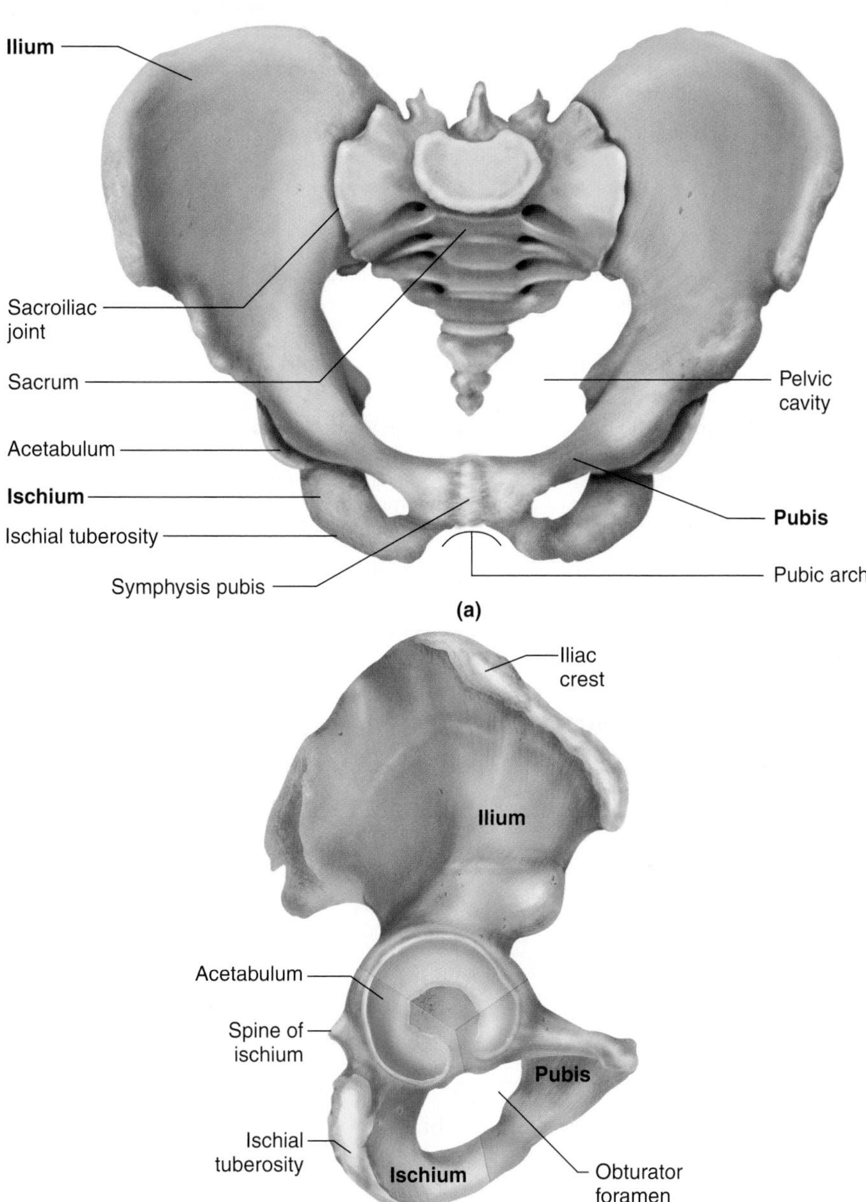

▲ **FIGURE 9.10 Pelvic Girdle.** (a) Front view. (b) Side view.

The Pelvic Girdle

The pelvic girdle is the two hip bones that articulate anteriorly with each other at the **symphysis pubis** and posteriorly with the sacrum to form the bowl-shaped **pelvis.** The two joints between the hip bones and the sacrum are the **sacroiliac joints.**

The pelvic girdle has the following functions:

1. Support the axial skeleton.
2. Transmit the upper body's weight through to the lower limbs.
3. Provide attachments for the lower limbs.
4. Protect the internal reproductive organs, urinary bladder, and distal segment of the large intestine.

Each hip bone is a fusion of three bones, the **ilium, ischium,** and **pubis** (*Figure 9.10a*). The fusion takes place in the region of the **acetabulum,** a cup-shaped cavity on the lateral surface of the hip bone, which receives the head of the **femur,** or thigh bone (*Figure 9.10b*).

The lower part of the pelvis is formed by the lower ilium, ischium, and pubic bones that surround a short canal-like pelvic cavity, through which the rectum, vagina, and urethra pass. The infant passes down this canal during childbirth.

Disorders of the Pelvic Girdle

Sacroiliac (SI) joint strain is a common cause of low-back pain. Unlike most joints, the SI joint is designed to move only one-quarter of an inch during weight bearing and forward flexion. Its main function is to provide shock absorption for the spine.

During pregnancy, hormones enable connective tissue to relax so that the pelvis can expand enough to allow birth.

WORD ANALYSIS AND DEFINITION

S/ = Suffix P/ = Prefix R/ = Root R/CF = Combining Form

WORD	PRONUNCIATION	ELEMENTS		DEFINITION
arthritis	ar-**THRI**-tis	S/ R/	**-itis** *inflammation* **arthr-** *joint*	Inflammation of a joint or joints.
carpus	**KAR**-pus		Greek *wrist*	The eight carpal bones of the wrist.
carpal (adj)	**KAR**-pal	S/ R/	**-al** *pertaining to* **carp-** *wrist bones*	Pertaining to the wrist.
metacarpal	**MET**-ah-**KAR**-pal	P/	**meta-** *after, subsequent to*	The five bones between the carpus and the fingers.
Colles fracture	**KOL**-ez **FRAK**-chur		Abraham Colles, Irish surgeon, 1773–1843	Fracture of the distal radius at the wrist.
cyst	SIST		Greek *fluid-filled sac*	An abnormal, fluid-containing sac.
dorsum	**DOOR**-sum		Latin *back*	The back of any part of the body, including the hand.
dorsal (adj)	**DOOR**-sal	S/ R/	**-al** *pertaining to* **dors-** *back*	Pertaining to the back of any part of the body.
ventral (opposite of *dorsal*)	**VEN**-tral	R/	**ventr-** *belly*	Pertaining to the belly or situated nearer to the surface of the belly.
ganglion	**GANG**-lee-on		Greek *swelling*	Fluid-containing swelling attached to the synovial sheath of a tendon.
Heberden node	**HEH**-ber-den NOHD		William Heberden, English physician, 1710–1801	Bony lump on the terminal phalanx of the fingers in osteoarthritis.
metacarpophalangeal	**MET**-ah-**KAR**-poh-fay-**LAN**-jee-al	S/ P/ R/CF R/CF	**-al** *pertaining to* **meta-** *after, subsequent to* **-carp/o-** *bones of the wrist* **-phalang/e-** *phalanx, finger or toe*	The joints between the metacarpal bones and the phalanges.
osteoarthritis	**OSS**-tee-oh-ar-**THRI**-tis	S/ R/CF R/	**-itis** *inflammation* **oste/o-** *bone* **arthr-** *joint*	Chronic inflammatory disease of joints.
phalanx **phalanges (pl)**	**FAY**-lanks fay-**LAN**-jeez		Latin *bone of finger or toe*	One of the bones of the digits (fingers or toes).
rheumatism	**RU**-mat-izm	S/ R/	**-ism** *condition* **rheumat-** *a flow*	Pain in various parts of the musculoskeletal system.
rheumatic (adj)	ru-**MAT**-ik	S/	**-ic** *pertaining to*	Pertaining to or characterized by rheumatism.
rheumatoid arthritis	**RU**-mah-toyd ar-**THRI**-tis	S/	**-oid** *resembling*	Systemic disease affecting many joints.
stenosis	steh-**NOH**-sis		Greek *narrowing*	Narrowing of a passage.
tenosynovitis	**TEN**-oh-sin-oh-**VIE**-tis	S/ R/CF R/	**-itis** *inflammation* **ten/o-** *tendon* **-synov-** *synovial membrane*	Inflammation of both a tendon and its synovial sheath.
thenar eminence	**THAY**-nar **EM**-in-ens	R/	eminence Latin *stand out* **thenar** *palm*	The fleshy mass at the base of the thumb.
hypothenar eminence	high-poh-**THAY**-nar **EM**-in-ens	P/	**hypo-** *below, smaller*	The fleshy mass at the base of the little finger.

EXERCISES

***Elements** can have more than one meaning and must be applied differently to specific medical terms. Fill in the blanks.*

1. Give an example of a medical term (other than *hypothenar*), from any chapter, that begins with *hypo* and in which *hypo* means *below* (in the sense of location).

   Term: ______________________ means ______________________.

2. Give an example of another medical term, from any chapter, that begins with *hypo* and in which *hypo* means *smaller.* Smaller can also be meant in the sense of "less than" or "deficient."

   Term: ______________________ means ______________________.

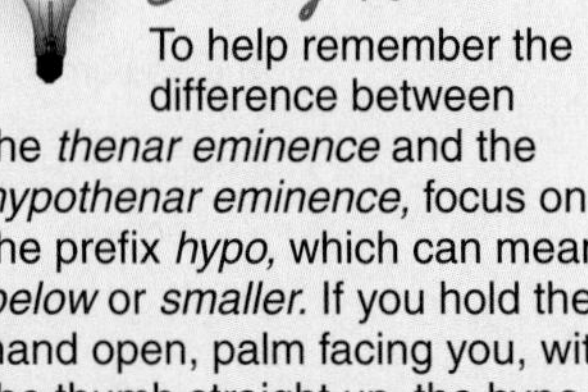

Study Hint

To help remember the difference between the *thenar eminence* and the *hypothenar eminence,* focus on the prefix *hypo,* which can mean *below* or *smaller.* If you hold the hand open, palm facing you, with the thumb straight up, the hypothenar eminence is the fleshy mass at the base of the **little** finger, which is the **smallest** one on the hand.

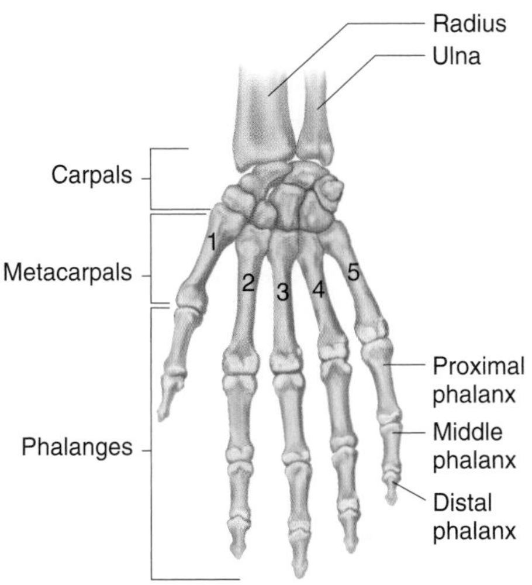

▲ **FIGURE 9.5 Bones of the Wrist and Hand.**

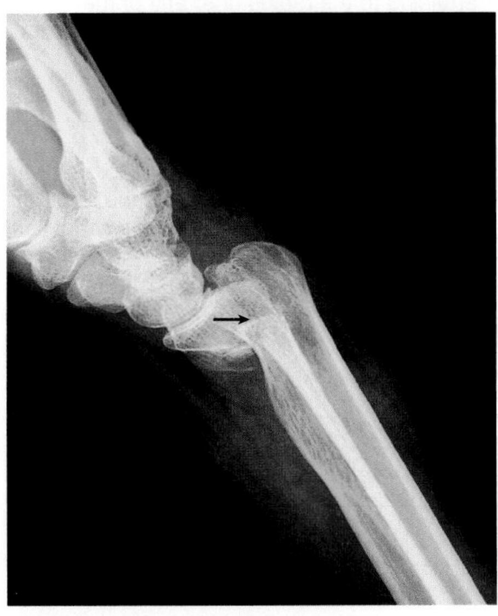

▲ **FIGURE 9.6 X-Ray of Colles Fracture.**

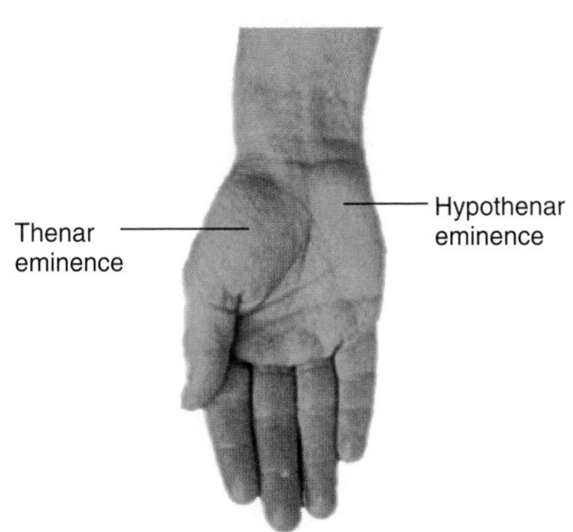

▲ **FIGURE 9.7 Palmar Surface of the Hand.**

Abbreviations	
CTS	carpal tunnel syndrome
OA	osteoarthritis
RA	rheumatoid arthritis

WRIST AND HAND

The Wrist

In your forearm, the radius on the thumb side and the ulna on the little-finger side articulate at the wrist joint with the small **carpal** bones (*Figure 9.5*). The muscles of the forearm supinate and pronate the forearm, flex and extend the wrist joint and hand, and move the hand medially and laterally. All these terms of movement are detailed in *Chapter 2.*

Your forearm is bigger near the elbow because the fleshy bellies of the forearm muscles are bulky. Your wrist is much thinner because the muscles have become tendons that pass over the wrist on the way to being inserted into the bones of the fingers.

Common Disorders of the Wrist

Ganglion cysts are fluid-filled cysts that arise when the synovial tendon sheaths running over the back of the wrist are irritated or inflamed. They often disappear spontaneously.

Stenosing tenosynovitis is inflammation of the synovial sheaths on the back of the wrist, producing pain in the wrist.

Carpal tunnel syndrome (CTS) develops similarly on the front of the wrist as a result of inflammation and swelling of tendon sheaths arising from overuse of repetitive movements, for example, in computer keyboard operation. This causes "pins and needles" or pain and loss of muscle power in the thumb side of the hand.

Colles fracture is a common fracture of the radius just above the wrist joint (*Figure 9.6*). It occurs when a person tries to break a fall with an outstretched hand.

The Hand

The five fingers of one hand have 14 bones called **phalanges.** The thumb has two phalanges. The remaining four fingers each has three (*Figure 9.5*). In the palm of the hand, the five bones proximal to the fingers are **metacarpals** that connect to the phalanges at the **metacarpophalangeal** joints. The metacarpals connect at the wrist to eight small **carpal** bones. These, in turn, connect the hand to the bones of the forearm. All these bones require numerous joints with ligaments to connect and stabilize them.

When you look at the palmar surface of your hand, at the base of the thumb is a prominent pad of muscles called the **thenar eminence.** A smaller pad of muscles at the base of the little finger is called the **hypothenar eminence** (*Figure 9.7*). The back of the hand is called the **dorsum.**

Disorders of the Hand

Osteoarthritis (OA) in the hand joints occurs from wear and tear leading to degeneration and breakdown of joint cartilage. Small bony spurs called **Heberden nodes** form over the joint (*Figure 9.8*).

Rheumatoid arthritis (RA), with destruction of joint surfaces, joint capsule, and ligaments, leads to marked deformity and joint instability (*Figure 9.9*). It occurs mostly in women, with onset between ages 40 and 60. The disease affects the synovial membrane that lines joints and tendons. Lumps known as **rheumatic** nodules form over the small joints of the hand and wrist.

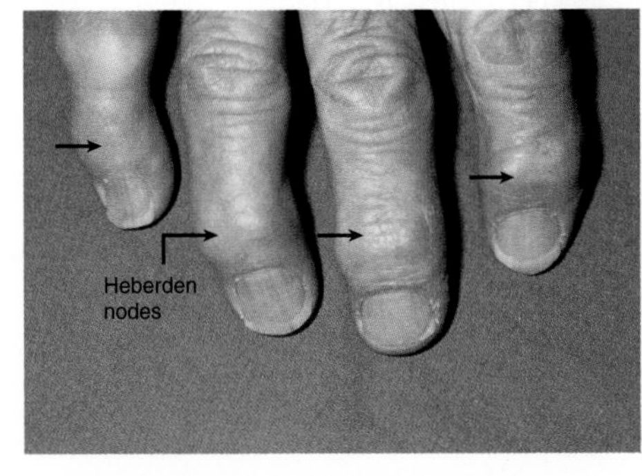

▲ **FIGURE 9.8 Hand with Osteoarthritis.**

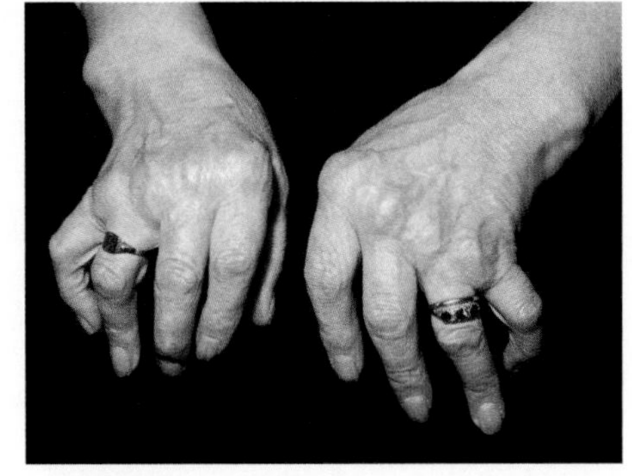

▲ **FIGURE 9.9 Hands with Rheumatoid Arthritis.**

WORD ANALYSIS AND DEFINITION

S/ = Suffix P/ = Prefix R/ = Root R/CF = Combining Form

WORD	PRONUNCIATION		ELEMENTS	DEFINITION
biceps brachii	**BYE**-sepz **BRAY**-key-eye	P/ R/ R/CF	**bi-** *two* **-ceps** *head* **brachi/i** *of the arm*	A muscle of the arm that has two heads or points of origin on the scapula.
brachialis	**BRAY**-kee-al-is	S/ R/	**-alis** *pertaining to* **brachi-** *arm*	Muscle that lies underneath the biceps and is the strongest flexor of the forearm.
brachioradialis	**BRAY**-kee-oh-**RAY**-dee-al-is	S/ R/CF R/	**-alis** *pertaining to* **brachi/o-** *arm* **-radi-** *radius*	Muscle that helps flex the forearm.
deltoid	**DEL**-toyd	S/ R/	**-oid** *resembling* **delt-** *triangle*	Large, fan-shaped muscle connecting the scapula and clavicle to the humerus.
latissimus dorsi	lah-**TISS**-ih-muss **DOOR**-sigh	S/ R/ R/	**-imus** *most* **latiss-** *wide* **dorsi** *of the back*	The widest (broadest) muscle in the back.
radius **radial (adj)**	**RAY**-dee-us **RAY**-dee-al	 S/ R/	Latin *spoke of a wheel* **-al** *pertaining to* **radi-** *radius*	The forearm bone on the thumb side. Pertaining to the radius or to any of the structures (artery, vein, nerve) named after it.
triceps brachii	**TRY**-sepz **BRAY**-key-eye	P/ R/ R/CF	**tri-** *three* **-ceps** *head* **brachi/i** *of the arm*	Muscle of the arm that has three heads or points of origin.
ulna **ulnar (adj)**	**UL**-nah **UL**-nar	 S/ R/	Latin *elbow, arm* **-ar** *pertaining to* **uln-** *ulna*	The medial and larger bone of the forearm. Pertaining to the ulna or to any of the structures (artery, vein, nerve) named after it.

EXERCISES

Elements *will assist you in determining the difference among these muscle terms. Reduce the terminology of these muscles to the basic elements. Fill in the chart.*

Muscle	Prefix	Root/Combining Form	Suffix
biceps brachii			
brachialis			
brachioradialis			
deltoid			
latissimus dorsi			
triceps brachii			

Using the terms from the chart, write the name of the correct muscle on the line next to its description. Use the elements as your guide.

1. three heads or points of origin: ____________________
2. strongest flexor of the forearm: ____________________
3. fan-shaped muscle (resembles a triangle): ____________________
4. helps flex the forearm: ____________________
5. two points of origin on the scapula: ____________________
6. broadest (most wide) muscle of the back: ____________________

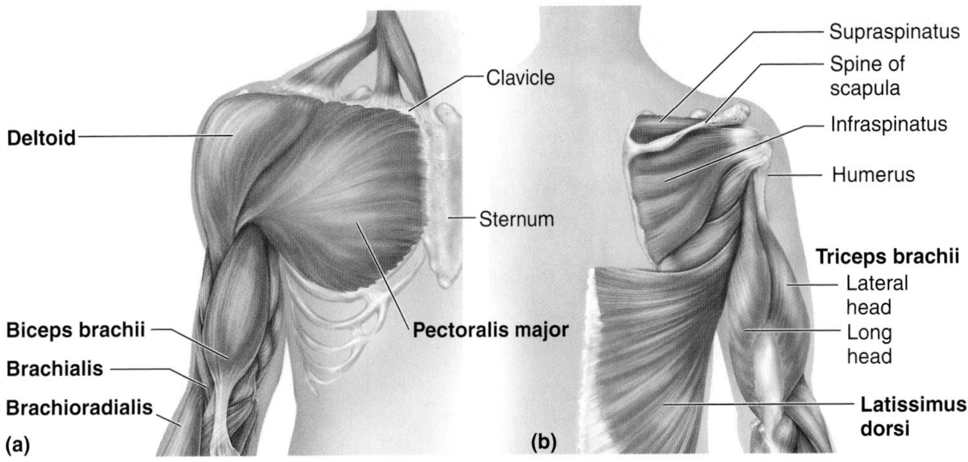

▲ **FIGURE 9.3 Muscles Joining Arm to Body.** (a) Anterior view. (b) Posterior view.

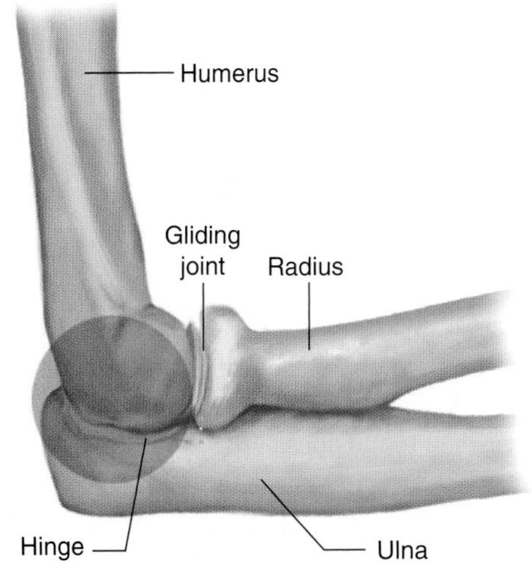

▲ **FIGURE 9.4 Elbow Joint.**

UPPER ARM AND ELBOW JOINT

Muscles connect the humerus to the pectoral girdle, vertebral column, and ribs. These muscles enable the arm to be freely movable at the shoulder joint. The major anterior muscles are the **deltoid** and **pectoralis major** (*Figure 9.3a*), and among the major posterior muscles is the **latissimus dorsi** (*Figure 9.3b*).

Elbow Joint

The elbow joint (*Figure 9.4*) has two articulations:

1. A hinge joint between the humerus and **ulna** bone of the forearm, which allows flexion and extension of the elbow.
2. A gliding joint between the humerus and **radius** bone of the forearm, which allows pronation and supination of the forearm and hand.

A joint capsule and ligaments hold the two articulations together.

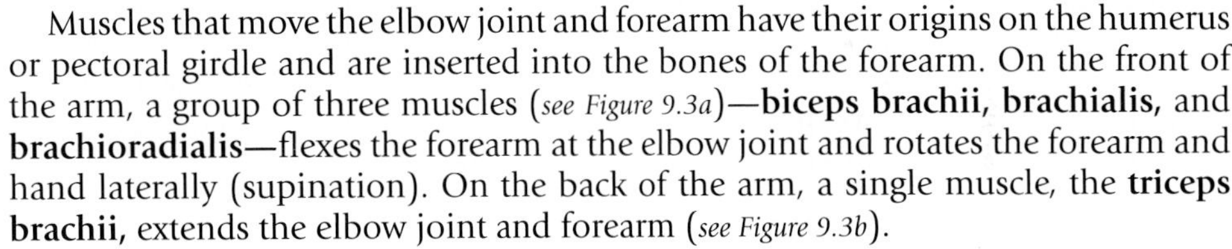

Muscles that move the elbow joint and forearm have their origins on the humerus or pectoral girdle and are inserted into the bones of the forearm. On the front of the arm, a group of three muscles (*see Figure 9.3a*)—**biceps brachii, brachialis,** and **brachioradialis**—flexes the forearm at the elbow joint and rotates the forearm and hand laterally (supination). On the back of the arm, a single muscle, the **triceps brachii,** extends the elbow joint and forearm (*see Figure 9.3b*).

Common Disorders of the Elbow Joint

Tennis elbow is a common joint injury. Tendons of upper-arm and forearm muscles are inserted into the humerus just above the elbow joint. Small tears in the tendons at their attachments are caused by trauma; or by overuse of the elbow joint, such as in weight lifting; or by poor techniques playing tennis or golf. Pain occurs when straightening the elbow or opening and closing the fingers. Treatment is rest, ice, pain medication, massage, and stretch exercises.

Ligament strains and bone fractures due to a heavy fall or a blow to the elbow are also common injuries.

WORD ANALYSIS AND DEFINITION

S/ = Suffix P/ = Prefix R/ = Root R/CF = Combining Form

WORD	PRONUNCIATION		ELEMENTS	DEFINITION
acromion	ah-**CROW**-mee-on		Greek *tip of the shoulder*	Lateral end of the scapula, extending over the shoulder joint.
acromioclavicular	ah-**CROW**-mee-oh-klah-**VICK**-you-lar	S/ R/CF R/	**-ar** *pertaining to* **acromi/o-** *acromion* **-clavicul-** *clavicle*	The joint between the acromion and the clavicle.
ambulatory	am-byu-**LAY**-tor-ee	S/ R/	**-ory** *having the function of* **ambulat-** *walking*	Surgery or any other care provided without an overnight stay in a medical facility.
articulate	ar-**TIK**-you-late	S/ R/	**-ate** *composed of* **articul-** *joint*	Two separate bones have formed a joint.
articulation	ar-tik-you-**LAY**-shun	S/	**-ation** *process*	A joint.
clavicle	**KLAV**-ih- kul		Latin *collarbone*	Curved bone that forms the anterior part of the pectoral girdle.
clavicular (adj)	klah-**VICK**-you-lah	S/ R/	**-ar** *pertaining to* **clavicul-** *clavicle*	Pertaining to the clavicle.
dislocation	dis-low-**KAY**-shun	S/ P/ R/	**-ion** *action, condition* **dis-** *apart, away from* **-locat-** *place*	Completely out of joint.
humerus	**HYU**-mer-us		Latin *shoulder*	Single bone of the upper arm.
insertion **insert (verb)**	in-**SIR**-shun in-**SIRT**	S/ R/	**-ion** *action, condition* **insert-** *put together*	The insertion of a muscle is the attachment of a muscle to a more movable part of the skeleton, as distinct from the origin.
musculoskeletal system	**MUSS**-kyu-loh-**SKEL**-eh-tal **SIS**-tem	S/ R/CF R/	**-al** *pertaining to* **muscul/o-** *muscle* **-skelet-** *skeleton* system, Greek *an organized whole*	Pertaining to the muscles and the bony skeleton, functioning as an organized whole.
origin	**OR**-ih-gin		Latin *source of*	Fixed source of a muscle at its attachment to bone.
pectoral	**PEK**-tor-al	S/ R/	**-al** *pertaining to* **pector-** *chest*	Pertaining to the chest.
pectoral girdle	**PEK**-tor-al **GIR**-del		girdle, Old English *encircle*	Incomplete bony ring that attaches the upper limb to the axial skeleton.
rotator cuff	roh-**TAY**-tor CUFF	S/ R/	**-or** *one who does* **rotat-** *rotate* cuff, Old English *band*	Part of the capsule of the shoulder joint.
scapula **scapulae (pl)**	**SKAP**-you-lah **SKAP**-you-lee		Latin *shoulder blade*	Shoulder blade.
scapular (adj)	**SKAP**-you-lar	S/ R/	**-ar** *pertaining to* **scapul-** *scapula*	Pertaining to the shoulder blade.
subluxation	sub-luck-**SAY**-shun	S/ P/ R/	**-ion** *action, condition* **sub-** *under, below, slightly* **-luxat-** *dislocate*	An incomplete dislocation when some contact between the joint surfaces remains.

EXERCISES

Build *medical terms using the language of orthopedics to complete this exercise. Each term is defined and partially complete. Add the rest of the elements to complete the term, and write under the line the element(s) you have used. The first one is done for you. Fill in the blanks.*

1. incomplete dislocation ___sub___ / ___luxat___ / ___ion___
   P R S

2. joint between acromion and clavicle ________ / ________ / ___ar___

3. a joint ________ / ________ / ___ation___

4. pertaining to the shoulder blade ________ / ________ / ___ar___

WORD ANALYSIS AND DEFINITION

S/ = Suffix P/ = Prefix R/ = Root R/CF = Combining Form

WORD	PRONUNCIATION		ELEMENTS	DEFINITION
acetabulum	ass-eh-**TAB**-you-lum		Latin *vinegar cup*	The cup-shaped cavity of the hip bone that receives the head of the femur to form the hip joint.
arthrodesis	ar-**THROW**-dee-sis	S/ R/CF	**-desis** *to fuse together* **arthr/o-** *joint*	Fixation or stiffening of a joint by surgery.
brace	BRACE		Old English *to fasten*	Appliance to support a part of the body in its correct position.
diastasis	die-**ASS**-tah-sis		Greek *separation*	Separation of normally joined parts.
femur **femoral (adj)**	**FEE**-mur **FEM**-oh-ral	 S/ R/	Latin *thigh* **-al** *pertaining to* **femor-** *femur*	The thigh bone. Pertaining to the femur.
ilium	**ILL**-ee-um		Latin *groin*	Large wing-shaped bone at the upper and posterior part of the pelvis.
ischium **ischia (pl)** **ischial (adj)**	**IS**-kee-um **IS**-kee-ah **IS**-kee-al	 S/ R/	Greek *hip* **-al** *pertaining to* **ischi-** *ischium, hip bone*	Lower and posterior part of the hip bone. Pertaining to the ischium.
pelvis **pelvic (adj)**	**PEL**-viss **PEL**-vik	 S/ R/	Latin *basin* **-ic** *pertaining to* **pelv-** *pelvis*	Basin-shaped ring of bones, ligaments, and muscles at the base of the spine. Also, any basin-shaped cavity, like the pelvis of the kidney (*Chapter 13*). Pertaining to the pelvis.
pubis **pubic (adj)**	**PYU**-bis **PYU**-bik	 S/ R/	Latin *pubis* **-ic** *pertaining to* **pub-** *pubis*	Alternative name for the pubic bone. Pertaining to the pubic bone.
radiology **radiologist**	ray-dee-**OL**-oh-jee ray-dee-**OL**-oh-jist	S/ R/CF S/	**-logy** *study of* **radi/o-** *radiation, x-rays* **-logist** *one who studies, specialist*	The study of medical imaging. Medical specialist in the use of x-rays and other imaging techniques.
sacroiliac joint	say-kroh-**ILL**-ih-ak JOINT	S/ R/CF R	**-ac** *pertaining to* **sacr/o-** *sacrum* **-ili-** *ilium*	The joint between the sacrum and the ilium.
symphysis **symphyses (pl)**	**SIM**-feh-sis **SIM**-feh-sees		Greek *grow together*	Two bones joined by fibrocartilage; in this case, the two pubic bones.

The stretching in the SI joint ligaments makes it "overmobile," susceptible to wear and tear and generation of painful arthritis.

Another cause of pain in the SI joint is trauma when tearing of the joint ligaments allows too much motion and pain results.

A diagnosis of sacroiliac pain can be made by clinical examination, joint x-ray **(radiology)**, and CT scan. An injection of local anesthetic into the joint is used to relieve the pain temporarily. Treatment is usually stabilization of the joint with a **brace** and physical therapy to strengthen the low-back muscles. Occasionally, **arthrodesis** of the joint is necessary.

Diastasis symphysis pubis is another result of the stretching of pelvic ligaments during pregnancy. This stretches the joint between the two pubic bones and leads to pain and difficulty in walking, climbing stairs, and turning over in bed.

Abbreviation

SI	sacroiliac

EXERCISES

Demonstrate *your knowledge of the precise medical term to answer the following questions. Every answer can be found in the above WAD.*

Write the term that is:

1. a surgical procedure ______________________
2. a medical occupation ______________________
3. another name for the thigh bone ______________________
4. basin-shaped ring of bones ______________________

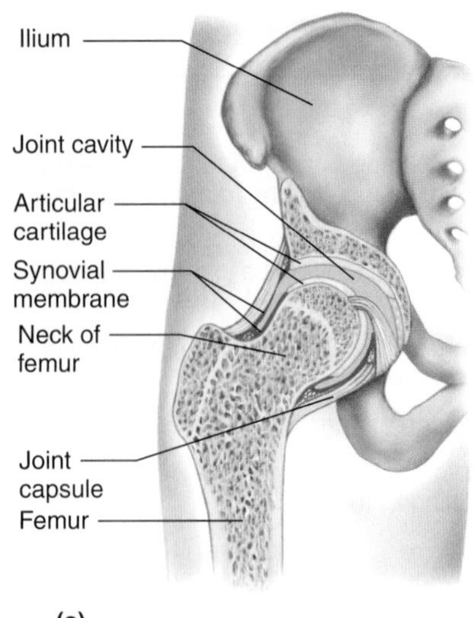

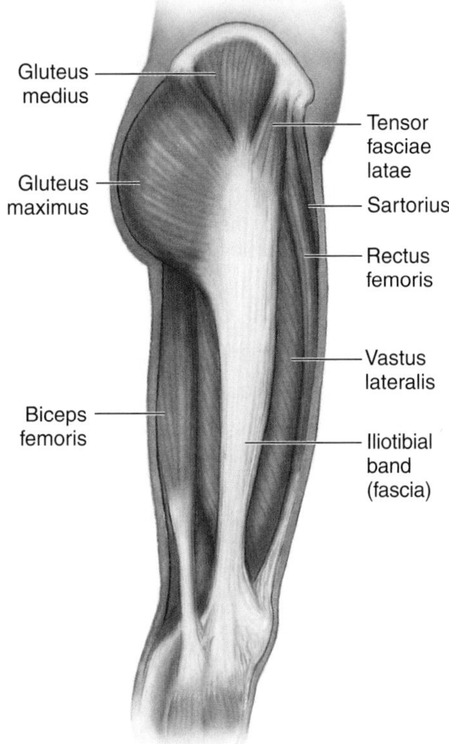

▲ **FIGURE 9.11 Hip Joint.**
(a) Right frontal view of a section of the hip joint. (b) Muscles of the hip and thigh, lateral view.

BONES, JOINTS, AND MUSCLES OF THE HIP AND THIGH

The hip joint is a ball-and-socket synovial joint between the head of the femur (thigh bone) and the cup-shaped acetabulum of the hip bone (*Figure 9.11a*). The joint is held in place by a thick joint capsule reinforced by strong ligaments that connect the neck of the femur to the rim of the acetabulum.

Powerful muscles that support the hip joint and move the thigh have their origins on the pelvic girdle and their insertions into the femur. Prominent among them are the three **gluteus** muscles—**maximus, medius,** and **minimus** (*Figure 9.11b*)—and the **adductor** muscles that run down the inner thigh.

Disorders and Injuries of the Hip Joint

Hip pointer, usually a football-related injury, is a blow to the rim of the pelvis that leads to bruising of the bone and surrounding tissues.

Osteoarthritis is common in the hip as a result of aging, weight bearing, and repetitive use of the joint. The cartilage on both the acetabulum and the head of the femur degenerates. The resulting friction between the bones of the head of the femur and the acetabulum leads to pain and loss of mobility.

Rheumatoid arthritis can also affect the hip, beginning in the synovial membrane and progressing to destroy cartilage and bone.

Avascular necrosis of the femoral head is the death (necrosis) of bone tissue when the blood supply is cut off (avascular), usually as a result of trauma.

Fractures of the neck of the femur occur as a result of a fall, most commonly in elderly women with osteoporosis.

Surgical Procedures of the Hip Joint

Arthroplasty, a total replacement of the hip joint with a metal **prosthesis,** is the most common hip surgery today. The diseased parts of the joint are removed and replaced with artificial parts made of titanium, other metals, ceramics, and, recently, plastics (*Figure 9.12*).

Arthrodesis is the fixation or stiffening of a joint by surgery.

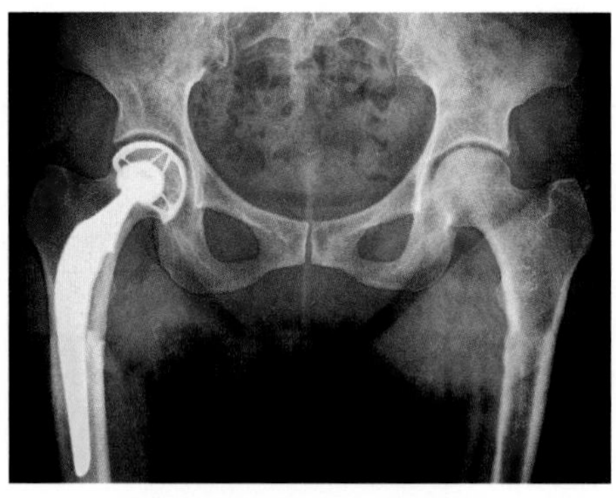

▲ **FIGURE 9.12 Total Hip Replacement.**
Colored x-ray of prosthetic hip.

WORD ANALYSIS AND DEFINITION

S/ = Suffix P/ = Prefix R/ = Root R/CF = Combining Form

WORD	PRONUNCIATION	ELEMENTS		DEFINITION
abductor **abduction** **adductor** **adduction** *(See the Study Hint in Exercises.)*	ab-**DUCK**-tor ab-**DUCK**-shun ah-**DUCK**-tor ah-**DUCK**-shun	S/ P/ R/ S/ P/	**-or** *thing that does* **ab-** *away from* **-duct-** *to lead* **-ion** *action, condition* **ad-** *toward*	Muscle that moves the thigh away from the midline. Action of moving away from the midline. Muscle that moves the thigh toward the midline. Action of moving toward the midline.
arthroplasty	**AR**-throw-plas-tee	S/ R/CF	**-plasty** *reshaping by surgery* **arthr/o-** *joint*	Surgery to repair, as far as possible, the function of a joint.
avascular	a-**VAS**-cue-lar	S/ P/ R/	**-ar** *pertaining to* **a-** *without* **-vascul-** *blood vessel*	Without a blood supply.
gluteus **gluteal (adj)**	**GLUE**-tee-us **GLUE**-tee-al	 S/ R/	Greek *buttocks* **-eal** *pertaining to* **glut-** *buttocks*	Refers to one of three muscles in the buttocks. Pertaining to the buttocks.
maximus	**MAKS**-ih-mus		Latin *the biggest or the greatest*	The gluteus maximus muscle is the largest muscle in the body, covering a large part of each buttock.
medius	**ME**-dee-us		Latin *middle*	The gluteus medius muscle is partly covered by the gluteus maximus.
minimus	**MIN**-ih-mus		Latin *smallest*	The gluteus minimus is the smallest of the gluteal muscles and lies under the gluteus medius.
necrosis **necrotic (adj)**	neh-**KROH**-sis neh-**KROT**-ik	 S/ R/CF	Greek *death* **-tic** *pertaining to* **necr/o-** *death*	Pathologic death of cells or tissue. Pertaining to or affected by necrosis.
prosthesis **prosthetic (adj)**	**PROS**-thee-sis pros-**THET**-ik		Greek *addition*	An artificial part to remedy a defect in the body. Pertaining to a prosthesis.

EXERCISES

Word association *can help you remember the meaning of a term. Relate the term to English: If someone is* **ab***ducted, that person is taken* away *from his or her family. The* **ab***ductor muscle moves the thigh* away *from the midline.*

Precision is a necessary component of every day on the job. Patient safety is at stake.

Continue working with the language of orthopedics. Demonstrate your knowledge of the terms by circling the correct answers below.

1. *Maximus, medius,* and *minimus* are all:
   bones   muscles   tendons   ligaments   phalanges
2. *Gluteus* refers to the:
   wrist   hand   buttocks   hip   spine
3. If tissue is dead, it is:
   necrotic   sacral   rheumatic   degenerative   fibrotic
4. *Avascular* means no:
   living tissue   tendon support   movement   blood supply   defect
5. A *prosthesis* is:
   a procedure   a cast   a diagnostic test   a brace   an artificial body part

Study Hint

A**b**ductor and a**d**ductor are easy to confuse. *Be very careful writing or speaking these terms because they have opposite meanings.*

Operative Report. Fulwood Medical Center.

Gail Griffith, aged 17.

Preoperative Diagnosis: Traumatic ACL tear and tear of medial **meniscus** right knee.

CASE REPORT 9.2

Postoperative Diagnosis: Same.

Procedure Performed: **Arthroscopy**, repair of medial **collateral** ligament, ACL reconstruction, repair of torn medial meniscus, right knee.

Operative Findings: An avulsed anterior *cruciate* ligament (ACL) of the femur with a tear of the posterior horn of the medial meniscus and tear of medial collateral ligament.

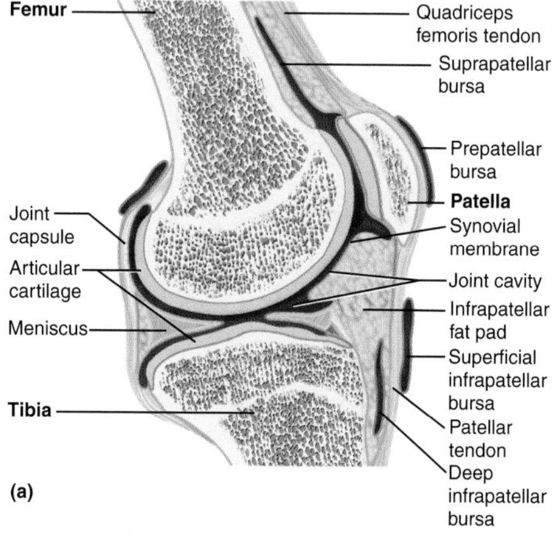

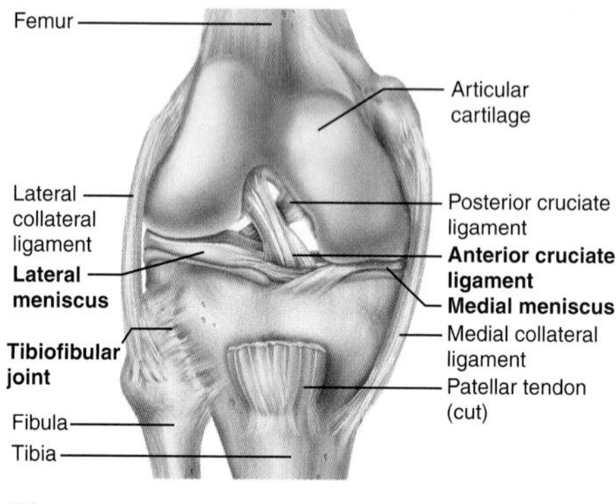

▲ **FIGURE 9.13 Knee Joint.**
(a) Section of knee joint. (b) Right knee joint, anterior view.

BONES, JOINTS, AND MUSCLES OF THE KNEE AND THIGH

The Knee Joint

The knee is a hinged joint formed with four bones:

1. The lower end of the **femur,** shaped like a horseshoe.
2. The flat upper end of the **tibia.**
3. The flat triangular bone, the **patella** (kneecap), which is embedded in the **patellar** tendon and articulates with the femur (*Figure 9.13a*).
4. The **fibula,** which forms a separate joint by articulating with the tibia. This is called the **tibiofibular joint** (*Figure 9.13b*).

Mechanically, the role of the patella is to provide an increase of about 30% in the strength of extension of the knee joint.

Within the knee joint, two crescent-shaped pads of cartilage lie on top of the tibia and articulate with the femur. They are the **medial** and **lateral menisci.** Their function is to distribute weight more evenly across the joint surface to minimize wear and tear.

The knee joint has a fibrous capsule, lined with synovial membrane that secretes synovial fluid, which provides lubrication for the joint. The joint is held together by ligaments. Two are outside the joint and are called the **medial** and **lateral collateral ligaments.** Two other ligaments are located inside the joint cavity and are called the **anterior cruciate ligament** (**ACL**) and **posterior cruciate ligament** (**PCL**). They cross over each other to form an "X" (*see Figure 9.13b*).

Thigh Muscles

The thigh muscles move the knee joint and lower leg. The anterior thigh is composed of the large **quadriceps femoris** muscle, which has four heads and is the most powerful muscle in the body. The four muscles converge into the **quadriceps tendon,** which contains the patella, and continue as the patellar tendon to be inserted into the tibia (*Figure 9.14*). The quadriceps muscle **extends** (straightens) the knee joint and, because of the weight of the lower leg, has to be a powerful muscle.

The posterior thigh is composed mostly of the three **hamstring muscles**—the **biceps femoris, semimembranosus,** and **semitendinosus** (*Figure 9.15*). They **flex** (bend) the knee joint and rotate the leg. The hollow at the back of the knee between the hamstring tendons is called the **popliteal fossa.**

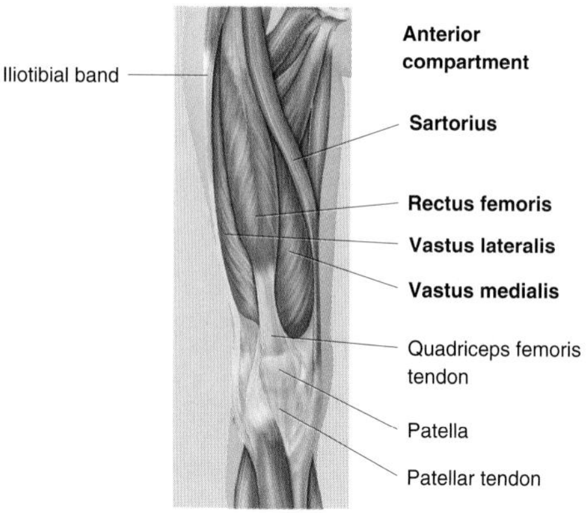

▲ **FIGURE 9.14 Anterior Muscles of the Thigh.**

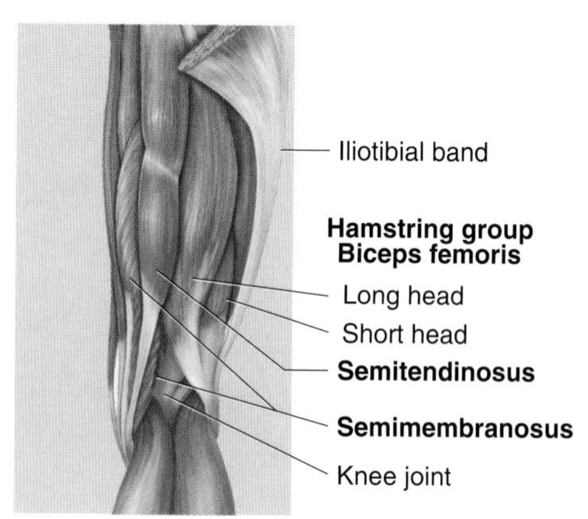

▲ **FIGURE 9.15 Posterior Thigh Muscles.**

WORD ANALYSIS AND DEFINITION

S/ = Suffix P/ = Prefix R/ = Root R/CF = Combining Form

WORD	PRONUNCIATION		ELEMENTS	DEFINITION
arthroscopy	ar-**THROS**-koh-pee	S/	**-scopy** *the process of using an instrument to examine visually*	Visual examination of the interior of a joint.
		R/CF	**arthr/o-** *joint*	
arthroscope	**AR**-thro-skope	S/	**-scope** *instrument to examine visually*	Endoscope used to examine the interior of a joint.
collateral (***Note:*** An extra "l" has been inserted.)	koh-**LAT**-er-al	S/ P/ R/	**-al** *pertaining to* **co-** *together* **-later-** *side*	Situated at the side, often to bypass an obstruction.
cruciate	**KRU**-she-ate		Latin *cross*	Shaped like a cross. In this case, the two internal ligaments of the knee joint cross over each other to form an "X."
fibula	**FIB**-you-lah		Latin *clasp or buckle*	The smaller of the two bones of the lower leg.
fibular (adj)	**FIB**-you-lar	S/ R/	**-ar** *pertaining to* **fibul-** *fibula*	Pertaining to the fibula.
meniscus **menisci (pl)**	meh-**NISS**-kuss meh-**NISS**-key		Greek *crescent*	Disc of cartilage between the bones of a joint, in this case, the knee joint.
patella (kneecap) **patellae (pl)**	pah-**TELL**-ah pah-**TELL**-ee		Latin *small plate*	Thin, circular bone in front of the knee joint, embedded in the patellar tendon.
patellar (adj)	pah-**TELL**-ar	S/ R/	**-ar** *pertaining to* **patell-** *patella*	Pertaining to or embedded in the patellar tendon.
popliteal fossa	pop-**LIT**-ee-al **FOSS**-ah	S/ R/CF	**-al** *pertaining to* **poplit/e-** *ham, back of knee* fossa, Latin *trench, ditch*	The hollow at the back of the knee.
quadriceps femoris	**KWAD**-rih-seps **FEM**-or-is	P/ R/ S/ R/	**quadri-** *four* **-ceps** *head* **-is** *belonging to, pertaining to* **femor-** *femur*	An anterior thigh muscle with four heads (origins).
tibia	**TIB**-ee-ah		Latin *large shinbone*	The larger bone of the lower leg.
tibial (adj)	**TIB**-ee-al	S/ R/	**-al** *pertaining to* **tibi-** *tibia*	Pertaining to the tibia.

Abbreviations

ACL	anterior cruciate ligament
PCL	posterior cruciate ligament

EXERCISES

Spelling: *A medical term must be spelled correctly. Work with a fellow student on this exercise. Cover the WAD box with a sheet of paper while your partner dictates any 7 terms in the above WAD box to you. Write them in the "Test" column, and do your best to spell them correctly. When you are finished, remove the cover and check your spelling. If you have made any errors, rewrite the correct spelling in the "Corrections" column. After you have corrected your exercise, dictate the remaining 8 terms to your partner.*

Test	Corrections
1. ____________	____________
2. ____________	____________
3. ____________	____________
4. ____________	____________
5. ____________	____________
6. ____________	____________
7. ____________	____________

Case Report 9.2 (continued)

Gail Griffith, a 17-year-old, landed awkwardly after jumping for a ball during a basketball game. "My knee kinda popped as I landed." She had to be assisted off the court. In the Emergency Department, the knee was swollen and unstable. An MRI showed a partial tear of the medial collateral ligament, a complete **rupture** of the anterior cruciate ligament, and a partial tear of the medial meniscus. Gail decided to have surgery, even though full recovery would take 6 months to 1 year of rehabilitation.

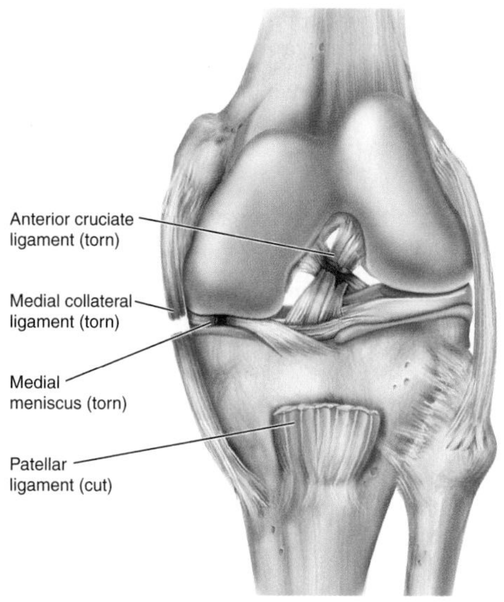

▲ **FIGURE 9.16 Gail Griffith's Knee Injuries.**

THE KNEE JOINT

Injuries to the Knee Joint

The anterior cruciate ligament (ACL) is the most commonly injured ligament in the knee (*Figure 9.16*), particularly in football players and female athletes. The injury is often caused by a sudden **hyperflexion** of the knee joint when landing awkwardly on flat ground, as in Gail's injury. Because of its poor vascular (blood) supply, once torn the ligament does not heal and has to be stitched back together surgically.

Other major ligaments that are commonly injured are the medial and lateral collateral ligaments and the posterior cruciate ligament.

Meniscus injuries result from a twist to the knee. Pain and locking are the result of the torn meniscus flipping in and out of the joint as it moves. Because loss of a meniscus leads to arthritic changes, repair of the meniscus, as in Gail's case, rather than removal is preferred. Removal of a meniscus is a **meniscectomy.**

Patellar problems produce pain that is noticed particularly when descending stairs. The force on the patella when descending stairs is about seven times body weight compared to about two times body weight when ascending stairs.

Patellar subluxation or dislocation produces an unstable, painful kneecap.

Prepatellar bursitis ("housemaid's knee") produces painful swelling over the **bursa** at the front of the knee and is seen in people who kneel for extended periods of time, such as carpet layers.

Tendinitis of the patellar tendon is produced by overuse during such activities as cycling, running, or dancing. Pain is felt where the tendon is inserted into the tibia. It is treated by R.I.C.E (**r**est, **i**ce, **c**ompression, **e**levation).

Surgical Procedures of the Knee Joint

Arthrocentesis, aspiration of fluid from the knee joint, is used to establish a diagnosis by laboratory examination of the fluid, drain off infected fluid, or insert medication such as local corticosteroids.

Arthrography is an x-ray of a joint after injection of contrast medium into the joint to make the inside details of the joint visible.

Diagnostic arthroscopy is an exploratory procedure performed using an arthroscope to examine the internal compartments of the knee joint.

Surgical arthroscopy, performed through an arthroscope, can be a **debridement** or removal of torn tissue such as a meniscus or a ligament. It can also be repair of a torn ligament by suturing or tendon autograft or repair of a torn meniscus.

Arthroplasty involves a total replacement of the knee joint (*Figure 9.17*), usually because of osteoarthritis of the joint. The damaged cartilage and bone from the surfaces of the knee joint are removed and replaced with metal and plastic.

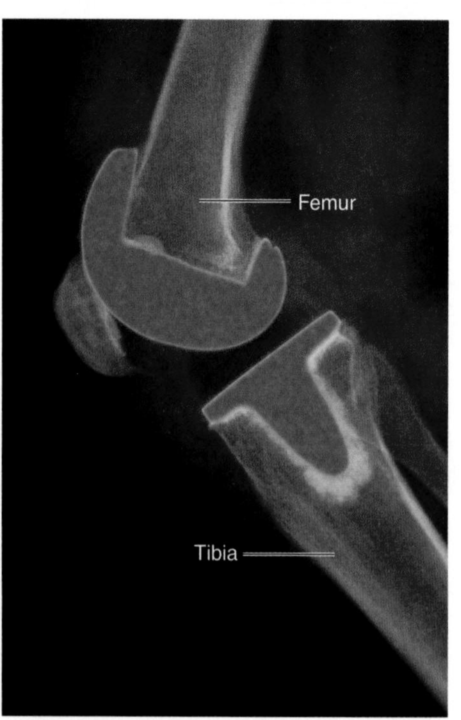

◀ **FIGURE 9.17 Total Knee Replacement.** Colored x-ray of total knee replacement, left knee.

WORD ANALYSIS AND DEFINITION

S/ = Suffix P/ = Prefix R/ = Root R/CF = Combining Form

WORD	PRONUNCIATION		ELEMENTS	DEFINITION
arthrocentesis	**AR**-throw-sen-**TEE**-sis	S/ R/CF	**-centesis** *puncture* **arthr/o-** *joint*	Aspiration of fluid from a joint.
arthrography	ar-**THROG**-rah-fee	S/ R/CF	**-graphy** *process of recording* **arthr/o-** *joint*	X-ray of a joint taken after the injection of a contrast medium into the joint.
bursa **bursitis**	**BURR**-sah burr-**SIGH**-tis	 S/ R/	Latin *purse* **-itis** *inflammation* **burs-** *bursa*	A closed sac containing synovial fluid. Inflammation of a bursa.
debridement	day-**BREED**-mon ("Mon" is the French pronunciation of *ment.*)	S/ P/ R/	**-ment** *action* **de-** *removal, out of* **-bride-** *rubble, rubbish*	The removal of injured or necrotic tissue.
hyperflexion	high-per-**FLEK**-shun	S/ P/ R/	**-ion** *action, condition* **hyper-** *excessive* **-flex-** *bend*	Flexion of a limb or part beyond the normal limits.
meniscectomy	men-ih-**SEK**-toh-me	S/ R/	**-ectomy** *surgical excision* **menisc-** *crescent, meniscus*	Excision (cutting out) of all or part of a meniscus.
prepatellar	pree-pah-**TELL**-ar	S/ P/ R/	**-ar** *pertaining to* **pre-** *before, in front of* **-patell-** *patella*	In front of the patella.
rupture	**RUP**-tyur		Latin *break, fracture*	Break or tear of any organ or body part.
tendinitis (also spelled *tendonitis*)	ten-dih-**NYE**-tis	S/ R/	**-itis** *inflammation* **tendin-** *tendon*	Inflammation of a tendon.

EXERCISES

Elements: *Recognition of word elements will help you understand a medical term. For each of the following terms, identify the type of the element shown in bold italic, and then define that element. Fill in the chart, and answer the questions below it.*

Remember: An element that begins a term is not necessarily a prefix!

Medical Term	Type of Element (P, R/CF, or S)	Meaning of Element
tendin***itis***		
prepatellar		
de***bride***ment		
bursitis		
arthro***centesis***		
hyperflexion		
meniscectomy		

1. List every term that is a surgical procedure:

____

2. List every term that could be a diagnosis:

____

3. Which term contains a prefix that describes where something is located?

____

BONES, MUSCLES, AND JOINTS OF THE LOWER LEG, ANKLE, AND FOOT

The two bones of the lower leg are the larger and medial tibia and the thinner and lateral fibula. The lower end of the tibia on its medial border forms a prominent process called the medial malleolus. The lower end of the fibula forms the lateral malleolus (*Figure 9.18a*). You can palpate (feel) both these prominences at your own ankle.

The muscles of the lower leg move the ankle, foot, and toes. Those on the front of the leg are in a compartment between the tibia and the fibula (*Figure 9.18b*). They dorsiflex the foot at the ankle and extend the toes. Those on the lateral side of the leg **evert** (turn outward) the foot.

Those on the back of the leg plantar-flex the foot at the ankle, flex the toes, and **invert** (turn inward) the foot. The **gastrocnemius** muscle forms a large part of the calf. The distal end of it joins with the tendon of the **soleus** muscle to form the **Achilles (calcaneal) tendon,** which is attached to the heel bone **(calcaneus)** (*Figure 9.18c*). The gastrocnemius muscle and the Achilles tendon enable you to "push off" and start running or jumping. (The terms describing movements of the foot are detailed in *Chapter 2*.)

The ankle has two joints:

- One between the lateral malleolus of the fibula and the talus.
- One between the medial malleolus of the tibia and the talus.

The **talus** is the most superior of the seven **tarsal** bones of the ankle and proximal foot (*Figure 9.18a*), and its upper surface articulates with the tibia. The tarsal bones help the ankle bear the body's weight. Strong ligaments on both sides of the ankle joint hold the joint together.

Attached to the tarsal bones are the five parallel **metatarsal** bones that form the instep and then fan out to form the ball of the foot, where they bear weight. Each toe has three phalanges, except for the big toe, which has only two. This is identical to the thumb and its relation to the hand. The tendons of the leg muscles are inserted into the phalanges (*Figure 9.18a*).

Disorders and Injuries of the Ankle and Foot

Podiatry is a health care specialty concerned with the diagnosis and treatment of disorders and injuries of the foot and toenails. A **podiatrist** is not an MD.

Bunions occur usually at the base of the big toe and are swellings of the bones that cause the metatarsophalangeal joint to be misaligned and stick out medially. This deformity is called **hallux valgus.**

Strains and **sprains** are more common in the ankle than in any other joint in the body. A strain is an acute injury resulting from overstretching or overcontraction of a muscle or tendon. A sprain is the result of an abnormal stretch or tear of a ligament. Some severe sprains with tearing of the ligament may require surgical repair.

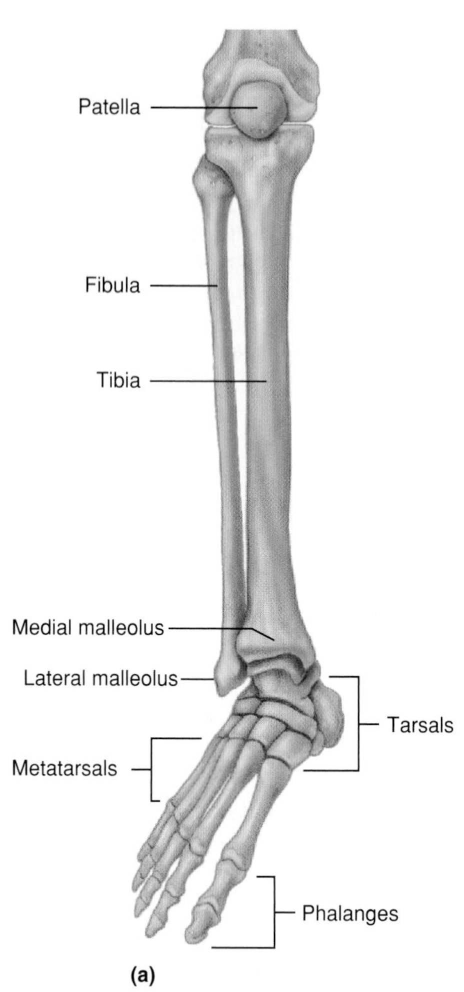

Patella
Patellar ligament
Tibia
Fibularis longus
Gastrocnemius
Soleus
Fibularis brevis
Tibialis anterior
Extensor digitorum longus
Extensor retinacula
(b)

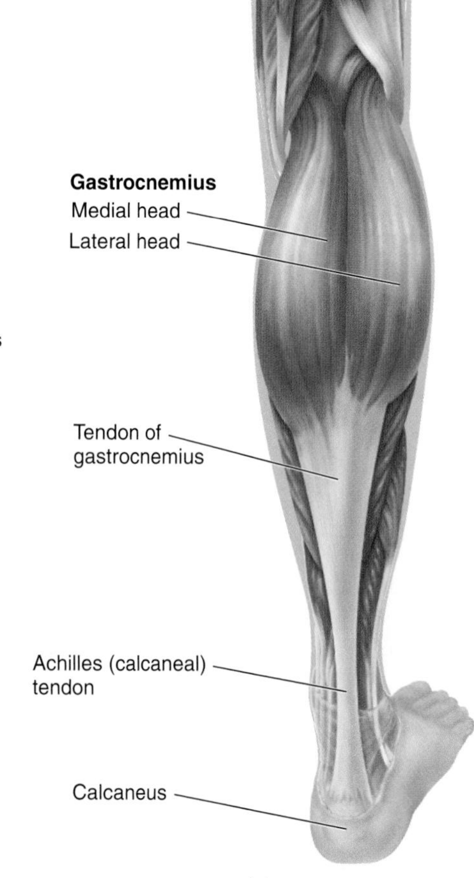

▲ **FIGURE 9.18 Bones and Muscles of Lower Leg and Foot.** (a) Bones of the lower leg and foot. (b) Muscles of the front of the right leg. (c) Muscles of the back of the right leg.

WORD ANALYSIS AND DEFINITION

S/ = Suffix P/ = Prefix R/ = Root R/CF = Combining Form

WORD	PRONUNCIATION		ELEMENTS	DEFINITION
bunion	**BUN**-yun		French *bump*	A swelling at the base of the big toe.
calcaneus **calcaneal (adj)**	kal-**KAY**-knee-us kal-**KAY**-knee-al	 S/ R/	Latin *the heel* **-eal** *pertaining to* **calcan-** *calcaneus*	Bone of the tarsus that forms the heel. Pertaining to the calcaneus.
calcaneal tendon (same as **Achilles tendon**)	ah-**KILL**-eeze		mythical Greek warrior	A tendon formed from gastrocnemius and soleus muscles and inserted into the calcaneus.
fascia **fasciitis** (note spelling)	**FASH**-ee-ah fash-ee-**I**-tis	 S/ R/	Latin *a band* **-itis** *inflammation* **fasci-** *fascia*	Sheet of fibrous connective tissue. Inflammation of the fascia.
gastrocnemius	gas-trok-**KNEE**-me-us	S/ R/	**-ius** *pertaining to* **gastrocnem-** *calf of leg*	Major muscle in back of the lower leg (the calf).
hallux valgus	**HAL**-uks **VAL**-gus	R/ R/	**hallux** *big toe* **valgus** *turn out*	Deviation of the big toe toward the lateral side of the foot.
metatarsus **metatarsal (adj)**	**MET**-ah-**TAR**-sus **MET**-ah-**TAR**-sal	S/ P/ R/ S/	**-us** *pertaining to* **meta-** *after, subsequent to* **-tars-** *ankle* **-al** *pertaining to*	The five parallel bones of the foot between the tarsus and the phalanges. Pertaining to the metatarsus.
podiatry **podiatrist**	poh-**DIE**-ah-tree poh-**DIE**-ah-trist	S/ R/ S/	**-iatry** *treatment* **pod-** *foot* **-iatrist** *practitioner*	The diagnosis and treatment of disorders and injuries of the foot. Practitioner of podiatry.
Pott fracture	POT **FRAK**-shur		Percival Pott, London surgeon, 1714–1788	Fracture of the lower end of the fibula, often with fracture of the tibial malleolus.
sprain **strain**	SPRAIN STRAIN		origin unknown Latin *to bind*	A wrench or tear in a ligament. Overstretch or tear in a muscle or tendon.
talus	**TAY**-luss		Latin *heel bone*	The tarsal bone that articulates with the tibia to form the ankle joint.
tarsus **tarsal (adj)**	**TAR**-sus **TAR**-sal	 S/ R/	Latin *ankle* **-al** *pertaining to* **tars-** *ankle*	The collection of seven bones in the foot that form the ankle and instep. Pertaining to the tarsus.

Pott fracture is a fracture of the fibula near the ankle, often accompanied by a fracture of the medial malleolus of the tibia.

Achilles tendinitis is a small stretch injury that causes the tendon to become swollen and painful. A complete tear needs surgical repair.

Plantar fasciitis is tearing of the sheet of fascia that supports the arch of the foot.

EXERCISES

Identify *the medical term, based on the brief meaning given to you. Be sure to check your spelling when you are finished! Fill in the blanks.*

1. swelling at the base of the big toe __________
2. another name for the calcaneal tendon __________
3. sheet of fibrous connective tissue __________
4. bone of the tarsus that forms the heel __________
5. wrench or tear in a ligament __________

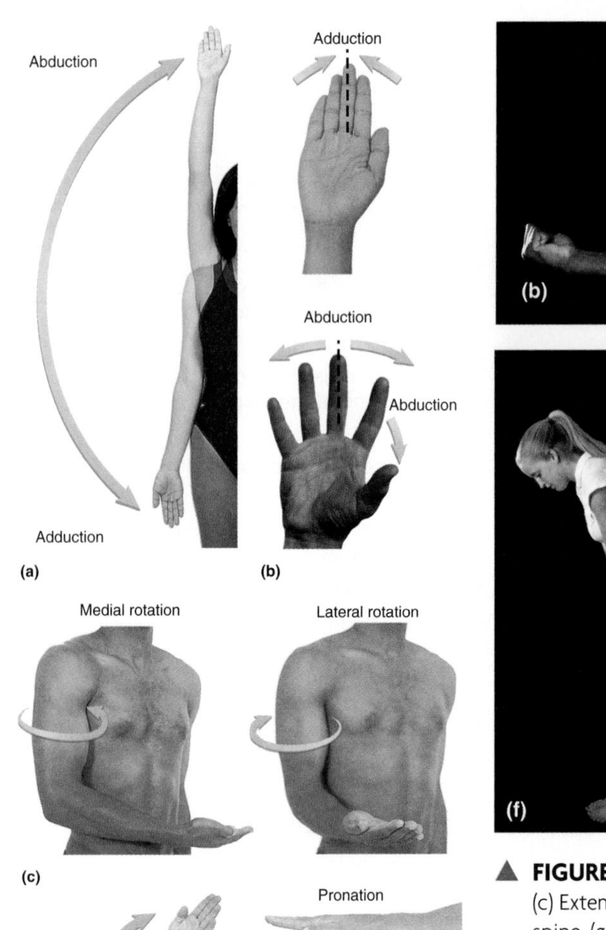

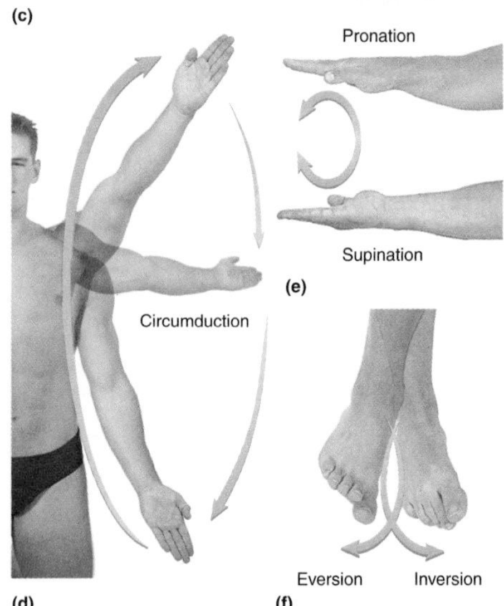

FIGURE 9.20 (a) Abduction and adduction of the upper limb. (b) Abduction and adduction of the fingers. (c) Medial and lateral rotation of the arm. (d) Circumduction. (e) Pronation and supination of the Hand. (f) Eversion and inversion of the Foot.

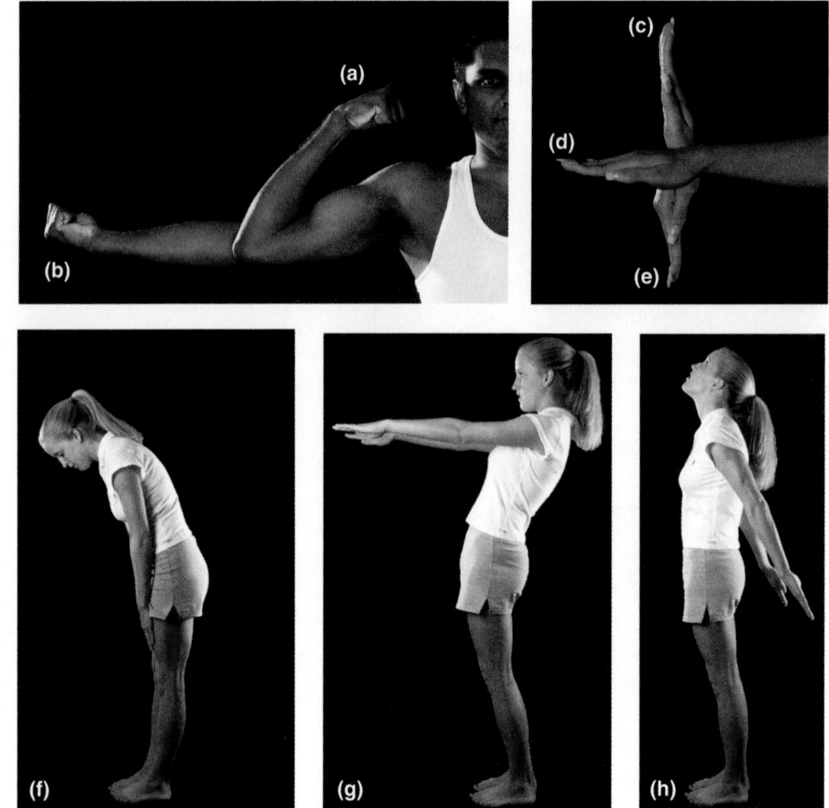

FIGURE 9.19 Joint Flexion and Extension. (a) Flexion of the elbow. (b) Extension of the elbow. (c) Extension of the wrist. (d) Neutral position of the wrist. (e) Flexion of the wrist. (f) Flexion of the spine. (g) Flexion of the shoulder. (h) Extension of the shoulder.

JOINT MOVEMENT

Flexion and Extension of Joints

Flexion (bending) and **extension** (straightening) are shown in the elbow joint (*Figure 9.19a and b*), in the wrist joint (*Figure 9.19c, d, and e*), in the spine (*Figure 9.19f*), and in the shoulder joint (*Figure 9.19g and h*).

For most of the rest of the body, flexion is movement of a body part anterior to the coronal plane (*Chapter 2*). Extension is movement posterior to the coronal plane (*Figure 9.19g*). For example, when you bend your trunk forward, that is flexion (*Figure 9.19f*). When you bend your trunk backward, that is extension (*Figure 9.19g*). When you bend your trunk sideways to the right or left, that is called lateral flexion.

Abduction and Adduction of Joints

Abduction is movement away from the midline. **Adduction** is movement toward the midline. Abduction of your arm is moving it sideways away from your trunk; adduction is bringing it back to the side of your trunk (*Figure 9.20a*). Abduction of your fingers is spreading them apart, away from the middle finger; adduction is bringing them back together (*Figure 9.20b*).

Rotation of Joints

Rotation means ***to turn around an axis.*** Medial rotation of the upper-arm bone, the humerus, with the elbow flexed brings the palm of the hand toward the body. Lateral rotation moves the palm away from the body (*Figure 9.20c*).

Pronation and Supination

When you lie flat on the ground face down on your belly with your palms touching the ground, you are **prone.** When you lie flat on your back with your spine on the floor and your palms facing up, you are **supine.**

WORD ANALYSIS AND DEFINITION

S/ = Suffix P/ = Prefix R/ = Root R/CF = Combining Form

WORD	PRONUNCIATION		ELEMENTS	DEFINITION
circumduction **circumduct (verb)**	ser-kum-**DUCK**-shun ser-kum-**DUCKT**	S/ P/ R/	**-ion** *action, condition* **circum-***around* **-duct-** *lead*	Movement of an extremity in a circular motion. Move an extremity in a circular motion.
eversion **evert (verb)**	ee-**VER**-shun ee-**VERT**		Latin *overturn*	Turning outward. Turn outward.
inversion **invert (verb)**	in-**VER**-shun in-**VERT**		Latin *to turn about*	Turning inward. Turn inward.
pronation **pronate (verb)** **prone**	pro-**NAY**-shun **PRO**-nate PRONE	S/ R/	**-ion** *action, condition* **pronat-** *bend down* Latin, *lying down*	Process of lying face down or of turning a hand or foot with the volar (palm or sole) surface down. Lying face down, flat on your belly.
supination **supine**	soo-pih-**NAY**-shun soo-**PINE**	S/ R/	**-ion** *action, condition* **supinat-** *bend backward* Latin, *bend backward*	Process of lying face upward or of turning a hand or foot so that the palm or sole is facing up. Lying face up, flat on your spine.

When you rotate your forearm so that your palm faces the floor, that is **pronation.** When you rotate the forearm so that your palm is facing upward, that is **supination** (*Figure 9.20c and e*).

When you turn your ankle so that the sole of your foot faces toward the opposite foot, that is supination or **inversion.** When you turn your ankle so that the sole of the foot faces laterally away from the other foot, that is pronation or **eversion** (*Figure 9.20f*).

Circumduction of Joints

Circumduction of the shoulder is moving it in a circular movement so that it forms a cone with the shoulder joint as the apex of the cone (*Figure 9.20d*).

EXERCISES

Parts of speech: *Some medical terms can act as both a* **verb** *(action) and a* **noun** *(person, place, or thing). Each pair of terms below includes a verb and a noun. Write the correct form of the term* on the line *to complete the sentence. Then* under the line *write the part of speech you have used for the answer (noun or verb).*

1. **circumduct** **circumduction**

   The baseball pitcher was unable to ______________________ his arm to wind up his pitch.

2. **invert** **inversion**

   A clubfoot would be an ______________________ of the foot.

3. **abduct** **abduction**

   Moving away from the midline of the body is called ______________________.

4. **evert** **eversion**

   The patient was unable to ______________________ his ankle due to great pain.

5. **adduct** **adduction**

   The patient was asked to ______________________ his arms from a horizontal plane toward the center of his chest.

THE APPENDICULAR SKELETON, JOINTS, AND MUSCLES

CHALLENGE YOUR KNOWLEDGE

A. **Roots/combining forms:** The meaning of the R/CF element is given in the following chart. Identify the R/CF, and list a medical term containing that element. Define the term as well. Fill in the chart.

Meaning of R/CF	R/CF	Term Containing This R/CF	Definition of the Medical Term
ankle			
arm			
blood vessel			
chest			
dislocate			
foot			
ham			
head			
joint			
side			
to lead			
walking			

B. **Word attack:** Don't jump to conclusions. Follow the instructions below the question.

Question:

Circle the term that means *no blood supply:*

a. hematemesis

b. hemoptysis

c. avascular

d. hemophilia

e. hematuria

1. Read the question and all the answers twice.
2. At first glance, you might discount *avascular* because it is the only answer choice without *hem/hemat* (which means *blood*) as a root. Do not do this before you systematically analyze each term.
3. Consider a, b, d, and e as a group since they all have the same R/CF, which you know means *blood.* The only way to tell them apart is to analyze each suffix. Analyze the suffixes here:

   emesis ______________________________

   ptysis ______________________________

   philia ______________________________

   uria ______________________________

4. What is your conclusion after answering question 3?

   ____

5. Analyze the remaining choice: *avascular.* Write the meaning of the elements under the line:

   ____________/____________/____________

6. The correct answer is ______________________________.

C. **Terminology challenge:** Think back to elements you have learned throughout the book so far. Fill in the blanks.

1. The element *pector* means ______________________________.

   A term using this element is ______________________________.

   Use this term in a sentence of your choice:

   ____

   ____

   ____

2. Another element that means the same thing is ______________________________.

   A term using this element is ______________________________.

   Use this term in a sentence of your choice:

   ____

   ____

   ____

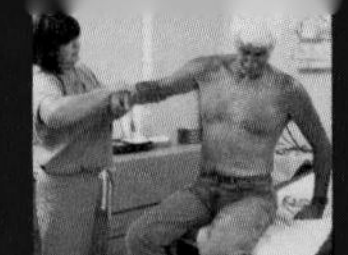

CHAPTER 9 REVIEW

THE APPENDICULAR SKELETON, JOINTS, AND MUSCLES

D. Prefixes: Match the correct medical term in column 1 with the meaning of its prefix in column 2.

	Column 1	Column 2
_____	1. subluxate	A. two
_____	2. hypothenar	B. apart
_____	3. triceps	C. toward
_____	4. adductor	D. after
_____	5. quadriceps	E. without
_____	6. dislocation	F. under
_____	7. avascular	G. three
_____	8. abductor	H. away from
_____	9. biceps	I. four
_____	10. metacarpal	J. below, smaller

Study Hint

To help you focus, first underline or highlight only the prefixes in the terms in column 1. Then read all the answer choices in column 2. Finally, go back and do the matching to the highlighted or underlined prefixes. Fill in the blanks.

E. Patient education: Patients will ask for a layperson's explanation of medical terms. If your patient asks, could you briefly explain in nonmedical language the following terms?

1. shoulder separation:

____

____

____

2. shoulder dislocation:

____

____

____

3. shoulder subluxation:

____

____

____

F. **Look similar/sound similar**—but not the same. Train your eye and ear to know the difference between the terms below.

1. *ilium*

   Definition: ______________________________

2. *ischium*

   Definition: ______________________________

3. *sprain*

   Definition: ______________________________

4. *strain*

   Definition: ______________________________

5. *evert*

   Definition: ______________________________

6. *invert*

   Definition: ______________________________

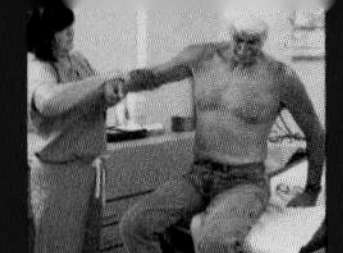

CHAPTER 9 REVIEW

THE APPENDICULAR SKELETON, JOINTS, AND MUSCLES

G. **Recall and review:** This exercise on word elements contains elements from throughout the book. Try to recall the elements without turning back in your book. Indicate the type of element by placing a ✓ in the appropriate column; then write its meaning.

	Type of Element			
Element	**Prefix**	**Root/CF**	**Suffix**	**Meaning of Element**
ary				
chondro				
chiro				
epi				
gen				
ortho				
path				
peri				
plasia				
sarc				

H. **Create** your own test question for this chapter. It must be a "translate into layperson's terms" or "translate into medical terms" type of question. You may need to draft (practice writing) the question several times before you are satisfied with it. Write the draft of your question below. Then write the finished question, and the answer, on a separate piece of paper, and *hand it in to your instructor.*

___

___

___

___

___

Do not forget to write this question and your answer on a separate piece of paper and hand it in to your instructor.

I. The language of orthopedics: Because the body has so many bones, there is an extensive amount of orthopedic vocabulary. Build your knowledge of the language of orthopedics with this exercise. Circle the best answer.

1. The term *dorsum* means:
   front   back   middle   end

2. The medical term for *collarbone* is:
   humerus   ulna   scapula   clavicle

3. An incomplete dislocation is a:
   separation   subluxation   fracture   malunion

4. Small bony spurs in osteoarthritis are:
   Pott   Heberden   rheumatoid   Colles

5. The term *metacarpal* refers to the:
   arm   elbow   forearm   hand

6. The narrowing of a passage is:
   rheumatic   stenosis   tenosynovitis   arthrodesis

7. A bone of a finger *or* toe is a(n):
   carpus   fascia   eminence   phalanx

8. The forearm bone on the thumb side is the:
   radius   ulna   humerus   calcaneus

9. Flesh at the base of the thumb is:
   thenar   metacarpal   hypothenar   phalanges

10. A higher place or part is an:
   emmenince   emminence   eminence   emenince

J. Spelling demons: The following terms from this chapter are particularly difficult to spell and pronounce. Correct pronunciation and spelling of medical terms is the mark of an educated professional. Circle the correct spelling; then check (✔) that you have practiced the pronunciation. (Remember that pronunciations are on the Student Online Learning Center [www.mhhe.com/AllanEssMedLanguage].) WWW

				Pronunciation ✔
1.	menisci	meniskci	menisschi	________
2.	poplitial	popliteal	poplitteal	________
3.	meniskectomy	menissectomy	meniscectomy	________
4.	debridement	debridment	debredment	________
5.	fasceitis	fasscitis	fasciitis	________
6.	humerus	humourous	humorus	________
7.	rhumatism	rumatism	rheumatism	________
8.	sympasis	synphysis	symphysis	________
9.	acetabulum	ascetabulum	acetabullum	________

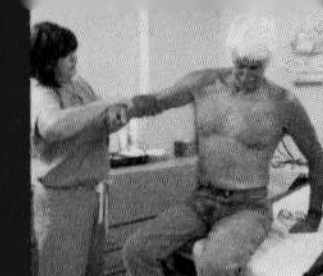

CHAPTER 9 REVIEW

THE APPENDICULAR SKELETON, JOINTS, AND MUSCLES

K. **Seek and find:** Some terms in this book are defined in context and do not always appear in a WAD box. Here is a sample of such terms that have appeared in this chapter. Challenge yourself to answer without turning back to check in the book. *Make sure you pay attention to this type of information, as some of it may be a test question!* Fill in the blanks.

Also Known As:

1. housemaid's knee ______________________
2. thigh bone ______________________
3. to bend ______________________
4. R.I.C.E. ______________________
5. shoulder blades ______________________
6. Achilles tendon ______________________
7. collarbone ______________________
8. to straighten ______________________
9. kneecap ______________________
10. heel bone ______________________

L. **Partner exercise:** Ask your study partner to close his or her text. Dictate the following sentences to your partner so that he or she can write them down and then hand them back to you. Check your partner's sentences against the text below. The sentences are not correct unless every word is present and spelled correctly. When you have finished checking your partner's answers, switch places—your partner dictates, and you write.

1. On the front of the arm, a group of three muscles—(1) biceps brachii, (2) brachialis, and (3) brachioradialis—flexes the forearm at the elbow joint and rotates the forearm and hand laterally in supination.
2. The muscles of the forearm supinate and pronate the forearm, flex and extend the wrist joint and hand, and move the hand medially and laterally.
3. The acromion is attached to the clavicle at the acromioclavicular joint, which provides a connection between the axial skeleton, the pectoral girdle, and the arm.

M. **Translate** the following sentences into language your patient can understand. Rewrite the sentences below.

1. The talus is the most superior of the seven tarsal bones of the ankle and proximal foot, and its upper surface articulates with the tibia.

____

____

____

____

2. Flexion is movement of a body part anterior to the coronal plane.

____

____

____

3. The pelvic girdle is the two hip bones that articulate anteriorly with each other at the symphysis pubis and posteriorly with the sacrum to form the bowl-shaped pelvis.

____

____

____

N. **Greek and Latin:** Many medical terms come directly from Greek or Latin. Test your knowledge of these terms with this exercise. Fill in the chart with a brief definition of each term.

Medical Term	**Definition**
acetabulum	
carpus	
diastasis	
eminence	
ganglion	
gluteus	
humerus	
ilium	
ischium	
pelvis	
scapula	
symphysis	
ulna	
cyst	

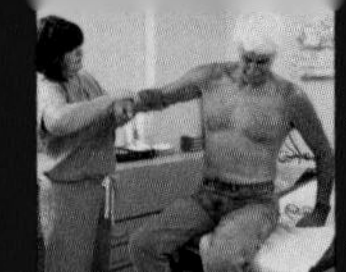

CHAPTER 9 REVIEW

THE APPENDICULAR SKELETON, JOINTS, AND MUSCLES

O. Suffixes: The beauty of suffixes, especially surgical suffixes, is that they can be applied to various body system roots. The root *arthr* and R/CF *arthr/o* appear in many orthopedic terms. Build the terms using *arthr/arthro* and the suffixes listed below. There are more suffixes than you will need.

desis	plasty	scopy	scope	graphy
centesis	itis	osis	itic	ary

Patient is scheduled for:

1. x-ray of a joint after contrast medium injection:

   ______________________ / ______________________

2. withdrawal of fluid from the joint with a needle:

   ______________________ / ______________________

3. surgery to restore/repair joint function:

   ______________________ / ______________________

4. surgical fixation of the joint:

   ______________________ / ______________________

5. visual examination of the interior of the joint:

   ______________________ / ______________________

6. The procedure in question 5 will use an

   ______________________ / ______________________.

7. The diagnosis for all these patients could be

   ______________________ / ______________________.

8. The diagnosis in question 7 would cause the patient to have:

   ______________________ / ______________________ joints.

P. **Proofread** the following sentences for errors in fact and/or spelling. Rewrite each sentence with the corrections made. There is only one sentence that is totally correct.

1. The two glueteus muscles (maximus and minimus) are powerful muscles that support the knee joint.

2. Dislocations of the neck of the femur occur as a result of a fall, most commonly in women with osteopenia.

3. Arthodesis, the aspiration of fluid from the hip joint and replacement of the fluid with a steroid solution, is performed frequently for osteoarthritis.

4. When you rotate your forearm so that your palm faces the floor, that is suppination.

5. The bones of the pectoral girdle are the scapulum and the claviccle.

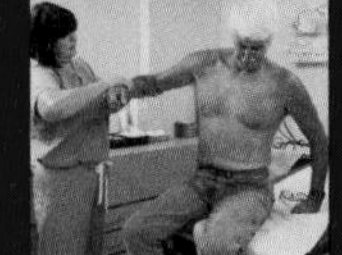

CHAPTER 9 REVIEW

THE APPENDICULAR SKELETON, JOINTS, AND MUSCLES

Q. Chapter challenge: Circle the correct answer.

> ***Study Hint***
> Immediately cross off any answer you know is not correct. In your remaining choices, there is only *one best answer.*

1. The correct pair of singular and plural terms is:

   a. scapula — scapulum
   b. clavical — clavicals
   c. phalanx — phalanges
   d. synphysis — synpheses
   e. patela — patellae

2. Which of the following is *not* a bone?

   a. humerus
   b. brachioradialis
   c. ulna
   d. radius
   e. metacarpal

3. Which of the following is a true statement about the terms *bursitis, tenosynovitis, fasciitis,* and *tendonitis?*

   a. They all involve inflammation.
   b. None of them involves bone.
   c. Every one can be a diagnosis.
   d. Only a and c are true.
   e. Answers a, b, and c are all true.

4. Which group of terms are the *hamstring muscles?*

   a. biceps, triceps, quadriceps
   b. biceps femoris, semimembranosus, semitendinosus
   c. deltoid, pectoralis major, latissimus dorsi
   d. gastrocnemius, soleus, tarsus
   e. biceps brachii, brachialis, brachioradialis

5. Which abbreviation stands for a disease condition?

   a. SI
   b. MRI
   c. CMA
   d. RA
   e. AC

6. What comes together at an *articulation?*

   a. bone and muscle
   b. muscle and tendon
   c. bone and bone
   d. bone and cartilage
   e. tendon and ligament

7. Four muscles that originate on the scapula wrap around the shoulder joint and fuse to form one large tendon called the:

   a. gastrocnemius
   b. rotator cuff
   c. clavicle
   d. gluteus maximus
   e. humerus

8. What do these terms all have in common: *brachialis, deltoid, latissimus,* and *dorsi?*

a. They are all tendons.
b. They are all bones.
c. They are all muscles.
d. They are all ligaments.
e. They have nothing in common.

9. Choose the pair that has both terms spelled correctly:

a. fasciitis arthritis
b. ruhmatoid rheumatic
c. phalanx acetabum
d. arthrodisis diastasis
e. sacroilliac ischiel

10. Based on the surgical suffix in the term *arthroplasty,* what is being done?

a. incision into
b. removal of
c. repair of
d. puncture of
e. fixation of

11. Which two terms relate to muscles?

a. prone supine
b. origin insertion
c. carpal metacarpal
d. thenar hypothenar
e. ilium ischium

12. Where is the *popliteal fossa* located?

a. in the crook of the elbow
b. at the back of the neck
c. at the base of the spine
d. at the back of the knee
e. in the pelvic girdle

13. Which two terms contain elements that mean a number:

a. anterior posterior
b. thenar hypothenar
c. biceps triceps
d. metacarpal carpal
e. maximum medius

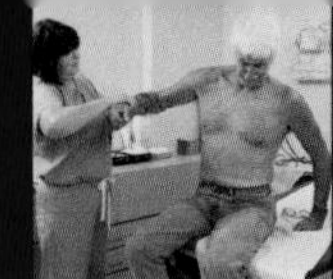

THE APPENDICULAR SKELETON, JOINTS, AND MUSCLES

R. **Case Report challenge:** Now that you are more comfortable with the terms in this chapter, you can apply that knowledge and briefly answer the questions about the Case Report. If you read the report through and underline all the medical terminology, this will make it easier to answer the questions.

Operative Report. Fulwood Medical Center.

Gail Griffith, aged 17.

Preoperative Diagnosis: Traumatic ACL tear and tear of medial *meniscus* right knee.

Postoperative Diagnosis: Same.

Procedure Performed: **Arthroscopy,** repair of medial *collateral* ligament, ACL reconstruction, repair of torn medial meniscus, right knee.

Operative Findings: An avulsed anterior **cruciate** ligament (ACL) of the femur with a tear of the posterior horn of the medial meniscus and tear of medial collateral ligament.

1. Find the two pairs of opposite terms in the above report:

   (a) ______________________ means ______________________

   ______________________ means ______________________

   (b) ______________________ means ______________________

   ______________________ means ______________________

2. What type of scope would be used in this procedure?

   ____

3. Use the Glossary or a dictionary to define these two terms:

   (a) *avulsion:*

   ____

   ____

   (b) *traumatic:*

   ____

   ____

4. *Medial* describes a location on the body—where is it?

____

____

____

5. What medical term is the opposite of *medial*?

Term: ______________________________

Definition: ______________________________________

____

____

6. What is the function of a *meniscus*?

____

____

____

7. What is the plural of the term above? ______________________________

8. What is the *femur*? ______________________________________

____

____

Endocrine System

The Essentials of the Language of Endocrinology

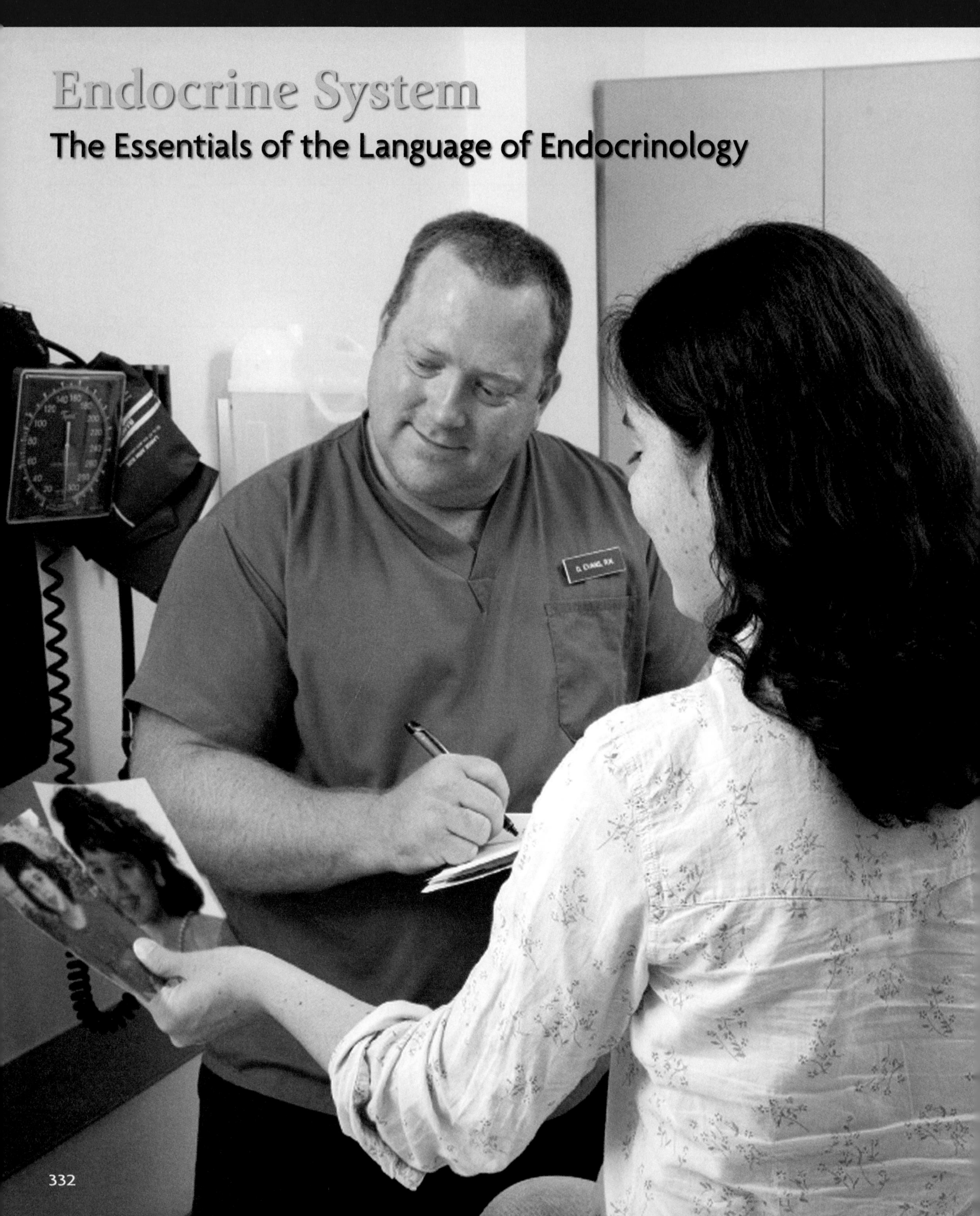

CHAPTER 10

CASE REPORT 10.1

You are

. . . a registered nurse working with endocrinologist Sabina Khalid, MD, in the Endocrinology Clinic at Fulwood Medical Center.

You are communicating with

. . . Mrs. Gina Tacher, a 33-year-old schoolteacher. She complains of coarsening of her facial features and enlargement of the bones of her hands. Over the past 10 years her nose and jaw have increased in size and her voice has become husky. She has brought photos of herself at ages 9 and 16 and as she is now. She has no other health problems.

Keynote

A hormone is secreted by an endocrine gland or cell and is carried by the bloodstream to act at distant target sites.

Learning Outcomes

The **endocrine** system is a communication system. The **hormones** produced by the system are blood-borne messengers secreted by endocrine glands; they circulate in the bloodstream, gaining access to all other cells of the body. They are distributed anywhere that blood goes, but they affect only target cells that have receptors for them. The hormones have intracellular effects to alter the metabolism of the target cells. The information in this chapter enables you to:

- 10.1 Apply the language of endocrinology to the structures and functions of the endocrine system.
- 10.2 Comprehend, analyze, spell, and write the medical terms of endocrinology.
- 10.3 Recognize and pronounce the medical terms of endocrinology.
- 10.4 Use medical terminology to explain the effects of common endocrine disorders on health.

LESSON 10.1 ENDOCRINE SYSTEM

Endocrine System, Hypothalamus, and Pituitary and Pineal Glands

OBJECTIVES

The endocrine system is a network of ductless glands that secrete **hormones** and discharge them directly into the bloodstream. The information in this lesson will enable you to:

- **Name the glands that make up the endocrine system.**
- **List the hormones produced by the hypothalamus and pituitary gland.**
- **Identify medical terminology for the control that the hypothalamic and pituitary hormones exert over other endocrine glands.**
- **Specify the roles of the pineal gland.**
- **Use correct medical terminology to describe disorders of the hypothalamus, pituitary gland, and pineal gland.**

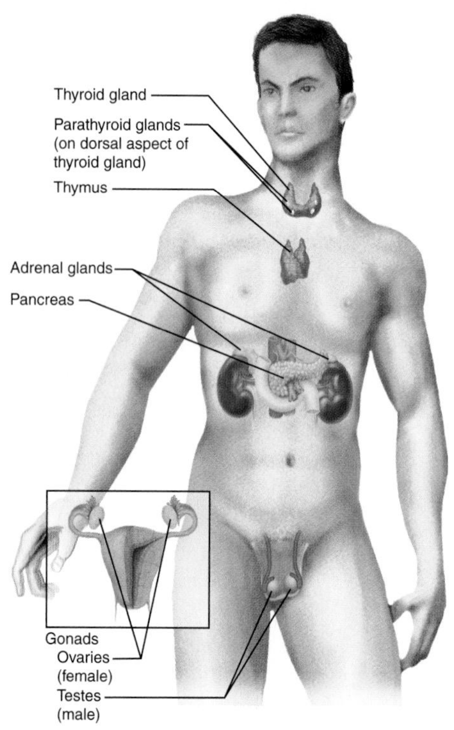

▲ **FIGURE 10.1 Major Endocrine Glands.**

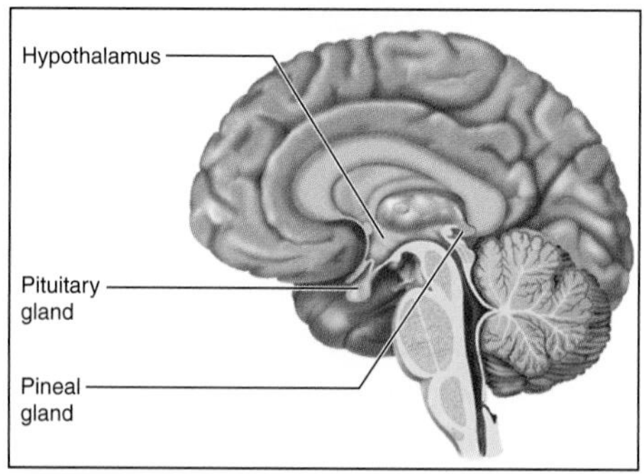

▲ **FIGURE 10.2 Hypothalamus, Pituitary Gland, and Pineal Gland.**

Abbreviations

ADH	antidiuretic hormone
SAD	seasonal affective disorder

THE ENDOCRINE SYSTEM

The endocrine system comprises several major organs (*Figures 10.1 and 10.2*):

- **Pituitary gland (1)** and the nearby **hypothalamus (1).**
- **Pineal gland (1).**
- **Thyroid gland (1).**
- **Parathyroid glands (4).**
- **Thymus gland (1).**
- **Adrenal glands (2).**
- **Pancreas (1).**
- **Gonads: testes (2) in the male; ovaries (2) in the female** (The male gonads are discussed in *Chapter 13* and the female in *Chapter 14.*)

In addition, endocrine cells found in tissues all over the body secrete hormones. For example:

- **Cells in the upper GI tract** secrete hormones that include gastrin, which stimulates gastric secretions.
- **Cells in the kidney** secrete erythropoietin, which stimulates erythrocyte production.
- **Fat cells** secrete leptin, which helps suppress appetite. Lack of leptin can lead to overeating and obesity.
- **Cells in tissues throughout the body** secrete prostaglandins, which act locally to dilate blood vessels, relax airways, stimulate uterine contractions in menstrual cramps or labor, and lower acid secretion in the stomach. When tissues are injured, prostaglandins promote an inflammatory response.

Hypothalamus

The hypothalamus (*Figure 10.2*) forms the floor and walls of the brain's third ventricle (*Chapter 12*) and produces eight hormones. Six of them are local hormones that regulate the production of hormones by the anterior pituitary gland. Two of them, **oxytocin** and **antidiuretic hormone (ADH),** are transported to the posterior pituitary, where they are stored until they are needed elsewhere in the body.

Pineal Gland

The pineal gland is located on the roof of the third ventricle of the brain, posterior to the hypothalamus. It secretes **serotonin** by day and converts it to **melatonin** at night. The gland reaches its maximum size in childhood and may regulate the timing of puberty. It may also play a role in **seasonal affective disorder (SAD)**, in which people are depressed in the dark days of winter.

WORD ANALYSIS AND DEFINITION

S/ = Suffix P/ = Prefix R/ = Root R/CF = Combining Form

WORD	PRONUNCIATION		ELEMENTS	DEFINITION
antidiuretic (***Note:*** This term has two prefixes.)	**AN**-tih-die-you-**RET**-ik	S/ P/ P/ R/	**-ic** *pertaining to* **anti-** *against* **-di-** *complete* **-uret-** *urination*	An agent that decreases urine production.
endocrine	**EN**-doh-krin	P/ R/	**endo-** *within* **-crine** *secrete*	A gland that produces an internal or hormonal secretion.
endocrinology (***Note:*** The "e" in *crine* changes to "o" for better flow.)	**EN**-doh-krih-**NOL**-oh-jee	S/	**-logy** *study of*	Medical specialty concerned with the production and effects of hormones.
endocrinologist	**EN**-doh-krih-**NOL**-oh-jist	S/	**-logist** *one who studies, specialist*	A medical specialist in endocrinology.
hormone	**HOR**-mohn		Greek *to set in motion*	Chemical formed in one tissue or organ and carried by the bloodstream to stimulate or inhibit a function of another tissue or organ.
hormonal (adj)	hor-**MOHN**-al	S/ R/	**-al** *pertaining to* **hormon-** *hormone*	Pertaining to hormones.
hypothalamus	high-poh-**THAL**-ah-muss	P/ R/	**hypo-** *below* **-thalamus**	An endocrine gland in the floor and wall of the third ventricle of the brain.
hypothalamic (adj)	high-poh-thah-**LAM**-ik	S/ R/	**-ic** *pertaining to* **-thalam-** *thalamus*	Pertaining to the hypothalamus.
melatonin	mel-ah-**TONE**-in	S/ R/ R/	**-in** *substance* **mela-** *black* **-ton-** *tension, pressure*	Hormone formed by the pineal gland.
oxytocin	**OCK**-see-toe-sin	S/ R/ R/	**-in** *substance* **oxy-** oxygen **-toc-** *labor and childbirth*	Pituitary hormone that stimulates the uterus to contract.
pineal	**PIN**-ee-al		Latin *like a pine cone*	Pertaining to the pineal gland.
pituitary	pih-**TYU**-ih-tary	S/ R/	**-ary** *pertaining to* **pituit-** *pituitary*	Pertaining to the pituitary gland.
serotonin	ser-oh-**TOE**-nin	S/ R/CF R/	**-in** *substance* **ser/o-** *serum* **-ton-** *tension, pressure*	Neurotransmitter in central and peripheral nervous systems.

EXERCISES

Elements *remain your best tool for understanding medical terms. The elements are listed in column 1. Identify the type of element in column 2, its meaning in column 3, and an example of a term containing that element in column 4. Fill in the chart.*

Element	Type of Element (P, R, CF, or S)	Meaning of Element	Medical Term Containing This Element
anti			
di			
endo			
hypo			
logist			
mela			
oxy			
sero			
toc			
uret			

1. Choose any term from column 4 and use it in a sentence of your choice that is not a definition:

____

____

____

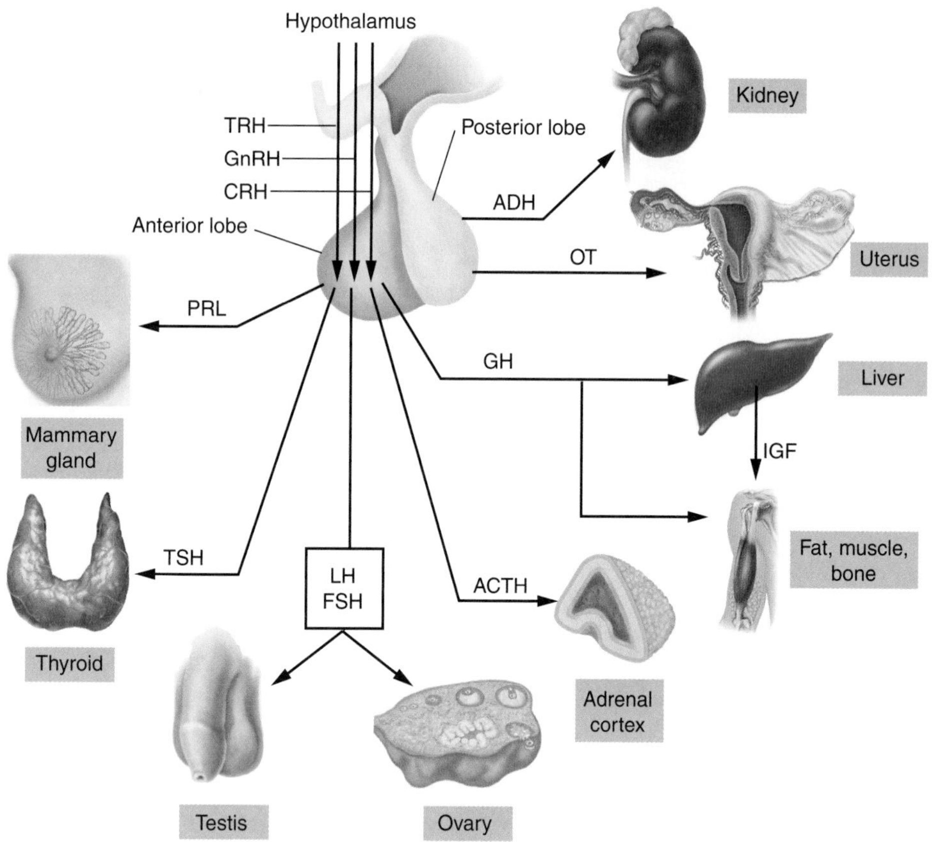

▲ **FIGURE 10.3 Hormones of the Pituitary Gland and Their Target Organs.**

PITUITARY GLAND

Each hormone plays its part in maintaining homeostasis, but the pituitary gland and the hypothalamus work together and often influence the production of hormones in the other endocrine glands. The pituitary gland is suspended from the hypothalamus (*Figure 10.3*). The pituitary gland has two components:

- A large anterior lobe
- A smaller, posterior lobe

Anterior-lobe hormones are six in number:

1. **Follicle-stimulating hormone (FSH)** stimulates target cells in the ovaries to develop eggs and stimulates sperm production in the testes.
2. **Luteinizing hormone (LH)** stimulates ovulation and the formation of a corpus luteum in the ovary (*Chapter 14*) to secrete estrogen and progesterone. In the male, LH stimulates production of testosterone (*Chapter 13*).
3. **Thyroid-stimulating hormone (TSH)**, or **thyrotropin,** stimulates the growth of the thyroid gland and the production of thyroxine.
4. **Adrenocorticotropic hormone (ACTH)**, or **corticotropin,** stimulates the adrenal glands to produce hormones called **corticosteroids.**
5. **Prolactin (PRL)** stimulates the mammary glands after pregnancy to produce milk. In the male it sensitizes the testes to LH to enhance production of testosterone.
6. **Growth hormone (GH)**, or **somatotropin,** is produced in quantities at least a thousand times as great as any other pituitary hormone. It stimulates cells to enlarge and divide.

Tropic hormones are hormones that stimulate other endocrine glands to produce their hormones. FSH and LH are called **gonadotropins** because they stimulate **gonadal** functions.

Posterior-lobe hormones are produced by nuclei in the hypothalamus and stored and released in the pituitary posterior lobe (*Figure 10.4*). There are two types:

1. **Oxytocin (OT)** in childbirth stimulates uterine contractions and in lactation forces milk to flow down ducts to the nipple. In both sexes, its production increases during sexual intercourse to help give the feelings of satisfaction and emotional bonding.
2. **Antidiuretic hormone (ADH)** reduces the volume of urine produced by the kidneys. It is also called **vasopressin.**

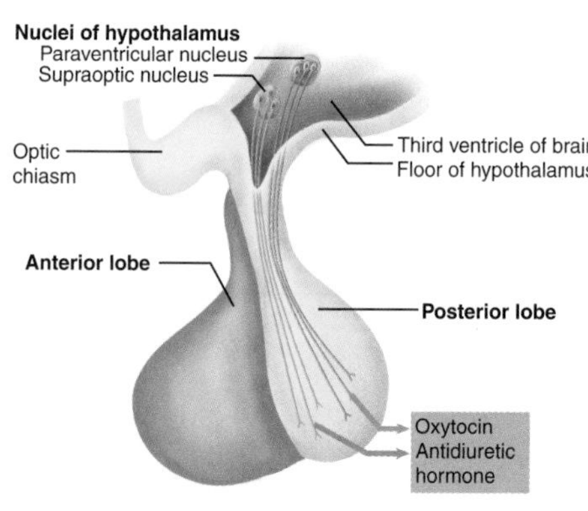

▲ **FIGURE 10.4 Hormones of the Posterior Lobe of the Pituitary Gland.**

Abbreviations

ACTH	adrenocorticotropic hormone
DI	diabetes insipidus
FSH	follicle-stimulating hormone
GH	growth hormone
LH	luteinizing hormone
OT	oxytocin
PRL	prolactin
TSH	thyroid-stimulating hormone

Case Report 10.1 (continued)

Dr. Khalid's examination of Mrs. Tacher shows a protruding mandible and an enlarged, deeply grooved tongue. Her feet and hands are enlarged, her ribs are thickened, and her heart is enlarged. X-rays show a thickened skull and enlarged nasal sinuses. CT and MRI scans show a tumor in the pituitary gland. Blood tests show high levels of growth hormone. She is scheduled for surgery to remove the tumor.

WORD ANALYSIS AND DEFINITION

S/ = Suffix P/ = Prefix R/ = Root R/CF = Combining Form

WORD	PRONUNCIATION	ELEMENTS		DEFINITION
acromegaly	ak-roe-**MEG**-ah-lee	S/ R/CF	**-megaly** *enlargement* **acr/o-** *peak, highest point*	Enlargement of head, face, hands, and feet due to excess growth hormone in an adult.
adrenocorticotropic	ah-**DREE**-noh-**KOR**-tih-koh-**TROH**-pik	S/ R/CF R/CF	**-tropic** *a turning, change* **adren/o-** *adrenal gland* **-cortic/o-** *cortex, cortisone*	Hormone of the anterior pituitary that stimulates the cortex of the adrenal gland to produce its own hormones.
corticosteroid	**KOR**-tih-koh-**STEHR**-oyd	S/ R/CF	**-steroid** *steroid* **cortic/o-** *cortex, cortisone*	A hormone produced by the adrenal cortex.
corticotropin	**KOR**-tih-koh-**TROH**-pin	S/ R/CF	**-tropin** *stimulation* **cortic/o-** *cortisone, cortex*	Pituitary hormone that stimulates the cortex of the adrenal gland to secrete cortisone.
diabetes insipidus (DI)	dye-ah-**BEE**-teez in-**SIP**-ih-dus	 S/ P/ R/	diabetes, Greek siphon **-us** *pertaining to* **in-** *not* **-sipid-** *flavor*	Excretion of large amounts of dilute urine as a result of inadequate ADH production.
gonadotropin **gonad**	**GO**-nad-oh-**TROH**-pin **GO**-nad	S/ R/CF	**-tropin** *stimulation* **gonad/o-** *testis, ovary* gonad, Latin *seed*	Any hormone that stimulates gonadal function. An organ that produces sex cells; a testis or an ovary.
panhypopituitarism (Note: Two prefixes.)	pan-**HIGH**-poh-pih-**TYU**-ih-tah-rizm	S/ P/ P/ R/	**-ism** *condition* **-hypo-** *deficient* **pan-** *all* **-pituitar-** *pituitary*	Deficiency of all the pituitary hormones.
prolactin	pro-**LAK**-tin	S/ P/ R/	**-in** *substance* **pro-** *before* **-lact-** *milk*	Pituitary hormone that stimulates the production of milk.
somatotropin (also called **growth hormone, GH**)	**SO**-mah-toh-**TROH**-pin	S/ R/CF	**-tropin** *stimulation* **somat/o-** *the body*	Hormone of the anterior pituitary that stimulates the growth of body tissues.
thyrotropin	thigh-roe-**TROH**-pin	S/ R/CF	**-tropin** *stimulation* **thyr/o-** *thyroid*	Hormone from the anterior pituitary gland that stimulates function of the thyroid gland.

Overproduction of growth hormone in children produces gigantism. In adults, the overproduction produces **acromegaly** (*Figure 10.5*). This is the condition Mrs. Gina Tacher has.

Underproduction of growth hormone, present at birth, leads to dwarfism.

Diabetes insipidus (DI) results from a decreased production of antidiuretic hormone (ADH). Symptoms are excessive urine production leading to excessive thirst.

▲ **FIGURE 10.5**
Woman with Acromegaly, Age 52.

EXERCISES

Abbreviations *will be your answers in this matching exercise. Match the correct abbreviation to its description.*

____ 1. stimulates ovulation A. LH

____ 2. stimulates production of corticosteroids B. GH

____ 3. stimulates uterine contractions C. ACTH

____ 4. stimulates ovaries to develop eggs D. TSH

____ 5. reduces volume of urine E. OT

____ 6. stimulates production of thyroxin F. FSH

____ 7. stimulates cells to enlarge and divide G. ADH

LESSON 10.2 Thyroid, Parathyroid, and Thymus Glands

ENDOCRINE SYSTEM

OBJECTIVES

In order to understand what is happening to Ms. Leary, you need to be able to:

- **Describe the location and anatomy of the thyroid gland.**
- **Identify medical terms to explain how the three thyroid hormones are produced and secreted.**
- **Specify the medical terminology for the functions of the thyroid hormones.**
- **Discuss common disorders of the thyroid gland.**

In addition, information in this lesson will enable you to:

- **Locate the positions of the parathyroid and thymus glands.**
- **Name the hormones produced by the parathyroid and thymus glands and state their functions.**

You are

. . . an EMT working in the Emergency Department at Fulwood Medical Center at 0200 hours in the morning.

You are communicating with

. . . the parents of Ms. Norma Leary, a 22-year-old college student living with her parents for the summer.

CASE REPORT 10.2

Ms. Leary is **emaciated**, extremely agitated, and at times disoriented and confused. Her parents tell you that in the past 3 or 4 days she has been coughing and not feeling well. In the past 12 hours she has become feverish and been complaining of a left-sided chest pain. With questioning, the parents reveal that prior to this acute illness Ms. Leary had lost about 20 pounds in weight, although she was eating voraciously. Her **VS** are T 105.2, P 180 and irregular, R 24, BP 160/85.

You call for Dr. Hilinski STAT. On his initial examination he believes that the patient is in thyroid storm. This is a medical emergency. There are no immediate laboratory tests that can confirm this diagnosis.

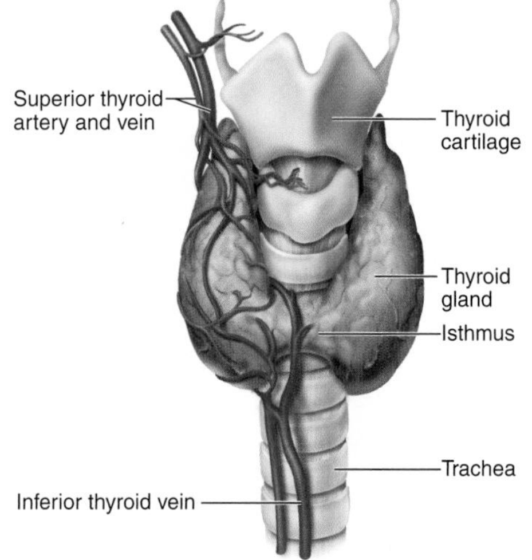

▲ **FIGURE 10.6**
Anatomy of the Thyroid Gland.

Abbreviations

PTH	parathyroid hormone
T3	triiodothyronine
T4	tetraiodothyronine (thyroxine)
VS	vital signs: measurement of temperature (T), pulse (P), respiration (R), and blood pressure (BP)

THYROID GLAND

The **thyroid** gland lies just beneath the skin of the neck and below the thyroid cartilage ("Adam's apple"). It is about 2 inches across and shaped like a bow tie. Two lobes extend up on either side of the trachea and are joined by an isthmus (*Figure 10.6*).

Cells in the gland secrete the two thyroid hormones **T3** and **T4.** The latter is known as **thyroxine.** The term **thyroid hormone** refers to T3 and T4 collectively. The thyroid hormone acts in three interrelated ways:

- **Stimulates** almost every tissue in the body to produce proteins.
- **Increases** the amount of oxygen that cells use.
- **Controls** the speed at which the body's chemical functions proceed **(metabolic rate).**

The thyroid also produces the hormone **calcitonin,** which promotes calcium deposition and bone formation.

PARATHYROID GLANDS

The **parathyroid** glands are usually four in number and are partially embedded in the posterior surface of the thyroid gland. They secrete **parathyroid hormone (PTH).** PTH and calcitonin are **antagonistic:** PTH stimulates bone resorption to bring calcium back into the blood, whereas calcitonin takes calcium from the blood and stimulates bone deposition (*see Chapter 8*).

LESSON 10.3 Adrenal Glands and Hormones

ENDOCRINE SYSTEM

OBJECTIVES

The information in this lesson will enable you to:

- **Locate the adrenal glands.**
- **Differentiate between the adrenal cortex and the medulla.**
- **Identify the medical terminology for the functions of the hormones produced by the cortex and medulla.**
- **Detail how the body adapts to stress.**
- **Use correct medical terminology to explain common disorders of the adrenal glands.**

▲ **FIGURE 10.11 John F. Kennedy.**

CASE REPORT 10.3

John Fitzgerald Kennedy (JFK) (1917–1963) was elected President of the United States of America in 1960 at the age of 43 (*Figure 10.11*). From the age of 13, when he was diagnosed as having colitis, he had had health problems. At age 27 he had low-back pain that necessitated low-back surgery. He was then diagnosed as having adrenal gland insufficiency (Addison disease) with osteoporosis of his lumbar spine. This required low-back surgery on three more occasions. JFK received adrenal hormone replacement therapy for the rest of his life, together with pain medication for his low-back pain, until his assassination in Dallas, Texas, in 1963.

In medical retrospect, instead of colitis, JFK probably had celiac disease (*see Chapter 5*), which has strong associations with Addison disease.

ADRENAL GLANDS

An **adrenal (suprarenal)** gland is anchored like a cap on the upper pole of each kidney (*Figure 10.12 inset*).

The outer layer of the gland, the adrenal cortex (*Figure 10.12*), synthesizes more than 25 **steroid** hormones known collectively as **adrenocortical** hormones, or corticosteroids. These hormones include:

1. **Glucocorticoids,** mainly **hydrocortisone (cortisol),** which help regulate blood glucose levels, particularly in response to stress. They also have an anti-inflammatory effect, and are often found in dermatologic preparations.
2. **Mineralocorticoids,** mostly **aldosterone,** which promote sodium retention and potassium secretion by the kidneys.
3. **Sex steroids,** which include mainly a weak androgen that is converted to testosterone (*see Chapter 13*) and an estrogen (*see Chapter 14*).

The inner layer of the adrenal gland, the adrenal **medulla** (*Figure 10.12*), secretes hormones called **catecholamines,** principally **epinephrine (adrenaline)** and **norepinephrine (noradrenaline).** These hormones prepare the body for physical activity and are responsible for the "flight or fight" response.

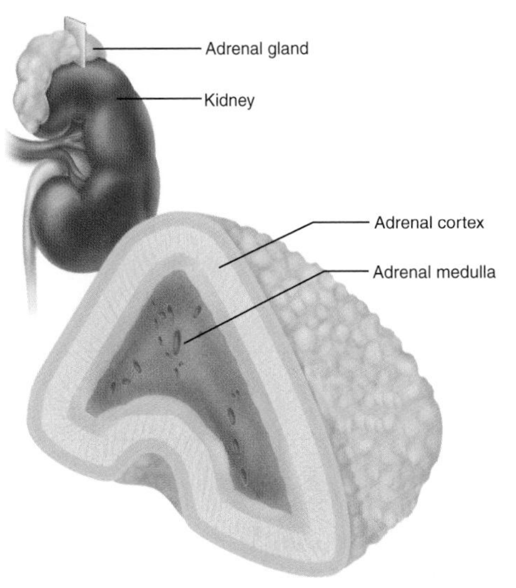

▲ **FIGURE 10.12 Adrenal Gland.**

Disorders of the Adrenal Glands

Adrenocortical hypofunction is most commonly seen as **Addison disease,** caused by the **idiopathic** atrophy (wasting away) of the adrenal cortex. Symptoms are weakness, fatigue, increased susceptibility to infection, and diminished resistance to stress. Hormone replacement therapy is given, and John F. Kennedy received replacement therapy until he died.

Adrenocortical hyperfunction is most commonly seen as **Cushing syndrome.** Excess production of the steroid hormones produces "moon" **facies,** muscle wasting and weakness, kidney stones, and reduced resistance to infection. Most cases are due to a pituitary tumor secreting too much ACTH, causing the adrenal glands to produce an excess of steroids. Sometimes, Cushing syndrome can be produced by administration of excess steroid medications.

WORD ANALYSIS AND DEFINITION

S/ = Suffix P/ = Prefix R/ = Root R/CF = Combining Form

WORD	PRONUNCIATION	ELEMENTS		DEFINITION
cretin **cretinism**	**KREH**-tin **KREH**-tin-izm	 S/ R/	French *cretin* **-ism** *condition, process* **cretin-** *cretin*	Severe congenital hypothyroidism. Condition of severe congenital hypothyroidism.
exophthalmos	ek-sof-**THAL**-mos	P/ R/	**ex-** *out* **-ophthalmos** *eye*	Protrusion of the eyeball.
goiter	**GOY**-ter		Latin *throat*	Enlargement of the thyroid gland.
Graves disease	GRAVZ diz-**EEZ**		Robert Graves, Irish physician, 1796–1853	Hyperthyroidism with toxic goiter.
Hashimoto disease	hah-shee-**MOH**-toe diz-**EEZ**		Hakaru Hashimoto, Japanese surgeon, 1881–1934	Autoimmune disease of the thyroid gland.
hyperparathyroidism (***Note:*** Contains 2 suffixes and 2 prefixes.) **hypoparathyroidism**	**HIGH**-per-par-ah-**THIGH**-royd-izm **HIGH**-poh-par-ah-**THIGH**-royd-izm	S/ S/ P/ P/ R/ P/	**-ism** *condition* **-oid-** *resembling* **hyper-** *excessive* **-para-** *adjacent* **-thyr-** *thyroid* **hypo-** *deficient, below*	Excessive levels of parathyroid hormone. Deficient levels of parathyroid hormone.
hyperpyrexia	**HIGH**-per-pie-**REK**-see-ah	S/ P/ R/	**-ia** *condition* **hyper-** *excessive* **-pyrex-** *fever*	Extremely high body temperature or fever.
hyperthyroidism (Note 2 suffixes.) (Also called **thyrotoxicosis.**) **hypothyroidism** (Note 2 suffixes.)	high-per-**THIGH**-royd-izm high-poh-**THIGH**-royd-izm	S/ S/ P/ R/ S/ S/ P/ R/	**-ism** *condition* **-oid-** *resembling* **hyper-** *excessive* **-thyr-** *thyroid* **-ism** *condition* **-oid-** *resembling* **hypo-** *deficient, below* **-thyr-** *thyroid*	Excessive production of thyroid hormones. Deficient production of thyroid hormones.
myxedema	miks-eh-**DEE**-muh	P/ R/	**myx-** *mucus* **-edema** *swelling*	Nonpitting, waxy edema of the skin in hypothyroidism.
tetany	**TET**-ah-nee		Greek *convulsive tension*	Severe muscle twitches, cramps, and spasms.
thyroidectomy **thyroiditis**	thigh-roy-**DEK**-toe-me thigh-roy-**DIE**-tis	S/ S/ R/ S/	**-ectomy** *surgical excision* **-oid-** *resembling* **thyr-** *thyroid* **-itis** *inflammation*	Surgical removal of the thyroid gland. Inflammation of the thyroid gland.
thyrotoxicosis (Also called **hyperthyroidism.**)	**THIGH**-roe-toks-ih-**KOH**-sis	S/ R/CF R/CF	**-sis** *condition* **thyr/o-** *thyroid* **-toxic/o-** *poison*	Disorder produced by excessive thyroid hormone production.

EXERCISES

Elements: *Make the elements work for you! One word in each of the descriptions below is bolded. This is your clue to finding the correct medical term in the word bank. When you write the term on the line, circle or highlight the element in that term that matched the clue of the bolded term. There are more answers than questions. Fill in the blanks.*

Word Bank:

thyroiditis	hyperthyroidism	exophthalmos
hypothyroidism	thyroidectomy	hyperpyrexia

1. **excessive** production of thyroid hormones: ____________________

2. **removal** of the thyroid gland: ____________________

3. protrusion of the **eyeball**: ____________________

4. extremely high **fever:** ____________________

5. **deficient** production of thyroid hormones: ____________________

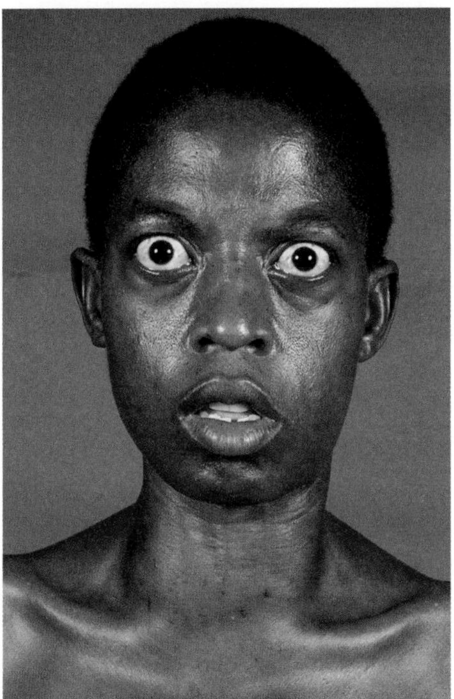

▲ **FIGURE 10.8 Hyperthyroidism May Cause the Eyes to Protrude (Exophthalmos).**

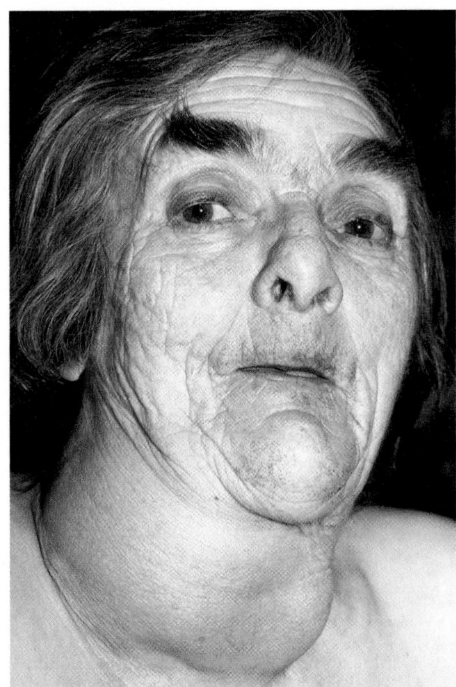

▲ **FIGURE 10.9 Elderly Woman with Hypothyroidism and Goiter.**

Case Report 10.2 (continued)

Thyroid storm is the condition Ms. Norma Leary presented with in the Emergency Department. It is the most extreme state of hyperthyroidism, with severely exaggerated effects of the thyroid hormones causing **hyperpyrexia**, tachycardia, agitation, and delirium. The weight loss prior to her illness becoming acute was part of her undiagnosed **hyperthyroidism**.

DISORDERS OF THE THYROID AND PARATHYROID GLANDS

Hyperthyroidism (Thyrotoxicosis)

The symptoms of **hyperthyroidism** (excessive thyroid hormone production) are those of increased body metabolism. These include tachycardia, hypertension, sweating, shakiness, anxiety, weight loss despite increased appetite, and diarrhea.

Graves disease is an autoimmune disorder (*see Chapter 15*) in which an antibody stimulates the thyroid to produce and secrete excessive amounts of thyroid hormone into the blood. It is associated with one or more symptoms of a **goiter,** which is an enlarged thyroid gland; **exophthalmos,** or bulging outward of the eyes (*Figure 10.8*); and nonpitting, waxy edema of the lower leg.

Hypothyroidism

Hypothyroidism results from an inadequate production of thyroid hormone, leading to a slowing of the body's metabolism. Primary hypothyroidism, in which no specific cause is found, affects 10% of older women. Symptoms develop gradually. They include loss of hair; dry, scaly skin; puffy face and eyes; slow, hoarse speech; weight gain; constipation; and inability to tolerate cold (*Figure 10.9*).

Severe hypothyroidism is called **myxedema.** In developing countries, a common cause is lack of iodine in the diet. In this country, iodine is added to table salt to prevent hypothyroidism and iodine is also found in dairy products and seafood.

Thyroiditis is an inflammation of the thyroid gland. It presents most commonly as **Hashimoto thyroiditis,** an autoimmune disease with lymphocytic infiltration of the gland. Hypothyroidism results, necessitating lifelong thyroid hormone replacement therapy.

Cretinism (*Figure 10.10*) is a congenital form of thyroid deficiency that severely retards mental and physical growth. If it is diagnosed and treated early with thyroid hormones, significant improvement can be achieved.

Thyroid cancer usually presents as a symptomless nodule in the thyroid gland. It can metastasize (spread) to cervical and mediastinal lymph nodes and to liver, lungs, and bones. Total **thyroidectomy** with local lymph node dissection is the initial step in treatment.

Disorders of the Parathyroid Glands

Hypoparathyroidism is a deficiency of parathyroid hormone (PTH) that lowers levels of blood calcium. Most symptoms are neuromuscular, ranging from tingling in the fingers to muscle cramps and the painful muscle spasms of **tetany** (not **tetanus,** which is caused by a toxin acting on the central nervous system).

Hyperparathyroidism is an excess of PTH and is seen more often than hypoparathyroidism. It is usually caused by one of the four glands enlarging and working out of pituitary control. It leads to bones being depleted of calcium and becoming brittle, high blood calcium levels, and kidney stones.

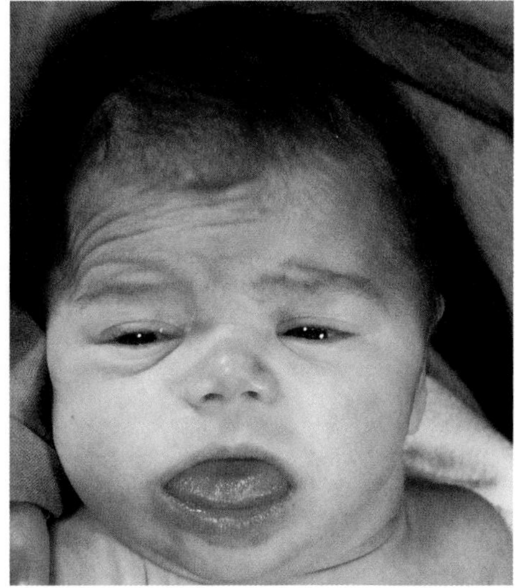

▲ **FIGURE 10.10 Infant with Cretinism.**

WORD ANALYSIS AND DEFINITION

S/ = Suffix P/ = Prefix R/ = Root R/CF = Combining Form

WORD	PRONUNCIATION		ELEMENTS	DEFINITION
antagonist	an-**TAG**-oh-nist	S/ P/ R/	**-ist** *agent* **ant-** *against* **-agon-** *to fight*	An opposing structure, agent, disease, or process.
antagonistic (adj) (***Note:*** two suffixes)	an-**TAG**-oh-nist-ik	S/	**-ic** *pertaining to*	Having an opposite function.
calcitonin	kal-sih-**TONE**-in	S/ R/CF R/	**-in** *substance* **calc/i-** *calcium* **-ton-** *tension, pressure*	Thyroid hormone that moves calcium from blood to bones.
emaciation	ee-may-see-**AY**-shun	S/ R/CF	**-ation** *process* **emac/i-** *make thin*	Abnormal thinness.
emaciated (adj)	ee-may-see-**AY**-ted	S/	**-ated** *pertaining to a condition*	Pertaining to or suffering from emaciation.
iodine	**EYE**-oh-dine *or* **EYE**-oh-deen	S/ R/	**-ine** *pertaining to* **iod-** *violet*	Chemical element, the lack of which causes thyroid disease.
parathyroid	par-ah-**THIGH**-royd	S/ P/ R/	**-oid** *resembling* **para-** *adjacent, beside* **-thyr-***thyroid*	Endocrine glands embedded in the back of the thyroid gland.
thymus	**THIGH**-mus		Greek *sweetbread*	Endocrine gland located in the mediastinum.
thyroid	**THIGH**-royd		Greek *an oblong shield*	Endocrine gland in the neck; or a cartilage of the larynx.
thyroxine	thigh-**ROCK**-sin	S/ R/ R/	**-ine** *pertaining to* **thyr-** *thyroid gland* **-ox-** *oxygen*	Thyroid hormone T4, tetraiodothyronine.

THYMUS GLAND

The **thymus** gland is located in the mediastinum behind the sternum between the lungs and above the heart (*Figure 10.7*). It is large in children and then decreases in size until in the elderly it is mostly fibrous tissue. It secretes a group of hormones that stimulate the production of T lymphocytes (*see Chapter 11*).

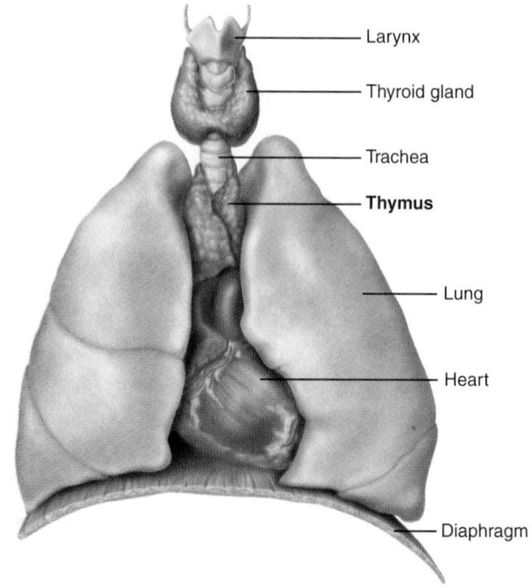

FIGURE 10.7 Position of Thymus Gland in Mediastinum.

EXERCISES

Precision in documentation: *Strive always to have your documentation error-free, both in fact and in spelling. Documentation with errors is unacceptable and unprofessional. Proofread the following statements carefully to determine the errors. Any sentence containing an error in fact or spelling (or both) should be rewritten correctly on the line below it. There is only one sentence that is entirely correct.*

1. Thyroid cartiledge in the neck is known as the Adam's apple.

____

2. This patient is emaciated, extremely agitated, and at times disorientation and confused.

____

3. The term *thyroid hormone* refers to T3 and T4 collectively.

____

4. The thymus gland is located in the mediastenum behind the stermun between the lungs and behind the heart.

____

LESSON 10.2 ENDOCRINE SYSTEM

Thyroid, Parathyroid, and Thymus Glands

OBJECTIVES

In order to understand what is happening to Ms. Leary, you need to be able to:

- **Describe the location and anatomy of the thyroid gland.**
- **Identify medical terms to explain how the three thyroid hormones are produced and secreted.**
- **Specify the medical terminology for the functions of the thyroid hormones.**
- **Discuss common disorders of the thyroid gland.**

In addition, information in this lesson will enable you to:

- **Locate the positions of the parathyroid and thymus glands.**
- **Name the hormones produced by the parathyroid and thymus glands and state their functions.**

You are

. . . an EMT working in the Emergency Department at Fulwood Medical Center at 0200 hours in the morning.

You are communicating with

. . . the parents of Ms. Norma Leary, a 22-year-old college student living with her parents for the summer.

CASE REPORT 10.2

Ms. Leary is **emaciated**, extremely agitated, and at times disoriented and confused. Her parents tell you that in the past 3 or 4 days she has been coughing and not feeling well. In the past 12 hours she has become feverish and been complaining of a left-sided chest pain. With questioning, the parents reveal that prior to this acute illness Ms. Leary had lost about 20 pounds in weight, although she was eating voraciously. Her **VS** are T 105.2, P 180 and irregular, R 24, BP 160/85.

You call for Dr. Hilinski STAT. On his initial examination he believes that the patient is in thyroid storm. This is a medical emergency. There are no immediate laboratory tests that can confirm this diagnosis.

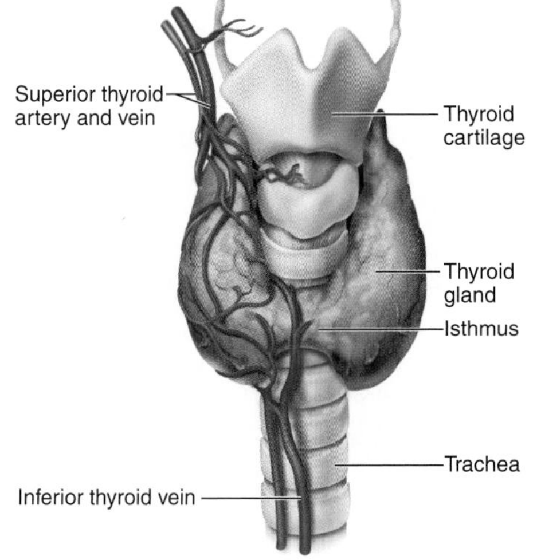

▲ **FIGURE 10.6**
Anatomy of the Thyroid Gland.

Abbreviations

PTH	parathyroid hormone
T3	triiodothyronine
T4	tetraiodothyronine (thyroxine)
VS	vital signs: measurement of temperature (T), pulse (P), respiration (R), and blood pressure (BP)

THYROID GLAND

The **thyroid** gland lies just beneath the skin of the neck and below the thyroid cartilage ("Adam's apple"). It is about 2 inches across and shaped like a bow tie. Two lobes extend up on either side of the trachea and are joined by an isthmus (*Figure 10.6*).

Cells in the gland secrete the two thyroid hormones **T3** and **T4.** The latter is known as **thyroxine.** The term **thyroid hormone** refers to T3 and T4 collectively. The thyroid hormone acts in three interrelated ways:

- **Stimulates** almost every tissue in the body to produce proteins.
- **Increases** the amount of oxygen that cells use.
- **Controls** the speed at which the body's chemical functions proceed **(metabolic rate).**

The thyroid also produces the hormone **calcitonin,** which promotes calcium deposition and bone formation.

PARATHYROID GLANDS

The **parathyroid** glands are usually four in number and are partially embedded in the posterior surface of the thyroid gland. They secrete **parathyroid hormone (PTH).** PTH and calcitonin are **antagonistic:** PTH stimulates bone resorption to bring calcium back into the blood, whereas calcitonin takes calcium from the blood and stimulates bone deposition (*see Chapter 8*).

WORD ANALYSIS AND DEFINITION

S/ = Suffix P/ = Prefix R/ = Root R/CF = Combining Form

WORD	PRONUNCIATION	ELEMENTS		DEFINITION
acromegaly	ak-roe-**MEG**-ah-lee	S/ R/CF	**-megaly** *enlargement* **acr/o-** *peak, highest point*	Enlargement of head, face, hands, and feet due to excess growth hormone in an adult.
adrenocorticotropic	ah-**DREE**-noh-**KOR**-tih-koh-**TROH**-pik	S/ R/CF R/CF	**-tropic** *a turning, change* **adren/o-** *adrenal gland* **-cortic/o-** *cortex, cortisone*	Hormone of the anterior pituitary that stimulates the cortex of the adrenal gland to produce its own hormones.
corticosteroid	**KOR**-tih-koh-**STEHR**-oyd	S/ R/CF	**-steroid** *steroid* **cortic/o-** *cortex, cortisone*	A hormone produced by the adrenal cortex.
corticotropin	**KOR**-tih-koh-**TROH**-pin	S/ R/CF	**-tropin** *stimulation* **cortic/o-** *cortisone, cortex*	Pituitary hormone that stimulates the cortex of the adrenal gland to secrete cortisone.
diabetes insipidus (DI)	dye-ah-**BEE**-teez in-**SIP**-ih-dus	 S/ P/ R/	diabetes, Greek siphon **-us** *pertaining to* **in-** *not* **-sipid-** *flavor*	Excretion of large amounts of dilute urine as a result of inadequate ADH production.
gonadotropin **gonad**	**GO**-nad-oh-**TROH**-pin **GO**-nad	S/ R/CF	**-tropin** *stimulation* **gonad/o-** *testis, ovary* gonad, Latin *seed*	Any hormone that stimulates gonadal function. An organ that produces sex cells; a testis or an ovary.
panhypopituitarism (Note: Two prefixes.)	pan-**HIGH**-poh-pih-**TYU**-ih-tah-rizm	S/ P/ P/ R/	**-ism** *condition* **-hypo-** *deficient* **pan-** *all* **-pituitar-** *pituitary*	Deficiency of all the pituitary hormones.
prolactin	pro-**LAK**-tin	S/ P/ R/	**-in** *substance* **pro-** *before* **-lact-** *milk*	Pituitary hormone that stimulates the production of milk.
somatotropin (also called **growth hormone, GH**)	**SO**-mah-toh-**TROH**-pin	S/ R/CF	**-tropin** *stimulation* **somat/o-** *the body*	Hormone of the anterior pituitary that stimulates the growth of body tissues.
thyrotropin	thigh-roe-**TROH**-pin	S/ R/CF	**-tropin** *stimulation* **thyr/o-** *thyroid*	Hormone from the anterior pituitary gland that stimulates function of the thyroid gland.

Overproduction of growth hormone in children produces gigantism. In adults, the overproduction produces **acromegaly** (*Figure 10.5*). This is the condition Mrs. Gina Tacher has.

Underproduction of growth hormone, present at birth, leads to dwarfism.

Diabetes insipidus (DI) results from a decreased production of antidiuretic hormone (ADH). Symptoms are excessive urine production leading to excessive thirst.

▲ **FIGURE 10.5 Woman with Acromegaly, Age 52.**

EXERCISES

Abbreviations *will be your answers in this matching exercise. Match the correct abbreviation to its description.*

_____ 1. stimulates ovulation — A. LH

_____ 2. stimulates production of corticosteroids — B. GH

_____ 3. stimulates uterine contractions — C. ACTH

_____ 4. stimulates ovaries to develop eggs — D. TSH

_____ 5. reduces volume of urine — E. OT

_____ 6. stimulates production of thyroxin — F. FSH

_____ 7. stimulates cells to enlarge and divide — G. ADH

WORD ANALYSIS AND DEFINITION

S/ = Suffix P/ = Prefix R/ = Root R/CF = Combining Form

WORD	PRONUNCIATION		ELEMENTS	DEFINITION
Addison disease	**ADD**-ih-son diz-**EEZ**		Thomas Addison, English physician, 1793–1860	An autoimmune disease leading to decreased production of adrenocortical steroids.
adrenal (same as *suprarenal*)	ah-**DREE**-nal	S/ P/ R/ P/	**-al** *pertaining to* **ad-** *to, toward* **-ren-** *kidney* **supra-** *above*	Endocrine gland on the upper pole of each kidney.
adrenaline (also called **epinephrine**)	ah-**DREN**-ah-lin ep-ih-**NEF**-rin	S/ P/ R/	**-ine** *pertaining to* **epi-** *above* **-nephr-** *kidney*	Main catecholamine produced by the adrenal medulla.
adrenocortical	ah-dree-noh-**KOR**-tih-kal	S/ R/CF R/	**-al** *pertaining to* **adren/o-** *adrenal* **-cortic-** *cortex*	Pertaining to the cortex of the adrenal gland.
aldosterone	al-**DOS**-ter-own	S/ R/CF R/	**-one** *hormone* **ald/o-** *organic compound* **-ster-** *steroid*	Mineralocorticoid hormone of the adrenal cortex.
catecholamine	kat-eh-**COAL**-ah-meen	S/ R/	**-amine** *nitrogen-containing substance* **catechol-** *tyrosine containing*	Major elements in the stress response; include epinephrine and norepinephrine.
Cushing syndrome	**KUSH**-ing **SIN**-drohm		Harvey Cushing, U.S. neurosurgeon, 1869–1939	Hypersecretion of cortisol (hydrocortisone) by the adrenal gland.
facies	**FASH**-eez		Latin *appearance*	Facial features and expressions.
glucocorticoid	glu-co-**KOR**-tih-koyd	S/ R/ R/CF	**-oid** *resembling* **-cortic-** *cortisone* **gluc/o-** *glucose*	Hormone of the adrenal cortex that helps regulate glucose metabolism.
hydrocortisone (also called **cortisol**)	high-droh-**KOR**-tih-sohn	S/ R/CF R/	**-one** *hormone* **hydr/o-** *water* **-cortis-** *cortisone*	Potent glucocorticoid with anti-inflammatory properties.
idiopathic	id-ih-oh-**PATH**-ik	S/ R/CF R/	**-ic** *pertaining to* **idi/o-** *unknown* **-path-** *disease*	Pertaining to a disease of unknown etiology.
mineralocorticoid	**MIN**-er-al-oh-**KOR**-tih-koyd	S/ R/ R/CF	**-oid** *resemble* **-cortic-** *cortex* **mineral/o-** *inorganic material*	Hormone of the adrenal cortex that influences sodium and potassium metabolism.
norepinephrine (***Note:*** two prefixes) (also called **noradrenaline**)	**NOR**-ep-ih-**NEFF**-rin	S/ P/ P/ R/	**-ine** pertaining to **nor-** *normal* **-epi-** *above* **-nephr-** *kidney*	Catecholamine hormone of the adrenal gland that is a parasympathetic neurotransmitter.
steroid	**STER**-oyd	S/ R/	**-oid** *resembling* **ster-** *solid*	Large family of chemical substances found in many drugs, hormones, and body components.

EXERCISES

Analyze *the elements of medical terms as a clue to understanding the meaning of the term. Match the element in column 1 to its correct meaning in column 2. Review the WAD box before you begin the exercise. Fill in the blanks.*

	Column 1	Column 2
_____	1. hydro	A. resemble
_____	2. oid	B. sugar
_____	3. ster	C. water
_____	4. gluco	D. solid
_____	5. nephr	E. kidney

LESSON 10.4 The Pancreas

ENDOCRINE SYSTEM

OBJECTIVES

The information provided in this lesson will enable you to:

- **Distinguish between the different cells of the pancreas and their secretions.**
- **Identify medical terminology for the functions of the hormones produced by the pancreas.**
- **Use medical terminology to explain common disorders of the pancreatic hormones.**

You are

. . . a certified medical assistant working with Susan Lee, MD, in her Primary Care Clinic at Fulwood Medical Center.

You are communicating with

. . . Mrs. Martha Jones, who is here for her monthly checkup. She is a 53-year-old type 2 diabetic on insulin.

CASE REPORT 10.4

Mrs. Jones has diabetic retinopathy and diabetic neuropathy of her feet. Bariatric surgery has enabled her to reduce her weight from 275 to 156 pounds. The time is 0930 hrs.

She is complaining of having a cold and cough for the past few days. Now she is feeling drowsy and nauseous and has a dry mouth. As you talk with her, you notice that her speech is slurred. She cannot remember if she gave herself her morning insulin. Examination of her lungs reveals rales at her right base.

Her VS: T 97.8, P 120, R 20, BP 100/50. You perform her blood **glucose** measurement. The reading is 600 milligrams per deciliter (mg/dL). A recommended value 2 hours after breakfast is < 145 mg/dL.

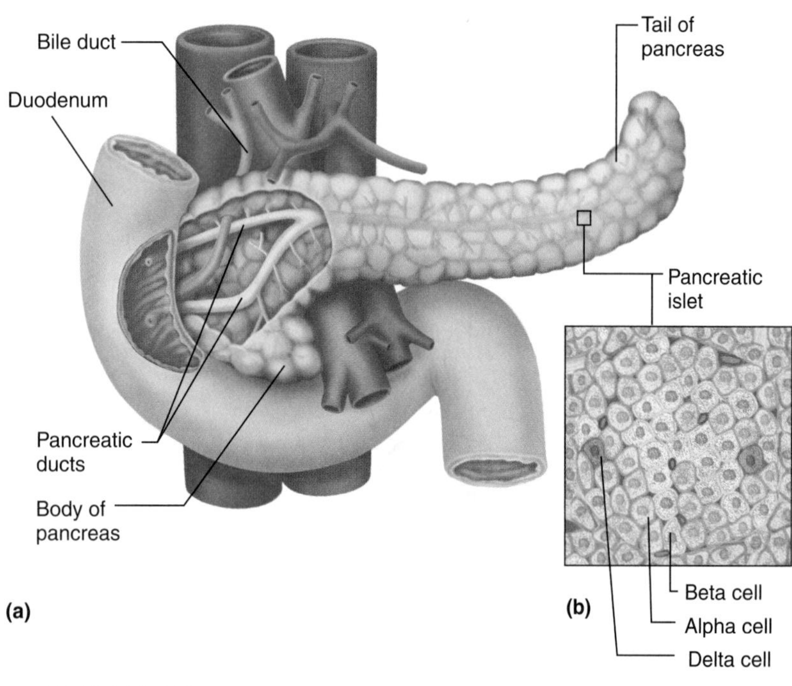

▲ **FIGURE 10.13 Pancreas.**
(a) Anatomy of the pancreas. (b) Alpha, beta, and delta cells.

THE PANCREAS

The location and structure of the pancreas is detailed in *Chapter 5*. Most of the pancreas is an **exocrine** gland that secretes digestive juices through a duct (*Figure 10.13a*). Scattered throughout the pancreas are clusters of endocrine cells grouped around blood vessels. These clusters are called **pancreatic islets (Islets of Langerhans).** Within the islets are three distinct cell types (*Figure 10.13b*):

1. **Alpha cells:** Secrete the hormone **glucagon** in response to a low blood **glucose.** Glucagon's actions are:
   a. In the liver, to stimulate **gluconeogenesis, glycogenolysis,** and the release of glucose into the bloodstream.
   b. In adipose tissue, to stimulate fat catabolism and the release of free fatty acids.
2. **Beta cells:** Secrete **insulin** in response to a high blood glucose. Insulin has the opposite effects to those of glucagon:
   a. In muscle and fat cells, to absorb glucose and to store glycogen and fat.
   b. In the liver, to stimulate the conversion of glucose to glycogen.
3. **Delta cells:** Secrete **somatostatin,** which acts within the pancreas to inhibit the secretion of glucagon and insulin.

WORD ANALYSIS AND DEFINITION

S/ = Suffix P/ = Prefix R/ = Root R/CF = Combining Form

WORD	PRONUNCIATION	ELEMENTS		DEFINITION
exocrine	**EK**-soh-krin	P/ R/	**exo-** *outward* **-crine** *secrete*	A gland that secretes outwardly through excretory ducts.
glucagon	**GLU**-kah-gon	S/ R/	**-agon** *to fight* **gluc-** *sugar, glucose*	Pancreatic hormone that supports blood glucose levels.
gluconeogenesis	**GLU**-koh-nee-oh-**JEN**-eh-sis	S/ P/ R/CF	**-genesis** *creation* **-neo-** *new* **gluc/o-** *sugar, glucose*	Formation of glucose from noncarbohydrate sources.
glucose	**GLU**-kose	S/ R/	**-ose** *full of* **gluc-** *sugar, glucose*	The final product of carbohydrate digestion and the main sugar in the blood.
glycogenolysis	**GLYE**-koh-jen-oh-**LYE**-sis	S/ R/CF R/CF	**-lysis** *separate, dissolve* **glyc/o-** *glycogen* **-gen/o-** *create*	Conversion of glycogen to glucose.
insulin	**IN**-syu-lin	S/ R/	**-in** *a substance* **insul-** *an island*	Hormone produced by the islet cells of the pancreas.
islets of Langerhans	**EYE**-lets of **LAHNG**-er-hahnz		Paul Langerhans, German anatomist, 1847–1888	Areas of pancreatic cells that produce insulin and glucagon.
somatostatin	**SOH**-mah-toh-**STAT**-in	S/ R/CF	**-statin** *inhibit* **somat/o-** *body*	Hormone that inhibits release of growth hormone and insulin.

Glucagon is not the only hormone that raises blood glucose; epinephrine, hydrocortisone, and growth hormone also have that effect. Insulin is the only hormone that lowers blood glucose.

EXERCISES

Build *the correct medical term that matches the definition. Insert each missing element* on the line, *and label what type of element it is* under the line. *Then answer questions 4 and 5.*

1. conversion of glycogen to glucose:

   ______________________/______________________/lysis

2. main sugar in the blood:

   ______________________/______________________/ose

3. hormone that inhibits release of GH and insulin:

   ______________________/______________________/statin

4. the formation of glucose from noncarbohydrate sources:

   gluco/______________________/genesis

5. What is unusual about the elements in the term above?

   ____

DIABETES MELLITUS (DM)

Diabetes mellitus is a syndrome characterized by hyperglycemia resulting from an absolute or relative impairment of insulin secretion and/or insulin action. This leads to a disruption of carbohydrate, fat, and protein metabolism. It is the world's most prevalent metabolic disease and the leading cause of blindness, renal failure, and gangrene. There are four categories of diabetes mellitus:

1. **Type 1 diabetes,** also called **insulin-dependent diabetes mellitus (IDDM),** accounts for 10% to 15% of all cases of DM, but it is the predominant type of DM occurring in patients under the age of 30. When symptoms become apparent, 90% of the pancreatic insulin-producing cells have been destroyed by **autoantibodies.**
2. **Type 2 diabetes,** also called **non-insulin-dependent diabetes mellitus (NIDDM),** accounts for 85% to 90% of all cases of DM. Almost 18 million U.S. residents are diagnosed with type 2 DM. Not only is there some impairment of insulin response, but there is decreased glucose uptake by tissues due to **insulin resistance.** In addition to contributing to type 2 DM, insulin resistance leads to other common disorders such as hypertension, hyperlipidemia, and coronary artery disease.
3. **Gestational diabetes** is seen in the latter half of 5% of pregnancies. While most cases of gestational diabetes resolve after the pregnancy, a woman who has this complication of pregnancy has a 30% chance of developing type 2 DM within 10 years.
4. **Mature-onset diabetes of the young (MODY)** is genetically inherited, occurs in thin individuals in their teens and twenties, and is comparable to type 2 in its severity.

Keynote

Type 2 diabetes used to occur over the age of 30, but it is now seen also in obese adolescents and young adults.

Abbreviations

DM	diabetes mellitus
IDDM	insulin-dependent diabetes mellitus
MODY	mature-onset diabetes of the young
NIDDM	non-insulin-dependent diabetes mellitus

Hypoglycemia

Hypoglycemia is present when blood glucose is below 70 mg/dL. Because brain metabolism depends primarily on glucose, the brain is the first organ affected by hypoglycemia. The disorder presents clinically as impaired mental efficiency, followed by shakiness, anxiety, confusion, tremor, seizures, and, if untreated, loss of consciousness. Symptomatic hypoglycemia is sometimes called **insulin shock.** Low blood glucose can be raised to normal in minutes by taking 10 to 20 grams of carbohydrate (3 to 4 ounces), such as orange, apple, or grape juice or a sugar-containing soft drink.

Keynote

The brain is the first organ affected by hypoglycemia.

Hyperglycemia

In **hyperglycemia,** the classic symptoms are ***poly*uria** (excessive urination), ***poly*dipsia** (excessive thirst) and ***poly*phagia** (excessive hunger), with unexplained weight loss.

Symptomatic hyperglycemia is how type 1 DM usually presents. Type 2 can be symptomatic or asymptomatic and is often found during a routine health examination.

Because of high glucose levels, hyperglycemia damages capillary endothelial cells in the retina, renal glomerulus, and neurons and Schwann cells in peripheral nerves. Of all diabetics, 85% develop some degree of **diabetic retinopathy;** 30% develop diabetic nephropathy, which can progress to end-stage renal disease (*see Chapter 13*). Diabetic neuropathy causes sensory defects with numbness, tingling, and **paresthesias** in the stocking-glove (feet-hands) distribution.

In larger blood vessels the hyperglycemia contributes to endothelial cell-lining damage and atherosclerosis. Coronary artery disease and peripheral vascular disease with claudication (*see Chapter 6*) are complications. Hyperglycemia is the most common cause of foot ulcers and then gangrene of the lower extremity, sometimes necessitating **amputation.** The risk of infection is increased by the cellular hyperglycemia and the circulatory deficits.

The complications of hyperglycemia can be kept at bay by stringent control of blood glucose levels.

Keynote

Hyperglycemia causes damage to vascular endothelial cells at the microvascular and macrovascular levels.

Keynote

Untreated hyperglycemia can progress to diabetic **coma.**

WORD ANALYSIS AND DEFINITION

S/ = Suffix P/ = Prefix R/ = Root R/CF = Combining Form

WORD	PRONUNCIATION		ELEMENTS	DEFINITION
amputation	am-pyu-**TAY**-shun	S/ R/	**-ation** *a process* **amput-** *to prune, lop off*	Process of removing a limb, part of a limb, a breast, or other projecting part.
autoantibody (***Note:*** 2 prefixes)	awe-toe-**AN**-tee-bod-ee	P/ P/ R/	**auto-** *self, same* **-anti-** *against* **-body** *body*	Antibody produced in response to an antigen from the host's own tissue.
coma **comatose (adj)**	**KOH**-mah **KOH**-mah-toes	 S/ R/	Greek *deep sleep* **-ose** *full of* **comat-** *coma*	State of deep unconsciousness. In a state of coma.
diabetes mellitus **diabetic (adj)**	dye-ah-**BEE**-teez **MEL**-ih-tus dye-ah-**BET**-ik	 S/ R/	diabetes, Greek *a siphon* mellitus, Latin *sweetened with honey* **-ic** *pertaining to* **diabet-** *diabetes*	Metabolic syndrome caused by absolute or relative insulin deficiency and/or ineffectiveness. Pertaining to or suffering from diabetes.
hyperglycemia **hyperglycemic (adj)**	**HIGH**-per-gly-**SEE**-me-ah **HIGH**-per-gly-**SEE**-mik	S/ P/ R/ S/	**-emia** *a blood condition* **hyper-** *above* **-glyc-** *glucose* **-emic** *pertaining to a blood condition*	High level of glucose (sugar) in blood. Pertaining to or having hyperglycemia.
hypoglycemia **hypoglycemic (adj)**	**HIGH**-poh-gly-**SEE**-me-ah **HIGH**-poh-gly-**SEE**-mik	S/ P/ R/ S/	**-emia** *a blood condition* **hypo-** *below, deficient* **-glyc-** *glucose* **-ic** *pertaining to*	Low level of glucose (sugar) in the blood. Pertaining to or suffering from a low blood sugar.
paresthesia **paresthesias (pl)**	par-es-**THEE**-ze-ah	S/ P/ R/	**-ia** *condition* **par-** *abnormal* **-esthes-** *sensation*	An abnormal sensation; e.g., tingling, burning, pricking.
polydipsia	pol-ee-**DIP**-see-ah	S/ P/ R/	**-ia** *condition* **poly-** *many, excessive* **-dips-** *thirst*	Excessive thirst.
polyphagia	pol-ee-**FAY**-jee-ah	S/ P/ R/	**-ia** *condition* **poly-** *many, excessive* **-phag-** *eat*	Excessive eating.
polyuria	pol-ee-**YOU**-ree-ah	S/ P/ R/	**-ia** *condition* **poly-** *many, excessive* **-ur-** *urine*	Excessive production of urine.
retinopathy	ret-ih-**NOP**-ah-thee	S/ R/CF	**-pathy** *disease* **retin/o-** *retina of the eye*	Degenerative disease of the retina.

EXERCISES

Diabetes *is the world's most prevalent metabolic disease. Many patients have diabetes as a concurrent condition with other health problems, which always makes it a consideration in treatment and the prescribing of medications. Test your knowledge of this disease by answering the following questions. Circle the correct choice.*

1. Diabetes is the leading cause of:

   blindness   renal failure   gangrene   all of these   none of these

2. The first organ affected by hypoglycemia is the:

   kidney   heart   pancreas   brain   liver

3. The predominant type of DM under the age of 30 is:

   type 1   type 2   non-insulin-dependent DM   gestational diabetes

4. Impairment of insulin response and decreased insulin effectiveness is termed insulin:

   production   resistance   autoantibodies   control   conversion

5. Most cases of gestational diabetes resolve after:

   medication   treatment   testing   pregnancy   surgery

DIABETES MELLITUS (continued)

Diabetic ketoacidosis (DKA) is a state of marked hyperglycemia with dehydration, **metabolic acidosis,** and **ketone** formation. It is seen mostly in type 1 DM and is usually the result of a lapse in insulin treatment, an acute infection, or a trauma that renders the usual insulin treatment inadequate.

Keynote

Diabetic ketoacidosis is a medical emergency.

It presents with polyuria, vomiting, and lethargy and can progress to coma. The ketone **acetone** can be smelled on the breath. DKA is a medical emergency. There is a 2% to 5% mortality from circulatory collapse if the DKA is not promptly controlled.

Diabetic coma, a severe medical emergency, is caused by *hyper*glycemia.

Insulin shock, another severe medical emergency, is caused by *hypo*glycemia.

A blood glucose test will differentiate hypoglycemia from hyperglycemia.

Case Report 10.4 *(continued)*

Mrs. Martha Jones is in the early stages of a diabetic ketoacidosis coma, probably initiated by a right lower-lobe pneumonia. A urine specimen was obtained. Dr. Lee was notified. Blood was taken for a full chemistry panel, and arterial blood gases were drawn. Dr. Lee treated Mrs. Jones immediately with an IV infusion of **saline** solution and IV insulin. She was admitted to the hospital.

Treatment of Diabetes Mellitus

The basic principle of diabetes treatment is to avoid hyperglycemia and hypoglycemia. The following are the areas of treatment:

- **Diet and exercise** to achieve weight reduction of 2 pounds per week in overweight type 2 patients is essential.
- **Patient education** is necessary so that the patient understands the disease process, can recognize the indications for seeking immediate medical care, and will follow a regimen of foot care.
- **Plasma glucose monitoring** is an essential skill that all diabetics must learn. Patients on insulin must learn to adjust their insulin doses. Home glucose analyzers use a drop of blood obtained from the fingertip or forearm by a spring-powered lancet. The frequency of testing is varied individually.
- **Assessment of the patient** should be performed on routine physician visits for symptoms or signs of complications.
- **Periodic laboratory evaluation** includes **BUN** and serum creatinine (kidney function), lipid profile, ECG, and an annual complete ophthalmologic evaluation.

Keynote

Maintaining normal plasma glucose levels is the basis for good management of DM.

Glycosylated hemoglobin (Hb A1c) is used to monitor plasma glucose control during the preceding 1 to 3 months.

Oral antidiabetic drugs are used for type 2 DM but not type 1. These drugs include:

- *Metformin:* Acts by decreasing hepatic glucose production.
- *Acarbose:* Delays carbohydrate digestion and absorption in the intestine and can be used in combination with other drugs.
- *Thiazolidinediones* such as troglitazone: Improve insulin sensitivity in skeletal muscle and suppress hepatic glucose production. They are used for type 2 patients on insulin and allow them to reduce their insulin dose.

Keynote

Many insulin-dependent diabetics need multiple subcutaneous insulin injections each day.

Injectable insulin preparations are used in type 1 DM and sometimes in type 2. They are classified by their speed of action.

For some patients requiring frequent doses of insulin, continuous subcutaneous insulin infusion is given by an implanted battery-powered, programmable pump that provides continuous insulin through a small needle in the abdominal wall.

WORD ANALYSIS AND DEFINITION

S/ = Suffix P/ = Prefix R/ = Root R/CF = Combining Form

WORD	PRONUNCIATION	ELEMENTS		DEFINITION
acetone	**ASS**-eh-tone		Latin *vinegar*	Ketone that is found in blood, urine, and breath when diabetes mellitus is out of control.
ketoacidosis	**KEY**-toe-ass-ih-**DOE**-sis	S/ R/ R/CF	**-osis** *condition* **-acid-** *acid* **ket/o-** *ketone*	Excessive ketones in the blood, making it acid.
ketone	**KEY**-tone		Greek *acetone*	Chemical formed in uncontrolled diabetes or in starvation.
ketosis	key-**TOE**-sis	S/ R/	**-osis** *condition* **ket-** *ketone*	Excessive production of ketones.
metabolic acidosis	met-ah-**BOL**-ik ass-ih-**DOE**-sis	S/ R/ S/ R/	**-ic** *pertaining to* **metabol-** *change* **-osis** *condition* **acid-** *acid*	Decreased pH in blood and body tissues as a result of an upset in metabolism.
saline	**SAY**-leen		Latin *salt*	Salt solution, usually sodium chloride.

Abbreviations

BUN blood urea nitrogen
DKA diabetic ketoacidosis
Hb A1c glycosylated hemoglobin, hemoglobin A one-C

EXERCISES

Case Report (continued): *Read aloud the Case Report continuation on the opposite page. Read it a second time, and highlight or underline the medical terms. Answer the following questions based on this Case Report.*

1. "Diabetic ketoacidosis is a state of marked hyperglycemia with dehydration." Explain this in layperson's language.

____

____

____

____

____

2. Mrs. Jones's pneumonia is in her right lower lobe. How many lobes are in the right lung? ____________________

3. Define *pneumonia:*

____

____

____

4. Name another abbreviation that could have been used in this Case Report: ____________________, which stands for

____

5. What chemical is contained in a *saline* solution? ____________________

CHAPTER 10 REVIEW

ENDOCRINE SYSTEM

CHALLENGE YOUR KNOWLEDGE

A. **Build your knowledge** of the language of endocrinology by analyzing the elements in the following terms. Then write a sentence of your choice for any one term in the chart.

Medical Term	Meaning of Prefix	Meaning of Root/CF	Meaning of Suffix
acromegaly			
autoantibody			
endocrinology			
glycogenolysis			
hypoglycemia			
neuropathy			
panhypopituitarism			
paresthesia			
polyphagia			
serotonin			

Sentence:

____

____

____

____

____

B. **Symptoms:** Each of the documentations below presents a classic symptom of *hyperglycemia.* Write the correct medical terms for the symptoms on the blanks.

1. Patient reports he needs to urinate many times during the day and even gets up three or four times at night to urinate.

   Symptom: ______________________________

2. Patient states she has to carry water with her at all times because she is always very thirsty.

   Symptom: ______________________________

3. Patient says he is always hungry, despite eating three big meals and several smaller ones each day.

   Symptom: ______________________________

C. **Rewrite** the following sentences in language a patient can understand. Review any terms you need to in the Glossary or a dictionary before you start writing.

1. "PTH and calcitonin are antagonistic: PTH stimulates bone resorption, whereas calcitonin stimulates bone deposition."

___

___

___

2. "Thyroid cancer usually presents as a symptomless nodule in the thyroid gland, but it can metastasize to cervical and mediastinal lymph nodes and to liver, lungs, and bones."

___

___

___

D. **Precision in communication:** Because of an error in communication, this patient was sent to the wrong specialist! Find the error. *Underline the incorrect medical terminology in each sentence:*

1. Because of this patient's neuropathy, I am referring him to a kidney specialist.

This sentence *should* have read:

Because of ___

___

You are ultimately responsible for everything you communicate regarding patient care!

2. Because of this patient's nephropathy, I am referring her to an immunologist.

This sentence *should* have read:

Because of ___

___

ENDOCRINE SYSTEM

E. **Prefixes:** The following prefixes appear in this chapter and have also appeared in previous chapters. Give the meaning of the prefix, and list two different terms in which the prefix appears. Fill in the blanks.

1. *anti* means ______________________

   Term from this chapter: ______________________

   Term from a previous chapter: ______________________

2. *endo* means ______________________

   Term from this chapter: ______________________

   Term from a previous chapter: ______________________

3. *hyper* means ______________________

   Term from this chapter: ______________________

   Term from a previous chapter: ______________________

4. *ex* means ______________________

   Term from this chapter: ______________________

   Term from a previous chapter: ______________________

5. *ad* means ______________________

   Term from this chapter: ______________________

   Term from a previous chapter: ______________________

6. *epi* means ______________________

   Term from this chapter: ______________________

   Term from a previous chapter: ______________________

7. *neo* means ______________________

   Term from this chapter: ______________________

   Term from a previous chapter: ______________________

8. *poly* means ______________________

   Term from this chapter: ______________________

   Term from a previous chapter: ______________________

9. *pan* means ______________________

   Term from this chapter: ______________________

   Term from a previous chapter: ______________________

F. **Recall and review:** This exercise on word elements contains elements from a previous chapter. Try to recall the previous elements without turning back in your book. Check (✔) the type of element; then write its meaning. Fill in the chart.

	Type of Element			
Element	**Prefix**	**Root/CF**	**Suffix**	**Meaning of Element**
a				
ad				
ambulat				
centesis				
glut				
later				
menisc				
necrot				
quadri				
vascul				

G. **Spelling demons:** The following terms from this chapter are particularly difficult to spell and pronounce. Correct pronunciation and spelling of medical terms is the mark of an educated professional. Circle the correct spelling, and then check (✔) that you have practiced the pronunciation. *Remember:* Pronunciations are on the Student Online Learning Center (www.mhhe.com/AllanEssMedLanguage).

			Pronunciation ✔
1. emaciation	emmaciation	emacciation	
2. cretenism	creitenism	cretinism	
3. epinephrine	epineprine	epinephrin	
4. exofthalmos	exophalmus	exophthalmos	
5. thyrotoxicosis	tyrotoxicosis	thyrotoxicossis	
6. hyperpexia	hyperprexia	hyperpyrexia	
7. paresthesia	peresthesia	parestia	
8. facies	fascies	feces	
9. misedema	mixedema	myxedema	

CHAPTER 10 REVIEW

ENDOCRINE SYSTEM

H. Latin and Greek terms cannot be further deconstructed into prefix, root, or suffix. You must know them for what they are. Test your knowledge of these terms in this exercise. Match the meaning in column 1 with the correct medical term in column 2.

_____ 1. gland located in the mediastinum	A. thyroid
_____ 2. facial features and expression	B. acetone
_____ 3. salt solution	C. facies
_____ 4. severe muscle twitches	D. goiter
_____ 5. person with severe hypothyroidism	E. insulin
_____ 6. gland and cartilage in neck	F. coma
_____ 7. hormone produced by pancreas	G. tetany
_____ 8. deeply unconscious	H. saline
_____ 9. ketone found in blood	I. thymus
_____ 10. enlargement of thyroid gland	J. cretin

I. Terminology challenge: Find the two roots in this chapter that have the same meaning. (*Hint:* It is a body organ.)

1. The roots ______________________ and ______________________ both mean ______________________.

Think of two terms in which these roots appear (one term for each root). Write a sentence for each of these terms.

2. First term: ______________________

   Sentence: ______________________

   ____

   ____

3. Second term: ______________________

   Sentence: ______________________

   ____

   ____

J. Patient education: Briefly explain to your patient the difference between *tetany* and *tetanus.* Be sure to use nonmedical language the patient can understand. Consult the Glossary or a dictionary if you need to. Fill in the blanks.

1. *tetany:* ______________________

   ____

   ____

2. *tetanus:* ______________________

   ____

   ____

K. Elements are your clues in the following terms. You do not necessarily have to know what an entire term means, but if you recognize an element in the term, it will help you answer the questions. Choose the correct terms from among this group to fit the descriptions. Some blanks may need more than one term, and there are extra terms you will not use. Fill in the blanks.

nephropathy	comatose	adrenalectomy
retinopathy	polyuria	saline
hypoglycemia	antidiuretic	serotonin
thyroidectomy	endocrine	nephrectomy
prolactin	panhypopituitarism	ketoacidosis

1. terms connected to urine:

____

2. terms for blood conditions:

____

3. terms that are procedures:

____

4. terms connected to milk

____

5. terms describing a disease or condition:

____

L. Deconstruct the following terms into the meanings of their elements. First, slash the term into elements. Then write the appropriate elements *on the line,* and the meaning of each element *under the line.* Fill in the blanks.

1. endocrinology ____________ / ____________ / ____________
2. acromegaly ____________ / ____________ / ____________
3. panhypopituitarism ____________ / ____________ / ____________
4. polydipsia ____________ / ____________ / ____________
5. exophthalmos ____________ / ____________ / ____________
6. thyrotoxicosis ____________ / ____________ / ____________

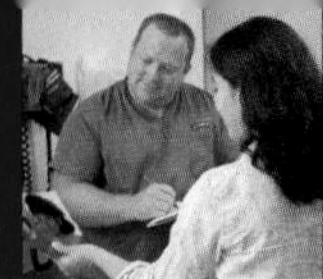

ENDOCRINE SYSTEM

M. **Short answer:** Write a brief description of each of the three terms below. Check your writing for correct spelling, and be ready to read your answers aloud in class. Fill in the blanks.

1. *insulin-dependent:* ______________________________

2. *insulin shock:* ______________________________

3. *insulin resistance:* ______________________________

N. **Brain teasers.**

1. In the term *thyrotoxicosis,* the element *toxic/o* means ______________________________

   What, then, is a *toxicologist?* ______________________________

   Where might this occupation be employed, and why? ______________________________

2. "Most diabetics develop some degree of *retinopathy, nephropathy,* and *neuropathy.*"

   Which body systems are affected by each of these diagnoses?

   retinopathy ______________________________

   nephropathy ______________________________

   neuropathy ______________________________

   Name the specialist a patient would consult for each of these conditions:

   retinopathy ______________________________

   nephropathy ______________________________

   neuropathy ______________________________

O. **Roots** are the fundamental core of every term. Test yourself on the roots below, found in the language of endocrinology. Write the root *on the line,* and write the meaning of just the root *under the line.* Based on the elements, you will not have to fill in every blank. Don't forget to answer the questions at the end of the exercise.

1. an agent that increases urine production — di/ __________ /ic
2. medical specialty concerned with hormones — endo/ __________ /logy
3. neurotransmitter in CNS — __________ / __________ /in
4. enlargement of extremities due to excess GH — acro/ __________ / __________
5. pituitary hormone that stimulates uterus to contract — __________ / __________ /in
6. hormone that stimulates secretion of milk — pro/ __________ /in
7. protrusion of the eyeball — ex/ __________ / __________
8. extremely high fever — hyper/ __________ / __________
9. severe or prolonged hypothyroidism — myx/ __________ / __________
10. disorder produced by excessive thyroid hormone production — __________ / __________ /osis
11. Based on what you see in the empty blanks above, notice that not every term needs a __________.
12. Sometimes, a __________ will start the term.

P. **Word attack.** Follow the instructions below the question.

Question:

Which is the only term that contains two prefixes?

a. endocrine
b. antidiuretic
c. hypothalamus
d. diabetic
e. neuropathy

> *Study Hint*
> Pay attention to small details!

1. Read the question and possible answers twice.
2. Notice that the question is asking for the term with *two* prefixes, not just one.
3. Slash each term into elements first. Since the question is asking about prefixes, analyze the answer choices starting at the beginning of the term, where the prefix would usually occur. *Remember that not every element at the beginning of a term has to be a prefix. Some roots or combining forms can start a term.*
4. Which term(s) can be eliminated because they have *no* prefix? __________

Remember to cross them out of the answer choices when you have eliminated them as possible correct answers.

5. Which terms among the remaining choices have a prefix? __________
6. Therefore, the only term that has *two* prefixes is __________.

ENDOCRINE SYSTEM

Q. Specialists: This chapter introduced you to an *endocrinologist.* Write below the names of five other specialists you have met in previous chapters, and give one disease or condition each would treat. Fill in the blanks.

	Specialist	Disease/Condition
1.	______________________	______________________
2.	______________________	______________________
3.	______________________	______________________
4.	______________________	______________________
5.	______________________	______________________

Have you checked your spelling?

R. Chapter challenge: Circle the correct answer.

Study Hint
Immediately cross off any answer you know is not correct. In your remaining choices, there is only *one best answer.*

1. Which term has an element that means *black?*
   - a. oxytocin
   - b. melatonin
   - c. serotonin
   - d. pineal
   - e. acromegaly

2. *Hyperpyrexia, tetany,* and *edema* can all be considered:
   - a. diseases
   - b. symptoms
   - c. diagnoses
   - d. a, b, and c
   - e. only b and c

3. In the abbreviation *BUN,* the "B" stands for:
   - a. blood
   - b. bile
   - c. bilirubin
   - d. bariatric
   - e. bone

4. Which term contains an element meaning *all:*
   - a. prolactin
   - b. panhypopituitarism
   - c. antagonist
   - d. parathyroid
   - e. exophthalmos

5. What is *serotonin* converted to at night?
   - a. antidiuretic
   - b. prostaglandin
   - c. oxytocin
   - d. prolactin
   - e. melatonin

6. Mrs. Tacher had an "enlarged heart"; the medical term for this is:
   - a. cardiopulmonary
   - b. cardiomyopathy
   - c. cardiomegaly
   - d. cardiography
   - e. carditis

7. This hormone reduces the volume of urine produced by the kidneys and is also called *vasopressin:*

a. FSH
b. LH
c. ACTH
d. PRL
e. ADH

8. Find the incorrectly spelled term:

a. cretenism
b. antagonist
c. myxedema
d. aldosterone
e. norepinephrine

9. *Idiopathic* means that a disease:

a. is just starting
b. has no known cause
c. is in the acute stage
d. is contagious
e. has no known cure

10. The symptoms of hyperthyroidism are *tachycardia* and *hypertension.* These are the same as:

a. increased heart rate and high blood pressure
b. decreased heart rate and low blood pressure
c. decreased heart rate and high blood pressure
d. increased heart rate and low blood pressure
e. none of these

11. Identify the only pair of terms that are spelled correctly:

a. aldostirone facies
b. hydrocortison mineralocorticoid
c. epinephrin somatostatin
d. glycogenolisis gluconeogenesis
e. mellitus paresthesia

12. What is the only hormone that can lower blood glucose?

a. GH
b. insulin
c. oxytocin
d. glucagon
e. ADH

13. In the term *hypothalamus,* the element *hypo* means:

a. excessive
b. deficient
c. below
d. beside
e. between

14. A term that contains a prefix in the middle of the word is:

a. gluconeogenesis
b. somatostatin
c. epinephrine
d. adrenocorticotropic
e. thyrotoxicosis

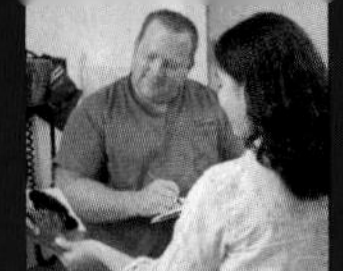

CHAPTER 10 REVIEW

ENDOCRINE SYSTEM

S. Case Report challenge: Now that you are more comfortable with the terms in this chapter, you can apply that knowledge and briefly answer the questions about the Case Report. If you read the report through and underline all the medical terminology, this will make it easier to answer the questions.

You are communicating with

. . . The parents of Ms. Norma Leary, a 22-year-old college student living with her parents for the summer. Norma is **emaciated,** extremely agitated, and at times disoriented and confused. Her parents tell you that in the past 3 or 4 days she has been coughing and not feeling well. In the past 12 hours she has become feverish and been complaining of a left-sided chest pain. With questioning, the parents reveal that prior to this acute illness Norma had lost about 20 pounds in weight, although she was eating voraciously. Her **VS** are T 105.2, P 180 and irregular, R 24, BP 160/85.

You call for Dr. Hilinski STAT. On his initial examination he believes that the patient is in thyroid storm. This is a medical emergency. There are no immediate laboratory tests that can confirm this diagnosis.

Thyroid storm is the condition Ms. Norma Leary presented with in the Emergency Department. It is the most extreme state of hyperthyroidism, with severely exaggerated effects of the thyroid hormones causing **hyperpyrexia,** tachycardia, agitation, and delirium. The weight loss prior to her illness becoming acute was part of her undiagnosed **hyperthyroidism.**

1. Norma is described as "emaciated"—how would she look?

____

____

2. List all of the patient's symptoms:

____

____

____

3. Use a dictionary to look up the word *voraciously.* Define it here:

____

____

4. After looking up the term above, what does *not* make sense about the patient's weight loss?

____

____

5. What is the opposite of an *acute* illness? _______________________________

6. What other specialist might be called in to consult regarding this patient's tachycardia? _______________________

7. Does this patient have an overactive or underactive thyroid condition?

____

Write your own study notes for this chapter.

The Blood, Lymphatic, and Immune Systems

The Essentials of the Languages of Hematology and Immunology

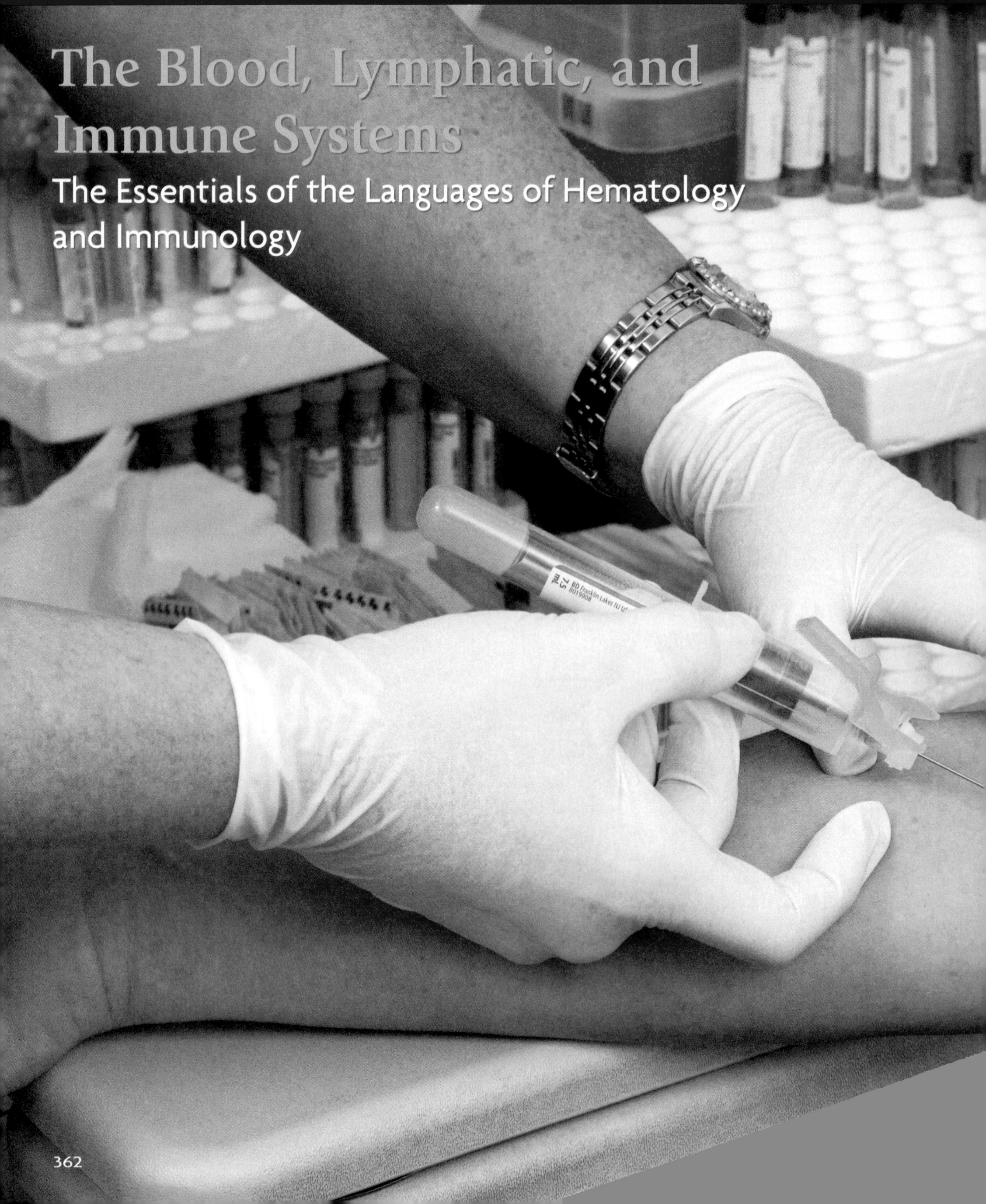

CHAPTER

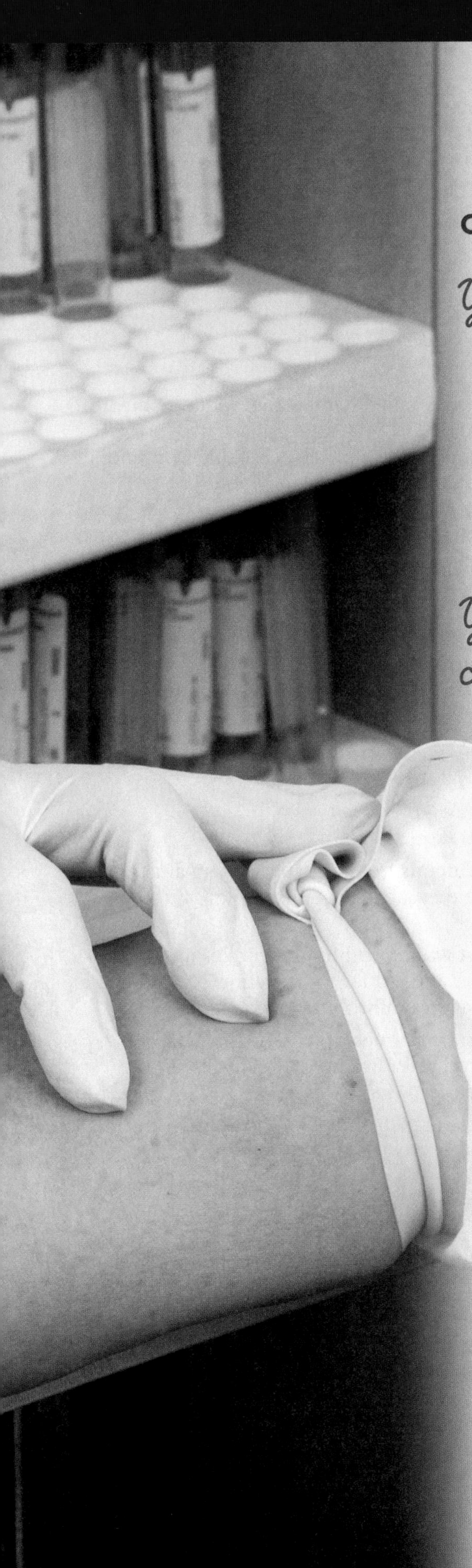

CHAPTER 11

CASE REPORT 11.1

You are

. . . a medical assistant employed by Susan Lee, MD, a primary care physician at Fulwood Medical Center.

You are communicating with

. . . Ms. Luisa Sosin, a 47-year-old woman who presented a week ago with fatigue, lethargy, and muscle weakness. Physical examination revealed pallor of her skin, a pulse rate of 90, and a respiratory rate of 20. Dr. Lee referred her for extensive blood work. She also determined that Ms. Sosin had been taking aspirin and NSAIDs (nonsteroidal anti-inflammatory drugs) for the past 6 months for low-back pain. Ms. Sosin's laboratory report shows an **iron-deficiency anemia**.

You are responsible for documenting her investigation and care. In order to be able to communicate intelligently with Dr. Lee about the patient, and to document Ms. Luisa Sosin's medical care, you need to have knowledge of the anatomy, physiology, and medical terminology of hematology.

Learning Outcomes

This chapter will provide you with information that enables you to:

11.1 Apply the language of hematology to the anatomy and physiology of the blood.

11.2 Comprehend, analyze, spell, and write the medical terms of hematology so that you can communicate and document accurately and precisely.

11.3 Recognize and pronounce the medical terms of hematology so that you can communicate verbally with accuracy and precision.

11.4 Use correct medical terminology to explain the effects of common disorders of the blood on health.

LESSON 11.1 Components and Functions of Blood

THE BLOOD, LYMPHATIC, AND IMMUNE SYSTEMS

OBJECTIVES

The information in this lesson relates to the composition, functions, and uses of blood and will enable you to:

- **Identify the medical terminology for the components of blood.**
- **Name the medical terms used to describe plasma and its functions.**
- **Apply correct medical terminology to explain the functions of blood.**

COMPONENTS OF BLOOD

The study of the blood and its disorders is called **hematology**, and the medical specialist in this area is called a **hematologist.**

Blood is a type of connective tissue and consists of cells contained in a liquid matrix. An average-sized adult has about 5 liters (10 pints) of blood.

If a specimen of blood is collected in a tube and centrifuged, the cells of the blood are packed into the bottom 45% of the tube (*Figure 11.1*). These cells consist of 99% red blood cells **(RBCs)**, together with white blood cells **(WBCs)** and **platelets.** The **hematocrit (Hct)** is the percentage of total blood volume composed of red blood cells.

Keynote

Serum is identical to plasma except for the absence of clotting proteins.

Plasma is the remaining 55% of the blood sample. It is a clear, yellowish liquid that is 91% water. Plasma is a **colloid,** a liquid that contains suspended particles, most of which are plasma proteins. **Nutrients,** waste products, hormones, and enzymes are dissolved in plasma for transportation. When blood is allowed to clot and the solid clot is removed, **serum** is left.

FUNCTIONS OF BLOOD

The functions of the blood are to:

1. **Maintain the body's homeostasis** (*see Chapter 2*).
2. **Transport nutrients, vitamins, and minerals** from digestive system and storage areas to organs and cells where they are needed. Examples of nutrients are glucose and amino acids (*see Chapter 5*).
3. **Transport waste products** from cells and organs to the liver and kidney for excretion. Examples are creatinine, urea, bilirubin, and lactic acid.
4. **Transport hormones** from endocrine glands to the target cells. Examples are insulin and thyroxine (*see Chapter 10*).
5. **Transport gases** to and from the lungs and cells. Examples are oxygen and carbon dioxide (*see Chapter 7*).
6. **Protect against foreign substances.** Cells and chemicals in the blood are an important part of the immune system to deal with microorganisms and toxins.
7. **Form clots.** This provides protection against blood loss and is the first step in tissue repair and restoration of normal function.

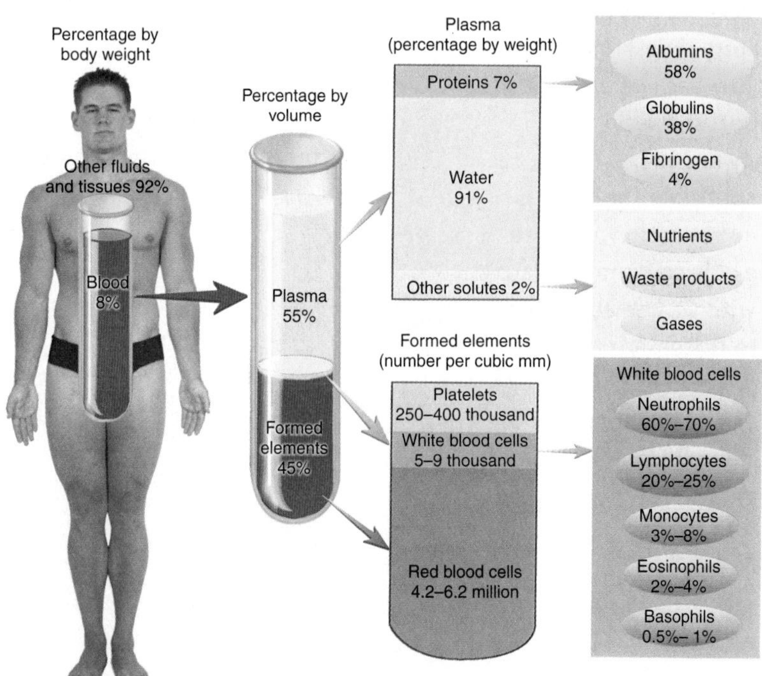

FIGURE 11.1 Components of Blood.

WORD ANALYSIS AND DEFINITION

S/ = Suffix P/ = Prefix R/ = Root R/CF = Combining Form

WORD	PRONUNCIATION	ELEMENTS		DEFINITION
anemia	ah-**NEE**-me-ah	P/ R/	**an-** *without* **-emia** *a blood condition*	Decreased number of red blood cells.
anemic (adj)	ah-**NEE**-mik	R/	**-emic** *pertaining to a condition of the blood*	Pertaining to or suffering from anemia.
colloid	**COLL**-oyd	S/ R/	**-oid** *resembling* **coll-** *glue*	Liquid containing suspended particles.
hematocrit (Hct)	**HE**-mat-oh-krit	S/ R/CF	**-crit** *to separate* **hemat/o-** *blood*	Percentage of red blood cells in the blood.
hematology	he-mah-**TOL**-oh-jee	S/ R/CF	**-logy** *study of* **hemat/o-** *blood*	Medical specialty of the blood and its disorders.
hematologist	he-mah-**TOL**-oh-jist	S/	**-logist** *one who studies, specialist*	Specialist in hematology.
nutrient	**NYU**-tree–ent		Latin *to nourish*	Constituent of food necessary for the body to function normally.
nutrition	nyu-**TRISH**-un	S/ R/	**-ion** *action, condition* **nutrit-** *nourishment*	The study of food and liquid requirements for normal function of the human body.
plasma	**PLAZ**-mah		Greek *something formed*	Fluid, noncellular component of blood.
platelet (Also called **thrombocyte.**)	**PLAYT**-let	S/ R/	**-let** *little, small* **plate-** *flat*	Small particle involved in the clotting process.
serum	**SEER**-um		Latin *whey*	Fluid remaining after removal of cells and fibrin clot.
vitamin (***Note:*** The duplicate letter "a" is omitted. It was originally thought that all vitamins were amines.)	**VYE**-tah-min	S/ R/	**-amin(e)** *nitrogen-containing substance* **vita-** *life*	Essential organic substance necessary in small amounts for normal cell function.

Abbreviations

Hct hematocrit
RBC red blood cell
WBC white blood cell

EXERCISES

Match *the definition in column 1 with the correct medical term in column 2. Fill in the blanks.*

_____ 1. suspension of particles in liquid — A. platelet
_____ 2. decreased number of RBCs — B. plasma
_____ 3. study of blood disorders — C. anemia
_____ 4. essential for normal cell function — D. hematocrit
_____ 5. fluid remaining after removal of clot — E. colloid
_____ 6. fluid noncellular component of blood — F. serum
_____ 7. also called a thrombocyte — G. vitamin
H. hematology

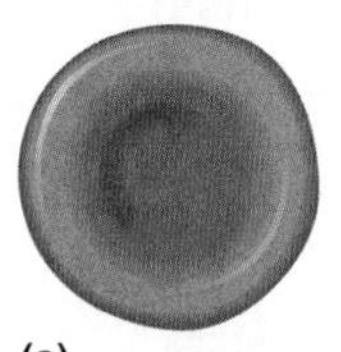

(a)

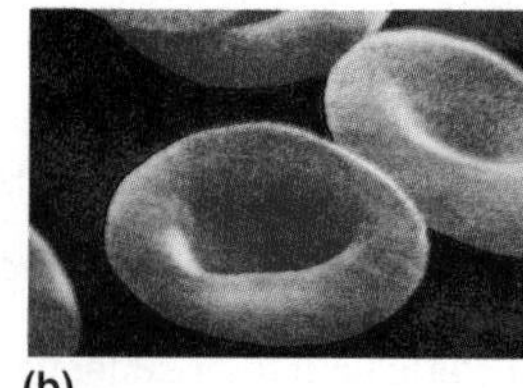

(b)

▲ **FIGURE 11.2 Red Blood Cells.**
(a) Top view. (b) Side view.

Keynote

RBCs are unable to move themselves and are dependent on the heart and blood flow to move them around the body.

Keynote

The average life span of an RBC is 120 days, during which time the cell circulates through the body about 75,000 times.

Keynote

Anemia produces **pallor** (pale color) because of the deficiency of the red-colored **oxyhemoglobin**, the combination of oxygen and hemoglobin.

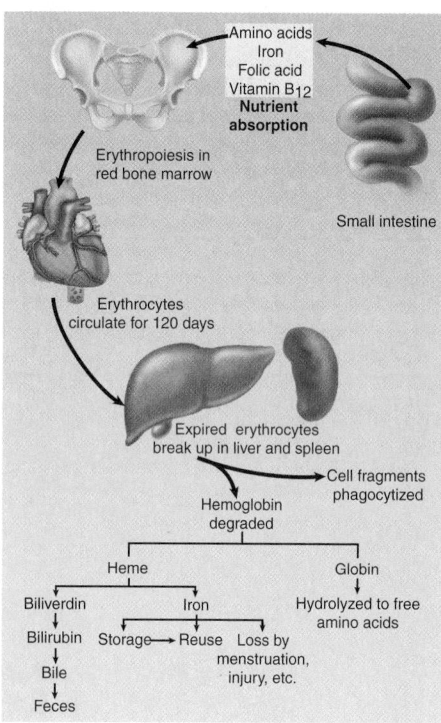

▲ **FIGURE 11.3 Life and Death of RBCs.**

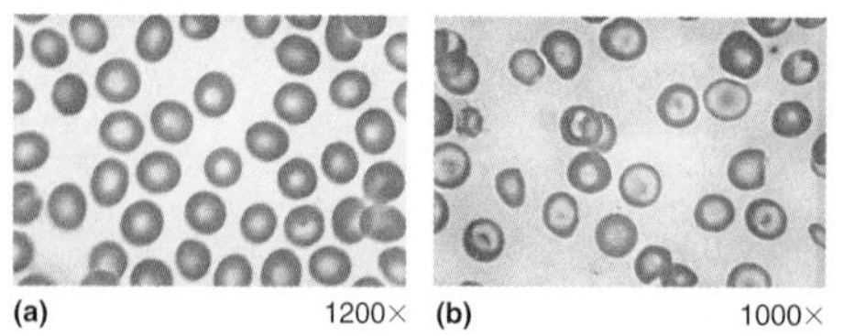

▲ **FIGURE 11.4 Red Blood Cells.**
(a) Normal RBCs. (b) Iron-deficient RBCs.

Keynote

Hemolysis liberates hemoglobin from RBCs.

STRUCTURE, FUNCTIONS, AND DISORDERS OF RED BLOOD CELLS (ERYTHROCYTES)

Structure of RBCs (Erythrocytes)

Each RBC is a disk with edges that are thicker and raised above the flattened center (*Figure 11.2*). This biconcave surface area enables a more rapid flow of gases into and out of the RBC.

The main component of RBCs is **hemoglobin (Hb),** which gives the cell and blood its red color. Hb is composed of the iron-containing pigment **heme** bound to a protein called **globin.** The rest of the red blood cell consists of the cell membrane, water, electrolytes, and enzymes. Mature RBCs do not have a nucleus.

Functions of RBCs (Erythrocytes)

The functions of the RBCs are to:

1. **Transport oxygen (O_2)** from the lungs to the cells all over the body. Oxygen is transported in combination with hemoglobin.
2. **Transport carbon dioxide (CO_2)** from the tissue cells to the lungs for excretion.
3. **Transport nitric oxide (NO),** a gas produced by the lining cells of blood vessels that signals smooth muscle to relax. Nitric oxide is also a transmitter of signals between nerve cells.

With age, RBCs become more fragile, and squeezing through tiny capillaries ruptures them. **Macrophages** in the liver and spleen take up the hemoglobin that is released and break it down into its components heme and the protein globin. The heme is broken down into iron and into an orange pigment called bilirubin (*Figure 11.3*).

Disorders of Red Blood Cells

Anemia is a reduction in the number of RBCs or in the amount of hemoglobin each RBC contains (*Figure 11.4*). Both of these reduce the oxygen-carrying capacity of the blood and produce the symptoms of shortness of breath **(SOB)** and fatigue.

There are several types of anemia.

Iron-deficiency anemia is the diagnosis for Ms. Luisa Sosin (*Case Report 11.1*). The cause was chronic bleeding from her gastrointestinal tract due to the aspirin and other painkillers she was taking. Her stools were positive for occult blood. Other causes of iron-deficiency anemia can be heavy menstrual bleeds or a diet deficient in iron.

Pernicious anemia (PA) is due to vitamin B_{12} deficiency. It is caused by a shortage of intrinsic factor, which is normally secreted by cells in the lining of the stomach and binds with vitamin B_{12}. This complex is absorbed into the bloodstream. Without vitamin B_{12}, hemoglobin cannot be formed; the red cells are decreased in number and in hemoglobin concentration and increased in size.

Sickle cell anemia is a genetic disorder found most commonly in African-Americans. It is due to production of an abnormal hemoglobin that causes the RBCs to form a rigid sickle cell shape (*Figure 11.5*). The abnormal cells **agglutinate** (clump together) and block small capillaries. This causes intense pain in the **hypoxic** tissues (a sickle cell crisis) and can cause stroke, kidney failure, and heart failure. There is a minor form of the disease, **sickle cell trait,** in which symptoms rarely occur.

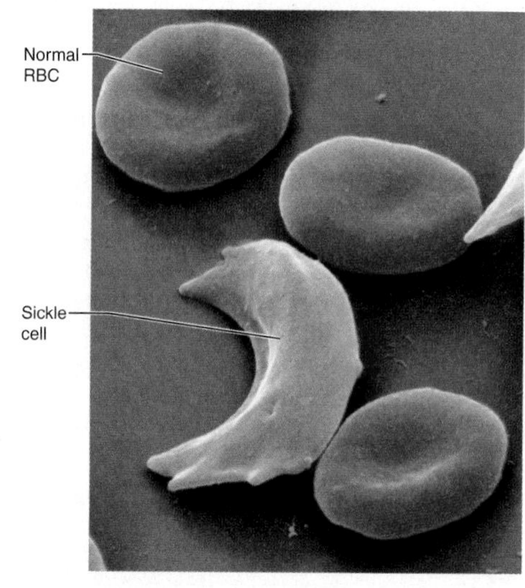

▲ **FIGURE 11.5 Sickle Cell Disease.**

Hemolytic anemia is due to excessive destruction of normal and abnormal RBCs. **Hemolysis,** the destruction

WORD ANALYSIS AND DEFINITION

S/ = Suffix P/ = Prefix R/ = Root R/CF = Combining Form

WORD	PRONUNCIATION	ELEMENTS		DEFINITION
agglutinate (verb)	ah-**GLUE**-tin-ate	S/ P/ R/	**-ate** *composed of, pertaining to* **ag- (ad-)** *to* **-glutin-** *stick*	Stick together to form clumps.
erythrocyte	eh-**RITH**-roh-site	S/ R/CF	**-cyte** *cell* **erythr/o-** *red*	Red blood cell (RBC).
heme	HEEM		Greek *blood*	The iron-based component of hemoglobin that carries oxygen.
hemoglobin **oxyhemoglobin** **hemoglobinopathy**	**HE**-moh-**GLOW**-bin **OCK**-see-he-moh-**GLOW**-bin **HE**-moh-**GLOW**-bih-**NOP**-ah-thee	R/ R/CF R/ S/ R/CF R/CF	**-globin** *protein* **hem/o-** *blood* **oxy-** *oxygen* **-pathy** *disease* **-hem/o-** *blood* **-globin/o-** *protein*	Red-pigmented protein that is the main component of red blood cells. Combination of hemoglobin and oxygen. Disease caused by the presence of an abnormal hemoglobin in red blood cells.
hemolysis **hemolytic (adj)**	he-**MOL**-ih-sis he-moh-**LIT**-ik	S/ R/CF S/ R/	**-lysis** *destruction* **hem/o-** *blood* **-ic** *pertaining to* **-lyt-** *destroy*	Destruction of red blood cells so that hemoglobin is liberated. Pertaining to the destruction of red blood cells.
hypoxia (***Note:*** The duplicate letter "o" is omitted.) **hypoxic (adj)**	high-**POCK**-see-ah high-**POCK**-sik	S/ P/ R/ S/	**-ia** *condition* **hypo-** *below, deficient* **-ox-** *oxygen* **-ic** *pertaining to*	Below-normal levels of oxygen in tissues, gases, or blood. Deficient in oxygen.
macrophage	**MACK**-roh-fayj	P/ R/	**macro-** *large* **-phage** *to eat*	Large white blood cell that removes bacteria, foreign particles, and dead cells.
pallor	**PAL**-or		Latin paleness	Paleness of the skin.
pernicious anemia	per-**NISH**-us ah-**NEE**-me-ah	S/ P/ R/ P/ R/	**-ous** *pertaining to* **per-** *through* **-nici-** *lethal* **an-** *without* **-emia** *a blood condition*	Chronic anemia due to lack of vitamin B_{12}.
trait	TRAYT		Latin *an extension*	A discrete characteristic that has a known quality.

of RBCs, can be caused by toxic substances such as snake and spider venoms, mushroom toxins, drug reactions, trauma to RBCs by hemodialysis or heart-lung machines, or an incompatible blood transfusion.

Aplastic anemia is a condition in which the bone marrow is unable to produce sufficient new cells of all types—red cells, white cells, and platelets. It can be associated with exposure to radiation, benzene, and certain drugs. The diagnosis is made by bone marrow aspiration. Treatment is by suppression of the immune system and, in severe cases, a bone marrow transplant.

Abbreviations

CO_2	carbon dioxide
Hb or **Hgb**	hemoglobin
NO	nitric oxide
O_2	oxygen
PA	pernicious anemia
SOB	short(ness) of breath

EXERCISES

Deconstruct *the medical term that appears in the first column. Separate with slashes the basic elements that form each term. Fill in the chart.*

Medical Term	Meaning of Prefix	Meaning of Root(s)/CF	Meaning of Suffix	Meaning of Term
agglutinate				
anemia				
erythrocyte				
hemolysis				
hypoxic				
macrophage				
oxyhemoglobin				

TYPES AND FUNCTIONS OF WHITE BLOOD CELLS (LEUKOCYTES)

Granulocytes

1. **Neutrophils** (*Figure 11.6*), also called **polymorphonuclear leukocytes (PMNLs),** are normally 55% to 65% of the total WBC count. These cells phagocytize bacteria, fungi, and some viruses. In **neutropenia** the number of neutrophils is decreased. In **neutrophilia** the number is increased.
2. **Eosinophils** (*Figure 11.7*) are normally 2% to 4% of the total WBC count. They leave the bloodstream to enter tissue undergoing an allergic response. In allergic reactions, the number and percentage of eosinophils increase.
3. **Basophils** (*Figure 11.8*) are normally less than 1% of the total WBC count. Basophils migrate to damaged tissues to release histamine (which increases blood flow) and heparin (which prevents blood clotting).

Because of their granular cytoplasm, the above three types of WBCs are called **granulocytes.** Their granules are sites for the production of enzymes and chemicals.

Agranulocytes

4. **Monocytes** (*Figure 11.9*) are the largest blood cell and are normally 3% to 8% of the total WBC count. Monocytes leave the bloodstream and become macrophages that phagocytize bacteria, dead neutrophils, and dead cells in the tissues.
5. **Lymphocytes** (*Figure 11.10*) are normally 25% to 35% of the total WBC count. They are the smallest WBC. Lymphocytes are produced in red bone marrow and migrate through the bloodstream to lymphatic tissues—lymph nodes, tonsils, spleen, and thymus—where they proliferate.

   There are two main types of lymphocyte:

   a. **B cells** differentiate into plasma cells. These are stimulated by bacteria or toxins to produce antibodies or immunoglobulins **(Ig)** (*see Chapter 11*).

   b. **T cells** attach directly to foreign antigen-bearing cells such as bacteria, which they kill with toxins they secrete.

In a laboratory report, a **differential white blood cell count (DIFF)** lists the percentages of the different leukocytes in a blood sample.

Disorders of White Blood Cells

Normally a cubic millimeter (mm^3) of blood contains 5,000 to 10,000 white blood cells.

In **leukocytosis** the total WBC count exceeds 10,000 per cubic millimeter.

- Allergic reactions increase the number of eosinophils.
- Typhoid fever, malaria, and tuberculosis increase the number of monocytes.
- Whooping cough and infectious mononucleosis increase the number of lymphocytes.

Infectious mononucleosis occurs in the 15- to 25-year-old population. Its cause, the **Epstein-Barr virus (EBV),** is a very common virus, a member of the herpes virus family. The EBV is transmitted by exchange of saliva, as in kissing.

Leukemia is cancer of the blood-forming tissues and produces a high number of leukocytes and their precursors. As the **leukemic** cells proliferate, they take over the bone marrow and cause a deficiency of normal red blood cells, white blood cells, and platelets. This makes the patient anemic and vulnerable to infection and bleeding.

In **leukopenia,** the WBC count drops below 5,000 cells per cubic millimeter of blood. Leukopenia is seen in viral infections such as measles, mumps, chickenpox, poliomyelitis, and AIDS.

In **pancytopenia,** the erythrocytes (red blood cells), leukocytes (white blood cells), and thrombocytes (platelets) in the circulating blood are all markedly reduced. This can occur with cancer chemotherapy.

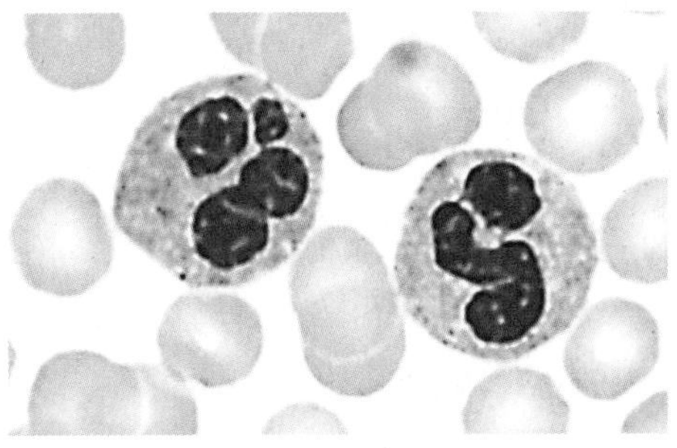

▲ **FIGURE 11.6 Neutrophils Are Granulocytes.**

Keynote

Neutrophils, eosinophils, and basophils are granulocytes.

Keynote

Because monocytes and lymphocytes have no granules in their cytoplasm, they are called **agranulocytes.**

Keynote

Leukocytosis is too many white blood cells and often indicates the presence of an infection.

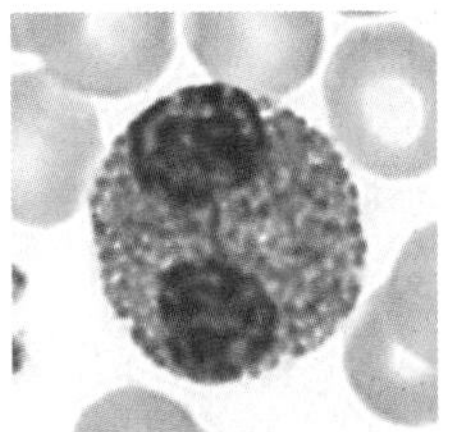

▲ **FIGURE 11.7 Eosinophils Are Granulocytes.**

▲ **FIGURE 11.8 Basophils Are Granulocytes.**

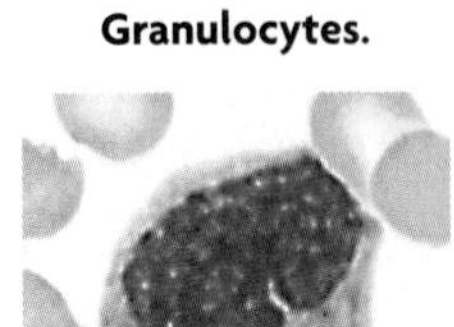

▲ **FIGURE 11.9 Monocytes Are Agranulocytes.**

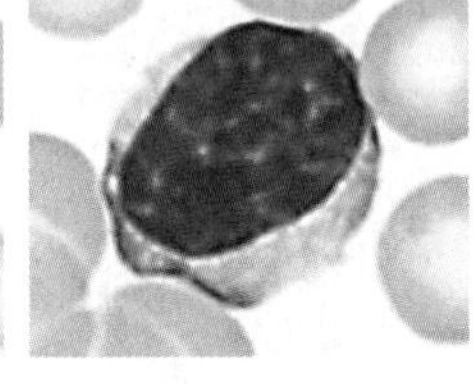

▲ **FIGURE 11.10 Lymphocytes Are Agranulocytes.**

Keynote

Leukopenia is too few white blood cells.

Keynote

Pancytopenia is too few RBCs, WBCs, and platelets.

Abbreviations

DIFF	differential white blood cell count
EBV	Epstein-Barr virus
Ig	immunoglobulin
PMNL	polymorphonuclear leukocyte

WORD ANALYSIS AND DEFINITION

S/ = Suffix   P/ = Prefix   R/ = Root   R/CF = Combining Form

WORD	PRONUNCIATION	ELEMENTS		DEFINITION
agranulocyte	a-**GRAN**-you-loh-site	S/ P/ R/CF	**-cyte** *cell* **a-** *without, not* **-granul/o-** *granule*	A white blood cell without any granules in its cytoplasm.
basophil	**BAY**-so-fill	S/ R/CF	**-phil** *attraction* **bas/o-** *base*	A basophil's granules attract a basic blue stain in the laboratory.
eosinophil	ee-oh-**SIN**-oh-fill	S/ R/CF	**-phil** *attraction* **eosin/o-** *dawn*	An eosinophil's granules attract a rosy-red color on staining.
granulocyte	**GRAN**-you-loh-site	S/ R/CF	**-cyte** *cell* **granul/o-** *small grain*	A white blood cell that contains multiple small granules in its cytoplasm.
leukemia **leukemic (adj)**	loo-**KEE**-mee-ah loo-**KEE**-mik	S/ R/ S/	**-emia** *a blood condition* **leuk-** *white* **-emic** *pertaining to a blood condition*	Disease when the blood is taken over by white blood cells and their precursors. Pertaining to or affected by leukemia.
leukocyte **leucocyte (syn)** (***Note:*** Either spelling is acceptable.)	**LOO**-koh-site	S/ R/CF	**-cyte** *cell* **leuk/o-** *white*	Another term for a white blood cell.
leukocytosis **leukopenia**	**LOO**-koh-sigh-**TOE**-sis loo-koh-**PEE**-nee-ah	S/ S/	**-osis** *condition* **-penia** *deficiency*	An excessive number of white blood cells. A deficient number of white blood cells.
lymphocyte	**LIM**-foh-site	S/ R/CF	**-cyte** *cell* **lymph/o-** *lymph*	Small white blood cell with a large nucleus.
monocyte	**MON**-oh-site	R/ P/	**-cyte** *cell* **mono-** *single*	Large white blood cell with a single nucleus.
mononucleosis	**MON**-oh-nyu-klee-**OH**-sis	S/ P/ R/	**-osis** *condition* **mono-** *single* **-nucle-** *nucleus*	Presence of large numbers of specific, diagnostic mononuclear leukocytes.
neutrophil **neutropenia** **neutrophilia**	**NEW**-troh-fill **NEW**-troh-**PEE**-nee-ah **NEW**-troh-**FILL**-ee-ah	S/ R/CF S/ S/	**-phil** *attraction* **neutr/o-** *neutral* **-penia** *deficiency* **-philia** *attraction*	Neutrophils' granules take up purple stain equally, whether the stain is acid or alkaline. A deficiency of neutrophils. An increase in neutrophils.
pancytopenia	**PAN**-site-oh-**PEE**-nee-ah	S/ P/ R/CF	**-penia** *deficiency* **pan-** *all* **-cyt/o-** *cell*	Deficiency of all types of blood cells.
polymorphonuclear	**POL**-ee-more-foh-**NEW**-klee-ah	S/ P/ R/CF R/	**-ar** *pertaining to* **poly-** *many* **-morph/o-** *shape* **-nucle-** *nucleus*	White blood cell with a multilobed nucleus.

EXERCISES

A **suffix** *can complete a medical term. The element* **leuko** *remains the same in each term below. Review the WAD before you start the exercise. Fill in the blanks.*

penia   **cyte**   **cytosis**

Leuko is a (type of element) ______________________ and means ______________________.

1. An excessive number of white blood cells is: ______________________
2. Another term for a white blood cell is: ______________________
3. A deficient number of white blood cells is: ______________________

CASE REPORT 11.2

You are

. . . an emergency medical technician (EMT) working in the Fulwood Medical Center Emergency Department.

You are communicating with

. . . Janis Tierney, a 17-year-old high school student, who presents with fainting at school. She is pale. Her pulse is 90, blood pressure 100/60. She tells you that she is having a menstrual period with excessive bleeding. Her physical examination is otherwise unremarkable. She has a past history of easy bruising and recurrent nosebleeds and an episode of severe bleeding after a tooth extraction.

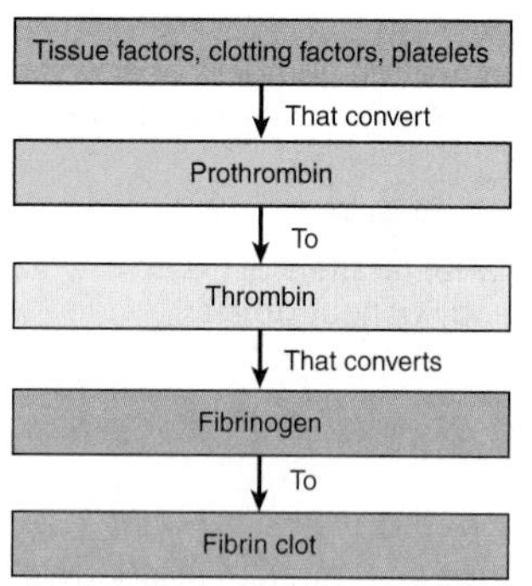

▲ **FIGURE 11.11 Blood Coagulation.**

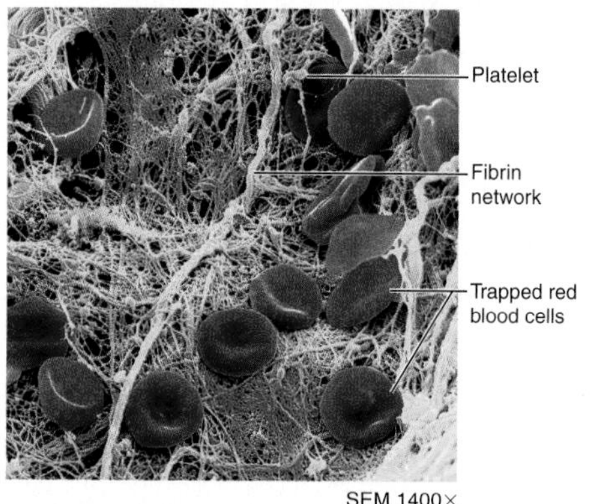

▲ **FIGURE 11.12 Blood Clot.**

Keynote

Hemostasis is the control of bleeding. *Homeostasis* is control of the state of equilibrium of the body.

Keynote

Thrombus is a clot. Embolus is a piece of thrombus that breaks off.

HEMOSTASIS

Hemostasis, the control of bleeding, is a vital issue in maintaining **homeostasis,** the state of equilibrium of the body. Uncontrolled bleeding can take the body out of balance by decreasing blood volume and lowering blood pressure.

Platelets (also called **thrombocytes**) play a key role in hemostasis. They are minute fragments of large bone marrow cells, and consist of a small amount of granular cytoplasm surrounded by a plasma membrane, and have no nucleus.

Hemostasis is achieved through a three-step mechanism:

1. **Vascular spasm,** an immediate but temporary constriction of the injured blood vessels.
2. **Platelet plug formation,** an accumulation of platelets that bind themselves together and adhere to surrounding tissues. The binding and adhesion of platelets is mediated through **von Willebrand factor (vWF),** a protein produced by the cells lining blood vessels.
3. **Blood coagulation,** the process of going through **prothrombin** and **thrombin** to the formation of a blood clot that traps blood cells, platelets, and tissue fluid in a network of **fibrin** (*Figures 11.11 and 11.12*).

After a blood clot forms, platelets adhere to strands of fibrin and contract to pull the fibers and the edges of the broken blood vessel together. **Fibroblasts** invade the clot to produce a fibrous connective tissue that seals the blood vessel.

Disorders of Coagulation (Coagulopathies)

Hemophilia in its classical form (hemophilia A) is a disease males inherit from their mothers and is due to a deficiency of a coagulation factor called **factor VIII.**

Von Willebrand disease (vWD) is a deficiency of a specific protein of the factor VIII complex different from the part involved in hemophilia and is the most common hereditary bleeding disorder.

Disseminated intravascular coagulation (DIC) occurs when a severe bacterial infection activates the clotting mechanism simultaneously throughout the cardiovascular system. Small clots form and obstruct blood flow into tissues and organs, particularly the kidney, leading to renal failure.

Thrombus formation **(thrombosis)** is a clot that forms attached to a diseased or damaged area on the walls of blood vessels or the heart. If part of the thrombus breaks loose and moves through the circulation, it is called an **embolus.**

Thrombocytopenia is a deficiency of platelets.

Purpura is bleeding into the skin from small arterioles that produces a larger individual lesion than **petechiae** from capillary bleeds (*Figure 11.13a and b*). **Bruises** (or **hematomas**) are leaks of blood from all types of blood vessels (*Figure 11.13c*).

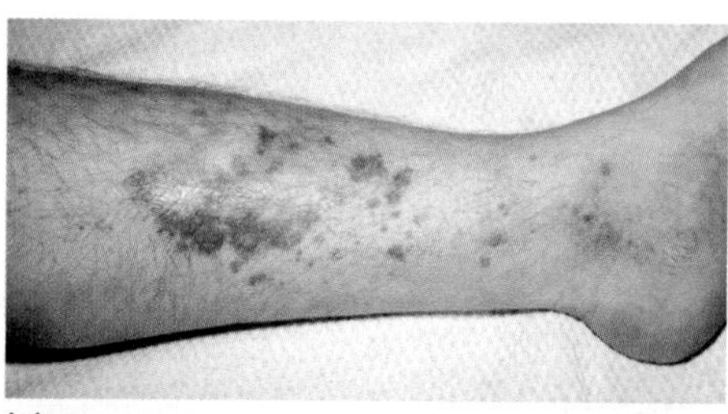

(a)

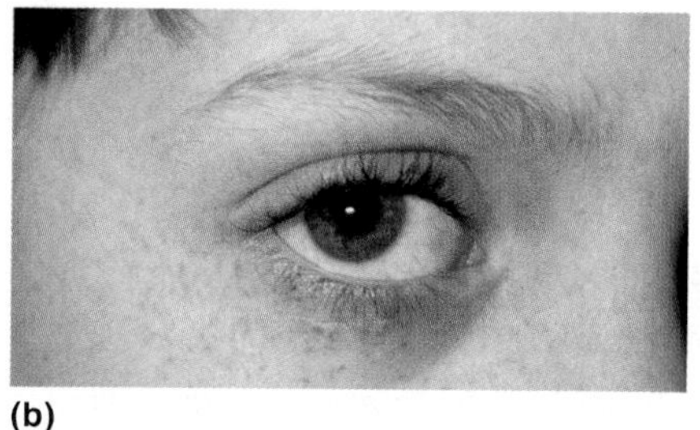

(b)

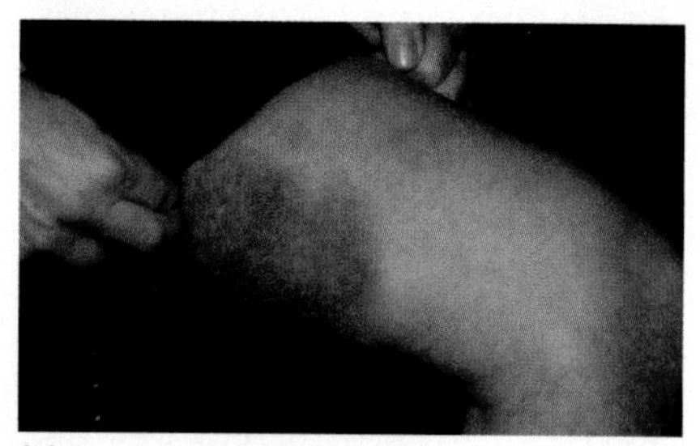

(c)

▲ **FIGURE 11.13 Subsurface Bleeding.** (a) Purpura. (b) Petechiae. (c) Bruises.

WORD ANALYSIS AND DEFINITION

S/ = Suffix   P/ = Prefix   R/ = Root   R/CF = Combining Form

WORD	PRONUNCIATION	ELEMENTS		DEFINITION
coagulant	koh-**AG**-you-lant	S/ R/	**-ant** *forming, pertaining to* **coagul-** *clot, clump*	Substance that causes clotting.
coagulation **anticoagulant**	koh-ag-you-**LAY**-shun **AN**-tee-koh-**AG**-you-lant	S/ P/	**-ation** *process* **anti-** *against*	Process of blood clotting. Substance that prevents clotting.
embolus	**EM**-boh-lus		Greek *plug, stopper*	Detached piece of thrombus, a mass of bacteria, quantity of air, or foreign body that blocks a blood vessel.
fibrin	**FIE**-brin		Latin *fiber*	Stringy protein fiber that is a component of a blood clot.
fibroblast	**FIE**-broh-blast	S/ R/CF	**-blast** *immature cell* **fibr/o-** *fiber*	Cell that forms collagen fibers.
hematoma (also called **bruise**)	he-mah-**TOE**-mah	S/ R/	**-oma** *mass, tumor* **hemat-** *blood*	Collection of blood that has escaped from vessels into surrounding tissues.
hemophilia	he-moh-**FILL**-ee-ah	S/ R/CF	**-philia** *attraction* **hem/o-** *blood*	An inherited disease from a deficiency of clotting factor VIII.
hemostasis (***Note:*** *Homeo*stasis has a very different meaning.)	he-moh-**STAY**-sis	S/ R/CF	**-stasis** *control, stop* **hem/o-** *blood*	Control of or stopping bleeding.
petechia **petechiae (pl)**	peh-**TEE**-kee-ah peh-**TEE**-kee-ee		Latin *spot on the skin*	Pinpoint capillary hemorrhagic spot in the skin.
prothrombin	pro-**THROM**-bin	S/ P/ R/	**-in** *substance* **pro-** *before* **-thromb-** *blood clot*	Protein formed by the liver and converted to thrombin in the blood-clotting mechanism.
purpura	**PUR**-pyu-rah		Greek *purple*	Skin hemorrhages that are red initially and then turn purple.
thrombocyte (also called **platelet**) **thrombocytopenia**	**THROM**-boh-site **THROM**-boh-site-oh-**PEE**-nee-uh	S/ R/CF S/ R/CF	**-cyte** *cell* **thromb/o-** *blood clot* **-penia** *deficiency* **-cyt/o-** *cell*	Another name for a platelet. Deficiency of platelets in circulating blood.

Case Report 11.2 (continued)

Janis Tierney, who presented in the ED with heavy menstrual bleeding, has a deficiency of von Willebrand factor (vWF). Her platelets are unable to stick together, and a platelet plug cannot form in the lining of her uterus to help end her menstrual flow.

Abbreviations

DIC	disseminated intravascular coagulation
vWD	von Willebrand disease
vWF	von Willebrand factor

EXERCISES

Review *the two pages open in front of you to determine the answers to the following questions. Circle the best choice.*

1. In the term *hemostasis,* the suffix means:
   blood   condition   control
2. Which term relates to a color?
   hemophilia   petechia   purpura
3. The process of blood clotting is called:
   homeostasis   coagulation   thrombocytopenia
4. A piece of thrombus that has broken off into the circulation is called:
   an embolus   a fibroblast   a coagulant
5. A substance used to prevent blood clotting is called:
   an anticoagulant   a hematoma   a thrombus
6. The suffix *ation* signifies:
   a procedure   a process   an excision

You are

... an emergency medical technician–paramedic (EMT-P) working in the Level One Trauma Unit at Fulwood Medical Center.

You are communicating with

... Ms. Joanne Rodi, an 18-year-old student, who has been admitted to the unit from the operating room.

CASE REPORT 11.3

Ms. Rodi has had surgery for multiple fractures in a car accident. She is receiving a blood **transfusion**. You document that her temperature has risen to 102.8°, her respirations have risen to 24 per minute, and she has chills. You take her blood pressure, and it has fallen to 90/60. What is your next step?

BLOOD GROUPS AND TRANSFUSIONS

Red Cell Antigens

On the surfaces of red blood cells are molecules called **antigens.** In the plasma, **antibodies** are present. Each antibody can combine with only a specific antigen. If the plasma antibodies combine with another red cell antigen, bridges are formed that connect the red cells together. This is called **agglutination,** or clumping, of the cells. Hemolysis (destruction) of the cells also occurs.

The antigens on the surfaces of the cells have been categorized into groups, of which two are the most important. These are the **ABO** and **Rh** blood groups.

ABO Blood Group

The two major antigens on the cell surface are antigen A and antigen B.

A person with only antigen A has ***type* A** blood.

A person with only antigen B has ***type* B** blood.

A person with both antigen A and antigen B has ***type* AB** blood.

A person with neither antigen has ***type* O** blood and is a universal donor, able to give blood to any other person.

Specific antibodies are synthesized in the plasma during the first 8 months after birth:

Whenever antigen A is absent, anti-A antibody is produced.

Whenever antigen B is absent, anti-B antibody is produced.

Figure 11.14 shows the different combinations of antigens and antibodies in the different blood types.

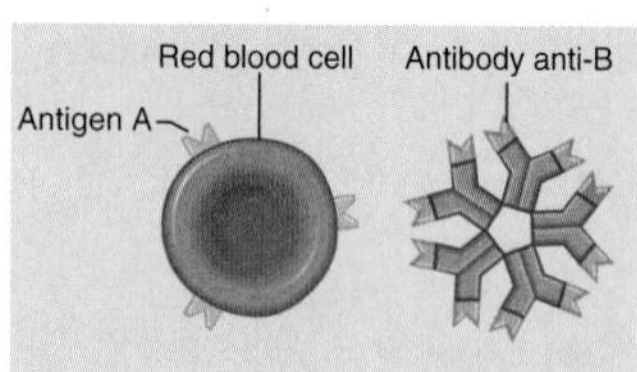

(a)

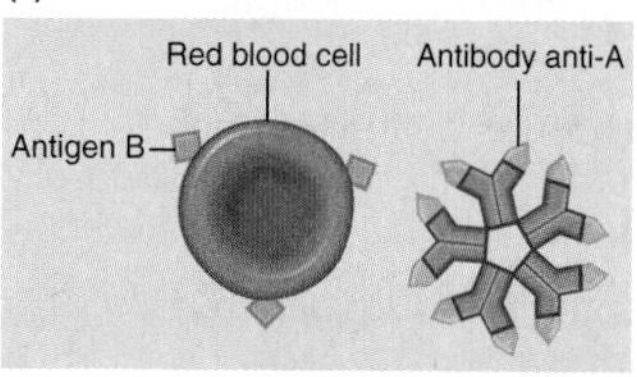

(b)

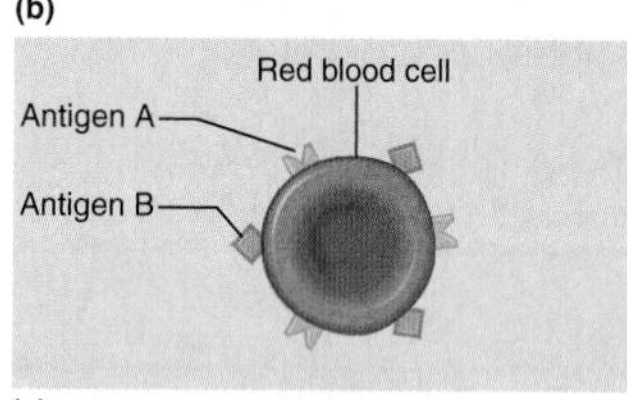

(c)

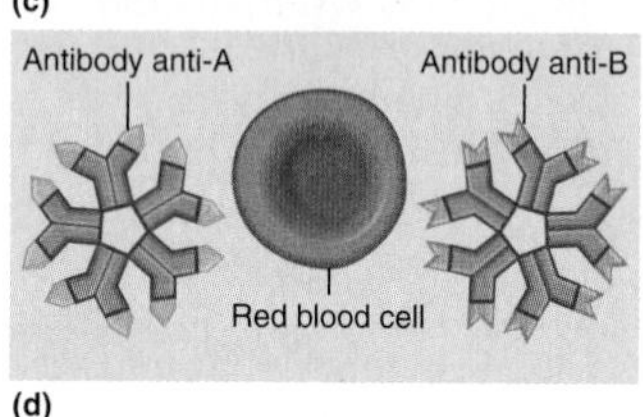

(d)

▲ **FIGURE 11.14 Blood Types.**
(a) Type A blood. (b) Type B blood. (c) Type AB blood. (d) Type O blood.

Case Report 11.3 (continued)

In Joann Rodi's case, she has type A blood and by mistake received blood of type AB, which agglutinated in the presence of her anti-B antibodies. Your immediate response is to stop the transfusion, replace it with a saline **infusion,** call your supervisor, and notify the doctor.

A **transfusion** of blood or packed red blood cells replaces lost red blood cells to restore the blood's oxygen-carrying capacity. **Autologous** donation and transfusion occurs when people donate their own blood ahead of time to be given to them if necessary during a surgical procedure.

Keynote

All blood groups are inherited.

Keynote

Rh factor is an antigen on the surface of a red blood cell. Its presence or absence is inherited.

Abbreviations

ABO	blood group system
HDN	hemolytic disease of the newborn
Rh	Rhesus
RhoGAM	Rhesus immune globulin

Rh Blood Group

If an Rh antigen is present on the red cell surface, the blood is said to be Rh-positive (Rh+). If there is no Rh antigen on the surface, the blood is Rh-negative (Rh−).

If an Rh-negative person receives a transfusion of Rh-positive blood, anti-Rh antibodies will be produced. This can cause RBC agglutination and hemolysis.

If an Rh-negative woman and an Rh-positive man conceive an Rh-positive child (*Figure 11.15a*), the placenta normally prevents maternal and fetal blood from mixing. However, at birth or during a miscarriage, fetal cells can enter the mother's bloodstream. These Rh-positive cells stimulate the mother's tissues to produce Rh antibodies (*Figure 11.15b*).

WORD ANALYSIS AND DEFINITION

S/ = Suffix P/ = Prefix R/ = Root R/CF = Combining Form

WORD	PRONUNCIATION	ELEMENTS		DEFINITION
agglutination (noun)	ah-glue-tih-**NAY**-shun	S/ P/ R/	**-ation** *process* **ag-** *to (same as ad-)* **-glutin-** *glue*	Process by which cells or other particles adhere to each other to form clumps.
antibody **antibodies (pl)**	**AN**-tee-body	P/ R/	**anti-** *against* **-body** *substance, body*	Protein produced in response to an antigen.
antigen	**AN**-tee-gen	P/ R/	**anti-** *against* **-gen** *produce, create*	Substance capable of triggering an immune response.
autologous	awe-**TOL**-oh-gus	P/ R/	**auto-** *self, same* **-logous** *relation*	Blood transfusion with the same person as donor and recipient—self-transfusion.
erythroblastosis fetalis	eh-**RITH**-roh-blast-oh-sis fee-**TAH**-lis	S/ R/CF R/ S/ R/	**-osis** *condition* **erythr/o-** *red* **-blast-** *immature cell* **-is** *belonging to* **fetal-** *fetus*	Erythroblastosis fetalis is a hemolytic disease of the newborn **(HDN).**
infusion	in-**FYU**-zhun	P/ R/	**in-** *in* **-fusion** *to pour*	Introduction intravenously of a substance other than blood.
transfusion	trans-**FYU**-zhun	P/ R/	**trans-** *across* **-fusion** *to pour*	Transfer of blood or a blood component from donor to recipient.

If the mother becomes pregnant with a second Rh-positive fetus, her Rh antibodies can cross the placenta and agglutinate and hemolyze the fetal red cells (*Figure 11.15c*). This causes hemolytic disease of the newborn **(HDN,** or **erythroblastosis fetalis)**.

Hemolytic disease of the newborn due to Rh incompatibility can be prevented. The Rh-negative mother giving birth to an Rh-positive child should be given Rh-immune globulin **(RhoGAM).**

Other causes of hemolytic disease in the newborn include ABO incompatibility, incompatibility in other blood group systems, hereditary spherocytosis, and some infections acquired before birth.

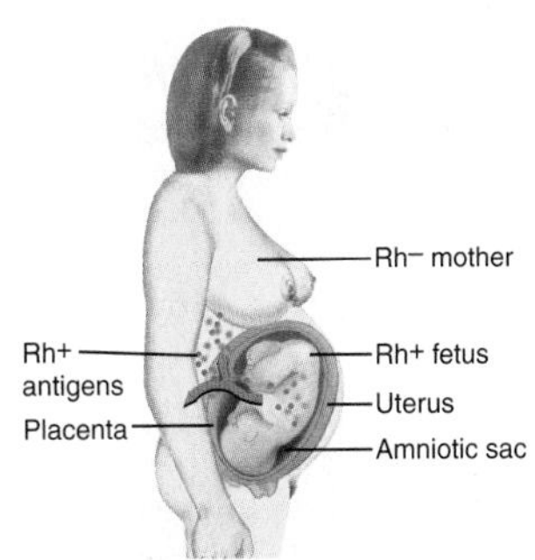

(a) First pregnancy

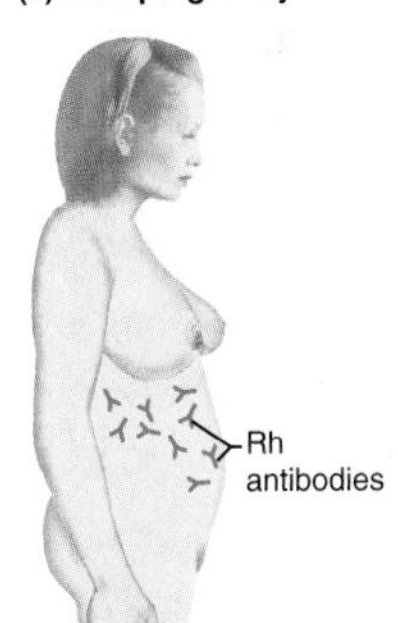

(b) Between pregnancy

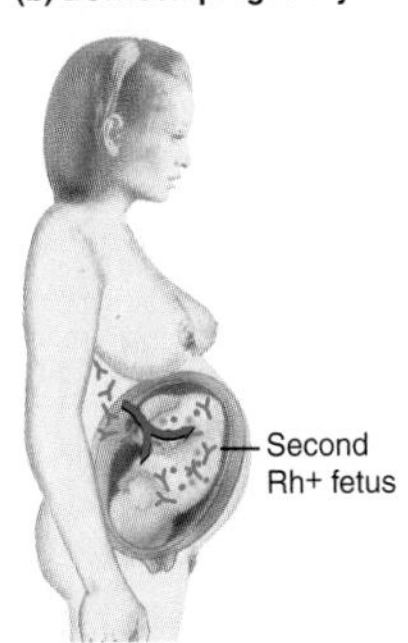

(c) Second pregnancy

▲ **FIGURE 11.15 Hemolytic Disease of the Newborn.** (a) First pregnancy. (b) Between pregnancies. (c) Second pregnancy.

EXERCISES

Review *the elements in the WAD before starting this exercise. Fill in the blanks.*

_____	1. trans	A. to pour
_____	2. anti	B. glue
_____	3. logous	C. immature cell
_____	4. auto	D. red
_____	5. osis	E. condition
_____	6. fusion	F. against
_____	7. erythro	G. relation
_____	8. glutin	H. across
_____	9. gen	I. produce
_____	10. blast	J. self

THE BLOOD, LYMPHATIC, AND IMMUNE SYSTEMS

LESSON 11.2 Lymphatic System

OBJECTIVES

You live in a world that surrounds you with chemicals and disease-causing organisms waiting for a chance to enter your body and harm you. Your body has **three lines of defense mechanisms** against foreign organisms **(pathogens),** cells **(cancer),** or molecules **(pollutants** and **allergens):**

1. **Physical,** including your skin and mucous membranes; chemicals in your perspiration, saliva, and tears; hairs in your nostrils; and cilia and mucus to protect your lungs.
2. **Cellular mechanisms,** based on defensive cells (lymphocytes) that directly attack suspicious cells such as cancer cells, transplanted tissue cells, or cells infected with viruses or parasites.
3. **Humoral defense mechanisms** *(see Lesson 3 in this chapter)*, based on antibodies **(Abs)** that are found in body fluids and bind to bacteria, toxins, and extracellular viruses, tagging them for destruction.

The physical mechanisms of defense are discussed in the individual body system chapters. As part of your defense mechanisms, the lymphatic system and its fluid provide surveillance and protection against foreign materials. In this lesson the information provided will enable you to:

- **Name the medical terminology for the anatomy of the** *lymphatic* **system.**
- **Detail the functions of the lymphatic system.**
- **Identify the major cells of the lymphatic system and their functions.**
- **Use correct medical terminology for the anatomy and functions of the lymph** *nodes, tonsils, thymus,* **and** *spleen.*
- **Recognize the common disorders of the lymphatic system.**

You are

. . . a medical assistant working with Susan Lee, MD, in her primary care clinic.

You are communicating with

. . . Ms. Anna Clemons, a 20-year-old waitress, who is a new patient. She has noticed a lump in her left neck.

CASE REPORT 11.4

On questioning, you elicit that Ms. Clemons has lost about 8 pounds in weight in the past couple of months, has felt tired, and has had some night sweats.

Her vital signs are normal. There are two firm, enlarged **lymph** nodes in her left neck in front of the sternocleidomastoid muscle. Physical examination is otherwise unremarkable.

WORD ANALYSIS AND DEFINITION

S/ = Suffix P/ = Prefix R/ = Root R/CF = Combining Form

WORD	PRONUNCIATION	ELEMENTS		DEFINITION
allergen (*Note:* The duplicate letter "g" is omitted.) **allergic (adj)** **allergy**	**AL**-er-jen ah-**LER**-jik **AL**-er-jee	S/ R/ R/ S/ S/	**-gen** *create* **all-** *other, strange* **-erg-** *work* **-ic** *pertaining to* **-ergy** *process of working*	Substance creating a hypersensitivity (allergic) reaction. Pertaining to or suffering from an allergy. Hypersensitivity to a particular allergen.
lymph	LIMF		Latin *clear, spring water*	A clear fluid collected from tissues and transported by lymph vessels to the venous circulation.
lymphatic (adj) **lymphoid (adj)**	lim-**FAT**-ik **LIM**-foyd	S/ R/ S/	**-atic** *pertaining to* **lymph-** *lymph* **-oid** *resembling*	Pertaining to lymph or the lymphatic system. Resembling lymphatic tissue.
node	NOHD		Latin *a knot*	A circumscribed mass of tissue.
pathogen	**PATH**-oh-jen	S/ R/CF	**-gen** *produce, create, form* **path/o-** *disease*	A disease-causing microorganism.
pollutant	poh-**LOO**-tant	S/ R/	**-ant** *pertaining to* **pollut-** *unclean*	Substance that makes an environment unclean or impure.
spleen	SPLEEN		Greek *spleen*	Vascular, lymphatic organ in left upper quadrant of abdomen.

Abbreviation

Ab	antibody

EXERCISES

Build *the correct medical terms that match the definitions given. Fill in the blanks.*

1. substance that makes the environment unclean or impure:

_______________ / _______________ / _______________
P R/CF S

2. substance creating a hypersensitivity reaction:

_______________ / _______________ / _______________
P R/CF S

3. resembling lymphatic tissue:

_______________ / _______________ / _______________
P R/CF S

4. a disease-causing microorganism:

_______________ / _______________ / _______________
P R/CF S

5. pertaining to lymph:

_______________ / _______________ / _______________
P R/CF S

6. hypersensitivity to a particular allergen:

_______________ / _______________ / _______________
P R/CF S

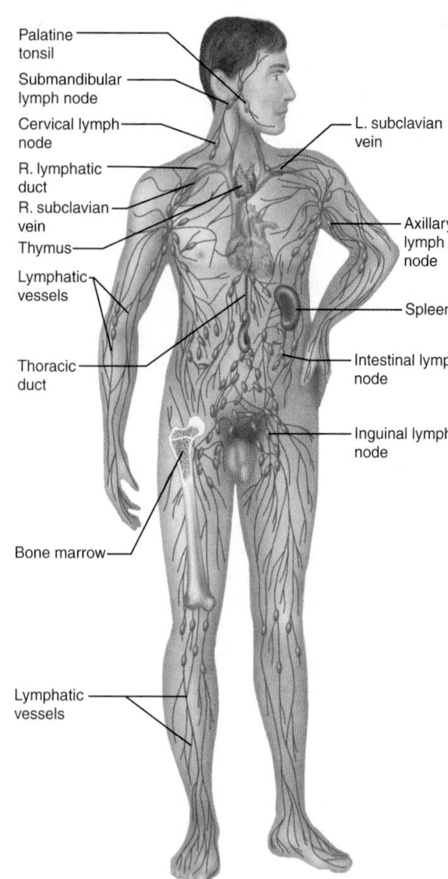

▲ **FIGURE 11.16 The Lymphatic System.** (R. = right; L. = left)

LYMPHATIC SYSTEM

The lymphatic system (*Figure 11.16*) has three components:

1. A network of thin **lymphatic capillaries and vessels,** similar to blood vessels, that penetrates the **interstitial** spaces of nearly every tissue in the body except cartilage, bone, red bone marrow, and the CNS.
2. A group of tissues and organs that produce **immune cells.**
3. **Lymph,** a clear colorless fluid similar to blood plasma but whose composition varies from place to place in the body. It flows through the network of lymphatic capillaries and vessels.

The lymphatic system has three functions:

1. **Absorb** excess interstitial fluid and return it to the bloodstream.
2. **Remove** foreign chemicals, cells, and debris from the tissues.
3. **Absorb** dietary lipids from the small intestine (*see Chapter 5*).

The **lymphatic network** begins with lymphatic capillaries that are closed-ended tubes nestled among **blood capillary networks** (*Figure 11.17*). The lymphatic capillaries are designed to let interstitial fluid enter, and the interstitial fluid becomes lymph. In addition, bacteria, viruses, cellular debris, and traveling cancer cells can enter the lymphatic capillaries with the interstitial fluid. The lymphatic capillaries converge to form the larger **lymphatic collecting vessels.** These resemble small veins and have one-way valves in their lumen. They travel alongside veins and arteries.

Lymph Nodes

At irregular intervals, the collecting vessels enter into the part of the lymphatic network called *lymph nodes* (*Figure 11.18*). There are hundreds of lymph nodes stationed all over the body (*Figure 11.16*). They are concentrated in the neck, axilla, and groin. Their functions are to filter impurities from the lymph and alert the immune system to the presence of pathogens.

The lymph moves slowly through the node, which filters the lymph and removes any foreign matter. Macrophages in the lymph nodes ingest and break down the foreign matter and display fragments of it to T cells. This alerts the immune system to the presence of an invader. Lymph leaves the nodes when it enters into the **efferent** collecting vessels. All these lymph vessels move lymph toward the thoracic cavity.

Collecting vessels merge into **lymphatic trunks** that drain lymph from a major body region. In turn, these lymphatic trunks merge into two large **lymphatic ducts,** the thoracic duct on the left and the right lymphatic duct, which empty into the subclavian veins (*Figure 11.16*).

▼ **FIGURE 11.17 Lymphatic Flow.**

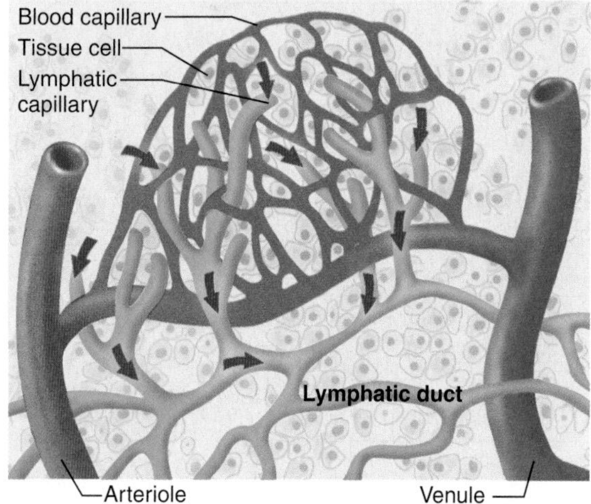

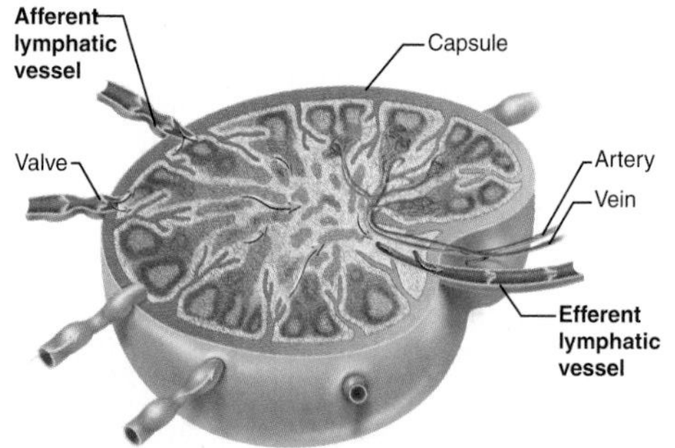

▲ **FIGURE 11.18 Lymph Node.**

Case Report 11.4 (continued)

Anna Clemons has cancerous nodes in her neck. They were caused not by metastatic cancer but by a cancer of the lymph nodes called **Hodgkin lymphoma,** which is detailed later on this page.

DISORDERS OF THE LYMPHATIC SYSTEM

Physicians routinely palpate accessible lymph nodes in the neck **(cervical nodes),** axilla (armpit) **(axillary nodes),** and groin **(inguinal nodes)** for enlargement and tenderness, which indicate disease in the tissues drained by the lymph nodes.

Cancerous lymph nodes are enlarged, firm, and usually painless. Infections in the lymph nodes cause them to be swollen and tender to the touch, a condition called **lymphadenitis.** All lymph node enlargements are collectively called **lymphadenopathy.** When lymph nodes are removed, the process is called **lymphadenectomy.**

Lymphoma is a malignant neoplasm of the lymphatic organs, usually the lymph nodes. Associated symptoms can be fever, night sweats, fatigue, and weight loss. Lymphomas are grouped into two categories by microscopic examination of affected lymphatic tissues:

Keynote

Treatment options for lymphomas include radiation, chemotherapy, and bone marrow transplant.

1. **Hodgkin lymphoma,** or **Hodgkin disease:** The cancer spreads in an orderly manner to adjoining lymph nodes. This enables the disease to be staged depending on how far it has spread.
2. **Non-Hodgkin lymphomas:** These occur much more frequently than Hodgkin lymphoma. They include some 30 different disease entities in 10 different subtypes.

Diagnostic procedures used include biopsy of an enlarged node, x-rays, CT and MRI scans, **lymphangiogram,** and bone marrow biopsy.

Tonsillitis, inflammation of the tonsils and adenoids, occurs mostly in infancy and childhood. The infection can be viral or bacterial, usually streptococcal. It produces enlarged, tender lymph nodes under the jaw. The infection can be recurrent, and **tonsillectomy** is sometimes performed (*see Chapter 7*). If the adenoids also show recurrent infection, adenoidectomy can be performed, often at the same time as tonsillectomy.

Keynote

The body can function without a spleen but is somewhat more vulnerable to infection.

Splenomegaly, an enlarged spleen, is not a disease in itself but the result of an underlying disorder. But, when the spleen enlarges, it traps and stores an excessive number of blood cells and platelets **(hypersplenism)** and reduces the number of blood cells and platelets in the bloodstream. Occasionally splenectomy is necessary.

Ruptured spleen is a common complication from car accidents or other trauma when the abdomen and rib cage are damaged. Intra-abdominal bleeding from the ruptured spleen can be extensive, with a dramatic fall in blood pressure. It is a surgical emergency requiring splenectomy.

Keynote

Edema due to water retention, when pressed with a finger, leaves a depression (pit). Lymphedema is brawny and pits minimally.

Lymphedema is localized, brawny, nonpitting fluid retention caused by a compromised lymphatic system, often after surgery or radiation therapy. It can also be primary, where the cause is unknown.

WORD ANALYSIS AND DEFINITION

S/ = Suffix P/ = Prefix R/ = Root R/CF = Combining Form

WORD	PRONUNCIATION		ELEMENTS	DEFINITION
adenoid	**ADD**-eh-noyd	S/ R/	**-oid** *resemble* **aden-** *gland*	Single mass of lymphoid tissue in the midline at the back of the throat.
adenoidectomy	**ADD**-eh-noy-**DEK**-toh-me	S/	**-ectomy** *surgical excision*	Surgical removal of the adenoid tissue.
bacterium **bacteria (pl)**	bak-**TEER**-ee-um bak-**TEER**-ee-ah		Greek *a staff*	A unicellular, simple, microscopic organism.
follicle	**FOLL**-ih-kull		Latin *a small sac*	Spherical mass of cells containing a cavity; or a small cul-de-sac, such as a hair follicle
immunoglobulin	**IM**-you-noh-**GLOB**-you-lin	S/ R/CF R/	**-in** *chemical compound, substance* **immun/o-***immune* **-globul-** *globular, protein*	Specific protein evoked by an antigen. All antibodies are immunoglobulins.
spleen	SPLEEN		Greek *spleen*	Vascular, lymphatic organ in left upper quadrant of abdomen.
splenectomy (***Note:*** single "e" in *splen*)	sple-**NECK**-toe-me	S/ R/	**-ectomy** *surgical excision* **splen-** *spleen*	Surgical removal of the spleen.
splenomegaly	sple-noh-**MEG**-ah-lee	S/ R/CF	**-megaly** *enlargement* **-splen/o-** *spleen*	Enlarged spleen.

Abbreviation

Ig	immunoglobulin

EXERCISES

Review *all the terms and elements in the WAD box. Pay careful attention to spelling and pronunciation. Answer the following questions by circling the best choice.*

1. In the term *adenoid* one of the elements means:
   tissue organ gland
2. The correct plural of *bacterium* is:
   bacteria bacteriae bacteriums
3. In addition to referring to a lymph follicle, the term *follicle* can also apply to:
   digits bones hair
4. What do *tonsil* and *adenoid* have in common?
   lymphatic tissue connective tissue adipose tissue
5. Immunoglobulin can also be considered:
   tissue collection pathogens antibodies
6. The suffix in the term *adenoidectomy* implies:
   surgical excision surgical fixation surgical incision
7. A microscopic organism is:
   bacteria follicle bacterium
8. The spleen is located in the:
   ABG RLQ LUQ
9. The specific protein evoked by an antigen is:
   bacterium enzyme immunoglobulin
10. The element meaning *enlargement* is:
    megaly ectomy oid

Keynote

Tissues that are the first line of defense against pathogens—for example, the airway passages—have lymphatic tissue in the submucous layers to help protect against invasion.

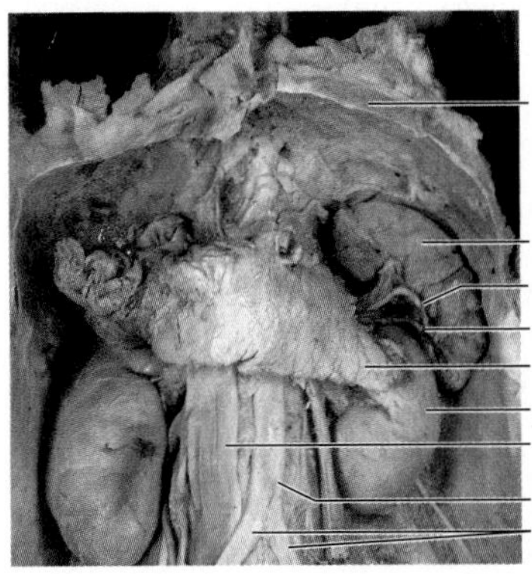

▲ **FIGURE 11.19 Position of Spleen.**

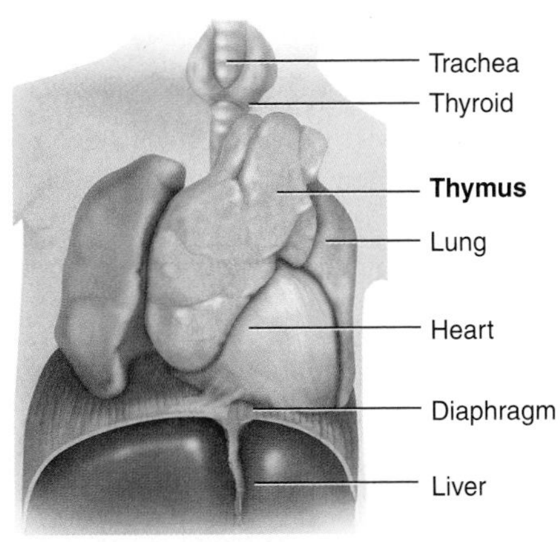

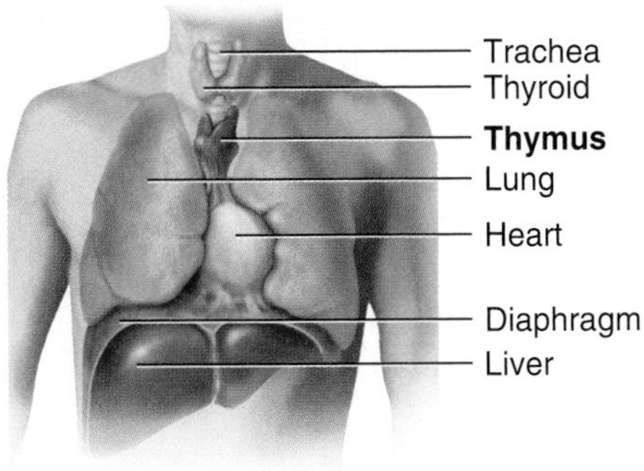

▲ **FIGURE 11.20 Thymus.**
(a) Large thymus in infant. (b) Adult thymus.

LYMPHATIC TISSUES AND CELLS

In some organs, lymphocytes and other cells form dense clusters called lymphatic **follicles.** These are constant features in lymph nodes and in the tonsils and the ileum.

Lymphatic tissues are composed of a variety of cells that include:

- **T lymphocytes (T cells):** The "T" stands for *thymus,* which is where they develop and mature. T lymphocytes make up 75% to 85% of body lymphocytes. There are several types of T-cells:
  - **Killer T cells** can recognize a specific antigen and destroy target cells.
  - **Helper T cells** begin the defense response against a specific antigen.
  - **Memory T cells** encounter and recognize a previously known antigen and quickly kill it.
  - **Suppressor T cells** suppress activation of the immune system. Failure of these cells to function properly may result in autoimmune diseases.
- **B lymphocytes (B cells):** These cells mature in the bone marrow. B lymphocytes make up 15% to 25% of lymphocytes. They respond to a specific antigen and become plasma cells to produce antibodies **(immunoglobulins, Ig)** that immobilize, neutralize, and prepare the specific antigen for destruction. Macrophages that have developed from monocytes (*see Chapter 6*) ingest and destroy antigens, tissue debris, **bacteria,** and other foreign matter **(phagocytosis).**

LYMPHATIC ORGANS

Spleen

The **spleen,** a highly vascular and spongy organ, is the largest lymphatic organ. It is located in the left upper quadrant of the abdomen, below the diaphragm and lateral to the kidney (*Figure 11.19*).

The functions of the spleen are to:

- **Phagocytose bacteria** and other foreign materials.
- **Initiate an immune response** when antigens are found in the blood.
- **Phagocytose old, defective erythrocytes** and platelets.
- **Serve as a reservoir** for erythrocytes and platelets.

Tonsils and Adenoids

The **tonsils** (*see Chapter 7*) are two masses of lymphatic tissue located at the entrance to the oropharynx that entrap inhaled and ingested pathogens. **Adenoids** are similar tissue on the posterior wall of the nasopharynx (*see Chapter 7*). The tonsils and adenoids form lymphocytes and antibodies, trap bacteria and viruses, and drain them into the tonsillar lymph nodes for elimination. They can become infected themselves.

Thymus

The thymus has both endocrine (*see Chapter 10*) and lymphatic functions. T lymphocytes develop and mature in the thymus and are released into the bloodstream. The thymus is largest in infancy (*Figure 11.20a*) and reaches its maximum size at puberty. It then shrinks (*Figure 11.20b*) and is eventually replaced by fibrous and adipose tissue.

WORD ANALYSIS AND DEFINITION

S/ = Suffix P/ = Prefix R/ = Root R/CF = Combining Form

WORD	PRONUNCIATION	ELEMENTS		DEFINITION
efferent	**EFF**-eh-rent	S/ R/	**-ent** *end result, pertaining to* **effer-** *move away from the center*	Moving away from a center.
afferent (***Note:*** These are opposite terms.)	**AFF**-eh-rent	R/	**affer-** *move toward the center*	Moving toward a center.
immune	im-**YUNE**		Latin *protected from*	Protected from an infectious disease.
immunity	im-**YOU**-nih-tee	S/ R/	**-ity** *condition* **immun-** *immune*	State of being protected.
immunology	im-you-**NOL**-oh-jee	S/ R/CF	**-logy** *study of* **immun/o-** *immune*	The science and practice of immunity and allergy.
immunologist	im-you-**NOL**-oh-jist	S/	**-logist** *one who studies, specialist*	Medical specialist in immunology.
immunize (verb)	**IM**-you-nize	S/ R/	**-ize** *affect in a specific way* **immun-** *immune*	Make resistant to an infectious disease.
immunization (noun)	**IM**-you-nih-**ZAY**-shun	S/	**-ization** *process of inserting or creating*	Administration of an agent to provide immunity.
interstitial	in-ter-**STISH**-al	S/ R/	**-al** *pertaining to* **interstiti-** *space between tissues*	Pertaining to spaces between cells in a tissue or organ.

EXERCISES

Precision *in usage is important if you want to communicate correct information. These six terms all contain a common root/combining form. Insert the correct term in each sentence, then match the elements correctly.*

immune **immunity** **immunization**

immunology **immunologist** **immunize**

1. One who specializes in (the study of the science of immunity and allergy) ______________________ is termed an (type of specialist) ______________________.

2. The ______________________ system is a group of specialized cells in different parts of the body that recognize foreign substances and neutralize them.

3. We need to ______________________ young children before they start school.

4. A prior (injection) ______________________ obtained before she went overseas boosted her (status of being immune) ______________________ to the disease.

____ 5. effer — A. pertaining to

____ 6. ity — B. space between tissues

____ 7. ization — C. away from center

____ 8. ent — D. process of inserting or creating

____ 9. affer — E. toward the center

____ 10. interstiti — F. condition

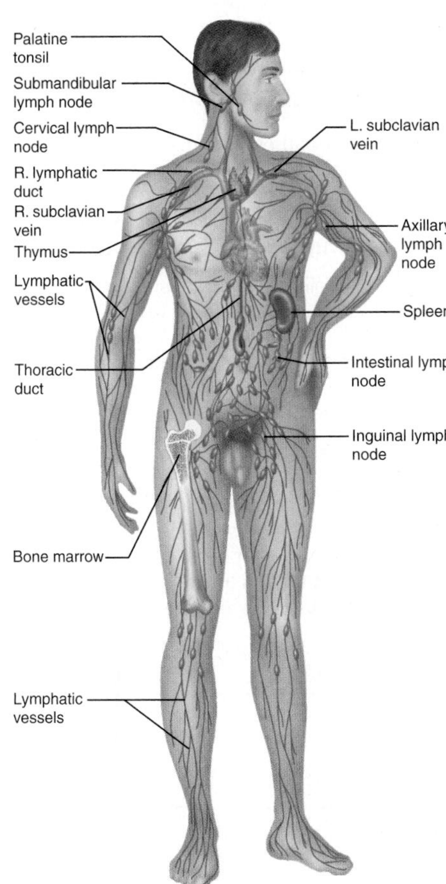

▲ **FIGURE 11.16 The Lymphatic System.** (R. = right; L. = left)

LYMPHATIC SYSTEM

The lymphatic system (*Figure 11.16*) has three components:

1. A network of thin **lymphatic capillaries and vessels,** similar to blood vessels, that penetrates the **interstitial** spaces of nearly every tissue in the body except cartilage, bone, red bone marrow, and the CNS.
2. A group of tissues and organs that produce **immune cells.**
3. **Lymph,** a clear colorless fluid similar to blood plasma but whose composition varies from place to place in the body. It flows through the network of lymphatic capillaries and vessels.

The lymphatic system has three functions:

1. **Absorb** excess interstitial fluid and return it to the bloodstream.
2. **Remove** foreign chemicals, cells, and debris from the tissues.
3. **Absorb** dietary lipids from the small intestine (*see Chapter 5*).

The **lymphatic network** begins with lymphatic capillaries that are closed-ended tubes nestled among **blood capillary networks** (*Figure 11.17*). The lymphatic capillaries are designed to let interstitial fluid enter, and the interstitial fluid becomes lymph. In addition, bacteria, viruses, cellular debris, and traveling cancer cells can enter the lymphatic capillaries with the interstitial fluid. The lymphatic capillaries converge to form the larger **lymphatic collecting vessels.** These resemble small veins and have one-way valves in their lumen. They travel alongside veins and arteries.

Lymph Nodes

At irregular intervals, the collecting vessels enter into the part of the lymphatic network called *lymph nodes* (*Figure 11.18*). There are hundreds of lymph nodes stationed all over the body (*Figure 11.16*). They are concentrated in the neck, axilla, and groin. Their functions are to filter impurities from the lymph and alert the immune system to the presence of pathogens.

The lymph moves slowly through the node, which filters the lymph and removes any foreign matter. Macrophages in the lymph nodes ingest and break down the foreign matter and display fragments of it to T cells. This alerts the immune system to the presence of an invader. Lymph leaves the nodes when it enters into the **efferent** collecting vessels. All these lymph vessels move lymph toward the thoracic cavity.

Collecting vessels merge into **lymphatic trunks** that drain lymph from a major body region. In turn, these lymphatic trunks merge into two large **lymphatic ducts,** the thoracic duct on the left and the right lymphatic duct, which empty into the subclavian veins (*Figure 11.16*).

▼ **FIGURE 11.17 Lymphatic Flow.**

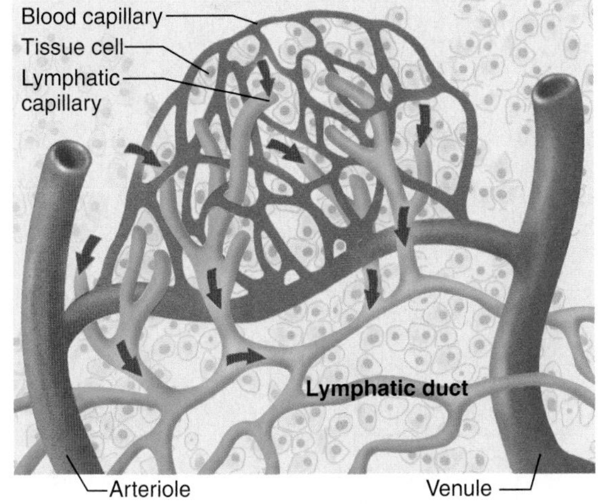

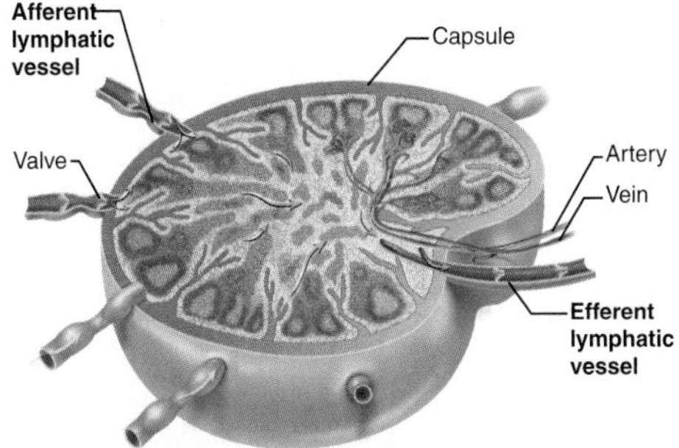

▲ **FIGURE 11.18 Lymph Node.**

WORD ANALYSIS AND DEFINITION

S/ = Suffix P/ = Prefix R/ = Root R/CF = Combining Form

WORD	PRONUNCIATION	ELEMENTS		DEFINITION
allergen (***Note:*** The duplicate letter "g" is omitted.) **allergic (adj)** **allergy**	**AL**-er-jen ah-**LER**-jik **AL**-er-jee	S/ R/ R/ S/ S/	**-gen** *create* **all-** *other, strange* **-erg-** *work* **-ic** *pertaining to* **-ergy** *process of working*	Substance creating a hypersensitivity (allergic) reaction. Pertaining to or suffering from an allergy. Hypersensitivity to a particular allergen.
lymph	LIMF		Latin *clear, spring water*	A clear fluid collected from tissues and transported by lymph vessels to the venous circulation.
lymphatic (adj) **lymphoid (adj)**	lim-**FAT**-ik **LIM**-foyd	S/ R/ S/	**-atic** *pertaining to* **lymph-** *lymph* **-oid** *resembling*	Pertaining to lymph or the lymphatic system. Resembling lymphatic tissue.
node	NOHD		Latin *a knot*	A circumscribed mass of tissue.
pathogen	**PATH**-oh-jen	S/ R/CF	**-gen** *produce, create, form* **path/o-** *disease*	A disease-causing microorganism.
pollutant	poh-**LOO**-tant	S/ R/	**-ant** *pertaining to* **pollut-** *unclean*	Substance that makes an environment unclean or impure.
spleen	SPLEEN		Greek *spleen*	Vascular, lymphatic organ in left upper quadrant of abdomen.

Abbreviation

Ab	antibody

EXERCISES

Build *the correct medical terms that match the definitions given. Fill in the blanks.*

1. substance that makes the environment unclean or impure:

_______________ / _______________ / _______________
P R/CF S

2. substance creating a hypersensitivity reaction:

_______________ / _______________ / _______________
P R/CF S

3. resembling lymphatic tissue:

_______________ / _______________ / _______________
P R/CF S

4. a disease-causing microorganism:

_______________ / _______________ / _______________
P R/CF S

5. pertaining to lymph:

_______________ / _______________ / _______________
P R/CF S

6. hypersensitivity to a particular allergen:

_______________ / _______________ / _______________
P R/CF S

WORD ANALYSIS AND DEFINITION

S/ = Suffix P/ = Prefix R/ = Root R/CF = Combining Form

WORD	PRONUNCIATION		ELEMENTS	DEFINITION
hypersplenism (***Note:*** one "e" in this term)	high-per-**SPLEN**-izm	S/ P/ R/	**-ism** *condition* **hyper-** *excessive* **-splen-** *spleen*	Condition in which the spleen removes blood components at an excessive rate.
lymphadenectomy	lim-**FAD**-eh-**NECK**-toe-me	S/ R/	**-ectomy** *surgical excision* **lymphaden-** *lymph node*	Surgical excision of a lymph node(s).
lymphadenitis **lymphadenopathy**	lim-**FAD**-eh-neye-tis lim-**FAD**-eh-**NOP**-ah-thee	S/ S/ R/CF	**-itis** *inflammation* **-pathy** *disease* **lymphaden/o-** *lymph node*	Inflammation of a lymph node(s). Any disease process affecting a lymph node(s).
lymphangiogram	lim-**FAN**-jee-oh-gram	S/ R/CF	**-gram** *recording* **lymphangi/o-** *lymphatic vessels*	Radiographic images of lymph vessels and nodes following injection of contrast material.
lymphedema	**LIMF**-eh-dee-mah	R/ R/	**lymph-** *lymph* **-edema** *edema*	Tissue swelling due to lymphatic obstruction.
lymphoma **Hodgkin**	lim-**FOH**-mah **HOJ**-kin	S/ R/	**-oma** *tumor, mass* **lymph-** *lymphatic system, lymph* Thomas Hodgkin, British physician, 1798–1866	Any neoplasm of lymphatic tissue. Hodgkin lymphoma is marked by chronic enlargement of lymph nodes spreading to other nodes in an orderly way.

EXERCISES

Elements *remain your best tool for understanding the meaning of a term. Match the meaning of the definition in column 1 with the correct element in column 2. Then use any term containing an element in column 2 in a sentence of patient documentation.*

	Column 1	Column 2
_____	1. recording	A. lymphangio
_____	2. excision	B. pathy
_____	3. excessive	C. inguin
_____	4. lymph node	D. oma
_____	5. inflammation	E. gram
_____	6. lymphatic vessels	F. itis
_____	7. tumor, mass	G. ectomy
_____	8. disease	H. hyper
_____	9. condition	I. lymphaden
_____	10. groin	J. ism

Sentence:

____

____

____

LESSON 11.3 Immune System

OBJECTIVES

The study of the immune system is called immunology. The medical specialist involved in the study and research of the immune system and in treating disorders of the immune system is called an immunologist. The information in this lesson will enable you to:

- **Define the medical terminology for the immune system and its characteristics.**
- **Contrast cellular and humoral immunity.**
- **Use correct medical terminology to explain the structure and actions of antibodies.**
- **Describe some common disorders of the immune system, including HIV and AIDS.**

You are

. . . a laboratory technician working the night shift at Fulwood Medical Center.

You are communicating with

. . . Mr. Michael Cowan, a 40-year-old homeless man and drug addict, who has presented to the Emergency Department with a high fever for which no cause is obvious on clinical examination.

CASE REPORT 11.5

You have been called to the emergency room to take blood from Mr. Cowan. You insert the needle into an **antecubital** vein, but he starts jerking his arm around and trying to get off the gurney. In the struggle, the needle comes out of the vein and pricks your hand through your glove.

As you immediately flush and clean the wound, report the incident, seek immediate medical attention, and go through your initial medical evaluation, it is essential that you have knowledge about your immune system and its response to the potential infection. Then you can make informed decisions about your treatment and future employment.

THE IMMUNE SYSTEM

Keynote

The immune system is not an organ system but a group of specialized cells in different parts of the body.

The immune system is a group of specialized cells in different parts of the body that recognize foreign substances and neutralize them. It is the **third line of defense** listed earlier in this chapter. When the immune system is weak, it allows pathogens, including the viruses that cause common colds and flu, and cancer cells to successfully invade the body.

Three characteristics distinguish immunity from the first two lines of defense of physical and cellular mechanisms:

1. **Specificity:** The immune response is directed against a specific pathogen. Immunity to one pathogen does not confer immunity on others. Specificity has one disadvantage. If a virus or a bacterium changes a component of its genetic code, it then becomes a new organism to the immune system. This **mutation** occurs, for example, with bacteria in response to antibiotics and in HIV's response to anti-HIV drugs **(development of resistance).**
2. **Memory:** When exposure to the same pathogen occurs again, the immune system recognizes the pathogen and has its responses ready to act quickly.
3. **Discrimination:** The immune system learns to recognize agents **(antigens)** that represent **"self"** and agents (antigens) that are **"non-self"** (foreign). Most of this recognition is developed prior to birth. A variety of disorders occur when this discrimination breaks down. They are known as **autoimmune disorders.**

Keynote

Some antigens are free molecules, such as **toxins.** Others are components of a cell membrane or a bacterial cell wall.

An **antigen** is any molecule that triggers an immune response. Most antigens are unique in their structure. It is this uniqueness that enables your body to distinguish its own (self) molecules from foreign (nonself) molecules.

WORD ANALYSIS AND DEFINITION

S/ = Suffix P/ = Prefix R/ = Root R/CF = Combining Form

WORD	PRONUNCIATION	ELEMENTS		DEFINITION
antecubital	an-teh-**KYU**-bit-al	S/ P/ R/	**-al** *pertaining to* **ante-** *in front of, before* **-cubit-** *elbow*	In front of the elbow.
autoimmune	awe-toe-im-**YUNE**	P/ R/	**auto-** *self, same.* **-immune** *protected*	Immune reaction directed against a person's own tissue.
discrimination	**DIS**-krim-ih-**NAY**-shun	S/ P/ R/	**-ation** *process* **dis-** *away from* **-crimin-** *distinguish*	Ability to distinguish between different things.
mutation	myu-**TAY**-shun		Latin *to change*	Change in the chemistry of a gene.
specific **specificity** (***Note:*** Has two suffixes.)	speh-**SIF**-ik spes-ih-**FIS**-ih-tee	S/ R/ S/	**-ic** pertaining to **specif-** *species* **-ity** *condition, state*	Relating to a particular entity. State of having a specific, fixed relation to a particular entity.
toxin **toxicity**	**TOK**-sin toks-**ISS**-ih-tee	 S/ R/	Greek *poison* **-ity** *state, condition* **toxic-** *poison*	Poisonous substance formed by a cell or organism. The state of being poisonous.

EXERCISES

Review *the two pages spread open before you. Provide a brief answer to each of the questions below.*

1. What happens when the immune system becomes weak?

____

____

____

2. Does the immune system contain any specific organs?

____

____

____

3. Describe an *autoimmune* disorder:

____

____

____

4. Briefly describe the development of resistance:

____

____

5. What occurs when discrimination of the immune system breaks down?

____

____

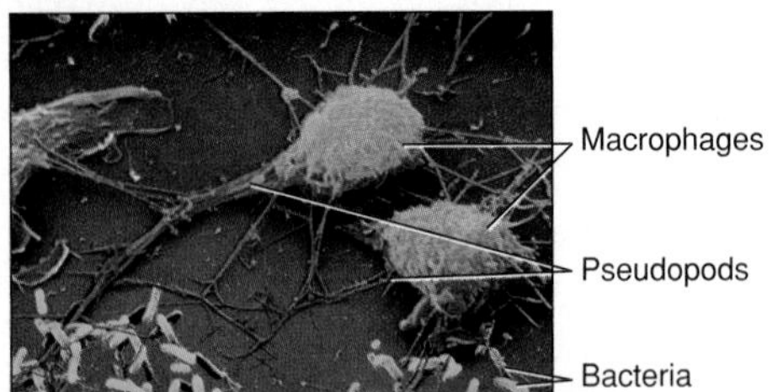

 **FIGURE 11.21 Macrophages Phagocytose Bacteria.**
Filamentous extensions of the macrophage snare the rod-shaped bacteria and draw them to the cell surface, where they are engulfed.

Keynote

Antibodies do not actively destroy an antigen. They render it harmless and mark it for destruction by phagocytes.

IMMUNITY

Immunity is the state of being able to resist a specific infectious disease. It is classified biologically into two types, though these two mechanisms often respond to the same antigen:

1. **Cellular (cell-mediated) immunity:** This is a direct-form of defense based on the actions of lymphocytes to attack foreign and diseased cells and destroy them. The many different types of T cells, B cells, and macrophages described in the previous lesson of this chapter are involved in this style of attack (*Figure 11.21*).
2. **Humoral (antibody-mediated) immunity:** This is an indirect form of attack that employs **antibodies** produced by plasma cells, which have been developed from B cells. The antibodies bind to an antigen and tag it for destruction. These antibodies are called **immunoglobulins,** present in blood plasma and body secretions.

Complement Fixation

The **complement** system is a group of 20 or more proteins continually present in blood plasma. Immunoglobulins bind to foreign cells, initiating the binding of complement to the cell and leading to its destruction.

Immunization

Immunization is the preventive method of stimulating the immune system without exposing the body to an infection. An agent is used that is composed of the antigenic components of a killed or **attenuated** microorganism or its inactivated toxins. This agent is called a **vaccine. Vaccination** has eradicated smallpox worldwide. However, if we stop vaccinating against smallpox, we will have a population that is susceptible to smallpox and outbreaks will occur. The same concept applies to the diseases in childhood immunizations (*Table 11.1*).

Keynote

The immune system is thought to be able to produce some 2 million different antibodies.

TABLE 11.1 Recommended Immunizations for Persons Aged 0 to 6 Years

Hepatitis A (HepA)	Hepatitis B (HepB)
Rotavirus (Rota)	Influenza
Inactivated poliovirus (IPV)	Varicella (chickenpox)
Diphtheria, tetanus, pertussis (DTaP)	Pneumococcal (PCV)
Measles, mumps, rubella (MMR)	Meningococcal (MPSV4)
Hemophilus influenza type b (Hib)	

Centers for Disease Control and Prevention, 2007.

The process that generates your own antibodies in response to an outside agent is called **artificial active immunity.** If you are unfortunate enough to acquire an infection, the antibodies you produce during that infection will provide you with a **natural active immunity** for some time in the future.

Artificial passive immunity is a temporary immunity from the injection of an **immune serum** obtained from another individual or an animal. Immune serum is used to treat snakebite, tetanus, and rabies.

WORD ANALYSIS AND DEFINITION

S/ = Suffix P/ = Prefix R/ = Root R/CF = Combining Form

WORD	PRONUNCIATION	ELEMENTS		DEFINITION
attenuate	ah-**TEN**-you-ate	S/ R/	**-ate** *composed of, pertaining to* **attenu-** *weaken*	Weaken the ability of an organism to produce disease.
attenuated (adj)	ah-**TEN**-you-a-ted	S/	**-ated** *pertaining to a condition*	Weakened.
complement	**KOM**-pleh-ment		Latin *that which completes*	Group of proteins in serum that finish off the work of antibodies to destroy bacteria and other cells.
humoral immunity	**HYU**-mor-al im-**YOU**-nih-tee	S/ R/ S/ R/	**-al** *pertaining to* **humor-** *fluid* **-ity** *condition* **immun-** *immune*	Defense mechanism arising from antibodies in the blood.
immune serum	im-**YUNE SEER**-um		immune, Latin *protected from* serum, Latin *whey*	Serum taken from another human or animal that has antibodies to a disease.
(also called **antiserum**)	an-tee-**SEER**-um	P/ R/	**anti-** *against* **-serum** *serum*	
vaccinate (verb)	**VAK**-sin-ate	S/ R/	**-ate** *pertaining to, composed of* **vaccin-** *vaccine, giving a vaccine*	To administer a vaccine.
vaccination	vak-sih-**NAY**-shun	S/	**-ation** *process*	Administration of a vaccine.
vaccine	**VAK**-seen		Latin, *related to a cow*	Preparation to generate active immunity.

EXERCISES

Patient education: *Translate the following sentence into language your patient can understand. First read the sentence. Go back and read it a second time, and underline or highlight all the terms that will need explaining. Rewrite the sentence on the lines below.*

1. "An agent is used that is composed of the antigenic components of a killed or attenuated microorganism or its inactivated toxins."

____

____

____

2. Now explain to your patient the difference between:

Artificial active immunity:

____

____

____

Artificial passive immunity:

____

____

____

3. Explain to your patient the mechanism of immunization.

____

____

____

Keynote

Common food and drug allergens are peanuts, milk, eggs, wheat, shellfish, penicillin and related antibiotics, and sulfa drugs.

▲ **FIGURE 11.22 Boy with Combined Immunodeficiency Disease in Protective Sterile Enclosure.**

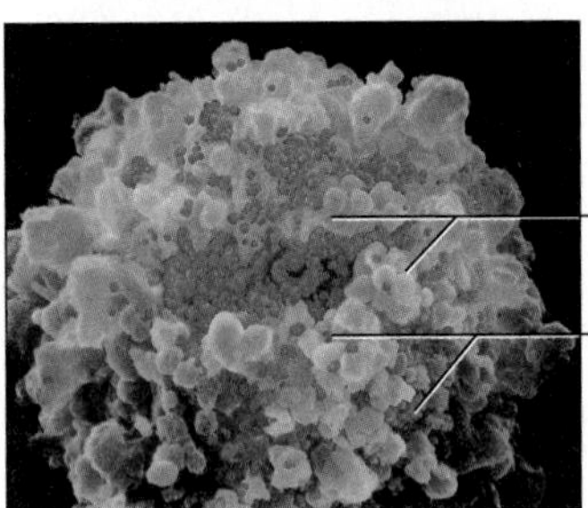

▲ **FIGURE 11.23 Viruses Emerging from a Dying Helper T Cell.**

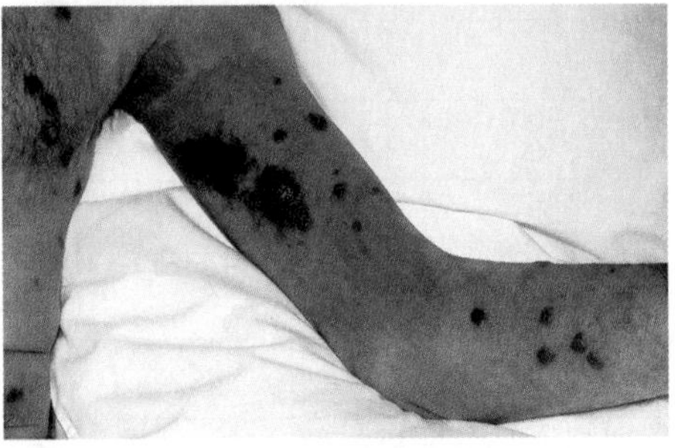

▲ **FIGURE 11.24 Lesions of Kaposi Sarcoma.**

DISORDERS OF THE IMMUNE SYSTEM

Hypersensitivity is an excessive immune response to an antigen that would normally be tolerated. Hypersensitivity includes:

- **Allergies,** which are reactions to environmental antigens such as pollens, molds, dusts, foods, and drugs.
- Abnormal reactions to your own tissues **(autoimmune disorders).**
- Reactions to tissues transplanted from another person **(alloimmune disorders).**

In most **allergic** (hypersensitivity) reactions, **allergens** (antigens) stimulate the cells to produce **histamine.** The symptoms produced by these changes include edema, mucus hypersecretion and congestion, watery eyes, and hives **(urticaria).**

Anaphylaxis is an acute, immediate, and severe allergic reaction. It can be relieved by **antihistamines.**

Anaphylactic shock is more severe and is characterized by dyspnea due to bronchiole constriction, circulatory shock, and sometimes death. It is a life-threatening medical emergency.

Asthma is triggered by allergens, listed above, and by air pollutants, drugs, and emotions. Bronchioles constrict spasmodically **(bronchospasm)** leading to the wheezing and coughing of asthma.

Autoimmune disorders are an overvigorous response of the immune system in which the immune system fails to distinguish self-antigens from foreign antigens. These self-antigens produce autoantibodies that attack the body's own tissues. This type of response occurs, for example, in lupus erythematosus, type 1 diabetes, multiple sclerosis, rheumatoid arthritis, and psoriasis.

Immunodeficiency disorders are a deficient response of the immune system in which it fails to respond vigorously enough. Disorders are classified in three categories:

1. **Congenital** (inborn) disorders are caused by a genetic abnormality that is often sex-linked, with boys affected more often than girls. An example is **inherited combined immunodeficiency disease,** in which there is an absence of both T cells and B cells (*Figure 11.22*). These children are very susceptible to **opportunistic** infections and must live in protective sterile enclosures.
2. **Immunosuppression** is a common side effect of corticosteroids used in treatment to prevent transplant rejection and in chemotherapy treatment for cancer.
3. **Acquired immunodeficiency** results from diseases such as **acquired immunodeficiency syndrome (AIDS),** which involves a severely depressed immune system from infection with the **human immunodeficiency virus (HIV).**

HIV and AIDS

HIV is one of a group of viruses known as **retroviruses.** Like other viruses, it can replicate only inside a living host cell and it invades helper T cells and cells in the upper respiratory tract and CNS. Inside the cell, the virus can stay **dormant** for months or years. When it is activated (AIDS), the new viruses emerge from the dying host cell and attack more cells (*Figure 11.23*). This dormant phase **(incubation)** can range from a few months to 12 years.

As the virus destroys more and more cells, antibodies cannot be produced. Symptoms appear, including chills, fever, night sweats, fatigue, weight loss, and lymphadenitis. Opportunistic infections by bacteria, viruses, and fungi can occur. These infections include toxoplasmosis, pneumocystitis, tuberculosis, herpes simplex, cytomegalovirus, and candidiasis. Cancers can also invade, and a form of skin cancer called **Kaposi sarcoma** (*Figure 11.24*) is often seen.

HIV survives poorly outside the human body. It is destroyed by laundering, dishwashing, chlorination, disinfectants, alcohol, and germicidal skin cleansers.

WORD ANALYSIS AND DEFINITION

S/ = Suffix P/ = Prefix R/ = Root R/CF = Combining Form

WORD	PRONUNCIATION	ELEMENTS		DEFINITION
alloimmune	**AL**-oh-im-**YUNE**	P/ R/	**all/o-** *other, strange* **-immune** *immunity*	Immune reaction directed against foreign tissue.
anaphylaxis	**AN**-ah-fih-**LAK**-sis	P/ R/	**ana-** *excessive* **-phylaxis** *protection*	Immediate severe allergic response.
anaphylactic (adj)	**AN**-ah-fih-**LAK**-tik	S/ R/	**-tic** *pertaining to* **-phylac-** *protect*	Pertaining to anaphylaxis.
asthma	**AZ**-mah		Greek *asthma*	Episodes of breathing difficulty due to narrowed or obstructed airways.
asthmatic (adj)	az-**MAT**-ik	S/ R/	**-atic** *pertaining to* **asthm-** *asthma*	Suffering from or pertaining to asthma.
dormant	**DOR**-mant	S/ R/	**-ant** *forming* **dorm-** *sleep*	Inactive.
histamine	**HIS**S-tah-mean	R/ R/	**hist-** *derived from histidine* **-amine** *nitrogen-containing substance*	Compound liberated in tissues as a result of injury or an immune response.
antihistamine	an-tee-**HIS**-tah-mean	P/	**anti -** *against*	Drug used to treat allergic symptoms because of its action antagonistic to histamine.
hypersensitivity	**HIGH**-per-sen-sih-**TIV**-ih-tee	S/ P/ R/	**-ity** *condition* **hyper-** *excessive* **-sensitiv-** *feeling*	Exaggerated abnormal reaction to an allergen.
immunodeficiency	**IM**-you-noh-dee-**FISH**-en-see	S/ R/CF R/	**-ency** *quality, state of* **immun/o-** *immune response* **-defici-** *failure, lacking*	Failure of the immune system.
immunosuppression	**IM**-you-noh-suh-**PRESH**-un	S/ R/	**-ion** *action, condition* **-suppress-** *press under*	Failure of the immune system caused by an outside agent.
incubation	in-kyu-**BAY**-shun	S/ R/	**-ation** *process* **incub-** *lie on, hatch*	Process to develop an infection.
Kaposi sarcoma	kah-**POH**-see sar-**KOH**-mah		Moritz Kaposi, Hungarian dermatologist, 1837–1902	A skin cancer often seen in AIDS patients.
opportunistic	**OP**-or-tyu-**NIS**-tik	S/ R/	**-istic** *pertaining to* **opportun-** *take advantage of*	An organism or a disease in a host with lowered resistance.
retrovirus	**REH**-troh-vie-rus	P/ R/	**retro-** *backward* **-virus** *poison*	Virus with an RNA core.
urticaria	ur-tee-**KARE**-ee-ah		Latin *nettle*	Rash of itchy wheals (hives).

HIV is found in blood, semen, vaginal secretions, saliva, tears, and breast milk of infected mothers. The most common means of transmission are:

- **Sexual intercourse** (vaginal, oral, anal).
- **Shared needles** for drug use.
- **Contaminated blood products.** (All donated blood is now tested for HIV.)
- **Transplacental transmission,** from an infected mother to her fetus.

Abbreviations

AIDS	acquired immunodeficiency syndrome
HIV	human immunodeficiency virus

EXERCISES

Build *more medical vocabulary for the language of immunology. Complete the construction of each term by using the following elements to fill in the blanks. There are more answers than you need.*

defici allo suppress sensitiv ion hyper incub osis phylaxis ic ency dorm

1. exaggerated, abnormal reaction to an allergen: ____________/ ____________/ ____________ ity
2. inactive: ____________/ ____________/ ____________ ant
3. process to develop an infection: ____________/ ____________/ ____________ ation

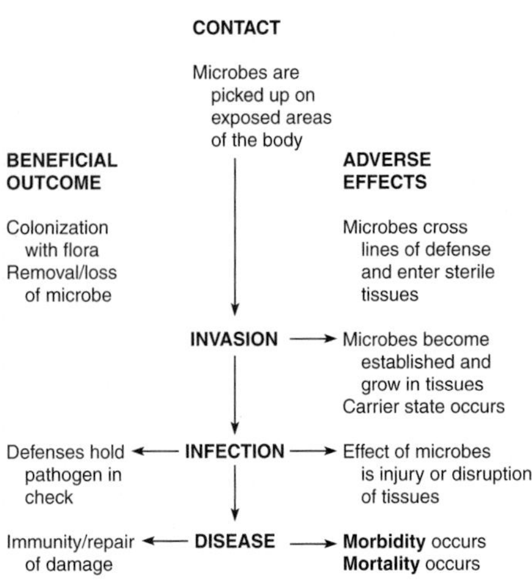

▲ **FIGURE 11.25 Associations between Microbes and Humans.**

INFECTION

Microbes (microorganisms) are everywhere—in the air, water, and soil and all over our bodies, where they are called **normal flora.** These normal microorganisms are found on your skin, in your nose and respiratory tract, and in your mouth and digestive tract. Areas such as your brain and cardiovascular system remain microbe-free **(sterile).**

If microorganisms other than the normal flora invade the body, they become pathogens and an infection occurs. If the infection causes harm to the body, an **infectious** disease is produced (*Figure 11.25*). Pathogens include bacteria, viruses, fungi, and parasites.

Bacterial Infections

The thousands of different bacteria are single-celled microorganisms that reproduce by dividing. Frequently seen bacteria include the **staphylococcus ("staph"),** which can be present harmlessly on the skin but causes infections in wounds or any normally sterile place such as a joint or the peritoneum; the **streptococcus (strep),** which is a cause of sore throats; the **pneumococcus,** which is a cause of pneumonia; and **coliform** bacteria, which live normally in the gut but cause infections elsewhere, for example, in the urinary tract.

Methicillin-resistant *Staphylococcus aureus* **(MRSA)** is a type of bacteria that is resistant to the antibiotics normally used to treat staph infections. MRSA infections occur most frequently in hospitals and health care facilities. This form of staph is known as community-associated MRSA **(CA-MRSA).**

Viral Infections

Viruses are the smallest of the microorganisms. They cannot be seen under an ordinary light **microscope** but are visible through electron **microscopy.** They spread from person to person through coughs and sneezes or via hands that are not properly washed.

Viruses cause specific diseases of childhood such as measles (rubeola), German measles (rubella), chickenpox (varicella), and mumps. They cause upper respiratory infections (*see Chapter 7*) and such modern respiratory infections as **severe acute respiratory syndrome (SARS), avian influenza** (bird flu), and **West Nile virus (WNV)**, a seasonal **epidemic** in North America flaring up in the summer and fall.

Fungal Infections

Many fungi are "good fungi," for example, the mushrooms that you eat and the yeasts that ferment beer and bread. Penicillin was derived from a fungus. The most common pathogenic fungi are those that cause skin infections (*see Chapter 3*).

Opportunistic fungi are normally harmless, but cause disease in people who are on prolonged doses of antibiotics, are receiving chemotherapy or immunosuppressive therapy, or have diabetes mellitus or AIDS.

Parasitic Infections

Parasites are organisms that live on or in another organism and draw nourishment from their host. In many rural areas of the world, parasites are **endemic. Malaria** is caused by a parasite, which is transmitted from person to person by the bite of a mosquito.

Pinworms are the most common parasite in America. Pinworm eggs are ingested, they hatch in the human intestine, and the young worms migrate to the anus, where the female deposits her eggs. The eggs can be transferred by fingers from the anus or from infected bedding to the mouth of the same child or another.

Abbreviations

CA-MRSA	community-associated methicillin-resistant *Staphylococcus aureus*
MRSA	methicillin-resistant *Staphylococcus aureus*
SARS	severe acute respiratory syndrome
WNV	West Nile virus

Keynote

When viruses spread from person to person, they are said to be **contagious.**

Keynote

Viral diseases do not respond to antibiotics.

Keynote

Nosocomial infections (hospital-acquired infections) are becoming increasingly common and lethal.

Keynote

Handwashing is the most important factor in preventing the transmission of infections.

WORD ANALYSIS AND DEFINITION

S/ = Suffix P/ = Prefix R/ = Root R/CF = Combining Form

WORD	PRONUNCIATION	ELEMENTS		DEFINITION
contagious	kon-**TAY**-jus	S/ P/ R/	**-ious** *pertaining to* **con-** *with, together* **-tag-** *touch*	Infection can be transmitted from person to person or from a person to a surface to a person.
endemic	en-**DEM**-ik	S/ P/ R/	**-ic** *pertaining to* **en-** *in* **-dem-** *the people*	Pertaining to a disease always present in a community.
epidemic	ep-ih-**DEM**-ik	P/	**epi-** *above, upon*	Pertaining to an outbreak in a community of a disease or a health-related behavior.
pandemic	pan-**DEM**-ik	P/	**pan-** *all*	Pertaining to a disease attacking the population of a very large area.
flora	**FLO**-rah		Latin *flower*	Microorganisms covering the exterior and interior surfaces of a healthy animal.
infect (verb)	in-**FEKT**		Latin *invade internally*	To invade an organism by a microorganism.
infection (noun)	in-**FEK**-shun	S/ R/	**-ion** *condition, action* **infect-** *internal invasion*	Invasion of the body by disease-producing microorganisms.
infectious (adj)	in-**FEK**-shus	S/	**-ious** *pertaining to*	Capable of being transmitted to a person; or a disease caused by the action of a microorganism.
microbe	**MY**-krohb	P/ R/	**micro-** *small* **-be** *life*	Short for *microorganism.*
microorganism	**MY**-kroh-**OR**-gan-izm	S/ R/	**-ism** *process* **-organ-** *organ, instrument*	Any organism too small to be seen by the naked eye.
microscope	**MY**-kroh-skope	P/ R/	**micro-** *small* **-scope** *instrument for viewing*	Instrument for viewing something small that cannot be seen in detail by the naked eye.
microscopic	**MY**-kroh-**SKOP**-ik	S/	**-ic** *pertaining to*	Visible only with the aid of a microscope.
nosocomial	noh-soh-**KOH**-mee-al	S/ R/CF R/	**-ial** *pertaining to* **nos/o-** *disease* **-com-** *take care of*	Acquired while in the hospital.

EXERCISES

Correct usage *of the appropriate grammatical form of a medical term is the mark of an educated professional. Practice your language of immunology in the following sentences. Fill in the blanks.*

1. **infect** **infection** **infectious**

   This patient has a rarely seen ____________________. Please refer her to the ____________________ disease specialist.

2. **bacterium** **bacteria** **bacterial**

   The ____________________ streptococcus causes ____________________ infections in the throat.

3. **sterile** **sterility** **sterilize**

   An autoclave is used to ____________________ instruments.

4. If the ____________________ of an instrument is in question, it should not be used.

5. The term ____________________ can also mean *unable to reproduce.*

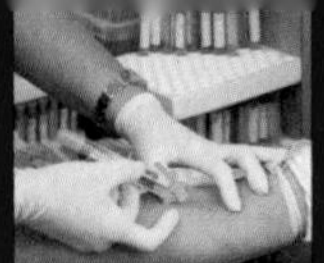

CHAPTER 11 REVIEW

THE BLOOD, LYMPHATIC, AND IMMUNE SYSTEMS

CHALLENGE YOUR KNOWLEDGE

A. **Prefixes:** Each of the following terms is lacking its prefix. After you have entered the prefix *on* the line, write the meaning of the prefix next to it. The first answer is done for you. Fill in the blanks.

1. protein produced in response to an antigen ______ anti/body; against ______
2. spleen removes blood components at an excessive rate ______/splen/ism; ______
3. in front of the elbow ______ /cubit/al; ______
4. directed against the person's own tissues ______/immune; ______
5. immediate, severe, allergic response ______/phylaxis; ______
6. immune reaction against transplanted tissue ______/immune; ______
7. virus with an RNA core ______/virus; ______
8. *Anti* means ______.

   Ante means ______.

   My study hint:

   ______

   ______

> **Study Hint**
> The elements *anti* and *ante* can be pronounced the same, but the difference of only one letter makes them two different prefixes. Create for yourself a study hint for how to tell them apart.

B. **What am I?** Each of the following medical terms is a noun, which is the name of a person, place, or thing. Match the meaning in column 1 to the correct medical term in column 2.

	Column 1	Column 2
____	1. poisonous substance	A. vaccine
____	2. nonpitting fluid retention	B. antibody
____	3. substance producing an allergic reaction	C. retrovirus
____	4. agent intended to prevent disease	D. immunity
____	5. lymphatic tissue in the nasopharynx	E. pathogen
____	6. malignant neoplasm	F. lymphedema
____	7. produced in response to an antigen	G. lymphoma
____	8. has an RNA core	H. allergen
____	9. state of being protected	I. adenoid
____	10. disease-causing microorganism	J. toxin

C. **Suffixes:** Work with the seven terms below from the language of immunology. Their roots/combining forms are similar, and their suffixes help define them. First, deconstruct each of the terms in the chart, and then define the medical term.

Medical Term	Meaning of Prefix	Meaning of Root/CF	Meaning of Suffix	Meaning of Medical Term
lymphatic				
lymphadenectomy				
lymphadenitis				
lymphadenopathy				
lymphangiogram				
lymphedema				
lymphoma				

Using the terms from this chart, answer the following questions.

1. List the terms that can be billed as a diagnosis:

   ___

2. Write the term that is a surgical procedure:

   ___

3. Write the term that is a radiological procedure:

   ___

4. List the terms that relate to lymph nodes:

   ___

5. Name a term that concerns lymphatic vessels (*hint:* check the elements):

   ___

6. Write an explanation to a fellow student giving the difference between *lymphoma* and *lymphedema:*

   ___

   ___

   ___

   ___

7. What have you noticed about all the terms in the chart?

   ___

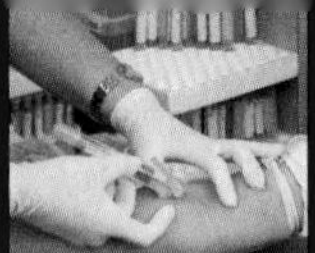

THE BLOOD, LYMPHATIC, AND IMMUNE SYSTEMS

D. Analyze and discover the difference. Medical language has many terms that appear similar but have unique meanings all their own. If you can analyze similar terms, you will understand the difference and be able to explain it to your patient.

1. *Alloimmune*

   Prefix: ______________________ Means: ______________________

   Root: ______________________ Means: ______________________

   Autoimmune

   Prefix: ______________________ Means: ______________________

   Root: ______________________ Means: ______________________

   Write an explanation to your patient of the difference between *alloimmune* and *autoimmune:*

   ____

   ____

   ____

2. *Immunodeficiency*

   Root: ______________________ Means: ______________________

   CF: ______________________ Means: ______________________

   Suffix: ______________________ Means: ______________________

   Immunosuppression

   Root: ______________________ Means: ______________________

   CF: ______________________ Means: ______________________

   Suffix: ______________________ Means: ______________________

   Write an explanation to your patient of the difference between *immunodeficiency* and *immunosuppression:*

   ____

   ____

   ____

   ____

E. Review: Previously, you have studied directional terms that aid you in locating some part of the body. Apply those terms to the following sentence, and briefly explain, in nonmedical terms, the location of the spleen.

"The spleen, a highly vascular and spongy organ, is the largest lymphatic organ and is located in the left upper quadrant of the abdomen, below the diaphragm and lateral to the kidney."

Read the sentence; then read it again and underline all the medical terms. Be sure you also know the position of the diaphragm and the kidneys. You may be asked by the instructor to illustrate them on the classroom skeleton.

____

____

____

____

____

F. **Spelling demons:** Some terms pose difficulty because they may or may not double some letters (or drop some letters) in various forms of the term. *Spleen* and *splenectomy* are an example. Test yourself on the correct spelling of the following terms.

1. lymphatic tissue in oropharynx ______________________
2. removal of this tissue ______________________
3. inflammation of this tissue ______________________
4. vascular, lymphatic organ ______________________
5. removal of this organ ______________________
6. enlargement of this organ ______________________
7. organ removes blood components at an excessive rate ______________________

G. **Immunology terminology:** Increase your knowledge of the language of immunology. The element's meaning is given to you in column l. Name the element in column 2, and identify the type of element it is in column 3. In the last column, give an example of a medical term containing that element.

Meaning of Element	Element	Type of Element (P, R, CF, or S)	Medical Term Containing This Element
disease			
enlargement			
excision			
gland			
groin			
in front of			
lymph node			
lymphatic vessel			
sleep			
spaces within tissues			
to produce			
weaken			

H. **Where am I?** Several terms in this chapter denote a place or direction on the body. Challenge yourself to insert the correct term on the line.

1. pertaining to the groin area ______________________
2. in front of the elbow ______________________
3. pertaining to the neck ______________________
4. in spaces between cells in a tissue or organ ______________________
5. pertaining to the armpit ______________________
6. moving outward from an organ or part ______________________

Challenge: Can you remember any other similar term from a previous chapter? [*Hint:* Think of prefixes that denote a location (*within, behind,* etc.) or direction (*toward, across,* etc.).] Write the term below, and use it in a sentence that is not a definition.

7. Term: ______________ means ______________________.

   Sentence:

   ____

   ____

I. **Recall and review:** How well do you remember these word elements from the previous chapter? Try to answer without looking back to check. Fill in the chart.

Element	Type of Element (P, R, CF, or S)	Meaning of Element	Term Using Element
acro			
lact			
para			
myx			
toxico			
lysis			
gluc			
idio			
hydro			
nephr			

J. **Dictionary exercise:** Look up the word *reservoir.* Define it, and then give a brief explanation of how the spleen functions as a reservoir.

Definition of *reservoir:*

____

Spleen as a reservoir:

____

____

____

K. **Similar but different:** The following terms all contain the word *edema,* but additional words or elements add new meaning to the term. Give a brief answer that describes each form of *edema.* Use the Glossary or a dictionary if you are unsure of any term.

1. *edema:* _______________
2. *peripheral edema:* _______________
3. *pitting edema:* _______________
4. *lymphedema:* _______________

L. **Seek and find:** The following terms were defined in context (within the text) and may not always appear in WAD boxes. You are given a brief meaning as the term appeared in the text—write the medical term on the line. *Remember to pay attention to these terms—they could be a test question!*

1. clot _______________
2. ingest and destroy _______________
3. lymph nodes in the neck _______________
4. lymph nodes in the armpit _______________
5. lymph nodes in the groin _______________
6. inborn _______________
7. dormant phase _______________
8. spasmodic constriction of bronchioles _______________
9. hives _______________
10. hypersensitivity reactions _______________

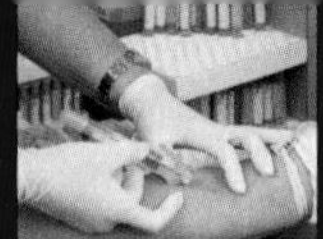

CHAPTER 11 REVIEW

THE BLOOD, LYMPHATIC, AND IMMUNE SYSTEMS

M. Translation: Reduce the sentences below to the most basic language that a nonmedical person could understand. First, use your knowledge of medical terminology to understand the statement. Then organize your thoughts and formulate your answer.

1. "Intra-abdominal hemorrhage from the ruptured spleen can be extensive, with dramatic hypotension, and is a surgical emergency requiring splenectomy."

____

____

2. "Hemolysis can be caused by toxic substances or an incompatible blood transfusion."

____

____

3. "The abnormal cells agglutinate and block small capillaries, which causes intense pain in the hypoxic tissues."

____

____

N. Roots and combining forms are the foundation of every medical term. Slash the term into elements in the first column, identify the R/CF in the second column, and provide the meaning for the R/CF in the third column. Complete the chart with the meaning of the medical term in the last column.

Medical Term (slash first)	Root/CF	Meaning of Root/CF	Meaning of Medical Term
anaphylactic			
immunologist			
anemic			
hypoxic			
pernicious			
autoimmune			
attenuate			
interstitial			
hypersplenism			
pancytopenia			
anaphylaxis			
nosocomial			

O. **Latin and Greek terms** cannot be further deconstructed into prefix, root, or suffix. You must know them for what they are. Test your knowledge of these terms with this exercise. Match the meaning in column 1 with the correct medical term in column 2.

	Column 1	Column 2
_____	1. clear fluid	A. asthma
_____	2. mass of tissue	B. follicle
_____	3. change in gene chemistry	C. urticaria
_____	4. that which completes	D. lymph
_____	5. rash of itchy wheals	E. mutation
_____	6. spherical cluster of cells	F. complement
_____	7. hypersensitive lung disorder	G. node
_____	8. heme	H. paleness
_____	9. embolus	I. spot on skin
_____	10. lymph	J. clear spring water
_____	11. pallor	K. blood
_____	12. petechia	L. plug, stopper

P. **Terminology challenge.**

The challenge element from this chapter is *megaly,* which means

____.

1. Find a term from this chapter containing this element.

Term: ______________________________

Means: ______________________________

2. There are two other terms with this element that have appeared in previous chapters. Can you name them and define them?

Term: ______________________________

Means: ______________________________

Term: ______________________________

Means: ______________________________

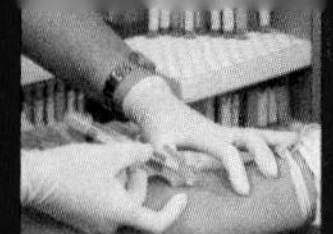

CHAPTER 11 REVIEW

THE BLOOD, LYMPHATIC, AND IMMUNE SYSTEMS

Q. Chapter challenge: Circle the correct answer.

> **Study Hint**
> Immediately cross off any answer you know is not correct. In your remaining choices, there is only *one best answer.*

1. Where is a *nosocomial* infection acquired?
   a. in a school
   b. in an airplane
   c. on a cruise ship
   d. in the hospital
   e. where you work

2. *Dyspnea* due to bronchiole constriction and circulatory shock are symptoms of:
   a. asthma
   b. anaphylaxis
   c. Kaposi sarcoma
   d. anaphylactic shock
   e. urticaria

3. "A" in the abbreviation *AIDS* stands for:
   a. acute
   b. acquired
   c. active
   d. attenuated
   e. asthmatic

4. The medical term for leaks of blood from all types of blood vessels is:
   a. lymphoma
   b. hematoma
   c. bruise
   d. a and b
   e. b and c

5. Choose the pair of correct spellings:
   a. tonsel tonselectomy
   b. tonsil tonsillectomy
   c. tonssil tonsilectomy
   d. tonsill tonsilectomy
   e. tonnsil tonsillectomy

6. Identify the term in which a root appears at the end of the term:
   a. allergen
   b. antibody
   c. efferent
   d. pathogen
   e. interstitial

7. What can be said about the terms *cervical, axillary,* and *inguinal:*
   a. They are all areas accessible to palpation.
   b. Lymph nodes are located in these areas.
   c. Their suffixes all mean *pertaining to.*
   d. All of these statements are correct.
   e. None of these statements is correct.

8. Which set consists of terms that only contain lymphatic tissue?
   a. spleen, tonsil, adenoid
   b. pancreas, tonsil
   c. spleen, adenoid, adrenal
   d. pancreas, adenoid, tonsil
   e. spleen, pancreas

9. What is the medical term for *hives*?

a. thrombocytopenia
b. urticaria
c. hemolysis
d. hematoma
e. embolus

10. The correct plural of *bacterium* is:

a. bacteriums
b. bacteriae
c. bacteria
d. bacteriaes
e. bacterias

11. What is the dormant phase of a developing infection called?

a. incubation
b. hemoglobinopathy
c. agglutination
d. colloid
e. differential

12. Which disease has been virtually eliminated worldwide by the use of vaccination?

a. hemophilia
b. leucopenia
c. leukemia
d. smallpox
e. EBV

13. *Purpura* is bleeding into the skin from small arterioles. What is bleeding from capillary beds called?

a. thrombosis
b. embolus
c. petechiae
d. hemophilia
e. thrombocytopenia

14. *Immunosuppressive* drugs are given after:

a. organ transplant
b. anaphylactic shock
c. retrovirus
d. opportunistic infection
e. Hodgkin lymphoma

15. *Avian influenza* is another name for what type of flu?

a. WNV
b. bird
c. stomach
d. respiratory
e. SARS

16. An *allergic reaction* is one of:

a. hypoglycemia
b. hypersensitivity
c. hypotension
d. hypersplenism
e. hypertension

17. *Kaposi sarcoma* is a skin cancer often seen in patients with:

a. DJD
b. CF
c. AIDS
d. RA
e. BPH

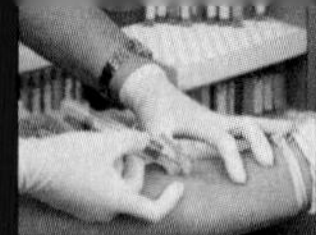

THE BLOOD, LYMPHATIC, AND IMMUNE SYSTEMS

R. Case Report challenge: Now that you are more comfortable with the terms in this chapter, you can apply that knowledge and briefly answer the questions about the Case Report. If you read the report through and underline all the medical terminology, this will make it easier to answer the questions.

CASE REPORT 11.4

You are

. . . a medical assistant working with Susan Lee, MD, in her primary care clinic.

You are communicating with

. . . Ms. Anna Clemons, a 20-year-old waitress, who is a new patient. She has noticed a lump in her left neck. On questioning, you elicit that she has lost about 8 pounds in weight in the past couple of months, has felt tired, and has had some night sweats.

Her vital signs are normal. There are two firm, enlarged lymph nodes in her left neck in front of the sternocleidomastoid muscle. Physical examination is otherwise unremarkable.

Ms. Clemons has cancerous nodes in her neck. They were not caused by metastatic cancer but by a cancer of the lymph nodes called Hodgkin lymphoma.

1. Where in this Case Report could you substitute an abbreviation for something that has been spelled out?

   ________________ means ________________.

2. The medical term for *enlarged lymph nodes* is

   ________________.

3. The element *oid* in *sternocleidomastoid* means

   ________________.

4. Explain what is meant by *metastatic cancer:*

   ____

   ____

   ____

5. The suffix in the term *lymphoma* means

   ________________.

6. What are Ms. Clemons' presenting symptoms?

   ____

   ____

   ____

Write your own study notes for this chapter.

The Nervous System

The Essentials of the Language of Neurology

12

CASE REPORT 12.1

You are

. . . an **electroneurodiagnostic** technologist working with Gregory Solis, MD, a **neurosurgeon** in Fulwood Medical Center.

You are communicating with

. . . Ms. Roberta Gaston, a 39-year-old woman, who has been referred by Raul Cardenas, MD, a **neurologist,** for evaluation for possible **neurosurgery.** Ms. Gaston has had **epileptic** seizures since the age of 16. She also has daily minor spells in which she stops interacting and blinks rhythmically for about 20 seconds, after which she returns to normal. She is not able to work and is cared for by her parents. Her neurologic examination is normal. Her **electroencephalogram (EEG)** shows diffuse epileptic discharges in the left frontal region. Her CT scan is normal. An MRI shows a 20-mm-diameter mass adjacent to her left ventricle.

Learning Outcomes

Your roles are to perform electroneurodiagnostic evaluations, communicate with Ms. Gaston and her parents, communicate with other health professionals, and maintain and document Ms. Gaston's history. To perform these roles, you must be able to:

12.1 Apply the language of neurology to the structures and functions of the nervous system.

12.2 Comprehend, analyze, spell, and write the medical terms of neurology and mental health.

12.3 Recognize and pronounce the medical terms of neurology and mental health.

12.4 Explain the effects of common disorders of the nervous system on health.

LESSON 12.1 Functions and Structure of the Nervous System

THE NERVOUS SYSTEM

OBJECTIVES

The trillions of cells in your body must communicate and work together for you to function effectively. This is done through your **nervous system.** It is essential that you understand how this system operates. In this lesson, you will learn to:

- **Apply medical terminology to the functions of the nervous system.**
- **Relate the functions of the nervous system to the structures of its components.**
- **List the medical terminology for the subdivisions of the nervous system and their basic cells.**

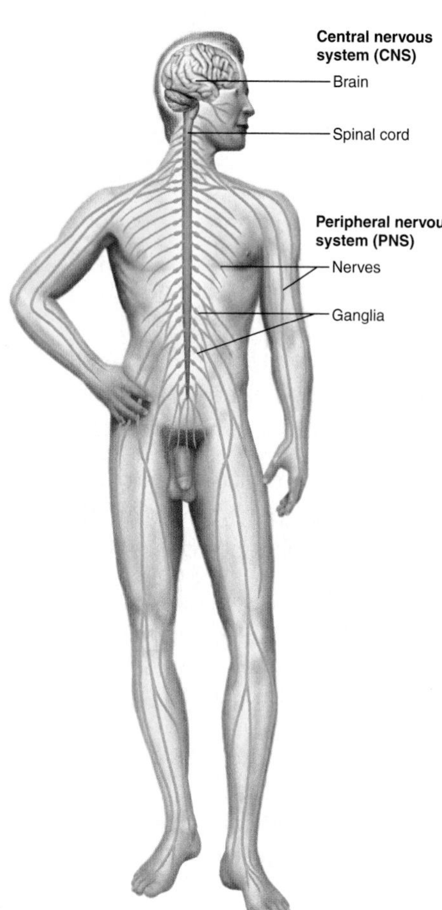

▲ **FIGURE 12.1**
The Nervous System.

FUNCTIONS OF THE NERVOUS SYSTEM

1. **Sensory input** to the brain comes from receptors all over your body at both the conscious and subconscious levels (*Figure 12.1*). You are conscious of external stimuli that you receive from your body as it interacts with your environment. Inside your body, internal stimuli about the amount of oxygen and carbon dioxide in your blood and other homeostatic variables are being processed continually at the subconscious level.
2. **Motor output** from your brain stimulates the skeletal muscles to contract and enables you to move in any way. The production of sweat, saliva, and digestive enzymes is controlled by the nervous system without active input from you.
3. **Evaluation and integration** occur in the brain and spinal cord to process the **sensory** input, initiate a response, and store the event in memory.
4. **Homeostasis** is maintained by the nervous system taking in internal sensory input. For example, the nervous system responds by stimulating the heart to deliver the correct volume of blood for oxygenation of and removal of waste products from cells.
5. **Mental activity** occurs in the brain so that you can think, feel, understand, respond, and remember.

Abbreviations

CT	computed tomography
EEG	electroencephalogram
MRI	magnetic resonance imaging

WORD ANALYSIS AND DEFINITION

S/ = Suffix P/ = Prefix R/ = Root R/CF = Combining Form

WORD	PRONUNCIATION	ELEMENTS		DEFINITION
electroencephalogram (EEG)	ee-**LEK**-troh-en-**SEF**-ah-low-gram	S/ R/CF R/CF	**-gram** *recording* **electr/o-** *electricity* **-encephal/o-** *brain*	Record of the electrical activity of the brain.
electroencephalograph	ee-**LEK**-troh-en-**SEF**-ah-low-graf	S/	**-graph** *to write, record*	Device used to record the electrical activity of the brain.
electroencephalography	ee-**LEK**-troh-en-**SEF**-ah-**LOG**-rah-fee	S/	**-graphy** *process of recording*	The process of recording the electrical activity of the brain.
electroneurodiagnostic (adj)	ee-**LEK**-troh-**NYUR**-oh-die-ag-**NOS**-tik	S/ R/CF R/CF R/	**-ic** *pertaining to* **electr/o-** *electricity* **-neur/o-** *nerve* **-diagnost-** *decision*	Pertaining to the use of electricity in the diagnosis of a neurologic disorder.
epilepsy	**EP**-ih-**LEP**-see		Greek *seizure*	Chronic brain disorder due to paroxysmal excessive neuronal discharges.
epileptic (adj) (***Note:*** An epileptic episode is called a **seizure.**)	**EP**-ih-**LEP**-tik **SEE**-zhur	S/ R/	**-ic** *pertaining to* **epilept-** *seizure*	Pertaining to or suffering from epilepsy.
motor	**MOH**-tor		Latin *to move*	Structures of the nervous system that send impulses out to cause muscles to contract or glands to secrete.
nerve	NERV		Latin *nerve*	A cord of nerve fibers bound together by connective tissue.
nervous system	**NER**-vus **SIS**-tem	S/ R/	**-ous** *pertaining to* **nerv-** *nerve* system, Greek *an organized whole*	The whole, integrated nerve apparatus.
neurology	nyu-**ROL**-oh-jee	S/ R/CF	**-logy** *study of* **neur/o-** *nerve*	Medical specialty of disorders of the nervous system.
neurologist	nyu-**ROL**-oh-jist	S/	**-logist** *one who studies, specialist*	Medical specialist in disorders of the nervous system.
neurologic (adj)	**NYUR**-oh-**LOJ**-ik	S/ R/	**-ic** *pertaining to* **-log-** *to study*	Pertaining to the nervous system.
neurosurgeon	**NYU**-roh-**SUR**-jun	S/ R/	**-eon** *one who does* **-surg-** *operate*	One who operates on the nervous system.
neurosurgery	**NYU**-roh-**SUR**-jer-ee	S/	**-ery** *process of*	Operating on the nervous system.
sensory	**SEN**-soh-ree	S/ R/	**-ory** *having the function of* **sens-** *feel*	Pertaining to sensation; structures of the nervous system that carry impulses to the brain.

EXERCISES

Documentation: *Fill in the following paragraph with the appropriate language of neurology.*

electroencephalography
epilepsy
neurologist
neurosurgeon
electroneurodiagnostic
electroencephalograph
electroencephalogram

Roberta Gaston and her parents were sent to this office by her ____________________. Dr. Solis has ordered some ____________________ tests because of her ____________________. The particular type of test is an ____________________, which will produce an EEG on the (device) ____________________.

After the results of the ____________________, if Ms. Gaston needs surgery, it will be performed by a(n) ____________________.

FIGURE 12.2 Components of the Nervous System.

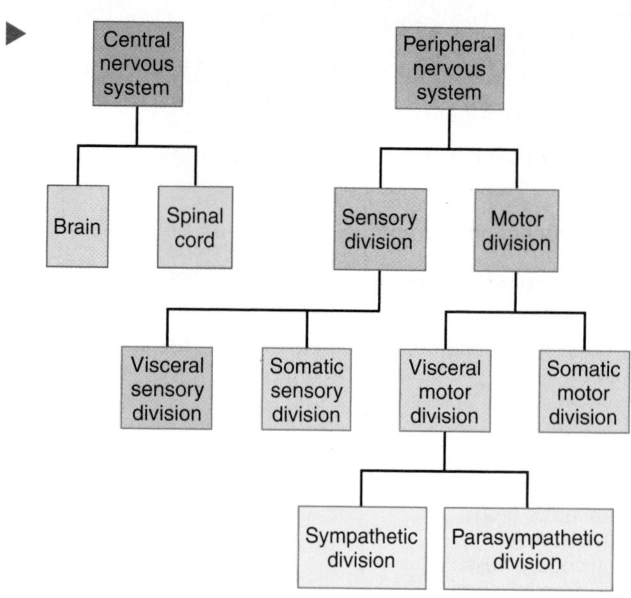

COMPONENTS OF THE NERVOUS SYSTEM

The nervous system has two major anatomical subdivisions (*Figure 12.2*):

1. The **central nervous system (CNS),** consisting of the brain and spinal cord.
2. The **peripheral nervous system (PNS),** consisting of all the **neurons** and nerves outside the central nervous system. It includes 12 pairs of **cranial nerves** originating from the brain and 31 pairs of **spinal nerves** originating from the **spinal cord.**

The peripheral nervous system, in turn, is further subdivided into:

i. **Sensory division,** in which sensory nerves **(afferent nerves)** carry messages toward the spinal cord and brain from sense organs. **Visceral nerves** carry signals from the **viscera** in the thoracic and abdominal cavities, for example, the heart and lungs and the stomach and intestines. **Somatic nerves** carry signals from the skin, muscles, bones, and joints.

ii. **Motor division,** in which motor nerves **(efferent nerves)** carry messages away from the brain and spinal cord to muscles and organs.

   a. The **visceral motor division** is called the **autonomic nervous system (ANS).** It carries signals to glands and to cardiac and smooth muscle. It operates at a subconscious level outside your voluntary control. It, in turn, has two subdivisions:

      - The **sympathetic division** arouses the body for action, for example, by increasing the heart and respiratory rates to increase oxygen supply to brain and muscles.
      - The **parasympathetic division** calms the body, slowing down the heartbeat but stimulating digestion.

   b. The **somatic motor division** carries signals to the skeletal muscles and is under your voluntary control.

Study Hint

Remember the acronym **SAME:**

Sensory nerves are **A**fferent (toward the brain).

Motor nerves are **E**fferent (toward the skeleton).

Abbreviations

ANS	autonomic nervous system
CNS	central nervous system
PNS	peripheral nervous system

WORD ANALYSIS AND DEFINITION

S/ = Suffix P/ = Prefix R/ = Root R/CF = Combining Form

WORD	PRONUNCIATION	ELEMENTS		DEFINITION
afferent (***Note:*** also called *sensory;* opposite of *efferent*)	**AFF**-eh-rent		Latin *to bring to*	Moving toward a center; for example, nerve fibers conducting impulses to the spinal cord and brain.
autonomic	awe-toh-**NOM**-ik	S/ P/ R/	**-ic** *pertaining to* **auto-** *self* **-nom-** *law*	Self-governing visceral motor division of the peripheral nervous system.
efferent (***Note:*** also called *motor;* opposite of *afferent*)	**EFF**-eh-rent		Latin *to bring away from*	Moving away from a center; for example, conducting nerve impulses away from the brain or spinal cord.
neuron	**NYUR**-on		Greek *nerve*	Technical term for a nerve cell; consists of cell body with its dendrites and axons.
parasympathetic (***Note:*** This term contains two prefixes.)	par-ah-sim-pah-**THET**-ik	S/ P/ P/ R/	**-ic** *pertaining to* **para-** *beside* **-sym-** *together* **-pathet-** *suffering*	Division of the autonomic nervous system; has opposite effects to the sympathetic division.
somatic	soh-**MAT**-ik	S/ R/	**-ic** *pertaining to* **somat-** *body*	A division of the peripheral nervous system serving the skeletal muscles.
sympathetic	sim-pah-**THET**-ik	S/ P/ R/	**-ic** *pertaining to* **sym-** *together* **-pathet-** *suffering*	Division of the autonomic nervous system operating at an unconscious level.
visceral (adj)	**VISS**-er-al	S/ R/	**-al** *pertaining to* **viscer-** *internal organs*	Pertaining to the internal organs.
viscus **viscera (pl)**	**VISS**-kus **VISS**-er-ah		Latin *an internal organ*	Any single internal organ.

EXERCISES

Elements: *Solid knowledge of elements is the key to learning medical terminology. Match each element in column 1 with its correct meaning in column 2.*

	Column 1		Column 2
_____	1. ic	A.	self
_____	2. sym	B.	pertaining to
_____	3. viscer	C.	suffering
_____	4. auto	D.	law
_____	5. pathet	E.	body
_____	6. para	F.	together
_____	7. somat	G.	internal organ
_____	8. nom	H.	beside

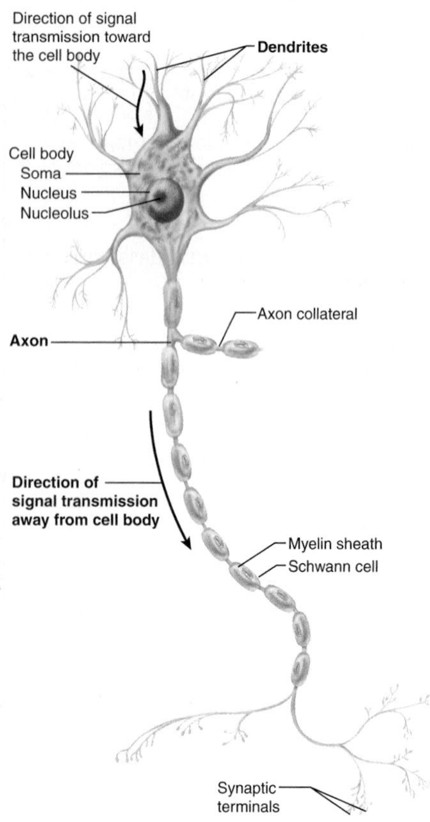

▲ **FIGURE 12.3 Neuron.**

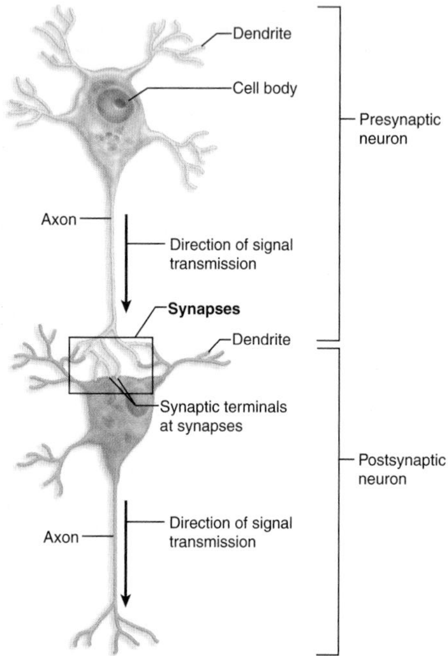

▲ **FIGURE 12.4 Synapse.**

CELLS OF THE NERVOUS SYSTEM

Neurons (nerve cells) receive stimuli and transmit impulses to other neurons or to receptors in organs. Each neuron consists of a **cell body** and two types of processes, or extensions, called **axons** and **dendrites** (*Figure 12.3*).

Dendrites are short, multiple, highly branched extensions of the neuron's cell body. They conduct impulses toward the cell body. The more dendrites a neuron has, the more impulses it can receive from other neurons.

A single axon, or nerve fiber, arises from the cell body, is covered in a fatty **myelin** sheath, and carries the impulse away from the cell body. Each axon can range in length from a few millimeters to a meter.

Bundles of these axons appear white and create the **white matter** of the brain and spinal cord. Neuron cell bodies, dendrites, and synapses appear gray and create the **gray matter.**

The axon terminates in a network of small branches that ends at a **synapse** (junction) with a dendrite from another neuron or with a receptor on a muscle cell or gland cell (*Figure 12.4*). **Neurotransmitters** cross the synapse to stimulate or inhibit another neuron or the cell of a muscle or gland. Examples of neurotransmitters are norepinephrine, serotonin, and **dopamine.**

Groups of cell bodies cluster together to form ganglia, and groups of cell bodies and axons collect together to form nerves.

The trillion neurons in the nervous system are outnumbered 50 to 1 by the supportive **glial** cells **(neuroglia).**

The blood-brain barrier is a physical barrier composed of glial cells and the capillary blood vessel walls that prevents foreign substances, toxins, and infections from leaving the bloodstream and affecting the brain cells.

WORD ANALYSIS AND DEFINITION

S/ = Suffix P/ = Prefix R/ = Root R/CF = Combining Form

WORD	PRONUNCIATION		ELEMENTS	DEFINITION
axon	**ACK**-son		Greek *axis*	Single process of a nerve cell carrying nervous impulses away from the cell body.
dendrite	**DEN**-dright		Greek *looking like a tree*	Branched extension of the nerve cell body that receives nervous stimuli.
dopamine	**DOH**-pah-meen		Precursor of norepinephrine	Neurotransmitter in some specific small areas of the brain.
glia	**GLEE**-ah		Greek *glue*	Connective tissue that holds a structure together.
glial (adj) **neuroglia**	**GLEE**-al nyu-roh-**GLEE**-ah	S/ R/ R/CF	**-al** *pertaining to* **-glia** *glue* **neur/o-** *nerve*	Pertaining to glia or neuroglia. Connective tissue holding nervous tissue together.
myelin	**MY**-eh-lin	S/ R/	**-in** *substance, chemical compound* **myel-** *spinal cord*	Material of the sheath around the axon of a nerve.
neurotransmitter (***Note:*** *Transmit* is a word itself, so the prefix *trans* is in the middle of the overall word.)	**NYUR**-oh-trans-**MIT**-er	S/ R/CF P/ R/	**-er** *agent* **neur/o-** *nerve* **-trans-** *across* **-mitt-** *send*	Chemical agent that relays messages from one nerve cell to the next.
synapse	**SIN**-aps	P/ R/	**syn-** *together* **-apse** *clasp*	Junction between two nerve cells, or a nerve fiber and its target cell, where electrical impulses are transmitted between the cells.

EXERCISES

Spelling *is very important for every medical term. Choose the correct spelling in this language of neurology. Circle the best choice in the first half of the exercise; then follow the instructions below for the second half.*

1. material of the sheath around a nerve axon:

   myelin   mieline   myeline

2. branched extension of the nerve cell body that receives nervous stimuli:

   denderite   dendrite   dendryite

3. chemical agent that relays messages from one nerve cell to the next:

   neutrontransmitter   neurontransmiter   neurotransmitter

4. junction between two nerve cells:

   synnapse   synaps   synapse

Choose the correct terms from the WAD box above, based on the brief definition given to you. Fill in the blanks.

5. neurotransmitter in specific areas of the brain ______________________

6. connective tissue that holds a structure together ______________________

7. single process of a nerve cell carrying impulses away from the body ______________________

Check—have you spelled answers 5 through 7 correctly?

LESSON 12.2 The Brain and Cranial Nerves

OBJECTIVES

The sensations of smelling the roses, seeing them, and touching them are recognized and interpreted in the brain, as are all sensations. The actions of kneeling down, cutting the rose stem, walking into the house and placing it in a vase originate in the brain, as do all our voluntary actions. The information in this lesson will enable you to:

- **Use correct medical terminology to describe the essential structures of the brain and spinal cord.**
- **Identify the medical terminology of the major sensory and motor areas of the brain.**
- **Select the correct medical terminology to describe how the brain and spinal cord are protected and supported.**

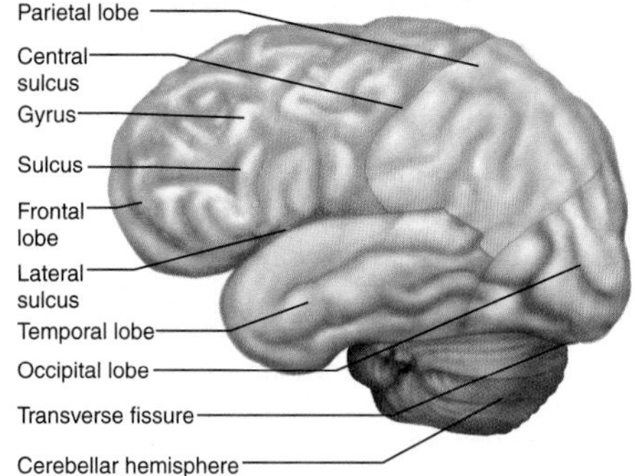

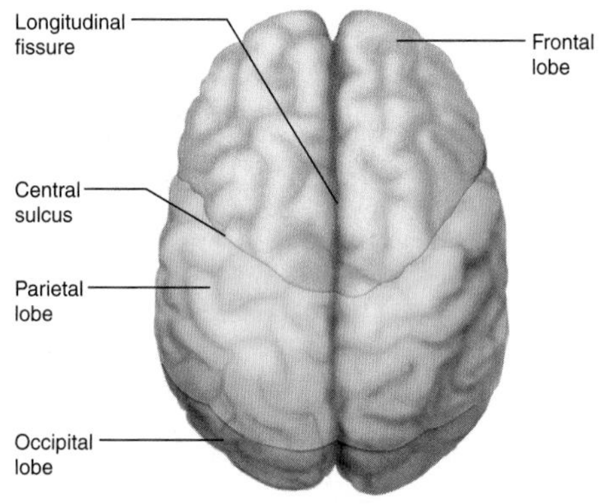

▲ **FIGURE 12.5 Brain.**
(a) View from the left side. (b) View from above.

Abbreviation	
CSF	cerebrospinal fluid

THE BRAIN

The adult brain weighs about 3 pounds. Its size and weight are proportional to body size, not intelligence. The brain is divided into three major regions: 1. the **cerebrum,** 2. the **brain stem,** and 3. the **cerebellum.**

Cerebrum

The cerebrum is about 80% of the brain and consists of two **cerebral** hemispheres, which are mirror images of each other. They are separated by a deep longitudinal fissure, at the bottom of which they are connected by a bridge of nerve fibers called the **corpus callosum.**

On the surface of the cerebrum, numerous ridges, **gyri,** are separated by fissures called **sulci** (*Figure 12.5*). The cerebral hemispheres are covered by a thin layer of gray matter (nerve cells and dendrites) called the cerebral **cortex.** It is folded into the gyri, and sulci, and contains 70% of all the neurons in the nervous system. Below the cerebral cortex is a mass of white matter, in which bundles of myelinated nerve fibers connect the neurons of the cortex to the rest of the nervous system.

Functional Cerebral Regions

Each cerebral hemisphere is divided into four lobes:

1. The **frontal** lobe is located behind the forehead. It forms the anterior part of the hemisphere. It is responsible for intellect, concentration, planning, problem solving, and the voluntary motor control of muscles.
2. The **parietal** lobe is posterior to the frontal lobe. The parietal lobe receives and interprets sensory information, such as receiving and interpreting spoken words.
3. The **temporal** lobe is below the frontal and parietal lobes. The temporal lobe is involved in interpreting sensory experiences.
4. The **occipital** lobe forms the posterior part of the hemisphere. It interprets visual images and the written word.

Deep inside each cerebral hemisphere are spaces called ventricles. They contain a watery fluid called **cerebrospinal fluid (CSF)** that circulates through the ventricles and around the brain and spinal cord. The CSF helps protect, cushion, and provide nutrition for the brain and spinal cord.

Underneath the cerebral hemispheres and ventricles are located the:

- **Thalamus** (*Figure 12.5*), which receives all sensory impulses and channels them to the appropriate region of the cortex for interpretation.
- **Hypothalamus,** which regulates blood pressure, body temperature, and water and electrolyte balance.

WORD ANALYSIS AND DEFINITION

S/ = Suffix P/ = Prefix R/ = Root R/CF = Combining Form

WORD	PRONUNCIATION	ELEMENTS		DEFINITION
cerebellum	ser-eh-**BELL**-um	S/ R/	**-um** *structure* **cerebell-** *little brain*	The most posterior area of the brain located between the midbrain and the cerebral hemispheres.
cerebrospinal (adj)	**SER**-ee-broh-**SPY**-nal	S/ R/CF R/	**-al** *pertaining to* **cerebr/o-** *brain* **-spin-** *spinal cord*	Pertaining to the brain and spinal cord.
cerebrospinal fluid (CSF)				Fluid formed in the ventricles of the brain; surrounds the brain and spinal cord.
cerebrum **cerebral (adj)**	**SER**-ee-brum **SER**-ee-bral	 S/ R/	Latin *brain* **-al** *pertaining to* **cerebr-** *brain*	Cerebral hemispheres. Pertaining to the cerebral hemispheres or the brain.
corpus callosum	**KOR**-pus kah-**LOW**-sum	R/ S/ R/	**corpus** *body* **-um** *structure* **callos-** *thickening*	Bridge of nerve fibers connecting the two cerebral hemispheres.
gyrus **gyri (pl)**	**JI**-rus **JI**-ree		Greek *circle*	Rounded elevation on the surface of the cerebral hemispheres.
hypothalamus	high-poh-**THAL**-ah-muss	S/ P/ R/	**-us** *pertaining to* **hypo-** *below* **-thalam-** *thalamus,*	An area of gray matter lying below the thalamus.
medulla oblongata	meh-**DULL**-ah ob-lon-**GAH**-tah	R/ S/ R/	**medulla** *middle* **-ata** *place* **oblong-** *elongated*	Most posterior subdivision of the brainstem, continuation of the spinal cord.
occipital lobe	ock-**SIP**-it-al LOBE	S/ R/	**-al** *pertaining to* **occipit-** *back of head*	Posterior area of the cerebral hemispheres.
parietal lobe	pah-**RYE**-eh-tal LOBE	S/ R/	**-al** *pertaining to* **pariet-** *wall*	Area of the brain under the parietal bone.
pons	PONZ		Latin *bridge*	Part of the brainstem.
sulcus **sulci (pl)**	**SUL**-cuss **SUL**-sigh		Latin *furrow, ditch*	Groove on the surface of the cerebral hemispheres that separates gyri.
temporal lobe	**TEM**-pore-al LOBE	S/ R/	**-al** *pertaining to* **tempor-** *time, temple*	Posterior two-thirds of the cerebral hemispheres.
thalamus	**THAL**-ah-mus		Greek *inner room*	Mass of gray matter under the ventricle in each cerebral hemisphere.

Brainstem and Cerebellum

The **brainstem** contains two major areas:

- The **pons,** which relays sensory impulses from peripheral nerves to higher brain centers.
- The **medulla oblongata,** which has centers to control vital cardiovascular and respiratory activities.

The **cerebellum**, the most posterior area of the brain, coordinates skeletal muscle activity to maintain posture and balance.

EXERCISES

Roots: *The lobes of the brain share the suffix* al, *meaning* pertaining to. *It is the roots that describe their exact location in the brain. Use your knowledge of roots to understand the anatomical location of the lobes. Fill in the blanks.*

1. *front/al* root means: ____________________; lobe is located: ____________________
2. *occipit/al* root means: ____________________; lobe is located: ____________________
3. *pariet/al* root means: ____________________; lobe is located: ____________________
4. *tempor/al* root means: ____________________; lobe is located: ____________________

Oh	(olfactory-I)
once	(optic-II)
one	(oculomotor-III)
takes	(trochlear-IV)
the	(trigeminal-V)
anatomy	(abducens-VI)
final	(facial-VII)
very	(vestibulocochlear-VIII)
good	(glossopharyngeal-IX)
vacations	(vagus-X)
are	(accessory-XI)
heavenly!	(hypoglossal-XII)

▲ **FIGURE 12.6**

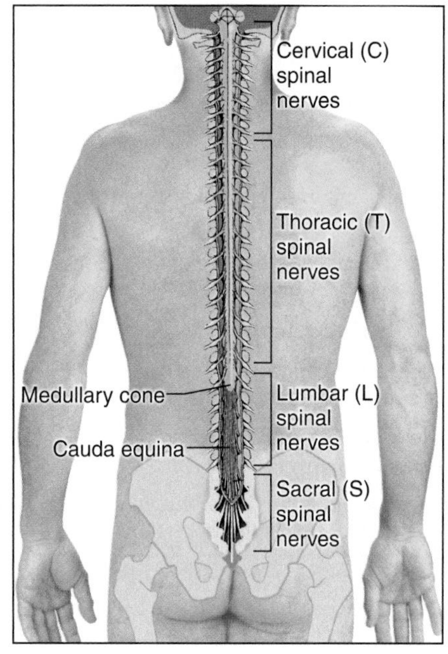

▲ **FIGURE 12.7**
Spinal Cord Regions.

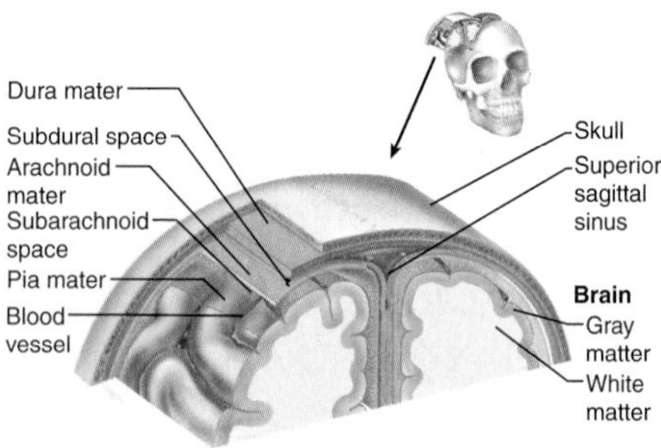

▲ **FIGURE 12.8**
Meninges of Brain.

CRANIAL NERVES, SPINAL CORD, AND MENINGES

Cranial Nerves

The brain communicates with the rest of the body through the cranial nerves and spinal cord. The cranial nerves are part of the peripheral nervous system and originate on the inferior surface of the brain. They have both names and numbers, the latter written in Roman numerals (*Figure 12.6*).

The cranial nerves provide sensory and motor functions for the head and neck areas except for the vagus nerve, which supplies the organs of the thorax and abdomen.

Spinal Cord

The spinal cord lies within and is protected by the vertebral canal. It has 31 segments, each of which gives rise to a pair of spinal nerves (*Figure 12.7*). These are the major link between the brain and the peripheral nervous system and are a pathway for sensory and motor impulses.

The spinal cord is divided into four regions (*Figure 12.7*):

1. The **cervical region** is continuous with the medulla oblongata. It contains the motor neurons that supply the neck, shoulders, and upper limbs through 8 pairs of cervical spinal nerves **(C1–C8).**
2. The **thoracic region** contains the motor neurons that supply the thoracic cage, rib movement, vertebral column movement, and postural back muscles through 12 pairs of thoracic spinal nerves **(T1–T12).**
3. The **lumbar** region supplies the hips and the front of the lower limbs through 5 pairs of lumbar nerves **(L1–L5).**
4. The **sacral** region supplies the buttocks, genitalia, and backs of the legs through 5 sacral nerves **(S1–S5)** and 1 coccygeal nerve.

Meninges

The brain and spinal cord are protected by the cranium and the vertebrae, cushioned by the CSF, and covered by the **meninges** (*Figure 12.8*). The meninges have three layers:

1. **Dura mater,** the outermost layer of tough connective tissue attached to the inner surface of the cranium but separated from the vertebral canal by the **epidural space,** into which epidural injections are introduced.
2. **Arachnoid mater,** a thin web over the brain and spinal cord. The CSF is contained in the **subarachnoid space** between the arachnoid and pia mater.
3. **Pia mater,** the innermost layer of the meninges, attached to the surface of the brain and spinal cord. It supplies nerves and blood vessels that nourish the outer cells of the brain and spinal cord.

WORD ANALYSIS AND DEFINITION

S/ = Suffix P/ = Prefix R/ = Root R/CF = Combining Form

WORD	PRONUNCIATION	ELEMENTS		DEFINITION
arachnoid mater	ah-**RACK**-noyd **MAY**-ter	S/ R/ R/	**-oid** *resembling* **arachn-** *cobweb, spider* **mater** *mother*	Weblike middle layer of the three meninges.
dura mater (***Note:*** Both terms are stand-alone roots.)	**DYU**-rah **MAY**-ter	R/ R/	**dura** *hard* **mater** *mother*	Hard, fibrous outer layer of the meninges.
epidural **epidural space**	ep-ih-**DYU**-ral	S/ P/ R/	**-al** *pertaining to* **epi-** *above* **-dur-** *dura*	Above the dura. Space between the dura mater and the wall of the vertebral canal.
meninges **meningitis**	meh-**NIN**-jeez men-in-**JIE**-tis	 S/ R/	Greek *membrane* **-itis** *inflammation* **mening-** *meninges*	Three-layered covering of the brain and spinal cord. Inflammation of the meninges.
pia mater (stand-alone roots)	**PEE**-ah **MAY**-ter	R/ R/	**pia** *delicate* **mater** *mother*	Delicate inner layer of the meninges.
subarachnoid space	sub-ah-**RACK**-noyd SPASE	S/ P/ R/	**-oid** *resembling* **sub-** *under* **-arachn-** *cobweb, spider*	Space between the pia mater and the arachnoid membrane.

EXERCISES

Construct *the following medical terms by filling in the missing elements. After you have inserted the correct element, circle the word in the statement that was your clue to the missing element. Fill in the blanks.*

1. under the arachnoid mater: ____________ /arachn/oid
2. inflammation of the meninges: mening/ ____________
3. weblike middle layer of meninges: ____________ /oid mater
4. pertaining to above the dura: ____________ /dur/al
5. hard, fibrous outer layer of meninges: ____________ mater
6. Which two prefixes in this WAD denote opposite locations?

   ____________ means ____________.

   ____________ means ____________.
7. Which term is a diagnosis? ____________
8. CSF is found in the sub/ ____________ / ____________.

LESSON 12.3 Disorders of the Brain, Cranial Nerves, and Meninges

OBJECTIVES

When patients communicate with you as a health professional, they will often enhance your continual, ongoing learning by informing and educating you and reinforcing your knowledge about a specific disorder. Information in this lesson will enable you to:

- **Describe common disorders of the brain and cranial nerves.**
- **Select the correct terminology to communicate about the brain and cranial nerves and their disorders with physicians and other health professionals.**

You are

. . . a medical assistant working with Dr. Raul Cardenas, a neurologist at Fulwood Medical Center.

You are communicating with

. . . Mr. Lester Rood, a 75-year-old man who was diagnosed a year ago as having dementia. He lives with his daughter, Judy, and she is with him today.

CASE REPORT 12.2

Patient interview:

Lester Rood: How am I feeling? Scared stiff. Sometimes I don't know where I am. I get so messed up. I can't cook anymore. I forget what I'm doing, can't get things straight. I find myself in the street and don't know how I got there. Judy has to help me shower and remind me to go to the bathroom. And it's only going to get worse. I don't want to be a burden. I used to have 100 people work for me. It's so frustrating, so frightening.

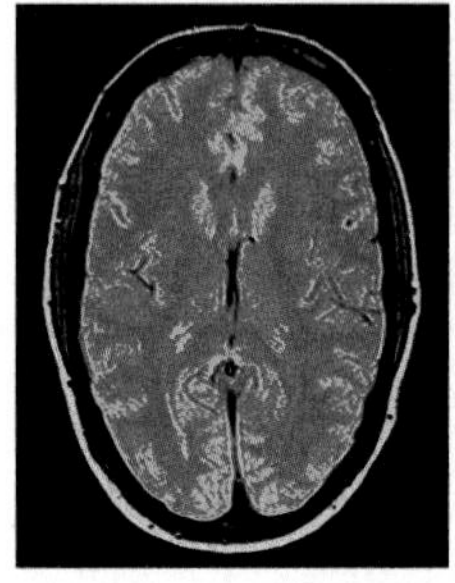

(a)

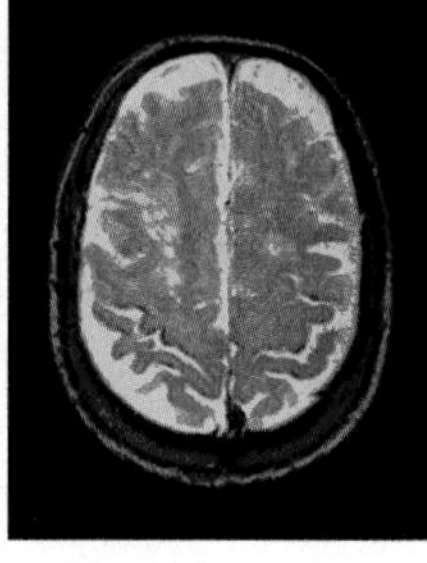

(b)

▲ **FIGURE 12.9 Brain Sections.** (a) MRI scan of normal brain. (b) MRI scan of Alzheimer disease, showing cerebral atrophy *(yellow)*.

Abbreviations

AMS	altered mental state
CVA	cerebrovascular accident

DISORDERS OF THE BRAIN

Dementia

Your **empathy** allowed Mr. Rood to talk without interruption. He reminded you that the symptoms of **dementia** include irreversible short-term memory loss, inability to solve problems, confusion, inappropriate behavior such as wandering away, and impaired intellectual function that interferes with normal activities and relationships. Dementia requires a lot of **sympathy** from family and caregivers.

Senile dementia is not a normal part of aging and is not a specific disease. It is a term used for a collection of symptoms that can be caused by a number of disorders affecting the brain.

Alzheimer disease (*Figure 12.9*) is the most common form of dementia. It affects 10% of the population over 65, and 50% of the population over 85. Nerve cells in the areas of the brain associated with memory and **cognition** are replaced by abnormal clumps and tangles of a protein.

Vascular dementia is the second most common form of dementia. It can come on gradually when arteries supplying the brain become arteriosclerotic (narrowed or blocked), depriving the brain of oxygen. It can also occur suddenly after a stroke (*see below*).

Confusion is used to describe people who cannot process information normally. For example, they cannot answer questions appropriately, understand where they are, or remember important facts, such as their name and address.

Delirium is the sudden onset of disorientation, an inability to think clearly or pay attention. There is a change in the level of consciousness, varying from increased wakefulness to drowsiness. It is an altered mental state **(AMS)**, not a disease, and it is reversible. It can be part of dementia or a **stroke.**

Other conditions causing dementia include reactions to medications, such as **sedatives** and antiarthritics; depression in the elderly; and infections, such as AIDS or encephalitis.

WORD ANALYSIS AND DEFINITION

S/ = Suffix P/ = Prefix R/ = Root R/CF = Combining Form

WORD	PRONUNCIATION	ELEMENTS		DEFINITION
Alzheimer disease	**AWLZ**-high-mer **DIZ**-eez		Dr. Alois Alzheimer, German physician, 1864–1915	Common form of dementia.
cognition	kog-**NIH**-shun		Latin *knowledge*	Process of acquiring knowledge through thinking, learning, and memory.
confusion	kon-**FEW**-zhun	S/ R/	**-ion** *condition, action* **confus-** *bewildered*	Mental state in which environmental stimuli are not processed appropriately.
delirium	de-**LIR**-ee-um	S/ R/	**-um** *structure* **deliri-** *disorientation, confusion*	Acute altered state of consciousness with agitation and disorientation.
dementia	dee-**MEN**-she-ah	S/ P/ R/	**-ia** *condition* **de-** *removal, without* **-ment-** *mind*	Chronic, progressive, irreversible loss of the mind's cognitive and intellectual functions.
empathy	**EM**-pah-thee	P/ R/	**em-** *into* **-pathy** *emotion, disease*	Ability to place yourself into the feelings, emotions, and reactions of another person.
sympathy	**SIM**-pah-thee	P/ R/	**sym-** *together* **-pathy** *emotion, disease*	Appreciation and concern for another person's mental and emotional state.
sedative	**SED**-ah-tiv	S/ R/	**-ive** *pertaining to, quality of* **sedat-** *to calm*	Agent that calms nervous excitement.
sedation	seh-**DAY**-shun	S/	**-ion** *condition, action*	State of being calmed.
senile	**SEE**-nile	S/ R/	**-ile** *pertaining to* **sen-** *old age*	Characteristic of old age.
stroke (same as *cerebrovascular accident*, **CVA**)	STROHK		Old English *to strike*	Acute clinical event caused by impaired cerebral circulation.

EXERCISES

Identify *the elements in each medical term, and unlock the meaning of the word. Fill in the chart; then use two of the terms in patient documentation of your choice.*

Medical Term	Meaning of Prefix	Meaning of Root/CF	Meaning of Suffix	Meaning of the Term
dementia				
sympathy				
delirium				
confusion				
empathy				
sedation				

Patient documentation:

1. ______________________________

____

____

2. ______________________________

____

____

Case Report 12.1 *(continued from the chapter opening)*

For Ms. Roberta Gaston, who was seen in Dr. Solis' neurosurgery clinic, the EEG did not localize an epileptic source. Hence, deep brain electrodes were inserted into the region of the suspicious mass that showed on MRI. Seizures were recorded arising in the mass itself. Dr. Solis performed a surgical resection of the mass, which was a glioma. Ms. Gaston has been seizure-free since the surgery a year ago.

Keynote

Epilepsy affects 1 in 200 people, and 50% of cases develop before age 10.

EPILEPSY

Epilepsy is a chronic disorder in which clusters of neurons (nerve cells) discharge their electrical signals in an abnormal rhythm. This disturbed electrical activity (a **seizure** or a **convulsion**) can cause strange sensations and behavior, convulsions, and loss of consciousness. The causes of epilepsy are numerous, from abnormal brain development to brain damage.

An accepted classification of seizures is that from the International League Against Epilepsy:

1. **Partial seizures** occur when the epileptic activity is in one localized area of the brain only, causing, for example, involuntary jerking movements of a single limb.
2. **Generalized seizures:**
   a. **Absence seizures,** previously known as **"petit mal,"** begin between ages 5 and 10. The child stares vacantly for a few seconds. The child may be accused of daydreaming.
   b. **Tonic-clonic seizures,** previously called **"grand mal,"** are dramatic. The person experiences a **loss of consciousness (LOC),** breathing stops, eyes roll up, and the jaw is clenched. This "tonic" phase lasts for 30 to 60 seconds. It is followed by the "clonic" phase, in which the whole body shakes with a series of violent, rhythmic jerkings of the limbs. The seizures last for a couple of minutes, and then consciousness returns.
   c. **Febrile seizures** are triggered by a fever in infants aged 6 months to 5 years. Very few of these infants go on to develop epilepsy.

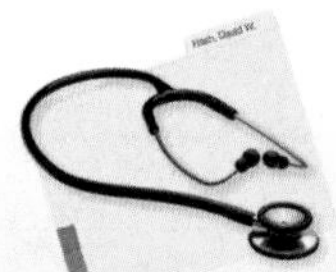

Case Report 12.1 *(continued)*

Ms. Gaston suffered from both absence and tonic-clonic seizures.

Status epilepticus is a medical emergency and requires maintenance of the airway, breathing, and circulation and intravenous administration of anticonvulsant drugs.

Keynote

The first-aid treatment during a seizure is to place the person in a reclining position, cushion the head, and turn the person on his or her side. Do not try to hold down the person's arms or legs or restrain the tongue.

Keynote

There is no cure for tic disorders, but they can be treated pharmacologically with haloperidol or clonidine.

Status epilepticus is defined as having one continuous seizure or recurrent seizures without regaining consciousness for 30 minutes or more.

Any seizure may be followed by a period of diminished function in the area of the brain surrounding the seizure focal origin. This transient neurologic deficit is called a **postictal** state.

Tourette syndrome and other **tic disorders** are characterized by episodes of involuntary, rapid, repetitive, fixed movements of individual muscle groups in the face or the limbs. They are associated with meaningless vocal sounds or meaningful words and phrases. The tics are probably genetic.

Narcolepsy is a chronic disorder in which patients fall asleep during the day for a few seconds or up to an hour. There is no cure.

WORD ANALYSIS AND DEFINITION

S/ = Suffix P/ = Prefix R/ = Root R/CF = Combining Form

WORD	PRONUNCIATION	ELEMENTS		DEFINITION
grand mal	GRAHN MAL	R/ R/	**grand** *big* **mal** *bad*	Old name for generalized tonic-clonic seizure.
narcolepsy	**NAR**-koh-lep-see	S/ R/CF	**-lepsy** *seizure* **narc/o-** *stupor*	Involuntary falling asleep.
petit mal	peh-**TEE** MAL	R/ R/	**petit** *small* **mal** *bad*	Old name for absence seizures.
postictal (adj)	post-**IK**-tal	S/ P/ R/	**-al** *pertaining to* **post-** *after* **-ict-** *seizure*	Transient neurologic deficit after a seizure.
ictal (adj)	**ICK**-tal	S/	**-al** *pertaining to*	Pertaining to, or a condition caused by, a stroke or epilepsy.
tic	TIK		French *tic*	Sudden, involuntary, repeated contraction of muscles.
tonic	**TON**-ik	S/ R/	**-ic** *pertaining to* **ton-** *pressure, tension*	State of muscular contraction.
tonic-clonic seizure	**TON**-ik **KLON**-ik **SEE**-zhur	R/ S/ R/	**clon-** *tumult* **-ure** *process* **seiz-** *to grab*	The body alternates between excessive muscular rigidity (tonic) and jerking muscular contractions (clonic).
Tourette syndrome	tur-**ET SIN**-drome		Gilles de la Tourette, French neurologist, 1857–1904	Disorder of multiple motor and vocal tics.

EXERCISES

Elements—recall and review: *Match the meaning in column 1 to the correct element in column 2. Fill in the blanks.*

____ 1. condition A. oid

____ 2. after B. narco

____ 3. one C. ict

____ 4. stupor D. para

____ 5. resembling E. mono

____ 6. abnormal F. post

____ 7. seizure G. osis

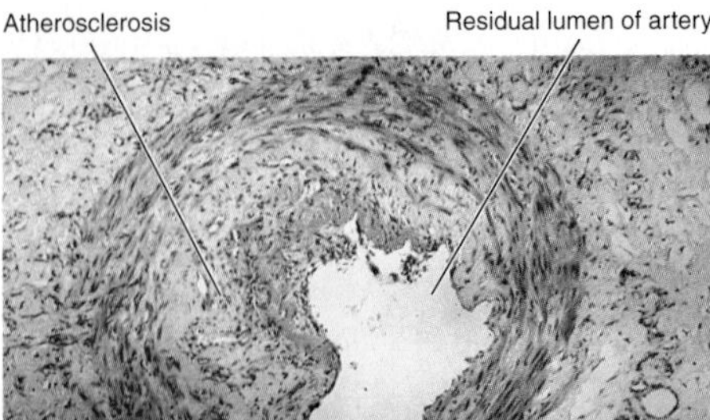

(a)

(b)

▲ **FIGURE 12.10 Causes of Ischemic Strokes.**
(a) Atherosclerosis in a cerebral artery, leaving a small, residual lumen. (b) Embolus blocking an artery. Healthy tissue is on the left *(pink)*; blood-starved tissue is on the right *(blue)*.

Keynote

Risk factors for ischemic strokes are hypertension, diabetes mellitus, high cholesterol levels, cigarettes, and obesity.

Keynote

Risk factors for hemorrhagic strokes are hypertension, cerebral arteriovenous malformations, and cerebral aneurysms.

Keynote

One-third of people with TIAs have subsequent TIAs, and one-third have a full-blown stroke later.

Abbreviations

BSE	bovine spongiform encephalopathy
CJD	Creutzfeldt-Jakob disease
TIA	transient ischemic attack
tPA	tissue plasminogen activator

CEREBROVASCULAR ACCIDENTS (CVAs) OR STROKES

A **stroke** (also known as a **cerebrovascular accident** or **CVA**) occurs when the blood supply to a part of the brain is suddenly interrupted and thus brain cells are deprived of oxygen. Some cells die; others are left badly damaged. With timely treatment, the damaged cells can be saved. There are two types of stroke:

1. **Ischemic strokes** account for 90% of all strokes and are caused by:
   a. **Atherosclerosis:** Plaque in the wall of a cerebral artery (*Figure 12.10a*).
   b. **Embolism:** Blood clot in a cerebral artery originating from elsewhere in the body (*Figure 12.10b*).

   Treatment of ischemic strokes is by thrombolysis using clot busters such as **tissue plasminogen activator (tPA)** within 3 hours of the stroke, with supportive measures followed by rehabilitation.
2. **Hemorrhagic strokes (intracranial hemorrhage)** occur when a blood vessel in the brain bursts or when a cerebral **aneurysm** ruptures.

Cerebral arteriography can determine the site of bleeding in hemorrhagic strokes, enabling surgery to be performed to stop the bleed or to clip off the aneurysm.

Transient Ischemic Attack (TIA)

Transient ischemic attacks **(TIAs)** are short-term, small strokes with symptoms lasting for less than 24 hours. If neurologic symptoms persist for more than 24 hours, the condition is a full-blown stroke with brain cell damage and death.

The most frequent cause is a small embolus that occludes (blocks) a small artery in the brain. Often, the embolus arises from a clot in the atrium in atrial fibrillation or from an atherosclerotic plaque in a carotid artery.

Treatment is directed at the underlying cause. **Carotid endarterectomy** may be necessary if a carotid artery is significantly occluded with plaque.

OTHER BRAIN DISORDERS

Parkinson disease is caused by the degeneration of neurons in the basic ganglia that produce a neurotransmitter called dopamine. Motor symptoms of abnormal movements, **tremor** of the hands, rigidity, a shuffling gait, and weak voice appear. The symptoms gradually increase in severity. There is no cure.

Creutzfeld-Jakob disease (CJD) produces a rapid deterioration of mental function, with difficulty in coordination of muscle movement. Some cases are linked to the consumption of beef from cattle with mad cow disease **(bovine spongiform encephalopathy, BSE).**

Syncope (fainting or passing out) is a temporary loss of consciousness and posture. It is usually due to hypotension and the associated deficient oxygen supply (hypoxia) to the brain.

Migraine produces an intense throbbing, pulsating pain in one area of the head, often with nausea and vomiting. It can be preceded by an aura, visual disturbances such as flashing lights, or temporary loss of vision. Prevention is difficult.

Encephalitis is inflammation of the parenchyma of the brain. It is usually caused by a virus such as HIV, West Nile virus, herpes simplex, or the childhood diseases of measles, mumps, chickenpox, and rubella (*see Chapter 20*).

Brain abscess is most often a direct spread of infection from sinusitis, otitis media, or mastoiditis. It can also be a result of blood-borne pathogens from lung or dental infections.

Brain tumors are most often secondary tumors that have metastasized from cancers in the lung, breast, skin, or kidney. Primary brain tumors arise from any of the glial cells and are called gliomas. In Case Report 10.1, Ms. Roberta Gaston has a glioma.

WORD ANALYSIS AND DEFINITION

S/ = Suffix   P/ = Prefix   R/ = Root   R/CF = Combining Form

WORD	PRONUNCIATION	ELEMENTS		DEFINITION
aneurysm	**AN**-yur-izm		Greek *dilation*	Small, circumscribed dilation of an artery or cardiac chamber.
arteriography	ar-teer-ee-**OG**-rah-fee	S/ R/	**-graphy** *recording* **arteri/o-** *artery*	X-ray visualization of an artery after injection of contrast material.
bovine spongiform encephalopathy (BSE)	**BO**-vine **SPON**-jee-form en-sef-ah-**LOP**-ah-thee	S/ R/ S/ R/CF S/ P/ R/CF	**-ine** *pertaining to* **bov-** *cattle* **-form** *appearance of* **spong/i-** *sponge* **-pathy** *disease* **en-** *in* **-cephal/o-** *head*	Disease of cattle ("mad cow disease") that can be transmitted to humans, causing Creutzfeldt-Jakob disease.
carotid endarterectomy	kah-**ROT**-id **END**-ar-ter-**EK**-toe-me	 S/ P/ R/	carotid, Greek *large neck artery* **-ectomy** *surgical excision* **end-** *inside* **-arter-** *artery*	Surgical removal of diseased lining from the carotid artery to leave a smooth lining and restore blood flow.
Creutzfeldt-Jakob disease (CJD)	**KROITS**-felt **YAK**-op **DIZ**-eez		Hans Creutzfeldt, 1885–1964, and Alfons Jakob, 1884–1931, German psychiatrists	Progressive, incurable, neurologic disease caused by infectious prions.
encephalitis	en-**SEF**-ah-**LIE**-tis	S/ R/	**-itis** *inflammation* **encephal-** *brain*	Inflammation of brain cells and tissues.
migraine	**MY**-grain	P/ R/	**mi-** *derived from* hemi, *half* **-graine** *head pain*	Paroxysmal severe headache confined to one side of the head.
Parkinson disease	**PAR**-kin-son **DIZ**-eez		James Parkinson, British physician, 1755–1824	Disease of muscular rigidity, tremors, and a masklike facial expression.
syncope	**SIN**-koh-peh		Greek *cutting short*	Temporary loss of consciousness and postural tone due to diminished cerebral blood flow.
tremor	**TREM**-or		Latin *to shake*	Small, shaking, involuntary, repetitive movements of hands, extremities, neck, or jaw.

EXERCISES

Diagnosis: *Patient documentation is given to you below. Circle the correct language of neurology for the diagnosis.*

1. Patient is experiencing shaking, involuntary movements of her extremities.
   Diagnosis: syncope prion tremor
2. Radiologic studies show a small, circumscribed dilation of the cerebral artery.
   Diagnosis: Parkinson disease aneurysm migraine
3. The patient's paroxysmal headache is confined to the left temporal region.
   Diagnosis: syncope tremor migraine
4. Tests and studies have confirmed inflammation of the brain cells and tissues for this patient.
   Diagnosis: meningitis fasciitis encephalitis
5. Unfortunately, this patient has a progressive, incurable neurologic disease caused by infectious prions.
   Diagnosis: TIA CJD BSE
6. This patient has experienced loss of consciousness due to diminished cerebral blood flow.
   Diagnosis: syncope tremor migraine

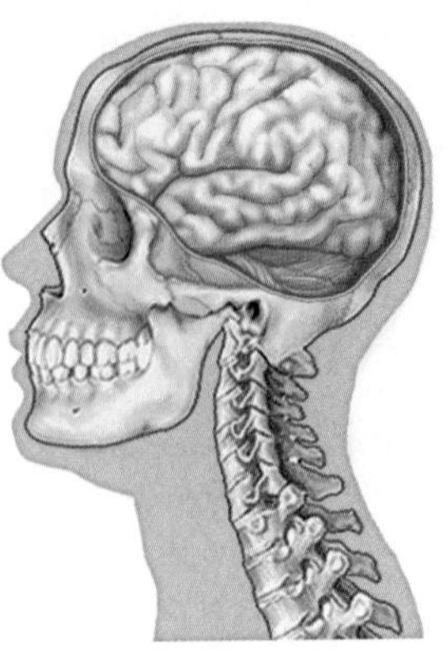

1. Position prior to impact.

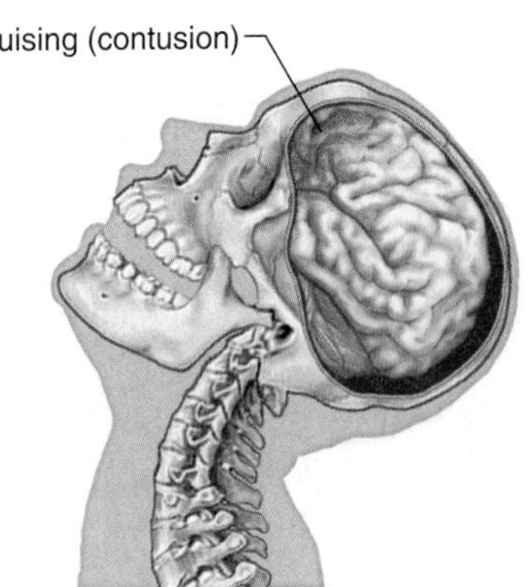

2. Impact (coup).

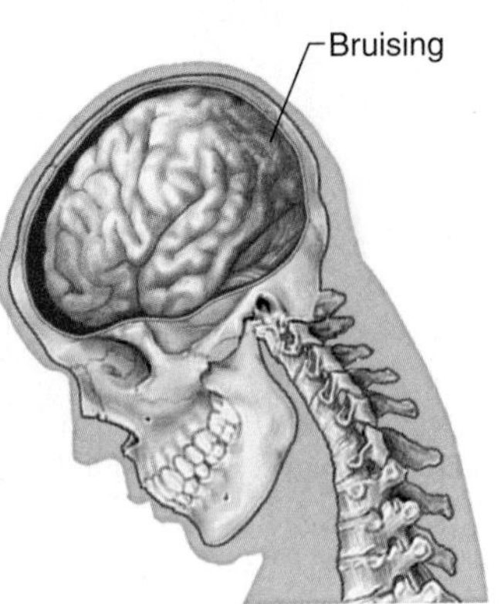

3. Contrecoup action.

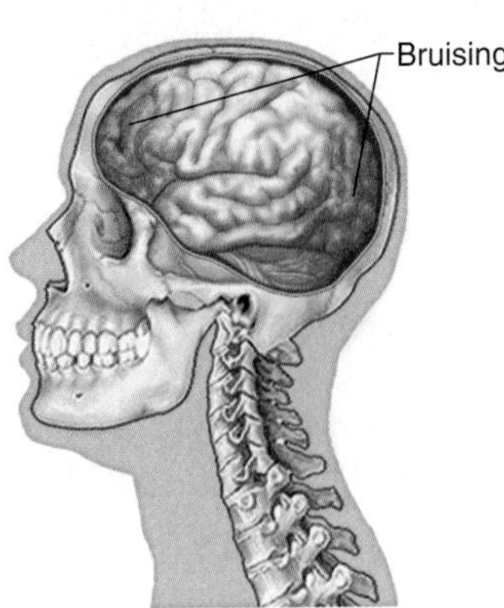

4. Subsequent coup-contrecoup injury.

▲ **FIGURE 12.11 Contusions Caused by Back-and-Forth Movement of Brain in Skull.**

Keynote

About 5.3 million Americans (2% of the population) are living with disabilities from TBI.

TRAUMATIC BRAIN INJURY (TBI)

Traumatic brain injury (TBI) causes damage to the brain. Over 1 million people are seen by medical doctors each year following a blow on the head. Of these, 50,000 to 100,000 will have prolonged problems affecting their work and their **activities of daily living (ADLs).**

If you are driving your car at 50 miles per hour and are hit head-on, your brain goes from 50 miles per hour to zero instantly. Your soft brain tissue is propelled forward and squished against the front of your hard skull **(coup).** Then the brain and the rest of your body rebound backward. The soft brain is then squished against the back of your rigid skull **(contracoup).** The squished front and back areas are at least bruised. This bruise is called a **contusion** (*Figure 12.11*).

If the process is more severe, blood vessels tear and blood flows into the brain. In addition, the brain itself can tear and cut brain signals and connections. In any injury, brain swelling can occur. Because the skull is hard and rigid, it cannot expand to cope with this extra volume. So the soft brain tissue is compressed, and some areas can stop working.

A mild head injury is called a **concussion.** You may feel dazed or have a period of confusion and not recall the event that caused the concussion. Repeated concussions have a cumulative effect, with loss of mental ability and/or traumatically induced Parkinson disease (as happened with professional boxer Muhammad Ali).

Shaken baby syndrome (SBS) is a type of TBI produced when a baby is violently shaken. A baby has weak neck muscles and a heavy head. Shaking makes the brain bounce back and forth in the skull, leading to severe brain damage.

DISORDERS OF THE MENINGES

Subdural hematoma is bleeding into the subdural space outside the brain. Most subdural hematomas are associated with closed head injuries and bleeding from broken veins caused by violent rotational movement of the head.

Epidural hematoma is a pooling of blood in the epidural space outside the brain. Most epidural hematomas are associated with a fractured skull and bleeding from an artery that lies in the meninges.

Meningitis is inflammation of the meninges covering the brain and spinal cord. Viral meningitis is the most common form and occurs at all ages. Bacterial meningitis is more common in the very young or very old. **Meningococcal meningitis** is contagious through droplet infection by coughing and sneezing and through close living conditions as in dormitories. Vaccines are available to prevent most causes of meningitis.

▲ **FIGURE 12.12 Bell Palsy of Left Side of Face.**

DISORDERS OF THE CRANIAL NERVES

Bell palsy is a disorder of the seventh cranial nerve (facial nerve) causing a sudden onset of weakness or paralysis of facial muscles on one side of the face. Common symptoms are inability to smile, whistle, or grimace, drooping of the mouth with drooling of saliva, and inability to close the eye (*Figure 12.12*).

WORD ANALYSIS AND DEFINITION

S/ = Suffix P/ = Prefix R/ = Root R/CF = Combining Form

WORD	PRONUNCIATION		ELEMENTS	DEFINITION
analgesia **analgesic (adj)**	an-al-**JEE**-zee-ah an-al-**JEE**-zik	S/ P/ R/ S/	**-ia** *condition* **an-** *without* **-alges-** *sensation of pain* **-ic** *pertaining to*	State in which pain is reduced. Substance that produces analgesia.
Bell palsy	BELL **PAWL**-zee		Charles Bell, Scottish surgeon, 1774–1842	Paresis, or paralysis, of one side of the face.
concussion	kon-**KUSH**-un	S/ R/	**-ion** *action, condition* **concuss-** *shake or jar violently*	Mild brain injury.
contrecoup	**KON**-treh-koo		French *counterblow*	Injury to the brain at a point directly opposite the point of injury.
contusion	kon-**TOO**-zhun	S/ R/	**-ion** *action, condition* **contus-** *bruise*	Hemorrhage into a tissue (bruising), including the brain.
coup	KOO		French *a blow*	Injury to the brain directly under the skull at the point of contact.
fibromyalgia	fie-broh-my-**AL**-jee-ah	S/ R/CF R/	**-algia** *pain* **fibr/o-** *fiber* **-my-** *muscle*	Pain in the muscle fibers.
meningococcal	meh-nin-goh-**KOK**-al	S/ R/CF R/	**-al** *pertaining to* **mening/o-** *meninges* **-cocc-** *round bacterium*	Pertaining to the *meningococcus* bacterium.
subdural space	sub-**DYU**-ral SPACE	S/ P/ R/	**-al** *pertaining to* **sub-** *below* **-dur-** *dura, hard*	Space between the arachnoid and dura mater layers of the meninges.

PAIN MANAGEMENT

More than 6 million Americans are affected by the pain of **fibromyalgia,** 5 million are disabled by back pain, and 40 million suffer from chronic, recurrent headaches.

Medications are prescribed based on the severity of the pain:

- For mild pain, **analgesics** (such as acetaminophen) and nonsteroidal anti-inflammatory drugs **(NSAIDs)** are used.
- For moderate pain, **opiates** (codeine, hydrocodone, and oxycodone) in combination with analgesics are used.
- For severe pain, higher doses of opiates are used, usually on their own. These include **morphine** and fentanyl.

Abbreviations

ADL activity of daily living
NSAID nonsteroidal anti-inflammatory drug
SBS shaken baby syndrome
TBI traumatic brain injury

EXERCISES

Deconstruct *the following language of neurology. First, slash the term in column 1 into its elements. Then write the meaning of each element to see how it will aid you in understanding the meaning of the complete term. Fill in the chart.*

Medical Term	Meaning of Prefix	Meaning of Root/CF	Meaning of Suffix	Meaning of Term
concussion				
contusion				
fibromyalgia				
subdural				

Remember: Every term does not need a prefix!

LESSON 12.4 Disorders of the Spinal Cord and Peripheral Nerves

THE NERVOUS SYSTEM

OBJECTIVES

There are many disorders of the nervous system that affect the spinal cord and the peripheral nerves without affecting the brain. In this lesson, the information will enable you to:

- **Discuss disorders of the spinal cord and peripheral nerves.**
- **Select the correct terminology to communicate about the nervous system and its disorders with patients and other health professionals.**
- **Name common congenital disorders of the nervous system.**

You are

... Tanisha Colis, an electroneurodiagnostic technologist working for Raul Cardenas, MD, a neurologist at Fulwood Medical Center.

You are communicating with

... Mrs. Suzanne Kalish, a 42-year-old social worker employed by the medical center.

CASE REPORT 12.3

She has recently had an exacerbation of her symptoms due to multiple sclerosis (MS). She is going to have a visual evoked potential (VEP) test, followed by an MRI of her brain and spinal cord.

Patient interview:

Tanisha: Good morning, Mrs. Kalish, I'm Tanisha Colis, the technologist who'll be performing your visual evoked potential test. How are you feeling?

Mrs. Kalish: I've been doing OK for the last 4 or 5 years. Then, a few weeks ago, I started dragging my right foot like a wounded witch. I've got to hang onto the walls to stay vertical. I'm tired out, can't come to work. It's a struggle to walk the few yards just to pick up the mail.

Tanisha: The MRI you are going to have today will give us a lot of information about what's going on.

Mrs. Kalish: My mind is going "wheelchair, wheelchair, wheelchair." Especially since in the last couple of days the vision in my right eye has gotten all blurred.

Tanisha: That's the reason you are having the visual evoked potential test.

Mrs. Kalish: I hate this disease. If I had cancer, I've got a chance. I'd fight it to the end, whatever that would be. Nobody's ever beat MS. You can only lose. It can be kind and leave you for a while, but it's never far away.

Tanisha: Let me help you up, and we'll go get this test done.

Mrs. Kalish: I can manage, thank you.

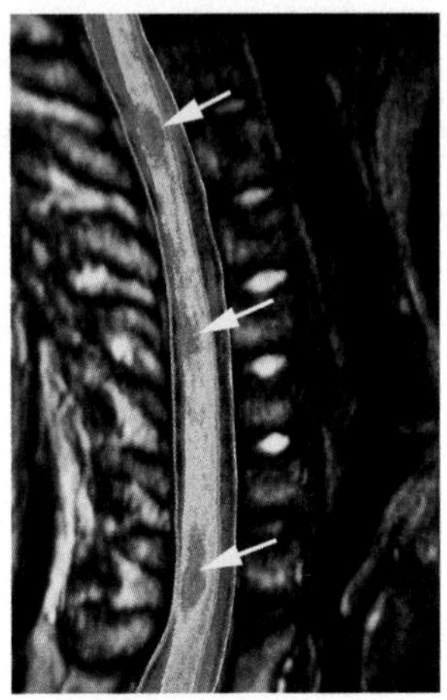

▲ **FIGURE 12.13**
Multiple Sclerosis.
Note areas of spinal cord where myelin sheath has been destroyed *(arrows)*. Normal spinal cord is yellow.

Keynote

Other causes of demyelination in the CNS are injury, ischemia, toxic agents such as chemotherapy or radiotherapy, and congenital disorders such as **Tay-Sachs disease.**

DISORDERS OF THE MYELIN SHEATH OF NERVE FIBERS

When the myelin sheath surrounding nerve fibers is damaged, nerves do not conduct impulses normally. In newborns, many of their nerves have immature myelin sheaths, which is why some of their movements are jerky and uncoordinated.

Demyelination, the destruction of an area of the myelin sheath, can occur in the PNS and be caused by inflammation, vitamin B_{12} deficiency, poisons, and some medications.

Multiple sclerosis (MS) is the most common disorder in which demyelination of nerve fibers in the brain, spinal cord, and optic nerves can occur (*Figure 12.13*). MS is a chronic, progressive disorder. **Intermittent** myelin damage and scarring slows nerve impulses. This leads to muscle weakness, pain, funny sensations **(paresthesias),** numbness, and vision loss. Because different nerve fibers are affected at different times, MS symptoms often worsen **(exacerbations)** or show partial or complete reduction **(remissions).** Suzanne is now in an exacerbation.

MS has an average age of onset between 18 and 35 years and is more common in women. Its cause is unknown, but it is thought to be an autoimmune disease. There is no known cure for MS.

WORD ANALYSIS AND DEFINITION

S/ = Suffix P/ = Prefix R/ = Root R/CF = Combining Form

WORD	PRONUNCIATION	ELEMENTS		DEFINITION
demyelination	dee-**MY**-eh-lin-**A**-shun	S/ P/ R/	**-ation** *process* **de-** *without* **-myelin-** *myelin*	Process of losing the myelin sheath of a nerve fiber.
exacerbation (contrast remission)	ek-zas-er-**BAY**-shun	S/ R/	**-ation** *process* **exacerbat-** *increase, aggravate*	Period when there is an increase in the severity of a disease.
intermittent	**IN**-ter-**MIT**-ent	S/ P/ R/	**-ent** *end result* **inter-** *between* **-mitt-** *send*	Alternately ceasing and beginning again.
modify	**MOD**-ih-fie		Latin *to limit*	Change the form or qualities of something; here, the pathology of MS.
paresthesia **paresthesias (pl)**	par-es-**THEE**-ze-ah	S/ P/ R/	**-ia** *condition* **par(a)-** *abnormal* **-esthes-** *sensation*	Abnormal sensation; e.g., tingling, burning, pricking.
remission (contrast exacerbation)	ree-**MISH**-un	S/ P/ R/	**-ion** *condition, action* **re-** *back* **-miss-** *send*	Period when there is a lessening or absence of the symptoms of a disease.
sclerosis	skleh-**ROH**-sis		Greek *hardness*	Thickening and hardening of a tissue; in the nervous system, hardening of nervous tissue by fibrous and glial connective tissue.
Tay-Sachs disease	TAY-SAKS **DIZ**-eez		Warren Tay, British ophthalmologist, 1843–1927; Bernard Sachs, U.S. neurologist, 1858–1944	Congenital fatal disorder of fat metabolism.

Abbreviations

MS multiple sclerosis
VEP visual evoked potential

EXERCISES

Test *yourself on the elements and terms in the above WAD box.* Practice pronouncing the terms before you start the exercise. Listen to the pronunciations on the Student Online Learning Center (www.mhhe.com/AllanEssMedLanguage). *Circle the best answer.*

1. The term *intermittent* contains:
   prefix, root, and suffix   prefix and root   combining form and suffix
2. Which term is a congenital, fatal disorder of fat metabolism?
   Tay-Sachs disease   demyelination   multiple sclerosis
3. *Remission* is the opposite of:
   intermittent   exacerbation   demyelination
4. To *modify* something is to:
   dispose of it   change it   renew it
5. In the term *demyelination* the prefix means:
   without   in front of   half
6. If a disease is *exacerbated,* it:
   is more severe   has not changed   is less severe
7. The term with a root that means *to increase or aggravate* is:
   intermittent   remission   exacerbation

Have you pronounced each word correctly?

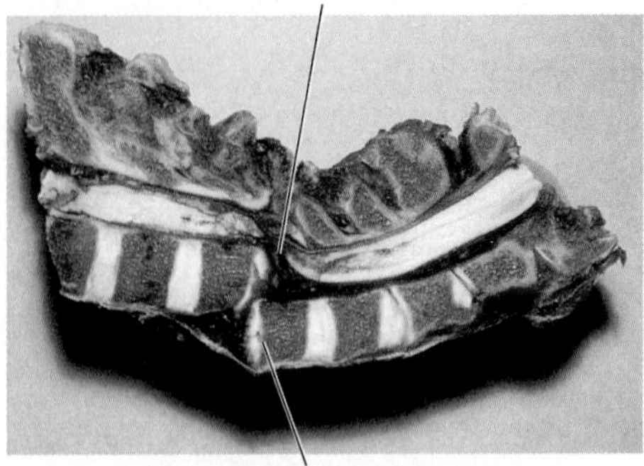

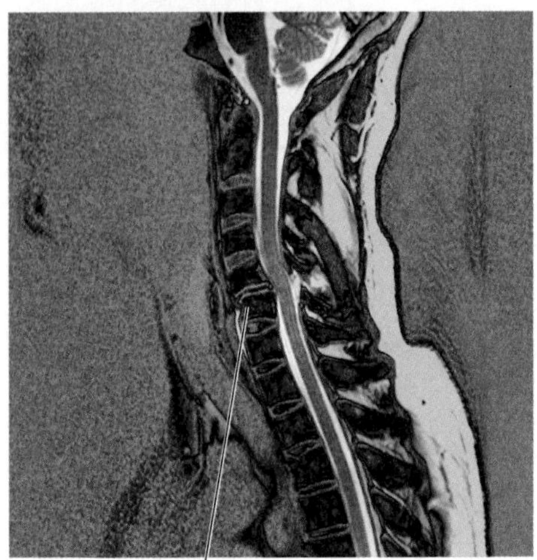

▲ **FIGURE 12.14**
Spinal Cord Injuries.
(a) Severed spinal cord from fracture-dislocation of vertebra. (b) Compressed spinal cord with vertebral fracture.

Keynote

Approximately half a million people in the United States have spinal cord injuries (SCIs). There are approximately 8000 new injuries each year, of which 82% involve males between the ages of 16 and 30.

Abbreviations

ALS	amyotrophic lateral sclerosis
Polio	poliomyelitis
PPS	postpolio syndrome
SCI	spinal cord injuries

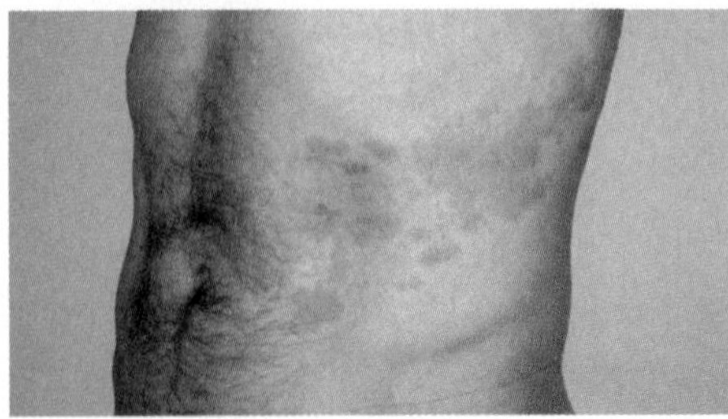

▲ **FIGURE 12.15**
Herpes Zoster (Shingles).

DISORDERS OF THE SPINAL CORD

Trauma

The spinal cord is injured in three ways:

1. **Severed,** by a fractured vertebra (*Figure 12.14a*).
2. **Contused,** as in a sudden, violent jolt to the spine.
3. **Compressed,** by a dislocated vertebra, bleeding, or swelling (*Figure 12.14b*).

Because of its anatomy with nerve fibers and tracts going up and down, to and from the brain, injury to the spinal cord results in loss of function below the injury. For example, if the cord is injured in the thoracic region, the arms function normally but the legs can be **paralyzed.** Both muscle control and sensation are lost. Partial paralysis is called **paresis.**

If the spinal cord is severed, the loss of function is permanent. Contusions can cause temporary loss lasting days, weeks, or months.

Compression of the cord can also occur from a tumor in the cord or spine or from a **herniated** disc. Cancer or osteoporosis can cause a vertebra to collapse and compress the cord.

In **syringomyelia,** fluid-filled cavities grow in the spinal cord and compress nerves that detect pain and temperature. There is no specific cure.

Other Disorders of the Spinal Cord

Acute transverse myelitis is a localized disorder of the spinal cord that blocks transmission of impulses up and down the spinal cord.

Subacute combined degeneration of the spinal cord is due to a deficiency of vitamin B_{12}. The sensory nerve fibers in the spinal cord degenerate, producing weakness, clumsiness, tingling, and loss of the position sense as to where limbs are. Treatment is injections of vitamin B_{12}.

Poliomyelitis is an acute infectious disease, occurring mostly in children, due to the poliovirus. The virus destroys motor neurons. Poliomyelitis is preventable by vaccination and has almost been eradicated in the world.

Postpolio syndrome (PPS) is when people develop tired, painful, and weak muscles many years after recovery from polio.

Amyotrophic lateral sclerosis (ALS), or "Lou Gehrig disease," occurs when motor nerves in the spinal cord progressively deteriorate. This leads to muscle weakness that can progress to paralysis. There is no cure.

DISORDERS OF PERIPHERAL NERVES

Neuropathy is used here as any disorder affecting one or more peripheral nerves.

Mononeuropathy is damage to a single peripheral nerve. Examples are:

- *Carpal tunnel syndrome,* in which the median nerve at the wrist is compressed between the wrist bones and a strong overlying ligament.
- *Ulnar nerve palsy,* which results from nerve damage as the ulnar nerve crosses close to the surface over the humerus at the back of the elbow.
- *Peroneal nerve palsy,* which arises from nerve damage as the peroneal nerve passes close to the surface near the back of the knee. Compression of the nerve occurs in people who are bedridden or strapped in a wheelchair.

Polyneuropathy is damage to, and the simultaneous malfunction of, many motor and/or sensory peripheral nerves throughout the body.

Herpes zoster, or **shingles** (*Figure 12.15*), is an infection of peripheral nerves arising from a reactivation of the dormant primary virus infection in childhood chickenpox (varicella).

WORD ANALYSIS AND DEFINITION

S/ = Suffix P/ = Prefix R/ = Root R/CF = Combining Form

WORD	PRONUNCIATION	ELEMENTS		DEFINITION
amyotrophic	a-my-oh-**TROH**-fik	S/ P/ R/CF R/	**-ic** *pertaining to* **a-** *without* **-my/o-** *muscle* **-troph-** *nourishment, development*	Pertaining to muscular atrophy.
compression	kom-**PRESH**-un	S/ P/ R/	**-ion** *action, condition* **com-** *together* **-press-** *squeeze*	Squeeze together to increase density and/or decrease a dimension of a structure.
herniation **hernia (noun)** **herniate (verb)**	**HER**-nee-ay-shun **HER**-nee-ah **HER**-nee-ate	S/ R/	**-ation** *process* **herni-** *rupture*	Protrusion of an anatomical structure from its normal location. Protrusion of a structure through the tissue that normally contains it.
neuropathy **mononeuropathy** **polyneuropathy**	nyu-**ROP**-ah-thee **MON**-oh-nyu-**ROP**-ah-thee **POL**-ee-nyu-**ROP**-ah-thee	S/ R/CF P/ P/	**-pathy** *disease* **neur/o-** *nerve* **mono-** *one* **poly-** many	Any disorder of the nervous system. Disorder affecting a single nerve. Disorder affecting many nerves.
paralyze (verb) **paralysis (noun)** **paralytic (adj)**	**PAR**-ah-lyze pah-**RAL**-ih-sis par-ah-**LYT**-ik	P/ R/ R/ S/ R/	**para-** *beside, abnormal* **-lyze** *destroy* **-lysis** *destruction* **-ic** *pertaining to* **-lyt-** *destroy*	To make incapable of movement. Loss of voluntary movement. Suffering from paralysis.
paresis **hemiparesis**	par-**EE**-sis **HEM**-ee-pah-**REE**-sis	 P/ R/	Greek *weakness* **hemi-** *half* **-paresis** *weakness*	Partial paralysis (weakness). Weakness of one side of the body.
poliomyelitis **polio (abbreviation)** **postpolio syndrome (PPS)**	**POE**-lee-oh-**MY**-eh-lie-tis post-**POE**-lee-oh **SIN**-drome	S/ R/ R/ P/ R/ P/ R/	**-itis** *inflammation* **polio-** *gray matter* **-myel-** *spinal cord* **post-** *after* **-polio** *gray matter* **syn-** *together* **-drome** *running*	Inflammation of the gray matter of the spinal cord, leading to paralysis of the limbs and muscles of respiration. Progressive muscle weakness in a person previously affected by polio.
syringomyelia	sih-**RING**-oh-my-**EE**-lee-ah	S R/ R/CF	**-ia** *condition* **-myel-** *spinal cord* **syring/o-** *tube, pipe*	Abnormal longitudinal cavities in the spinal cord cause paresthesias and muscle weakness.

EXERCISES

Identify *the element, and give its meaning. Complete the exercise by listing a medical term that contains this element. Review the WAD box above before you start the exercise. Fill in the chart.*

Element	Element Identity (P, R, CF, or S)	Meaning of Element	Medical Term Containing This Element
myo			
herni			
pathy			
troph			
lysis			
poly			

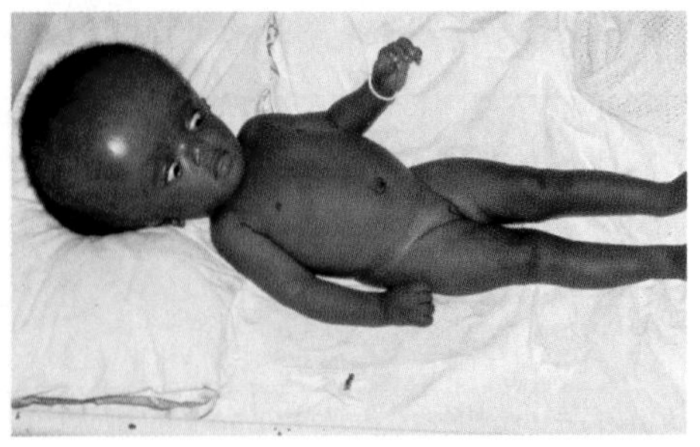

▲ **FIGURE 12.16**
Infant with Hydrocephalus.

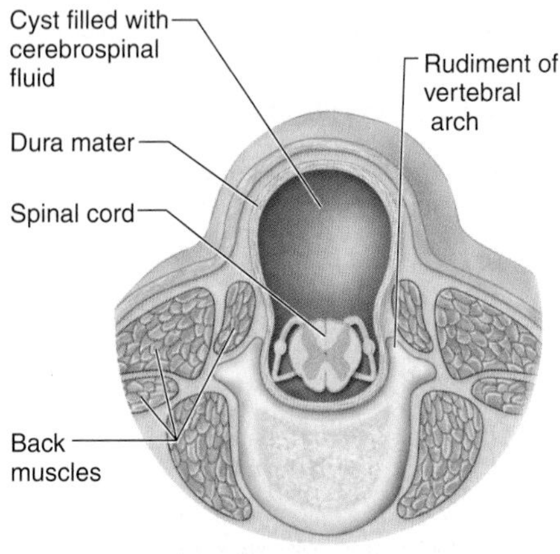

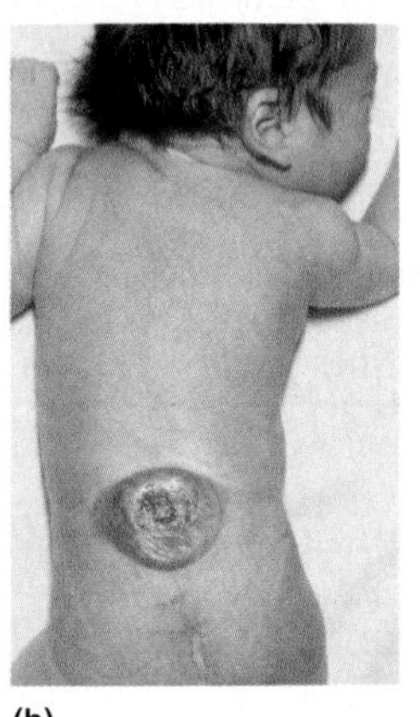

▲ **FIGURE 12.17 Spina Bifida Cystica.**
(a) Cross section of spinal cord. (b) Child with spina bifida cystica.

Keynote

Classification of CP by number of limbs impaired:

- **Quadriplegia:** All four limbs are involved.
- **Paraplegia:** Both lower extremities are affected.
- **Hemiplegia:** The arm and leg of one side of the body are affected.
- **Monoplegia:** Only one limb is affected, usually an arm.

Abbreviations

FAS	fetal alcohol syndrome
CP	cerebral palsy

CONGENITAL ANOMALIES OF THE NERVOUS SYSTEM

Some of the most devastating congenital neurologic abnormalities develop in the first 8 to 10 weeks of gestation (pregnancy), when the nervous system is in its early stages of formation. These malformations can be detected using ultrasonography and amniocentesis (*see Chapter 15*). Many can be prevented by the mother taking 4 mg/day of folic acid before conception and during early pregnancy.

A **teratogen** is an agent that can cause **anomalies** of an embryo or fetus (*see Chapter 15*). It can be a chemical, a virus, or radiation. Some teratogens are encountered in the workplace and include textile dies, photographic chemicals, semiconductor materials, and the metals lead, mercury, and cadmium.

Anencephaly is absence of the cerebral hemispheres and is incompatible with life. **Microcephaly,** decreased head size, is associated with small cerebral hemispheres and with moderate to severe motor and mental retardation.

Hydrocephalus (*Figure 12.16*) is ventricular enlargement in the cerebral hemispheres with excessive CSF; it is usually due to an obstruction that prevents the CSF from exiting the ventricles to circulate around the spinal cord. Treatment involves placing a tube (shunt) into the ventricle to divert the excess fluid into the abdominal cavity or a neck vein.

Spina bifida occurs mostly in the lumbar and sacral regions. It is very variable in its presentation and symptoms. **Spina bifida occulta** has a small partial defect in the vertebral arch. The spinal cord or meninges do not protrude. Often the only sign is a tuft of hair on the skin overlying the defect.

In **spina bifida cystica** there is no vertebral arch formed. The spinal cord and meninges protrude through the opening and may or may not be covered with a thin layer of skin (*Figure 12.17a and b*). Protrusion of only the meninges is called a **meningocele.** Protrusion of the meninges and spinal cord is called a **meningomyelocele.** Paralysis of the lower limbs may be present.

Fetal alcohol syndrome (FAS) can occur when a pregnant woman drinks alcohol. The child born with FAS has a small head, narrow eyes, and a flat face and nose. Intellect and growth are impaired. FAS is the third most common cause of mental retardation in newborns.

CEREBRAL PALSY

Cerebral palsy (CP) is the term used to describe the motor impairment resulting from brain damage in an infant or young child regardless of the cause or the effect on the child. It is not hereditary. In congenital CP the cause is often unknown but can be a brain malformation or maternal use of cocaine. CP developed at birth or in the neonatal period is usually related to an incident causing hypoxia of the brain.

Cerebral palsy causes delay in the development of normal milestones in infancy and childhood (*see Chapter 16*). The affected limbs can be **spastic** (muscles are tight and resistant to stretch). They can show **athetoid** movements, in which the limbs have involuntary writhing movements and are in constant motion. There may be present a poor sense of balance and coordination leading to **ataxia.**

WORD ANALYSIS AND DEFINITION

S/ = Suffix P/ = Prefix R/ = Root R/CF = Combining Form

WORD	PRONUNCIATION	ELEMENTS		DEFINITION
anencephaly	**AN**-en-**SEF**-ah-lee	S/ P/ R/	**-aly** *condition* **an-** *without* **-enceph-** *brain*	Born without cerebral hemispheres.
microcephaly	**MY**-kroh-**SEF**-ah-lee	P/ R/	**micro-** *small* **-ceph-** *head*	An abnormally small head.
ataxia	a-**TAK**-see-ah	S/ P/ R/	**-ia** *condition* **a-** *without* **-tax-** *coordination*	Inability to coordinate muscle activity, leading to jerky movements.
ataxic (adj)	a-**TAK**-sik	S/	**-ic** *pertaining to*	Pertaining to or suffering from ataxia.
athetosis	ath-eh-**TOE**-sis	S/ R/	**-osis** *condition* **athet-** *without position, uncontrolled*	Slow, writhing involuntary movements.
athetoid (adj)	**ATH**-eh-toyd	S/	**-oid** *resembling*	Resembling or suffering from athetosis.
hemiplegia	hem-ee-**PLEE**-jee-ah	S/ P/ R/	**-ia** *condition* **hemi-** *half* **-pleg-** *paralysis*	Paralysis of one side of the body.
hemiplegic (adj)	hem-ee-**PLEE**-jik	S/	**-ic** *pertaining to*	Pertaining to or suffering from hemiplegia.
hemiparesis	**HEM**-ee-pah-**REE**-sis	R/	**-paresis** *weakness*	Weakness of one side of the body.
meningocele	meh-**NING**-oh-seal	S/ R/CF	**-cele** *hernia* **mening/o-***meninges*	Protrusion of the meninges from the spinal cord or brain through a defect in the vertebral column or cranium.
meningomyelocele	meh-nin-goh-**MY**-el-oh-seal	S/ R/CF	**-cele** *hernia* **-myel/o-** *spinal cord*	Protrusion of the spinal cord and meninges through a defect in the vertebral arch of one or more vertebrae.
monoplegia	**MON**-oh-**PLEE**-jee-ah	S/ P/ R/	**-ia** *condition* **mono-** *one* **-pleg-** *paralysis*	Paralysis of one limb.
monoplegic (adj)	**MON**-oh-**PLEE**-jik	S/	**-ic** *pertaining to*	Pertaining to or suffering from monoplegia.
palsy	**PAWL**-zee		Latin *paralysis*	Paralysis or paresis from brain damage.
paraplegia	par-ah-**PLEE**-jee-ah	S/ P/ R/	**-ia** *condition* **para-** *abnormal* **-pleg-** *paralysis*	Paralysis of both lower extremities.
paraplegic (adj)	par-ah-**PLEE**-jik	S/	**-ic** *pertaining to*	Pertaining to or suffering from paraplegia.
quadriplegia	kwad-rih-**PLEE**-jee-ah	S/ P/ R/	**-ia** *condition* **quadri-** *four* **-pleg-** *paralysis*	Paralysis of all four limbs.
quadriplegic (adj)	kwad-rih-**PLEE**-jik	S/	**-ic** *pertaining to*	Pertaining to or suffering from quadriplegia.
spina bifida	**SPY**-nah **BIH**-fih-dah	R/CF P/ R/	**spin/a** *spine* **bi-** *two* **-fida** *split*	Failure of one or more vertebral arches to close during fetal development.
spina bifida cystica	**SIS**-tik-ah	S/ R/	**-ica** *pertaining to* **cyst-** *cyst*	Meninges and spinal cord protruding through the absent vertebral arch and having the appearance of a cyst.
spina bifida occulta	**OH**-kul-tah	R/CF	**occult/a** *hidden*	The deformity of the vertebral arch is not apparent from the surface.

EXERCISES

Search and find *the correct term for the element you are given. Circle the best answer.*

1. Find the term with the prefix meaning *without:*
   hydrocephalus anencephaly bifida
2. Find the term with the suffix meaning *hernia:*
   meningocele cystica teratogen
3. Find the term with the root meaning *hidden:*
   spina bifida spina bifida cystica spina bifida occulta

LESSON 12.5 Mental Health

OBJECTIVES

As a health professional, you will find that many of your patients have a mental health problem in addition to their physical ailments. This lesson will enable you to:

- **Comprehend, analyze, spell, and write the essential terms of psychology and psychiatry to communicate and document accurately and precisely.**
- **Recognize and pronounce the essential terms of psychology and psychiatry to communicate verbally with accuracy and precision.**

Keynote

A licensed specialist in psychology is a **psychologist.** Psychologists are not licensed to prescribe medications.

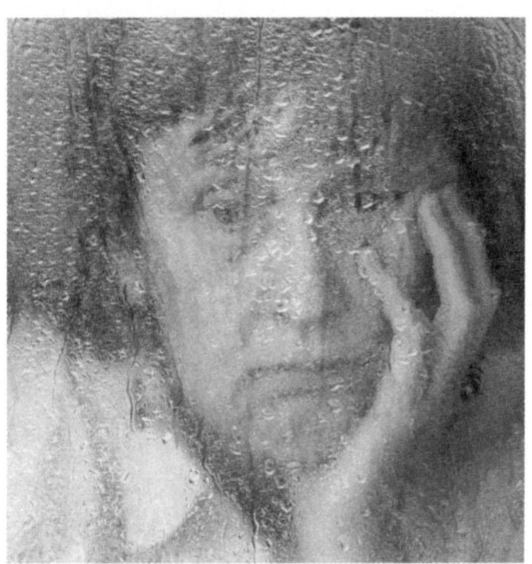

▲ **FIGURE 12.18 Depressed Woman at Window.**

Keynote

Insanity is a legal term for a severe mental illness that impairs a defendant's ability to understand the moral wrong of the act he or she committed. *It is not a medical diagnosis.*

MENTAL HEALTH

Mental health is defined as emotional, behavioral, and social well-being that enables an individual to cope with internal and external events.

Psychology is the scientific study of behavior—talking, reading, sleeping, interacting with others—and mental processes—thinking, feeling, remembering, dreaming.

Psychiatry is the medical specialty concerned with the origin, diagnosis, prevention, and treatment of mental, emotional, and behavioral disorders.

Mood Disorders

Major depression, also called **unipolar disorder** (*Figure 12.18*), occurs when a person is so deeply sad for at least 2 weeks that he or she feels hopeless, sees nothing but sorrow and despair in the future, and may not want to live anymore.

Bipolar disorder (which used to be called **manic-depressive disorder**) is the alternation of episodes of major depression with excessive overexcitement and impulsive behavior called **mania.**

Anxiety Disorders

Anxiety disorders are characterized by unreasonable anxiety or fear that is so intense and chronic that it disrupts the person's life.

- **Generalized anxiety disorder (GAD)** consists of unreasonable anxiety that is not focused on one particular situation or event.
- **Posttraumatic stress disorder (PTSD)** arises after significant trauma such as a life-threatening incident, loss of a loved one, abuse, or combat in war.
- **Panic disorder** is characterized by sudden, brief attacks of intense fear.
- **Phobias** differ from generalized anxiety and panic attacks in that a specific situation or object brings on the strong fear response. Examples are **claustrophobia** (fear of being trapped in a confined space), **acrophobia** (fear of heights), and **agoraphobia** (fear of crowded places).
- **Obsessive-compulsive disorder (OCD)** patients have both obsessions and compulsions. The obsessions are recurrent thoughts, fears, doubts, images, or impulses. The compulsions are recurrent, irresistible actions such as hand washing, counting, and checking. The recurrent actions can be violent or sexual.

WORD ANALYSIS AND DEFINITION

S/ = Suffix   P/ = Prefix   R/ = Root   R/CF = Combining Form

WORD	PRONUNCIATION		ELEMENTS	DEFINITION
anxiety	ang-**ZI**-eh-tee		Greek *distress, anxiety*	Distress and dread caused by fear.
bipolar disorder	bi-**POH**-lar dis-**OR**-der	S/ P/ R/	**-ar** *pertaining to* **bi-** *two* **-pol-** *pole*	A mood disorder with alternating periods of depression and mania.
delusion	dee-**LOO**-shun	S/ R/	**-ion** *action, condition* **delus-** *deceive*	Fixed, unyielding false belief held despite strong evidence to the contrary.
hallucination	hah-loo-sih-**NAY**-shun	S/ R/	**-ation** *process* **hallucin-** *imagination*	Perception of an object or event when there is no such thing present.
mania **manic (adj)**	**MAY**-nee-ah **MAN**-ik	 S/ R/	Greek *frenzy* **-ic** *pertaining to* **man-** *frenzy*	Mood disorder with hyperactivity, irritability, and rapid speech. Pertaining to or suffering from mania.
mute	MYUT		Latin *silent*	Unable or unwilling to speak.
paranoia **paranoid**	par-ah-**NOY**-ah **PAR**-ah-noyd	P/ R/ S/	**para-** *abnormal, beside* **-noia** *to think* **-oid** *resembling*	Presence of persecutory delusions. Having delusions of persecution.
phobia	**FOH**-bee-ah		Greek *fear*	Pathologic fear or dread.
psychiatry **psychiatrist** **psychiatric (adj)**	sigh-**KIGH**-ah-tree sigh-**KIGH**-ah-trist sigh-kee-**AH**-trik	S/ R/ S/ S/	**-iatry** *treatment* **psych-** *mind* **-iatrist** *one who treats* **-ic** *pertaining to*	Diagnosis and treatment of mental disorders. Licensed medical specialist in psychiatry. Pertaining to psychiatry.
psychology **psychologist**	sigh-**KOL**-oh-jee sigh-**KOL**-oh-jist	S/ R/CF S/	**-logy** *study of* **psych/o-** *mind* **-logist** *one who studies, specialist*	Science concerned with the behavior of the human mind. Licensed specialist in psychology.
psychosis **psychotic (adj)**	sigh-**KOH**-sis sigh-**KOT**-ik	S/ R/ S/ R/CF	**-osis** *condition* **psych-** *mind* **-tic** *pertaining to* **psych/o-** *mind*	Disorder causing mental disruption and loss of contact with reality. Pertaining to or affected by psychosis.
sociopath	**SOH**-see-oh-path	S/ R/CF	**-path** *disease* **soci/o-** *society, social*	Person with antisocial personality disorder.

Schizophrenia

Schizophrenia is a form of **psychosis** in which there is a loss of contact with reality. People with schizophrenia do *not* have a split personality, but their words are separated from the meaning, their perceptions are separated from reality, and their behaviors are separated from their thought processes. They perceive things without a stimulation **(hallucinations).** They suffer from **delusions** (mistaken beliefs that are contrary to facts). The delusions can be **paranoid** (with pervasive distrust and suspicion of others). Their speech is disorganized and can be incoherent, or they refuse or are unable to speak **(mute).**

Abbreviations

GAD	generalized anxiety disorder
MPD	multiple-personality disorder
OCD	obsessive-compulsive disorder
PTSD	posttraumatic stress disorder

EXERCISES

Elements *remain your best tool for deconstructing and understanding the meaning of a medical term. Each of the following terms has an element bolded. Identify the type of that element, and write the meaning of the element.*

Medical Term	Type of Element (P, R, CF, or S)	Meaning of the Element
1. socio**path**	______________	______________
2. **man**ic	______________	______________
3. psych**osis**	______________	______________
4. **para**noia	______________	______________
5. **hallucin**ation	______________	______________
6. **mut**ism	______________	______________

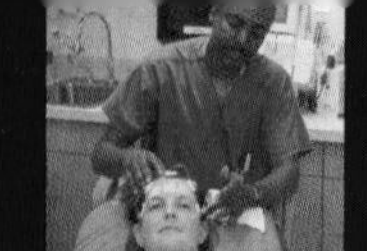

CHAPTER 12 REVIEW

NERVOUS SYSTEM

CHALLENGE YOUR KNOWLEDGE

A. **Prefixes:** This chapter contains a large number of important prefixes that you have also seen in previous chapters and will see again. Knowing these prefixes well will help increase your medical vocabulary. S-T-R-E-T-C-H your memory to recall previous terms with the same prefixes. *You must be prepared to define each medical term!* Fill in the chart.

Prefix	Meaning of Prefix	Medical Term in This Chapter	Medical Term from a Previous Chapter
a			
anti			
auto			
de			
epi			
mono			
para			
post			
re			
sub			
syn			
trans			

B. **Brief answer:** If you really understand the meanings of terms, you can explain the difference between terms to a patient or a fellow student. Explain the difference between:

1. *delusion:*

   hallucination:

2. *concussion:*

   contusion:

C. **Word attack:** Find your clues to the correct answer in the question.

Question:

Endarterectomy means:

a. surgical excision of an organ

b. surgical excision inside the skull

c. surgical excision of a gland

d. surgical excision inside an artery

e. surgical excision of a bone

1. Slash the term in question. On the line, write the elements. Below the line, write the meaning of the elements.

_______________ / _______________ / _______________

2. Since all the answers refer to "surgical excision" (which you know is the suffix), use another element for a better clue.

Prefix means: _______________

Root means: _______________

3. Cross off all the choices that do not have any relation to either the prefix or the root.

I've crossed off choices _______________.

4. My remaining choices are _______________.

5. What is the *only* choice that contains the meanings of *both* the prefix and the root? _______________

6. Therefore, the correct answer is _______________.

D. **Abbreviations:** The following abbreviations could be present in many types of clinical documentation. You must be able to interpret them correctly. The abbreviation is given to you—write out in words the meaning of the abbreviation, and check (✔) whether it is a diagnosis, procedure, test, or "other." If it is "other," assign it a category. The first one is done for you. Fill in the chart.

Abbreviation	Meaning of Abbreviation	Diagnosis	Procedure/Test	Other
tPA	tissue plasminogen activator			✔ drug
CP				
LOC				
CJD				
MRI				
EEG				
MS				
UA				
ALS				
PPS				
OCD				

NERVOUS SYSTEM

E. Analyze: Slash the following medical terminology into basic elements. Write the meaning of each element in the space beneath the line. The first one is done for you. Fill in the blanks. *Remember: Every term may not have a prefix.*

anencephaly	an /	enceph /	aly
	without	brain	condition

1. spastic ______ / ______ / ______
2. epidural ______ / ______ / ______
3. hydrocephalus ______ / ______ / ______
4. teratogen ______ / ______ / ______
5. meningocele ______ / ______ / ______
6. psychiatrist ______ / ______ / ______
7. sympathetic ______ / ______ / ______
8. synapse ______ / ______ / ______
9. dementia ______ / ______ / ______
10. postictal ______ / ______ / ______

F. Suffixes can have more than one meaning, but only one of the meanings will apply to a particular term. In this exercise, practice using the suffix *oma,* and then use it to construct new terms.

1. The suffix *oma* has two meanings:

   ______ or ______

   Recall from a previous chapter:

2. In the term *hematoma,* the meaning of *oma* is ______.
3. Describe a hematoma: ______
4. In the term *glioma,* the meaning of *oma* is ______.
5. Describe a glioma: ______

   ______

Construct more terms for the following tumors:

6. A nerve tumor ______ /oma
7. A tumor of the meninges ______ /oma
8. Any tumor implying "cancerous" ______ /oma
9. A tumor arising in a gland ______ /oma

G. **Deconstruct** the following medical terms into their basic elements, and provide a meaning for each element. Remember to answer the final questions.

Medical Term	Prefix	Meaning of Prefix	Root/CF	Meaning of Root/CF	Suffix	Meaning of Suffix
anesthesiologist						
ataxia						
endarterectomy						
exacerbation						
fibromyalgia						
glioma						
postictal						
remission						
synapse						
subdural						

1. Which two terms in the above table are opposites, and what do they mean? ____________________ and ____________________ are opposites.

   Term 1 means ____________________.

   Term 2 means ____________________.

Study Hint
Always study opposite terms in pairs—they are easier to remember that way.

2. Which term in the table is a medical specialty? ____________________

   ____

3. Which term is a surgical procedure? ____________________

4. Which terms can be a diagnosis? ____________________

   ____

5. Which term refers to the acuteness of an illness? ____________________

6. Which term refers to the meninges? ____________________

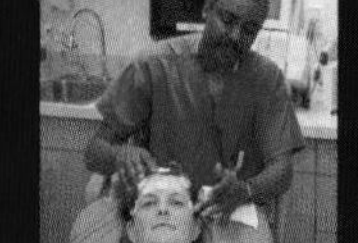

NERVOUS SYSTEM

H. **Spelling demons:** You choose the 10 most difficult terms to spell in this chapter. Write them below; then dictate them to a fellow student, and see who has more of them spelled correctly. Then have your exercise partner dictate his or her 10 terms to you. Good luck—the terms you find most difficult will probably be ones that will appear on a test.

Your 10 Most Difficult Terms	**Your Partner's 10 Terms**
1. ______________________	______________________
2. ______________________	______________________
3. ______________________	______________________
4. ______________________	______________________
5. ______________________	______________________
6. ______________________	______________________
7. ______________________	______________________
8. ______________________	______________________
9. ______________________	______________________
10. ______________________	______________________

My score for my 10 terms: ______________________

My score for my partner's dictated terms: ______________________

Write here any terms you have spelled incorrectly from either test:

____

____

____

____

Can you correctly pronounce each term? Can you define each term?

1. What is the difference between a *neurologist* and a *neurosurgeon?*

2. Why is Ms. Gaston unable to work?

3. What diagnostic tests did Ms. Gaston undergo?

4. Define *glioma:*

5. What type of surgery did the patient have?

6. Explain the difference in the types of seizures Ms. Gaston had:

7. What other body system has a *ventricle,* and where is it located?

CUMULATIVE REVIEW

These exercises are built around terms and elements from this, and the preceding three chapters.

They are an excellent source of review for tests—be sure to do every exercise!

A. Suffixes: Medical terms are given in the first column—begin by slashing each term into its elements. Write the suffix in the second column and its meaning in the third column. Challenge yourself to think of another medical term from any previous chapter with this suffix—write that term in the fourth column, and write its meaning in the last column.

Your goal is to complete the entire last column!

Medical Term	Suffix	Meaning of Suffix	Another Medical Term with This Suffix	Meaning of This Medical Term
arthroscopy				
pituitary				
articulation				
adenoid				
acromegaly				
electroencephalogram				
tenosynovitis				
cerebellum				
autonomic				
tonsillectomy				
endocrinology				
interstitial				

B. Abbreviations: The abbreviations are given. Write out in words the meaning of each abbreviation. Fill in the blanks.

The answer is not correct unless all the words are spelled correctly!

1. ADLs

____

2. GH

____

3. Ig

____

4. CNS

____

5. OA

____

6. BBB

____

7. HIV

____

8. ACL

____

9. AIDS

____

10. SAD

____

11. EEG

____

12. VS

____

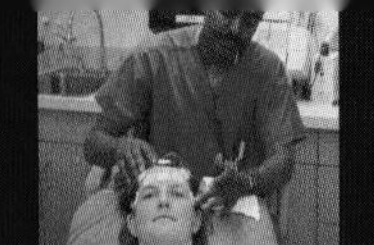

C. **Diagnosis/procedure:** Each of the following medical terms is either a diagnosis or a procedure. Place a check mark (✓) in the appropriate column for diagnosis or procedure. In the last column, write the name of the specialist who would be using that medical term in his or her dictation and billing. Fill in the chart.

Medical Term	Diagnosis	Procedure	Specialist
myxedema			
plantar fasciitis			
tonsillitis			
arthrocentesis			
urticaria			
epilepsy			
hypersplenism			
thyroidectomy			
arteriography			
polyphagia			
lymphadenopathy			
arthrodesis			
meningitis			
lymphangiogram			
diabetes insipidus			

D. Roots/combining forms are always your best clue for understanding the meaning of a term. The R/CF is given to you in column 1. Define its meaning in column 2. In column 3, give an example of a term in which it appears. Write in the final column the meaning of the medical term.

Root/CF	Meaning of R/CF	Medical Term Containing This R/CF	Meaning of Medical Term
tars			
effer			
toc			
inguin			
acromio			
neuro			
patho			
pod			
cerebro			
sero			
encephalo			
crine			
myel			
hallux			
aden			
lact			

CUMULATIVE REVIEW

E. **Prefixes** generally add description to a medical term—color, location, size, and so on. Build the following terms by completing them with the appropriate prefix *on the line.* Then write the meaning of the prefix *under the line.*

an	syn	ad	ante	pro	epi	ex	ana
anti	endo	de	a	para	dis	auto	re

1. gland that produces an internal secretion ______________________/crine
2. backward flow ______________________/flux
3. in front of the elbow ______________________/cubit/al
4. state in which pain is reduced ______________________/alges/ia
5. an endocrine gland adjacent to the thyroid gland ______________________/thyroid
6. immediate, severe allergic response ______________________/phylaxis
7. space between the dura mater and the vertebral canal ______________________/dur/al
8. junction between two nerve cells ______________________ /apse
9. completely out of joint ______________________ /loc/ation
10. loss of the mind's cognitive and mental functions ______________________ /ment/ia
11. hormone that stimulates milk secretion ______________________/lact/in
12. directed against a person's own tissues ______________________/immune
13. protein produced in response to an antigen ______________________/body
14. muscle that moves the thigh toward the midline ______________________/duct/or
15. protrusion of the eyeball ______________________/ophthalmos
16. without a blood supply ______________________/vascul/ar

F. **Spelling demons** Some medical terms are particularly difficult to pronounce and spell. Develop your ability to recognize a misspelling when you see it, and know how to correct it. *Only one term in this exercise is spelled correctly*—for each incorrectly spelled term, write the correct spelling on the line next to it.

1. serotonnin ______
2. imunization ______
3. acromoeoclavicular ______
4. electronurodiagnostic ______
5. lymphedima ______
6. panhypopituitarism ______
7. metatarsis ______
8. interstital ______
9. dopeamine ______
10. adrenocorticotrophic ______
11. tenosinovitis ______
12. tonsilectomy ______
13. ocipital ______
14. oxitocin ______
15. gastroknemius ______

G. **Monitor My Progress:** An honest self-assessment of your progress with this material will make it easier for you to learn from your mistakes. Jot down below any wrong answers you have corrected; terms, elements, or spellings you find difficult to remember; and so on. The more times you see, speak, or write the terms, the better you will come to know them.

Based on these review exercises, I need to spend more time on:

______

______

______

______

______

______

______

Urinary System and Male Reproductive System

The Essentials of the Language of Urology

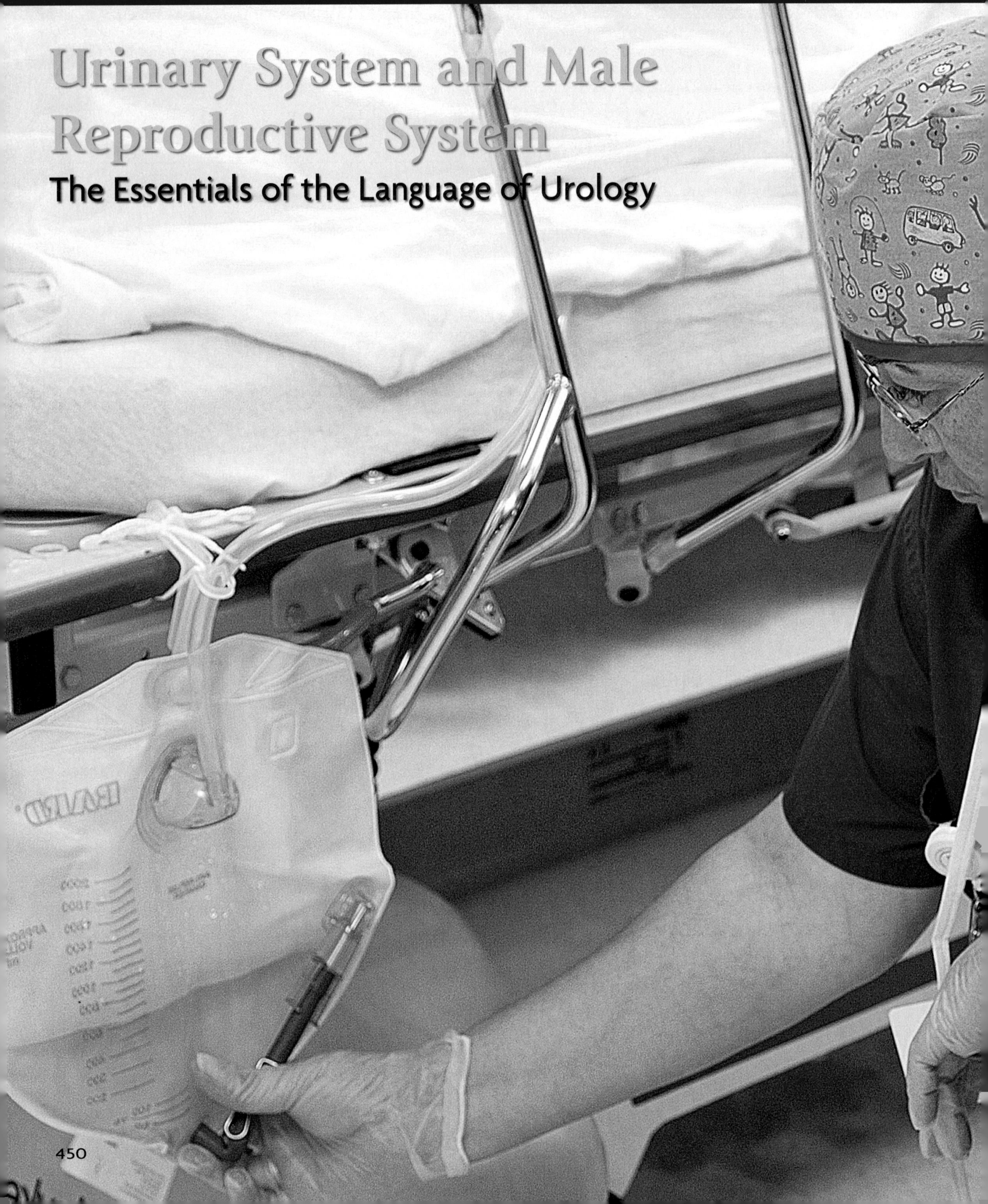

CHAPTER

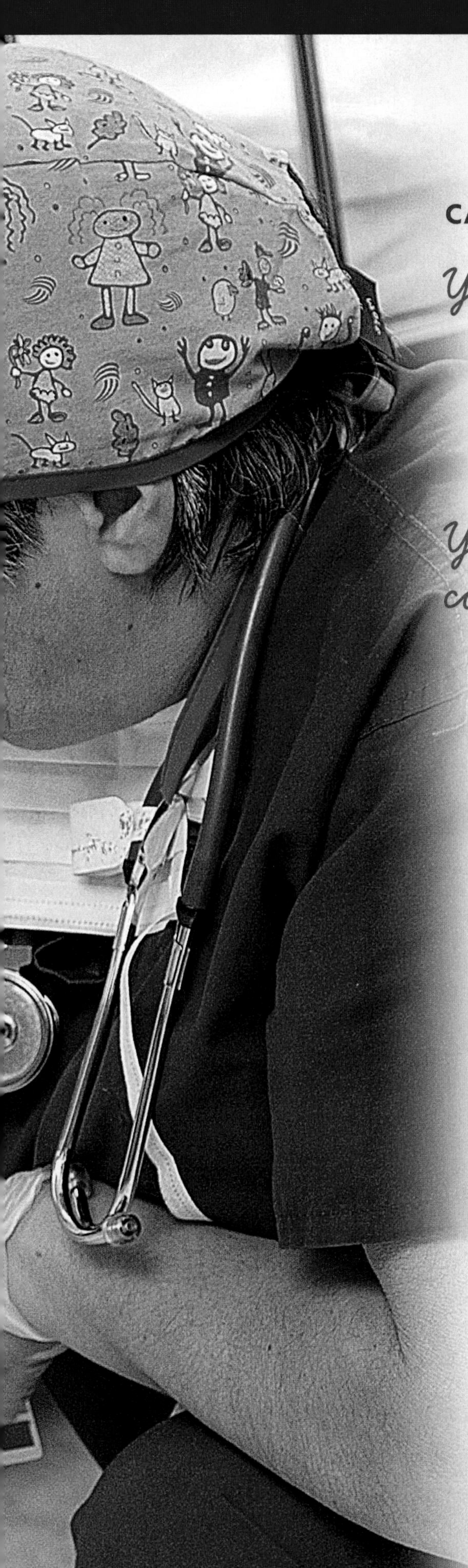

CASE REPORT 13.1

You are

. . . a surgical physician assistant working with **urologist** Phillip Johnson, MD, at Fulwood Medical Center.

You are communicating with

. . . Mr. Nelson Hughes, a 58-year-old school principal. You are making your afternoon hospital visits to Dr. Johnson's patients. Earlier today you assisted at Mr. Hughes's surgery. A laparoscopic radical nephrectomy for a TNM Stage II **renal** cell carcinoma (cancer) with no evidence of local invasion or lymph node involvement (metastasis) was performed.

Your job is to assess Mr. Hughes's postoperative state and determine whether postoperative complications exist.

Learning Outcomes

In order to understand Case Report 13.1, define and recognize areas of concern, communicate with Dr. Johnson and the patient, and document the patient's progress, you need to be able to:

13.1 Apply the language of urology to the structures and functions of the urinary system and the male reproductive system.

13.2 Comprehend, analyze, spell, and write the medical terms of urology.

13.3 Recognize and pronounce the medical terms of urology.

13.4 Explain the effects of common urinary disorders on health.

13.5. Use correct medical terminology to describe the process of spermatogenesis.

13.6 Identify the medical terminology of common disorders of the male reproductive system.

LESSON 13.1 Urinary System, Kidneys, and Ureters

URINARY SYSTEM AND MALE REPRODUCTIVE SYSTEM

OBJECTIVES

The body's metabolism continually produces metabolic waste products that poison the body if not eliminated. The **urinary** system carries the major burden of excretion of these waste products. Within the urinary system, the kidney is the agent that eliminates the metabolic waste products. Therefore, the kidney is a vital organ to understand, and it brings with it a whole new set of terminology.

In this lesson, the information will enable you to:

- **Name and locate the organs of the urinary system.**
- **Identify the medical terminology for the structures and functions of the kidneys and ureters.**
- **Apply correct medical terminology to the disorders of the kidneys and ureters.**

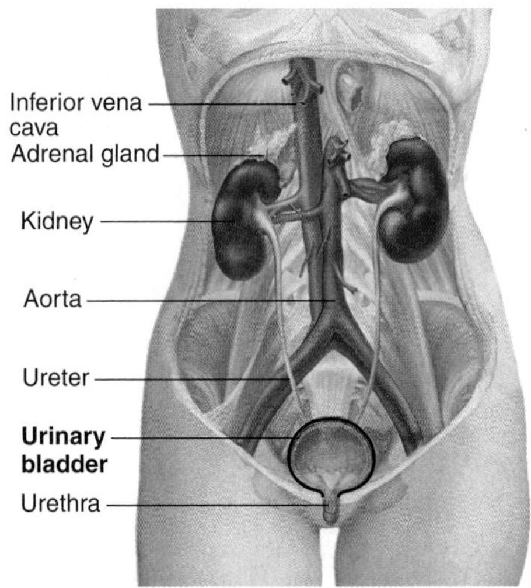

▲ **FIGURE 13.1 Urinary System.**

URINARY SYSTEM

The **urinary system** (*Figure 13.1*) consists of **six organs:**

- Two **kidneys**
- Two **ureters**
- A single **urinary bladder**
- A single **urethra**

The process of removing metabolic waste is called excretion. It is an essential process in maintaining homeostasis (*see Chapter 2*). The metabolic wastes include carbon dioxide, excess water and electrolytes, **nitrogenous** compounds including **ammonia** (from the breakdown of proteins), and **urea.** If these wastes are not eliminated, they poison the whole body.

THE KIDNEYS

Each kidney is a bean-shaped organ about the size of a clenched fist. One kidney is located on each side of the vertebral column and lies against the deep muscles of the back.

Waste-laden blood enters the kidney at its hilum (*Figure 13.2*) through the renal artery. Excess water, urea, and other waste products are **filtered** from the blood by the **nephrons** in the cortex of the kidney. The filtrate, now called **urine,** is carried off from the kidneys by the **ureters** to the urinary **bladder.**

The **excretion functions of the kidneys** are to:

- **Filter** blood to eliminate wastes.
- **Regulate** blood volume and pressure by eliminating or conserving water as necessary.
- **Maintain homeostasis** by controlling the amounts of water and electrolytes that are eliminated.

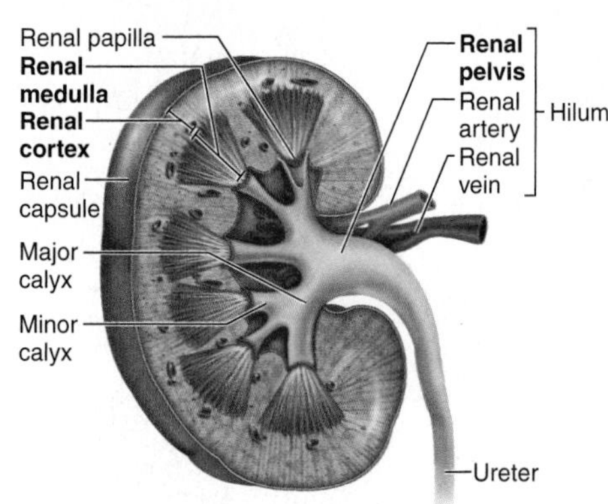

▲ **FIGURE 13.2 Section of a Kidney.**

Keynote

The kidney removes waste products from the blood by a process of **filtration.**

Keynote

Each renal cortex contains about 1 million nephrons, the functional filtration unit of the kidney.

Keynote

The material that remains in the tubule is urine. It consists of excess water, electrolytes, and urea.

Abbreviation

TNM tumor, node, metastasis (tumor staging method)

WORD ANALYSIS AND DEFINITION

S/ = Suffix P/ = Prefix R/ = Root R/CF = Combining Form

WORD	PRONUNCIATION		ELEMENTS	DEFINITION
ammonia	ah-**MOAN**-ih-ah	S/ R/	**-ia** *condition* **ammon-** *ammonia*	Toxic breakdown product of amino acids (proteins).
bladder	**BLAD**-er		Old English *bladder*	Hollow sac that holds fluid; e.g., urine or bile.
filter (noun or **verb)**	**FIL**-ter		Latin *strain through material*	Porous substance used to separate liquids and gases from particulate matter; *or* to subject a substance to the action of a filter.
filtrate	**FIL**-trate	S/ R/	**-ate** *composed of, pertaining to* **filtr-** *strain through*	Liquid that has passed through a filter.
filtration	fil-**TRAY**-shun	S/	**-ation** *process*	Process of passing liquid through a filter.
kidney	**KID**-nee		Old English *kidney*	Organ of excretion.
nephron	**NEF-ron**		Greek *kidney*	Filtration unit of the kidney; glomerulus + renal tubule.
nephrology	neh-**FROL**-oh-jee	S/ R/CF	**-logy** *study of* **nephr/o-** kidney	Medical specialty of diseases of the kidney.
nephrologist	neh-**FROL**-oh-jist	S/	**-logist** *one who studies, specialist*	Medical specialist in disorders of the kidney.
nitrogenous (adj)	ni-**TROJ**-en-us	S/ R/CF R/	**-ous** *pertaining to* **nitr/o-** *nitrogen* **-gen-** *create*	Containing or generating nitrogen.
renal (adj)	**REE**-nal	S/ R/	**-al** *pertaining to* **ren-** kidney	Pertaining to the kidney.
urea	you-**REE**-ah		Greek *urine*	End product of nitrogen metabolism.
ureter (***Note:*** Two "e's" = two tubes.)	you-**REE**-ter		Greek *urinary canal*	Tube that connects each kidney to the urinary bladder.
ureteral (adj)	you-ree-**TER**-al	S/ R/	**-al** *pertaining to* **ureter-** *ureter*	Pertaining to a ureter.
urine	**YUR**-in		Latin *urine*	Fluid and dissolved substances excreted by kidney.
urinary (adj)	**YUR**-in-ary	S/ R/	**-ary** *pertaining to* **urin-** *urine*	Pertaining to urine.
urinate (verb)	**YUR**-in-ate	S/	**-ate** *composed of, pertaining to*	To pass urine.
urination	yur-ih-**NAY**-shun	S/	**-ation** *process*	Process of passing urine.
urology	yur-**ROL**-oh-jee	S/ R/CF	**-logy** *study of* **ur/o-**, *urinary system*	Medical specialty that studies the urinary system.
urologist	yur-**ROL**-oh-jist	S/	**-logist** *one who studies, specialist*	Specialist in urology.
urological (adj)	yur-roh-**LOJ**-ik-al	S/	**-ical** *pertaining to*	Pertaining to urology.

EXERCISES

Construct *the following terms by inserting the correct elements. After you insert an element, write the identity (P, R, CF, or S) of the element below the line. Review the terms in the WAD box first. Fill in the blanks.*

1. specialist in the urinary system:

_______________ / _______________ / _______________

2. containing or generating nitrogen:

_______________ / _______________ / _______________

3. pertaining to the ureter:

_______________ / _______________ / _______________

Case Report 13.1 (continued)

Mr. Nelson Hughes had been well until a few months before his surgery, when he noticed a vague, aching pain in his left loin. One week prior to surgery, he suddenly passed bright red urine. Urinalysis showed red blood cells **(hematuria).** Physical examination revealed an enlarged left kidney. Intravenous **pyelogram** (IVP) and other imaging tests showed a tumor 3 inches in diameter in the center of the left kidney. Bone scan was normal, indicating no metastases to bone.

DISORDERS OF THE KIDNEYS

Kidney Cancer

Keynote

25% to 30% of all renal cancers relate directly to smoking.

Renal cell carcinoma is the most common form of kidney cancer and occurs twice as often in men as in women. The cancer develops in the lining cells of the renal tubules, which is why Mr. Hughes had hematuria. Radical **nephrectomy** is the most common treatment for renal cell carcinoma.

Wilms tumor, or **nephroblastoma,** is a malignant kidney tumor of childhood, occurring usually between ages 3 and 8 years. It is treated effectively with a combination of surgery and chemotherapy.

Renal adenomas, benign kidney tumors, are usually asymptomatic (they produce no symptoms), are discovered incidentally, and are not life threatening.

Keynote

As little as 1 milliliter of blood will turn the urine red.

Hematuria, blood in the urine, can be caused by lesions anywhere in the urinary system, including trauma, infections, and congenital diseases such as sickle cell anemia. In microscopic hematuria, the urine is not red, and red blood cells can be seen only under a microscope or be identified by a urine dipstick.

Keynote

The acute form of glomerulonephritis has 100% recovery.

Acute glomerulonephritis is an inflammation of the filtration unit of the kidney (the nephron). It damages the glomerular capillaries and allows protein and red blood cells to leak into the urine. It can develop rapidly after an episode of strep throat infection, most often in children.

Chronic glomerulonephritis can occur with no history of kidney disease and present as kidney failure. It also occurs in **diabetic nephropathy** and can be associated with autoimmune diseases such as lupus erythematosus.

Nephrotic syndrome causes large amounts of protein to leak out into the urine, so the level of protein in the blood falls. In children it nearly always responds to treatment with steroids. The most obvious symptom is fluid retention, with edema of the ankles and legs. This is treated with **diuretics** (to increase fluid excretion), by restricting salt in the diet, and by reducing fluid intake.

WORD ANALYSIS AND DEFINITION

S/ = Suffix P/ = Prefix R/ = Root R/CF = Combining Form

WORD	PRONUNCIATION	ELEMENTS		DEFINITION
diuretic (adj) (***Note:*** The "a" is dropped from *dia* to enable the word to flow.)	die-you-**RET**-ik	S/ P/ R/	**-etic** *pertaining to* **di(a)-** *complete* **-ur-** *urine*	Agent that increases urine output.
diuresis (noun)	die-you-**REE**-sis	S/	**-esis** *condition*	Excretion of large volume of urine.
glomerulonephritis	glo-**MER**-you-low-nef-**RYE**-tis	S/ R/CF R/	**-itis** *inflammation* **glomerul/o-** *glomerulus* **-nephr-** *kidney*	Infection of the glomeruli of the kidney.
hematuria	he-mah-**TYU**-ree-ah	S/ R/	**-uria** *urine* **hemat-** *blood*	Blood in the urine.
nephrectomy	neh-**FREK**-toe-me	S/ R/	**-ectomy** *surgical excision* **nephr-** *kidney*	Surgical removal of a kidney.
nephroblastoma (Wilms tumor) **Wilms tumor**	**NEF**-roh-blas-**TOE**-mah WILMZ **TOO**-mor	S/ R/CF R/	**-oma** *tumor, mass* **nephr/o-** *kidney* **-blast-** *immature cell* Max Wilms, German surgeon, 1867–1918	Cancerous kidney tumor of childhood.
nephropathy	neh-**FROP**-ah-thee	S/ R/CF	**-pathy** *disease* **nephr/o-** *kidney*	Any disease of the kidney.
nephrotic syndrome **nephrosis** (same as *nephrotic syndrome*)	neh-**FROT**-ik **SIN**-drohm neh-**FROH**-sis	S/ R/CF S/ R/	**-tic** *pertaining to* **nephr/o-** *kidney* **-osis** *condition* **nephr-** *kidney*	Glomerular disease with marked loss of protein.
pyelogram	**PIE**-el-oh gram	S/ R/CF	**-gram** *record, recording* **pyel/o-** *renal pelvis*	X-ray image of renal pelvis and ureters.

Abbreviation

IVP intravenous pyelogram

EXERCISES

Unscramble *the terms at the right to match them with their definitions. Check the correct spelling in the WAD box. Fill in the blanks.*

1. inflammation of the glomeruli of the kidney: eeopilrnrtohisugml
____
2. also known as Wilms tumor: hbsaelmoanprto
____
3. any disease of the kidney: teopyahnprh
____
4. production of large volume of urine: iessudri
____
5. x-ray image of renal pelvis and ureters: lrgeyoapm
____
6. blood in the urine: arameithu
____

DISORDERS OF THE KIDNEYS (continued)

Keynote

Acute interstitial nephritis causes 15% of all cases of acute renal failure.

Interstitial nephritis is an inflammation of the kidney tissue between the renal tubules. Most often it is acute and temporary. It can be an allergic reaction to or a side effect of drugs such as penicillin or ampicillin, **NSAIDs,** and diuretics.

Pyelonephritis is an infection of the renal pelvis. Most often it occurs as part of a total urinary tract infection **(UTI),** commencing in the urinary bladder (*see the next lesson in this chapter*). It has a high mortality rate in the elderly and in people with a compromised immune system.

Polycystic kidney disease (PKD) is an inherited disease. Large fluid-filled cysts grow within the kidneys and press against the kidney tissue. Eventually, the kidneys cannot function effectively.

Acute renal failure (ARF) makes the kidneys suddenly stop filtering waste products from the blood. Initially **oliguria** and then **anuria** are associated with confusion, seizures, and coma.

Keynote

Acute renal failure is usually reversible.

The causes of acute renal failure include severe burns, trauma, septicemia, toxins such as mercury and excess alcohol, and drugs such as aspirin, ibuprofen, and antibiotics such as streptomycin and gentamycin.

In the treatment of ARF, the goal is to treat the underlying disease. **Dialysis** may be necessary while the kidneys are healing.

Keynote

Chronic renal failure has no cure.

Chronic renal failure (CRF), or **chronic kidney disease (CKD),** is a gradual loss of renal function. Symptoms and signs may not appear until kidney function is less than 25% of normal. The causes of chronic renal failure include diabetes, hypertension, and kidney disease including chronic glomerulonephritis, nephrotic syndrome, and lead poisoning.

Uremia is the complex of symptoms resulting from excess nitrogenous waste products in the blood, as seen in renal failure.

End-stage renal disease (ESRD) means the kidneys are functioning at less than 10% of their normal capacity. At this point, life cannot be maintained, and either dialysis or kidney transplant is needed.

Abbreviations

ARF	acute renal failure
CKD	chronic kidney disease
CRF	chronic renal failure
ESRD	end-stage renal disease
NSAID	nonsteroidal anti-inflammatory drug
PKD	polycystic kidney disease
UTI	urinary tract infection

WORD ANALYSIS AND DEFINITION

S/ = Suffix P/ = Prefix R/ = Root R/CF = Combining Form

WORD	PRONUNCIATION	ELEMENTS		DEFINITION
anuria	an-**YOU-**ree-ah	S/ P/ R/	**-ia** *condition* **an-** *lack of* **-ur-** urine	Absence of urine production.
dialysis	die-**AL**-ih-sis	P/ R/	**dia-** *complete* **-lysis** *destruction*	An artificial method of filtration to remove excess waste materials and water from the body.
hemodialysis	**HE**-moh-die-**AL**-ih-sis	R/CF	**hem/o-** *blood*	An artificial method of filtration to remove excess waste materials and water directly from the blood.
nephritis	neh-**FRY**-tis	S/ R/	**-itis** *inflammation* **nephr-** *kidney*	Inflammation of the kidney.
oliguria	ol-ih-**GYUR**-ee-ah	S/ R/ R/	**-ia** *condition* **olig-** *scanty* **-ur-** *urine*	Scanty production of urine.
polycystic	pol-ee-**SIS**-tik	S/ P/ R/	**-ic** *pertaining to* **poly-** *many* **-cyst-** *sac, bladder, cyst*	Composed of many cysts.
pyelonephritis	**PIE**-eh-loh-neh-**FRY**-tis	S/ R/CF R/	**-itis** *inflammation* **pyel/o-** *renal pelvis* **-nephr-** *kidney*	Inflammation of the kidney and renal pelvis.
uremia	you-**REE**-me-ah	S/ R/	**-emia** *a blood condition* **ur-** *urine*	The complex of symptoms arising from renal failure.

EXERCISES

Build your knowledge *of the elements contained in the language of urology. First slash the term into elements. A specific element in each term is bolded. Write the identity (P, R, CF, or S) of the bolded element under it, and write the meaning of that element on the line next to the term.*

1. **hemo**dialysis ____________
2. **an**uria ____________
3. **olig**uria ____________
4. **poly**cystic ____________
5. uremia ____________
6. **pyelo**nephritis ____________

Remember: Every element at the beginning of a medical term is not necessarily a prefix!

7. Give an example from this exercise: ______________________________

Documentation.

CASE REPORT 13.2

Emergency Department, Fulwood Medical Center

9/12/08

Mr. Justin Leandro, a 37-year-old construction worker, presented at 1520 hrs. He complained of a sudden onset of excruciating pain in his right abdomen and back an hour previously, while at work. The pain is spasmodic and radiates down into his *groin.*

He has vomited once, and keeps having the urge to urinate. He has no previous medical history of significance.

VS: T 99.4°F, P 92, R 20, BP 130/86.

Abdomen slightly distended, with tenderness in the right upper and lower quadrants and **flank.** A dipstick test showed blood in his urine.

Provisional diagnosis by Mark Eagle, MD: stone in right ureter.

An IV line was started, 2 mg morphine sulfate given by IV push at 1540 hrs. He is going to x-ray stat for KUB and IVP.

Andrea Facundo, EMT-P. 1555 hrs.

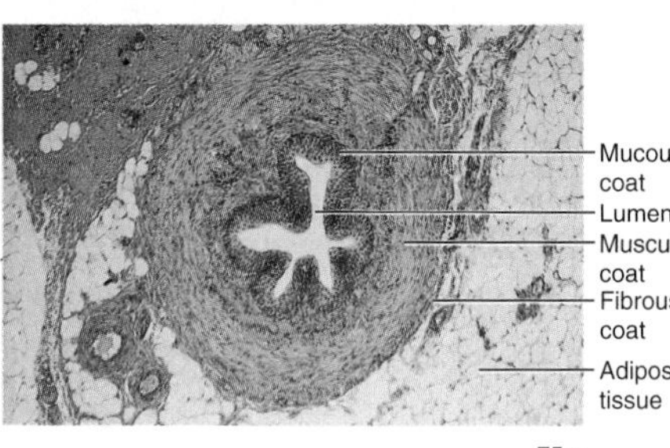

▲ **FIGURE 13.3 Cross-Section of a Ureter.**

Keynote

The muscle wall of the bladder acts as a sphincter around the ureters to prevent reflux of urine.

Keynote

The presence of urine in the renal pelvis initiates peristalsis of the ureters.

THE URETERS

Each ureter (*Figure 13.3*) is a muscular tube, about 10 inches long and ¼ inch wide, that carries urine from the renal pelvis to the urinary bladder. It lies on the posterior abdominal wall, which is why Mr. Leandro had pain in his back.

Each ureter passes obliquely through the muscle wall of the bladder. As pressure builds in a filling bladder, the muscle wall compresses the ureter and prevents urine from being forced back up the ureter to the kidneys **(reflux).**

In addition to gravity, muscular **peristaltic** waves, originating in the renal pelvis, squeeze urine down the ureter and squirt it into the bladder. The peristaltic waves are intermittent (come and go), which is why Mr. Leandro's pain was spasmodic.

Case Report 13.2 *(continued)*

Mr. Leandro's KUB (x-ray of *K*idney, *U*reter and *B*ladder) showed a suspicious lesion halfway down his right ureter. IVP confirmed that this was a stone (renal **calculus**) blocking the ureter and showed the pelvis of the right kidney to be slightly dilated. Mr. Leandro's stone was large enough to be lodged in the ureter, blocking the flow of urine, with the backflow pressure leading to **hydronephrosis** of the kidney.

Kidney and Ureteral Stones (Nephrolithiasis)

Stones **(calculi)** begin in the pelvis of the kidney as a tiny grain of undissolved material, usually a mineral called *calcium oxalate* (*Figure 13.4*). When the urine flows out of the kidney, the grain of material is left behind. Over time, more material is deposited and a stone is formed.

Most stones enter the ureter while they are still small enough to pass down the ureter into the bladder and out of the body in urine.

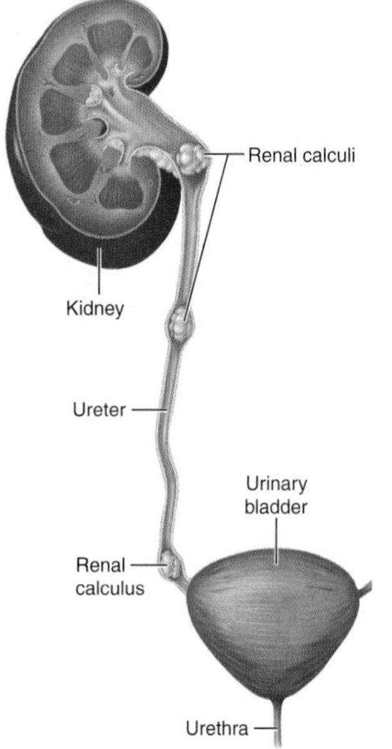

▲ **FIGURE 13.4 Renal Calculi.** Calculi can become lodged at sites along the urinary tract.

Case Report 13.2 *(continued)*

Mr. Leandro was kept in the hospital overnight with IV pain medication but did not pass the stone. **Extracorporeal** shock wave **lithotripsy** (ESWL) was successful in crumbling the stone. He urinated through a strainer so that the stone fragments could be recovered and chemically analyzed.

WORD ANALYSIS AND DEFINITION

S/ = Suffix P/ = Prefix R/ = Root R/CF = Combining Form

WORD	PRONUNCIATION		ELEMENTS	DEFINITION
calculus **calculi (pl)**	**KAL**-kyu-lus **KAL**-kyu-lie		Latin *pebble*	Small stone.
extracorporeal	**EKS**-tra-kor-**POH**- ree-al	S/ P/ R/CF	**-al** *pertaining to* **extra-** *outside* **-corpor/e-** *body*	Outside the body.
flank	FLANK		Latin *broad*	Side of the body between the pelvis and the ribs.
groin	GROYN		Old English *groin*	Crease where the thigh joins the abdomen.
hydronephrosis	**HIGH**-droh-neh-**FRO**-sis	S/ R/CF R/CF	**-osis** *condition* **hydr/o-** *water* **-nephr/o-** *kidney*	Dilation of the pelvis and calyces of a kidney.
hydronephrotic (adj)	**HIGH**-droh-neh-**FROT**-ik	S/	**-tic** *pertaining to*	Pertaining to or suffering from the dilation of the pelvis and calyces of the kidney.
lithotripsy	**LITH**-oh-trip-see	S/ R/CF	**-tripsy** *crushing* **lith/o-** *stone*	Crushing stones by sound waves.
lithotripter	**LITH**-oh-trip-ter	S/	**-tripter** *crusher*	Instrument that generates sound waves.
nephrolithiasis	**NEF**-roe-lih-**THIGH**-ah-sis	S/ R/CF R/	**-iasis** *condition, state of* **nephr/o-** *kidney* **-lith-** *stone*	Presence of a kidney stone.
peristalsis	per-ih-**STAL**-sis	P/ R/	**peri-** *around* **-stalsis** *constrict*	Waves of alternate contraction and relaxation of the muscle wall of a tube.

Abbreviations

ESWL	extracorporeal shock wave lithotripsy
IV	intravenous
KUB	x-ray of abdomen to show ***K***idneys, ***U***reters, and ***B***ladder

EXERCISES

Match *the element in column 1 with the correct meaning in column 2.*

____	1. extra	A. around
____	2. lith	B. crushing
____	3. corpore	C. constriction
____	4. tripsy	D. stone
____	5. hydro	E. kidney
____	6. stalsis	F. outside
____	7. iasis	G. pertaining to
____	8. peri	H. body
____	9. nephro	I. water
____	10. tic	J. abnormal condition

LESSON 13.2 Urinary Bladder and Urethra

URINARY SYSTEM AND MALE REPRODUCTIVE SYSTEM

OBJECTIVES

The urinary bladder is a temporary storage place for urine before it is **voided** through the **urethra.** A moderately full bladder contains about 500 mL (1 pint) of urine. The maximum capacity of the bladder is around 750 to 800 mL (1½ pints). **Urination,** emptying of the bladder, is also called **micturition.** The information in this lesson will enable you to:

- **Use correct medical terminology to describe the structure and functions of the urinary bladder.**
- **Contrast the differences in structure of the male and female urethras and the incidence of urinary tract infections in the two sexes.**
- **Discuss common disorders of the bladder and urethra, using correct medical terminology.**

You are

... a medical assistant working in the office of Dr. Susan Lee, a primary care physician, at Fulwood Medical Center.

You are communicating with

... Mrs. Caroline Dobson, a 32-year-old housewife. You have asked her the reason for her visit to the office today.

CASE REPORT 13.3

Patient Interview:

Mrs. Dobson: Since yesterday afternoon I've had a lot of pain low down in my belly and in my lower back. I keep having to go to the bathroom every hour or so to pee. It's often difficult to start, and it burns as it comes out. I've had this problem twice before when I was pregnant with my two kids, so I've started drinking cranberry juice. I've been shivering since I woke up this morning, and the last urine I passed was pink. Was that due to the cranberry juice?

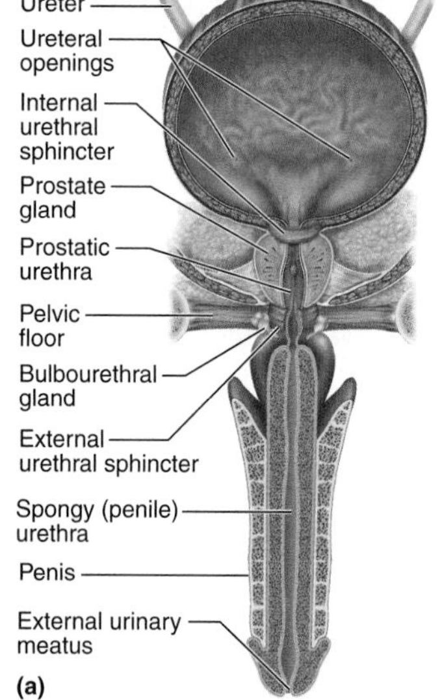

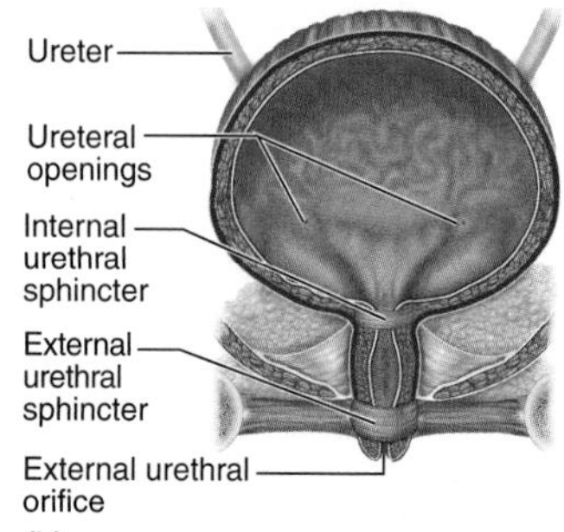

◀ **FIGURE 13.5 Urinary Bladder.** (a) Male anatomy. (b) Female anatomy.

URINARY BLADDER AND URETHRA

Urinary Bladder

The urinary bladder is a hollow, muscular organ on the floor of the pelvic cavity, posterior to the pubic symphysis (*Figure 13.5*).

Urethra

The urethra, a thin-walled tube, takes urine from the floor of the bladder to the outside. At the base of the bladder, the muscular wall of the bladder is thickened to form the **internal urethral sphincter.** As the urethra passes through the skeletal muscles of the pelvic floor, the **external urethral sphincter** provides voluntary control of micturition.

In the male (*Figure 13.5a*), the urethra is 7 to 8 inches in length and passes through the penis. In the female (*Figure 13.5b*), the urethra is only about 1½ inches long, and it opens to the outside just above the vagina.

In both the male and female, the opening of the urethra to the outside is called the **external urinary meatus.**

Micturition

When the bladder contains about 200 mL of urine, stretch receptors in its wall trigger the **micturition reflex.** Parasympathetic nerves stimulate the muscle wall of the bladder to contract and the internal sphincter to relax, and the need to urinate is sensed as urgent. However, voluntary control of the external sphincter can keep that sphincter contracted and can hold urine in the bladder until you decide to urinate. Involuntary micturition during sleep in older children or adults is called **enuresis.**

WORD ANALYSIS AND DEFINITION

S/ = Suffix P/ = Prefix R/ = Root R/CF = Combining Form

WORD	PRONUNCIATION	ELEMENTS		DEFINITION
enuresis	en-you-**REE**-sis	S/ R/	**-esis** *condition* **enur-** *urinate*	Involuntary bedwetting.
meatus	me-**AY**-tus		Latin *a passage*	The external opening of a passage.
micturition (noun) **micturate (verb)**	mik-choo-**RISH**-un **MIK**-choo-rate	S/ R/ S/	**-ition** *process* **mictur-** *pass urine* **-ate** *pertaining to*	Act of passing urine. Pass urine.
reflex	**REE**-fleks		Latin *to bend back*	An involuntary response to a stimulus.
urethra (***Note:*** One "e" = one tube.)	you-**REE**-thra		Greek *passage for urine*	Tube that carries urine from bladder to outside.
void	VOYD		Latin *to empty*	To evacuate urine or feces.

EXERCISES

Reread *the Case Report on the opposite page and review all the terms and elements in the WAD box above. Fill in the blanks.*

1. List here all of Mrs. Dobson's symptoms:

____

____

____

____

2. List here only the symptoms from question 1 that are specific to the urinary system:

____

____

____

____

3. If Dr. Lee feels that Mrs. Dobson might need a referral to a specialist, what kind of specialist would she recommend?

____

4. From the terms in the WAD, what symptom does Mrs. Dobson *not* have?

____

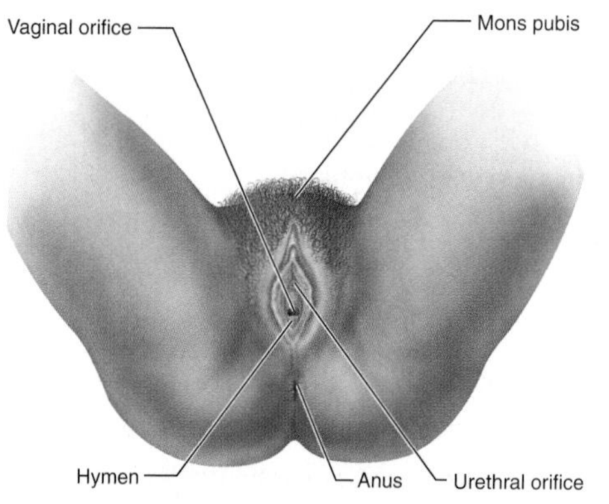

▲ **FIGURE 13.6 Female External Genitalia.**

DISORDERS OF URINARY BLADDER AND URETHRA

Urinary Tract Infection (UTI)

A urinary tract infection occurs when bacteria invade and multiply in the urinary tract. The point of entry for the bacteria is through the urethra. Because the female urethra is shorter than that of the male and opens to the surface near the anus (*Figure 13.6*), bacteria from the GI tract, such as *E. coli,* can more easily invade the female urethra. This is why women are more prone to UTIs than men. Once UTIs have occurred, they often recur.

Infection of the urethra is called **urethritis;** infection of the urinary bladder, **cystitis.** If cystitis is untreated, infection can spread up the ureters to the renal pelvis, causing **pyelitis,** and carry on to reach the renal cortex and nephrons, causing pyelonephritis.

Keynote

Ten million doctor visits each year are for UTIs.

Case Report 13.3 (continued)

Mrs. Dobson described many of the symptoms of cystitis. She had **suprapubic** and low-back pain. She had increased frequency of micturition with **dysuria** and difficulty in and burning on micturition. Her pink urine is probably hematuria.

The diagnosis can be made through urinalysis (*see later in this chapter*). Culture of the organism causing the infection and testing of its sensitivity to different antibiotics enables appropriate antibiotic therapy to be prescribed. Cranberry juice makes the urine more acid and resistant to infection.

Urinary Incontinence

Keynote

Incontinence is not a way of life. It can be helped.

Loss of control of the bladder is called **urinary incontinence.** The result is wet clothes and bedding.

About 12 million adults in America have urinary incontinence. It is most common in women over the age of 50 years, and it is frequent in elderly men.

Aging itself is not a cause of urinary incontinence.

Urinary Retention

Urinary retention is the abnormal, involuntary holding of urine in the bladder. **Acute retention** can be caused by an obstruction in the urinary system, for example, an enlarged prostate in the male or neurologic problems such as multiple sclerosis. **Chronic retention** can be caused by an untreated obstruction in the urinary tract, such as an enlarged prostate gland.

Keynote

Cigarette smoking contributes to more than 50% of bladder cancers.

Bladder Cancer

Bladder cancer is more common in men than women. It is the fourth most common cancer in men and the eighth in women.

WORD ANALYSIS AND DEFINITION

S/ = Suffix P/ = Prefix R/ = Root R/CF = Combining Form

WORD	PRONUNCIATION	ELEMENTS		DEFINITION
cystitis	sis-**TIE**-tis	S/ R/	**-itis** *inflammation* **cyst-** *bladder*	Inflammation of the urinary bladder.
cystoscope	**SIS**-toh-skope	S/ R/CF	**-scope** *instrument for viewing* **cyst/o-** *bladder*	An endoscope inserted to view the inside of the bladder.
cystoscopy	sis-**TOS**-koh-pee	S/	**-scopy** *to examine*	The process of using a cystoscope.
dysuria	dis-**YOU**-ree-ah	S/ P/ R/	**-ia** *condition* **dys-** *bad, difficult* **-ur-** *urine*	Difficulty or pain with urination.
incontinence	in-**KON**-tin-ence	S/ P/ R/	**-ence** *state of* **in-** *not* **-contin-** *hold together*	Inability to prevent discharge of urine or feces.
incontinent	in-**KON**-tin-ent	S/	**-ent** *pertaining to*	Denoting incontinence.
pyelitis	pie-eh-**LYE**-tis	S/ R/	**-itis** *inflammation* **pyel-** *renal pelvis*	Inflammation of the renal pelvis.
retention	ree-**TEN**-shun		Latin *hold back*	Holding back in the body what should normally be discharged (e.g., urine).
suprapubic	**SOO**-prah-pyu-bik	S/ P/ R/	**-ic** *pertaining to* **supra-** *above* **-pub-** *pubis*	Above the symphysis pubis.
urethritis	you-ree-**THRI**-tis	S/ R/	**-itis** *inflammation* **urethr-** *urethra*	Inflammation of the urethra.

EXERCISES

Deconstruct *the language of urology found in the WAD above. Knowledge of elements is your best key to understanding the meaning of a term. Slash the terms in column 1 into elements, and fill in the chart. Then use any one term in a sentence of your choice that is not a definition.*

Medical Term	Meaning of Prefix	Meaning of Root/CF	Meaning of Suffix
cystoscope			
urethritis			
cystitis			
dysuria			
incontinence			
pyelitis			
suprapubic			

Sentence:

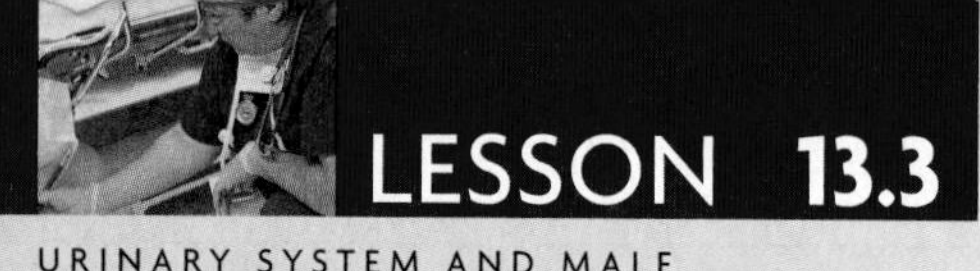

URINARY SYSTEM AND MALE REPRODUCTIVE SYSTEM

LESSON 13.3 Male Reproductive System and Testes

OBJECTIVES

As you set up the care of any urological patient, there are immediate clinical decisions to be made. You will have to communicate clearly with Dr. Helinski and other health professionals and with the patient and his close family. You must also document the patient's care. In order to participate effectively in all this, you must be able to:

- **Apply the language of urology to the structure and functions of the male reproductive system.**
- **Discuss the medical terminology for the process of *spermatogenesis.***
- **List the functions of testosterone.**
- **Identify the medical terminology of common disorders of the male reproductive system.**

You are

. . . an EMT-P working in the Emergency Department at Fulwood Medical Center.

You are communicating with

. . . Joseph Davis, a 17-year-old high school senior, who is brought in by his mother at 0400 hrs.

CASE REPORT 13.4

He is complaining of **(c/o)** sudden onset of pain in his left **testicle** 3 hours earlier that woke him up. The pain is intense and has made him vomit. VS: T 99.2°F, P 88, R 15, BP 130/70. Examination reveals his left testicle to be enlarged, warm, and tender. His abdomen is normal to palpation.

At your request, Dr. Helinski, the emergency physician on duty, examines Joseph immediately. He diagnoses a **torsion** of the patient's left testicle.

MALE REPRODUCTIVE SYSTEM

The **male reproductive organ system** (*Figure 13.7*) consists of:

1. **Primary sex organs,** or **gonads:** the two **testes.**
2. **Secondary sex organs:**
   a. **Penis.**
   b. **Scrotum.**
   c. System of ducts, including the **epididymis, ductus (vas) deferens,** and **urethra.**
3. **Accessory glands:**
   a. **Prostate.**
   b. **Seminal vesicles.**
   c. **Bulbourethral glands.**

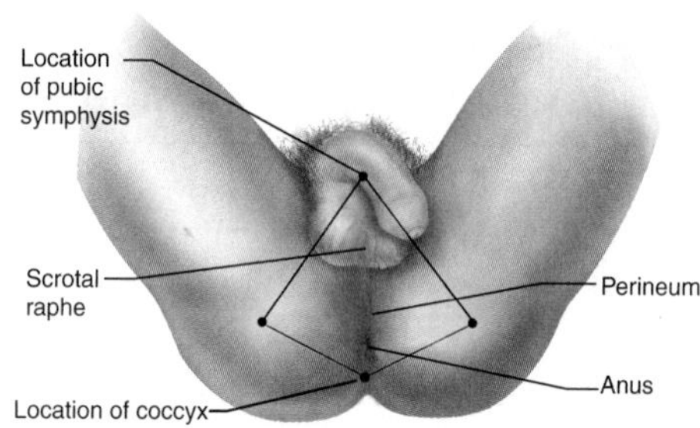

FIGURE 13.8 Male Perineum.

Perineum

The **external genitalia** (the penis, scrotum, and testes) occupy the **perineum,** a diamond-shaped region between the thighs. Its border is at the pubic symphysis anteriorly and the coccyx posteriorly (*Figure 13.8*). The anus is also in the perineum.

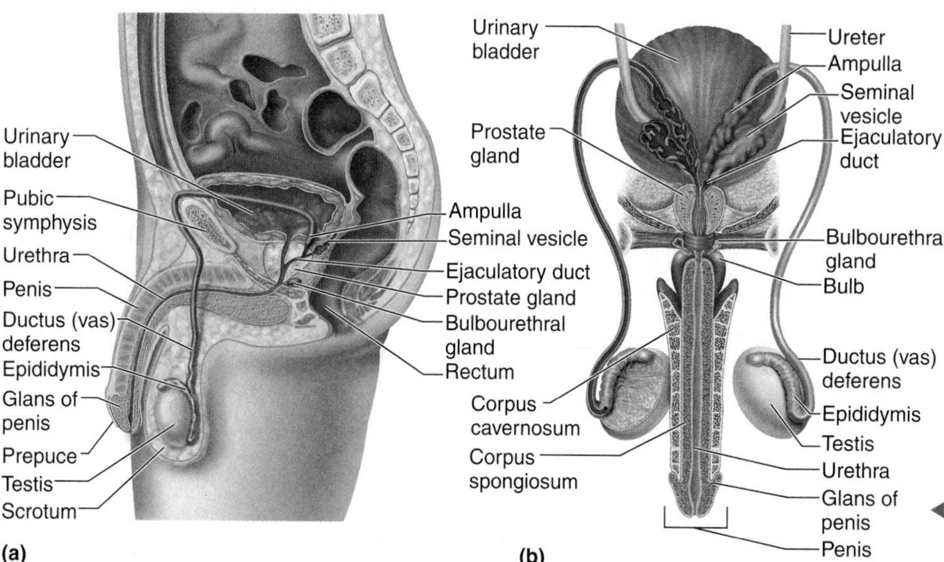

FIGURE 13.7 Male Reproductive System.
(a) Male pelvic cavity, midsaggital section.
(b) Male reproductive organs.

WORD ANALYSIS AND DEFINITION

S/ = Suffix P/ = Prefix R/ = Root R/CF = Combining Form

WORD	PRONUNCIATION		ELEMENTS	DEFINITION
bulbourethral	**BUL**-boh-you-**REE**-thral	S/ R/CF R/	**-al** *pertaining to* **bulb/o-** *bulb* **-urethr-** *urethra*	Pertaining to the bulbous penis and urethra.
ductus deferens (same as **vas deferens**)	**DUK**-tus **DEH**-fuh-renz VAS		ductus, Latin *to lead* deferens, Latin *carry away* vas, Latin *vessel, canal*	Tube that receives sperm from the epididymis.
epididymis	**EP**-ih-**DID**-ih-miss	P/ R/	**epi-** *above* **-didymis** *testis*	Coiled tube attached to testis.
genitalia (***Note:*** This term contains two suffixes.) **genital (adj)**	**JEN**-ih-**TAY**-lee-ah **JEN**-ih-tal	S/ R/ S/	**-ia** *condition* **genit-** *primary male or female sex organs* **-al-** *pertaining to*	External and internal organs of reproduction. Pertaining to reproduction or to the male or female sex organs.
gonad **gonads (pl)**	**GO**-nad **GO**-nads		Greek *seed*	Testis or ovary.
penis **penile (adj)**	**PEE**-nis **PEE**-nile	 S/ R/	Latin *tail* **-ile** *pertaining to* **pen-** *penis*	Conveys urine and semen to the outside. Pertaining to the penis.
perineum **perineal (adj)**	**PER**-ih-**NEE**-um **PER**-ih-**NEE**-al	 S/ R/	Greek *perineum* **-al** *pertaining to* **perine-** *perineum*	Area between the thighs, extending from the coccyx to the pubis. Pertaining to the perineum.
scrotum **scrotal (adj)**	**SKRO**-tum **SKRO**-tal	 S/ R/	Latin *scrotum* **-al** *pertaining to* **scrot-** *scrotum*	Sac containing the testes. Pertaining to the scrotum.
sperm **spermatozoa (pl)** **spermatic (adj)** **spermatogenesis**	SPERM **SPER**-mat-oh-**ZOH**-ah **SPER**-mat-ik **SPER**-mat-oh-**JEN**-eh-sis	 S/ R/CF S/ S/	Greek *seed* **-zoa** *animal* **spermat/o-** *sperm* **-ic** *pertaining to* **-genesis** *creation, formation*	Mature male sex cell. Sperm (plural). Pertaining to sperm. The process by which male germ cells differentiate into sperm.
testicle **testicular (adj)** **testis** **testes (pl)**	**TES**-tih-kul tes-**TICK**-you-lar **TES**-tis **TES**-tez	 S/ R/	Latin *small testis* **-ar** *pertaining to* **testicul-** *testicle* Latin *testis*	One of the male reproductive glands. Pertaining to the testicle. Same as testicle.
torsion	**TOR**-shun		Latin *to twist*	The act or result of twisting.

Scrotum

The **scrotum** is a skin-covered sac between the upper thighs. It is divided into two compartments. Each compartment contains a testis. The scrotum's function is to provide a cooler environment for the testes than that in the body. **Sperm** are best produced and stored at a few degrees cooler than the internal body temperature.

Abbreviation

c/o complaining of

EXERCISES

Documentation: *You could be working in the Emergency Department when this patient comes in. Take the information presented in Case Report 13.4 (opposite page), and document the case in the patient's record. Use the following terms to fill in the blanks.*

testicular **testes** **testis** **testicle** **torsion**

This patient presented to the ED because of pain in his left ______________. Both ______________ were examined, but the left ______________ was enlarged, warm, and tender. The emergency physician on duty diagnosed ______________. Patient will be scheduled for surgery immediately.

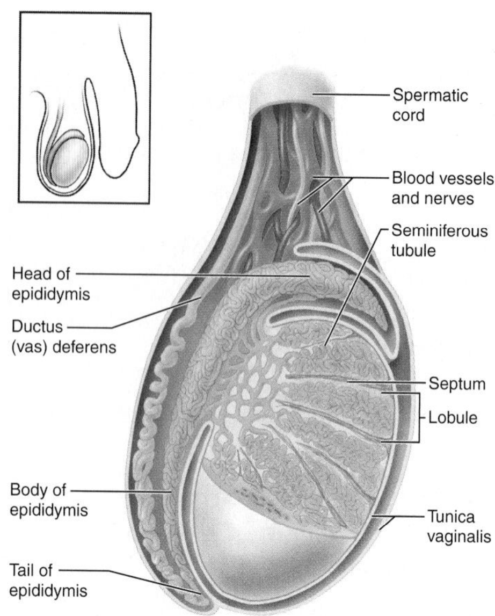

▲ **FIGURE 13.9 The Testis and Associated Structures.**

Keynote

The sperm count of a 65-year-old man is approximately one-third of the count when he was 20.

Keynote

The spermatic cord can be palpated through the skin of the scrotum.

TESTES AND SPERMATIC CORD

Testes

In the adult male, each testis is a small, oval organ about 2 inches long and ¾ inch in width (*Figure 13.9*). Each testis is covered by a serous membrane, the **tunica vaginalis,** which has outer and inner layers separated by serous fluid.

Inside the testis some 250 lobules each contains three or four **seminiferous tubules** in which several layers of germ cells are in the process of becoming sperm. Between the seminiferous tubules are the interstitial cells. They produce hormones called **androgens.**

Testosterone is the major androgen produced by the interstitial cells of the testes. Its effects include:

- **Stimulates spermatogenesis.** Testosterone levels peak at age 20 and then decline steadily to one-fifth of that level at age 80.
- **Stimulates** the development of male secondary sex characteristics at puberty:
  - Enlargement of testes, scrotum, penis.
  - Development of pubic, axillary, body, and facial hair.
  - Secretion of sebum in skin. Acne can result (*Chapter 3*).
- **Stimulates** a burst of growth at puberty, including muscle mass, higher basal metabolic rate **(BMR),** and larger larynx. This last effect deepens the voice.
- **Stimulates** the brain to increase **libido** (sex drive) in the male.

Spermatic Cord

The blood vessels and nerves to the testis arise in the abdominal cavity. They pass through the inguinal canal, where they join with connective tissue to form a **spermatic cord** that suspends each testis in the scrotum (*Figure 13.9*). The left testis is suspended lower than the right. Within the cord are an artery, a plexus of veins, nerves, a thin muscle, and the ductus (vas) deferens into which sperm go when they leave the testis.

WORD ANALYSIS AND DEFINITION

S/ = Suffix   P/ = Prefix   R/ = Root   R/CF = Combining Form

WORD	PRONUNCIATION		ELEMENTS	DEFINITION
androgen	**AN**-droh-jen	S/ R/CF	**-gen** *create, produce* **andr/o-** *masculine*	Hormone that promotes masculine characteristics.
libido	lih-**BEE**-doh		Latin *lust*	Sexual desire.
semen	**SEE**-men		Latin *seed*	Penile ejaculate containing sperm and seminal fluid.
seminal vesicle	**SEM**-in-al **VES**-ih-kull	S/ R/ S/ R/	**-al** *pertaining to* **semin-** *semen* **-le** *small* **vesic-** *sac containing fluid*	Sac of the ductus deferens that produces seminal fluid.
seminiferous (adj)	sem-ih-**NIF**-er-us	S/ R/CF R/	**-ous** *pertaining to* **semin/i-** *semen* **-fer-** *to bear, carry*	Pertaining to carrying semen.
testosterone	tes-**TOSS**-ter-own	S/ R/CF	**-sterone** *steroid* **test/o-** *testis*	Powerful androgen produced by testes.
tunica vaginalis	**TYU**-nih-kah vaj-ih-**NAHL**-iss	 S/ R/	tunica, Latin *coat* **-alis** *pertaining to* **vagin-** sheath, *vagina*	Covering, particularly of a tubular structure. The tunica vaginalis is the sheath of the testis and epididymis.

Abbreviation

BMR basal metabolic rate

EXERCISES

Deconstruction *of medical terms is a tool for analyzing their meaning. In the following chart, you are given medical terms. Deconstruct each term into its root (or combining form) and suffix.* Notice that none of these terms has a prefix. *Write the element in the appropriate column and the meaning of each element in the last column. The first one is done for you. Then use any term or terms from the WAD in a sentence of your choice that is not a definition.*

Term	Root/CF	Suffix	Meaning of Element
vaginalis	vagin		sheath
		alis	pertaining to
testosterone			
androgen			
seminal			
vesicle			

Sentence:

____

____

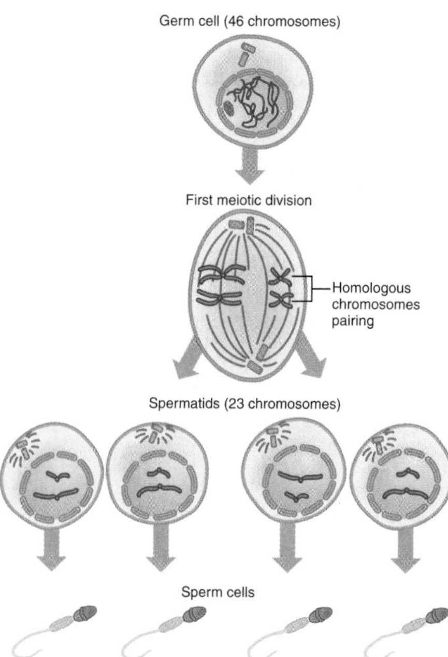

▲ **FIGURE 13.10 Spermatogenesis.**

SPERMATOGENESIS

Spermatogenesis is the process in which the germ cells of the seminiferous tubules mature and divide into two daughter cells with identical copies of their DNA **(mitosis).** The germ cells have 23 pairs of chromosomes (a total of 46). They then undergo two divisions called **meiosis.** The four daughter cells produced by meiosis differentiate into **spermatids** and then spermatozoa (sperm) (*Figure 13.10*). Because of meiosis, each sperm has only 23 chromosomes that can combine at fertilization with the 23 chromosomes in a female oocyte (egg).

The mature sperm has a pear-shaped head and a long tail. The head contains three segments (*Figure 13.11*):

- The **nucleus,** which contains 23 chromosomes.
- The **cap,** which contains enzymes used to penetrate the egg if the sperm is successful.
- The **basal body** of the tail.

The tail is divided into three segments and is responsible for movement as the sperm swims up the female reproductive tract.

Case Report 13.4 (continued)

Joseph Davis presented with typical symptoms and signs of testicular torsion. The affected testis rapidly became painful, tender, swollen, and inflamed. Emergency surgery was performed, and the testis and cord were manually untwisted through an incision in the scrotum. The testis was stitched to surrounding tissues to prevent a recurrence.

Keynote

The testis in testicular torsion will die in approximately 6 hours unless the blood supply is restored.

DISORDERS OF THE TESTES

Testicular torsion is the twisting of a testis on its spermatic cord. The testicular artery in the twisted cord becomes blocked, and the blood supply to the testis is cut off. The condition occurs in men between puberty and age 25. In half the cases, it starts in bed at night.

Varicocele is a condition in which the veins in the spermatic cord become dilated and tortuous as varicose veins. If it is uncomfortable, it can be treated by surgically tying off the affected veins.

Hydrocele is a collection of excess fluid in the space between the visceral and parietal layers of the tunica vaginalis of the testis. It is most common after age 40.

Spermatocele is a collection of sperm in a sac formed in the epididymis. It occurs in about 30% of men, is benign, and rarely causes symptoms. It does not require treatment unless it becomes bothersome.

Cryptorchism occurs when a testis fails to descend from the abdomen into the scrotum before a boy is 12 months old. As undescended testicles have a higher risk of infertility and cancer, **orchiopexy** is performed to bring the testis into the scrotum.

Epididymitis is inflammation of the epididymis; **epididymoorchitis (orchitis)** is inflammation of the epididymis and testis. Orchitis, inflammation of the testis, is usually a consequence of epididymitis. They are most commonly caused by a bacterial infection spreading from a urinary tract infection or infection of the prostate. They can also be caused by sexually transmitted diseases **(STDs),** such as gonorrhea or chlamydia.

A viral cause of orchitis is **mumps.** In males past puberty who develop mumps, 30% will develop orchitis, and 30% of those will develop resulting testicular atrophy. If the infection is bilateral, **infertility** can result. Mumps is avoidable by immunization in childhood.

Testicular cancer usually develops in men younger than 40.

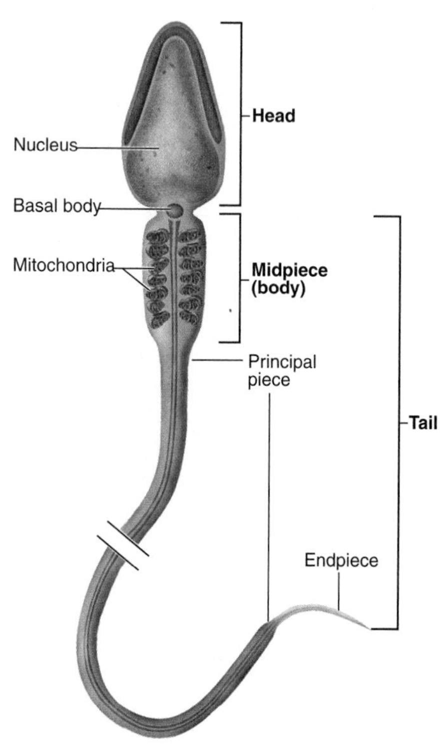

▲ **FIGURE 13.11 Mature Sperm.**

WORD ANALYSIS AND DEFINITION

S/ = Suffix P/ = Prefix R/ = Root R/CF = Combining Form

WORD	PRONUNCIATION	ELEMENTS		DEFINITION
cryptorchism	krip-**TOR**-kizm	S/ P/ R/	**-ism** *condition* **crypt-** *hidden* **-orch-** *testicle*	Failure of one or both testes to descend into the scrotum.
epididymitis	**EP**-ih-did-ih-**MY**-tis	S/ P/ R/	**-itis** *inflammation* **epi-** *above* **-didym-** *testis*	Inflammation of the epididymis.
epididymoorchitis (same as *orchitis*)	ep-ih-**DID**-ih-moh-or-**KIE**-tis	S/ P/ R/CF R/	**-itis** *inflammation* **epi-** *above* **-didym/o-** *testis* **-orch-** *testicle*	Inflammation of the epididymis and testicle.
hydrocele	**HIGH**-droh-seal	S/ R/CF	**-cele** *swelling* **hydr/o-** *water*	Collection of fluid in the space of the tunica vaginalis.
meiosis	my-**OH**-sis	S/ R/	**-osis** *condition* **mei-** *lessening*	Two rapid cell divisions, resulting in half the number of chromosomes (23).
mitosis	my-**TOE**-sis	S/ R/	**-osis** *condition* **mit-** *threadlike structure*	Cell division that creates two identical cells, each with 46 chromosomes.
orchiectomy	or-key-**ECK**-toe-me	S/ R/	**-ectomy** *surgical excision* **orchi-** *testicle*	Removal of one or both testes.
orchiopexy	**OR**-key-oh-**PEK**-see	S/ R/CF	**-pexy** *surgical fixation* **orchi/o-** *testicle*	Surgical fixation of a testis in the scrotum.
orchitis	or-**KIE**-tis	S/ R/	**-itis** *inflammation* **orchi-** *testicle*	Inflammation of the testis.
spermatid	**SPER**-mat-id	S/ R/	**-id** *having a particular quality* **spermat-** *sperm*	A cell late in the development process of sperm.
spermatocele	**SPER**-mat-oh-seal	S/ R/CF	**-cele** *swelling* **spermat/o-** *sperm*	Cyst of the epididymis that contains sperm.
varicocele	**VAIR**-ih-koh-seal	S/ R/CF	**-cele** *swelling* **varic/o-** *varicosity*	Varicose veins of the spermatic cord.

Case Report 13.5

Lance Armstrong, the world-renowned cyclist, presented with hemoptysis (coughing up blood). He had ignored a slight swelling of one testis, and the cancer in that testis had already metastasized to his lungs and brain. He required extensive chemotherapy, with brain surgery to remove several metastases.

Abbreviations

SET	self-examination of the testis
STD	sexually transmitted disease

The initial treatment for testicular cancer is surgical removal of the affected testis **(orchiectomy)**, followed by chemotherapy and sometimes radiation therapy.

Keynote

Self-examination of the testes (**SET**) for swelling or tenderness should be performed monthly by all men.

EXERCISES

Build your knowledge *of the language of urology by correctly answering the questions regarding the elements in the following terms. Circle the best answer.*

1. In the term *hydrocele,* the R/CF means:
   testis water sperm
2. In the term *cryptorchism,* the prefix means:
   outside of behind hidden
3. In the term *orchiopexy,* the suffix means:
   removal incision fixation
4. In the term *epididymitis,* which element means *above?*
   epi didym itis
5. In the term *epididymoorchitis,* the element *orch* means:
   threadlike testicle hidden

Study Hint

Before you attempt to answer the question, divide each term into its components with a slash. This will help you identify which element is referred to in the question.

LESSON 13.4 Spermatic Ducts, Accessory Glands, and Penis

URINARY SYSTEM AND MALE REPRODUCTIVE SYSTEM

OBJECTIVES

The male prostate and urethra have both **urological** and **reproductive** functions, as the flow of urine and semen goes through both organs. Disorders of the prostate and urethra produce symptoms and signs that arise in both areas. This makes it essential to have knowledge of their anatomy, physiology, and terminology to be able to understand both functions. The information provided in this lesson will enable you to:

- **Trace the pathway taken by a sperm cell from a testis to the sperm cell's ejaculation.**
- **Identify the medical terminology for the structure and functions of the prostate and other male accessory glands.**
- **Describe the origins, structure, and functions of semen.**
- **Apply correct medical terminology to the anatomy and physiology of the spermatic ducts, the prostate, and other accessory glands.**
- **Discuss the medical terminology of common disorders of the prostate gland.**
- **Apply correct medical terminology to the structures and functions of the penis and its disorders.**

You are

... a surgical technologist (LCC-ST) working for urologist Phillip Johnson, MD, in the Urology Clinic at Fulwood Medical Center.

You are communicating with

... Mr. Ronald Detrick, a 60-year-old man, who has been referred to the Urology Clinic.

CASE REPORT 13.6

Mr. Detrick complains of (c/o) having to get out of bed to urinate four or five times at night. He has difficulty starting urination, has a weak stream, and feels he is not emptying his bladder completely. His symptoms have been gradually worsening over the past year. He has lost interest in sex. His physical examination is unremarkable except that digital rectal examination (DRE) reveals a diffusely enlarged prostate with no nodules.

SPERMATIC DUCTS

As the sperm cells mature in the testes over a 60-day period, they move down the seminiferous tubules into the epididymis for storage. To be **ejaculated,** the sperm move into the ductus (vas) deferens, the **ejaculatory** duct, and finally the urethra to reach the outside of the body (*Figure 13.12*).

The epididymis adheres to the posterior side of the testis. It is a single-coiled duct in which the sperm are stored for 12 to 20 days until they mature and become **motile.**

The **ductus (vas) deferens** is a muscular duct that travels up from the epididymis in the scrotum, passes behind the urinary bladder, and widens into a terminal ampulla, which joins with the duct of the **seminal vesicle** (*Figure 13.12*).

The **ejaculatory duct,** formed by the ductus deferens and seminal vesicle, is a short (¾-inch) duct that passes through the prostate gland and empties its contents of sperm and seminal fluid **(semen)** into the urethra.

ACCESSORY GLANDS

The five **accessory glands** are:

1. The two **seminal vesicles,** located on the posterior surface of the urinary bladder. They produce a viscous (sticky), yellowish, alkaline fluid that contains fructose and **prostaglandins.** Fructose is a sugar that provides energy for the sperm.
2. The single **prostate gland,** located immediately below the bladder and anterior to the rectum. It surrounds the urethra and the ejaculatory duct. It is composed of 30 to 50 glands that open directly into the urethra. The glands secrete a slightly milky fluid that contains citric acid, a nutrient for sperm, and **prostate-specific antigen (PSA),** an enzyme that helps liquefy the **ejaculate.**

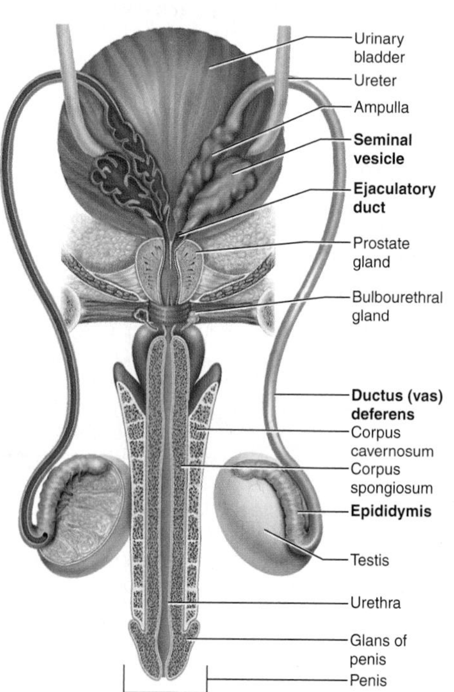

FIGURE 13.12 Components of Male Reproductive Ducts.

WORD ANALYSIS AND DEFINITION

S/ = Suffix P/ = Prefix R/ = Root R/CF = Combining Form

WORD	PRONUNCIATION	ELEMENTS		DEFINITION
ejaculate (***Note:*** can be a verb or a noun)	ee-**JACK**-you-late	S/	**-ate** *composed of, pertaining to*	To expel suddenly; *or* the semen expelled in ejaculation.
		R/	**ejacul-** *shoot out*	
ejaculation (noun)	ee-**JACK**-you-**LAY**-shun	S/	**-ation** *process*	Process of expelling semen suddenly.
ejaculatory (adj)	ee-**JACK**-you-**LAY**-tor-ee	S/	**-atory** *pertaining to*	Pertaining to ejaculation.
motile	**MOH**-til	S/ R/	**-ile** *pertaining to* **mot-** *to move*	Capable of spontaneous movement.
motility	moh-**TILL**-ih-tee	S/	**-ity** *condition, state*	The ability for spontaneous movement.
prostaglandin	**PROS**-tah-**GLAN**-din	S/ R/ R/	**-in** *chemical* **prosta-** *prostate* **-gland-** *gland*	Hormone present in many tissues, but first isolated from the prostate gland.
prostate	**PROS**-tate (***Note:*** not **PROS**-t*r*ate, which means *exhausted*)		Greek *one standing before*	Organ surrounding the beginning of the urethra.
prostatic (adj)	pros-**TAT**-ik	S/ R/	**-tic** *pertaining to* **prosta-** *prostate*	Pertaining to the prostate.
residual	reh-**ZID**-you-al	S/ R/CF	**-al** *pertaining to* **resid/u-** *what is left over*	Pertaining to anything left over.

3. The two **bulbourethral glands,** located one on each side of the membranous urethra. Each gland has a short duct leading into the spongy (penile) urethra. The glands secrete a clear, slippery, alkaline mucus that protects the sperm as they pass through the urethra by neutralizing any **residual** acid urine.

Seminal fluid contains components produced by all five accessory glands. The functions of seminal fluid are to:

- Provide sperm with **nutrients** as they pass through the urethra and female reproductive tract.
- **Neutralize** the acid secretions of the vagina, in which sperm cannot survive.
- Provide hormones **(prostaglandins)** that widen the opening of the cervix to enable sperm to enter more easily.
- Provide the **fluid** vehicle to deliver the sperm.

Combined with sperm, the seminal fluid becomes **semen.**

Keynote

Semen is derived from the secretions of several glands:

- 5% comes from the testicles and epididymis (sperm).
- 50% to 80%, from the seminal vesicles.
- 15% to 33%, from the prostate gland.
- 2% to 5%, from the bulbourethral glands.

Keynote

A normal sperm count is in the range of 75 to 150 million sperm per milliliter **(mL)** of semen. A normal ejaculation consists of 2 to 5 mL of semen.

Abbreviations

DRE	digital rectal examination
LCC-ST	Liaison Council on Certification for the Surgical Technologist
mL	milliliter
PSA	prostate-specific antigen

EXERCISES

Writing practice: *Write a sentence that is not a definition for each of the following terms. Be sure to employ the* correct form of the term *in your sentence. Fill in the blanks.*

1. *ejaculate* (as a noun—thing): ______________________________

2. *ejaculate* (as a verb—action): ______________________________

3. *ejaculatory* (adjective): ______________________________

Study Hint

SEVEN UP is used to remember the pathway of sperm:

S = seminiferous tubules
E = epididymis
V = vas (ductus) deferens
E = ejaculatory duct
N = nothing
U = urethra
P = penis

DISORDERS OF THE PROSTATE GLAND

Benign prostatic hyperplasia (BPH), also known as **benign prostatic hypertrophy** or **benign enlargement of the prostate (BEP)**—a noncancerous enlargement of the prostate—can begin to cause symptoms from the age of 45, and by age 80 some 90% of men have symptoms. The enlargement is one of hyperplasia rather than hypertrophy; it compresses the prostatic urethra to produce symptoms of:

- Difficulty starting and stopping the urine stream.
- **Nocturia, polyuria,** and dysuria.

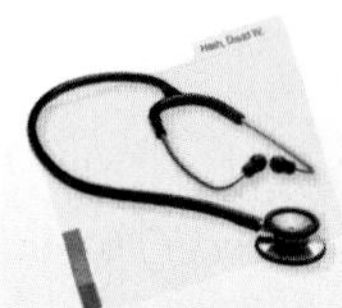

Case Report 13.6 (continued)

These were the symptoms that Ronald Detrick was describing when he was referred to the Urology Clinic.

In some patients, surgical treatment by **transurethral** resection **(TURP)** relieves the symptoms. An endoscope called a **resectoscope** is inserted into the urethra and used to remove the tissue surrounding and compressing the urethra.

Prostatic cancer affects 10% of men over the age of 50, and its incidence is increasing. It forms hard nodules in the periphery of the gland and is often asymptomatic in its early stages, as it does not compress the urethra.

Keynote

Survival from prostate cancer is up to 80% if it is detected before it spreads outside the gland.

Screening for prostatic cancer is performed by:

- **Digital rectal exam (DRE):** The size and texture of the prostate are palpated by a finger inserted into the rectum.
- **Prostate-specific antigen (PSA)** level in the blood: Even though cancer can be present when the level is zero, the benefit of the test is that it can show whether levels rise rapidly over time.
- **Early prostate cancer antigen (EPCA-2):** This is more accurate than PSA and is in clinical trials.

Several treatment options involving radiotherapy are available for prostate cancer. **Brachytherapy,** in which small radioactive rods are inserted directly into the tumor, is also used. Sometimes, a **radical prostatectomy,** with complete surgical removal of the prostate and surrounding tissues, is performed.

Prostatitis is inflammation of the prostate gland. It occurs in three main types:

Type I: An acute bacterial infection with fever, chills, frequency, dysuria, and hematuria.

Type II: A chronic bacterial infection, with less severe symptoms.

Type III: A chronic nonbacterial prostatitis in which the urinary symptoms are present but no bacteria can be detected. This is the most common type. Its etiology is unknown, and treatment is difficult.

MALE INFERTILITY

Keynote

Male infertility is involved in 40% of the 2.6 million infertile married couples in the United States. [Source: National Institutes of Health (**NIH**)]

Keynote

Vasectomy is almost 100% successful in producing male sterility.

Male infertility is the inability to conceive after at least one year of unprotected intercourse. The primary causes of infertility are:

- Impaired sperm production: cryptorchidism; anorchism (absence of one or both testes); testicular trauma; testicular cancer; orchitis after puberty.
- Impaired sperm delivery: infections and blockage of spermatic ducts.
- Testosterone deficiency **(hypogonadism):** medications to treat hypertension or high cholesterol; environmental endocrine disrupters that adversely affect the endocrine system. Examples are phthalates in plastics and dioxins in paper.

WORD ANALYSIS AND DEFINITION

S/ = Suffix P/ = Prefix R/ = Root R/CF = Combining Form

WORD	PRONUNCIATION	ELEMENTS		DEFINITION
brachytherapy	brah-kee-**THAIR**-ah-pee	P/ R/	**brachy-** *short* **-therapy** *treatment*	Radiation therapy in which the source of radiation is implanted in the tissue to be treated.
hyperplasia	**HIGH**-per-**PLAY**-zee-ah	S/ P/ R/	**-ia** *condition* **hyper-** *excessive* **-plas-** *molding, formation*	Increase in the *number* of the cells in a tissue or organ.
hypertrophy (***Note:*** See Study Hint below.)	high-**PER**-troh-fee	R/	**-trophy** *development*	Increase in the *size* of the cells in a tissue or organ.
hypogonadism	**HIGH**-poh-**GOH**-nad-izm	S/ P/ R/	**-ism** *condition* **hypo-** *deficient* **-gonad-** *testis or ovary*	Deficient gonad production of sperm or eggs or hormones.
infertility	in-fer-**TIL**-ih-tee	S/ P/ R/	**-ity** *condition* **in-** *not* **-fertil-** *able to conceive*	Failure to conceive.
nocturia	nok-**TYU**-ree-ah	S/ P/ R/	**-ia** *condition* **noct-** *night* **-ur-** *urine*	Excessive urination at night.
polyuria	pol-ee-**YOU**-ree-ah	S/ P/ R/	**-ia** *condition* **poly-** *excessive* **-ur-** *urine*	Excessive production of urine.
prostatectomy	pros-tah-**TEK**-toe-me	S/ R/	**-ectomy** *surgical excision* **prostat-** prostate	Surgical removal of the prostate.
prostatitis	pros-tah-**TIE**-tis	S/	**-itis** *inflammation*	Inflammation of the prostate.
resectoscope	ree-**SEK**-toe-skope	R/ R/CF	**-scope** *instrument for viewing* **resect/o-** *cut off*	Endoscope for the transurethral removal of lesions.
transurethral	**TRANS**-you-**REE**-thral	S/ P/ R/	**-al** *pertaining to* **trans-** *across, through* **-urethr-** *urethra*	Procedure performed through the urethra.
vasectomy	vah-**SEK**-toe-me	S/ R/	**-ectomy** *surgical excision* **vas-** *duct*	Excision of a segment of the ductus deferens.
vasovasostomy (vasectomy reversal)	**VAY**-soh-vay-**SOS**-toe-me	S/ R/CF	**-stomy** *new opening* **vas/o-** *duct*	Reanastomosis of the ductus deferens to restore the flow of sperm.

In the United States each year, 500,000 men choose to be made infertile **(sterile)** by having a **vasectomy.** Under local anesthesia, the ductus deferens is cut in two places, a 1-centimeter segment is removed, and the ends are cauterized and tied. The site of the surgery makes the man infertile but still able to produce and ejaculate seminal fluid.

Vasovasostomy is a microsurgical procedure to suture back together the cut ends of the ductus deferens and reverse a vasectomy.

Abbreviations

BEP	benign enlargement of the prostate
BPH	benign prostatic hyperplasia
EPCA-2	early prostate cancer antigen-2
TURP	transurethral resection of the prostate
NIH	National Institutes of Health

EXERCISES

Recognition: *As you become more familiar with elements (especially suffixes), you will be able to recognize medical terms that are procedures and diagnoses. Practice with the terms in the WAD box above. Fill in the blanks.*

1. List all the terms in the WAD that could be a possible diagnosis for a patient:

____

____

____

2. List all the terms in the WAD that are procedures:

____

____

Study Hint

Word association for *hypertrophy:*

*Hyper**trophy*** relates to an increase in *size.*

Remember that sports *trophies* come in all *sizes.*

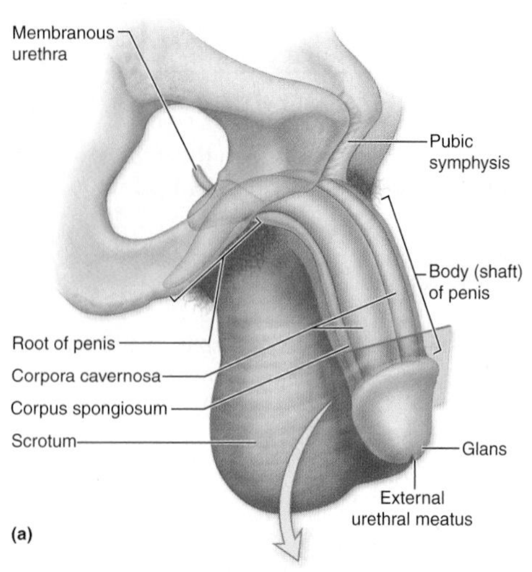

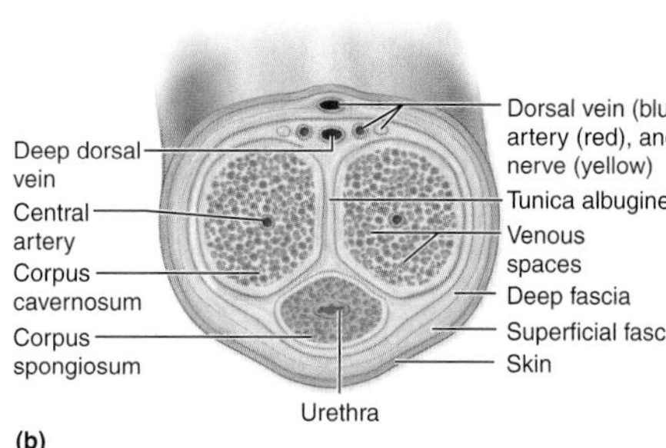

▲ **FIGURE 13.13 Anatomy of the Penis.** (a) External anatomy. (b) Cross-sectional view.

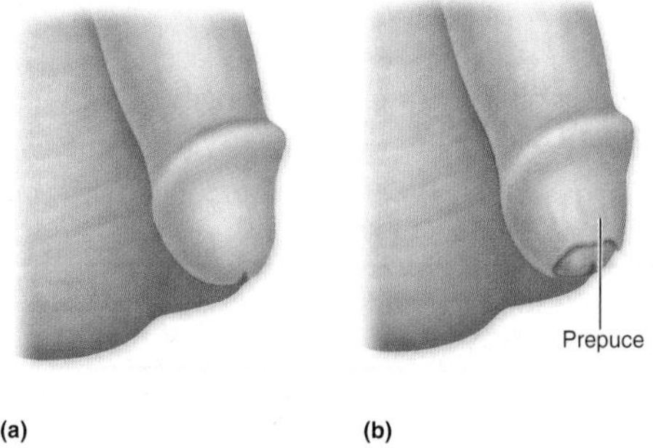

▲ **FIGURE 13.14 Prepuce.** (a) Circumcised penis. (b) Uncircumcised penis.

Keynote

Erectile dysfunction occurs in some 20 million American men.

Keynote

Erectile dysfunction can be associated with diabetes, stroke, multiple sclerosis, hypertension, cigarette smoking, radiation therapy, drugs such as antidepressants and cholesterol-lowering medications, and loss of interest in one's sexual partner.

PENIS

The **penis** (*Figure 13.13*) is specifically designed to meet its two main functions:

- Enable urine to flow to the outside.
- Deposit semen in the female vagina around the cervix.

The external, visible part of the penis is composed of the **shaft** and the **glans,** at the tip of which the external urethral meatus is located. The skin of the penis continues over the glans as the **prepuce** (foreskin). A ventral fold of tissue, the **frenulum,** attaches the skin to the glans.

The shaft of the penis contains three **erectile** vascular bodies (*Figure 13.13a*):

- Paired **corpora cavernosa** are located dorsolaterally.
- A single **corpus spongiosum** is located inferiorly. It contains the urethra and goes on to form the glans.

Erection occurs when the corpora cavernosa fill with blood, causing the erectile bodies to distend and become rigid. It is a parasympathetic nervous system response to stimulation.

Ejaculation occurs when the sympathetic nervous system stimulates the smooth muscle of the ductus deferens, ejaculatory ducts, and the prostate gland to contract.

The Prepuce (Foreskin) and Urethra

The functions of the **prepuce** (foreskin) are to cover and protect the glans (*Figure 13.14b*) and produce **smegma,** a lubricant containing lipids and cell debris and some natural antibiotics. Removal of the foreskin is called **circumcision** (*Figure 13.14a*).

Disorders of the Penis

Trauma to the penis can vary from being caught in a pants' zipper to being fractured while erect during vigorous sexual intercourse.

Priapism is a persistent, painful erection that occurs when blood cannot escape from the erectile tissue. It can be caused by drugs such as epinephrine, by blood clots, or by spinal cord injury.

Cancer of the penis occurs most commonly on the glans and is rare in circumcised men.

Sexually transmitted diseases (STDs) are discussed in detail in *Chapter 14*.

Erectile dysfunction (ED), or **impotence,** is the inability to have a satisfactory erection. Treatment is addressed to any underlying disease.

Premature ejaculation is more common than erectile dysfunction. It occurs when a man ejaculates so quickly during intercourse that it causes distress or embarrassment to one or both partners.

Disorders of the Prepuce

- **Balanitis** is infection of the glans and foreskin with bacteria or yeast.
- **Phimosis** is a condition in which the foreskin is tight because of a small opening and cannot be retracted over the glans for cleaning. It can lead to balanitis.
- **Paraphimosis** is a condition in which the retracted foreskin cannot be pulled forward to cover the glans.

Disorders of the Penile Urethra

Urethritis is an inflammation of the urethra. It can be caused by bacteria, STDs, viruses, and chemical irritants from **spermicides** and contraceptive gels.

Urethral stricture is scarring that narrows the urethra. It results from infection or injury.

Hypospadias is a congenital defect in which the opening of the urethra is on the undersurface of the penis instead of at the head of the glans. It can be corrected surgically.

Epispadias is a congenital defect in which the opening of the urethra is on the dorsum of the penis.

WORD ANALYSIS AND DEFINITION

S/ = Suffix P/ = Prefix R/ = Root R/CF = Combining Form

WORD	PRONUNCIATION	ELEMENTS		DEFINITION
balanitis	bal-ah-**NIE**-tis	S/ R/	**-itis** *inflammation* **balan-** *glans penis*	Inflammation of the glans and prepuce of the penis.
cavernosa	kav-er-**NOH**-sah	S/ R/	**-osa** *like* **cavern-** *cave*	Resembling a cave.
circumcision	ser-kum-**SIZH**-un	S/ P/ R/	**-ion** *action, condition* **circum-** *around* **-cis-** *to cut*	To remove part or all of the prepuce.
corpus **corpora (pl)**	**KOR**-pus kor-**POR**-ah		Latin *body*	Major part of a structure.
epispadias	ep-ih-**SPAY**-dee-as	S/ P/ R/	**-ias** *condition* **epi-** *above* **-spad-** *tear or cut*	Condition in which the urethral opening is on the dorsum of the penis.
erectile **erection**	ee-**REK**-tile ee-**REK**-shun	S/ R/ S/	**-ile** *pertaining to* **erect-** *to set up, straight* **-ion** *action, condition*	Capable of erection or being distended with blood. Distended and rigid state of an organ.
frenulum	**FREN-**you-lum		Latin *small bridle*	Fold of mucous membrane between the glans and the prepuce.
glans	GLANZ		Latin *acorn*	Head of the penis or clitoris.
hypospadias	high-poh-**SPAY**-dee-as	S/ P/ R/	**-ias** *condition* **hypo-** *below* **-spad-** *tear or cut*	Urethral opening is more proximal than normal on the ventral surface of the penis.
impotence	**IM-**poh-tence		Latin *inability*	Inability to achieve an erection.
paraphimosis	**PAR**-ah-fih-**MOH**-sis	S/ P/ R/	**-osis** *condition* **para-** *abnormal* **-phim-** *muzzle*	Condition in which a retracted prepuce cannot be pulled forward to cover the glans.
phimosis	fih-**MOH**-sis	S/ R/	**-osis** *condition* **phim-** *muzzle*	Prepuce cannot be retracted.
prepuce (same as *foreskin*)	**PREE**-puce		Latin *foreskin*	Fold of skin that covers the glans penis.
priapism	**PRY-**ah-pizm		Priapus, mythical Roman god of procreation	Persistent erection of the penis.
smegma	**SMEG-**mah		Greek *ointment*	Oily material produced by the glans and prepuce.
spermicide **spermicidal (adj)**	**SPER**-mih-side sper-mih-**SIGH**-dal	S/ R/CF S/	**-cide** *destroy* **sperm/i-** *sperm* **-al** *pertaining to*	Agent that destroys sperm. Pertaining to the killing of sperm; *or* destructive to sperm.
spongiosum	spun-jee-**OH-**sum	S/ R/	**-um** *tissue* **spongios-** *sponge*	Spongelike tissue.

Abbreviation

ED erectile dysfunction

EXERCISES

Language of urology: *Refine your knowledge of urological terminology by choosing the correct term to complete the statement. Circle the best choice.* Be precise, and watch the spelling!

1. Condition in which a retracted prepuce cannot be pulled forward to cover the glans:
   hypospadias phimosis spongiosum paraphimosis
2. To remove all or part of the prepuce:
   circumcision circumscion circummcision circumsion
3. Skin that covers the glans penis:
   forskin fourskin forksin foreskin
4. Fold of mucous membrane:
   frennulum freeulum freenulum frenulum

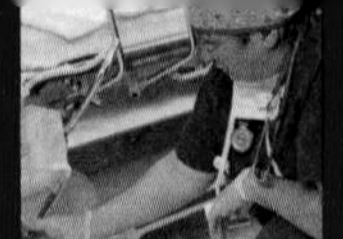

CHAPTER 13 REVIEW

URINARY SYSTEM AND MALE REPRODUCTIVE SYSTEM

CHALLENGE YOUR KNOWLEDGE

A. **Roots/combining forms** The root/combining form *nephr/o* will be the basis for each of the defined terms below. Use the appropriate suffix to construct the correct term. Elements you have seen in previous chapters will aid you in constructing the appropriate term. Fill in the blanks.

logist	iasis	osis	pathy	scope	
itis	tomy	scopy	ectomy	logy	tic

1. study of the kidney and its diseases:

   ______________/______________/______________

2. instrument used to examine the kidney:

   ______________/______________/______________

3. surgical removal of a kidney:

   ______________/______________/______________

4. any disease of the kidney:

   ______________/______________/______________

5. examination of the kidney with a scope:

   ______________/______________/______________

6. specialist in the study of the kidney:

   ______________/______________/______________

7. inflammation of a kidney:

   ______________/______________/______________

8. incision into the kidney:

   ______________/______________/______________

9. pertaining to the kidney:

   ______________/______________/______________

10. incision for removal of a kidney stone:

    ______________/______________/______________

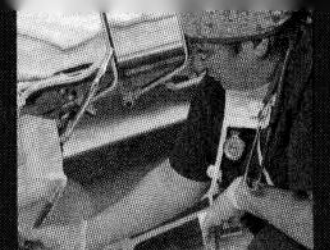

CHAPTER 13 REVIEW

URINARY SYSTEM AND MALE REPRODUCTIVE SYSTEM

CHALLENGE YOUR KNOWLEDGE

A. **Roots/combining forms** The root/combining form *nephr/o* will be the basis for each of the defined terms below. Use the appropriate suffix to construct the correct term. Elements you have seen in previous chapters will aid you in constructing the appropriate term. Fill in the blanks.

logist	iasis	osis	pathy	scope	
itis	tomy	scopy	ectomy	logy	tic

1. study of the kidney and its diseases:

   ______________/______________/______________

2. instrument used to examine the kidney:

   ______________/______________/______________

3. surgical removal of a kidney:

   ______________/______________/______________

4. any disease of the kidney:

   ______________/______________/______________

5. examination of the kidney with a scope:

   ______________/______________/______________

6. specialist in the study of the kidney:

   ______________/______________/______________

7. inflammation of a kidney:

   ______________/______________/______________

8. incision into the kidney:

   ______________/______________/______________

9. pertaining to the kidney:

   ______________/______________/______________

10. incision for removal of a kidney stone:

   ______________/______________/______________

WORD ANALYSIS AND DEFINITION

S/ = Suffix P/ = Prefix R/ = Root R/CF = Combining Form

WORD	PRONUNCIATION	ELEMENTS		DEFINITION
balanitis	bal-ah-**NIE**-tis	S/ R/	**-itis** *inflammation* **balan-** *glans penis*	Inflammation of the glans and prepuce of the penis.
cavernosa	kav-er-**NOH**-sah	S/ R/	**-osa** *like* **cavern-** *cave*	Resembling a cave.
circumcision	ser-kum-**SIZH**-un	S/ P/ R/	**-ion** *action, condition* **circum-** *around* **-cis-** *to cut*	To remove part or all of the prepuce.
corpus **corpora (pl)**	**KOR**-pus kor-**POR**-ah		Latin *body*	Major part of a structure.
epispadias	ep-ih-**SPAY**-dee-as	S/ P/ R/	**-ias** *condition* **epi-** *above* **-spad-** *tear or cut*	Condition in which the urethral opening is on the dorsum of the penis.
erectile **erection**	ee-**REK**-tile ee-**REK**-shun	S/ R/ S/	**-ile** *pertaining to* **erect-** *to set up, straight* **-ion** *action, condition*	Capable of erection or being distended with blood. Distended and rigid state of an organ.
frenulum	**FREN-**you-lum		Latin *small bridle*	Fold of mucous membrane between the glans and the prepuce.
glans	GLANZ		Latin *acorn*	Head of the penis or clitoris.
hypospadias	high-poh-**SPAY**-dee-as	S/ P/ R/	**-ias** *condition* **hypo-** *below* **-spad-** *tear or cut*	Urethral opening is more proximal than normal on the ventral surface of the penis.
impotence	**IM**-poh-tence		Latin *inability*	Inability to achieve an erection.
paraphimosis	**PAR**-ah-fih-**MOH**-sis	S/ P/ R/	**-osis** *condition* **para-** *abnormal* **-phim-** *muzzle*	Condition in which a retracted prepuce cannot be pulled forward to cover the glans.
phimosis	fih-**MOH**-sis	S/ R/	**-osis** *condition* **phim-** *muzzle*	Prepuce cannot be retracted.
prepuce (same as *foreskin*)	**PREE**-puce		Latin *foreskin*	Fold of skin that covers the glans penis.
priapism	**PRY-**ah-pizm		Priapus, mythical Roman god of procreation	Persistent erection of the penis.
smegma	**SMEG-**mah		Greek *ointment*	Oily material produced by the glans and prepuce.
spermicide **spermicidal (adj)**	**SPER**-mih-side sper-mih-**SIGH**-dal	S/ R/CF S/	**-cide** *destroy* **sperm/i-** *sperm* **-al** *pertaining to*	Agent that destroys sperm. Pertaining to the killing of sperm; *or* destructive to sperm.
spongiosum	spun-jee-**OH**-sum	S/ R/	**-um** *tissue* **spongios-** *sponge*	Spongelike tissue.

Abbreviation

ED	erectile dysfunction

EXERCISES

Language of urology: *Refine your knowledge of urological terminology by choosing the correct term to complete the statement. Circle the best choice.* Be precise, and watch the spelling!

1. Condition in which a retracted prepuce cannot be pulled forward to cover the glans:
   hypospadias phimosis spongiosum paraphimosis
2. To remove all or part of the prepuce:
   circumcision circumscion circummcision circumsion
3. Skin that covers the glans penis:
   forskin fourskin forksin foreskin
4. Fold of mucous membrane:
   frennulum freeulum freenulum frenulum

B. **Latin and Greek terms** cannot be further deconstructed into prefix, root, or suffix. You must know them for what they are. Test your knowledge of these terms with this exercise. Match the meaning in column 1 with the correct medical term in column 2.

_____	1. small stone	A. flank
_____	2. external opening of a passage	B. retention
_____	3. to empty	C. groin
_____	4. to twist	D. gonad
_____	5. between pelvis and ribs	E. scrotum
_____	6. involuntary response to stimulus	F. calculus
_____	7. to hold back	G. void
_____	8. crease where thigh joins abdomen	H. reflex
_____	9. testis or ovary	I. meatus
_____	10. sac containing testes	J. torsion

C. **Abbreviations:** Test your knowledge of this chapter's abbreviations by circling the correct answer.

1. A nephrologist would be consulted for:
   NSAID ARF IV
2. Which of the following is *not* a diagnosis?
   UTI CRF c/o
3. What should be performed monthly by all men?
   SET PSA EPCA-2
4. Increase in the number of cells in an organ:
   TURP BPH ED
5. TNM is a:
   diagnosis tumor staging method procedure
6. Which abbreviation would be found in medication orders?
   mL BEP NIH
7. Which of the following is a procedure?
   ESWL ESRD PKD
8. Used to diagnose an enlarged prostate:
   PKD CKD DRE
9. Which one is *not* a radiological procedure?
   KUB IVP UA
10. IVP is a:
    diagnostic test sexually transmitted disease diagnosis

URINARY SYSTEM AND MALE REPRODUCTIVE SYSTEM

D. **Deconstruct** the following terms into their basic elements. First slash the term; then write the meaning of each element under the line. Give the definition on the line below. Know the elements—know the term:

1. *vaginalis* __________/__________/__________

   Definition:

   ____

2. *cryptorchism* __________/__________/__________

   Definition:

   ____

3. *hydrocele* __________/__________/__________

   Definition:

   ____

4. *transurethral* __________/__________/__________

   Definition:

   ____

5. *vasovasostomy* __________/__________/__________

   Definition:

   ____

6. *hypospadias* __________/__________/__________

   Definition:

   ____

7. *hematuria* __________/__________/__________

   Definition:

   ____

E. **Translate** the following documentation into layperson's language your patient can understand. Rewrite each sentence, and do not use any abbreviations.

1. "The IVP and KUB confirmed a renal calculus occluding the ureter, and a dilated pelvis of the right kidney, leading to hydronephrosis."

___

___

___

___

2. "Mrs. Dobson described many of the symptoms of cystitis: suprapubic and low-back pain and increased frequency of micturition with dysuria and hematuria."

___

___

___

___

F. **Terminology challenge.**

1. Recall this term from a previous chapter: *occult.*

   Meaning of *occult:* ___

   Use this term in a sentence:

   ___

   ___

   ___

2. What *element* from this chapter means the same thing?

   ___

3. List a term with this element: ___

   Use this term in a sentence:

   ___

   ___

   ___

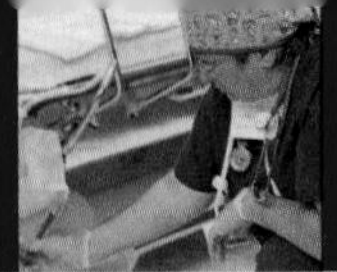

CHAPTER 13 REVIEW

URINARY SYSTEM AND MALE REPRODUCTIVE SYSTEM

G. Spelling demons: The following terms from this chapter are particularly difficult to spell and pronounce. Correct pronunciation and spelling of medical terms is the mark of an educated professional. Circle the correct spelling, and then check (✓) that you have practiced the pronunciation. Remember that pronunciations are on the Student Online Learning Center (www.mhhe.com/AllanEssMedLanguage).

				Pronunciation ✓
1.	genitallia	genitalea	genitalia	________
2.	perineal	pereneal	parineal	________
3.	libydo	lybido	libido	________
4.	incontienence	incontinence	incontenance	________
5.	epidydymis	epididymis	epidydimis	________
6.	beenign	benine	benign	________
7.	gomerulonephritis	glomerulonephritis	glomerulonepritis	________
8.	circumcision	cercumcision	cirrcumsion	________
9.	parapymosis	paraphimosis	parapimosis	________
10.	priaprism	preaprism	priapism	________

H. Prefixes: The following terms all end with *uria,* but it is the prefix that makes each term different. First match the prefix in column 1 with its meaning in column 2. Then construct the new terms to match the definitions in the rest of the exercise. Fill in the blanks.

_____ 1. dys	A. blood
_____ 2. olig	B. bad, difficult
_____ 3. hemat	C. night
_____ 4. an	D. lack of
_____ 5. noct	E. scanty

6. blood in the urine ________/________/________
7. lack of urine production ________/________/________
8. scanty production of urine ________/________/________
9. difficulty or pain with urination ________/________/________
10. excessive urination at night ________/________/________

I. **Difference between:** If you really understand a term, you can explain it to someone else. Write an explanation to a classmate of the difference between:

1. *ureter:*

____

____

____

2. *urethra:*

____

____

____

J. **Elements** remain your best clue to understanding the medical term. Write the meaning of the element in the second column, give an example of a medical term containing this element in the third column, and give the meaning of the medical term in the fourth column.

Element	Meaning of Element	Medical Term Containing This Element	Meaning of the Medical Term
lysis			
supra			
cyst			
emia			
tomy			
testo			
pexy			
orchi			
lith			

K. **Patient education:** Your patient has just been told by the doctor that he has *nephrolithiasis* and has the option of *ESWL* for a procedure. Explain this in language your patient can understand.

____

____

____

____

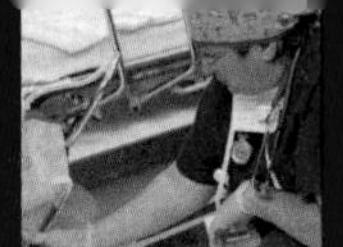

CHAPTER 13 REVIEW

URINARY SYSTEM AND MALE REPRODUCTIVE SYSTEM

L. *Scope/scopy:* Change the root, and you have the name of a different "scope" and a different procedure ("scopy"). Recall the roots/combining forms from previous chapters to complete the table. Don't forget to answer the question at the end.

Root/CF	Meaning of Root	Body System	Term for Instrument Used for This Examination	Term for Process of Examining with This Scope	Name of Specialist Who Performs This Procedure
arthro					
broncho					
cysto					
gastro					
nephro					
oto					
uretero					

1. Generally speaking, any scope inserted within the body can be called an ____________________.
(*Clue:* "within")

M. **Proofread** the following sentences for errors in fact or spelling. If a sentence is incorrect, rewrite it correctly on the lines below. There is only one sentence that is entirely correct.

1. At the base of the bladder, the muscular wall of the bladder is thickened to form the external ureteral spinchter.

2. The urinary bladder is a hollow, muscular organ on the floor of the abdominal cavity, anterior to the public symphysis.

3. Involuntary micturition during sleep in older children or adults is called *enuresis.*

4. Loss of control of the kidney is called *urinary retention.*

5. Most stones enter the uretter while they are still small enough to pass down the urethra into the bladder and out of the body in urea.

N. Recall and review: How well do you remember these word elements from the previous chapter? Try to answer without looking back to check. Fill in the chart.

Element	Type of Element (P, R, CF, or S)	Meaning of Element
arachn		
cerebell		
encephalo		
myel		
occipit		
para		
pleg		
plexy		
sym		
um		

O. Study hint: Keep a list in the back of your book of various elements like *nephr/o* and *ren* that both have the same meaning but generate their own vocabulary.

1. *Nephr/o* means ______________________.

   Example of a term using this element: ______________________

   Ren means ______________________.

   Example of a term using this element: ______________________

Recall other elements from previous chapters that both have the same meaning.

2. In the *language of ophthalmology* there are two examples. Write them here:

   ______________________ means ______________________.

   Example of a term using this element: ______________________

   ______________________ means ______________________.

   Example of a term using this element: ______________________

3. In the *language of pulmonology* there are two examples. Write them here:

   ______________________ means ______________________.

   Example of a term using this element: ______________________

   ______________________ means ______________________.

   Example of a term using this element: ______________________

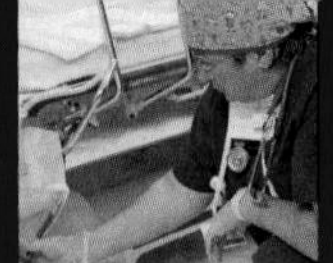

CHAPTER 13 REVIEW

URINARY SYSTEM AND MALE REPRODUCTIVE SYSTEM

P. **Seek and find:** Some terms in this book are defined in context and do not appear in a WAD. Below is a sample of terms and definitions that have appeared in this chapter. Challenge yourself to answer without turning back to check in the book. Fill in the blanks.

What Is Another Name For:

1. produces no symptoms ____________________
2. coughing up blood ____________________
3. vas deferens ____________________
4. nephroblastoma ____________________
5. infertile ____________________
6. foreskin ____________________

Q. **Chapter challenge:** Circle the correct answer.

1. A *suprapubic, transabdominal* needle aspiration is:
   a. below the pubis, through the abdomen
   b. above the pelvis, below the pubis
   c. above the pubis, through the abdomen
   d. inside the pelvis, below the abdomen
   e. below the pelvis, through the abdomen

> *Study Hint*
> Immediately cross off any answer you know is not correct. In your remaining choices, there is only *one best answer.*

2. *Renal calculus* is another term for:
   a. kidney stone
   b. vas deferens
   c. tunica vaginalis
   d. seminiferous tubules
   e. malignant kidney tumor

3. What surgical procedure reverses sterility?
   a. vasectomy
   b. orchiopexy
   c. vasovasostomy
   d. prostatectomy
   e. cystectomy

4. Which pair of terms is spelled correctly?
   a. hypogonadism vassovasostomy
   b. segma prepuce
   c. frennulum forskin
   d. errectile dysfunction
   e. paraphimosis priapism

5. If something is described as *residual,* it is:
   a. flowing backward
   b. in temporary remission
   c. whatever is left over
   d. an involuntary action
   e. capable of spontaneous movement

6. Undescended testicles present a higher risk for:
   a. infertility
   b. cancer
   c. spermatocele
   d. a and b
   e. a, b, and c
7. An agent that increases urine output is:
   a. diuresis
   b. dysuria
   c. polyuria
   d. diuretic
   e. anuria
8. Within the urinary system, which organ is the agent that filtrates the metabolic waste products?
   a. ureter
   b. kidney
   c. urethra
   d. penis
   e. pancreas
9. Inflammation of the kidney and renal pelvis would be:
   a. cystitis
   b. pyelitis
   c. urethritis
   d. pyelonephritis
   e. nephritis
10. In both the male and the female, the opening of the urethra to the outside is called the:
    a. internal urethral sphincter
    b. external urethral sphincter
    c. external urinary meatus
    d. vas deferens
    e. pubic symphysis
11. What is true of the abbreviations *BPH* and *BEP?*
    a. Both pertain to the male.
    b. Neither one is malignant.
    c. Both can be a diagnosis.
    d. Answers a, b, and c are true.
    e. Only a and c are true.
12. Which of these is an inherited disease?
    a. PKD
    b. ESRD
    c. BPH
    d. CRF
    e. ARF
13. The area between the thighs, extending from the coccyx to the pubis, is the:
    a. periosteum
    b. perineum
    c. symphis pubis
    d. vas deferens
    e. scrotum
14. What can repair *cryptorchism?*
    a. micturition
    b. ochitis
    c. orchiectomy
    d. orchiopexy
    e. spermatogenesis

URINARY SYSTEM AND MALE REPRODUCTIVE SYSTEM

R. Case Report challenge: Now that you are more comfortable with the terms in this chapter, you can apply that knowledge and briefly answer the questions about the Case Report. If you read the report through and underline all the medical terminology, this will make it easier to answer the questions.

CASE REPORT 13.4

You are

. . . an EMT-P working in the Emergency Department at Fulwood Medical Center.

You are communicating with

. . . Joseph Davis, a 17-year-old high school senior, who is brought in by his mother at 0400 hrs. He is complaining of (c/o) sudden onset of pain in his left testicle 3 hours earlier that woke him up. The pain is intense and has made him vomit. VS: T 99.2°F, P 88, R 15, BP 130/70. Examination reveals his left testicle to be enlarged, warm, and tender. His abdomen is normal to palpation.

At your request, Dr. Helinski, the emergency physician on duty, examines Joseph immediately. He diagnoses a torsion of the patient's left testicle. Joseph is healthy otherwise.

Joseph Davis presented with typical symptoms and signs of testicular torsion. The affected testis rapidly became painful, tender, swollen, and inflamed.

Dr. Helinski, the emergency physician on duty, has referred Joseph Davis to a surgeon for repair of his condition, and it is your duty to obtain from the patient's insurance company preapproval for the surgery. You must be prepared to furnish answers to the questions the insurance company will ask.

You make the telephone call to the 800 number listed on the patient's insurance card for preauthorization of hospitalization or surgery. You have answered all the usual insurance questions regarding the patient's subscriber ID number, group number, name of policyholder, and so on. The insurance company needs answers to the following clinical questions in order to preapprove the surgery. Fill in the blanks.

1. What is the patient's diagnosis?

____

2. What symptoms did the patient present with to the Emergency Department?

____

____

____

____

3. Is this condition the result of an accident? ______________________________

Emergency surgery was performed, and the testis and cord were manually untwisted through an incision in the scrotum. The testis was stitched to surrounding tissues to prevent a recurrence.

4. What procedure is the surgeon going to perform?

____

5. Can you briefly describe this procedure?

____

____

____

6. Why is this procedure a surgical emergency?

____

____

____

On the basis of your answers, the insurance company representative has preapproved the surgery and one overnight stay in the hospital for the patient. Be sure to write down in the patient's chart the preapproval reference number, the name of the person who gave it to you, and the date and time of your conversation.

7. Does the patient have any preexisting conditions?

____

____

____

Female Reproductive System

The Essentials of the Language of Gynecology

14

CASE REPORT 14.1

You are

. . . a licensed practical nurse (LPN) working in the Emergency Department at Fulwood Medical Center.

You are communicating with

. . . Ms. Lara Baker, a 32-year-old single mother who works in the Billing Department of the medical center. You have been asked to take her vital signs. For the past couple of days she has had muscle aches and a feeling of general uneasiness that she thought were due to her heavy menstrual period. In the past 3 hours, she has developed a severe headache with nausea and vomiting. A diffuse rash over her trunk that looks like sunburn is now spreading to her upper arms and thighs.

VS: T 104.2°F, P 120 and irregular, R 20, BP 86/50. As you took her VS, you noted that she did not seem to understand where she was. She was unable to pass a urine specimen.

For this patient, the treatment she receives in the next few minutes is vital for her survival. You have your supervising nurse and the emergency physician come to see her immediately. As you participate in this patient's care, clear communication among the team members is essential.

Learning Outcomes

You will need to be able to:

14.1 Apply the language of gynecology to the structures and functions of the female reproductive system.

14.2 Comprehend, analyze, spell, and write the medical terms of gynecology as they relate to the female reproductive system.

14.3 Recognize and pronounce the medical terms of gynecology as they relate to the female reproductive system.

14.4 Use correct medical terminology to describe common disorders of the female reproductive system.

LESSON 14.1 External Genitalia and Vagina

FEMALE REPRODUCTIVE SYSTEM

OBJECTIVES

The female reproductive system is dormant until puberty, when the **ovaries** begin to secrete significant amounts of the sex hormones **estrogen** and **progesterone.** Then, the external genitalia become more prominent, pubic hair develops, the **vagina** becomes lubricated, and breast enlargement occurs. The information in this lesson will enable you to:

- **Describe the essential structures and functions of the external genitalia, vagina, and accessory glands.**
- **Apply correct medical terminology to the structures and functions of the external genitalia and vagina.**
- **Name the medical terminology of common disorders of the external genitalia and vagina.**

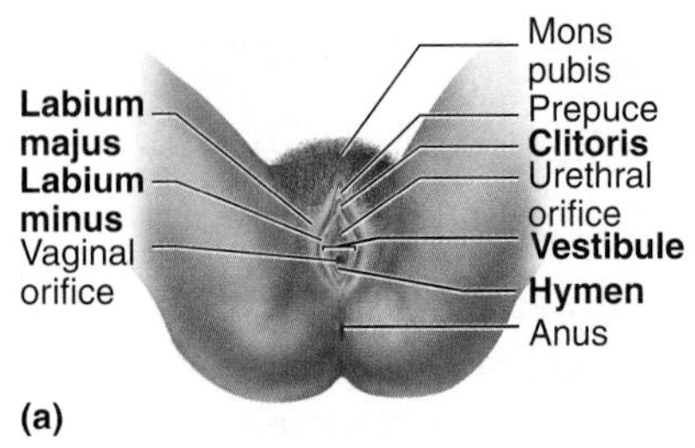

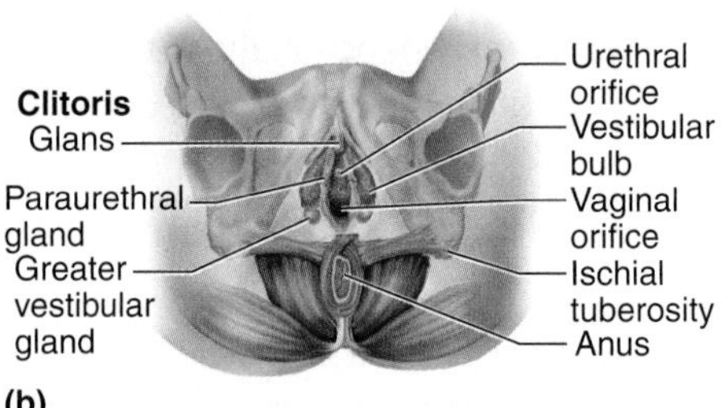

▲ **FIGURE 14.1 Female Perineum and Vulva.** (a) Surface anatomy. (b) Subcutaneous structures.

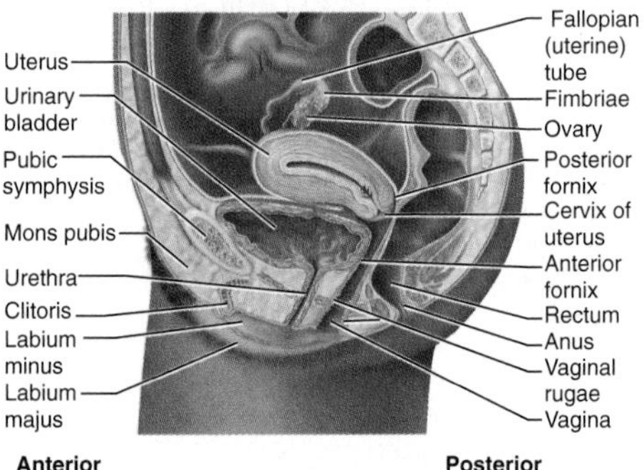

▲ **FIGURE 14.2 Female Reproductive Organs.**

Abbreviation	
LPN	licensed practical nurse

EXTERNAL GENITALIA

The female external genitalia occupy most of the **perineum** and are collectively called the **vulva.** The structures of the vulva (*Figure 14.1a*) include the:

- **Mons pubis,** a mound of skin and adipose tissue overlying the symphysis pubis.
- **Labia majora,** a pair of thick folds of skin, connective tissue, and adipose tissue.
- **Labia minora,** a pair of thin folds of hairless skin immediately internal to the labia majora. Anteriorly, the labia minora join together to form the **prepuce** (hood) of the **clitoris,** a small erectile body capped with a glans. Posteriorly, they merge with the labia majora.
- **Vestibule,** the area enclosed by the labia minora. It contains the urinary and vaginal openings.

Deep to the labia majora on each side of the vaginal orifice (opening) is a pea-sized greater vestibular **(Bartholin)** gland (*Figure 14.1b*). Both of these glands secrete mucin that lubricates the vagina. Secretion is increased during sexual intercourse.

VAGINA

The **vagina,** or birth canal, is a fibromuscular tube, 4 to 5 inches in length. It connects the vulva with the uterus (*Figure 14.2*). It has three main functions:

- Discharge menstrual fluid.
- Receive the penis and semen.
- Deliver a baby.

It is located between the rectum and the urethra. The urethra is embedded in its anterior wall.

At its posterior end, the vagina extends beyond the **cervix** of the uterus and forms blind spaces called the anterior and posterior **fornices.** The lower end of the vagina contains numerous transverse folds. The folds project into the vaginal opening to form the **hymen,** which stretches across the opening. The hymen contains one or two openings to allow the escape of menstrual fluid.

WORD ANALYSIS AND DEFINITION

S/ = Suffix P/ = Prefix R/ = Root R/CF = Combining Form

WORD	PRONUNCIATION		ELEMENTS	DEFINITION
cervix **cervical (adj)** (***Note:*** This term also means *pertaining to the neck region.*)	**SER**-viks **SER**-vih-kal	 S/ R/	Latin *neck* **-al** *pertaining to* **cervic-** *neck*	Lower part of the uterus. Pertaining to the cervix.
clitoris	**KLIT**-oh-ris		Greek *clitoris*	Erectile organ of the vulva.
estrogen	**ES**-troh-jen	S/ R/CF	**-gen** *produce* **estr/o** *woman*	Generic term for hormones that stimulate female secondary sex characteristics.
fornix **fornices (pl)**	**FOR**-niks **FOR**-nih-seez		Latin *a vault*	Arch-shaped, blind-ended part of the vagina behind and around the cervix.
hymen	**HIGH**-men		Greek *membrane*	Thin membrane partly occluding the vaginal orifice.
labium **labia (pl)**	**LAY**-bee-um **LAY**-bee-ah		Greek *lip*	Fold of the vulva.
majus **majora (pl)**	**MAY**-jus **MAY**-jora		Latin *greater*	Bigger or greater; e.g., the labia majora.
minus **minora (pl)**	**MY**-nus **MY**-nora		Latin *smaller*	Smaller or lesser; e.g., the labia minora.
mons pubis	MONZ **PYU**-bis		mons, Latin *mountain* pubis, Latin *pubic bone*	Fleshy pad with pubic hair, overlying the pubic bone.
ovary **ovaries (pl)** **ovarian (adj)**	**OH**-vah-ree **OH**-vah-rees oh-**VAIR**-ee-an	 S/ R/	Latin *egg* **-an** *pertaining to* **ovari-** *ovary*	One of the paired female reproductive glands. Pertaining to the ovary(ies).
progesterone (***Note:*** two suffixes)	pro-**JESS**-ter-own	S/ S/ P/ R/	**-one** *hormone, chemical substance* **-er-** *agent, one who does* **pro-** *before* **-gest-** *pregnancy*	Hormone that prepares the uterus for pregnancy.
vagina **vaginal (adj)**	vah-**JIE**-nah **VAJ**-ih-nal	 S/ R/	Latin *sheath* **-al** *pertaining to* **vagin-** *vagina*	The female genital canal extending from the uterus to the vulva. Pertaining to the vagina.
vulva **vulvar (adj)**	**VUL**-vah **VUL**-var	 S/ R/	Latin *a wrapper or covering* **-ar** *pertaining to* **vulv-** *vulva*	Female external genitalia. Pertaining to the vulva.

EXERCISES

Latin and Greek terms *cannot be further deconstructed into prefix, root, or suffix. You must know them for what they are. Match the meaning in column 1 with the correct medical term in column 2.*

	Column 1	Column 2
_____	1. membrane	A. vagina
_____	2. a covering or wrapper	B. ovary
_____	3. pubic bone	C. minora
_____	4. mountain	D. majora
_____	5. a vault	E. vulva
_____	6. smaller	F. fornix
_____	7. lip	G. hymen
_____	8. greater	H. mons
_____	9. egg	I. pubis
_____	10. sheath	J. labia

Keynote

Bacterial vaginosis is associated with increased risk of gonorrhea and HIV infection.

DISORDERS OF THE VULVA AND VAGINA

Bacterial vaginosis is the most common cause of **vaginitis** in women of child-bearing age. Different types of invading bacteria outnumber the normal bacteria of the vagina. The main symptom is an abnormal vaginal discharge with a fishlike odor. Diagnosis is made by laboratory examination of a specimen taken by vaginal swab. Treatment is with appropriate antibiotics.

Toxic shock syndrome is a life-threatening illness caused by toxins (poisons) circulating in the bloodstream. Bacteria in the vagina are encouraged to grow by the presence of a superabsorbent **tampon** that is not changed frequently (*see Case Report 14.1*); the bacteria produce toxins that are absorbed into the bloodstream. Other risk factors for toxic shock syndrome include skin wounds and surgery.

Case Report 14.1 (continued)

Ms. Lara Baker presented to the Emergency Department with toxic shock syndrome. Because of her heavy period, she was using a superabsorbent tampon.

She was admitted to intensive care. The tampon was removed and cultured. IV fluids and antibiotics were administered. Her kidney and liver functions were monitored. The causative organism was *Staphylococcus aureus*. She recovered well but had a second episode 6 months later.

Keynote

Of all adult women, 75% have at least one genital yeast infection in their lifetimes.

Keynote

Ten million office visits annually are for vulvodynia.

Vulvovaginal candidiasis is a common cause of genital itching or burning with a "cottage-cheese" vaginal discharge. It is caused by an overgrowth of the yeast fungus called ***Candida***. It can occur after taking antibiotics. Treatment is with anti-fungal drugs.

Vulvovaginitis can be caused by allergic and irritative agents found in vaginal hygiene products, spermicides, detergents, and synthetic underwear.

Vulvodynia is a chronic, lasting, severe pain around the vaginal orifice, which feels raw. Painful intercourse **(dyspareunia)** is common. The vulva may look normal or be slightly swollen. The etiology (cause) is unknown. Treatment varies from local anesthetics and creams to biofeedback therapy with exercises for the muscles of the pelvic floor. Surgical removal of the affected area **(vestibulectomy)** has been tried with variable results.

Vaginal cancers are uncommon, comprising 1% to 2% of gynecologic malignancies. They can be effectively treated with surgery and radiation therapy.

WORD ANALYSIS AND DEFINITION

S/ = Suffix P/ = Prefix R/ = Root R/CF = Combining Form

WORD	PRONUNCIATION	ELEMENTS		DEFINITION
dyspareunia	dis-pah-**RUE**-nee-ah	S/ P/ R/	**-ia** *condition* **dys-** *painful* **-pareun-** *lying beside, sexual intercourse*	Pain during sexual intercourse.
tampon	**TAM**-pon		French *plug*	Plug or pack in a cavity to absorb or stop bleeding.
vaginosis **vaginitis**	vah-jih-**NOH**-sis vah-jih-**NIE**-tis	S/ R/ S/	**-osis** *condition* **vagin-** *vagina* **-itis** *inflammation*	A disease of the vagina. Inflammation of the vagina.
vestibulectomy	ves-tib-you-**LEK**-toe-me	S/ R/	**-ectomy** *surgical excision* **vestibul-** *entrance*	Surgical excision of the vulva.
vulvodynia	vul-voh-**DIN**-ee-uh	S/ R/CF	**-dynia** *pain* **vulv/o-** *vulva*	Chronic vulvar pain.
vulvovaginal **vulvovaginitis**	**VUL**-voh-**VAJ**-ih-nal **VUL**-voh-vaj-ih-**NIE**-tis	S/ R/CF R/ S/	**-al** *pertaining to* **vulv/o-** *vulva* **-vagin-** *vagina* **-itis** *inflammation*	Pertaining to the vulva and vagina. Inflammation of the vulva and vagina.

EXERCISES

Suffixes *in this WAD box have appeared in many terms in previous chapters. Slash the medical term in the first column into elements to isolate the suffix. Define the meaning of the suffix, define the term in which it appears, write another term with a similar ending, and define that term. Fill in the chart.*

Medical Term from WAD Above	Meaning of Suffix	Meaning of This Medical Term	Term from a Previous Chapter, with the Same Suffix	Meaning of That Medical Term
dyspareunia				
vaginosis				
vestibulectomy				
vulvovaginal				
vulvovaginitis				

SEXUALLY TRANSMITTED DISEASES (STDs)

Keynote

Three million cases of chlamydia are recognized annually in the United States and can be prevented by abstinence or by using a condom.

Keynote

Infection with gonorrhea can be prevented by abstinence or by using a condom.

According to the Centers for Disease Control **(CDC)**, 15 million new cases of sexually transmitted diseases **(STDs)** are reported annually in the United States, with adolescents and young adults being at the greatest risk.

Chlamydia is known as the "silent" disease because up to 75% of infected women and men have no symptoms. When there are signs, a vaginal or penile discharge and irritation with dysuria are common. Highly accurate urine tests and DNA probes are available for diagnosis. Treatment is with oral antibiotics. Untreated, chlamydia can spread higher into the female reproductive tract and cause **pelvic inflammatory disease (PID).** It can be passed on to a newborn during childbirth and cause eye infections or pneumonia. This is a reason for giving antibiotic eyedrops to newborns.

Gonorrhea is spread by unprotected sex and can be passed on to a baby in childbirth, causing a serious eye infection. This is another reason for giving newborns eyedrops. Symptoms include a vaginal discharge, bleeding, and dysuria. Laboratory testing on a swab taken from the surface of the infected area can confirm the diagnosis. DNA probes are also available. Gonorrhea can be treated with a single dose of an antibiotic, but it is developing resistance to antibiotics.

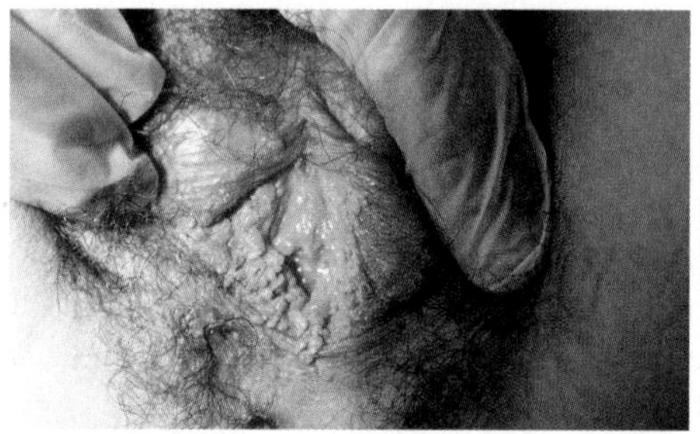

▲ **FIGURE 14.3 HPV in the Female Vulva.**

Syphilis is transmitted sexually and can then spread through the bloodstream to every organ in the body. **Primary syphilis** begins 10 to 90 days after infection as an ulcer, a **chancre**, at the place of infection. Four to ten weeks later, if the primary syphilis is not treated, **secondary syphilis** appears as a rash on the palms of the hands and the soles of the feet, with swollen glands and muscle and joint pain. **Tertiary syphilis** can occur years after the primary infection and cause permanent damage to the brain, with dementia.

Keynote

Three million cases of trichomoniasis occur annually in the United States and could be prevented by abstinence or by using a condom.

Trichomoniasis ("trich") is caused by the parasite *Trichomonas vaginalis.* In women it can produce a frothy yellow-green vaginal discharge with irritation and itching of the vulva. Because it is a "ping-pong" infection that goes back and forth between partners, both individuals should be treated.

Genital herpes simplex is a disease caused by the virus herpes simplex **(HSV)** 2. The genital sores are painful and can recur throughout life. There is no cure for genital herpes. Antiviral medications can provide clinical benefit by limiting the **replication** of the virus.

Human papilloma virus (HPV) causes genital warts in both men and women (*Figure 14.3*). HPV can also cause changes to the cells in the cervix. Some strains of the virus can increase a woman's risk for cervical cancer. More than 90% of abnormal **Pap** smears are caused by HPV infections. A vaccine is available that can prevent lasting infections with strains that cause cervical cancers and genital warts. The vaccine can be given to females aged 9 to 26, before they are active sexually.

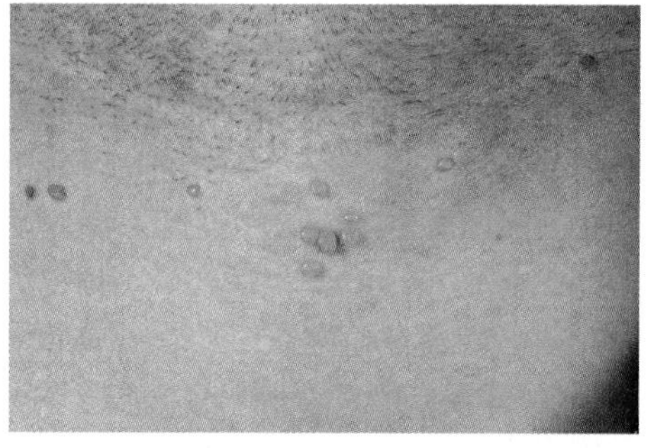

▲ **FIGURE 14.4 Molluscum Contagiosum.**

Molluscum contagiosum is a virus that can be sexually transmitted and produces small, shiny bumps that have a milky-white fluid inside (*Figure 14.4*). They can disappear and reappear anywhere on the body.

Human immunodeficiency virus (HIV) is a virus that attacks the immune system and usually leads to **acquired immune deficiency syndrome (AIDS).** HIV is carried in body fluids and transmitted during unprotected sex. Sharing needles can spread the virus. The virus can also pass from an infected pregnant woman to her unborn child, so she must take medications to protect the baby.

There is no cure for HIV or AIDS, but combinations of anti-HIV medications are taken to stop the replication of the virus in the cells of the body and stop the progression of the disease. The development of resistance to the drugs is a problem.

HIV damages the immune system, allowing infections to develop that the body would normally cope with easily. These are **opportunistic infections** and include herpes simplex, candidiasis, syphilis, and tuberculosis **(TB)**.

Abbreviations

AIDS	acquired immunodeficiency syndrome
CDC	Centers for Disease Control and Prevention
HIV	human immunodeficiency virus
HPV	human papilloma virus
HSV	herpes simplex virus
Pap	Papanicolaou (test, stain)
PID	pelvic inflammatory disease
STD	sexually transmitted disease
TB	tuberculosis

WORD ANALYSIS AND DEFINITION

S/ = Suffix P/ = Prefix R/ = Root R/CF = Combining Form

WORD	PRONUNCIATION		ELEMENTS	DEFINITION
acquired immunodeficiency syndrome (AIDS)	ah-**KWIRED** **IM**-you-noh-dee-**FISH**-en-see	S/ R/CF R/	acquired, Latin *obtain* **-ency** *condition* **immun/o** *immune response* **-defici-** *lacking, inadequate*	Infection with the HIV virus.
	SIN-drohm	P/ R/	**syn-** *together* **-drome** *running*	Combination of signs and symptoms associated with a particular disease process.
chancre	**SHAN**-ker		Latin *cancer*	Primary lesion of syphilis.
chlamydia	klah-**MID**-ee-ah		Latin *cloak*	An STD caused by infection with *Chlamydia*, a species of bacteria.
condom	**KON**-dom		Old English *sheath or cover*	A sheath or cover for the penis or vagina to prevent conception and infection.
gonorrhea	gon-oh-**REE**-ah	R/ R/CF	**-rrhea** *flow, discharge* **gon/o-** *seed*	Specific contagious sexually transmitted infection.
human immunodeficiency virus (HIV)	**HYU**-man **IM**-you-noh-dee-**FISH**-en-see **VIE**-rus	R/ S/ R/CF R/	**human** *human being* **-ency** *condition* **immun/o** *immune response* **-defici-** *lacking, inadequate* virus, Latin *poison*	Etiologic agent of acquired immunodeficiency syndrome (AIDS).
human papilloma virus (HPV)	**HYU**-man pap-ih-**LOW**-mah **VIE**-rus	R/ S/ R/	**human** *human being* **-oma** *tumor* **papill-** *pimple* virus, Latin *poison*	Causes warts on the skin and genitalia and can increase the risk for cervical cancer.
molluscum contagiosum (***Note:*** "s" in "sum" added to enable word to flow.) (modern word *contagious)*	moh-**LUS**-kum kon-**TAY**-jee-oh-sum	S/ R/ R/CF	**-um** *structure* **mollusc-** *soft* **contagi/o-** *transmissible by contact*	STD caused by a virus.
opportunistic infection (***Note:*** This term contains two suffixes.)	**OP**-or-tyu-**NIS**-tik in-**FEK**-shun	S/ S/ R/	**-ic** *pertaining to* **-ist-** *agent, specialist* **opportun-** *take advantage of*	An infection that causes disease when the immune system is compromised for other reasons.
replication	rep-lih-**KAY**-shun	S/ R/	**-ation** *process* **replic-** *reply*	Reproduction to produce an exact copy.
syphilis	**SIF**-ih-lis		Principal character in a Latin poem	Sexually transmitted disease caused by a spirochete.
Trichomonas	trik-oh-**MOH**-nas	R/CF R/	**trich/o-** *hair* **-monas** *single unit*	A parasite causing an STD.
trichomoniasis	**TRIK**-oh-moh-**NIE**-ah-sis	S/ R/	**-iasis** *condition* **-mon-** *single*	Infection with *Trichomonas vaginalis.*

EXERCISES

Construct *the correct medical term to match the definition that is given. Write the appropriate* element *on the line and the* meaning *of the element* below *the line.*

1. infection with *Trichomonas* __________ / __________ / __________

2. reproduction of an exact copy __________ / __________ / __________

3. infection that takes advantage of a compromised immune system __________ / __________ / __________

LESSON 14.2 Ovaries, Fallopian (Uterine) Tubes, and Uterus

FEMALE REPRODUCTIVE SYSTEM

OBJECTIVES

The female primary sex organs are the **ovaries.** The female **internal accessory organs** include a pair of uterine tubes, a uterus, and a vagina. Women are born with all the eggs **(ova)** that they will release in their lifetimes, but it is not until puberty that the eggs mature and start to leave the ovary. The ovarian hormones, estrogen and progesterone, are involved in **menstruation** and **pregnancy.** These complex interactions are the core of the human reproductive system and an essential part of understanding the human body. The information in this lesson will enable you to:

- **Describe the structure and functions of an ovary.**
- **List the functions of estrogen and progesterone.**
- **Apply correct medical terminology to the structures and functions of the ovaries, fallopian (uterine) tubes, and uterus.**
- **Name the medical terminology of common disorders of the ovaries, fallopian tubes, and uterus.**

You are

. . . a certified health education specialist (CHES) employed by Fulwood Medical Center.

You are communicating with

. . . Ms. Claire Marcos, a 21-year-old student referred to you by Anna Rusak, MD, a gynecologist.

CASE REPORT 14.2

Ms. Marcos has been diagnosed with polycystic ovarian syndrome, and your task is to develop a program of self-care as part of her overall plan of therapy.

From her medical record, you see that she presented with irregular, often missed menstrual periods since the beginning of puberty, persistent acne, patches of dark skin on the back of her neck and under her arms, loss of hair from the front of her scalp, and inability to control her weight. She is 5 feet 4 inches tall and weighs 150 pounds.

You are to counsel her about her self-care program involving exercise, diet, and use of birth control medications.

Abbreviations

CHES	certified health education specialist
GYN	gynecology

ANATOMY OF THE FEMALE REPRODUCTIVE TRACT

Ovaries

Each **ovary** is an almond-shaped organ about 1 inch long and ½ inch in diameter. It is held in place by ligaments that attach it to the pelvic wall and uterus. The ovaries' main functions are to:

- Produce and release eggs.
- Secrete hormones that affect puberty, menstruation, and pregnancy.

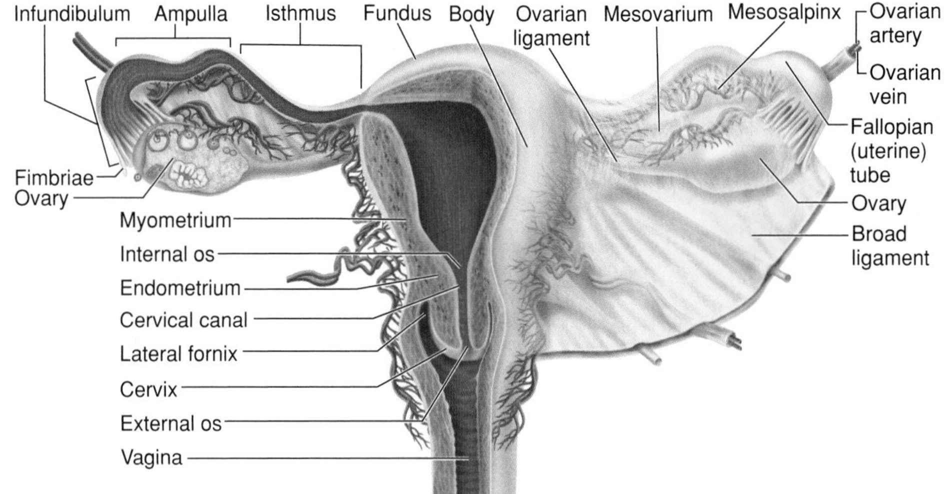

▲ **FIGURE 14.5 Female Reproductive Tract.**

Fallopian (Uterine) Tubes

Each fallopian tube is a canal about 4 inches long extending from the uterus and opening to the abdominal cavity near an ovary. At the ovarian end, the outer third of the tube flares out into finger-like folds, each of which is called a **fimbria.** At ovulation the **fimbriae** enclose the ovary (*Figure 14.5*). The tube's main functions are to enable sperm and eggs to meet and fertilize.

Uterus

The **uterus** is a thick-walled, muscular organ in the pelvic cavity. Anatomically it is divided into three regions:

WORD ANALYSIS AND DEFINITION

S/ = Suffix P/ = Prefix R/ = Root R/CF = Combining Form

WORD	PRONUNCIATION		ELEMENTS	DEFINITION
endometrium	en-doh-**ME**-tree-um	S/ P/ R/CF	**-um** *tissue* **endo-** *within, inside* **-metr/i-** *uterus*	Inner lining of the uterus.
endometrial (adj)	en-doh-**ME**-tree-al	S/	**-al** *pertaining to*	Pertaining to the inner lining of the uterus.
fallopian tubes (also called *uterine tubes*)	fah-**LOW**-pee-an		Gabrielle Fallopio, Italian anatomist, 1523–1562	Uterine tubes connected to the fundus of the uterus.
fimbria **fimbriae (pl)**	**FIM**-bree-ah **FIM**-bree-ee		Latin *fringe*	Fringelike structure.
gynecology (GYN)	guy-nih-**KOL**-oh-jee	S/ R/CF	**-logy** *study of* **gynec/o-** *woman, female*	Medical specialty of diseases of the female.
gynecologist	guy-nih-**KOL**-oh-jist	S/	**-logist** *one who studies, specialist*	Specialist in gynecology.
gynecologic (adj)	**GUY**-nih-koh-**LOJ**-ik	S/ R/CF R/	**-ic** *pertaining to* **gynec/o-** *woman, female* **-log-** *to study*	Pertaining to gynecology.
os	OS		Latin *mouth*	Opening into a canal; e.g., the cervix.
ovum **ova (pl)**	**OH**-vum **OH**-vah		Latin *egg*	Egg.
uterus **uterine (adj)**	**YOU**-ter-us **YOU**-ter-ine	 S/ R/	Latin *womb* **-ine** *pertaining to* **uter-** *uterus*	Organ in which an egg develops into a fetus. Pertaining to the uterus.

- The **fundus** is the broad, curved upper region between the lateral attachments of the fallopian (uterine) tubes.
- The **body** is the midportion.
- The **cervix** is the cylindrical inferior portion that projects into the vagina.

The main functions of the uterus are to cradle and nourish the fetus from conception to birth and to produce a woman's monthly menstrual flow (period).

The lower end of the uterine cavity communicates with the vagina through the cervical canal that has an external **os** into the vagina. The inner lining of the uterus is called the **endometrium.**

EXERCISES

Recall: *The prefix* endo and *the suffix* um *have appeared in a previous chapter, the one on the heart. Recall terms in that chapter that have this prefix and suffix. Fill in the chart.*

Prefix/Suffix	Meaning of P/S	Term in This Chapter	Meaning of This Term	Term in Previous Chapter	Meaning of That Term
endo					
um					

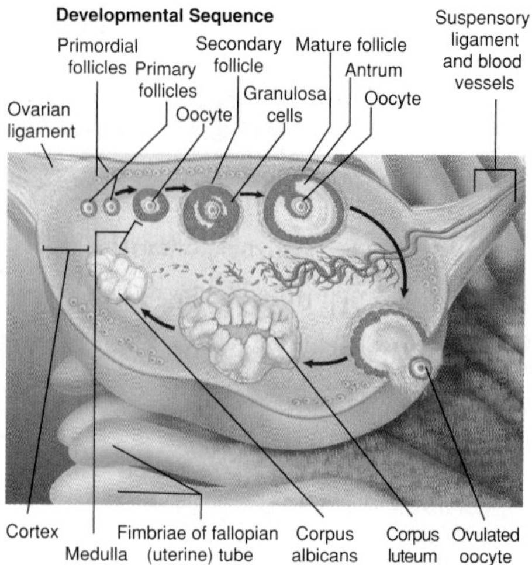

▲ **FIGURE 14.6 Oogenesis.**
The development of the oocyte begins with the primordial follicles.

▲ **FIGURE 14.7 Ovulation.**

Keynote

Menarche cannot occur until a girl has at least 17% body fat.

SEXUAL CYCLE OOGENESIS (EGG FORMATION) AND THE FEMALE SEXUAL CYCLE

Oogenesis

During **prenatal** development, small groups of cells in the ovarian cortex form some 2 million follicles. These are a single large cell, the **primary oocyte.** Many of these degenerate, a process called **atresia.**

At puberty, some 400,000 primary **oocytes** (egg cells) remain, each containing 23 pairs of chromosomes. They undergo meiosis to produce **secondary oocytes** containing 23 single chromosomes. The oocyte is surrounded by, and the fluid-filled **antrum** of the follicle is lined by, **granulosa** cells that secrete estrogen (*Figure 14.6*).

At the beginning of each menstrual cycle, as many as 20 primary follicles can start the maturing process, but only one develops fully. The remainder degenerate. By the midpoint of the menstrual cycle, the mature follicle bulges out on the surface of the ovary and ruptures (**ovulation,** *Figure 14.7*). The oocyte and lining cells from the follicle can either be taken into the fallopian tube or fall into the pelvic cavity and degenerate.

Ovarian Hormones

The ovaries of the sexually mature female secrete estrogens and progesterone. **Estrogens** are produced in the ovarian follicles. Their sexual functions are to:

1. Convert girls into sexually mature women through **thelarche, pubarche,** and **menarche.**
2. Regulate the menstrual cycle (*Figure 14.8*).
3. Be involved in pregnancy when that occurs (*see Chapter 15*).

Progesterone is produced by the corpus luteum of the ovary and also by the adrenal glands (*see Chapter 10*). Its sexual functions are to:

1. Prepare the lining (endometrium) of the uterus for implantation of the egg (*Figure 14.8*).
2. Inhibit lactation during pregnancy.
3. Produce menstrual bleeding if pregnancy does not occur.

The ovaries also secrete small amounts of androgens, male hormones.

The Sexual Cycle

The sexual cycle averages 28 days in length:

- Days 1 to 14: This is the **follicular phase**. **Menstruation** occurs during the first 3 to 5 days.
- Day 14: The developing ovarian follicles mature, and one of them **ovulates** around day 14.
- Days 15 to 24: After menstruation ends, the uterus starts to replace the endometrial tissue lost during menstruation. After **ovulation,** in the **postovulatory phase,** the endometrial lining continues to grow. The residual ovarian follicle becomes a **corpus luteum** containing **lutein** cells that are involved in progesterone production.
- Day 24: The corpus luteum **involutes.**
- Day 26: The corpus luteum is an inactive scar called a **corpus albicans.** At this time, the arteries supplying the endometrium of the uterus contract. This leads to ischemia, tissue necrosis, and the start of menstruation.

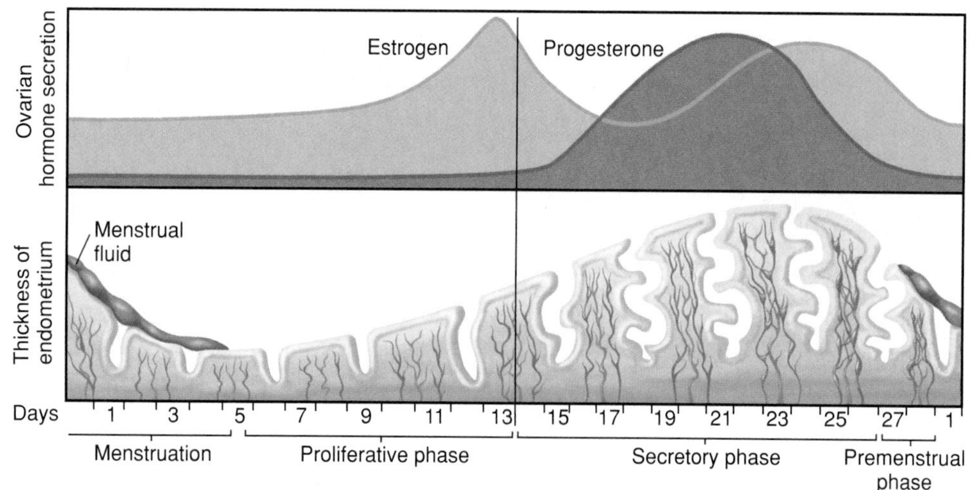

▲ **FIGURE 14.8 Menstrual Cycle.**

WORD ANALYSIS AND DEFINITION

S/ = Suffix P/ = Prefix R/ = Root R/CF = Combining Form

WORD	PRONUNCIATION		ELEMENTS	DEFINITION
antrum	**AN**-trum		Greek *cave*	A closed cavity.
corpus albicans	**KOR**-pus **AL-**bih-kanz		corpus, Latin *body* albicans, Latin *white*	An atrophied corpus luteum.
corpus luteum	**KOR**-pus **LOO**-teh-um	 S/ R/	corpus, Latin *body* **-um** *tissue* **lute-** *yellow*	Yellow structure formed at the site of a ruptured ovarian follicle.
luteal (adj)	**LOO**-teh-al	S/	**-al** *pertaining to*	Pertaining to a corpus luteum.
granulosa cell	gran-you-**LOW**-sah SELL	S/ R/	**-osa** *like* **granul-** *small grain* cell, Latin *storeroom*	Cell lining the ovarian follicle.
involution **involute (verb)**	in-voh-**LOO**-shun in-voh-**LUTE**	S/ P/ R/	**-ion** *action, condition* **in-** *in* **-volut-** *roll up, shrink*	Decrease in size.
lutein	**LOO**-teh-in		Latin *saffron yellow*	Yellow pigment.
menarche	meh-**NAR**-key	S/ R/	**-arche** *beginning* **men-** *month*	First menstrual period.
menses (noun)	**MEN**-seez		Latin *month*	Monthly uterine bleeding.
menstruation (noun)	men-stru-**AY**-shun	S/ R/CF	**-ation** *process* **menstr/u-** *menses*	Same as *menses.*
menstruate (verb)	**MEN**-stru-ate	S/	**-ate** *composed of, pertaining to*	Act of menstruation.
menstrual (adj)	**MEN**-stru-al	S/	**-al** *pertaining to*	Pertaining to menstruation.
ovulation	**OV**-you-**LAY**-shun	S/ R/	**-ation** *process* **ovul-** *ovum, egg*	Release of an oocyte from a follicle.
ovulate (verb)	**OV**-you-late	S/	**-ate** *pertaining to, composed of*	To release the oocyte.
prenatal	pree-**NAY**-tal	S/ P/ R/	**-al** *pertaining to* **pre-** *before* **-nat-** *birth*	Before birth.
pubarche	pyu-**BAR**-key	S/ R/	**-arche** *beginning* **pub-** *pubis*	Onset of development of pubic and axillary hair.
thelarche	thee-**LAR**-key	S/ R/	**-arche** *beginning* **thel-** *breast, nipple*	Onset of breast development.

EXERCISES

Deconstruct *the following medical terms into the meanings of their basic elements. This will help you learn the meaning of the term. Fill in the chart.*

Medical Term	Meaning of Prefix	Meaning of Root/CF	Meaning of Suffix	Meaning of Medical Term
pubarche				
menstruation				
ovulation				
menarche				
thelarche				
involution				

1. Put the three terms of female sexual maturity in the correct order of their occurrence:

a. ______________________________

b. ______________________________

c. ______________________________

LESSON 14.3

FEMALE REPRODUCTIVE SYSTEM

Disorders of the Female Reproductive Tract

OBJECTIVES

Information provided in this lesson will enable you to:

- **Use correct medical terminology for common disorders of the ovaries, fallopian tubes, and uterus.**
- **Select medical terminology for the causes of and treatments for infertility.**
- **Discuss different methods of contraception and their failure rates.**

Case Report 14.2 *(continued)*

When Ms. Claire Marcos first presented in the gynecology clinic, Dr. Rusak examined her abdomen and pelvis. The doctor was able to palpate both enlarged ovaries on vaginal examination. A vaginal ultrasound scan showed multiple small cysts in each ovary. Dr. Rusack diagnosed polycystic ovarian syndrome. She prescribed birth control pills because they contain estrogen and progesterone. They can correct the hormone imbalance, regulate Ms. Marcos' menses, and lower the level of testosterone to diminish her acne and hair problems.

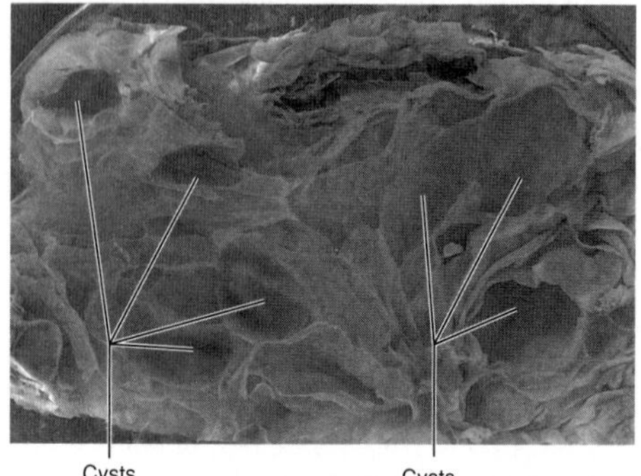

▲ **FIGURE 14.9 Polycystic Ovary.**

DISORDERS OF THE OVARIES

Ovarian Cysts

Polycystic ovarian syndrome (PCOS), in which multiple follicular cysts form in both ovaries (*Figure 14.9*), is the disorder Ms. Marcos presented with. Because of the repeated cyst formation, no egg matures and is released, so ovulation does not occur and progesterone is not produced. Without progesterone, her menstrual cycle is irregular or absent.

The cysts produce androgens, which prevent ovulation and produce acne, the male-pattern hair loss from the front of the scalp, and weight gain. Women with this syndrome are also at increased risk for endometrial cancer, type 2 diabetes, high blood cholesterol, hypertension, and heart disease.

Keynote

The peak incidence of ovarian cancer is in the 50- and 60-year age groups.

Ovarian cancer is the second most common gynecologic cancer after endometrial cancer, but it accounts for more deaths than any other gynecologic malignancy. Symptoms develop late in the disease process and are usually vague. A mass in the abdomen may be detected during routine pelvic examination. Treatment is to surgically remove the tumor and administer chemotherapy. The five-year survival rate is below 20%.

Primary amenorrhea occurs when a girl has not menstruated by age 16. This can occur with or without other signs of puberty. There are numerous possible causes:

- Drastic weight loss from malnutrition, dieting, bulimia, or anorexia nervosa.
- Extreme exercise, as in some young gymnasts.
- Extreme obesity.
- Chronic illness.

Treatment is directed at the basic cause.

Secondary amenorrhea occurs when a woman who has menstruated normally misses three or more periods in a row and she is not pregnant or in her **menopause.** The causes include:

- Ovarian disorders, such as polycystic ovarian syndrome.
- Excessive weight loss, low body fat percentage (e.g., as in gymnasts), or excessive exercise (e.g., as in marathon runners).
- Certain drugs, including antidepressants.
- Stress.

WORD ANALYSIS AND DEFINITION

S/ = Suffix P/ = Prefix R/ = Root R/CF = Combining Form

WORD	PRONUNCIATION	ELEMENTS		DEFINITION
amenorrhea	a-men-oh-**REE**-ah	R/ P/ R/CF	**-rrhea** *flow, discharge* **a-** *without* **-men/o-** *menses*	Absence or abnormal cessation of menstrual flow.
dysmenorrhea	dis-men-oh-**REE**-ah	R/ P/ R/CF	**-rrhea** *flow, discharge* **dys-** *painful or difficult* **men/o-** *menses*	Painful and difficult menstruation.
menopause (***Note:*** This term has no suffix, only a combining form and root.)	**MEN**-oh-paws	R/ R/CF	**-pause** *cessation* **men/o-** *menses*	Permanent ending of menstrual periods.
menopausal (adj)	**MEN**-oh-paws-al	S/	**-al** *pertaining to*	Pertaining to the menopause.
premenstrual	pree-**MEN**-stru-al	S/ P/ R/	**-al** *pertaining to* **pre-** *before* **-menstru-** *menses*	Pertaining to the time immediately before the menses.
primary	**PRY**-mah-ree		Latin *first*	The first disease or symptom, after which others may occur as complications.
secondary	**SEK**-ond-ah-ree		Latin *following or second*	Diseases or symptoms following a primary disease or symptom.

Primary dysmenorrhea, or **premenstrual syndrome (PMS),** refers to pain or discomfort associated with menstruation. The pain can begin 1 or 2 days before menses, peak on the first day of flow, and then slowly subside. The pain can be severe enough to prevent normal functioning for 2 or 3 days. Treatment is with **NSAIDs.** If these are ineffective, oral contraceptives are 80% to 90% effective.

Secondary dysmenorrhea is pain associated with disorders such as infection in the genital tract or endometriosis.

Keynote

Oral contraceptives are 80% to 90% effective in relieving symptoms of PMS.

Abbreviations

GYN	gynecology
NSAID	nonsteroidal anti-inflammatory drug
PCOS	polycystic ovarian syndrome
PMS	premenstrual syndrome

EXERCISES

Apply *your knowledge of the medical terminology for the female reproductive system by choosing the correct answer to each of the following questions. Circle the best choice.*

1. Another name for *primary dysmenorrhea* is:
   PID PMS HPV
2. *Premenstrual* happens:
   before menses after menses in the middle of menses
3. The gynecologic malignancy that accounts for the most deaths is:
   cervical cancer breast cancer ovarian cancer
4. Pick the term that contains a prefix, root, and suffix:
   menopause secondary premenstrual
5. In the medical term *dysmenorrhea,* the prefix *dys* means:
   without dead first before painful

You are

... a women's health nurse practitioner working with Anna Rusak, MD, a gynecologist at Fulwood Medical Center.

You are communicating with

... Mrs. Carol Isbell, a 29-year-old woman c/o severe dysmenorrhea since the age of 15.

CASE REPORT 14.3

Mrs. Isbell has been unable to conceive after 2 years of unprotected intercourse. She has severe, cramping lower abdominal pain for 2 days before, during, and 2 days after her periods, which are very heavy. She also has lower abdominal pain on intercourse. Physical examination is unremarkable except that pelvic examination shows several tender masses on each side of a normal-sized uterus. Dr. Rusak has also examined her and agreed with your diagnosis of endometriosis. Mrs. Isbell is to have an ultrasound examination.

Keynote

Residual scarring of the fallopian tube from salpingitis is a common cause of infertility.

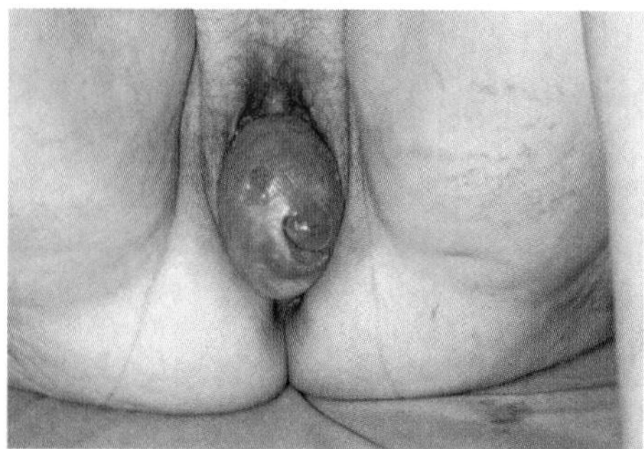

▲ **FIGURE 14.10 Prolapsed Uterus Protruding from the Vagina.**

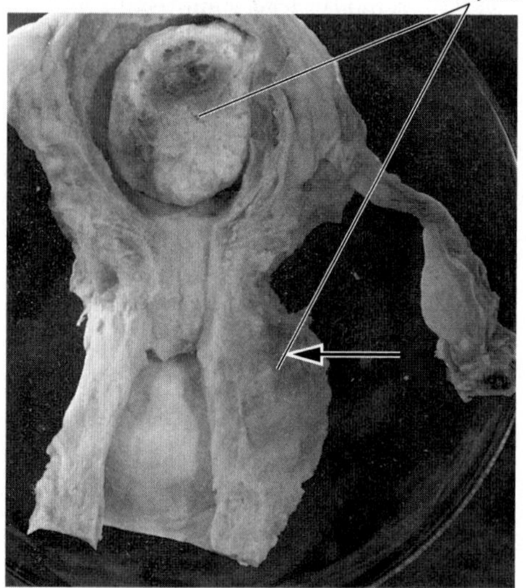

▲ **FIGURE 14.11 Leiomyomas (Fibroids).** In this sectioned uterus a smaller, rounded leiomyoma is present, causing a bulge in the uterine wall. The larger mass at the top is another leiomyoma, projecting from the surface into the uterine cavity.

DISORDERS OF THE FEMALE REPRODUCTIVE TRACT

Disorders of the Uterus and Fallopian (Uterine) Tubes

Endometriosis is said to affect 1 in 10 American women of childbearing age. The endometrium becomes implanted outside the uterus on the fallopian tubes, the ovaries, and the pelvic peritoneum. The displaced endometrium continues to go through its monthly cycle. It thickens and bleeds, leads to cysts and scar tissue, and produces pain. The etiology (cause) of endometriosis is unknown. Treatment is to remove the abnormal tissue through laparoscopy.

Salpingitis is an inflammation of the fallopian (uterine) tubes and is part of pelvic inflammatory disease **(PID)**. A bacterial infection, often from an **STD**, spreads from the vagina through the cervix and uterus. Symptoms are lower abdominal pain, fever, and a vaginal discharge. Treatment is with appropriate antibiotics. If a pelvic abscess has developed, it may be necessary to remove the damaged tube **(salpingectomy)**.

Uterine Prolapse

The uterus is normally supported by the muscles, ligaments, and connective tissue of the pelvic floor. Difficult childbirth, aging, obesity, lack of exercise, chronic coughing, and chronic constipation can weaken these tissues, causing the uterus to descend into the vaginal canal (*Figure 14.10*). Uterine **prolapse** can be accompanied by prolapse of the bladder and anterior vaginal wall (a disorder called a **cystocele**) or by prolapse of the rectum and posterior wall of the vagina (a disorder called a **rectocele**).

Treatment can be an individually fitted vaginal **pessary** inserted into the vagina to support the uterus. Surgical procedures such as sacral **colpopexy,** in which a mesh patch is inserted into the pelvic floor, or vaginal **hysterectomy,** in which the uterus is removed through the vagina, achieve good results.

Retroversion of the uterus is a normal variation, found in 20% of women, in which the body of the uterus is tipped backward instead of forward **(anteversion)**. It can also be caused by lax pelvic muscles and ligaments, pelvic adhesions (scar tissue in the pelvis following salpingitis), or pelvic inflammatory disease (PID). Retroversion by itself does not cause symptoms, and treatment is not usually necessary.

Uterine Fibroids

Fibroids are noncancerous growths that appear during childbearing years. Three out of four women have them, but only one out of four women has symptoms that include **menorrhagia, metrorrhagia, polymenorrhea,** low-back pain, and pelvic pain.

Uterine fibroids are also called **fibromyomas, leiomyomas,** or **myomas** (*Figure 14.11*). They arise in the **myometrium** and produce a pale, firm, rubbery mass separate from the surrounding tissue. They vary in size from seedlings to large masses that distort the uterus. They can protrude into the uterine cavity, causing menorrhagia, or project outside the uterus and press on the bladder or rectum, thereby producing symptoms.

Treatment options are numerous and include watchful waiting, **myomectomy,** and **focused ultrasound surgery (FUS). Hysterectomy** is major surgery and is considered by many gynecologists as a last resort.

WORD ANALYSIS AND DEFINITION

S/ = Suffix P/ = Prefix R/ = Root R/CF = Combining Form

WORD	PRONUNCIATION		ELEMENTS	DEFINITION
anteversion	an-teh-**VER**-shun	S/ P/ R/	**-ion** *action, condition* **ante-** *forward* **-vers-** *turn*	Forward displacement or tilting of a structure (in this case, the uterus).
colpopexy	**KOL**-poh-peck-see	S/ R/CF	**-pexy** *surgical fixation* **colp/o-** *vagina*	Surgical fixation of the vagina.
cystocele	**SIS**-toh-seal	S/ R/CF	**-cele** *hernia* **cyst/o-** *bladder*	Hernia of the bladder into the vagina.
endometriosis	**EN**-doh-me-tree-**OH**-sis	S/ P/ R/CF	**-osis** *condition* **endo-** *within, inside* **-metr/i-** *uterus*	Endometrial tissue outside the uterus.
fibroid	**FIE**-broyd	S/ R/	**-oid** *resembling* **fibr-** *fiber*	Uterine tumor resembling fibrous tissue.
fibromyoma	**FIE**-broh-my-**OH**-mah	S/ R/CF R/	**-oma** *tumor, mass* **fibr/o-** *fiber* **-my-** *muscle*	Benign neoplasm derived from smooth muscle and containing fibrous tissue.
hysterectomy	his-ter-**EK**-toe-me	S/ R/	**-ectomy** *surgical excision* **hyster-** *uterus*	Surgical removal of the uterus.
leiomyoma (also called *fibroid*)	**LIE**-oh-my-**OH**-mah	S/ R/ R/CF	**-oma** *tumor, mass* **-my-** *muscle* **lei/o-** *smooth*	Benign neoplasm derived from smooth muscle.
menorrhagia	men-oh-**RAY**-jee-ah	R/ R/CF	**-rrhagia** *flow, discharge* **men/o-** *menses*	Excessive menstrual bleeding.
metrorrhagia	**MEH**-troh-**RAY**-jee-ah	R/ R/CF	**-rrhagia** *flow, discharge* **metr/o-** *uterus*	Irregular uterine bleeding between menses.
myoma	my-**OH**-mah	S/ R/	**-oma** *tumor, mass* **my-** *muscle*	Benign tumor of muscle.
myomectomy	my-oh-**MEK**-toe-me	S/ R/ R/	**-ectomy** *surgical excision* **my-** *muscle* **-om-** *tumor*	Surgical removal of a myoma.
myometrium	my-oh-**MEE**-tree-um	S/ R/CF R/CF	**-um** *tissue* **my/o-** *muscle* **-metr/i-** *uterus*	Muscle wall of the uterus.
pessary	**PES**-ah-ree		Greek *an oval stone*	Appliance inserted into the vagina to support the uterus.
polymenorrhea	**POL**-ee-men-oh-**REE**-ah	R/ P/ R/CF	**-rrhea** *flow* **poly-** *many, excessive* **men/o-** *menses*	More than normal frequency of menses.
rectocele	**REK**-toe-seal	S/ R/CF	**-cele** *hernia* **rect/o-** *rectum*	Hernia of the rectum into the vagina.
retroversion	reh-troh-**VER**-shun	S/ P/ R/	**-ion** *action, condition* **retro-** *backward* **-vers-** *turn*	Tipping backward of the uterus.
salpingitis	sal-pin-**JIE**-tis	S/ R/	**-itis** *inflammation* **salping-** *tube*	Inflammation of the fallopian (uterine) tube.
salpingectomy	sal-pin-**JEK**-toe-me	S/	**-ectomy** *surgical excision*	Surgical removal of a fallopian tube.

EXERCISES

Review *all the terms in the WAD and their meanings* before *you answer the following questions. Fill in the blanks.*

1. List all the terms that are surgical procedures: ____________________

____

2. List all the terms that are opposites: ____________________

____

3. List all the terms that are diagnoses: ____________________

____

DISORDERS OF THE FEMALE REPRODUCTIVE TRACT (continued)

Keynote

More than 600,000 hysterectomies are performed each year in the United States.

Endometrial Cancer

Endometrial cancer is the fourth most common cancer in women, after lung, breast, and colon cancer. Each year, 40,000 new cases are diagnosed in the United States, mostly in women between ages 60 and 70. The most common symptom is vaginal bleeding after the menopause. Surgery is the most common treatment. It can be a **total hysterectomy,** in which the uterus and cervix are removed, or a **radical hysterectomy,** in which the fallopian tubes and ovaries are removed as well as the uterus. If the cancer has spread to other parts of the body, progesterone therapy, radiation therapy, and chemotherapy are used.

Cervical cancer is less common than endometrial cancer, but 50% of cases occur between ages 35 and 55. Some 10,000 new cases are diagnosed in the United States each year. Early cervical cancer produces no symptoms or signs and may be found on a routine **Pap** test. In the **precancerous** stage, abnormal cells **(dysplasia)** are found only in the outer layer of the cervix. Thirteen types of human papilloma virus (HPV) can convert these **dysplastic** cells to cancer cells (*Figure 14.12*). A vaccine has been developed that appears to make people immune to two of the most common types of HPV. It has been licensed by the FDA for use in girls and women aged 9 to 26.

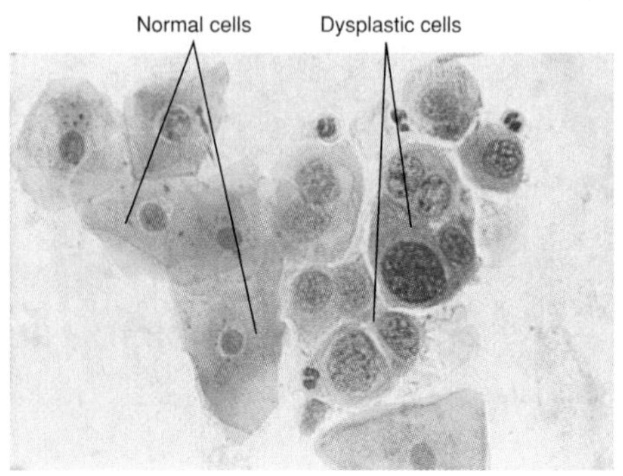

▲ **FIGURE 14.12 Abnormal Pap Smear with Dysplastic Cells.**

Treatment depends on the stage of the cancer. In **preinvasive** cancer, when the cancer is only in the outer layer of the lining of the cervix, the abnormal cells can be removed with a scalpel **(conization),** electrical current, or laser beam. In an **invasive** stage, when the cancer has invaded the cervix and beyond, treatment can include simple or radical hysterectomy, chemotherapy, and radiation therapy.

Pap Test

In a Pap test, the doctor brushes cells from the cervix (*Figure 14.13*). The cells are smeared onto a slide or rinsed into a special liquid and sent to the laboratory for examination. The test enables abnormal cells, precancerous or cancerous, to be detected. It is the most successful and accurate test for early detection of cervical abnormalities.

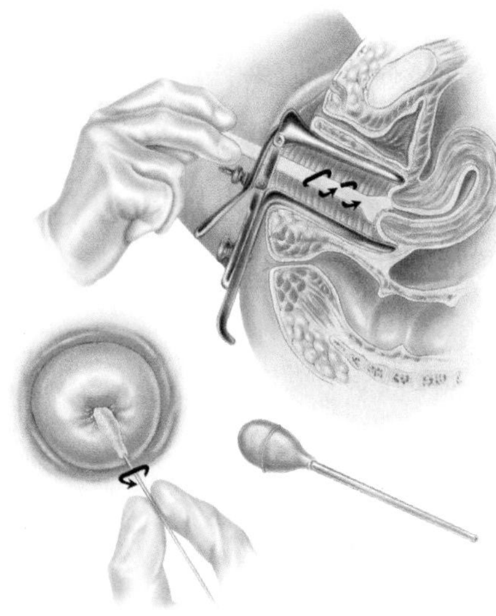

▲ **FIGURE 14.13 Pap Smear Being Performed.**

Other Causes of Uterine Bleeding

Dysfunctional uterine bleeding is a term applied when no cause can be found for a patient's menorrhagia. Treatment is with oral contraceptives. If that fails, **dilation and curettage (D & C)** may be effective. This procedure involves dilating the entrance to the uterus through the cervix so that a thin instrument can be inserted to scrape or suction away the lining of the uterus and take tissue samples. An alternative is endometrial **ablation,** in which a heat-generating tool or a laser removes or destroys the lining of the uterus and prevents or reduces menstruation. Endometrial ablation and hysterectomy are used in women who have finished childbearing.

WORD ANALYSIS AND DEFINITION

S/ = Suffix P/ = Prefix R/ = Root R/CF = Combining Form

WORD	PRONUNCIATION	ELEMENTS		DEFINITION
ablation	ab-**LAY**-shun	S/ P/ R/	**-ion** *action, condition* **ab-** *away from* **-lat-** *to take*	Removal of tissue to destroy its function.
conization	koh-ni-**ZAY**-shun	S/ R/	**-ation** *process* **coniz-** cone	Surgical excision of a cone-shaped piece of tissue.
curettage	kyu-reh-**TAHZH**	S/ R/	**-age** *pertaining to* **curett-** *to cleanse*	Scraping the interior of a cavity.
dysfunctional	dis-**FUNK**-shun-al	S/ P/ R/	**-al** *pertaining to* **dys-** *painful, difficult* **-function-** *perform*	Difficulty in performing.
dysplasia	dis-**PLAY**-zee-ah	S/ P/ R/	**-ia** *condition* **dys-** *painful, difficult* **-plas-** *molding, formation*	Abnormal tissue formation.
dysplastic (adj)	dis-**PLAS**-tik	S/	**-tic** *pertaining to*	Pertaining to or showing abnormal tissue formation.
Pap test	PAP TEST		George Papanicolaou, Greek-U.S. physician, 1883–1962	Examination of cells taken from the cervix.

Abbreviations

Pap	Papanicolaou (Pap test, Pap smear)
D & C	dilation and curettage

EXERCISES

Deconstruct *the following medical terms into their elements with slashes. Be sure to write the identity of the element (P, R, CF, or S) under the line.* Remember: Every element may not be present in every term. *Space is provided for you to briefly define the term. Fill in the blanks.*

1. *ablation* __________/__________/__________

   Definition: ____________________

2. *dysfunctional* __________/__________/__________

   Definition: ____________________

3. *curettage* __________/__________/__________

   Definition: ____________________

4. *conization* __________/__________/__________

   Definition: ____________________

Keynote

In postmenopausal women, the risk of heart disease becomes almost the same as that of men.

Keynote

The risks and benefits of hormone replacement therapy in menopause are an ongoing debate.

Keynote

In women, fertility begins to decrease as early as age 30, and **pregnancy** rates are very low after age 44.

Keynote

Approximately 20% of women now have their first child at age 35 or older.

Keynote

In 20% to 30% of female infertility problems, no identifiable cause is found.

Keynote

The success rate for IVF is approximately 30% for each egg retrieval.

▲ **FIGURE 14.14** **Female Condom.**

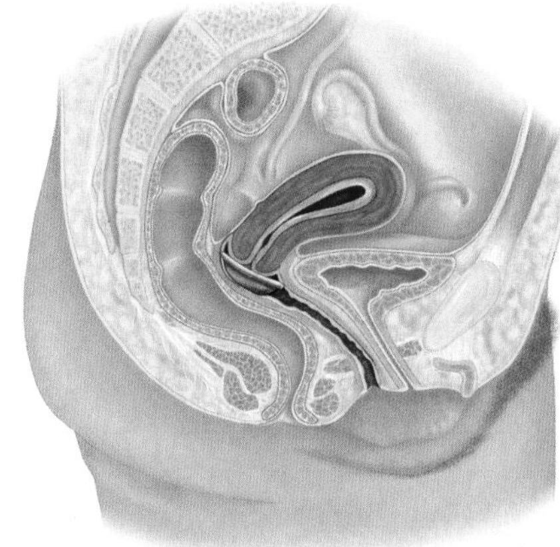

▲ **FIGURE 14.15** **Diaphragm.**

Keynote

Unprotected sex has a failure rate of 85%, with resulting pregnancy.

MENOPAUSE, FEMALE INFERTILITY, AND CONTRACEPTION

Menopause

Menopause occurs when a woman has not menstruated for a year and is not **pregnant**; she is in the "change of life." For most women, menstruation ceases between the ages of 45 and 55.

Without estrogen and progesterone, the uterus, vagina, and breasts atrophy (waste away), and more bone is lost than is replaced. Blood vessels constrict and dilate in response to changing hormone levels and can cause "hot flashes."

Female Infertility

Infertility is the inability to become pregnant after 1 year of unprotected intercourse. It affects 10% to 15% of all couples. The causes of infertility are due to:

- The female factor alone in 35% of cases.
- The male factor alone in 30%.
- Male and female factors in 20%.
- Unknown factors in 15%.

Causes of infertility in a woman include scarring of the fallopian tubes, structural abnormalities of the uterus, and infrequent ovulation, all of which were addressed earlier in this chapter. Treatment is of any underlying cause arising from the results of the infertility evaluation.

Surgical procedures to initiate pregnancy include:

- **Intrauterine insemination,** in which sperm are inserted directly into the uterus via a special catheter.
- **In vitro fertilization (IVF),** in which eggs and sperm are combined in a laboratory dish and two to four resulting embryos are placed inside the uterus. This can result in twins or triplets.

Contraception

Contraception is the prevention of pregnancy. Common methods of contraception include:

- **Behavioral methods:** These include **abstinence, coitus interruptus,** and the **rhythm method.** In the latter two methods there is a 20% failure rate.
- **Barrier methods:**
  - **Condoms** are available for males and females (*Figure 14.14*). They have a 5% to 10% failure rate.
  - **Diaphragms** (*Figure 14.15*) and **cervical caps** consist of a latex or rubber dome that is inserted into the vagina and placed over the cervix. When used with a spermicide, they have a 5% to 10% failure rate.
  - **Spermicidal foams and gels** are inserted into the vagina. Used on their own, they have a 25% failure rate.
- **Intrauterine devices (IUDs):** These are T-shaped flexible plastic or copper devices inserted into the uterus and left in place for 1 to 4 years. The failure rate is less than 3%.
- **Hormonal methods:**
  - **Oral contraceptives** (birth control pills) utilize a mixture of estrogen and progesterone to prevent follicular development and ovulation. They have a 5% failure rate, usually due to inconsistent pill taking.
  - **Estrogen/progestin patches** deliver the hormones transdermally (through the skin). Their failure rate is less than 1%.
  - **Injected progestins,** such as Depo-Provera, are given by injection every 3 months. Their failure rate is less than 1%.
  - **Implanted progestins,** such as Norplant, are contained in porous silicone tubes that are inserted under the skin and slowly release the progestin for up to 5 years. Their failure rate is less than 1%.

WORD ANALYSIS AND DEFINITION

S/ = Suffix P/ = Prefix R/ = Root R/CF = Combining Form

WORD	PRONUNCIATION		ELEMENTS	DEFINITION
coitus	**KOH**-it-us		Latin *come together*	Sexual intercourse
postcoital (adj)	post-**KOH**-ih-tal	S/ P/ R/	**-al** *pertaining to* **post-** *after* **-coit-** *sexual intercourse*	After sexual intercourse
contraception	kon-trah-**SEP**-shun	S/ R/ R/	**-ion** *action, condition* **contra-** *against* **-cept-** *receive*	Prevention of pregnancy.
contraceptive (adj)	kon-trah-**SEP**-tiv	S/	**-ive** *quality of*	An agent that prevents conception.
diaphragm (***Note:*** also the term for the muscle that separates the thoracic and abdominal cavities)	**DIE**-ah-fram		Greek *partition or wall*	A ring and dome-shaped material inserted into the vagina to prevent pregnancy.
insemination	in-sem-ih-**NAY**-shun	S/ P/ R/	**-ation** *process* **in-** *in* **-semin-** *semen*	Introduction of semen into the vagina.
inseminate (verb)	in-**SEM**-ih-nate	S/	**-ate** *pertaining to*	To introduce semen into the vagina.
in vitro fertilization (IVF)	IN **VEE**-troh **FER**-til-ih-**ZAY**-shun	P/ S/ R/	**in-** *in* vitro, Latin *glass* **-ization** *process of creating* **fertil-** *able to conceive*	Process of combining sperm and egg in a laboratory dish and placing resulting embryos inside the uterus.
pregnant	**PREG**-nant	S/ R/	**-ant** *pertaining to* **pregn-** *with child*	Having conceived.
pregnancy	**PREG**-nan-see	S/	**-ancy** *state of*	State of being pregnant.
progestin	pro-**JESS**-tin	S/ P/ R/	**-in** *substance* **pro-** *before* **-gest-** *produce, gestation*	A synthetic form of progesterone.

- **Morning-after pills,** such as Plan B, contain large doses of progestins to inhibit or delay ovulation. They are a backup when taken within 72 hours of unprotected intercourse. Their failure rate is around 10%.
- **Mifepristone (RU486),** when taken with a prostaglandin, induces a miscarriage. It has an 8% failure rate.

Abbreviations

IUD intrauterine device
IVF in vitro fertilization

Surgical Methods

- **Tubal ligation** ("getting your tubes tied") is performed with laparoscopy. Both fallopian tubes are cut, a segment is removed, and the ends are tied off and cauterized shut. The contraception failure rate is less than 1%.

EXERCISES

Match *the definitions in column 1 with the correct medical term in column 2.* Hint: Be sure to use the correct form (noun, verb, adjective) of the term. *Fill in the blanks.*

_____	1. synthetic form of progesterone	A. diaphragm
_____	2. agent that prevents conception	B. pregnant
_____	3. sexual intercourse	C. insemination
_____	4. introduction of semen into vagina	D. coitus
_____	5. having conceived	E. progestin
_____	6. egg and sperm in a lab dish	F. contraceptive
_____	7. ring and dome-shaped contraceptive device	G. IVF

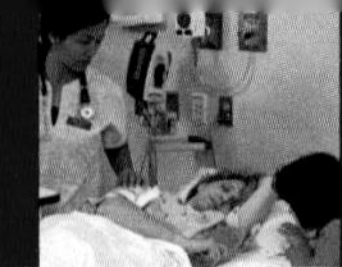

CHAPTER 14 REVIEW

FEMALE REPRODUCTIVE SYSTEM

CHALLENGE YOUR KNOWLEDGE

A. **Language of gynecology:** Understanding the terminology for the anatomy and physiology of the genitalia and reproductive organs is an important introduction to the female reproductive system. Identify each of the following terms based on its root or combining form and its meaning, and then give the meaning of the medical term. Fill in the chart.

Medical Term	Root/CF	Meaning of Root/CF	Meaning of Term
progesterone			
dyspareunia			
gonorrhea			
endometrial			
hysterectomy			
gynecologic			
menarche			
amenorrhea			
colpopexy			
menstruation			

Use any *one* of the terms above in a sentence of your choice that could be patient documentation:

____

____

____

B. **Diseases:** Test your knowledge of diseases. Anyone working in a GYN clinic or internal medicine practice will have patients diagnosed with sexually transmitted diseases. Make the correct association between the name of the STD and its description by matching the statement in column 1 with the term in column 2.

	Column 1	Column 2
_____	1. "silent disease"	A. HPV
_____	2. begins as a chancre	B. trichomoniasis
_____	3. can be treated with single dose of antibiotic	C. chlamydia
_____	4. untreated chlamydia can cause this	D. primary syphilis
_____	5. can cause dementia if untreated	E. genital herpes simplex
_____	6. appears as a rash with joint pain	F. tertiary syphilis
_____	7. "ping-pong infection"	G. PID
_____	8. caused by HSV 2	H. gonorrhea
_____	9. produces shiny bumps with milky fluid inside	I. secondary syphilis
_____	10. causes genital warts	J. molluscum contagiosum

Spelling note: The diseases in the above exercise can be particularly difficult to spell.

Work with a fellow student: Ask your partner to close his or her book and write on a separate sheet of paper the above terms, which you dictate. Check your partner's spelling. Then close your book and ask your partner to dictate the same terms to you. Check your spelling.

Master these terms now—they will surely be in a test question.

C. **Abbreviations:** The following abbreviations have appeared in this chapter. Write out the words each abbreviation stands for, and give a brief definition for it. Refer to your dictionary and/or the Glossary for help. Fill in the blanks.

1. *LPN* means ______.
   Definition: ______
2. *STD* means ______.
   Definition: ______
3. *HSV* means ______.
   Definition: ______
4. *NSAID* means ______.
   Definition: ______
5. *IVF* means ______.
   Definition: ______
6. *c/o* means ______.
   Definition: ______
7. *FUS* means ______.
   Definition: ______
8. *PCOS* means ______.
   Definition: ______
9. *D & C* means ______.
   Definition: ______
10. *IUD* means ______.
    Definition: ______
11. *PMS* means ______.
    Definition: ______

D. Elements: Knowing the meanings of word elements is your best tool for building and analyzing medical terms. Choose from among the following word elements, and insert the correct element on the blank to build the medical term.

1. pertaining to many cysts: ____________________ cystic
   (pick one: mono poly cyano endo)

2. one who studies female diseases: ____________________ logist
   (pick one: neuro dermato cardio gyneco)

3. term synonymous with *menstruation:* ____________________
   (pick one: gynecology gynecologic menses menstrual)

4. egg cell: ____________________ cyte
   (pick one: ovari gyneco granul oo)

5. hormone produced by the ovary: ____________________ gen
   (pick one: menstru estro ovari proges)

6. surgical excision of the vulva: ____________________ ectomy
   (pick one: vulv hystero myo vestibul)

7. pain during sexual intercourse: ____________________ pareunia
   (pick one: dys a trans mal)

8. inner lining of the uterus: ____________________ metrium
   (pick one: endo myo peri pre)

9. surgical fixation of the vagina: ____________________ pexy
   (pick one: vagino colpo vestibul hystero)

10. inflammation of the fallopian tube: ____________________ itis
    (pick one: vagin salping my hyster)

E. Dictate the following sentences to your class partner. Have your partner write them on a separate piece of paper. After he or she has written the sentences, have your partner spell the medical terms for you. (**Note:** For the answer to be correct, *every word* has to be spelled correctly.) If either of you is having difficulties with the pronunciations, check the term on the Student Online Learning Center (www.mhhe.com/AllanEssMedLanguage). Be prepared to define any of these terms if the instructor should ask!

WWW

1. Some means of contraception are diaphragms, spermicidal foams and gels, intrauterine devices, and oral contraceptives.

2. Tubal ligation is performed with laparoscopy.

3. An alternative to dilation and curettage is endometrial ablation, in which a laser removes or destroys the lining of the uterus and prevents or reduces menstruation.

Write any misspelled or mispronounced terms on the lines below:

____________________ ____________________

____________________ ____________________

____________________ ____________________

F. **Spelling demons:** This chapter contains some very difficult spelling demons. Challenge your ability to spell all these terms correctly. Circle the best choice, and then fill in the blanks.

1.	dyspareuria	dyspareunia	dyspariunia
2.	clamidia	clammydia	chlamydia
3.	chancre	cankcre	cranchre
4.	gonnorhea	gonorrea	gonorrhea
5.	trichomonosis	trichomoniasis	trichomonis
6.	syphillis	syphilis	syphylis
7.	dismenorhea	dysmenorhea	dysmenorrhea
8.	liomyoma	leomyoma	leiomyoma
9.	menorrhagia	mennoragia	menorhagia
10.	diaphragm	diaphram	diaphramm

11. Write a definition for any two terms in the above list:

    Term: ______________________________

    Definition: ______________________________

    Term: ______________________________

    Definition: ______________________________

12. Use any one term in a sentence that is *not* a definition:

    ____

    ____

    ____

13. Which term is the most difficult for you to spell correctly?

    Term: ______________________________

    Definition: ______________________________

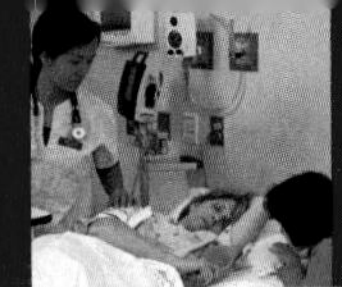

FEMALE REPRODUCTIVE SYSTEM

G. Surgical suffixes: Each of the following terms is an "ectomy," that is, the surgical removal of some body part. What is the doctor removing in each procedure? Fill in the first chart.

Medical Term	Meaning of Medical Term (What Is Being Removed?)
hysterectomy	
salpingectomy	
myomectomy	
vestibulectomy	
cystectomy	

Remembering roots and terms from previous chapters, complete the second chart:

Medical Term	Meaning of Medical Term (What Is Being Removed?)
vasectomy	
gastrectomy	
pancreatectomy	
nephrectomy	
laryngectomy	
pneumonectomy	
cholecystectomy	

The principle is the same: Know one suffix, and it is your clue to a dozen different terms. Your knowledge of the roots will help you with the specifics of each term.

H. Train your eye and ear: There is no room in medicine for "almost got it right." Precision in communication is critical for patient safety. Explain the difference between these very similar looking (and sounding) medical terms.

1. *pericardium:*

___

___

2. *perineum:*

___

___

3. *metrorrhagia:*

____

____

4. *menorrhagia:*

____

____

I. **Terminology challenge:** The terms *diaphragm* and *cervical* appear in this chapter, but they have also appeared previously in other chapters, where they have different meanings. Define all these terms, including their meanings from previous chapters. Fill in the blanks.

1. *Cervical* in this chapter means ______________________.

   Where is this located? ______________________.

2. *Cervical* in a previous chapter means ______________________.

   Where is this located? ______________________.

3. *Diaphragm* in this chapter means ______________________.

4. *Diaphragm* in a previous chapter means ______________________.

J. **Prefixes:** The prefix is given in the first column. Identify the meaning of the prefix, give an example of a medical term containing this prefix (the term does not have to be from this chapter), and give the meaning of the term in the last column. Fill in the chart.

Prefix	Meaning of Prefix	Medical Term Containing This Prefix	Meaning of This Term
a			
ab			
contra			
dys			
peri			
poly			
pre			
pro			
re			
retro			
syn			

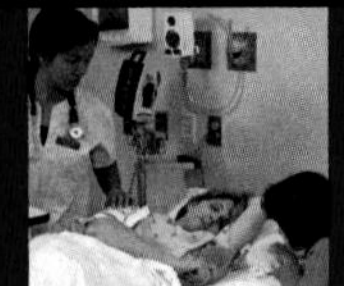

FEMALE REPRODUCTIVE SYSTEM

K. Study hint: Write yourself a study hint for remembering the tricky spelling in terms like *dysmenorrhea* and *metrorrhagia*. The "rrh" combination appears in several terms in other chapters as well.

Always check your spelling when these words are answers to test questions.

Study Hint:

List as many terms as you can (from this or other chapters) that have the "rrh" combination of letters:

L. Dictionary: The terms below have appeared in the text but are not defined in a WAD box. Look them up in a dictionary or online. Identify the part of speech (noun, verb, adjective), and demonstrate that you can use it correctly in a sentence.

1. *diffuse*

   Definition:

   Part of speech:

   Sentence:

2. *puberty*

   Definition:

   Part of speech:

   Sentence:

M. **Seek and find:** How carefully do you pay attention to what you are reading? The answers to all the following questions can be found in this chapter. Fill in the blanks. Try to score a perfect 10!

1. Medical terms can be very long, but some are very short. What medical term appears in this chapter and only has two letters? ____________________

2. Name one set of opposite terms that appear in this chapter, and define them.

   a. Term: ____________________

      Means: ____________________

   b. Term: ____________________

      Means: ____________________

3. List and define all the medical terms in this chapter that mean something is *painful or difficult:*

   ____

   ____

   ____

4. What other terms can also mean *uterine fibroids?*

   ____

   ____

5. Where is the patient when FUS is performed?

   ____

6. Medication patches deliver drugs *transdermally.* What does this mean?

   ____

7. What does *prepuce* mean? ____________________

8. What type of fungus is *Candida?* ____________________

N. **Translate** the following sentence into language your patient can understand. Be brief but accurate. Rewrite the sentence below.
"On day 26 of the sexual cycle the arteries supplying the endometrium of the uterus contract. This leads to ischemia, tissue necrosis, and the start of menstruation."

Rewrite:

____

____

____

____

____

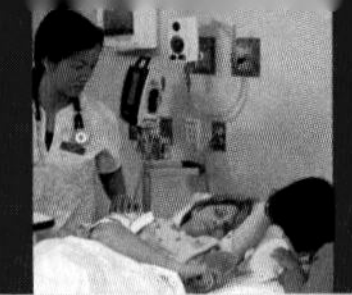

CHAPTER 14 REVIEW

FEMALE REPRODUCTIVE SYSTEM

O. Latin and Greek terms cannot be further deconstructed into prefix, root, or suffix. You must know them for what they are. Test your knowledge of these terms with this exercise. Match the meaning in column 1 with the correct medical term in column 2. There are more answers than you need.

_____ 1. cave

_____ 2. partition or wall

_____ 3. lip

_____ 4. greater

_____ 5. tampon

_____ 6. month

_____ 7. egg

_____ 8. sheath

_____ 9. membrane

_____ 10. come together

A. menses

B. ovary

C. vagina

D. coitus

E. minora

F. hymen

G. majora

H. antrum

I. labium

J. diaphragm

K. plug

L. oocyte

P. Same but different: More than one body part can have the same term but obviously different meanings.

1. The *vestibule* is the area enclosed by the labia minora and contains the urinary and vaginal orifices. Do you remember what other chapter you have read that talks about a different vestibule? Briefly describe that vestibule, and what body system it is in, on the lines below.

   Vestibule:

   ____

   Body system:

   ____

2. The cervical canal *dilates* in labor. What else *dilates* in a different body system?

   Dilate:

   ____

   Body system:

   ____

Q. **Deconstruct terms:** One element in each of the following medical terms has been bolded. First slash the term to isolate the elements. Identify the bolded element as a prefix, root, combining form, or suffix under the line, and write the meaning of the element on the line. Then write another term in which this same element appears. Fill in the blanks.

1. ***dys**pareunia* Meaning: ______________________

   Name another term in which this same element appears: ______________________

2. *gono**rrhea*** Meaning: ______________________

   Name another term in which this same element appears: ______________________

3. ***gyneco**logy* Meaning: ______________________

   Name another term in which this same element appears: ______________________

4. corpus ***luteum*** Meaning: ______________________

   Name another term in which this same element appears: ______________________

5. ***ovulation*** Meaning: ______________________

   Name another term in which this same element appears: ______________________

6. *recto**cele*** Meaning: ______________________

   Name another term in which this same element appears: ______________________

7. *hyster**ectomy*** Meaning: ______________________

   Name another term in which this same element appears: ______________________

8. ***salping**itis* Meaning: ______________________

   Name another term in which this same element appears: ______________________

FEMALE REPRODUCTIVE SYSTEM

R. Elements: The following medical terms all have the same prefix, but the rest of their elements make them entirely different terms. Deconstruct each medical term into its basic elements, write their meanings, and then use each term in a sentence of your own making. Fill in the chart, and create the sentences.

Medical Term	Meaning of Prefix	Meaning of Root/CF	Meaning of Suffix	Meaning of Term
dysplasia				
dysmenorrhea				
dysplastic				
dysfunctional				
dyspareunia				

Sentences:

1. ______________________________

2. ______________________________

3. ______________________________

4. ______________________________

5. ______________________________

S. Brain teaser: In this chapter there are three roots/combining forms that all mean the same thing. (*Hint:* The elements refer to a body part.) List those elements, with their meanings; then give an example of each one in a different term, and give the meaning of each term.

Remember: Even though these elements all mean the same thing, they are not interchangeable, and only certain elements go with specific terms.

1. (element) ______________________ means ______________________.
   Term containing this element:
   ____
   Meaning of this term:
   ____
   ____

2. (element) ______________________ means ______________________.
   Term containing this element:
   ____
   Meaning of this term:
   ____
   ____

3. (element) ______________________ means ______________________.
   Term containing this element:
   ____
   Meaning of this term:
   ____
   ____

Now find two suffixes in this chapter that also have the same meaning.

4. (element) ______________________ means ______________________.
   Term containing this element:
   ____
   Meaning of this term:
   ____
   ____

5. (element) ______________________ means ______________________.
   Term containing this element:
   ____
   Meaning of this term:
   ____
   ____

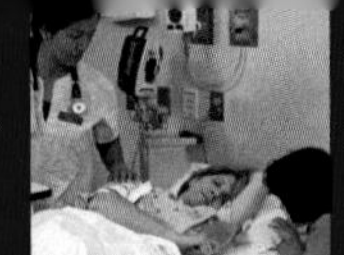

FEMALE REPRODUCTIVE SYSTEM

T. **Chapter challenge:** Circle the correct answer.

> **Study Hint**
> Immediately cross off any answer you know is not correct. In your remaining choices, there is only *one best answer.*

1. Which is the only correct statement regarding the external female genitalia?
   a. The female external genitalia occupy most of the perimetrium.
   b. The female external genitalia do not include the mons pubis.
   c. The female external genitalia are collectively called the *vulva.*
   d. The female external genitalia include the vagina.
   e. The female external genitalia do not include the vestibule.

2. Which of the following abbreviations is *not* a diagnosis?
   a. HIV
   b. PID
   c. AIDS
   d. CDC
   e. PMS

3. Which pair of terms contains a disease and a surgical procedure:

a.	chancre	chlamydia
b.	chlamydia	replication
c.	gonorrhea	opportunistic infection
d.	PID	FUS
e.	fimbriae	fallopian tubes

4. The terms *fundus, body,* and *cervix* are associated with which body organ?
   a. ovary
   b. bladder
   c. uterus
   d. fallopian tube
   e. vagina

5. Which term contains two suffixes?
   a. trichomoniasis
   b. menstruation
   c. endometriosis
   d. opportunistic
   e. anteversion

6. Painful intercourse is:
   a. dysphagia
   b. dysmenorrhea
   c. dysphoric
   d. dyspareunia
   e. dysuria

7. The correct plural of the term *fornix* is:
   a. forniae
   b. fornixes
   c. fornices
   d. fornium
   e. fornia

8. Which term has a suffix that means *condition?*
   a. dyspareunia
   b. endometriosis
   c. trichomoniasis
   d. none of the above
   e. all of the above

9. What is the diagnosis when a woman who has menstruated normally misses three or more periods in a row and she is not pregnant or in her menopause?

a. dysmenorrhea
b. tertiary syphilis
c. secondary amenorrhea
d. gonorrhea
e. primary amenorrhea

10. Another name for the greater vestibular gland is:

a. Bartholin
b. labia minora
c. labia majora
d. a and b
e. a and c

11. *Fibromyomas, leiomyomas,* or *myomas* are all terms for:

a. ovarian cysts
b. uterine fibroids
c. hernias
d. rectoceles
e. spirochetes

12. Which of the following terms has no prefix?

a. trichomoniasis
b. endometrium
c. prenatal
d. amenorrhea
e. dysphoria

13. Which term can be found in two different body systems?

a. follicle
b. cervix
c. fundus
d. a and c
e. a, b, and c

14. What does the "I" in *PID* stand for?

a. inflamed
b. infected
c. impaired
d. inflammatory
e. involution

15. **Brain teaser:** Which choice is the correct order of these terms?

a. menstruation, menopause, involution, ovulation
b. ovulation, involution, menstruation, menopause
c. menstruation, ovulation, involution, menopause
d. involution, ovulation, menopause, menstruation
e. menstruation, involution, ovulation, menopause

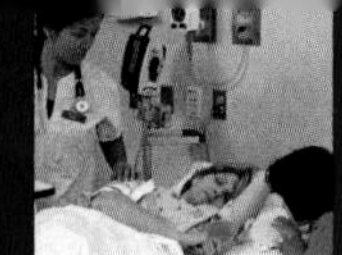

FEMALE REPRODUCTIVE SYSTEM

U. Case Report challenge: Now that you are more comfortable with the terms in this chapter, you can apply that knowledge and briefly answer the questions about the Case Report. If you read the report through and underline all the medical terminology, this will make it easier to answer the questions.

CASE REPORT 14.2

You are

. . . a certified health education specialist (CHES) employed by Fulwood Medical Center.

You are communating with

. . . Ms. Claire Marcos, a 21-year-old student referred to you by Anna Rusak, MD, a gynecologist. Ms. Marcos has been diagnosed with polycystic ovarian syndrome, and your task is to develop a program of self-care as part of her overall plan of therapy.

From her medical record, you see that she presented with irregular, often missed menstrual periods since the beginning of puberty, persistent acne, patches of dark skin on the back of her neck and under her arms, loss of hair from the front of her scalp, and inability to control her weight. She is 5 feet 4 inches tall and weighs 150 pounds.

You are to counsel her about her self-care program involving exercise, diet, and use of birth control medications.

When Ms. Claire Marcos first presented in the gynecology clinic, Dr. Rusak examined her abdomen and pelvis. The doctor was able to palpate both enlarged ovaries on vaginal examination. A vaginal ultrasound scan showed multiple small cysts in each ovary. Dr. Rusak diagnosed polycystic ovarian syndrome. She prescribed birth control pills because they contain estrogen and progesterone. These can correct the hormone imbalance, regulate Ms. Marcos' menses, and lower the level of testosterone to diminish her acne and hair problems.

1. Which body cavity contains the abdomen and the pelvis?

____

2. What is the meaning of the term *palpate?*

____

____

3. What type of specialist would Dr. Rusak refer Ms. Marcos to for her acne? _______________

4. What is the singular form of *ovaries?* _______________

5. What is the abbreviation for Ms. Marcos' diagnosis? _______________

6. What is another way of saying "multiple cysts"?

____

7. What diagnostic test did Ms. Marcos have? _______________

8. What are the hormone terms mentioned in this Case Report?

____

9. Identify each of the hormones as either male or female hormones:

Male: ______________________________

Female: ______________________________

10. What symptoms did Ms. Marcos have when she saw Dr. Rusak?

____

____

11. Why did Dr. Rusak prescribe birth control pills for Ms. Marcos?

____

____

____

12. What is another term for *menses?*

____

13. What is the effect of testosterone on Ms. Marcos?

____

____

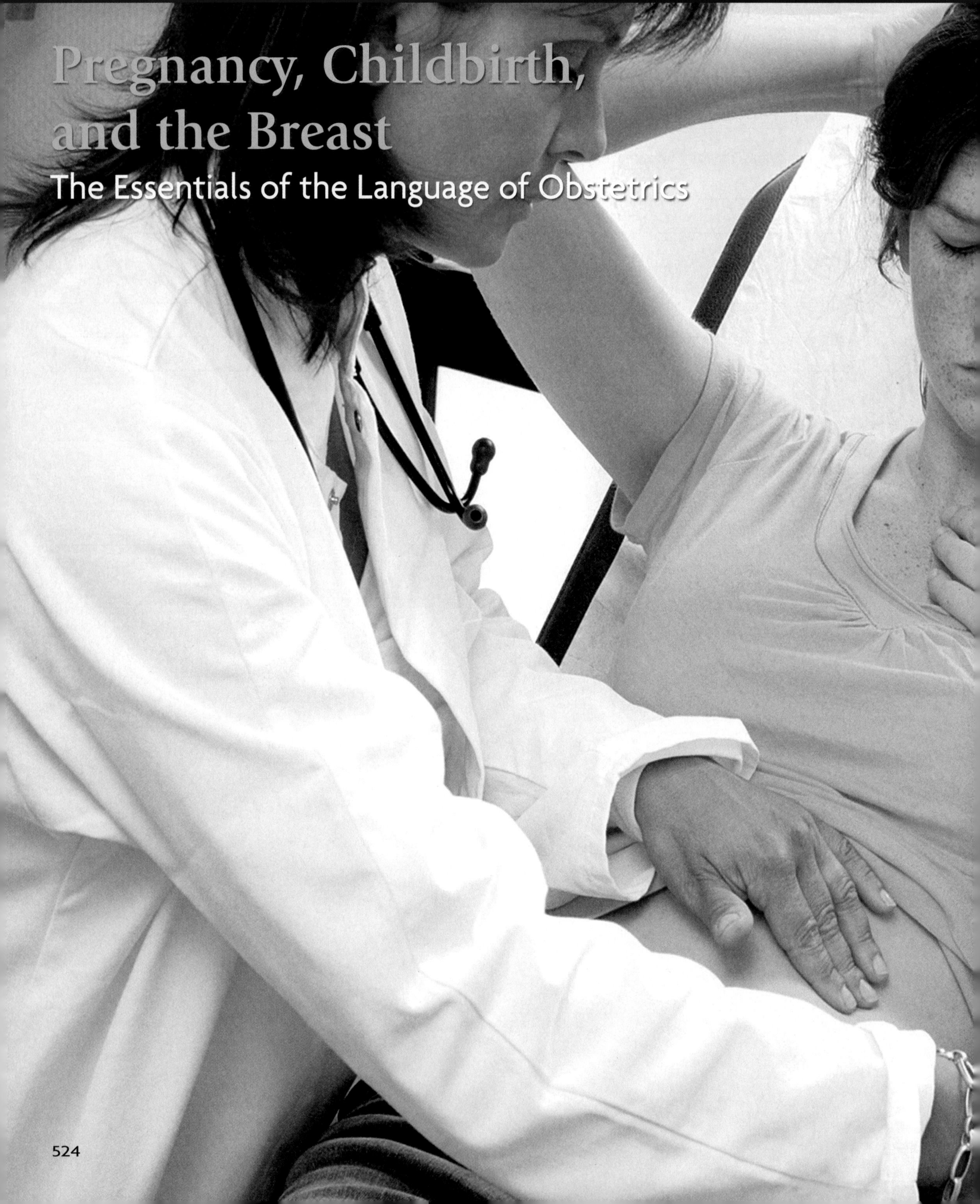

Pregnancy, Childbirth, and the Breast

The Essentials of the Language of Obstetrics

CHAPTER

CASE REPORT 15.1

You are

. . . an obstetric assistant (OA) employed by Garry Joiner, MD, an obstetrician at Fulwood Medical Center.

You are communicating with

Mrs. Gloria Maggay, a 29-year-old housekeeper. Her last menstrual period was 8 weeks ago, and she has a positive home pregnancy test. This is her first pregnancy. She has breast tenderness and mild nausea. For the past 2 days she has had some cramping, right-sided, lower abdominal pain and this morning had vaginal spotting. Her VS are T 99°F, P 80, R 14, BP 130/70.

While you are waiting for Dr. Joiner to come and examine her, Mrs. Maggay complains of feeling faint and has a sharp, severe pain in her lower abdomen on the right side. Her pulse rate has increased to 92. You need to recognize what is happening and utilize the correct medical terminology as you talk to Dr. Joiner.

Learning Outcomes

When the nuclei of the sperm and the egg unite, their chromosomes combine. Fertilization **(conception)** is complete, and a zygote is formed. This process will be described in this lesson to enable you to:

15.1 Specify the correct medical terminology for the stages of embryonic development, the implantation of the embryo in the uterus, and fetal development.

15.2 Recognize and use appropriately the medical terminology for the functions of the placenta and for childbirth.

15.3 Discuss some of the most common problems of fetal development and childbirth using correct medical terminology.

15.4 Apply correct medical terminology to the structure and functions of the breast and its common disorders.

LESSON 15.1 Conception and Development

PREGNANCY, CHILDBIRTH, AND THE BREAST

OBJECTIVES

The information in this lesson will enable you to:

- **Apply medical terminology to describe conception and implantation.**
- **Identify the terminology for the characteristics of an embryo and fetus.**
- **List the functions of the placenta.**
- **Use correct medical terminology to discuss the mechanisms of pregnancy (PGY) and childbirth and their common disorders.**

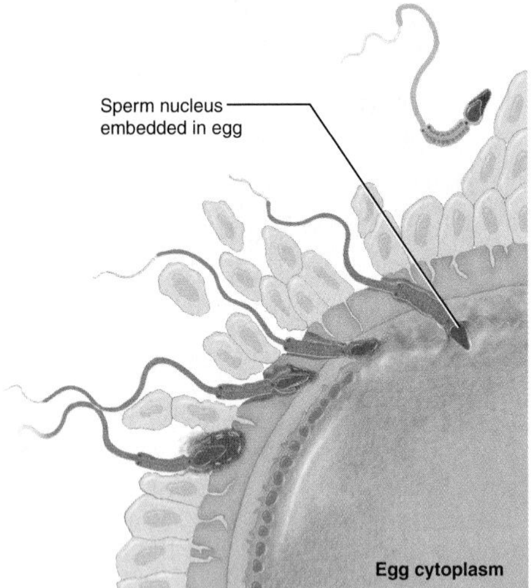

▲ **FIGURE 15.1 Fertilization.**

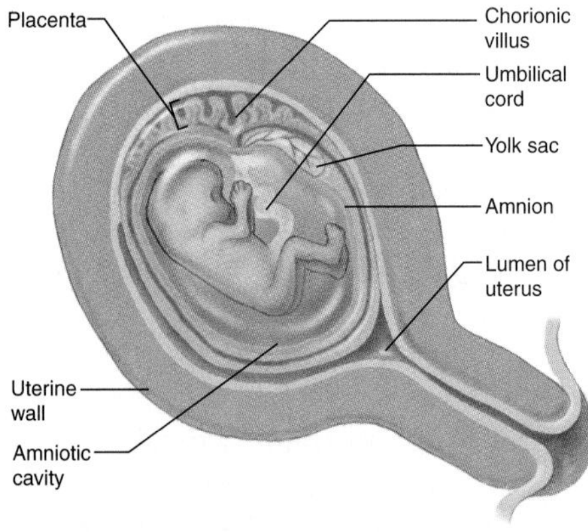

▲ **FIGURE 15.2 Embryo and Placenta at 13½ weeks.**

Abbreviations

OA	obstetric assistant
PGY	pregnancy

CONCEPTION AND DEVELOPMENT

Conception

When released from the ovary, an egg takes 72 hours to reach the uterus—but it must be **fertilized** within 12 to 24 hours to survive. Therefore, **fertilization** must take place in the distal third of the fallopian tube.

Between 200 million and 600 million sperm are deposited in the vagina near the cervix. Many are destroyed by the acidity in the vagina or just drain out. Others fail to get through the cervical mucus. Half the survivors will go up the wrong tube. The journey through the uterus into the fallopian tube takes about an hour. Some 2,000 to 3,000 sperm reach the egg. Several of them penetrate the outer layers of the egg to clear the path for the one sperm that will penetrate all the way into the egg to fertilize it (*Figure 15.1*). The egg becomes a **zygote.**

Implantation

While still in the fallopian tube, the zygote divides; a fluid-filled cavity develops, and the zygote becomes a **blastocyst.** A week after fertilization, the blastocyst enters the uterine cavity and burrows into the endometrium, and **implantation** occurs. A group of cells in the blastocyst differentiate into the **embryo.** Other cells from the blastocyst, together with endometrial cells, form the **placenta.**

Embryo

From week 2 until week 8 is the **embryonic period,** in which most of the external structures and internal organs of the embryo are formed, together with the placenta, umbilical cord, **amnion,** and **chorion**. The amnion grows to envelop the embryo and is filled with amniotic fluid. At the eighth week all the organ systems are present. The embryo is just over 1 inch long and is now called a **fetus.**

Placenta

The placenta is a disc of tissue that increases in size as pregnancy proceeds (*Figure 15.2*). The surface facing the fetus is smooth and gives rise to the **umbilical cord.** The surface attached to the uterine wall consists of treelike structures called **chorionic villi.** The cells of the villi keep the maternal and fetal circulations separate, but they are very thin and allow an exchange of gases, nutrients, and waste products to occur.

The functions of the placenta are to:

- **Transport** nutrients and oxygen from the mother to the fetus.
- **Transport** nitrogenous wastes and carbon dioxide from the fetus to the mother, who can excrete them.
- **Transport** maternal antibodies and hormones to the fetus.
- **Secrete** hormones such as estrogen and progesterone.

WORD ANALYSIS AND DEFINITION

S/ = Suffix P/ = Prefix R/ = Root R/CF = Combining Form

WORD	PRONUNCIATION	ELEMENTS		DEFINITION
amnion	**AM**-nee-on		Greek *membrane around fetus*	Membrane around the fetus that contains amniotic fluid.
amniotic (adj)	am-nee-**OT**-ic	S/ R/CF	**-tic** *pertaining to* **amni/o-** *amnion, fetal membrane*	Pertaining to the amnion.
amniocentesis	**AM**-nee-oh-sen-tee-sis	S/	**-centesis** *to puncture*	Removal of amniotic fluid for diagnostic purposes.
blastocyst	**BLAS**-toe-sist	S/ R/CF	**-cyst** *cyst, bladder* **blast/o-** *germ cell*	First 2 weeks of the developing embryo.
chorion **chorionic (adj)**	**KOH**-ree-on koh-ree-**ON**-ick	S/ R/	Greek *membrane* **-ic** *pertaining to* **chorion-** *chorion*	The fetal membrane that forms the placenta. Pertaining to the chorion.
chorionic villus	**VILL**-us		villus, Latin *shaggy hair*	Vascular process of the embryonic chorion to form the placenta.
conception	kon-**SEP**-shun		Latin *something received*	Fertilization of the egg by sperm to form a zygote.
embryo	**EM**-bree-oh		Greek *a young one*	Developing organism from conception until the end of the second month.
embryonic (adj)	em-bree-**ON**-ic	S/ R/	**-ic** *pertaining to* **embryon-** *embryo*	Pertaining to the embryo.
fertilize (verb) **fertilization (noun)**	**FER**-til-ize **FER**-til-eye-**ZAY**-shun	S/ R/	Latin *make fruitful* **-ation** *process* **fertiliz-** *make fruitful*	Penetration of the oocyte by sperm. Union of a male sperm and a female egg.
fetus	**FEE**-tus		Latin *offspring*	Human organism from the end of the eighth week after conception to birth.
fetal (adj)	**FEE**-tal	S/ R/	**-al** *pertaining to* **fet-** *fetus*	Pertaining to the fetus.
implantation	im-plan-**TAY**-shun	S/ P/ R/	**-ation** *process* **im-** *in* **-plant-** *to plant, insert*	Attachment of a fertilized egg to the endometrium.
placenta	plah-**SEN**-tah		Latin *a cake*	Organ that allows metabolic exchange between the mother and the fetus.

Unfortunately, some undesirable items and many medications can cross the placenta. These include the HIV and rubella viruses, bacteria that cause syphilis, alcohol, nicotine and carbon monoxide from smoking, and drugs ranging from aspirin to heroin and cocaine. All these have bad effects on the fetus.

Amniocentesis and **chorionic villus sampling** are performed to test for chromosomal abnormalities and genetic birth defects.

EXERCISES

Precision in documentation *includes using the correct form (noun, verb, adjective) of the medical term. Practice precision in this* written language of obstetrics. *Fill in the blanks.*

amniocentesis **amnion** **amniotic**

1. The ____________________ will be punctured in the procedure ____________________ in order to withdraw the ____________________ fluid.

embryo **embryonic**

2. In the ____________________ stage of gestation, the ____________________ forms in the first 8 weeks of human development.

gestation **gestational**

3. ____________________ is divided into trimesters. It is not uncommon for some women to develop ____________________ diabetes during pregnancy.

Practice your precision in pronunciation please!

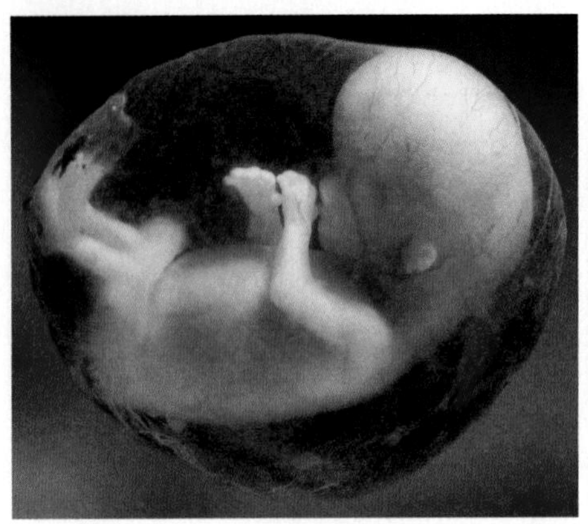

▲ **FIGURE 15.3 Developing Fetus at 20 Weeks.**

CONCEPTION AND DEVELOPMENT (continued)

Fetus

The fetal period lasts from the eighth week until birth. At the eighth week, the heart is beating. By the twelfth week, the bones have begun to calcify, and the external genitalia can be differentiated as male or female. In the fourth month, downy hair called **lanugo** appears on the body. In the fifth month, skeletal muscles become active, and the baby's movements are felt between 16 and 22 weeks of **gestation** (*Figure 15.3*). A protective substance called **vernix caseosa** covers the skin. In the sixth and seventh months, weight gain is increased and body fat is deposited.

At 38 weeks, the baby is at full-term and ready for birth.

Gestation is divided into **trimesters.** The first trimester is up to week 12, the second from weeks 13 to 24, and the third from week 25 to birth.

DISORDERS OF PREGNANCY

Ectopic Pregnancy

If the fallopian tube is obstructed, the fertilized egg will be prevented from moving into the uterus and will continue its development in the fallopian tube. This is called an **ectopic pregnancy.** Tubal disorders that cause ectopic pregnancy include previous salpingitis, pelvic inflammatory disease (PID), and endometriosis.

Abbreviations

OR	operating room
POC	products of conception

Case Report 15.1 (continued)

Mrs. Gloria Maggay's symptoms are those of an ectopic pregnancy. The sudden increase in the pain and the rise in pulse rate can indicate that the tube had ruptured and was hemorrhaging into the abdominal cavity. The gynecologist should see Mrs. Maggay immediately and take her to the operating room **(OR)** for laparoscopic surgery to stop the bleeding and evacuate the products of conception **(POC)**.

Keynote

Preeclampsia and eclampsia threaten the life of both mother and fetus.

Preeclampsia and Eclampsia

Preeclampsia is a sudden, abnormal increase in blood pressure after the 20th week of pregnancy, with edema (swelling) of the face, hands, and feet and proteinuria. Severe preeclampsia can lead to stroke, bleeding disorders, and death of the mother and fetus.

Eclampsia is a life-threatening condition, characterized by the signs and symptoms of preeclampsia, with the addition of convulsions. Management involves immediate admission to the hospital and control of the mother's blood pressure. The baby is delivered as soon as the mother is stabilized, regardless of maturity.

Amniotic Fluid Abormalities

Amniotic fluid abnormalities occur in the second trimester, when the fetus breathes in and swallows amniotic fluid. This promotes development of the gastrointestinal tract and lungs. In the third trimester, the amount of amniotic fluid is about 1 quart and is mostly fetal urine.

Oligohydramnios is too little amniotic fluid. It is associated with an increase in the risk of birth defects and poor fetal growth. Its etiology (cause) is unknown.

Polyhydramnios is too much amniotic fluid. It causes abdominal discomfort and breathing difficulties for the mother. It is associated with **preterm delivery,** placental problems, and poor fetal growth.

WORD ANALYSIS AND DEFINITION

S/ = Suffix P/ = Prefix R/ = Root R/CF = Combining Form

WORD	PRONUNCIATION	ELEMENTS		DEFINITION
eclampsia	ek-**LAMP**-see-uh		Greek *a shining forth*	Convulsions in a patient with preeclampsia.
ectopic	ek-**TOP**-ik	S/ R/	**-ic** *pertaining to* **ectop-** *on the outside, displaced*	Out of place, not in a normal position.
gestation **gestational (adj)**	jes-**TAY**-shun jes-**TAY**-shun-al	S/ R/ S/	**-ion** *action, condition* **gestat-** *pregnancy* **-al** *pertaining to*	From conception to birth. Pertaining to gestation.
lanugo	la-**NYU**-go		Latin *wool*	Fine, soft hair on the fetal body.
oligohydramnios **polyhydramnios**	**OL**-ih-goh-high-**DRAM**-nee-os **POL**-ee-high-**DRAM**-nee-os	P/ R/ R/ P/	**oligo-** *scanty, too little* **-hydr-** *water* **-amnios** *amnion* **poly-** *many, excessive*	Too little amniotic fluid. Too much amniotic fluid.
preeclampsia	pree-eh-**KLAMP**-see-uh	S/ P/ R/	**-ia** *condition* **pre-** *before* **-eclamps** *shining forth*	Hypertension, edema, and proteinuria during pregnancy.
preterm (same as *premature*)	pree-**TERM**	P/ R/	**pre-** *before* **-term** *limit, end*	Baby delivered before 37 weeks of gestation.
trimester	**TRY**-mes-ter		Latin *of 3 months' duration*	One-third of the length of a full-term pregnancy.
vernix caseosa	**VER**-nicks kay-see-**OH**-sah		vernix, Latin *varnish* caseosa, Latin *cheese*	Cheesy substance covering the skin of the fetus.

EXERCISES

Prefixes: *The WAD box contains prefixes that have appeared in previous chapters. In the chart below, the meaning of the prefix is given to you in the first column. List a term with that prefix from the WAD above, and give the meaning of the term. Then add another term with the same prefix from* any *previous chapter, and be sure to write its meaning as well. Fill in the chart.*

Meaning of Prefix	WAD Term with This Prefix	Meaning of This Term	Term from Previous Chapter	Meaning of Term from Previous Chapter
before				
excessive				
many				
new				
too little (scanty)				

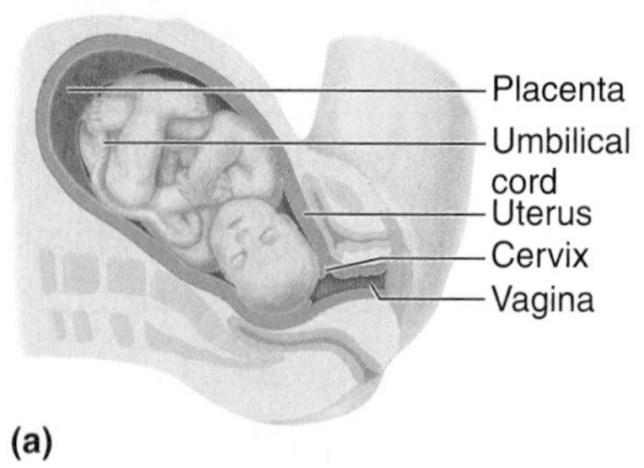

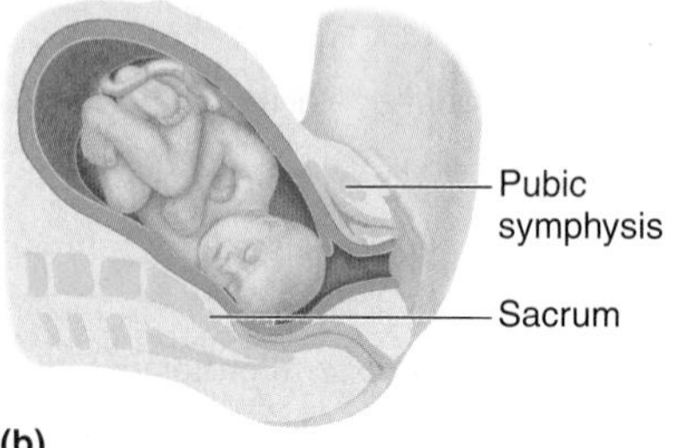

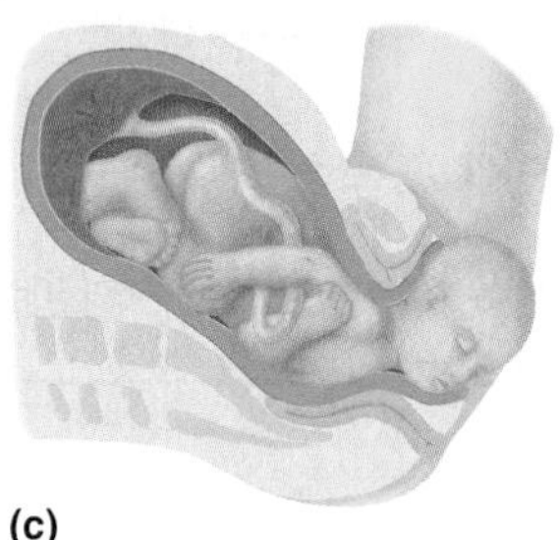

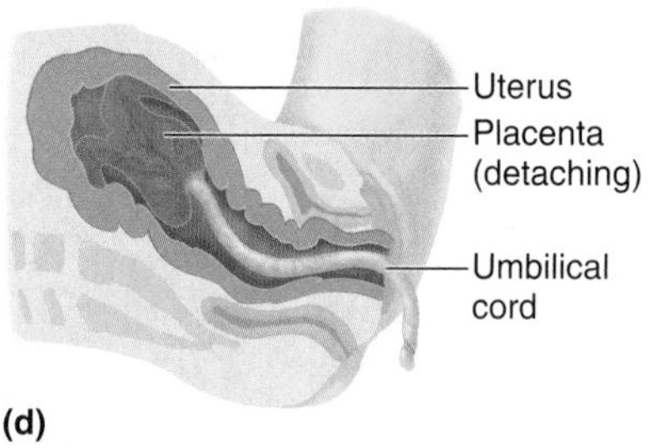

▲ **FIGURE 15.4 The Stages of Childbirth.** (a) First stage: Early dilation. (b) First stage: Late dilation. (c) Second stage: Expulsion of the fetus. (d) Third stage: Expulsion of the placenta.

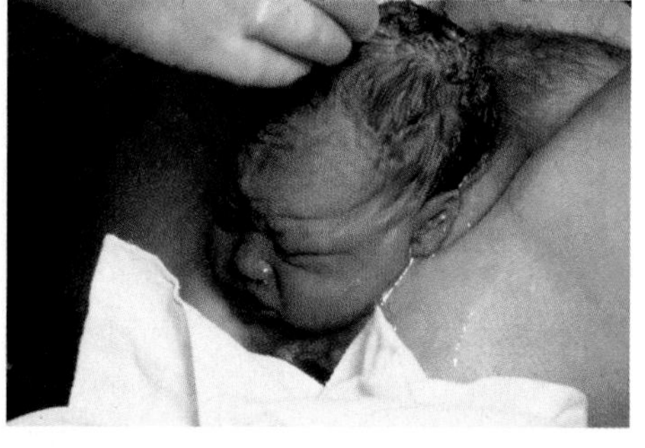

▲ **FIGURE 15.5 Delivery of the Head.**

Abbreviation	
GDM	gestational diabetes mellitus

DISORDERS OF PREGNANCY (continued)

Gestational Diabetes Mellitus (GDM)

In some pregnant women, the amount of insulin they can produce decreases. This leads to hyperglycemia. For the mother, gestational diabetes mellitus **(GDM)** increases the risk of preeclampsia and future type 2 diabetes. For the **neonate,** it increases the risk of **perinatal** mortality, birth trauma, and **neonatal** hypoglycemia. Later in life, both mother and child are at increased risk for developing type 2 diabetes and obesity.

Hyperemesis Gravidarum

Eighty percent of pregnant women experience some degree of "morning sickness." It is at its worst between 2 and 12 weeks and resolves in the second trimester. For a few women, nausea is persistent, and vomiting is extreme and can lead to dehydration. This is **hyperemesis gravidarum.** Severe cases may have to be admitted to the hospital for IV fluids.

Teratogenesis

Teratogenesis is the production of fetal abnormalities—**congenital malformations**—by a chemical agent affecting the mother during the early development of the fetus, while organs and structures are being formed. All medications readily cross the placenta. **Teratogens** include alcohol, isoretinoin (acne medication), valproic acid (anticonvulsant), and the rubella virus.

CHILDBIRTH

Labor contractions begin about 30 minutes apart. They have to be intermittent because each contraction shuts down the maternal blood supply to the placenta and to the fetus. Labor pains are due to ischemia of the uterine muscle during contractions.

Labor is divided into three stages (*Figure 15.4*), each of which is usually longer in a **primipara** (first-time birth) than in a **multipara** (two or more births). Another term for pregnant is **gravid,** and a pregnant woman can be called **gravida.** A first pregnancy can be called a **primigravida.**

First Stage—Dilation of the Cervix

This is the longest stage of labor. It can be a few minutes in a multipara to more than 1 day in a primipara. **Dilation** is the widening of the cervical canal to the same diameter as that of the baby's head (*Figure 15.4b*). At the same time the wall of the cervix becomes thinner, a process called **effacement.** During dilation, the fetal membranes rupture, and the "waters break" as amniotic fluid is released.

Second Stage—Expulsion of the Fetus

As the uterus continues to contract, additional pain is generated by the stretching of the cervix and vagina by the baby's head. When the head reaches the vaginal opening and stretches the vulva, the head is said to be **crowning** and can be delivered (*Figure 15.5*). This process is sometimes helped by performing an **episiotomy,** making an incision in the perineum.

After the baby is delivered, blood in the placental vein is drained into the baby, and the umbilical cord is clamped in two places and cut between the two clamps.

Third Stage—Expulsion of the Placenta

After the baby is delivered, the uterus continues to contract. It pushes the placenta off the uterine wall and expels it out the vagina (*Figure 15.4d*).

Puerperium

The 6 weeks **postpartum** (after the birth) are called the **puerperium.** The uterus shrinks **(involution)** through self-digestion **(autolysis)** of uterine cells by their own enzymes. This generates a vaginal discharge called **lochia** that lasts about 10 days.

WORD ANALYSIS AND DEFINITION

S/ = Suffix P/ = Prefix R/ = Root R/CF = Combining Form

WORD	PRONUNCIATION	ELEMENTS		DEFINITION
autolysis	awe-**TOL**-ih-sis	P/ R/	**auto-** *self, same* **-lysis** *destruction*	Destruction of cells by enzymes within the cells.
dilation	die-**LAY**-shun	S/ R/	**-ion** *process* **dilat-** *open out*	Stretching or enlarging an opening.
effacement	ee-**FACE**-ment	S/ R/	**-ment** *resulting state* **efface-** *wipe out*	Thinning of the cervix in relation to labor.
episiotomy	eh-piz-ee-**OT**-oh-me	S/ R/CF	**-tomy** *surgical incision* **episi/o-** *vulva*	Surgical incision of the vulva.
gravid **gravida**	**GRAV**-id **GRAV**-ih-dah		Latin *pregnant* Latin *pregnant woman*	Pregnant. A pregnant woman.
hyperemesis	high-per-**EM**-ee-sis	P/ R/	**hyper-** *excessive* **-emesis** *vomiting*	Excessive vomiting.
labor	**LAY**-bore		Latin *toil, suffering*	Process of expulsion of the fetus.
lochia	**LOW**-kee-uh		Greek *relating to childbirth*	Vaginal discharge following childbirth.
multipara	mul-**TIP**-ah-ruh	P/ R/	**multi-** *many* **-para** *to bring forth*	Woman who has given birth to two or more children.
neonate	**NEE**-oh-nate	P/ R/	**neo-** *new* **-nate** *born*	A newborn infant.
neonatal (adj) (***Note:*** The "e" in the root is dropped because of the following vowel.)	**NEE**-oh-**NAY**-tal	S/	**-al** *pertaining to*	Pertaining to the newborn infant or the newborn period.
perinatal	per-ih-**NAY**-tal	S/ P/ R/	**-al** *pertaining to* **peri-** around **-nat-** birth	Around the time of birth.
postpartum	post-**PAR**-tum	P/ R/	**post-** *after* **-partum** *childbirth, to bring forth*	After childbirth.
primigravida	pree-mih-**GRAV**-ih-dah	P/ R/	**primi-** *first* **-gravida** *pregnancy*	First pregnancy.
primipara	pry-**MIP**-ah-ruh	P/ R/	**primi-** *first* **-para** *to bring forth*	Woman giving birth for the first time.
puerperium (***Note:*** This term is composed of roots only.)	pyu-er-**PER**-ee-um	R/ R/	**puer-** *child* **-perium** *a bringing forth*	Six-week period after birth in which the uterus involutes.
teratogen	**TER**-ah-toe-jen	S/ R/CF	**-gen** *create, produce* **terat/o** *monster, malformed fetus*	Agent that produces fetal deformities.
teratogenic (adj) **teratogenesis**	**TER**-ah-toe-**JEN**-ik **TER**-ah-toe-**JEN**-eh-sis	S/ S/	**-ic** *pertaining to* **-esis** *condition*	Capable of producing fetal deformities. Process involved in producing fetal deformities.

EXERCISES

Elements: *Several of these elements you have seen before, and you will certainly see them again in other terms. Learn an element once, and recognize it all the time. Circle the best answer to each question.*

1. The term that contains the prefix meaning *many* is:
   primipara lochia multipara
2. The term that contains the suffix meaning *incision* is:
   episiotomy effacement dilation
3. The term that contains the root meaning *to bring forth* is:
   primipara involution effacement
4. The term that contains the root meaning *child* is:
   gravid multipara puerperium
5. The term that relates to *after childbirth* is:
   primipara postpartum gavida
6. The term that contains the prefix meaning *first* is:
   multipara puerperium primipara
7. The term that contains the prefix meaning *self* is:
   multipara autolysis lochia

DISORDERS OF CHILDBIRTH

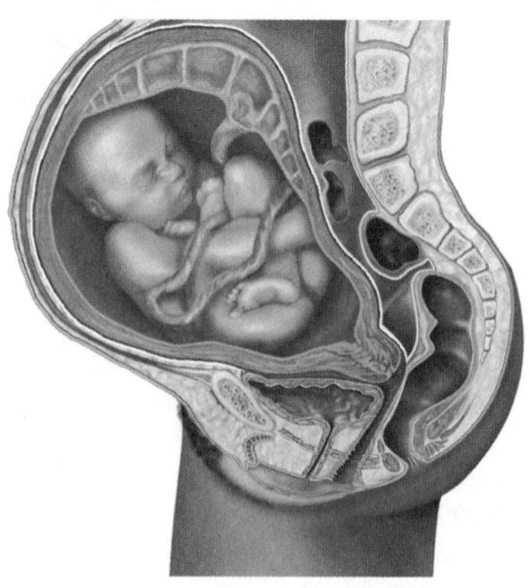

▲ **FIGURE 15.6 Breech Presentation.**

Fetal distress due to lack of oxygen is an uncommon complication of labor, but it is detrimental if not recognized. During labor, there is electronic fetal heart monitoring. Treatment is to give the mother oxygen or increase IV fluids. If distress persists, the baby is delivered as quickly as possible by forceps extraction, vacuum extractor, or **cesarean section (C-section).**

An **abnormal position of fetus** occurs at the beginning of labor if the baby is not a head-first **(vertex)** presentation facing rearward. Abnormal positions include:

- **Breech:** The buttocks present (*Figure 15.6*).
- **Face:** The face, instead of the top of the head, presents.
- **Shoulder:** The shoulder and upper back are trying to exit the uterus first.

If the baby cannot be turned into a vertex presentation, a C-section is usually performed.

In **prolapsed umbilical cord,** the cord precedes the baby down the birth canal. Pressure on the cord can cut off the baby's blood supply, which is still being provided through the umbilical arteries.

In **nuchal cord,** the cord is wrapped around the baby's neck during delivery. This occurs in 20% of deliveries.

Premature rupture of the membranes occurs in 10% of normal pregnancies and increases the risk of infection of the uterus and fetus.

Gestational Classification

Every newborn ***(neonate)*** is either:

- **Premature,** less than 37 weeks gestation.
- **Full-term,** between 37 and 42 weeks gestation.
- **Postterm,** longer than 42 weeks gestation.

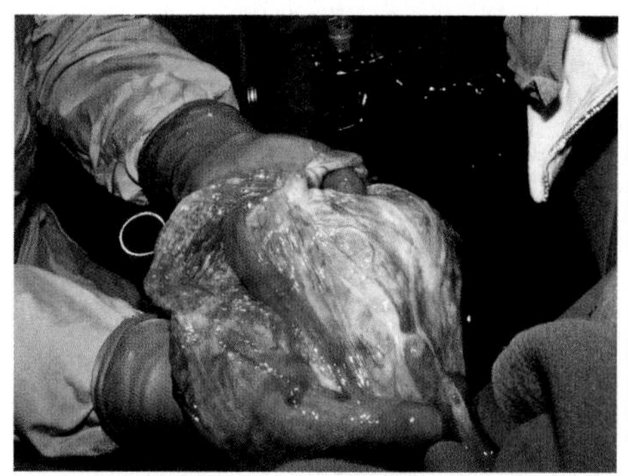

▲ **FIGURE 15.7 Placenta (Afterbirth).**

Prematurity occurs in about 8% of newborns. The earlier the baby is born, the more life-threatening problems occur.

Because their lungs are underdeveloped, premature babies can develop **respiratory distress syndrome (RDS),** also called **hyaline membrane disease.** Their lungs are not mature enough to produce **surfactant,** a mixture of lipids and proteins that keeps the alveoli from collapsing. Respiratory exchange of oxygen and carbon dioxide cannot occur.

An immature liver can impair the excretion of bilirubin (*see Chapter 5*), and premature babies become jaundiced. High levels of bilirubin can produce **kernicterus,** in which deposits of bilirubin in the brain cause brain damage.

Postmaturity is much less common than prematurity. Its etiology is unknown, but the placenta begins to shrink and is less able to supply sufficient nutrients to the baby. This leads to hypoglycemia; loss of subcutaneous fat; dry, peeling skin; and, if oxygen is lacking, fetal distress. The baby can pass stools **(meconium)** into the amniotic fluid. In its distress, the baby can take deep gasping breaths and inhale the meconium fluid. This leads to **meconium aspiration syndrome** and respiratory difficulty at birth.

An **abortion** is the expulsion from the uterus of an embryo or fetus before the 20th week of gestation. It can be spontaneous (occurring from natural causes) or be induced medically or surgically.

Placental Disorders

Placenta abruptio is separation of the placenta from the uterine wall before delivery of the baby. The baby's oxygen supply is cut off, and fetal distress appears quickly. It is an **obstetric (OB)** emergency, and usually a **C-section** is indicated.

Placenta previa is a low-lying placenta between the baby's head and the internal os (opening) of the cervix. It can cause severe bleeding during labor, and a C-section may be necessary.

Abbreviations

C-section	cesarean section
OB	obstetrics
PPH	postpartum hemorrhage
RDS	respiratory distress syndrome

WORD ANALYSIS AND DEFINITION

S/ = Suffix P/ = Prefix R/ = Root R/CF = Combining Form

WORD	PRONUNCIATION	ELEMENTS		DEFINITION
abortion	ah-**BOR**-shun	S/ R/	**-ion** *action* **abort-** *fail at onset*	Spontaneous or induced expulsion of the embryo or fetus from the uterus at 20 weeks or less.
breech	BREECH		Old English *trousers*	Buttocks-first presentation of the fetus at delivery.
cesarean section **C-section (abbrev)**	seh-**ZAH**-ree-an **SEK**-shun		Roman law under the Caesars required that pregnant women who died be cut open and the fetus be extracted.	Extraction of the fetus through an incision in the abdomen and uterine wall.
kernicterus	ker-**NICK**-ter-us	R/ R/	**kern-** *nucleus* **-icterus** *jaundice*	Bilirubin staining of the basal nuclei of the brain.
meconium	meh-**KOH**-nee-um		Greek *a little poppy*	The first bowel movement of the newborn.
nuchal cord	**NYU**-kul KORD		nuchal, French *the back (nape) of the neck*	Loop(s) of umbilical cord around the fetal neck.
placenta abruptio	plah-**SEN**-tah ab-**RUP**-she-oh		abruptio, Latin *to break off*	The premature detachment of the placenta.
placenta previa	plah-**SEN**-tah **PREE**-vee-ah	P/ R/	**pre-** *before, in front of* **-via** *the way*	Placenta obstructing the fetus during delivery.
postmature **postmaturity**	post-mah-**TYUR** post-mah-**TYUR**-ih-tee	P/ R/ S/	**post-** *after* **-mature** *ripe, ready* **-ity** *condition, state*	Infant born after 42 weeks of gestation. Condition of being postmature.
premature (slang term "preemie") **prematurity** (same as *preterm*)	pree-mah-**TYUR** pree-mah-**TYUR**-ih-tee	P/ R/ S/	**pre-** *before* **-mature** *ripe, fully developed* **-ity** *condition, state*	Occurring before the expected time. Condition of being premature.
surfactant	sir-**FAK**-tant		*surface active agent*	A protein and fat compound that creates surface tension to hold the lung alveolar walls apart.
vertex	**VER**-teks		Latin *whorl*	Topmost point of the vault of the skull.

In **retained placenta,** all or part of the placenta and/or membranes remain behind in the uterus 30 minutes to an hour after the baby has been delivered. The result of retained placenta is heavy uterine bleeding called **postpartum hemorrhage (PPH).** Manual removal of the retained product may be necessary under spinal, epidural, or general anesthesia (*Figure 15.7*).

EXERCISES

Patient documentation: *Pick any two terms from the WAD, and use them to create your own patient documentation.* You must be able to pronounce, spell, and define every term you use in your sentences as the instructor may ask you to read them aloud in class. *Fill in the blanks.*

1. ______________________________

____

2. ______________________________

____

LESSON 15.2 The Female Breast

PREGNANCY, CHILDBIRTH, AND THE BREAST

OBJECTIVES

Until the relatively recent introduction of bottles filled with liquid supplied by cows or soybeans, the milk produced by the female breast was essential for the survival of the human species. Nourishment of the infant remains the breast's major function, even though the breast is a visible, tangible, and beautiful symbol of femininity. It is important to complete your understanding of the female reproductive system by being able to:

- **Describe the medical terminology of the anatomy of the breast and the mammary gland.**
- **Explain the physiology and mechanisms of lactation using correct medical terminology.**
- **Apply correct medical terminology to common disorders of the breast.**

You are

. . . a surgical technologist working with Charles Walsh, MD, a surgeon at Fulwood Medical Center.

You are communicating with

. . . Mrs. Victoria Post, a 62-year-old woman who c/o finding a lump in her right breast.

CASE REPORT 15.2

Mrs. Post found the lump in her right breast together with some lumps in her right armpit on self-examination. She had not examined her breasts for about 6 months. She began her menopause at age 52. Her mother and a sister died of breast cancer. Physical examination shows a 2-cm firm, painless lesion in the upper outer quadrant of her right breast and several small, painless lymph nodes in her right axilla. Dr. Walsh has scheduled her for a biopsy of the breast lesion and for DNA testing.

Keynote

There is no relation between breast size and the ability to breastfeed.

Your roles are to keep Mrs. Post informed of the need for the diagnostic procedures, their results, and indications for future treatment and to document her care.

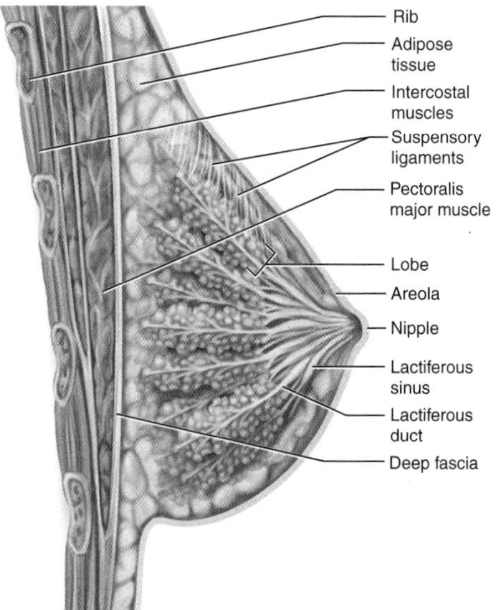

▲ **FIGURE 15.8 Anatomy of the Lactating Breast.**

THE BREAST

Anatomy of the Breast

The breasts of males and females are identical until puberty, when ovarian hormones stimulate the development of the breasts in females.

Each adult female breast has a **body** located over the pectoralis major muscle and an **axillary tail** extending toward the armpit. The **nipple** projects from the breast and contains multiple openings of the main milk ducts. The reddish-brown **areola** surrounds the nipple. The small bumps on its surface are **areolar glands.** These are sebaceous glands whose secretions prevent chapping and cracking during breastfeeding.

The **nonlactating breast** consists mostly of adipose and connective tissues. It has a system of ducts that branch through the connective tissue and converge on the nipple.

Mammary Gland

When the **mammary** gland develops during pregnancy, it is divided into 15 to 20 lobes that contain the secretory **alveoli** that produce milk. Each lobe is drained by the main milk ducts, called **lactiferous ducts** (*Figure 15.8*). Immediately before opening onto the nipple, each lactiferous duct dilates to form a **lactiferous sinus** in which milk is stored before being released from the nipple.

WORD ANALYSIS AND DEFINITION

S/ = Suffix P/ = Prefix R/ = Root R/CF = Combining Form

WORD	PRONUNCIATION	ELEMENTS		DEFINITION
apoptosis	**AP**-op-**TOE**-sis	P/ R/	**apo-** *separation from* **-ptosis** *drooping*	Programmed normal cell death.
areola **areolar (adj)**	ah-**REE**-oh-lah ah-**REE**-oh-lar	 S/ R/	Latin *small area* **-ar** *pertaining to* **areol-** *areolar*	Circular reddish area surrounding the nipple. Pertaining to the areola.
colostrum	koh-**LOSS**-trum		Latin *foremilk*	The first breast secretion at the end of pregnancy.
lactation (noun) **lactate (verb)** **lactiferous (adj)**	lak-**TAY**-shun **LAK**-tate lak-**TIF**-er-us	S/ R/ S/ S/ R/	**-ation** *a process* **lact-** *milk* **-ate** *composed of, pertaining to* **-ous** *pertaining to* **-ifer-** *to bear, carry*	Production of milk. To produce milk. Pertaining to or yielding milk.
mammary	**MAM**-ah-ree	S/ R/	**-ary** *pertaining to* **mamm-** *breast*	Pertaining to the lactating breast.
nipple	**NIP**-el		Old English *small nose*	Projection from the breast into which the lactiferous ducts open.
oxytocin	**OCK**-see-**TOE**-sin	S/ R/ R/	**-in** *substance* **oxy-** *oxygen* **-toc-** *labor*	Pituitary hormone that stimulates the uterus to contract.
prolactin	pro-**LAK**-tin	S/ P/ R/	**-in** *substance* **pro-** *before* **-lact-** *milk*	Pituitary hormone that stimulates the production of milk.

Lactation

In late pregnancy, the secretory alveoli and lactiferous ducts contain **colostrum.** This secretion contains more protein but less fat than human milk, but it also contains high levels of **immunoglobulins** (*see Chapter 11*) that give the infant protection from infections. Colostrum begins to be replaced by milk 2 or 3 days after the baby's birth, and this replacement is complete by day 5.

Milk production is mainly controlled by **prolactin,** a hormone from the pituitary gland. The other essential stimulus to milk production is the baby's sucking (*Figure 15.9*), which stimulates prolactin production. In addition, the **sucking reflex** stimulates the pituitary gland to produce **oxytocin** (*see Chapter 10*), which causes milk to be ejected from the alveoli into the duct system.

Milk production continues as long as the baby suckles at least twice daily. In the United States only 20% of babies are still breastfeeding at 6 months. The American Academy of Pediatrics recommends breastfeeding for 6 to 12 months.

After complete cessation of lactation, involution of the mammary gland occurs. The epithelial cells of the alveoli are lost through **apoptosis** (programmed cell death), the ducts shrink in size, and adipose and connective tissues return to be the major breast tissues.

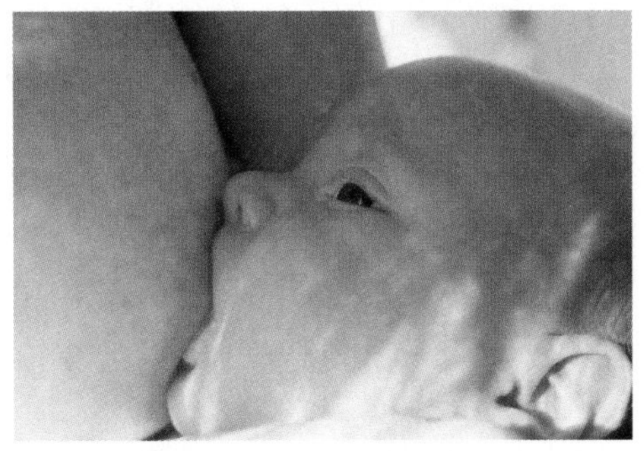

▲ **FIGURE 15.9 Breastfeeding.**

Keynote

Breastfeeding is not a reliable means of contraception. It has a failure rate of around 10%.

EXERCISES

Spelling *your documentation correctly is a mark of an educated professional. Read the following statements, and insert the correctly spelled term in the blanks.*

1. The circular, reddish area surrounding the nipple is the *(aireola/areola)* ____________________.
2. After complete cessation of lactation, involution of the *(mamery/mammary)* ____________________ gland occurs.
3. Programmed cell death is known as *(apotosis/apoptosis)* ____________________.
4. The first breast secretion at the end of pregnancy is known as *(colestrium/colostrum)* ____________________.

DISORDERS OF THE BREAST

Mastitis, inflammation of the breast, can occur in association with breastfeeding if the nipple or areola is cracked or traumatized. It is not an indication for stopping breastfeeding.

Mastalgia (breast pain) is the most common benign breast disorder. The pain can be associated with breast tenderness or be part of PMS.

Paget disease of the nipple presents as a scaling, crusting lesion of the nipple, sometimes with a discharge from the nipple (*Figure 15.10*). It is indicative of an underlying cancer that has to be the focus of diagnosis and treatment.

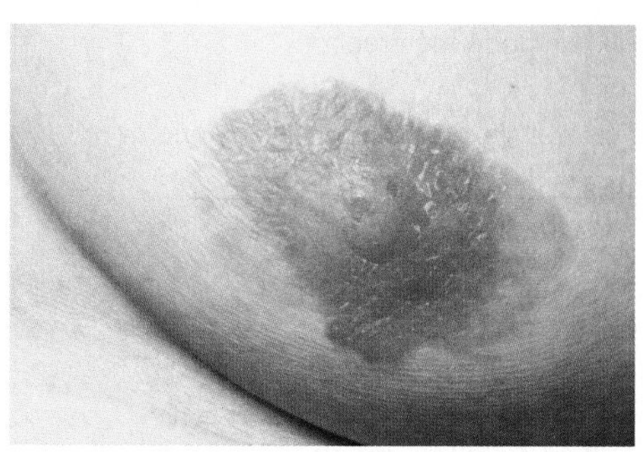

▲ **FIGURE 15.10 Paget Disease of the Nipple Is Associated with Breast Cancer.**

Nipple discharge, particularly if it is from one breast and bloody, is an indication of an underlying disorder such as breast cancer and warrants investigation.

Fibroadenomas are circumscribed, small, benign tumors that can be either cystic or solid and can be multiple. They can be excised surgically.

Fibrocystic breast disease presents as a dense, irregular, cobblestone consistency of the breast, often with intermittent breast discomfort. It occurs in over 60% of all women and is considered by many doctors as a normal variant.

Breast cancer affects one in eight women in their lifetimes. Risk factors include a family history, particularly if a woman carries either the **BRCA1** or **BRCA2 gene;** the use of postmenopausal estrogen therapy; and an early menarche and late menopause.

Most breast cancers are discovered as a lump by the patient, which is why **monthly breast self-examinations (BSEs)** are so important. Another 40% are discovered on routine **mammograms** (*Figure 15.11*). Routine **mammography** reduces breast cancer mortality by 25% to 30%. Breast cancer occurs rarely in the male.

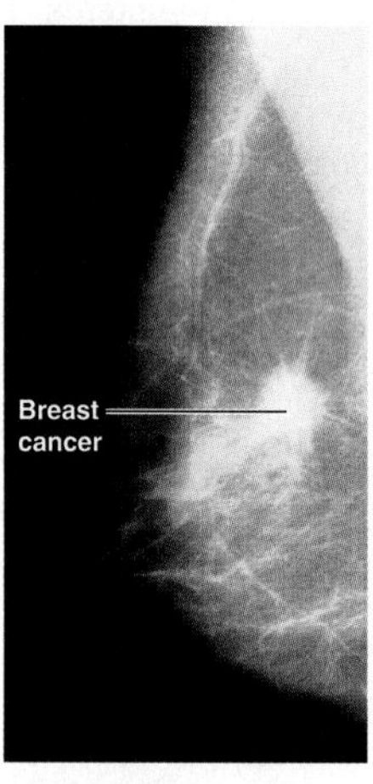

▲ **FIGURE 15.11 Mammogram Showing Breast Cancer.**

Most breast cancers occur in the upper and outer quadrant of the breast. If cancer is suspected, biopsy should be planned. This is being performed more and more often as a **stereotactic biopsy,** a needle biopsy performed during mammography. Breast cancer can metastasize to lymph nodes, lungs, liver, bone, brain, and skin. The surgical treatments for breast cancer include:

- **Lumpectomy:** Removal of the tumor with preservation of the surrounding breast tissue.
- Simple **mastectomy:** Removal of the breast with skin and nipple.
- Modified radical mastectomy: Simple mastectomy with lymph node dissection.
- Radical mastectomy: Modified mastectomy with removal of the pectoralis major muscle.

Abbreviations

BRCA1, BRCA2	breast cancer genes
BSE	breast self-examination

Case Report 15.2 (continued)

Mrs. Victoria Post's biopsy confirmed that the lump was cancer. DNA testing showed the presence of the BRCA1 gene, which explained the occurrence of breast cancer in her mother and sister. Mrs. Post is to attend a consultation with her surgeon Dr. Walsh and **geneticist** Ingrid Hughes, MD, PhD, to determine what treatment should be undertaken. For example, because of her high risk of cancer, should her left breast, currently cancer-free, be removed? Because with BRCA1 there is a 20% to 40% risk of ovarian cancer, should her ovaries also be removed?

Additional radiotherapy and chemotherapy are also used.

Galactorrhea occurs when a woman makes milk when she is not breastfeeding. Sometimes the cause cannot be found, but the disorder can occur in association with hormone therapy, antidepressants, tumor of the pituitary gland, and use of street drugs such as opiates and marijuana. In most cases, the milk production ceases with time.

Gynecomastia, enlargement of the breast, can be unilateral or bilateral and can occur in both sexes. It is usually associated with either liver disease, marijuana, or drug therapy such as estrogens, calcium channel blockers, and **antineoplastic** drugs. It remits or disappears after the drug is withdrawn. Occasionally cosmetic surgery is needed.

WORD ANALYSIS AND DEFINITION

S/ = Suffix P/ = Prefix R/ = Root R/CF = Combining Form

WORD	PRONUNCIATION	ELEMENTS		DEFINITION
antineoplastic	**AN**-tee-nee-oh-**PLAS**-tik	S/ P/ R/CF R/	**-tic** *pertaining to* **anti-** *against* **-ne/o-** *new* **-plas-** *growth*	Pertaining to the prevention of the growth and spread of cancer cells.
fibroadenoma	**FIE**-broh-ad-en-**OH**-muh	S/ R/CF R/	**-oma** *tumor* **fibr/o-** *fiber* **-aden-** *gland*	Benign tumor containing much fibrous tissue.
fibrocystic disease	fie-broh-**SIS**-tik **DIZ**-eez	S/ R/CF R/	**-ic** *pertaining to* **fibr/o-** *fiber* **-cyst-** *cyst*	Benign breast disease with multiple tiny lumps and cysts.
galactorrhea	gah-**LAK**-toe-**REE**-ah	R/ R/CF	**-rrhea** *flow* **galact/o-** *milk*	Abnormal flow of milk from the breasts.
gene **genetics** **geneticist**	JEEN jeh-**NET**-iks jeh-**NET**-ih-sist	 S/ R/ S/	Greek *origin, birth* **-ics** *knowledge* **genet-** *origin* **-ist** *specialist*	Functional segment of DNA molecule. Science of the inheritance of characteristics. A specialist in genetics.
gynecomastia	**GUY**-nih-koh-**MAS**-tee-ah	S/ R/CF R/	**-ia** *condition* **gynec/o-** *female* **-mast-** *breast*	Enlargement of the breast.
lumpectomy	lump-**ECK**-toe-me	S/ R/	**-ectomy** *surgical excision* **lump-** *piece*	Removal of a lesion with preservation of surrounding tissue.
mammogram **mammography**	**MAM**-oh-gram mah-**MOG**-rah-fee	S/ R/CF S/	**-gram** *a record* **mamm/o-** *breast* **-graphy** *process of recording*	The record produced by x-ray imaging of the breast. Process of x-ray examination of the breast.
mastalgia	mass-**TAL**-jee-uh	S/ R/	**-algia** *pain* **mast-** *breast*	Pain in the breast.
mastectomy	mass-**TECK**-toe-me	S/ R/	**-ectomy** *surgical excision* **mast-** *breast*	Surgical excision of the breast.
mastitis	mass-**TIE**-tis	S/ R/	**-itis** *inflammation* **mast-** *breast*	Inflammation of the breast.
stereotactic	**STER**-ee-oh-**TAK**-tic	S/ R/ R/CF	**-ic** *pertaining to* **-tact-** *orderly arrangement* **stere/o-** *three-dimensional*	A precise three-dimensional method to locate a lesion.

EXERCISES

Suffixes *always provide a big clue to the meaning of a medical term. Review all the terms in the WAD above, and then answer the following questions. All of these questions can be answered by analyzing the suffix in each term.*

Which of the terms in this WAD means:

1. a condition of enlargement of the breast? ____________________
2. a glandular tumor? ____________________
3. process of an x-ray examination of the breast? ____________________
4. knowledge of the inheritance of characteristics? ____________________
5. an inflammation of the breast? ____________________
6. breast pain? ____________________

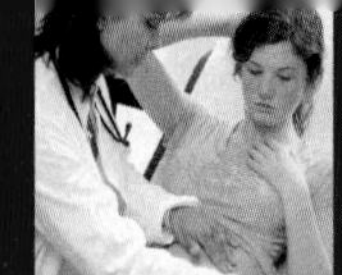

CHAPTER 15 REVIEW

PREGNANCY, CHILDBIRTH, AND THE BREAST

CHALLENGE YOUR KNOWLEDGE

A. **Patient documentation:** The following mammogram report for a patient contains terminology you should understand after reading this and previous chapters in this text. Answer the questions about the report by filling in the blanks.

> **Exam:** Diagnostic bilateral mammogram
>
> **Reason for exam:** Fibrocystic breasts
>
> The glandular tissue is heterogeneously dense. There are scattered fibroglandular densities bilaterally. There is a benign-appearing curvilinear area of glandular tissue in the right lower inner breast. There are a few faint calcifications in the lower inner left breast and in the central left breast. Magnification views were performed of the left breast calcifications in two positions, and they have an appearance most consistent with a benign process. There is no associated mass or distortion.
>
> **Impression:** Heterogeneously dense glandular tissue with focal glandular density in the medial aspect of the right breast and faint calcifications in the left breast. Probably benign.
>
> **Recommendation:** Comparison with prior films. If prior films are not available, recommend follow-up mammogram in 6 months. Findings were submitted to the patient in writing.

1. Define *bilateral:* ______________________________

____

2. Deconstruct the term *fibroglandular* by putting slashes between each element:

______________ / ______________ / ______________
P R/CF S

Define *fibroglandular:*

____

3. Describe where the "*medial* aspect of the right breast" would be:

____

____

4. Look up the term *calcification* in the Glossary or a dictionary and describe it here: ______________________________

____

5. What is a *benign* process? ______________________________

6. Were any malignancies found in this report? ______________________________

7. What makes this a *diagnostic* procedure? ______________________________

____

Discussion questions on this report:

8. Why does the radiologist recommend follow-up in 6 months if no previous films are available?

____

____

9. Although this was a routinely scheduled mammogram, what breast symptoms might also require a mammogram?

____

____

10. Why is a new film always compared to a previous one?

____

____

B. **Word attack.** This exercise teaches you to think logically about analyzing a medical term.

Which term means *too little* of something?

a. polyhydramnios

b. fibroadenoma

c. endometriosis

d. multipara

e. oligohydramnios

1. Read the question and the possible answers; then read them again.

2. The words in the question—*too little*—are describing the quantity of something. *Words* that describe are adjectives, but *elements* that describe how big, how little, the color of something, where something is located, and the like, are usually prefixes.

3. Here is where your knowledge of elements is important. Slash each term into elements. Start by analyzing just the prefixes.

4. Does every term in this question have a prefix? ______________________ If not, which terms do not have prefixes?

____

5. Analyze the *prefix only* in each of the remaining terms.

6. Which term has the prefix that means *too little?* ____

7. Therefore, the answer is ____.

Note: Remember that another way of saying "too little" would be to say "deficient" *(hypo)*, so if there had been an answer choice beginning with the element *hypo*, it could have been the correct answer.

Elements can have similar meanings, but they are not interchangeable in terms.

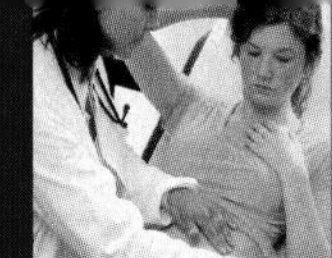

CHAPTER 15 REVIEW

PREGNANCY, CHILDBIRTH, AND THE BREAST

C. **Roots/combining forms** are the core foundation of every term. Circle just the roots or combining forms in the following terms, and give their meanings on the lines provided.

1. *blastocyst:* ____________________
2. *gestation:* ____________________
3. *ectopic:* ____________________
4. *oligohydramnios:* ____________________
5. *teratogenic:* ____________________
6. *neonatal:* ____________________
7. *episiotomy:* ____________________
8. *amniocentesis:* ____________________
9. *obstetrician:* ____________________
10. *apoptosis:* ____________________
11. *fibrocystic:* ____________________
12. *gynecologist:* ____________________
13. *mastectomy:* ____________________
14. *postpartum:* ____________________
15. *neonatal:* ____________________

D. Language of obstetrics: Pregnancy has its own associated set of obstetric terms. Apply your knowledge of obstetric terms to answering the following questions. Write the correct elements on the line, and the identity (P, R, CF, or S) of the element below the line. *Remember: Some terms may not contain every element.*

1. disorder in which a fertilized egg continues its development in the fallopian tube instead of the uterus:

   ______________________ / ______________________ / ______________________

2. woman giving birth for the first time:

   ______________________ / ______________________ / ______________________

3. medical term that relates to measured time of a pregnancy:

   ______________________ / ______________________ / ______________________

4. removal of amniotic fluid for diagnostic purposes:

   ______________________ / ______________________ / ______________________

5. from conception to birth:

   ______________________ / ______________________ / ______________________

6. first 2 weeks of the developing embryo:

   ______________________ / ______________________ / ______________________

7. too much amniotic fluid:

   ______________________ / ______________________ / ______________________

8. hypertension, edema, and proteinuria during pregnancy:

   ______________________ / ______________________ / ______________________

9. birth occurring before 37 weeks of gestation:

   ______________________ / ______________________ / ______________________

10. destruction of cells by enzymes within the cells:

    ______________________ / ______________________ / ______________________

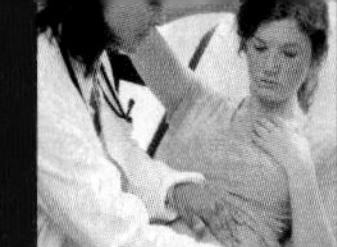

PREGNANCY, CHILDBIRTH, AND THE BREAST

E. **Spelling demons:** Every chapter contains spelling demons. Challenge your ability to spell all these terms correctly. Circle the best choice.

1.	choirionic	khorionic	chorionic
2.	pre-eclampsia	preeclampsia	preclampsia
3.	oligohydramnios	oligohydromios	oligohydrammios
4.	puerperpium	puerperium	purperpium
5.	cesarian	cesarrian	cesarean
6.	obstrician	obstetrican	obstetrician
7.	apotosis	apoptossis	apoptosis
8.	galactorrhea	galactorhea	galactorea
9.	hyperemessis	hyperemmesis	hyperemesis
10.	gynecomastia	gyncomastia	gynecomasstia

Write the other terms from this chapter that you found the most difficult to spell and give their definitions.

Term: ______________________ Definition: ______________________

____

Term: ______________________ Definition: ______________________

____

F. **Deconstruct** the following medical terms into their basic word elements, and write the meaning of each element in the last column. The first one is done for you. Fill in the chart.

Medical Term	Prefix	Root(s)/CF	Suffix	Meaning of Element
prematurity	pre			before
		matur		ripe
			ity	condition
galactorrhea				
lactiferous				
mammary				
mastalgia				
amniocentesis				
fertilization				
ectopic				
gynecomastia				
multipara				

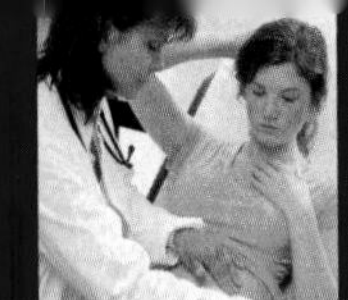

CHAPTER 15 REVIEW

PREGNANCY, CHILDBIRTH, AND THE BREAST

G. **Definitions:** The following medical terms do not deconstruct into elements, so there is no clue as to their meaning. You must know them for what they are. Define the following terms on the lines below.

1. *chorion:* ______________________

2. *vertex:* ______________________

3. *embryo:* ______________________

4. *fetus:* ______________________

5. *lochia:* ______________________

6. *breech:* ______________________

7. *placenta:* ______________________

8. *meconium:* ______________________

9. *gravida:* ______________________

10. *areola:* ______________________

H. Terminology challenge: *Autolysis* is a new term you learned in this chapter. It is composed of two elements you have met before in previous chapters. Recall other terms you have already learned that have those elements. Fill in the blanks.

1. Slash the term *autolysis,* and write the meaning of each element below the line.

   ______________________ / ______________________ / ______________________

Other terms with the same elements:

2. *Auto:* ______________________

3. *Lysis:* ______________________

Brain teaser: Why is the term *autolysis* like the term *apoptosis?*

4. ______________________

   ____

   ____

I. Translate the following sentences from medical language to layperson's language and from layperson's language to medical language.

Remember to check your spelling, and you should be able to pronounce everything you write. You may be called on to read your sentences in class.

1. From medical language to layperson's language:

   "For the neonate, it increases the risk of perinatal mortality, birth trauma, and neonatal hypoglycemia."

   ____

   ____

   ____

2. From layperson's language to medical language:

   "The signs of this condition are a sudden, abnormal increase in blood pressure after the 20th week of pregnancy, with swelling of the face, hands, and feet and protein in the urine."

   ____

   ____

   ____

   ____

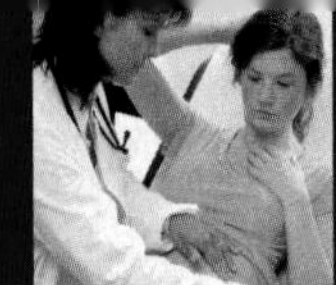

PREGNANCY, CHILDBIRTH, AND THE BREAST

J. **Prefixes:** Some of the same prefixes continue to appear throughout these chapters and always have the same meaning. Compare the meanings of medical terms from this chapter with those of medical terms that are from previous chapters and have the same prefix. Fill in the blanks.

Element/Term	Meaning of Element and Term
1. hyper	______________________
hyperemesis	______________________
hypertension	______________________
2. neo	______________________
neonatal	______________________
neoplasm	______________________
3. oligo	______________________
oligohydramnios	______________________
oliguria	______________________
4. poly	______________________
polycystic	______________________
polyphagia	______________________
5. pre	______________________
preterm	______________________
prepatellar	______________________
6. auto	______________________
autolysis	______________________
autologous	______________________
7. post	______________________
postpartum	______________________
postprandial	______________________
8. pro	______________________
prolactin	______________________
prognosis	______________________

9. anti ____________________

antineoplastic ____________________

anticoagulant ____________________

10. peri ____________________

perinatal ____________________

pericardium ____________________

K. Short answer: If you understand the questions below and the terminology well enough, you can explain the concepts to someone else, as you will often have to do on the job. Write your explanation on the lines provided.

1. Explain how the difference of one word element changes the meanings of these words: *conception* *contraception*

____

____

____

2. A new patient's chart says she is "gravida 4 para 3." What does that mean?

____

____

____

____

L. Recall and review: How well do you remember these word elements from the previous chapter? Try to answer without looking back to check. Write your explanation on the lines provided.

Element	Type of Element	Meaning of Element
1. metri	______	______
2. pareun	______	______
3. oma	______	______
4. syn	______	______
5. dys	______	______

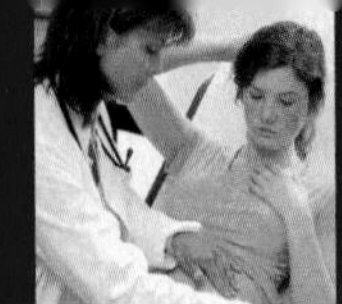

CHAPTER 15 REVIEW

PREGNANCY, CHILDBIRTH, AND THE BREAST

M. **Build** the following medical terms using the definitions as clues. Write the correct elements on the line, and the meaning of each element under the line. Fill in the blanks.

1. thinning of cervix in labor ____________/____________/____________
2. self-digestion ____________/____________/____________
3. giving birth for the first time ____________/____________/____________
4. after childbirth ____________/____________/____________
5. 6-week period after birth ____________/____________/____________
6. surgical incision to enlarge the opening of the vulva ____________/____________/____________
7. newborn infant ____________/____________/____________
8. occurring before the expected time ____________/____________/____________
9. stimulates production of milk ____________/____________/____________
10. inflammation of the breast ____________/____________/____________
11. removal of amniotic fluid ____________/____________/____________
12. pertaining to pregnancy ____________/____________/____________
13. too much amniotic fluid ____________/____________/____________
14. excessive vomiting ____________/____________/____________

N. **Abbreviations** must be used carefully to convey the correct information. Circle the best choice to complete the sentence.

1. This is the patient's first (OB/PGY/RDS).
2. The patient has developed (OR/POC/GDM) in her last trimester.
3. The emergency (C-section/STAT/OB) will be scheduled for the first available (OR/OB/OA).
4. Order 3 units of blood for the patient having the (POC/RDS/PPH).
5. (GDM/RDS/BSE) can save a life.

O. **Partner exercise:** Write a multiple-choice question using terms from this chapter. You must have five possible answers. Give your question to your partner, and you answer his or her question. Check your spelling of the terms before submitting your question.

1. Question:

____

____

____

____

____

a. ______________________________

b. ______________________________

c. ______________________________

d. ______________________________

e. ______________________________

2. Ask your partner to write a multiple-choice question using five different terms from this chapter for you to answer, spell, and pronounce to your partner.

Question:

____

____

____

____

a. ______________________________

b. ______________________________

c. ______________________________

d. ______________________________

e. ______________________________

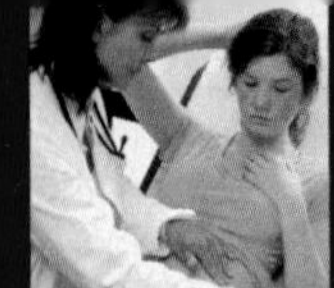

PREGNANCY, CHILDBIRTH, AND THE BREAST

P. **Chapter challenge:** Circle the correct answer.

> **Study Hint**
> Immediately cross off any answer you know is not correct. In your remaining choices, there is only *one best answer.*

1. When does the embryo cease being called an embryo and start being called a fetus?
   a. after delivery
   b. after the eighth week
   c. after the third trimester
   d. just before delivery
   e. after the first month
2. Fine, soft hair on the fetal body is called:
   a. lochia
   b. meconium
   c. lanugo
   d. kernicterus
   e. vernix caseosa
3. What is a vertex presentation?
   a. Head presents first, facing rearward.
   b. Shoulder and upper back present first.
   c. Fetus presents face first.
   d. Head presents first, facing downward.
   e. Buttocks present first.
4. Identify the pair of incorrect spellings:
   a. galactorrhea gynecomastia
   b. antineoplastic excisional
   c. fibroadenoma mammogram
   d. mammography stereotactic
   e. falopian corionic
5. The letter "H" in the abbreviation *PPH* stands for:
   a. hemorrhage
   b. hyperemesis
   c. hypoglycemia
   d. hyaline
   e. hysterecomy
6. When the head of the fetus reaches the vaginal opening, the condition is termed:
   a. crowning
   b. effacement
   c. apoptosis
   d. autolysis
   e. involution
7. Previous salpingitis, PID, or endometriosis can sometimes be a cause of:
   a. mastalgia
   b. mastitis
   c. cervical cancer
   d. ectopic pregnancy
   e. eclampsia

8. The protective substance covering the skin of the fetus is called:

a. vernix caseosa
b. chorion
c. lanugo
d. lochia
e. kernicterus

9. Which term has embedded within it another complete term (a word within a word that can stand by itself)?

a. lactiferous
b. apoptosis
c. mammography
d. mastectomy
e. stereotactic

10. Which diagnosis is considered a life-threatening condition?

a. endometriosis
b. hyperemesis gravidarum
c. mastalgia
d. eclampsia
e. mastitis

11. *Nuchal cord* is:

a. present only in ectopic pregnancy
b. wrapped around the baby's neck during delivery
c. a congenital malformation
d. only present in breech births
e. present in 50% of births

12. What is usually performed if a fetus cannot be turned into vertex presentation?

a. hysterectomy
b. C-section
c. amniocentesis
d. mastectomy
e. episiotomy

13. The first bowel movement of the newborn is:

a. kernicterus
b. meconium
c. oxytocin
d. surfactant
e. colostrum

14. One abbreviation does *not* belong in the following group: OB, PGY, RDS, PPH, POC. Why?

a. It represents a different body system.
b. Everything else is a diagnosis; this is a procedure.
c. It's an abbreviation for a class of drugs.
d. Everything else is a procedure; this is a diagnosis.
e. It's an abbreviation for a radiological procedure.

15. Which disease/condition is indicative of an underlying cancer that has to be the focus of diagnosis and treatment?

a. fibrocystic breast disease
b. galactorrhea
c. Paget disease
d. ectopic pregnancy
e. fibroadenoma

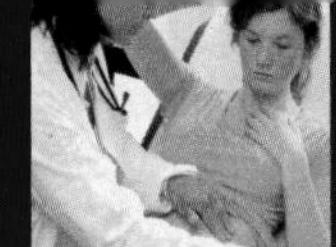

CHAPTER 15 REVIEW

PREGNANCY, CHILDBIRTH, AND THE BREAST

Q. Case Report challenge: Now that you are more comfortable with the terms in this chapter, you can apply that knowledge and briefly answer the questions about the Case Report. If you read the report through and underline all the medical terminology, this will make it easier to answer the questions.

CASE REPORT 15.1

You are

. . . an obstetric assistant employed by Garry Joiner, MD, an obstetrician at Fulwood Medical Center.

You are communicating with

Mrs. Gloria Maggay, a 29-year-old housekeeper. Her last menstrual period was 8 weeks ago, and she has a positive home pregnancy test. This is her first pregnancy. She has breast tenderness and mild nausea. For the past 2 days she has had some cramping, right-sided, lower abdominal pain and this morning had vaginal spotting. Her vital signs are T 99°F, P 80, R 14, BP 130/70.

While you are waiting for Dr. Joiner to come and examine her, Mrs. Maggay complains of feeling faint and has a sharp, severe pain in her lower abdomen on the right side. Her pulse rate has increased to 92. You need to recognize what is happening.

Mrs. Gloria Maggay's symptoms are those of an ectopic pregnancy. The sudden increase in the pain and the rise in pulse rate can indicate that the tube had ruptured and was hemorrhaging into the abdominal cavity. The obstetrician should see Mrs. Maggay immediately and take her to the operating room for laparoscopic surgery to stop the bleeding and evacuate the products of conception.

1. List all of Mrs. Maggay's symptoms:

____

____

____

____

2. There are abbreviations that could be substituted for medical terms in the above case report. Write them here:

____

____

____

3. What body quadrant has the "sharp, severe pain"? Use the abbreviation:

____

4. Mrs. Maggay is hemorrhaging into the abdominal cavity—what does this mean?

____

____

5. Name one other body cavity: ____________________

Name one organ located in this cavity: ____________________

6. What "tube" has ruptured? ____________________

7. In this case, what happened when the tube ruptured?

____

____

8. Define *ectopic pregnancy:*

____

____

____

9. Why, then, has the tube ruptured?

____

____

10. What type of instrument will be used in Mrs. Maggay's surgery?

____

11. What medical term can be applied to Mrs. Maggay since this is her first pregnancy?

____

12. In Vital Signs, what do the following abbreviations stand for?

T ____________________

P ____________________

R ____________________

BP ____________________

Life Span

The Essentials of the Languages of Pediatrics and Gerontology

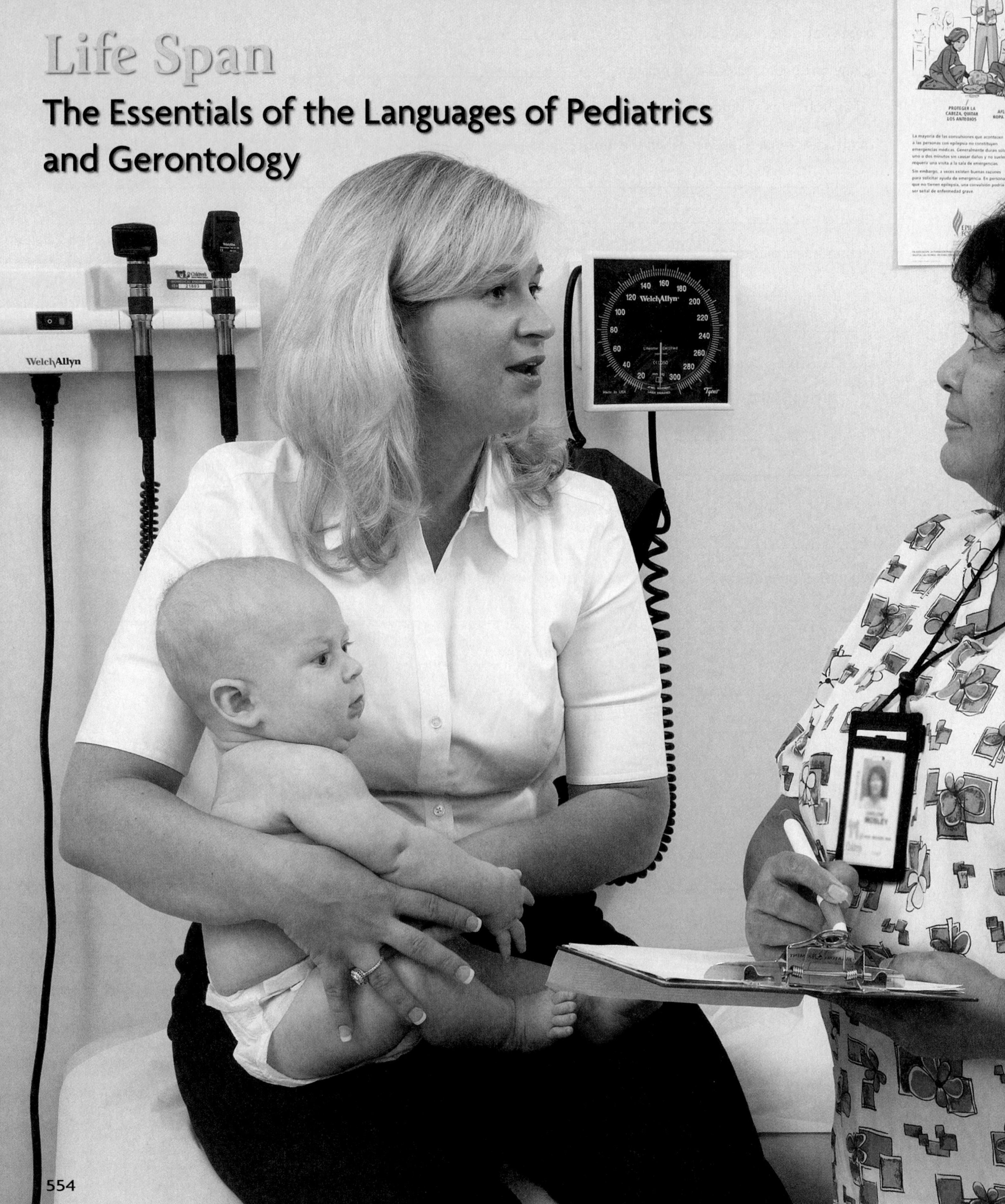

NEONATAL ADAPTATIONS (continued)

Cardiovascular System (CVS) Adaptations

The fetus is dependent on the **placenta** and umbilical cord to provide oxygen and nutrients and to remove carbon dioxide and fetal wastes. At birth, the two major divisions of the cardiovascular system (**CVS**), the pulmonary and systemic circulations (*see Chapter 6*), become separate and operational.

Congenital cardiovascular defects are present in about 1% of births. Before birth a **patent** (open) vessel (the ductus arteriosus) connects the aorta and pulmonary artery. Normally this closes within a few hours of birth. When it doesn't close, the patent ductus arteriosus **(PDA)** allows blood that should flow through the aorta and nourish the body to return to the lungs. Children with a PDA grow slowly, tire easily, and develop pneumonia. A small patent ductus can close spontaneously. A ductus can be closed with medication, by surgically tying it, or by inserting a plug.

Septal defects occur when the baby is born with an opening ("hole in the heart") in the septum (wall) that separates the right and left sides of the heart. An opening between the two upper chambers is called an *atrial septal defect* **(ASD).** An opening between the two lower chambers is called a *ventricular septal defect* **(VSD).** Small defects often close spontaneously during the first year of life. If not, the defects can be closed by surgery.

Cyanosis occurs when insufficient blood is being pumped to the lungs and thus the blood being pumped to the body contains less oxygen than the tissues need. Neonates with cyanosis are called "blue babies" because of their **cyanotic** blue skin (*Figure 16.3*). One type of cyanotic heart disease is the **tetralogy of Fallot (TOF),** in which four heart defects all shunt blood away from the lungs. Children with TOF do not grow, are **dyspneic,** and require open-heart surgery to correct the defects.

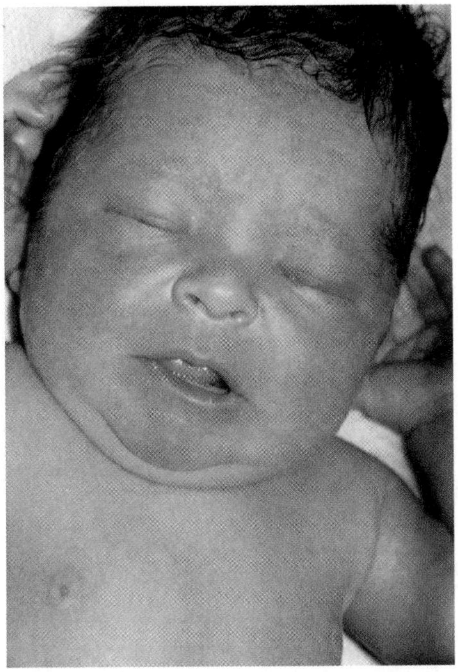

▲ **FIGURE 16.3**
Cyanotic ("Blue") Baby.

Respiratory Adaptations

Respiratory adaptations occur immediately at birth when, unless the baby's respiratory function is depressed by too much sedation or anesthesia in the mother, the first breath is taken spontaneously.

The development of the chemical surfactant is a key to the development of fully functioning lungs. When surfactant is deficient, the lung alveoli collapse and the exchange of gases cannot take place. This deficiency occurs in premature infants born before the 37th week of gestation and produces **respiratory distress syndrome (RDS),** previously called **hyaline membrane disease (HMD).** The more premature, the greater the risk of RDS. Infants with RDS are at risk for cerebral ischemia, hemorrhage, and neonatal death.

Keynote

Meconium is the first stool a baby passes.

Meconium aspiration syndrome occurs in newborns who are stressed **in utero** or at the time of delivery. The distress causes the fetus to expel **meconium** (stool) into the amniotic fluid. Then deep gasping for breath by the distressed fetus causes the aspiration of meconium into the lungs. Treatment is by **intubation** and **suction** to remove the meconium-stained fluid.

Transient tachypnea of the newborn (TTN) occurs most often in C-section and **precipitate** (unduly rapid) vaginal deliveries. Amniotic fluid remains in the infant's lungs and causes a self-limiting respiratory distress.

Sudden infant death syndrome (SIDS) is thought to be caused by a failure of cardiorespiratory control mechanisms to mature. It is the sudden death of an infant for which no cause is found after a thorough investigation and autopsy. It occurs during sleep and peaks between 2 and 3 months. It can be prevented by placing infants on their backs to sleep (*Figure 16.4*).

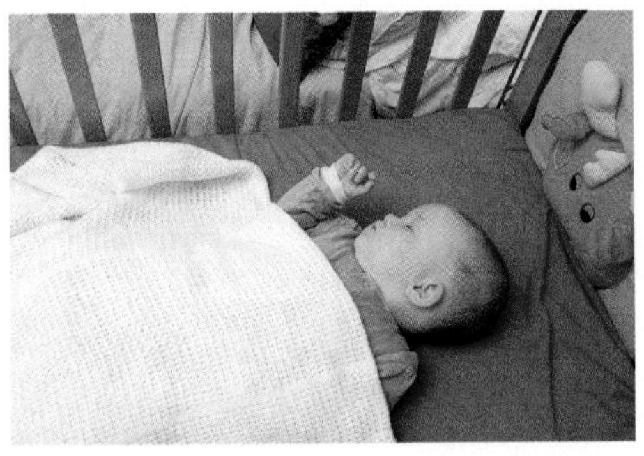

▲ **FIGURE 16.4**
Infant Sleeping on Back.

WORD ANALYSIS AND DEFINITION

S/ = Suffix   P/ = Prefix   R/ = Root   R/CF = Combining Form

WORD	PRONUNCIATION		ELEMENTS	DEFINITION
adaptation **adapt (verb)**	ad-ap-**TAY**-shun a-**DAPT**	S/ R/	**-ation** *process* **adapt-** *to adjust*	Change in the function or structure of an organ to meet new conditions.
anomaly **anomalies (pl)**	ah-**NOM**-ah-lee		Greek *irregularity*	Structural abnormality present at birth.
Apgar score	**AP-**gar SKOR		Virginia Apgar, U.S. anesthesiologist, 1909–1974	Evaluation of a newborn's status.
fetus **fetal (adj)**	**FEE**-tus **FEE**-tal	 S/ R/	Latin *offspring* **-al** *pertaining to* **fet-** *fetus*	Human organism from the end of the eighth week after conception to birth. Pertaining to the fetus.
neonate **neonatal (adj)** (***Note:*** The "e" is dropped as it is followed by another vowel, "a.")	**NEE**-oh-nate **NEE**-oh-**NAY**-tal	P/ R/CF S/ R/	**neo-** *new* **-nat/e** *born* **-al** *pertaining to* **-nat-** *born*	A newborn infant. Pertaining to the newborn infant or the newborn period.
pediatrics **pediatrician**	pee-dee-**AT**-riks **PEE**-dee-ah-**TRISH**-an	S/ R/ R/ S/	**-ics** *knowledge* **-iatr-** *medical treatment* **ped-** *child* **-ician** *expert*	Medical specialty of treating children during development from birth through adolescence. Medical specialist in pediatrics.

Abbreviation

SIDS sudden infant death syndrome

EXERCISES

Match *the meaning in column 1 with the correct element in column 2. The elements are ones you will see in terms throughout this chapter. Knowledge of these elements will help to increase your medical vocabulary. Fill in the blanks.*

	Column 1	Column 2
_____	1. new	A. nate
_____	2. expert	B. ics
_____	3. medical treatment	C. ped
_____	4. knowledge	D. ician
_____	5. to adjust	E. neo
_____	6. child	F. al
_____	7. born	G. iatr
_____	8. pertaining to	H. ation
_____	9. process	I. adapt

10. Use any one term from the above WAD in a sentence of patient documentation:

Sentence: ____

____

____

LESSON 16.1 Neonatal Period

LIFE SPAN

OBJECTIVES

In order to be able to understand where **neonates** are in their development and to talk to mothers and other health professionals about this, you need to be able to:

- **Name the anatomic and physiologic *adaptations* that occur at birth and in the neonatal period.**
- **Apply correct medical terminology to common congenital *anomalies* that interfere with normal anatomical and physiologic development.**
- **Describe the medical terminology of disorders in growth and development in the neonatal period.**

Case Report 16.1 (continued)

Mrs. Anna Hotteling, the mother, has several areas of concern. A friend's baby recently died of **SIDS** (sudden infant death syndrome). How can she prevent this for her child? Carol wants to feed every 2 to 3 hours day and night, and she (Anna) is exhausted. How long will this go on, and for how long should she continue breastfeeding? Mrs. Hotteling also wants to know what her baby can see and how she should communicate with her.

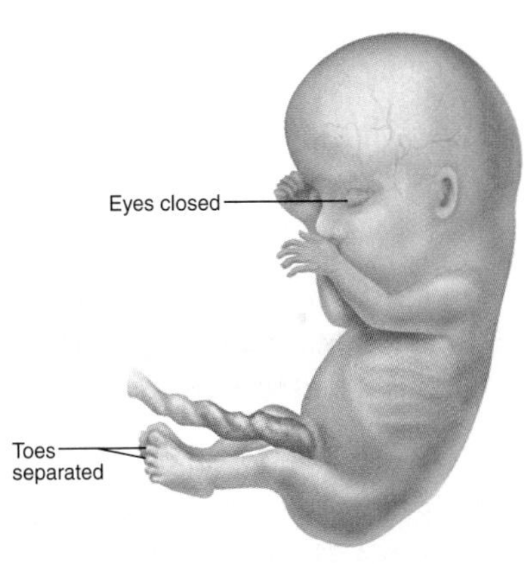

▲ **FIGURE 16.1**
Fetus at 8 Weeks (56 Days).

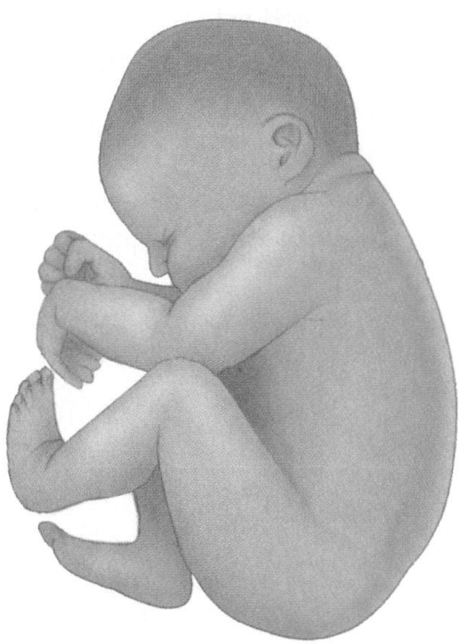

▲ **FIGURE 16.2**
Full-Term Infant (38 Weeks).

NEONATAL ADAPTATIONS

Fetal life is a preparation for birth. At the end of the first 8 weeks of fetal life, all the organ systems are in place (*Figure 16.1*). From then until birth, the organs grow and acquire the functional capabilities to support life outside the mother. Sometimes a part of this process will fail, and the fetus can have a developmental abnormality.

At birth, normal organ development is not yet complete, but the **neonate** (*Figure 16.2*) suddenly has to **adapt** to a totally different environment. Each of the organ systems has to adapt to the new environment and then go on to complete its development during childhood. This developmental process is why children are not just "little adults," and why there is a specialty of **pediatrics** practiced by **pediatricians.**

Immediately after birth, the neonate is evaluated for her **Apgar score** at 1 minute and again at 5 minutes of life. The Apgar score gives health care providers an immediate assessment of the baby's condition at birth. The five measurements of Activity, Pulse, Grimace, Appearance, and Respiration are each scored on a 3-point scale: 0 (poor), 1, or 2 (normal). The total score obtainable is between 0 and 10 (*Table 16.1*). A score of 7 or above is normal. Below 7, the baby needs special immediate care including oxygen and further airway suctioning. The Apgar score does not predict long-term health, intellectual status, or outcome.

TABLE 16.1 Apgar Scoring

Apgar Sign		0	1	2
A	**Activity (muscle tone)**	Limp, no movement	Limbs flexed, little movement	Active spontaneous movement
P	**Pulse**	No pulse	Below 100 beats per minute	Above 100 beats per minute
G	**Grimace (responsiveness)**	No response to airway suction	Grimace only to airway suction	Pulls away from airway suction
A	**Appearance (skin color)**	Blue-gray all over	Pink, except hands and feet are bluish	Pink all over
R	**Respiration**	No breathing	Weak cry, gasping	Strong cry, normal breathing effort and rate

CASE REPORT 16.1

You are

. . . a pediatric medical assistant employed by Sandra Mendes, MD, a **pediatrician** at Fulwood Medical Center. You are working in her Well-Baby Clinic.

You are communicating with

. . . Mrs. Anna Hotteling, a 35-year-old mother, who has brought her 10-week-old daughter, Carol, to the Well-Baby Clinic. Carol, Mrs. Hotteling's first baby, was born via a normal vaginal delivery at term. She weighed 7 pounds 2 ounces. She had normal **Apgar scores.** Carol had persistent jaundice when seen at 2 weeks of age. She is being breastfed. You ask Mrs. Hotteling if she has any concerns.

Learning Outcomes

In order to understand the growth and development of children, adults, and the elderly and communicate about this with your employer, other health professionals, and patients, you need to be able to:

16.1 **Apply the languages of pediatrics and gerontology to normal human growth and development.**

16.2 **Comprehend, analyze, spell, and write the medical terms of pediatrics and gerontology.**

16.3 **Recognize and pronounce the medical terms of pediatrics and gerontology.**

16.4 **Use correct medical terminology to specify common disorders in growth and development.**

Life Span

The Essentials of the Languages of Pediatrics and Gerontology

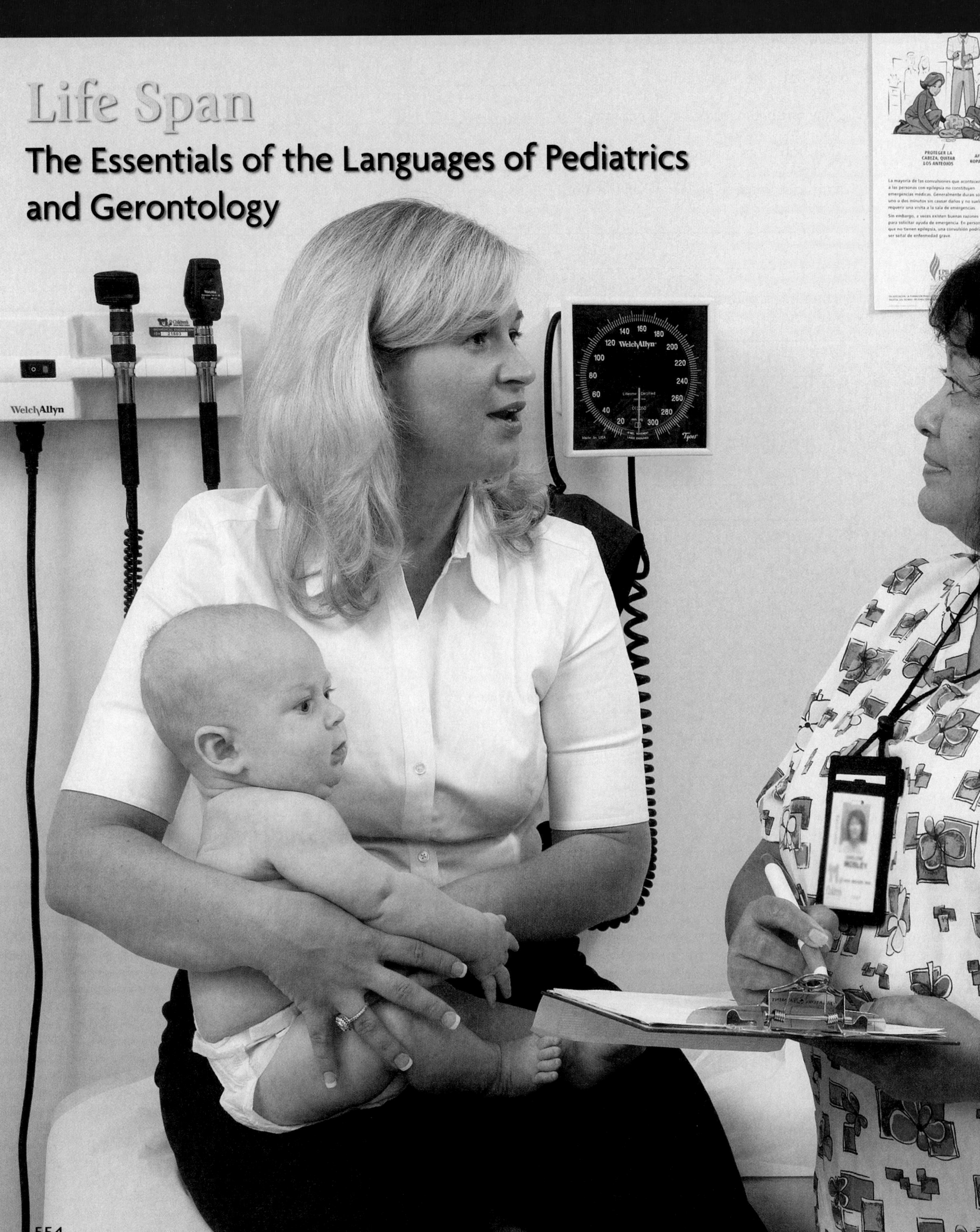

5. Name one other body cavity: ______________________________

Name one organ located in this cavity: ______________________________

6. What "tube" has ruptured? ______________________________

7. In this case, what happened when the tube ruptured?

____

____

8. Define *ectopic pregnancy:*

____

____

____

9. Why, then, has the tube ruptured?

____

____

10. What type of instrument will be used in Mrs. Maggay's surgery?

____

11. What medical term can be applied to Mrs. Maggay since this is her first pregnancy?

____

12. In Vital Signs, what do the following abbreviations stand for?

T ______________________________

P ______________________________

R ______________________________

BP ______________________________

WORD ANALYSIS AND DEFINITION

S/ = Suffix P/ = Prefix R/ = Root R/CF = Combining Form

WORD	PRONUNCIATION	ELEMENTS		DEFINITION
cyanosis **cyanotic (adj)**	sigh-ah-**NO**-sis sigh-ah-**NOT**-ik	S/ R/ S/ R/CF	**-osis** *condition* **cyan-** *blue* **-tic** *pertaining to* **cyan/o-** *blue*	Blue discoloration of skin, lips, and nail beds due to low levels of oxygen in the blood. Pertaining to or marked by cyanosis.
dyspnea **dyspneic (adj)**	disp-**NEE**-ah disp-**NEE**-ik	P/ R/ S/	**dys-** *bad, difficult* **-pnea** *breathe* **-ic** *pertaining to*	Difficulty breathing. Pertaining to or suffering from difficulty in breathing.
Fallot	fah-**LOW**		Etienne Louis Arthur Fallot, French physician, 1850–1911	First described the tetralogy of the four heart defects.
hyaline membrane disease	**HIGH**-ah-line **MEM**-brain **DIZ**-eez	S/ R/ R/ P/ R/	**-ine** *pertaining to* **hyal-** *glass* **membrane** *cover* **dis-** *apart from* **-ease** *normal function*	Respiratory distress syndrome of the newborn.
in utero	IN **YOU**-ter-oh	R/CF	**uter/o** *uterus*	Within the womb; not yet born.
meconium	meh-**KOH**-nee-um		Greek *a little poppy*	The first bowel movement of the newborn.
patent	**PAY**-tent		Latin *lie open*	Open.
precipitate labor	pree-**SIP**-ih-tate **LAY**-bore		precipitate, Latin *to throw down headlong* labor, Latin *toil, suffering*	A very rapid labor and delivery.
suction	**SUK**-shun		Latin *sucking*	Use of a catheter to clear the upper airway or other tubes.
tetralogy	te-**TRAL**-oh –jee	P/ R/	**tetra-** *four* **-logy** *study of*	A set of four congenital heart defects occurring together.

Abbreviations

ASD	atrial septal defect
CVS	cardiovascular system
HMD	hyaline membrane disease
PDA	patent ductus arteriosus
RDS	respiratory distress syndrome
SIDS	sudden infant death syndrome
TOF	tetralogy of Fallot
TTN	transient tachypnea of the newborn
VSD	ventricular septal defect

EXERCISES

Abbreviations: *Incorporate your knowledge of abbreviations from this lesson into the following patient documentation. Fill in the blanks, using the choices below; there are more choices than answers.*

ASD SIDS TOF TTN VSD PDA

1. Diagnostic testing confirms ____________________—an opening between the two lower chambers of the infant's heart. Patient will be scheduled for open-heart surgery early next week.
2. Due to a precipitate delivery, and amniotic fluid remaining in the lungs, this infant has ____________________.
3. Because this infant's cardiorespiratory mechanisms have failed to mature, he is at great risk for ____________________.
4. The septal defect in this infant has been confirmed on testing as being between the two upper heart chambers— ____________________.
5. One type of cyanotic heart disease is ____________________.

Keynote

A baby is born with more than 100 billion brain cells. In the first months and years of life, connections are made (wiring) to form complex circuits that allow for thinking, feelings, and behaviors.

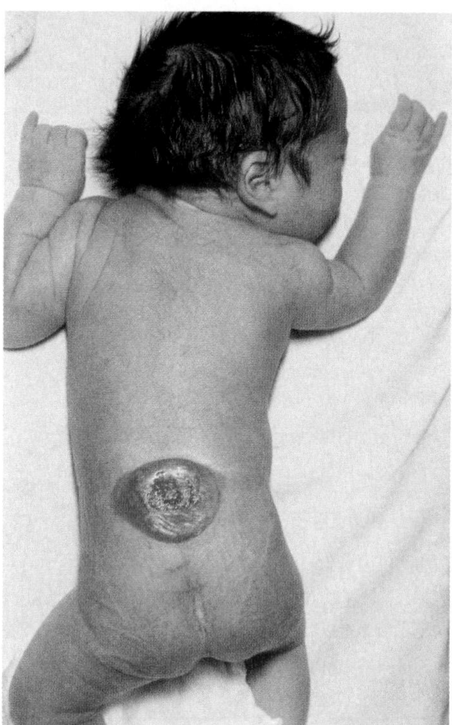

▲ **FIGURE 16.5**
Child with Spina Bifida Cystica.

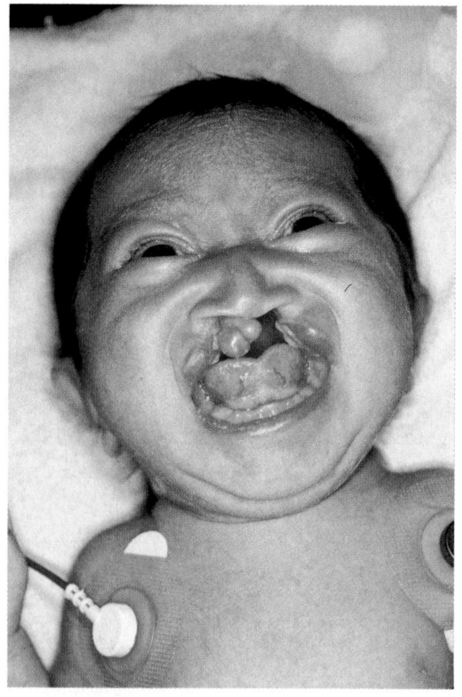

▲ **FIGURE 16.6**
Infant with Cleft Lip and Palate.

NEONATAL ADAPTATIONS (continued)

Thermoregulation and Adaptations

The infant has a larger ratio of surface area to body volume than an adult. Therefore, the infant loses heat more easily, particularly if the body surface is wet. This is why the newborn is dried, wrapped, and placed in a warmer after birth. **Hypothermia** is more likely to occur in premature or small-for-date **(SFD)** neonates.

Brain and Neurologic Adaptations

A newborn baby's brain is one-quarter of its adult size. The brain's growth is monitored by charting increases in head circumference. At birth, only the spinal cord and brainstem are well developed. The cortex is primitive. All of the newborn's kicking, grasping, rooting (searching for the nipple), and crying behaviors are functions of the brainstem, which is why they are involuntary and/or not well coordinated.

Congenital Neurologic Abnormalities

Anencephaly is the absence of the cerebral hemispheres. This is incompatible with life.

Microcephaly is characterized by small cerebral hemispheres, leading to motor and mental retardation.

An **encephalocele** is a protrusion of nervous tissue and meninges through a defect in the skull.

Hydrocephalus is an enlargement of the ventricles with excessive CSF, and is the most common cause of a large head in the neonate (*see Chapter 12*).

Spina bifida is a failure of the vertebral column to close over the spinal cord in the lumbar and sacral regions. **Spina bifida occulta** occurs when the vertebral arches fail to unite and there is no neurologic involvement (*see Chapter 8*). When the vertebral defect is more open, nervous tissue can protrude through it in a sac. In **spina bifida cystica** (*Figure 16.5*), the sac can contain **meninges (meningocele)**, part of the spinal cord **(myelocele)**, or both **(myelomeningocele)** (*see Chapter 8*).

Neonatal seizures are a common and sometimes serious neonatal disorder. They can be primary, from an intracranial process such as meningitis or a cerebral hemorrhage from a difficult birth, or can be secondary, from a systemic or metabolic problem such as hypoxia, hypoglycemia, or hypocalcemia. Treatment is directed to the underlying pathology.

Skeletal Adaptations

The skeletal system can fail in utero and produce congenital abnormalities in different parts of the newborn's body. **Craniofacial** malformations are most often **cleft lip** and **cleft palate,** which occur once in 800 births (*Figure 16.6*). A cleft palate interferes with feeding and speech development. The end treatment is surgical closure.

Developmental dysplasia of the hip (congenital dislocation of the hip) occurs more commonly in female infants and following a breech delivery. Hip ultrasound can confirm the diagnosis. Treatment is with padded diapers to keep the affected femur abducted. Surgery may be necessary.

WORD ANALYSIS AND DEFINITION

S/ = Suffix P/ = Prefix R/ = Root R/CF = Combining Form

WORD	PRONUNCIATION	ELEMENTS		DEFINITION
anencephaly	**AN-en-SEF-ah-lee**	P/ R/	**an-** *without* **-encephaly** *condition of the brain*	Born without cerebral hemispheres.
bifid **spina bifida**	**BI**-fid **SPY**-nah **BIH**-fih-dah		bifid, Latin *split into two* spina, Latin *backbone*	Separated into two parts. Failure of one or more vertebral arches to close during fetal development.
cleft lip **cleft palate**	KLEFT LIP KLEFT **PAL**-ate		cleft, Latin *fissure, separation of parts*	Congenital defect of the upper lip. Congenital defect of the palate.
craniofacial	**KRAY**-nee-oh-**FAY**-shal	S/ R/CF R/	**-al** *pertaining to* **crani/o-** *cranium* **-faci-** *face*	Pertaining to both the face and the cranium.
dysplasia	dis-**PLAY**-zee-ah	S/ P/ R/	**-ia** *condition* **dys-** *bad, difficult* **-plas-** *molding, formation*	Abnormal tissue formation.
dysplastic (adj)	dis-**PLAS**-tik	S/	**-tic** *pertaining to*	Pertaining to or showing abnormal tissue formation.
encephalocele	en-**SEF**-ah-loh-seal	S/ R/CF	**-cele** *swelling, hernia* **encephal/o-** *brain*	Congenital defect of the cranium with herniation of brain tissue.
hydrocephalus	high-droh-**SEF**-ah-lus	P/ R/CF	**hydro-** *water* **encephal/o-** *brain*	Enlarged head due to excess CSF in the cerebral ventricles.
hypothermia	high-poh-**THER**-me-ah	S/ P/ R/	**-ia** *condition* **hypo-** *below* **-therm-** *heat*	Very low core body temperature.
microcephaly	**MY**-kroh-**SEF**-ah-lee	P/ R/	**micro-** *small* **-cephaly** *condition of the head*	Small head.
microcephalic (adj)	**MY**-kroh-**SEF**-ah-lik	S/	**-ic** *pertaining to*	Pertaining to or suffering from a small head.
myelocele	**MY**-eh-low-seal	S/ R/CF	**-cele** *hernia, swelling* **myel/o-** *spinal cord*	Protrusion of the spinal cord through a defect in the vertebral arch.
myelomeningocele	**MY**-eh-low-meh-**NING**-oh-seal	S/ R/CF R/CF	**-cele** *hernia, swelling* **myel/o-** *spinal cord* **-mening/o-** *meninges*	Protrusion of the spinal cord *and* meninges through a defect in the vertebral arch of one or more vertebrae.

Abbreviation

SFD small for date

EXERCISES

Build *the correct medical term based on the statement given. Write the element on the line to build the term, and write the meaning of each element below the line.*

1. pertaining to both the face and the cranium: ____________ / ____________ / ____________

2. small head: ____________ / ____________ / ____________

3. very low core body temperature: ____________ / ____________ / ____________

4. congenital defect of the cranium with herniation of brain tissue: ____________ / ____________ / ____________

NEONATAL ADAPTATIONS (continued)

Growth Adaptations

Failure to grow fully in the uterus (in utero) has two main causes:

1. **Inadequate nutrition** (caused by poor placental function). The newborn is below the 10th percentile (*see below*) of babies of the same gestational age. This is called **small for gestational age (SGA).** Good nutrition after delivery will enable growth to accelerate to normal.
2. **Premature labor** (the infant is born before 37 weeks of gestation). The premature infant (*Figure 16.7*) weighs less than 5½ pounds, has little subcutaneous fat or hair, and has low spontaneous activity. Immature development of the CNS leads to poor sucking and swallowing. Premature infants may have to be fed by IV or by **gavage (stomach tube).**

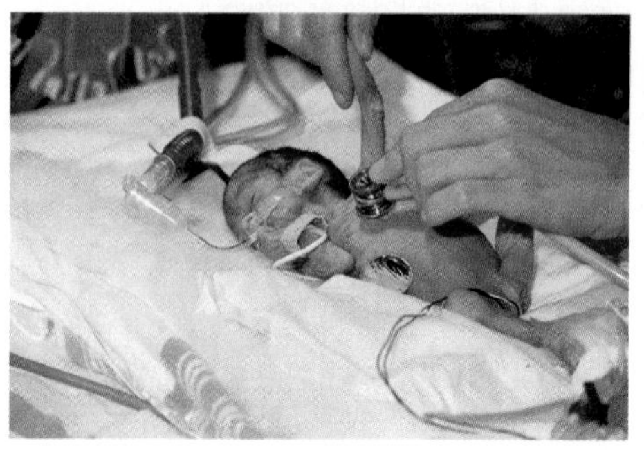

▲ **FIGURE 16.7 Premature Baby.**

Occasionally, labor does not start until after 42 weeks' gestation and produces a **postmature infant** (*Figure 16.8*). Past term, the placenta atrophies and may calcify, so the fetus receives insufficient nutrition in the days after term. As a result, postmature infants have decreased subcutaneous fat and dry, peeling skin.

Normal full-term babies lose up to 10% of their birth weight during the first few days of life. By 1 month, infants should be gaining 2/3 to 1 ounce per day, and they will grow 1 to 1½ inches per month. These changes will be plotted on growth charts that indicate the size and growth patterns of individual children compared to other children in the United States.

Failure to thrive (FTT) is the term used for an infant or young child who is not growing and developing as expected. There are two main reasons for the failure:

1. **Organic disorders,** such as chronic illness (e.g., celiac disease), genetic (e.g., Down syndrome), metabolic (e.g., fetal alcohol syndrome, or **FAS**), and hormone disorders (e.g., pituitary dwarfism).
2. **Psychosocial disorders,** including poverty, lack of education about feeding, neglect or abuse, and parental mental illness or substance abuse.

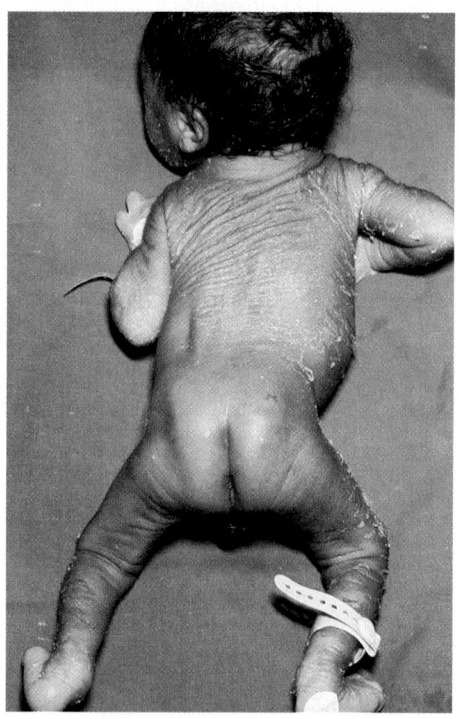

▲ **FIGURE 16.8**
Postmature Infant with Skin Changes.

Urinary System Adaptations

Kidneys at birth are not fully developed. Infants cannot concentrate their urine, so they have a high rate of water loss. They require more fluid intake, relative to body weight, than adults. In the first month of life, an infant will produce five to six wet diapers daily.

Congenital urinary tract disorders include renal **agenesis,** in which one or both kidneys are absent; blockage of urinary flow in utero, causing hydronephrosis of the kidneys; polycystic kidney disease, one of the most common genetic disorders; and hypospadias of the penis, in which the urethra does not extend to the end of the penis and opens along the underside of the penis (*see Chapter 13*).

Digestive System Adaptations

Until 6 months of age, the baby needs only breast milk or commercially prepared formula. Around 1 month of age, a feeding routine will be established.

Esophageal **atresia,** incomplete formation of the esophagus, is often associated with a **fistula** between the esophagus and the trachea. This leads to feeding difficulties and respiratory distress in the first few days of life.

An immature digestive system may be the cause of **colic,** in which the neonate cries for hours at a time until you feel like joining in. Of all infants, 25% have colic. It begins between the third and sixth weeks and goes away on its own by the twelfth week. There is no known treatment.

Food intolerance, totally different from food allergy (*see below*), is an adverse digestive system reaction to food. It does not involve the immune system. It can be a metabolic reaction to a digestive enzyme deficiency such as **lactase deficiency** (*see Chapter 5*), leading to intolerance to cows' milk.

Keynote

Breast milk provides valuable antibodies, digestive enzymes, and hormones that infants need. It is nonallergenic.

WORD ANALYSIS AND DEFINITION

S/ = Suffix P/ = Prefix R/ = Root R/CF = Combining Form

WORD	PRONUNCIATION	ELEMENTS		DEFINITION
agenesis	a-**JEN**-eh-sis	P/ R/	**a-** *without* **-genesis** *creation, production*	Failure to develop any organ or any part.
atresia	a-**TREE**-zee-ah	P/ R/	**a-** *without* **-tresia** *a hole*	Congenital absence of a normal opening or lumen.
colic	**KOL**-ik	S/ R/	**-ic** *pertaining to* **col-** *colon*	Spasmodic, crampy pains in the abdomen.
diphtheria	dif-**THEER**-ee-ah		Greek *leather*	Disease with a thick (leathery) coating of the pharynx.
fistula	**FIS**-tyu-lah		Latin *pipe, tube*	Abnormal passage.
gavage	guh-**VAHZH**		French *to force-feed geese (to make pate de foie gras)*	To feed by a stomach tube.
intolerance	in-**TOL**-er-ance		Latin *impatient*	Inability of the small intestine to digest and dispose of a particular dietary substance.
lactase	**LAK**-tase	S/ R/	**-ase** *enzyme* **lact-** *milk*	Enzyme that breaks down lactose (milk sugar) to glucose and galactose.
pertussis (also known as *whooping cough*)	per-**TUSS**-is **WHO**-ping KOFF	P/ R/	**per-** *intense* **-tussis** *cough*	Infectious disease with a spasmodic, intense cough ending with a whoop (stridor).
phototherapy	foh-toe-**THAIR**-ah-pee	S/ R/CF	**-therapy** *treatment* **phot/o-** *light*	Treatment using light rays.

Abbreviations

FAS	fetal alcohol syndrome
FTT	failure to thrive
SGA	small for gestational age

Hematologic Adaptations

The newborn infant has an excess of RBCs. When these are broken down, bilirubin is produced (*see Chapter 6*). In addition, the neonate's liver is immature and cannot process bilirubin quickly. The excess bilirubin is deposited in the tissues, producing jaundice. In the first 3 days after birth, neonatal jaundice affects 60% of full-term and 80% of premature infants. In most infants, no specific treatment is needed. More severe cases respond to **phototherapy** (*Figure 16.9*), which employs blue wavelengths of light to convert bilirubin to less toxic chemicals that can be excreted in bile or urine.

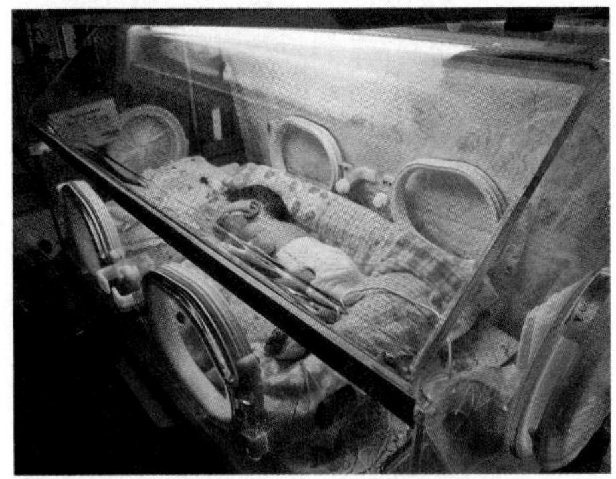

▲ **FIGURE 16.9**
One-Day-Old Premature Baby Undergoing Phototherapy.

Immunologic Adaptations

The baby is born with IgG levels (*see Chapter 11*) near those of an adult, having acquired them from the mother through the placenta. These levels remain high enough in the first 6 months to protect the baby against some infectious diseases but not against others, including **whooping cough (pertussis)** and **diphtheria.** This is why immunization against these diseases takes place at 2, 4, and 6 months of age.

EXERCISES

Documentation: *Use the terminology* and *the abbreviations found on these two pages to correctly fill in the following patient documentation.*

1. This infant's mother has been an alcoholic for the past year and a half. Infant was born suffering from ____________________.
2. This preemie falls below the l0th percentile and is ____________________.
3. Because of throat cancer, this patient is being fed by ____________________.
4. Patient was born with only the right kidney present. Diagnosis: renal ____________________.
5. Because of jaundice, this infant will receive 8 hours of ____________________ each day until discharge.

CASE REPORT 16.2

True Story

A singer in a Master Chorale during her seventh and eighth months of pregnancy was in rehearsal for Handel's *Messiah* for a Christmas concert. At home, she frequently practiced her solo, which began "I know that my redeemer liveth." When her daughter was 6 months old, the mother was changing the baby's diaper and started to sing "I know that my . . ." The child turned her head and chimed in "redeemer." Now, at 10 years old, the daughter has absolute pitch. She has the rare ability to replicate on the piano the note for any sound made in nature.

Immunologic Adaptations (continued)

In **food allergies,** the body's immune system reacts as though a particular food is harmful (an allergen) and creates IgE antibodies to it (*see Chapter 11*). This, in turn, generates chemicals such as histamine that produce symptoms of a runny nose, itchy skin rash, swelling of the lips, or wheezing. The most common allergens in the neonate are milk, eggs, wheat, soy, and peanuts. Between 25% and 50% of affected children outgrow their food allergies by age 3.

Keynote

In the first 2 years of life, most children have 8 to 10 colds.

Keynote

Six percent of children under the age of 3 have food allergies.

Hearing Adaptations

The fetus begins to hear loud noises around the beginning of the third trimester (24 weeks). By the seventh month of pregnancy, the fetus can hear **maternal** speech and remember what is heard after birth. **Congenital malformations** of the external auditory canal and middle ear can result in a conductive hearing loss (*see Chapter 4*). Congenital malformations of the inner ear can result in a sensorineural hearing loss (*see Chapter 4*).

Keynote

All newborn babies should receive hearing screening tests before they go home from the hospital.

Visual Adaptations

At birth, a nonanesthetized baby is able to see and focus on an object between 8 and 12 inches away. This distance probably relates to the distance between the mother's and baby's faces during breastfeeding (*Figure 16.10*). For color vision, the baby will see only very brightly colored objects. Full color vision comes at between 3 and 4 months of age.

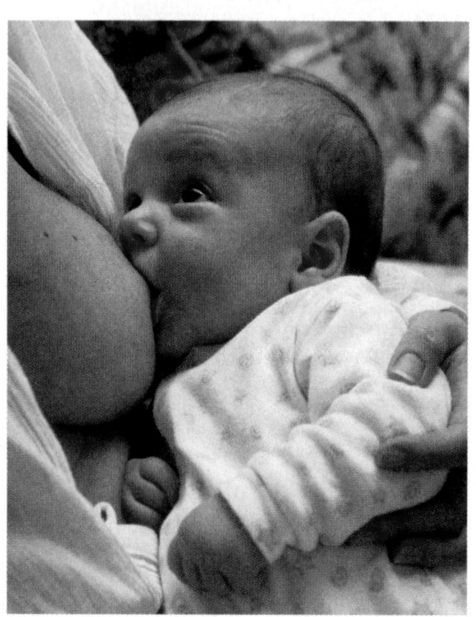

▲ **FIGURE 16.10 One-Month-Old Baby Breastfeeds and Interacts with Mother.**

Social Adaptations

Identity. The building blocks of a baby's identity began in the last trimester in utero when she played with her fingers and toes and responded to sounds, music, and voices. Genes and her inborn **temperament** play their roles. At 1 month of age, her interaction with her caregiver's love, attention, and caring skills is the **stimulus** for how she reacts to other people. By 2 to 3 months, as she plays with her fingers and toes, she will be starting the process of being aware of her own physical identity.

Communication. Crying is the baby's primary communication method. She cries in different ways to say "I'm wet, hungry, lonely, or just overwhelmed by the sights and sounds of this world around me." She'll start to "coo" and gurgle, and she will enjoy hearing the caregiver respond with the same sounds. When she turns away or closes her eyes, she may just need her own space.

Curiosity. At 1 month, baby can show interest only in what is in front of her, such as a parent's face or a brightly colored toy. As her eye and neck muscles develop, she can turn her head to see any object that catches her eye. At 3 months, she'll take swipes at a mobile of shapes hanging over her crib, and she'll look into a child-safe mirror in her crib near her head.

Keynote

By 3 months of age, the baby will smile at the sound of a known voice and turn her head to the direction of the sound.

WORD ANALYSIS AND DEFINITION

S/ = Suffix P/ = Prefix R/ = Root R/CF = Combining Form

WORD	PRONUNCIATION	ELEMENTS		DEFINITION
congenital	kon-**JEN**-ih-tal	S/ P/ R/	**-al** *pertaining to* **con-** *together, with* **-genit-** *birth, bring forth*	Present at birth, either inherited or due to an event during gestation up to the moment of birth.
malformation	**MAL**-for-**MAY**-shun	S/ P/ R/	**-ion** *condition* **mal-** *bad,* **-format-** *to form*	Failure of proper or normal development.
maternal	mah-**TER**-nal	S/ R/	**-al** *pertaining to* **matern-** *mother*	Pertaining to or derived from the mother.
stimulus **stimuli (pl)** **stimulant**	**STIM**-you-lus **STIM**-you-lie **STIM**-you-lant	 S/ R/	Latin *goad, incite* **-ant** *forming* **stimul-** *excite, strengthen*	Something that excites or strengthens the functional activity of an organ or part. An agent that excites or strengthens.
temperament	**TEM**-per-ah-ment		Latin *disposition*	Predisposition to character or personality.

EXERCISES

Demonstrate *that you can use the correct form of the medical term.*

1. Write three different sentences that are not definitions—and not directly out of the text—using the terms.

stimulus stimuli stimulant

a. ______________________________

____

____

b. ______________________________

____

____

c. ______________________________

____

____

2. Write a sentence of patient documentation using the term *congenital malformation.*

____

____

____

LESSON 16.2 Infant to Teenager

LIFE SPAN

OBJECTIVES

To be both a pediatric medical assistant and a parent, you need to be able to use correct terminology to:

- **Identify the medical terminology used in different stages of development in childhood.**
- **Apply the medical terminology used for common disorders of development in childhood.**

You are

. . . a pediatric medical assistant working with Dr. Sarah Mendes, a pediatrician, at Fulwood Medical Center.

You are communicating with

7-year-old Carol Hotteling and her mother.

CASE REPORT 16.3

You have watched Carol grow over the 7 years that you have worked in Dr. Mendes' Well-Baby Clinic. Her mother, Mrs. Anna Hotteling, is telling you about her current concerns:

"As you know, she's always been a handful. The neighbors call her a tomboy. She's trashed every room in the house. The other kids' mothers stopped inviting her to parties. Nothing holds her attention for very long. I've lost her a couple of times in the shopping mall when she's darted away. I always thought she'd grow out of it. But now she's in second grade. The teacher's complaining that Carol can't sit still, she interferes with the other kids, disrupts the class, and she's having trouble reading and staying focused. It's got to be due to something I did or didn't do, hasn't it?"

STAGES 1–3, INFANCY THROUGH EARLY CHILDHOOD

Around 1-year-old, your child can be introduced to an optimal nutritional environment that reflects your own good eating habits. This can ensure a normal growth pattern that can be disturbed only by illness or **psychosocial** disorders, such as poverty or abuse, as she goes through the transformation from a **dependent** infant into a **competent, independent** adult. This transformation to adulthood has been broken down by **psychologist** Erik Erikson into eight stages.

Stage 1: Infancy—Age 0 to 1 Year

Keynote

Stage 1 issue: trust versus mistrust.

In the first year of life, your infant is **dependent** on parents and caregivers for food, affection, and warmth. Consistently meeting these needs will result in your infant's secure, trusting attachment to parents and their environment. If those needs are not met, your infant can develop distrust toward people and their surroundings. At the end of the first year, your infant will be starting to show **independence** by pulling herself up to stand, walking holding on to furniture, and perhaps saying a first word.

Stage 2: Toddler—Age 1 to 2 Years

Keynote

Stage 2 issue: independence versus self-doubt.

Independence is the issue for your toddler, who learns to walk and talk and wants to do things for herself like use the toilet and put on her own clothes. Out of these abilities and desires begins the development of **self-confidence** and **self-control.** If your child's initiatives are encouraged and mistakes are accepted, she will be able to make choices and achieve self-control and independence. If parents are overprotective or disapproving, the child can have doubts about her abilities. By the end of this year, she will run, climb on furniture, scribble artwork, say "I love you," and be able to cope with other children.

Stage 3: Early Childhood—Age 2 to 6 Years

Keynote

Stage 3 issue: initiative versus impulse.

Your child finds power with motor skills and her ability to engage in social interaction. She wants to be the center of attention and part of an adult world for which she is not ready. Encouragement with boundaries and consistent **discipline** help her define her role in the family. The word *discipline* has the same root as *disciple,* a learner of the responsible way of living. **Impulse control** and **responsibility** are learned as a key part of her independence.

WORD ANALYSIS AND DEFINITION

S/ = Suffix P/ = Prefix R/ = Root R/CF = Combining Form

WORD	PRONUNCIATION	ELEMENTS		DEFINITION
competent	**KOM**-peh-tent		Latin *in harmony*	Capable of performing a task or function.
dependent	dee-**PEN**-dent	S/ R/	**-ent** *end result* **depend-** *relying on*	Having to rely on someone else.
discipline	**DIS**-ih-plin	S/ R/	**-ine** *pertaining to* **discipl-** *understand*	Training for proper conduct or action.
independent	in-dee-**PEN**-dent	S/ P/ R/	**-ent** *end result* **in-** *not* **-depend-** *relying on*	Able to fend for oneself.
independence	in-dee-**PEN**-dense	S/	**-ence** *state of, quality of*	The state of being able to think and act for oneself.
infant **infancy**	**IN**-fant **IN**-fan-see		Latin *not speaking*	Child in first year of life. The first year of life.
psychosocial	sigh-koh-**SOH**-shal	S/ R/CF R/	**-ial** *pertaining to* **psych/o-** *the mind* **-soc-** *partner*	Involving both the mind and various social and community aspects of life.
psychologist	sigh-**KOL**-oh-jist	S/	**-logist** *specialist*	Licensed specialist in psychology.

EXERCISES

Lesson objective: *Meet this lesson's objective of being able to identify stages of development in childhood by answering the following questions. Use the vocabulary on these two pages to fill in the blanks.*

1. Stage 1 of childhood development encompasses what ages? ____________________

2. What is the main issue in stage 1 of childhood development? ____________________

3. An infant must transform from a ____________________ infant to an ____________________ adult.

4. What is the main issue for toddlers? ____________________

5. What must toddlers begin to develop at that age?

   Self-____________________ and self-____________________

6. What age range is considered "early childhood"? ____________________

7. What does this psychologist believe are key parts of a child's independence? ____________ and ____________

8. Who is the famous psychologist who formulated the eight stages of childhood development? ____________________

STAGES 4 AND 5—GRADE SCHOOL YEARS AND ADOLESCENCE

Stage 4: Grade School Years—Age 6 to 12 Years

Keynote

Stage 4 issue: competence versus failure.

Transition takes place from the world of home to the world of **peers.** The parents' role is to stimulate curiosity and pleasure in intellectual success so that the child develops a sense of competence, not feelings of inferiority and expectations of failure.

Stage 5: Adolescence—Age 12 to 18 Years

Keynote

Stage 5 issue: identity versus confusion.

This is the time of identity crisis. "Who am I?"—the most powerful and critical question in anyone's lifetime. If your **adolescent** is able to integrate the trust learned in stage 1 with the independence learned in stage 2, with the impulse control and responsibility from stage 3, with the competence and intellectual curiosity of stage 4, your adolescent can integrate these values into a good self-image. As a result, there will be a strong identity and readiness to be an adult. If not, your adolescent can sink into confusion and fear and be unable to make choices about roles in life.

Depression, **bipolar** disorder, anxiety disorders, and the eating disorders of **anorexia nervosa, bulimia,** and **binge eating** arise out of this confusion.

DEVELOPMENTAL DISORDERS OF CHILDHOOD AND ADOLESCENCE

Enuresis occurs in the child who is dry by day but wets the bed at night without waking up. Fifteen percent of children aged 5 have enuresis.

Encopresis is persistent fecal soiling beyond the age at which toilet training should be complete (i.e., at 3 to 4 years of age).

Eating disorders in early childhood are part of the syndrome **failure to thrive:**

- **Pica** is the persistent ingestion of nonnutritive substances such as dried lead-containing paint. It is associated with iron deficiency anemia and developmental delay.
- **Rumination** disorder occurs between 3 and 12 months and is characterized by the persistent **regurgitation** and rechewing of food after feeding. It goes away spontaneously.
- **Feeding disorder of infancy** occurs when a child under 6 years refuses to eat adequately and fails to gain his expected weight. There is no physical explanation found.

WORD ANALYSIS AND DEFINITION

S/ = Suffix P/ = Prefix R/ = Root R/CF = Combining Form

WORD	PRONUNCIATION		ELEMENTS	DEFINITION
adolescence (state of)	ad-oh-**LESS**-ents	S/ R/	**-ence** *state of, quality of* **adolesc-** *beginning of adulthood*	Stage that begins with puberty and ends with physical maturity.
adolescent (person)	ad-oh-**LESS**-ent	S/	**-ent** *end result, pertaining to*	Pertaining to adolescence or a person in that stage.
anorexia	an-oh-**RECK**-see-ah	S/ P/ R/	**-ia** *condition* **an-** *without* **-orex-** *appetite*	Without an appetite; or having an aversion to food.
binge eating	BINJ **EE**-ting		binge, Old English *to soak*	Eating with periods of excessive intake.
bipolar disorder	bi-**POH**-lar dis-**OR**-der	S/ P/ R/	**-ar** *pertaining to* **bi-** *two* **-pol-** *pole*	A mood disorder with alternating periods of depression and mania.
bulimia	byu-**LEEM**-ee-ah		Greek *hunger*	Episodic bouts of excessive eating with compensatory throwing up.
encopresis	en-koh-**PREE**-sis		Greek *full of manure*	Repeated soiling with feces.
enuresis	en-you-**REE**-sis	P/ R/	**en-** *in, within* **-uresis** *to urinate*	Bedwetting; urinary incontinence.
peer	PEER		Latin *equal*	A person at the same level or standing.
pica	**PIE**-kah		Latin *magpie*	Eating substances not considered to be food.
regurgitation	ree-gur-jih-**TAY**-shun	S/ P/ R/	**-ation** *process* **re-** *back, backward* **-gurgit-** *flood*	Expel contents of the stomach into the mouth, short of vomiting.
rumination	**ROO**-min-ay-shun	S/ R/	**-ation** *process* **rumin-** *throat*	To bring back food into the mouth to chew over and over.

EXERCISES

Lesson objective: *Continue to meet the lesson objective of identifying Erickson's stages of development. Fill in the chart.*

Stage	Ages	Major Issue in Development
4		
5		

Pick any one of these stages, and write a two-sentence description of what might happen to a child in this stage.

Always aim for neat and legible handwriting!

DEVELOPMENTAL DISORDERS OF CHILDHOOD AND ADOLESCENCE (continued)

Attention deficit hyperactivity disorder (ADHD) is the problem that Carol Hotteling and her mother are wrestling with. It affects 4% to 8% of school-age children, boys three times as often as girls. It has three major components:

1. **Inattention,** with an inability to pay attention to details, carelessness, distractibility, and problems with organization.
2. **Hyperactivity,** shown by fidgeting, squirming, always being on the go, and talking excessively.
3. **Impulsiveness,** shown by interrupting, intruding, and acting inappropriately.

The etiology (cause) is unknown, though some areas of the brain are known to be smaller in ADHD. There is no evidence that parenting plays any role in the development of ADHD. Treatment is with stimulants, such as Ritalin, or antidepressants, such as Prozac, in combination with special educational interventions, parent training to support the child, and **cognitive-behavioral therapy.** This form of therapy identifies the thinking that is causing unwanted feelings and behaviors and tries to replace this thinking with thoughts that lead to more desirable behaviors.

Learning disability (LD) refers to a group of disorders with difficulties in the acquisition and use of listening, speaking, reading, writing, and reasoning abilities. LD is not associated with obvious problems such as bad eyesight and lack of intellectual ability but refers to a discrepancy between a child's capacity to learn and the level of achievement. The term **dyslexia** is used for a learning disability in which there is primarily difficulty with reading and spelling.

Oppositional defiant disorder (ODD) shows a persistent pattern of defiance, disobedience, and hostility to parents and teachers. Children with ODD blame others for their mistakes and frequently lose their tempers.

Conduct disorder (CD) occurs between 9 and 17 years and is expressed by fighting, bullying, cruelty, and early substance abuse. Running away from home is common, and some young women can go into prostitution. Of all children with CD, 25% to 50% become antisocial adults.

Therapy for these disruptive disorders is difficult. Behavioral therapy and psychotherapy can help patients express and control their anger. The only two therapeutic programs shown to be effective both involved parent training to help their children.

Obsessive compulsive disorder (OCD) is an anxiety disorder with recurrent unwanted thoughts **(obsessions)** and repetitive behaviors **(compulsions or rituals)** that include hand washing, counting, and cleaning. Treatment involves finding the right medication for the child from the many antianxiety and antidepressant medications and using cognitive-behavioral therapy.

Autism spectrum disorders (ASDs) are characterized by varying degrees of impairment in social interactions and communication skills associated with repetitive and **stereotyped** patterns of behavior. This spectrum varies from a severe form called **autism** to a milder and higher-functioning form called **Asperger syndrome.** If a child has symptoms of these disorders, but does not meet specific criteria for their diagnosis, the term **pervasive developmental disorder, not otherwise specified (PDD-NOS)** is used.

Abbreviations

ADHD	attention deficit hyperactivity disorder
ASD	autism spectrum disorder
CD	conduct disorder
LD	learning disability
OCD	obsessive compulsive disorder
ODD	oppositional defiant disorder
PDD-NOS	pervasive developmental disorder, not otherwise specified

Keynote

Autism is a spectrum of disorders.

Eating Disorders in Adolescence

- **Anorexia nervosa** is a refusal to maintain a normal body weight and occurs more frequently in adolescent females. The young women have an inaccurate perception of their body size, weight, and shape.
- **Bulimia (bulimia nervosa)** occurs in females who "binge-eat" two or three times per week and then make themselves throw up **(purge).** They tend to fast (not eat) between the binges.
- **Binge-eating disorder** is characterized by binge eating two or three times per week with no compensatory vomiting.

Treatment for eating disorders requires nutritional support and intense psychiatric care.

Keynote

Adolescent eating disorders occur 90% of the time in females.

WORD ANALYSIS AND DEFINITION

S/ = Suffix P/ = Prefix R/ = Root R/CF = Combining Form

WORD	PRONUNCIATION		ELEMENTS	DEFINITION
Asperger syndrome (disorder)	**AHS**-per-ger **SIN**-drohm	 P/ R/	Hans Asperger, Austrian pediatrician, 1906–1980 **syn-** *together* **-drome** *running*	Developmental disorder of children. Combination of signs and symptoms associated with a particular disease process.
autism	**AWE**-tizm		Greek *self*	Developmental disorder of children.
cognitive-behavioral therapy	**KOG**-nih-tiv be-**HAYV**-yur-al **THAIR**-ah-pee	S/ R/ S/ S/ R/ R/	**-ive** *quality of* **cognit-** *thinking* **-al** *pertaining to* **-ior-** *pertaining to* **behav-** *mental or motor activity* **therapy** *treatment*	Psychotherapy that emphasizes thoughts and attitudes in one's behavior.
compulsion **compulsive**	kom-**PULL**-shun kom-**PULL**-siv	S/ R/ S/	**-ion** *action, condition* **compuls-** *to drive, compel* **-ive** *quality of*	Uncontrollable impulses to perform an act repetitively. Possessing uncontrollable impulses to perform an act repetitively.
dyslexia (noun) **dyslexic (adj)**	dis-**LEK**-see-ah dis-**LEK**-sik	S/ P/ R/ S/	**-ia** *condition* **dys-** *difficult, painful* **-lex-** *word* **-ic** *pertaining to*	Impaired reading and writing ability below the person's level of intelligence. Pertaining to or suffering from dyslexia.
hyperactivity	**HIGH**-per-ac-**TIV**-ih-tee	S/ P/ R/	**-ity** *state, condition* **hyper-** *excessive* **-activ-** *movement*	Excessive restlessness and movement.
impulsive	im-**PUL**-siv	S/ P/ R/	**-ive** *quality of* **im-** *not* **-puls-** *to drive*	Inability to resist performing inappropriate actions.
inattention	**IN**-ah-**TEN**-shun	S/ P/ R/	**-ion** *condition* **in-** *not* **-attent-** *awareness*	Lack of concentration and direction.
obsession (noun) **obsessive (adj)**	ob-**SESH**-un ob-**SES**-iv	 S/ R/	Latin *to besiege* **-ive** *quality of* **obsess-** *besieged by thoughts*	Persistent, recurrent, uncontrollable thoughts or impulses. Possessing persistent, recurrent, uncontrollable thoughts or impulses.
purge	PURJ		Latin *to cleanse*	Consciously throw up or cause bowel evacuation.
stereotype	**STER**-ee-oh-tipe	S/ R/CF	**-type** *particular kind, model* **stere/o-** *three-dimensional*	An image held in common by members of a group.

EXERCISES

Build *the correct medical terms based on the statements given. Write the elements on the line and the meaning of the elements below the line.*

1. excessive restlessness and movement: ______________ / ______________ / ______________

2. possessing persistent, recurrent, unwanted thoughts that cannot be suppressed: ______________ / ______________ / ______________

3. impaired reading and writing ability below the person's level of intelligence: ______________ / ______________ / ______________

4. uncontrollable impulses to perform an act repetitively: ______________ / ______________ / ______________

LESSON 16.3 Adulthood, Aging, Senescence, and Death

LIFE SPAN

OBJECTIVES

The information in this lesson will enable you to:

- **Distinguish aging from senescence.**
- **Use correct medical terminology to describe changes that occur with senescence in major organ systems.**
- **Discuss the medical terminology involved in preparations for death.**

You are

. . . a medical assistant working in the Geriatric Clinic at Fulwood Medical Center.

You are communicating with

. . . 85-year-old Mr. Mathew Hickman, who has an early **dementia.**

CASE REPORT 16.4

Mr. Hickman also has a slow-growing prostate cancer, for which he has opted to have no treatment. With the help of his two daughters and day care providers, he is struggling to stay at home. His daughter, Sandra Hotteling, is with him.

Mr. Hickman: I'm still here, you know, somewhere inside this frail old body. I can remember yesteryear, though I'm a bit hazy about today. I can't hear like I used to. But I can still tie my shoes and put my own socks on. I'm not frightened of death, but the process of getting there scares the heck out of me. I've lived my life with dignity. I want to live my death the same way, you know, and leave with my dignity intact.

Keynote

Stage 6 issue: intimacy versus isolation.

Keynote

Stage 7 issue: others versus self.

Keynote

Stage 8 issue: fulfillment versus despair.

DEFINITIONS

- **Life span** is the age to which individual humans aspire to live and the process of getting there.
- **Life expectancy** is the average length of life for any given population.
- **Longevity** is living beyond the normal life expectancy.
- **Aging** is the gradual, spontaneous change resulting in maturation through childhood, adolescence, and young adulthood. Changes then decline, rather than mature, through late adulthood and old age.
- **Senescence** is the loss over time of the capacity of cells to divide, grow, and function, a process that terminates in death.

ADULTHOOD

Stage 6: Young Adulthood—Age 19 to 40 Years

The sense of identity that you acquired in adolescence enables you to achieve intimacy with spouse and children and to view your parents in a different light. Without identity you may well be successful at work, but you will fear commitment, be isolated emotionally, and be unable to depend on anyone else.

Stage 7: Middle Adulthood—Age 40 to 70 Years

Your sense of identity enables you to look outside yourself and care for others through parenting or social commitment. You want to create a living legacy. Without this identity, you remain self-centered, stop growing emotionally, wonder why nobody loves you, and become addicted to work or a substance.

Stage 8: Late Adulthood—Age 70 to Death

When you look back, has life been filled with satisfaction or disappointment? Do you feel good about yourself? "Just as a healthy child does not fear life, a healthy adult will not fear death" (Erikson). Otherwise, old age is filled with despair, fear, and sickness.

The medical study, care, and treatment of the elderly is called **geriatrics,** and the medical specialist in geriatrics is called a **geriatrician.**

SENESCENCE OF ORGAN SYSTEMS

Organ systems begin to show signs of senescence at very different ages and do not degenerate at the same speed. Most physiologic studies show general peak physical performance in the twenties.

Integumentary system changes begin in the forties. Melanocytes (*see Chapter 3*) die, and hair becomes gray and thinner. The skin becomes paper thin, loses elasticity, hangs loose, and wrinkles (*Figure 16.11*). Flat brown-black spots, **senile lentigines (age spots),** appear on the back of the hands and areas exposed to sunlight.

WORD ANALYSIS AND DEFINITION

S/ = Suffix P/ = Prefix R/ = Root R/CF = Combining Form

WORD	PRONUNCIATION		ELEMENTS	DEFINITION
aging **aged**	**A**-jing **A**-jid		Latin *aging*	The process of human maturation and decline. Having lived to advanced age.
dementia	dee-**MEN**-she-ah	S/ P/ R/	**-ia** *condition* **de-** *removal, take away* **-ment-** *mind*	Chronic, progressive, irreversible loss of the mind's cognitive and intellectual functions.
geriatrics **geriatrician**	jer-ee-**AT**-riks jer-ee-ah-**TRISH**-an	R/ R/ S/	**-iatrics** *medical knowledge* **ger-** *old age* **-ian** *one who does*	Medical specialty that deals with the problems of old age. Medical specialist in geriatrics.
gerontology **gerontologist**	jer-on-**TOL**-oh-jee jer-on-**TOL**-oh-jist	S/ R/CF S/	**-logy** *study of* **geront/o-** *old age* **-logist** *specialist*	Study of the process and problems of aging. Specialist in gerontology.
lentigo **lentigines (pl)**	len-**TIE**-go len-**TIHJ**-ih-neez		Greek *lentil*	Age spot; small, flat, brown-black spot in the skin of older people.
life expectancy	LIFE eck-**SPEK**-tan-see	S/ R/	**-ancy** *state of* **expect-** *await*	Statistical determination of the number of years an individual is expected to live.
life span	LIFE SPAN		life, Old English *life* span, Old English *reach*	The age that a person reaches.
longevity	lon-**JEV**-ih-tee	S/ R/	**-ity** *condition* **longev-** *long life*	Duration of life beyond the normal expectation.
senescence **senescent**	seh-**NES**-ens seh-**NES**-ent	 S/ R/	Latin *to grow old* **-ent** *end result, pertaining to* **senesc-** *growing old*	The state of being old. Growing old.
senile **senility**	**SEE**-nile seh-**NIL**-ih-tee	S/ R/ S/ R/	**-ile** *pertaining to* **sen-** *old age* **-ity** *condition* **senil-** *senile*	Characteristic of old age. Mental disorders occurring in old age.

Special senses start to decline in the twenties. Visual acuity declines at that time. In the forties, presbyopia (*see Chapter 4*) appears, and many people develop cataracts in old age. Hearing loss occurs as the ossicles become stiffer and the number of cochlear hair cells declines (*see Chapter 4*). This was the cause of Mr. Hickman's hearing loss. Taste and smell are blunted late in life as taste cells and olfactory buds decline in number.

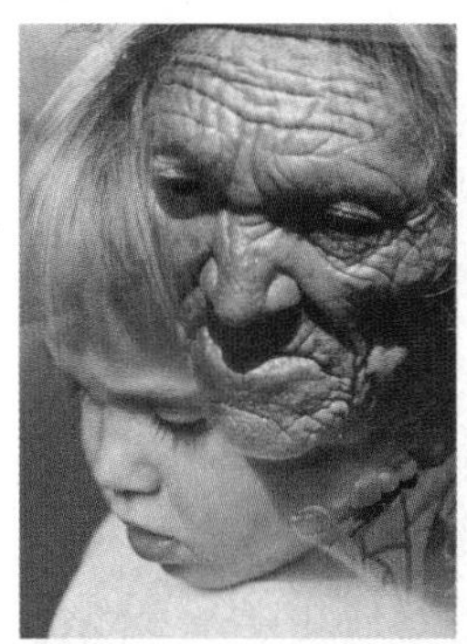

FIGURE 16.11
Senescence of the Skin.

EXERCISES

Lesson objective: *Complete the lesson objective by identifying the final stages of development in Erickson's theory. Fill in the chart.*

Stage	Ages	Major Issues
6		
7		
8		

Pick any one of these stages, and write a short description of what might happen to a person in that stage.

Keynote

Exercise and good nutrition help prevent osteopenia.

Keynote

Exercise and good nutrition help prevent muscle degeneration.

Keynote

Exercising your brain enhances the quality of life in old age.

Keynote

Exercise and good nutrition extend longevity and enhance the quality of life.

Keynote

Because of lowered immunity, vaccinations against **influenza** and other seasonal infections are recommended for the elderly.

SENESCENCE OF ORGAN SYSTEMS (continued)

Skeletal system changes appear in the thirties, when osteoblasts become less active than osteoclasts. The result is osteopenia, which goes on to become osteoporosis, particularly in postmenopausal women. Joints in the older age groups have less synovial fluid and thinner articular cartilage. Osteoarthritis results (*see Chapter 9*).

Muscular system changes occur with age as muscle mass and strength are lost **(sarcopenia).** As muscle atrophies, there are fewer muscle fibers to do the work and the available blood supply is decreased.

Nervous system changes begin around age 30, when the brain weighs twice as much as it does at age 75. Motor coordination, balance, intellectual function, and short-term memory decline more quickly than long-term memory and language skills.

Cardiovascular systems always show coronary artery atherosclerosis from an early age. As a result, when aging myocardial cells die, the heart wall gets thinner and weaker and cardiac output declines. This causes the decline in physical capabilities with aging. The atherosclerotic plaques narrow arteries and trigger thrombosis, leading to strokes and heart attacks (*see Chapter 6*). In veins, valves become weaker and blood flows back and pools in the legs, leading to poor venous return to the heart and heart failure.

Respiratory system changes are noticeable in the thirties, when pulmonary ventilation declines, a factor in the gradual loss of stamina. The rib cage becomes less flexible; the lungs, less elastic, with fewer alveoli. Respiratory function declines. As respiratory health declines, hypoxic degenerative changes occur in all the other organ systems.

Urinary system changes begin in the twenties, when the number of nephrons starts to decline. Later in life, many of the remaining glomeruli become atherosclerotic. The **glomerular filtration rate (GFR)** decreases, and the kidneys become less efficient. For example, drug doses in the elderly need to be reduced because drugs cannot be cleared from the blood as rapidly.

Immune system function declines in the elderly as the amounts of lymphatic tissue and red bone marrow decrease with age. This leads to a reduction in both cellular and humoral (antibody) immunity. This means that the elderly have less protection against infectious diseases and cancer.

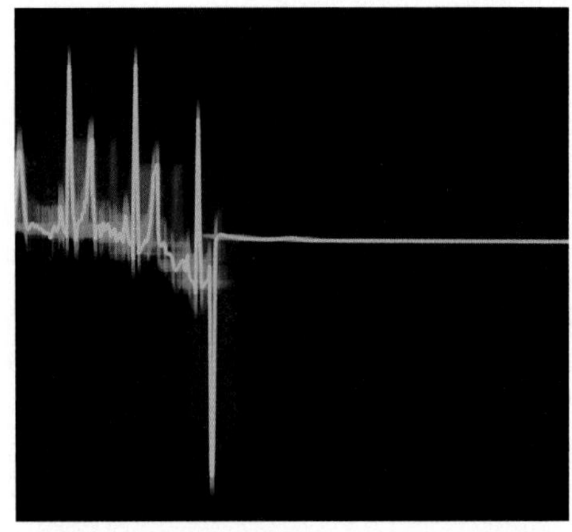

▲ **FIGURE 16.12 Electroencephalogram Shows Brain Death.**

DYING AND DEATH

Death is inevitable. Just as fetal life in the womb is a preparation for birth, so living and aging are a preparation for death. The process of dying, rather than death itself, is of concern to most elderly people. Dying should be dignified and free from physical and emotional pain. A **hospice** provides **palliative care** and provides for the emotional and spiritual needs of terminally ill patients and their loved ones at an in-patient facility or in the patient's home. Palliative care is designed to provide pain and symptom management to maintain the highest quality of life for as long as life remains.

There is no universally accepted moment of biological death. In most states in the United States, death is now defined in terms of **brain death (BD),** in which there is no cerebral or brainstem activity and the EEG is flat for a specific length of time (*Figure 16.12*). Two other conditions involving brain damage and loss of brain function cause medical difficulty:

1. **Persistent vegetative state (PVS)** occurs in people who suffer enough brain damage, usually from trauma, that they are unaware of themselves or their surroundings, even though their eyes are open. Yet they still have certain reflexes and can breathe and pump blood because the brainstem still functions. With medical care and artificial feeding, patients can survive for decades.
2. **Minimally conscious state (MCS)** is a condition of severely altered **consciousness** in which minimal, inconsistent evidence of awareness of self or surroundings is demonstrated. PET scans of MCS patients show cortical function, for example, when the patients' loved ones speak to them. They are more likely to improve than are PVS patients. Trauma is a common cause of MCS.

Keynote

PVS and MCS differ from **coma**, in which the person is unresponsive and his or her eyes are closed.

If the cause of death is uncertain or a crime is suspected, an **autopsy (postmortem)** is performed.

WORD ANALYSIS AND DEFINITION

S/ = Suffix P/ = Prefix R/ = Root R/CF = Combining Form

WORD	PRONUNCIATION	ELEMENTS		DEFINITION
autopsy (same as postmortem) **postmortem**	**AWE**-top-see post-**MOR**-tem	 S/ P/ R/	Greek *see with one's own eyes* **-em** *condition* **post-** *after* **-mort-** *death*	Examination of the body and organs of a dead person to determine the cause of death.
coma	**KOH**-mah		Greek *deep sleep*	State of deep unconsciousness.
consciousness	**KON**-shus-ness	S/ R/	**-ness** *quality, state* **conscious-** *to be aware of*	The state of being aware of and responsive to the environment.
death	DETH		Old English *to die*	Total and permanent cessation of all vital functions.
hospice	**HOS**-pis		Latin *lodging*	Provides care to the dying and their families.
influenza	in-flew-**EN**-zah		Latin *caused by the influence of the heavenly bodies*	Acute viral infection of upper and lower respiratory tracts.
palliative care	**PAL-**ee-ah-tiv KAIR	S/ R/ R/	**-ive** *quality of* **palliat-** *reduce suffering* **care** *be responsible for*	To relieve symptoms and pain without curing.
sarcopenia	sar-koh-**PEE**-nee-ah	S/ R/CF	**-penia** *deficiency* **sarc/o-** *flesh*	Progressive loss of muscle mass and strength in aging.
vegetative	**VEJ**-eh-tay-tiv	S/ R/	**-ive** *quality of* **vegetat-** *growth*	Functioning unconsciously, as plant life is assumed to do.

Abbreviations

BD	brain death
GFR	glomerular filtration rate
MCS	minimally conscious state
PVS	persistent vegetative state

EXERCISES

Practice *using your medical vocabulary. Write a brief scenario about a patient, using as many of the terms in the above WAD box as you reasonably can. Be sure to proofread your scenario for correct spelling when you are finished writing. Be prepared to read the scenario aloud in class if called on.*

Scenario:

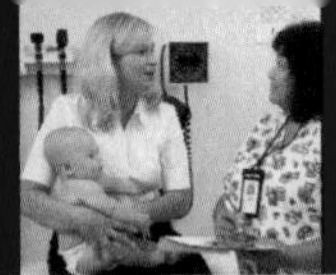

CHAPTER 16 REVIEW

LIFE SPAN

CHALLENGE YOUR KNOWLEDGE

A. **Roots/combining forms** are the foundation of every term. Add prefixes and suffixes to the root/combining form *encephal/o* to build the terms for the definitions given.

1. absence of cerebral hemispheres:

   ______________/______________/______________

2. enlargement of the ventricles with excessive CSF:

   ______________/______________/______________

3. protrusion of nervous tissue and meninges through a defect in the skull:

   ______________/______________/______________

4. small head:

   ______________/______________/______________

Using this same root, and *prefixes and suffixes you have learned from previous chapters,* build the terms for the following definitions.

5. record of the electrical activity of the brain:

   ______________/______________/______________

6. inflammation of the brain:

   ______________/______________/______________

7. inflammation of the brain and spinal cord:

   ______________/______________/______________

8. disease of the brain:

   ______________/______________/______________

B. **Correct form** of the medical term: In addition to knowing how to spell and pronounce medical terms, you must also know how to use the correct form of the term (singular, plural, noun, verb, adjective). The terms are given to you—use the correct form in the appropriate sentence. You will not use all the terms. *Hint:* Once you have used a term, cross it off the list. Fill in the blanks.

cyanotic	infant	adolescents	cyanosis
stimulant	senile	microcephaly	senescent
compulsion	adapt	senescence	stimuli
senility	adaptation	infancy	longevity
microcephalic	adolescence	compulsive	stimulus

1. The body will ______________ to the loss of a limb.
2. Because of the blue color of the patient's nail beds, the doctor is considering a diagnosis of ______________.
3. This child has a ______________ skull.
4. Something that excites an organ or part is called a ______________. The plural of this term is ______________. Caffeine can be termed a ______________.

5. Her baby is in his first year of ______________________________.

6. Her ______________________________ to set fires is called *pyromania.*

7. ______________________________ is a state that begins with puberty and ends with physical maturity.

8. The patient's ______________________________ hand washing can be relieved with intense therapy.

9. Some ______________________________ prefer to see a family practitioner instead of a pediatrician.

10. Please place the ______________________________ on the scale.

11. Her grandfather suffers from ______________________________.

12. A ______________________________ organ is aged, with loss of function.

C. **Make your own study hint:** Association with English words sometimes makes medical terms easier to remember.

1. You are given the English word *thermometer:*

   Therm/o means ______________________________.

   Meter means ______________________________.

2. What is the purpose of a thermometer?

   ____

   ____

3. What medical term in this chapter means *low core body temperature?*

   ____________________ / ____________________ / ____________________

4. Deconstruct the above term:

   P: ______________________________ means ______________________________

   R: ______________________________ means ______________________________

   S: ______________________________ means ______________________________

5. Write here your own study hint for helping you remember the meaning of the medical term by associating the English term with it.

   ____

   ____

   ____

   ____

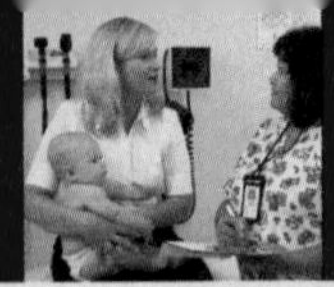

LIFE SPAN

D. Review the *language of pediatrics and gerontology,* and choose the correct medical terms to write in the blanks. *Watch your spelling!*

1. An artery that is open:

   ____

   (occluded patent)

2. Predisposition to character and personality:

   ______________________________.

   (parameter temperament)

3. Treatment using light rays:

   ____

   (fototherapy phototherapy)

4. Result of excess bilirubin in tissues:

   ____

   (jaundice pertussis)

5. Another name for whooping cough:

   ____

   (diphtheria pertussis)

6. Agent that reduces surface tension in lung alveoli:

   ____

   (surfactant ventilator)

7. Pertaining to or derived from the mother:

   ____

   (paternal maternal)

8. Capable of performing a task or function:

   ____

   (independent competent)

9. Eating substances not considered to be food:

   ____

   (encopresis pica)

E. **Prefixes:** Match the meaning of the term in column 1 with the correct prefix in column 2. Fill in the blanks.

	Column 1	Column 2
_____	1. apart from	A. hypo
_____	2. not	B. re
_____	3. without	C. syn
_____	4. two	D. a
_____	5. together	E. micro
_____	6. below	F. post
_____	7. back	G. dis
_____	8. through	H. in
_____	9. after	I. bi
_____	10. small	J. per

F. **Terminology challenge:** This is a three-part exercise. Medical terms can sound the same but can have different spellings and different meanings, like *ileum* and *ilium.* (Remember them?)

An example from this chapter is:

adolescence meaning *the state (or time of) being an adolescent*

adolescents plural of the term *adolescent,* which means *a person who is at this stage of life*

Part One: In the following sentences, use the correct form of either of the two terms above.

1. ______________________________ are sometimes confused about their feelings.

2. ______________________________ is an especially difficult time to lose a parent.

Part Two: You compose a sentence for each of the above terms.

3. *adolescence:*

____

____

____

4. *adolescents:*

____

____

____

Part Three: **Dictate** your sentences (numbers 3 and 4) to a classmate, and then see if he or she used the correct spellings (and meanings) of the terms. *How did your classmate do?*

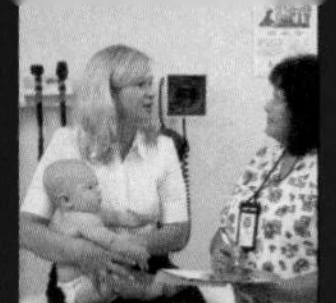

LIFE SPAN

G. **Discussion question or short answer:** You have completed your study of all the body systems. Explain the difference between *food allergy* and *food intolerance.* Be sure to explain which body systems are affected in each process and how they react. Keep your answer brief but to the point. *You may be called on in class to read your explanation.*

H. **Seek and find:** Some English words or medical terms in this book are defined in context and do not always appear in a WAD box. Other terms have a more common name by which they are known. Here is a sample of words and terms that have appeared in this chapter. Challenge yourself to answer without turning back to check in the book. Fill in the chart, and then check your answers in the text, a dictionary, or the Glossary.

Word	**Meaning/Common Name**
septum	
precipitate	
in utero	
gavage	
pertussis	
etiology	
ritual	
lentigines	
visual acuity	
terminally ill	

I. **Spelling demons:** Every chapter contains spelling demons. Challenge your ability to spell all these terms correctly. Circle the best choice.

1.	hyaline	hyeline	hialine
2.	gavage	garage	gavahge
3.	diptheria	diphtheria	diphteria
4.	adolescence	adolesence	adolessence
5.	bulemia	bulimia	bullemia
6.	remmination	rumenation	rumination
7.	lentiggines	lentegines	lentigines
8.	senescence	senessence	seniscence
9.	conciousness	consciousness	consiousness
10.	paliative	palliative	paliateve

J. **Deconstruct** the following medical terms into their basic elements. Fill in the chart. Complete the exercise by *defining* each of the terms on the lines provided.

Medical Term	Meaning of Prefix	Meaning of Root/CF	Meaning of Suffix
agenesis			
anorexia			
cyanosis			
dementia			
dyslexia			
enuresis			
gerontologist			
microcephalic			
pertussis			
tetralogy			

Definitions:

agenesis ______________________________

anorexia ______________________________

cyanosis ______________________________

dementia ______________________________

dyslexia ______________________________

enuresis ______________________________

gerontologist ______________________________

microcephalic ______________________________

pertussis ______________________________

tetralogy ______________________________

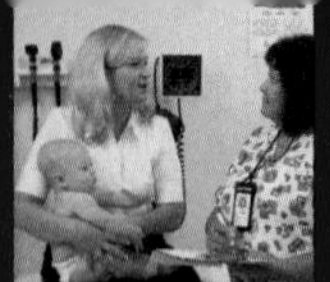

CHAPTER 16 REVIEW

LIFE SPAN

K. Translate the following sentences from medical language to nonmedical language your patient can understand.

1. "Infants with RDS are at risk for cerebral ischemia, hemorrhage, and neonatal death."

2. "In senescence of the cardiovascular system, the atherosclerotic plaques occlude narrow arteries and trigger thrombosis, leading to CVAs and MIs."

3. "TTN occurs most often in C-section and precipitate vaginal deliveries."

L. Suffixes: Knowledge of suffixes provides an important clue to the meaning of a medical term. The suffixes are given to you—fill in the meaning of each suffix, give an example of a medical term *from this or a previous chapter* that ends with this suffix, and give the meaning of the medical term.

Suffix	Meaning of Suffix	Medical Term with This Suffix	Meaning of Medical Term
al			
ar			
ase			
ia			
ician			
ics			
ion			
ive			
logy			
ness			
osis			

M. **Latin and Greek terms** cannot be further deconstructed into prefix, root, or suffix for a clue as to the meaning of the term. You simply must know them for what they are. Every one of the answer choices is either a Latin or a Greek term. Choose the correct term to write on the line.

1. to lie open:

   ______________________________ (patent precipitate)

2. same as *postmortem:*

   ______________________________ (autopsy coma)

3. forms a leathery coating on the pharynx:

   ______________________________ (encopresis diphtheria)

4. bouts of eating and then throwing up:

   ______________________________ (bulimia pica)

5. developmental disorder of children:

   ______________________________ (autism obsession)

6. to cause bowel evacuation:

   ______________________________ (senescence purge)

7. deep sleep:

   ______________________________ (coma death)

8. fissure, separation of parts:

   ______________________________ (bifid cleft)

9. first bowel movement of newborn:

   ______________________________ (meconium hyaline)

10. abnormal passage:

    ______________________________ (fissure fistula)

11. bedwetting:

    ______________________________ (encopresis enuresis)

CHAPTER 16 REVIEW

LIFE SPAN

N. Abbreviations: Choose the correct abbreviation to match the description given. You have more answers than you need. Fill in the blanks.

ASD	CPAP	SIDS	CVS	TTN	BD	LD
ADD	TOF	RDS	MCS	VSD	FTT	COPD

1. sudden death of an infant for which no cause can be found on autopsy __________
2. opening between two upper chambers of the heart __________
3. amniotic fluid remains in the infant's lungs and causes a self-limiting respiratory distress __________
4. also called *hyaline membrane disease* __________
5. infant or child who is not growing as expected __________
6. opening between the two lower chambers of the heart __________
7. body system made up of the heart and blood vessels __________
8. condition indicated by a flat EEG for a specific length of time __________
9. four heart defects all shunting blood away from the lungs __________
10. more likely to improve than PVS patients __________
11. difficulty in understanding words or sounds __________

O. Chapter challenge: Circle the correct answer.

1. What do these two terms have in common: *bifida* and *tetralogy?*
   a. Neither one has a prefix.
   b. Each one is a surgical procedure.
   c. They both refer to a number.
   d. Their suffixes both mean *condition.*
   e. They have nothing in common.
2. The purpose of *palliative* care is to:
   a. make the patient comfortable
   b. relieve pain
   c. reduce suffering
   d. not cure
   e. all of the above
3. *Phototherapy* is a specific treatment for:
   a. hypothermia
   b. senescence
   c. jaundice
   d. celiac disease
   e. diphtheria
4. Which statement applies to *agenesis?*
   a. constantly chewing food
   b. born with one kidney instead of two
   c. an unresponsive state of consciousness
   d. repeated soiling with feces
   e. refusal to maintain normal body weight

> *Study Hint*
> Immediately cross off any answer you know is not correct. In your remaining choices, there is only *one best answer.*

5. Which term contains only two roots?

a. lentigo
b. dementia
c. geriatrics
d. hospice
e. influenza

6. Repeated soiling with feces is:

a. pica
b. anorexia
c. bipolar disorder
d. enuresis
e. encopresis

7. In the acronym *APGAR,* what does the "R" stand for?

a. rooting
b. regurgitation
c. reflux
d. respiration
e. rumination

8. Loss of cell function and being old is:

a. longevity
b. senescence
c. senility
d. dementia
e. geriatrics

9. If something is *postmortem,* it has occurred after:

a. eating
b. surgery
c. death
d. fainting
e. illness

10. The term *cyanosis* has an element meaning:

a. glass
b. surface
c. blue
d. heat
e. brain

11. If your patient has very low core body temperature, he or she has:

a. hypoglycemia
b. hypotension
c. hypothermia
d. hypochondriasis
e. hypophysis

12. An 8-year-old would see a *pediatrician,* and an 80-year-old would see a:

a. dermatologist
b. orthopedist
c. cardiologist
d. gerontologist
e. otologist

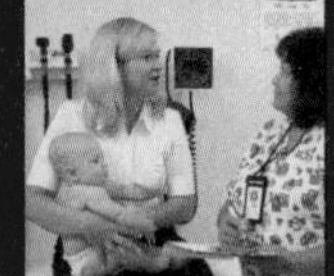

LIFE SPAN

P. **Case Report challenge:** Now that you are more comfortable with the terms in this chapter, you can apply that knowledge and briefly answer the questions about the Case Reports. If you read the reports through and underline all the medical terminology, this will make it easier to answer the questions.

CASE REPORT 16.1

You are

. . . a pediatric medical assistant employed by Sandra Mendes MD, a pediatrician at Fulwood Medical Center. You are working in her Well-Baby Clinic.

You are communicating with

. . . Mrs. Anna Hotteling, a 35-year-old mother, who has brought her 10-week-old daughter, Carol, to the Well-Baby Clinic. Carol, Mrs Hotelling's first baby, was born via a normal vaginal delivery at term. She weighed 7 pounds 2 ounces. She had normal Apgar scores. Carol had persistent jaundice when seen at 2 weeks of age. She is being breastfed. You ask Mrs. Hotteling if she has any concerns.

Mrs. Anna Hotteling, the mother, has several areas of concern. A friend's baby recently died of SIDS (sudden infant death syndrome). How can she prevent this for her child? Carol wants to feed every 2 to 3 hours day and night, and she (Anna) is exhausted. How long will this go on, and for how long should she continue breastfeeding? Mrs. Hotteling also wants to know what her baby can see and how she should communicate with her.

CASE REPORT 16.3

You are

. . . a pediatric medical assistant working with Dr. Sarah Mendes at Fulwood Medical Center. You are working in her Well-Baby Clinic.

You are communicating with

. . . 7-year-old Carol Hotteling and her mother. You have watched Carol grow over the 7 years that you have worked in Dr. Mendes' Well-Baby Clinic. Her mother, Mrs. Anna Hotteling, is telling you about her current concerns:

"As you know, she's always been a handful. The neighbors call her a tomboy. She's trashed every room in the house. The other kids' mothers stopped inviting her to parties. Nothing holds her attention for very long. I've lost her a couple of times in the shopping mall when she's darted away. I always thought she'd grow out of it. But now she's in second grade. The teacher's complaining that Carol can't sit still, she interferes with the other kids, disrupts the class, and she's having trouble reading and staying focused. It's got to be due to something I did or didn't do, hasn't it?"

1. What does it mean to be born normally "at term"?

____

____

____

2. What does the *Apgar* score measure?

3. Carol had *jaundice* as a baby. Describe what that is and a treatment for it.

4. What does the abbreviation *SIDS* stand for?

5. Describe SIDS:

6. What does medical science think is a possible cause of SIDS?

7. At 7 years of age, Carol is in which of Erickson's stages?

8. What is the main issue at this stage of development?

9. List all of Carol's behavioral issues:

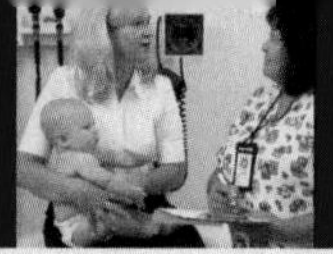

CUMULATIVE REVIEW

These exercises are built around terms and elements from this and the preceding three chapters. They are an excellent source of review for tests—be sure to do every exercise!

A. Suffixes: Medical terms are given in the first column. Begin by slashing each term into its elements. Write the suffix in the second column and its meaning in the third column. Challenge yourself to think of another medical term with the same suffix from any previous chapter. Write that term in the fourth column, and write its meaning in the last column. *Your goal is to complete the entire chart!*

Medical Term	Suffix	Meaning of Suffix	Another Medical Term with This Suffix	Meaning of This New Medical Term
cyanosis				
urologist				
episiotomy				
glomerulonephritis				
replication				
mammary				
nephrectomy				
myelocele				
amniocentesis				
gonorrhea				
dyslexic				
nephropathy				
implantation				

B. Abbreviations: The abbreviations are given in this exercise. Write out in words the meaning of each abbreviation. Fill in the blanks.

The answer is not correct unless all the words are spelled correctly!

1. PID: ____________________
2. GDM: ____________________
3. PCOS: ____________________
4. IVP: ____________________
5. ARF: ____________________
6. OCD: ____________________
7. PPH: ____________________
8. STD: ____________________
9. PGY: ____________________
10. ESRD: ____________________

C. Diagnosis/procedure: The following medical terms are either a diagnosis or a procedure. Place a check mark (✔) in the appropriate column for diagnosis or procedure. In the last column, write the name of the specialist who would be using that medical term in dictation and billing.

orthopedist	geriatrician	neurologist	immunologist
pediatrician	endocrinologist	urologist	gynecologist

Medical Term	**Diagnosis**	**Procedure**	**Specialist**
gynecomastia			
vestibulectomy			
mastalgia			
hyaline membrane disease			
meconium aspiration syndrome			
colpopexy			
mammography			
lithotripsy			
encopresis			
dysmenorrhea			
dementia			
fibroadenoma			
pertussis			
senility			
herpes simplex			
enuresis			

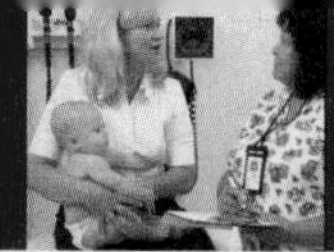

D. **Roots/combining forms** are always your best clue for understanding the meaning of a term. The R/CF is given to you in column 1; define its meaning in column 2. In column 3, give an example of a term in which it appears. Write in the final column the meaning of the medical term.

Root/CF	Meaning of R/CF	Medical Term Containing This R/CF	Meaning of Medical Term
cyst			
gestat			
cyan			
hyster			
encephalo			
pyel			
photo			
mast			
meno			
radio			
genesis			
hydr			
metri			
emesis			
didymis			

E. **Prefixes** generally add description to a medical term—color, location, size, and the like. Build the following terms by completing them with the appropriate prefix. Then write the meaning of the prefix.

in	contra	noct	poly
ab	crypt	neo	an
pre	oligo	retro	auto
hyper	syn	en	com

1. tilting backward of the uterus: ______________________ /vers/ion

   Meaning: ______________________

2. too little amniotic fluid: ______________________ /hydr/amnios

   Meaning: ______________________

3. failure of one or both testes to descend into the scrotum: ______________________ /orch/ism

   Meaning: ______________________

4. an aversion to food or no appetite for food: ______________________ /orex/ia

   Meaning: ______________________

5. whole body response to a deficiency: ______________________________ /drome

Meaning: ______________________________

6. destruction of cells by enzymes within the cells: ______________________________ /lysis

Meaning: ______________________________

7. more than normal frequency of menses: ______________________________ /meno/rrhea

Meaning: ______________________________

8. increase in the size of cells in a tissue or organ: ______________________________ /trophy

Meaning: ______________________________

9. uncontrollable impulse to perform an act repeatedly: ______________________________ /puls/ion

Meaning: ______________________________

10. birth occurring before 37 weeks of gestation: ______________________________ /term

Meaning: ______________________________

11. failure to conceive: ______________________________ /fertility

Meaning: ______________________________

12. bedwetting: ______________________________ /uresis

Meaning: ______________________________

13. removal of tissue to destroy its function: ______________________________ /lat/tion

Meaning: ______________________________

14. excessive urination at night: ______________________________ /uria

Meaning: ______________________________

15. a newborn infant: ______________________________ /nate

Meaning: ______________________________

16. prevention of pregnancy: ______________________________ /cept/ion

Meaning: ______________________________

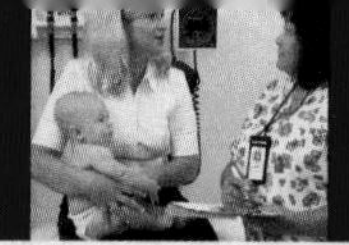

CUMULATIVE REVIEW

F. **Spelling demons** Some medical terms are particularly difficult to pronounce and spell. Develop your ability to recognize a misspelling when you see it, and know how to correct it. *Only one term in this exercise is spelled correctly.* For each incorrectly spelled term, write the correct spelling on the line next to it.

1. syphilus ____________________
2. efacement ____________________
3. senesence ____________________
4. vasovasostomy ____________________
5. microencephaly ____________________
6. puerperiam ____________________
7. transuretral ____________________
8. paliative ____________________
9. menarch ____________________
10. apotosis ____________________
11. circumcicion ____________________
12. obstetrican ____________________
13. chankre ____________________
14. mylomeningocele ____________________
15. gerantology ____________________
16. adeptation ____________________
17. pedeatrics ____________________
18. dysneic ____________________
19. tetralegy ____________________
20. microsephalic ____________________

Monitor My Progress: An honest self-assessment of your progress with this material will make it easier for you to learn from your mistakes. Jot down below any wrong answers you have corrected; terms, elements, or spellings you find difficult to remember; and so on. The more times you see, speak, or write the terms, the better you will come to know them.

Based on these review exercises, I need to spend more time on:

___

___

___

___

___

___

___

Appendices

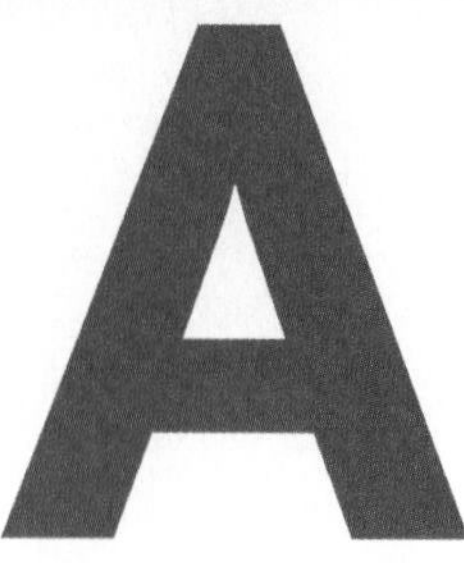

End-of-Book Exercises

A. Roots/combining forms: Each of the elements in the first column refers to a specific body part. Test your knowledge of roots/combining forms by filling in the chart.

	Root/Combining Form	**Meaning of R/CF**	**Medical Term with This R/CF**	**Definition of This Term**
1.	articul			
2.	blephar			
3.	brachi			
4.	cephal			
5.	cerebr			
6.	chondro			
7.	crani			
8.	entero			
9.	lymphadeno			
10.	metri			
11.	myel			
12.	myo			
13.	naso			
14.	nephr			
15.	neuro			
16.	ophthalmo			
17.	pector			
18.	pneum			
19.	pulmono			
20.	pyelo			
21.	ren			
22.	rhin			
23.	thorac			
24.	vascul			
25.	occul			
26.	os			
27.	axill			
28.	proct			

B. Spelling demons: Precision in communication and professionalism requires correct spelling of all medical terms. Circle the correct spelling of each term, and write the term on the line. Then write a brief definition of the medical term on the line below.

1. akromion acrommion acromion ______________________

   Definition: ______________________

2. teratogen terratogin terattogen ______________________

   Definition: ______________________

3. urticaria erticarria urtticaria ______________________

   Definition: ______________________

4. cirrosis sirosis cirrhosis ______________________

   Definition: ______________________

5. anurism aneurysm anerysm ______________________

   Definition: ______________________

6. systocele cystocele cystosele ______________________

   Definition: ______________________

7. osteomyelitis osteomylitis osteomyeletis ______________________

   Definition: ______________________

8. allopesia alopecia alopecsia ______________________

   Definition: ______________________

9. preeclampsia preclampsia preklampsia ______________________

   Definition: ______________________

10. epilepsy epelepsy epilepsy ______________________

    Definition: ______________________

11. pneumonthorax pneumothorax pnemonthorax ______________________

    Definition: ______________________

12. cholesteatoma colesteatoma colestiatoma ______________________

    Definition: ______________________

13. umbillical umbilical umbilikal ______________________

    Definition: ______________________

14. ketoacidosis ketoneacidosis ketoasidosis ______________________

    Definition: ______________________

15. epididimis   epididymus   epididymis   ____________________

Definition: ____________________

16. encefalocele   encephalosele   encephalocele   ____________________

Definition: ____________________

C. **Suffixes** provide an important clue to the meaning of a medical term. Always begin to decipher a medical term at the back element (suffix), and then look at the front element (prefix). The root/combining form will fill in the rest of the meaning of the term. Complete the chart.

	Suffix	Meaning of Suffix	Medical Term with This Suffix	Definition of This Term
1.	pathy			
2.	ptosis			
3.	oid			
4.	itis			
5.	scope			
6.	logy			
7.	megaly			
8.	fusion			
9.	opsy			
10.	gram			
11.	oma			
12.	stasis			
13.	pnea			
14.	emia			
15.	rrhea			
16.	um			

D. **Abbreviations:** Reading patient documentation to extract or abstract information for coding purposes requires that you understand what you are reading. Identify the abbreviations in the chart below. Use a check mark (✓) to indicate whether the abbreviation is a diagnosis, a procedure, or something else. In the last column, write the meaning of the abbreviation.

	Abbreviation	Diagnosis	Procedure	Other	Meaning of Abbreviation
1.	SLE				
2.	AKA				
3.	GI				
4.	CXR				
5.	EEG				
6.	AMI				

	Abbreviation	Diagnosis	Procedure	Other	Meaning of Abbreviation
7.	c/o				
8.	WNV				
9.	PID				
10.	CABG				
11.	PKD				
12.	MRI				
13.	RUQ				
14.	CVS				
15.	DKA				
16.	URI				

E. Prefixes: Insert the correct prefix to complete each medical term.

1. an agent that decreases urine production ______________ diuretic
2. high blood sugar without ketoacidosis ______________ osmolar
3. decreased number of RBCs ______________ emia
4. junction between two nerve cells ______________ apse
5. deficiency of all types of blood cells ______________ cytopenia
6. excessive vomiting ______________ emesis
7. stick together to form clumps ______________ glutinate
8. absence of urine production ______________ uria
9. chemical agent that relays message to nerve cells ______________ mitter
10. large WBC that removes bacteria ______________ phage

F. Grammatical forms—noun, verb, adjective: Write *on the line* the correct form of the medical term in the parentheses. *Under the line* identify the term as a noun, verb, or adjective.

1. The in vitro *(fertilize/fertilization)* ______________ was successful after the third attempt.
2. Breast cancer can *(metastasize/metastasis)* ______________ to the brain.
3. The patient's *(retinal/retina)* ______________ is in danger of detaching.
4. The *(digestive/digest)* ______________ tract is also called the *alimentary canal.*
5. Arteries can become *(occlusion/occluded)* ______________ with plaque.
6. A *(resection/resect)* ______________ is a cutting back or removal of a body part.
7. Because of his injury, the patient could not *(circumduct/circumduction)* ______________ his left arm.
8. Her high level of blood glucose made her *(hyperglycemia/hyperglycemic)* ______________.

G. **Translate** the following medical statements into language your patient can understand. Rewrite the sentences below.

1. Dr. Lee believes that the third lesion is a squamous cell carcinoma and wants to perform a biopsy-removal of the cutaneous lesion.

2. Mrs. Hughes, your conjunctivitis involves your conjunctiva and eyelids, two of the periorbital accessory structures of the eye.

3. Bariatric surgery is used for the treatment of obesity.

4. The heart is located in the thoracic cavity between the two lungs in an area called the mediastinum.

5. Intervertebral discs consist of fibrocartilage and inhabit the intervertebral space between the bodies of adjacent vertebrae.

6. Mr. Jacobs has a history of COPD. Last evening he had increased SOB and a cough that produced yellow sputum.

7. The muscles on the lateral side of the leg evert the foot.

8. Maintaining normal plasma glucose levels is the basis for good management of DM.

H. Suffixes: One of the most prevalent meanings of a suffix in a medical term is "pertaining to." Write below all the suffixes you can find in the text that mean "pertaining to."

I. Proofread the following sentences, and make the necessary corrections on the lines below. There are errors in fact and/or spelling. Only one sentence is entirely correct.

1. Short bones, like the tibia, carpals, and patela have a length nearly equal to their width.

2. Shoulder subluxation occurs when the ball of the radius slips out of the socket of the scapulla, usually posteriorly.

3. The pituitary gland secretes serotonin by day and converts it to melatonin at night.

4. Platelets that have developed from monocytes ingest and destroy antigens, tissue debris, bacteria, and other foreign matter by the process of photocytosis.

5. Visceral nerves carry signals from the viscera in the cranial and spinal cavities.

6. Excretion is an essential process in maintaining hemostasis.

7. Vulvovaginitis is a chronic, lasting, severe pain around the vaginal orifice; dispareunia is common.

____

____

____

8. Cells from the blastocyst, together with endometrial cells, form the placenta.

____

____

____

J. Plurals: A lot of medical terms do not form their plurals with a simple "s" at the end of the term. You are given the singular term; write the plural form of the term in the next column and a brief definition in the final column.

	Singular	Plural	Definition
1.	varix		
2.	phalanx		
3.	fungus		
4.	sperm		
5.	patella		
6.	bacterium		
7.	bronchus		
8.	vertebra		
9.	nevus		
10.	foramen		

K. Roots/combining forms: Roots and combining forms are the foundation of all medical terms. Each of the elements below is the basis for a specialist's title. Complete the chart.

	R/CF	Meaning of R/CF	Medical Specialist Title with This R/CF	Definition of This Term
1.	cardio			
2.	chiro			
3.	dermato			
4.	gastr			
5.	gyneco			
6.	hemat			
7.	nephro			
8.	neuro			
9.	opthalmo			
10.	ortho			

	R/CF	Meaning of R/CF	Medical Specialist Title with This R/CF	Definition of This Term
11.	osteo			
12.	oto			
13.	ped			
14.	psycho			
15.	pulmono			
16.	uro			

L. Multiple choice: After you have read the question and answer choices, start by eliminating the answers you know are incorrect. Then circle the correct answer. *Remember:* There is only *one best* answer.

1. Which term contains an element meaning "water"?
   - a. hematuria
   - b. arthroscopy
   - c. dehydration
   - d. bilateral
   - e. dermatologist
2. In the term *arthroscopy,* the element *scopy:*
   - a. is a root
   - b. indicates a procedure
   - c. is a suffix
   - d. is a root that functions as a suffix
   - e. is a suffix that indicates a procedure
3. In the term *bilateral,* the element *bi* means:
   - a. before
   - b. two
   - c. black
   - d. many
   - e. deficient
4. "Pink eye" is associated with which medical term?
   - a. periorbital
   - b. lacrimal
   - c. conjunctivitis
   - d. stye
   - e. ptosis
5. The tube linking the pharynx and the stomach is the:
   - a. larynx
   - b. esophagus
   - c. trachea
   - d. bronchus
   - e. colon
6. The term *thoracic* has a root that means:
   - a. lung
   - b. stomach
   - c. tube
   - d. chest
   - e. cavity
7. Which term has no prefix?
   - a. cyanosis
   - b. respiration
   - c. decongestant
   - d. epistaxis
   - e. paranasal

8. The term with an element meaning "around" is:
   a. medullary
   b. osteogenesis
   c. sarcoma
   d. malunion
   e. periosteum
9. Which of the following terms is an adjective?
   a. dislocation
   b. clavicular
   c. scapulae
   d. radius
   e. dorsum
10. Which term has two prefixes?
    a. endocrinology
    b. hypothalamus
    c. adrenocorticotropic
    d. antidiuretic
    e. polyphasia
11. An immediate, severe, allergic response is:
    a. anaphylaxis
    b. homeostasis
    c. hemastasis
    d. hormonal
    e. autoimmune
12. The term with an element meaning "to record" is:
    a. electroneurodiagnostic
    b. electroencephalograph
    c. electroencephalogram
    d. electroencephalography
    e. none of the above
13. A hollow sac that holds fluid is the:
    a. hilum
    b. urethra
    c. nephron
    d. ureter
    e. bladder
14. Which medical term can be applied to more than one place in the body?
    a. fornix
    b. cervical
    c. majora
    d. ovarian
    e. ova
15. Which of the following is (are) performed to test for chromosomal abnormalities?
    a. amniocentesis
    b. chorionic villus sampling
    c. thoracentesis
    d. both a and b
    e. both b and c
16. Which of these terms is incompatible with life?
    a. hydrocephalus
    b. anencephaly
    c. hypothermia
    d. microcephaly
    e. encephalocele

M. Suffixes: The meaning of each of the following suffixes is the same as the meaning of another suffix. Fill in the blanks.

Suffix	Meaning of Suffix	Another Suffix with Same Meaning
1. osis		
2. ation		
3. elle		
4. ism		
5. dynia		

N. Difference between elements: Many elements sound and look similar but have very different meanings. Be precise in your communication—patient safety depends on you! Fill in the chart.

	Element	Type of Element (P, R/CF, or S)	Meaning of Element	Medical Term (from Any Chapter) Containing This Element
1.	stalsis			
2.	stasis			
3.	uretero			
4.	urethro			
5.	acro			
6.	acromio			
7.	homeo			
8.	hemo			
9.	bi			
10.	bio			
11.	inter			
12.	intra			
13.	colono			
14.	colpo			
15.	echo			
16.	ecto			
17.	auto			
18.	allo			
19.	metacarpo			
20.	metatarso			
21.	vaso			
22.	veno			
23.	thymo			
24.	thyro			

	Element	Type of Element (P, R/CF, or S)	Meaning of Element	Medical Term (from Any Chapter) Containing This Element
25.	necro			
26.	narco			
27.	lipo			
28.	litho			
29.	sacro			
30.	sarco			
31.	oro			
32.	orchio			
33.	brady			
34.	brachy			
35.	plexy			
36.	pexy			

O. Prefixes generally appear at the beginning of a term but occasionally will be in the middle of a term. A term may also have more than one prefix. Fill in the chart.

	Prefix	Meaning of Prefix	Medical Term with This Prefix	Definition of This Term
1.	allo			
2.	an			
3.	anti			
4.	auto			
5.	con			
6.	eu			
7.	dia			
8.	dys			
9.	hetero			
10.	homo			
11.	post			
12.	pre			
13.	pro			
14.	re			
15.	syn			
16.	xeno			

P. Latin and Greek terms do not deconstruct into elements. You must know these terms for what they are, and you must be able to use them correctly. Fill in the chart. *The first term is done for you.*

	Medical Term	Meaning in Latin or Greek	Brief Definition
	ileum	*to twist or roll up*	*third portion of the small intestine*
1.	thorax		
2.	bolus		
3.	diaphysis		
4.	ischium		
5.	eschar		
6.	follicle		
7.	esophagus		
8.	pons		
9.	paresis		
10.	coitus		
11.	femur		
12.	vertex		
13.	ilium		
14.	embolus		
15.	hospice		
16.	melanin		
17.	trachea		
18.	toxin		
19.	coccyx		
20.	diaphragmatic		
21.	coma		
22.	sulcus		
23.	ptosis		
24.	calculus		
25.	meatus		
26.	edema		
27.	tetany		
28.	pharynx		
29.	septum		
30.	purge		
31.	areola		

32. Use any one term from the chart above in a sentence of patient documentation:

____

____

____

____

____

Q. Chief-complaint medical terms: The patients of Fulwood Medical Center have come into the various clinics with the following complaints. Use the correct medical terminology to enter the chief complaint in each patient's chart. *The first one is done for you.*

	Patient Complains Of:	Correct Medical Terminology for Chart
	no appetite or aversion to food	*anorexia*
1.	mouth ulcer	
2.	blood in urine	
3.	ringing in ears	
4.	difficulty breathing	
5.	bedwetting at night	
6.	nosebleed	
7.	"pink eye"	
8.	wart	
9.	hair loss	
10.	inability to urinate	
11.	fainting	
12.	painful muscle spasms	

R. Transcription exercises: The following sentences from patient documentation and office forms contain errors. Circle the errors, and then write the correct medical terminology on the lines below.

1. This 65-year-old male with a prior history of an appendectomy awoke on the day of admission with crampy addominal pain that was accompanied by destention, nausea, and vomitting.

____

____

____

2. The patient is a 58-year-old black female with an ostosarcoma of the hemipelvis who underwent successful wide resuction and reconstruction; however, 2 weeks later her wound became infected and perforated the rectim, which resulted in gross contamination and her death.

____

____

____

3. Under general anesthesia it was clear that the patient had a grade II disruption of the latteral colateral ligament as well as an anterior and posterior cruxiate ligament disruption.

4. PA/lateral CXR: The heart size is normal. There is bilateral atellectasis noted. There is density within the overlying soft tissues of the right chest consistent with recent right-sided breast biopsie. There is no evidence of plural efusion or pneumothorax.

5. Numerous enlarged, abnormal-appearing limph nodes in the right axila, with markedly thickened cortices effacement. Concerning for lympoma or metastaic disease. Further followup is recommended.

6. The gastriscope was advanced into the distal esopagus, which was essentially normal. Advancement of the scope into the stomach showed evidence of erytema and gasritis. The pylorus was intubated and the duodenal bulb was visualized.

7. Past medical history:

punmonia	plurisy	arthralgia	migrain headaches
malaria	sleep apnia	broncitis	cardiomyopathy
conjunctivetis	gout	otitus media	anemmia

8. There is diffuse fatty infiltration of the liver. The spleen is unremarkable as well as kidneys and adrennal glands. There is no retroperitonial adenopathy or abdominal aortic aneursm. There are multiple calcifications within the galbladder consistent with choleithiasis.

9. Increased activity at the ankles and feet is seen bilateraly, likely related to degenerative/hypertopic changes. There is minimal increase activity at both shoulders and at the right sternocavicular joint, likely related to arthretic changes.

10. Past medical history:

diphtheria	typhoid	rubela	rhumatic fever
tonsilitis	chickenbox	conjunctivetis	pertusis

____

____

____

S. Prefixes: Match the prefix in column l to its correct meaning in column 2.

_____	1. brady	A. rapid
_____	2. ultra	B. water
_____	3. myx	C. not
_____	4. media	D. normal, good
_____	5. hydro	E. slow
_____	6. eu	F. mucus
_____	7. tachy	G. beyond
_____	8. meta	H. together
_____	9. non	I. middle
_____	10. con	J. after, subsequent to

T. Elements by grouping: Studying terms or elements together by group makes them easier to remember. Fill in the chart below.

Chart 1: Identify each element (prefix or root) that denotes a color, and write the color in the final column.

	Element	Identify Element (P, R/CF, S)	Color
1.	cyano		
2.	jaundice/cirrh		
3.	leuko		
4.	erthryo		
5.	melan		
6.	violet		
7.	purpura		

Chart 2: Identify each element that denotes a number or size, and write the number or size in the final column.

	Element	Identify Element (P, R/CF, S)	Number or Size
1.	tetra		
2.	poly		
3.	macro		
4.	mono		
5.	uni		
6.	bi		
7.	multi		
8.	tri		
9.	micro		
10.	pan		
11.	semi		
12.	quadri		

Chart 3: Identify each element that denotes a location or direction (above, below, etc.), and write the meaning of the element in the next column; then write the location or direction in the final column. (More than one element may have the same meaning.)

	Element	Meaning of Element	Location or Direction
1.	ectop		
2.	effer		
3.	endo		
4.	epi		
5.	ex		
6.	hyper		
7.	hypo		
8.	inter		
9.	intra		
10.	occip		
11.	para		
12.	peri		
13.	sub		
14.	super		
15.	trans		
16.	eso		
17.	exo		

	Element	Meaning of Element	Location or Direction
18.	re		
19.	dis		
20.	ab		
21.	ad		
22.	meta		
23.	circum		
24.	acro		
25.	supra		

U. Suffixes: The following suffixes are all associated with surgical and diagnostic procedures. CPT codes are specific as to type of procedure—you can't code and bill a **nephrotomy** if the patient actually had a **nephrectomy.** You must code what is documented in the medical record. Know your suffixes! In the chart below, give the meaning of the suffix, an example of a term with this suffix, and the definition of the term. *The first one is done for you.*

	Suffix	Meaning of Suffix	Term with This Suffix	Meaning of the Term
	ectomy	*removal of*	*nephrectomy*	*removal of the kidney*
1.	centesis			
2.	cision			
3.	desis			
4.	pexy			
5.	plasty			
6.	rrhaphy			
7.	tomy			
8.	tripsy			
9.	graphy			
10.	lysis			
11.	stomy			
12.	scopy			

V. Application exercise: Choose the correct term, form of the term, spelling of the term, plural, or abbreviation from this word bank to fill in the blanks of the following patient documentation. Be precise!

Word Bank:

electroencephalogram	lithotripsy	mastalgia	coma
menarche	ABGs	blepharitis	pneumonia
urologist	aplastic anemia	mastitis	menopause
PID	amniocentesis	gastroenterologist	comma
varicose	laparoscopy	renal cell carcinoma	varices

1. The hematologist diagnosed the patient as having ____________________.
2. This breastfeeding mother has a severe inflammation of the breast— ____________________ —due to a cracked areola.
3. The neurologist has ordered an ____________________ for the patient.
4. This patient has progressed to a deep state of unconsciousness—her ____________________ is not reversible.
5. This patient has not menstruated for over a year and is not pregnant. She has entered ____________________.
6. The patient has multiple ____________________ in her right leg that will require surgery.
7. Because this patient's hematuria is not clearing up with antibiotics, I am referring him to a ____________________ for further tests.
8. Schedule this patient for a ____________________ to check for internal bleeding.
9. The laparoscopic nephrectomy was performed for ____________________.
10. The pulmonologist ordered ____________________ for the patient with chronic obstructive pulmonary disease.

W. Roots/combining forms: Roots and combining forms are the foundation of all medical terms. Each of the elements below is the basis for a medical term that identifies a treatment or procedure. Fill in the chart.

Root/Combining Form	Meaning of R/CF	Medical Term with This R/CF	Definition of This Term
cryo			
laparo			
electro			
photo			
stereo			

B

Word Parts

Note: For easy identification, the word parts in this appendix appear in the same colors as in the Word Analysis and Definition boxes: suffix, prefix, root, root/combining form. Any term that is used in the text in both root and combining form is shown only in this appendix as a combining form.

WORD PART	DEFINITION
a-	not, without
a-	variant of ad
ab-	away from
abdomin/o	abdomen
ability	competence
ablat	take away
-able	capable
abort	fail at onset, expel nonviable fetus
abras	scrape off
absorpt	to swallow, take in
ac-	toward
-ac	pertaining to
-acea	condition, remedy
acid/o	acid, low pH
acin	grape
acous	hearing
acr/o	peak, topmost, extremity, highest point
acromi/o	acromion
act	to do, perform, performance
activ	movement
acu	needle
acu-	sharp
acumin	to sharpen
ad-	to, toward, into
adapt	to adjust

WORD PART	DEFINITION
-ade	process
aden/o	gland
adenoid	adenoid
adip	fat
adjust	alter
adjuv	give help
adolesc	beginning of adulthood
adren/o	adrenal gland
aer/o	air, gas
affer	move toward the center
ag-	to
-age	pertaining to
agglutin	sticking together, clumping
-ago	disease
agon	to fight
-agon	to fight
agor/a	marketplace
-agra	severe pain
-al	pertaining to
alanine	an amino acid, protein synthesized in muscle
albicans	white
albino/o	white
albumin	albumin
ald/o	organic compound

S = Suffix P = Prefix R = Root R/CF = Combining Form

WORD PART	DEFINITION
-ale	pertaining to
alges	sensation of pain
-algia	pain, painful condition
aliment	nourishment
-alis	pertaining to
alkal	base
all/o	strange, other
allo-	strange, other
alopec-	baldness, mange
alpha-	first letter in the Greek alphabet
alveol	alveolus, air sac
-aly	condition
ambly-	dull
ambulat	to walk, walking
amin/o	nitrogen containing
-amine	nitrogen-containing substance
ammon	ammonia
amni/o	amnion, fetal membrane
amnios	amnion
amph-	around
ampull	bottle-shaped
amput	to prune, cut off
amyl	starch
an-	not, lack of, without
-an	pertaining to
an/o	anus
ana-	away from, excessive
anabol	build up
analysis	process to study whole in terms of its parts
analyst	one who separates
anastom	provide a mouth
-ance	condition, state of
-ancy	state of
andr/o	male, masculine
aneurysm	dilation
angi/o	blood vessel, lymph vessel
angina	sore throat, chest pain radiating to throat
ankyl	stiff

WORD PART	DEFINITION
-ant	forming, pertaining to
ant-	against
ante-	before, forward, in front of
anter	before, front part
anthrac	coal
anti-	against
aort	aorta
apo-	different from, separation from
appendic	appendix
apse	clasp
aqu-	water
-ar	pertaining to
arachn	cobweb, spider
-arche	beginning
areol	areolar
aria	air
-arian	one who is
-aris	pertaining to
aroma	smell, sweet herb
array	place in order
arter	artery
arteri/o	artery
arteriosus	like an artery
arthr/o	joint
articul	joint
-ary	pertaining to
asbest	asbestos
ascit	fluid in the belly
-ase	enzyme
aspartate	an amino acid
aspergill	aspergillus
aspirat	to breathe on
assay	evaluate
assist	aid, help
asthm	asthma
astr/o	star
-ata	action, place, use
-ate	composed of, pertaining to
-ated	pertaining to a condition
atel	incomplete

S = Suffix P = Prefix R = Root R/CF = Combining Form

WORD PART	DEFINITION
ather/o	porridge, gruel, fatty substance
athet	without position, uncontrolled
-atic	pertaining to
-ation	a process
-ative	pertaining to, quality of
-ator	agent, instrument, person or thing that does something
-atory	pertaining to
atri/o	entrance, atrium
-atric	treatment
attent	awareness
attenu	to weaken
audi/o	hearing
audit	hearing
aur-	ear
auscult	listen to
auto-	self, same
avail	useful
axill	armpit, axilla
ayur-	life
azot	nitrogen
-back	back, toward the starting point
bacteri/o	bacteria
balan	glans penis
bar	pressure
bari	weight
bas/o	base, opposite of acid
basal	deepest part
basil-	base, support
be	life
behav	mental or motor activity
beta	second letter of Greek alphabet
bi-	two, twice, double
bi/o	life
-bil	able
bil/i	bile
-blast	germ cell, immature cell
blast/o	germ cell, immature cell
blephar/o	eyelid
body	body, mass, substance

WORD PART	DEFINITION
bov	cattle
brachi/o	arm
brachii	of the arm
brachy-	short
brady-	slow
bride	rubbish, rubble
bronch/i	bronchus
bronch/o	bronchus
brucell	pathologist David Bruce
buccinat	cheek
bulb/o	bulb
burs	bursa
calc/i	calcium
calcan	calcaneous
calcul	stone, little stone
callos	thickening
calor	heat
canal	duct or channel
cancer	cancer
candid	*Candida*
capill	hairlike, capillary
capit	head
capn	carbon dioxide
caps	box, cover, shell
capsul	box
carb/o	carbon
carboxy	group of organic compounds
carcin/o	cancer
card	heart
cardi/o	heart
care	be responsible for
carotene	yellow-red pigment
carotid	large neck artery
carp/o	bones of the wrist
cartilag	cartilage
cata-	down
catabol	break down
catechol	tyrosine-containing
catheter	insert, catheter
caud	tail

S = Suffix P = Prefix R = Root R/CF = Combining Form

WORD PART	DEFINITION
cav	hollow space
cava	cave
cavern	cave
cec	cecum
-cele	cave, hernia, swelling
celi	abdomen
cellul	cell, small cell
cent-	hundred
-centesis	to puncture
centr/o	central
ceph	head
cephal/o	head
-cephalus	head
cephaly	condition of the head
ceps	head
cept	to receive
cerebell	little brain, cerebellum
cerebr/o	brain
cerumin	cerumen
cervic	neck
cess	going forward
chancr	chancre
chem/o	chemical
chemic	chemical
-chete	hair
-chezia	pass a stool
chir/o	hand
chlor	green
chol/e	bile
cholangi	bile duct
cholecyst	gallbladder
choledoch/o	common bile duct
chondr/o	cartilage, rib, granule
chorion	chorion, membrane
chrom/o	color
chromat	color
chron/o	time
chym/o	chyme
-cidal	pertaining to killing
cide	to kill

WORD PART	DEFINITION
cili	hairlike structure
circulat	circular route
circum-	around
cirrh	yellow
cis	to cut
cit/i	cell
-clast	break, break down
claudic	limp
claustr/o	confined space
clav	clavicle
clave	lock
clavicul	clavicle
-cle	small
clitor	clitoris
clon	cutting used for propagation, tumult
-clonus	violent action
co-	with, together
coagul/o	clot, clump
coarct	press together, narrow
cobal	cobalt
cocc	round bacterium
coccus	berry, spherical bacterium
cochle	cochlear
code	information system
cognit	thinking
coit	sexual intercourse
col-	with, together
col	colon
coll	collect, glue
coll/a	glue
colon/o	colon
coloniz	form a colony
colp/o	vagina
com	take care of
com-	with, together
comat	coma
combin	combine
comminut	break into pieces
commodat	adjust
compat	tolerate

S = Suffix P = Prefix R = Root R/CF = Combining Form

WORD PART	DEFINITION
compet	strive together
complex	woven together
compli	fulfill
compress	press together
compuls	drive, compel
con-	with, together
concav	arched, hollow
concept	become pregnant
concuss	shake or jar violently
condyl	knuckle
confus	bewildered
congest	accumulation of fluid
coni	dust
coniz	cone
conjunctiv	conjunctiva
connect	join together
conscious	awareness
constip	press together
constrict	to narrow
contagi/o	transmissible by contact
contaminat/o	to corrupt, make unclean
contin	hold together
contra-	against
contract	draw together, pull together
contus	bruise
convalesc	recover
cor	heart
cori	skin
corne/o	cornea
coron	crown, coronary
corpor/e	body
corpus	body
cortic/o	cortex, cortisone
cortis	cortisone
cost	rib
crani/o	cranium, skull
crease	groove
creat	flesh
creatin	creatine
crete	to separate

WORD PART	DEFINITION
cretin	cretin
crimin	distinguish
crine	secrete
crista	crest
-crit	to separate
cry/o	cold
crypt-	hidden
cub	cube
cubit	elbow
cubitus	lying down
cune/i	wedge
cur	cleanse, cure
curat	to care for
curett	to cleanse
cursor	run
cusp	point
cutan/e	skin
cyan/o	dark blue
-cyst	cyst, bladder
cyst/o	bladder, sac, cyst
cysteine	an amino acid
cyt/o	cell
-cyte	cell
cyte	cell
dacry/o	tears, lacrimal duct
dai	day
de-	without, out of, removal, from
defec	clear out waste
defici	failure, lacking, inadequate
degenerat	deteriorate
deglutit	to swallow
del	visible
deliri	confusion, disorientation
delt	triangle
delus	deceive
dem	the people
demi-	half
dendr/o	treelike
dent	tooth
depend	rely on

S = Suffix P = Prefix R = Root R/CF = Combining Form

WORD PART	DEFINITION
depress	press down
derm/a	skin
-derma	skin
dermat/o	skin
dermis	skin
-desis	bind together, fixation of bone or joint
di-	two
dia-	complete
diabet	diabetes
diagnost	decision
dialectic	argument
dialy	separate
diaphor	sweat
diaphragm/a	diaphragm
diastol	diastole, relaxation
dict	consent, surrender
didym/o	testis
didymis	testis
diet	a way of life
different	not identical
digest	to break down food
digit	finder or toe
dilat	open up, expand, widen
dips	thirst
dis-	apart, away from
discipl	understand
disciplin	disciple, instruction
dist	away from the center
-dium	appearance
diuret	increase urine output
diverticul	byroad
dorm	sleep
dors	back
dorsi	of the back
drome	running
drop	liquid globule
duce	to lead
ducer	to lead, leader
duct	to lead, lead
ductus	leading

WORD PART	DEFINITION
duoden	twelve, duodenum
dur	dura
dura	hard
dwarf	miniature
dynam/o	power
-dynia	pain
dys-	bad, difficult, painful
e-	out of, from
-eal	pertaining to
ease	normal function, freedom from pain
ec-	out, outside
ech/o	sound wave
echin	hedgehog
eclamps	shining forth
eco-	environment
-ectasis	dilation
-ectomy	excision, surgical excision
ectop	on the outside, displaced
eczem/a	eczema
-ed	pertaining to
edema	edema, swelling
-ee	person who is the object of an action
efface	wipe out
effer	move out from the center
effus	pour out
ejacul	shoot out
ejaculat	shoot out
elasma	plate
elect	choice
electr/o	electric, electricity
elimin	throw away, expel
-elle	small
-em	condition
em-	in, into
-ema	result
emac/i	make thin
embol	plug
embryon	embryo, fertilized egg
emesis	vomiting
-emesis	to vomit, vomiting

S = Suffix P = Prefix R = Root R/CF = Combining Form

WORD PART	DEFINITION
emet	to vomit
emia	a blood condition
-emia	a blood condition
-emic	pertaining to a blood condition
emic	pertaining to a blood condition
emmetr-	measure
emuls	suspend in a liquid
en-	in
-ence	forming, quality of, state of
enceph	brain
encephal/o	brain
encephaly	condition of the brain
-ency	condition, state of, quality of
end-	inside, within
endo-	inside, within
-ent	end result, pertaining to
enter/o	intestine
entery	condition of the intestine
enur	urinate
environ	surroundings
-eon	one who does
eosin/o	dawn
ependym	lining membrane
epi-	above, upon, over
epilept	seizure
epiphys/i	growth
episi/o	vulva
equi-	equal
equin	horse
equip	to fit out
-er	agent, one who does
erect	straight, to set up
erg/o	work
-ergy	process of working
-ery	process of
erysi-	red
erythemat	redness
erythr/o	red
-escent	process
-esis	condition

WORD PART	DEFINITION
eso-	inward
esophag/e	esophagus
essent	existence
esthes	sensation, perception
esthet	sensation, perception
estr/o	woman
ethm	sieve
eti/o	cause
-etic	pertaining to
-etics	pertaining to
-ette	little
eu-	good, normal
ex-	away from, out, out of
exacerbat	increase, aggravate
examin	test, examine
excis	cut out
excret	separate, discharge
exo-	outside, outward
expect	await
expir	breathe out
extra-	out of, outside
faci	face
factor	maker
farct	area of dead tissue
fasc/i	fascia
febr	fever
fec	feces
feed	to give food, nourish
femor	femur
fer	to bear, to carry
ferrit	iron
fertil	able to conceive
fertiliz	to make fruitful
fet/o	fetus
fibr/o	fiber, fibrous
fibrill	small fiber
fibrin/o	fibrin
fibul	fibula
-fication	remove
fida	split

S = Suffix P = Prefix R = Root R/CF = Combining Form

WORD PART	DEFINITION
field	definite area
filar	roundworm
filtr	strain through
fiss	split
fistul	tube, pipe
flammat	flame
flat	flatus
flatul	excessive gas
flavin	yellow
flex	bend
fluid	flowing
fluo-	fluorine
fluor/o	flux, flow
flux	flow
foc	center, focus
follicul	follicle
foramin	opening, foramen
fore-	in front
-form	appearance of, resembling
format	to form
fract	break
fraction	small amount
free	free
frequ	repeated, often
front	front, forehead
fructos	fruit sugar
function	perform
fund/o	fundus
fung/i	fungus
fusion	to pour
galact/o	milk
gall	bile
gastr/o	stomach
gastrin	stomach hormone
gastrocnem	calf of leg
gemin	twin, double
gen/o	produce, create
-gen	create, produce, form
gen-	birth
-gene	production, give birth

WORD PART	DEFINITION
gener	create, produce
genesis	origin, creation, production
-genesis	creation, origin, formation, source
genet	origin
-genic	creation, producing
genit	bring forth, birth, primary male or female sex organ
genitor	offspring
ger	old age
geront/o	old age
gest	gestation, pregnancy, produce
gestat	gestation, pregnancy, to bear
gigant	giant
gingiv	gums
gland	gland
glauc	lens opacity, grey
gli/o	glue, supportive tissue of nervous system
-glia	glue, supportive tissue of nervous system
globin/o	protein
globul	globular, protein
glomerul/o	glomerulus
gloss/o	tongue
glott	mouth of windpipe
glottis	mouth of windpipe
gluc/o	glucose, sugar
glut	buttocks
glutin	glue, stick
glyc/o	glycogen, glucose, sugar
glycer	glycerol, sweet
gnath	jaw
gnose	use knowledge
gnosis	knowledge
gomph	bolt, nail
gon/o	seed
gonad/o	gonads, testes, or ovaries
gong	daily practice
-grade	going
graft	splice, transplant

S = Suffix P = Prefix R = Root R/CF = Combining Form

WORD PART	DEFINITION
-graft	tissue for transplant
graine	head pain
-gram	a record, recording
grand-	big
grand	big
granul/o	granule, small grain
-graph	to record, write
-grapher	one who records
-graphy	process of recording
gravida	pregnant
gravis	serious
gru	to move
guan	dung
gurgit	flood
gynec/o	woman, female
habilitat	restore
hale	breathe
halit	breath
hallucin	imagination
hallux	big toe
hem/o	blood
hemangi/o	blood vessel
hemat/o	blood
heme	red iron-containing pigment
hemi-	half
hepar	liver
hepat/o	liver
herb/i	plant
herni/o	hernia, rupture
herp	blister
hetero-	different
hiat	opening
hist	derived from histidine
hist/o	tissue
holist	entire, whole
hom/i	man
home/o	the same
homo-	same, alike
hormon	chemical messenger, hormone
human	human being

WORD PART	DEFINITION
humor	fluid
hyal	glass
hydr/o	water
hyp-	below
hyper-	above, beyond, excess, excessive
hypn/o	sleep
hypo-	below, deficient, smaller, low, under
hyster/o	uterus
-ia	condition
-iac	pertaining to
-ial	pertaining to
-ian	one who does, specialist
-ias	condition
-iasis	condition, state of
iatr	medical treatment, physician
-iatric	relating to medicine, medical knowledge
iatrics	medical knowledge
-iatrist	practitioner, one who treats
-iatry	treatment, field of medicine
-ible	can do, able to
-ic	pertaining to
-ica	pertaining to
-ical	pertaining to
-ician	expert
-ics	knowledge
ict	seizure
icterus	jaundice
-id	having a particular quality, pertaining to
-ide	having a particular quality
idi/o	unknown, personal
ifer	to bear, carry
-ify	to become
-il	a thing
-ile	pertaining to
ile/o	ileum
ili/o	ilium (hip bone)
im-	in, not
imag	likeness
immun/o	immune, immune response, immunity
immune	protected from

S = Suffix P = Prefix R = Root R/CF = Combining Form

WORD PART	DEFINITION
immuniz	make immune
impair	worsen
impede	obstruct
-imus	most
-in	substance, chemical compound
in-	not, into, in
incis	cut into
incub	sit on, lie on, hatch
index	to declare
-ine	pertaining to
infant	infant
infect	internal invasion, infection
infer	below, beneath
infest	invade, attack
inflammat	set on fire
inflat	blow up
infra-	below, beneath
-ing	quality of, doing
ingest	carry in
inguin	groin
inhal	breathe in
inhibit	repress
inject	force in
ino	sinew
insect/i	insect
insert	put together
inspir	breathe in
insul	island
integr	whole
integument	covering of the body
inter-	between
interstiti	space between tissues
intestin	gut, intestine
intra-	inside, within
intrins	on the inside
intus-	within
iod	violet, iodine
-ion	action, condition
-ior	pertaining to
-iosum	pertaining to

WORD PART	DEFINITION
-ious	pertaining to
irrig	to water
-is	belonging to, pertaining to
isch	to block
ischi	ischium
-ism	condition, process
-ismus	take action
iso-	equal
-ist	agent, specialist
-istic	pertaining to
-isy	inflammation
-ites	associated with
-ition	process
-itis	inflammation, infection
-ity	condition, state
-ium	structure
-ius	pertaining to
-ive	nature of, quality of, pertaining to
-iz	subject to
-ization	process of inserting or creating
-ize	action, affect in a specific way, policy
-ized	affected in a specific way
-izer	affects in a particular way, line of action
jejun	jejunum
jugul	throat
junct	joining together
juxta-	beside, near, close to
kal	potassium
kary/o	nucleus
kel/o	tumor
kerat	keratin, hard protein
kerat/o	cornea
kern	nucleus
ket/o	ketone
keton	ketone
ketone	organic compound
kin	motion
kinase	enzyme
kinesi/o	movement
kinet	motion

S = Suffix P = Prefix R = Root R/CF = Combining Form

WORD PART	DEFINITION
-kinin	move in
klept/o	to steal
kyph/o	bent, humpback
labi	lip
labyrinth	inner ear
lacer	to tear
lacrim	tears, tear duct
lact	milk
lactat	secrete milk
lapar/o	abdomen in general
lapse	clasp, fall together
-lapse	fall together, slide
laryng/o	larynx
lash	end of whip
lat	to take
lateral	at the side
latiss	wide
-le	small
lei/o	smooth
-lemma	covering
-lepsy	seizure
lept	thin, small
-let	small
leuk/o	white
lex	word
librium	balance
ligament	ligament
ligat	tie up, tie off
lign	line
-ling	small
lingu	tongue
lip/o	fat
lipid	fat
lith/o	stone
liv	life, live
load	to carry
lob	lobe
locat	a place
log	to study
-logist	one who studies, specialist

WORD PART	DEFINITION
logous	relation
-logy	study of
logy	study of
longev	long life
lord/o	curve, swayback
lubric	make slippery
lucid	bright clear
lumb	lower back, loin
lump	piece
lun	moon
lupus	wolf
-lus	small
lute	yellow
luxat	dislocate
ly	break down, separate
-ly	every
lymph/o	lymph, lymphatic system
lymphaden/o	lymph node
lymphangi/o	lymphatic vessels
lys/o	decompose, dissolve
lysis	destruction
-lysis	destruction, dissolve, separation
lyt	dissolve, destroy
-lyte	soluble
-lytic	relating to destruction
lyze	destruct, dissolve
macro-	large
macul	spot
magnet	magnet
mak	makes
-maker	one who makes
mal-	bad, difficult
mal	bad, difficult, inadequate
-malacia	abnormal softness
malign	harmful, bad
malleol	small hammer, malleolus
mamm/o	breast
man	frenzy, madness
man/o	pressure
mandibul	mandible

S = Suffix P = Prefix R = Root R/CF = Combining Form

WORD PART	DEFINITION
-mania	frenzy, madness
manic	affected by frenzy
manipul	handful, use of hands
marker	sign
mast	breast
mastic	chew
mastoid	mastoid process
mater	mother
matern	mother
matur(e)	ripe, ready, fully developed
maxilla	maxilla
medi	middle
media	middle
mediastin/o	mediastinum
medic	medicine
medulla	middle
mega-	enormous
-megaly	enlargement
mei	lessening
mela	black
melan/o	melanin, black pigment
mellit	sweetened with honey
membran/o	cover, skin
men/o	menses, monthly, month
mening/o	meninges, membranes
menisc	crescent, meniscus
menstr/u	menses, occurring monthly
ment	mind, chin
-ment	action, state, resulting state
mere	part
mero-	partial
meso-	middle
meta-	after, beyond, subsequent to
metabol	change
metacarp	bones of the hand
metatars	bones of the feet
-meter	measure, instrument to measure
metr/o	uterus
-metric	pertaining to measurement
-metrist	skilled in measurement

WORD PART	DEFINITION
-metry	process of measuring
mi-	derived from *hemi*, half
micr/o	small
micro-	small
mictur	pass urine
mid-	middle
mileusis	lathe
milli-	one-thousandth
miner	mines
mineral/o	inorganic material
miss	send
mit	thread
mito-	thread
mitr-	having two points
mitt	send
mod	nature, form, method
molec	mass
mollusc	soft
mon	single
monas	single unit
monil	type of fungus
mono-	one, single
morbid	disease
morph/o	shape
mort	death
mot	move
motiv	move
muc/o	mucus, mucous membrane
mucosa	lining of a cavity
multi-	many
mune	in service
muscul/o	muscle
mut	silent
muta	genetic change
mutil	to maim
my/o	muscle
myc/o	fungus
myel/o	spinal cord, bone marrow
myelin	in the spinal cord, myelin
myo-	to blink

S = Suffix P = Prefix R = Root R/CF = Combining Form

WORD PART	DEFINITION
myop-	to blink
myos	muscle
myring/o	tympanic membrane, eardrum
myx-	mucus
narc/o	stupor
nas	nose
nat	born, birth
nate	born, birth
natr/i	sodium
natur/o	nature
ne/o	new
nebul	cloud
necr/o	death
neo-	new
nephr/o	kidney
nerv	nerve
-ness	quality, state
neur/o	nerve, nervous tissue
neutr/o	neutral
nici	lethal
nitr/o	nitrogen
noct-	night
noia	to think
nom	law
non-	no, not
nor	normal
norm-	normal
nos/o	disease
nucle/o	nucleus
nucleol	small nucleus
nutri	nourish
nutrit	nourishment
o/o	egg
oblong	elongated
obsess	besieged by thoughts
obstetr	pregnancy and childbirth
occipit	back of head
occulta	hidden
ocul/o	eye
-ode	way, road, path

WORD PART	DEFINITION
odont	tooth
odyn/o	pain
-oid	resembling
-ol	alcohol, chemical, substance
-ola	small
-ole	small
olfact	smell
oligo-	scanty, too little
om/o	body, tumor
-oma	tumor, mass
onc/o	tumor
-one	chemical substance, hormone
onych/o	nail
ophthalm/o	eye
ophthalmos	eye
-opia	sight
opportun	take advantage of
-opsis	vision
-opsy	to view
opt/o	vision
optic	eye
-or	a doer, one who does, that which does something
or/o	mouth
orbit	orbit
orchi/o	testicle
ordin	arrange
orex	appetite
organ	organ, tool, instrument
orth/o	straight
orthot	correct
-orum	function of
-ory	having the function of
os	mouth
-osa	full of, like
-ose	full of
-osis	condition
osmo	push
osmol	concentration
oss/e	bone

S = Suffix P = Prefix R = Root R/CF = Combining Form

WORD PART	DEFINITION
oste/o	bone
-osus	condition
ot/o	ear
-otomy	incision
-ous	pertaining to
ov/i	egg
ovari	ovary
ovul	ovum, egg
ox	oxygen
-oxia	oxygen condition
oxid	oxidize
oxy	oxygen
pace	step
palat	palate
palliat	reduce suffering
palm	palm
palpat	touch, stroke
palpit	throb
pan-	all
pancreat	pancreas
panto-	entire
papill/o	pimple
par-	abnormal, beside
-para	to bring forth
para-	adjacent to, alongside, beside, abnormal
paresis	weakness
parasit	parasite
pareun	lying beside, sexual intercourse
pariet	wall
paroxysm	irritate; sudden, sharp attack
particul	little piece
partum	childbirth, to bring forth
pat	lie open
patell	patella
patent	lie open
-path	disease
path/o	disease
pathet	suffering
-pathic	pertaining to a disease

WORD PART	DEFINITION
pathy	disease, emotion
-pathy	disease
paus	cessation
pause	cessation
pector	chest
ped	child, foot
pedicul	louse
pelas	skin
pelv	pelvis
pen	penis
-penia	deficient, deficiency
peps	digestion
pepsin/o	pepsin
pept	digest
per-	through, intense
perforat	bore through
perfus	to pour
peri-	around
perine	perineum
peripher	external boundary, outer part, outer edge
periton/e	stretch over, peritoneum
perium	a bringing forth
perm/e	pass through
pes	foot
pesti	pest
petit	small
petit-	small
-pexy	fixation, surgical fixation
phaco-	lens
phag/o	to eat
phage	to eat
-phage	to eat
-phagia	swallowing, eating
phagia	swallowing
phalang/e	phalanx, finger, toe
pharmac/o	drug
pharyng/o	pharynx
pharynx	pharynx, throat
phenol	benzene derivative

S = Suffix P = Prefix R = Root R/CF = Combining Form

WORD PART	DEFINITION
phenyl	chemical group
pheo-	gray
pher/o	to carry
-pheresis	removal
-phil	attraction
-phile	attraction
-philia	attraction
phim	muzzle
phleb/o	vein
phob	fear
-phobia	fear
phon/o	sound, voice
phor	bear, carry
phosphat	phosphorus
phot/o	light
phren	mind
phylac	protect
phylaxis	protection
-phyll	leaf
physema	blowing
physi/o	body
physis	growth
phyt/o	plant
pia	delicate
pituit	pituitary
pituitar	pituitary
plak	plate, plaque
plant	insert, plant
planus	flat surface
plas	molding, formation, growth
-plasia	formation
-plasm	something formed
plasm/o	to form
-plasty	formation, repair, surgical repair
plate	flat
pleg	paralysis
plete	filled
pleur	pleura
plexy	stroke
-pnea	breathe

WORD PART	DEFINITION
pneum/o	air, lung
pneumat	structure filled with air
pneumon	air, lung
pod	foot
-poiesis	to make
-poiet	the making
-poietin	the maker
poikilo-	irregular
point	to pierce
pol	pole
polio-	gray matter
pollut	unclean
poly-	excessive, many, much
polyp	polyp
poplit/e	ham, back of knee
por/o	opening
post-	after
poster	back part
pract	efficient, practical
prand/i	breakfast
pre-	before, in front of
precis	accurate
pregn	with child, pregnant
presby	old man
press	press close, press down, squeeze
prevent	prevent
primi-	first
pro-	before, in front, projecting forward
proct/o	anus and rectum
product	lead forth
prolifer	bear offspring
pronat	bend down
prosta	prostate
prosthet	artificial part
prot/e	first
protein	protein
proto-	first
provision	provide
proxim	nearest to the center
prurit	itch

S = Suffix P = Prefix R = Root R/CF = Combining Form

WORD PART	DEFINITION
pseudo-	false
psych/o	mind, soul
psyche	mind, soul
pteryg	wing
ptosis	drooping, falling
-ptosis	drooping
ptysis	spit
pub	pubis
puer	child
pulmon/o	lung
puls	to drive
pump	pump
punct	puncture
pupill-	pupil
pur	pus
purific	make pure
purul	pus
py/o	pus
pyel/o	renal pelvis
pylor	gate, pylorus
pyr/o	fire, heat
pyrex	fever
pyrid	heat
qi	vital force
quadrant	quadrant
quadri-	four
radi/o	radius, x-ray, radiation
radic	root
re-	again, back, backward
recept	receive
rect/o	rectum
reflex	bend back
refract-	bend
regul	to rule, control
remiss	send back, give up
ren	kidney
replic	reply
rescein	resin
resect/o	cut off
resid/u	left over, what is left over

WORD PART	DEFINITION
resist-	to withstand
respire/a	to breathe
restor	renew
resuscit	revive from apparent death
reticul	fine net, network
retin/o	retina
retinacul	hold back
retro-	backward
rhabd/o	rod shaped, striated
rheumat	a flow, rheumatism
rhin/o	nose
rhythm	rhythm
rib/o	like a rib
ribo	a sugar, pentose
ribo-	from ribose, a sugar
rigid	stiff
rose	rose
rotat	rotate
-rrhagia	excessive flow, discharge
-rrhaphy	suture
rrhea	flow, discharge
-rrhoid	flow
rrhyth	rhythm
rrhythm	rhythm
-rubin	rust colored
rumin	throat
sacchar	sugar
sacr/o	sacrum
sagitt	arrow
saliv	saliva
salping/o	fallopian tube, uterine tube
salpinx	trumpet
san	sound, healthy
sanit	health
sapon	soap
sarc/o	flesh, muscle, sarcoma
satur	to fill
scapul	scapula
schiz/o	to split, cleave
scintill	spark

S = Suffix P = Prefix R = Root R/CF = Combining Form

WORD PART	DEFINITION
scler/o	hardness, white of eye
scoli/o	crooked
scope	instrument for viewing
-scope	instrument for viewing, instrument to examine
-scopy	to examine, to view
scorb	scurvy
scrot	scrotum
seb/o	sebum
sebac/e	wax
sebum	wax
secret	secrete, produce, separate
sect	cut off
sedat	to calm
sedent	sitting
segment	section
seiz	to grab
self	me, own individual
semi-	half
semin/i	semen
seminat	scatter seed
sen	old age
senesc	growing old
senil	senile
sens	feel
sensitiv	feeling
sensor/i	sensation, sensory
separat	move apart
seps	decay, infection
sept/o	septum, partition
ser/o	serum
serum	serum
sib	relative
-side	glycoside
sigm	Greek letter "S"
silic	silicon, glass
simi	ape, monkey
simul	imitate
sin/o	sinus
sinus	sinus

WORD PART	DEFINITION
sipid	flavor
-sis	abnormal condition, process
sit/u	place
skelet	skeleton
smear	spread
soc	partner
soci/o	society, social
soma	body
somat/o	body
-some	body
somn/o	sleep
son/o	sound
sorbit	fruit of a tree
sorpt	swallow
spad	tear or cut
spasm	spasm, sudden involuntary tightening
spast	tight
specif	species
sperm/i	sperm
spermat/o	sperm
sphen	wedge
spher/o	sphere
sphygm/o	pulse
spin/a	spine
spin/o	spine, spinal cord
spir/o	to breathe
spirat	breathe
spirit/u	spirit
spiro-	spiral, coil
splen/o	spleen
spongios	sponge
spor	spore
stable	steady
stag	standing place
stalsis	constrict, constriction
staphyl/o	bunch of grapes
-stasis	stop, stand still, control
stasis	stagnate, to stand still
stat	stationary
-static	stopped, standing still

S = Suffix P = Prefix R = Root R/CF = Combining Form

WORD PART	DEFINITION
-statin	inhibit
stax	fall in drops
steat	fat
stein	stone
sten/o	narrow, contract
ster	solid, steroid
stere/o	three-dimensional
steril	sterile, make sterile
stern	chest, breastbone
-steroid	steroid
-sterol	steroid
-sterone	steroid
steth/o	chest
sthen	strength
stigmat	focus
stimul	excite, strengthen
stin	partition
stip	press
stiti	space
stoma	mouth
-stomy	new opening
stone	stone, pebble
storm	crisis
strab	squint
strat	layer
strept/o	twisted
strict	narrow
study	inquiry
su/i	self
sub-	below, under, underneath
suct	suck
suffic/i	enough
sulf	sulfur
super	above, excessive
super-	above, excessive
supinat	bend backward
supplement	supply to remedy a deficiency
suppress	pressed under, push under
supra-	above
surfact	surface

WORD PART	DEFINITION
surg	operate
suscept	to take up
-sylated	linked
sym-	together
symptomat	collection of symptoms
syn-	together
syndesm	bind together
synov	synovial membrane
syring/o	tube, pipe
system	the body as a whole
systol/e	contraction, systole
tachy-	rapid
tact	orderly arrangement
tag	touch
tain	hold
talip	ankle bone
tamin	touch
tampon	plug
tangent	touch
tars	ankle
tax	coordination
tempor/o	time, temple
ten/o	tendon
tendin	tendon
tens	pressure
-tensin	tense, taut
terat/o	monster, malformed fetus
term	limit, end
test/o	testis, testicle
testicul	testicle, testis
tetra-	four
thalam	thalamus
thalamus	thalamus
thalass	sea
thel	breast, nipple
then	motion
thenar	palm
therap/o	healing, treatment
therapeut	healing, treatment
-therapist	one who treats

S = Suffix P = Prefix R = Root R/CF = Combining Form

WORD PART	DEFINITION
therapy	treatment
-therapy	treatment
therm/o	heat
thesis	arrange, place, organize
thet	arrange, place, organize
thi	sulfur
thora	chest
thorac/o	chest
thorax	chest
thromb/o	blood clot, clot
thym	thymus gland
thyr/o	thyroid
tibi	tibia
-tic	pertaining to
-tion	process, being
-tiz	pertaining to
toc	labor, birth
toler	endure
tom/o	cut, slice, layer
-tome	instrument to cut
-tomy	surgical incision
ton/o	pressure, tension
tonsil	tonsil
tonsill/o	tonsil
tope	part, location
topic	local
-tous	pertaining to
tox	poison
-toxic	able to kill
toxic/o	poison
trache/o	trachea, windpipe
tract	draw, pull
tranquil	calm
trans-	across, through
traumat	wound, injury
tresia	a hole
tri-	three
trich/o	hair
-tripsy	crushing
-tripter	crusher

WORD PART	DEFINITION
trochle	pulley
trop	turn, turning
troph	development, nourishment
trophy	development, nourishment
-tropic	a turning, change
-tropin	nourishing, stimulation
tryps	friction
tub	tube
tubercul	swelling, nodule, tuberculosis
tubul	small tube
tussis	cough
tympan/o	eardrum, tympanic membrane
-type	model, particular kind
typh	typhus
ulcer	a sore
-ule	little, small
-ulent	abounding in
uln	ulnar
ultra-	higher, beyond
-um	tissue, structure
umbilic	navel, umbilicus
un	one
un-	not
uni-	one
ur/o	urine, urinary system
-ure	process, result of
uresis	to urinate
uret	ureter, urine, urination
ureter/o	ureter
urethr/o	urethra
-uria	urine
urin/a	urine
-us	pertaining to
uter/o	uterus
uve	uvea
uvul	uvula
vaccin	vaccine, giving a vaccine
vag	vagus nerve
vagin	sheath, vagina
valgus	turn out

S = Suffix P = Prefix R = Root R/CF = Combining Form

WORD PART	DEFINITION
valv	valve
varic/o	varicosity; dilated, tortuous vein
vas/o	blood vessel, duct
vascul	blood vessel
ved	knowledge
veget	plants
vegetat	growth
ven/a	vein
ven/o	vein
ventil	wind
ventr	belly
ventricul	ventricle
vers	turn
-version	change
vert	to turn
vertebr	vertebra
vesic	sac containing fluid
vestibul/o	vestibule, entrance
via	the way
violet	bluish purple
viril	masculine
virus	poison

WORD PART	DEFINITION
viscer	internal organs
viscos	viscous, sticky
visu	sight
vita	life
voc	voice
vol	volume
volunt	willing
volut	shrink, roll up
-volut	rolled up
vuls	tear pull
vulv/o	vulva
whip	to swing
xanth	yellow
xeno-	foreign
-xis	condition
-yl	substance
zea-	to live
-zoa	animal
zyg	zygote
zygomat	cheekbone
zyme	fermenting, enzyme, transform

Abbreviations

ABBREVIATION	DEFINITION
µg	microgram; one-millionth of a gram
↑	increase/ above
↓	decrease/ below
1°	primary
2°	secondary
Ab	antibody
ABGs	arterial blood gases
ABO	agents of biologic origin
ABO	a blood group system
AC	acromioclavicular
ACL	anterior cruciate ligament
ACLS	advanced cardiac life support
ACTH	adrenocorticotropic hormone
AD	right ear
ADD	attention deficit disorder
ADH	antidiuretic hormone
ADHD	attention deficit hyperactivity disorder
ADL	activity of daily living
AED	automatic external defibrillator
Afib	atrial fibrillation
Ag	antigen
AIDS	acquired immunodeficiency syndrome
AKA	above-knee amputation
ALL	acute lymphocytic leukemia
ALS	amyotrophic lateral sclerosis
AMI	acute myocardial infarction
AMS	altered mental state
ANS	autonomic nervous system
AOM	acute otitis media
AP	anteroposterior

ABBREVIATION	DEFINITION
ARDS	adult respiratory distress syndrome
ARF	acute respiratory failure
ARF	acute renal failure
AROM	active range of motion
AS	left ear
ASD	atrial septal defect
ASD	autism spectrum disorder
ASHD	arteriosclerotic heart disease
AU	both ears
AV	atrioventricular
AVM	arteriovenous malformation
BBB	blood brain barrier
BD	brain death
BEP	benign enlargement of the prostate
BKA	below-knee amputation
BM	bowel movement
BMD	bone mineral density
BMR	basal metabolic rate
BOM	bilateral otitis media
BP	blood pressure
BPD	borderline personality disorder
BPH	benign prostatic hyperplasia
BPPV	benign paroxysmal positional vertigo
BRCA1	genetic mutation responsible for breast and ovarian cancer (**br**east **ca**ncer 1)
BRCA2	genetic mutation responsible for breast cancer (**br**east **ca**ncer 2)
BSE	bovine spongiform encephalopathy
BSE	breast self-examination

APPENDIX C

ABBREVIATION	DEFINITION
C1	first cervical vertebra
C5	fifth cervical vertebra or nerve
C7	seventh cervical vertebra
CA	cancer
CABG	coronary artery bypass graft
CAD	coronary artery disease
CAO	chronic airway obstruction
CAPD	continuous ambulatory peritoneal dialysis
CBC	complete blood count
CBT	cognitive-behavioral therapy
CD	conduct disorder
CDC	Centers for Disease Control and Prevention
CF	cystic fibrosis
CHD	congenital heart disease
CHES	certified health education specialist
CHF	congestive heart failure
CJD	Creutzfeldt-Jakob disease
CKD	chronic kidney disease
CMA	certified medical assistant
CMV	cytomegalovirus
CNA	certified nurse assistant
CNS	central nervous system
c/o	complains of
CO_2	carbon dioxide
COPD	chronic obstructive pulmonary disease
COT	certified occupational therapist
COTA	certified occupational therapist assistant
CP	cerebral palsy
CPAP	continuous positive airway pressure
CPR	cardiopulmonary resuscitation
CPT	cognitive processing therapy
CRF	chronic renal failure
CRP	C-reactive protein
C-section	cesarean section
CSF	cerebrospinal fluid
CT	computed tomography
CVA	cerebrovascular accident
CVP	central venous pressure
CVS	cardiovascular system
CVT	cardiovascular technologist
CXR	chest x-ray
D & C	dilation and curettage
DASH	dietary approaches to stop hypertension
DEXA	dual-energy x-ray absorptiometry
DI	diabetes insipidus
DIC	disseminated intravascular coagulation
DID	dissociative identity disorder
DIFF	differential white blood cell count
DJD	degenerative joint disease
DKA	diabetic ketoacidosis
dL	deciliter; one-tenth of a liter
DM	diabetes mellitus
DMD	Duchenne muscular dystrophy
DNA	deoxyribonucleic acid
DNR	do not resuscitate
DO	Doctor of Osteopathy
DRE	digital rectal examination
DSM-IV	*Diagnostic and Statistical Manual of Mental Disorders,* fourth edition
DVT	deep vein thrombosis
EBV	Epstein-Barr virus
ECG	electrocardiogram
ECT	electroconvulsive therapy
ED	erectile dysfunction
ED	emergency department
EEG	electroencephalogram
EKG	electrocardiogram
EMT	emergency medical technician
EMT-P	emergency medical technician–paramedic
EPCA-2	early prostate cancer antigen–2
ER	emergency room
ESR	erythrocyte sedimentation rate
ESRD	end-stage renal disease
ESWL	extracorporeal shock wave lithotripsy
FAS	fetal alcohol syndrome
FDA	U.S. Food and Drug Administration
FEV1	forced expiratory volume in 1 second
FSH	follicle-stimulating hormone

ABBREVIATION	DEFINITION
FTT	failure to thrive
FUS	focused ultrasound surgery
FVC	forced vital capacity
Fx	fracture
g	gram
GAD	generalized anxiety disorder
GDM	gestational diabetes mellitus
GERD	gastroesophageal reflux disease
GFR	glomerular filtration rate
GH	growth hormone, somatotrophin
GI	gastrointestinal
GI	glycemic index
GL	glycemic load
GTT	glucose tolerance test
GYN	gynecology
HAV	hepatitis A virus
Hb	hemoglobin
Hb A1c	glycosylated hemoglobin A one-C
HBOT	hyperbaric oxygen therapy
HBV	hepatitis B virus
HCG	human chorionic gonadotropin
HCl	hydrochloric acid
Hct	hematocrit
HCV	hepatitis C virus
HDL	high-density lipoprotein
HDN	hemolytic disease of the newborn
Hgb	hemoglobin
HIPAA	Health Insurance Portability and Accountability Act
HIV	human immunodeficiency virus
HMD	hyaline membrane disease
HPI	history of present illness
HPV	human papilloma virus
HRT	hormone replacement therapy
HSV	herpes simplex virus
HSV-1	herpes simplex virus, type 1
HTN	hypertension
HUS	hemolytic uremic syndrome
IBS	irritable bowel syndrome
ICD	implantable cardioverter/defibrillator

ABBREVIATION	DEFINITION
IDDM	insulin-dependent diabetes mellitus
Ig	immunoglobulin
IgA	immunoglobulin A
IgD	immunoglobulin D
IgE	immunoglobulin E
IgG	immunoglobulin G
IgM	immunoglobulin M
IM	intramuscular
INR	international normalized ratio
ITP	idiopathic (immunologic) thrombocytopenic purpura
IU	international unit(s)
IUD	intrauterine device
IV	intravenous
IVC	inferior vena cava
IVF	in vitro fertilization
IVP	intravenous pyelogram
JRA	juvenile rheumatoid arthritis
KUB	x-ray of abdomen to show **k**idneys, **ure**ters, and **b**ladder
LASER	**l**ight **a**mplification by **s**timulated **e**mission of **r**adiation
LCC-ST	certified surgical technologist
LD	learning disability
LDL	low-density lipoprotein
LFT	liver function test
LH	luteinizing hormone
LLQ	left lower quadrant
LOC	loss of consciousness
LPN	licensed practical nurse
LUQ	left upper quadrant
LVN	licensed vocational nurse
mcg	microgram; one-millionth of a gram
MCP	metacarpophalangeal
MCS	minimally conscious state
MD	Doctor of Medicine
mg	milligram
MI	myocardial infarction
mL	milliliter
mm^3	cubic millimeter

ABBREVIATION	DEFINITION
MOAB	monoclonal antibody
MODY	mature onset diabetes of the young
MONA	morphine, oxygen, nitroglycerine, and aspirin
MPD	multiple personality disorder
MRA	magnetic resonance angiography
MRI	magnetic resonance imaging
mRNA	messenger RNA
MRSA	methicillin-resistant *Staphylococcus aureus*
MS	multiple sclerosis
NCI	National Cancer Institute
NIDDM	non-insulin-dependent diabetes mellitus
NIH	National Institutes of Health
NKA	no known allergies
NO	nitric oxide
NRDS	neonatal respiratory distress syndrome
NSAID	nonsteroidal anti-inflammatory drug
O_2	oxygen
OA	osteoarthritis
OA	obstetric assistant
OB	obstetrics
OCD	obsessive compulsive disorder
OD	Doctor of Osteopathy
OD	right eye
ODD	oppositional defiant disorder
OGTT	oral glucose tolerance test
OME	otitis media with effusion
OS	left eye
OSHA	Occupational Safety and Health Administration
OT	ophthalmic technician
OT	oxytocin
OT	occupational therapy
OTC	over the counter
OU	both eyes
P	pulse rate
PA	posteroanterior
PA	pernicious anemia
PaO_2	partial pressure of arterial oxygen

ABBREVIATION	DEFINITION
Pap	Papanicolaou (Pap test, Pap smear)
PAT	paroxysmal atrial tachycardia
PCL	posterior cruciate ligament
PCOS	polycystic ovarian syndrome
PDA	patent ductus arteriosus
PDD-NOS	pervasive developmental disorder, not otherwise specified
PDT	postural drainage therapy
PE tube	pressure equalization tube
PEEP	positive end-expiratory pressure
PEFR	peak expiratory flow rate
PERRLA	pupils equal, round, reactive to light, and accommodation
PET	positron emission tomography
PFTs	pulmonary function tests
PGY	pregnancy
pH	hydrogen ion concentration
PhD	Doctor of Philosophy
PID	pelvic inflammatory disease
PIP	proximal interphalangeal
PKD	polycystic kidney disease
PMDD	premenstrual dysphoric disorder
PMNL	polymorphonuclear leukocyte
PMS	premenstrual syndrome
PNB	pulseless, nonbreather
PNS	peripheral nervous system
PO	by mouth
POC	products of conception
polio	poliomyelitis
PPH	postpartum hemorrhage
PPS	postpolio syndrome
p.r.n, PRN	when necessary
PSA	prostate-specific antigen
PT	physiotherapy
PT	prothrombin time
PT	physical therapy, physical therapist
PTA	physical therapy assistant
PTCA	percutaneous transluminal coronary angioplasty
PTH	parathyroid hormone

ABBREVIATION	DEFINITION
PTSD	posttraumatic stress disorder
PVC	premature ventricular contractions
PVD	peripheral vascular disease
PVS	persistent vegetative state
q.4.h.	every 4 hours
q.i.d.	four times each day
R	respiration rate
RA	rheumatoid arthritis
RBC	red blood cell
RDA	recommended dietary allowance
RDS	respiratory distress syndrome
Rh	Rhesus
Rho-GAM	Rhesus immune globulin
R.I.C.E.	rest, ice, compression, and elevation
RLQ	right lower quadrant
RN	registered nurse
RNA	ribonucleic acid
ROM	range of motion
RU-486	mifepristone
RUQ	right upper quadrant
SA	sinoatrial
SAD	seasonal affective disorder
SARS	severe acute respiratory syndrome
SBS	shaken baby syndrome
SC	subcutaneous
SCI	spinal cord injury
SET	self-examination of the testes
SFD	small for date
SG	specific gravity
SGA	small for gestational age
SI	sacroiliac
SIDS	sudden infant death syndrome
SLE	systemic lupus erythematosus
SOB	short(ness) of breath
SP	standard precautions
SSRI	selective serotonin reuptake inhibitor
STAT	immediately
STD	sexually transmitted disease
SVC	superior vena cava

ABBREVIATION	DEFINITION
T	temperature
T1	first thoracic vertebra or nerve
T_3	triiodothyronine
T_4	tetraiodothyronine (thyroxine)
TB	tuberculosis
TBI	traumatic brain injury
TENS	transcutaneous electrical nerve stimulation
THR	total hip replacement
TIA	transient ischemic attack
t.i.d.	(Latin *ter in die*) three times a day
TMJ	temporomandibular joint
TNM	**t**umor-**n**ode-**m**etastasis staging system for cancer
TOF	tetralogy of Fallot
tPA	tissue plasminogen activator
TSH	thyroid-stimulating hormone
TTM	trichotillomania
TTN	transient tachypnea of the newborn
TTP	thrombotic thrombocytopenic purpura
TURP	transurethral resection of the prostate
UA	urinalysis
UP	universal precautions
URI	upper respiratory infection
USDA	U.S. Department of Agriculture
UTI	urinary tract infection
UV	ultraviolet
VEP	visual evoked potential
Vfib	ventricular fibrillation
VS	vital signs
VSD	ventricular septal defect
V-tach	ventricular tachycardia
vWD	von Willebrand disease
vWF	von Willebrand factor
WAD	Word Analysis and Definition (box)
WBC	white blood cell; white blood (cell) count
WNL	within normal limits
WNV	West Nile virus

D

Diagnostic and Therapeutic Procedures

A compilation of the diagnostic and therapeutic procedural terms used in this book.

abdominoplasty Esthetic operation on the abdominal wall (tummy tuck).

ablation Removal of tissue to destroy its function.

activated partial thromboplastin time (APTT) Blood test used to monitor the dose of heparin, an anticoagulant.

adenoidectomy Surgical removal of the adenoid tissue.

alignment Process of bringing the ends of a fractured bone at the break back opposite each other so that they fit together as they did in the original bone.

ambulatory Surgery or any other care provided without an overnight stay in a medical facility.

ambulatory blood pressure monitor Device that provides a record of blood pressure readings over a 24-hour period as patients go about their daily activities.

amniocentesis Removal of amniotic fluid for diagnostic purposes.

amputation Process of removing a limb, part of a limb, a breast, or other projecting part.

anastomosis Surgically made union between two tubular structures.

angiogram Radiographic image of arteries or veins after injection of contrast material.

angiography The process of obtaining an angiogram.

angioplasty Reopening of a blood vessel by surgery.

anoscopy Examination of the anus and lower rectum with a rigid instrument.

Apgar score Evaluation of a newborn's status.

appendectomy Surgical removal of the appendix.

arterial blood gases The measurement of the levels of oxygen and carbon dioxide in the blood—a good indicator of respiratory function.

arteriography X-ray visualization of an artery after injection of contrast material.

arthrocentesis Aspiration of fluid from a joint; used to establish a diagnosis by laboratory examination of the fluid, drain off infected fluid, or insert medication such as local corticosteroids.

arthrodesis Fixation or stiffening of a joint by surgery.

arthrography X-ray of a joint after injection of a contrast medium into the joint to make the inside details of the joint visible.

arthroplasty Replacement of a joint with a prosthesis.

arthroscopy Procedure performed using an arthroscope to examine the internal compartments of a joint or perform a surgical procedure such as debridement, removal of damaged tissue, or repair of torn ligaments.

aspiration Removal by suction of fluid or gas from a body cavity.

atherectomy Surgical removal of atheroma from a blood vessel.

audiometer Electronic device that generates sounds in different frequencies and intensities to test for hearing loss.

auscultation Diagnostic method of listening to body sounds with a stethoscope.

autograft Graft removed from the patient's own skin.

automatic external defibrillator (AED) Device that sends an electric shock to the heart to stop the heart and allow a normal contraction rhythm to resume.

bariatric surgery Surgical treatment of obesity.

barium meal Ingestion of barium sulfate to study the distal esophagus, stomach, and duodenum on x-ray.

barium swallow Ingestion of barium sulfate, a contrast material, to show details of the pharynx and esophagus on x-ray.

biopsy Removal of tissue from a living person for laboratory examination.

blepharoplasty Correction of defects in the eyelids.

bone marrow aspiration or biopsy Use of a needle to remove bone marrow cells.

bone mineral density (BMD) Screening test for osteoporosis using a dual-energy x-ray absorptiometry (DEXA) scan.

brace Appliance to support a part of the body in its correct position.

brachytherapy Radiation therapy in which the source of irradiation is implanted in the tissue to be treated.

bronchoscopy Examination of the interior of the tracheobronchial tree with an endoscope.

cannula Tube inserted into a blood vessel or cavity as a channel for fluid or gas.

cardiac catheterization Procedure that detects patterns of pressures and blood flows in the heart. A thin tube is guided into the heart under x-ray guidance after being inserted into a vein or artery.

cardiac stress testing Exercise tolerance test that raises the heart rate and monitors the effect on cardiac function.

cardiopulmonary resuscitation Attempt to restore cardiac and pulmonary function.

cardioversion Restoration of a normal heart rhythm by electrical shock. Also called *defibrillation.*

catheterization Introduction of a catheter.

cerebral angiography Injection of a radiopaque dye into the blood vessels of the neck and brain to detect blood vessels that are partially or completely blocked, aneurysms, or arteriovenous malformations.

cerebral arteriography Procedure used to determine the site of bleeding in hemorrhagic strokes, enabling surgery to be performed to stop the bleed or to clip off the aneurysm.

chest x-ray Radiograph image of the chest that can be taken in anteroposterior (AP), posteroanterior (PA), lateral, and sometimes oblique and lateral decubitus positions.

cholangiography Use of a contrast medium to radiographically visualize the bile ducts.

cholecystectomy Surgical removal of the gallbladder.

cholelithotomy Surgical removal of a gallstone(s).

circumcision Removal of part or all of the prepuce of the penis.

clean-catch, midstream urine specimen Sample collected after the external urethral meatus is cleaned. The first part of the urine stream is not collected, and the sterile collecting vessel is introduced into the urinary stream to collect the last part.

clot-busting drugs Drugs injected within a few hours of an MI or thrombotic stroke to dissolve the thrombus. Also called *thrombolytic drugs.*

colonoscopy Examination of the inside of the colon by endoscopy.

colostomy Artificial opening from the colon to the outside of the body.

colpopexy Surgical fixation of a relaxed and prolapsed vagina to the anterior abdominal wall.

computed tomography (CT) Scan in which images of sections of the body are generated by a computer synthesis of x-rays obtained in many different directions in a given plane.

conization Surgical excision of a cone-shaped piece of tissue, e.g., from the outer lining of the cervix.

constant positive airway pressure (CPAP) Attempt to keep alveoli open by maintaining a positive pressure in the airways. A mask is fitted over the nose and mouth and attached to a ventilator.

continuous ambulatory peritoneal dialysis (CAPD) Dialysis performed by the patient at home through an implanted peritoneal catheter, usually 4 times a day, 7 days a week.

continuous cycling peritoneal dialysis Use of a machine to automatically infuse dialysis solution into and out of the abdominal cavity through a peritoneal catheter during sleep.

coronary angiogram Injection of a contrast dye during cardiac catheterization to identify coronary artery blockages.

coronary artery bypass surgery (CABG) Procedure in which healthy blood vessels harvested as a graft from the leg, chest, or arm are used to bypass (detour) the blood around blocked coronary arteries.

cryosurgery Use of liquid nitrogen or argon gas in a probe to freeze and kill abnormal tissue.

curette Scoop-shaped instrument for scraping or removing new growths (or earwax).

cystoscopy Insertion of a pencil-thin, flexible, tubelike telescope through the urethra into the bladder to examine directly the lining of the bladder and to take a biopsy if needed.

cystourethrogram X-ray image during voiding to show the structure and function of the bladder and urethra.

debridement Removal of injured or necrotic tissue.

defibrillation Restoration of uncontrolled twitching of cardiac muscle fibers to a normal rhythm.

dermabrasion Removal of upper layers of the skin using a high-powered rotating brush.

dialysis Artificial method of removing waste materials and excess fluid from blood.

digital rectal examination Palpation of the rectum and prostate gland with an index finger.

dilation and curettage (D&C) Dilation of the cervix so that a thin instrument can be inserted to scrape away the lining of the uterus and take tissue samples.

dipstick Plastic strip bearing paper squares of reagent—the most cost-effective method of screening urine. After the stick is dipped in the urine specimen, the color change in each segment of the dipstick is compared to a color chart on the container. Dipsticks can screen for pH, specific gravity, protein, blood, glucose, ketones, bilirubin, nitrite, and leukocyte esterase.

Doppler ultrasound Diagnostic instrument that sends an ultrasonic beam into the body.

early morning urine collection Process used to determine the ability of the kidneys to concentrate urine following overnight dehydration.

echocardiography Ultrasound recording of heart function.

echoencephalography Use of ultrasound in the diagnosis of intracranial lesions.

electrocardiogram Record of the electrical signals of the heart.

electrocardiography Interpretation of electrocardiograms.

electroconvulsive therapy (ECT) Passage of electric current through the brain to produce convulsions and treat persistent depression.

electroencephalography Recording of the electrical activity of the brain.

electromyography Recording of electrical activity in muscle.

endarterectomy Surgical removal of plaque from an artery.

endometrial ablation Use of a heat-generating tool or a laser to remove or destroy the lining of the uterus and prevent or reduce menstruation.

endoscope An instrument for the examination of the interior of a hollow or tubular organ.

endoscopy Use of an endoscope to examine the interior of a tubular or hollow organ and perform a biopsy, remove polyps (polypectomy), and coagulate bleeding lesions.

enema Injection of fluid into the rectum and lower bowel.

enteroscopy Examination of the lining of the digestive tract.

episiotomy Surgical incision in the perineum to dilate the opening of the vagina.

evoked responses Use of stimuli for vision, sound, and touch to activate specific areas of the brain and measure their responses with EEG. This provides information about how that specific area of the brain is functioning.

excision Surgical removal of part or all of a structure or organ.

excisional biopsy Removal of a tumor with a surrounding margin of normal tissue.

external fixation Method of maintaining the alignment of a fractured bone by immobilizing the bone through the use of plaster casts, splints, traction, and external fixators such as steel rods and pins.

extracorporeal shock wave lithotripsy (ESWL) Process in which a machine called a *lithotripter* produces shock waves that crumble renal or ureteral stones into small pieces that can pass down the ureter.

fasciectomy Surgical removal of fascia.

fecal occult blood test Diagnostic procedure that detects the presence of blood not visible to the naked eye. Trade name: *Hemoccult* test.

fistulectomy Surgical excision of a fistula.

fistulotomy Surgical enlargement or opening up of a fistula.

flexible endoscopy Use of a flexible, slim fiber-optic instrument that transmits light and sends back images to the observer.

forceps extraction Assisted delivery of a baby by an instrument that grasps the head of the baby.

fundoscopy Examination of the retina with an ophthalmoscope.

gastroscopy Endoscopic examination of the inside of the stomach.

gavage To feed by a stomach tube.

gingivectomy Surgical removal of diseased gum tissue.

heart transplant Surgery in which the heart of a recently deceased person (donor) is transplanted to the recipient after the recipient's diseased heart has been removed.

Hemoccult test Trade name for *fecal occult blood test.*

hemodialysis Process that filters blood through an artificial kidney machine (dialyzer).

hemorrhoidectomy Surgical removal of hemorrhoids.

herniorrhaphy Surgical repair of a hernia.

heterograft Graft from a nonhuman species. Also called *xenograft.*

Holter monitor Continuous ECG recorded on a tape cassette for at least 24 hours as a person works, plays, and rests.

homocysteine Amino acid in the blood. Elevated levels are related to a higher risk of CAD, stroke, and peripheral vascular disease.

homograft Skin graft from another person or a cadaver. Also called *allograft.*

hysterectomy Surgical removal of the uterus.

ileostomy Artificial opening from the ileum to the outside of the body.

implantable cardioverter/defibrillator (ICD) Implanted device that senses abnormal rhythms and gives the heart a small electrical shock to return the rhythm to normal.

incision Cut or surgical wound.

internal fixation Use of tissue-compatible materials such as stainless steel and titanium to stabilize fractured bony parts and enable the patient to return to function quicker and reduce the incidence of nonunion and malunion (improper healing). The types of internal fixation are wires used as sutures to "sew" the bone fragments together; plates that extend along both or all fragments of bone and are held in place by screws; rods inserted through the medullary cavity of both fragments to align the bones; and screws that can be used on their own as well as with plates.

intradermal injection Introduction of a short, thin needle into the epidermis, thus raising a small wheal. This site is used for allergy and tuberculosis (TB) testing.

intramuscular (IM) injection Use of a long needle that penetrates the epidermis, dermis, and hypodermis to reach into the muscles underneath. Some antibiotics and some immunizations are given by this route.

intrauterine insemination Insertion of sperm directly into the uterus via a special catheter to initiate pregnancy.

intravenous pyelogram (IVP). Procedure in which a contrast material containing iodine is injected intravenously and its progress through the urinary tract is then recorded on a series of rapid radiological images.

intubation Insertion of a tube into a canal, hollow organ, or cavity, e.g., into the trachea for anesthesia or control of ventilation.

in vitro fertilization (IVF) Process of combining sperm and egg in a laboratory dish and placing the resulting embryos inside the uterus.

Ishihara color system Test for color vision defects.

Jaeger reading card Chart containing type in different sizes of print for testing near vision.

keratomileusis Procedure that cuts and shapes the cornea.

keratotomy Incision through the cornea.

kidney transplant Surgery in which the kidney of a donor is transplanted to a recipient; provides a better quality of life than kidney dialysis, if a suitable donor can be found.

KUB X-ray of the abdomen to show **k**idneys, **u**reters, and **b**ladder.

laparoscopy Examination of the contents of the abdomen using an endoscope, which can also be used to perform surgery and take samples for biopsy.

laryngoscopy Use of a hollow tube with a light and camera to visualize or operate on the larynx.

laser surgery Use of a concentrated, intense narrow beam of electromagnetic radiation for surgery. (*laser:* **l**ight **a**mplification by **s**imulated **e**mission of **r**adiation)

lipectomy Surgical removal of fatty tissue.

lipid profile Group of blood tests that helps determine the risk of CAD and comprises total cholesterol; high-density lipoprotein

(HDL), or "good cholesterol"; low-density lipoprotein (LDL), or "bad cholesterol"; and triglycerides.

liposuction Surgical removal of fatty tissue using suction.

lobectomy Surgical removal of a lobe of a structure, for example, a lobe of a lung.

lumbar puncture Use of a hollow needle to remove CSF so that it can be examined in the laboratory. Also called *spinal tap.*

lumpectomy Removal of a lesion with preservation of surrounding tissue.

lymphadenectomy Surgical removal of a lymph gland(s).

lymphangiogram Radiographic images of lymph vessels and nodes following injection of contrast material.

magnetic resonance angiography (MRA) Method of visualizing vessels that contain flowing structures by producing a contrast between them and stationary structures.

magnetic resonance imaging (MRI) Diagnostic technique that creates detailed images of structures and tissues in various planes without exposing patients to radiation as in conventional radiography or computed tomography.

mammogram Record produced by x-ray imaging of the breast.

mammoplasty Surgical reshaping of the breasts.

mastectomy Surgical excision of the breast.

mechanical ventilation Process by which gases are moved into and out of the lungs via a ventilator that is set to meet the respiratory requirements of the patient.

mediastinoscopy Examination of the mediastinum using an endoscope inserted through an incision in the suprasternal notch.

myomectomy Surgical removal of a uterine myoma (fibroid).

myringotomy Incision through the tympanic membrane; e.g., for the placement of pressure equalization (PE) tubes to allow an effusion to drain.

nebulizer Device used to deliver liquid medicine in a fine mist.

nephrectomy Surgical removal of a kidney.

nephrolithotomy Incision into the kidney for removal of a stone.

nephroscopy Examination of the pelvis of the kidney.

nerve conduction studies Studies that measure the speed at which motor or sensory nerves conduct impulses.

nuclear imaging of the heart Use of an injection of a radioactive substance in association with a cardiac stress test to assess cardiac function.

ophthalmoscopy Examination of the retina using an ophthalmoscope.

orchiopexy Surgical fixation of a testis in the scrotum.

ostomy Artificial opening into a tubular structure, for example, ileostomy and colostomy.

otoscopy Examination of the ear using an otoscope.

pacemaker Device that regulates cardiac electrical activity. The device generates electronic signals carried along thin, insulated wires to the heart muscle.

palpation Examination with the fingers and hands.

panendoscopy Visual examination of the inside of the esophagus, stomach, and upper duodenum using a flexible fiber-optic endoscope.

Pap test Examination of cells taken from the cervix.

parathyroidectomy Surgical removal of the parathyroid glands.

peak flow meter Instrument used to record the greatest flow of air that can be sustained for 10 milliseconds on forced expiration, the peak expiratory flow rate (PEFR). It is of value in following the course of asthma and in postoperative care to monitor the return of lung function after anesthesia.

percutaneous nephrolithotomy Insertion of a nephroscope through the skin to locate and remove a renal pelvic or ureteral stone.

percutaneous transluminal coronary angioplasty (PTCA) Procedure in which a balloon-tipped catheter is guided to the site of the blockage and inflated to expand the artery from the inside by compressing the plaque against the walls of the artery.

percutaneous transthoracic needle aspiration Insertion of a needle with a cutting chamber through an intercostal space to hook a specimen of parietal pleura for laboratory examination.

peritoneal dialysis Procedure in which a dialysis solution is infused into and drained out of the abdominal cavity through a small, flexible, implanted catheter.

phlebotomy Process of taking blood from a vein.

photocoagulation Use of a laser beam to form a clot or destroy abnormal capillaries. In the eye, this slows the pace of the visual loss in macular degeneration

phototherapy Treatment using light rays.

pneumonectomy Surgical removal of a lung.

polypectomy Excision or removal of a polyp.

polysomnography Test to monitor brain waves, muscle tension, eye movement, and oxygen levels in the blood as the patient sleeps.

positive end expiratory pressure (PEEP) Technique in ventilation to keep the alveoli from collapsing in adult and neonatal respiratory distress syndromes.

positron emission tomography (PET) Scan that shows the uptake and distribution of substances such as sugar in tissues to locate abnormal, often malignant, structures.

postural drainage therapy Treatment that involves positioning and tilting the patient so that gravity promotes drainage of secretions from lung segments. Chest percussion (tapping) on the chest wall can help loosen, mobilize, and drain the retained secretions.

proctoscopy Examination of the inside of the anus and rectum by endoscopy.

prostatectomy Surgical removal of part or all of the prostate.

prosthesis Manufactured substitute for a missing part of the body.

prothrombin time (PT) Test used to monitor the dose of Coumadin, an anticoagulant. It is reported as an *international normalized ratio (INR)* instead of in seconds.

pulmonary rehabilitation Therapeutic restoration of lung function that includes education, breathing exercises and retraining, exercises for the upper and lower extremities, and psychosocial support.

pulse oximeter Sensor placed on the finger to measure the oxygen saturation of the blood.

pyelogram X-ray image of the renal pelvis and ureters.

quadrantectomy Surgical excision of a quadrant of the breast.

radical hysterectomy Surgical removal of the fallopian tubes and ovaries as well as the uterus.

radical mastectomy Complete surgical removal of all breast tissue, the pectoralis major muscle, and associated lymph nodes.

radical prostatectomy Complete surgical removal of the prostate and surrounding tissues.

random urine collection Process in which a sample is taken with no precautions regarding contamination. It is often used for collecting samples for drug testing. "Pee into a cup."

reduction Procedure in which the distal segment of a fractured bone is pulled back into alignment with the proximal segment. Anesthesia may be used.

rehabilitation Therapeutic restoration of an ability to function as before after disease, illness, or injury.

renal angiogram X-ray with contrast material used to assess blood flow to the kidneys.

resection Removal of a specific part of an organ or structure.

retrograde pyelogram Injection of contrast material through a urinary catheter into the ureters to locate stones and other obstructions.

rhinoplasty Surgical procedure to alter the size or shape of the nose.

Rinne test Test for a conductive hearing loss.

salpingectomy Surgical removal of a fallopian tube.

sclerotherapy Injection of a solution into a vein to thrombose it.

segmentectomy Surgical excision of a segment of a tissue or organ.

sigmoidoscopy Endoscopic examination of the sigmoid colon.

Snellen letter chart Test for acuity of distant vision.

sphygmomanometer Instrument for measuring arterial blood pressure.

spinal tap Placement of a needle through an intervertebral space into the subarachnoid space to withdraw CSF.

spirometer Device used to measure the volume of air that a patient moves in and out of the respiratory system.

splenectomy Surgical removal of the spleen.

staging Process of determination of the extent of the distribution of a neoplasm. The TNM (tumor-node-metastasis) staging system can be used.

stent placement Procedure in which a wire mesh tube, or stent, is placed inside the vessel to reduce the likelihood that an occluded artery will close up again. Some stents (drug-eluting stents) are covered with a special medication to help keep the artery open.

sterilization Process of making sterile.

stethoscope Instrument for listening to cardiac, respiratory, and other sounds.

stoma Artificial opening.

subcutaneous (SC) injection Injection in which a needle pierces the epidermis and dermis to reach the hypodermis (subcutaneous) layer. This site is used for insulin injections and for some immunizations.

suprapubic transabdominal needle aspiration of the bladder Procedure used with newborns and small infants to obtain a pure sample of urine.

suture Process or material that brings together the edges of a wound to enhance tissue healing. Also, a form of fibrous joint to unite two bones.

thoracentesis Insertion of a needle through an intercostal space to remove fluid from a pleural effusion for laboratory study or to relieve pressure. Also called *pleural tap.*

thoracoscopy Examination of the pleural cavity with an endoscope.

thoracotomy Incision through the chest wall.

thymectomy Surgical removal of the thymus gland.

thyroidectomy Surgical removal of the thyroid gland.

tomography Radiographic image of a selected slice of tissue.

tonometry Measurement of intraocular pressure.

tonsillectomy Surgical removal of the tonsils.

tracheal aspiration Procedure in which a soft catheter is passed in the trachea to allow brushings and washings to remove cells and secretions from the trachea and main bronchi for diagnostic study.

tracheostomy Insertion of a tube into the windpipe to assist breathing.

tracheotomy The process of making an incision into the trachea.

traction Gentle but continuous application of a pulling force that can align a fracture, reduce muscle spasm, and relieve pain.

transdermal application Administration of some medications through the skin by an adhesive transdermal patch that is applied to the skin. The medication diffuses across the epidermis and enters the blood vessels in the dermis. Contraceptive hormones, analgesics, and antinausea/seasickness medications are examples.

transthoracic Going through the chest wall.

tubal ligation Surgery, using laparoscopy, in which both fallopian tubes are cut, a segment is removed, and the ends are tied off and cauterized shut.

24-hour urine collection Process that determines the amount of protein being excreted daily and estimates the kidneys' filtration ability.

tympanostomy Surgically created new opening in the tympanic membrane to allow fluid to drain from the middle ear.

ultrasonography Delineation of deep structures using sound waves.

ureteroscopy Examination of the ureter. A small flexible ureteroscope is passed through the urethra and bladder into the ureter. Devices can be passed through the endoscope to remove or fragment stones.

urinalysis (U/A) Examination of urine to separate it into its elements and define their kind and/or quantity. A routine urinalysis in the laboratory can include tests for color, clarity, pH, specific gravity, protein, glucose, ketones, and leukocyte esterase (indicator of infection).

urinalysis (microscopic) Analysis of the solids deposited by centrifuging a specimen of urine. It can reveal RBCs, WBCs, and renal tubular epithelial cells stuck together to form casts (cylindrical molds of cells) in nephrotic syndrome.

urine culture Culture taken from a clean-catch urine specimen. It is the definitive test for a urinary tract infection.

vasectomy Excision of a segment of the ductus deferens to interrupt the flow of sperm.

vasovasostomy Microsurgical procedure to suture back together the cut ends of the ductus deferens to restore the flow of sperm. Also called *vasectomy reversal.*

venogram Radiograph of veins after injection of radiopaque contrast material.

vestibulectomy Surgical excision of the vulva.

voiding cystourethrogram (VCUG) Imaging in which a contrast material is inserted into the bladder through a catheter and X-rays are taken during voiding.

Weber test Test for sensorineural hearing loss.

xenograft Graft from a nonhuman species. Also called *heterograft.*

E

Pharmacology

A compilation of pharmacologic terms used in this book.

acetaminophen Analgesic (reduces response to pain) and antipyretic (reduces fever).

adrenaline (1) Hormone produced by the adrenal medulla that boosts the supply of oxygen and glucose to the brain and increases heart rate and output. (2) Drug used to treat cardiac arrest and dysrhythmias and relieve bronchospasm in asthma. Also called *epinephrine.*

allergen Substance producing a hypersensitivity (allergic) reaction. Examples are animal fur and dander, penicillins, and foods such as eggs, milk, and wheat.

analgesic Substance that reduces or relieves the response to pain without producing loss of consciousness. Examples are aspirin and other NSAIDs, acetaminophen, and codeine.

androgen Hormone that promotes masculine characteristics. An example is testosterone.

anesthetic Agent that causes absence of feeling or sensation. Examples of local anesthetics are lidocaine and novocaine; examples of general anesthetics are nitrous oxide, thiopental, and ketamine.

antacid Agent that neutralizes the acidity of stomach contents. Examples are aluminum hydroxide, magnesium hydroxide, and calcium carbonate.

antibiotic Substance that has the capacity to inhibit the growth of or destroy bacteria and other microorganisms. Examples are penicillin, erythromycin, cefotaxime, and flucloxacillin.

anticoagulant Substance that prevents clotting. Examples are heparin and Coumadin (warfarin).

antidiabetic drugs Medications used to treat diabetes. Those given orally include metformin, acarbose, and thiazolidinediones, such as troglitazone. Insulin is given by injection or inhaled.

antidiuretic Agent that decreases urine production. Examples are vasopressin, amiloride, and chlorpropamide.

antiepileptic Agent capable of preventing or arresting epilepsy, Examples are phenobarbitol, phenytoin, and valproate.

antihistamine Agent used to treat allergic symptoms because of its action antagonistic to histamine. Examples are benadryl, diphenhydramine, and cimetidine.

anti-inflammatory Agent that reduces inflammation by acting on the body's responses, without affecting the causative agent. Examples are corticosteroids and aspirin.

antimicrobial Agent used to destroy or prevent multiplication of organisms. (See antibiotic.)

antineoplastic Agent that prevents the growth and spread of cancer cells. Examples are methotrexate, fluorouracil, and cyclophosphamide.

antipruritic Medication against itching. Examples are calamine lotion, hydrocortisone cream applied topically, and diphenhydramine (Benadryl) taken orally.

antipyretic Agent that reduces fever. Examples are aspirin and acetaminophen.

antiseptic Agent that reduces the number of microorganisms in different situations. Examples are alcohol, chlorhexidine, and providone-iodine.

atropine Agent used to dilate the pupils.

beta blocker Agent used in the treatment of a variety of cardiovascular diseases. Examples are propanalol and acebutolol.

bronchodilator Agent that relaxes the smooth muscles of the bronchioles. Examples are theophylline; beta-2 agonists, such as salbutamol and terbutaline; and anticholinergics, such as ipratropium bromide.

calcium channel blocker Agent that decreases the force of contraction of the myocardium, dilates coronary arteries, and reduces blood pressure. Examples are amlodipine and verapamil.

chemotherapy Treatment using chemical agents, usually in relation to neoplastic disease. Examples are platinum compounds such as cisplatin or paraplatin.

coagulant Substance that causes clotting. Thrombin and fibrin glue are used surgically to treat bleeding.

contraceptive Agent that prevents conception. Examples are condoms, diaphragms, and birth control pills using a mixture of estrogen and progesterone.

corticosteroids Hormones produced by the adrenal cortex. Examples are cortisol and aldosterone.

cortisol One of the glucocorticoids produced by the adrenal cortex; has anti-inflammatory effects. Also called *hydrocortisone.*

decongestant An agent that reduces the swelling and fluid in the nose and sinuses. Examples are pseudoephedrine and phenylephrine.

depressant Substance that diminishes activity, sensation, or tone, particularly in relation to the nervous system. Examples are alcohol, barbiturates, and benzodiazepines.

disease-modifying drugs Agent that has partial success in slowing down the accumulation of disabilities in a specific disease process. Examples in multiple sclerosis (MS) include interferons and mitoxantrone.

disinfectant Agent used to destroy pathogenic and other microorganisms on nonliving surfaces. Examples are alcohol, hydrogen peroxide, and hypochlorites.

diuretic Agent that increases urine output. Examples are furosemide, hydrochlorthiazide, spironolactone, and mannitol.

dopamine Chemical neurotransmitter in some specific areas of the brain.

epinephrine (1) Hormone produced by the adrenal medulla that boosts the supply of oxygen and glucose to the brain and increases heart rate and output. (2) Drug used to treat cardiac arrest and dysrhythmias and relieve bronchospasm in asthma. Also called *adrenaline.*

estrogen Generic term for hormones that stimulate female secondary sex characteristics.

fluorescein Dye that produces a vivid green color under a blue light; used to diagnose corneal abrasions and foreign bodies in the eye.

histamine Compound liberated in tissues as a result of injury or an immune response.

hydrocortisone Potent glucocorticoid with anti-inflammatory properties. Also called *cortisol.*

immunization Treatment with an agent designed to protect susceptible people from a communicable disease, such as agents that protect against the childhood diseases of measles, rubella, and pertussis.

insulin Hormone produced by the islet cells of the pancreas that promotes glucose use. Injectable insulin preparations are classified by their speed of action.

melatonin Hormone formed and secreted by the pineal gland during darkness. Serotonin is a precursor. It assists in the control of daily body rhythms, stimulates the immune system, and is an antioxidant.

morphine Derivative of opium used as an analgesic or sedative.

mucolytic Agent that attempts to break up mucus to allow it to be cleared more effectively from the airways. Examples are guaifenesin (common in over-the-counter cough medications), potassium iodide, and *N*-acetylcysteine (taken through a nebulizer).

narcotic Drug derived from opium. Examples are heroin, morphine, codeine, and demerol.

neurotransmitter Chemical that crosses a synapse to stimulate or inhibit another neuron or the cell of a muscle or gland. Examples are norepinephrine, serotonin, and dopamine.

opiate Drug derived from opium. Examples are morphine, codeine, heroin, and demerol.

oxygen Gas given by nasal cannula or by mask and intubation to relieve hypoxia. Patients with severe, chronic COPD can be attached to a portable cylinder of oxygen.

pharmacist Person licensed by the state to prepare and dispense drugs.

pharmacology Science of the preparation, uses, and effects of drugs.

pharmacy Facility licensed to prepare and dispense drugs.

placebo Inert, medicinally inactive compound with no intrinsic therapeutic value.

progesterone Hormone used to correct abnormalities of menstruation, and as a contraceptive.

psychoactive Agent able to alter mood, behavior, and/or cognition. Examples include narcotics, stimulants, antidepressants, and hallucinogens.

saline Salt solution, usually sodium chloride.

somatotropin Hormone of the anterior pituitary gland that stimulates the growth of body tissues. Also called *growth hormone (GH).*

spermicide Agent that destroys sperm. Examples are nonoxynol-9 and benzalkonium chloride.

sterilization Elimination of all microorganisms by high-pressure steam (autoclave), dry heat (oven), or radiation.

steroid Large family of chemical substances found in many drugs, hormones, and body components.

stimulant Agent that excites or strengthens. Examples include caffeine, nicotine, and cocaine.

surfactant Protein and fat compound that creates surface tension to hold the lung alveolar walls apart.

teratogen Agent that produces fetal abnormalities—congenital malformations—while organs and structures are being formed. (All medications readily cross the placenta.) Examples include alcohol, isoretinoin (acne medication), valproic acid (anticonvulsant), and the rubella virus.

testosterone The major androgen that promotes development of male sex characteristics.

thrombolytic Agent injected within a few hours of a myocardial infarction (MI) or stroke to dissolve the thrombus causing the arterial blockage. Examples are streptokinase and tissue plasminogen activator (tPA). Also called *clot-busting drug.*

thyroxine Thyroid hormone T4, tetraiodothyronine.

topical Medication applied to the skin to obtain a local effect. Examples are ointments, creams, gels, lotions, patches, and sprays.

toxin Poisonous substance formed by a living cell or organism. Examples are venom from bee stings, snake bites, and jellyfish stings.

vaccine Agent used to generate immunity and composed of the antigenic components of a killed or attenuated microorganism or its inactivated toxins. See *immunization.*

vitamin Essential organic substance necessary in small amounts for normal cell function.

warfarin Anticoagulant; also used as rat poison. Trade name: *Coumadin.*

A

abdomen (**AB**-doh-men) Part of the trunk between the thorax and the pelvis.

abdominal (ab-**DOM**-in-al) Pertaining to the abdomen.

abdominopelvic (ab-**DOM**-ih-no-**PEL**-vik) Pertaining to the abdomen and pelvis.

abdominoplasty (ab-**DOM**-ih-noh-plas-tee) Surgical removal of excess subcutaneous fat from the abdominal wall (tummy tuck).

abduction (ab-**DUCK**-shun) Action of moving away from the midline.

abductor (ab-**DUCK**-tor) Muscle that moves the thigh away from the midline.

ablation (ab-**LAY**-shun) Removal of tissue to destroy its function.

abortion (ah-**BOR**-shun) Spontaneous or induced expulsion of the fetus from the uterus at 20 weeks or less.

abrasion (ah-**BRAY**-shun) Area of skin or mucous membrane that has been scraped off.

abruptio (ab-**RUP**-she-oh) Placenta abruptio is the premature detachment of the placenta.

absorption (ab-**SORP**-shun) Uptake of nutrients and water by cells in the GI tract.

accommodate (ah-**KOM**-oh-date) To adapt to meet a need.

accommodation (ah-kom-oh-**DAY**-shun) The act of adjusting something to make it fit the needs; for example, the lens of the eye adjusts itself.

accomodative (ah-kom-oh-**DAY**-tiv) Pertaining to accommodation.

acetabulum (ass-eh-**TAB**-you-lum) The cup-shaped cavity of the hip bone that receives the head of the femur to form the hip joint.

acetaminophen (ah-seat-ah-**MIN**-oh-fen) Medication that is an analgesic and an antipyretic.

acetone (**ASS**-eh-tone) Ketone that is found in blood, urine, and breath when diabetes mellitus is out of control.

Achilles tendon (ah-KILL-eeze) A tendon formed from gastrocnemius and soleus muscles and inserted into the calcaneus bone. Also called *calcaneal tendon.*

achondroplasia (a-kon-droh-**PLAY**-zee-ah) Condition with abnormal conversion of cartilage into bone, leading to dwarfism.

acne (**AK**-nee) Inflammatory disease of sebaceous glands and hair follicles.

acquired immunodeficiency syndrome (**AIDS**) (ah-**KWIRED IM**-you-noh-dee-**FISH**-en-see **SIN**-drohm) Infection with the HIV virus.

acromegaly (ak-roe-**MEG**-ah-lee) Enlargement of the head, face, hands, and feet due to excess growth hormone in an adult.

acromioclavicular (AC) (ah-**CROW**-mee-oh-klah-**VICK**-you-lar) The joint between the acromion and the clavicle.

acromion (ah-**CROW**-mee-on) Lateral end of the scapula, extending over the shoulder joint.

activities of daily living (ADLs) (ak-**TIV**-ih-tees of **DAY**-lee **LIV**-ing) Daily routines for mobility, personal care, bathing, dressing, eating, and moving.

acute (ah-**KYUT**) Disease of sudden onset.

adapt (a-**DAPT**) To adjust to different conditions.

adaptation (ad-ap-**TAY**-shun) Change in the function or structure of an organ to meet new conditions.

addict (**ADD**-ikt) One who cannot live without a substance or practice.

addiction (ah-**DIK**-shun) Habitual psychologic and physiologic dependence on a substance or practice.

Addison disease (**ADD**-ih-son diz-**EEZ**) An autoimmune disease leading to decreased production of adrenocortical steroids.

adduction (ah-**DUCK**-shun) Action of moving toward the midline.

adductor (ah-**DUCK**-tor) Muscle that moves the thigh toward the midline.

adenocarcinoma (**ADD**-eh-noh-kar-sih-**NOH**-mah) A cancer arising from glandular epithelial cells.

adenoid (**ADD**-eh-noyd) Single mass of lymphoid tissue in the midline at the back of the throat.

adenoidectomy (**ADD**-eh-noy-**DEK**-toh-me) Surgical removal of the adenoid tissue.

adipose (**ADD**-ih-pose) Containing fat.

adolescence (ad-oh-**LESS**-ence) Stage that begins with puberty and ends with physical maturity.

adolescent (ad-oh-**LESS**-ent) Pertaining to adolescence or a person in that stage.

adrenal gland (ah-**DREE**-nal GLAND) The suprarenal gland on the upper pole of each kidney.

adrenaline (ah-**DREN**-ah-lin) One of the catecholamines. Also called *epinephrine.*

adrenocortical (ah-**DREE**-noh-**KOR**-tih-kal) Pertaining to the cortex of the adrenal gland.

adrenocorticotropic (ah-**DREE**-noh-**KOR**-tih-koh-**TROH**-pik) Hormone of the anterior pituitary that stimulates the cortex of the adrenal gland to produce its own hormones.

afferent (**AFF**-eh-rent) Moving toward a center; for example, nerve fibers conducting impulses to the spinal cord and brain.

aged (**A**-jid) Having lived to an advanced age.

agenesis (a-**JEN**-eh-sis) Failure to develop any organ or any part.

agglutinate (ah-**GLUE**-tin-ate) Stick together to form clumps.

agglutination (ah-glue-tih-**NAY**-shun) Process by which cells or other particles adhere to each other to form clumps.

aging (**A**-jing) The process of human maturation and decline.

agranulocyte (a-**GRAN**-you-loh-site) A white blood cell without any granules in its cytoplasm.

aldosterone (al-**DOS**-ter-own) Mineralocorticoid hormone of the adrenal cortex.

alignment (a-**LINE**-ment) Having a structure in its correct position relative to other structures.

alimentary (al-ih-**MEN**-tar-ee) Pertaining to the digestive tract.

alimentary canal (al-ih-**MEN**-tar-ee kah-**NAL**) Digestive tract.

allergen (AL-er-jen) Substance producing a hypersensitivity (allergic) reaction.

allergenic (al-er-JEN-ik) Pertaining to the capacity to produce an allergic reaction.

allergic (ah-LER-jik) Pertaining to or suffering from an allergy.

allergy (AL-er-jee) Hypersensitivity to an allergen.

alloimmune (AL-oh-im-YUNE) Immune reaction directed against foreign tissue.

alopecia (al-oh-PEE-shah) Partial or complete loss of hair, naturally or from medication.

alveolus (al-VEE-oh-lus) Terminal element of the respiratory tract where gas exchange occurs. Plural *alveoli.*

Alzheimer disease (AWLZ-high-mer DIZ-eez) Common form of dementia.

amblyopia (am-blee-OH-pee-ah) Failure or incomplete development of the pathways of vision to the brain.

ambulatory (am-byu-LAY-tor-ee) Surgery or any other care provided without an overnight stay in a medical facility.

amenorrhea (a-men-oh-REE-ah) Absence or abnormal cessation of menstrual flow.

amino acid (ah-ME-no ASS-id) The basic building block for protein.

ammonia (ah-MOAN-ih-ah) Toxic breakdown product of amino acids (proteins).

amniocentesis (AM-nee-oh-sen-TEE-sis) Removal of amniotic fluid for diagnostic purposes.

amnion (AM-nee-on) Membrane around the fetus that contains amniotic fluid.

amniotic (am-nee-OT-ic) Pertaining to the amnion.

ampulla (am-PULL-ah) Dilated portion of a canal or duct.

amputation (am-pyu-TAY-shun) Process of removing a limb, a part of a limb, a breast, or some other projecting part.

amputee (AM-pyu-tee) A person with an amputation.

amylase (AM-il-aze) One of a group of enzymes that breaks down starch.

amyotrophic (a-my-oh-TROH-fik) Pertaining to muscular atrophy.

anabolism (an-AB-oh-lizm) The buildup of complex substances in the cell from simpler ones as a part of metabolism.

anal (A-nal) Pertaining to the anus.

analgesia (an-al-JEE-zee-ah) State in which pain is reduced.

analgesic (an-al-JEE-zic) Substance that produces analgesia.

anaphylactic (AN-ah-fih-LAK-tik) Pertaining to anaphylaxis.

anaphylaxis (AN-ah-fih-LAK-sis) Immediate severe allergic response.

anastomosis (ah-NAS-toh-MOH-sis) A surgically made union between two tubular structures. Plural *anastomoses.*

androgen (AN-droh-jen) Hormone that promotes masculine characteristics.

anemia (ah-NEE-me-ah) Decreased number of red blood cells.

anemic (ah-NEE-mik) Pertaining to or suffering from anemia.

anencephaly (AN-en-SEF-ah-lee) Born without cerebral hemispheres.

anesthesia (an-es-THEE-zee-ah) Complete loss of sensation.

anesthesiologist (AN-es-thee-zee-OL-oh-jist) Medical specialist in anesthesia.

anesthesiology (AN-es-thee-zee-OL-oh-jee) Medical specialty related to anesthesia.

anesthetic (an-es-THET-ik) Agent that causes absence of feeling or sensation.

aneurysm (AN-yur-izm) Circumscribed dilation of an artery or cardiac chamber.

angiogram (AN-jee-oh-gram) Radiograph obtained after injection of radiopaque contrast material into blood vessels.

angiography (an-jee-OG-rah-fee) Radiography of vessels after injection of contrast material.

angioplasty (AN-jee-oh-PLAS-tee) Recanalization of a blood vessel by surgery.

anomaly (ah-NOM-ah-lee) Structural abnormality present at birth.

anorectal junction (A-no-RECK-tal JUNK-shun) Junction between the anus and the rectum.

anorexia (an-oh-RECK-see-ah) Without an appetite; *or* having an aversion to food.

anoxia (an-OCK-see-ah) Without oxygen.

anoxic (an-OCK-sik) Pertaining to or suffering from a lack of oxygen.

antacid (ant-ASS-id) Agent that neutralizes the acidity of stomach contents.

antagonist (an-TAG-oh-nist) An opposing structure, agent, disease, or process.

antagonistic (an-TAG-oh-nist-ik) Having an opposite function.

antecubital (an-teh-KYU-bit-al) In front of the elbow.

anterior (an-TEER-ee-or) Front surface of body; situated in front.

anteversion (an-teh-VER-shun) Forward displacement or tilting of a structure.

anthracosis (an-thra-KOH-sis) Lung disease caused by the inhalation of coal dust.

anthrax (AN-thraks) A severe, malignant infectious disease.

antibiotic (AN-tih-bye-OT-ik) A substance that has the capacity to inhibit growth of and destroy bacteria and other microorganisms.

antibody (AN-tee-body) Protein produced in response to an antigen. Plural *antibodies.*

anticoagulant (AN-tee-koh-AG-you-lant) Substance that prevents clotting.

antidiuretic (AN-tih-die-you-RET-ik) An agent that decreases urine production.

antidiuretic hormone (ADH) (AN-tih-die-you-RET-ik HOR-mohn) Posterior pituitary hormone that decreases urine output by acting on the kidney. Also called *vasopressin.*

antiepileptic (AN-tee-epih-LEP-tik) A pharmacologic agent capable of preventing or arresting epilepsy.

antigen (AN-tee-jen) Substance capable of triggering an immune response.

antihistamine (an-tee-HIS-tah-meen) Drug used to treat allergic symptoms because of its action antagonistic to histamine.

antimicrobial (AN-tee-my-KROH-bee-al) Agent to destroy or prevent multiplication of organisms.

cecum (SEE-kum) Blind pouch that is the first part of the large intestine.

celiac (SEE-lee-ack) Relating to the abdominal cavity.

celiac disease (SEE-lee-ak DIZ-eez) Disease caused by sensitivity to gluten.

cell (SELL) The smallest unit of the body capable of independent existence.

cellular (SELL-you-lar) Pertaining to a cell.

cellulitis (SELL-you-LIE-tis) Infection of subcutaneous connective tissue.

cephalic (seh-FAL-ik) Pertaining to or nearer to the head.

cerebellum (ser-eh-BELL-um) The most posterior area of the brain, located between the midbrain and the cerebral hemispheres.

cerebral (SER-ee-bral) Pertaining to the cerebral hemispheres or the brain.

cerebrospinal (SER-ee-broh-SPY-nal) Pertaining to the brain and spinal cord.

cerebrospinal fluid (CSF) (SER-ee-broh-SPY-nal FLU-id) Fluid formed in the ventricles of the brain that surrounds the brain and spinal cord.

cerebrum (SER-ee-brum) Cerebral hemispheres.

cerumen (seh-ROO-men) Waxy secretion of the ceruminous glands of the external ear.

ceruminous (seh-ROO-mih-nus) Pertaining to cerumen.

cervical (SER-vih-kal) Pertaining to the cervix or to the neck region.

cervix (SER-viks) The lower part of the uterus.

cesarean section (seh-ZAH-ree-an SEK-shun) Extraction of the fetus through an incision in the abdomen and uterine wall. Also called *C-section.*

chancre (SHAN-ker) Primary lesion of syphilis.

chemotherapy (KEY-moh-THAIR-ah-pee) Treatment using chemical agents.

chiasm (KYE-asm) X-shaped crossing of the two optic nerves at the base of the brain. Alternative term *chiasma.*

chickenpox (CHICK-en-pocks) Acute, contagious viral disease. Also called *varicella.*

chiropractic (kye-roh-PRAK-tik) Diagnosis, treatment, and prevention of mechanical disorders of the musculoskeletal system.

chiropractor (kye-roh-PRAK-tor) Practitioner of chiropractic.

chlamydia (klah-MID-ee-ah) An STD caused by an infection with *Chlamydia,* a species of bacteria

cholangiography (KOH-lan-jee-OG-rah-fee) Use of a contrast medium to radiographically visualize the bile ducts.

cholecystectomy (KOH-leh-sis-TECK-toe-me) Surgical removal of the gallbladder.

cholecystitis (KOH-leh-sis-TIE-tis) Inflammation of the gallbladder.

choledocholithiasis (KOH-leh-DOH-koh-li-THIGH-ah-sis) Presence of a gallstone in the common bile duct.

cholelithiasis (KOH-leh-lih-THIGH-ah-sis) Condition of having bile stones (gallstones).

cholelithotomy (KOH-leh-lih-THOT-oh-me) Surgical removal of a gallstone(s).

cholesteatoma (KOH-less-tee-ah-TOE-mah) Yellow, waxy tumor arising in the middle ear.

cholesterol (koh-LESS-ter-ol) Formed in liver cells; is the most abundant steroid in tissues and circulates in the plasma attached to proteins of different densities.

chorea (kor-EE-ah) Involuntary, irregular spasms of limb and facial muscles.

choreic (kor-EE-ik) Pertaining to or suffering from chorea.

chorion (KOH-ree-on) The fetal membrane that forms the placenta.

chorionic (koh-ree-ON-ick) Pertaining to the chorion.

chorionic villus (koh-ree-ON-ik VILL-us) Vascular process of the embryonic chorion to form the placenta.

choroid (KOR-oid) Region of the retina and uvea.

chromosome (KROH-moh-sohm) Body in the nucleus that contains DNA and genes.

chronic (KRON-ik) A persistent, long-term disease.

chyle (KYLE) A milky fluid that results from the digestion and absorption of fats in the small intestine.

chyme (KYME) Semifluid, partially digested food passed from the stomach into the duodenum.

cilium (SILL-ee-um) Hairlike motile projection from the surface of a cell. Plural *cilia.*

circulation (SER-kyu-LAY-shun) Continuous movement of blood through the heart and blood vessels.

circumcision (ser-kum-SIZH-un) To remove part or all of the prepuce.

circumduct (ser-kum-DUCKT) To move an extremity in a circular motion.

circumduction (ser-kum-DUCK-shun) Movement of an extremity in a circular motion.

cirrhosis (sir-ROE-sis) Extensive fibrotic liver disease.

claudication (klaw-dih-KAY-shun) Intermittent leg pain and limping.

clavicle (KLAV-ih-kul) Curved bone that forms the anterior part of the pectoral girdle.

clavicular (klah-VICK-you-lah) Pertaining to the clavicle.

cleft lip (KLEFT LIP) Congenital defect of the upper lip.

cleft palate (KLEFT PAL-ate) Congenital defect of the palate.

clitoris (KLIT-oh-ris) Erectile organ of the vulva.

clonic (KLON-ik) State of rapid successions of muscular contractions and relaxations.

closed fracture (KLOSD FRAK-chur) A bone is broken but the skin over it is intact.

Clostridium botulinum (klos-TRID-ee-um bot-you-LIE-num) Bacterium that causes food poisoning.

clot (KLOT) The mass of fibrin and cells that is produced in a wound.

coagulant (koh-AG-you-lant) Substance that causes clotting.

coagulate (koh-AG-you-late) Form a clot.

coagulation (koh-ag-you-LAY-shun) The process of blood clotting.

coagulopathy (koh-ag-you-LOP-ah-thee) Disorder of blood clotting. Plural *coagulopathies.*

coarctation (koh-ark-TAY-shun) Constriction, stenosis, particularly of the aorta.

bronchiolitis (brong-key-oh-**LYE**-tis) Inflammation of the small bronchioles.

bronchitis (bron-**KI**-tis) Inflammation of the bronchi.

bronchoconstriction (**BRONG**-koh-kon-**STRIK**-shun) Reduction in the diameter of a bronchus.

bronchodilator (**BRONG**-koh-die-**LAY**-tor) Agent that increases the diameter of a bronchus.

bronchogenic (**BRONG**-koh-**JEN**-ik) Arising from a bronchus.

bronchopneumonia (**BRONG**-koh-new-**MOH**-nee-ah) Acute inflammation of the walls of smaller bronchioles with spread to lung parenchyma.

bronchoscope (**BRONG**-koh-skope) Endoscope used for bronchoscopy.

bronchoscopy (brong-**KOS**-koh-pee) Examination of the interior of the tracheobronchial tree with an endoscope.

bronchus (**BRONG**-kuss) One of two subdivisions of the trachea. Plural *bronchi.*

bulbourethral (**BUL**-boh-you-**REE**-thral) Pertaining to the bulbous penis and urethra.

bulimia (byu-**LEEM**-ee-ah) Episodic bouts of excessive eating with compensatory throwing up.

bulla (**BULL**-ah) Bubble-like dilated structure. Plural *bullae.*

bundle of His (**BUN**-del of HISS) Pathway for electrical signals to be transmitted to the ventricles.

bunion (**BUN**-yun) A swelling at the base of the big toe.

bursa (**BURR**-sah) A closed sac containing synovial fluid.

bursitis (burr-**SIGH**-tis) Inflammation of a bursa.

C

cadaver (kah-**DAV**-er) A dead body or corpse.

calcaneal (kal-**KAY**-knee-al) Pertaining to the calcaneus.

calcaneus (kal-**KAY**-knee-us) Bone of the tarsus that forms the heel.

calcitonin (kal-sih-**TONE**-in) Thyroid hormone that moves calcium from blood to bones.

calculus (**KAL**-kyu-lus) Small stone. Plural *calculi.*

calyx (**KAY**-licks) Funnel-shaped structure. Plural *calyces.*

cancer (**KAN**-ser) General term for a malignant neoplasm.

Candida (**KAN**-did-ah) A yeastlike fungus.

Candida albicans (**KAN**-did-ah **AL**-bih-kanz) The most common form of *Candida.*

candidiasis (can-dih-**DIE**-ah-sis) Infection with the yeastlike fungus *Candida.* Also called *thrush.*

canker (**KANG**-ker) Nonmedical term for an aphthous ulcer. Also called *mouth ulcer.*

cannula (**KAN**-you-lah) Tube inserted into a blood vessel or cavity as a channel for fluid.

capillary (**KAP**-ih-lair-ee) Minute blood vessel between the arterial and venous systems.

capsular (**KAP**-syu-lar) Pertaining to a capsule.

capsule (**KAP**-syul) Fibrous tissue layer surrounding a joint or other structure.

carbohydrate (kar-boh-**HIGH**-drate) Group of organic food compounds that includes sugars, starch, glycogen, and cellulose.

carbuncle (**KAR**-bunk-ul) Infection of many furuncles in a small area, often on the back of the neck.

carcinogen (kar-**SIN**-oh-jen) Cancer-producing agent.

carcinogenesis (kar-**SIN**-oh-**JEN**-eh-sis) Origin and development of cancer.

carcinoma (kar-sih-**NOH**-mah) A malignant and invasive epithelial tumor.

carcinoma in situ (kar-sih-**NOH**-mah in **SIGH**-too) Carcinoma that has not invaded surrounding tissues.

cardiac (**KAR**-dee-ak) Pertaining to the heart.

cardiogenic (**KAR**-dee-oh-**JEN**-ik) Of cardiac origin.

cardiologist (**KAR**-dee-**OL**-oh-jist) A medical specialist in the diagnosis and treatment of disorders of the heart (cardiology).

cardiology (**KAR**-dee-**OL**-oh-jee) Medical specialty of diseases of the heart.

cardiomegaly (**KAR**-dee-oh-**MEG**-ah-lee) Enlargement of the heart.

cardiomyopathy (**KAR**-dee-oh-my-**OP**-ah-thee) Disease of the heart muscle, the myocardium.

cardiopulmonary resuscitation (**KAR**-dee-oh-**PUL**-moh-nary ree-sus-ih-**TAY**-shun) The attempt to restore cardiac and pulmonary function.

cardiovascular (**KAR**-dee-oh-**VAS**-kyu-lar) Pertaining to the heart and blood vessels.

cardioversion (**KAR**-dee-oh-**VER**-shun) Restoration of a normal heart rhythm by electric shock. Also called *defibrillation.*

caries (**KARE**-eez) Bacterial destruction of teeth.

carotid (kah-**ROT**-id) Main artery of the neck.

carotid endarterectomy (kah-**ROT**-id **END**-ar-ter-**EK**-toe-me) Surgical removal of diseased lining from the carotid artery to leave a smooth lining and restore blood flow.

carpal (**KAR**-pal) Pertaining to the wrist.

carpus (**KAR**-pus) Collective term for the eight carpal bones of the wrist.

cartilage (**KAR**-tih-lage) Nonvascular, firm connective tissue found mostly in joints.

catabolism (kah-**TAB**-oh-lizm) Breakdown of complex substances into simpler ones as a part of metabolism.

cataplexy (**KAT**-ah-plek-see) Sudden loss of muscle tone with brief paralysis.

cataract (**KAT**-ah-ract) Complete or partial opacity of the lens.

catecholamine (kat-eh-**COAL**-ah-meen) Major element in the stress response; includes epinephrine and norepinephrine.

catheter (**KATH**-eh-ter) Hollow tube to allow passage of fluid into or out of a body cavity, organ, or vessel.

catheterization (**KATH**-eh-ter-ih-**ZAY**-shun) Introduction of a catheter.

catheterize (**KATH**-eh-teh-**RIZE**) To introduce a catheter.

caudal (**KAW**-dal) Pertaining to or nearer to the tailbone.

cautery (**KAW**-ter-ee) Agent or device used to burn or scar a tissue.

cavernosa (kav-er-**NOH**-sah) Resembling a cave.

cavity (**KAV**-ih-tee) Hollow space or body compartment. Plural *cavities.*

cecal (**SEE**-kal) Pertaining to the cecum.

audiologist (aw-dee-**OL**-oh-jist) Specialist in evaluation of hearing function.

audiology (aw-dee-**OL**-oh-jee) Study of hearing disorders.

audiometer (aw-dee-**OM**-ee-ter) Instrument to measure hearing.

audiometric (**AW**-dee-oh-**MET**-rik) Pertaining to the measurement of hearing.

auditory (**AW**-dih-tor-ee) Pertaining to the sense or the organs of hearing.

aura (**AWE**-rah) Sensory experience preceding an epileptic seizure or a migraine headache.

auricle (**AW**-ri-kul) The shell-like external ear.

auscultation (aws-kul-**TAY**-shun) Diagnostic method of listening to body sounds with a stethoscope.

autism (**AWE**-tizm) Developmental disorder of children.

autoantibody (awe-toe-**AN**-tee-bod-ee) Antibody produced in response to an antigen from the host's own tissue.

autograft (**AWE**-toe-graft) A graft using tissue taken from the same individual who is receiving the graft.

autoimmune (awe-toe-im-**YUNE**) Immune reaction directed against a person's own tissue.

autologous (awe-**TOL**-oh-gus) Blood transfusion with the same person as donor and recipient.

autolysis (awe-**TOL**-ih-sis) Destruction of cells by enzymes within the cells.

autonomic (awe-toh-**NOM**-ik) Self-governing visceral motor division of the peripheral nervous system.

autopsy (**AWE**-top-see) Examination of the body and organs of a dead person to determine the cause of death.

avascular (a-**VAS**-cue-lar) Without a blood supply.

avulsion (a-**VUL**-shun) Forcible separation or tearing away, often of a tendon from bone.

axilla (**AK**-sill-ah) Medical name for the armpit. Plural *axillae.*

axillary (**AK**-sill-air-ee) Pertaining to the armpit.

axon (**ACK**-son) Single process of a nerve cell carrying nervous impulses away from the cell body.

azotemia (azo-**TEE**-me-ah) Excess nitrogenous waste products in the blood.

B

bacterial (bak-**TEER**-ee-al) Pertaining to bacteria.

bacterium (bak-**TEER**-ee-um) A unicellular (single-cell), simple, microscopic organism. Plural *bacteria.*

balanitis (bal-ah-**NIE**-tis) Inflammation of the glans and prepuce of the penis.

bariatric (bar-ee-**AT**-rik) Treatment of obesity.

basilar (**BAS**-ih-lar) Pertaining to the base of a structure.

basophil (**BAY**-so-fill) A basophil's granules attract basic blue stain in the laboratory.

Bell palsy (BELL **PAWL**-zee) Paresis, or paralysis, of one side of the face.

biceps brachii (**BYE**-sepz **BRAY**-key-eye) A muscle of the arm that has two heads or points of origin on the scapula.

bicuspid (by-**KUSS**-pid) Having two points. A bicuspid heart valve has two flaps; a bicuspid (premolar) tooth has two points.

bifid (**BI**-fid) Separated into two parts.

bilateral (by-**LAT**-er-al) On two sides; for example, in both ears.

bile (BILE) Fluid secreted by the liver into the duodenum.

bile acids (BILE **AH**-sids) Steroids synthesized from cholesterol.

biliary (**BILL**-ee-air-ree) Pertaining to bile or the biliary tract.

bilirubin (bill-ee-**RU**-bin) Bile pigment formed in the liver from hemoglobin.

binge eating (BINJ **EE**-ting) Eating with periods of excessive intake.

biopsy (**BI**-op-see) Removing tissue from a living person for laboratory examination.

bipolar disorder (bi-**POH**-lar dis-**OR**-der) A mood disorder with alternating episodes of depression and mania.

bladder (**BLAD**-er) Hollow sac that holds fluid, for example, urine or bile.

blastocyst (**BLAS**-toe-sist) First 2 weeks of the developing embryo.

blepharitis (blef-ah-**RYE**-tis) Inflammation of the eyelid.

blepharoplasty (**BLEF**-ah-roh-**PLAS**-tee) Surgical repair of the eyelid.

blepharoptosis (**BLEF**-ah-**ROP**-toe-sis) Drooping of the upper eyelid.

blood-brain barrier (BBB) (BLUD BRAYN **BAIR**-ee-er) A selective mechanism that protects the brain from toxins and infections.

bolus (**BOH**-lus) Single mass of a substance.

botulism (**BOT**-you-lizm) Food poisoning caused by the neurotoxin produced by *Clostridium botulinum.*

bovine spongiform encephalopathy (**BO**-vine **SPON**-jee-form en-sef-ah-**LOP**-ah-thee) Disease of cattle that can be transmitted to humans, causing Creutzfeldt-Jakob disease. Also called *mad cow disease.*

bowel (**BOUGH**-el) Another name for *intestine.*

brace (BRACE) Appliance to support a part of the body in its correct position.

brachial (**BRAY**-kee-al) Pertaining to the arm.

brachialis (**BRAY**-kee-al-is) Muscle that lies underneath the biceps and is the strongest flexor of the forearm.

brachioradialis (**BRAY**-kee-oh-**RAY**-dee-al-is) Muscle that helps flex the forearm.

brachytherapy (brah-kee-**THAIR**-ah pee) Radiation therapy in which the source of radiation is implanted in the tissue to be treated.

bradycardia (brad-ee-**KAR**-dee-ah) Slow heart rate (below 60 beats per minute).

bradypnea (brad-ip-**NEE**-ah) Slow breathing.

brainstem (BRAYNSTEM) Comprises the thalamus, pineal gland, pons, fourth ventricle, and medulla oblongata.

breech (BREECH) Buttocks-first presentation of the fetus at delivery.

bronchiectasis (brong-key-**ECK**-tah-sis) Chronic dilation of the bronchi following inflammatory disease and obstruction.

bronchiole (**BRONG**-key-ole) Increasingly smaller subdivisions of bronchi.

antineoplastic (AN-tee-nee-oh-**PLAS**-tik) Pertaining to the prevention of the growth and spread of cancer cells.

antipruritic (AN-tee-pru-**RIT**-ik) Medication against itching.

antipyretic (AN-tee-pie-**RET**-ik) Agent that reduces fever.

antisepsis (an-tih-**SEP**-sis) Inhibiting the growth of infectious agents.

antiseptic (an-tee-**SEP**-tik) An agent or substance capable of affecting antisepsis.

antiserum (an-tee-**SEER**-um) Serum taken from another human or animal that has antibodies to a disease. Also called *immune serum.*

antrum (**AN**-trum) A closed cavity.

anuria (an-**YOU**-ree-ah) Absence of urine production.

anus (**A**-nus) Terminal opening of the digestive tract through which feces are discharged.

anxiety (ang-**ZI**-eh-tee) Distress and dread caused by fear.

aorta (a-**OR**-tuh) Main trunk of the systemic arterial system.

aortic (a-**OR**-tik) Pertaining to the aorta.

apex (**A**-peks) Tip or end; for example, of the cone-shaped heart.

Apgar score (**AP**-gar SKOR) Evaluation of a newborn's status.

aphthous (**AF**-thus) Painful small oral ulcers (canker sores).

apnea (**AP**-nee-ah) Absence of spontaneous respiration.

apoptosis (**AP**-op-**TOE**-sis) Programmed normal cell death.

appendicitis (ah-pen-dih-**SIGH**-tis) Inflammation of the appendix.

appendix (ah-**PEN**-dicks) Small blind projection from the pouch of the cecum.

aqueous humor (**ACHE**-we-us **HEW**-mor) Watery liquid in the anterior and posterior chambers of the eye.

arachnoid mater (ah-**RACK**-noyd **MAY**-ter) Weblike middle layer of the three meninges.

areola (ah-**REE**-oh-luh) Circular reddish area surrounding the nipple.

areolar (ah-**REE**-oh-lar) Pertaining to the areola.

arrhythmia (a-**RITH**-me-ah) Condition when the heart rhythm is abnormal.

arteriography (ar-teer-ee-**OG**-rah-fee) X-ray visualization of an artery after injection of contrast material.

arteriole (ar-**TIER**-ee-ole) Small terminal artery leading into the capillary network.

arteriosclerosis (ar-**TIER**-ee-oh-skler-**OH**-sis) Hardening of the arteries.

arteriosclerotic (ar-**TIER**-ee-oh-skler-**OT**-ik) Pertaining to or suffering from arteriosclerosis.

artery (**AR**-ter-ee) Thick-walled blood vessel carrying blood away from the heart.

arthritis (ar-**THRI**-tis) Inflammation of a joint or joints.

arthrocentesis (**AR**-throw-sen-**TEE**-sis) Withdrawal of fluid from a joint through a needle.

arthrodesis (ar-**THROW**-dee-sis) Fixation or stiffening of a joint by surgery.

arthrography (ar-**THROG**-rah-fee) X-ray of a joint taken after the injection of a contrast medium into the joint.

arthroplasty (**AR**-throw-plas-tee) Surgery to restore, as far as possible, the function of a joint.

arthroscope (**AR**-thro-skope) Endoscope used to examine the interior of a joint.

arthroscopy (ar-**THROS**-koh-pee) Visual examination of the interior of a joint.

articulate (ar-**TIK**-you-late) Two separate bones have formed a joint.

articulation (ar-tik-you-**LAY**-shun) A joint.

asbestosis (as-bes-**TOE**-sis) Lung disease caused by the inhalation of asbestos particles.

ascites (ah-**SIGH**-teez) Accumulation of fluid in the abdominal cavity.

asepsis (a-**SEP**-sis) Absence of living pathogenic organisms.

Asperger syndrome (**AHS**-per-ger **SIN**-drohm) Developmental disorder of children.

aspiration (**AS**-pih-**RAY**-shun) Removal by suction of fluid or gas from a body cavity.

asthma (**AZ**-mah) Episodes of breathing difficulty due to narrowed or obstructed airways.

asthmatic (az-**MAT**-ik) Suffering from or pertaining to asthma.

astigmatism (ah-**STIG**-mah-tism) Inability to focus light rays that enter the eye in different planes.

asymptomatic (**A**-simp-toe-**MAT**-ik) Without any symptoms experienced by the patient.

asystole (a-**SIS**-toe-lee) Absence of contractions of the heart.

ataxia (a-**TAK**-see-ah) Inability to coordinate muscle activity, leading to jerky movements.

ataxic (a-**TAK**-sik) Pertaining to or suffering from ataxia.

atelectasis (at-el-**ECK**-tah-sis) Collapse of part of a lung.

atherectomy (ath-er-**EK**-toe-me) Surgical removal of atheroma.

atheroma (ath-er-**ROE**-mah) Lipid deposit in the lining of an artery.

atherosclerosis (**ATH**-er-oh-skler-**OH**-sis) Atheroma in arteries.

athetoid (**ATH**-eh-toyd) Resembling or suffering from athetosis.

athetosis (ath-eh-**TOE**-sis) Slow, writhing involuntary movements.

atonic (a-**TOHN**-ik) Without normal muscular tone.

atopic (ay-**TOP**-ik) Pertaining to an allergy.

atopy (**AT**-oh-pee) State of hypersensitivity to an allergen—allergic.

atresia (a-**TREE**-zee-ah) Congenital absence of a normal opening or lumen.

atrial (**A**-tree-al) Pertaining to the atrium.

atrioventricular (AV) (**A**-tree-oh-ven-**TRICK**-you-lar) Pertaining to both the atrium and the ventricle.

atrium (**A**-tree-um) Chamber where blood enters the heart on both right and left sides. Plural *atria.*

atrophy (**AT**-roh-fee) Wasting or diminished volume of a tissue or organ.

atropine (**AT**-ro-peen) Pharmacologic agent used to dilate pupils.

attenuate (ah-**TEN**-you-ate) Weaken the ability of an organism to produce disease.

allergen (AL-er-jen) Substance producing a hypersensitivity (allergic) reaction.

allergenic (al-er-JEN-ik) Pertaining to the capacity to produce an allergic reaction.

allergic (ah-LER-jik) Pertaining to or suffering from an allergy.

allergy (AL-er-jee) Hypersensitivity to an allergen.

alloimmune (AL-oh-im-YUNE) Immune reaction directed against foreign tissue.

alopecia (al-oh-PEE-shah) Partial or complete loss of hair, naturally or from medication.

alveolus (al-VEE-oh-lus) Terminal element of the respiratory tract where gas exchange occurs. Plural *alveoli*.

Alzheimer disease (AWLZ-high-mer DIZ-eez) Common form of dementia.

amblyopia (am-blee-OH-pee-ah) Failure or incomplete development of the pathways of vision to the brain.

ambulatory (am-byu-LAY-tor-ee) Surgery or any other care provided without an overnight stay in a medical facility.

amenorrhea (a-men-oh-REE-ah) Absence or abnormal cessation of menstrual flow.

amino acid (ah-ME-no ASS-id) The basic building block for protein.

ammonia (ah-MOAN-ih-ah) Toxic breakdown product of amino acids (proteins).

amniocentesis (AM-nee-oh-sen-TEE-sis) Removal of amniotic fluid for diagnostic purposes.

amnion (AM-nee-on) Membrane around the fetus that contains amniotic fluid.

amniotic (am-nee-OT-ic) Pertaining to the amnion.

ampulla (am-PULL-ah) Dilated portion of a canal or duct.

amputation (am-pyu-TAY-shun) Process of removing a limb, a part of a limb, a breast, or some other projecting part.

amputee (AM-pyu-tee) A person with an amputation.

amylase (AM-il-aze) One of a group of enzymes that breaks down starch.

amyotrophic (a-my-oh-TROH-fik) Pertaining to muscular atrophy.

anabolism (an-AB-oh-lizm) The buildup of complex substances in the cell from simpler ones as a part of metabolism.

anal (A-nal) Pertaining to the anus.

analgesia (an-al-JEE-zee-ah) State in which pain is reduced.

analgesic (an-al-JEE-zic) Substance that produces analgesia.

anaphylactic (AN-ah-fih-LAK-tik) Pertaining to anaphylaxis.

anaphylaxis (AN-ah-fih-LAK-sis) Immediate severe allergic response.

anastomosis (ah-NAS-toh-MOH-sis) A surgically made union between two tubular structures. Plural *anastomoses*.

androgen (AN-droh-jen) Hormone that promotes masculine characteristics.

anemia (ah-NEE-me-ah) Decreased number of red blood cells.

anemic (ah-NEE-mik) Pertaining to or suffering from anemia.

anencephaly (AN-en-SEF-ah-lee) Born without cerebral hemispheres.

anesthesia (an-es-THEE-zee-ah) Complete loss of sensation.

anesthesiologist (AN-es-thee-zee-OL-oh-jist) Medical specialist in anesthesia.

anesthesiology (AN-es-thee-zee-OL-oh-jee) Medical specialty related to anesthesia.

anesthetic (an-es-THET-ik) Agent that causes absence of feeling or sensation.

aneurysm (AN-yur-izm) Circumscribed dilation of an artery or cardiac chamber.

angiogram (AN-jee-oh-gram) Radiograph obtained after injection of radiopaque contrast material into blood vessels.

angiography (an-jee-OG-rah-fee) Radiography of vessels after injection of contrast material.

angioplasty (AN-jee-oh-PLAS-tee) Recanalization of a blood vessel by surgery.

anomaly (ah-NOM-ah-lee) Structural abnormality present at birth.

anorectal junction (A-no-RECK-tal JUNK-shun) Junction between the anus and the rectum.

anorexia (an-oh-RECK-see-ah) Without an appetite; *or* having an aversion to food.

anoxia (an-OCK-see-ah) Without oxygen.

anoxic (an-OCK-sik) Pertaining to or suffering from a lack of oxygen.

antacid (ant-ASS-id) Agent that neutralizes the acidity of stomach contents.

antagonist (an-TAG-oh-nist) An opposing structure, agent, disease, or process.

antagonistic (an-TAG-oh-nist-ik) Having an opposite function.

antecubital (an-teh-KYU-bit-al) In front of the elbow.

anterior (an-TEER-ee-or) Front surface of body; situated in front.

anteversion (an-teh-VER-shun) Forward displacement or tilting of a structure.

anthracosis (an-thra-KOH-sis) Lung disease caused by the inhalation of coal dust.

anthrax (AN-thraks) A severe, malignant infectious disease.

antibiotic (AN-tih-bye-OT-ik) A substance that has the capacity to inhibit growth of and destroy bacteria and other microorganisms.

antibody (AN-tee-body) Protein produced in response to an antigen. Plural *antibodies*.

anticoagulant (AN-tee-koh-AG-you-lant) Substance that prevents clotting.

antidiuretic (AN-tih-die-you-RET-ik) An agent that decreases urine production.

antidiuretic hormone (ADH) (AN-tih-die-you-RET-ik HOR-mohn) Posterior pituitary hormone that decreases urine output by acting on the kidney. Also called *vasopressin*.

antiepileptic (AN-tee-epih-LEP-tik) A pharmacologic agent capable of preventing or arresting epilepsy.

antigen (AN-tee-jen) Substance capable of triggering an immune response.

antihistamine (an-tee-HIS-tah-meen) Drug used to treat allergic symptoms because of its action antagonistic to histamine.

antimicrobial (AN-tee-my-KROH-bee-al) Agent to destroy or prevent multiplication of organisms.

A

abdomen (AB-doh-men) Part of the trunk between the thorax and the pelvis.

abdominal (ab-DOM-in-al) Pertaining to the abdomen.

abdominopelvic (ab-DOM-ih-no-PEL-vik) Pertaining to the abdomen and pelvis.

abdominoplasty (ab-DOM-ih-noh-plas-tee) Surgical removal of excess subcutaneous fat from the abdominal wall (tummy tuck).

abduction (ab-DUCK-shun) Action of moving away from the midline.

abductor (ab-DUCK-tor) Muscle that moves the thigh away from the midline.

ablation (ab-LAY-shun) Removal of tissue to destroy its function.

abortion (ah-BOR-shun) Spontaneous or induced expulsion of the fetus from the uterus at 20 weeks or less.

abrasion (ah-BRAY-shun) Area of skin or mucous membrane that has been scraped off.

abruptio (ab-RUP-she-oh) Placenta abruptio is the premature detachment of the placenta.

absorption (ab-SORP-shun) Uptake of nutrients and water by cells in the GI tract.

accommodate (ah-KOM-oh-date) To adapt to meet a need.

accommodation (ah-kom-oh-DAY-shun) The act of adjusting something to make it fit the needs; for example, the lens of the eye adjusts itself.

accomodative (ah-kom-oh-DAY-tiv) Pertaining to accommodation.

acetabulum (ass-eh-TAB-you-lum) The cup-shaped cavity of the hip bone that receives the head of the femur to form the hip joint.

acetaminophen (ah-seat-ah-MIN-oh-fen) Medication that is an analgesic and an antipyretic.

acetone (ASS-eh-tone) Ketone that is found in blood, urine, and breath when diabetes mellitus is out of control.

Achilles tendon (ah-KILL-eeze) A tendon formed from gastrocnemius and soleus muscles and inserted into the calcaneus bone. Also called *calcaneal tendon*.

achondroplasia (a-kon-droh-PLAY-zee-ah) Condition with abnormal conversion of cartilage into bone, leading to dwarfism.

acne (AK-nee) Inflammatory disease of sebaceous glands and hair follicles.

acquired immunodeficiency syndrome (AIDS) (ah-KWIRED IM-you-noh-dee-FISH-en-see SIN-drohm) Infection with the HIV virus.

acromegaly (ak-roe-MEG-ah-lee) Enlargement of the head, face, hands, and feet due to excess growth hormone in an adult.

acromioclavicular (AC) (ah-CROW-mee-oh-klah-VICK-you-lar) The joint between the acromion and the clavicle.

acromion (ah-CROW-mee-on) Lateral end of the scapula, extending over the shoulder joint.

activities of daily living (ADLs) (ak-TIV-ih-tees of DAY-lee LIV-ing) Daily routines for mobility, personal care, bathing, dressing, eating, and moving.

acute (ah-KYUT) Disease of sudden onset.

adapt (a-DAPT) To adjust to different conditions.

adaptation (ad-ap-TAY-shun) Change in the function or structure of an organ to meet new conditions.

addict (ADD-ikt) One who cannot live without a substance or practice.

addiction (ah-DIK-shun) Habitual psychologic and physiologic dependence on a substance or practice.

Addison disease (ADD-ih-son diz-EEZ) An autoimmune disease leading to decreased production of adrenocortical steroids.

adduction (ah-DUCK-shun) Action of moving toward the midline.

adductor (ah-DUCK-tor) Muscle that moves the thigh toward the midline.

adenocarcinoma (ADD-eh-noh-kar-sih-NOH-mah) A cancer arising from glandular epithelial cells.

adenoid (ADD-eh-noyd) Single mass of lymphoid tissue in the midline at the back of the throat.

adenoidectomy (ADD-eh-noy-DEK-toh-me) Surgical removal of the adenoid tissue.

adipose (ADD-ih-pose) Containing fat.

adolescence (ad-oh-LESS-ence) Stage that begins with puberty and ends with physical maturity.

adolescent (ad-oh-LESS-ent) Pertaining to adolescence or a person in that stage.

adrenal gland (ah-DREE-nal GLAND) The suprarenal gland on the upper pole of each kidney.

adrenaline (ah-DREN-ah-lin) One of the catecholamines. Also called *epinephrine*.

adrenocortical (ah-DREE-noh-KOR-tih-kal) Pertaining to the cortex of the adrenal gland.

adrenocorticotropic (ah-DREE-noh-KOR-tih-koh-TROH-pik) Hormone of the anterior pituitary that stimulates the cortex of the adrenal gland to produce its own hormones.

afferent (AFF-eh-rent) Moving toward a center; for example, nerve fibers conducting impulses to the spinal cord and brain.

aged (A-jid) Having lived to an advanced age.

agenesis (a-JEN-eh-sis) Failure to develop any organ or any part.

agglutinate (ah-GLUE-tin-ate) Stick together to form clumps.

agglutination (ah-glue-tih-NAY-shun) Process by which cells or other particles adhere to each other to form clumps.

aging (A-jing) The process of human maturation and decline.

agranulocyte (a-GRAN-you-loh-site) A white blood cell without any granules in its cytoplasm.

aldosterone (al-DOS-ter-own) Mineralocorticoid hormone of the adrenal cortex.

alignment (a-LINE-ment) Having a structure in its correct position relative to other structures.

alimentary (al-ih-MEN-tar-ee) Pertaining to the digestive tract.

alimentary canal (al-ih-MEN-tar-ee kah-NAL) Digestive tract.

coccyx (KOK-sicks) Small tailbone at the lower end of the vertebral column.

cochlea (KOK-lee-ah) An intricate combination of passages; used to describe the inner ear.

cochlear (KOK-lee-ar) Pertaining to the cochlea.

cognition (kog-NIH-shun) Process of acquiring knowledge through thinking, learning, and memory.

cognitive-behavioral therapy (CBT) (KOG-nih-tiv be-HAYV-yur-al THAIR-ah-pee) Psychotherapy that emphasizes thoughts and attitudes in one's behavior.

coitus (KOH-it-us) Sexual intercourse.

colic (KOL-ik) Spasmodic, crampy pains in the abdomen.

colitis (koh-LIE-tis) Inflammation of the colon.

collagen (KOLL-ah-jen) Major protein of connective tissue, cartilage, and bone.

collateral (koh-LAT-er-al) Situated at the side, often to bypass an obstruction.

Colles fracture (KOL-ez FRAK-chur) Fracture of the distal radius at the wrist.

colloid (COLL-oyd) Liquid containing suspended particles.

colon (KOH-lon) The large intestine, extending from the cecum to the rectum.

colostomy (koh-LOSS-toe-me) Artificial opening from the colon to the outside of the body.

colostrum (koh-LOSS-trum) The first breast secretion at the end of pregnancy.

colpopexy (KOL-poh-peck-see) Surgical fixation of the vagina.

coma (KOH-mah) State of deep unconsciousness.

comatose (KOH-mah-toes) In a state of coma.

comedo (KOM-ee-doh) A whitehead or blackhead caused by too much sebum and too many keratin cells blocking the hair follicle. Plural *comedones.*

comminuted fracture (KOM-ih-nyu-ted FRAK-chur) A fracture in which the bone is broken into small pieces.

competent (KOM-peh-tent) Capable of performing a task or function.

complement (KOM-pleh-ment) Group of proteins in the serum that finish off the work of antibodies to destroy bacteria and other cells.

complete fracture (kom-PLEET FRAK-chur) A bone is fractured into two separate pieces.

compliance (kom-PLY-ance) Measure of the capacity of a chamber or hollow viscus (e.g., the lungs) to expand; *or* consistency and accuracy with which a patient follows a treatment regimen.

compression (kom-PRESH-un) Squeeze together to increase density and/or decrease a dimension of a structure.

compression fracture (kom-PRESH-un FRAK-chur) Fracture of a vertebra causing loss of height of the vertebra.

compulsion (kom-PULL-shun) Uncontrollable impulses to perform an act repetitively.

compulsive (kom-PULL-siv) Possessing uncontrollable impulses to perform an act repetitively.

conception (kon-SEP-shun) Fertilization of the egg by sperm to form a zygote.

concha (KON-kah) Shell-shaped bone on the medial wall of the nasal cavity. Plural *conchae.*

concussion (kon-KUSH-un) Mild brain injury.

condom (KON-dom) A sheath or cover for the penis or vagina to prevent conception and infection.

conductive hearing loss (kon-DUK-tiv) Hearing loss caused by lesions in the outer ear or middle ear.

condyle (KON-dile) Large, smooth, rounded expansion of the end of a bone where it forms a joint with another bone.

confusion (kon-FEW-zhun) Mental state in which environmental stimuli are not processed appropriately.

congenital (kon-JEN-ih-tal) Present at birth, either inherited or due to an event during gestation up to the moment of birth.

conization (koh-ni-ZAY-shun) Surgical excision of a cone-shaped piece of tissue.

conjunctiva (kon-junk-TIE-vah) Inner lining of the eyelids.

conjunctival (kon-junk-TIE-val) Pertaining to the conjunctiva.

conjunctivitis (kon-junk-tih-VI-tis) Inflammation of the conjunctiva.

connective tissue (koh-NECK-tiv TISH-you) The supporting tissue of the body.

consciousness *(*KON-shus-ness) The state of being aware of and responsive to the environment.

constipation (kon-stih-PAY-shun) Hard, infrequent bowel movements.

constrict (kon-STRIKT) To become or make narrow.

constriction (kon-STRIK-shun) A narrowed portion of a structure.

contagious (kon-TAY-jus) Infection can be transmitted from person to person or from a person to a surface to a person.

contaminate (kon-TAM-in-ate) To cause the presence of an infectious agent to be on any surface or in any substance.

contamination (KON-tam-ih-NAY-shun) Presence of an infectious agent on a surface or in substances.

contraception (kon-trah-SEP-shun) Prevention of pregnancy.

contraceptive (kon-trah-SEP-tiv) An agent that prevents conception.

contract (kon-TRAKT) Draw together or shorten.

contracture (kon-TRAK-chur) Muscle shortening due to spasm or fibrosis.

contrecoup (KON-treh-koo) Injury to the brain at a point directly opposite the point of original impact.

contusion (kon-TOO-zhun) Hemorrhage into a tissue (bruising), including the brain.

convulsion (kon-VUL-shun) Alternative name for seizure.

cor pulmonale (KOR pul-moh-NAH-lee) Right-sided heart failure arising from chronic lung disease.

cornea (KOR-nee-ah) The central, transparent part of the outer coat of the eye covering the iris and pupil.

corneal (KOR-nee-al) Pertaining to the cornea.

coronal (KOR-oh-nal) Pertaining to the vertical plane dividing the body into anterior and posterior portions.

coronal plane (KOR-oh-nal PLAIN) Vertical plane dividing the body into anterior and posterior portions.

coronary circulation (KOR-oh-nair-ee SER-kyu-LAY-shun) Blood vessels supplying the heart.

corpus (KOR-pus) Major part of a structure. Plural *corpora.*

corpus albicans (**KOR**-pus **AL**-bih-kanz) An atrophied corpus luteum.

corpus callosum (**KOR**-pus kah-**LOW**-sum) Bridge of nerve fibers connecting the two cerebral hemispheres.

corpus luteum (**KOR**-pus **LOO**-teh-um) Yellow structure formed at the site of a ruptured ovarian follicle.

corpuscle (**KOR**-pus-ul) A blood cell.

cortex (**KOR**-teks) Outer portion of an organ, such as bone; gray covering of cerebral hemispheres. Plural *cortices.*

cortical (**KOR**-tih-cal) Pertaining to a cortex.

corticosteroid (**KOR**-tih-koh-**STEHR**-oyd) A hormone produced by the adrenal cortex.

corticotropin (**KOR**-tih-koh-TROH-pin) Pituitary hormone that stimulates the cortex of the adrenal gland to secrete corticosteroids.

cortisol (**KOR**-tih-sol) One of the glucocorticoids produced by the adrenal cortex; has anti-inflammatory effects. Also called *hydrocortisone.*

coryza (koh-**RYE**-zah) Viral inflammation of the mucous membrane of the nose. Also called *acute rhinitis.*

coup (KOO) Injury to the brain occurring directly under the skull at the point of impact.

coxa (**COCK**-sah) Hipbone. Plural *coxae.*

cranial (**KRAY**-nee-al) Pertaining to the cranium.

craniofacial (**KRAY**-nee-oh-**FAY**-shal) Pertaining to both the face and the cranium.

cranium (**KRAY**-nee-um) The skull.

cretin (**KREH**-tin) A person with severe congenital hypothyroidism.

cretinism (**KREH**-tin-izm) Condition of severe congenital hypothyroidism.

Creutzfeldt-Jakob disease (**KROITS**-felt **YAK**-op **DIZ**-eez) Progressive incurable neurologic disease caused by infectious prions.

cricoid (**CRY**-koyd) Ring-shaped cartilage in the larynx.

crista ampullaris (**KRIS**-tah am-**PULL**-air-is) Mound of hair cells and gelatinous material in the ampulla of a semicircular canal.

Crohn disease (KRONE diz-**EEZ**) Narrowing and thickening of terminal small bowel. Also called *regional enteritis.*

croup (KROOP) Infection of the upper airways in children, characterized by a barking cough. Also called *laryngotracheobronchitis.*

cruciate (**KRU**-she-ate) Shaped like a cross.

cryosurgery (cry-oh-**SUR**-jer-ee) Use of liquid nitrogen or argon gas in a probe to freeze and kill abnormal tissue.

cryptorchism (krip-**TOR**-kizm) Failure of one or both testes to descend into the scrotum.

curettage (kyu-reh-**TAHZH**) Scraping the interior of a cavity.

curette (kyu-**RET**) Scoop-shaped instrument for scraping the interior of a cavity or removing new growths.

Cushing syndrome (**KUSH**-ing **SIN**-drohm) Hypersecretion of cortisol (hydrocortisone) by the adrenal cortex.

cutaneous (kyu-**TAY**-nee-us) Pertaining to the skin.

cuticle (**KEW**-tih-cul) Nonliving epidermis at the base of the fingernails and toenails.

cyanosis (sigh-ah-**NO**-sis) Blue discoloration of the skin, lips, and nail beds due to low levels of oxygen in the blood.

cyanotic (sigh-ah-**NOT**-ik) Pertaining to or marked by cyanosis.

cyst (SIST) An abnormal, fluid-containing sac.

cystic (**SIS**-tik) Relating to a cyst.

cystic fibrosis (CF) (**SIS**-tik fie-**BROH**-sis) Genetic disease in which excessive viscid mucus obstructs passages, including bronchi.

cystitis (sis-**TIE**-tis) Inflammation of the urinary bladder.

cystocele (**SIS**-toh-seal) Hernia of the bladder into the vagina.

cystoscope (**SIS**-toh-skope) An endoscope inserted to view the inside of the bladder.

cystoscopy (sis-**TOS**-koh-pee) Using a cystoscope to examine the inside of the urinary bladder.

cystourethrogram (sis-toh-you-**REETH**-roe-gram) X-ray image during voiding to show the structure and function of the bladder and urethra.

cytologist (**SIGH**-tol-oh-jist) Specialist in the structure, chemistry, and pathology of the cell.

cytology (**SIGH**-tol-oh-jee) Study of the cell.

cytoplasm (**SIGH**-toh-plazm) Clear, gelatinous substance that forms the substance of a cell except for the nucleus.

cytotoxic (sigh-toh-**TOX**-ik) Destructive to cells.

D

dandruff (**DAN**-druff) Scales in hair from shedding of the epidermis.

death (DETH) Total and permanent cessation of all vital functions.

debridement (day-**BREED**-mon) The removal of injured or necrotic tissue.

decongestant (dee-con-**JESS**-tant) Agent that reduces the swelling and fluid in the nose and sinuses.

decubitus ulcer (de-**KYU**-bit-us **UL**-ser) Sore caused by lying down for long periods of time.

defecation (def-eh-**KAY**-shun) Evacuation of feces from the rectum and anus.

defect (**DEE**-fect) An absence, malformation, or imperfection.

defective (dee-**FEK**-tiv) Imperfect.

defibrillation (de-fib-rih-**LAY**-shun) Restoration of uncontrolled twitching of cardiac muscle fibers to normal rhythm.

defibrillator (de-fib-rih-**LAY**-tor) Instrument for defibrillation.

deformity (de-**FOR**-mih-tee) A permanent structural deviation from the normal.

degenerative (dee-**JEN**-er-a-tiv) Relating to the deterioration of a structure.

deglutition (dee-glue-**TISH**-un) The act of swallowing.

dehydration (dee-high-**DRAY**-shun) Process of losing body water.

delirium (de-**LIR**-ee-um) Acute altered state of consciousness with agitation and disorientation; condition is reversible.

deltoid (**DEL**-toyd) Large, fan-shaped muscle connecting the scapula and clavicle to the humerus.

delusion (dee-**LOO**-shun) Fixed, unyielding false belief held despite strong evidence to the contrary.

dementia (dee-**MEN**-she-ah) Chronic, progressive, irreversible loss of intellectual and mental functions.

demyelination (dee-**MY**-eh-lin-**A**-shun) Process of losing the myelin sheath of a nerve fiber.

dendrite (**DEN**-dright) Branched extension of the nerve cell body that receives nervous stimuli.

dental (**DEN**-tal) Pertaining to the teeth.

dentine (**DEN**-tin) Dense, ivory-like substance located under the enamel in a tooth. (Also spelled *dentin.*)

dentist (**DEN**-tist) Legally qualified specialist in dentistry.

dentistry (**DEN**-tis-tree) Evaluation, diagnosis, prevention, and treatment of conditions of the oral cavity and associated structures.

deoxyribonucleic acid (DNA) (dee-**OCK**-see-**RYE**-boh-noo-**KLEE**-ik **ASS**-id) Source of hereditary characteristics found in chromosomes.

dependent (dee-**PEN**-dent) Having to rely on someone else.

depressant (de-**PRESS**-ant) Substance that diminishes activity, sensation, or tone.

depression (de-**PRESS**-shun) Mental disorder with feelings of deep sadness and despair.

dermabrasion (der-mah-**BRAY**-shun) Removal of upper layers of skin by rotary brush.

dermal (**DER**-mal) Pertaining to the skin.

dermatitis (der-mah-**TYE**-tis) Inflammation of the skin.

dermatologic (der-mah-toh-**LOJ**-ik) Pertaining to the skin and dermatology.

dermatologist (der-mah-**TOL**-oh-jist) Medical specialist in diseases of the skin.

dermatology (der-mah-**TOL**-oh-jee) Medical specialty concerned with disorders of the skin.

dermatomyositis (**DER**-mah-toe-**MY**-oh-site-is) Inflammation of the skin and muscles.

dermis (**DER**-miss) Connective tissue layer of the skin beneath the epidermis.

detoxification (dee-**TOKS**-ih-fih-**KAY**-shun) Remove poison from a tissue or substance.

diabetes insipidus (dye-ah-**BEE**-teez in-**SIP**-ih-dus) Excretion of large amounts of dilute urine as a result of inadequate ADH production.

diabetes mellitus (dye-ah-**BEE**-teez **MEL**-ih-tus) Metabolic syndrome caused by absolute or relative insulin deficiency and/or insulin ineffectiveness.

diabetic (dye-ah-**BET**-ik) Pertaining to or suffering from diabetes.

diagnose (die-ag-**NOSE**) To make a diagnosis.

diagnosis (die-ag-**NO**-sis) The determination of the cause of a disease. Plural *diagnoses.*

diagnostic (die-ag-**NOS**-tik) Pertaining to or establishing a diagnosis.

dialysis (die-**AL**-ih-sis) An artificial method of filtration to remove excess waste materials and water from the body.

diaphoresis (**DIE**-ah-foh-**REE**-sis) Sweat or perspiration.

diaphoretic (**DIE**-ah-foh-**RET**-ic) Pertaining to sweat or perspiration.

diaphragm (**DIE**-ah-fram) A ring and dome-shaped material inserted into the vagina to prevent pregnancy; *or* the muscular sheet separating the abdominal and thoracic cavities.

diaphragmatic (**DIE**-ah-frag-**MAT**-ik) Pertaining to the diaphragm.

diaphysis (die-**AF**-ih-sis) The shaft of a long bone.

diarrhea (die-ah-**REE**-ah) Abnormally frequent and loose stools.

diastasis (die-**ASS**-tah-sis) Separation of normally joined parts.

diastole (die-**AS**-toe-lee) Dilation of heart cavities, during which they fill with blood.

diastolic (die-as-**TOL**-ik) Pertaining to diastole.

differential (dif-er-**EN**-shal) A differential white blood cell count lists percentages of the different leukocytes in a blood sample.

diffuse (di-**FUSE**) To disseminate or spread out.

diffusion (di-**FYU**-zhun) The means by which small particles move between tissues.

digestion (die-**JEST**-shun) Breakdown of food into elements suitable for cell metabolism.

digestive (die-**JEST**-iv) Pertaining to digestion.

digital (**DIJ**-ih-tal) Pertaining to a finger or toe.

dilate (**DIE**-late) To perform or undergo dilation.

dilation (die-**LAY**-shun) Stretching or enlarging an opening or a structure.

diphtheria (dif-**THEER**-ee-ah) Disease with a thick, membranous (leathery) coating of the pharynx.

diplegia (die-**PLEE**-jee-ah) Paralysis of all four limbs, with the two legs affected most severely.

disability (dis-ah-**BILL**-ih-tee) Diminished capacity to perform certain activities or functions.

discipline (**DIS**-ih-plin) Training for proper conduct or action.

disease (**DIZ**-eez) A disorder of body functions, systems, or organs.

disinfectant (dis-in-**FEK**-tant) Agent that disinfects.

disinfection (dis-in-**FEK**-shun) Process of destruction of microorganisms by chemical agents.

dislocation (dis-low-**KAY**-shun) Completely out of joint.

displaced fracture (dis-**PLAYSD FRAK**-chur) A fracture in which the fragments are separated and are not in alignment.

disseminate (dih-**SEM**-in-ate) Widely scattered throughout the body or an organ.

dissociative identity disorder (di-**SO**-see-ah-tiv eye-**DEN**-tih-tee dis-**OR**-der) Part of an individual's personality is separated from the rest, leading to multiple personalities.

distal (**DISS**-tal) Situated away from the center of the body.

diuresis (die-you-**REE**-sis) Excretion of large volumes of urine.

diuretic (die-you-**RET**-ik) Agent that increases urine output.

diverticulitis (**DIE**-ver-tick-you-**LIE**-tis) Inflammation of the diverticula.

diverticulosis (**DIE**-ver-tick-you-**LOW**-sis) Presence of a number of small pouches in the wall of the large intestine.

diverticulum (die-ver-**TICK**-you-lum) A pouchlike opening or sac from a tubular structure (e.g., gut). Plural *diverticula.*

dopamine (**DOH**-pah-meen) Neurotransmitter in specific small areas of the brain.

Doppler (**DOP**-ler) Diagnostic instrument that sends an ultrasonic beam into the body.

Doppler ultrasonography (**DOP**-ler **UL**-trah-soh-**NOG**-rah-fee) Detects direction, velocity, and turbulence of blood flow; used in workup of stroke patients.

dormant (**DOOR**-mant) Inactive.

dorsal (**DOOR**-sal) Pertaining to the back or situated behind.

dorsum (**DOOR**-sum) Upper, posterior, or back surface.

Down syndrome (DOWN **SIN**-drome) A syndrome with variable abnormalities associated with three chromosomes 21.

Duchenne muscular dystrophy (**DOO**-shen **MUSS**-kyu-lar **DISS**-troh-fee) Symmetrical weakness and wasting of pelvic, shoulder, and proximal limb muscles.

ductus arteriosus (**DUK**-tus ar-**TEER**-ih-**OH**-sus) Fetal vessel that connects the descending aorta with the left pulmonary artery.

ductus deferens (**DUK**-tus **DEH**-fuh-renz) Tube that receives sperm from the epididymis. Also known as *vas deferens.*

duodenal (du-oh-**DEE**-nal) Pertaining to the duodenum.

duodenum (du-oh-**DEE**-num) The first part of the small intestine; approximately 12 finger-breadths (9 to 10 inches) in length.

dura mater (**DYU**-rah **MAY**-ter) Hard, fibrous outer layer of the meninges.

dysentery (**DIS**-en-tare-ee) Disease with diarrhea, bowel spasms, fever, and dehydration.

dysfunctional (dis-**FUNK**-shun-al) Difficulty in functioning.

dyslexia (dis-**LEK**-see-ah) Impaired reading and writing ability below the person's level of intelligence.

dyslexic (dis-**LEK**-sik) Pertaining to or suffering from dyslexia.

dysmenorrhea (dis-men-oh-**REE**-ah) Painful and difficult menstruation.

dyspareunia (dis-pah-**RUE**-nee-ah) Pain during sexual intercourse.

dyspepsia (dis-**PEP**-see-ah) "Upset stomach," epigastric pain, nausea, and gas.

dysphagia (dis-**FAY**-jee-ah) Difficulty in swallowing.

dysphoria (dis-**FOR**-ee-ah) Psychiatric mood disorder.

dysplasia (dis-**PLAY**-zee-ah) Abnormal tissue formation.

dysplastic (dis-**PLAS**-tik) Pertaining to or showing abnormal tissue formation.

dyspnea (disp-**NEE**-ah) Difficulty breathing.

dyspneic (disp-**NEE**-ik) Pertaining to or suffering from difficulty in breathing.

dysrhythmia (dis-**RITH**-me-ah) An abnormal heart rhythm.

dysuria (dis-**YOU**-ree-ah) Difficulty or pain with urination.

E

echocardiography (**EK**-oh-kar-dee-**OG**-rah-fee) Ultrasound recording of heart function.

echoencephalography (**EK**-oh-en-sef-ah-**LOG**-rah-fee) Use of ultrasound in the diagnosis of intracranial lesions.

eclampsia (ek-**LAMP**-see-uh) Convulsions in a patient with preeclampsia.

ectopic (ek-**TOP**-ik) Out of place, not in a normal position.

eczema (**EK**-zeh-mah) Inflammatory skin disease, often with a serous discharge.

eczematous (**EK**-zem-ah-tus) Pertaining to or marked by eczema.

edema (ee-**DEE**-mah) Excessive accumulation of fluid in cells and tissues.

edematous (ee-**DEM**-ah-tus) Pertaining to or marked by edema.

effacement (ee-**FACE**-ment) Thinning of the cervix in relation to labor.

efferent (**EFF**-eh-rent) Moving away from a center; for example, conducting nerve impulses away from the brain or spinal cord.

effusion (eh-**FYU**-shun) Collection of fluid that has escaped from blood vessels into a cavity or tissues.

ejaculate (ee-**JACK**-you-late) To expel suddenly; *or* the semen expelled in ejaculation.

ejaculation (ee-**JACK**-you-**LAY**-shun) Process of expelling semen suddenly.

ejaculatory (ee-**JACK**-you-**LAY**-tor-ee) Pertaining to ejaculation.

elective (e-**LEK**-tiv) Surgery or a procedure that is not urgent or vital.

electrocardiogram (ECG or **EKG)** (ee-**LEK**-troh-**KAR**-dee-oh-gram) Record of the electrical signals of the heart.

electrocardiograph (ee-**LEK**-troh-**KAR**-dee-oh-graf) Machine that makes the electrocardiogram.

electrocardiography (ee-**LEK**-troh-kar-dee-**OG**-rah-fee) Interpretation of electrocardiograms.

electroconvulsive therapy (ECT) (ee-**LEK**-troh-kon-**VUL**-siv **THAIR**-ah-pee) Passage of electric current through the brain to produce convulsions and treat persistent depression.

electrode (ee-**LEK**-trode) A device for conducting electricity.

electroencephalogram (EEG) (ee-**LEK**-troh-en-**SEF**-ah-low-gram) Record of the electrical activity of the brain.

electroencephalograph (ee-**LEK**-troh-en-**SEF**-ah-low-graf) Device used to record the electrical activity of the brain.

electroencephalography (ee-**LEK**-troh-en-**SEF**-ah-**LOG**-rah-fee) The process of recording the electrical activity of the brain.

electrolyte (ee-**LEK**-troh-lite) Substance that, when dissolved in a suitable medium, forms electrically charged particles.

electromyogram (ee-**LEK**-troh-**MY**-oh-gram) Recording of electric currents associated with muscle action.

electromyography (ee-**LEK**-troh-my-**OG**-rah-fee) Recording of electrical activity in muscle.

electroneurodiagnostic (ee-**LEK**-troh-**NYUR**-oh-die-ag-**NOS**-tik) Pertaining to the use of electricity in the diagnosis of a neurologic disorder.

elimination (e-lim-ih-**NAY**-shun) Removal of waste material from the digestive tract.

emaciated (ee-may-see-**AY**-ted) Pertaining to or suffering from emaciation.

emaciation (ee-may-see-**AY**-shun) Abnormal thinness.

embolus (**EM**-boh-lus) Detached piece of thrombus, a mass of bacteria, quantity of air, or foreign body that blocks a blood vessel.

embryo (**EM**-bree-oh) Developing organism from conception until the end of the second month.

embryology (em-bree-**OL**-oh-jee) Science of the origin and early development of an organism.

embryonic (em-bree-**ON**-ic) Pertaining to the embryo.

emesis (**EM**-eh-sis) Vomit.

emmetropia (emm-eh-**TROH**-pee-ah) Normal refractive condition of the eye.

empathy (**EM**-pah-thee) Ability to place yourself into the feelings, emotions, and reactions of another person.

emphysema (em-fih-**SEE**-mah) Dilation of respiratory bronchioles and alveoli.

empyema (**EM**-pie-**EE**-mah) Pus in a body cavity, particularly in the pleural cavity.

emulsify (ee-**MUL**-sih-fye) Break up into very small droplets to suspend in a solution (emulsion).

emulsion (ee-**MUL**-shun) The system that contains small droplets suspended in a liquid.

enamel (ee-**NAM**-el) Hard substance covering a tooth.

encephalitis (en-**SEF**-ah-**LIE**-tis) Inflammation of brain cells and tissues.

encephalocele (en-**SEF**-ah-loh-seal) Congenital defect of the cranium with herniation of brain tissue.

encephalomyelitis (en-**SEF**-ah-loh-**MY**-eh-lie-tis) Inflammation of the brain and spinal cord.

encephalopathy (en-sef-ah-**LOP**-ah-thee) Any disorder of the brain.

encopresis (en-koh-**PREE**-sis) Repeated soiling with feces.

endarterectomy (**END**-ar-ter-**EK**-toe-me) Surgical removal of plaque from an artery.

endemic (en-**DEM**-ik) Pertaining to a disease always present in a community.

endocardial (**EN**-doh-kar-**DEE**-al) Pertaining to the endocardium.

endocarditis (**EN**-doh-kar-**DIE**-tis) Inflammation of the lining of the heart.

endocardium (**EN**-doh-kar-**DEE**-um) The inside lining of the heart.

endocrine (**EN**-doh-krin) A gland that produces an internal or hormonal substance.

endocrinologist (**EN**-doh-krih-**NOL**-oh-jist) A medical specialist in endocrinology.

endocrinology (**EN**-doh-krih-**NOL**-oh-jee) Medical specialty concerned with the production and effects of hormones.

endometrial (en-doh-**ME**-tree-al) Pertaining to the inner lining of the uterus.

endometriosis (**EN**-doh-me-tree-**OH**-sis) Endometrial tissue outside the uterus.

endometrium (en-doh-**ME**-tree-um) Inner lining of the uterus.

endoscope (**EN**-doh-skope) Instrument to examine the inside of a tubular or hollow organ.

endoscopy (en-**DOS**-koh-pee) The use of an endoscope.

endotracheal (en-doh-**TRAY**-kee-al) Pertaining to being inside the trachea.

enema (**EN**-eh-mah) An injection of fluid into the rectum.

enteric (en-**TEHR**-ik) Pertaining to the intestine.

enteroscope (**EN**-ter-oh-**SKOPE**) Slender, tubular instrument with light source and camera to visualize the digestive tract.

enteroscopy (en-ter-**OSS**-koh-pee) The examination of the lining of the digestive tract.

enuresis (en-you-**REE**-sis) Bedwetting; urinary incontinence.

enzyme (**EN**-zime) Protein that induces changes in other substances.

eosinophil (ee-oh-**SIN**-oh-fill) An eosinophil's granules attract a rosy-red color on staining.

epicardial (**EP**-ih-kar-**DEE**-al) Pertaining to the epicardium.

epicardium (**EP**-ih-kar-**DEE**-um) The outer layer of the heart wall.

epicondyle (**EP**-ih-**KON**-dile) Projection above the condyle for attachment of a ligament or tendon.

epidemic (ep-ih-**DEM**-ik) Pertaining to an outbreak in a community of a disease or a health-related behavior.

epidermal (ep-ih-**DER**-mal) Pertaining to the epidermis.

epidermis (ep-ih-**DER**-miss) Top layer of the skin.

epididymis (**EP**-ih-**DID**-ih-miss) Coiled tube attached to the testis.

epididymitis (**EP**-ih-did-ih-**MY**-tis) Inflammation of the epididymis.

epididymoorchitis (**EP**-ih-**DID**-ih-moh-or-**KIE**-tis) Inflammation of the epididymis and testicle. Also called *orchitis.*

epidural (ep-ih-**DYU**-ral) Above the dura.

epidural space (ep-ih-**DYU**-ral SPASE) Space between the dura mater and the wall of the vertebral canal.

epigastric (ep-ih-**GAS**-trik) Pertaining to the abdominal region above the stomach.

epigastrium (ep-ih-**GAS**-tri-um) The abdominal region above the stomach.

epiglottis (ep-ih-**GLOT**-is) Leaf-shaped plate of cartilage that shuts off the larynx during swallowing.

epiglottitis (ep-ih-**GLOT**-eye-tis) Inflammation of the epiglottis.

epilepsy (**EP**-ih-**LEP**-see) Chronic brain disorder due to paroxysmal excessive neuronal discharges.

epileptic (**EP**-ih-**LEP**-tik) Pertaining to or suffering from epilepsy.

epinephrine (ep-ih-**NEF**-rin) Main catecholamine produced by the adrenal medulla. Also called *adrenaline.*

epiphysial (eh-**PIF**-ih-see-al) Pertaining to an epiphysis.

epiphysial plate (eh-**PIF**-ih-see-al PLATE) Layer of cartilage between the epiphysis and the metaphysis where bone growth occurs.

epiphysis (eh-**PIF**-ih-sis) Expanded area at the proximal and distal ends of a long bone to provide increased surface area for attachment of ligaments and tendons.

episiotomy (eh-piz-ee-**OT**-oh-me) Surgical incision of the vulva.

epispadias (ep-ih-**SPAY**-dee-as) Condition in which the urethral opening is on the dorsum of the penis.

epistaxis (ep-ih-**STAK**-sis) Nosebleed.

epithelium (ep-ih-**THEE**-lee-um) Tissue that covers surfaces or lines cavities.

equilibrium (ee-kwi-**LIB**-ree-um) Being evenly balanced.

erectile (ee-**REK**-tile) Capable of erection or being distended with blood.

erection (ee-**REK**-shun) Distended and rigid state of an organ.

erosion (ee-**ROE**-shun) Form a shallow ulcer in the lining of a structure.

erythroblast (eh-**RITH**-ro-blast) Precursor to a red blood cell.

erythroblastosis (eh-**RITH**-roh-blast-oh-sis) Condition of many immature red cells in blood.

erythrocyte (eh-**RITH**-roh-site) A red blood cell.

erythropoiesis (eh-**RITH**-roh-poy-**EE**-sis) The formation of red blood cells.

erythropoietin (eh-**RITH**-roh-**POY**-ee-tin) Protein secreted by the kidney that stimulates red blood cell production.

eschar (**ESS**-kar) The burnt, dead tissue lying on top of third-degree burns.

Escherichia coli (esh-eh-**RIK**-ee-ah **KOH**-lie) Organism in the intestine; releases an exotoxin that can cause diarrhea.

esophageal (ee-**SOF**-ah-**JEE**-al) Pertaining to the esophagus.

esophagitis (ee-**SOF**-ah-**JI**-tis) Inflammation of the lining of the esophagus.

esophagus (ee-**SOF**-ah-gus) Tube linking the pharynx to the stomach.

esotropia (es-oh-**TROH**-pee-ah) Turning the eye inward toward the nose.

estrogen (**ES**-troh-jen) Generic term for hormones that stimulate female secondary sex characteristics.

ethmoid (**ETH**-moyd) Bone that forms the back of the nose and encloses numerous air cells.

etiology (ee-tee-**OL**-oh-jee) The study of the causes of a disease.

eupnea (yoop-**NEE**-ah) Normal breathing.

eustachian tube (you-**STAY**-shun TYUB) Tube that connects the middle ear to the nasopharynx. Also called *auditory tube.*

euthyroid (you-**THIGH**-royd) Normal thyroid function.

eversion (ee-**VER**-shun) Turning outward.

evert (ee-**VERT**) To turn outward.

evolve (ee-**VOLV**) To develop gradually.

exacerbation (ek-zas-er-**BAY**-shun) Period when there is an increase in the severity of a disease.

exanthem (ek-**ZAN**-them) Skin eruption or rash occurring as the outward sign of a viral or bacterial disease.

excision (ek-**SIZH**-un) Surgical removal of part or all of a structure.

excoriate (eks-**KOR**-ee-ate) To scratch.

excoriation (eks-**KOR**-ee-**AY**-shun) Scratch mark.

excrete (eks-**KREET**) To pass waste products of metabolism out of the body.

excretion (eks-**KREE**-shun) Removal of waste products of metabolism out of the body.

exhale (**EKS**-hail) Breathe out.

exocrine (**EK**-soh-krin) A gland that secretes substances outwardly through excretory ducts.

exophthalmos (ek-sof-**THAL**-mos) Protrusion of the eyeball.

exotropia (ek-soh-**TROH**-pee-ah) Turning the eye outward away from the nose.

expectorate (ek-**SPEK**-toh-rate) Cough up and spit out mucus from the respiratory tract.

expiration (**EKS**-pih-**RAY**-shun) Breathe out.

extension (eks-**TEN**-shun) Straighten a joint to increase its angle.

extracorporeal (**EKS**-tra-kor-**POH**-ree-al) Outside the body.

extravasate (eks-**TRAV**-ah-sate) To ooze out from a vessel into the tissues.

extrinsic (eks-**TRIN**-sik) Any muscle located entirely on the outside of the structure under consideration; for example, the eye.

F

facies (**FASH**-eez) Facial features and expressions.

fallopian tubes (fah-**LOW**-pee-an) Uterine tubes connected to the fundus of the uterus.

Fallot (fah-**LOW**) First described the tetralogy of congenital heart defects.

fascia (**FASH**-ee-ah) Sheet of fibrous connective tissue

fasciectomy (fash-ee-**EK**-toe-me) Surgical removal of fascia.

fasciitis (fash-ee-**EYE**-tis) Inflammation of fascia.

fasciotomy (fash-ee-**OT**-oh-me) An incision through a band of fascia, usually to relieve pressure on underlying structures.

febrile (**FEB**-ril or **FEB**-rile) Pertaining to or suffering from fever.

fecal (**FEE**-kal) Pertaining to feces.

feces (**FEE**-sees) Undigested, waste material discharged from the bowel.

femoral (**FEM**-oh-ral) Pertaining to the femur.

femur (**FEE**-mur) The thigh bone.

fertilization (**FER**-til-eye-**ZAY**-shun) Union of a male sperm and a female egg.

fertilize (**FER**-til-ize) Penetration of the egg by sperm.

fetal (**FEE**-tal) Pertaining to the fetus.

fetalis (fee-**TAH**-lis) Erythroblastosis fetalis is a hemolytic disease of the newborn.

fetus (**FEE**-tus) Human organism from the end of the eighth week after conception to birth.

fever (**FEE**-ver) Increased body temperature that is a physiologic response to disease.

fibrillation (fi-brih-**LAY**-shun) Uncontrolled quivering or twitching of the heart muscle.

fibrin (**FIE**-brin) Stringy protein fiber that is a component of a blood clot.

fibrinogen (fie-**BRIN**-oh-jen) Precursor of fibrin in blood-clotting process.

fibroadenoma (**FIE**-broh-ad-en-**OH**-muh) Benign tumor containing much fibrous tissue.

fibroblast (**FIE**-broh-blast) Cell that forms collagen fibers.

fibrocystic disease (fie-broh-**SIS**-tik **DIZ**-eez) Benign breast disease with multiple tiny lumps and cysts.

fibroid (**FIE**-broyd) Uterine tumor resembling fibrous tissue.

fibromyalgia (fie-broh-my-**AL**-jee-ah) Pain in the muscle fibers.

fibromyoma (**FIE**-broh-my-**OH**-mah) Benign neoplasm derived from smooth muscle and containing fibrous tissue.

fibrosis (fie-**BROH**-sis) Repair of dead tissue cells by formation of fibrous tissue.

fibrotic (fie-**BROT**-ik) Pertaining to or affected by fibrosis.

fibula (**FIB**-you-lah) The smaller of the two bones of the lower leg.

fibular (**FIB**-you-lar) Pertaining to the fibula.

filter (**FIL**-ter) Porous substance used to separate liquids or gases from particulate matter; *or* to subject a substance to the action of a filter.

filtrate (**FIL**-trate) That which has passed through a filter.

filtration (fil-**TRAY**-shun) Process of passing liquid through a filter.

fimbria (**FIM**-bree-ah) A fringelike structure on the surface of a cell or microorganism. Plural *fimbriae.*

fissure (**FISH**-ur) Deep furrow or cleft. Plural *fissures.*

fistula (**FIS**-tyu-lah) Abnormal passage. Plural *fistulae* or *fistulas.*

flank (FLANK) Side of the body between pelvis and ribs.

flatulence (**FLAT**-you-lence) Excessive amount of gas in the stomach and intestines.

flatus (**FLAY**-tus) Gas or air expelled through the anus.

flex (FLEKS) To bend a joint so that the two parts come together.

flexion (**FLEK**-shun) Bend a joint to decrease its angle.

flexor (**FLEK**-sor) Muscle or tendon that flexes a joint.

flexure (**FLEK**-shur) A bend in a structure.

flora (**FLO**-rah) Microrganisms covering the exterior and interior surfaces of a healthy animal.

fluorescein (flor-**ESS**-ee-in) Dye that produces a vivid green color under a blue light to diagnose corneal abrasions and foreign bodies.

follicle (**FOLL**-ih-kull) Spherical mass of cells containing a cavity; *or* a small cul-de-sac, such as a hair follicle.

follicular (fo-**LIK**-you-lar) Pertaining to a follicle.

foramen (fo-**RAY**-men) An opening through a structure. Plural *foramina.*

forceps extraction (**FOR**-seps ek-**STRAK**-shun) Assisted delivery of the baby by an instrument that grasps the head of the baby.

foreskin (**FOR**-skin) Skin that covers the glans penis.

fornix (**FOR**-niks) Arch-shaped, blind-ended part of the vagina behind and around the cervix. Plural *fornices.*

fovea centralis (**FOH**-vee-ah sen-**TRAH**-lis) Small pit in the center of the macula that has the highest visual acuity.

frenulum (**FREN**-you-lum) Fold of mucous membrane between the glans and the prepuce.

frontal (**FRON**-tal) Vertical plane dividing the body into anterior and posterior portions.

function (**FUNK**-shun) The ability of an organ or tissue to perform its special work.

fundoscopic (fun-doh-**SKOP**-ik) Pertaining to fundoscopy.

fundoscopy (fun-**DOS**-koh-pee) Examination of the fundus (retina) of the eye.

fundus (**FUN**-dus) Part farthest from the opening of a hollow organ.

fungicide (**FUN**-jee-side) Agent to destroy fungi.

fungus (**FUN**-gus) General term used to describe yeasts and molds. Plural *fungi.*

G

galactorrhea (gah-**LAK**-toe-**REE**-ah) Abnormal flow of milk from the breasts.

gallbladder (**GAWL**-blad-er) Receptacle on the inferior surface of the liver for storing bile.

gallstone (**GAWL**-stone) Hard mass of cholesterol, calcium, and bilirubin that can be formed in the gallbladder and bile duct.

ganglion (**GANG**-lee-on) Collection of nerve cells outside the brain and spinal cord; *or* a fluid-filled cyst. Plural *ganglia.*

gastric (**GAS**-trik) Pertaining to the stomach.

gastrin (**GAS**-trin) Hormone secreted in the stomach that stimulates secretion of HCl and increases gastric motility.

gastritis (gas-**TRY**-tis) Inflammation of the lining of the stomach.

gastrocnemius (gas-trok-**NEE**-me-us) Major muscle in back of the lower leg (the calf).

gastroenteritis (**GAS**-troh-en-ter-**I**-tis) Inflammation of the stomach and intestines.

gastroenterologist (**GAS**-troh-en-ter-**OL**-oh-jist) Medical specialist in gastroenterology.

gastroenterology (**GAS**-troh-en-ter-**OL**-oh-gee) Medical specialty of the stomach and intestines.

gastroesophageal (**GAS**-troh-ee-sof-ah-**JEE**-al) Pertaining to the stomach and esophagus.

gastrointestinal (GI) (**GAS**-troh-in-**TESS**-tin-al) Pertaining to the stomach and intestines.

gastroscope (**GAS**-troh-skope) Endoscope for examining the inside of the stomach.

gastroscopy (gas-**TROS**-koh-pee) Endoscopic examination of the stomach.

gavage (guh-**VAHZH**) To feed by a stomach tube.

gene (JEEN) Functional segment of DNA molecule.

geneticist (jeh-**NET**-ih-sist) A specialist in genetics.

genetics (jeh-**NET**-iks) Science of the inheritance of characteristics.

genital (**JEN**-ih-tal) Relating to reproduction or to the male or female sex organs.

genitalia (**JEN**-ih-**TAY**-lee-ah) External and internal organs of reproduction.

geriatrician (jer-ee-ah-**TRISH**-an) Medical specialist in geriatrics.

geriatrics (jer-ee-**AT**-riks) Medical specialty that deals with the problems of old age.

gerontologist (jer-on-**TOL**-oh-jist) Medical specialist in gerontology.

gerontology (jer-on-**TOL**-oh-jee) Study of the process and problems of aging.

gestation (jes-**TAY**-shun) From conception to birth.

gestational (jes-**TAY**-shun-al) Pertaining to gestation.

gigantism (**JI**-gan-tizm) Abnormal height and size of the entire body.

gingiva (**JIN**-jih-vah) Tissue surrounding the teeth and covering the jaw.

gingival (**JIN**-jih-val) Pertaining to the gingiva.

gingivectomy (jin-jih-**VEC**-toe-me) Surgical removal of diseased gum tissue.

gingivitis (jin-jih-**VI**-tis) Inflammation of the gums.

glans (GLANZ) Head of the penis or clitoris.

glaucoma (glau-**KOH**-mah) Increased intraocular pressure.

glia (**GLEE**-ah) Connective tissue that holds a structure together.

glial (**GLEE**-al) Pertaining to glia or neuroglia.

glioma (gli-**OH**-mah) Tumor of a glial cell.

glomerulonephritis (glo-**MER**-you-low-nef-**RYE**-tis) Infection of the glomeruli of the kidney.

glomerulus (glo-**MER**-you-lus) Plexus of capillaries; part of a nephron. Plural *glomeruli.*

glossopharyngeal (**GLOSS**-oh-fah-**RIN**-jee-al) Ninth (IX) cranial nerve, supplying the tongue and pharynx.

glottis (**GLOT**-is) Vocal apparatus of the larynx.

glucagon (**GLU**-kah-gon) Pancreatic hormone that supports blood glucose levels.

glucocorticoid (glu-co-**KOR**-tih-koyd) Hormone of the adrenal cortex that helps regulate glucose metabolism.

gluconeogenesis (**GLU**-koh-nee-oh-**JEN**-eh-sis) Formation of glucose from noncarbohydrate sources.

glucose (**GLU**-kose) The final product of carbohydrate digestion and the main sugar in the blood.

gluteal (**GLU**-tee-al) Pertaining to the buttocks.

gluten (**GLU**-ten) Insoluble protein found in wheat, barley, and oats.

gluteus (**GLU**-tee-us) Refers to one of three muscles in the buttocks.

glycogen (**GLYE**-koh-gen) The body's principal carbohydrate reserve, stored in the liver and skeletal muscle.

glycogenolysis (**GLYE**-koh-jen-oh-**LYE**-sis) Conversion of glycogen to glucose.

glycosuria (**GLYE**-koh-**SYU**-ree-ah) Presence of glucose in urine.

glycosylated hemoglobin (Hb A1c) (**GLYE**-koh-sih-lay-ted **HE**-moh-**GLOW**-bin) Hemoglobin A fraction linked to glucose; used as an index of glucose control.

goiter (**GOY**-ter) Enlargement of the thyroid gland.

gomphosis (gom-**FOE**-sis) Joint formed by a peg and socket. Plural *gomphoses.*

gonad (**GO**-nad) Testis or ovary. Plural *gonads.*

gonadotropin (**GO**-nad-oh-**TROH**-pin) Any hormone that stimulates gonad function.

gonorrhea (gon-oh-**REE**-ah) Specific contagious sexually transmitted infection.

grade (GRAYD) In cancer pathology, a classification of the rate of growth of cancer cells.

graft (GRAFT) Transplantation of living tissue.

grand mal (GRAHN MAL) Old name for generalized tonic-clonic seizure.

granulation (gran-you-**LAY**-shun) New fibrous tissue formed during wound healing.

granulocyte (**GRAN**-you-loh-site) A white blood cell that contains multiple small granules in its cytoplasm.

granulosa cell (gran-you-**LOW**-sah SELL) Cell lining the ovarian follicle.

Graves disease (GRAVZ diz-**EEZ**) Hyperthyroidism with toxic goiter.

gravid (**GRAV**-id) Pregnant.

gravida (**GRAV**-ih-dah) A pregnant woman.

gray matter (GRAY **MATT**-er) Regions of the brain and spinal cord occupied by cell bodies and dendrites.

greenstick fracture (**GREEN**-stik **FRAK**-chur) A fracture in which one side of the bone is partially broken and the other side is bent. Occurs mostly in children.

groin (GROYN) Crease where the thigh joins the abdomen.

Guillain-Barré syndrome (GEE-yan bah-**RAY SIN**-drom) Disorder in which the body makes antibodies against myelin, disrupting nerve conduction.

gynecologic (**GUY**-nih-koh-**LOJ**-ik) Pertaining to gynecology.

gynecologist (guy-nih-**KOL**-oh-jist) Specialist in gynecology.

gynecology (guy-nih-**KOL**-oh-jee) Medical specialty of diseases of the female.

gynecomastia (**GUY**-nih-koh-**MAS**-tee-ah) Enlargement of the breast.

gyrus (**JI**-rus) Rounded elevation on the surface of the cerebral hemispheres. Plural *gyri.*

H

hairline fracture (**HAIR**-line **FRAK**-chur) A fracture without separation of the fragments.

halitosis (hal-ih-**TOE**-sis) Bad odor of the breath.

hallucination (hah-loo-sih-**NAY**-shun) Perception of an object or event when there is no such thing present.

hallux valgus (**HAL**-uks **VAL**-gus) Deviation of the big toe toward the lateral side of the foot.

Hashimoto disease (hah-shee-**MOH**-toe diz-**EEZ**) Autoimmune disease of the thyroid gland. Also called *Hashimoto thyroiditis.*

Haversian canals (hah-**VER**-shan ka-**NALS**) Vascular canals in bone. Also called *central canals.*

Heberden node (**HEH**-ber-den NOHD) Bony lump on the terminal phalanx of the fingers in osteoarthritis.

hemangioma (he-**MAN**-jee-oh-mah) Abnormal mass of proliferating blood vessels.

hematemesis (he-mah-**TEM**-eh-sis) Vomiting of red blood.

hematocrit (Hct) (**HE**-mat-oh-krit) Percentage of red blood cells in the blood.

hematologist (he-mah-**TOL**-oh-jist) Specialist in hematology.

hematology (he-mah-**TOL**-oh-jee) Medical specialty of disorders of the blood.

hematoma (he-mah-**TOE**-mah) Collection of blood that has escaped from the blood vessels into surrounding tissues. Also called *bruise.*

hematuria (he-mah-**TYU**-ree-ah) Blood in the urine.

heme (HEEM) The iron-based component of hemoglobin that carries oxygen.

hemiparesis (**HEM**-ee-pah-**REE**-sis) Weakness of one side of the body.

hemiplegia (hem-ee-**PLEE**-jee-ah) Paralysis of one side of the body.

hemiplegic (hem-ee-**PLEE**-jik) Pertaining to or suffering from hemiplegia.

Hemoccult test (HEEM-o-kult TEST) Trade name for a fecal occult blood test.

hemodialysis (HE-moh-die-AL-ih-sis) An artificial method of filtration to remove excess waste materials and water directly from the blood.

hemodynamics (HE-moh-die-NAM-iks) The science of the flow of blood through the circulation.

hemoglobin (HE-moh-GLOW-bin) Red-pigmented protein that is the main component of red blood cells.

hemoglobinopathy (HE-moh-GLOW-bih-NOP-ah-thee) Disease caused by the presence of an abnormal hemoglobin in the red blood cells.

hemolysis (he-MOL-ih-sis) Destruction of red blood cells so that hemoglobin is liberated.

hemolytic (he-moh-LIT-ik) Pertaining to the process of destruction of red blood cells.

hemophilia (he-moh-FILL-ee-ah) An inherited disease from a deficiency of clotting factor VIII.

hemoptysis (he-MOP-tih-sis) Bloody sputum.

hemorrhage (HEM-oh-raj) To bleed profusely.

hemorrhoid (HEM-oh-royd) Dilated rectal vein producing painful anal swelling. Plural *hemorrhoids.*

hemorrhoidectomy (HEM-oh-royd-ECK-toh-me) Surgical removal of hemorrhoids.

hemostasis (he-moh-STAY-sis) Control of or stopping bleeding.

hemothorax (he-moh-THOR-ax) Blood in the pleural cavity.

heparin (HEP-ah-rin) An anticoagulant secreted particularly by liver cells.

hepatic (hep-AT-ik) Pertaining to the liver.

hepatitis (hep-ah-TIE-tis) Inflammation of the liver.

hernia (HER-nee-ah) Protrusion of a structure through the tissue that normally contains it.

herniate (HER-nee-ate) To protrude.

herniation (HER-nee-ay-shun) Protrusion of an anatomical structure from its normal location.

herniorrhaphy (HER-nee-OR-ah-fee) Repair of a hernia.

herpes simplex virus (HSV) (HER-peez SIM-pleks VIE-rus) Manifests with painful, watery blisters on the skin and mucous membranes.

herpes zoster (HER-pees ZOS-ter) Painful eruption of vesicles that follows a nerve root on one side of the body. Also called *shingles.*

heterograft (HET-er-oh-graft) A graft using tissue taken from another species. Also called *xenograft.*

hiatal (high-AY-tal) Pertaining to a hernia.

hiatus (high-AY-tus) An opening through a structure.

hilum (HIGH-lum) The site where the nerves and blood vessels enter and leave an organ. Plural *hila.*

histamine (HISS-tah-mean) Compound liberated in tissues as a result of injury or an allergic response.

histologist (his-TOL-oh-jist) Specialist in histology.

histology (his-TOL-oh-jee) Study of the structure and function of cells, tissues, and organs.

Hodgkin lymphoma (HOJ-kin lim-FOH-mah) Marked by chronic enlargement of lymph nodes spreading to other nodes in an orderly way.

holistic (ho-LIS-tik) Pertaining to the care of the whole person in physical, mental, emotional, and spiritual dimensions.

homeostasis (hoh-mee-oh-STAY-sis) Maintaining the stability of a system or the body's internal environment.

homograft (HOH-moh-graft) Skin graft from another person or a cadaver.

hordeolum (hor-DEE-oh-lum) Abscess in an eyelash follicle. Also called *stye.*

hormonal (hor-MOHN-al) Pertaining to a hormone.

hormone (HOR-mohn) Chemical formed in one tissue or organ and carried by the blood to stimulate or inhibit a function of another tissue or organ.

Horner syndrome (HOR-ner SIN-drome) Disorder of the sympathetic nerves to the face and eye.

hospice (HOS-pis) Provides care to the dying and their families.

human immunodeficiency virus (HIV) (HYU-man IM-you-noh-dee-FISH-en-see VIE-rus) Etiologic agent of acquired immunodeficiency syndrome (AIDS).

human papilloma virus (HPV) (HYU-man pap-ih-LOW-mah VIE-rus) Causes warts on the skin and genitalia and can increase the risk for cervical cancer.

humerus (HYU-mer-us) Single bone of the upper arm.

humoral immunity (HYU-mor-al ihm-YOU-nih-tee) Defense mechanism arising from antibodies in the blood.

Huntington disease (HUN-ting-ton diz-EEZ) Progressive inherited, degenerative, incurable neurologic disease. Also called *Huntington chorea.*

hyaline (HIGH-ah-line) Cartilage that looks like frosted glass and contains fine collagen fibers.

hyaline membrane disease (HIGH-ah-line MEM-brain DIZ-eez) Respiratory distress syndrome of the newborn.

hydrocele (HIGH-droh-seal) Collection of fluid in the space of the tunica vaginalis.

hydrocephalus (high-droh-SEF-ah-lus) Excess CSF in the cerebral ventricles; may cause enlarged head.

hydrochloric acid (HCl) (high-droh-KLOR-ic ASS-id) The acid of gastric juice.

hydrocortisone (high-droh-KOR-tih-sohn) Potent glucocorticoid with anti-inflammatory properties. Also called *cortisol.*

hydronephrosis (HIGH-droh-neh-FROH-sis) Dilation of the pelvis and calyces of a kidney.

hydronephrotic (HIGH-droh-neh-FROT-ik) Pertaining to or suffering from hydronephrosis.

hymen (HIGH-men) Thin membrane partly occluding the vaginal orifice.

hyperactivity (HIGH-per-ac-TIV-ih-tee) Excessive restlessness and movement.

hypercalcemia (HIGH-per-cal-SEE-me-ah) Excessive level of calcium in the blood.

hypercapnia (HIGH-per-KAP-nee-ah) Abnormal increase of carbon dioxide in the arterial bloodstream.

hyperemesis (high-per-EM-eh-sis) Excessive vomiting.

hyperflexion (high-per-FLEK-shun) Flexion of a limb or part beyond the normal limits.

hyperglycemia (HIGH-per-gly-SEE-me-ah) High level of glucose (sugar) in blood.

hyperglycemic (**HIGH**-per-gly-**SEE**-mik) Pertaining to or having hyperglycemia.

hyperimmune globulin (**HIGH**-per-im-**YUNE GLOB**-you-lin) Immunoglobulin prepared from serum of people with a high antibody titer to a specific antigen.

hyperkalemia (**HIGH**-per-kah-**LEE**-me-ah) High level of potassium in the blood.

hypernatremia (**HIGH**-per-nah-**TREE**-me-ah) High level of sodium in the blood.

hyperopia (high-per-**OH**-pee-ah) Able to see distant objects but unable to see close objects.

hyperosmolar (**HIGH**-per-os-**MOH**-lar) Marked hyperglycemia without ketoacidosis.

hyperparathyroidism (**HIGH**-per-para-**THIGH**-royd-izm) Excessive production of parathyroid hormone.

hyperplasia (**HIGH**-per-**PLAY**-zee-ah) Increase in the number of cells in a tissue or organ.

hyperpnea (high-perp-**NEE**-ah) Deeper and more rapid breathing than normal.

hyperpyrexia (**HIGH**-per-pie-**REK**-see-ah) Extremely high body temperature or fever.

hypersecretion (**HIGH**-per-seh-**KREE**-shun) Excessive secretion (of mucus or enzymes or waste products).

hypersensitivity (**HIGH**-per-sen-sih-**TIV**-ih-tee) Exaggerated abnormal reaction to an allergen.

hypersplenism (high-per-**SPLEN**-izm) Condition in which the spleen removes blood components at an excessive rate.

hypertension (**HIGH**-per-**TEN**-shun) Persistent high arterial blood pressure.

hypertensive (**HIGH**-per-**TEN**-siv) Pertaining to or suffering from high blood pressure.

hyperthyroidism (high-per-**THIGH**-royd-izm) Excessive production of thyroid hormones.

hypertrophy (high-**PER**-troh-fee) Increase in size, but not in number, of an individual tissue element.

hypochondriac (high-poh-**KON**-dree-ack) A person who exaggerates the significance of symptoms.

hypochromic (high-poh-**CROW**-mik) Pale in color, as in RBCs when hemoglobin is deficient.

hypodermic (high-poh-**DER**-mik) Pertaining to the hypodermis.

hypodermis (high-poh-**DER**-miss) Tissue layer of skin below the dermis.

hypogastric (high-poh-**GAS**-trik) Abdominal region below the stomach.

hypoglossal (high-poh-**GLOSS**-al) Twelfth (XII) cranial nerve, supplying muscles of the tongue.

hypoglycemia (**HIGH**-poh-gly-**SEE**-me-ah) Low level of glucose (sugar) in the blood.

hypoglycemic (**HIGH**-poh-gly-**SEE**-mik) Pertaining to or suffering from low blood sugar.

hypogonadism (**HIGH**-poh-**GOH**-nad-izm) Deficient gonad production of sperm or eggs or hormones.

hypokalemia (**HIGH**-poh-kah-**LEE**-me-ah) Low level of potassium in the blood.

hyponatremia (**HIGH**-poh-nah-**TREE**-me-ah) Low level of sodium in the blood.

hypoparathyroidism (**HIGH**-poh-par-ah-**THIGH**-royd-izm) Deficient production of parathyroid hormone.

hypophysis (high-**POF**-ih-sis) Another name for *pituitary gland*.

hypopituitarism (**HIGH**-poh-pih-**TYU**-ih-tah-rizm) Condition of one or more deficient pituitary hormones.

hypospadias (high-poh-**SPAY**-dee-as) Urethral opening more proximal than normal on the ventral surface of the penis.

hypotension (**HIGH**-poh-**TEN**-shun) Persistent low arterial blood pressure.

hypotensive (**HIGH**-poh-**TEN**-siv) Pertaining to or suffering from low blood pressure.

hypothalamic (high-poh-thah-**LAM**-ik) Pertaining to the hypothalamus.

hypothalamus (high-poh-**THAL**-ah-muss) An area of gray matter lying below the thalamus.

hypothenar eminence (high-poh-**THAY**-nar **EM**-in-nens) The fleshy mass at the base of the little finger.

hypothermia (high-poh-**THER**-me-ah) Very low core body temperature.

hypothyroidism (high-poh-**THIGH**-royd-izm) Deficient production of thyroid hormones.

hypovolemic (**HIGH**-poh-vo-**LEE**-mick) Decreased blood volume in the body.

hypoxia (high-**POCK**-see-ah) Below-normal levels of oxygen in tissues, gases, or blood.

hypoxic (high-**POCK**-sik) Deficient in oxygen.

hysterectomy (his-ter-**EK**-toe-me) Surgical removal of the uterus.

I

ictal (**ICK**-tal) Pertaining to, or a condition caused by, a stroke or epilepsy.

idiopathic (**ID**-ih-oh-**PATH**-ik) Pertaining to a disease of unknown etiology.

ileocecal (**ILL**-ee-oh-**SEE**-cal) Pertaining to the junction of the ileum and cecum.

ileocecal sphincter (**ILL**-ee-oh-**SEE**-cal **SFINK**-ter) A band of muscle that encircles the junction of the ileum and cecum.

ileoscopy (ill-ee-**OS**-koh-pee) Endoscopic examination of the ileum.

ileostomy (ill-ee-**OS**-toe-me) Artificial opening from the ileum to the outside of the body.

ileum (**ILL**-ee-um) Third portion of the small intestine.

iliac (**ILL**-ee-ack) A structure related to the ilium (pelvic bone).

ilium (**ILL**-ee-um) Large wing-shaped bone at the upper and posterior part of the pelvis. Plural *ilia*.

immune (im-**YUNE**) Protected from an infectious disease.

immune serum (im-**YUNE SEER**-um) Serum taken from another human or animal that has antibodies to a disease. Also called *antiserum*.

immunity (im-**YOU**-nih-tee) State of being protected.

immunization (**IM**-you-nih-**ZAY**-shun) Administration of an agent to provide immunity.

immunize (**IM**-you-nize) To make resistant to an infectious disease.

immunodeficiency (**IM**-you-noh-dee-**FISH**-en-see) Failure of the immune system.

immunoglobulin (**IM**-you-noh-**GLOB**-you-lin) Specific protein evoked by an antigen. All antibodies are immunoglobulins.

immunologist (im-you-**NOL**-oh-jist) Medical specialist in immunology.

immunology (im-you-**NOL**-oh-jee) The science and practice of immunity and allergy.

immunosuppression (**IM**-you-noh-suh-**PRESH**-un) Failure of the immune system caused by an outside agent.

impacted (im-**PAK**-ted) Immovably wedged, as with earwax blocking the external canal.

impacted fracture (im-**PAK**-ted **FRAK**-chur) A fracture in which one bone fragment is driven into the other.

impairment (im-**PAIR**-ment) The state of being worse, weaker, or damaged.

impetigo (im-peh-**TIE**-go) Infection of the skin producing thick, yellow crusts.

implant (im-**PLANT**) To insert material into tissues; *or* the material inserted into tissues.

implantable (im-**PLAN**-tah-bul) A device that can be inserted into tissues.

implantation (im-plan-**TAY**-shun) Attachment of a fertilized egg to the endometrium.

impotence (**IM**-poh-tence) Inability to achieve an erection.

impulsive (im-**PUL**-siv) Inability to resist performing inappropriate actions.

in situ (IN **SIGH**-tyu) In the correct place.

in utero (in **YOU**-ter-oh) Within the womb; not yet born.

in vitro fertilization (IVF) (IN **VEE**-troh **FER**-til-ih-**ZAY**-shun) Process of combining sperm and egg in a laboratory dish and placing the resulting embryos inside the uterus.

inattention (**IN**-ah-**TEN**-shun) Lack of concentration and direction.

incision (in-**SIZH**-un) A cut or surgical wound.

incompetence (in-**KOM**-peh-tense) Failure of valves to close completely.

incomplete fracture (in-kom-**PLEET FRAK**-chur) A fracture that does not extend across the bone, as in a hairline fracture.

incontinence (in-**KON**-tin-ence) Inability to prevent discharge of urine or feces.

incontinent (in-**KON**-tin-ent) Denoting incontinence.

incubation (in-kyu-**BAY**-shun) Process to develp an infection.

incus (**IN**-cuss) Middle one of the three ossicles in the middle ear; shaped like an anvil.

independence (in-dee-**PEN**-dense) The state of being able to think and act for oneself.

independent (in-dee-**PEN**-dent) Pertaining to the ability to think and act for oneself.

indigestion (in-dee-**JESS**-chun) Symptoms resulting from difficulty in digesting food.

infancy (**IN**-fan-see) The first year of life.

infant (**IN**-fant) Child in the first year of life.

infarct (in-**FARKT**) Area of cell death resulting from blockage of its blood supply.

infarction (in-**FARK**-shun) Sudden blockage of an artery.

infect (in-**FEKT**) To invade an organism by a microorganism.

infection (in-**FEK**-shun) Invasion of the body by disease-producing microorganisms.

infectious (in-**FEK**-shus) Capable of being transmitted to a person; *or* a disease caused by the action of a microorganism.

inferior (in-**FEE**-ree-or) Situated below.

infertility (in-fer-**TIL**-ih-tee) Failure to conceive.

infestation (in-fes-**TAY**-shun) Act of being invaded on the skin by a troublesome other species, such as a parasite.

inflammation (in-flah-**MAY**-shun) A complex of cell and chemical reactions in response to an injury or a chemical or biologic agent.

inflammatory (in-**FLAM**-ah-tor-ee) Causing or affected by inflammation.

influenza (in-flew-**EN**-zah) An acute, viral infection of upper and lower respiratory tracts.

infusion (in-**FYU**-zhun) Introduction intravenously of a substance other than blood.

ingestion (in-**JEST**-shun) Intake of food, either by mouth or through a nasogastric tube.

inguinal (**ING**-gwin-ahl) Pertaining to the groin.

inhale (**IN**-hail) Breathe in.

insanity (in-**SAN**-ih-tee) Nonmedical term for a person unable to be responsible for his or her actions.

insecticide (in-**SEK**-tih-side) Agent to destroy insects.

inseminate (in-**SEM**-ih-nate) To introduce semen into the vagina.

insemination (in-sem-ih-**NAY**-shun) The introduction of semen into the vagina.

insertion (in-**SIR**-shun) The insertion of a muscle is the attachment of a muscle to a more movable part of the skeleton, as distinct from the origin.

inspiration (in-spih-**RAY**-shun) Breathe in.

instability (in-stah-**BIL**-ih-tee) Abnormal tendency of a joint to partially or fully dislocate.

insufficiency (in-suh-**FISH**-en-see) Lack of completeness of function; for example, for a heart valve to fail to close properly.

insulin (**IN**-syu-lin) A hormone produced by the islet cells of the pancreas.

integument (in-**TEG**-you-ment) Organ system that covers the body, the skin being the main organ within the system.

integumentary (in-**TEG**-you-**MENT**-ah-ree) Pertaining to the covering of the body.

interatrial (**IN**-ter-**AY**-tree-al) Between the atria of the heart.

intercostal (**IN**-ter-**KOS**-tal) The space between two ribs.

intermittent (**IN**-ter-**MIT**-ent) Alternately ceasing and beginning again.

interosseous (in-ter-**OSS**-ee-us) A structure between bones; for example, muscles.

interphalangeal (**IN**-ter-fay-**LAN**-jee-al) Finger or toe joint between two phalanges.

interstitial (in-ter-**STISH**-al) Pertaining to spaces between cells in a tissue or organ.

interventricular (**IN**-ter-ven-**TRIK**-you-lar) Between the ventricles of the heart.

intervertebral (**IN**-ter-**VER**-teh-bral) The space between two vertebrae.

intestinal (in-**TESS**-tin-al) Pertaining to the intestine.

intestine (in-**TESS**-tin) The digestive tube from stomach to anus.

intolerance (in-**TOL**-er-ance) Inability of the small intestine to digest and dispose of a particular dietary substance.

intracellular (in-trah-**SELL**-you-lar) Within the cell.

intracranial (in-trah-**KRAY**-nee-al) Within the cranium (skull).

intradermal (in-trah-**DER**-mal) Within the epidermis.

intramuscular (in-trah-**MUSS**-kew-lar) Within the muscle.

intraocular (in-trah-**OCK**-you-lar) Pertaining to the inside of the eye.

intrathecal (**IN**-trah-**THEE**-kal) Within the subarachnoid or subdural space.

intrauterine (**IN**-trah-**YOU**-ter-ine) Inside the uterine cavity.

intravenous (**IN**-trah-**VEE**-nus) Through a vein.

intrinsic (in-**TRIN**-sik) Any muscle located entirely within (inside) the structure under consideration; for example, muscles inside the vocal cords or the eye.

intrinsic factor (in-**TRIN**-sik **FAK**-tor) Makes the absorption of vitamin B_{12} happen.

intubation (**IN**-tyu-**BAY**-shun) Insertion of a tube into the trachea.

intussusception (**IN**-tuss-sus-**SEP**-shun) The slipping of one part of bowel inside another to cause obstruction.

inversion (in-**VER**-shun) Turning inward.

invert (in-**VERT**) Turn inward.

involuntary (in-**VOL**-un-tah-ree) Not under control of the will.

involute (in-voh-**LUTE**) Regressive changes in a tissue.

involution (in-voh-**LOO**-shun) Decrease in size.

iodine (**EYE**-oh-dine or **EYE**-oh-deen) Chemical element, the lack of which causes thyroid disease.

iris (**EYE**-ris) Colored portion of the eye with the pupil in its center.

irrigation (ih-rih-**GAY**-shun) Use of water to remove wax out of the external ear canal.

ischemia (is-**KEY**-me-ah) Lack of blood supply to tissue.

ischemic (is-**KEY**-mik) Pertaining to or affected by the lack of blood supply to tissue.

ischial (**IS**-key-al) Pertaining to the ischium.

ischium (**IS**-key-um) Lower and posterior part of the hip bone. Plural *ischia.*

Ishihara color system (ish-ee-**HAR**-ah) Test for color vision defects.

islet cells (**EYE**-let SELLS) Hormone-secreting cells of the pancreas.

islets of Langerhans (**EYE**-lets of **LAHNG**-er-hahnz) Areas of pancreatic cells that produce insulin and glucagon.

isotope (**I**-so-tope) Radioactive element used in diagnostic procedures.

J

Jaeger reading cards (**YA**-ger) Type of different sizes of print for testing near vision.

jaundice (**JAWN**-dis) Yellow staining of tissues with bile pigments, including bilirubin.

jejunal (je-**JEW**-nal) Pertaining to the jejeunum.

jejunum (je-**JEW**-num) Segment of small intestine between the duodenum and the ileum.

K

Kaposi sarcoma (kah-**POH**-see sar-**KOH**-mah) A skin cancer seen in AIDS patients.

keloid (**KEY**-loyd) Raised, irregular, lumpy scar due to excess collagen fiber production during healing of a wound.

keratin (**KER**-ah-tin) Protein found in the skin, nails, and hair.

keratomileusis (ker-ah-**TOE**-mill-oo-sis) Cuts and shapes the cornea.

keratotomy (ker-ah-**TOT**-oh-mee) Incision in the cornea.

kernicterus (ker-**NICK**-ter-us) Bilirubin staining of the basal nuclei of the brain.

ketoacidosis (**KEY**-toe-ass-ih-**DOE**-sis) Excessive production of ketones, making the blood acid.

ketone (**KEY**-tone) Chemical formed in uncontrolled diabetes or in starvation.

ketosis (key-**TOE**-sis) Excess production of ketones.

kidney (**KID**-nee) Organ of excretion.

kyphosis (ki-**FOH**-sis) A normal posterior curve of the thoracic spine that can be exaggerated in disease.

kyphotic (ki-**FOT**-ik) Pertaining to or suffering from kyphosis.

L

labium (**LAY**-bee-um) Fold of the vulva. Plural *labia.*

labor (**LAY**-bore) Process of expulsion of the fetus.

labyrinth (**LAB**-ih-rinth) The inner ear.

labyrinthitis (**LAB**-ih-rin-**THI**-tis) Inflammation of the inner ear.

laceration (lass-eh-**RAY**-shun) A tear of the skin.

lacrimal (**LAK**-rim-al) Pertaining to tears; *or* bone that forms the medial wall of the orbit.

lactase (**LAK**-tase) Enzyme that breaks down lactose (milk sugar) to glucose and galactose.

lactate (**LAK**-tate) To produce milk.

lactation (lak-**TAY**-shun) Production of milk.

lacteal (**LAK**-tee-al) A lymphatic vessel carrying chyle away from the intestine.

lactiferous (lak-**TIF**-er-us) Pertaining to or yielding milk.

lactose (**LAK**-toes) The disaccharide found in cow's milk.

lanugo (la-**NYU**-go) Fine, soft hair on the fetal body.

laparoscope (**LAP**-ah-roh-skope) Instrument (endoscope) used for viewing the abdominal contents.

laparoscopic (**LAP**-ah-roh-**SKOP**-ik) Pertaining to laparoscopy.

laparoscopy (lap-ah-**ROS**-koh-pee) Examination of the contents of the abdomen using an endoscope.

laryngeal (lah-**RIN**-jee-al) Pertaining to the larynx.

laryngitis (lah-rin-**JEYE**-tis) Inflammation of the larynx.

laryngopharynx (lah-**RIN**-go-**FAH**-rinks) Region of the pharynx below the epiglottis that includes the larynx.

laryngoscope (lah-**RING**-oh-skope) Hollow tube with a light and camera used to visualize or operate on the larynx.

laryngotracheobronchitis (lah-**RING**-oh-**TRAY**-kee-oh-brong-**KIE**-tis) Inflammation of the larynx, trachea, and bronchi. Also called *croup*.

larynx (**LAH**-rinks) Organ of voice production.

laser surgery (**LAY**-zer **SUR**-jer-ee) Use of a concentrated, intense narrow beam of electromagnetic radiation for surgery.

lateral (**LAT**-er-al) Situated at the side of a structure.

latissimus dorsi (lah-**TISS**-ih-muss **DOOR**-sigh) The widest (broadest) muscle in the back.

leiomyoma (**LIE**-oh-my-**OH**-mah) Benign tumor derived from smooth muscle.

lens (LENZ) Transparent refractive structure behind the iris.

lentigo (len-**TIE**-go) Age spot; small, flat, brown-black spot in the skin of older people. Plural *lentigines*.

leptin (**LEP**-tin) Hormone secreted by adipose tissue.

lesion (**LEE**-zhun) Pathologic change or injury in a tissue.

leukemia (loo-**KEE**-mee-ah) Disease when the blood is taken over by white blood cells and their precursors.

leukemic (loo-**KEE**-mik) Pertaining to or affected by leukemia.

leukocyte (**LOO**-koh-site) Another term for a white blood cell. Alternative spelling *leucocyte*.

leukocytosis (**LOO**-koh-sigh-**TOE**-sis) An excessive number of white blood cells.

leukopenia (loo-koh-**PEE**-nee-ah) A deficient number of white blood cells.

libido (lih-**BEE**-doh) Sexual desire.

life expectancy (LIFE eck-**SPEK**-tan-see) Statistical determination of the number of years an individual is expected to live.

life span (LIFE SPAN) The age that a person reaches.

ligament (**LIG**-ah-ment) Band of fibrous tissue connecting two structures.

limbic (**LIM**-bic) Array of nerve fibers surrounding the thalamus.

linear fracture (**LIN**-ee-ar **FRAK**-chur) A fracture running parallel to the length of the bone.

lipase (**LIE**-paze) Enzyme that breaks down fat.

lipectomy (lip-**ECK**-toe-me) Surgical removal of adipose tissue.

lipid (**LIP**-id) General term for all types of fatty compounds; for example, cholesterol, triglycerides, and fatty acids.

lipoprotein (**LIP**-oh-pro-teen) Bonding of molecules of fat and protein.

liposuction (**LIP**-oh-suck-shun) Surgical removal of adipose tissue using suction.

lithotripsy (**LITH**-oh-trip-see) Crushing stones by sound waves.

lithotripter (**LITH**-oh-trip-ter) Machine that generates sound waves.

liver (**LIV**-er) Body's largest organ, located in the right upper quadrant of the abdomen.

lobar (**LOW**-bar) Pertaining to a lobe.

lobe (LOBE) Subdivision of an organ or other part.

lobectomy (low-**BECK**-toe-me) Surgical removal of a lobe.

lochia (**LOW**-kee-uh) Vaginal discharge following childbirth.

longevity (lon-**JEV**-ih-tee) Duration of life beyond the normal expectation.

lordosis (lore-**DOH**-sis) A normal forward curvature of the lumbar spine that can be exaggerated in disease.

lordotic (lore-**DOT**-ik) Pertaining to or suffering from lordosis.

louse (LOWSE) Parasitic insect. Plural *lice*.

lumbar (**LUM**-bar) Region in the back and sides between the ribs and pelvis.

lumen (**LOO**-men) The interior space of a tubelike structure.

lumpectomy (lump-**ECK**-toe-me) Removal of a lesion with preservation of surrounding tissue.

luteal (**LOO**-teh-al) Pertaining to a corpus luteum.

lutein (**LOO**-tee-in) Yellow pigment.

luteum (**LOO**-tee-um) Corpus luteum is the yellow (lutein) body formed after an ovarian follicle ruptures.

lymph (LIMF) A clear fluid collected from body tissues and transported by lymph vessels to the venous circulation.

lymphadenectomy (lim-**FAD**-eh-**NECK**-toe-me) Surgical excision of a lymph node(s).

lymphadenitis (lim-**FAD**-eh-neye-tis) Inflammation of a lymph node(s).

lymphadenopathy (lim-**FAD**-eh-**NOP**-ah-thee) Any disease process affecting a lymph node.

lymphangiogram (lim-**FAN**-jee-oh-gram) Radiographic images of lymph vessels and nodes following injection of contrast material.

lymphatic (lim-**FAT**-ik) Pertaining to lymph or the lymphatic system.

lymphedema (**LIMF**-eh-dee-mah) Tissue swelling due to lymphatic obstruction.

lymphocyte (**LIM**-foh-site) Small white blood cell with a large nucleus.

lymphoid (**LIM**-foyd) Resembling lymphatic tissue.

lymphoma (lim-**FOH**-mah) Any neoplasm of lymphatic tissue.

M

macrocyte (**MACK**-roh-site) Large red blood cell.

macrocytic (mack-roh-**SIT**-ik) Pertaining to a macrocyte.

macrophage (**MACK**-roh-fayj) Large white blood cell that removes bacteria, foreign particles, and dead cells.

macula lutea (**MACK**-you-lah **LOO**-tee-ah) Yellowish spot on the back of the retina; contains the fovea centralis.

macule (**MACK**-yul) Small, flat spot or patch on the skin.

majus (**MAY**-jus) Bigger or greater; for example, labium majus. Plural *majora*.

malabsorption (mal-ab-**SORP**-shun) Inadequate gastrointestinal absorption of nutrients.

malformation (MAL-for-MAY-shun) Failure of proper or normal development.

malfunction (mal-FUNK-shun) Inadequate or abnormal function.

malignancy (mah-LIG-nan-see) State of being malignant.

malignant (mah-LIG-nant) Tumor that invades surrounding tissues and metastasizes to distant organs.

malleus (MAL-ee-us) Outer (lateral) one of the three ossicles in the middle ear; shaped like a hammer.

malnutrition (mal-nyu-TRISH-un) Inadequate nutrition from poor diet or inadequate absorption of nutrients.

malunion (mal-YOU-nee-un) The two bony ends of a fracture fail to heal together in the correct position.

mammary (MAM-ah-ree) Relating to the lactating breast.

mammogram (MAM-oh-gram) The record produced by x-ray imaging of the breast.

mammography (mah-MOG-rah-fee) The process of x-ray examination of the breast.

mammoplasty (MAM-oh-plas-tee) Surgical reshaping of the breast.

mandible (MAN-di-bel) Lower jawbone.

mandibular (man-DIB-you-lar) Pertaining to the mandible.

mania (MAY-nee-ah) Mood disorder with hyperactivity, irritability, and rapid speech.

manic (MAN-ik) Pertaining to or suffering from mania.

marrow (MAH-roe) Fatty, blood-forming tissue in the cavities of long bones.

mastalgia (mass-TAL-jee-uh) Pain in the breast.

mastectomy (mass-TECK-toe-me) Surgical excision of the breast.

masticate (MASS-tih-kate) To chew.

mastication (mass-tih-KAY-shun) The process of chewing.

mastitis (mass-TIE-tis) Inflammation of the breast.

mastoid (MASS-toyd) Small bony protrusion immediately behind the ear.

maternal (mah-TER-nal) Pertaining to or derived from the mother.

matrix (MAY-triks) Substance that surrounds and protects cells, is manufactured by the cells, and holds them together.

maturation (mat-you-RAY-shun) Process to achieve full development.

mature (mah-TYUR) Fully developed.

maxilla (mak-SILL-ah) Upper jawbone, containing right and left maxillary sinuses.

maxillary (mak-SILL-ah-ree) Pertaining to the maxilla.

maximus (MAKS-ih-mus) The gluteus maximus muscle is the largest muscle in the body, covering a large part of each buttock.

meatal (me-AY-tal) Pertaining to a meatus.

meatus (me-AY-tus) The external opening of a passage.

meconium (meh-KOH-nee-um) The first bowel movement of the newborn.

medial (ME-dee-al) Nearer to the middle of the body.

mediastinal (ME-dee-ass-TIE-nal) Pertaining to the mediastinum.

mediastinoscopy (ME-dee-ass-tih-NOS-koh-pee) Examination of the mediastinum using an endoscope.

mediastinum (ME-dee-ass-TIE-num) Area between the lungs containing the heart, aorta, venae cavae, esophagus, and trachea.

medius (ME-dee-us) The gluteus medius muscle is partly covered by the gluteus maximus.

medulla (meh-DULL-ah) Central portion of a structure surrounded by cortex.

medulla oblongata (meh-DULL-ah ob-lon-GAH-tah) Most posterior subdivision of the brainstem; continuation of the spinal cord.

medullary (meh-DULL-ah-ree) Pertaining to a medulla.

meiosis (my-OH-sis) Two rapid cell divisions, resulting in half the number of chromosomes.

melanin (MEL-ah-nin) Black pigment found in the skin, hair, and retina.

melanoma (mel-ah-NO-mah) Malignant neoplasm formed from cells that produce melanin.

melatonin (mel-ah-TONE-in) Hormone formed by the pineal gland.

melena (mel-EN-ah) The passage of black, tarry stools.

membrane (MEM-brain) Thin layer of tissue covering a structure or cavity.

membranous (MEM-brah-nus) Pertaining to a membrane.

menarche (meh-NAR-key) First menstrual period.

Ménière disease (men-YEAR DIZ-eez) Disorder of the inner ear with acute attacks of tinnitus, vertigo, and hearing loss.

meninges (meh-NIN-jeez) Three-layered covering of the brain and spinal cord.

meningitis (men-in-JIE-tis) Inflammation of the meninges.

meningocele (meh-NING-oh-seal) Protrusion of the meninges from the spinal cord or brain through a defect in the vertebral column or cranium.

meningococcal (meh-nin-goh-KOK-al) Pertaining to the *meningococcus* bacterium.

meningomyelocele (meh-nin-goh-MY-el-oh-seal) Protrusion of the spinal cord and meninges through a defect in the vertebral arch of one or more vertebrae.

meniscectomy (men-ih-SEK-toh-me) Excision (cutting out) of all or part of a meniscus.

meniscus (meh-NISS-kuss) Disc of cartilage between the bones of a joint; for example, in the knee joint. Plural *menisci.*

menopausal (MEN-oh-paws-al) Pertaining to the menopause.

menopause (MEN-oh-paws) Permanent ending of menstrual periods.

menorrhagia (men-oh-RAY-jee-ah) Excessive menstrual bleeding.

menses (MEN-seez) Monthly uterine bleeding.

menstrual (MEN-stru-al) Pertaining to menstruation.

menstruate (MEN-stru-ate) The act of menstruation.

menstruation (men-stru-AY-shun) Synonym of *menses.*

mesentery (MESS-en-ter-ree) A double layer of peritoneum enclosing the abdominal viscera.

metabolic (met-ah-BOL-ik) Pertaining to metabolism.

metabolic acidosis (met-ah-BOL-ik ass-ih-DOE-sis) Decreased pH in the blood and body tissues as a result of an upset in metabolism.

metabolism (meh-**TAB**-oh-lizm) The constantly changing physical and chemical processes occurring in the cell that are the sum of anabolism and catabolism.

metacarpal (MET-ah-**KAR**-pal) The five bones between the carpus and the fingers.

metacarpophalangeal (**MET**-ah-**KAR**-poh-fay-**LAN**-jee-al) The articulations (joints) between the metacarpal bones and the phalanges.

metastasis (meh-**TAS**-tah-sis) Spread of a disease from one part of the body to another. Plural *metastases.*

metastasize (meh-**TAS**-tah-size) To spread to distant parts.

metastatic (meh-tah-**STAT**-ik) Pertaining to the character of cells that can metastasize.

metatarsal (MET-ah-**TAR**-sal) Pertaining to the metatarsus.

metatarsus (MET-ah-**TAR**-sus) The five parallel bones of the foot between the tarsus and the phalanges.

metrorrhagia (MEH-troh-**RAY**-jee-ah) Irregular uterine bleeding between menses.

microbe (**MY**-krohb) Short for *microorganism.*

microcephalic (**MY**-kroh-**SEF**-ah-lik) Pertaining to or suffering from a small head.

microcephaly (**MY**-kroh-**SEF**-ah-lee) An abnormally small head.

microcyte (**MY**-kroh-site) Small red blood cell.

microcytic (my-kroh-**SIT**-ik) Pertaining to a small cell.

microorganism (**MY**-kroh-**OR**-gan-izm) Any organism too small to be seen by the naked eye.

microscope (**MY**-kroh-skope) Instrument for viewing something small that cannot be seen in detail by the naked eye.

microscopic (**MY**-kroh-**SKOP**-ik) Visible only with the aid of a microscope.

micturate (**MIK**-choo-rate) Pass urine.

micturition (mik-choo-**RISH**-un) Act of passing urine.

migraine (**MY**-grain) Paroxysmal severe headache confined to one side of the head.

mineral (**MIN**-er-al) Inorganic compound usually found in the earth's crust.

mineralocorticoid (**MIN**-er-al-oh-**KOR**-tih-koyd) Hormone of the adrenal cortex that influences sodium and potassium metabolism.

minimus (**MIN**-ih-mus) The gluteus minimus is the smallest of the gluteal muscles and lies under the gluteus medius.

minus (**MY**-nus) Smaller or lesser; for example, labium minus. Plural *minora.*

mitochondrion (my-toe-**KON**-dree-on) Organelle that generates, stores, and releases energy for cell activities. Plural *mitochondria.*

mitosis (my-**TOE**-sis) Cell division that creates two identical cells, each with 46 chromosomes.

mitral (**MY**-tral) Shaped like the headdress of a Catholic bishop.

modify (**MOD**-ih-fie) Change the form or qualities of something.

molar (**MO**-lar) One of six teeth in each jaw that grind food.

mole (MOLE) Benign localized area of melanin-producing cells.

molecule (**MOLL**-eh-kyul) Very small particle.

molluscum contagiosum (moh-**LUS**-kum kon-**TAY**-jee-oh-sum) STD caused by a virus.

monocyte (**MON**-oh-site) Large white blood cell with a single nucleus.

mononeuropathy (MON-oh-nyu-**ROP**-ah-thee) Disorder affecting a single nerve.

mononucleosis (MON-oh-nyu-klee-**OH**-sis) Presence of large numbers of specific, diagnostic mononuclear leukocytes.

monoplegia (**MON**-oh-**PLEE**-jee-ah) Paralysis of one limb.

monoplegic (**MON**-oh-**PLEE**-jik) Pertaining to or suffering from monoplegia.

mons pubis (MONZ **PYU**-bis) Fleshy pad with pubic hair, overlying the pubic bone.

morbidity (mor-**BID**-ih-tee) The frequency of the appearance of a disease.

morphine (**MOR**-feen) Derivative of opium used as an analgesic or sedative.

mortality (mor-**TAL**-ih-tee) Death rate.

motile (**MOH**-til) Capable of spontaneous movement.

motility (moh-**TILL**-ih-tee) The ability for spontaneous movement.

motor (**MOH**-tor) Structures of the nervous system that send impulses out to cause muscles to contract or glands to secrete.

mouth (MOWTH) External opening of a cavity or canal.

mucin (**MYU**-sin) Protein element of mucus.

mucocutaneous (**MYU**-koh-kyu-**TAY**-nee-us) Junction of skin and mucous membrane; for example, the lips.

mucolytic (**MYU**-koh-**LIT**-ik) Agent capable of dissolving or liquefying mucus.

mucosa (myu-**KOH**-sah) Lining of a tubular structure that secretes mucus. Another name for *mucous membrane.*

mucous (**MYU**-kus) Pertaining to mucus or the mucosa.

mucus (**MYU**-kus) Sticky secretion of cells in mucous membranes.

multipara (mul-**TIP**-ah-ruh) Woman who has given birth to two or more children.

murmur (**MUR**-mur) Abnormal heart sound heard with a stethoscope when a valve closes or opens abnormally.

muscle (**MUSS**-el) A tissue consisting of contractile cells.

musculoskeletal (**MUSS**-kyu-loh-**SKEL**-eh-tal) Pertaining to the muscles and the bony skeleton.

mutation (myu-**TAY**-shun) Change in the chemistry of a gene.

mute (MYUT) Unable or unwilling to speak.

mutism (**MYU**-tizm) Absence of speech.

myasthenia gravis (my-as-**THEE**-nee-ah **GRA**-vis) Disorder of fluctuating muscle weakness.

myelin (**MY**-eh-lin) Material of the sheath around the axon of a nerve.

myelitis (**MY**-eh-**LIE**-tis) Inflammation of the spinal cord.

myelocele (**MY**-eh-low-seal) Protrusion of the spinal cord through a defect in the vertebral arch.

myelomeningocele (**MY**-eh-low-meh-**NING**-oh-seal) Protrusion of the spinal cord and meninges through a defect in the vertebral arch of one or more vertebrae.

myocardial (my-oh-**KAR**-dee-al) Pertaining to heart muscle.

myocarditis (**MY**-oh-kar-**DIE**-tis) Inflammation of the heart muscle.

myocardium (MY-oh-**KAR**-dee-um) All the heart muscle.

myoma (my-**OH**-mah) Benign tumor of muscle.

myomectomy (my-oh-**MEK**-toe-me) Surgical removal of a myoma (fibroid).

myometrium (my-oh-**MEE**-tree-um) Muscle wall of the uterus.

myopia (my-**OH**-pee-ah) Able to see close objects but unable to see distant objects.

myringotomy (mir-in-**GOT**-oh-me) Incision in the tympanic membrane.

myxedema (miks-eh-**DEE**-muh) Nonpitting, waxy edema of the skin in hypothyroidism.

N

narcolepsy (**NAR**-koh-lep-see) Involuntary falling asleep.

narcotic (nar-**KOT**-ik) Drug derived from opium or any drug with effects similar to those of opium derivatives.

naris (**NAH**-ris) Nostril. Plural *nares.*

nasal (**NAY**-zal) Pertaining to the nose.

nasogastric (**NAY**-zoh-**GAS**-trik) Pertaining to the nose and stomach.

nasolacrimal duct (**NAY**-zoh-**LAK**-rim-al DUKT) Passage from the lacrimal sac to the nose.

nasopharyngeal (**NAY**-zoh-fah-**RIN**-jee-al) Pertaining to the nasopharynx.

nasopharynx (**NAY**-zoh-**FAH**-rinks) Region of the pharynx at the back of the nose and above the soft palate.

natal (**NAY**-tal) Pertaining to birth.

nebulizer (**NEB**-you-liz-er) Device used to deliver liquid medicine in a fine mist.

necrosis (neh-**KROH**-sis) Pathologic death of cells or tissue.

necrotic (neh-**KROT**-ik) Pertaining to or affected by necrosis.

necrotizing fasciitis (neh-kroh-**TIZE**-ing fash-ee-**EYE**-tis) Inflammation of fascia producing death of the tissue.

neonatal (**NEE**-oh-**NAY**-tal) Pertaining to the newborn infant or the newborn period.

neonate (**NEE**-oh-nate) A newborn infant.

nephrectomy (nef-**REK**-toe-me) Surgical removal of a kidney.

nephritis (nef-**RY**-tis) Inflammation of the kidney.

nephroblastoma (**NEF**-roh-blas-**TOE**-mah) Cancerous kidney tumor of childhood. Also known as *Wilms tumor.*

nephrolithiasis (**NEF**-roe-lih-**THIGH**-ah-sis) Presence of a kidney stone.

nephrolithotomy (**NEF**-roe-lih-**THOT**-oh-me) Incision for removal of a stone.

nephrologist (nef-**ROL**-oh-jist) Medical specialist in disorders of the kidney.

nephrology (nef-**ROL**-oh-jee) Medical specialty of diseases of the kidney.

nephron (**NEF**-ron) Filtration unit of the kidney; glomerulus + renal tubule.

nephropathy (nef-**ROP**-ah-thee) Any disease of the kidney.

nephroscope (**NEF**-roe-skope) Endoscope to view the inside of the kidney.

nephroscopy (nef-**ROS**-koh-pee) To examine the kidney.

nephrosis (nef-**ROH**-sis) Same as *nephrotic syndrome.*

nephrotic syndrome (nef-**ROT**-ik **SIN**-drome) Glomerular disease with marked loss of protein. Also called *nephrosis.*

nerve (NERV) A cord of nerve fibers bound together by connective tissue.

nervous (**NER**-vus) Pertaining to a nerve or the nervous system; *or* easily excited or agitated.

nervous system (**NER**-vus **SIS**-tem) The whole, integrated nerve apparatus.

neural (**NYU**-ral) Pertaining to nervous tissue.

neuralgia (nyu-**RAL**-jee-ah) Pain in the distribution of a nerve.

neuroglia (nyu-roh-**GLEE**-ah) Connective tissue holding nervous tissue together.

neurohypophysis (**NYU**-roh-high-**POF**-ih-sis) Posterior lobe of the pituitary gland.

neurologic (**NYU**-roh-**LOJ**-ik) Pertaining to the nervous sytem.

neurologist (nyu-**ROL**-oh-jist) Medical specialist in disorders of the nervous system.

neurology (nyu-**ROL**-oh-jee) Medical specialty of disorders of the nervous system.

neuromuscular (**NYU**-roh-**MUSS**-kyu-lar) A junction where a nerve supplies muscle tissue.

neuron (**NYU**-ron) Technical term for a nerve cell; consists of the cell body with its dendrites and axons.

neuropathy (nyu-**ROP**-ah-thee) Any disorder affecting the nervous system.

neurosurgeon (**NYU**-roh-**SUR**-jun) One who operates on the nervous system.

neurosurgery (**NYU**-roh-**SUR**-jer-ee) Operating on the nervous system.

neurotoxin (**NYU**-roh-tock-sin) Agent that poisons the nervous system.

neurotransmitter (**NYU**-roh-trans-**MIT**-er) Chemical agent that relays messages from one nerve cell to the next.

neutropenia (**NEW**-troh-**PEE**-nee-ah) A deficiency of neutrophils.

neutrophil (**NEW**-troh-fill) A neutrophil's granules take up (purple) stain equally, whether the stain is acid or alkaline.

neutrophilia (**NEW**-troh-**FILL**-ee-ah) An increase in neutrophils.

nevus (**NEE**-vus) Congenital lesion of the skin. Plural *nevi.*

nipple (**NIP**-el) Projection from the breast into which the lactiferous ducts open.

nitrite (**NI**-trite) Chemical formed in urine by *E. coli* and other microorganisms.

nitrogenous (ni-**TROJ**-en-us) Containing or generating nitrogen.

nocturia (nok-**TYU**-ree-ah) Excessive urination at night.

node (NOHD) A circumscribed mass of tissue.

nonunion (non-**YOU**-nee-un) Total failure of healing of a fracture.

norepinephrine (**NOR**-ep-ih-**NEFF**-rin) Catecholamine hormone of the adrenal gland that is a parasympathetic neurotransmitter. Also called *noradrenaline.*

nosocomial (noh-soh-**KOH**-mee-al) Acquired while in the hospital.

phagocyte (FAG-oh-site) Blood cell that ingests and destroys foreign particles and cells.

phagocytic (fag-oh-SIT-ik) Pertaining to a phagocyte.

phagocytosis (FAG-oh-sigh-TOE-sis) Process of ingestion and destruction.

phalanx (FAY-lanks) A bone of a finger or toe. Plural *phalanges.*

pharmacist (FAR-mah-sist) Person licensed by the state to prepare and dispense drugs.

pharmacology (far-mah-KOLL-oh-jee) Science of the preparation, uses, and effects of drugs.

pharmacy (FAR-mah-see) Facility licensed to prepare and dispense drugs.

pharyngeal (fah-RIN-jee-al) Pertaining to the pharynx.

pharyngitis (fah-rin-JIE-tis) Inflammation of the pharynx.

pharynx (FAH-rinks) Tube from the back of the nose to the larynx.

phimosis (fih-MOH-sis) Prepuce cannot be retracted.

phlebitis (fleh-BIE-tis) Inflammation of a vein.

phlebotomist (fleh-BOT-oh-mist) Person skilled in taking blood from veins.

phlebotomy (fleh-BOT-oh-me) Taking blood from a vein.

phlegm (FLEM) Abnormal amounts of mucus expectorated from the respiratory tract.

phobia (FOH-bee-ah) Pathologic fear or dread.

photocoagulation (foh-toe-koh-ag-you-LAY-shun) Using light (laser beam) to form a clot.

photophobia (foh-toe-FOH-bee-ah) Fear of the light because it hurts the eyes.

photophobic (foh-toe-FOH-bik) Pertaining to or suffering from photophobia.

photoreceptor (foh-toe-ree-SEP-tor) A photoreceptor cell receives light and converts it into electrical impulses.

photosensitive (FOH-toe-SEN-sih-tiv) Abnormally sensitive to light.

photosensitivity (FOH-toe-SEN-sih-tiv-ih-tee) Light produces pain in the eye.

phototherapy (FOH-toe-THAIR-ah-pee) Treatment using light rays.

pia mater (PEE-ah MAY-ter) Delicate inner layer of the meninges.

pica (PIE-kah) Eating substances not considered to be food.

pineal (PIN-ee-al) Pertaining to the pineal gland.

pink eye (PINK EYE) Conjunctivitis.

pinna (PIN-ah) Another name for *auricle.* Plural *pinnae.*

pitting edema (ee-DEE-mah) An indentation made by a finger in an edematous area persists for a long time.

pituitary (pih-TYU-ih-tary) Pertaining to the pituitary gland.

placebo (plah-SEE-boh) An inert compound with no innate therapeutic value.

placenta (plah-SEN-tah) Organ that allows metabolic interchange between the mother and the fetus.

placenta abruptio (plah-SEN-tah ab-RUP-she-oh) Premature detachment of the placenta.

placenta previa (plah-SEN-tah PREE-vee-ah) Placenta obstructing the fetus during delivery.

plaque (PLAK) Patch of abnormal tissue.

plasma (PLAZ-mah) Fluid, noncellular component of blood.

platelet (PLAYT-let) Cell fragment involved in the clotting process. Also called *thrombocyte.*

pleura (PLUR-ah) Membrane covering the lungs and lining the ribs in the thoracic cavity. Plural *pleurae.*

pleural (PLUR-al) Pertaining to the pleura.

pleurisy (PLUR-ih-see) Inflammation of the pleura.

plexus (PLEK-sus) A weblike network of joined nerves. Plural *plexuses.*

pneumoconiosis (NEW-moh-koh-nee-OH-sis) Fibrotic lung disease caused by the inhalation of different dusts.

pneumonectomy (NEW-moh-NECK-toe-me) Surgical removal of a lung.

pneumonia (new-MOH-nee-ah) Inflammation of the lung parenchyma (tissue).

pneumonitis (new-moh-NI-tis) Synonym for *pneumonia.*

pneumothorax (new-moh-THOR-ax) Air in the pleural cavity of the chest.

podiatrist (poh-DIE-ah-trist) Practitioner of podiatry.

podiatry (poh-DIE-ah-tree) The diagnosis and treatment of disorders and injuries of the foot.

poliomyelitis (POE-lee-oh-MY-eh-lie-tis) Inflammation of the gray matter of the spinal cord, leading to paralysis of the limbs and muscles of respiration. Abbreviation *polio.*

pollutant (poh-LOO-tant) Substance that makes an environment unclean or impure.

pollution (poh-LOO-shun) Condition that is unclean, impure, and a danger to health.

polycystic (pol-ee-SIS-tik) Composed of many cysts.

polycythemia vera (POL-ee-sigh-THEE-me-ah) Chronic disease with bone marrow hyperplasia, increase in number of RBCs, and increase in blood volume.

polydipsia (pol-ee-DIP-see-ah) Excessive thirst.

polyhydramnios (POL-ee-high-DRAM-nee-os) Too much amniotic fluid.

polymenorrhea (POL-ee-men-oh-REE-ah) More than normal frequency of menses.

polymorphonuclear (POL-ee-more-foh-NEW-klee-ah) White blood cell with a multilobed nucleus.

polyneuropathy (POL-ee-nyu-ROP-ah-thee) Disorder affecting many nerves.

polyp (POL-ip) Any mass of tissue that projects outward.

polypectomy (pol-ip-ECK-toh-mee) Excision or removal of a polyp.

polyphagia (pol-ee-FAY-jee-ah) Excessive eating.

polyposis (pol-ih-POH-sis) Presence of several polyps.

polysomnography (POLL-ee-som-NOG-rah-fee) Test to monitor brain waves, muscle tension, eye movement, and oxygen levels in the blood as the patient sleeps.

polyuria (pol-ee-YOU-ree-ah) Excessive production of urine.

pons (PONZ) Part of the brainstem.

popliteal (pop-LIT-ee-al) Pertaining to the back of the knee.

popliteal fossa (pop-LIT-ee-al FOSS-ah) The hollow at the back of the knee.

parenteral (pah-**REN**-ter-al) Giving medication by any means other than the gastrointestinal tract.

paresis (par-**EE**-sis) Partial paralysis (weakness).

paresthesia (par-es-**THEE**-ze-ah) An abnormal sensation; for example, tingling, burning, prickling. Plural *parasthesias.*

parietal (pah-**RYE**-eh-tal) Pertaining to the outer layer of the pericardium and the wall of any body cavity; *or* the two bones forming the sidewalls and roof of the cranium.

parietal lobe (pah-**RYE**-eh-tal LOBE) Area of the brain under the parietal bone.

Parkinson disease (**PAR**-kin-son **DIZ**-eez) Disease of muscular rigidity, tremors, and a masklike facial expression.

paronychia (par-oh-**NICK**-ee-ah) Infection alongside the nail.

parotid (pah-**ROT**-id) Parotid gland is the salivary gland beside the ear.

paroxysmal (par-ock-**SIZ**-mal) Occurring in sharp, spasmodic episodes.

particle (**PAR**-tih-kul) A small piece of matter.

particulate (par-**TIK**-you-late) Relating to a fine particle.

patella (pah-**TELL**-ah) Thin, circular bone in front of the knee joint and embedded in the patellar tendon. Also called *kneecap.* Plural *patellae.*

patellar (pah-**TELL**-ar) Pertaining to the patella.

patent (**PAY**-tent) Open.

patent ductus arteriosus (**PAY**-tent **DUK**-tus ar-**TER**-ee-oh-sus) An open, direct channel between the aorta and the pulmonary artery.

pathogen (**PATH**-oh-jen) A disease-causing microorganism.

pathogenic (path-oh-**JEN**-ik) Causing disease.

pathologic fracture (path-oh-**LOJ**-ik **FRAK**-chur) Fracture occurring at a site already weakened by a disease process, such as cancer.

pathologist (pa-**THOL**-oh-jist) A specialist in pathology.

pathology (pa-**THOL**-oh-jee) Medical specialty dealing with the structural and functional changes of a disease process or the cause, development, and structural changes in disease.

pectoral (**PEK**-tor-al) Pertaining to the chest.

pectoral girdle (**PEK**-tor-al **GIR**-del) Incomplete bony ring that attaches the upper limb to the axial skeleton.

pectoralis (**PEK**-tor-ah-lis) Pertaining to the chest.

pedal (**PEED**-al) Pertaining to the foot.

pediatrician (**PEE**-dee-ah-**TRISH**-an) Medical specialist in pediatrics.

pediatrics (pee-dee-**AT**-riks) Medical specialty of treating children during development from birth through adolescence.

pediculosis (peh-dick-you-**LOH**-sis) An infestation with lice.

peer (PEER) A person at the same level or standing.

pelvic (**PEL**-vik) Pertaining to the pelvis.

pelvis (**PEL**-viss) A basin-shaped ring of bones, ligaments, and muscles at the base of the spine; *or* a basin-shaped cavity, as in the pelvis of the kidney.

penile (**PEE**-nile) Pertaining to the penis.

penis (**PEE**-nis) Conveys urine and semen to the outside.

pepsin (**PEP**-sin) Enzyme produced by the stomach that breaks down protein.

pepsinogen (pep-**SIN**-oh-jen) Converted by HCl in stomach to pepsin.

peptic (**PEP**-tik) Relating to the stomach and duodenum.

percutaneous (**PER**-kyu-**TAY**-nee-us) Passage through the skin.

perforated (**PER**-foh-ray-ted) Punctured with one or more holes.

perforation (per-foh-**RAY**-shun) A hole through the wall of a structure.

perfuse (per-**FYUSE**) To force blood to flow through a lumen or a vascular bed.

perfusion (per-**FYU**-shun) The act of perfusing.

pericardial (per-ih-**KAR**-dee-al) Pertaining to the pericardium.

pericarditis (**PER**-ih-kar-**DIE**-tis) Inflammation of the pericardium, the covering of the heart.

pericardium (per-ih-**KAR**-dee-um) A double layer of membranes surrounding the heart.

perimetrium (per-ih-**ME**-tree-um) The covering of the uterus; part of the peritoneum.

perinatal (per-ih-**NAY**-tal) Around the time of birth.

perineal (**PER**-ih-**NEE**-al) Pertaining to the perineum.

perineum (**PER**-ih-**NEE**-um) Area between the thighs, extending from the coccyx to the pubis.

periodontal (**PER**-ee-oh-**DON**-tal) Around a tooth.

periodontics (**PER**-ee-oh-**DON**-tiks) Branch of dentistry specializing in disorders of tissues around the teeth.

periodontist (**PER**-ee-oh-**DON**-tist) Specialist in periodontics.

periodontitis (**PER**-ee-oh-don-**TIE**-tis) Inflammation of tissues around a tooth.

periorbital (per-ee-**OR**-bit-al) Pertaining to tissues around the orbit.

periosteal (**PER**-ee-**OSS**-tee-al) Pertaining to the periosteum.

periosteum (**PER**-ee-**OSS**-tee-um) Fibrous membrane covering a bone.

peripheral (peh-**RIF**-er-al) Pertaining to the periphery or external boundary.

peripheral vision (peh-**RIF**-er-al **VIZH**-un) Ability to see objects as they come into the outer edges of the visual field.

peristalsis (per-ih-**STAL**-sis) Waves of alternate contraction and relaxation of the muscle wall of a tube; for example, of the intestinal wall to move food along the digestive tract.

peritoneal (**PER**-ih-toe-**NEE**-al) Pertaining to the peritoneum.

peritoneum (**PER**-ih-toe-**NEE**-um) Membrane that lines the abdominal cavity.

peritonitis (**PER**-ih-toe-**NIE**-tis) Inflammation of the peritoneum.

pernicious anemia (per-**NISH**-us ah-**NEE**-me-ah) Chronic anemia due to lack of vitamin B_{12}.

pertussis (per-**TUSS**-is) Infectious disease with a spasmodic, intense cough ending on a whoop (stridor). Also called *whooping cough.*

pes planus (PES **PLAY**-nuss) A flat foot with no plantar arch.

pessary (**PES**-ah-ree) Appliance inserted into the vagina to support the uterus.

petechia (peh-**TEE**-kee-ah) Pinpoint capillary hemorrhagic spot in the skin. Plural *petechiae.*

petit mal (peh-**TEE** MAL) Old name for absence seizures.

osteocyte (OSS-tee-oh-site) A bone-maintaining cell.

osteogenesis imperfecta (OSS-tee-oh-JEN-eh-sis im-per-FEK-tah) Inherited condition when bone formation is incomplete, leading to fragile, easily broken bones.

osteogenic sarcoma (OSS-tee-oh-JEN-ik sar-KOH-mah) Malignant tumor originating in bone-producing cells.

osteomalacia (OSS-tee-oh-mah-LAY-she-ah) Soft, flexible bones lacking in calcium (rickets).

osteomyelitis (OSS-tee-oh-my-eh-LIE-tis) Inflammation of bone tissue.

osteopath (OSS-tee-oh-path) Practitioner of osteopathy.

osteopathy (OSS-tee-OP-ah-thee) Medical practice based on maintaining the balance of the body.

osteopenia (OSS-tee-oh-PEE-nee-ah) Decreased calcification of bone.

osteoporosis (OSS-tee-oh-poh-ROE-sis) Condition in which the bones become more porous, brittle, and fragile and more likely to fracture.

ostomy (OSS-toe-me) Artificial opening into a tubular structure.

otitis media (oh-TIE-tis ME-dee-ah) Inflammation of the middle ear.

otolith (OH-toe-lith) A calcium particle in the vestibule of the inner ear.

otologist (oh-TOL-oh-jist) Medical specialist in diseases of the ear.

otology (oh-TOL-oh-jee) Study of the function and diseases of the ear.

otorhinolaryngologist (OH-toe-RHINO-lah-rin-GOL-oh-jist) Ear, nose, and throat medical specialist.

otosclerosis (oh-toe-sklair-OH-sis) Hardening at the junction of the stapes and oval window that causes loss of hearing.

otoscope (OH-toe-skope) Instrument for examining the ear.

otoscopic (oh-toe-SKOP-ik) Pertaining to examination with an otoscope.

otoscopy (oh-TOS-koh-pee) Examination of the ear.

ovarian (oh-VAIR-ee-an) Pertaining to the ovary(ies).

ovary (OH-vah-ree) One of the paired female egg-producing glands. Plural *ovaries.*

ovulate (OV-you-late) Release the oocyte from a follicle.

ovulation (OV-you-LAY-shun) Release of an oocyte from a follicle.

ovum (OH-vum) Egg. Also called *oocyte.* Plural *ova.*

oxygen (OCK-see-jen) The gas essential for life.

oxyhemoglobin (OCK-see-he-moh-GLOW-bin) Hemoglobin in combination with oxygen.

oxytocin (OCK-see-TOE-sin) Pituitary hormone that stimulates the uterus to contract.

P

pacemaker (PACE-may-ker) Device that regulates cardiac electrical activity.

palate (PAL-ate) Roof of the mouth.

palatine (PAL-ah-tine) Bone that forms the hard palate and parts of the nose and orbits.

palliative care (PAL-ee-ah-tiv KAIR) To relieve symptoms and pain without curing.

pallor (PAL-or) Paleness of the skin.

palm (PAHLM) Flat or anterior surface of the hand.

palmar (PAHL-mar) Pertaining to the palm.

palpate (PAL-pate) To examine with the fingers and hands.

palpation (pal-PAY-shun) Examination with the fingers and hands.

palpitation (pal-pih-TAY-shun) Forcible, rapid beat of the heart felt by the patient.

palsy (PAWL-zee) Paralysis or paresis from brain damage.

pancreas (PAN-kree-as) Lobulated gland, the head of which is tucked into the curve of the duodenum.

pancreatic (PAN-kree-AT-ik) Pertaining to the pancreas.

pancreatitis (PAN-kree-ah-TIE-tis) Inflammation of the pancreas.

pancytopenia (PAN-site-oh-PEE-nee-ah) Deficiency of all types of blood cells.

pandemic (pan-DEM-ik) Pertaining to a disease attacking the population of a very large area.

panendoscopy (pan-en-DOS-koh-pee) A visual examination of the inside of the esophagus, stomach, and upper duodenum using a flexible fiber-optic endoscope.

panhypopituitarism (pan-HIGH-poh-pih-TYU-ih-tah-rizm) Deficiency of all the pituitary hormones.

Pap test (PAP TEST) Examination of cells taken from the cervix.

papilla (pah-PILL-ah) Any small projection. Plural *papillae.*

papilledema (pah-pill-eh-DEE-mah) Swelling of the optic disc in the retina.

papillomavirus (pap-ih-LOH-mah-vi-rus) Virus that causes warts and is associated with cancer.

papule (PAP-yul) Small, circumscribed elevation on the skin.

para (PAH-rah) Abbreviation for number of deliveries.

paralysis (pah-RAL-ih-sis) Loss of voluntary movement.

paralytic (par-ah-LYT-ik) Suffering from paralysis.

paralyze (PAR-ah-lyze) To make incapable of movement.

paranasal (PAR-ah NAY-zal) Adjacent to the nose.

paranoia (par-ah-NOY-ah) Presence of persecutory delusions.

paranoid (PAR-ah-noyd) Having delusions of persecution.

paraphimosis (PAR-ah-fih-MOH-sis) Condition in which a retracted prepuce cannot be pulled forward to cover the glans.

paraplegia (par-ah-PLEE-jee-ah) Paralysis of both lower extremities.

paraplegic (par-ah-PLEE-jik) Pertaining to or suffering from paraplegia.

parasite (PAR-ah-site) An organism that attaches itself to, lives on or in, and derives its nutrition from another species.

parasitic (par-ah-SIT-ik) Pertaining to a parasite.

parasympathetic (par-ah-sim-pah-THET-ik) Division of the autonomic nervous system; has opposite effects of the sympathetic division.

parathyroid (par-ah-THIGH-royd) Endocrine glands embedded in the back of the thyroid gland.

paraurethral (PAR-ah-you-REE-thral) Situated around the urethra.

parenchyma (pah-RENG-kih-mah) Characteristic functional cells of a gland or organ that are supported by the connective tissue framework.

nuchal cord (**NYU**-kul KORD) Loop(s) of umbilical cord around the fetal neck.

nuclear (**NYU**-klee-ar) Pertaining to a nucleus.

nucleolus (nyu-**KLEE**-oh-lus) Small mass within the nucleus.

nucleus (**NYU**-klee-us) Functional center of a cell or structure.

nutrient (**NYU**-tree-ent) A substance in food required for normal physiologic function.

nutrition (nyu-**TRISH**-un) The study of food and liquid requirements for normal function of the human body.

O

obesity (oh-**BEE**-sih-tee) Excessive amount of fat in the body.

oblique fracture (ob-**LEEK FRAK**-chur) A diagonal fracture across the long axis of the bone.

obsession (ob-**SESH**-un) Persistent, recurrent, uncontrollable thoughts or impulses.

obsessive (ob-**SES**-iv) Possessing persistent, recurrent, uncontrollable thoughts or impulses.

obstetrician (ob-steh-**TRISH**-un) Medical specialist in obstetrics.

obstetrics (OB) (ob-**STET**-ricks) Medical specialty for the care of women during pregnancy and the postpartum period.

occipital (ock-**SIP**-it-al) The back of the skull.

occipital lobe (ock-**SIP**-it-al LOBE) Posterior area of the cerebral hemispheres.

occlude (oh-**KLUDE**) To close, plug, or completely obstruct.

occlusion (oh-**KLU**-zhun) A complete obstruction.

occult (oh-**KULT**) Not visible on the surface, hidden.

occult blood (oh-**KULT** BLUD) Blood that cannot be seen in the stool but is positive on a fecal occult blood test.

ocular (**OCK**-you-lar) Pertaining to the eye.

olfaction (ol-**FAK**-shun) Sense of smell.

olfactory (ol-**FAK**-toh-ree) Related to the sense of smell.

oligohydramnios (**OL**-ih-goh-high-**DRAM**-nee-os) Too little amniotic fluid.

oliguria (ol-ih-**GYUR**-ee-ah) Scanty production of urine.

omentum (oh-**MEN**-tum) Membrane that encloses the bowels.

onychomycosis (oh-nih-koh-my-**KOH**-sis) Condition of a fungus infection in a nail.

oocyte (**OH**-oh-site) Female egg cell.

oogenesis (oh-oh-**JEN**-eh-sis) Development of a female egg cell.

open fracture (**OH**-pen **FRAK**-chur) The skin over the fracture is broken.

ophthalmia neonatorum (off-**THAL**-me-ah ne-oh-nay-**TOR**-um) Conjunctivitis of the newborn.

ophthalmic (off-**THAL**-mick) Pertaining to the eye.

ophthalmologist (off-thal-**MALL**-oh-jist) Medical specialist in ophthalmology.

ophthalmology (off-thal-**MALL**-oh-jee) Diagnosis and treatment of diseases of the eye.

ophthalmoscope (off-**THAL**-moh-skope) Instrument for viewing the retina.

ophthalmoscopic (**OFF**-thal-**MOS**-koh-pik) Pertaining to the use of an ophthalmoscope.

ophthalmoscopy (**OFF**-thal-**MOS**-koh-pee) The process of viewing the retina.

opiate (**OH**-pee-ate) A drug derived from opium.

opportunistic (**OP**-or-tyu-**NIS**-tik) An organism or a disease in a host with lowered resistance.

opportunistic infection (**OP**-or-tyu-**NIS**-tik in-**FEK**-shun) An infection that causes disease when the immune system is compromised for other reasons.

optic (**OP**-tick) The eye or vision; *or* second (II) cranial nerve, which carries visual information.

optical (**OP**-tih-kal) Pertaining to the eye or vision.

optometrist (op-**TOM**-eh-trist) Someone skilled in the measurement of vision but who cannot treat eye diseases or prescribe medication.

optometry (op-**TOM**-eh-tree) The profession of the measurement of vision.

oral (**OR**-al) Pertaining to the mouth.

orbit (**OR**-bit) The bony socket that holds the eyeball.

orbital (**OR**-bit-al) Pertaining to the orbit.

orchiectomy (or-key-**ECK**-toe-me) Removal of one or both testes.

orchiopexy (**OR**-key-oh-**PEK**-see) Surgical fixation of a testis in the scrotum.

orchitis (or-**KIE**-tis) Inflammation of the testis. Also called *epididymoorchitis.*

organ (**OR**-gan) Structure with specific functions in a body system.

organelle (**OR**-gah-nell) Part of a cell having specialized function(s).

organism (**OR**-gan-izm) Any whole, living individual animal or plant.

orifice (**OR**-ih-fis) Any opening or aperture.

origin (**OR**-ih-gin) Fixed source of a muscle at its attachment to bone.

oropharyngeal (**OR**-oh-fah-**RIN**-jee-al) Pertaining to the oropharynx.

oropharynx (**OR**-oh-fah-rinks) Region at the back of the mouth between the soft palate and the tip of the epiglottis.

orthopedic (or-tho-**PEE**-dik) Pertaining to the correction and cure of deformities and diseases of the musculoskeletal system; originally, most of the deformities treated were in children. Also spelled *orthopaedic.*

orthopedist (or-tho-**PEE**-dist) Specialist in orthopedics.

orthopnea (or-**THOP**-nee-ah) Difficulty in breathing when lying flat.

orthopneic (or-**THOP**-nee-ik) Pertaining to or affected by orthopnea.

orthotic (or-**THOT**-ik) Orthopedic appliance to correct an abnormality.

orthotist (or-**THOT**-ist) Maker and fitter of orthopedic appliances.

os (OSS) Opening into a canal; for example, the cervix.

ossicle (**OSS**-ih-kel) A small bone, particularly relating to the three bones in the middle ear.

osteoarthritis (**OSS**-tee-oh-ar-**THRI**-tis) Chronic inflammatory disease of the joints, with pain and loss of function.

osteoblast (**OSS**-tee-oh-blast) Bone-forming cell.

myocardium (MY-oh-KAR-dee-um) All the heart muscle.

myoma (my-OH-mah) Benign tumor of muscle.

myomectomy (my-oh-MEK-toe-me) Surgical removal of a myoma (fibroid).

myometrium (my-oh-MEE-tree-um) Muscle wall of the uterus.

myopia (my-OH-pee-ah) Able to see close objects but unable to see distant objects.

myringotomy (mir-in-GOT-oh-me) Incision in the tympanic membrane.

myxedema (miks-eh-DEE-muh) Nonpitting, waxy edema of the skin in hypothyroidism.

N

narcolepsy (NAR-koh-lep-see) Involuntary falling asleep.

narcotic (nar-KOT-ik) Drug derived from opium or any drug with effects similar to those of opium derivatives.

naris (NAH-ris) Nostril. Plural *nares.*

nasal (NAY-zal) Pertaining to the nose.

nasogastric (NAY-zoh-GAS-trik) Pertaining to the nose and stomach.

nasolacrimal duct (NAY-zoh-LAK-rim-al DUKT) Passage from the lacrimal sac to the nose.

nasopharyngeal (NAY-zoh-fah-RIN-jee-al) Pertaining to the nasopharynx.

nasopharynx (NAY-zoh-FAH-rinks) Region of the pharynx at the back of the nose and above the soft palate.

natal (NAY-tal) Pertaining to birth.

nebulizer (NEB-you-liz-er) Device used to deliver liquid medicine in a fine mist.

necrosis (neh-KROH-sis) Pathologic death of cells or tissue.

necrotic (neh-KROT-ik) Pertaining to or affected by necrosis.

necrotizing fasciitis (neh-kroh-TIZE-ing fash-ee-EYE-tis) Inflammation of fascia producing death of the tissue.

neonatal (NEE-oh-NAY-tal) Pertaining to the newborn infant or the newborn period.

neonate (NEE-oh-nate) A newborn infant.

nephrectomy (nef-REK-toe-me) Surgical removal of a kidney.

nephritis (nef-RY-tis) Inflammation of the kidney.

nephroblastoma (NEF-roh-blas-TOE-mah) Cancerous kidney tumor of childhood. Also known as *Wilms tumor.*

nephrolithiasis (NEF-roe-lih-THIGH-ah-sis) Presence of a kidney stone.

nephrolithotomy (NEF-roe-lih-THOT-oh-me) Incision for removal of a stone.

nephrologist (nef-ROL-oh-jist) Medical specialist in disorders of the kidney.

nephrology (nef-ROL-oh-jee) Medical specialty of diseases of the kidney.

nephron (NEF-ron) Filtration unit of the kidney; glomerulus + renal tubule.

nephropathy (nef-ROP-ah-thee) Any disease of the kidney.

nephroscope (NEF-roe-skope) Endoscope to view the inside of the kidney.

nephroscopy (nef-ROS-koh-pee) To examine the kidney.

nephrosis (nef-ROH-sis) Same as *nephrotic syndrome.*

nephrotic syndrome (nef-ROT-ik SIN-drome) Glomerular disease with marked loss of protein. Also called *nephrosis.*

nerve (NERV) A cord of nerve fibers bound together by connective tissue.

nervous (NER-vus) Pertaining to a nerve or the nervous system; *or* easily excited or agitated.

nervous system (NER-vus SIS-tem) The whole, integrated nerve apparatus.

neural (NYU-ral) Pertaining to nervous tissue.

neuralgia (nyu-RAL-jee-ah) Pain in the distribution of a nerve.

neuroglia (nyu-roh-GLEE-ah) Connective tissue holding nervous tissue together.

neurohypophysis (NYU-roh-high-POF-ih-sis) Posterior lobe of the pituitary gland.

neurologic (NYU-roh-LOJ-ik) Pertaining to the nervous sytem.

neurologist (nyu-ROL-oh-jist) Medical specialist in disorders of the nervous system.

neurology (nyu-ROL-oh-jee) Medical specialty of disorders of the nervous system.

neuromuscular (NYU-roh-MUSS-kyu-lar) A junction where a nerve supplies muscle tissue.

neuron (NYU-ron) Technical term for a nerve cell; consists of the cell body with its dendrites and axons.

neuropathy (nyu-ROP-ah-thee) Any disorder affecting the nervous system.

neurosurgeon (NYU-roh-SUR-jun) One who operates on the nervous system.

neurosurgery (NYU-roh-SUR-jer-ee) Operating on the nervous system.

neurotoxin (NYU-roh-tock-sin) Agent that poisons the nervous system.

neurotransmitter (NYU-roh-trans-MIT-er) Chemical agent that relays messages from one nerve cell to the next.

neutropenia (NEW-troh-PEE-nee-ah) A deficiency of neutrophils.

neutrophil (NEW-troh-fill) A neutrophil's granules take up (purple) stain equally, whether the stain is acid or alkaline.

neutrophilia (NEW-troh-FILL-ee-ah) An increase in neutrophils.

nevus (NEE-vus) Congenital lesion of the skin. Plural *nevi.*

nipple (NIP-el) Projection from the breast into which the lactiferous ducts open.

nitrite (NI-trite) Chemical formed in urine by *E. coli* and other microorganisms.

nitrogenous (ni-TROJ-en-us) Containing or generating nitrogen.

nocturia (nok-TYU-ree-ah) Excessive urination at night.

node (NOHD) A circumscribed mass of tissue.

nonunion (non-YOU-nee-un) Total failure of healing of a fracture.

norepinephrine (NOR-ep-ih-NEFF-rin) Catecholamine hormone of the adrenal gland that is a parasympathetic neurotransmitter. Also called *noradrenaline.*

nosocomial (noh-soh-KOH-mee-al) Acquired while in the hospital.

metabolism (meh-**TAB**-oh-lizm) The constantly changing physical and chemical processes occurring in the cell that are the sum of anabolism and catabolism.

metacarpal (**MET**-ah-**KAR**-pal) The five bones between the carpus and the fingers.

metacarpophalangeal (**MET**-ah-**KAR**-poh-fay-**LAN**-jee-al) The articulations (joints) between the metacarpal bones and the phalanges.

metastasis (meh-**TAS**-tah-sis) Spread of a disease from one part of the body to another. Plural *metastases.*

metastasize (meh-**TAS**-tah-size) To spread to distant parts.

metastatic (meh-tah-**STAT**-ik) Pertaining to the character of cells that can metastasize.

metatarsal (**MET**-ah-**TAR**-sal) Pertaining to the metatarsus.

metatarsus (**MET**-ah-**TAR**-sus) The five parallel bones of the foot between the tarsus and the phalanges.

metrorrhagia (**MEH**-troh-**RAY**-jee-ah) Irregular uterine bleeding between menses.

microbe (**MY**-krohb) Short for *microorganism.*

microcephalic (**MY**-kroh-**SEF**-ah-lik) Pertaining to or suffering from a small head.

microcephaly (**MY**-kroh-**SEF**-ah-lee) An abnormally small head.

microcyte (**MY**-kroh-site) Small red blood cell.

microcytic (my-kroh-**SIT**-ik) Pertaining to a small cell.

microorganism (**MY**-kroh-**OR**-gan-izm) Any organism too small to be seen by the naked eye.

microscope (**MY**-kroh-skope) Instrument for viewing something small that cannot be seen in detail by the naked eye.

microscopic (**MY**-kroh-**SKOP**-ik) Visible only with the aid of a microscope.

micturate (**MIK**-choo-rate) Pass urine.

micturition (mik-choo-**RISH**-un) Act of passing urine.

migraine (**MY**-grain) Paroxysmal severe headache confined to one side of the head.

mineral (**MIN**-er-al) Inorganic compound usually found in the earth's crust.

mineralocorticoid (**MIN**-er-al-oh-**KOR**-tih-koyd) Hormone of the adrenal cortex that influences sodium and potassium metabolism.

minimus (**MIN**-ih-mus) The gluteus minimus is the smallest of the gluteal muscles and lies under the gluteus medius.

minus (**MY**-nus) Smaller or lesser; for example, labium minus. Plural *minora.*

mitochondrion (my-toe-**KON**-dree-on) Organelle that generates, stores, and releases energy for cell activities. Plural *mitochondria.*

mitosis (my-**TOE**-sis) Cell division that creates two identical cells, each with 46 chromosomes.

mitral (**MY**-tral) Shaped like the headdress of a Catholic bishop.

modify (**MOD**-ih-fie) Change the form or qualities of something.

molar (**MO**-lar) One of six teeth in each jaw that grind food.

mole (MOLE) Benign localized area of melanin-producing cells.

molecule (**MOLL**-eh-kyul) Very small particle.

molluscum contagiosum (moh-**LUS**-kum kon-**TAY**-jee-oh-sum) STD caused by a virus.

monocyte (**MON**-oh-site) Large white blood cell with a single nucleus.

mononeuropathy (**MON**-oh-nyu-**ROP**-ah-thee) Disorder affecting a single nerve.

mononucleosis (**MON**-oh-nyu-klee-**OH**-sis) Presence of large numbers of specific, diagnostic mononuclear leukocytes.

monoplegia (**MON**-oh-**PLEE**-jee-ah) Paralysis of one limb.

monoplegic (**MON**-oh-**PLEE**-jik) Pertaining to or suffering from monoplegia.

mons pubis (MONZ **PYU**-bis) Fleshy pad with pubic hair, overlying the pubic bone.

morbidity (mor-**BID**-ih-tee) The frequency of the appearance of a disease.

morphine (**MOR**-feen) Derivative of opium used as an analgesic or sedative.

mortality (mor-**TAL**-ih-tee) Death rate.

motile (**MOH**-til) Capable of spontaneous movement.

motility (moh-**TILL**-ih-tee) The ability for spontaneous movement.

motor (**MOH**-tor) Structures of the nervous system that send impulses out to cause muscles to contract or glands to secrete.

mouth (MOWTH) External opening of a cavity or canal.

mucin (**MYU**-sin) Protein element of mucus.

mucocutaneous (**MYU**-koh-kyu-**TAY**-nee-us) Junction of skin and mucous membrane; for example, the lips.

mucolytic (**MYU**-koh-**LIT**-ik) Agent capable of dissolving or liquefying mucus.

mucosa (myu-**KOH**-sah) Lining of a tubular structure that secretes mucus. Another name for *mucous membrane.*

mucous (**MYU**-kus) Pertaining to mucus or the mucosa.

mucus (**MYU**-kus) Sticky secretion of cells in mucous membranes.

multipara (mul-**TIP**-ah-ruh) Woman who has given birth to two or more children.

murmur (**MUR**-mur) Abnormal heart sound heard with a stethoscope when a valve closes or opens abnormally.

muscle (**MUSS**-el) A tissue consisting of contractile cells.

musculoskeletal (**MUSS**-kyu-loh-**SKEL**-eh-tal) Pertaining to the muscles and the bony skeleton.

mutation (myu-**TAY**-shun) Change in the chemistry of a gene.

mute (MYUT) Unable or unwilling to speak.

mutism (**MYU**-tizm) Absence of speech.

myasthenia gravis (my-as-**THEE**-nee-ah **GRA**-vis) Disorder of fluctuating muscle weakness.

myelin (**MY**-eh-lin) Material of the sheath around the axon of a nerve.

myelitis (**MY**-eh-**LIE**-tis) Inflammation of the spinal cord.

myelocele (**MY**-eh-low-seal) Protrusion of the spinal cord through a defect in the vertebral arch.

myelomeningocele (**MY**-eh-low-meh-**NING**-oh-seal) Protrusion of the spinal cord and meninges through a defect in the vertebral arch of one or more vertebrae.

myocardial (my-oh-**KAR**-dee-al) Pertaining to heart muscle.

myocarditis (**MY**-oh-kar-**DIE**-tis) Inflammation of the heart muscle.

portal vein (**POR**-tal) The vein that carries blood from the intestines to the liver.

postcoital (post-**KOH**-ih-tal) After sexual intercourse.

posterior (pos-**TEER**-ee-or) Pertaining to the back surface of the body; situated behind.

postictal (post-**IK**-tal) Transient neurologic deficit after a seizure.

postmature (post-mah-**TYUR**) Infant born after 42 weeks of gestation.

postmaturity (post-mah-**TYUR**-ih-tee) Condition of being postmature.

postmortem (post-**MOR**-tem) Examination of the body and organs of a dead person to determine the cause of death.

postnatal (post-**NAY**-tal) After the birth.

postpartum (post-**PAR**-tum) After childbirth.

postpolio syndrome (PPS) (post-**POE**-lee-oh **SIN**-drome) Progressive muscle weakness in a person previously affected by polio.

postprandial (post-**PRAN**-dee-al) Following a meal.

posttraumatic (post-traw-**MAT**-ik) Occurring after and caused by trauma.

Pott fracture (POT **FRAK**-shur) Fracture of the lower end of the fibula, often with fracture of the tibial malleolus.

precancerous (pree-**KAN**-ser-us) Lesion from which cancer can develop.

precipitate labor (pree-**SIP**-ih-tate **LAY**-bore) A very rapid labor and delivery.

precursor (pree-**KUR**-sir) Cell or substance formed earlier in the development of the cell or substance.

preeclampsia (pree-eh-**KLAMP**-see-uh) Hypertension, edema, and proteinuria during pregnancy.

preemie (**PREE**-me) Slang for *premature baby.*

pregnancy (**PREG**-nan-see) State of being pregnant.

pregnant (**PREG**-nant) Having conceived.

prehypertension (pree-**HIGH**-per-**TEN**-shun) Precursor to hypertension.

premature (pree-mah-**TYUR**) Occurring before the expected time; for example, an infant born before 37 weeks of gestation.

prematurity (pree-mah-**TYUR**-ih-tee) Condition of being premature.

premenstrual (pree-**MEN**-stru-al) Pertaining to the time immediately before the menses.

prenatal (pree-**NAY**-tal) Before birth.

prepatellar (pree-pah-**TELL**-ar) In front of the patella.

prepuce (**PREE**-puce) Fold of skin that covers the glans penis. Same as *foreskin.*

presbyopia (prez-bee-**OH**-pee-ah) Difficulty in nearsighted vision occurring in middle and old age.

preterm (**PREE**-term) Baby delivered before 37 weeks of gestation. Also called *premature.*

previa (**PREE**-vee-ah) Anything blocking the fetus during its birth; for example, an abnomally situated placenta, *placenta previa.*

priapism (**PRY**-ah-pizm) Persistent erection of the penis.

primary (**PRY**-mah-ree) The first of a disease or symptom, after which others may occur as complications arise.

primigravida (pry-mih-**GRAV**-ih-dah) First pregnancy.

primipara (pry-**MIP**-ah-ruh) Woman giving birth for the first time.

prion (**PREE**-on) Small infectious protein particle.

proctitis (prok-**TIE**-tis) Inflammation of the lining of the rectum.

proctoscopy (prok-**TOSS**-koh-pee) Examination of the inside of the anus by endoscopy.

progesterone (pro-**JESS**-ter-own) Hormone that prepares the uterus for pregnancy.

progestin (pro-**JESS**-tin) A synthetic form of progesterone.

prognathism (**PROG**-nah-thizm) Condition of a forward-projecting jaw.

prognosis (prog-**NO**-sis) Forecast of the probable future course and outcome of a disease.

prolactin (pro-**LAK**-tin) Pituitary hormone that stimulates the production of milk.

prolapse (pro-**LAPS**) A sinking down of an organ or tissue.

proliferate (pro-**LIF**-eh-rate) To increase in number through reproduction.

pronate (**PRO**-nate) Rotate the forearm so that the surface of the palm faces posteriorly in the anatomical position.

pronation (pro-**NAY**-shun) Process of lying face down or of turning a hand or foot with the volar (palm or sole) surface down.

prone (PRONE) Lying face down, flat on your belly.

prophylactic (pro-fih-**LAK**-tik) The act or the agent that prevents a disease.

prophylaxis (pro-fih-**LAX**-is) Prevention of disease.

prostaglandin (**PROS**-tah-**GLAN**-din) Hormone present in many tissues but first isolated from the prostate gland.

prostate (**PROS**-tate) Organ surrounding the urethra at the base of the male urinary bladder.

prostatectomy (pros-tah-**TEK**-toe-me) Surgical removal of the prostate.

prostatic (pros-**TAT**-ik) Pertaining to the prostate.

prostatitis (pros-tah-**TIE**-tis) Inflammation of the prostate.

prosthesis (**PROS**-thee-sis) Manufactured substitute for a missing part of the body.

prosthetic (pros-**THET**-ik) Pertaining to a prosthesis.

prostrate (pros-**TRAYT**) To lay flat or to be overcome by physical weakness and exhaustion.

prostration (pros-**TRAY**-shun) To be lying flat or to be overcome by physical weakness and exhaustion.

protein (**PRO**-teen) Class of food substances based on amino acids.

proteinuria (pro-tee-**NYU**-ree-ah) Presence of protein in urine.

prothrombin (pro-**THROM**-bin) Protein formed by the liver and converted to thrombin in the blood-clotting mechanism.

provisional diagnosis (pro-**VISH**-un-al die-ag-**NO**-sis) A temporary diagnosis pending further examination or testing.

proximal (**PROK**-sih-mal) Situated nearest the center of the body.

pruritic (proo-**RIT**-ik) Itchy.

pruritus (proo-**RYE**-tus) Itching.

psoriasis (so-**RYE**-ah-sis) Rash characterized by reddish, silver-scaled patches.

psychiatric (sigh-kee-**AH**-trik) Pertaining to psychiatry.

psychiatrist (sigh-**KIGH**-ah-trist) Licensed medical specialist in psychiatry.

psychiatry (sigh-**KIGH**-ah-tree) Diagnosis and treatment of mental disorders.

psychologic (sigh-koh-**LOJ**-ik) Pertaining to psychology.

psychological (sigh-koh-**LOJ**-ik-al) Pertaining to psychology.

psychologist (sigh-**KOL**-oh-jist) One who studies and becomes a specialist in psychology.

psychology (sigh-**KOL**-oh-jee) Study of the behavior of the human mind.

psychopath (**SIGH**-koh-path) Person with antisocial personality disorder.

psychosis (sigh-**KOH**-sis) Disorder causing mental disruption and loss of contact with reality.

psychosocial (**SIGH**-koh-**SOH**-shal) Involving both the mind and various social and community aspects of life.

psychosomatic (**SIGH**-koh-soh-**MAT**-ik) Disorders of the body influenced by the mind.

psychotic (sigh-**KOT**-ik) Pertaining to or affected by psychosis.

ptosis (**TOE**-sis) Sinking down of the upper eyelid or an organ.

pubarche (pyu-**BAR**-key) Development of pubic and axillary hair.

puberty (**PYU**-ber-tee) Process of maturing from child to young adult.

pubic (**PYU**-bik) Pertaining to the pubis.

pubis (**PYU**-bis) Alternative name for *pubic bone.*

puerperium (pyu-er-**PER**-ee-um) Six-week period after birth in which the uterus involutes.

pulmonary (**PULL**-moh-**NAR**-ee) Pertaining to the lungs.

pulmonologist (**PULL**-moh-**NOL**-oh-jist) Specialist in treating disorders of the lungs.

pulmonology (**PULL**-moh-**NOL**-oh-jee) Study of the lungs, or the medical specialty of disorders of the lungs.

pulp (PULP) Dental pulp is the connective tissue in the cavity in the center of the tooth.

pupil (**PYU**-pill) The opening in the center of the iris that allows light to reach the lens. Plural *pupillae.*

pupillary (PYU-pill-ah-ree) Pertaining to the pupil.

purge (PURJ) Consciously throw up or cause bowel evacuation.

Purkinje fibers (per-**KIN**-jee fi-**BERS**) Network of nerve fibers in the myocardium.

purpura (**PUR**-pyu-rah) Skin hemorrhages that are red initially and then turn purple.

purulent (**PURE**-you-lent) Showing or containing a lot of pus.

pustule (**PUS**-tyul) Small protuberance on the skin that contains pus.

pyelitis (pie-eh-**LYE**-tis) Inflammation of the renal pelvis.

pyelogram (**PIE**-el-oh-gram) X-ray image of renal pelvis and ureters.

pyelonephritis (**PIE**-eh-loh-neh-**FRY**-tis) Inflammation of the kidney and renal pelvis.

pyloric (pie-**LOR**-ik) Pertaining to the pylorus.

pylorus (pie-**LOR**-us) Exit area of the stomach.

pyorrhea (pie-oh-**REE**-ah) Purulent discharge.

pyrexia (pie-**REK**-see-ah) An abnormally high body temperature or fever.

Q

quadrant (**KWAD**-rant) One-quarter of a circle; *or* one of four regions of the surface of the abdomen.

quadrantectomy (kwad-ran-**TEK**-toe-me) Surgical excision of a quadrant of the breast.

quadriceps femoris (**KWAD**-rih-seps **FEM**-or-is) An anterior thigh muscle with four heads.

quadriplegia (kwad-rih-**PLEE**-jee-ah) Paralysis of all four limbs.

quadriplegic (kwad-rih-**PLEE**-jik) Pertaining to or suffering from quadriplegia.

R

radial (**RAY**-dee-al) Pertaining to the forearm or to any of the structures (artery, vein, nerve) named after it; *or* diverging in all directions from any given center.

radiation (ray-dee-**AY**-shun) To spread out.

radiologic (**RAY**-dee-oh-**LOJ**-ik) Pertaining to radiology.

radiologist (ray-dee-**OL**-oh-jist) Medical specialist in the use of x-rays and other imaging techniques.

radiology (ray-dee-**OL**-oh-jee) Study of medical imaging.

radius (**RAY**-dee-us) The forearm bone on the thumb side.

rale (RAHL) Crackle heard through a stethoscope when air bubbles through liquid in the lungs. Plural *rales.*

rash (RASH) Skin eruption.

rectocele (**RECK**-toe-seal) Hernia of the rectum into the vagina.

rectum (**RECK**-tum) Terminal part of the colon from the sigmoid to the anal canal.

reduction (ree-**DUCK**-shun) Restore a structure to its normal position.

reflex (**REE**-fleks) An involuntary response to a stimulus.

reflux (**REE**-fluks) Backward flow.

refract (ree-**FRACT**) Make a change in the direction of, or bend, a ray of light.

refraction (ree-**FRAK**-shun) The bending of light

regenerate (ree-**JEN**-eh-rate) Reconstitution of a lost part.

regeneration (ree-**JEN**-eh-**RAY**-shun) The process of reconstitution.

regulate (reg-you-**LATE**) To control the way in which a process progresses.

regulation (reg-you-**LAY**-shun) Control of the way in which a process progresses.

regurgitate (ree-**GUR**-jih-tate) To flow backward; for example, blood through a heart valve.

regurgitation (ree-gur-jih-**TAY**-shun) Expel contents of the stomach into the mouth, short of vomiting.

rehabilitation (**REE**-hah-bill-ih-**TAY**-shun) Therapeutic restoration of an ability to function as before.

remission (ree-**MISH**-un) Period when there is a lessening or absence of the symptoms of a disease.

renal (**REE**-nal) Pertaining to the kidney.

replication (rep-lih-**KAY**-shun) Reproduction to produce an exact copy.

reproduction (ree-pro-**DUK**-shun) The process by which organisms produce offspring.

reproductive (ree-pro-**DUK**-tiv) Pertaining to reproduction.

resection (ree-**SEK**-shun) Removal of a specific part of an organ or structure.

resectoscope (ree-**SEK**-toe-skope) Endoscope for transurethral removal of lesions of the prostate.

residual (reh-**ZID**-you-al) Pertaining to anything left over.

resistance (reh-**ZIS**-tants) Ability of an organism to withstand the effects of an antagonistic agent.

resistant (reh-**ZIS**-tant) Able to resist.

respiration (**RES**-pih-**RAY**-shun) Process of breathing; fundamental process of life used to exchange oxygen and carbon dioxide.

respirator (**RES**-pih-**RAY**-tor) Another name for *ventilator.*

respiratory (**RES**-pih-rah-tor-ee) Pertaining to respiration.

retention (ree-**TEN**-shun) Holding back in the body what should normally be discharged (e.g., urine).

retina (**RET**-ih-nah) Light-sensitive innermost layer of the eyeball.

retinal (**RET**-ih-nal) Pertaining to the retina.

retinoblastoma (**RET**-in-oh-blas-**TOE**-mah) Malignant neoplasm of primitive retinal cells.

retinopathy (ret-ih-**NOP**-ah-thee) Any disease of the retina.

retrograde (**RET**-roh-grade) Reversal of a normal flow; for example, back from the bladder into the ureters.

retroversion (reh-troh-**VER**-shun) Tipping backward of the uterus.

retroverted (**REH**-troh-vert-ed) Tilted backward.

retrovirus (**REH**-troh-vie-rus) Virus with an RNA core.

rheumatic (ru-**MAT**-ik) Pertaining to or affected by rheumatism.

rheumatism (**RU**-mat-izm) Pain in various parts of the musculoskeletal system.

rheumatoid arthritis (RA) (**RU**-mah-toyd ar-**THRI**-tis) Disease of connective tissue, with arthritis as a major manifestation.

rhinitis (rye-**NIE**-tis) Inflammation of the nasal mucosa. Also called *coryza.*

rhinoplasty (**RYE**-no-plas-tee) Surgical procedure to change size or shape of the nose.

rhoncus (**RONG**-kuss) Wheezing sound heard on auscultation of the lungs; made by air passing through a constricted lumen. Plural *rhonci.*

ribonucleic acid (RNA) (**RYE**-boh-nyu-**KLEE**-ik **ASS**-id) Information carrier from DNA in the nucleus to an organelle to produce protein molecules.

ribosome (**RYE**-boh-sohm) Structure in the cell that assembles amino acids into protein.

rickets (**RICK**-ets) Disease due to vitamin D deficiency, producing soft, flexible bones.

rigidity (ri-**JID**-ih-tee) Increased muscle tone at rest.

Rinne test (**RIN**-eh TEST) Test for conductive hearing loss.

root (ROOT) Fundamental or beginning part of a structure.

rooting (rue-**TING**) A neonatal reflex to turn toward the nipple and open the mouth when a nipple is placed on the cheek.

rosacea (roh-**ZAY**-she-ah) Persistent erythematous rash of the central face.

rotator cuff (roh-**TAY**-tor CUFF) Part of the capsule of the shoulder joint.

rumination (**ROO**-min-ay-shun) To bring back food into the mouth to chew over and over.

rupture (**RUP**-tyur) Break or tear of any organ or body part.

S

sacral (**SAY**-kral) Pertaining to or in the neighborhood of the sacrum.

sacroiliac joint (say-kroh-**ILL**-ih-ak JOINT) The joint between the sacrum and the ilium.

sacrum (**SAY**-crum) Segment of the vertebral column that forms part of the pelvis.

sagittal (**SAJ**-ih-tal) Vertical plane through the body dividing it into right and left portions.

saline (**SAY**-leen) Salt solution, usually sodium chloride.

saliva (sa-**LIE**-vah) Secretion in the mouth from salivary glands.

salivary (**SAL**-ih-var-ee) Pertaining to saliva.

salpingectomy (sal-pin-**JEK**-toe-me) Surgical excision of a fallopian tube.

salpingitis (sal-pin-**JIE**-tis) Inflammation of the uterine tube.

saphenous (**SAPH**-ih-nus) Relating to the saphenous vein in the thigh.

sarcoidosis (sar-koy-**DOH**-sis) Granulomatous lesions of the lungs and other organs; cause is unknown.

sarcoma (sar-**KOH**-mah) A malignant tumor originating in connective tissue.

sarcopenia (sar-koh-**PEE**-nee-ah) Progressive loss of muscle mass and strength with aging.

saturated fatty acid (satch-you-**RAY**-ted **FAT**-ee **ASS**-id) Lipid that is incapable of absorbing any more hydrogen.

scab (SKAB) Crust that forms over a wound or sore during healing.

scabies (**SKAY**-bees) Skin disease produced by mites.

scald (SKAWLD) Burn from contact with hot liquid or steam.

scapula (**SKAP**-you-lah) Shoulder blade. Plural *scapulae.*

scapular (**SKAP**-you-lar) Pertaining to the scapula.

scar (SKAR) Fibrotic seam that forms when a wound heals.

schizophrenia (skitz-oh-**FREE**-nee-ah) Disorder of perception, thought, emotion, and behavior.

Schwann cell (SHWANN SELL) Connective tissue cell of the peripheral nervous system that forms a myelin sheath.

sclera (**SKLAIR**-ah) Fibrous outer covering of the eyeball and the white of the eye.

scleral (**SKLAIR**-al) Pertaining to the sclera.

scleritis (sklair-**RI**-tis) Inflammation of the sclera.

scleroderma (sklair-oh-**DERM**-ah) Thickening and hardening of the skin due to new collagen formation.

sclerose (skleh-**ROSE**) To harden or thicken.

sclerosis (skleh-**ROH**-sis) Thickening or hardening of a tissue; in the nervous system, hardening of nervous tissue by fibrous and glial connective tissue.

sclerotherapy (**SKLAIR**-oh-**THAIR**-ah-pee) Injection of a solution into a vein to thrombose it.

scoliosis (skoh-lee-**OH**-sis) An abnormal lateral curvature of the vertebral column.

scoliotic (**SKOH**-lee-**OT**-ik) Pertaining to or suffering from scoliosis.

scrotal (**SKRO**-tal) Pertaining to the scrotum.

scrotum (**SKRO**-tum) Sac containing testes.

seasonal affective disorder (see-**ZON**-al af-**FEK**-tiv dis-**OR**-der) Depression that occurs at the same time every year, often in winter.

sebaceous glands (seh-**BAY**-shus GLANZ) Glands in the dermis that open into hair follicles and secrete an oily fluid called *sebum*.

seborrhea (seb-oh-**REE**-ah) Excessive amount of sebum.

seborrheic (seb-oh-**REE**-ik) Pertaining to seborrhea.

sebum (**SEE**-bum) Waxy secretion of the sebaceous glands.

secondary (**SEK**-ond-ah-ree) Diseases or symptoms following a primary disease or symptom.

secrete (seh-**KREET**) To produce a chemical substance in a cell and release it from the cell.

secretion (seh-**KREE**-shun) The production of a chemical substance in a cell and its release from the cell.

sedation (seh-**DAY**-shun) State of being calmed.

sedative (**SED**-ah-tiv) Agent that calms nervous excitement.

segment (**SEG**-ment) A section of an organ or structure.

segmentectomy (seg-men-**TEK**-toe-me) Surgical excision of a segment of a tissue or organ.

seizure (**SEE**-zhur) Event due to excessive electrical activity in the brain.

self-examination (**SELF**-ek-zam-ih-**NAY**-shun) Conduct an examination of one's own body.

self-mutilation (self-myu-tih-**LAY**-shun) Injury or disfigurement made to one's own body.

semen (**SEE**-men) Penile ejaculate containing sperm and seminal fluid.

semilunar (sem-ee-**LOO**-nar) Appears like a half moon.

seminal vesicle (**SEM**-in-al **VES**-ih-kull) Sac of the ductus deferens that produces seminal fluid.

seminiferous (sem-ih-**NIF**-er-us) Pertaining to carrying semen.

seminiferous tubule (sem-ih-**NIF**-er-us **TU**-byul) Coiled tubes in the testes that produce sperm.

seminoma (sem-ih-**NO**-mah) Neoplasm of germ cells of a testis.

senescence (seh-**NES**-ens) The state of being old.

senescent (seh-**NES**-ent) Growing old.

senile (**SEE**-nile) Characteristic of old age.

senility (seh-**NIL**-ih-tee) Mental disorders occurring in old age.

sensation (sen-**SAY**-shun) The conscious feeling of the effects of a stimulation.

sensorineural hearing loss (**SEN**-sor-ih-**NYUR**-al) Hearing loss caused by lesions of the inner ear or the auditory nerve.

sensory (**SEN**-soh-ree) Having the function of sensation; structures of the nervous system that carry impulses to the brain.

sepsis (**SEP**-sis) Presence of pathogenic organisms or their toxins in blood or tissues.

septicemia (sep-tih-**SEE**-mee-ah) Microorganisms circulating in, and infecting, the blood (blood poisoning).

septum (**SEP**-tum) A thin wall separating two cavities or tissue masses. Plural *septa*.

serotonin (ser-oh-**TOE**-nin) A neurotransmitter in the central and peripheral nervous systems.

serum (**SEER**-um) Fluid remaining after removal of cells and fibrin clot from blood.

shock (SHOCK) Sudden physical or mental collapse or circulatory collapse.

shunt (SHUNT) A bypass or diversion of fluid; for example, blood.

sigmoid (**SIG**-moyd) Sigmoid colon is shaped like an "S."

sigmoidoscopy (sig-moi-**DOS**-koh-pee) Endoscopic examination of the sigmoid colon.

sign (SINE) Physical evidence of a disease process.

silicosis (sil-ih-**KOH**-sis) Fibrotic lung disease from inhaling silica particles.

sinoatrial (SA) node (sigh-noh-**AY**-tree-al NODE) The center of modified cardiac muscle fibers in the wall of the right atrium that acts as the pacemaker for the heart rhythm.

sinus (**SIGH**-nus) Cavity or hollow space in a bone or other tissue.

sinus rhythm (**SIGH**-nus **RITH**-um) The normal (optimal) heart rhythm arising from the sinoatrial node.

sinusitis (sigh-nyu-**SIGH**-tis) Inflammation of the lining of a sinus.

skeletal (**SKEL**-eh-tal) Pertaining to the skeleton.

skeleton (**SKEL**-eh-ton) The bony framework of the body.

Skene glands (SKEEN GLANZ) Paraurethral glands in the anterior wall of the vagina. Also called *paraurethral glands*.

smegma (**SMEG**-mah) Oily material produced by the glans and prepuce.

Snellen letter chart (**SNEL**-en) Test for acuity of distant vision.

snore (SNOR) Noise produced by vibrations in the structures of the nasopharynx.

sociopath (**SOH**-see-oh-path) Person with antisocial personality disorder.

somatic (soh-**MAT**-ik) Relating to the body in general; *or* a division of the periperal nervous system serving the skeletal muscles.

somatostatin (**SOH**-mah-toh-**STAT**-in) Hormone that inhibits release of growth hormone and insulin.

somatotropin (**SOH**-mah-toh-**TROH**-pin) Hormone of the anterior pituitary that stimulates the growth of body tissues. Also called *growth hormone*.

spasm (SPASM) Sudden involuntary contraction of a muscle group.

spasmodic (spaz-**MOD**-ik) Intermittent contractions.

spastic (**SPAZ**-tik) Increased muscle tone on movement.

specific (speh-**SIF**-ik) Relating to a particular entity.

specificity (spes-ih-**FIS**-ih-tee) State of having a fixed relation to a particular entity.

sperm (SPERM) Mature male sex cell. Also called *spermatozoon.*

spermatic (**SPER**-mat-ik) Pertaining to sperm.

spermatid (**SPER**-mat-id) A cell late in the development process of sperm.

spermatocele (**SPER**-mat-oh-seal) Cyst of the epididymis that contains sperm.

spermatogenesis (**SPER**-mat-oh-**JEN**-eh-sis) The process by which male germ cells differentiate into sperm.

spermatozoa (**SPER**-mat-oh-**ZOH**-ah) Sperm (plural of *spermatozoon*).

spermicidal (sper-mih-**SIGH**-dal) Pertaining to the killing of sperm; *or* destructive to sperm.

spermicide (**SPER**-mih-side) Agent that destroys sperm.

sphenoid (**SFEE**-noyd) Wedge-shaped bone at the base of the skull.

sphincter (**SFINK**-ter) Band of muscle that encircles an opening; when it contracts, the opening squeezes closed.

sphygmomanometer (**SFIG**-moh-mah-**NOM**-ih-ter) Instrument for measuring arterial blood pressure.

spina bifida (**SPY**-nah **BIH**-fih-dah) Failure of one or more vertebral arches to close during fetal development.

spina bifida cystica (**SIS**-tik-ah) Meninges and spinal cord protruding through the absent vertebral arch and having the appearance of a cyst.

spina bifida occulta (**OH**-kul-tah) The deformity of the vertebral arch is not apparent from the surface.

spinal (**SPY**-nal) Pertaining to the spine.

spinal tap (**SPY**-nal TAP) Placement of a needle through an intervertebral space into the subarachnoid space to withdraw CSF.

spine (SPINE) The vertebral column; *or* a short bony projection.

spiral fracture (**SPY**-ral **FRAK**-chur) A fracture in the shape of a coil.

spirochete (**SPY**-roh-keet) Spiral-shaped bacterium causing a sexually transmitted disease (syphilis).

spirometer (spy-**ROM**-eh-ter) An instrument used to measure respiratory volumes.

spirometry (spy-**ROM**-eh-tree) Use of a spirometer.

spleen (SPLEEN) Vascular, lymphatic organ in the left upper quadrant of the abdomen.

splenectomy (sple-**NECK**-toe-me) Surgical removal of the spleen.

splenomegaly (sple-noh-**MEG**-ah-lee) Enlarged spleen.

spondylosis (spon-dih-**LOH**-sis) Degenerative osteoarthritis of the spine.

spongiosum (spun-jee-**OH**-sum) Spongelike tissue.

sprain (SPRAIN) A wrench or tear in a ligament.

sputum (**SPYU**-tum) Matter coughed up and spat out by individuals with respiratory disorders.

squamous cell (**SKWAY**-mus SELL) Flat, scalelike epithelial cell.

stage (STAYJ) Definition of the extent and dissemination of a malignant neoplasm.

staging (**STAY**-jing) Process of determination of the extent of the distribution of a neoplasm.

stapes (**STAY**-peas) Inner (medial) one of the three ossicles of the middle ear; shaped like a stirrup.

Staphylococcus (**STAF**-ih-loh-**KOK**-us) Genus of gram-positive bacteria that divide in more than one plane to form clusters. Plural *staphylococci.*

starch (STARCH) Complex carbohydrate made of multiple units of glucose attached together.

stasis (**STAY**-sis) Stagnation in the flow of any body fluid.

status (**STAT**-us) A state or condition.

status epilepticus (**STAT**-us ep-ih-**LEP**-tik-us) Latin phrase for being in a prolonged or recurrent seizure for longer than a specific time frame.

stem cell (STEM SELL) Undifferentiated cell found in a differentiated tissue that can divide to yield the specialized cells in that tissue.

stenosis (steh-**NOH**-sis) Narrowing of a canal or passage.

stent (STENT) Wire-mesh tube used to keep arteries open.

stereopsis (ster-ee-**OP**-sis) Three-dimensional vision.

stereotactic (**STER**-ee-oh-**TAK**-tic) A precise three-dimensional method to locate a lesion.

stereotype (**STER**-ee-oh-tipe) An image held in common by members of a group.

sterile (**STER**-isle) Free from all living organisms and their spores; *or* unable to fertilize or reproduce.

sterility (ster-**RIL**-ih-tee) Inability to reproduce.

sterilization (**STER**-ih-lih-**ZAY**-shun) Process of making sterile.

sterilize (**STER**-ih-lize) To make sterile.

sternum (**STIR**-num) Long, flat bone forming the center of the anterior wall of the chest.

steroid (**STER**-oyd) Large family of chemical substances found in many drugs, hormones, and body components.

stethoscope (**STETH**-oh-skope) Instrument for listening to cardiac and respiratory sounds.

stimulant (**STIM**-you-lant) Agent that excites or strengthens.

stimulation (stim-you-**LAY**-shun) Arousal to increased functional activity.

stimulus (**STIM**-you-lus) Something that excites or strengthens the functional activity of an organ or part. Plural *stimuli.*

stoma (**STOW**-mah) Artificial opening.

strabismus (strah-**BIZ**-mus) Turning of an eye away from its normal position.

strain (STRAIN) Overstretch or tear in a muscle or tendon.

stratum basale (**STRAH**-tum bay-**SAL**-eh) Deepest layer of the epidermis, from which the other cells originate and migrate.

Streptococcus (strep-toe-**KOK**-us) Genus of gram-positive bacteria that grow in chains. Plural *streptococci.*

striated muscle (**STRI**-ay-ted **MUSS**-el) Another term for *skeletal muscle.*

stricture (**STRICK**-shur) Narrowing of a tube.

stridor (**STRY**-door) High-pitched noise made when there is a respiratory obstruction in the larynx or trachea.

stroke (STROHK) Acute clinical event caused by impaired cerebral circulation.

subarachnoid space (sub-ah-**RACK**-noyd SPASE) Space between the pia mater and the arachnoid membrane.

subclavian (sub-**CLAY**-vee-an) Underneath the clavicle.

subcutaneous (sub-kew-**TAY**-nee-us) Below the skin. Also called *hypodermic.*

subdural space (sub-**DYU**-ral SPACE) Space between the arachnoid and dura mater layers of the meninges.

sublingual (sub-**LING**-wal) Underneath the tongue.

subluxation (sub-luck-**SAY**-shun) An incomplete dislocation when some contact between the joint surfaces remains.

submandibular (sub-man-**DIB**-you-lar) Underneath the mandible.

submucosa (sub-mew-**KOH**-sa) Tissue layer underneath the mucosa.

substernal (sub-**STER**-nal) Under (behind) the sternum or breastbone.

suction (**SUK**-shun) Use of a catheter to clear the upper airway or other tubes.

sulcus (**SUL**-cuss) Groove on the surface of the cerebral hemispheres that separates gyri. Plural *sulci.*

superficial (soo-per-**FISH**-al) Situated near the surface.

superior (soo-**PEE**-ree-or) Situated above.

supinate (**SOO**-pih-nate) Rotate the forearm so that the surface of the palm faces anteriorly in the anatomical position.

supination (soo-pih-**NAY**-shun) Process of lying face upward or of turning a hand or foot so that the palm or sole is facing up.

supine (soo-**PINE**) Lying face up, flat on your spine.

suprapubic (**SOO**-prah-pyu-bik) Above the symphysis pubis.

surfactant (sir-**FAK**-tant) A protein and fat compound that creates surface tension to hold the lung alveolar walls apart.

suture (**SOO**-chur) Two bones are joined together by a fibrous band continuous with their periosteum, as in the skull; *or* a stitch to hold the edges of a wound together; *or* to stitch the edges of a wound together. Plural *sutures.*

sympathetic (sim-pah-**THET**-ik) Division of the autonomic nervous system operating at an unconscious level.

sympathy (**SIM**-pah-thee) Appreciation and concern for another person's mental and emotional state.

symphysis (**SIM**-feh-sis) Two bones joined by fibrocartilage. Plural *symphyses.*

symptom (**SIMP**-tum) Departure from the normal experienced by a patient.

symptomatic (simp-toe-**MAT**-ik) Pertaining to the symptoms of a disease.

synapse (**SIN**-aps) Junction between two nerve cells or a nerve fiber and its target cell, where electrical impulses are transmitted between the cells.

syncope (**SIN**-koh-peh) Temporary loss of consciousness and posture due to diminished cerebral blood flow.

syndesmosis (sin-dez-**MOH**-sis) Binding together of two bones with ligaments. Plural *syndesmoses.*

syndrome (**SIN**-drohm) Combination of signs and symptoms associated with a particular disease process.

synovial (sin-**OH**-vee-al) Pertaining to synovial fluid and the synovial membrane.

synthesis (**SIN**-the-sis) The process of building a compound from different elements.

synthetic (sin-**THET**-ik) Built up or put together from simpler compounds.

syphilis (**SIF**-ih-lis) Sexually transmitted disease caused by a spirochete.

syringomyelia (sih-**RING**-oh-my-**EE**-lee-ah) Abnormal longitudinal cavities in the spinal cord cause paresthesias and muscle weakness.

systemic (sis-**TEM**-ik) Relating to the entire organism.

systemic lupus erythematosus (sis-**TEM**-ik **LOO**-pus er-ih-**THEE**-mah-toe-sus) Inflammatory connective tissue disease affecting the whole body.

systole (**SIS**-toe-lee) Contraction of the heart muscle.

systolic (sis-**TOL**-ik) Pertaining to systole.

T

tachycardia (tack-ih-**KAR**-dee-ah) Rapid heart rate (above 100 beats per minute).

tachypnea (tack-ip-**NEE**-ah) Rapid breathing.

talipes (**TAL**-ip-eze) Deformity of the foot involving the talus.

talus (**TAY**-luss) The tarsal bone that articulates with the tibia to form the ankle joint.

tampon (**TAM**-pon) Plug or pack in a cavity to absorb or stop bleeding.

tamponade (tam-poh-**NAID**) Pathologic compression of an organ, such as the heart.

tapeworm (TAPEWORM) Intestinal parasitic worm.

tarsal (**TAR**-sal) Pertaining to the tarsus

tarsus (**TAR**-sus) The collection of seven bones in the foot that form the ankle and instep; *or* the flat fibrous plate that gives shape to the outer edges of the eyelids.

tartar (**TAR**-tar) Calcified deposit at the gingival margin of the teeth.

Tay-Sachs disease (TAY SAKS **DIZ**-eez) Congenital fatal disorder of fat metabolism.

temperament (**TEM**-per-ah-ment) Predisposition to character or personality.

temporal (**TEM**-pore-al) Bone that forms part of the base and sides of the skull.

temporal lobe (**TEM**-pore-al LOBE) Posterior two-thirds of the cerebral hemispheres.

temporomandibular joint (TMJ) (**TEM**-pore-oh-man-**DIB**-you-lar JOYNT) The joint between the temporal bone and the mandible.

tendinitis (ten-dih-**NYE**-tis) Inflammation of a tendon. Also spelled *tendonitis.*

tendon (**TEN**-dun) Fibrous band that connects muscle to bone.

tenosynovitis (**TEN**-oh-sin-oh-**VIE**-tis) Inflammation of a tendon and its surrounding synovial sheath.

teratogen (**TER**-ah-toe-jen) Agent that produces fetal deformities.

teratogenesis (**TER**-ah-toe-**JEN**-eh-sis) Process involved in producing fetal deformities.

teratogenic (**TER**-ah-toe-**JEN**-ik) Pertaining to or capable of producing fetal deformities.

teratoma (ter-ah-**TOE**-mah) Neoplasm of a testis or ovary containing multiple tissues from other sites in the body.

testicle (TES-tih-kul) One of the male reproductive glands. Also called *testis.*

testicular (tes-TICK-you-lar) Pertaining to the testicle.

testis (TES-tis) A synonym for *testicle.* Plural *testes.*

testosterone (tes-TOSS-ter-own) Powerful androgen produced by the testes.

tetany (TET-ah-nee) Severe muscle twitches, cramps, and spasms.

tetralogy (te-TRAL-oh-jee) A set of four congenital heart defects occurring together.

tetralogy of Fallot (TOF) (te-TRAL-oh-jee of fah-LOW) Set of four congenital heart defects occurring together.

thalamus (THAL-ah-mus) Mass of gray matter underneath the ventricle in each cerebral hemisphere.

thalassemia (thal-ah-SEE-mee-ah) Group of inherited blood disorders that produce a hemolytic anemia.

thelarche (thee-LAR-key) Onset of breast development.

thenar (THAY-nar) The thenar eminence is the fleshy mass at the base of the thumb.

therapeutic (THAIR-ah-PYU-tik) Pertaining to the treatment of a disease or disorder.

therapy (THAIR-ah-pee) Systematic treatment of a disease, dysfunction, or disorder.

thoracentesis (THOR-ah-sen-TEE-sis) Insertion of a needle into the pleural cavity to withdraw fluid or air. Also called *pleural tap.*

thoracic (THOR-ass-ik) Pertaining to the chest (thorax).

thoracic cavity (THOR-ass-ik KAV-ih-tee) Space within the chest containing the lungs, heart, aorta, venae cavae, esophagus, trachea, and pulmonary vessels.

thoracoscopy (thor-ah-KOS-koh-pee) Examination of the pleural cavity with an endoscope.

thoracotomy (thor-ah-KOT-oh-me) Incision through the chest wall.

thorax (THOR-acks) The part of the trunk between the abdomen and the neck.

thrombin (THROM-bin) Enzyme that forms fibrin.

thrombocyte (THROM-boh-site) Another name for *platelet.*

thrombocytopenia (THROM-boh-site-oh-PEE-nee-uh) Deficiency of platelets in circulating blood.

thromboembolism (THROM-boh-EM-boh-lizm) A piece of detached blood clot (embolus) blocking a distant blood vessel.

thrombolysis (throm-BOL-ih-sis) Dissolving of a thrombus (clot).

thrombolytic (throm-boh-LIT-ik) Able to dissolve a thrombus.

thrombophlebitis (THROM-boh-fleh-BY-tis) Inflammation of a vein with clot formation.

thrombosis (throm-BOH-sis) Formation of a thrombus.

thrombus (THROM-bus) A clot attached to a diseased blood vessel or heart lining. Pleural *thrombi.*

thrush (THRUSH) Infection with *Candida albicans.*

thymectomy (thigh-MEK-toe-me) Surgical removal of the thymus gland.

thymoma (thigh-MOH-mah) Benign tumor of the thymus.

thymus (THIGH-mus) Lymphoid and endocrine gland located in the mediastinum.

thyroid (THIGH-royd) Endocrine gland in the neck; *or* a cartilage of the larynx.

thyroid hormone (THIGH-royd HOR-mohn) Collective term for the two thyroid hormones, T_3 and T_4.

thyroid storm (THIGH-royd STORM) Medical crisis and emergency due to excess thyroid hormones.

thyroidectomy (thigh-royd-ECK-toe-me) Surgical removal of the thyroid gland.

thyroiditis (thigh-royd-EYE-tis) Inflammation of the thyroid gland.

thyrotoxicosis (THIGH-roe-toks-ih-KOH-sis) Disorder produced by excessive thyroid hormone production.

thyrotropin (thigh-roe-TROH-pin) Hormone from the anterior pituitary gland that stimulates function of the thyroid gland.

thyroxine (thigh-ROCK-sin) Thyroid hormone T_4, tetraiodothyronine.

tibia (TIB-ee-ah) The larger bone of the lower leg.

tibial (TIB-ee-al) Pertaining to the tibia.

tic (TIK) Sudden, involuntary, repeated contraction of muscles.

tic douloureux (tik duh-luh-RUE) Painful, sudden, spasmodic involuntary contractions of the facial muscles supplied by the trigeminal nerve. Also called *trigeminal neuralgia.*

tinea (TIN-ee-ah) General term for a group of related skin infections caused by different species of fungi.

tinnitus (TIN-ih-tus) Persistent ringing, whistling, clicking, or booming noise in the ears.

tissue (TISH-you) Collection of similar cells.

tomography (toe-MOG-rah-fee) Radiographic image of a selected slice of tissue.

tone (TONE) Tension present in resting muscles.

tongue (TUNG) Mobile muscle mass in the mouth; bears the taste buds.

tonic (TON-ik) State of muscular contraction.

tonic-clonic (TON-ik KLON-ik) The body alternates between excessive muscular rigidity (tonic) and jerking muscular contractions (clonic).

tonic-clonic seizure (TON-ik KLON-ik SEE-zhur) Generalized seizure due to epileptic activity in all or most of the brain.

tonometer (toe-NOM-eh-ter) Instrument for determining intraocular pressure.

tonometry (toe-NOM-eh-tree) The measurement of intraocular pressure.

tonsil (TON-sill) Mass of lymphoid tissue on either side of the throat at the back of the tongue.

tonsillectomy (ton-sih-LEK-toh-me) Surgical removal of the tonsils.

tonsillitis (ton-sih-LIE-tis) Inflammation of the tonsils.

topical (TOP-ih-kal) Medication applied to the skin to obtain a local effect.

torsion (TOR-shun) The act or result of twisting.

Tourette syndrome (tur-ET SIN-drome) Disorder of multiple motor and vocal tics.

toxic (TOK-sick) Pertaining to a toxin.

toxicity (toks-ISS-ih-tee) The state of being poisonous.

toxin (TOK-sin) Poisonous substance formed by a cell or organism.

trachea (TRAY-kee-ah) Air tube from the larynx to the bronchi.

trachealis (tray-kee-AY-lis) Pertaining to the trachea.

tracheostomy (tray-kee-**OST**-oh-me) incision into the windpipe into which a tube can be inserted to assist breathing.

tracheotomy (tray-kee-**OT**-oh-me) Incision made into the trachea to create a tracheostomy.

tract (TRACKT) Bundle of nerve fibers with a common origin and destination.

traction (**TRAK**-shun) Pulling or dragging force.

trait (TRAYT) A discrete characteristic that has a known quality.

tranquilizer (**TRANG**-kwih-lie-zer) Agent that calms without sedating or depressing.

transdermal (tranz-**DER**-mal) Going across or through the skin.

transfusion (tranz-**FYU**-zhun) Transfer of blood or a blood component from donor to recipient.

transplant (**TRANZ**-plant) The tissue or organ used; *or* the act of transferring tissue from one person to another.

transplantation (**TRANZ**-plan-**TAY**-shun) The moving of tissue or an organ from one person or place to another.

transthoracic (tranz-thor-**ASS**-ik) Going through the chest wall.

transurethral (**TRANZ**-you-**REE**-thral) Procedure performed through the urethra.

transverse (tranz-**VERS**) Horizontal plane dividing the body into upper and lower portions.

transverse fracture (tranz-**VERS FRAK**-chur) A fracture perpendicular to the long axis of the bone.

tremor (**TREM**-or) Small, shaking, involuntary, repetitive movements of hands, extremities, neck, or jaw.

triceps brachii (**TRY**-sepz **BRAY**-key-eye) Muscle of the arm that has three heads or points of origin.

Trichomonas (trik-oh-**MOH**-nas) A parasite causing an STD.

trichomoniasis (**TRIK**-oh-moh-**NIE**-ah-sis) Infection with *Trichomonas vaginalis.*

tricuspid (try-**KUSS**-pid) Having three points; a tricuspid heart valve has three flaps.

triglyceride (try-**GLISS**-eh-ride) Lipid containing three fatty acids.

trimester (**TRY**-mes-ter) One-third of the length of a full-term pregnancy.

triplegia (try-**PLEE**-jee-ah) Paralysis of three limbs.

triplegic (try-**PLEE**-jik) Pertaining to or suffering from triplegia.

trochanter (troh-**KAN**-ter) One of two bony prominences near the head of the femur.

tropic (**TROH**-pik) Tropic hormones stimulate other endocrine glands to produce hormones.

tuberculosis (too-**BER**-kyu-**LOW**-sis) Infectious disease that can infect any organ or tissue.

tumor (**TOO**-mor) Any abnormal swelling.

tunica (**TYU**-nih-kah) A covering layer in the wall of a blood vessel or other tubular structure.

tunica vaginalis (**TYU**-nih-kah vaj-ih-**NAHL**-iss) The sheath of the testis and epididymis.

tympanic (tim-**PAN**-ik) Pertaining to the tympanic membrane (eardrum) or tympanic cavity.

tympanostomy (tim-pan-**OS**-toe-me) Surgically created new opening in the tympanic membrane to allow fluid to drain from the middle ear.

U

ulna (**UL**-nah) The medial and larger of the bones of the forearm.

ulnar (**UL**-nar) Pertaining to the ulna or to any of the structures (artery, vein, nerve) named after it.

ultrasonography (**UL**-trah-soh-**NOG**-rah-fee) Delineation of deep structures using sound waves.

ultraviolet (ul-trah-**VIE**-oh-let) Light rays at a higher frequency than the violet end of the spectrum.

umbilical (um-**BILL**-ih-kal) Pertaining to the umbilicus or the center of the abdomen.

umbilicus (um-**BILL**-ih-kuss) Pit in the abdomen where the umbilical cord entered the fetus.

unilateral (you-nih-**LAT**-er-al) Pertaining to one side.

urea (you-**REE**-ah) End product of nitrogen metabolism.

uremia (you-**REE**-me-ah) The complex of symptoms arising from renal failure.

ureter (you-**REE**-ter) Tube that connects a kidney to the urinary bladder.

ureteral (you-ree-**TER**-al) Pertaining to the ureter.

ureteroscope (you-**REE**-ter-oh-scope) Endoscope to view the inside of the ureter.

ureteroscopy (you-**REE**-ter-os-koh-pee) To examine the ureter.

urethra (you-**REE**-thra) Canal leading from the bladder to the outside.

urethritis (you-ree-**THRI**-tis) Inflammation of the urethra.

urinalysis (yur-ih-**NAL**-ih-sis) Examination of urine to separate it into its elements and define their kind and/or quantity.

urinary (**YUR**-in-ary) Pertaining to urine.

urinate (**YUR**-in-ate) To pass urine.

urination (yur-ih-**NAY**-shun) The act of passing urine.

urine (**YUR**-in) Fluid and dissolved substances excreted by the kidney.

urological (yur-oh-**LOJ**-ih-kal) Pertaining to urology.

urologist (you-**ROL**-oh-jist) Medical specialist in disorders of the urinary system.

urology (you-**ROL**-oh-jee) Medical specialty of disorders of the urinary system.

urticaria (ur-tee-**KARE**-ee-ah) Rash of itchy wheals (hives).

uterine (**YOU**-ter-ine) Pertaining to the uterus.

uterus (**YOU**-ter-us) Organ in which an egg develops into a fetus.

uvea (**YOU**-vee-ah) Middle coat of the eyeball—includes the iris, ciliary body, and choroid.

uveitis (you-vee-**EYE**-tis) Inflammation of the uvea.

uvula (**YOU**-vyu-lah) Fleshy projection of the soft palate.

V

vaccinate (**VAK**-sin-ate) To administer a vaccine.

vaccination (vak-sih-**NAY**-shun) Administration of a vaccine.

vaccine (**VAK**-seen) Preparation to generate active immunity.

vagina (vah-**JIE**-nah) Female genital canal extending from the uterus to the vulva.

vaginal (**VAJ**-ih-nal) Pertaining to the vagina.

vaginitis (vah-jih-**NIE**-tis) Inflammation of the vagina.

vaginosis (vah-jih-**NOH**-sis) A disease of the vagina.

vagus (**VAY**-gus) Tenth (X) cranial nerve; supplies many different organs throughout the body.

varicocele (**VAIR**-ih-koh-seal) Varicose veins of the spermatic cord.

varicose (**VAIR**-ih-kos) Characterized by or affected with varices.

varicosities (vair-ih-**KOS**-ih-tees) Collection of varicose veins.

varix (**VAIR**-iks) Dilated, tortuous vein. Plural *varices.*

vasectomy (vah-**SEK**-toe-me) Excision of a segment of the ductus deferens.

vasoconstriction (**VAY**-soh-con-**STRIK**-shun) Reduction in the diameter of a blood vessel.

vasodilation (**VAY**-soh-dih-**LAY**-shun) Increase in the diameter of a blood vessel.

vasovasostomy (**VAY**-soh-vay-**SOS**-toe-me) Reanastomosis of the ductus deferens to restore the flow of sperm. Also called *vasectomy reversal.*

vegetative (**VEJ**-eh-tay-tiv) Functioning unconsciously, as plant life is assumed to do.

vein (VANE) Blood vessel carrying blood toward the heart.

vena cava (**VEE**-nah **KAY**-vah) One of the two largest veins in the body. Plural *venae cavae.*

venogram (**VEE**-noh-gram) Radiograph of veins after injection of radiopaque contrast material.

venous (**VEE**-nuss) Pertaining to a vein.

ventilation (ven-tih-**LAY**-shun) Movement of gases into and out of the lungs.

ventilator (**VEN**-tih-lay-tor) Device that breathes for the patient.

ventral (**VEN**-tral) Pertaining to the belly or situated nearer to the surface of the belly.

ventricle (**VEN**-trih-kel) Chamber of the heart (pumps blood) or brain (produces cerebrospinal fluid).

venule (**VEN**-yule or **VEEN**-yule) Small vein leading from the capillary network.

vernix caseosa (**VER**-nicks kay-see-**OH**-sah) Cheesy substance covering the skin of the fetus.

verruca (ver-**ROO**-cah) Wart caused by a virus.

vertebra (**VER**-teh-brah) One of the bones of the spinal column. Plural *vertebrae.*

vertebral (**VER**-teh-bral) Pertaining to a vertebra.

vertex (**VER**-teks) Topmost point of the vault of the skull.

vertigo (**VER**-tih-go) Sensation of spinning or whirling.

vesicle (**VES**-ih-kull) Small sac containing liquid; for example, a blister.

vestibular (ves-**TIB**-you-lar) Pertaining to the vestibule.

vestibule (**VES**-tih-byul) Space at the entrance to a canal.

vestibulectomy (ves-tib-you-**LEK**-toe-me) Surgical excision of the vulva.

villus (**VILL**-us) Thin, hairlike projection, particularly of a mucous membrane lining a cavity. Plural *villi.*

virus (**VIE**-rus) Group of infectious agents that require living cells for growth and reproduction.

viscera (**VISS**-er-ah) Internal organs, particularly in the abdomen.

visceral (**VISS**-er-al) Pertaining to the internal organs.

viscosity (viss-**KOS**-ih-tee) The resistance of a fluid to flow.

viscous (**VISS**-kus) Sticky fluid that is resistant to flow.

viscus (**VISS**-kus) Any single internal organ.

visual acuity (**VIH**-zhoo-al ah-**KYU**-ih-tee) Sharpness and clearness of vision.

vitamin (**VYE**-tah-min) Essential organic substance necessary in small amounts for normal cell function.

vitreous (**VIT**-ree-us) Vitreous humor is a gelatinous liquid in the posterior cavity of the eyeball with the appearance of glass.

vocal (**VOH**-kal) Pertaining to the voice.

void (VOYD) To evacuate urine or feces.

voluntary muscle (**VOL**-un-tare-ee **MUSS**-el) Muscle that is under the control of the will.

vomer (**VOH**-mer) Lower nasal septum.

vulva (**VUL**-vah) Female external genitalia.

vulvar (**VUL**-var) Pertaining to the vulva.

vulvodynia (vul-voh-**DIN**-ee-uh) Chronic vulvar pain.

vulvovaginal (**VUL**-voh-**VAJ**-ih-nal) Pertaining to the vulva and vagina.

vulvovaginitis (**VUL**-voh-vaj-ih-**NIE**-tis) Inflammation of the vulva and vagina.

W

warfarin (**WAR**-fuh-rin) Anticoagulant; also used as rat poison; trade name *Coumadin.*

Weber test (**VA**-ber TEST) Test for sensorineural hearing loss.

wheal (WHEEL) Small, itchy swelling of the skin. Wheals raised by an injection do not itch. Also called *hives.*

whiplash (**HWIP**-lash) Symptoms caused by sudden, uncontrolled extension and flexion of the neck, often in an automobile accident.

white matter (WITE **MATT**-er) Regions of the brain and spinal cord occupied by bundles of axons.

whooping cough (**WHO**-ping KAWF) Infectious disease with spasmodic, intense cough ending on a whoop (stridor). Also called *pertussis.*

Wilms tumor (WILMZ **TOO**-mor) Cancerous kidney tumor of childhood. Also known as *nephroblastoma.*

wound (WOOND) Any injury that interrupts the continuity of skin or a mucous membrane.

X

xenograft (**ZEN**-oh-graft) A graft from another species. Also called *heterograft.*

Y

yeast (YEEST) Microscopic fungus.

Z

zygoma (zye-**GO**-mah) Bone that forms the prominence of the cheek.

zygomatic (zye-go-**MAT**-ik) Pertaining to the zygoma.

zygote (**ZYE**-goat) Cell resulting from the union of the sperm and egg.

PHOTOS

FRONT MATTER

Page iii (top): © Autumn Paul; **p. iii (bottom):** © Feragne Portrait Studio; **p. xiv:** © Royalty-Free Corbis; **p. xv:** © Vol. 69 PhotoDisc/Getty; **p. xvi:** © Vol. 107 PhotoDisc/Getty; **p. xvii:** © Banana Stock.

WELCOME CHAPTER

Opener, W.1: © The McGraw-Hill Companies/Rick Brady, photographer ; **W.2:** © Corbis/RF; **W.3, W.4:** © The McGraw-Hill Companies/Rick Brady, photographer; **W.5:** © Vol. 115 PhotoDisc/Getty.

CHAPTER 1

Opener: © The McGraw-Hill Companies, Inc./Rick Brady, photographer.

CHAPTER 2

Opener: © The McGraw-Hill Companies, Inc./Rick Brady, photographer; **2.1:** © The McGraw-Hill Companies, Inc./Joe DeGrandis; **2.2:** © The McGraw-Hill Companies, Inc./Eric Wise, photographer; **2.3, 2.5:** © The McGraw-Hill Companies, Inc./Joe DeGrandis; **2.6:** © Francis Leroy, BIOCOSMOS/Photo Researchers, Inc.

CHAPTER 3

Opener: © The McGraw-Hill Companies, Inc./Rick Brady, photographer; **3.5:** © Dr. P. Marazzi/SPL/Photo Researchers, Inc.; **3.6:** © BioPhoto Associates/Photo Researchers, Inc.; **3.7:** © James Stevenson/Photo Researchers, Inc.; **3.8:** © Medical-on-line/Alamy; **3.9:** © Kenneth Greer/Visuals Unlimited; **3.10:** © Logical Images; **3.11:** © SPL/Photo Researchers, Inc.; **3.12:** © Dr. P. Marazzi/SPL/Photo Researchers, Inc.; **3.13:** © A. M. Siegelman/Visuals Unlimited; **3.14:** © Eye of Science/Photo Researchers, Inc.; **3.15:** Farrar, W, E., Woods, M. J., Innes, J. S.: *Infectious Diseases: Text and Color Atlas,* ed. 2. London, Mosby, Europe, 1993; **3.17:** Courtesy Dr. Maureen Mayes; **3.18:** © Mediscan/Visuals Unlimited; **3.19:** © Medical-on-line/Alamy; **3.21:** © Kenneth Greer/Visuals Unlimited; **3.25:** Reprinted from J. Walter Wilson, *Fungous Diseases of Man,* Plate 42 (middle right), © 1965, The Regents of the University of California; **3.26:** © Logical Images; **3.27a:** © Sheila Terry/Photo Researchers, Inc.; **3.27b:** © Dr. P. Marazzi/Photo Researchers, Inc.; **3.27c:** © John Radcliffe Hospital/Photo Researchers, Inc.; **3.32:** © Ken Geer/Visuals Unlimited.

CHAPTER 4

Opener: © The McGraw-Hill Companies, Inc./Rick Brady, photographer; **4.1:** © The McGraw-Hill Companies, Inc./Joe DeGrandis, photographer; **4.3:** © Phototake, Inc./Alamy; **4.4:** © Western Ophthalmic Hospital/ Photo Researchers Inc.; **4.5:** © Mediscan; **4.6:** © ISM/Phototake; **4.8:** © BioPhoto Associates/Photo Researchers Inc.; **4.9:** © Matt Harris Photography/Alamy; **4.16:** © Phototake Inc./Alamy; **4.17:** National Eye Institute, National Institutes of Health; **4.18:** © Dr. P. Marazzi/Photo Researchers, Inc.; **4.19:** National Eye Institute, National Institutes of Health; **4.20:** © Volume 58/PhotoDisc/Getty; **4.21:** © Paul Parker/Photo Researchers, Inc.; **4.22:** © Chris Barry/Phototake; **4.23:** National Eye Institute, National Institutes of Health; **4.25, Page 118:** © The McGraw-Hill Companies, Inc./Rick Brady, photographer; **4.31:** © ISM/Phototake; **4.33:** © Lester V. Bergman/Corbis; **4.34:** © ISM/Phototake; **4.35:** © Collection CNRI/Phototake.

CHAPTER 5

Opener: © The McGraw-Hill Companies, Inc./Rick Brady, photographer; **5.6:** © Photo Network Stock/Grant Heilman Photography Inc.; **5.7:** © ISM/Phototake; **5.8:** © Medical-on-line/Alamy; **5.9:** © Mediscan/Visuals Unlimited; **5.15:** © CNRI/SPL/Photo Researchers Inc.; **5.19:** © SIU Biomedical/Photo Researchers, Inc.; **5.23:** © CNRI/Photo Researchers, Inc.; **5.25:** © Susan Leavine/Photo Researchers Inc.; **5.26:** © Dr. P. Marazzi/SPL/Photo Researchers, Inc.; **5.27:** © Phototake Inc./Alamy.

CHAPTER 6

Opener, 6.10: © The McGraw-Hill Companies, Inc./Rick Brady, photographer; **6.12:** © Ed Reschke; **6.13:** © Pasieka/SPL/Photo Researchers, Inc.; **6.19:** © Dr. P. Marazzi/Photo Researchers, Inc.

CHAPTER 7

Opener: © The McGraw-Hill Companies, Inc./Rick Brady, photographer; **7.7:** © Phototake; **7.10, 7.13:** © Ralph Hutchings/Visuals Unlimited; **7.15:** © Collection CNRI/Phototake; **7.16:** © Wellcome Photo Library; **7.17:** © Garo/Photo Researchers, Inc.

CHAPTER 8

Opener: © The McGraw-Hill Companies, Inc./Rick Brady, photographer; **8.2a:** © The McGraw-Hill Companies, Inc./Christine Eckel, photographer; **8.4:** © Dr. Michael Klein/Peter Arnold, Inc.; **8.5:** © The McGraw-Hill Companies, Inc./Joe DeGrandis, photographer; **8.6:** Dr. Francis Collins.

CHAPTER 9

Opener: © The McGraw-Hill Companies, Inc./Rick Brady, photographer; **9.6:** © Zephyr/Photo Researchers Inc.; **9.7:** © The McGraw-Hill Companies, Inc./Eric Wise, photographer; **9.8:** © Dr. P. Marazzi/Science Photo Library; Photo Researchers, Inc.; **9.9:** © Ralph Hutchings/Visuals Unlimited; **9.12:** © AJPhoto/Photo Researchers Inc.; **9.17:** © Charles McRae, M.D./Visuals Unlimited; **9.19:** © The McGraw-Hill Companies, Inc./Tim Vacula, photographer.

CHAPTER 10

Opener: © The McGraw-Hill Companies, Inc./Rick Brady, photographer; **10.5:** Reprinted by permission of publisher from Albert Mendeloff, "Acromegaly, diabetes, hypermetabolism, proteinuria, and heart failure," *American Journal of Medicine:* 20:1, 01-56, p. 135, © by Excerpta Medica, Inc.; **10.8:** © Dr. M. A. Ansary/Photo Researchers, Inc.; **10.9:** © Dr. P. Marazzi/SPL/Photo Researchers, Inc.;**10.10:** © Mediscan; **10.11:** © Bettmann/Corbis.

CHAPTER 11

Opener: © Vol. 40/PhotoDisc/Getty; **11.1:** © The McGraw-Hill Companies, Inc./Eric Wise, photographer; **11.2b:** © Bill Longcore/Photo Researchers, Inc.; **11.4:** © Ed Reschke; **11.5:** © Meckes/Ottawa/Photo Researchers, Inc.; **11.6–11.10:** © Ed Reschke; **11.12:** © Oliver Meckes/Photo Researchers, Inc.; **11.13a:** © Medical-on-line/Alamy; **11.13b:** © Dr. P. Marazzi/SPL/Photo Researchers, Inc.; **11.13c:** © Paul Cox/Alamy; **11.19:** © The McGraw-Hill Companies,

Inc./Dennis Strete, photographer; **11.21:** © Peter Arnold, Inc.; **11.22:** © Visuals Unlimited; **11.23:** © NIBSC/SPL/Photo Researchers, Inc.; **11.24:** © SPL/Photo Researchers, Inc.

CHAPTER 12

Opener: © The McGraw-Hill Companies, Inc./Rick Brady, photographer; **12.9:** © Phototake Inc./Alamy; **12.10:** © Wellcome Photo Library; **12.12:** © NIH/Phototake Inc.; **12.13:** © James Cavallini/Photo Researchers, Inc.; **12.14a:** © Wellcome Photo Library; **12.14b:** © Simon Fraser/Newcastle Hospitals NHS/Science Photo Library; Photo Researchers, Inc.; **12.15:** © Bart's Medical Library/Phototake Inc.; **12.16:** © Dr. M. A. Ansary/Custom Medical Stock Photography; **12.17:** © NMSB/Custom Medical Stock; **12.18:** © Vol. 15 PhotoDisc/Getty.

CHAPTER 13

Opener: © The McGraw-Hill Companies, Inc./Rick Brady, photographer; **13.3:** © Per H. Kjeldsen.

CHAPTER 14

Opener: © The McGraw-Hill Companies, Inc./Rick Brady, photographer; **14.3:** © Kenneth Greer/Visuals Unlimited; **14.4:** © Biophoto Associates/Photo Researchers, Inc.; **14.7:** © Dr. Landrum Shettles; **14.9:** The William Boyd Museum, Dept. of Pathology and Laboratory Medicine, The University of British Columbia; **14.10:** © Wellcome Photo Library; **14.11:** The William Boyd Museum, Dept. of Pathology and Laboratory Medicine, The University of British Columbia; **14.12:** © Parviz M. Pour/Photo Researchers Inc.; **14.14:** © Corbis RF.

CHAPTER 15

Opener: © Phanie/Photo Researchers, Inc.; **15.3:** © BrandX Pictures/Punchstock; **15.5:** © Photo Researchers; **15.7:** © Visuals Unlimited; **15.9:** © EP Vol. 90/Getty Images; **15.10:** © Wellcome Photo Library; **15.11:** ALIX/Phanie/Photo Researchers, Inc.

CHAPTER 16

Opener: © The McGraw-Hill Companies, Inc./Rick Brady, photographer; **16.3:** © St. Bartholomew's Hospital, London/Photo Researchers, Inc.; **16.4:** © Sally and Richard Greenhill/Alamy; **16.5:** © NMSB/Custom Medical Stock Photo; **16.6:** © Scott Camazine/Alamy; **16.7:** © Susan Leavines/Photo Researchers, Inc.; **16.8:** © Wellcome Medical Trust Library; **16.9:** © Barry Slaven, MD, PhD/Phototake; **16.10:** © David Parker/Photo Researchers, Inc.; **16.12:** © Pasieka/Science Photo Library, Photo Researchers, Inc.

ILLUSTRATIONS

CHAPTER 2

2.4, 2.7, 2.8, 2.9, 2.10a, 2.11, 2.12: Saladin, *Anatomy & Physiology: The Unity of Form and Function,* 3/E, McGraw-Hill © 2004; **2.10b:** Shier, Butler, Lewis, *Hole's Human Anatomy and Physiology,* 10/E, McGraw-Hill © 2004.

CHAPTER 3

3.1, 3.4, 3.20, 3.22, 3.24, 3.29, 3.30, 3.31: Saladin, *Anatomy & Physiology: The Unity of Form and Function,* 3/E, McGraw-Hill © 2004; **3.2, 3.28:** Shier, Butler, Lewis, *Hole's Human Anatomy and Physiology,* 10/E, McGraw-Hill © 2004; **3.3:** Saladin, *Human Anatomy,* 1/E, McGraw-Hill © 2005; **3.23:** McKinley/O'Loughlin, *Human Anatomy,* McGraw-Hill © 2006; **3.27:** Saladin, *Anatomy & Physiology: The Unity of Form and Function,* 3/E, McGraw-Hill © 2004 and McKinley/O'Loughlin, *Human Anatomy,* McGraw-Hill © 2006.

CHAPTER 4

4.2, 4.12: Saladin, *Human Anatomy,* 1/E, McGraw-Hill © 2005; **4.7, 4.30:** Shier, Butler, Lewis, *Hole's Human Anatomy and Physiology,* 10/E, McGraw-Hill © 2004; **4.10, 4.11, 4.13, 4.14, 4.37:** McKinley/O'Loughlin, *Human Anatomy,* McGraw-Hill © 2006; **4.15:** Seeley et al., *Anatomy and Physiology,* 7/E, McGraw-Hill © 2006; **4.24, 4.26, 4.27, 4.32, 4.36, 4.38, 4.39:** Saladin, *Anatomy & Physiology: The Unity of Form and Function,* 3/E, McGraw-Hill © 2004; **4.28:** Allan, *Medical Language for Modern Health Care,* 1/E, McGraw-Hill © 2008.

CHAPTER 5

5.1, 5.13: Shier, Butler, Lewis, *Hole's Human Anatomy and Physiology,* 10/E, McGraw-Hill © 2004; **5.2, 5.10, 5.11, 5.12, 5.16, 5.17, 5.18, 5.21, 5.22, 5.24:** Saladin, *Anatomy & Physiology: The Unity of Form and Function,* 3/E, McGraw-Hill © 2004; **5.3:** Nowak/Handford, *Pathophysiology: Concepts and Applications for Healthcare Professionals,* 3/E, McGraw-Hill © 2004; **5.4, 5.5:** Saladin, *Human Anatomy,* 1/E, McGraw-Hill © 2005; **5.14:** McKinley/O'Loughlin, *Human Anatomy,* McGraw-Hill © 2006; **5.20, 5.23a:** Seeley et al., *Anatomy and Physiology,* 7/E, McGraw-Hill © 2006; **5.28:** Allan, *Medical Language for Modern Health Care,* 1/E, McGraw-Hill © 2008.

CHAPTER 6

6.1, 6.2, 6.15, Consultation Request and Report Form, page 216, 6.16: Allan, *Medical Language for Modern Health Care,* 1/E, McGraw-Hill © 2008; **6.3, 6.14:** McKinley/O'Loughlin, *Human Anatomy,* McGraw-Hill © 2006; **6.4, 6.5, 6.6, 6.9, 6.17, 6.18:** Saladin, *Anatomy & Physiology: The Unity of Form and Function,* 3/E, McGraw-Hill © 2004; **6.7, 6.8:** Shier, Butler, Lewis, *Hole's Human Anatomy and Physiology,* 10/E, McGraw-Hill © 2004; **6.11:** Seeley et al., *Anatomy and Physiology,* 7/E, McGraw-Hill © 2006.

CHAPTER 7

7.1, 7.2, 7.3, 7.4, 7.5, 7.6, 7.8, 7.9, 7.11: Saladin, *Anatomy & Physiology: The Unity of Form and Function,* 3/E, McGraw-Hill © 2004; **7.12, 7.14, 7.18:** McKinley/O'Loughlin, *Human Anatomy,* McGraw-Hill © 2006.

CHAPTER 8

8.1: Saladin, *Anatomy & Physiology: The Unity of Form and Function,* 3/E, McGraw-Hill © 2004; **8.2, 8.3:** McKinley/O'Loughlin, *Human Anatomy,* McGraw-Hill © 2006; **8.7, 8.9, 8.10, 8.11:** Saladin, *Human Anatomy,* 1/E, McGraw-Hill © 2005; **8.8:** Seeley et al., *Anatomy and Physiology,* 7/E, McGraw-Hill © 2006.

CHAPTER 9

9.1, 9.4, 9.5, 9.11, 9.18: Shier, Butler, Lewis, *Hole's Human Anatomy and Physiology,* 10/E, McGraw-Hill © 2004; **9.2, 9.10, 9.13, 9.14, 9.15, 9.16:** Saladin, *Human Anatomy,* 1/E, McGraw-Hill © 2005; **9.3:** Saladin, *Anatomy & Physiology: The Unity of Form and Function,* 3/E, McGraw-Hill © 2004.

CHAPTER 10

10.1: Allan, *Medical Language for Modern Health Care,* 1/E, McGraw-Hill © 2008; **10.2, 10.3, 10.4, 10.6, 10.7, 10.12, 10.13:** Saladin, *Anatomy & Physiology: The Unity of Form and Function,* 3/E, McGraw-Hill © 2004.

CREDITS

CHAPTER 11

11.1: Seeley et al., *Anatomy and Physiology,* 7/E, McGraw-Hill © 2006; **11.2a, 11.14:** Shier, Butler, Lewis, *Hole's Human Anatomy and Physiology,* 10/E, McGraw-Hill © 2004; **11.3, 11.15, 11.16, 11.18:** Saladin, *Anatomy & Physiology: The Unity of Form and Function,* 3/E, McGraw-Hill © 2004; **11.11, 11.20:** Allan, *Medical Language for Modern Health Care,* 1/E, McGraw-Hill © 2008; **11.17:** Saladin, *Human Anatomy,* 1/E, McGraw-Hill © 2005; **11.25:** Cowan, *Microbiology,* 1/E, McGraw-Hill, © 2006.

CHAPTER 12

12.1, 12.2, 12.3, 12.7, 12.8: Saladin, *Anatomy & Physiology: The Unity of Form and Function,* 3/E, McGraw-Hill © 2004; **12.4, 12.11:** Allan, *Medical Language for Modern Health Care,* 1/E, McGraw-Hill © 2008; **12.5:** Shier, Butler, Lewis, *Hole's Human Anatomy and Physiology,* 10/E, McGraw-Hill © 2004; **12.6, 12.17a:** McKinley/O'Loughlin, *Human Anatomy,* McGraw-Hill © 2006.

CHAPTER 13

13.1, 13.2, 13.5, 13.6, 13.7, 13.8, 13.9, 13.11, 13.12: Saladin, *Anatomy & Physiology: The Unity of Form and Function,* 3/E, McGraw-Hill © 2004; **13.4, 13.13, 13.14:** McKinley/O'Loughlin, *Human Anatomy,* McGraw-Hill © 2006; **13.10:** Shier, Butler, Lewis, *Hole's Human Anatomy and Physiology,* 10/E, McGraw-Hill © 2004.

CHAPTER 14

14.1, 14.2, 14.5, 14.6, 14.8: Saladin, *Anatomy & Physiology: The Unity of Form and Function,* 3/E, McGraw-Hill © 2004; **14.13:** Allan, *Medical Language for Modern Health Care,* 1/E, McGraw-Hill © 2008; **14.15:** McKinley/O'Loughlin, *Human Anatomy,* McGraw-Hill © 2006.

CHAPTER 15

15.1, 15.2, 15.4: Saladin, *Anatomy & Physiology: The Unity of Form and Function,* 3/E, McGraw-Hill © 2004; **15.6:** Allan, *Medical Language for Modern Health Care,* 1/E, McGraw-Hill © 2008.

CHAPTER 16

16.1, 16.2: Saladin, *Anatomy & Physiology: The Unity of Form and Function,* 3/E, McGraw-Hill © 2004.

A

INDEX

B

D

E

F

G

H

I

J

K

L

M

N

O

P

T

U

V

W

X

Y

Z

P9-ECZ-074

THE STATE OF FOOD AND AGRICULTURE 1990

THE STATE OF FOOD AND AGRICULTURE 1990

FOOD AND AGRICULTURE ORGANIZATION OF THE UNITED NATIONS
Rome, 1991

The statistical material in this publication has been prepared from the information available to FAO up to October 1990.

The designations employed and the presentation of the material in this publication do not imply the expression of any opinion whatsoever on the part of the Food and Agriculture Organization of the United Nations concerning the legal status of any country, territory, city or area, or of its authorities, or concerning the delimitation of its frontiers or boundaries. In some tables, the designations "developed" and "developing" economies are intended for statistical convenience and do not necessarily express a judgement about the stage reached by a particular country or area in the development process.

Part Three, "Structural adjustment and agriculture", was based on the work of Diane Elton, Lawrence D. Smith, Marco Spinedi, John Weeks and Trevor Young, Consultants.

David Lubin Memorial Library Cataloguing in Publication Data

FAO, Rome (Italy)
The state of food and agriculture 1990.
(FAO Agriculture Series, no. 23)
ISBN 92-5-102989-X

1. Agriculture. 2. Food production. 3. Trade.

I. Title II. Series

FAO code: 70 AGRIS: E16 E70

Printed in Italy

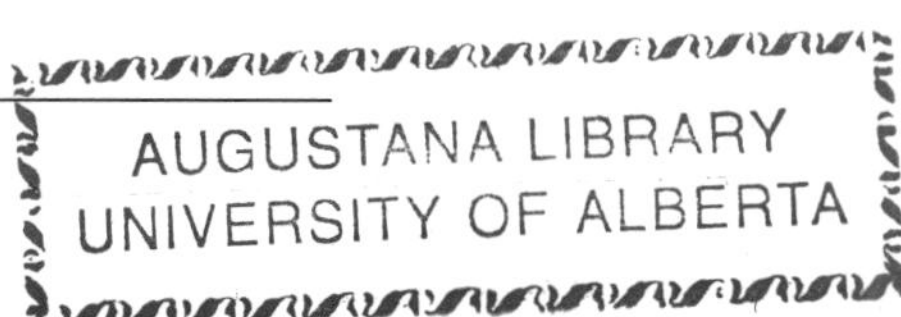

Foreword

After the extraordinary sequence of momentous events that made 1989 such a turning point in the course of history, 1990 has been a year of optimism and hope but also, in several respects, a year of frustration.

Optimism and hope because barriers between countries and political and economic systems continued to fall. A new atmosphere of *rapprochement* developed in international relationships that replaced ideological and political confrontation. Radical changes in Europe, in particular, transformed, and continue to transform, the world political and economic scene, but we should also not forget such significant events in the developing world as the unification of Yemen and the attainment of independence by Namibia. Despite the complexity of the problems involved in the process of economic and political reform, I believe its direction is essentially toward improved prospects for peace, international cooperation and prosperity.

This has also been a year of frustration, however, because old problems and injustices persist. Events in the Near East have somewhat shrouded our optimism. After only a few months of hopeful expectation, the benefits of the long-awaited "peace dividend" already appear illusory. Prospects still remain uncertain for global economic growth, trade, and the solution of the grave economic and social problems affecting much of the developing world.

Another reason for frustration is the continuation of the disappointing economic and agricultural performance of many developing countries, more particularly in 1990 in Africa and Latin America and the Caribbean.

Deteriorating terms of trade and weak markets for their non-oil commodity exports, critical financial problems linked to external debt and mounting inflation have all combined to depress incomes and wages, and thus to impoverish the populations of many developing countries. Sharply higher oil prices have only further darkened the economic and social prospects for most of them.

On external debt, I welcome the progress achieved in implementing a strengthened debt strategy and the various promising new initiatives reported in this document. Nonetheless, the problem remains critical. Only slow and partial progress has been made in reducing debt and debt-service costs and the economies of many indebted developing countries are still far from reaching a stage of sustained recovery. As in every year since 1984, net transfers continue to flow from developing to creditor industrialized countries, an anomaly that derides both the principle of equity and the process of development. These net transfers have amounted to almost $43 billion in 1989, and totalled $168 billion between 1984 and 1989.

Whatever the value of the ongoing efforts to help developing countries reduce their economic imbalances and debt, no viable solution to these problems can exist in the absence of an international economic environment providing adequate opportunities to expand their export trade. Of equal importance is to provide developing countries with remunerative prices for their agricultural commodity exports. However, it is disturbing that, as the deadline approaches, the necessary political will seems to be still lacking to maintain the momentum of the Uruguay Round of multilateral trade negotiations through to their successful conclusion, which is essential to improve the agricultural trade outlook in the 1990s.

Turning to the current world food and agricultural situation, I must express once again my deep concern that one out of five persons of the populations of nearly 100 developing countries remains undernourished. Also, we cannot be oblivious of, nor indifferent to, the fate of the millions of people who are threatened by starvation as soon as a season of insufficient rainfall, or any other natural or man-made event, disrupts the normal supply of food or hinders their access to it. It is this concern that has prompted FAO, together with the World Health Organization, to sponsor a major International Conference on Nutrition, to be held in Rome in December 1992. This important intergovernmental conference should define the terms of a coordinated strategy for national and international action to eradicate hunger and malnutrition, mobilize the necessary resources and set up a world information system to monitor human nutrition status.

Recent performance of the global food and agriculture sector has been uneven and prospects remain somewhat uncertain. Growth in world agricultural production in 1990 is expected to be lower than in 1989 but significantly above the depressed growth rates of 1986-88. However, such global expansion would still be unevenly distributed. As in 1989, the increase is expected to occur mainly in developed countries, particularly North America, Eastern Europe and the USSR. Agricultural performances are likely to be less satisfactory for much of the developing world.

Indeed, in both Africa and Latin America and the Caribbean, food production is expected to fall below population growth in 1990, a highly unsatisfactory feature in view of the already disappointing performance of these two regions in recent years.

World cereal output is expected to reach record levels in 1990. As in the previous year, developed countries, in particular North America and the USSR, would account for most of the increase, although production is also forecast to rise in developing countries in the aggregate. With world output of cereals exceeding global utilization for the first time in four years, cereal stocks by the end of 1990/91 are likely to increase a little and, as a consequence, the outlook for world food security to improve slightly.

Growth in agricultural trade slowed in 1989, not only well below that of 1988, but also below the average growth of the 1980s. Moreover, the modest global expansion of agricultural exports reflected largely the stronger performance of the developed countries. Agricultural export earnings of developing countries in 1989 rose only marginally, and at an even slower pace than the sluggish average growth rate of the 1980s.

The continuation of such unsatisfactory trends for developing countries is of great concern because of the serious balance-of-payments difficulties already faced by many of them, particularly those that predominantly export agricultural products. In both Africa and Latin America and the Caribbean, the two developing regions currently facing the largest economic and financial imbalances, and depending more on agricultural exports for correcting them, export earnings from agriculture fell in 1989.

The deterioration in the terms of trade of agricultural exports of developing countries in recent years is another disquieting feature. A further strong decline is also to be feared in 1990 and possibly in 1991 as well, in view of the higher prices of energy and oil-related production costs. Against this trend, prices of some of the developing countries' main export commodities such as sugar, tea and rubber have fallen sharply in the past year, and those of others such as coffee and cocoa have continued to be severely depressed.

At this juncture, when the globalization of interests is becoming more widely appreciated, the need for enhanced solidarity and partnership between North and South, rich and poor, is greater than ever before. The need for closer cooperation and collaboration has been reiterated at the 18th Special Session of the UN General Assembly devoted to international economic cooperation, in particular to the revitalization of economic growth and development of the developing countries. Yet the record shows that net disbursements of official development assistance from developed countries declined in 1989, both in real terms and in relation to the aggregate size of their economies. Commitments to agriculture have recently increased in real terms, but thanks to the effort of one country alone — Japan. Furthermore, multilateral commitments on concessional terms have continued their downward trend in 1989, as did the proportion allocated to agriculture out of total multilateral lending. I must therefore once again urge donor countries to contribute fully to the common objective of spurring and sustaining growth in developing countries, with particular attention to the least developed among them. However serious the budgetary constraints of developed countries, and however necessary the financial effort required to help Eastern European countries overcome their current difficulties, the development needs of the poorest countries must remain as a first priority. The Programme of Action for the 1990s defined at the Second United Nations Conference on the Least Developed Countries in September 1990 is a valuable instrument to translate awareness of the special requirements of the poorest countries into concrete results.

Against this mixed picture of accomplishments, uncertainties and frustrations, the magnitude of the development task in front of us as we enter a new decade needs no highlighting. Nevertheless, a valuable framework for action is provided by the International Development Strategy (IDS) for the Fourth United Nations International Development Decade, which examines the challenges and opportunities for the 1990s and defines the main objectives to be pursued. The IDS identifies food and agriculture as one of the main priority aspects of development, and policies for achieving objectives in food and agriculture are closely interlinked with those appropriate for the other priority areas.

As a contribution to the IDS, FAO has defined the elements of a long-term strategy for the food and agricultural sector which, if successfully implemented, can be instrumental in achieving the primary objectives of growth with equity, poverty alleviation, improved nutrition, health and education, enhanced status of women and people's participation. Among the matters of most direct interest to FAO is the additional objective of environmentally sound and sustainable agricultural development, the theme of the FAO/Netherlands Conference on Agriculture and Environment which is scheduled to be held in April 1991 and which will provide important inputs for the United Nations Conference on Environment and Development in 1992.

This year's special chapter in *The State of Food and Agriculture* is devoted to structural adjustment and agriculture. Although the process of structural adjustment has been at the centre of the attention of analysts, international institutions and policy-makers for at least a decade, it has gained, rather than lost, prominence among the main current issues of the global economic environment.

What emerges from this review is an uneven, and often disappointing, picture of accomplishments in structural adjustment so far. What also emerges, however, is the generally positive and in many cases irreplaceable role of agriculture in helping economies overcome external shocks and providing a basis for recovery and sustained growth. With the proper incentives and relatively small investment, agriculture can play a major positive role at whatever stage of a country's economic cycle — growth, crisis, adjustment or recovery. Its inherent qualities enable it to act as a buffer to external shocks as well as an engine for economic recovery.

This document shows that agriculture has been, and continues to be an important net earner of foreign exchange and a positive contributor to the debt financing of developing countries throughout the 1980s. Also, unlike other sectors, agriculture has maintained a positive net transfer position. It has also helped hold down inflation in many cases, and remained an important contributor to growth and employment in periods when economic activity was slackening and labour opportunities dwindling. Such a positive contribution of the sector to the overall recovery effort and to welfare renders all the more unjustified a neglect of agriculture on the part of some governments.

While the past experience with structural adjustment programmes indicates possible areas for improving some of their conceptual and operational aspects, structural adjustment remains an inescapable necessity for many countries to foster economic recovery and growth. At the same time, however, the effort and sacrifice involved in structural adjustment programmes has fallen on the debtor countries and their populations in a disproportionate way, emphasizing the need for a more balanced sharing of the burden. Also within debtor countries, there is now a well-documented tendency for the incidence of the programme of financial austerity to fall unevenly on the economically weak segments of the population. Thus, an extremely careful evaluation is needed of the social effects of these programmes, that must be designed, phased and sequenced in a way that would minimize hardship for the poorer and more vulnerable population groups, and involve them actively in the development process. In this context, the poorest of the poor are the rural poor in many countries undergoing structural adjustment. FAO is actively participating with the countries concerned, as well as other technical agencies within the UN system, in providing the benefit of its expertise and experience in food and agriculture to ensure that structural adjustment programmes attain their primary economic objectives while protecting the welfare needs of the most destitute population groups.

Edouard Saouma
DIRECTOR-GENERAL

Main events relating to food and agriculture

March 1989
Inter-American Development Bank (IDB) funding

In March 1989 the IDB Board of Governors finally approved the proposal for the Seventh General Increase in Resources by $26.5 billion to a total of $61 billion. Based on past trends, about 30 percent of these resources may be expected to be allocated to agriculture over the coming four years.

10 March 1989
Brady Plan initiative (Washington)

The United States Secretary of Treasury, Mr N. Brady, in a statement to the Brookings Institute and the Bretton Woods Committee Conference on Third World debt, launched a plan that encouraged debt and debt-service reduction on a voluntary basis, while recognizing the importance of continued new lending. The IMF and World Bank were called upon to provide support, as part of their policy-based lending programmes, for debt or debt-service reduction purposes.

3-4 April 1989
IMF spring meeting (Washington)

The meeting reached a decision to provide additional resources for debt-reduction operations to member countries undertaking "sound" economic reforms, by setting aside a portion of members' purchases under fund-supported arrangements.

5-8 April 1989
Mid-Term Review of Uruguay Round (Geneva)

The meeting of the Trade Negotiations Committee (TNC) completed the Mid-term Review of the Uruguay Round begun in Montreal in December 1988. Participants adopted four decisions: on agriculture; textiles and clothing; safeguards; and trade-related aspects of intellectual property rights. Specifically in the area of agriculture, a "framework approach" was agreed, covering interrelated long- and short-term elements and arrangements on sanitary and phytosanitary regulations, tropical products, natural resource-based products, and stabilization of markets and export earnings.

19 April 1989
Agreement on the Global System of Trade Preferences (GSTP) among Developing Countries

The agreement set out a framework of rules based on "most favoured nation treatment" among participants covering tariff, paratariff and non-tariff measures on all types of products, direct trade measures and sectoral agreements. Following the signature of the agreement by 15 countries representative of the Group of 77, discussions began on the issues of widening and deepening the scope of the System, by expanding the accession of a larger number of Group of 77 countries and by expanding the trade coverage.

22-25 May 1989
Meeting of World Food Council (Cairo)

At their 15th ministerial session, the member states of WFC adopted the Cairo Declaration and Programme of Cooperative Action. While recognizing that each country must take its own initiatives in the fight against hunger and poverty, the Cairo Declaration places a premium on cooperative actions among countries.

20 June 1989
Agreement establishing the Common Fund for Commodities (New York)

The Common Fund for Commodities finally entered into force on 19 June 1989, about 13 years after the proposal was made. Since July 1989, it has been functioning as an independent international organization with its headquarters in Amsterdam. The Fund's general objective is to enhance the stability and growth of commodity export earnings of developing countries through two accounts: one for financing buffer-stocking under international commodity agreements, unlikely to be used in the near future; and one for financing commodity development measures aimed at improving structural conditions in commodity markets.

27 June 1989
Group of Ten Industrialized Countries (G-10) Meeting of Finance Ministers (Spain)

The Dini Report endorsed the G-10's support for the new approach to the debt crisis, which places greater emphasis on accelerating the reduction of debt and the debt-service burden in the developing world.

14-16 July 1989
15th World Economic Summit of the Group of Seven (G-7) (Paris)

The leaders of the Group of Seven cited the main challenges to the current world economic situation to be: the maintenance of growth, the development and integration of developing countries into the world economy and the safeguarding of the environment for future generations. Countries with fiscal and current account deficits were urged to take action to reduce these deficits, while countries with external surpluses were advised to pursue policies conducive to non-inflationary growth of domestic demand.

22-23 July 1989
Agreement Mexico-creditor banks (Washington)

A major package agreement was reached, between Mexico and representatives from 500 creditor banks, applying to $53 billion of Mexico's $107 billion total debt. Creditor banks had three options: cut the principal of old loans; swap old for new loans at fixed and interest rates lower than current market rates; or provide new loans. Creditworthiness of new bonds were to be enhanced by additional resources from the IMF, the World Bank, the Government of Mexico and Japan.

24-27 July 1989
OAU Summit (Cairo)

At the OAU summit meeting, the African Alternative Framework for Structural Adjustment Programmes was considered and adopted by the African states. The resolution recommended that African governments use the framework for preparing their country programmes and for negotiating assistance.

11 September 1989
Conference on Global Environment (Tokyo)

The Conference, sponsored by the World Bank, announced a series of measures by the Bank which would make available $1.3 billion in loans to the developing countries, over the next three years, to combat pollution and protect the environment. Moreover a new Gas Development Unit would be established to promote production, consumption and export of natural gas. The Bank was also to triple its lending to forestry in the context of the Tropical Forestry Action Plan.

26 September 1989
Annual Meeting of IMF/World Bank (Washington)

The meeting acknowledged the spread of market-oriented policies throughout the world and a new industrial revolution linked to the use of computer technology, leading to a sustained growth of 3 percent for the developed world, but a growing gap between North and South. The developed countries were unwilling to extend the so-called Toronto Terms of official debt relief to poor countries outside Africa, and the United States blocked a replenishment of funds for the IDA, a soft loan affiliate of the World Bank.

15 October 1989
Meeting of Antarctic Treaty Signatories (Paris)

The 39 signatories of the Treaty reached agreement at the 15th meeting of the consultative parties to hold a special round of negotiations in 1990 to set up a "comprehensive" protection system for the continent's environment. Parallel negotiations on liability for environmental damage under the Wellington Convention are also to be held.

11-30 November 1989
25th Session of the FAO Conference (Rome)

At the 25th Session of the FAO Conference, the major trends and policies in food and agriculture were discussed including: the world food and agriculture situation; progress report on the GATT multilateral trade negotiations; International Code of Conduct on the Distribution and Use of Pesticides; a plan of action for the integration of women into agriculture and development; and preparation for and FAO's contribution to the International Development Strategy for the Fourth UN Development Decade.

22 November 1989
The European Community (EC) and the European Free Trade Association (EFTA) meet to try to establish Common Economic Zone (Brussels)

The European Community announced that negotiations for a common economic zone between the European Community and the six states of EFTA will commence in 1990. The aim would be to extend to EFTA the Community's single market benefits of free movements of goods, resources, capital and people, without jeopardizing the autonomy of EC decision-making.

15 December 1989
Signature of Fourth Lomé Convention (Lomé IV)

The new Convention was concluded between the 12 countries of the European Economic Community, 68 countries of Africa, the Caribbean and the Pacific (ACP) (including Haiti and the Dominican Republic, which also acceded to the new pact). Namibia was also to join the Convention later in 1990. Significant features included an increase in the financial resources from ECU 8.5 billion in Lomé III to ECU 12 billion for the 1990-95 period; more favourable conditions for the STABEX and SYSMIN regulatory mechanisms, made available as grants and thus no longer debt-creating; and aid provisions for structural adjustment programmes in associate countries.

8 January 1990
New funding for IDA

Thirty-two nations agreed to provide SDR 11.68 billion over the next three years to the world's poorest countries. Negotiations were completed for the ninth replenishment of the resources of the International Development Agency (IDA), a World Bank affiliate that provides concessional loans to low-income developing countries. Priority areas for IDA funds were poverty reduction; support for sound macro-economic and sectoral policies and programmes, including further emphasis on institutional strengthening; and environmental programmes. The funds covered operations in the three years from 1 July 1990 to 30 June 1993.

5 February 1990
Meeting of EC Foreign Affairs Ministers (Brussels)

The future of relations between the European Community and countries in Eastern Europe was the main theme of the meeting of EC Foreign Affairs Ministers on 5 February 1990 in Brussels. Member countries came out in favour of a European Community proposal on establishing agreements of association with East European countries, which implement programmes for political and economic reform and which were geared toward a market economy.

9 February 1990
US Government unveils 1990 Farm Bill

The United States Government's latest proposals on the 1990 Farm Bill were sent to Congress as the 1985 Farm Bill was due to expire at the end of 1990. The Administration's new proposals aimed at giving the forthcoming bill more flexibility in adapting to changes on the world market, while pursuing the positive aspects of the 1985 bill.

23 February 1990
EC proposal to GATT on trade, monetary and financial policies

The EC submitted a proposal for a joint declaration on coherence between trade, monetary and financial policies, to be adopted at the ministerial level by GATT. The proposed joint declaration set out a number of guiding principles as regards trade policy, the international monetary system and finance linked to development.

12-16 March 1990
20th FAO Regional Conference for the Near East (Tunis)

The conference discussed, *inter alia*, regional economic cooperation for agricultural development in the Near East, a balanced diet as a way to good nutrition and the status of agricultural research in the region.

3 April 1990
17th FAO European Regional Conference (Venice)

The possible socio-economic impacts on agriculture of environmental policies, food quality, and the consequences of the changing situation in Eastern Europe were the main topics for discussion. FAO has offered to contribute to the reform process in Eastern Europe through technical assistance, specialized advice and experience gained from agricultural and rural development and modernization in other countries.

7 April 1990
Group of Seven Meeting of Finance Ministers (Paris)

The finance ministers and Central Bank governors of Canada, France, the Federal Republic of Germany, Italy, Britain, Japan and the United States met for an exchange of views on current global economic issues. The ministers and governors expressed the need for continued close coordination of their macro-economic and structural policies to obtain sustained growth, low inflation and greater stability of exchange rates.

9 April 1990
Formation of the European Bank for Reconstruction and Development

The EC, 14 other developed countries and eight East European states formally established a European Bank for Reconstruction and Development with its headquarters in London. The bank will concentrate its lending on the private sector and on productive investment.

17 April 1990
Trade Ministers meet to discuss progress of GATT (Mexico)

Trade ministers from 30 major trading countries met in Puerto Vallarta, Mexico, to examine the progress of GATT Uruguay Round negotiations. A deadline of July has been set for establishing a package of agreements to help complete the Round on schedule in December 1990.

23-27 April 1990
20th FAO Regional Conference for Asia and the Pacific (Beijing, China)

The conference discussed, *inter alia*, action programmes to overcome specific nutritional deficiencies in the region, and progress and prospects of biotechnology for crop and livestock production in Asia and the Pacific.

28 April - 1 May 1990
18th Special Session of the UN General Assembly (New York)

The session was devoted to international economic cooperation, in particular to the revitalization of economic growth and development of developing countries.

2 May 1990
Conference on the Environment (Bergen, Norway)

The conference, gathering participants from 35 industrialized countries, sought agreement on economic measures to reduce pollution and avert global warming — the greenhouse effect. The conference, sponsored by the UN Economic Commission for Europe, provided a forum for future East-West environmental cooperation. Western countries agreed to set measures to stabilize emissions of carbon dioxide and to bring the Third World into international environmental negotiations.

2-4 May 1990
European Agricultural Forum (Austria)

The role of the agricultural sector in the united and integrated Europe of tomorrow, the problems facing the farming industries in the new democracies of Central and Eastern Europe and the search for a common European approach in the GATT Uruguay Round of multilateral trade negotiations, were the main themes of the forum.

21 May 1990
World Bank and OECD release studies on effects of subsidies to farmers

A series of studies released by the World Bank and the Organisation for Economic Cooperation and Development state that cutting subsidies to farmers would save a considerable part of the $200 billion they cost consumers and taxpayers in 24 industrial countries every year. The studies assessed the damage inflicted on the developing countries by the cost of these farm programmes — in terms of distorted markets, depressed commodity prices and underdeveloped farm economies.

11-15 June 1990
16th FAO Regional Conference for Africa (Marrakech)

The main themes of the conference were strategies for combating malnutrition in Africa and a programme for soil preservation.

27 June 1990
Proposal for economic partnership between the United States and Latin America

President George Bush launched proposals for a new economic parternership between the United States and Latin America. The proposals included the launching of a process aimed at a hemisphere-wide free trade zone; measures for enhancing capital flows toward Latin American countries involved in IMF adjustment programmes and receiving World Bank loans, through the creation of a five-year multilateral investment fund run by the IDB (and providing grants to assist specific reforms toward a market-oriented system); and a reduction of the $12 billion debt owed to the United States Government.

28 June 1990
IMF quotas increased

Following recommendations from the Interim Committee, the IMF's Board of Governors approved a 50 percent increase in quotas. Member countries have until 31 December 1991 to ratify the decision according to national legislative or other procedures.

9-11 July 1990
Summit Meeting of the seven main western industrial countries (Houston, USA)

The communiqué rejected trade protectionism and stated that the successful outcome of the GATT negotiations on agriculture required substantial and progressive reductions in support and protection to agriculture in all its forms. It stated that help would be provided to Central and Eastern European nations that were firmly committed to economic and political reform, but reiterated that the commitment to assist developing countries, particularly the poorest of them, would not be weakened by support to Central and Eastern Europe. It encouraged commercial banks to conclude promptly agreements on financial packages including debt and debt-service reduction and new money, in favour of indebted countries implementing courageous reforms. It agreed that, in the face of threats of irreversible environmental damage, lack of full scientific certainty was no excuse to postpone adequate action.

9-13 July 1990
21st FAO Regional Conference for Latin America and the Caribbean (Santiago, Chile)

The conference discussed, *inter alia,* sustainable rural development in fragile ecosystems in Latin America and the Caribbean and causes and prevention of malnutrition in the region.

23-26 July 1990
Meeting of Trade Negotiations Committee of the Uruguay Round of Multilateral Trade Negotiations within GATT (Geneva)

The Trade Negotiations Committee (the governing body of the Uruguay Round) set down a timetable for final negotiations leading to the planned conclusion of the Uruguay Round at a ministerial meeting in Brussels from 3 to 7 December 1990. The deadline for submission of offers concerning negotiations on agriculture was set at 15 October 1990.

2 August 1990
Persian Gulf crisis

The Iraqi invasion of Kuwait and the ensuing trade embargo of Iraq decided upon by the UN Security Council caused substantial increases in world oil prices. Countries most directly affected by the crisis are those in the Near East and Asia with a large number of migrant workers in Iraq and Kuwait. But the increase in oil prices seriously circumscribed growth prospects particularly of non-oil exporting developing countries and reform-pursuing countries in Eastern Europe.

3-14 September 1990
Second UN Conference on the Least Developed Countries (LDCs) (Paris)

The conference discussed new development strategies for the 1990s. Industrialized countries undertook to increase development assistance to the poorest developing countries centred on a target of 0.15 percent of their GDP. This was matched by a commitment from the least-developed countries to implement appropriate national development policies.

19 September 1990
Commonwealth Finance Ministers' Meeting

The United Kingdom's Chancellor of the Exchequer launched the "Trinidad and Tobago Terms" which are a development of the existing Toronto debt-relief terms for poor countries undergoing economic adjustment programmes. The plan, which concerns middle- and lower-middle-income countries, would cut the stock of debt of 17 African and two South American debtors already benefiting from the Toronto Terms arrangements.

24-28 September 1990
Tenth Session of FAO's Committee on Forestry (Rome)

The session considered, *inter alia,* biodiversity and the review of the Tropical Forestry Action Plan.

27 September 1990
Annual Meeting of IMF/World Bank (Washington)

A major theme of the meeting was the response to the negative economic effects of the events in the Near East, including aid strategies for the countries hard hit by the crisis. But the World Bank president also urged the international community to make as big an effort combating poverty as it had in dealing with the Gulf crisis and outlined ways in which countries could cooperate in reducing poverty.

29 October-7 November 1990
Second World Climate Conference (Geneva)

The conference considered the report of the Intergovernmental Panel on Climate Change (IPCC). The scientific and technical session recommended that although uncertainties remained, nations should take steps toward reducing emissions of greenhouse gases through national and regional actions and negotiations for a global convention on climate change. The ministerial session was not willing to adopt quantitative targets to stabilize emissions of greenhouse gases, but welcomed the commitments of many developed countries to do so.

Contents

TABLES

FIGURES

BOXES

ANNEX TABLES

Glossary

AAF	African Alternative Framework
ACC	Arab Cooperation Council
ACP	African, Caribbean and Pacific States
AfDB	African Development Bank
AMS	Aggregate measure of support
AMU	Arab Maghreb Union
ARPs	Acreage Reduction Programmes
AsDB	Asian Development Bank
BMR	Basal metabolic rate
CAP	Common Agricultural Policy
CARICOM	Caribbean Common Market
CBI	Caribbean Basin Initiative
CGEs	Computable general equilibrium models
COCOBOD	Cocoa Marketing Parastatal
COMECON	(CMEA) Council for Mutual Economic Assistance
CPEs	Centrally planned economies
DAC	Development Assistance Committee
DES	Dietary energy supply
DRS	Debt Reporting System
EC	European Communities
ECA	Economic Commission for Africa
ECLAC	Economic Commission for Latin America and the Caribbean
ECOSOC	Economic and Social Council of the United Nations
ECU	European Currency Unit
EEC	European Economic Community
EEP	Export Enhancement Programme
EMS	European Monetary System
FOR	Farmer-owned reserve
FSAS	Food Security Assistance Scheme
FY	Fiscal year
GATT	General Agreement on Tariffs and Trade
GDP	Gross domestic product
GNP	Gross national product
GOSAGRO-PROM	State Agro-industrial Committee (USSR)
GOSPLAN	State Planning Committee (USSR)
GSP	Generalized System of Preferences
HL	Hybrid lending
IBRD	International Bank for Reconstruction and Development
IDA	International Development Association
IDB	Inter-American Development Bank
IEFR	International Emergency Food Reserve
IFAD	International Fund for Agricultural Development
ILO	International Labour Organisation
IMF	International Monetary Fund
IPCC	International Panel on Climate Change
ITTO	International Tropical Timber Organization
LDCs	Least-developed countries
LIBOR	London inter-bank borrowing rate
MCAs	Monetary compensatory amounts

MFA	Multifibre arrangement
MLAR	Ministry of Lands, Agriculture and Resettlement
MTN	Multilateral trade negotiations
NAEs	Newly agro-industrialized economies
NAMBOARD	National Agricultural Marketing Board
NCA	Normal Crop Acreage
NERP	New Economic Recovery Programme
NIEs	Newly industrialized economies
OAU	Organization of African Unity
OECD	Organisation for Economic Cooperation and Development
OPEC	Organization of the Petroleum Exporting Countries
PAAERD	UN Programme of Action for African Economic Recovery and Development
PAMSCAD	Programmes of Action to Mitigate the Social Costs of Adjustment
PSE	Producer subsidy equivalent
SADCC	Southern African Development Coordination Conference
SALs	Structural adjustment loans
SAPs	Structural adjustment programmes
SDA	Social dimensions of adjustment
SDR	Special drawing right
SECALs	Sectoral adjustment loans
SECAP	Sectoral adjustment programme
SSAPs	Sector structural adjustment programmes
STABEX	Système de stabilisation des recettes d'exportation
SYSMIN	Système de stabilisation des recettes d'exportation des produits minéraux
TFAP	Tropical Forestry Action Plan
UDI	Unilateral Declaration of Independence
UNCTAD	United Nations Conference on Trade and Development
UNDP	United Nations Development Programme
UNICEF	United Nations Children's Fund
VAT	Value added taxes
WFC	World Food Council
WFP	World Food Programme
WHO	World Health Organization
ZCF	Zambian Cooperative Federation

Explanatory note

The following symbols are used in the tables:

–	=	none or negligible
...	=	not available
1988/89	=	a crop, marketing or fiscal year running from one calendar year to the next
1987-89	=	average for three calendar years.

Figures in statistical tables may not add up because of rounding. Annual changes and rates of change have been calculated from unrounded figures. Unless otherwise indicated, the metric system is used throughout. The dollar sign ($) refers to US dollars. "Billion", used throughout, is equal to 1 000 million.

Production index numbers
FAO index numbers have *1979-81* as the base period. The production data refer to primary commodities (e.g. sugar cane and sugar beet instead of sugar) and national average producer prices are used as weights. The indices for food products exclude tobacco, coffee, tea, inedible oil-seeds, animal and vegetable fibres and rubber. They are based on production data presented on a calendar-year basis.[1]

Trade index numbers
The indices of trade in agricultural products also are based on 1979-81. They include all the commodities and countries shown in the *FAO Trade Yearbook 1989*. Indices of total food products include those edible products generally classified as "food".

All indices represent changes in current values of exports (f.o.b.) and imports (c.i.f.), all expressed in US dollars. If some countries report imports valued at f.o.b. (free on board), these are adjusted to approximate c.i.f. (cost, insurance, freight) values. This method of estimation shows a discrepancy whenever the trend of insurance and freight diverges from that of the commodity unit values.

Volume and unit value indices represent the changes in the price-weighted sum of quantities and of the quantity-weighted unit values of products traded between countries. The weights are respectively the price and quantity averages of 1979-81, which is the base reference period used for all the index number series currently computed by FAO. The Laspeyres formula is used in the construction of the index numbers.[2]

Definitions of "narrow" and "broad"
The OECD definitions of agriculture are generally used in reporting on external assistance to agriculture. The *narrow* definition of agriculture, now referred to as "directly to the sector" includes the following items:

Appraisal of natural resources
Development and management of natural resources
Research
Supply of production inputs
Fertilizers
Agricultural services
Training and extension
Crop production
Livestock development
Fisheries
Agriculture (subsector unallocated)

The *broad* definition includes, in addition to the above items, activities that are defined as "indirectly to the sector". These activities are:

Forestry
Manufacturing of inputs
Agro-industries
Rural infrastructure
Rural development
Regional development
River development

Regional coverage
Developing countries include: i) developing market economies (Africa, Latin America, Near East,[3] Far East and other) and ii) Asian centrally planned economies or Asian CPEs (China, Cambodia, Democratic People's Republic of Korea, Mongolia and Viet Nam).

Developed countries include[4]: i) developed market economies (North America, western Europe including Yugoslavia, Oceania, Israel, Japan and South Africa) and ii) Centrally planned economies of Eastern Europe and the USSR (Bulgaria, Czechoslovakia, German Democratic Republic, Hungary, Poland, Romania and the USSR).[5]

Country designations used in this publication remain those current during the period in which the data were prepared.

---

[1] For full details, see *FAO Production Yearbook 1989*.

[2] For full details, see *FAO Trade Yearbook 1989*.

[3] The *Near East* includes: Egypt, Libyan Arab Jamahiriya, the Sudan, Afghanistan, Bahrain, Cyprus, Islamic Republic of Iran, Iraq, Jordan, Kuwait, Lebanon, Oman, Qatar, Kingdom of Saudi Arabia, Syrian Arab Republic, Turkey, United Arab Emirates and Yemen Arab Republic and Democratic Yemen.

[4] Note that "industrial countries", as defined by the International Monetary Fund (IMF) (see Figures 1.1, 1.3 and 1.4), include: Australia, Austria, Belgium, Canada, Denmark, Finland, France, Germany (Fed. Rep. of), Iceland, Ireland, Italy, Japan, Luxembourg, the Netherlands, New Zealand, Norway, Spain, Sweden, Switzerland, the United Kingdom and the United States. (They do not include Yugoslavia, Greece, Israel, South Africa, the centrally planned economies and some other smaller countries.)

[5] Albania is omitted in this report for insufficient data.

PART ONE

WORLD REVIEW

WORLD REVIEW

OVERALL ECONOMIC ENVIRONMENT

Recent events in the Near East have made prospects more uncertain for the global economic environment surrounding agriculture. What is certain is that prices of oil-based products and the costs of energy in all its forms, will significantly rise and negatively affect most countries for an indefinite period of time.[1] The likely consequences for oil-importing countries are a slowdown in economic activity, reinforced inflationary pressures and a worsening in fiscal and external imbalances. Related effects are the likelihood of more restrictive monetary and fiscal policies in many countries and tighter conditions for external financing in international markets. Within the OECD countries, the risk of introducing potentially recessionary measures — in particular raising interest rates — is politically more acceptable than promoting economic growth at the potential cost of higher rates of inflation. Indeed international interest rates already were rising in several industrial countries during the first half of 1990 in response to inflationary fears; following the events in the Near East, the increase has tended to be more accentuated.

Equity prices have fallen sharply in all the major markets, particularly those of countries more dependent on imported oil. By contrast, exchange rate movements have been relatively orderly, with a continuining depreciation of the US dollar only briefly interrupted in early August.

The extent of these influences can only be tentatively quantified in the current fluid situation, although they are likely to affect the short- rather than the medium-term outlook. Barring the outbreak of a major conflict, and assuming no further major oil price increases, most OECD member countries should be able to absorb the shock without entering into what can be technically defined as a recession. Several tentative estimates have been made for the short term, pointing to variable degrees of reduction in the rate of economic growth in industrial countries, higher inflation rates, and a deterioration in their external balances. IMF economic estimates for 1982-90 and forecasts for 1991 are shown in Figures 1.1 to 1.4.[2]

For industrial countries, the main features in the IMF's estimates for 1990/91 are:

- Real GDP would rise 2.5 percent in both 1990 and 1991, sharply down from an average of nearly 4 percent in 1988-89. Relatively high growth is projected in the Federal Republic of Germany and Japan, but in North America and the United Kingdom growth in 1990 and 1991 would barely exceed 1 percent.
- Employment growth in 1990 is projected to slow to 1.2 percent, the smallest increase since 1985. Should the crisis be protracted and lead to increased unemployment, stronger protectionism is to be feared.
- With higher energy prices and rates of industrial capacity utilization still high, inflation would rise to 4.8 percent in 1990 and slightly fall in 1991.
- Lower levels of economic activity and higher interest rates would worsen the fiscal position of most industrial countries.
- Adjustment in the current account imbalances among the main trading countries is expected to continue, although such imbalances are still unsustainably large.
- The overall trade deficit of developed countries is expected to decline, despite the negative effect of higher oil prices on the trade balance of net oil importers. For the

[1] From a low of $16 per barrel in July 1990, oil prices were fluctuating around $30-40 per barrel in September/October.

[2] *World Economic Outlook.* IMF, October 1990. Projections presented in this report assume oil prices averaging $26 per barrel for the remainder of 1990, and declining gradually to $21 per barrel in the fourth quarter of 1991. In the light of September/October prices ranging up to $40 per barrel, these projections may be optimistic.

major industrial countries the trade balance is expected to shift from a deficit in 1989 to a surplus in 1990. However, higher oil prices are expected to cause a worsening in the trade deficit of other industrial countries in 1990.

- Exports and imports of industrial countries are estimated to expand by approximately 6 percent in 1990, about one percentage point less than previously estimated. The terms of trade of these countries are now expected to deteriorate slightly in 1990.

In sum, unlike the situation that followed the oil crisis of the mid-1970s, or that of 1979-82, the industrial world would be spared recession and growth could resume at a relatively strong pace in 1992. What makes the difference between the current and previous oil price shocks, and permits a relatively optimistic

Figure 1.1

WORLD ECONOMIC OUTPUT
(Percentage change over preceding year)

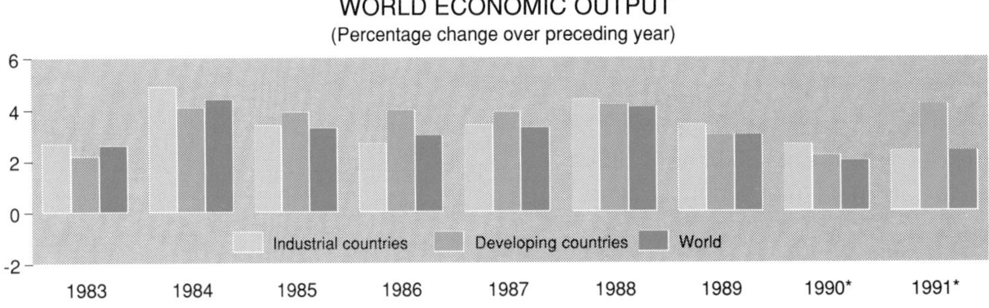

ECONOMIC GROWTH, DEVELOPING COUNTRY REGIONS
(Percentage change over preceding year)

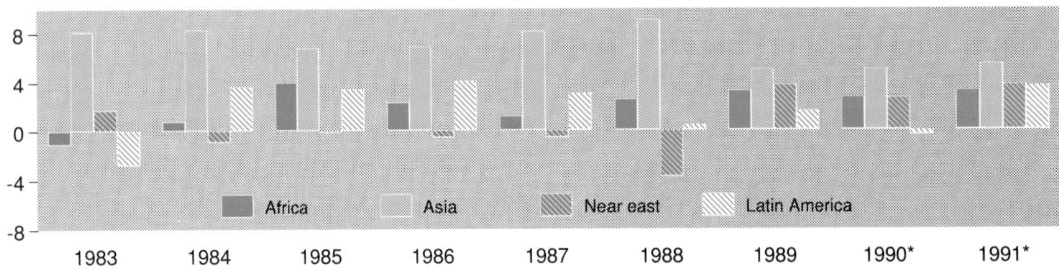

Source: IMF

* 1990: Estimate - 1991: Forecast

Figure 1.2

WORLD OUTPUT AND VOLUME OF WORLD TRADE 1983-91
(Percentage change over preceding year)

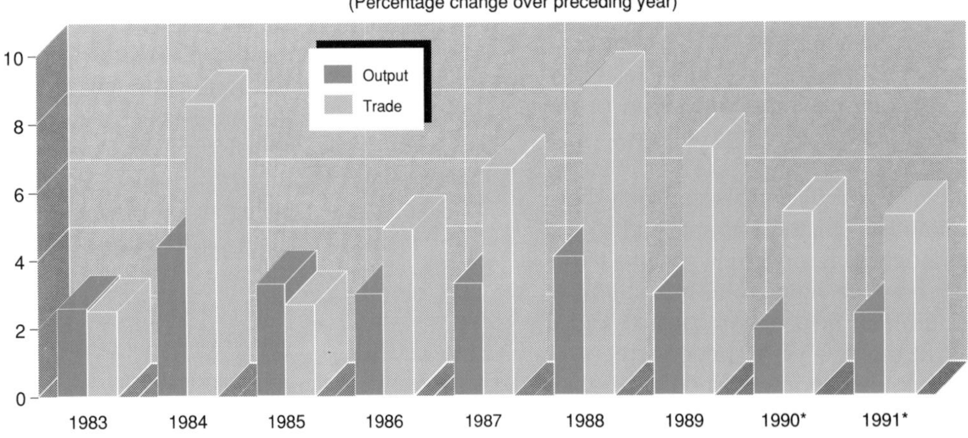

Source: IMF

* 1990: Estimate - 1991: Forecast

forecast for OECD countries, is the remarkable reduction in these countries' dependence on imported oil.[3] Another reason for optimism is that, unlike the situation in the early 1980s when the major OECD economies were in a similar position *vis-à-vis* their business cycles — close to full employment — and adopting a similar policy stance — restrictive measures to control inflation — the situation in recent years has been characterized by a diversity in cyclical positions. This makes the generalized adoption of potentially recessive policies unlikely. Also, countries which are still growing fast, such as

[3] Conservation efforts and the development of alternative sources of energy enabled OECD countries to reduce oil consumption as a share of GDP by almost one-third since the early 1980s. Moreover, current OECD emergency reserves are ample, their volume being sufficient to cover demand for three months.

Figure 1.3

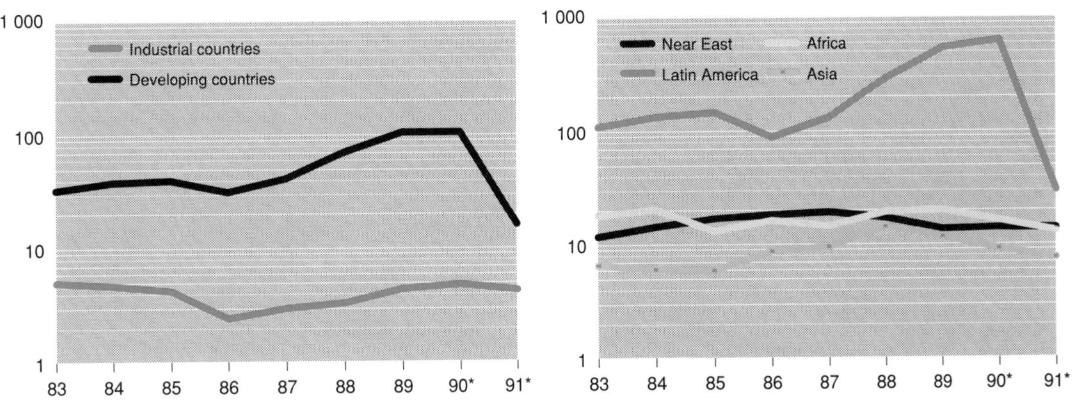

Source: IMF

* 1990: Estimate - 1991: Forecast

Figure 1.4

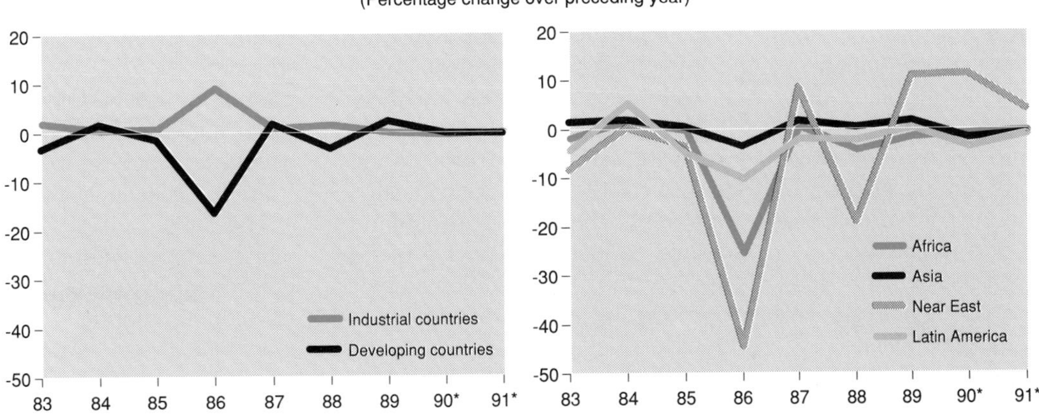

Source: IMF

* 1990: Estimate - 1991: Forecast

the Federal Republic of Germany and Japan, should provide strong demand for the exports of slower-growing countries.

A particularly difficult period is to be feared for the Eastern European countries. They were already facing grave problems in restructuring their economies, they are generally highly dependent on imported sources of energy, and they will have to pay hard currency for their oil imports from the USSR following the breakdown of barter arrangements of COMECON. They are also relatively inefficient users of energy, their energy use per unit of output being about twice the average levels of the OECD countries. Finally, trade sanctions against Iraq imply significant market losses for some Eastern European countries. Thus, for Yugoslavia and Romania, exports to Iraq and Kuwait accounted for about 2.5 percent of their total exports. For the USSR, on the other hand, the oil price rise could provide major economic benefits as long as the windfall gain is well managed and allocated.

Contrasting situations would be faced by developing countries according to their net trade position in oil. Varying gains from higher oil prices would be realized by the relatively small number of net oil exporters, some of which have also announced a substantial expansion in the volume of their oil exports.[4] Among the main highly indebted countries of this group, only Venezuela, Ecuador, Algeria,

[4] The World Bank estimates that only 11 non-Gulf oil-exporting developing countries would directly benefit from higher oil prices.

BOX 1.1

Recent developments in debt reduction and debt relief

Slow and uneven progress has been achieved in recent years in debt and debt-service reductions due to the complex negotiation process required, the divergence of interests among creditors and uncertainties about the adjustment efforts of some debtors. Such uncertainties appear even greater now. For many developing countries the deteriorating economic environment may undermine their capacity and resolve to pursue efforts toward stabilization and structural reform. However, the recent past has been fertile in promising new initiatives and developments, including those described below.

The "Toronto Terms", announced at the Toronto summit of the seven most industrialized countries in June 1988, are targeted on rescheduling official debt of sub-Saharan Africa. Since late 1988, 18 countries have obtained rescheduling on these terms through the Paris Club. Under the "Toronto Terms", creditors can choose from a menu of rescheduling options, including a significant writing-off of the amount rescheduled, a reduction in the interest rate on the rescheduled amount and an extension of the grace period and maturity.

The Brady Plan, launched in March 1989, is aimed at lowering the debt burden of 19 middle-income debtor countries with predominantly commercial debt. The main mechanisms are various forms of debt and debt-service reduction, including debt buy-backs, exchange of old debts at a discount for new collateralized bonds, policies to encourage repatriation of capital flight and foreign direct investment. Commercial banks are expected to provide debt reductions and new loans. The IMF and the World Bank are expected to provide up to $20 to 15 billion of financial resources. These measures should reduce annual debt service of the severely indebted middle-income countries by approximately $6 billion per year between 1990 and 1993. By mid-1990, four countries (Mexico, the Philippines, Costa Rica and Venezuela) reached agreement with creditor banks. The Moroccan Government and creditor banks have also agreed in principle to reschedule Morocco's bank debt under the Brady Plan. A major problem in the implementation of the Brady Plan has been the inadequacy of funds available to the World Bank and IMF for including debt reduction in their programme of operations — although the agreement reached in May 1990, that IMF quotas should be increased by 50 percent, was a positive development in this context. Moreover, the use of World Bank and IMF funds to replace commercial bank debt envisaged in the Brady Plan means less flexibility for developing countries involved because loans from the IMF and World Bank cannot be rescheduled. Also, debt-reduction borrowing reduces a country's access to other loans from these agencies.

The Paris Club of Creditors, which brings together creditor countries for the rescheduling of developing countries' official debt, launched a new strategy of offering more favourable terms for lower middle-income countries. Four countries (Morocco, Honduras,

Indonesia, Nigeria and Mexico should benefit, although in the case of Mexico, the expected gains would be largely offset by the negative effects of the reduction in economic activity in the United States. As regards net oil importers, many of them are likely to suffer serious economic hardship, according to their dependency on external sources of energy. Other than the direct impact of higher oil prices on their balance of payments and domestic production costs, the economies of the oil-importing developing countries are also likely to face indirect depressing effects from higher petroleum prices. Their trade and growth prospects, already grim for many, particularly those predominantly exporting agricultural products, would suffer further from a slow-down in economic activity in industrial countries. For the many indebted countries among them, tighter financial and monetary policies, involving in particular higher interest rates in creditor countries, is a potentially disastrous perspective. Finally, several countries in the Near East and Asia are expected to face major difficulties in view of the massive repatriation of workers from Iraq and Kuwait, which represented for some of them major sources of income in the form of remittances; and in view of the reduced trade and trans-shipment activities in the zone more directly affected by the trade embargo: Jordan, Turkey and Egypt. Jordan's losses from higher oil prices, lost remittances and lost trade with Iraq and Kuwait could amount to 35 percent or more of its GDP in 1991. The losses should represent less than 5 percent for Egypt and Turkey.

El Salvador and the Congo) have benefited from rescheduling packages which will be spread over a period of 15 years, five years longer than was previously allowed by the Paris Club.

Besides the Toronto and the Brady measures, certain other steps have been taken to assist low-income countries. There has been a sharp increase in the forgiveness of official development assistance (ODA) loans to sub-Saharan Africa. The United States announced that it would forgive $1 billion of these loans beginning 1 October 1989; France has indicated that it will forgive $2.3 billion, the Federal Republic of Germany has already cancelled $1.5 billion and Canada plans to write off more than $425 million.

In June 1990, the United States President announced a plan to boost economic reform and development in Latin America; the "Enterprise for the Americas Initiative" including a reduction in part of the $12 billion debt owed to the United States Government, conditional to undergoing reforms with the IMF/IBRD.

At the Commonwealth Finance Ministers' meeting in September 1990, Mr John Major, the then UK Chancellor of the Exchequer, launched the "Trinidad and Tobago Terms", which are a development of the existing Toronto debt-relief terms for poor countries undergoing economic adjustment programmes. The plan, which concerns middle-income and lower-middle income countries, will cut the stock of debt of 17 African and two South American debtors already benefiting from the Toronto Terms arrangements. The plan provides for writing off up to $18.3 billion of official debts, rescheduling the countries' whole stock of debt; providing additional cash flow benefits of $2.7 billion by capitalizing interest during its first year of operation; and extending repayment periods to 25 years. The Trinidad Terms would eventually be extended to other low-income countries not included in the original Toronto Terms.

Concerns about the possible negative effects of events in the Near East and the changed economic environment on efforts to resolve the debt problem and the capacity of many developing countries to pursue stabilization and structural reform policies were voiced at the second United Nations Conference on Trade and Development (UNCTAD) Conference on the Least Developed Countries (LDCs). The conference, held in Paris from 3 to 14 September 1990, considered the poorest countries to be the first victims of the crisis. It is to be hoped that the welcome aid initiatives announced at the conference in favour of LDCs were signs of a resolve by richer countries to pursue, and even strengthen, their overall aid efforts in favour of less developed countries despite the current difficulties in the global economic environment.

Such worsened prospects for a majority of developing countries can be seen from the following IMF estimates.

- Growth in real GDP in the developing countries as a group is estimated to fall from 3 percent in 1989 to 2.2 percent in 1990. At the regional level, growth in 1990 would be: Africa 2.7 percent; Asia 5 percent; Near East 2.6 percent; and Latin America and the Caribbean −0.4 percent. However, all regions are forecast to achieve significantly higher growth rates in 1991.
- The increase in consumer prices in 1990 is now estimated as follows: Africa 16 percent; Asia 9 percent; Near East 14 percent; and Latin America and the Caribbean 626 percent. Much lower rates are expected in 1991 in all regions except the Near East.
- For non-oil exporting developing countries, the growth in the volume of exports is estimated to fall to about 6 percent in 1990, the lowest rate since 1985, and decelerate further to 5.6 percent in 1991. Their terms of trade are expected to deteriorate 3.3 percent in 1990, and by a further 1.5 percent in 1991, reflecting a sharp fall in the prices of non-fuel primary commodities (−8 percent in 1990).
- The aggregate current account deficit of non-oil exporting developing countries is estimated to increase by $12 billion to $22 billion in 1991. The financing of these deficits is expected to be provided largely by official creditors.
- Total external debt of developing countries (excluding IMF credit) is expected to increase by 9 percent in 1990-91 and reach $1,354 billion by the end of 1991. Latin America and the Caribbean is the only region where external debt is projected to remain broadly unchanged (at $415 billion).

The possible disruptive effect of the situation in the Near East on the continuing efforts to resolve the problem of debt is cause for concern.

The debt-service ratio (payments on interest and amortization as a share of exports of goods and services) of developing countries is forecast to increase from 19 percent in 1990 to 20 percent in 1991. The situation, however, differs between regions. Thus in Latin America and the Caribbean, the debt-service ratio is expected to increase from 37 percent in 1990 to 43 percent in 1991 (compared to 33 percent in 1989). In Asia the debt-service ratio is forecast to decline slightly to 9 percent in 1991, down from 10 percent in 1990 and 11 percent in 1989. In Africa it should stabilize at 27 percent in 1990 and 1991, up from 25 percent in 1989, whereas in the Near East a ratio of 26 percent is forecast for both 1990 and 1991, compared to a ratio of 25 percent in 1989.

Consequences for agriculture

Agriculture's sensitivity to the macro-economic environment varies according to the sector's position in the different countries and its exposure to external influences.

In industrial countries, while the importance of agriculture is limited in terms of its direct contribution to income and employment,[5] its indirect impact on economic activity is considerable — as a provider of inputs for downstream activities such as food manufacturing, processing, retailing, etc., and as an outlet for upstream industries such as agricultural inputs, machinery and services. However, policy-induced rigidities and distortions have prevented economic signals from being fully transmitted to the agricultural sector in many of these countries. These policies have: attracted resources into the sector that could have been more cost-effectively used elsewhere; artificially raised domestic prices, negatively affecting consumers and distorting competitive relationships throughout downstream and upstream activities; insulated agriculture from world markets by border protection; and rendered excess production spuriously competitive in world markets through widespread use of export subsidies. To the extent that the exchange rate is also influenced by agricultural support, these policies have also indirectly affected the competitive position of non-agricultural activities. Despite awareness of the undesirable effects of these policies and their heavy costs to budgets and taxpayers, political and social considerations have only permitted partial and uneven progress toward greater market orientation.[6]

---

[5] Gross value added in agriculture accounts for about 3 percent of GDP in the OECD area. Employment in agriculture accounts for about 8 percent of civilian employment in Japan and the EEC, and 3 percent in the United States.

[6] According to a recent World Bank/OECD compendium of studies, the current annual cost of farmer subsidies to consumers and taxpayers in 24 industrial countries is about $200 billion.

For developing countries' agriculture, the overall economic environment and policy developments in industrial countries assume a particular significance. Non-oil developing countries typically rely on agriculture for 20 to 40 percent of their GDP, 60 to 80 percent of their employment and 50 to 70 percent of their total export earnings. At the same time, developing countries offer the highest potential for demand-based agricultural development in a long-term worldwide perspective.

However, recent economic and market developments have not moved toward the realization of this potential. At the current levels of per caput income growth, domestic demand for food cannot be expected to exceed 2 to 2.5 percent in 1989-90 in much of Africa and Latin America and the Caribbean — hardly the kind of impulse required to stimulate production growth. Nor can agricultural trade be expected to contribute significantly to economic recovery given current prospects for import demand from industrial countries, low prices for several key agricultural export commodities, high oil prices and market instability caused by higher interest rates. This is shown for the 40 developing countries defined by the IMF as predominantly agricultural exporters, by the following estimates:

- For the predominantly agricultural exporter developing countries, per caput GDP fell 1.1 percent in 1989 and was expected to fall 0.8 percent in 1990. By comparison, per caput GDP is estimated to have increased 1.1 percent in 1989 and 0.2 percent in 1990 in developing countries with a developed — more diversified — export base.
- Gross capital formation for these countries fell from about 17 percent of GDP in the early 1980s to less than 14 percent in 1989 and 1990 — compared to a relatively constant 20 percent of GDP for developing countries with a diversified export base.
- Inflation was significantly higher in predominantly agricultural exporter developing countries than in those with diversified exports throughout the 1980s.
- Export volumes rose at a significantly lower rate in predominantly agricultural exporter developing countries than in manufactured goods exporters during most of the 1980s, with the terms of trade of the former estimated to have declined by almost one-third between 1985 and 1990.
- Import volumes declined by an aggregate 11 percent during 1982-90 with prospects for a further contraction in 1991.

These developments raise a number of important issues. Trade is assumed to be the transmission belt by which economic growth in industrial countries produces growth elsewhere. In the 1980s, however, despite the long period of expansion in industrial countries, the impact of this transmission has been uneven between different groups of developing countries. On the one hand, a number of countries, particularly in Southeast Asia, have not only succeeded in overcoming external shocks and constraints on their markets but have succeeded in basing their rapid growth on export expansion. On the other hand, a large number of other countries that primarily rely on agriculture as a source of foreign exchange and growth have recorded poorer performances and are facing grimmer prospects than most other developing countries on almost every economic measure. What has differentiated the former group of countries has been their ability, or inherent capacity, to turn into labour-intensive, export-oriented, and increasingly broad-based manufacturing economies.

For the latter group the fact of being primarily agricultural exporters in a hostile economic environment may not totally explain their negative experience. There have been cases when such inimical external influences have coincided with unsound growth and trade-supportive government policies, and with unsustained domestic efforts as a consequence.

It can also be argued that, as development proceeds and market prospects for agriculture are reduced, the sector must lose relative importance as a primary source of foreign exchange and should, instead, be increasingly regarded as a generator of domestic supply, a source of labour and as a market for local industry. Whatever the merit of this argument in a long-term perspective, many countries cannot afford to let agricultural exports become but a marginal source of growth before exports from other sectors have grown sufficiently. For most of them the sector must remain the main source of export earnings for a long time to come and so the external environment will remain a dominant influence in their development performance, domestic policies and efforts. Often the success of domestic policy reforms depends on improvements in the external environment, a fact which calls for

specific action. Measures should include providing these countries with better access to international markets; granting differential and more favourable treatment for their exports; ensuring non-automatic reciprocity of concessions granted by industrial countries; and ensuring continuity and improvement in existing preferential trade and aid arrangements.

TRENDS IN FOOD AND AGRICULTURAL PRODUCTION IN 1986-89

At the world level, the second part of the 1980s was a period of faltering agricultural output growth. This was especially the case for developed countries, North America in particular, where output declined by a yearly average of 0.7 percent during 1986-89, reflecting three consecutive years of decline not fully compensated by a recovery in 1989 (Figure 1.5). Output also declined during 1986-89 in Oceania, virtually stagnated in western Europe and rose only moderately in Eastern Europe and the USSR amid large year-to-year fluctuations.

Growth in food and agricultural production varied widely in developing countries in the latter half of the 1980s. At the regional level, the overall pattern was one of sustained growth in Asia, although the pace of growth slowed from the 1981-85 average; mediocre performances in Latin America and the Caribbean where production lagged behind population growth; marked instability in the Near East with virtually no progress overall; and a disquietingly poor record in Africa where production growth fell below the already inadequate levels of the early 1980s.

The commodity composition of the main groups of food products is shown in Table 1.1. Outstanding features are:

- World *cereal* production only marginally increased in 1986-89 reflecting an overall decline in developed countries and a slow-down in growth in developing countries to half the average rate of 1981-85. The shortfall in developed countries resulted from the prolonged drought of 1988 in North America, a sharp decline in production in Oceania, particularly of coarse grain, and overall stagnation in western Europe. On the other hand, cereal production reportedly rose nearly 3 percent yearly in Eastern Europe and the USSR, as a result of bountiful harvests in 1986 and a production recovery in the USSR and Bulgaria in 1989 after two years of setbacks.
- Among developing country regions, cereal production was favourable only in the Far East, despite the poor monsoon in 1986 and 1987 which caused growth in paddy output in 1986-89 to fall markedly below previous

TABLE 1.1. **Agricultural production, by commodity**

Item	Developed countries: Average annual change 1981-85	Developed countries: Average annual change 1986-89	Developing countries: Average annual change 1981-85	Developing countries: Average annual change 1986-89	World: Average annual change 1981-85	World: Average annual change 1986-89
	(	....	%	....	....	)
Total cereals[1]	3.3	−0.5	3.8	1.6	3.4	0.4
Wheat	0.4	1.3	6.4	2.3	2.5	1.7
Rice, paddy	2.3	−0.9	3.6	1.9	3.4	1.7
Coarse grains	5.5	−1.1	2.6	0.6	4.1	−1.0
Root crops	3.5	−1.4	0.6	1.6	1.6	0.5
Pulses	9.6	6.8	2.9	2.0	4.6	3.3
Vegetable oils	3.9	2.2	4.8	6.8	4.4	4.9
Sugar, centrifugal (raw)	−0.4	1.1	2.8	1.7	1.4	1.4
Cocoa beans	—	—	3.5	6.3	2.9	5.9
Coffee	—	—	−0.8	3.8	−0.8	3.8
Tea	1.3	−0.9	5.8	3.0	5.2	2.6
Cotton lint	3.9	4.1	6.9	—	5.3	1.5
Tobacco	1.1	−2.5	9.4	3.1	6.3	1.4
Total meat	1.3	1.8	4.4	4.3	2.4	2.8
Total milk	1.4	0.3	3.8	3.3	2.0	1.1
Hen eggs	0.9	0.5	8.2	4.4	3.6	2.2

NOTE: Percentage changes have been calculated from unrounded figures.
[1] Including rice in terms of paddy.
Source: FAO, Statistics Division (ESS), Commodities and Trade Division (ESC).

high rates. For all other developing regions, the period was characterized by generally mediocre or poor cereal harvests, both in terms of their domestic requirements — none of them succeeded in increasing per caput cereal production — and in relation to the average levels of 1981-85.

- The *vegetable oils* sector has been among the most dynamic despite a slow-down in production growth in developed countries in the second half of the 1980s which reflected, on the one hand, policy measures to reduce the incentive to production resulting from high support prices in some importing countries and, on the other hand, farmers' response to depressed world market prices in some exporting countries. Faster growth in developing countries more than offset the slow-down in developed countries and resulted in the former increasing their stake in the sector to nearly two-thirds of the total. At the regional level, growth was particularly strong in the exporting countries of Southeast Asia and Latin America as well as, among importing countries, in China and, more recently, in India, reflecting increased policy incentives. Production growth continued to lag well behind population growth in Africa and the Near East.
- Among other food products, growth of *root crop* production was inadequately low in Latin America and Africa, but moderately accelerated in the other major producing and consuming region, the Asian CPEs. Production of *pulses* rose close to population growth in the Far East, with dry bean and chickpea output expanding significantly in India in 1986 and 1989. Pulse production also increased strongly in the Near East and Africa but rose only slightly in Latin America and actually fell in the Asian CPEs. Sugar production moderately expanded in 1986-89 in both developed and developing countries, with declines in Latin America and the Caribbean, the Asian CPEs and the Near East more than offset by increases in the other regions.
- Growth in output of *animal products,* particularly milk, slowed in developed countries in recent years reflecting relatively high feed prices as well as efforts to reduce structural surpluses in the EC and some

other European countries. However, developing countries, more markedly the Asian CPEs, achieved significant increases in both milk and meat production.

- Tropical beverages showed disparate trends, with *cocoa bean* production strongly recovering in some main producing countries in Africa, but remaining stagnant in Latin America and the Caribbean since 1986; *coffee* production showing large fluctuations in Latin America and the Caribbean, mainly as a consequence of weather-based shortfalls in 1986 and 1988 in Brazil, but no actual trend in either direction worldwide; and *tea* expanding at a sustained rate on a global level with strong growth in Kenya and the Asian CPEs.

Food production lagged behind population growth in 49 out of 72 developing countries (68 percent) that recorded significant changes in per caput food production (above or below 1 percent) in 1985-89 (Figure 1.6). Out of these 72 countries, per caput food production fell significantly in about 80 percent of the total number of countries in Africa, about 65 percent of those in the Near East and Asia and the Pacific, while in Latin America and the Caribbean about the same number of countries experienced increases and declines in per caput production. However, food production exceeded population growth in several of the most populous developing countries, e.g. China, India, Indonesia, Brazil and Colombia. By contrast in Africa, countries with declining per caput production accounted for an overwhelming majority of the region's population.

Food and agricultural production in 1990

The first estimates of world agricultural production in 1990 indicate an increase of around 2.4 percent (2.6 percent for food), a smaller expansion than in 1989, but significantly above the depressed average growth rates of 1986-88 (Table 1.2).

As in 1989, though less pronouncedly, the global increase in food and agricultural production in 1990 is expected to occur mainly in developed countries, particularly North America and Eastern Europe and the USSR. Production in Oceania and the EEC may only moderately increase.

For much of Africa and Latin America and the Caribbean, 1990 is likely to be another disappointing agricultural year. In both regions food production is expected to fall below population growth for the second consecutive year. In Latin America and the Caribbean, the overall shortfall largely reflected severely reduced crops in Brazil and some Andean and Caribbean countries which more than offset the average or good harvests in most other countries in the region. In Africa, particularly poor food output levels are expected in Liberia, Zambia, Morocco, Botswana and Senegal. Furthermore, while food output in the Sudan is forecast to significantly rise from the depressed 1989 levels, the increase will be inadequate to compensate for the heavy production setbacks of earlier years. A similar situation is faced by

Figure 1.5

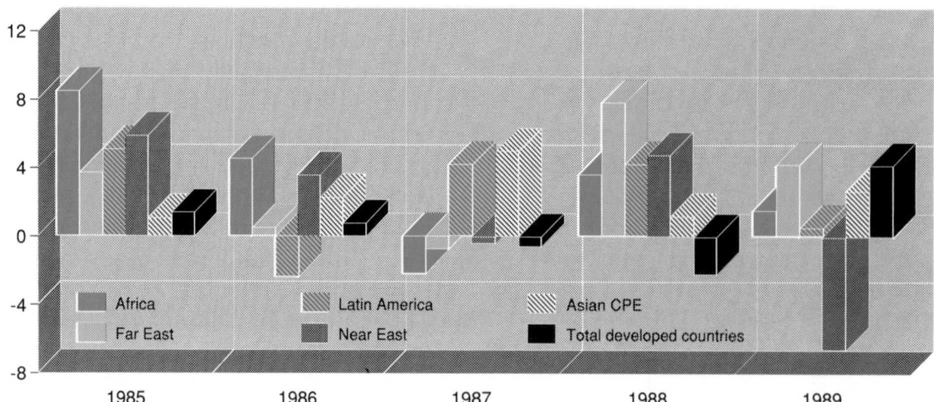

Source: FAO, Statistics Division (ESS)

TABLE 1.2. **Changes in world and regional food, agricultural, crop and livestock production, 1988-1990**

Country groups	Food		Agriculture		Crops		Livestock	
	1988 to 1989	1989 to 1990*	1988 to 1989	1989 to 1990*	1988 to 1989	1989 to 1990*	1988 to 1989	1989 to 1990*
	(........				%			........)
Developing market economies	1.6	1.4	1.5	1.4	1.3	1.7	2.7	1.7
Africa	1.1	1.4	1.5	1.5	1.4	—	2.7	2.1
Far East	4.4	1.1	4.2	1.0	5.2	0.7	4.1	3.5
Latin America and Caribbean	0.9	0.6	0.5	0.8	−0.1	1.2	2.1	—
Near East	−6.9	5.3	−6.6	5.4	−9.6	9.3	1.2	1.7
Asian centrally planned economies	2.9	4.5	2.7	4.2	2.4	3.3	4.0	1.2
All developing countries	2.0	2.3	1.9	2.3	1.7	2.2	3.1	1.5
Developed market economies	4.6	2.2	4.1	2.3	9.5	2.5	−0.7	1.2
North America	11.6	3.4	10.3	3.6	18.9	6.1	0.3	1.3
Oceania	−2.7	0.9	−2.6	1.2	1.5	2.5	−3.0	2.3
EEC (12)	—	1.8	—	1.8	1.8	−0.5	−1.5	1.1
Other western Europe	2.0	1.1	2.0	1.1	8.1	−1.1	−1.5	0.5
Eastern Europe and the USSR	2.7	4.5	4.1	2.9	4.6	5.3	1.2	—
All developed countries	4.1	2.9	4.2	2.5	7.8	3.4	—	0.7
World	2.9	2.6	2.8	2.4	3.9	2.7	1.1	1.1

* Preliminary.
Source: FAO. Based on information available up to 31 August 1990.

Ethiopia, where domestic food supply prospects give rise to serious concern (see the section, Food shortages and emergencies).

Expectations for Asia are for a significant decline in the growth rate of food and agricultural production relative to recent trends. Food production is expected to slightly exceed population growth in India but may significantly fall in per caput terms in Nepal, Bangladesh, Thailand and the Republic of Korea. On the other hand, growth in production is forecast to be well above both population growth and recent trends in Myanmar, Laos, Sri Lanka, Malaysia, Viet Nam, and Pakistan.

In China, prospects for 1990 are varied, ranging from bountiful crops for some commodities, in particular oil crops and sugar, some recovery in pulse output, but slow growth in livestock production mainly caused by feed shortage. In the case of cereals, the outlook is generally favourable and the record output level achieved in 1989 is likely to be exceeded.

As regards individual commodities, world *cereal* output is currently forecast at a record level of 1 941 million tonnes in 1990, approximately 66 million more than in 1989 (Table 1.3). Developed countries, in particular North America and the USSR, would account for most of the increase, although production is also forecast to rise in developing countries in the aggregate. The most significant increase is in *wheat,* expected to rise 27 percent in North America and reach record levels in the USSR. Among the main producer developing countries, wheat crops are likely to be significantly larger than in 1989 in China, Pakistan, Egypt and Argentina; close to the 1989 peak in India; markedly improved from the severely drought-reduced 1989 harvest in Turkey; but smaller in Mexico and Brazil. World production of *coarse grain* is forecast to rise 12.7 million tonnes (1.5 percent), the bulk of the increase being again in North America, the USSR and Turkey. Coarse grain production is estimated to decline in both western and Eastern Europe, Oceania, Africa and South America, however. The overall expansion in developing country coarse grain production would mainly reflect larger crops in Asia, particularly China and India, and in Turkey. At over 515 million tonnes, world *rice* (paddy) production in 1990 would exceed the extremely high level of 1989 by about 1 percent, with bountiful crops in Asia again accounting for most of the increase.

Among other commodities the following features stand out:

- *Vegetable oil* production in 1990 is expected to fall slightly in industrial countries, and grow at a lower, if still appreciable, rate in

Figure 1.6

ANNUAL CHANGES IN PER CAPUT FOOD PRODUCTION, 1985-89

	DEVELOPING COUNTRIES	DEVELOPED COUNTRIES
MORE THAN 5 %	CAPE VERDE, CYPRUS (2)	
3.01 % TO 5 %	BHUTAN, CHILE, MALAYSIA, PARAGUAY, TRINIDAD AND TOBAGO (5)	
1.01 % TO 3 %	BRAZIL, CHINA, COLOMBIA, BENIN, ECUADOR, INDIA, INDONESIA, MOROCCO, NAMIBIA, NEPAL, PERU, GUINEA-BISSAU, SAUDI ARABIA, URUGUAY, YEMEN ARAB REPUBLIC, ZAMBIA (16)	BELGIUM+LUXEMBOURG, CZECHOSLOVAKIA, HUNGARY, NORWAY, SOUTH AFRICA, USSR (6)
0.01 % TO 1 %	BOLIVIA, CENTRAL AFRICAN REPUBLIC, CUBA, DOMINICAN REPUBLIC, EGYPT, GHANA, GUATEMALA, HONDURAS, KENYA, KAMPUCHEA DEM., KOREA DEM., MACAU, MALI, NIGER, PAKISTAN, PUERTO RICO, THAILAND, TOGO, VIET NAM, SAMOA (20)	AUSTRIA, BULGARIA, NETHERLANDS, SPAIN (4)
-0.01 % TO -1 %	ARGENTINA, BANGLADESH, BELIZE, CAMEROON, CHAD, CONGO, COSTA RICA, ETHIOPIA GAMBIA, HONG KONG, CÔTE D'IVOIRE, LIBYA, MALAWI, MAURITANIA, MEXICO, PHILIPPINES, REUNION, SOMALIA, TONGA, TURKEY, BURKINA FASO (21)	YUGOSLAVIA, ALBANIA, DENMARK, FRANCE, GERMANY DEM.REP., GERMANY FED.REP., ITALY, POLAND, PORTUGAL, ROMANIA (10)
-1.01 % TO -3 %	ANGOLA, BARBADOS, BOTSWANA, MYANMAR, BURUNDI, EL SALVADOR, GABON, GUINEA, JORDAN, KOREA REP., LAOS, LESOTHO, LIBERIA, MADAGASCAR, MAURITIUS, MONGOLIA, MOZAMBIQUE, GUADALOUPE, PAPUA NEW GUINEA, SENEGAL, SIERRA LEONE, ZAIRE, SINGAPORE, SUDAN, SWAZILAND, TANZANIA, UGANDA, YEMEN DEM. (28)	AUSTRALIA, CANADA, FINLAND, GREECE, IRELAND, ISRAEL, JAPAN, NEW ZEALAND, SWITZERLAND, UNITED KINGDOM, UNITED STATES OF AMERICA (11)
-3.01 % TO -5 %	ALGERIA, SOLOMON ISLANDS, BRUNEI, SRI LANKA, FIJI, GUYANA, HAITI, MARTINIQUE, VANUATU, NIGERIA, PANAMA, ZIMBABWE, VENEZUELA (13)	MALTA, SWEDEN (2)
BELOW -5 %	IRAN, IRAQ, JAMAICA, NICARAGUA, RWANDA, SURINAME, SYRIA, TUNISIA (8)	ICELAND (1)

Source: FAO

developing countries. Production is likely to change little from 1989 levels in North America and western Europe, and fall about 4 percent in Eastern Europe and the USSR. By contrast, all developing country regions may be expected to raise their oil crop production, more significantly Asia (large gains in China, Malaysia and the Philippines more than offsetting reductions in India and Thailand). In Latin America, bumper crops are expected in Argentina and Paraguay. In contrast, output in Brazil is likely to fall significantly below the exceptionally high levels of 1989, though still representing a second all-time record.

- Prospects for world *sugar* production are for an increase to 108 million tonnes in 1990 (beet sugar 39.4 million tonnes, up 4.5 percent; and cane sugar 68.9 million tonnes, up 1.7 percent), a record for the third year in succession. Most of the increase in production would occur in Asia, crop expectations being particularly favourable in India and China and, among developed countries, in Eastern Europe and the USSR. In Latin America, a decline of nearly 6 percent in sugar output in 1990 may result from production shortfalls in Brazil, Mexico and the Dominican Republic which more than offset larger crops in Cuba, Colombia and Argentina. The global increase in sugar production would be inadequate to satisfy consumption, however, with stocks expected to fall in 1989/90 for the fifth consecutive year. Nevertheless, with another record crop anticipated in 1990/91, consumption and production could come into closer balance.
- Production of *roots and tubers* may rise over 5 percent in Latin America and the Caribbean, a significantly improved performance compared to recent trends for the region. In Africa, however, production is expected to continue stagnating.
- A significant expansion in world *pulse* production is expected in 1990, mainly reflecting continued favourable performances in Asia where production is expected to hit a new record. India's production of pulses is forecast to exceed 14 million tonnes, implying a cumulative 20 percent increase during 1989 and 1990. Pulse production may also largely recover in the Near East from the 1989 setback, though remaining below previous years' levels. On the other hand, performances are likely to remain at best mediocre in most pulse producing and consuming countries in Africa. In Latin America, severely reduced crops in Brazil are expected to be more than offset by bumper harvests in several other countries including Mexico and Argentina.
- Production prospects for *animal products* are for a continuation of recent trends, with slow growth in the main producing industrial countries, but fairly dynamic performance in developing countries. *Milk* production would remain virtually unchanged from the previous year in western Europe and only slightly increase in North America, Eastern Europe and the USSR. By contrast milk production in developing countries is likely to rise 4 percent, close to 1980s trends, although much of the increase would accrue to a limited number of large producing countries in Latin America (in particular Mexico, Brazil and Argentina) and the Far East (mainly India). *Meat* production may only slightly increase in North America, remain broadly unchanged in western Europe (although EEC countries have been facing market difficulties arising from supply in excess of sluggish demand and mounting stocks of meat) and fall below 1988 and 1989 levels in Eastern Europe and the USSR. Among developing regions, meat production in Africa may rise 3.7 percent, the best performance in many years, but fall slightly short of the high levels of the previous year in Latin America.
- *Green coffee* production in Africa may fall by 2.5 percent to its lowest level since 1985, with all main producers except Ethiopia and Uganda sharing in the decline. In Central America and the Caribbean and Brazil, slight increases are expected in coffee production, while in Colombia coffee output is expected to expand to the highest levels since 1984. World *cocoa bean* production is forecast to change little from 1989, despite significantly reduced output levels in Côte d'Ivoire, Nigeria and Brazil. While the cocoa crop in Ghana will remain stable, a spectacular growth in production is expected in Malaysia where cocoa output in 1990 is likely to increase to about twice the levels of the mid-1980s. Cocoa production is also likely to expand in Indonesia. *Tea* production is forecast to continue rapidly rising in Kenya, Sri Lanka and India, with all three countries achieving record harvests in 1990.

Supply/demand trends for cereals

The period 1986-90 was marked by a sharp

turnaround in the world market situation for cereals. In a short period of time after the abnormal drought that hit North America in 1988, world markets switched from a situation of abundance, with depressed prices and stocks reaching embarrassingly high levels, to one where stocks have fallen close to the global food security threshold. With world cereal production falling 3.5 percent in 1987 and further 3.2 percent in 1988, consumption requirements could only be met through a heavy drawdown in stocks (Figure 1.7). While production substantially recovered in 1989, growing by an estimated 7 percent, this was still insufficient to prevent a further reduction of stocks which fell to an equivalent of 17 percent of global consumption.

The favourable prospects for world output of cereals in 1990, estimated to exceed global utilization for the first time in four years, indicate that cereal stocks by the end of 1990/91 are likely to increase by more than 8 percent above their opening level (Table 1.4). The increase in global cereal stocks would result from a forecast 20 percent expansion in *wheat*, a 1.7 percent increase in *rice*, but a 2.5 percent fall in *coarse grain* stocks to their lowest level since 1983/84. At the forecast level of 324 million tonnes in 1990/91, cereal stocks would represent 18 percent of estimated world consumption, the same share as in 1988/89, slightly above the minimum level FAO considers necessary for safeguarding world food security.

While in a longer perspective the underlying production and consumption trends may be conducive to new surpluses in the main producing countries — and vice versa current conditions may be favourable for their budgets and producers — it is the immediate situation that calls for attention. The margin of world food security offered by current cereal stocks would be inadequate to prevent supply shortages and escalating prices in the event of significant production shortfalls in 1991, even if production in 1990 is relatively favourable.

Almost two-thirds of the total increase in global wheat stocks in 1990/91 is projected to be in the major exporter countries. However, the ratio of their forecast stocks (currently 54 million tonnes) to their trend utilization (domestic use plus exports) in 1991/92 would still only be 24 percent, compared to a level of around 30 percent before surplus stocks started accumulating during 1982/83-1986/87.

For coarse grains, the forecast volume of stocks in 1990/91 would only represent 14 percent of the trend level of global utilization in 1991/92. The ratio was on average 22 percent in the early 1980s. In addition, the stocks held by major exporters (currently forecast at 61.5 million tonnes) would be the second lowest since the world food crisis in the early 1970s, representing only 16 percent of their total trend utilization for 1991/92, compared with an average of 22 percent during the more typical period of the early 1980s.

Although forecasts for rice carryover stocks are still highly tentative, a marginal increase may be expected in 1991 ending stocks of rice after the strong recovery of the previous year.

Figure 1.7

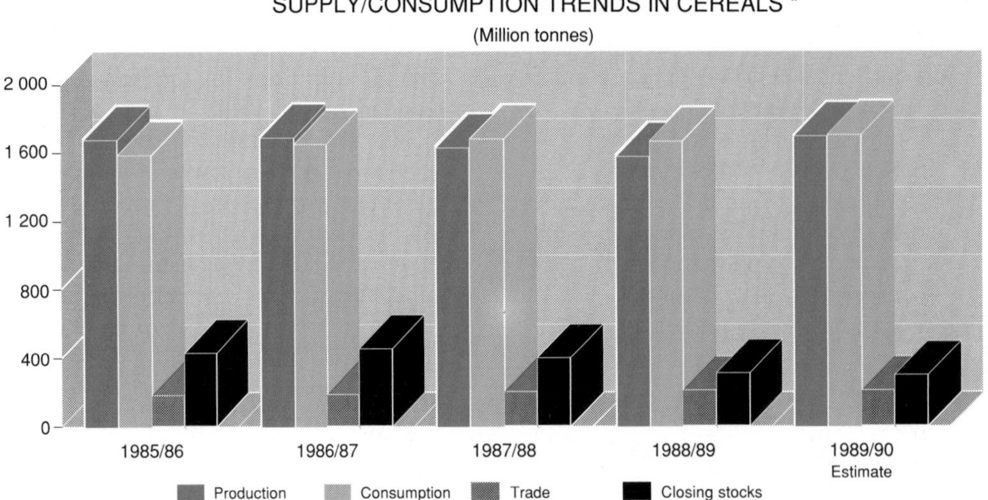

Source: FAO, Commodities and Trade Division (ESC)

* Rice, milled

As regards developments in the distribution of cereal stocks, major exporting developed countries accounted for most of the decline between 1986/87 and 1989/90. The slight forecast increase in 1990/91 would mainly stem from higher inventories in countries other than major exporters, including the USSR. Stocks held by developing countries are estimated to have recovered somewhat from levels of the previous year but are not forecast to change significantly in 1990/91. Nevertheless, inventories held in many developing countries continue to remain at unsatisfactorily low levels particularly when viewed in the light of rising consumption requirements.

Fertilizers

The severe economic and financial difficulties faced by many developing countries, adversely affecting their ability to import and subsidize the prices of fertilizers for low-income farmers, had resulted in a sharp slow-down in the growth of their fertilizer consumption in the early 1980s. The negative impact of this trend on output is difficult to assess but is bound to be considerable. On average, fertilizers have contributed 50 percent of the increase in crop yields. It has been estimated, for example, that in the Philippines the value of cereal production foregone as a result of reduced fertilizer imports may be ten times greater than the cost of these imports. Since 1985 the trend has been for a relative recovery in the use of fertilizers, particularly potash and phosphate, although the situation varied according to regions. China and the other Asian centrally planned economies, the developing region where fertilizer use was already highest relative to arable land (140 kg/ha in the early 1980s) recorded the highest and more sustained increases in fertilizer use throughout the 1980s. As a consequence, the Asian CPEs' consumption rose

BOX 1.2

Cereals and food security

Recent developments in cereal markets should be assessed in the light of the prominent position of these staples in developing countries' food production, consumption and trade — and hence food security. No less than 30 percent, and up to 75 percent, of total calorie supply is cereal-based in developing countries. Cereals account for about one-third of their total volume of food production (16 percent in Africa and Latin America, over 40 percent in Asia); and cereal imports in many of these countries have gained importance in domestic diets and total import expenditure, accounting for over 20 percent of the total value of agricultural imports of developing countries in recent years. On the other hand the weight of developing countries in world cereal exports is relatively small (about 14 percent), with no more than ten of these countries being normally net cereal exporters.

For a large majority of developing countries a global tightening in cereal markets and higher prices are therefore bad news. Admittedly, some of them have become excessively — and possibly unnecessarily — dependent on cereal imports, not having adequately exploited their domestic potential. From this perspective, high prices may be a blessing in disguise if they are to promote domestic farming and reduce import dependency in the long term. But there are also short-term considerations that impose priority attention. Among the many net cereal-importing developing countries, a particularly vulnerable group has been identified by FAO: the low-income food-deficit countries. It has been estimated that the increase in cereal prices in 1988/89 added about $3 billion to the annual cereal bill of these countries — twice the value of food aid they have received. For many of these countries even a small reduction in cereal supplies — domestic, imported or in the form of food aid — may have devastating repercussions. An increase in international prices can have a severe impact on retail markets, resulting in reduced or qualitatively impoverished diets, unless the government is prepared — and financially able — to protect consumers through subsidies. Underlying the FAO Secretariat estimate of minimum cereal stock levels required to safeguard global food security is the sharp rise in prices that would be expected to follow a fall in stocks to significantly below 17 percent of world consumption.

TABLE 1.3. **Agricultural production, by commodity, 1989 and 1990**

Item	Developed countries			Developing countries			World		
			Change			Change			Change
	1989	1990[1]	1989-1990	1989	1990[1]	1989-1990	1989	1990[1]	1989-1990
	(. . . .million tonnes. . . .)		%	(. . . .million tonnes. . . .)		%	(. . . .million tonnes. . . .)		%
Total cereals[2]	878.6	928.7	5.7	995.9	1 012.1	1.6	1 874.5	1 940.8	3.5
Wheat	316.7	357.4	12.9	225.5	232.8	3.2	542.2	590.2	8.8
Rice, paddy	25.4	26.1	2.8	484.3	489.2	1.0	509.7	515.3	1.1
Coarse grains	536.5	545.1	1.6	286.0	290.1	1.4	822.5	835.2	1.5
Root crops	203.9	206.2	1.1	387.6	389.9	0.6	591.5	596.1	0.8
Pulses	20.5	21.0	2.4	35.9	37.9	5.6	56.4	58.9	4.4
Oil-seeds	106.9	106.1	−0.7	172.5	178.7	3.6	279.3	284.8	2.0
Sugar, centrifugal (raw)	44.4	44.5	0.2	61.0	63.6	4.3	105.4	108.1	2.6
Cocoa beans	—	—	—	2.4	2.4	—	2.4	2.4	—
Coffee	—	—	—	5.8	5.8	—	5.8	5.8	—
Tea	0.2	0.2	—	2.3	2.4	4.3	2.5	2.6	4.0
Cotton lint	6.0	6.5	8.3	11.0	12.0	9.1	17.0	18.5	8.8
Tobacco	1.7	1.8	5.9	5.3	5.1	−3.8	7.1	6.9	−2.8
Total meat	103.1	103.3	0.2	66.2	66.8	0.9	169.3	170.1	0.5
Total milk	384.7	388.0	0.9	150.5	155.7	3.4	535.2	543.6	1.6
Hen eggs	19.1	19.0	−0.5	16.4	16.9	3.0	35.5	35.9	1.1

[1] Preliminary.
[2] Including rice in terms of paddy.
Source: FAO, Based on information available up to 31 August 1990.

TABLE 1.4. **Carryover stocks of cereals**

Country and country groups	Crop year ending in					
	1986	1987	1988	1989	1990	1991
	(million tonnes)					
Developed countries	289.6	319.7	276.7	184.2	165.5	185.9
Canada	14.4	18.5	13.5	9.7	10.8	14.3
United States	181.2	203.8	169.4	86.1	61.5	69.6
Australia	6.1	4.1	3.1	3.2	3.1	3.7
EEC	36.1	31.6	28.7	28.5	28.5	28.3
Japan	5.2	5.8	5.6	5.4	5.1	5.1
USSR	31.0	38.0	39.0	36.0	39.0	46.0
Developing countries	136.9	135.9	123.6	122.7	133.5	138.1
Asia	113.1	108.5	97.7	97.0	108.0	118.3
Bangladesh	1.0	0.7	1.5	1.2	1.1	1.1
China	52.0	46.3	47.5	42.8	42.3	44.8
India	17.0	15.0	5.4	4.4	11.1	13.6
Pakistan	2.0	3.1	1.6	2.5	2.5	2.6
Turkey	0.4	0.9	1.0	0.8	0.7	0.9
Africa	11.8	14.6	10.4	12.3	12.0	10.2
Latin America and the Caribbean	11.8	12.6	15.1	13.2	13.2	9.4
Argentina	0.7	0.7	1.3	1.3	1.1	1.1
Brazil	3.0	4.6	6.5	5.3	5.5	2.4
World	426.5	455.6	400.3	306.9	299.0	324.0
Wheat	160.2	167.1	142.3	114.5	118.9	143.0
Rice	58.0	54.5	45.0	46.2	52.2	56.2
Coarse grains	208.3	234.0	213.0	146.3	127.8	124.7
World stock as % consumption	26	27	24	18	17	18

Source: FAO, Commodities and Trade Division (ESC).

to over 200 kg/ha in recent years, second only to western Europe. Far East countries also strongly expanded nutrient use, bringing their consumption levels to 60 kg/ha. While the increase was widespread across the region it was to a large extent spearheaded by Indonesia, Thailand and India, although consumption in the latter has tended to level off in recent years. Fertilizer use in 1985-88 picked up strongly in Latin America and the Caribbean after the near stagnation of the early 1980s, and has shown a steady increase, albeit at a declining pace, in the Near East. By contrast, plant nutrient application has only very slowly increased in Africa overall, and there was even a decline in consumption in 1986-88 after the exceptionally high rise of 1985 following the breaking of the drought.

The rates of increase in fertilizer use in the 1980s fall well below those of the 1970s, when growth in consumption in all developing countries averaged nearly 10 percent a year.

Fertilizer markets were generally calm by mid-1990 with prices at levels close to those a year earlier (a notable exception being ammonium sulphate, f.o.b. western Europe, for which prices had fallen by almost one-third). However, recent events in the Near East caused a tightening in markets and upward pressure on prices, particularly of ammonia and urea. Such tightening was caused, on the one hand, by reduced supply in world markets resulting from the trade embargo on Iraq (Iraq and Kuwait being large suppliers, producing together about 1.140 million tonnes of ammonia and 785 000 tonnes of urea); and, on the other hand, from the impact of higher prices of oil and related products on fertilizer production costs (over 90 percent of world ammonia production being derived from natural gas).

Recent increases in fertilizer prices have varied according to the market, but have been significant in some cases. For instance, increases in ammonia f.o.b. export prices between 26 July and 27 September 1990 were about 44 percent for US Gulf, 30 percent for Caribbean and 22 percent for Near East markets. For urea, price increases were of the order of 35 percent for US Gulf, 8 percent for Near East and 39 percent for Caribbean markets.

AGRICULTURAL TRADE, TRENDS IN 1985-88

The period 1985-88 was characterized by considerable buoyancy in world agricultural trade in relation to the earlier part of the decade. World trade in crops and livestock rose by a cumulative 28 percent during 1985-88, compared to an overall decline of over 5 percent in the previous four years (Figure 1.8). The upsurge in trade was particularly pronounced in developed countries in 1987 and 1988, on the side of both exports and imports. By contrast, of all developing regions, only the Far East and Asian CPEs recorded rapidly expanding trade. Africa and Latin America and the Caribbean, the two regions for which agricultural exports matter most in an overall economic context, were those where such exports were more depressed.

Even more buoyant than trade in crops and livestock was that in fishery and forestry products in recent years (see the Fisheries and Forestry sections). Also, unlike trade in crops and livestock, exports of fishery and forestry products in 1985-88 showed more dynamism in developing than developed countries.

The strong expansion in agricultural trade in developed market economies mainly reflected an intensification of intra-OECD country trade. On the contrary, developing countries lost relative importance as outlets for agricultural exports from the OECD. For instance, developing countries accounted for 40 percent of total agricultural exports of the EEC in recent years compared to about 50 percent in the early 1980s. OECD imports from developing countries barely increased throughout the 1980s (0.7 percent yearly in current value terms) after having expanded 14 percent yearly during the 1970s.

Agricultural trade balances in developing countries showed significant shifts between 1981-84 and 1985-88. Export/import ratios were as follows (percentages):

	1981-84	1985-88
Africa	**85.8**	**107.3**
Latin America	**250.7**	**293.3**
Far East	**110.3**	**109.9**
Near East	**26.1**	**28.4**
Asian CPEs	**71.3**	**121.8**

Latin America and the Caribbean consolidated its surplus position, and the Asian CPEs became significant net exporters in the latter period. However, while increasing surpluses in Latin America during the 1980s were based on import contraction rather than export expansion, the opposite was true in the Asian CPEs. Despite the heavy agricultural component of their economies, African countries alternated net deficit and surplus situations. The improvement in African trade balances in 1985-88 mainly reflected a sharp reduction in food imports in 1986 and 1987. The Far East maintained its moderate surplus position while Near East countries remained heavily dependent on food imports.

Although agricultural export prices generally firmed during 1985-88 in relation to the period 1981-84, this trend was not shared by many of the products chiefly exported by developing countries. Moreover, prices of other major traded products rose significantly faster than agricultural prices in recent years. As a result, the net barter terms of trade of agricultural products against manufactured goods and crude petroleum in developing countries as a whole deteriorated by nearly 7 percent yearly between 1985-88 after having fallen 0.3 percent yearly in the period 1981-84. All developing country regions shared in the negative trend, with Latin America and the Near East hardest hit throughout the whole 1981-88 period.

Many developing countries attempted to compensate for depressed agricultural prices through an increased export effort. However, expanded export volumes were generally insufficient to prevent a decline in the purchasing power of agricultural exports (income terms of trade) in 1985-88. Thus, while net barter terms of trade for developing market economies fell 6.1 percent yearly during this period, income terms of trade fell 4.5 percent yearly. Among developing country regions, only the Asian CPEs achieved an expansion of export volumes that more than offset the deterioration in net barter terms of trade, thereby permitting significant gains in the purchasing power of agricultural exports.

Agricultural trade in 1989

In 1989 the value of world exports of agricultural, fishery and forestry products is estimated to have increased by about 3 percent (Table 1.5). The increase was not only well below that of 1988, it was also lower than the average growth of the 1980s.

Figure 1.8

TRADE IN CROPS AND LIVESTOCK, 1986-89

(US$ billion)

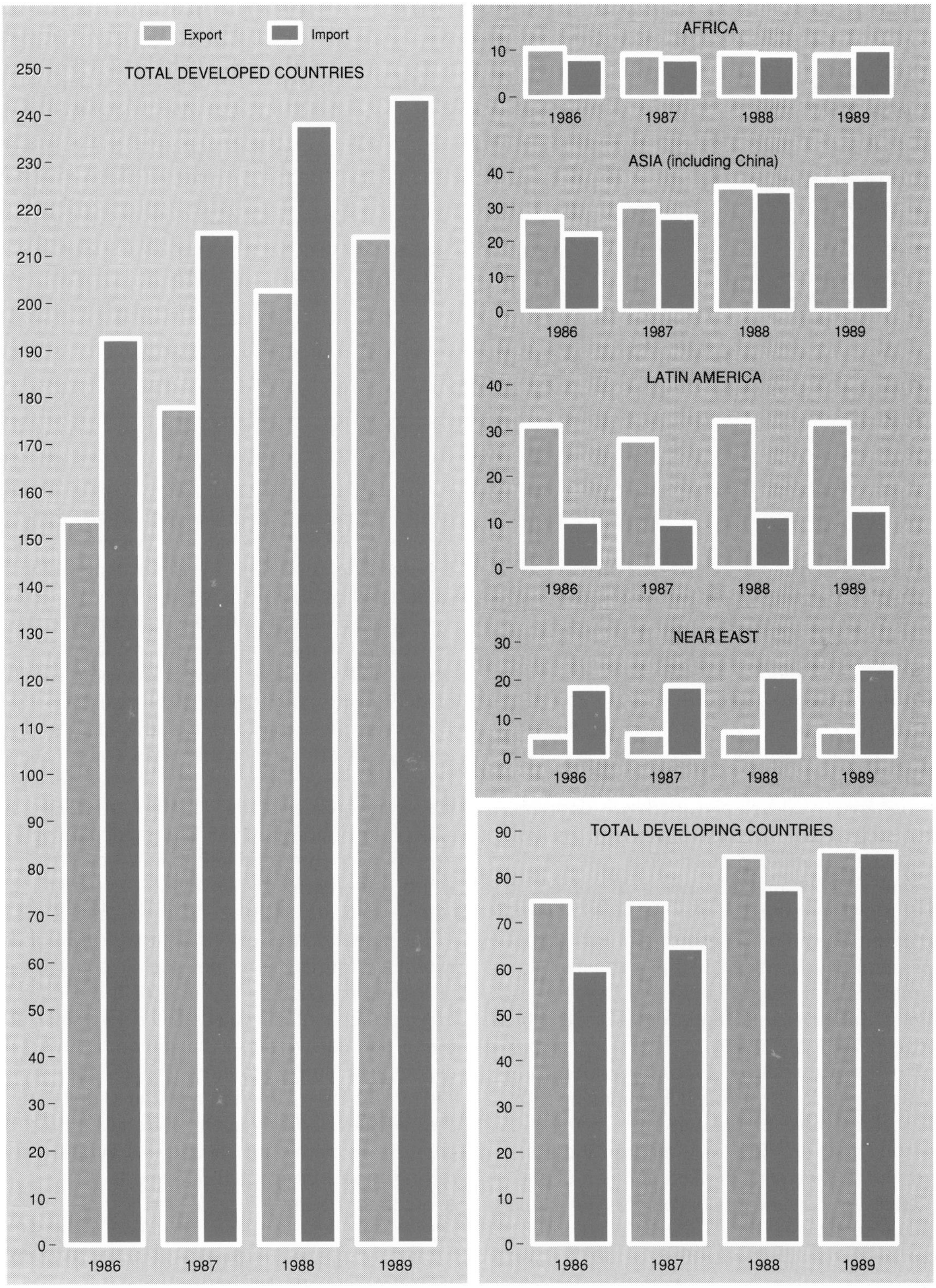

Source: FAO

TABLE 1.5. **Value of world exports of agricultural (crops and livestock), fishery and forest products, at current prices, 1987-89**

Item	1987	1988	1989	Change 1987-1988	Change 1988-1989	Average of annual changes 1980-1988
	(.......... $ '000 million..........)			(.............. % ....)		
Agricultural products	**252.11**	**287.3**	**300.0**	**13.8**	**4.4**	**3.0**
Total developing countries	**74.3**	**84.5**	**85.7**	**13.7**	**1.5**	**2.1**
Total developed countries	**178.0**	**202.8**	**214.2**	**13.9**	**5.7**	**3.5**
Fishery products	**28.2**	**32.2**	**32.1**	**14.4**	**−0.5**	**9.1**
Total developing countries	**12.9**	**15.0**	**15.0**	**16.6**	**−0.1**	**11.1**
Total developed countries	**15.3**	**17.2**	**17.1**	**12.6**	**−0.7**	**7.6**
Forest products	**73.3**	**85.8**	**85.9**	**17.2**	–	**5.5**
Total developing countries	**10.2**	**12.1**	**12.1**	**18.6**	–	**4.0**
Total developed countries	**63.0**	**73.7**	**73.7**	**16.9**	–	**5.8**
Total	**353.8**	**405.4**	**417.9**	**14.6**	**3.1**	**3.8**
Total developing countries	**97.4**	**111.7**	**112.9**	**14.6**	**1.1**	**3.0**
Total developed countries	**256.4**	**293.7**	**305.1**	**14.6**	**3.9**	**4.2**
	(.............. % ....)					
Share of developing countries	**27.5**	**27.5**	**27.0**			

NOTE: Figures may not add up because of rounding.
Source: FAO.

The slow-down of growth reflected mainly a stagnation in exports of fishery and forestry products, the two subsectors that had experienced more dynamic growth in previous years, in both developed and developing countries. By contrast, export earnings of crop and livestock products expanded at a faster pace than the average of the 1980s, even though prices of several commodities declined.

The global expansion of crop and livestock exports largely reflected the stronger performance of developed market economies. Most striking was the performance of Australia where agricultural export earnings rose 16 percent in 1989 after expanding 28 percent in 1988. The United States also achieved a further significant expansion in export earnings (about 9 percent) which, added to the booming performance of the previous year, resulted in a cumulative increase of approximately 38 percent between 1987 and 1989.

The main agricultural exporting countries in western Europe also expanded shipments in 1989, albeit more moderately than in the previous year (the EEC registered an increase in trade of 9 percent in 1988 and 5 percent in 1989). Among the major exporter developed market economies, only Canada failed to improve further over the 1988 export performance. Agricultural export earnings fell below the 1988 levels in all Eastern European countries except Poland, where exports rose nearly 19 percent, and Hungary where they remained virtually unchanged. Exports fell about 5 percent in the USSR, to approximately the same level as two years earlier.

In sharp contrast with the performance of agricultural markets in most industrial countries, agricultural (crops and livestock) export earnings of developing countries in 1989 rose only 1.5 percent, less than the already sluggish average growth rate of the 1980s (Table 1.6). This slow growth resulted in a further decline in the share of developing countries in world agricultural trade. In 1989 their agricultural exports accounted for less than 29 percent of world trade compared to over 30 percent in 1980 and nearly 40 percent in the mid-1960s.

The continuation of unsatisfactory agricultural export performances in developing countries is of increasing concern because of the serious external account difficulties faced by many of them, particularly those that predominantly export agricultural products. This is more so in Africa and Latin America and the Caribbean, the two developing regions currently facing the greatest economic and financial difficulties, and relying more on agricultural exports for overcoming them.[7]

[7] Agricultural exports currently account for about 21 percent of total exports in developing Africa and 29 percent in Latin America and the Caribbean compared to 14 percent for developing countries as a whole.

TABLE 1.6. **Value of world agricultural trade (crops and livestock), at current prices, by region, 1987-89**

Item	1987	1988	1989	Change 1987-1988	Change 1988-1989	Average of annual changes 1980-1989 current prices	Average of annual changes 1980-1989 volume
	(.......... $ '000 million..........)			(...................... % ....)			
Africa							
Export	**9.2**	**9.3**	**8.9**	**0.6**	**−3.4**	**−1.2**	**−0.2**
Import	**8.2**	**9.0**	**10.3**	**9.6**	**15.4**	**0.3**	**2.0**
Far East[1]							
Export	**30.5**	**36.0**	**37.8**	**18.3**	**4.2**	**5.4**	**5.7**
Import	**27.2**	**34.8**	**38.1**	**28.0**	**9.5**	**5.5**	**4.1**
Latin America							
Export	**28.2**	**32.2**	**31.7**	**14.4**	**−1.5**	**0.2**	**2.9**
Import	**9.9**	**11.6**	**12.9**	**17.7**	**10.5**	**−0.7**	**−1.5**
Near East							
Export	**6.0**	**6.5**	**6.7**	**8.5**	**3.7**	**3.4**	**7.7**
Import	**18.6**	**21.2**	**23.4**	**13.8**	**10.4**	**3.3**	**5.2**
Total developing countries							
Export	**74.3**	**84.5**	**85.7**	**13.7**	**1.5**	**2.1**	**3.8**
Import	**64.7**	**77.5**	**85.6**	**19.7**	**10.5**	**2.9**	**2.8**
Total developed countries							
Export	**178.0**	**202.8**	**214.2**	**13.9**	**5.7**	**3.5**	**1.2**
Import	**215.1**	**238.1**	**243.7**	**10.7**	**2.4**	**3.2**	**1.4**
World							
Export	**252.4**	**287.3**	**300.0**	**13.8**	**4.4**	**3.0**	**1.9**
Import	**279.8**	**315.6**	**229.3**	**12.8**	**4.4**	**3.1**	**1.9**
Share of developing countries in world agricultural trade	(................ % ....)						
Export	**29.4**	**29.4**	**28.6**				
Import	**23.1**	**24.6**	**26.0**				

NOTE: Figures may not add up because of rounding. Annual changes and their averages have been calculated from unrounded figures.
[1] Includes China.
***Source*: FAO.**

Agricultural exports fell by over 3 percent in 1989 in Africa (with a cumulative 11 percent decline in the current value, and 1.8 percent in the volume of exports, between 1980 and 1989) and by 1.5 percent in Latin America and the Caribbean (with exports only increasing 1.8 percent in current value despite a 26 percent increase in their volume during the same nine-year period). The situation appears more favourable in the Far East and the Near East regions, with regard to both their 1989 performance and their recent trends.

Behind these general trends and features the following regional developments can be highlighted.

Africa. The decline in exports in Africa mainly reflected a deterioration in trade of tropical beverages. Notwithstanding increased volume movements by several countries, the drop in prices — particularly for coffee and cocoa — curtailed earnings. Performances of other commodities were uneven. Sugar exports expanded in Swaziland but remained relatively unchanged in Mauritius, the largest exporter of the region. Cotton exports from the Sudan declined but this was offset by increased earnings in Côte d'Ivoire and Mali. Both Nigeria and Cameroon increased their export of rubber while earnings from this commodity remained almost unchanged in Liberia.

Latin America and the Caribbean. The overall decline in the value of agricultural exports resulted mainly from lower trade in coffee, sugar and soybean cake, three important sources of agricultural earnings in the region, which was not offset by gains in other commodities. All the major coffee exporters except Mexico recorded sharp declines in export earnings from this commodity. Sugar exports also fell in Cuba and Brazil. While

exports of soybean cake increased in Brazil, they declined in Argentina — the net result being a slightly negative growth rate for this commodity. Banana export earnings are estimated to have remained broadly unchanged at the previous year's level.

Far East (including China). A strong increase in rice export earnings in Thailand and, to a lesser extent, Viet Nam, more than offset stagnant or reduced shipments in most other rice-exporting countries. Sugar exports also expanded sharply in Thailand. Exports of rubber, which had shown considerable dynamism in 1988, deteriorated markedly in Malaysia. While export earnings for coffee remained relatively favourable, they were depressed for cocoa and for palm oil, although palm oil exports continued to expand. Exports of cotton lint from Pakistan again increased significantly.

Near East. The export performance of cotton lint compared unfavourably to that of other agricultural products (in particular hazelnuts, animal products and pulses), and the share of this commodity in total agricultural exports of the Near East continued to fall, reaching an estimated 12 percent (compared to 23 percent in the early 1980s). Cotton lint exports fell in both Egypt and Turkey in 1989.

Expenditure on food and agricultural imports in 1989 followed divergent patterns. The growth in food and agricultural imports of developed countries significantly declined in value terms compared with the average for the 1980s; but the opposite occurred in developing countries.

Among developing country regions, Africa registered the largest increase in food imports (over 15 percent in 1989, and an aggregate 25 percent during 1987-89). Such a large recourse to imports reflected the deteriorating domestic food supply conditions in several African countries. A similar consideration applied to Latin America and the Caribbean where imports expanded 10.5 percent in 1989, despite continuing financial difficulties (the region's per caput food production stagnated in 1988 and declined in 1989). In the Far East, the strong growth in agricultural imports in 1989 mainly reflected larger food purchases by China, which surpassed Egypt as the largest food-importing developing country. In the Near East, while Egypt expanded food imports only moderately, other large importing countries, including Saudi Arabia, significantly expanded their purchases.

With the growth in imports exceeding that in exports in all developing regions, agricultural trade balances markedly deteriorated in 1989, particularly in Africa and Latin America and the Caribbean. Africa and the Far East became net agricultural importers and the export/import ratio fell to 245 percent in Latin America and the Caribbean (against an average 293 percent during 1985-88).

World trade in cereals in 1990/91

The preliminary FAO forecast of world trade in cereals in 1990/91 is 202 million tonnes, down from the estimated 209 million tonnes of 1989/90. Almost all the decline would be in coarse grains. The forecast reflects a reduction in total cereal imports by developed countries to about 81 million tonnes in 1990/91 from 88 million tonnes in 1989/90, whereas the aggregate cereal imports of the developing countries are expected to remain virtually unchanged at 120 million tonnes.

World imports of wheat are expected to remain virtually unchanged, at 96 million tonnes in 1990/91. The lack of growth in wheat imports is mainly a result of good 1990 cereal crops in a number of importing countries, stagnation in the economies of the developing countries, and no expected increase in food aid shipments. Among developed countries, a marginal reduction in wheat imports is expected for the USSR (although the forecast for this country remains highly tentative) alongside a decline in Eastern Europe. Total wheat purchases by western Europe and Japan are estimated to remain basically unchanged. A slight increase in imports is projected for developing countries resulting from an expected increase in Latin America, a reduction in Asia and an unchanged level in Africa.

For coarse grains, a major reduction in imports is forecast, from 101 million tonnes in 1989/90 to 94 million tonnes in 1990/91. The decline mainly reflects the excellent 1990 cereal crop outlook in the USSR and the competitive prices of wheat on international markets, making coarse grain imports less attractive for animal feeding. A reduction in coarse grain imports primarily reflects sharply reduced purchases by the USSR as well as lower imports by Eastern Europe. Aggregate coarse grain imports by other developed countries are expected to remain at about the same level as in 1989/90. Total imports by developing countries are expected to remain unchanged as the net result of slight declines in Asia,

Figure 1.9

TERMS OF TRADE OF AGRICULTURAL EXPORTS FOR MANUFACTURED GOODS AND CRUDE PETROLEUM, 1986-89

(Percentages)

NET BARTER TERMS OF TRADE

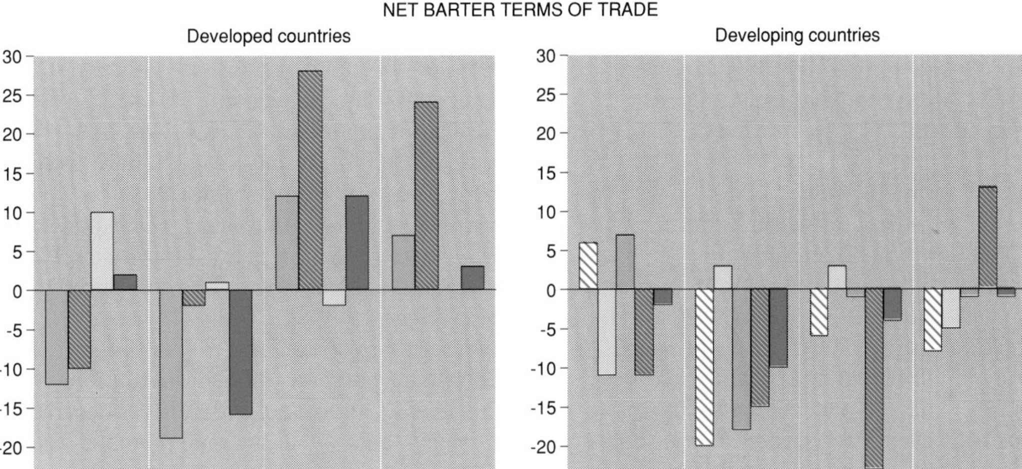

INCOME TERMS OF TRADE

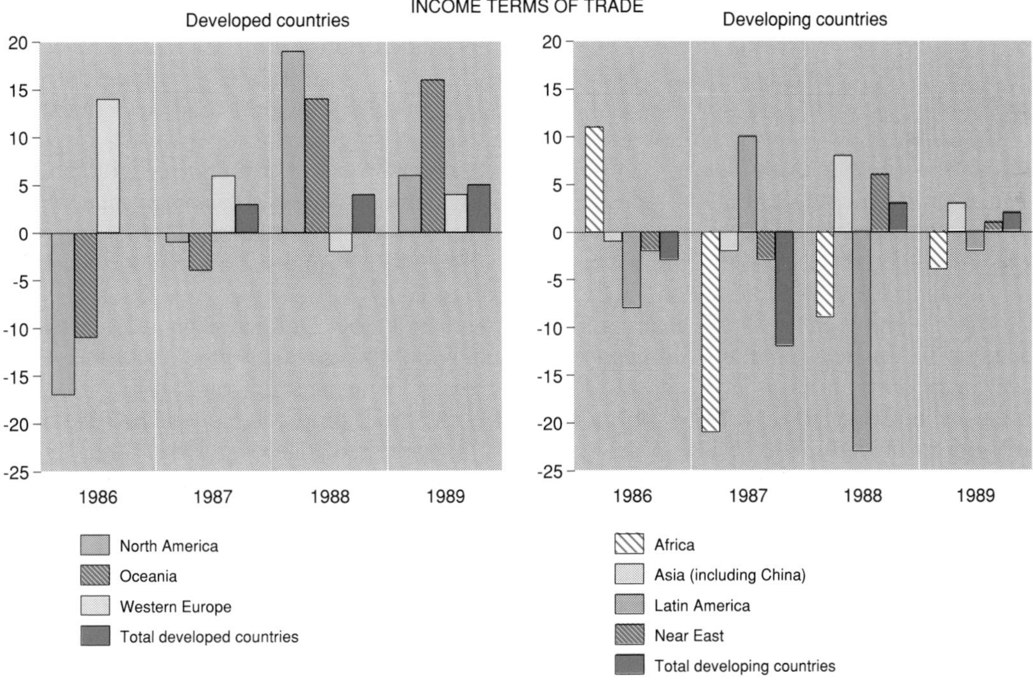

Net barter terms of trade refer to the relative unit prices of products.
Income terms of trade refer to the actual purchasing power of countries' exports, since they also take into account changing export volumes, and hence the total amount of foreign exchange earned by exports and available to pay for imports.

Source: FAO

Africa and Central America, and an increase in South America.

The world rice market has entered into a phase of uncertainty with structural shifts taking place. A decline in the global rice trade is expected in the 1990 calendar year, settling at 11 million tonnes against a level of 13 million tonnes in 1989. Lower rice imports are expected in both Asia and Africa. For 1991, however, rice exports are tentatively estimated at 12 million tonnes.

Agricultural terms of trade

Trends in agricultural terms of trade for the period 1986-89 are shown in Figure 1.9. Main features include: i) for developing countries, a pronounced deterioration in the terms of trade of agricultural exports *vis-à-vis* trade-weighted prices of manufactured goods and crude petroleum. By 1989, their net barter terms of trade were at their lowest level in the 1980s; ii) for developed countries, benefiting from relatively better market conditions for temperate product exports, a comparatively more stable pattern of change, and a less pronounced deterioration of the agricultural terms of trade overall; iii) varying degrees of compensation for depressed agricultural export prices in the form of increasing volumes of exports. This is observed in most groups of countries by the comparatively better performance of the income terms of trade (or purchasing power of agricultural exports), which takes into account changes in both prices and quantities of agricultural exports; and iv) uneven performances among regions. Once again, it is Africa and Latin America and the Caribbean that have fared worst in an area of particular importance for both regions. By 1989, the index of agricultural income terms of trade (base 1979-81 = 100) had fallen to 72 for Africa and 82 for Latin America and the Caribbean. In other words, Africa's earnings from agricultural exports in 1989 would enable the region to purchase 28 percent less manufactures and crude petroleum than in the period 1979-81; and Latin America's, 18 percent less.

While data for 1990 do not yet permit a full assessment of changes in terms of trade and purchasing power of agricultural exports, recent events do not provide ground for optimism.

Export prices of most agricultural commodities have generally tended to weaken in 1990 compared with the average levels of the previous year (Figure 1.10). For cereals, export prices of *wheat* fell about 30 percent between January and August 1990. The price of *maize* significantly strengthened during the first half of 1990 but weakened in July/August to about the same level as in late 1989. With improved supplies and sharply reduced import demand for *rice* in the main importing countries, particularly in Asia, rice prices in mid-1990 were about 10 percent lower than in early 1990. Prices of the main groups of *oil-seeds* and *meals* were also significantly lower in 1990 than in the previous year. On average, quotations for *soybeans* in January/August 1990 were about 10 percent lower than in January/December 1989; and 17-20 percent lower in the case of palm oil and soybean meal. Soybean oil quotations remained close to the average of 1989.

Prices for *tropical beverages* (coffee, tea and cocoa) showed pronounced fluctuations in 1990 with average quotations during January/July remaining below the average levels of the previous year. Particularly striking was the overall decline in coffee prices in 1990, which were 23 percent below the average levels of 1989 despite some strengthening during the first quarter of the year. *Cotton* prices sharply strengthened during the first half of 1990. Prices of *sugar* further strengthened during the first quarter of 1990 as world markets remained tighter and stocks continued to fall. However, with expectations of a record output in 1990/91 and a better balance between production and consumption, prices began to fall in the second quarter and by August had reached levels of a year earlier.

Against a background of generally weakening agricultural prices, the outlook suggests increases in the cost of imports of manufactured goods and crude petroleum in world markets. While the US dollar unit value index of manufactured goods exported by developed market economies had remained unchanged in 1989 from the level of the previous year, it increased by almost 6 percent in the first quarter of 1990 compared to a year earlier. This tendency could be accentuated. In particular, recent events in the Near East have given oil prices an immediate jolt and, while it is not possible to foresee at which levels prices will stabilize, prospects for at least the short term are for significantly higher levels than in 1989 and the first half of 1990. Higher oil prices are also likely to affect the price of manufactured goods because of their impact on the energy and petrol-related component of production costs. Therefore, a

Figure 1.10

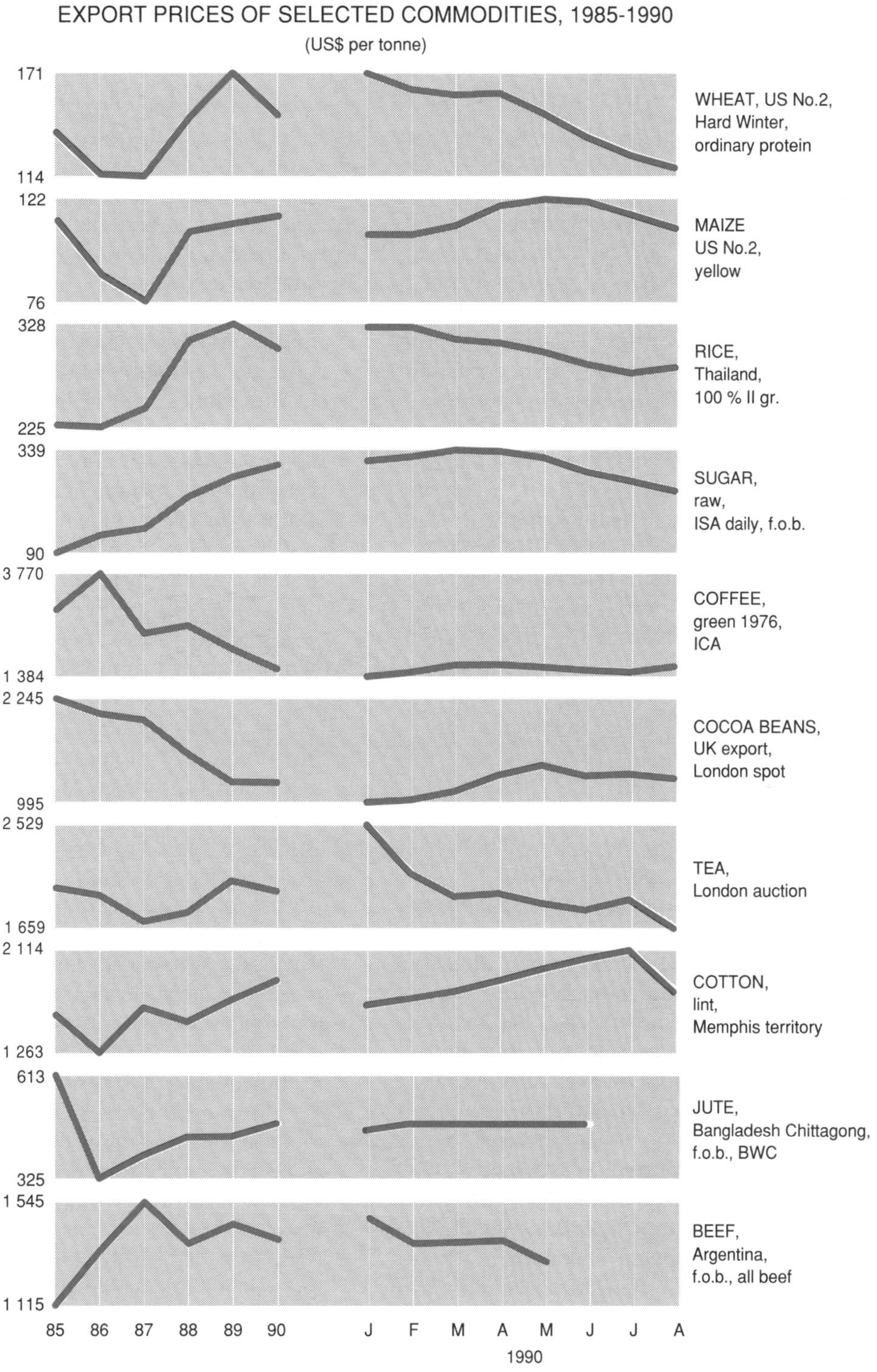

EXPORT PRICES OF SELECTED COMMODITIES, 1985-1990

(US$ per tonne)

171
114
WHEAT, US No.2,
Hard Winter,
ordinary protein

122
76
MAIZE
US No.2,
yellow

328
225
RICE,
Thailand,
100 % II gr.

339
90
SUGAR,
raw,
ISA daily, f.o.b.

3 770
1 384
COFFEE,
green 1976,
ICA

2 245
995
COCOA BEANS,
UK export,
London spot

2 529
1 659
TEA,
London auction

2 114
1 263
COTTON,
lint,
Memphis territory

613
325
JUTE,
Bangladesh Chittagong,
f.o.b., BWC

1 545
1 115
BEEF,
Argentina,
f.o.b., all beef

85 86 87 88 89 90
J F M A M J J A
1990

Source: FAO

* eight-month average except for jute (six months) and beef (five months)

further significant deterioration may be expected in 1990 in agricultural terms of trade *vis-à-vis* other major traded products.

Agriculture in the Uruguay Round of multilateral trade negotiations

The eighth round of GATT multilateral trade negotiations (MTN) — the Uruguay Round — is due to conclude in December 1990. Its success or failure, critically dependent on the last stage of negotiations, will be of the highest importance for overall and agricultural trade in the years to come. What is at stake is either a freer and more equitable trading system granting fair participation of all countries in the benefits of expanded trade; or an accentuation of competing trading blocks, of short-term benefit to a few, but potentially disastrous for many trade-dependent economies, particularly in the developing world.

A number of particular features have characterized the Uruguay Round. Prominent among these is the central role played by agriculture — unlike previous negotiations in which the sector was regarded as the preserve of domestic policy. Indeed, the very success of the Round is now seen to hinge on the ability of negotiators to reach an agreement on the liberalization of trade in farm products and the reduction of farm subsidies.

Another important feature of the current Round is the broad scope of the negotiations which cover not only border measures of protection, i.e. tariffs, import quotas and export subsidies, but also measures of support for domestic production. Thus, all policies affecting agricultural trade are under negotiation.

The record of achievements has been mixed and positions remain far apart on a number of key issues. Wide differences over agricultural policies and farm reform in particular had been at the basis of a disappointingly inconclusive mid-term review in Montreal in November 1988. Reconvened in Geneva in April 1989, the review agreed on a set of interim measures which set the framework for renewed negotiations. These covered long-term and short-term elements for reform as well as sanitary and phytosanitary regulations. It also recognized the special position of developing countries and the need for their special treatment, including ways and means to compensate for the negative effects of reform on net food importers. Agreement was also reached on further negotiations to liberalize trade in tropical products, including their processed and semi-processed form.

Several proposals have been tabled since then and intensive discussions have continued between the main negotiators: the United States; the European Communities; the Cairns Group of 13 agricultural trading nations; North European countries, Japan and Switzerland, sharing concern for domestic food security considerations; and food-importing developing countries. The discussions have been dominated by dissent between the two main protagonists, the United States and the European Community, on a number of issues. These include the following:

- "Tariffication" of *import barriers,* i.e. conversion of part of non-tariff barriers into customs duties that would initially provide equivalent protection. The United States, supported by the Cairns Group, suggested converting such barriers into bound tariff quotas and then reducing them to zero or very low levels over a period of ten years. The EC was opposed to such an approach unless special provisions are made, including a "corrective factor" to protect farmers' incomes against abrupt changes in world prices or fluctuations in exchange rates. Technical difficulties, linked to the measurement of domestic and world market prices and the choice of a reference year, further complicate the issue.
- Crucial to the issue of *domestic support* is the way of measuring it. The EC suggested the use of an aggregate measure of support (AMS), including all the government support provided to a specific farm product; and the freezing of all AMS at their current levels and their successive reduction, by agreed amounts, on a yearly schedule. There would be, however, a "rebalancing" of support; that is, for some products subsidies/border tariffs could actually increase provided the overall support to agriculture declined. The United States' suggestion is that internal support measures should be categorized as "permitted" (i.e. those not distorting trade), "disciplined" (for which the AMS approach could be applied) and "prohibited". On this basis, all domestic support directly affecting farm production and prices could be phased out over ten years. This proposal is supported by the Cairns Group but opposed by the EC.
- Positions between the EC and the United States on *export subsidies* are even more difficult to bridge. The United States

advocates phasing out these subsidies over five years, while the EC considers export refunds as an integral part of its current variable levy and dual pricing system. Thus, export refunds would be subject to the corrective factor the EC proposes for import charges.

A number of additional concerns were raised by other countries. Several North European countries, Switzerland and Japan expressed the view that food security considerations and other non-trade concerns such as environmental protection should also be taken into account in discussing a reform process that would affect agricultural production. Several of these countries had serious reservations regarding the use of the AMS approach as a means for reducing agricultural support.

A number of developing countries emphasized the need to take into account the possible negative effects of agricultural reform on net food-importing developing countries. They suggested several measures to counter such effects, including improved access to markets, provision of financial and technical assistance for agricultural development, food aid and concessional sales. The need for special and differentiated treatment for the developing countries in implementing reform was also emphasized, with particular reference to a proposal presented by Brazil.

In April 1990, trade ministers from 30 countries met in Mexico in an attempt to summon the political impetus needed to complete successfully the MTN by the end of the year. While the meeting did not result in any breakthrough on the key issues, it set a deadline of July for establishing an outline package of agreements. An encouraging feature of this package is that it covers all 15 items in the Uruguay Round including reduction of farm support in all three main areas — import barriers, export subsidies and domestic support.

At their July 1990 economic summit meeting in Houston, leaders of the seven largest western industrial countries reached a compromise agreement on farm support. The agreement called for cuts — but not outright elimination — in all categories of farm support, within a framework that includes a common instrument for measuring such support and taking into account food security considerations. It is hoped that this compromise agreement will provide the political impetus needed for reaching basic agreements on farm reform and other trade issues and so pave the way to a successful conclusion of the Uruguay Round in December 1990.

At the subsequent meeting of the Trade Negotiations Committee (the governing body of the Uruguay Round) in late July an accelerated timetable was laid down for negotiations, leading to the conclusion of the Uruguay Round at a ministerial meeting in Brussels from 3 to 7 December 1990.

FOOD AVAILABILITY AND NUTRITION

Calorie supply

Food availability at the country/regional level can be assessed on the basis of dietary energy supply (DES), generally measured in terms of kilocalories per caput per day. Being primarily the result of changes in domestic food production and net trade, DES reveals not only the capability of a country's food productive system in relation to its population's requirements, but also the extent to which domestic supply deficiencies have been compensated through importing food, either commercially or as food aid.

The following features can be observed from the early 1970s to recent years (Table 1.7):

- A wide gap of over 900 calories per caput/day currently separates average calorie supply in developed and developing countries. However, since calorie supply rose faster in developing regions, the gap has narrowed significantly since the early 1970s. Thus, DES in developing countries was equivalent to 65 percent of that in developed countries in 1969-71, compared to 72 percent by 1986-88.
- While the growth in DES increased in North America and western Europe between the 1970s and 1980s, it decelerated significantly in most developing country regions during the 1980s, a period of particularly unfavourable conditions for economic and agricultural growth for many of them. China was the major exception.
- Among developing country regions, a polarization of extreme situations occurred. On the one hand, the Near East, the developing region with the highest DES levels since the 1970s, continued to show an expansion in calorie supply during the first half of the 1980s — an expansion exceeded only by that of China during this period. On the other hand, Africa, where DES growth and levels were already the lowest during the 1970s, experienced an actual decline in average DES (0.2 percent yearly) during 1979-81 to 1986-88.

The obvious two determinants in a country's food supply situation are its capacity to expand domestic production or finance a net import position. For developing countries as a whole, there is a clear relation between DES and production changes, with exports and imports broadly netting out. At the individual region's level, however, the relative importance of production and net imports greatly varies. Thus, the two developing regions that achieved the fastest expansion in DES (Asia and the Near East) did so through opposing means. The Asian CPEs — overriden by the weight of China — relied almost entirely on domestically produced food, their imports only accounting for a minor and stable share of total DES. A similar phenomenon occurred in the Far East (non-CPE Asia), although the growth in production and DES was less dramatic. By contrast, the steep increase in DES in the Near East was primarily import-based, as per caput calorie production has declined since the mid-1970s.

In the case of Latin America, the very slow growth in DES during the 1980s was a consequence of a slow-down in production expansion and a sharp reduction in imports — both trends reflecting the severe crisis that affected the region's economic and agricultural performances since 1981-82.

In Africa, dismal food production performances since the early 1970s had to be compensated for by increasing imports. The latter, however, only marginally expanded in calorie terms during the 1980s, mainly because the lack of foreign exchange limited imports on commercial terms. As a result, Africa experienced a fall in DES during the 1980s — the only region to do so — after a long period of slow growth.

Incidence of undernourishment

Given assumptions on calorie intake distribution and minimum requirements, DES also allows the incidence of malnutrition to be estimated at the country and regional levels. Estimates of the incidence of undernourishment, based on 1983-85 data,[8] were last presented to the FAO

---

[8] *The Current World Food Situation.* CL 95/2. June 1989. This exercise was based on the methodology and assumptions used for the *Fifth FAO World Food Survey.* It incorporated the application of the cut-off point (minimum per caput requirement) of 1.4 BMR as the minimum energy requirement for adults and adolescents. The application of a given cut-off point obviously determines the overall picture of absolute undernourishment estimates. However, since the purpose was to provide indications of trends rather than absolute numbers, the choice of the cut-off point using 1.4 BMR, *vis-à-vis* the more conservative 1.2 BMR, was not crucial.

TABLE 1.7. **Dietary energy supplies (DES) and annual rates of change, by region/country group 1969-71, 1979-81 and 1986-88**

Region/country group	DES			Average of annual changes	
	1969-71	1979-81	1986-88	1969-71 to 1979-81	1979-81 to 1986-88
	(.......... kcal/per caput/day..........)			*(............ % ....)*	
Developed market economies	3 186	3 300	3 389	0.3	0.5
North America	3 371	3 487	3 626	0.3	0.7
Western Europe	3 233	3 371	3 445	0.4	0.4
Oceania	3 200	3 287	3 369	0.4	0.4
Other developed market economies	2 735	2 810	2 860	0.2	0.3
Developing market economies	2 158	2 317	2 352	0.7	0.2
Africa	2 046	2 148	2 119	0.6	0.2
Latin America	2 514	2 675	2 732	0.7	0.3
Near East	2 399	2 794	2 914	1.6	0.5
Far East	2 049	2 185	2 220	0.6	0.3
Other developing market economies	2 222	2 347	2 379	0.4	0.2
Centrally planned economies	2 378	2 606	2 824	0.9	1.1
Asian CPEs	2 006	2 323	2 620	1.4	1.6
Eastern Europe and USSR	3 318	3 401	3 418	0.3	0.1
Total developed countries	3 229	3 333	3 399	0.3	0.3
Total developing countries	2 106	2 319	2 434	1.0	0.7
World	2 435	2 587	2 671	0.6	0.5

Source: FAO.

Council in June 1989. The main conclusions were as follows:

- The number of undernourished persons in developing market economies rose by an estimated 15 million in the ten-year period 1969-71 and 1979-81, and by 37 million (to a total 512 million) in the following five years, an increase of 11 percent. Despite this increase in the absolute number of undernourished persons, their proportion in relation to total population declined during the 1980s, to an estimated 21 percent in 1984-86, because population grew faster overall — by 40 percent.
- The Far East region held the majority of undernourished persons in developing market economies — expectedly, given the region's massive population. The share of this region in the total number of malnourished persons in developing market economies tended to decline, however, from 61 percent in 1969-71 to 56 percent in 1983-85 (Figure 1.11).
- In relative terms — as a proportion of the respective populations — the incidence of undernourishment was considerably higher in Africa than in other regions. Furthermore, unlike other regions, this proportion rose during the 1980s and the absolute number increased by 27 percent. It is estimated that nearly one in every three Africans was undernourished in 1983-85, broadly the same proportion as in the early 1970s. All other regions recorded varying degrees of improvement between 1969-71 and 1984-86, the most notable being that in the Near East, where the proportion of undernourished population was halved.
- The Near East was also the sole region where the absolute numbers of undernourished declined, although here again such progress was arrested in the 1980s.

The fact that more than 20 percent of the population of 98 developing countries was estimated to be undernourished by the mid-1980s is a major indictment of virtually all aspects of development. No contemporary problem compares in gravity to the human devastation caused by persisting hunger and malnutrition. Recognition of this fact has prompted FAO and WHO to sponsor jointly an International Conference on Nutrition, to be held in Rome in December 1992. It is expected that the conference will result in an agreement on a coordinated strategy for national and international action to eradicate hunger and malnutrition.

Food shortages and emergencies

Serious problems of food supply shortages continue to persist in Africa. In *Ethiopia* and

the Sudan the 1990 harvest outlook is poor. Inadequate early season rainfalls in northern and western Ethiopia have jeopardized crop prospects in areas already facing serious food shortages. Although later rainfall brought some relief to several areas, the situation remains critical. International emergency relief programmes have so far averted widespread famine but the situation remains precarious and could still deteriorate in the period leading up to harvest at the end of the year. Insufficient early season rainfall has also hit areas of the Sudan already confronted by food shortages following droughts last year, and the food supply situation in the affected areas continues to deteriorate. In both countries, the outcome of the harvest at the end of the year will be crucial. In the south of the Sudan, Operation Lifeline Sudan Phase II, aimed at delivering more than 100 000 tonnes of food, is being implemented amid logistical difficulties. Thus there are fears that the food aid pledged under the programme may not be delivered to all at-risk populations because of the late start of the operation and the deterioration of the roads.

The food supply situation also remains highly critical in *Angola* and *Mozambique* with food aid deliveries falling well short of requirements. In both countries continued civil strife is seriously constraining agricultural activity and marketing as well as impeding the distribution of relief supplies to at-risk populations. In Mozambique, despite an increase in the recent harvest as a result of better weather and improving security conditions in some areas, cereal output will still cover only half of requirements. Moreover, logistic constraints will continue to hamper food distribution in deficit areas. Food shortages are widespread and serious, with current food distributions insufficient to ward off the threat of malnutrition and starvation. In Angola a decline in the recent harvest of cereal crops due to poor rainfall in southern and central provinces has further aggravated the situation. Cereals output in 1990 covers less than 40 percent of domestic consumption requirements. The needs of urban, displaced and drought-affected populations have to be met entirely by imports while pledges and deliveries of food aid to date are well below minimum needs. In both countries, as of August 1990, increased international assistance is required, in particular special measures to expedite delivery of food aid and logistic support for its internal distribution, including airlifts to otherwise inaccessible areas.

In *Liberia* civil strife has led to a rapid deterioration of the food supply situation, impeding the marketing and distribution of basic food supplies and disrupting agricultural

Figure 1.11

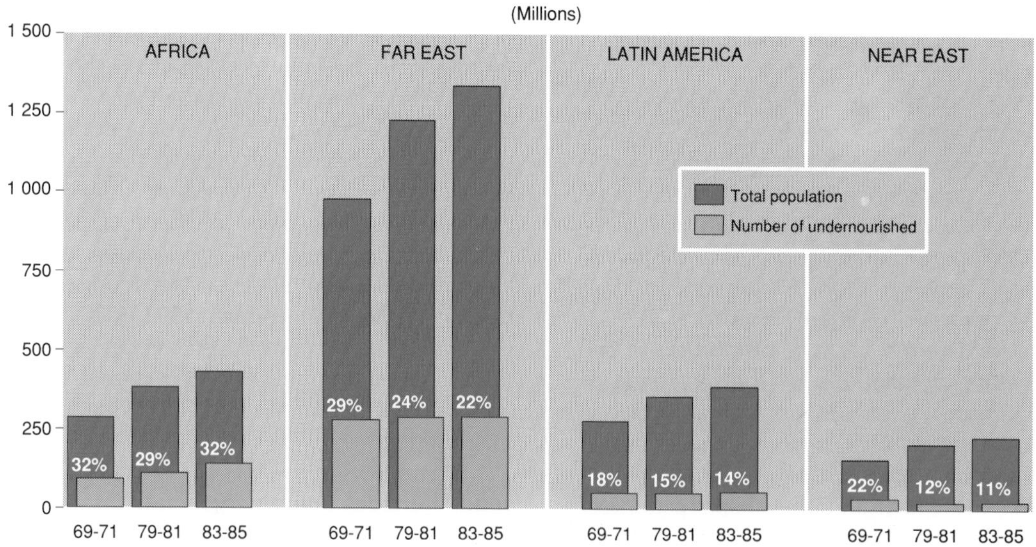

Source: FAO

activity. Severe food shortages are reported throughout the country while the number of refugees fleeing to neighbouring countries continues to grow. Substantial emergency food relief for the war-affected and displaced population is needed.

Other African countries facing shortfalls in food supplies and requiring exceptional assistance in the current marketing year, as of September 1990, were: Burundi, Cape Verde, Malawi, the Niger, Rwanda, Sierra Leone and Somalia. At the same time, a number of countries were experiencing situations of exceptional surpluses in some areas while requiring external assistance for transportation to deficit areas in the same or neighbouring countries. Examples are Benin, Burkina Faso, Chad, Côte d'Ivoire, Ethiopia, Ghana, Kenya, Mali, Mauritania, the Niger, the Sudan, Togo, Uganda and Zimbabwe.

Against the background of a difficult food supply situation in several African countries, there are four main areas which require immediate attention: i) priority to accelerating the delivery of relief supplies to at-risk populations of northern Ethiopia and the Sudan as well as monitoring the development of crops for harvest at the end of the year in both countries; ii) further pledges and logistic support for internal food distribution are needed to ensure supplies for strife-afflicted populations of Angola, Mozambique and Liberia; iii) further pledges of food aid for 1990/91 are needed for 12 countries; and iv) additional donor support is required for the purchase and internal distribution of local surpluses as well as for the disposal of unutilized exportable surpluses.

Serious food supply shortfalls are currently reported in nine other developing countries outside Africa. In Asia, despite a further anticipated improvement in the overall situation this year, difficult food supply situations are reported in Afghanistan and Sri Lanka as well as Jordan and Lebanon in the Near East. Shortfalls requiring exceptional and/or emergency assistance are furthermore reported in four Latin American and Caribbean countries — Bolivia, Haiti, Nicaragua and Peru — as well in Western Samoa, where emergency assistance is being provided to victims of the cyclone, Ofa, experienced in early February.

Another problem, hopefully temporary but currently causing major concern, is the massive outflow of people from the Near East following recent events in the region. In the absence of adequate transportation facilities to repatriate these refugees, tens of thousands of them have gathered in desert tent camps in the Iraq-Jordan border area in conditions of extreme hardship. Serious shortages of food and water as well as medical and other services are reported. The situation in these camps remains critical despite some relief in the form of external aid and a recent improvement in the ratio of arrivals to departures. However, a large number of people are still reported to be waiting to cross the border from Iraq.

EXTERNAL ASSISTANCE TO AGRICULTURE AND FOOD AID

Official external assistance to agriculture

Overall commitments of external assistance to agriculture reached about $16 billion in 1988, which is the latest year for which complete estimates are available. At current prices this amount was 10 percent more than in 1987 and 39 percent more than in 1985 — the latter level being virtually the same as in 1980-81. However, if these figures are deflated by prices of manufactured goods, in real terms there was a 4 percent decrease in 1988 commitments from the levels of 1985 (Figure 1.12). Nevertheless, there has been a steady increase in the estimated share of grants in total commitments, from 19 percent in 1980-81 to 24 percent in 1984-85 and 31 percent in 1987-88.

A notable feature since 1986 has been the sustained increase in bilateral commitments from the OECD's Development Assistance Committee (DAC) which more than offset a decline in commitments from multilateral institutions. Thus, bilateral commitments accounted for over half of the total in 1988 compared to 40 percent in 1985. The increase in bilateral commitments from DAC/EEC was particularly pronounced in 1988 (29 percent more than 1985 and 19 percent more than 1987 at constant 1985 prices). This increase mainly reflected the very high commitment figure of $2.3 billion from Japan in 1988 compared to only $930 million in 1987. An increase of a similar magnitude was expected for 1989 since Japan intends to raise the share of its official development assistance in DAC's total, in line with its share in DAC countries' GNP. EEC lending also rose noticeably in 1987 and 1988 ($1.2 and 1.3 billion respectively compared to a little over $500 million in 1986).

Estimates for 1989 are only available for multilateral lending. At an estimated $7.2 billion in 1989 — $5 billion at constant 1985 prices — multilateral commitments of external assistance to agriculture fell by almost one-third in real terms from the high figure of 1986. World Bank

Figure 1.12

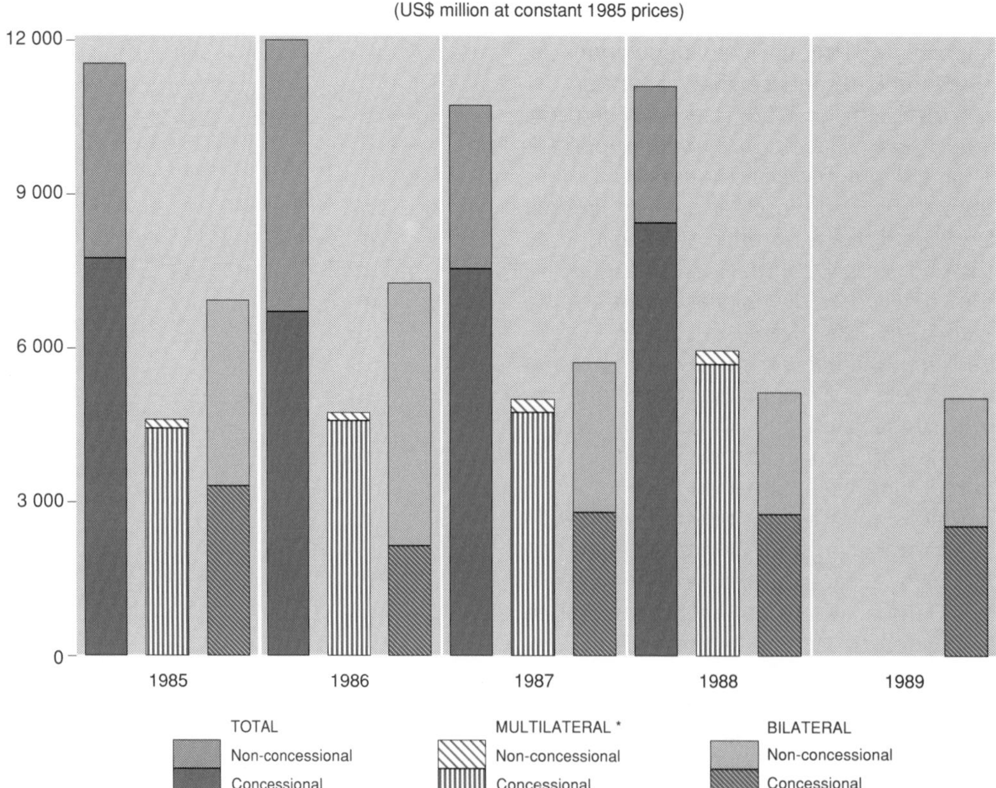

Source: FAO, Policy Analysis Division and OECD

* Not available for 1989

lending has remained stagnant at about $4 million a year since 1987, despite an increase in lending to other sectors. The share of agriculture in both IBRD and IDA commitments has continued to fall. IBRD loans to agriculture accounted for 27 percent of the total in 1980/82 compared to only 17 percent in 1987/89. One reason for the relative decline of agriculture has been the growing importance of World Bank structural adjustment loans which often have an agricultural component.

Regional banks' lending sharply fell in 1989, mainly reflecting a very low level of commitments by the Inter-American Development Bank (IDB) (which does not include a number of large IDB loans that were approved in the course of 1989 but not signed by the end of the year). By contrast the African Development Bank (AfDB) considerably expanded its commitments to agriculture in 1989.

Disbursements of external lending to agriculture are estimated to have reached $14.2 billion in 1988, 15 percent more than in 1987. ($7.5 billion from multilateral, and $6.7 billion from bilateral, sources.)

The World Bank and the Asian Development Bank (AsDB) accounted for much of the increase in multilateral disbursements in 1988. The World Bank's disbursements of $4.7 billion reflected the high levels of commitments during 1986 to 1988 (totalling $4.3 billion) for fast disbursing loans for agricultural adjustment programmes, credit and inputs. Commitments made in 1989 for credit and agricultural adjustment projects reached $1.5 billion with $267 million, or 18 percent of the total, disbursed in the same year. The remainder is expected to be disbursed in 1990 and 1991.

The sharp increase in the transfer of resources from the Asian Development Bank (from $435 million in 1987 to $885 million in 1988) was mainly owing to the expansion of quick disbursing programme loans and also to more effective delivery operations.

Multilateral disbursements of external lending to agriculture in 1989 are tentatively estimated at $6.2 billion, approximately 17 percent less than the amount disbursed the previous year. Information for bilateral flows in 1989 is not yet available.

Recent developments in agency funding

The latest information for *IFAD's* Third Replenishment indicates a total amount of $566.3 million. This is above the Second Replenishment figure (1985-87) of $540 million, but falls short of the target of $750 million. The Third Replenishment is expected to be effective by 30 June 1990 and covers the period to June 1992.

The target of $300 million for IFAD's Special Programme for sub-Saharan Africa was reached in January 1988. However, because of fluctuations in the dollar exchange rate, this total is currently about $276 million. Most of this amount has been committed and the Programme is due to terminate by the end of 1990 unless a decision is taken on its continuation.

In December 1989, 32 countries agreed to provide SDR 11.68 ($15 billion) to *IDA* (IDA-9) for the period 1 July 1990 to 30 June 1993. In addition, the repayments of funds lent in IDA's early years will provide an estimated SDR 1.6 billion in new commitments in the same period, bringing the total resources available to SDR 13.3 billion ($17 billion). In comparison, IDA-8 totalled SDR 10.5 billion ($13.4 billion). Priority areas for IDA funding will be: poverty reduction; support for sound macro-economic and sectoral policies and programmes, including further emphasis on institutional strengthening; and environmental programmes. Conventional investment should represent about 75 percent of all IDA lending, leaving around 25 percent for quick disbursing adjustment programmes.

The share of IDA-9 resources going to sub-Saharan Africa is likely to continue to be between 45 and 50 percent.

In March 1989 the *IDB* Board of Governors finally approved the proposal for the seventh General Increase in Resources by $26.5 billion to a total of $61 billion. The increase of resources will permit a lending programme of $22.5 billion from 1990 to 1993. If the trend of allocating roughly 30 percent to agriculture is maintained, approximately $6.5 billion would be available to agriculture in Latin America and the Caribbean during the next four years. The amount allocated to agriculture from 1986 to 1989 was $2.5 billion.

Food aid

In 1989/90, food aid in cereals is estimated at 11.5 million tonnes compared to 10.2 million tonnes in 1988/89. The increase largely reflects additional allocations to Eastern Europe (mainly Poland and Romania) from the EEC, the United States and other donors. Total shipments to low-income food-deficit countries are expected

to be 8.6 million tonnes, only marginally above the 1988/89 level and about three million tonnes below the quantities shipped to them in 1987/88.

Contributions to the International Emergency Food Reserve (IEFR) in 1989 amounted to 390 000 tonnes of cereals and 34 000 tonnes of non-cereal food commodities — well short of the 500 000 tonne target. About 94 percent of the contributions in total cereals and all of those in non-cereals were channelled through the World Food Programme (WFP). As of May 1990 contributions to the IEFR amounted to 158 000 tonnes of cereals and 6 322 tonnes of non-cereal food commodities.

There were 46 WFP emergency operations in 1989 of which 29 were in Africa, five in Asia, eight in the Near East and four in Latin America for a total cost of $93 million. More than two-thirds of the cost or $63.5 million were from the IEFR and $29.5 million from the WFP annual emergency allocation of $45 million. About 88 percent of the total WFP emergency operations from all sources in 1989 were directed to refugees, returnees and displaced persons. As of end-April 1990, 11 WFP operations have been approved (nine in Africa, one in Asia and one in the Near East) at a total cost of $23.9 million.

As of June 1990 pledges to WFP for 1989-90, by 77 donors, amounted to $1 152.1 million or 82 percent of the pledging target of $1 400 million, approximately the same proportion as during the same period in 1987-88. The total amount pledged includes $875.2 million in the form of commodities and $276.9 million in cash, or 24 percent compared with an expected cash component of at least one-third of total contributions, as stipulated by the WFP general regulations.

The preliminary FAO estimate of total food aid in cereals in 1990/91 is 9.8 million tonnes, about 1.7 million tonnes lower than the previous year, when exceptionally high shipments were made to Eastern European countries. No allocations to these countries are currently envisaged in 1990/91. Food aid in cereals to low-income food-deficit countries are expected to increase by about 4 percent from the previous year, though still remaining well below the levels provided to these countries during 1984/85-1987/88. Food aid in cereals would represent about 17 percent of total cereal imports by low-income food-deficit countries, the same as in the previous year, compared to an average of about 20 percent in the previous five years. In 1990/91 a higher proportion of shipments could be in the form of wheat and rice, in view of the decline in world prices of these commodities *vis-à-vis* those of coarse grains.

Budgetary allocations for food aid programmes of major donors for 1990/91 are expected to remain at about the same level as in the previous year. Funding for the United States' PL 480 programme for fiscal year (FY) 1990/91 is set at about $1.5 billion, ($846 million under Title II and the remainder under Title I), slightly lower than for FY 1989/90, but the volume of cereals to be shipped could be higher because of lower commodity prices. However, the United States' food aid in maize and sorghum under Section 416, which totalled over 1.6 million tonnes in FY 1989/90, is likely to be 1.5 million tonnes (500 000 tonnes of maize and one million tonnes of sorghum) during FY 1990/91. Thus, total shipments by the United States from both PL 480 funding and Section 416 would amount to about six million tonnes in 1990/91 — roughly the same level as in 1989/90.

Food aid in cereals by the EEC during 1990/91, financed by the regular food aid budget, is also expected to be maintained at the previous season's level of about 1.7 million tonnes (grain equivalent). In 1989/90 an additional 1.65 million tonnes were extraordinarily shipped to Eastern European countries, which were financed outside the food aid budget. Substantial amounts of EEC food aid in cereals will continue to be in the form of triangular transactions.

Shipments of food aid in cereals in 1990/91 from Canada, Japan, and other donor countries are also expected to be maintained at the level of the previous year.

As of July 1990, contributions from 13 donors to the IEFR amounted to 168 000 tonnes of cereals and 9 000 tonnes of non-cereal food commodities. Of these contributions, 120 000 tonnes of cereals and 9 000 tonnes of non-cereal commodities have been pledged for multilateral channelling through the WFP. In addition to IEFR contributions, 417 000 tonnes of cereals and 42 000 tonnes of other food commodities have been pledged under the subset of the WFP regular resources for meeting the requirements of protracted refugee situations.

As of July 1990, 64 000 tonnes of cereals and 5 000 tonnes of other food commodities have been allocated to 22 WFP emergency

operations in 19 countries. Of these, 32 000 tonnes were allocated to assist refugees and displaced persons, 21 000 tonnes to assist victims of drought and crop failure, while the balance was for victims of sudden natural disasters. In addition, to date in 1990, 391 000 tonnes of cereals and 38 000 tonnes of other food commodities have been allocated to 22 protracted refugee operations against resources which have been made available for that purpose.

FISHERIES

Production

In 1988, world fish production continued the strong growth which has characterized the period since the mid-1970s. The world fish harvest in 1988 was almost 98 million tonnes, or nearly 5 percent higher than the combined catch and aquaculture production of the previous year (Figure 1.13). While recorded higher catches were widespread, larger landings of small pelagic species in the southeastern Pacific and increased production by China were the principal contributing factors to the overall increase.

Catches of small pelagics by Chile, Ecuador and Peru had been unfavourably affected in 1987 by the change in the El Niño current. In 1988, however, catches by these countries increased by over 2.5 million tonnes. Peru was the principal beneficiary of the fishery, its catches increasing by almost 45 percent to over 6.6 million tonnes. These increased catches enabled Peru to increase its fish-meal production by over 50 percent, matching the Chilean production of 1.1 million tonnes. In Chile, catches were adversely affected by the movement of fish stocks from the northern to the southern part of the country and, consequently, increased by the relatively small margin of 8 percent. Nevertheless, catches reached the second highest level ever of 5.2 million tonnes.

Fish production in China increased by a further one million tonnes, a consequence of the continued development of aquaculture as in the previous year. Elsewhere in Asia, Japan's catches of 11.9 million tonnes were only slightly higher than in 1987. The country nevertheless maintained its position as the leading fishing nation. Several other Asian countries reached record levels of production.

Although total output by African countries showed a further increase, catches off southern and southwest Africa fell by over 9 percent to 1.3 million tonnes. As a result, production of both fish-meal and canned fish declined.

The USSR maintained its catch at about the level of the previous year, i.e. 11.3 million tonnes. The stability of United States production at 5.9 million tonnes was in marked contrast to that of the previous year when it had increased by 16 percent as a result of increased operations in the northwest Pacific. New Zealand production, however, boosted by

expansion of its national operations and through joint venture enterprises, expanded by a further 17 percent. Canadian catches increased by some 34 000 tonnes to 1.6 million tonnes, but catches of the higher-priced ground fish declined slightly. Output in the Scandinavian and EC countries remained, in aggregate, at about the same level as in the previous year. Countries where there was a fall in production included Norway, where there was decline in cod catches, and the United Kingdom; countries recording increases included Denmark, Iceland, France and Spain.

Primarily, as a result of the higher catches of small pelagic species, the greater part of the overall increase in fish production in 1988 was used for reduction to fish-meal. Very favourable fish-meal/soymeal price ratios and low stock levels encouraged a marked growth in fish-meal production, which rose by some 400 000 tonnes. Therefore, although supplies of fish for direct human consumption continued to grow in 1988, the rate of increase (4.3 percent) was less than the overall expansion in total world production (Table 1.8).

As in the previous year, aquaculture continued to make an important contribution to fish and shellfish supplies, especially for carp, tilapia, eel, trout, salmon, molluscs, shrimps and prawns. Aquacultural production grew particularly rapidly in Asia.

Trade

Short supplies of some important species and sustained demand further boosted the value of fish prices and international fish trade (Table 1.9). Available data indicate that the value of fish trade in 1988 rose by some 12 percent to reach well over $31 billion. Exports by developing countries exceeded $14 billion, double those achieved in 1984.

The United States became the world's leading exporter of fishery products in 1988; taking into account the $220 million received for trans-shipments to foreign vessels in the North Pacific, the United States achieved a remarkable 33 percent growth in fish exports, the total value of which exceeded $2.4 billion. Other major exporters also significantly expanded their sales (Canada +5.5 percent, Denmark +5.9 percent, Republic of Korea +18.5 percent, Thailand +29.3 percent and Norway +9 percent).

Japan was again by far the world's major importer of fishery products, with its purchases rising by a further 28 percent in value to a total representing almost one-third of world fish trade.

Imports by the United States declined in both quantity and value but all other major importing countries increased their purchases significantly. Notable among these were France (+10.5 percent) whose imports reached a level more than twice their value of the early 1980s, Spain (+30 percent) and the United Kingdom (+14 percent). Thailand (+70 percent) and Taiwan, Province of China (+18 percent), also merit mention because their fish imports have increased rapidly in recent years as an input for export-oriented fish processing industries.

Figure 1.13

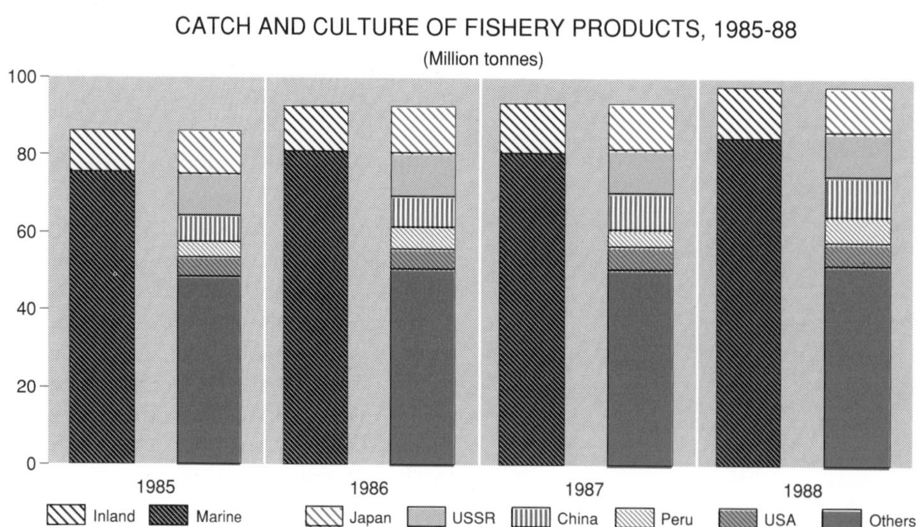

Source: FAO

BOX 1.3

Shrimp farming — a success for aquaculture?

One of the undoubted highlights in the recent history of aquaculture is the growth in production of shrimp, mainly the Peneus species. In tropical coastal land, using simple earth ponds, tidal or pumped water supplies, local or hatchery-reared seed, and simple fertilization or feeding, a marketable crop of 15 to 30 g shrimp can be reared in as little as three or four months. Annual yields range from 200 to 500 kg/ha in simple extensive tidal ponds using natural productivity, to upwards of ten tonnes/ha in modern intensive systems with complete feeds, aeration and regular water exchange. Sold on the international markets of Japan, the United States and western Europe, shrimp is almost a perfect export, using basic local resources to bring in foreign exchange. Recognizing its potential, FAO has played a key role in initiating and supporting the industry with inputs in technical development, training, information services and institutional support.

In production terms, the result has been impressive. In 1981, cultured shrimp accounted for 2.1 percent of the total world shrimp harvest, while in 1988 this share was estimated to have risen to 22 percent. Output in 1988 reached some 450 000 tonnes, corresponding to a farm-gate market value of perhaps $2.5 billion. Much of this originates in the developing world, particularly in Southeast Asia and Latin America. In Bangladesh, shrimp is the second most important export commodity and much of it is from farming. In the Philippines, about 40 000 ha of ponds are producing shrimp; China was estimated to produce 100 000 tonnes in 1988, an increase of some 50 percent over the previous year. In Latin America, cultured shrimp accounts for almost half the total shrimp harvest in the region. Ecuador, now the leading producer of farmed shrimp in the Western Hemisphere, supplied 70 000 tonnes in 1988, more than 80 percent of the region's production.

As a result of this development, rural coastal land of limited agricultural potential, until recently regarded as of little value, has acquired a new status. There are also new prospects for income and employment; traditional brackish water fishpond operators have a new source of income; fishing communities can get involved in catching seed shrimp and broodstock; and farms and hatcheries in rural areas offer local employment and training for young people. Furthermore, there are new service industries; in Southeast Asia, "backyard" hatcheries have provided excellent opportunities for small family-based businesses whose earnings give stimulus to local economies.

Yet much of the potential, not least the current sustainability, of the industry is under question as, after a decade of increasingly rapid and profitable growth, shrimp farming faces the problems of its almost unrestrained expansion. Markets show increasing signs of saturation; in 1988 and 1989, high inventories and abundant supply pushed prices to levels at which many producers faced losses, and in some cases had to suspend production. In Taiwan, over 1987-88, production dropped sharply because of serious disease losses associated with degraded environmental conditions resulting from highly intensive production methods. Ecuador, with less intensive systems, had similar problems one year later. As yet there are few treatments available.

In many areas, rising demand for feed materials puts pressure on supplies, including low-value fish, which might be available for direct consumption. The widespread clearing of mangrove areas and the destruction of important habitats — not least for larval marine fish and shrimp — the erosion of coastal margins, and the salinization of coastal flood plain soils are cause for increasing concern. As production moves increasingly toward larger-scale entrepreneurs or corporate groups, traditional communities and their activities become displaced. While the industry was profitable and expanding, these factors might have been less noticeable; under present circumstances, however, the wider costs of the industry must be questioned.

If shrimp farming is to bring real and continuing benefits to producer countries, means will have to be found to stabilize the industry, increase its efficiency and reduce its adverse impacts. This will require the right kind of market and technical support, planning and regulatory advice, particular care over resource management, and social and environmental assessment. These concerns are already starting to be addressed in current FAO projects in Asia and Latin America; while support for production continues, the emphasis is shifting more strongly toward ensuring that shrimp farming has a sound and well-structured future.

TABLE 1.8. Disposition of world catch, 1985-88

Item	1985	1986	1987	1988	Change 1984-85	Change 1985-86	Change 1986-87	Change 1987-88
	(................. '000 tonnes ....)				(.................... % ....)			
World	**86 257**	**92 612**	**93 414**	**97 985**	**2.9**	**7.4**	**0.9**	**4.9**
For human consumption	**60 984**	**65 777**	**67 836**	**70 755**	**2.5**	**7.9**	**3.1**	**4.3**
Marketing fresh	16 381	19 498	20 908	23 015	5.3	19.0	7.2	10.1
Freezing	20 388	21 784	22 107	22 740	1.2	6.8	1.5	2.9
Curing	12 745	13 045	13 191	13 220	4.5	2.4	1.1	0.2
Canning	11 470	11 450	11 630	11 780	−1.2	−0.2	1.6	1.3
For other purposes	**25 273**	**26 835**	**25 578**	**27 230**	**4.0**	**6.2**	**−4.7**	**6.5**
Reduction	24 273	25 835	24 578	26 180	3.7	6.4	−4.9	6.5
Miscellaneous purposes	1 000	1 000	1 000	1 050	11.1	—	—	5.0

Source: FAO, Fisheries Department.

TABLE 1.9. Trade in fisheries, 1985-88

Item	1985	1986	1987	1988	Change 1984-85	Change 1985-86	Change 1986-87	Change 1987-88
	(................. $ million ....)				(.................... % ....)			
World								
Exports	17 382	23 072	28 127	31 597	7.1	32.7	21.9	12.3
Imports	18 602	24 226	30 508	35 096	8.2	30.2	25.9	15.0
Total developing countries								
Exports	7 698	10 514	12 819	14 365	7.3	36.6	21.9	12.1
Imports	2 467	3 028	3 708	4 596	0.5	22.7	22.5	23.9
Total developed countries								
Exports	9 684	12 558	15 308	17 232	6.8	29.7	21.9	12.6
Imports	16 135	21 198	26 800	30 500	9.5	31.4	26.4	13.8
Major exporters								
United States	1 162	1 481	1 836	2 442	15.9	27.5	24.0	33.0
Canada	1 359	1 752	2 092	2 207	6.8	28.9	19.4	5.5
Denmark	953	1 341	1 751	1 854	6.0	40.7	30.6	5.9
Korea, Rep.	797	1 171	1 506	1 784	1.9	46.9	28.6	18.5
China (Taiwan Province)	956	1 422	1 674	1 695	15.9	48.7	17.7	1.3
Thailand	675	1 012	1 261	1 631	6.6	49.9	24.6	29.3
Norway	922	1 171	1 475	1 608	2.1	27.0	26.0	9.0
Major importers								
Japan	4 744	6 594	8 308	10 658	12.8	39.0	26.0	28.3
United States	4 052	4 785	5 662	5 389	9.5	18.1	18.3	−4.8
France	1 040	1 510	2 022	2 234	6.6	45.2	33.9	10.5
Italy	985	1 265	1 738	1 894	32.7	28.4	37.4	9.0
Spain	412	722	1 322	1 716	5.6	75.2	83.1	29.8
UK	941	1 216	1 387	1 577	7.5	29.2	14.1	13.7
Germany, Fed. Rep.	820	1 113	1 270	1 411	2.4	35.7	14.1	11.1
	(.................... % ....)							
Export as share of catches								
Total developing countries	8.9	11.4	13.8	14.9				
Total developed countries	11.2	13.3	16.5	17.6				

Source: FAO, Fisheries Department.

As regards fish availability, there is an increasing contrast in the overall trends and patterns between the developed and developing countries, particularly in parts of Africa and Asia.

In general, for the developed countries, the situation is one of relative abundance of supply and stable or even declining real prices for a number of products. Although total allowable catches of the preferred demersal species in the northeast Atlantic have been severely reduced in recent years, low-price substitutes (for example, hake and Alaska pollack), have become increasingly available, except in countries where they are subject to high import barriers. Prices of aquaculture products, such as salmon and shrimp, are also becoming increasingly attractive to consumers, while oversupply and, consequently, falling prices of squid are benefiting such large importing countries as Italy and Japan.

For a number of developing countries, particularly in Africa and Asia where fish often is an important component in the diet and a major source of animal protein, per caput consumption of fish is declining in response to rising prices. The preferred species are becoming increasingly unaffordable by the lower-income groups. Inadequate domestic supplies of fish sometimes result from increased pressure on resources by industrial and semi-industrial vessels fishing for export to developed countries, directly or through feed fishing for aquaculture production oriented to export markets. Although the foreign exchange so earned can indirectly contribute to increasing food supply if used for financing food imports, the immediate effect is inevitably to reduce fish supplies for domestic consumers. Additionally, such fishing frequently diminishes the catches of the artisanal subsector, the overall economic benefit of which is often underestimated by governments. An important policy issue is, therefore, the proper evaluation of all economic and social aspects involved in promoting fish consumption as opposed to promoting fishery activities as a source of foreign exchange.

Outlook

Some slight expansion in the catches of marine fish can be expected provided that correct management practices are applied by the fishing nations. Strict management is also necessary to avoid loss of catch and overexploitation. Some reduction in fish losses is possible as a result of better utilization and improved fish handling and processing. The long-term and sustainable growth in overall fish production, however, appears to depend heavily upon the continued development of aquaculture. In particular, there appear to be opportunities for further significant increases in aquaculture output by China which already accounts for over one-half of total world production in this sector. There is clearly scope for additional growth in both yields and areas under cultivation. A major constraint may be the availability of feed. Aquaculture is becoming an increasingly significant consumer of fish-meal: it is estimated that some 650 000 tonnes of high-quality fish-meal involving the reduction of over three million tonnes (live weight) of fish were used in aquaculture operations in 1988.

Preliminary indications for 1989 suggest that there were further increases in production and trade. One of the most important positive influences on production in 1989 was aquaculture, because it is expected that China's rapid growth in freshwater aquaculture will have continued into 1989 and 1990. In 1989, however, most of the North European countries recorded lower catches while global catches of small pelagics for reduction to fish-meal and oil remained at about the same level as the previous year, a further decline in the production of fish-meal in southern Africa being offset by an increase in Chile.

Looking further ahead, the economic changes currently taking place in Eastern European countries may result in the development of new markets for fresh and frozen fish.

FORESTRY

Production and trade

In 1989 some 3 460 million m^3 of *roundwood* were produced in the world, 55 percent of which was produced in the developing world (Figure 1.14). *Fuelwood* is the largest single component accounting for 1 975 million m^3. About 1 520 million m^3 of this total is produced and consumed in the developing countries as an energy source for large parts of the rural population. It is estimated that fuelwood represents 18 percent of the energy needs of developing countries, with this share reaching 80 to 85 percent in some African countries. The production of fuelwood in the developed countries grew rapidly in the late 1970s, but has recently declined because of decreasing fossil fuel prices.

The *pulp and paper sector* continues to expand with all of the main producing countries setting record levels in production and capacity utilization in 1988. Trade was strong as domestic demand rose in some major importing countries, and prices for most pulp and paper grades significantly increased.

The sector continued to grow in 1989, although at a slower pace than previous years, with European production and consumption reaching new records for almost all grades. The European fibre market remained tight, however, as indicated by a 25 percent increase in imports of waste paper in 1989. Growth in the production of *pulp and paper* was particularly strong in Japan (9 percent). On the other hand, the rapid growth of the sector in North America during 1986-88 ended in 1989 with output decreasing slightly from the record level reached in 1988.

The pulp and paper industry also continued to expand in developing countries, although at a much slower rate than in previous years. China's production of paper and paperboard grew by only 1 percent in 1989.

The *mechanical wood products sector* recorded another year of expansion in 1989, following the good performance of 1988. Most of the growth occurred in western Europe where very mild winter conditions in 1988 and 1989 had a favourable effect on the level of housing starts and resulted in a considerable increase in production and consumption of mechanical wood products. Production of coniferous sawnwood, the main wood material used by the building industry, is estimated to have reached the historical record of 56.5 million m^3 in 1989, some 3 percent above the previous record of 1980.

Output in most East European countries was stagnant or declining because of the difficulties in introducing economic reforms and restructuring industries. In North America the steady growth of the mechanical wood product sector since 1983 ended in 1988, with building activity dropping 10 percent. The decline in output continued in 1989, with the main exceptions of two wood-based panel products, particle board and medium-density fibreboard. The decline of North American consumption was only partly compensated by the steady growth of exports from the United States in 1988 and 1989 as a result of the weakening of the US dollar. Japan experienced a marked increase in its wood housing activity in 1989, after a marginal decrease in 1988. Imports of mechanical wood products increased in 1989, reflecting notably expanded imports of sawnwood and wood-based panels, both from temperate and tropical origins.[9]

Japan's imports of tropical logs grew slightly in 1989, after declining in 1988, principally to replace a low level of inventories.

After years of sustained growth, developing country production of mechanical wood products decreased in 1988 but resumed expansion in 1989. Countries with modern export-oriented industries benefited from greater world demand for mechanical wood products.

Exports of tropical timber, an important source of foreign exchange for some tropical developing countries, expanded by about 2.5 percent in volume in 1988. Their total value reached $7.6 billion in 1988, 10 percent above 1987, with an increase in the share of the higher-value processed products. Early estimates for 1989 point to further expansion of this trade, with a strong growth in western Europe's imports of tropical sawnwood from Southeast Asia; and a further increase in exports of plywood from Indonesia, which may exceed the 8 million m^3 mark, most of which will be imported by Japan.

Current issues

Concern for the tropical forests. World public opinion is becoming increasingly concerned

---

[9] Japan's imports of sawnwood increased twofold in the period 1978-1989 but imports of plywood grew 20 times during the same period, indicating growing import demand for manufactured wood products.

Figure 1.14

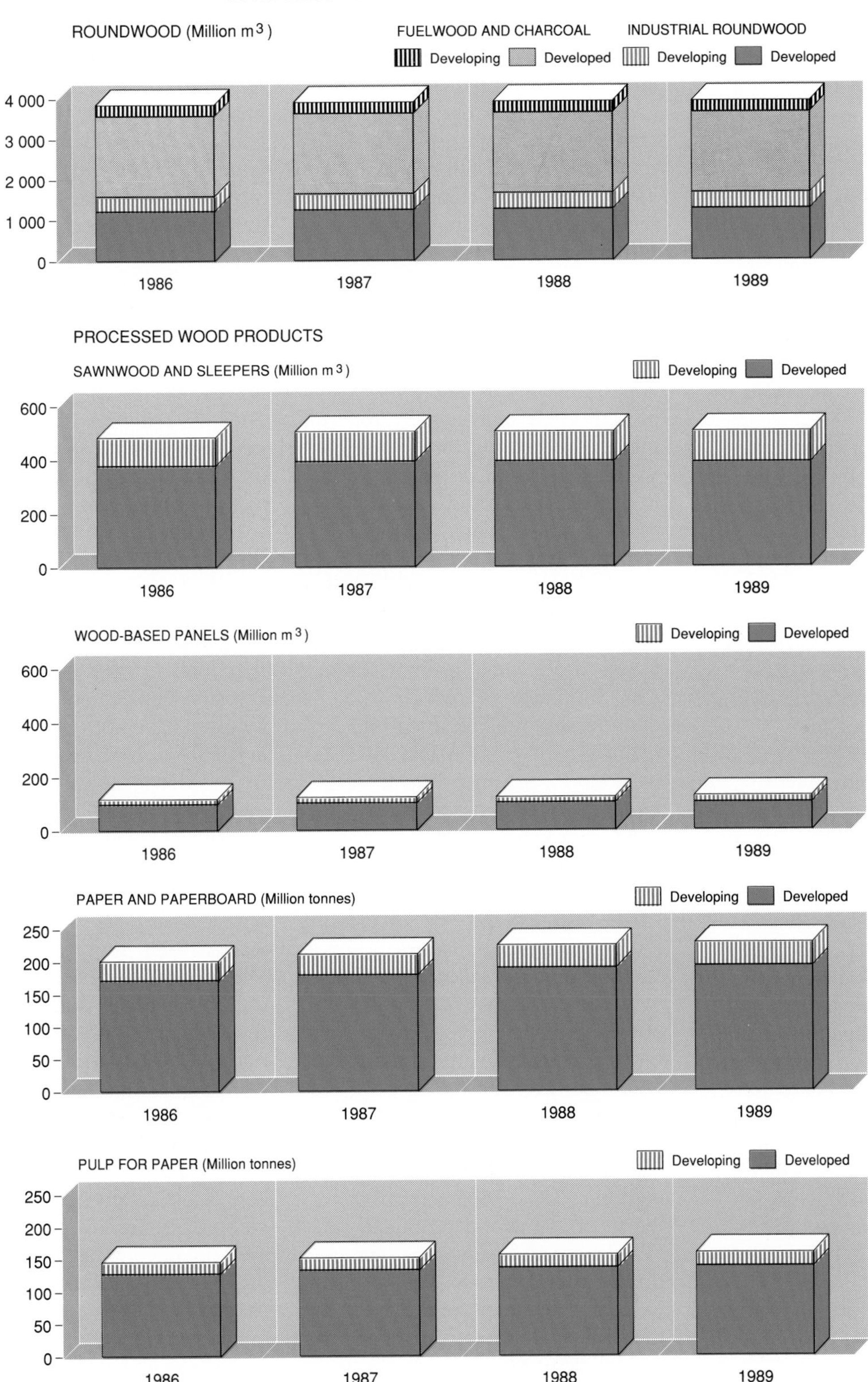

Source: FAO

BOX 1.4

Forestry and climate change

The problem, extent and causes

The so-called greenhouse gases, which are composed mainly of carbon dioxide (CO_2), methane (CH_4) and nitrogen compounds (N_2O, NO, NO_2), are major regulators of global temperature. A buildup in the atmosphere of these gases has been clearly documented over recent decades and a doubling of CO_2 has been predicted by the middle of the twenty-first century. This may cause the Earth's climate to become significantly warmer, with the main effects being felt at higher latitudes. Changes in rainfall and rainfall patterns may also be expected.

The modelling of these potential changes is very complex, and the rate and magnitude of actual changes have yet to be determined. However, it is thought that the average global temperature may rise between 1 and 5°C during the next century.

The primary source of CO_2 buildup is the combustion of fossil fuels, which account for approximately 5.2 billion[1] tonnes of carbon per year. However, changes in land use (shifting cultivation and other forms of conversion of forest to agriculture) which are eliminating considerable amounts of biomass, particularly in the tropical forest areas, also contribute to this buildup. The amount of CO_2 released during deforestation has been estimated at one to two billion tonnes per year; the amount released from soils in the course of this process is greater than that from the vegetative cover. Worldwide, atmospheric carbon has been estimated to be increasing by approximately three billion tonnes annually.

Global warming of the magnitude predicted above would also affect the geographic coverage of forests and woodlands, and the rate of growth and production of their component species. It may also lead to changes in the composition of species, shifts in genetic variation patterns and loss of biological diversity in forests and other natural ecosystems.

On the other hand, forests, plantations, and trees in non-forest areas act as an efficient sink for carbon. Considerable amounts of carbon may be absorbed by reforestation and afforestation by increasing the stocking and production of existing forests through sound forest management practices and by more extensive use of durable wood products.

Although it would seem feasible, in theory, to capture temporarily large parts of the yearly increase of carbon through accelerated tree planting, capturing the full excess would require an unrealistic input in terms of land and capital. The most optimistic calculations estimate that this would require the establishment of some 500 million ha of high-yielding plantations in the immediate future, at an initial cost of some \$200-500 billion, to which considerable maintenance and management costs would have to be added. As a measure of comparison, 1.1 million ha of plantation were established successfully per year in the 1980s in the tropical world (excluding China). Though the rate of planting has significantly increased during the 1980s, it is not commensurate with the total area mentioned above — which is the area needed to capture the excess carbon in the atmosphere.

Keeping the above facts in mind, however, it is clear that forestry can contribute substantially to capturing CO_2. For each cubic metre of stemwood produced or saved from destruction by fire or decay, approximately 0.26 tonnes of carbon are captured. It is therefore important to: diminish the rate of deforestation; increase the rate of afforestation/reforestation; improve the management of natural regeneration of trees and shrubs; improve the management and productivity of existing forests, including intensive protection against fire, diseases and insect pests; and increase the conversion of mature wood to durable wood products.

FAO's role and programme

FAO's contribution in this field resides first and foremost in its role as coordinating agency for the implementation of the Tropical Forestry Action Plan. TFAP, the general aim of which is to strengthen and harmonize international cooperation in tropical forestry, constitutes a forestry response to climate change at national as well as regional and global levels. It promotes the sustainable management of tropical forests, afforestation/reforestation and tree planting, all of which are key components in limiting the amounts of carbon dioxide present in the atmosphere.

In order to help foresters and development planners in FAO member countries gain a better understanding of the complex issue of global warming and the role that forestry may play, FAO's Forestry Department is preparing a Forestry Paper on Climate Change and Forests that reviews the status of knowledge as of 1989. The paper will be published during the autumn of 1990, with a shorter version to follow later.

A panel, consisting of nine high-level experts from forestry and related fields, met in Rome in March 1990 to review a draft of the above paper and to provide the Organization with guidelines on how best to serve its member countries in the field of climatic change and forests. The panel recognized that predictions on future climate change merited concerted action by the world community viz. dissemination of information on the role of forests as sinks and sources of carbon dioxide, and their potential role in mitigating climatic changes; protection and buffering of forests and trees against destruction and depletion caused by such changes; and adoption of global and

[1] 1 billion tonnes = 1 000 million tonnes = 1 gigaton = 1 petagram.

national strategies aimed at better scientific understanding of the processes involved. However, it was stressed that the present level of knowledge could not justify massive and costly action for the sole purpose of sequestering CO_2.

The Forestry Department also collaborates in the work of the International Panel on Climate Change (IPCC), and actively participated in meetings of its Working Group III on response strategies to the climate scare.

about the destruction of tropical forests. Deforestation may result in many forms of serious and far-reaching environmental damage, such as the loss of the genetic resources of unique ecosystems, the degradation of soils, the acceleration of soil erosion, decreased soil fertility and changes in water quality and waterlife habitats. In addition, the lives of indigenous forest dwellers may also be profoundly altered.

Alarm has been expressed about the possibility of deforestation contributing significantly to the global warming of our planet through the release of carbon dioxide into the atmosphere or by reducing the absorptive capacity of the woody biomass.

Governments of tropical countries have responded to this worldwide concern through a number of recent initiatives.

In May 1989, the governments of the eight countries situated in the Amazonian Basin adopted a joint declaration of the Amazon which aims to strengthen concerted action and cooperation within the context of the Amazon Cooperation Treaty. The declaration expresses strong commitment to the conservation and sustainable management of forests and other natural resources of the Basin, to the benefit of local communities.

The countries concerned have formed a Special Commission on the Environment for the Amazonia with the function of designing joint programmes to study measures for environmental management favouring the sustainability of Amazonian natural resources; to promote research; develop methodologies; evaluate environmental impact; obtain financial resources and technical cooperation; seek compatibility of environmental legislation; and exchange information on the protection of the environment in the Amazon region.

Some results of the recent efforts of the region's governments are already apparent. In 1989, the deforestation rate of the Brazilian Amazonia was slowed down considerably, mainly because of heavy rains, but also because of tighter restrictions placed on forest clearing by the relevant authorities, and the withdrawal of subsidies.

In 1989, an international mission was mounted under the auspices of the ITTO to assess the sustainable utilization and conservation of tropical forests and their genetic resources as well as the maintenance of the ecological balance in the state of Sarawak, Malaysia. On the basis of the latest assessment

of forest resources in the Philippines, which reveals the reduction of forest area to below 40 percent of the land area, that country's parliament is proposing a bill which would restrict logging to levels essential for domestic processing. In Thailand, where serious floods were attributed to excessive and uncontrolled logging, a ban on logging has been imposed by the government. In Laos, strict legislation has been introduced to eliminate uncontrolled logging and slash-and-burn practices.

Forestry in Europe. In May 1989, the Council of European Communities adopted a new Forestry Action Programme centred on four main objectives: protection of forests; development of forestry within the framework of the development of rural and backward regions; production and marketing of forest products; and support for afforestation of agricultural land as an economic alternative to agricultural surpluses. Coordination is to be organized through the creation of a Permanent Committee on Forestry and a system of forestry information. The forestry sector is important in the Community as it is estimated to employ more than two million workers. Forests cover some 20 percent of land area and produce about 115 million m^3 of roundwood annually. Domestic production covers approximately 50 percent of annual consumption by the member countries of the Community.

PART TWO

REGIONAL REVIEW

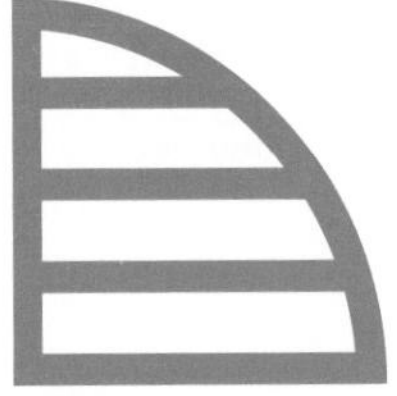

REGIONAL REVIEW

Developing country regions

This issue of *The State of Food and Agriculture* introduces a new approach to the coverage of the developing country regions. In order to review agricultural trends and issues with more insight and focus than an analysis at the aggregate regional level permits, the discussion centres on the experience of selected subregions: southern Africa, Pacific Asia, the Caribbean, and middle-income countries in the Near East. Future issues of this publication will successively cover other developing country subregions.

The subregional country groupings are used for analytical purposes only and do not necessarily refer to any territorial, political or institutional entity.

AFRICA

Overview

Africa's GDP growth accelerated somewhat in 1989 to 2.9 percent, compared with 2.4 percent in 1988. In per caput terms, the 1989 increase was negative but was nevertheless the second best performance since 1980. Indeed, the economic situation in sub-Saharan Africa appears to be desperate. Estimates vary as to the degree of decline undergone during the "lost" decade of the 1980s but, according to International Monetary Fund (IMF) estimates, real per caput GDP has fallen a cumulative 6 percent since 1981, following a previous percentage point decline during 1972-81. According to the African Development Bank (AfDB), Africa's real per caput income in 1989 was about 15 percent lower than in 1980. Whatever the figures are, the reality is significantly lower standards of living, a decaying infrastructure and smaller shares of world markets. Africa's total debt levelled off in 1988-89 at slightly less than $200 billion according to IMF estimates, but its debt service, having declined in 1987, rose again to 25 to 27 percent of its exports of goods and services — returning to the high levels of the mid-1980s. The Economic Commission for Africa (ECA) estimates that the net outflow of resources from Africa amounted to $5.5 billion in 1989.

Against this gloomy background, Africa's volume of export trade picked up in 1988 and 1989, permitting some growth in imports which had collapsed in the early to mid-1980s. Benefiting from more favourable weather conditions and economic circumstances in those countries determinedly carrying out policy reforms, Africa's agriculture has performed relatively better in the late 1980s, although growth in agricultural output still lags behind that of population.

Challenges and opportunities in southern Africa

Southern Africa provides a contrasting picture of formidable challenges and developmental opportunities. The challenges are a soaring population growth rate; continuing wars and civil strife; and widespread and grave economic problems. The opportunities are a rich natural resource base; a capricious but generally favourable climate for a wide range of agricultural activities; the presence of a relatively well developed transport infrastructure; and the makings of a functional institutional framework in some countries.

The southern African subregion has a unique opportunity for development arising from the gaining of independence of Namibia, the possible cessation of hostilities in Angola and Mozambique, and the general scaling down of the confrontation with the Republic of South Africa (RSA). The ECA estimates that the nine SADCC states suffered a loss in GDP of about $60 billion in 1980-88 owing to the destabilization activities of the Republic of South Africa, i.e. more than double these countries' GDP in 1988.[1] The opportunity for development would lie in the restoration of

[1] The nine states of the Southern African Development Coordination Conference (SADCC) are Angola, Botswana, Mozambique, Malawi, Lesotho, Swaziland, United Republic of Tanzania, Zambia, Zimbabwe. The source of these estimates is *South African destabilization: the economic cost of frontline resistance to apartheid.* UN-ECA, 1989.

stability in the subregion, the expected inflow of capital, the opening up of markets, and significantly reduced transport costs through the reopening of lower-cost rail links with the coast.

This section briefly reviews current agricultural and development issues in two countries which would be directly affected by such favourable events: Zambia and Zimbabwe.

Zambia. The economic and agricultural sector policy changes announced in mid-1989 marked a turning point in the problem-strewn path of Zambia's economic progress. A revaluation of the currency and price controls had been introduced as part of the May 1987 package of policies, the New Economic Recovery Programme (NERP), which marked the

BOX 2.1

Africa

"Out of Africa, always something new." So wrote Pliny, but some 2 000 years later, the publication of two reports has illustrated the vicissitudes and frustrations that Africa is enduring in modern times. In late 1989, the World Bank painted a bleak picture of sub-Saharan Africa which was threatened to become "marginalized" from the mainstream of global economic activities.[1] Sub-Saharan Africans were almost as poor as they were 30 years previously, the region's share of world markets had fallen by almost one-half and deindustrialization was becoming apparent. In April 1989, the joint meeting of African ministers of economic planning and development and ministers of finance had adopted a framework for analysis and strategy implementation which was a sober assessment of the ongoing economic crisis, a blueprint for future action and a basis for constructive dialogue between African countries and their development partners.[2]

A central theme of the World Bank report is the need for good governance, an efficient public service, a reliable judicial system and an administration which is accountable to the public. People must be given greater responsibility for improving their lives, grassroots organizations must be fostered and the informal sector promoted. As the world enters a new technological age, if sub-Saharan Africa is not to become more marginalized, it must improve science and technology and training and form new partnerships with the appropriate firms and research institutions, underlining the report's emphasis on the need to promote private sector activities in Africa. Savings and investment must be raised and, in the context of a rapidly growing population, a population policy must become an integral part of development. Increased productivity is needed in agriculture and elsewhere in the economy as well as improved infrastructure, stronger incentives and less, but better, government.

The genesis of AAF-SAP was the increasing concern among African policy-makers, not only with respect to the relevance of stabilization and structural adjustment programmes (SAPs) to Africa's long-term development objectives, but also with respect to its social, economic and financial impacts. Such programmes had been undertaken by more than 30 African countries since 1980. An assessment of them led to the conclusion that although they aimed at restoring economic growth, usually by achieving fiscal and external balances and permitting market forces to allocate resources, such objectives could not be achieved without

---

[1] *Sub-Saharan Africa: from crisis to sustainable growth.* World Bank, 1989.

[2] *African alternative framework to structural adjustment programmes for socio-economic recovery and transformation.* (AAF-SAP). ECA, 1989.

abandonment of the IMF programme of 1985. Since mid-1989, policies shifted back toward the IMF conditions, opened up the way for discussions with the Bretton Woods Institutions and led to renewed funding from bilateral and other multilateral institutions within the framework of the Fourth National Development Plan 1989-93, launched in January 1989. The new policy orientation involved a drastic devaluation of the kwacha to ZK16 = $1 from the ZK10 = $1 level set in November 1988, and the decontrolling of consumer prices. The introduction of a dual exchange rate system in February 1990 effectively resulted in a further massive devaluation of the kwacha. The market rate fell to ZK40 = $1 while the official rate

addressing the fundamental structural bottle-necks found in African economies. Hence, at the beginning of 1988, with the financial support of the United Nations Development Programme (UNDP), ECA began work on an alternative conceptual framework for SAPs that would address both the adjustment and structural problems facing African economies. The exercise was strongly supported by the 43rd Session of the UN General Assembly during the mid-term review of the UN Programme of Action for Africa's Economic Recovery and Development (UN-PAAERD).

The result of a lengthy process of consensus-building among African economic policy-makers, the main thrust of AAF-SAP is its holistic approach whereby the macro-economic framework, the policy measures and the strategies for their implementation take into account the dynamic relationship between economic adjustment and transformation. The aim is to remove any conflict between structural adjustment and long-term development, and to encourage the mobilization of domestic resources for investment as well as the creation of an appropriate environment to ensure a balance between the public and private sectors. The core of this alternative framework is the recognition of the human dimension in development which critics of orthodox SAPs claimed was not adequately covered by SAPs and thus led to increased marginalization of underprivileged groups of the population. This recognition implies the empowerment of people, the equitable distribution of income and wider democratization of the decision-making and implementation processes. AAF-SAP is not a rigid framework to be applied uniformly to all African countries, but could be used to design, implement and monitor national programmes of structural adjustment along with transformation and growth.

AAF-SAP was formally adopted by the ECA Conference of Ministers in April 1989.[3] It was subsequently submitted to the Council of Ministers of the Organization of African Unity (OAU), the Ninth Conference of Heads of State or Government of Non-aligned Countries, the Economic and Social Council of the United Nations (ECOSOC) and the 44th Session of the UN General Assembly.

With a view to making AAF-SAP operational, in-depth studies have been undertaken of three controversial instruments contained in the framework's proposals: production subsidies, multiple exchange rate systems and a differential interest rate policy. In addition, the general framework is being better defined by the elaboration of AAF-SAP model types that subsequently can be tailored to suit specific country situations. At the country level, Lesotho is the first country in which an AAF-SAP mission has been mounted.

Critics of AAF-SAP, while welcoming these attempts by Africa to develop its own development model to confront the region's grave economic problems, point out that certain elements of economic reform cannot be avoided and that the "alternative" tends to demand the very attribute which Africa most lacks: experienced and capable economic managers. It is agreed, however, that "social standards" must not regress and that vulnerable groups must be protected during the possibly protracted period of economic adjustment.

[3] Resolution 676 (XXIX).

continued to be linked to the special drawing right (SDR), at a rate equivalent to about ZK25 = $1. Although the new exchange rate policy provides an incentive to non-traditional exporters, tourism and private foreign investors, as well as balance-of-payments support, it has received criticism for its adverse impact on inflation. The annual inflation rate rose to more than 120 percent in early 1990, partly because all duties and sales taxes on imports are calculated on the value of goods at the market rate for the kwacha.

Nevertheless, the process of economic reform received a positive reception at the World Bank Consultative Group meeting for Zambia held in April 1990. Donors endorsed Zambia's economic restructuring programme and pledged $450 million of support. This financial support should go some way toward relieving the foreign exchange shortage which regularly imposes a severe brake on imports and which has resulted in a stagnating economy and deteriorating living standards. It is estimated that the programme requires about $1.9 billion of external aid, however. A major problem remaining is Zambia's external debt which was around $7.2 billion in mid-1990, about 36 percent of which is owed to multilateral donors and so is difficult to reschedule.

Alongside the radical changes in economic policy, introduced in mid-1989, the Zambian Government introduced some far-reaching sectoral reforms, particularly in agriculture. The core of these was the abolishment of the National Agricultural Marketing Board (NAMBOARD) with its produce marketing and fertilizer distribution functions taken over by the Zambian Cooperative Federation (ZCF) and Nitrogen Chemicals of Zambia respectively.

Ironically, it has been Zambia's recent success in expanding maize production in the small farm sector that has contributed to its maize marketing problems. There was a carryover of 5.5 million bags (90 kg) from the 1987/88 season, and supplies were 18 to 19 million bags in 1988-89 compared to the ten to 11 million bags required for consumption. Such large crops not only increased marketing costs but led to serious transportation and storage problems. Inadequate supplies of grain bags, lorry spares and tarpaulins have been a recurring problem for agricultural marketing, particularly of maize, during the past decade. As the government absorbs the greater part of both intra- and interprovincial transport costs as well as storage costs, the expansion of maize production away from the traditional commercial producing areas along the rail lines has led to increased marketing costs.

The other component of the maize subsidy is at the consumer level. In an attempt to hold down these costs, the government introduced a rationing scheme in 1989 whereby households qualifying for assistance (those earning at that time less than ZK20 500/month, assuming an average of six members) were limited to six coupons granting 14 kg of maize meal per month. At the then prevailing prices, the subsidy was about one-third of the selling price (which was already artificially low to the extent that transport and storage costs were also largely subsidized). Even with the introduction of the rationing scheme, the consumer subsidy was costing about ZK250 to 300 million a year. However, the total cost of the maize subsidy exceeded ZK1 billion, a significant share of the ZK3.5 billion budget deficit for 1989. In an attempt to reduce subsidy costs, the price of maize meal was more than doubled in mid-June 1990, the result of which has been violent protests and riots in the capital, Lusaka.

The issue of maize marketing in general, and the maize subsidy in particular, plays a key role in Zambia's agricultural policy. The country has considerable agricultural potential which was largely neglected until recent years and, indeed, a maize surplus has been achieved in the past two relatively favourable seasons. Producer prices of crop products other than maize have essentially been freed, with farmers able to negotiate with the buyers (millers and processors) for wheat, rice and oil-seeds. In April 1989, both Australia and the United States announced that they would end shipments of food aid in order to encourage domestic production of wheat. Shortages were expected because domestic production used to cover only about one-quarter of the annual consumption of 120 000 tonnes, but with the liberalization of prices, the price of bread increased significantly and demand has consequently fallen.

Future thrusts in the Government of Zambia's agricultural policy have become sharpened by the realization that the reserves of copper on which Zambia remains critically dependent will be almost exhausted by the end of the decade. Agriculture, therefore, will have to assume a much stronger role in economic development and, having been neglected in the past, it has received more attention in the current (Fourth) Plan, 1989-93. Non-traditional exports, including

agricultural products such as air-freighted horticultural products, are being actively promoted but there is still a long way to go as such exports comprise barely 10 percent of total merchandise exports. The agricultural sector itself also would have to more than double its present economic size to compensate for the eventual decline in Zambia's mining sector.

Zimbabwe. Within a difficult overall economic environment, Zimbabwe's agricultural sector has performed well, if erratically, since Independence. Output was depressed by drought in 1983/84 and again in 1987 but it has diversified considerably away from tobacco and maize. Increased production, particularly of maize and cotton from the small-scale or communal sector, has been notable and has more than compensated for the decline in the number of large-scale farms, witnessed immediately before and at Independence. With a diversified agricultural output and a land/population ratio in between that of Zambia and Malawi, Zimbabwe still has considerable agricultural potential. Nevertheless, there are several policy issues which have to be addressed.

The first of these is access to land, involving land reform. The ownership of land in Zimbabwe is extremely skewed:

- Approximately 4 500 large-scale farms occupy 13.5 million ha (3 000 ha/farm). Land ownership within the large-scale sector is also highly skewed; the sector covers a wide range of land types and supports an estimated 1.7 million people comprising farm labourers and their families.
- Approximately 8 500 small-scale commercial farms occupy 1.5 million ha (over 176 ha/farm); 800 000 family units, or about four million people, occupy 16 million ha (the communal lands) at an average of 20 ha/farm; since 1980, 40 000 families, about 250 000 people, have been resettled on 2.5 million ha at an average of 62.5 ha/farm.

Resettlement was of major political importance at Independence, with a programme envisaging the resettlement of about 162 000 families by mid-1985.[2] Because of difficulties in supporting resettled families — particularly during the drought years of 1983-84 — the marginal quality of some of the land which became available through the abandonment or voluntary sale of large-scale farms, and the subsequent shortage of land, only about 45 to 50 000 families have been resettled, either on individual holdings or on collective cooperative holdings.[3]

Pressure has been rising, particularly prior to the recent elections, for the fulfilment of earlier promises regarding access to land. In this context, several problems have to be resolved:

- Output and export earnings are still highly dependent on the large-scale sector. Investment is already low and a further loss of confidence could have serious consequences for food supply and exports. Therefore, the challenge is to obtain an adequate release of land without jeopardizing output and employment. The possibility of a land tax was raised in the budget statement for 1989 as a means of bringing underutilized land onto the market.
- Resettled families need to be supported during the first year at least, and maybe longer. Such expenditure will have to be borne by the government budget at a time when strenuous efforts are being made to halve its deficit.
- The land allocations per resettled family appear generous when account is taken of livestock. A wider range of farm systems incorporating livestock — but not just cattle — on extensively managed systems is necessary.
- Where possible, the integration of resettled smallholders within the existing large-scale sector would also seem desirable to avoid the continued dichotomy of the farming system and to enable the small-scale sector to benefit from the better infrastructure, transport and marketing facilities of the main commercial areas — without additional government expenditure — as well as from wider off-farm employment opportunities.

The second issue concerns agricultural marketing, pricing and input supply. In common with many other countries in Africa, Zimbabwe has a long history of intervention in agricultural

[2] The first five-year plan, 1986-90, foresaw the resettlement of 15 000 families per year.

[3] There are, in fact, four basic resettlement "models".

markets, promoted by its settler community and the need for food self-sufficiency following the Unilateral Declaration of Independence (UDI) in 1965. Price intervention is operated through four agricultural marketing parastatals (for beef cattle, maize and other food crops, dairy products and cotton) coordinated by an umbrella organization, the Agricultural Marketing Authority.[4] A controlled producer price system operates for wheat, maize and other coarse grains, soybeans, groundnuts, sunflower seed, cotton, beef and whole milk, and retail prices are fixed for maize meal, bread, cooking oil, beef, milk and stock feeds. The Grain Marketing Board is the sole buyer of maize in the commercial maize areas. Some intraregional trade is permitted within the communal areas, although not between them. Pan-territorial and pan-seasonal pricing systems inhibit the private sector from collecting maize from outlying areas and from performing a cereal storage function.

There are clear imperfections in agricultural markets in Zimbabwe, the most obvious sign being the very large subsidies required by the marketing boards. To what extent these result from the boards' being required to implement inconsistent or contradictory prices as opposed to being a consequence of their inefficiency, is difficult to establish but, in 1984/85, the agricultural marketing parastatal deficits amounted to 17 percent of the total government budgetary deficit and 1 to 2 percent of GDP. By 1989, as reported in the budget statement, the accumulated losses or subsidies of the agricultural marketing boards and the Agricultural Finance Corporation amounted to about $370 million, equivalent to 7 percent of total recurrent expenditure for 1988/89 or 37 percent of the budget deficit for that year.

While food subsidies have been reduced since 1980 as the government has abandoned its cheap food policy, it is notable that subsidies on beef continue and that beef has to be rationed because the Cold Storage Commission has insufficient throughput at existing prices.

[4] There are ten statutory boards under the Ministry of Lands, Agriculture and Resettlement (MLAR).

ASIA AND THE PACIFIC

Overview

Economic growth in Asia and the Pacific slowed in 1989, but the region remained one of the fastest growing areas in the world. Real growth of aggregate GDP fell to 5.3 percent in 1989 from a record 9.4 percent in 1988. The region's two largest economies, China and India, experienced rather sharp declines in GDP growth. In China, GDP growth fell from 11.2 percent in 1988 to 3.9 percent in 1989, while India's growth rate fell from a record high of 9.5 percent in 1988 to 4.5 percent in 1989. Agricultural GDP grew by 4.5 percent in China and 2.0 percent in India during 1989.

Throughout the region, only Nepal, Papua New Guinea and Western Samoa recorded GDP growth rates of less than 2 percent in 1989, and none were negative. On the other hand, the Republic of Korea, Singapore, Thailand, Indonesia, Malaysia, the Philippines, Pakistan, Fiji, Maldives, and Taiwan (Province of China) all had GDP growth rates over 5 percent. Thailand, with a GDP growth of 10.8 percent, was the world's fastest growing economy in 1989.

Similarly, several countries had strong agricultural sector performances during 1989. In Nepal, Pakistan, Myanmar (formerly Burma), Fiji and Viet Nam, agricultural GDP grew by around 6 percent or more in 1989. Indonesia, Laos, Malaysia, Thailand and the Philippines recorded agricultural growth rates between 3.5 percent and 4 percent.

The Asian region accounted for nearly 90 percent of 1989 world rice production, and 42 percent of total world cereal output. China, India, Bangladesh and Thailand all experienced record rice harvests resulting from a combination of good weather, expanded use of high-yielding inputs, and greater area planted.

Pacific Asia: the newly agro-industrialized economies

Pacific Asia includes a wide variety of countries that differ greatly from each other in natural resources, population size, geographic characteristics and economic systems. Japan, a leading economic power, is one of the largest and most stable industrial economies. China is the world's most populous country, with over one-fourth of the global population. The four natural resource-poor newly industrialized economies (NIEs) form a secondary industrial

base, together exporting manufactured products equal to around one-half of Japan's or the United States' exports.[5] The region also includes numerous resource-rich, low- and middle-income developing countries in Southeast Asia.

Over the past three decades, Pacific Asian countries have pursued widely divergent development strategies resulting in significantly different growth patterns. The exceptional economic performances recorded by the Asian NIEs since the early 1960s have been well documented. These four countries consistently applied macro-economic and trade policies that stimulated broad-based development through an export-led industrialization strategy. Rapid export expansion, initially founded on labour-intensive manufactured products, led to the strong industrial sector performance that has played such a dominant role in generating and sustaining their impressive growth rates.

In the 1980s, two additional Pacific Asian countries, Thailand and Malaysia, successfully established export-oriented, labour-intensive manufacturing sectors. Manufacturing exports grew by over 30 percent per year during the 1980s in Thailand, and by 20 percent per year in Malaysia. Both countries have become major exporters of textiles, clothing, footwear, semi-conductor devices, computer parts, furniture components, and plastic products.

But unlike the Asian NIEs, Thailand and Malaysia are building manufacturing sectors that also include agro-industries that take advantage of their abundant agricultural resources. Thailand's successful agro-export diversification strategy has included all agricultural subsectors. During the 1980s, annual rates of increase in the value of exports were frozen fowl 26 percent; canned fish and prawns 29 percent; wood products, including furniture parts, 21 percent; and canned pineapple 14 percent. Together, these products doubled their share of total export value to nearly 13 percent by the end of the decade. Similarly, in Malaysia, the yearly growth in the export value of rubber-based manufactured products was 33 percent, wood products 10 percent, and processed foods 10 percent during the 1980s.

Their success in increasing production and trade of traditional agricultural products, while diversifying into new agro-industries aimed at exporting higher-valued products explains why these two countries are frequently referred to as the newly agro-industrialized economies (NAEs).

The changing structure of trade and industry in Pacific Asia. The 1980s were characterized by record high interest rates, a prolonged recession in the industrial countries, rising protectionism, depressed primary commodity prices, shrinking supplies of development finance, increasing trade competition among developing countries, and the growing tendency of OECD countries to establish regional trading groups and to practise "trade bilateralism". These changing world economic conditions have contributed to a restructuring of trade and industry in Pacific Asia that has both benefited and constrained development efforts in Thailand and Malaysia.

For example, because of rising labour costs, currency appreciations, loss of Generalized System of Preferences (GSP) status, and tighter import quotas imposed by some OECD countries during the past decade, the NIEs have lost their comparative advantage in labour-intensive manufactured exports. The NIEs have responded by shifting toward more skill-intensive and capital-intensive manufacturing industries while investing heavily in relocating many labour-intensive export industries in the Southeast Asian countries where wages are lower.

Even though all countries faced the same changing world economic conditions, Thailand and Malaysia have been more successful in attracting, establishing, and expanding export manufacturing industries than the other Southeast Asian developing countries. Thailand and Malaysia have also diversified into new high-valued agricultural products and agro-industries. Finally, they have consistently increased domestic agricultural production and export earnings of traditional products in spite of declining terms of trade throughout most of the 1980s.

In the past, Thailand and Malaysia supplied the NIEs and Japan with raw materials and primary commodities. Today, both countries are not only steadily advancing toward newly-industrialized country status, they have also become major investors in commercial projects and importers of natural resources and raw materials from the least industrialized countries of the region — Myanmar, Laos, Cambodia and Viet Nam.

Beginning in the early 1980s, Thailand and Malaysia redirected their economic development

---

[5] The Asian NIEs include Singapore, Hong Kong Territory, Republic of Korea and Taiwan, Province of China.

policies toward a more outward-oriented strategy. Policy shifts included maintaining more competitive exchange rates, adjusting tariff structures, establishing export incentives, modernizing tax systems, providing investment incentives, changing public expenditure patterns and liberalizing financial sectors. The surging exports, continuing price stability and domestic and foreign investment booms experienced by Thailand and Malaysia are prompting more and more Southeast Asian countries to examine closely the development strategies and economic policies followed by these two NAEs.

Agricultural performance and policies: the cases of Thailand and Malaysia

Thailand. In 1985, textile products replaced rice as Thailand's major export. By 1987, textile products earned twice as much as rice. At present, Thailand's manufacturing sector accounts for about one-fourth of GDP and two-thirds of exports, but agriculture remains the major employer. Nearly 60 percent of the labour force, 17 percent of GDP and over 35 percent of exports originate in the agricultural sector. Storing, transporting and processing agricultural products provides even more jobs and export earnings in the rapidly expanding agro-based industries.

Thailand is one of the largest agricultural exporters in the world. The country typically supplies over one-third of the world's rice exports and is a leading exporter of tapioca, rubber, seafood, sugar, poultry and canned pineapple. The value of its agricultural exports increased by over 13 percent per year during the 1980s.

Over the past decade, the fishing industry has contributed significantly to agro-industrial exports. Thailand is the world's leading exporter of canned tuna, accounts for nearly 20 percent of the fresh and frozen prawn market and has the third largest fishing fleet in Asia. Seafood exports increased by 24 percent per year during the 1980s, and currently account for nearly 10 percent of total export earnings. Around 20 percent of agricultural imports consist of fish that is processed and canned for export.

Rice remains the country's most important food and export crop. In 1989, rice represented 60 percent of the cultivated area, 40 percent of total crop value, just over 25 percent of the average Thai's food budget and nearly 20 percent of agricultural export earnings.

Rice also helps to illustrate how changes in trade policies have promoted strong growth in Thai agriculture. For example, Thai policy-makers reduced the export duty on rice from 30 percent in 1980 to 5 percent in 1981 and then removed it and other taxes on rice exports completely in 1986. Exchange rate policy allowed the baht to depreciate 30 percent against the US dollar during the 1980s, maintaining the competitiveness of Thai exports. In addition, restructuring the import quota system, including lowering import tariffs, reduced the cost of yield-increasing inputs and encouraged greater use of fertilizers and pesticides. Finally, eliminating export licensing requirements, improved competition among buyers and traders, reducing their profit margins and passing price increases of rice to producers.

These and other policy changes created higher relative producer prices for commodities such as rice. The improved price incentives encouraged investments in small-scale irrigation works, the adoption of short-season varieties and increased fertilizer use on the dry season rice crop by about 20 percent in the last two years alone. The wet season rice crop is grown for home consumption so traditional varieties requiring fewer inputs are used. In addition, marginal increases in input costs have been offset by higher yields, resulting in lower variable costs per unit of production. Farm incomes increased by 8.8 percent in 1987, 19 percent in 1988, and 12.4 percent in 1989, while real agricultural GDP grew by 9.5 percent in 1988 and 4.1 percent in 1989.

Other important policy measures to promote agricultural and agro-based production include broader access to export financing; improved marketing support; special tax exemptions; and investments in roads, ports and related facilities. At the same time, government expenditure on production-related rural infrastructure such as large irrigation works, rural development projects, and research and extension services has increased. In 1988, the government announced special privileges and tax incentives to promote more rural industrialization and to encourage agro-processing firms to locate outside the crowded Bangkok area.

The growth process and the pace of industrialization are also presenting formidable development and environmental problems for Thai policy-makers. Since 1970, Thailand's forest cover has been reduced by around 38 percent. Over the past two decades, about 380 000 ha per year have been deforested and only 70 000 ha replanted. Government officials finally withdrew all logging concessions on national

forests early in 1989, shortly after devastating floods and landslides in the south had exposed the potential human and environmental dangers of deforestation. Officials remain concerned about how to enforce the logging ban while at the same time providing support for the wood-product industries. Thailand has been a net importer of tropical logs since the mid-1970s and recently arranged concessions for logging forests in Myanmar.

Likewise, extensive fishing in both the Gulf of Thailand and off the Isthmus of Kra has depleted fish stocks and the Thai Government has negotiated fishing rights in Myanmar's off-shore zones. Increased imports of fish are required to support the canned fish industry and the Thai fishing fleet is continually forced further from its traditional waters. The depletion of these and other resources needed to support the agro-processing industries remains a major problem for Thai policy-makers. Additional problems are arising from increased competition between industrial, urban and agricultural requirements. Despite adequate rainfall, reservoir levels are 17 percent below long-term averages because of the expanding need for electricity in Bangkok. This situation may threaten irrigation capacity in the near future as it did in 1987. Finally, the Bank of Thailand reports that rural poverty, land and income distribution problems, and inadequate infrastructure continue to hamper development efforts.

Malaysia. Like Thailand, Malaysia has pursued an outward-oriented, export-led development strategy over the past decade with a well-balanced emphasis on manufacturing, agro-industrial, and traditional exports. Since the early 1980s, Malaysian policy-makers have faced volatile oil prices, declines in prices of its major export commodities, slower import growth in the OECD countries and more limited access to international financial assistance.

Malaysia often experiences serious rural labour shortages because of cyclical demands for higher-paying jobs in urban industries and construction. In addition, the country suffered a recession in the period 1985-86, partly resulting from low commodity prices, although the economy made a strong recovery in the following years. GDP growth registered an 8.8 percent increase in 1988, and 7.6 percent in 1989, with the annual inflation rate never rising above 3 percent in the past three years. The agricultural sector grew by 7 percent in 1987 and nearly 4 percent in 1988.

The policy adjustments frequently credited with helping the country overcome these problems include tightening government spending; easing restrictions on industrial licensing and the foreign investment code; establishing new investment funds; introducing export credit refinancing programmes; maintaining a competitive exchange rate; and reforming the export duty structure. For example, export duties declined by 8 percent per year during the 1980s, falling from 18 percent of total government revenues in 1980 to 6.5 percent by 1988. The ringgit has depreciated 20 percent against the US dollar and nearly 100 percent against the yen since 1980.

In 1989, the agricultural sector accounted for about 16 percent of GDP and 30 percent of employment. Malaysia is the world's largest producer of palm fruit/kernel and rubber, as well as the largest exporter of vegetable oil, rubber and tropical timber and the third leading exporter of cocoa. Five natural resource-based commodities (rubber, tin, petroleum, palm oil and timber) accounted for 72 percent of export earnings in 1980, but only 41 percent by 1988.

Over 80 percent of Malaysia's agricultural land is in tree crops — primarily rubber with 36 percent and oil-palm with 34 percent. In the early 1980s, policy-makers began to recognize the environmental and economic value of the tropical forests. Research and extension efforts placed increasing emphasis on improving productivity to increase production instead of simply clearing forests to expand the planted area. During this period, development projects focused on rehabilitating crops, constructing irrigation facilities, expanding technical assistance capacity, introducing new varieties, and increasing the use of fertilizers and pesticides.

Both large estates and small farms were replanted with high-yielding rubber, oil-palm, and cocoa varieties. The area under high-yielding oil-palm expanded by over 60 percent during the 1980s; almost all cocoa areas are planted with high-yielding varieties, including plants developed by the Malaysian Agricultural Research and Development Institute; and nearly all the paddy grown in the country consists of high-yielding varieties. Input use has also increased substantially. Fertilizer consumption per hectare is up by over 50 percent since 1980 and specific projects aimed at improving fertilizer use on smallholdings have been implemented. Imports

of intermediate goods to support agricultural production have increased by around 6 percent per year in the 1980s.

The intensification strategy and policy shifts have helped Malaysia to compensate for low commodity prices in some cases. For example, while the price of palm oil decreased by 2.9 percent per year in the 1980s, export volume increased by 8.5 percent resulting in annual increases in the value of palm oil exports of 5.8 percent per year.

Malaysia's current policy efforts are focused on strengthening the agroprocessing industries of its traditional products to help alleviate poverty in rural areas, while developing new high-valued and agro-export products including fruits, vegetables, meats and fish to diversify the economic base and expand export potential. Recently, Malaysia became a net exporter of poultry and pork products, but high import duties on feed ingredients are inhibiting the expansion of the livestock sector.

Expanding agriculture's role in the industrialization process. Thailand and Malaysia have successfully promoted both manufactured and agro-industrial exports by practising an outward-oriented development strategy supported by fairly efficient import substitution policies. Stable social and political institutions combined with policy changes that encouraged agricultural growth have allowed these two countries to take advantage of the opportunities as well as to minimize the problems presented by changing external economic conditions.

At the same time, the rapid depletion of natural resources and increasingly unequal rural-urban and rural-rural income distribution patterns are a potential threat to long-term growth and social stability.

Both countries face the challenge of maintaining the growth process while finding ways to spread the benefits of growth to the rural poor. Development experiences in the NIEs demonstrate that increasing incomes of the rural and urban poor contributes significantly to the growth process by expanding domestic demand for locally produced products and by encouraging product diversification. Thailand and Malaysia appear to recognize the need to enhance rural incomes and to support the agricultural sector with a broad network of external support services along with trade and price policies that provide incentives and economic opportunities for the rural poor.

LATIN AMERICA AND THE CARIBBEAN: THE EXPERIENCE OF CARIBBEAN COUNTRIES

Overview[6]

As the decade came to a close, widespread evidence suggested that the regressive trend that had started in the early 1980s in the region was far from being arrested. GDP growth for Latin America and the Caribbean as a whole rose by a cumulative 11.7 percent between 1981-89, representing a 8.3 percent overall decline in per caput terms. Only five out of 24 countries (Barbados, Colombia, Cuba, Chile and the Dominican Republic) achieved increases in per caput GDP during 1981-89. After having fallen for the second consecutive year in 1989 (by 1 percent), per caput GDP was at the level of 1977-78.

Agricultural production rose 2.4 percent yearly during 1981-88, slightly above population growth, but only by 1.9 percent in 1989. Average inflation in 1989 rose for the third consecutive year to almost 1 000 percent, an all-time high. While a causation between inflation and GDP growth cannot be directly established, it is in countries with the highest inflation rates that growth tended to be more depressed (namely, Peru, Argentina, Venezuela and Nicaragua).

The external sector was less favourable in 1989 than in the two previous years. While exports rose by 9 percent, the rate was lower than in 1987-88 and — in the case of non-oil exporting countries — the increase in export earnings was largely offset by the increase in interest payments. Imports rose at a somewhat lower rate than exports in 1989, but the current account deficit remained at about $11 billion for the third consecutive year.

Agricultural trade stagnated overall on the side of exports, which rose by only 0.5 percent yearly between 1981-88, and declined by 2.7 percent yearly on the side of imports.

Total external debt declined by 5 percent in real terms in 1989, but debt/export and interest/export ratios remained very high — about 50 percent higher than the average levels

[6] The estimates of economic performance presented in this section are from the Economic Commission for Latin America and the Caribbean (ECLAC).

prior to the crisis of the early 1980s. The region continued to be a major net exporter of resources, estimated at about $17 billion in 1987, $29 billion in 1988 and $25 billion in 1989.

Caribbean countries

While presenting varying country situations, the Caribbean subregion generally shared in the major economic difficulties that reflected, and contributed to, the region's poor agricultural performance during the 1980s.

All but a few Caribbean countries — in particular Antigua, Dominica and St Vincent — recorded a marked slow-down in economic activity during the 1980s in relation to the previous decade, and several of them experienced pronounced declines in GDP. With sharply reduced export earnings and external financing flows in the early 1980s, the current account deficit of the subregion deteriorated to an equivalent of 17 percent of GDP in 1982 — over one-third of GDP in several countries. Although the deficit subsequently declined, it still averaged 10 percent of GDP in recent years. The problem of external debt was generally less acute than elsewhere in the Latin America region yet became critical in some countries: for instance, in 1989, external debt in Jamaica represented 120 percent of GDP, and debt service 40 percent of exports; in the Dominican Republic, the two ratios were over 70 percent and nearly 30 percent respectively; and in Trinidad and Tobago, 57 percent and 27 percent respectively.

Agriculture provided uneven support to economic growth and trade during the 1980s — the contribution of agriculture to overall economic expansion being in any event modest, given the sector's small, and in the long term, declining share in total GDP (about 8 percent of the total in CARICOM and 18 percent in non-CARICOM countries in recent years). Nevertheless, in most cases, agricultural growth exceeded that in other sectors while generally remaining well below population growth. Agricultural exports stagnated in value during the 1980s but with other trading sectors faring even worse, their importance in relation to total exports significantly increased. On average, agricultural export earnings contributed to financing over 40 percent of total import costs in recent years, although the share tended to fall in some countries, particularly in Guyana, the Dominican Republic and Haiti. Although food imports rose only moderately in volume and declined in value, they tended to absorb a greater share of foreign exchange earnings during the 1980s.

Against these trends, recent performances were relatively favourable in the subregion, particularly in the external sector. Favourable factors included the rebound of non-oil commodity prices, particularly sugar, and relative strength in those of petroleum; a booming tourism sector; and weakening of the US dollar, a factor of considerable importance given that several countries peg their currencies to the US dollar, and one implying significant gains in competitiveness since the dollar began weakening in 1988. The tendency was not uniform, however, as per caput GDP fell in 1989 in Jamaica, Trinidad and Tobago, Haiti and Guyana and only marginally increased in the Dominican Republic, despite that country's booming exports.

Underlying the generally faltering performances cited above have been the constraints and comparative disadvantages of most Caribbean countries — small open economies with a limited resource, manufacturing and exporting base and high costs of production. Being highly dependent on agricultural imports and exports, many of these countries are particularly sensitive to changes in the global economic and agricultural setting. Thus, the adverse external environment for agriculture during much of the 1980s (deteriorating agricultural terms of trade, falling international demand for most of the region's traditional export products, and reduced market access) had profound repercussions on their overall economic performance. These problems coincided with the collapse of several international commodity agreements as well as losses in international competitiveness — resulting from the strengthening of currencies pegged to the US dollar until 1987-88 and a disappointing trade performance despite preferential treatment by industrial countries. Performances in several countries also have been affected by natural disasters: more recently, hurricane Gilbert, which caused extensive damage in 1988, particularly in Jamaica, and another severe storm that devastated several northeastern Caribbean islands in September 1989.

Behind these temporary factors are a number of more fundamental constraints. Traditional export crops — sugar, bananas, tree crops — have become increasingly uneconomical, yet they still are the primary sources of foreign

exchange and employment.[7] While some of these crops, particularly sugar, could be rendered more viable by widespread mechanization, reducing labour intensity is not a feasible solution in some countries severely affected by unemployment problems. Export diversification (toward citrus, ornamentals, exotic fruits and products, aquaculture, etc.) has opened promising avenues in some cases, but leaving aside their risky, and initially costly exploitation, the potential for these products to substitute traditional exports remains limited. Despite austerity and import substitution efforts, food imports have become increasingly competitive in domestic markets — indeed, they have grown faster than food production for domestic consumption in many countries.

Recent developments in the external sector have presented some encouraging features for the short and mid-term (see section, Trade developments and issues). However, the above problems require more fundamental solutions in the longer term, including: improving productivity and external competitiveness; promoting export diversification — promoting the export of non-traditional and processed products — and reducing dependence on traditional markets; and further developing the domestic sector as a means of achieving food security, reducing rural poverty and lowering the costly and risky dependence on food imports. While these broad objectives are common to many other developing countries, they represent a particularly difficult challenge for the Caribbean, which must develop the sector from a more disadvantaged resource base and less competitive position than most.[8]

Trade developments and issues. The crucial importance of developed country markets for Caribbean countries sharply contrasts with the minor, and declining, economic weight of primary product trade with the Caribbean in industrial areas. Between 1980 and 1986, the value of Caribbean exports of primary products to the United States declined by 7 percent yearly; those to the EEC by 2.9 percent; while those to Canada, of comparatively lesser importance, rose by 7 percent.[9] Yet, the value of Caribbean (excluding Cuban) primary product exports to these industrial areas accounted for nearly one-fifth of their total merchandise exports during 1980-86. As regards market share in industrial countries, imports of primary products from the Caribbean accounted for about 3.5 percent of total United States imports in 1980-81, but only 1.8 percent of the total in 1985-86. The share also declined slightly from 0.6 percent to 0.5 percent in the EEC and rose only in Canada from 1.5 percent to 1.8 percent.

Market prospects are largely conditioned by developments in the preferential trade arrangements with industrial countries and areas: trade and aid agreement with the EEC under the Lomé Convention; the United States' Caribbean Basin Initiative (CBI); and Canada's Caribcan. In addition, Caribbean countries benefit from the industrial countries' GSP, granted to most developing countries. However the value of preferential agreements, at least in their past form, has been questioned in the light of the poor trade performance of Caribbean countries — in fact, poorer than that of most developing countries — and, more importantly, with regard to their long-term effects. Are these arrangements helping fragile agricultural industries become self-sustaining, or are they "addictive" mechanisms of indefinite dependence instead? What appears certain is that they constitute in the short term an irreplaceable life line to vital markets in industrial zones.

The recently-signed (December 1989) Lomé Convention (Lomé IV) contains provisions of considerable importance to the Caribbean as they directly address the concerns often expressed in the region: that the opportunities, including preferences, currently available to them may be jeopardized by the conclusion of an integrated market in the EEC in 1992; that the EC's opening toward Eastern Europe would lead to declining aid resources to Third World countries; and that EC concessions to all developing countries under the GATT principles would result in a dilution of aid and benefits for Asian, Caribbean and Pacific States (ACP) adhering to the convention.

On the issue of post-1992 developments, it is obviously impossible to assess the full

---

[7] For instance, in early 1989, the production cost of sugar in Jamaica was 19.5 US cents/lb compared with world market prices of 10 to 11 US cents/lb.

[8] A detailed framework for action at the regional level is provided by the *1989-91 Caribbean Community programme for agricultural development*, prepared by the CARICOM Secretariat.

[9] Caribbean exports of non-fuel primaries to the United States were in the order of $1 billion yearly during the 1980s, compared with about $700 billion to the EEC and $150 million to Canada.

consequences for the Caribbean until the process of EC market integration is complete. However, the fact that Lomé IV will have a life of ten years (it is due for renegotiation in the year 2000) gives reassurance at least for the mid-term. The increase in financial resources available under the convention is also an encouraging feature: the overall package rose from ECU 8.5 billion in Lomé III for the period 1985-90 to ECU 12 billion in Lomé IV for the period 1990 to 1995 — short of the minimum ECU 15.5 billion the ACP had demanded, however. STABEX and SYSMIN, the two instruments for assisting ACP commodity export revenue have both been strengthened financially and turned into non-reimbursable transfers — features of considerable interest for the region. In the past, Caribbean ACP have drawn from STABEX in the case of such agricultural commodities as sawnwood (Belize), coconuts (Dominica), nutmeg, mace and cocoa (Grenada), and bananas (Dominica, Grenada, Jamaica, St Lucia and St Vincent).

More important in terms of resource transfer than any other form of assistance to Caribbean ACP are the special trade protocols for traditional exports from the region. The sugar protocol, whereby the EC guarantees an annual purchase of up to 1.3 million tonnes of ACP sugar at EC prices, was not affected by the negotiations. In other words, while the protocol was maintained on an indefinite basis, it does not contemplate increased quotas, which can only occasionally change through the reallocation of country quotas. For bananas, the protocol resumed the provisions benefiting traditional suppliers. In particular, the Windward Islands, Jamaica, Belize and Suriname will continue their exports to the United Kingdom at unit prices significantly in excess of the world market level. It was also agreed that the banana protocol "does not prevent the community from establishing common rules for bananas in full consultation with the ACP as long as no ACP state, a traditional supplier to the Community, is placed as regards access to and advantages in the Community in a less favourable situation than in the past or the present".

The new rum protocol allows an increase in the quantities which may be imported free of customs duties and a progressive elimination of quotas by 1995. The previous protocol only allowed duty-free access within an overall ceiling, which had to be divided into member states' allocations. This may provide an opportunity to the Caribbean rum industry which, while accounting for over one-third of world rum production and trade, has experienced a decline in exports and has been operating below production capacity during much of the 1980s. The Caribbean currently supplies about 30 percent of the EC rum market.

Developments affecting trade with the United States and Canada are also of major importance for the region. The Caribbean Basin Economic Recovery Act, commonly referred to as the *Caribbean Basin Initiative* (CBI), provides tax incentives for foreign investments in beneficiary countries and duty-free access to the United States market for a range of products exported by the 23 eligible countries. The CBI began in January 1984 and remains effective until September 1995. The United States Congress is presently considering a 12-year extension known as CBI-II.

All eligible CBI exports can enter the United States duty-free if they are grown, produced or manufactured in a beneficiary country and meet the rules of origin requirements.[10] Participating Caribbean island countries have doubled their United States market share of CBI products since 1983. CBI exports have increased 33 percent per year between 1983 and 1987, although these exports were already growing at a similar rate in the four years prior to the CBI. More than 90 percent of CBI exports have included only seven product items: beef and veal, rum, tobacco, pharmaceutical products, ethyl alcohol, steel wire and bars, and electrical capacitors.

Several problems continue to inhibit CBI performance. In particular, the United States' GSP scheme has allowed duty-free access for many Caribbean exports since the 1970s, and an important purpose of the CBI was to broaden the product coverage. Nevertheless, many items of interest to Caribbean exporters are still excluded under the CBI, including textile and apparel articles subject to the multifibre arrangement (MFA), canned tuna, petroleum products, footwear, certain leather, rubber and plastic gloves, luggage and

[10] The three rules of origin are: i) the item must be imported directly from a beneficiary country; ii) at least 35 percent of the value of the item must be added in one or more beneficiary countries, but United States components may comprise 15 percent of the 35 percent value-added requirement; and iii) the product must be substantially transformed in one or more beneficiary countries.

handbags. However, in 1986, a special arrangement allowed textile products manufactured from 100 percent United States components to enter the United States outside the MFA.

The decline in the United States' *sugar quotas* and a sharp fall in sugar prices during most of the 1980s has severely reduced export earnings for many countries — although quotas have been raised in recent years as a consequence of reduced domestic production in the United States. Quotas from Caribbean countries[11] fell from about 594 000 short tonnes in 1984/85 to 410 000 short tonnes in 1985/86 and an average 230 000 short tonnes in 1987 and 1988. A significant increase was, however, announced for 1989/90, totalling about 634 000 short tonnes for the 1 January 1989/September 1990 21-month period. This increase only partially offsets the reductions operated between 1985 and 1987, however. Sugar exports by Caribbean countries to the United States dropped from $408 million in 1980 to $93 million by 1987. The surge in sugar prices in recent years has nevertheless resulted in considerable gains in export earnings from this commodity.

The *United States-Canada Free Trade Agreement* also presents some problems for the Caribbean Basin countries. Because each country has established different value-added levels to establish eligible products, Caribbean countries face two non-interchangeable markets. The Caribbean countries also fear losing market shares in Canada to United States exporters, and in the United States to Canadian exporters, as trade restrictions between the two countries are removed. Reduced access to the United States market would have greater consequences, given the high dependency of several countries in the region on these markets,[12] but the impact of possible losses in market share in Canada cannot be underestimated. Although Caribbean exports to Canada are relatively small and largely concentrated on minerals, Canada has been a fast growing market for the region. Exports of sugar, coffee, rum, fruits and vegetables account for about one-fourth of total Caribbean exports to Canada.

---

[11] Barbados, Belize, the Dominican Republic, Guyana, Haiti, Jamaica, St Kitts and Nevis and Trinidad and Tobago.

[12] For instance, the United States took almost 80 percent of the Dominican Republic's total exports, and supplied 56 percent of its imports in 1988.

NEAR EAST

Overview

GDP rose 3 percent yearly between 1980-1989 in the region as a whole, only slightly above population growth, but growth rates widely diverged within countries and country groups, ranging from 2.4 percent in the high-income countries to 3.9 percent in low-income countries. In 1989, GDP growth exceeded 5 percent in five countries while, in seven others, the increase in GDP fell below population growth.

Agricultural production rose by 2.8 percent between 1981 and 1988, a slightly slower rate than that of population. There were substantial increases in Saudi Arabia, the Libyan Arab Jamahiriya and Jordan, but production growth failed to increase in per caput terms in a majority of other countries. In 1989 agricultural production stagnated with decreases in Iran, Turkey and Jordan but good performances in most low-income countries of the region.

The inflation rate averaged 13.9 percent in 1989, the lowest rate since 1983, but was expected to accelerate slightly in 1990 to 14.2 percent.

The value of total merchandise exports declined 9.3 percent yearly between 1981 and 1988, mainly reflecting decreases in export revenues in oil-exporting countries. Merchandise imports only increased 1.3 percent yearly. In this period of severe external account constraints only a few countries — in particular Egypt, Afghanistan and Turkey — significantly expanded their imports.

Most non-oil exporters experienced stagnating or, in several cases, sharply falling export earnings from agriculture. The growth in imports was significantly reduced from the very high rate of the previous decade. However, despite some improvement in the agricultural trade balances in the second half of the decade, the value of agricultural exports only represented about 28 percent of that in agricultural imports in recent years.

The external debt of the Near East region approached $100 billion in both 1988 and 1989. Total debt has represented about 80 percent of GDP in recent years, this share being extremely high in some countries: in 1988 the debt/GDP ratio for Egypt was 142.5 percent; for Jordan 110 percent; for the Sudan 134 percent; and for the People's Democratic Republic of Yemen 212 percent. Debt service as a share of exports

steadily increased from 1982 until 1985 when it reached a peak of 13.6 percent. The ratio has broadly stabilized at 12 percent in recent years.

Middle-income countries in the Near East
This section focuses on five middle-income countries of the Near East — Cyprus, Egypt, Jordan, the Syriaan Arab Republic and Turkey.

Beyond their middle-income status, these countries present widely varying characteristics. The three largest economies — Egypt, Turkey and the Syrian Arab Republic — are the more agricultural oriented, with that sector accounting for about 18 to 20 percent of their GDP, employing 30 to 40 percent of their population and generating a varying but generally significant proportion of total export earnings (about 10 percent in the Syrian Arab Republic, 20 percent in Egypt and 25 percent in Turkey). The smaller economies are far less agriculture-based, although in Cyprus over one-fifth of the population is rural and farm product exports are important sources of foreign exchange.

Against this diverse structural background, the countries share major difficulties on the economic and financing fronts, with macro-economic imbalances reaching extreme proportions in some of them. Thus, the net trade deficit in recent years represented about 5 percent of GDP in Cyprus, 12 percent in Egypt and the Syrian Arab Republic, and as much as 40 to 45 percent in Jordan. Such imbalances could only be financed through massive external capital flows, mainly from oil-exporting neighbouring countries, in the form of aid and resident remittances. These flows, which represented over 20 percent of these countries' GDP in the late 1970s, enabled high and sustained investment levels, despite a very low domestic saving rate. However, with the end of the oil boom and large-scale repatriations, external financing flows sharply shrank, exposing the underlying weaknesses of these countries' economies.[13] Investment rates plunged from over 30 percent of GDP in the early 1980s to about 20 percent in 1987, despite some recovery in domestic savings; imports fell from an equivalent 50 percent of GDP to less than 40 percent of GDP during the same period; external debt approached $100 billion in 1988 and 1989, representing over 70 percent of the aggregate GNDP of these countries; and inflation accelerated, particularly in Egypt and Turkey.

These imbalances were accompanied by uneven economic performances, the positive exception being Turkey (which accounts for almost two-thirds of the combined output of the middle-income countries).[14] Agricultural production rose less than 2 percent yearly during 1981-89 (crops 1.8 percent and livestock 2.3 percent), compared with an average 3.2 percent during the 1970s — with Turkey again being the positive exception — and agriculture grew by 3.5 percent yearly. Agricultural exports, already a very small share of GDP by developing country standards (4 to 5 percent of GDP during the 1970s), fell to less than 3 percent of GDP in recent years following a collapse in agricultural terms of trade. On the other hand, agricultural import unit values fell considerably during the 1980s. However, as these countries' food imports also expanded strongly in volume, in financial terms they remained considerable — food imports accounted for about 20 percent of their total merchandise imports and 40 percent of their total export earnings in recent years. Egypt, the world's third largest importer of wheat and flour, currently imports over 40 to 50 percent of its total food supply; Cyprus over 70 percent; and Jordan as much as 80 to 90 percent.

Recent policy developments. Agricultural policy instances have been largely influenced by macro-economic policies to rectify internal and external imbalances. All countries have embarked on some form of stabilization and structural adjustment programme with the general objectives of reducing current account and fiscal deficits, and reducing government intervention in market mechanisms. While similar in approach, the implementation of these programmes has varied in severity: strong austerity has characterized recent policies in

[13] Flows from migrant workers remained considerable, however. Official remittances from migrants were $3.4 billion for Egypt (1989), $742 million for Jordan (1987) and $2 billion for Turkey (1987).

[14] Unlike other countries in the group, Turkey maintained a stable investment ratio (about 21 percent of GDP during 1981-87), largely financed through domestic savings (19 percent of GDP). Combined with stable terms of trade, the country's import capacity rose sharply, thus contributing to a 5 percent yearly growth in GDP during 1981-88. However, austerity measures introduced in 1988 to counter overheating and a severe drought reduced GDP growth to 1.1 percent in 1989.

the Syrian Arab Republic and Jordan, while a more gradual approach has been followed in Egypt and Cyprus. In Turkey, policy measures have been conditioned, on the one hand, by galloping inflation requiring fiscal and monetary restraint and, on the other hand, by the objective of resuming rapid growth which slowed down sharply in 1989.

Common to several countries has been the recognition of distortions introduced by currency overvaluation and measures to achieve more realistic exchange rates. Substantial devaluations were undertaken in 1989 in Egypt, Turkey and Jordan. Both Egypt and the Syrian Arab Republic are gradually moving toward a unified exchange rate, a move encouraged by the IMF.

Along with more realistic exchange rates, other measures have been introduced or strengthened, to promote exports and reduce imports, including agricultural products. Such measures have included revisions in the tariff structure for exports (Jordan, Syrian Arab Republic and Egypt) coupled with, in some cases, additional subsidies to exporters (Turkey). Private investment in agricultural export activities has been encouraged through generous tax holidays, for example, and other incentives (Turkey). The trade surplus recorded in the Syrian Arab Republic in 1989 — the first in over 30 years — was to a significant extent achieved through the successful promotion of private sector food and textile exports. In Cyprus, strong support is provided for the production of seasonally early crops for export.

Another primary concern remains the reduction of food import dependency through incentives to food crop production. However, adjustment-related policy changes have rendered the provision of such incentives increasingly difficult. All countries are making efforts to control public expenditure, a large part of which is in the form of producer and consumer subsidies. In Egypt, prices of many subsidized food items have been raised significantly and are expected to increase further as part of a recent agreement with the IMF. In Turkey, the cut in input subsidies which started in 1983 has been pursued, and the announced policy of Cyprus is to eliminate subsidies and reduce price control. On the other hand, food subsidies in Jordan's 1990 budget is estimated to remain at approximately the same level as in 1989; and the Syrian Arab Republic's pricing policy has been adjusted to offer higher profit margins to farmers, with continuing government intervention in agricultural markets of virtually all crop commodities and some livestock products.

Despite budgetary difficulties all five countries have strengthened efforts to expand land reclamation — seen in some cases as one of the most viable means of achieving food self-sufficiency — improve land productivity and increase water resources. Several projects of afforestation and soil conservation (e.g. in Jordan) and land reclamation in desert areas (Egypt) have been launched. Large irrigation projects are under way in Turkey (southeast Anatolia) and in Syria-Jordan (for the joint construction of the Al-Wenhad Dam), increasing irrigation water supplies for both countries. A significant feature of several recent programmes has been the importance given to the preservation of the natural environment. The Governments of Cyprus and Jordan have taken the most explicit steps in this direction.

For Cyprus and Turkey, the request to become members of the EEC has been another policy feature of far-reaching potential consequences for agriculture in the two countries.

Recent commodity-specific measures include the following:

- In Egypt, the government has allowed farmers more discretion in deciding what crops to plant. Controls have been relaxed on the production of wheat, rice, sugar cane and lentils. Cotton remains the most regulated crop, with 100 percent procured by the government. Wheat deliveries became voluntary in 1987 and ever since farmers have sold wheat on the open market and received higher prices than those set by the government.
- In Jordan, government intervention in agricultural/food marketing has remained pervasive with price controls on food staples, animal products, vegetables, etc. However, area restrictions under the Cropping Pattern Programme have been reduced and are now confined to tomatoes and eggplants in the Jordan Valley. Measures have been undertaken to improve quality control to gain share in the winter market in Europe and the EC has offered favourable import conditions. Plans are afoot to diversify cropping patterns to put the country's limited land and water resources to maximum use. The sheep stock is being built up to reduce reliance on imported meat.

- In the Syrian Arab Republic, the pervasive state controls on agricultural production are being gradually relaxed. To stimulate output, hefty increases have been effected in the procurement prices of major crops. The private sector has been allowed a greater role in internal trade and exports, although the government remains the only buyer of wheat, barley and chickpeas.
- As part of the structural adjustment programme in Turkey, support prices of major products have been increased to world market levels, although the number of commodities with support prices has been sharply reduced. In March 1989, the support price of wheat was raised by 80 percent and a 100 percent payment on the delivery of the crop enforced compared with 1988's staggered payment schedule. The sharp increase in the support price for wheat was intended to boost production of this crop which was cut down by a severe drought.

EASTERN EUROPE AND THE USSR

Overview

In 1989, and particularly in the last quarter of the year, political and economic reforms have swept over all countries of Eastern Europe, profoundly affecting their socio-economic and institutional systems. In this initial stage of reform, traditional and new problems have combined to render the current situation particularly difficult. Many countries in the region had been grappling for years with problems of "stagflation" and economic imbalances which, while unrevealed, were similar to those sometimes faced by developed market economies. Now, in addition, these countries must face the complexity of a transition from planned to market economies without a supporting theoretical background, which makes this experience a unique one. Policy reforms, particularly those for agriculture — a key sector for each of these economies — have greatly varied both in content and breadth. These differences reflect not only the economic weight and structure of agriculture in individual countries but also — perhaps mainly — the form and extent of their political reform.

It is still too early to assess the impact of the above reforms on the region's aggregate agricultural performance, generally characterized by inadequate and erratic growth. Regional production growth in 1989 was above the average for the 1980s, but individual country performances varied widely. The aggregate increase largely reflected a production recovery in the USSR and Bulgaria after two years of set-backs. For all other countries, 1989 was a poor agricultural year.

Recent policy changes in food and agriculture

Agriculture is a key sector in all East European economies and the USSR. Even in the German Democratic Republic (GDR), the most industrialized of these countries, about 9 percent of the population is engaged in farming, with the share rising to 10, 13 and 14 percent respectively in Czechoslovakia, Hungary and the USSR, and to around 22 percent in Poland and Romania. By comparison, only 7 percent of the population in western Europe and 2.5 percent of that in the United States are engaged in farming.

While regional agricultural trade is a relatively small share of total merchandise trade (5 percent in 1987 as compared with 10 percent in

western Europe and 15 percent in the United States), it is a vital source of foreign exchange earnings and hard currency. In Hungary, the most agricultural export-oriented country of the region, about one-fifth of total imports are financed by agricultural exports. In the other countries, this share ranges from 2 to 3 percent in the German Democratic Republic and the USSR to 11 percent in Bulgaria.

The economic importance of the sector justifies the special attention given it recently through policy reforms. A review of the main policy changes is given here by country.

On the broad economic front, reforms in *Hungary* included the establishment of an active stock market and further expansion of independent commercial banking. Joint venture laws allow a share of foreign ownership and repatriation of profits by foreign companies — despite the drain on hard currency reserves this represents. Since 1988, enterprises have been free to trade most products with partners in the convertible currency area. Measures are being introduced to render the forint partially convertible by 1992 or 1993 and limitations on the use of hard currency for the imports of a large group of products have been reduced. For a transitional period the forint will be successively devalued against convertible currencies and revalued against the currencies of other Eastern European countries.

In the case of agriculture, important reforms took place in Hungary regarding the legality of private land holdings and the establishment of markets for land. As of 1 January 1990, remaining government controls on most retail prices as well as farm subsidies were removed. This, however, had the immediate effect of fuelling price inflation which is expected to exceed 20 percent in 1990. A major objective is to expand agricultural exports to hard-currency countries, in view of the heavy debt burden (debt servicing absorbs 60 percent of annual export earnings in convertible currencies).

Among other problems, the new government elected in March 1990 will be facing:

- a debt of $20 billion, the highest in per caput terms in the COMECON, and increasing debt service;
- a huge current account deficit ($1.4 billion in 1989);
- high and rising price inflation;
- a decrease in total trade caused by reduced trade flows with the USSR, traditionally Hungary's main trading partner;
- an initial problem of increasing unemployment caused by looser cooperation with COMECON countries; and
- difficulties in expanding privatization, with 90 to 95 percent of enterprises being state-controlled.

Poland has also introduced significant market-oriented economic and agricultural policy reforms. Such reforms are interesting for a number of reasons. First, they have been proposed quickly, and introduced rapidly and with great determination. Second, the reforms are economy-wide and not just limited to agriculture. They send a clear message to economic agents that they must adjust their behaviour to the new policy conditions or face the consequences of non-adjustment with no help from the state. Third, the reforms maintain a significant support from the population which believes that current hardship is for a better future.

In the *USSR*, agricultural policies underwent numerous changes in 1989 in the attempt to revitalize the agricultural sector, but uneven progress was made toward greater alignment of USSR agriculture with world markets. Reforms continued in the agricultural administrative system. The State Agro-industrial Committee (GOSAGROPROM), established in November 1985, was abolished and replaced by the Council of Ministers' Commission for Food and Procurement, though some responsibilities were passed to other central organs such as the State Planning Committee (GOSPLAN). More authority was given to the republics, which now have control over the placement of state orders with farms, the setting of regional prices, and the use of investment funds at the regional level.

On 8 August 1989 a regulation was introduced allowing hard-currency payments to producers for deliveries of specific crops in excess of average annual sales during 1981-85 or 1986-88, depending on crops. The objective is to stimulate domestic production and sales to the state. Hard-currency payments were introduced in order to allow farms imported inputs or other production requisites they may need. The increased sales to the state would then decrease import needs and save foreign exchange. The programme, which was introduced on an experimental basis for two years, was not expected to result in any appreciable increases in output in the current season in view of its late announcement.

Since 1 January 1990 a law on leasing went into effect, superceding previous legislations, and a new legislation on land use was adopted on 28 February 1990. It contained the following provisions:

- rights of inheritance for certain land uses;
- land can be for industrial, cooperative or joint venture users;
- conditions for rescinding the right to use the land;
- system of land fees; and
- environmental land protection.

While these laws are aimed at providing incentives to production, their impact is expected to be limited because the negotiation of leases is dominated by state and collective farms; the distribution system for essential inputs is inadequate and, perhaps more importantly, leasees are reluctant to enter into a contract that would expose them to a greater risk than the largely guaranteed pay of state and collective farms.

The price reform of wholesale and intermediate goods that was expected to enter into force in January 1990 has been delayed until 1991 at least. While the planned revision does not intend to introduce market-determined prices, it would reduce some of the current drawbacks in Soviet pricing policy. Current retail prices for agricultural commodities are still heavily subsidized and are far out of line with world prices.

The pace and nature of agricultural policy reforms in the *German Democratic Republic, Czechoslovakia, Romania,* and *Bulgaria* will largely depend on the outcome of the planned elections in the spring and early summer of 1990. Some significant policy changes have already been operated in these countries, however.

In *Bulgaria,* the government has taken steps to introduce market forces into its agricultural sector. The previous government had already introduced wide-ranging reforms allowing prices of some commodities to be set freely; direct participation of enterprises in foreign trade; the formation of small, private enterprises, including agricultural enterprises; and providing for private sheep and cattle breeders to be paid partly in hard currency. The current government has recently abolished all limits on landholdings by private farmers and will now allow them to export their produce directly and receive hard currency for 20 percent of their exports.

In *Czechoslovakia,* the parliament is currently debating legislation to expand the private sector, break up state monopolies and liberalize foreign trade. It is also amending its laws on joint ventures to allow 100 percent foreign ownership. On 12 April 1990, a number of economic reforms were adopted including:

- internal convertibility of the koruna for companies by the end of 1990;
- freeing of prices in stages by the end of 1990 together with the abolition of subsidies; and
- elimination of central planning.

A sharply deflationary budget was proposed by the government in March. It was designed to force factories, farms and offices to reduce costs, cut jobs and to create favourable conditions for a market economy.

Spending on defence and security was cut by 12.5 percent while subsidies to industry, agriculture and food processing were reduced by 10.7 percent. Some "outside shocks" which could hit the economy in 1990 are feared, however. Supplies of oil and raw materials from the USSR are increasingly uncertain in view of the demand from hard-currency oil importers. Another uncertainty is represented by the currency and economic union between the German Democratic Republic and Federal Republic of Germany. The German Democratic Republic is Czechoslovakia's second largest trading partner after the USSR and, virtually overnight, will become part of the hard-currency trading area. Neighbouring Poland and Hungary are also expected to import fewer of industrial products from the country.

In the *German Democratic Republic,* a major influence for agriculture would be its monetary and economic union with the Federal Republic of Germany. Farm prices in the Federal Republic of Germany are subject to the Common Agricultural Policy of the EEC and, given the high support interventions, these prices are usually above world prices. German Democratic Republic farm prices have been even higher and would tend to decline if fully liberalized as recent events suggest. However, this potential threat to German Democratic Republic farm production units (i.e. combinats) could be avoided by raising retail food prices which have been kept unchanged since 1952. The decision in February 1990 to abolish the massive consumer subsidies (30 billion GDR marks) was a step in this direction.

The new Government of *Romania* immediately rescinded restrictive measures which had formerly affected private farmers by imposing mandatory delivery quotas. More recently, the government passed a decree giving complete freedom to both cooperatives and private farms to make all planting decisions and to sell their output to any buyer for whatever price they can negotiate. The government has tentatively endorsed the breakup of the largest socialized farms but intends to pursue that goal with great caution. The present government intends to maintain a substantial degree of state ownership although that policy may change after the elections.

In view of the acute problems of domestic food supply, the new government suspended all food exports and diverted stocks intended for export to the domestic market. Also, in the first quarter of 1990, the government authorized $150 million of food imports.

None of these four countries has yet confronted the issue of large consumer subsidies, as have Hungary and Poland. Such subsidies, on the one hand, severely strain these countries' budgets. However, with increasingly open borders and liberalization of foreign trade, pressure to slash these subsidies and align their consumer prices with world markets will certainly intensify. The governments all recognize the danger of inflation and would like to introduce reforms gradually.

Agricultural policy reforms: the case of Poland

Beginning in 1989, the Polish Government implemented a series of radical policy changes aimed at transforming Poland's state-run economy into one driven by market forces. Prices were liberalized, most subsidies were eliminated, and laws were passed calling for the breakup and privatization of the state-run monopolies. Immediate consequences for the consumers were a sharp increase in food prices, but also the disappearance of the shortages that had chronically characterized Poland. Farmers, on their side, found themselves in a very difficult position because procurement prices have lagged behind escalating input prices. In response, they are holding back output and thereby creating the risk of new supply shortfalls. Serious discord is developing between the farmers, who want a reinstatement of support, and the new government, which is determined to pursue its current market-oriented policy, involving the elimination of farm support — a course which is also prompted by budgetary constraints.

Approximately 80 percent of Poland's agricultural output comes from private farms. These farms are small, averaging 5 ha in size, and tend to be fragmented. Throughout most of Poland's post-war history, however, government policy discriminated against private farmers and favoured state farms. Reforms implemented in the early 1980s raised procurement prices received by farmers and provided somewhat better access to inputs. Nevertheless, at the beginning of 1989, farmers were still squeezed between the state monopoly on input supplies and the state purchase and distribution of farm output. Private farmers faced serious profitability problems and a steady erosion of their income.

In 1989, the following measures were passed, aimed at the commercialization of Poland's agriculture:

- In January, a decree abolished the monopoly of the state purchasing organization, allowing any individual or enterprise to compete with the state for the purchase of agricultural output. At the same time the system of state-controlled procurement prices was eliminated. Instead, the government set minimum prices for bulk commodities. These minimums were raised several times in 1989.
- In August, a decree removed all legal restrictions on retail food prices. Subsidies were frozen at their current level. A wage indexation plan was established which guaranteed 100 percent compensation for all food price rises in comparison with compensation for price rises of other goods which was set at 80 percent.
- In October, all consumer food subsidies were removed, except for those on a few basic items such as 2 percent fat milk and the lowest-quality bread.

Despite its expected long-term benefits, an immediate negative result of the August decree was an acceleration of inflation, which was almost 600 percent for the whole of 1989, with retail food prices rising nearly 1 000 percent. By October, however, the consumer shortages endemic to Poland had disappeared.

These consumer price rises did not translate into income gains for the farmers, however, who found themselves worse off than before. While their production costs rose dramatically, the prices received for their output failed to

increase proportionally. Despite the January 1989 decree, there was only a limited entry of new enterprises competing with the state purchasing monopoly. The main barrier was the lack of private capital markets to finance starting costs. State organizations also were very slow to bid up procurement prices. They claimed that extremely high processing and marketing costs, which were no longer covered by subsidies, and slack consumer demand prevented them from offering higher prices to farmers.

In response, the farmers refused to sell their output to the state. Livestock farmers eventually began to sell their herds in greater quantities, compelled by the high cost of feed. However, grain farmers continued to withhold output, seeing more value in their grain stocks than in the rapidly depreciating zloty. This created serious raw material shortages in the milling and baking industries, with the result that Poland was forced to request donations of wheat from the West.

In January 1990, a far-reaching programme was introduced in agreement with the IMF to combat what had become hyperinflation. The main elements were as follows:

- A balanced budget. Most remaining subsidies were abolished, state investment and other expenditures were curtailed, and the system of tax collection was tightened.
- Strict monetary control through the introduction of positive real interest rates.
- The devaluation and limited convertibility of the zloty and the removal of most restrictions on imports and exports.
- Strict wage controls. Wage increases were to be limited to 5 percent of the rate of inflation with heavy taxes to be imposed on enterprises granting excessive wage increases.
- More aggressive action to break up monopolies. The meat and sugar monopolies were especially targeted.
- A social safety net, including unemployment compensation and targeted food assistance.

The initial result of this programme was a 78.6 percent jump in prices during January 1990, as most remaining subsidies were removed. However, price rises fell to just 23 percent in February and less than 10 percent in March. By the end of the first quarter, inflation appeared to be under control though at the cost of a 30 percent drop in industrial sales and a growing threat of unemployment. Household purchasing power was down 24 percent in January and 17 percent in February.

While authorities had expected about 30 percent of the 2 620 state farms to go bankrupt, as of March only between 10 and 20 percent these farms had gone out of business and the authorities claim that most of the remainder are now well run and profitable. The farms managed to do this by specializing in crop production and eliminating unprofitable livestock operations.

But conditions have not improved for Poland's private farmers. Falling consumer demand continues to hold down prices for producers and virtually no progress has been made toward breaking up the state input monopoly. Grain farmers are still holding back output and livestock producers are cutting inventories because of the high cost of feed. As a result, red meat output is projected to decline 5 percent in 1990, following a 7 percent drop in 1989.

Because of the growing crisis in Poland's agriculture, the government has recently enacted measures partially reinstating support to producers. An Agency for Agricultural Marketing has recently been created. Its function will be to buy at market prices when supplies are high and to sell when there are shortages. It is already engaged in the purchase of hogs. Fertilizer subsidies have been reinstated, and the government is offering low-interest credit for the purchase of farmland. To give farmers more control over their affairs, the government has announced that elections will be held in the cooperatives which, under the previous regime, were largely coopted by the state. In addition, the APEX organizations governing the cooperatives will be eliminated.

The outlook for Polish agriculture is uncertain. In the long term, rising real prices for food products should stimulate supply while decreasing domestic demand and creating an exportable surplus of agricultural products. Poland has the potential to be a significant agricultural exporter (in contrast to the current non-competitive status of many Polish industrial products on the world market) but serious obstacles remain. With the fragmented farm structure and poor infrastructure, Polish farming remains a high-cost operation. These costs must be lowered for Polish agriculture to compete on world markets, but that will require huge amounts of capital, most of which must come from external sources.

Currently, a major debate is taking place among government officials as to the type of agricultural policy which Poland should pursue. The farmers are demanding substantial government support, warning that they will otherwise be forced to cut back output. But the question is who can pay for such support? The consumers are too poor, with a large share of their average income already being spent on food, and the resources that the state can devote are extremely limited.

DEVELOPED MARKET ECONOMIES

Overview

With generally favourable economic conditions sustaining demand, the current agricultural situation for developed market economies is characterized by tight markets. Despite a recovery in relation to earlier years in production of some key products like wheat, coarse grain and oil-seeds, prices tended to strengthen, thus allowing a reduction in subsidy intervention levels in several countries. Sugar prices sharply increased, though still generally failed to cover production costs. Overall, agricultural production rose by 4.1 percent in 1989, almost offsetting the cumulative 5 percent decline of the three previous years. However, this was entirely a result of North America's recovery from the 1988 drought, as production stagnated in western Europe and fell in Oceania. While prospects for global agricultural trade in 1989 were for an increase in volume, developed market economies were not expected to increase their import requirements.

A major feature of the market situation in these countries has been the reduction to very low levels of their cereal stocks.

While recent developments in agricultural markets in these countries have not arisen from any significant shift in their agricultural policies, they have presented positive short-term features — higher prices and farm incomes, lower subsidy payments and reduced storage costs. However, this situation needs to be assessed in a broader, longer-term perspective. There is a delicate balance between controlling the inherent excess production capacity in the area and safeguarding food security worldwide. This balance is all the more difficult to achieve considering the uncertain prospects for demand — stagnant, except for meat, in the industrial countries; potentially strong, but currently depressed by economic and financial difficulties in many developing countries; and unpredictable in centrally planned economies.

Selected policy issues affecting agriculture

Behind the serious imbalances and distortions existing in agricultural markets in most developed market economies are the excessive protectionist policies that have prevented an adequate transmission of market signals to farmers. Given the size of the countries involved, these policies are distorting markets worldwide. In recognition of this situation, a set

of market-oriented actions and principles for policy action were established by the OECD Council in May 1987. These constitute a reference framework against which agricultural market developments and the policies governing them are regularly monitored.

The third OECD annual report on agricultural policies, markets and trade in the area states that, for the second consecutive year, both the total and the rate of assistance to OECD agriculture declined in 1989.[15] As measured by the percentage producer subsidy equivalent (PSE), the rate of assistance fell from 45 percent in 1988 to 39 percent in 1989 — with PSEs ranging from 5 percent in highly market-oriented New Zealand to 72 percent in Japan. However, the significant decline in assistance in 1989 primarily resulted from the strengthening in world agricultural prices and the appreciation of the US dollar. Only a small part of the decline was a result of policy changes supporting greater market orientation. Agricultural trade remained distorted by high levels of farm support, import barriers and export aids, despite generally higher prices.

Prospects for agriculture in developed market economies — and, indeed, worldwide — are crucially dependent on three policy areas currently taking shape: the outcome of the Uruguay Round of GATT negotiations (see Part One, Agriculture in the Uruguay Round of multilateral trade negotiations); the orientations of the 1990 United States Farm Bill; and the process of integration in the European Community. The latter two issues are reviewed here.

The United States: the 1990 Farm Bill

Most farm programmes introduced in 1985 under the previous United States farm bill are due to expire in 1990, unless extended or modified. Thus in February 1990, after consultation with a large group of interested parties, the Administration presented Congress with a new set of wide-ranging proposals. The Administration hopes to have the new legislation signed into law before the autumn 1990 planting season, but it may not occur before early 1991.

The 1985 Food Security Act had been introduced at a time of excess supply in relation to demand for many agricultural products as well as a strong US dollar, and when the primary concerns were flagging United States exports and deteriorating farm financial conditions. Thus, its objectives were to make agriculture more competitive while maintaining direct income support until market conditions improved. Considering that United States agricultural exports expanded, farm income boomed (net farming income reached an all-time high of $48 billion in 1989) and farm financial conditions greatly improved (farm debt fell 28 percent between 1983 and 1989), the 1985 legislation was a successful one — although the depreciation of the US dollar also helped, as did the increase in prices following the 1988 drought. Thus, the new proposals build on the subsidy-backed, export-oriented foundations of the 1985 legislation, taking into account recent developments in the market place.

The set of proposals cover a wide range of issues such as price and income support, environment, international and crop disaster assistance programmes, food and consumer services, farm credit, science and education, marketing and inspection services as well as a variety of miscellaneous provisions. The proposed legislation aims to introduce greater flexibility to the existing programmes. In particular, farmers would be allowed to respond to market signals rather than to government support, thus permitting a further reduction in government spending on farm programmes.

- Greater production flexibility is expected to be introduced through a Normal Crop Acreage (NCA) concept. The NCA would define substitutable crops, authorize crop-specific acreage reduction programmes, and authorize planting on idled acres in exchange for giving up specified deficiency payments. This flexible approach would allow farmers to plant crops based on market signals without loss of farm programme benefits. It should also lead to higher production of crops in scarce supply, lower production of crops in surplus, and provide environmental benefits.
- On grain reserves and stocks policy, one undesirable effect of the current farmer-owned reserve (FOR) has been that farmers have been encouraged to hold stocks in government-financed storage at times when commodities have been needed by the market. The Administration's recommendation is to set a multiyear programme based on annual contracts and

[15] *Agricultural Policies and Markets and Trade: Monitoring and Outlook.* OECD, 1990.

characterized by greater simplicity and flexibility. It will also shift more of those decisions from the government to the farmer. In addition, the replenishment authority of the Food Security Wheat Reserve will be extended.

- The recommendations for triggered ARPs concerning wheat and feedgrains are that levels be triggered each marketing year on the basis of stocks-to-use ratios. If the ratio of ending stocks to total use is estimated to be more than 40 percent for wheat and 25 percent for maize, the ARP level would be 12.5 to 20 percent of the base acreage. If the stocks-to-use ratio is 40 percent or less for wheat and 25 percent or less for maize, the ARP level would be 0 to 12.5 percent. For cotton and rice, ARPs would be used to achieve, to the maximum extent practicable, a stocks-to-use ratio of 30 percent for cotton and 20 percent for rice.
- As regards loan rates for programme crops and soybeans, recommendations are to apply the loan rate formula specified in the 1985 Act for wheat and feedgrains also to upland cotton, extra longstaple cotton, and rice, and to extend the 1985 loan rate formula for soybeans. This would generally provide loan rates at 75 to 85 percent of previous market prices. All would be nine-month loans, but the secretary of agriculture could extend them if market conditions warranted it (except in the case of cotton which would have ten-month loans).

The proposed farm bill also envisages the continuation of export programmes. Thus, it proposes to extend the provisions of the Export Enhancement Programme (EEP) (although without mandatory programme levels or commodity programming requirements) as well as the Targeted Export Assistance Programme. One stated aim of the Administration is to maintain United States leverage during the current Uruguay Round of GATT negotiations, and not to give way on subsidies unilaterally in the absence of an agreement from competitors to follow suit. However, while efforts to increase agricultural exports would continue, a successful set of agreements on a reduction of trade distortions in late 1990 would open consideration to additional policy alternatives.

European Economic Community (EEC)

Economic integration: implications for agriculture. The European Economic Community (EEC) is in the process of completing the economic integration of its member countries as envisioned over 30 years ago. EC member countries signed the Single European Act in 1987 which committed them to remove all barriers to the free movement of goods, people, services and capital. In February of 1988 they agreed to remove these barriers by the end of 1992, and they have moved rapidly toward meeting that deadline.

Completion of a single EC market would make it the world's largest with 322 million people, a GNP of $4.6 trillion, and total food expenditure of around $450 billion. The political and technical complexity of such an undertaking is enormous but the political will to date seems to be sufficient to make the venture a successful one. Commercial interests have taken EC politicians at their word and are actively pushing for the creation of a single market. There have been thousands of joint business ventures in anticipation of increased commercial opportunities, including more than 150 joint ventures across borders within the EC's food and drink industry.

The creation of a single market will require the elimination of physical, technical, and fiscal trade barriers between the 12 member countries of the EC. These barriers will have to be abolished and rules, regulations, and standards will have to be harmonized in order for internal EC borders to be eliminated. The driving force behind the harmonization process is the removal of physical trade barriers which take the form of frontier controls. Removal of frontier controls requires the harmonization of fiscal and technical regulations that exist between the member countries. The EC has identified 279 directives which, if passed, would allow the removal of frontier controls.

Consequences for agriculture. The completion of the internal market has several implications for EC agricultural production and trade. The major implications are:

- removal of border taxes and subsidies (monetary compensatory amounts or MCAs) on agriculture and food products which should result in true common farm prices and more reliance on comparative advantage production in the EC;
- harmonization of animal and plant health and food safety regulations, resulting in more competitive farm input and output markets and simplified EC market access;

- harmonization in sectors related to agriculture such as transportation and financial services and value added and excise taxes which should lower farm supply prices;
- income and employment effects which should enhance EC farm income and provide increased employment for marginal farmers;
- pressure to reduce nationally based farm support measures which should result in more competitive agriculture; and
- increased focus on environmental issues in agriculture which should result in more production constraints.

The successful completion of the internal market could lead to a gradual but significant change in the pattern of agricultural production and trade in the EC through the above factors. There are also important links between the 1992 programme and the GATT negotiations as well as special implications for developing countries' agricultural trade, official development assistance, and foreign direct investment.

Reform of CAP prices. The lifting of internal EC frontier controls could result in elimination of the current internal economic distortion in the Common Agricultural Policy (CAP) — the agrimonetary system. Current functioning of the agrimonetary system leads to "uncommon" prices for agricultural products and prices which are higher than they would otherwise be. The EC Commission has attempted for over 20 years to return to a true common price for farm goods and the elimination of internal EC borders may provide the occasion.

The crux of the problem is that EC member countries have been allowed to establish an exchange rate for agriculture (commonly called the green rate) that differs from the official exchange rate. CAP prices are denominated in European Currency Units (ECU). Before the green rates were introduced in 1969 farmers were to be paid in national currencies at the official exchange rate. Since 1969, when exchange rate realignments occurred in the EC, member countries established a separate exchange rate for agricultural products and farmers were paid in national currencies at the green rate of exchange.

The effects of green rates are: i) farm prices differ between member states, and ii) member countries maintain some degree of control over national farm prices and, therefore, farm incomes and food prices. For example, the EC support price for the 1988/89 marketing year for feed wheat was $179 per tonne in Greece and $232 per tonne in the Federal Republic of Germany, while the so-called common ECU price was set at $201 per tonne. Price differentiation is accomplished by negotiating devaluations or revaluations of the green rate of exchange against the official exchange rate. There are even different green rates for commodities in the same country and there are currently 40 green rates in the EC.

When green rates appeared in 1969, first in France, then in the Federal Republic of Germany, and later in all EC member countries, it became necessary to establish a method of countering the price differentials between member countries created by the green rates of exchange. Without a price-equilibrating mechanism at the border, agricultural products would flow into the intervention system of the country with the highest price. Since the CAP guarantees that the intervention system must accept all quantities that qualify, there would have been a tremendous distortion of trade as price differences between member countries for some commodities reached over 60 percent.

The mechanism to prevent this trade distortion was a series of border taxes and subsidies that offset the price differences between the member countries. These taxes and subsidies are called monetary compensatory amounts (MCAs). MCAs change when green rates are devalued or revalued or when the official exchange rates change as a result, for instance, of a realignment in national currencies against the ECU in the European Monetary System (EMS).

In 1984 a special "correction factor" for green rates was introduced which effectively pegged all green rates to the revaluing currency in the EMS, thus giving birth to the green ECU. This was deemed necessary because the Deutschmark had been — and has continued to be until recently — the revaluing currency in the EMS, and revaluation of the German green rate meant price decreases for German farmers. Pegging the green rates to the Deutschmark solved that particular problem since German prices remain unchanged as the result of an EMS realignment but all of the devalued currencies receive farm price increases when their green rates are devalued. The correction factor had increased to 14.5 percent in early 1990 meaning that EC farm prices for CAP products have been increased by at least 14.5 percent since 1984 because of EMS realignments.

TABLE 2.1. **Annual changes in CAP prices in ECUs and national currencies**

Year	Common prices	National currencies
	(ECU)	*(from green rates)*
	(...........	%................)
1982/83	**10.4**	**12.4**
1983/84	**4.2**	**6.9**
1984/85	**−0.5**	**3.3**
1985/86	**0.1**	**1.8**
1986/87	**−0.3**	**2.2**
1987/88	**0.2**	**3.3**
1988/89	**—**	**1.6**
1989/90	**−0.2**	**1.3**
1990/91*	**−1.1**	**0.2**

*** Proposed.**

Source: The Agricultural Situation in the Community. **EC Commission, several issues.**

If economic integration is completed by the end of 1992 and borders are abolished, then it would not be possible to collect MCAs at the border. Alternative methods are technically feasible but are not considered acceptable because they would either be costly or susceptible to fraud. Hence it appears that the current system will have to be severely modified or abolished.

The move toward fixing exchange rates within such a narrow band that price differentials would be insufficient to cause trade movements based on green rate differentials could solve the problem. True common prices in agriculture would then be possible and no MCAs would be created. Production distortions created by the green rates would no longer occur if no compensating mechanisms were established. This would allow EC agricultural production to be based more on the principle of comparative advantage and would favour least cost producers throughout the EC.

Agricultural price negotiations in the EC would then be much more transparent and the problem created by the green rates and the correction factor would disappear. For example, the average weighted increase in CAP prices from 1982/83 to the proposed prices for the 1990/91 marketing year is 12.8 percent when measured in ECUs; however, when converted into national currencies, price increases to farmers had actually increased by 32.8 percent in the same period (Table 2.1).

In the medium to long term, abolition of the agrimonetary system would lead to lower average support prices and weaker intervention, thus resulting in a more market-oriented agriculture within the EEC. However, in the short term, prices are likely to increase because most countries would have to devalue their green rates, thereby increasing prices in national currencies, in order to return to official exchange rates. It is impossible to estimate how much lower EC price levels would have been in the past in the absence of the agrimonetary system because prices are negotiated.

Harmonization of plant and animal health and food safety. The EC intends to intervene in the food trade only when consumer health and safety are at risk and when the free movement of products is threatened. In the absence of harmonized rules, the principle of mutual recognition would apply meaning that products legally manufactured and sold in one member country could be sold in all member countries.

Completion of the internal market will have major effects on the food and drink industry because non-tariff barriers have long existed and have been increasing in recent years. The EC Commission, in a partial study of non-tariff barriers in the EC food and drink industry, estimated that these barriers were costing the EC industry over $1 billion annually. Harmonizing national standards so that one EC-wide standard is applied instead of 12 differing standards could allow EC food companies to realize greater economies of scale in food processing and distribution. In the event, EC food companies would be more competitive on the international market as well as within the EC market. The benefits of lower food prices, greater variety and food quality are expected to accrue to EC consumers.

Over 100 of the 279 directives required to eliminate internal borders are directed at the agricultural and food industry. Thirty-two of the directives will affect the food processing industry and food safety concerns in such areas as packaging, labelling, additives, nutrition, etc., and over 70 will have to harmonize rules pertaining to plant and animal health and food safety in areas such as pesticide residues, animal drug residues, inspection and certification controls, etc. Significant progress has been made in the harmonization process as nearly all of the agriculture-related directives have already been approved by the EC.

As regards food and agricultural imports, access to the EC could be enhanced if imports were treated in the same manner in all 12 countries because only one set of standards would have to be met instead of separate ones for each country. However, some standards may be set higher than previous levels in certain member countries which might constrain market access in some cases. Indeed,

it is the stated purpose of the EC to set standards at the highest level that negotiations will allow.

At this point it is unclear at what level this will be because there are enormous differences between countries. In the case of pesticide residues it appears that standards will be set at levels lower than the country with the strictest rules but considerably higher than countries with the highest tolerance levels for residues.

The result of the harmonization of plant and animal health rules will be a more competitive EC farm supply industry and, therefore, a reduction in farm input costs. Production patterns could then change as costs per unit of production are affected by geographical location in the EC. A more competitive food processing sector could also bid up the price of farm produce in some countries as food companies search for low-cost raw material throughout the EC. The food processing sector will likely push for lower CAP prices because of the need for least-cost sourcing of raw materials.

Third country exporters to the EC are concerned about how EC imports will be treated. The concern is particularly focused on how third country testing and certification procedures and results will be treated by the EC. The assessment of conformity to EC regulations and quality assurance will require testing and certification procedures that may result in the exclusive use of EC laboratories. The current round of GATT negotiations could play an important role in the EC's harmonization process. In the mid-term GATT review in April, the EC agreed to allow internationally recognized scientific bodies to provide the scientific evidence in trade disputes that involved sanitary and phytosanitary rules. The EC is more likely to accept internationally recognized standards if there is a GATT agreement. If not, the EC might then set standards for the EC which would become de facto standards for the rest of Europe, and this could result in trade diversion if they were to differ significantly from other internationally accepted standards.

Harmonization in sectors related to agriculture. Transportation. Elimination of border restrictions is expected to reduce transportation costs by 30 to 40 percent as average truck speeds would significantly increase from the current EC average of 12 km/h. When the transportation directive is fully implemented, which will facilitate return hauling, then even greater savings are expected. These savings may result in a relocation of food processing plants and should be particularly beneficial to the Mediterranean member countries of the EC.

Financial services. The banking industry will be subjected to a more competitive environment as national banks become EC banks and provide services throughout the EC. More competitive rates of interest should decrease credit and mortgage costs for farmers in some countries allowing for more capital investment. Freedom of capital movement was allowed on 1 July 1990 and is expected to result in a more open and competitive environment for investment and financial services.

Value added taxes (VAT). Current VAT rates for food vary widely among EC countries. Harmonizing VAT will be an extremely difficult task and it is still too early to tell what rates will apply after 1992. VAT rates will converge, but not necessarily be equalized, in such a way that the effects of VAT on trade flows and tax revenue will be minimized.

Excise taxes. There are enormous differences between excise taxes on cigarettes, alcohol and fuel throughout the member countries. For example, Greece levies a tax on cigarettes equal to 13 US cents per carton while Denmark's rate is $16.60 per carton; Italian fuel taxes are over twice the level of the Spanish tax and Ireland taxes pure alcohol at ten times the rate of Portugal. When these tax rates are harmonized, consumption patterns and farm input costs will be affected.

Income effects. The increase in GNP as a result of completion of the internal market is estimated by the EC Commission to be 5 to 7 percent over a five-year period. Many economists feel that this is an overestimate while others feel that it is underestimated by a factor of five. All agree that incomes will rise. An increase in income will be accompanied by a small but significant increase in food consumption, particularly of meat where EC per caput consumption is only two-thirds of North American consumption.

There are also income distributional effects that should increase incomes in economically disadvantaged regions of the EC where food consumption is relatively low. The agreements on CAP reform measures in February 1988 included a doubling of structural funds (from

$8 billion to $16 billion) destined for economically disadvantaged areas which are not able to withstand the competitive pressures created by the 1992 programme.

There is also a general expectation that a significant flow of investment to low-wage areas will occur as a result of the economic integration process. In the event, it would reinforce the redistributional effects of the increase in structural funds and the anticipated increase in GNP. Food costs are also expected to fall as a result of the 1992 programme and, in combination with the income distributional effects plus the increase in GNP, should result in a significant boost to food consumption in the EC. In turn, this may lead to a lower EC surplus food production and reduced agricultural exports.

Employment and farm structure effects. The EC Commission estimates that completion of the internal market will create a net addition of five million jobs. Many of these jobs will be created in the construction industry where farming skills are easily adapted. It is likely that some of the marginal farmers in the EC will be forced to abandon farming because of lower CAP prices in the future while others will abandon farming because alternative employment is available. In addition, many part-time farmers reside in economically disadvantaged areas where structural funds are to be doubled and where investment activity will increase.

The long-term macro-economic effects of the 1992 programme should thus have beneficial effects for food consumption and farm income. Of particular interest is the possibility of structural adjustments in the farming sector. With farmers abandoning farms for better-paying jobs made available by the 1992 programme, a consolidation of small farms into more economically-sized farms would help to alleviate the farm income problem in some areas of the EC. Such a development would reinforce the net farm income effects generated by lower input costs because of a more competitive demand by food processors for raw materials.

Nationally based programmes threatened. The elimination of internal borders will raise the question of nationally based production quotas for milk and sugar as well as import quotas allocated on a national basis such as those for bananas and beef. The problem of beef import quotas has already been solved but preferential treatment for banana imports by some EC member countries may have to be adjusted after 1992.

The milk quota regime is likely to be altered after 1992 because low-cost dairies will apply pressure for the transfer of quotas across national boundaries. Such a development would be in line with the objective of the 1992 programme to promote competition and could lead to lower milk prices. Such a development would also make the EC more competitive in the world dairy market.

Farm support derived from national treasuries accounted for at least one-third of all EC farm support from 1981-86 when CAP guarantee and guidance expenditures are included. The large affluent member countries spend more — as indicated by national treasury payments from France and the United Kingdom — accounting for 42 and 38 percent, respectively, of total support payments for their domestic farmers for the 1981-86 period.

It is difficult to estimate the production effects of national farm aids because a significant portion of them are direct transfer payments which do not enhance production. Social security payments in the form of pensions, health benefits and other payments to curb intensive production could not be objected to on competitive grounds but rebates, tax incentives, etc., which do provide advantages for farmers, would seem to be incompatible with a borderless economic market. Conversion of these production-enhancing payments to direct payments would be compatible both within the GATT framework and with the 1992 objective of promoting competition.

Environmental focus on agriculture. The 1987 Single European Act has a strong environmental component directed at agriculture. (Intensive agriculture has been associated with ground water pollution, soil erosion and exhaustion, and even acid rain, which is brought about by methane and ammonia emissions from intensive livestock operations.) The 1992 programme provides the legal framework and EC-wide institutions for environmental groups and governments to address environmental issues related to agriculture. It has been estimated that EC agricultural production will be lower ten years from now because of the environmental constraint.

The development of an EC environmental agency has been approved and subsequent

TABLE 2.2. **EC external trade, 1987**

Country	Exports to:	Imports from:
	(.............%.............	)
United States	**21**	**17**
Japan	**5**	**10**
Latin America	**5**	**7**
Africa	**7**	**17**
Asia	**10**	**10**
Near East	**10**	**7**
Other	**42**	**32**
Total	**100**	**100**

Source: **Ghosh, *European affairs.***

work will likely provide an even stronger rationale for transfers to farmers who lower the intensive use of inputs or who otherwise farm in a more environmentally safe manner.

Trade effects of the 1992 programme. From the above discussion it seems clear that EC agriculture could undergo a relatively dramatic change in production and trade patterns in the 1990s in the post-1992 integrated EC economy. The most obvious effect is a more competitive EC food processing and distribution sector. Economies of scale should accrue to the food companies that are able to focus on an EC-wide market without barriers. Economic efficiencies gained within the EC should result in more EC exports of processed food products to the world market.

It is difficult to judge whether there would be increased access to the EC market as a result of harmonization of standards. National and regional tastes will not disappear, which means there will be a premium on marketing. Higher standards for food safety could create conflicts with non-EC countries not able to meet the new standards. The EC will be particularly careful with tolerance levels for pesticides and animal drug residues. In addition, community preference may remain in place which means that the variable levy system for imports would continue protecting EC production from import competition.

An increase in food demand as a result of higher income levels and a redistribution of income should stimulate imports of animal feedstuffs. The amount imported will depend on the degree to which EC grain prices are reduced as a result of the elimination of the agrimonetary system.

A reduction in EC grain exports onto world markets should result from lower EC prices and slower production increases, as well as increased food consumption which reduces agricultural surpluses. This outcome depends on many other factors, not least of which is the rate of EC yield increases and the ability of the CAP budget to finance exports. Budget pressures should be alleviated by a more market-oriented EC agriculture and the emphasis on environmental degradation and food safety concerns should reduce intensive farming systems and thus slow yield increases for some crops.

Links between the 1992 programme and GATT. To some extent the outcome for agriculture under the 1992 programme will depend on the results of the GATT negotiations. The EC would benefit from a more liberalized world trade environment for industrial goods, services and intellectual property — all of which stand to gain from the 1992 programme. A GATT agreement on the major issue of agriculture would complement the trade liberalization effects of the 1992 programme. Without such agreement, the EC might turn inward and concentrate on the internal aspects of the integration effort, ignoring external effects. EC harmonization of animal and plant health and food safety regulations, for instance, might then become more problematical for non-EEC countries.

The potential for an agreement on plant and animal health and food safety is another important link to GATT. The Community's standard-setting process could be guided by the GATT framework as the EC countries have agreed with other GATT members to use recognized international standards for resolving trade disputes over food safety and plant and animal health.

GATT members are also considering the use of relevant international scientific organizations for determining scientific consensus on standards. The three organizations that GATT members have recognized are: the Codex Alimentarius Commission of FAO/WHO; the Plant Protection Convention of FAO; and the International Office of Epizootics. A GATT agreement on a dispute-settlement mechanism would be useful to the external dimension of the 1992 programme where disputes are likely to arise because of new standards.

Implications for developing countries. EC trade with developing countries is the largest of all the regions as they account for over 40 percent of EC exports and imports (Table 2.2). Closer economic integration within the EC could have important trade-creation and trade-diversion effects which developing countries will feel. For

example, it has been estimated that diversion could lead to a 9 percent decline in total non-fuel exports from the Near East (including Turkey) to the EC. Trade creation resulting from internal EC export growth is not expected to match the fall in the Near East region's exports.

The most important effects in agriculture are likely to be felt by the ACP countries which enjoy special treatment under the Lomé Convention. For example, the ACP countries export around 100 000 tonnes of bananas annually to the EC under preferential arrangements. Difficulties could arise from trying to maintain indefinitely this form of special treatment in a borderless EC. It is impossible to envisage precisely at this stage what the solution might be but the fact that ACP countries have increased in number from 44 in 1975 to 68 under Lomé IV (December 1989), while their share of EC imports has dropped from 8 percent to 3.8 percent during this period, is a cause for concern.

PART THREE
STRUCTURAL ADJUSTMENT AND AGRICULTURE

I. A review of the issues

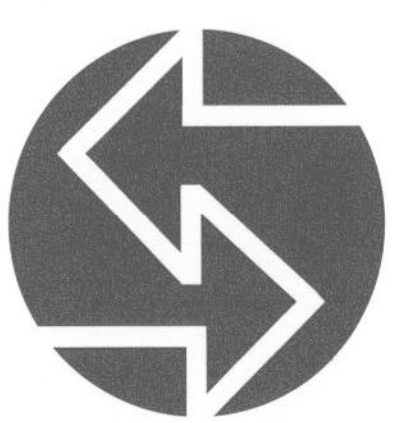

STRUCTURAL ADJUSTMENT AND AGRICULTURE

I. A review of the issues

THE NEED FOR ADJUSTMENT

A crisis unfolds

During the 1950s and 1960s developing countries experienced moderate to strong growth performances which, to a great extent, could be attributed to a relatively favourable world economic environment. While the terms of trade for most countries declined slightly during the 25 years after the Second World War, this was offset by inflows of private and official capital. The greater part of this financial inflow went to non-agricultural projects such as mining, manufacturing, and infrastructure. Yet agriculture performed reasonably well and, overall, developing countries benefited from a relatively stable world trading system of fixed exchange rates, preferential trading arrangements, and an increasing number of operative commodity agreements.

So sustained had been the growth performance of the Third World that, toward the end of the 1960s, issues regarding distribution of the gains from this growth received increasing emphasis, even to the point of making equity considerations integral to the definition of "economic development". While earlier development theory had tended to stress the likelihood of a positive relationship between income inequality and growth, by the early 1970s quite the opposite perspective gained respectability. This "basic needs" approach stressed the alleviation of poverty as the central concern of economic development. The emphasis on distribution tended in general to benefit agriculture for it was in the rural areas that the vast majority of the poor were found.

At the same time that a consensus emerged on the centrality of distribution to the development process, several shocks to the international system occurred, dramatically altering the international economic environment. In 1971, the stability of the international trading system was shaken by the United States Government's decision to suspend the fixed-price gold convertibility of the dollar. This measure ushered in a new set of international rules for trade and finance based on variable exchange rates (at least for the OECD countries). This new exchange rate regime proved particularly problematic for many developing countries because of the vulnerability of their economies to international price fluctuations and the relative lack of trained personnel with the skills necessary to manage increasingly complex economies. The collapse of the post-war system of international regulation coincided with a period of rising inflation as well as the petroleum price increases of late 1973-74 which generated substantial trade surpluses and deficits for developing country oil exporters and non-oil exporters, respectively.

Well before "structural adjustment" gained wide currency, a period of economic adjustment in the developing countries resulted from these two shocks. Accumulating foreign debt from both private and official lending sources to fill current account gaps became one manner of adjustment. The overwhelming consensus that emerged in the 1970s maintained that growing indebtedness did not represent a long-term danger. On the contrary, in the absence of indebtedness, the hard currency surpluses of the oil-exporting countries would not have been "recycled" into the expenditure stream, with the probable result being a world recession. Furthermore, as a result of the inflation of the late 1970s (itself fuelled by demand pressures arising from borrowing), real debt burdens of even the most indebted countries remained quite low except in a few cases. Most years, negative real interest rates characterized commercial bank loans and even more so official loans which were made on extremely concessionary terms. Along with the rise in petroleum prices, many other primary products enjoyed buoyant world market conditions which added to the impression that debts would be manageable (see Box 3.1).

The debt burden as such arose at the beginning of the 1980s when recession in the developed countries resulted in the desired and relatively sudden elimination of world

BOX 3.1

Debt accumulation and resource transfer

"With high inflation and a depreciating dollar, a developing country would be better off borrowing dollars than paying cash. In view of these financial inducements, a finance minister who abjured borrowing would be like a purchasing manager who preferred paying retail for goods that were being sold at wholesale prices. If he wasn't fired, he should have been."

Alfred J. Watkins[1]

Why countries borrowed

The two oil price shocks of 1973-74 and 1979 forced major adjustments by the world's economies. Oil-consuming developing nations, always in need of large amounts of capital to accelerate development, now required even more capital to finance rapidly rising fuel costs. To meet this need, they borrowed. By the mid-1970s, the depreciating US dollar, rising primary commodity prices, and real interest rates approaching zero made borrowing profitable. Finance officials turned to the private financial markets to obtain inexpensive capital.

Why banks lent

OPEC nations generally made large, short-term US dollar deposits in banks outside the United States, and outside the control of that country's banking authorities. Banks are not required to keep non-interest reserve accounts with the Federal Reserve on their off-shore deposits nor on their off-shore loans. They could lend out the entire amount, paying more for these deposits than for domestic deposits, but still make a higher profit by lending to developing countries.

To minimize the profit risks caused by variations between short-term deposit interest rates, long-term loan interest rates, and high rates of inflation, banks introduced floating rate loans. Unlike domestic borrowers in the United States who resisted floating rate loans, developing countries accepted them.

Banks had a huge supply of funds to lend. Developing countries had the commodities which were rising in price, a willingness to accept long-term floating rate loans and a need for funds. Inflation made the developing countries appear wealthier and their debt smaller.

How debt accumulated

As the world recession emerged in the early 1980s, the debt crisis matured. Oil-importing developing countries suddenly found their economies confronted by the most perverse concomitance of circumstances: economic recession and rising protectionism sharply reduced import demand for their products; their terms of trade collapsed as oil and oil-based energy prices initially soared and prices of other commodities fell; commercial bank eagerness to lend turned to eagerness for repayment; and interest rates sharply increased in real terms, resulting in higher service payments and an appreciation of the dollar. Policy-makers and international development agencies faced a profoundly different world economic environment where repaying loans dominated both discussions and decisions on how economic adjustment should proceed during the 1980s.

The debt picture in the 1980s

Overall, the Latin America and the Caribbean region has the largest amount of debt, peaking at over $440 billion in 1987 but remaining close to that level since. The region is also the largest borrower of loans from private financial institutions. Its debt-service/export ratio fluctuated around 35 percent between 1982 and 1986, but reached nearly 40 percent in 1988 and 1989.

Total debt as a percentage of GNP, however, is highest in Africa, south of the Sahara, where it reached 110 percent in 1989. Official lending dominates the African debt portfolio at a ratio of about two to one. The North

---

[1] A.J. Watkins. 1986. *Till debt do us part,* Landham, MD, University Press of America.

developments in the world economy, achieving these social and economic goals through adjustments in domestic policies poses a much more difficult challenge.

The most important external variables facing the developing countries are the terms of trade and trade barriers to their exports; interest rates on old debt and new borrowings; and the amount and terms of official development assistance (ODA).[3] These factors are influenced primarily by the actions of the developed countries, particularly the larger OECD countries. While the point is obvious, it is worth stressing that the economies holding the lion's share of world production and trade determine conditions in world markets and, therefore, the developing countries' circumstances. The greater part of developing country trade and borrowing is with developed countries, not among the developing countries themselves.

Changes in the international economic environment have an impact on each national economy. These changes eventually produce responses on the part of producers and consumers after reaching them in a form mediated by government policy. "Adjustment" as such is the process of producers and consumers responding to external changes, and "adjustment poilicies" are government actions to mediate those changes.

From the 1950s through the 1970s there was a strong, though certainly not unanimously supported, development strategy suggesting that countries could benefit from policies which

[3] Much of the commercial bank debt of developing countries has been contracted on the basis of variable interest rates, usually linked to the London inter-bank borrowing rate (LIBOR).

Nominal and real interest rates,[1] developing regions, 1971-80 and 1981 to 1988

Region	1971-80 Average	1981	1982	1983	1984	1985	1986	1987	1988
	(........	....	....	.... %.	....	....	....	....	........)
Sub-Saharan Africa									
Nominal	5.4	8.3	7.5	7.1	5.4	5.8	5.4	4.1	4.0
Real	−7.6	16.8	15.5	9.3	1.6	9.5	11.4	2.4	−0.2
Latin America & Caribbean									
Nominal	8.9	13.9	12.6	10.7	11.1	9.1	8.2	7.6	8.0
Real	−9.4	15.0	20.3	17.1	8.7	13.9	21.4	2.7	2.3
South Asia									
Nominal	3.2	4.9	5.8	5.0	5.5	5.0	5.3	4.6	5.3
Real	−9.4	3.3	10.7	8.8	3.7	9.5	10.6	−5.7	−2.8
East Asia & Pacific									
Nominal	7.7	10.2	9.9	8.8	8.8	8.0	6.8	6.3	6.2
Real	−4.9	8.6	14.8	12.6	7.0	12.5	12.1	−4.0	−1.9
North Africa & Near East									
Nominal	5.9	7.2	8.2	8.9	8.5	7.5	6.9	7.1	6.9
Real	−26.9	−4.3	10.5	20.7	10.4	13.5	46.1	25.7	20.3

[1] Average interest rates on new commitments, nominal and adjusted for changes in export unit values.
Source: World Economic Outlook. IMF, October 1989; *World Debt Tables 1989-90,* vol. 1. World Bank.

would reduce the impact of external changes on the domestic economy, either by delaying them — thus providing more time for adjustment — or by counteracting them. Import substitution, agricultural marketing boards and controls on foreign exchange mobility are examples of such "isolating" adjustment policies at the national level. Cooperative international actions by producer and consumer countries in the formulation of commodity agreements is another example and constitutes the general thrust of multilateral action within the United Nations Conference on Trade and Development (UNCTAD). These policy measures reflected a judgement that greater openness to the world economy carried with it costs as well as benefits. In general, the industrial economies, being more flexible and integrated as well as economically stronger, were better able to reap the benefits while minimizing the costs of openness than were the developing countries.

In the late 1970s this view, which stressed the "vulnerability" of developing economies to international fluctuations, lost currency in the policy debate. During the 1980s, "adjustment" came to refer to a particular way of responding to external changes, namely by facilitating their transmission through the domestic economy. Therefore, "adjustment" came to imply trade liberalization and a reduction of government intervention in markets. According to this approach, policy changes are seen as necessary not just to accommodate external shocks but also to reform ill-advised policy measures of the past. The extent to which the difficulties faced by developing countries in the 1980s were more the result of unsustainable national development policies and policy errors (such as excessive government intervention) than of external shocks (such as the debt burden, declining terms of trade and increased developed country protectionism) is a difficult matter to assess. On one hand, a number of countries succeeded in overcoming external shocks and maintaining or restoring some degree of macro-economic balance (see the section, Macro-economic overview, in Part Three, II). This suggests that sound national policies can make a difference. On the other hand, if the more inward-oriented policies of the 1960s and 1970s were detrimental to developing country growth, the fault does not lie with the implementing governments alone. Without development assistance from the international agencies and bilateral donors, the implementation of inward-oriented development plans would not have been possible. Indeed, in the 1960s and 1970s, it was not unusual for some of these agencies to recommend policies they now eschew as counterproductive.

The liberalization approach to adjustment had its antecedents in the stabilization programmes of the IMF, associated with the Stand-by Credit Arrangements established as early as 1952. These Stand-by Arrangements involved short-term lending by the IMF to countries suffering temporary balance-of-payments deficits. Quickly disbursed and having brief repayment periods, these "stabilization programmes" placed strict conditionality on the credit extended, usually involving currency devaluation and demand-contracting measures. In the 1960s and 1970s, the IMF initiated other balance-of-payments support programmes within the general framework of stabilization.[4] While in the view of the IMF these stabilization programmes served to foster recovery and improve the general economic health of a country, they are not growth or development programmes, nor are they intended to be. Their short time perspective alone restricted their role to stabilization. In the mid-1970s, the IMF began somewhat longer balance-of-payments support programmes (the Extended Fund Facility) and, in the mid-1980s, involved itself in three-year programmes (the Structural Adjustment Facility).

By the beginning of the 1980s, the overwhelming majority of developing countries required support for balance-of-payments problems that varied from acute to catastrophic. It was then that the World Bank entered into the field of policy-conditional lending for balance-of-payments support, following the orientation of the IMF. Since the institutional function of the World Bank is to foster development, the lending programmes were longer, more concessionary, and with a more formal emphasis on growth and stabilization than those of the IMF (see Box 3.3).

Structural adjustment policies were recommended in the framework of loan conditionality. These policies required changes in both short-term economic management as well as institutional reforms. Their aims were a

[4] Among these were the Buffer Stock Facility (for countries engaged in buffer stock arrangements under international commodity agreements); the Compensatory Financing Facility (for countries affected by price fluctuations of primary products); and the Oil Facility (for oil-importing countries).

more efficient allocation of resources and the enhancement of the economy's ability to withstand external shocks. These goals were to be achieved through devaluation, monetary and fiscal restraint (together with improved revenue performance, reduction of government intervention in markets, including labour markets) and liberalization of external sector policies. The expectation was that this set of policies would, *inter alia,* create incentives for the reallocation of resources toward internationally traded and/or tradable commodities (exportables and import substitutes) and would, therefore, enhance the foreign exchange position of countries implementing such policies — both through increased foreign exchange earnings and foreign exchange savings.

An important aspect of such loans and their associated policies is that they do not present a growth package as such. Conventional economic development plans involve a combination of more or less complex programmes and projects along with the means for their funding within an overall policy framework. Such programmes and projects, through their capacity-expanding potential and aggregate demand management, promote overall economic growth.

For the most part, the primary function of structural adjustment loans (SALs) is not to create assets directly, i.e. they are not project loans, nor are they restricted to the purchase of particular goods. Their primary role is to serve as a general balance-of-payments support. Such support relieves the foreign exchange

BOX 3.3

The World Bank and structural adjustment lending

As the terms of trade deteriorated, real interest rates increased and the debt crisis matured, the World Bank introduced structural adjustment loans (SALs) and sectoral adjustment loans (SECALs) in the 1980s. By linking these loans to policy changes and institutional reforms, adjustment lending is intended to help developing countries better absorb the impact of external shocks. In 1988, adjustment lending operations totalled $19.9 billion to 59 countries and represented 25 percent of the Bank's annual lending.

The World Bank's language of adjustment[1]

Adjustment. Policies to achieve changes in internal and external balances, changes in the structure of incentives and institutions, or both. A focus on internal and external balances is referred to as stabilization. A focus on incentives and institutions is referred to as structural adjustment.

Structural adjustment. Policy and institutional reforms aimed at improving resource allocation; increasing economic efficiency; expanding growth potential; and increasing resilience to future external shocks.

Stabilization programme. An economic programme, often supported by the IMF, aimed at achieving stabilization. Stabilization programmes involve particular fiscal and monetary policies.

Structural adjustment loans (SALs). World Bank lending that supports structural adjustment. The macro-economic framework in which these loans have been made has often included IMF stabilization targets, to which structural elements were added. SALs were introduced to address balance-of-payments needs, provided stabilization and structural adjustment policies were undertaken. All such loans have provided general import financing.

Sector adjustment loans (SECALs). World Bank lending for structural adjustment in a sector, focusing on major institutional and micro-economic policy changes. Some SECALs have provided general import financing, while others have confined financing to the import needs of particular sectors.

Hybrid lending (HL). Bank lending for adjustment in a sector, but with an investment component. Hybrid lending may also finance a part of the sectoral investment programme, or may finance recurrent expenditures if the budget is temporarily strained.

[1] *Source: Adjustment lending: an evaluation of ten years of experience.* World Bank, 1989.

constraint. The strategy is that the policy changes on which lending is conditional induce asset-creating behaviour on the part of investors, ranging from small-scale farmers up to large businesses, by altering the matrix of economic incentives. If the policy packages prove unsuccessful — and both the World Bank and the IMF have unilaterally or bilaterally ended adjustment programmes (i.e. suspended loan disbursements) — the recipient government may be left with an increased debt but no material assets on the other side of the ledger. In this way, unsuccessful adjustment programmes have a potentially high cost.

As mentioned above, major policy adjustments were required by developing countries as a result of unsustainable national development policies in the context of the greatly altered international environment. Adjustment by the developing countries included adapting to the conditions created by external developments beyond their control, i.e. the recession of the early 1980s, high real interest rates, and budgetary deficits in some developed countries that affected the availability of international liquidity and its distribution among countries. Given that the economies of the developed countries are generally much stronger and more flexible, the pressure on developing countries to adjust is striking for its asymmetry. Relevant adjustments by the developed countries could have been a reduction in the excessive levels of farm support, freer trade in primary products, and measures to achieve private and public debt relief. While initiatives emerged such as the Baker and Brady debt plans, they did not result in a systematic adjustment of developed country policies. This gap highlights the relative absence in the international system of any mechanism for encouraging adjustment in the industrial countries, particularly the larger ones.

Unfortunately, the structural adjustment policy packages provoked contentious debate over abstract and generalized issues of economic philosophy. Too often, the emphasis focused on justifying normative judgements rather than on the actual behaviour of economies in the developing world. From the point of view of the developing countries, the debate over adjustment appeared all the more frustrating because of the asymmetry in pressures for policy change. For example, while governments of developing countries came under heavy external pressure to liberalize trade, the OECD countries became increasingly protectionist with respect to many agricultural products and labour-intensive manufactures.

When one steps back from the heat of the debate over whether a particular set of policies brought improvement or deterioration in economic conditions, the most important lesson of the 1980s is the great diversity of developing country experiences. As shown in Part Three, II, in some detail, this diversity is quite striking with regard to agriculture, and correspondence between growth performance and policy regimes is not always easy to trace. In part, this may be because it is too soon to assess the impact of policy changes, particularly ones that act indirectly upon producers, such as deregulation and privatization. Furthermore, performance in agriculture is crucially dependent on the caprice of nature. This has been the case in Africa where apparent disaster cases have changed in a remarkably short time to become success stories as the result of favourable rains, then reversed again when drought returned. Others argued that unfavourable weather and falling primary product prices unveiled the underlying structural problems prevalent in some countries.

With the important exception of several Asian countries, the 1980s brought declining per caput incomes to most of the developing world. In terms of development, a "lost" decade has passed and both government and donor agency policy-makers look back to find little to inspire pride of accomplishment. After three post-war decades that involved grappling with the problems of economic growth, the 1980s introduced the grimmer task of confronting economic decline.

AGRICULTURE, ECONOMIC CRISIS AND STRUCTURAL ADJUSTMENT

Impact of economic crisis on agriculture

The previous section outlines how the combination of an adverse international economic environment and misguided domestic economic policies resulted in deep and widening economic imbalances in a number of developing countries. Experience has taught that, as empirically shown in Part Three, II, Macro-economic imbalances and agriculture, these developments and the accompanying economic crisis can adversely affect agriculture in several ways.

First, economic recession reduces the effective domestic or internal demand for food and other agricultural commodities, the magnitude of the reduction depending on the elasticities of demand with respect to income for these commodities. The effects of such price reductions depends on the ability of the country to channel the resulting surplus into export markets or, in the case of import substitutes, on the substitutability of imported and domestic commodities. To the extent that smallholders produce commodities primarily consumed in the domestic market, or commodities that have limited export potential, the reduction in effective demand results in proportionately larger price reductions. Thus, depressed demand may worsen the distribution of farm income.

Second, the general shortage of foreign exchange associated with an economic crisis also affects agriculture. Foreign exchange shortages reduce a developing a country's ability to acquire modern inputs such as machinery, fertilizers, chemicals, etc., the majority of which are usually imported. The overall impact of such a shortage includes a reduction in yields and, depending on the substitutability of specific inputs (e.g. hand labour for herbicides), a reduction in total production. Although larger commercial farmers are the principal users of imported inputs, a number of intermediate and small producers also use them. Thus, such shortages could affect a wide spectrum of farm producers.

Foreign exchange shortages may also lead to a deteriorating infrastructure. The maintenance of agricultural markets, and access to them, requires continuous outlays on roads, storage facilities, communications and inspection of graded exports like coffee. Reductions in government expenditure on such public goods and services as the result of a shortage of foreign exchange can negate the positive incentive effects arising from relative price changes.[5]

Another adverse consequence of foreign exchange shortages is the scarcity of consumer goods that it may entail. Since the limited amount of foreign exchange available is often allocated by the state in such a way that essential imports like food, investment goods or spare parts are given first priority, imports of consumer goods may be seriously curtailed. Hence, the quantity and variety of goods that can be acquired by the rural population may be restricted and may, therefore, constitute an additional (non-price) disincentive to the expansion or even maintenance of agricultural production.

Third, expansionary monetary and fiscal policies associated with an economic crisis, together with fixed nominal exchange rates, affect different sectors of the economy in different ways. The prices of goods and services produced mainly for the domestic market tend to increase under the pressure of the excess demand created by expansionary policies. Services in this case include public services, transportation and marketing services, housing and residential land, but also goods protected by import quotas, or goods with low values in relation to their transport costs. At the same time, nominal prices of exportable goods and import substitutes (i.e. tradables) do not change to the same extent since they are linked to those in world markets. Such goods include most agricultural commodities and a variety of manufactured products. Thus, tradables generally suffer a relative price loss *vis-à-vis* non-tradables. The effect on the prices of agricultural tradables relative to non-agricultural tradables depends, *inter alia*, on the differences in their levels of protection.

Thus, there is strong evidence that price incentives for agricultural producers generally

---

[5] "... cuts in the public expenditure programmes [as part of structural adjustment] are often indiscriminate... affecting the supply of critical public goods and services. The results of this survey show that if these cuts are large they can negate the expected supply response from improvements in price incentives...." Ajay Chhibber, The aggregate supply response. *In* S. Commander, ed. *Structural adjustment and agriculture: theory and practice in Africa and Latin America*, p. 66. London, Heinemann, 1989.

TABLE 3.1 **World Bank agricultural policy conditionality in 79 SALs and SECALs, financial years 1980-1987**

Policy Area	Africa	Asia	Near East & North Africa	Latin America & Caribbean	Total
Agricultural pricing (input & output)*	**30**	**3**	**3**	**9**	**45**
Trade liberalization	**16**	**8**	**5**	**23**	**49**
Institutional reform	**29**	**5**	**7**	**15**	**56**
Credit and banking	**11**	**3**	**1**	**9**	**24**
Public investment budget	**21**	**1**	**5**	**10**	**37**
Macro-economic	**20**	**2**	**6**	**10**	**38**
Environment	**3**	**1**	**1**	**2**	**7**
Total loans evaluated	**36**	**11**	**9**	**23**	**79**

*** Interest rate subsidies are included under "credit and banking".**

Source: **N. Ridley and C. Roberto. (Undated.)** ***Policy areas and trends in adjustment lending related to agriculture, FY 1980-87.*** **(mimeo) Washington, D.C., Agricultural Policies Division, World Bank.**

suffer during economic crises such as those widely experienced during the 1980s. Also, the degree to which production incentives are affected varies widely between countries and commodities.

Finally, the general climate of instability associated with an economic crisis can be quite debilitating for agriculture, particularly because of the long cycles of crop and livestock production processes. The combination of unanticipated external shocks and inappropriate policy responses — responses that magnify rather than dampen the effect of the shocks — can produce extreme instability in the form of rapid inflation and sharply fluctuating relative prices. For example, if marketing and processing margins remain relatively constant, changes in crop prices following changes in exchange rates can result in much greater short-term changes in producers' returns.

To the extent that structural adjustment policies succeed in reversing or reducing the disequilibria that are at the root of the economic crisis, they can be expected to have beneficial effects on the agricultural sector by:
i) reversing domestic demand contraction and increasing demand for agricultural products;
ii) increasing the return to tradables compared to non-tradables through devaluation;
iii) increasing foreign exchange availability;
iv) reducing the general climate of economic instability and policy uncertainty.

Agriculture under structural adjustment: agricultural policies and their effects

The general framework of agricultural policies. As noted above, structural adjustment programmes can have a positive impact on agriculture and hence on rural populations by reducing or eliminating disequilibria in the general economy. The total impact also depends on the measures that directly influence the agricultural sector through their effects on production, distribution, marketing and pricing of inputs and outputs.

In the 1980s, "structural adjustment" came to mean liberalization, deregulation, and privatization. In this approach, the key objective of adjustment is to improve resource allocation and to reduce or eliminate inefficiencies created by government intervention. Thus, in general, the programmes seek to reduce the government's role in the production, pricing and marketing of agricultural commodities.

Furthermore, agricultural adjustment lending aims at enhancing growth by changing the price and non-price incentive structure for agriculture in relation to other sectors, as well as within the agricultural sector itself, in ways that promote a more efficient use of resources. In practice, attempts to achieve this general objective have been made using a variety of policy instruments, with the release of the loan's funds being conditional on their implementation. A recent review of World Bank adjustment lending (SALs and SECALs) containing a degree of agricultural conditionality indicates that the major policy areas appearing in more than one-half of the 79 loans were institutional reform, trade, and agricultural input and output pricing (Table 3.1). There was considerable regional variation in the relative importance of the policy areas, with agricultural pricing being more important in Africa and trade policy more important in Latin America and the Caribbean.

Pricing policies: their objectives and possible impacts. Around 60 percent of all adjustment loans (80 percent in Africa) and virtually all agricultural SECALs included some form of agricultural pricing conditionality, with input and output pricing conditions assuming almost equal significance. According to the World Bank review, these pricing conditions have been considered critical to the success of loans in

almost one-third of the cases. However, successful implementation of agricultural pricing conditions occurred in only two-thirds of the cases reviewed and in less than one-half in the Latin America and the Caribbean region. The general thrust of pricing conditionalities has been to "get prices right", in the sense of adjusting domestic prices according to their border equivalents.

For the agricultural sector, the relative price is usually taken to be that between the sector as a whole and non-agricultural goods and services. Empirically, it is frequently expressed as the ratio of an index of prices received by farmers to an index of non-farm prices; that is, an indicator of the agricultural/non-agricultural terms of trade.[6] In the developing countries, four instruments have been used as the primary tools to affect the terms of trade between agricultural and non-agricultural sectors: i) direct price controls on output; ii) trade protection (tariff and non-tariff); iii) input pricing and allocation; and iv) different taxation regimes between sectors.

Many developing countries attempt to apply direct price controls on agricultural products, but they are often ineffective except in the case of export crops. Administrating comprehensive domestic price controls lies beyond the financial capability of a large number of developing countries. Even for export crops, the combination of taxation and low prices paid by marketing boards has often produced disincentives for these crops. In order to remedy this situation, adjustment programmes aim at raising producer prices to the levels of international prices, thereby also raising farmers' returns. The net effect of such policies on government tax revenue depends on the extent to which the reduction in revenues resulting from tax rate decreases is offset by the higher tax base brought about by increased exports. In addition, even if total export tax revenues decline, the net effect on the government budget depends on the changes in government outlays resulting from the reform or dismantling of inefficient parastatals as part of a general market liberalization policy.

Often, governments introduce price controls on major agricultural products to stabilize producer and consumer prices. Arguments for intervention and price stability are based on considerations such as the effect of risk and uncertainty on agricultural production induced by price movements. Although the debate on the effects of price stabilization on consumer and producer welfare is inconclusive, the importance governments of both developed and developing countries attach to stability suggests that most would be reluctant to forego it as part of a structural adjustment programme if it were conditional to complete price decontrol. The policy measures chosen by governments attempting to achieve stability, however, can and should minimize the inefficiencies and financial costs associated with such interventions.

Trade liberalization may benefit agriculture directly and indirectly. The lowering or abolition of trade taxes improves the relative prices of exportable commodities. In addition, a reduction in trade protection of manufactured goods, which generally receive higher rates of protection than agricultural tradables, improves the terms of trade between agriculture and the manufacturing sector. Deregulation and privatization of trade and marketing channels may also favour agricultural producers (see the section, Institutional reform).

However, several factors may contribute to reversing these benefits either in part or completely. In the case of agricultural products which had benefited from import protection tariffs, trade liberalization would tend to reduce their relative prices.[7] The net effect on the prices of such commodities will depend on the extent to which they were protected and on the degree of reduction in trade protection.

Existing evidence suggests that most policies preceding structural adjustment programmes (SAPs) have kept prices of agricultural commodities low compared to other sectors. This direct and indirect "taxation" (or negative protection) was usually stronger for exported commodities than for imported commodities (see Box 3.4). Thus, by eliminating the distortions induced by price incentives, structural adjustment policies improve the terms of trade of agricultural tradables; and within the agricultural sector, such adjustment would seem

[6] Governments throughout the developed and developing countries influence this ratio; indeed, it is doubtful that there exists a country in which this ratio exclusively reflects market forces.

[7] A thorough review of the issues arising from trade liberalization for agricultural commodities — sometimes called bringing domestic prices in line with "border" prices — is found in C. P. Timmer. 1986. *Getting prices right: the scope and limits of agricultural price policy*, particularly Chapter 4. Ithaca, Cornell University Press.

BOX 3.4

Discrimination against agriculture

It is by now commonly acknowledged that both economy-wide and sector-specific policies in many developing countries have discriminated against agriculture, resulting in a negative environment for agricultural production and investment opportunities.[1]

The sources of bias can be classified as direct or indirect. Direct discrimination includes those policies, widely implemented by parastatal marketing boards, that drive a "wedge" between prices received by farmers and border prices of tradable commodities. Indirect discrimination, on the other hand, is the outcome of policies not directly aimed at agriculture, including exchange rate overvaluation that tends to depress prices of exportables and import substitutes, and policies that protect industry and favour industrial import substitution, thereby raising the prices of non-agricultural goods and reducing agricultural prices in real terms. The effects of direct policies are readily observable and can be measured by the difference between domestic and border prices at the prevailing exchange rates. Indirect discrimination is harder to measure, as is the degree of overvaluation and the effects of trade protection on non-agricultural prices.

Evidence of effective protection of agriculture compared to the manufacturing sector in selected developing countries during a 25-year period shows that, with the exception of the Republic of Korea, all countries have ratios of less than 1.00. This is a clear indication of discrimination against agriculture. Nevertheless, there has been some improvement in the treatment of the agricultural sector during the latter period. The protection ratios reported incorporate policies directly affecting inputs and outputs but ignore indirect effects resulting from overvalued exchange rates.

An integrated approach that attempts to evaluate both direct and indirect protection or taxation of exported, as opposed to imported, agricultural commodities has been undertaken recently by Krueger, Schiff and Valdés.[2] The direct effect was calculated as the difference between the producer and border prices, while the total indirect effects reflected the discrepancy between the actual producer price and the one that would have prevailed in the absence of overvaluation and in the absence of all policies influencing trade. The results of this study are summarized as averages in two five-year periods for export and imported food crops. Negative numbers (percentages) denote negative protection; that is, the price that producers actually received was below what they would have received under a free-trade regime at realistic exchange rates with no direct intervention. For instance, producer prices for cocoa in Côte d'Ivoire in 1975-79 were about one-third, and in 1980-84 one-half, of what they would have been in the absence of their negative protection.

The degree of negative protection compared with that of imported food products, which in most countries are positively protected, is noteworthy. The widespread drive for self-sufficiency in food staples is reflected by the lower negative or even positive protection provided for these products, usually through the use of import tariffs or licensing. In addition, the effectiveness of

[1] For a wide range of examples and evidence see *African agriculture: the next 25 years.* FAO, 1986; *Potentials for agricultural and rural development in Latin America and the Caribbean.* FAO, 1988; *World Development Report.* World Bank. 1986.

[2] A.O. Krueger, M. Schiff and A. Valdés, Agricultural incentives in developing countries: measuring the effect of sectoral and economy-wide policies. 1989. In *The World Bank Economic Review,* 2:3(255-271).

Protection of agriculture compared with manufacturing in selected developing countries

Country and period	Year	Relative protection ratio[1]
In the 1960s		
Mexico	**1960**	**0.79**
Chile	**1961**	**0.40**
Malaysia	**1965**	**0.98**
Philippines	**1965**	**0.66**
Brazil	**1966**	**0.46**
Korea, Republic of	**1968**	**1.18**
Argentina	**1969**	**0.46**
Colombia	**1969**	**0.40**
In the 1970s and 1980s		
Philippines	**1974**	**0.76**
Colombia	**1978**	**0.49**
Brazil[2]	**1980**	**0.65**
Mexico	**1980**	**0.88**
Nigeria	**1980**	**0.35**
Egypt	**1981**	**0.57**
Peru[2]	**1981**	**0.68**
Turkey	**1981**	**0.77**
Korea, Republic of[2]	**1982**	**1.36**
Ecuador	**1983**	**0.65**

[1] Calculated as $(1 + EPR_a)/(1 + EPR_m)$, where EPR_a and EPR_m are the effective rates of protection for agriculture and the manufacturing sector respectively. A ratio of 1.00 indicates that effective protection is equal in both sectors, a ratio greater than 1.00 means that protection is in favour of agriculture; and less than 1.00, in favour of industry.
[2] Refers to primary sector.
Source: World development report. **World Bank, 1986.**

marketing boards is often limited, especially for those food crops involving large quantities traded through informal markets. Many export crops such as tropical beverages have only a small domestic market and generally have to be processed before consumption, thus making single-channel marketing less difficult.

The magnitude of negative protection is striking, especially for exportables, as is the fact that protection was greater in 1980-84 than in 1975-79 in most countries studied. The greater significance of indirect negative protection is also notable and implies that governments may be unwittingly penalizing agriculture.

Direct, indirect, and total nominal protection rates for exported products

Country	Product	1975-79 Direct	1975-79 Indirect	1975-79 Total	1980-84 Direct	1980-84 Indirect	1980-84 Total
		(%					)
Argentina	Wheat	−25	−16	−41	−13	−37	−50
Brazil	Soybeans	− 8	−32	−40	−19	−14	−33
Chile	Grapes	1	22	23	0	− 7	− 7
Colombia	Coffee	− 7	−25	−32	− 5	−34	−39
Côte d'Ivoire	Cocoa	−31	−33	−64	−21	−26	−47
Dominican Republic	Coffee	−15	−18	−33	−32	−19	−51
Egypt	Cotton	−36	−18	−54	−22	−14	−36
Ghana	Cocoa	26	−66	−40	34	−89	−55
Malaysia	Rubber	−25	− 4	−29	−18	−10	−28
Pakistan	Cotton	−12	−48	−60	− 7	−35	−42
Philippines	Copra	−11	−27	−38	−26	−28	−54
Portugal	Tomatoes	17	− 5	12	17	−13	4
Sri Lanka	Rubber	−29	−35	−64	−31	−31	−62
Thailand	Rice	−28	−15	−43	−15	−19	−34
Turkey	Tobacco	2	−40	−38	−28	−35	−63
Zambia	Tobacco	1	−42	−41	7	−57	−50
Average		−11	−25	−36	−11	−29	−40

Source: Krueger *et al.* 1989, op. cit.

Direct, indirect, and total nominal protection rates for imported food products

Country	Product	1975-79 Direct	1975-79 Indirect	1975-79 Total	1980-84 Direct	1980-84 Indirect	1980-84 Total
		(%					)
Brazil	Wheat	35	−32	3	− 7	−14	−21
Chile	Wheat	11	22	33	9	− 7	2
Colombia	Wheat	5	−25	−20	9	−34	−25
Côte d'Ivoire	Rice	8	−33	−25	16	−26	−10
Dominican Republic	Rice	20	−18	2	26	−19	7
Egypt	Wheat	−19	−18	−37	−21	−14	−35
Ghana	Rice	79	−66	13	118	−89	29
Korea, Republic of	Rice	91	−18	73	86	−21	74
Malaysia	Rice	38	− 4	34	68	−10	58
Morocco	Wheat	− 7	−12	−19	0	− 8	− 8
Pakistan	Wheat	−13	−48	−61	−21	−35	−56
Philippines	Corn	18	−27	− 9	26	−28	− 2
Portugal	Wheat	15	− 5	10	26	−13	13
Sri Lanka	Rice	18	−35	−17	11	−31	−20
Turkey	Wheat	28	−40	−12	− 3	−35	−38
Zambia	Corn	−13	−42	−55	− 9	−57	−66
Average		20	−25	− 5	21	−27	− 6

Source: Krueger *et al.* 1989, op. cit.

to benefit export more than import substitution commodities. However, the extent of improvement in relative agricultural prices depends also on the changes in non-agricultural prices, including input prices. Hence, if price signals for agriculture are to be fully effective they should become part of an overall strategy that includes prices of both agricultural commodities and non-agricultural goods and services.

Liberalization and accommodating external changes by opening an economy further are not without cost, however. By definition, such an opening reduces the institutional arrangements that cushion the impact of external shocks on domestic producers and consumers. In the 1980s, this increased exposure to external changes coincided with a worsening of the terms of agricultural trade in international markets. Furthermore, to a large extent, the disincentive signals sent by world markets with regard to developing country agriculture reflect market interventions or price distortions introduced by protectionist policies in most industrial countries. While the costs of ignoring international prices should be taken into account, an additional question can also be raised regarding the extent to which developing countries should align their price structures to distorted world market prices.

Supply response to price incentives. Bringing domestic prices closer to their border equivalents often implies increasing the prices of both agricultural inputs and outputs. The incentive to farmers will be reflected in the net return per unit of output. Even if the net return is positive, most experts agree that price liberalization can induce changes in the composition of commodities produced but, by itself, it cannot prompt a substantial increase in sectoral or aggregate growth in the short term.

Three major reasons account for this assessment of the role of prices. First, a rise in the *aggregate* supply of the agricultural sector in response to agricultural prices is limited by the relatively inelastic supply of resources in many countries, especially land. However, when policies and governmental intervention have been distorting land-use patterns, a move toward liberalization can lead to rapid gains in aggregate output. Furthermore, despite considerable seasonal underemployment of rural labour in developing countries, evidence suggests there is little or no idle labour during peak planting and harvesting periods. These shortages partly explain why various empirical studies have indicated that non-price factors are quite important in explaining agricultural growth. One recent study demonstrates that low levels of food stocks can influence both the area planted and use of non-household agricultural labour.[8] Estimates of supply elasticities are much higher when price and non-price factors are combined, for example, in irrigation projects. Investment in non-price factors should thus be viewed as complementary to a pricing policy.

In addition, public goods such as roads, irrigation systems, and research and extension services typically take years to construct or establish, so that their potential contribution to agricultural growth will only be measurable in the long term. The neglect and limitations of physical infrastructure and the human resource stock (scientists, technologists, planners, managers etc.) in many countries since the 1970s have created an enormous backlog of requirements. Increasing the supply of these public goods and services may be a necessary condition for output to respond to price incentives. Ironically, meeting macro-economic conditionality requirements has frequently worsened the scarcity of such public goods, because governments have had to severely curtail public-sector investment and recurrent expenditures in order to reduce fiscal deficits. Such situations represent one of the many potential conflicts in the design and implementation of structural adjustment programmes. Unless the agricultural sector is given a sufficiently high priority in the allocation of public-sector funds, agricultural growth may be slowed during the adjustment process, despite relative increases in producer prices.

Second, adjustment of agricultural prices by a reduction in direct taxation can be of relatively minor importance if protection of other sectors continues and the exchange rate remains overvalued. A recent analysis of 18 developing countries over the period 1975-84 shows that the indirect taxation burdens caused by exchange rate overvaluation and industrial protection policies were generally much more important than the difference between domestic and international agricultural prices (see Box 3.4).

---

[8] S.K. Kumar. 1988. Effect of seasonal food shortage on agricultural production in Zambia. In *World Development*, 16 : 9 (1051-1063).

Finally, the effects of price adjustments are likely to have a greater impact on larger farmers. Smallholders tend to produce food and consume most of it, particularly in the low-income developing countries. Moreover, empirical evidence suggests that smallholders are severely limited in their ability to expand output, owing to a lack of land in some cases and a lack of labour in others. Apart from land reform, overcoming these constraints requires access to credit and appropriate technology. However, adjustment programmes usually implement a policy of monetary austerity by raising real interest rates, making access to credit more difficult for smallholders.

Another important supply constraint is the time available to women for carrying out their tasks in crop production. The case of weeding is illustrative. Timely weeding is critical for raising yields in the short term without requiring complementary inputs. Weeding is sometimes done by men but more commonly by women, even on crops managed by men. The weeding period coincides both with the rainy season when health hazards are highest, and the "lean period" when food stocks are lowest. In order to spend more time on weeding, women have to reduce time spent on child care, cooking, collecting firewood and water, and other household tasks.[9] Women may be unable, or unwilling, to reduce further the time spent on these tasks, essential for the maintenance of health and nutrition standards, in order to devote more time to weeding.

The difficult trade-off between different uses of women's time may itself be exacerbated by non-price features of the adjustment package. Reductions in social expenditure may throw more health and nutrition maintenance burdens on rural women, whose extremely long working day has been well documented. Rural women usually have little or no spare time; they can devote more time to crop or livestock production only by reducing time spent on child care and the maintenance of family health and nutrition.[10]

Finally, when analysing possible factors that enhance or weaken the effects of price incentives on agricultural production, it is useful to keep in mind that agricultural producers face circumstances quite different in several respects from those of other producers. No doubt, manufacturers, miners, and commercial agents face uncertain demand; competition; changing regulations; and, if they are employers, interruptions as a result of strikes. However, these sources of risk and uncertainty, which are also present in agriculture, pale by comparison with the greatest source of shocks: nature. A flood can damage a factory or a drought can close down a paper mill for lack of water. These are unusual and insurable events for manufacturing; in agriculture, however, such events are so frequent as to haunt continuously the decisions of the producer.

Particularly for the small- and medium-sized cultivators, fishermen, and pastoralists in developing countries, the relationship with nature is a struggle for survival that requires constant vigilance. Therefore, one should distinguish between the rational behaviour of producers in non-agricultural pursuits where uncertainty and risk are important considerations, and in agriculture where they predominate. This difference makes the farmer more prone to risk-aversion and more likely to put greater emphasis on survival. These in turn affect the impact of short-term market signals on farmers' decisions. Hence, the results expected from policy reforms under structural adjustment programmes, insofar as they rely on changing price structures, may take appreciably longer to materialize in the agricultural sector than in other sectors.

Producer and consumer subsidies. Fifteen of the 21 SECALs analysed by the World Bank included input pricing conditions, although complete price decontrol or an end to government involvement was rare.[11] This situation perhaps reflects the reluctance of governments to eliminate producer subsidies on crucial inputs, particularly fertilizer, water and credit. The motives for removing producer

---

[9] R. Chambers *et al.* 1981. *Seasonal dimensions to rural poverty.* London, Francis Pinter.

[10] An example from a recent Zambian study shows how this may constrain agricultural supply response. As part of the stablization and structural adjustment measures, real per caput public expenditure on health provision was cut by 16 percent between 1983 and 1985. The result has been to shift more of the burden of health care on to women. Shortages of equipment and medical personnel in hospitals mean that women have to accompany other family members who are hospitalized to provide meals and care during their treatment. One woman reported missing the entire planting season for this reason (A. Evans and K. Young. 1958. *Gender issues in household labour allocation: the case of northern province, Zambia.* Research Report, ODA/ESCOR).

subsidies are usually to reduce government budget outlays and allocate resources more efficiently.

Fertilizer is a frequently subsidized input and the cost of the subsidy is usually small in relation to the total government budget. One argument in favour of a fertilizer subsidy is that, given the limitations of price or other incentives for expanding agricultural productivity/supply in many developing countries, it is justifiable to encourage the use of a specific input — such as fertilizers. Other arguments include lowering private costs to reduce the risk premium attached to innovations, and the complementary view that fertilizer is necessary to overcome land constraints. On the other hand, it has sometimes been claimed that subsidized fertilizers do not reach smaller farmers and give rise to black marketeering and illegal cross-border trade. In any event, the general effect of increases in fertilizer prices resulting from the removal or reduction of subsidies is to turn the terms of trade against farmers and, at a time of overall credit restraint, this may lead to significant reductions in fertilizer use. The result may be a conflict between structural adjustment objectives, between "correction" of relative prices and increasing agricultural output and exports. At the least, a removal of subsidies may need to be phased with targeted credit allocations to maintain input use, particularly fertilizers which, at the low levels of application commonly found in developing countries, are environmentally beneficial.

Credit subsidies are also often discouraged in adjustment programmes. Eliminating credit subsidies may hurt precisely those producers whom the adjustment package seeks to encourage: modernizing and innovative commercial farms. If removing credit subsidies leads to a more efficient use and allocation of the available credit, it is to be welcomed, but it involves more expensive credit to producers. If combined with a policy of monetary restraint that results in the reduction of the general supply of credit, the elimination of credit subsidies can be quite debilitating to agriculture. This is particularly the case for private and public marketing agencies that, in many countries, advance partial payment to farmers for the coming harvest. Higher real interest rates — frequently a goal of structural adjustment — can reduce the effect of higher crop prices at least in the short term. On the other hand, they can induce, over the longer term, the mobilization of national and rural savings and promote the creation of savings and credit organizations and cooperatives.

Several SALs and SECALs have called for the introduction of user charges for government services, previously supplied free of charge or at highly subsidized rates. Some of these services, such as the use of chemicals and drugs to treat and contain plant and animal pests and contagious diseases, might be considered public goods because of the externalities arising from non-treatment and the spread of the disease. In such cases, subsidies promote efficiency by bringing social benefits in line with social costs. Charging for other services might require administrative costs that could swamp the revenue generated by the charging scheme. Thus, a pragmatic approach is required with respect to user charges.

On the consumer side, food subsidies are a major item of government expenditure in some countries and thus are an important source of fiscal deficits. However, it would be naïve to ignore the danger that removing these subsidies can create severe political unrest. Food subsidies provide a clear example that all structural adjustment packages have implications for income distribution, for they are concerned with the reallocation and more effective use of a nation's resources. Adjustment brings "winners" and "losers" as does the period leading up to the crisis that provoked the need for adjustment. The policy challenge is to find methods of dealing with food subsidies that protect the most vulnerable people and promote economic efficiency, but that also minimize political instability. Steps taken in this direction could include targeted subsidies, food-for-work programmes and increased provision of educational and basic social services.

Institutional reform. In the original strategy for policy-based lending, SALs or broad-based SECALs were initially used to stabilize the economy and thereby to create a suitable environment in which sectoral policies would reinforce the policy changes of the stabilization phase. The intention was to "institutionalize" the policy process in order to create a foundation

[11] There have been cases of conditionality requiring the complete elimination of subsidies and price interventions for particular markets. Arrangements between the World Bank and some African governments are examples, sometimes as part of a pre-funding or "shadow" programme.

on which the economy could maintain stability, achieve long-term growth and respond more flexibly to external shocks. For the World Bank, a major aspect of policy regarding public institutions has not been their strengthening as such, but rather the transfer of their functions to the private sector. This approach would seem to be part of a more general aim to create a more favourable economic and political context for private investment.

In light of this approach, agricultural sector parastatals are frequently cited as causes of inefficiency in marketing and production. They often tax farmers indirectly by absorbing a large part of the prices of the commodities they handle. In some countries, the parastatal sector has generated large financial losses and is a major source of the government deficit. As a result, several SECALs have required the introduction of private-sector competition or even the closure of parastatals. With some exceptions, progress in improving efficiency or encouraging effective private-sector involvement has been limited to date, which can be attributed to a variety of economic, institutional and political causes. By their nature, state marketing agencies rarely concern themselves with commercial activities alone. Usually, they are also interventionist agencies through which government policies of resource allocation, employment absorption and income distribution are implemented. Very often, the inefficiencies and fiscal losses attributed to parastatal organizations are the outcome of implementing these wider objectives rather than performing marketing objectives. For example, food crop parastatals have been required to finance strategic reserves of basic staples from their own resources, or to operate within a marketing margin which would generate losses for the most efficiently run organization.

Four questions should be addressed to disentangle the controversial issues related to parastatals: i) what are the objectives of the marketing system; ii) which are the most cost-effective means of achieving these objectives; iii) what institutional structure, combining public and private agencies, is most appropriate for achieving these objectives in a cost-effective manner; and iv) how can the transition from the existing to the proposed structure be most effectively achieved?

Some governments have agreed to the reorganization, abolition, or divestiture of parastatals as conditionality for a SAL or SECAL. In many instances, however, there has been reluctance because of divergent views on the part of the potential lenders and the recipient government. In some cases, this may derive from political considerations, but it is important to recognize the genuine differences of views on the appropriate role of a marketing organization and the ability of private agents to perform their functions adequately. A lender would be loath to encourage parastatals if they hindered the overall adjustment and growth of the agricultural sector. At the same time, the reasons advanced for retaining a state monopoly marketing structure require careful consideration, particularly for small countries in which the private marketing structure itself is, or would be, monopolistic and promises few of the efficiency gains of competitive markets.

Major policy objectives of many governments include food security and price stability. Food security is frequently attempted through a parastatal marketing agency with monopolistic power over the import or export of major staple foods as well as being the sole domestic buyer. Such agencies typically operate a buffer stock, buying and selling domestically at predetermined prices. While parastatals have had some success in import-export trade, they have been much less effective in the marketing and processing of foodstuffs. The reasons for failure can be traced to lack of financial resources, lack of infrastructure, lack of clear policy directives, lack of internal and regional consistency in pricing and other policies that parastatals are called to implement, and involvement in non-commercial but socially desirable marketing activities. Continued informal trade in foodstuffs, both within as well as across borders, have contributed to their ineffectiveness.[12] Thus, privatization of those parastatals frequently involves little more than a formal acknowledgement of the private sector's dominant position. Sub-Saharan African and Central American countries, where rural markets are dispersed and transport infrastructure limited, are clear examples. On the other hand, the privatization of imports and exports can lead to monopolization by the private sector. This is particularly the case in small countries

[12] For further discussion see: FAO, *Food Security Assistance Programme: preparation of comprehensive national food security programmes: overall approach and issues.* Document prepared for the Second Ad Hoc Consultation with FSAS Donors. Rome, 27 October 1989.

with a legacy from the colonial period of a few dominant trading houses.

Ghana and Nigeria are examples of two different outcomes of parastatal reforms. In Ghana, the government drastically reformed the cocoa marketing parastatal (COCOBOD) by reducing staff and streamlining operations. In Nigeria, the abolition of the government cocoa marketing board led to numerous problems, the worst being a drastic drop in the quality of exported cocoa.

In terms of food security, experience in some developing countries indicates that efforts to abolish government involvement in agricultural services has in several cases worsened access to food as a result of artificial food scarcities created by hoarding and speculative activities of private traders. In some countries, especially in sub-Saharan Africa, some public intervention is deemed necessary for managing occasional surpluses to avoid negative effects on prices and production, and to assure distribution of food supplies to groups at risk in years of shortages.[13] Given the complexity of determining the correct mix of public agencies and private traders, a pragmatic approach is required in which future objectives are clearly defined and other forms of organization such as producer organizations and cooperatives are evaluated.

Another government motive for retaining control of marketing systems relates to credit, input supply, and maintenance of quality controls. With single-channel marketing, particularly of export crops, it is possible to lend seasonal credit against the security of crop deliveries with no other form of collateral — such as land titles — required. It is argued that this system would be unworkable with private traders. Experience suggests, however, that credit can be successfully channelled through traders who would be in personal contact with the farmers they are leading to. In the long term, a marketing system based on traders could prove to be more effective than relying on public agencies, but it is likely to require considerable time to evolve. There is also the argument that private traders would be unwilling or unable to distribute the limited and variable quantities of agricultural inputs required by farmers. However, input provision could be combined with an effective private credit system as suggested above. Similarly, introducing private marketing agents means that the nature of the government's regulatory functions changes. In principle, there is no reason why governments should liberalize quality standards: rather in many cases they should insist on higher standards.

[13] For instance, a SAL for Madagascar began a staged process aimed at encouraging the private sector to purchase and store rice, on the premise that a well-functioning storage market will stabilize prices optimally. As a first step, the government withdrew from domestic purchases and sales but operated a buffer stock of imported rice to prevent large speculative price increases.

PROBLEMS OF IMPLEMENTING STRUCTURAL ADJUSTMENT PROGRAMMES

Disappointing performance

After almost a decade of experience, it is difficult to avoid the conclusion that structural adjustment programmes (SAPs) have been associated with uneven economic performances within countries. Some argue that this inconsistency lies in the weakness of the theoretical underpinnings of the policy packages themselves.[14] Others, on a more practical level, have argued that various governments have implemented SAPs in very different ways and have lacked a thorough understanding of the consequences of enforcing policy measures in different orders or "sequencing".[15]

A recent FAO study of four sub-Saharan African countries reveals that in only one has careful consideration been given to policy sequencing before the implementation of the structural adjustment programme.[16]

Among others, the FAO policy review identified the following key issues: i) the sequencing of stabilization policies *vis-à-vis* policies of structural change; ii) the sequencing of liberalization measures across sectors; and iii) the sequencing of liberalization measures within agriculture itself.

While problems of sequencing are to a great extent specific to each country, an analysis of the general nature of the problem is both possible and fruitful.

Stabilization and structural sequencing

As working definitions, it may be recalled that "stabilization policy" refers to monetary and fiscal measures taken to restrict aggregate demand in order to control inflation and correct an unsustainable balance-of-payments situation; and that "structural adjustment policy" refers to measures that act on the supply side either directly, such as through agrarian reform or improving input supply, or indirectly — through market liberalization, for example. Broadly speaking, policy sequencing concerns the order in which these policy programmes are implemented. It is commonly perceived that stabilization measures should precede supply-side measures for a successful adjustment-with-growth programme.

Supply-side measures take time to produce changes in supply and, unless demand is restrained, such measures would result in an unmanageable current account deficit. Again, stabilization measures aim at restoring external and internal balances, and devaluation of the exchange rate is a major instrument toward this end. To sustain the exchange rate adjustment, appropriate monetary and fiscal policies have to be implemented. As this process proceeds, economic stability is essential for growth in savings and investment and for controlling inflation. Moreover, stabilization requires contractionary pressure on aggregate demand, and trade liberalization applies pressure on industries competing with imports. Simultaneous application of the two may increase business failure and unemployment and generate strong political opposition.

The sequencing debate has centred around the importance of stabilization programmes preceding structural adjustment measures and the need for stabilization policies to accompany supply-side measures during the adjustment process. In practice, there has been little disagreement over sequencing in this sense. Macro-economic reforms must be followed by certain elements of supply-side policies at the sector level, essential for ensuring an effective micro-economic response to the macro-economic incentive structure put in place. It is at this micro-economic level that major weaknesses have been identified in the sequencing of adjustment policies.

Sequencing of liberalization

The case for liberalization identifies government intervention in markets to be the cause of inefficiencies ("distortions"), and goes on to argue that ending these interventions improves resource allocation after the economy has fully adjusted. The sequencing issue arises because of the familiar problem that removing one intervention in an economy in which there are many others may worsen rather than improve

---

[14] The theoretical issues are discussed in A. Fishlow, The state of Latin American economics. In *Economic and social progress in Latin America, 1985 Report.* Washington, Inter-American Development Bank, 1985.

[15] Sequencing is treated in a non-technical way in J. Epstein, 1989. Old wine in new bottles: policy-based lending in the 1980s. *In* J. Weeks, ed., *Debt disaster?* New York, New York University Press.

[16] The study covered Ghana, the Gambia, Kenya and Malawi. N. Spooner and L.D. Smith, *Structural adjustment and policy sequencing in sub-Saharan Africa.* July 1989. Rome, FAO.

the allocation of resources. Economic theory offers little guidance on an optimal sequence for removing market distortions. Nevertheless, some broad conclusions can be derived from the general principle that the objective of structural adjustment is to achieve a reasonable and sustainable rate of growth.

Liberalization can be analysed under five categories: i) the domestic market for goods and services; ii) the external market for goods and services; iii) the domestic market for factors of production — land and labour; iv) domestic money markets; and v) the link between these markets and the international financial system.

Much of the analysis of sequencing focuses on the liberalization of external accounts. It is usually argued that the current account should be liberalized first, and the capital account later. A more practical explanation is that early liberalization of the capital account when the economy has yet to stabilize results in disruptive capital flight. For the same reason, the domestic money market should be liberalized before the external capital account in order to bring domestic interest rates in line with international rates.[17] However, liberalization of the domestic money market requires repressing inflationary pressures, which invariably means reducing the fiscal deficit. Thus, one arrives at a general rule for policy sequencing: liberalization of the domestic money market requires removing exchange controls on capital movements, which in turn requires establishing the fiscal deficit at the desired (non-inflationary) level.

This apparently commonsensical generalization is of little relevance for countries with very underdeveloped money markets, which is the case for virtually all of sub-Saharan Africa, many Asian countries and parts of Latin America and the Caribbean. In small countries particularly, but also in large low-income countries, money markets are very narrow and inefficient because of the limited range and quantity of financial assets and the small number of traders. In these countries, formal financial markets do not play a significant role in valuing the real capital stock of the country, most of which is not traded domestically because it is not actually a commodity (e.g. common property agricultural land in much of Africa) or because it is traded on foreign markets (e.g. local assets of international firms). Furthermore, the competitive efficiency of foreign exchange markets can be severely compromised by the presence of a few private financial institutions, usually with international links, that dominate trading.

More relevant to countries across the range of development is the liberalization of commodity markets. Comparatively little emphasis has been given to the sequencing of domestic market liberalization compared to liberalizing economic relations with the rest of the world. But some steps must be taken to liberalize key domestic markets and allow export-increasing incentives to change the structure of production. Without first generating export growth, trade liberalization will put pressure on foreign exchange reserves and the process will be unsustainable. However, increased export production initially requires increased imports of essential inputs.

FAO analyses have revealed contrasting experiences in the management of trade liberalization. In Kenya, for instance, trade liberalization was the major element of the first SAL but the foreign exchange requirements associated with it became unsustainable and the liberalization process had to be abandoned. Ghana, on the other hand, allocated foreign exchange according to availability through auction, and the types of goods eligible for entry to the auction expanded over time in line with the rising supply of foreign exchange.

Since the purpose of adjustment is the reallocation of resources, flexible factor markets must be a part of general liberalization. With regard to labour markets, this flexibility would seem to be present throughout the developing world beyond the needs of adjustment. In the 1980s, real wages in Africa, Asia and Latin America have been flexible downwards, in many cases to an extraordinary degree.[18]

Sequencing of agricultural liberalization

Given the importance of foreign exchange availability, and given that the agricultural sector is the most important sector for foreign exchange generation in many low-income economies, the stimulation of agricultural production is required at an early stage in the reform programme. This implies that

[17] S. Edwards, Sequencing and economic liberalization in developing countries. In *Finance and Development*, 24:1(1987)

[18] For Africa see V. Jamal and J. Weeks. 1987. Rural-urban income trends in sub-Saharan Africa. In *World Employment Programme Working Paper Series*, Working Paper 18. Geneva, Labour Market and Employment Planning Section.

stabilization measures must take due account of the sector's resource needs. If agriculture is the lead sector for export expansion, the subsectors where conditions favour an early expansion in production must be identified and their constraints to growth in production removed. For example, a large proportion of the higher domestic prices resulting from devaluation will have to be passed on to producers and be accompanied by other non-price policies which allow the agricultural (export) sector to respond to increased prices. At the same time, credit restrictions, which are also an essential element of the stabilization phase, must be well managed if private investment is not to be constrained.

If the assumptions about agricultural sector reforms are correct, then the benefits to agriculture should occur at a relatively early stage and precede trade liberalization in the sequencing of policy. However, FAO studies suggest that agricultural liberalization has not always been given sufficient priority. In Kenya, agricultural reforms were absent from the first SAL programme; and in Malawi, the adjustment programme focused on the smallholder subsector which accounted for only a minor proportion of foreign exchange earnings.

In practice, the benefits of liberalization and increased participation by private traders in agricultural input and output markets can only come about if, as discussed earlier, domestic financial markets are sufficiently developed. In fact, improvement in financial intermediation has been a relatively neglected element of adjustment programmes. Recognizing this, the World Bank has begun to emphasize financial reforms in the latest SAL programmes developed in West Africa. But this emphasis clearly represents a suboptimal pattern of sequencing, particularly given the long time frame required for some of the financial sector reforms. In the meantime, it can be expected that lack of finance will constrain the response of private agents to the new incentive frameworks established in Africa.

On the basis of the foregoing discussion, a recommended sequencing can be formulated:[19] i) relax controls on imported inputs essential to expanding production; ii) devalue to achieve a competitive exchange rate, which simultaneously requires monetary and fiscal restraint to curb inflation and convert a nominal devaluation to a real devaluation; iii) allocate foreign exchange to the maintenance and repair of infrastructure required to facilitate production increases; iv) initiate policies to make credit markets more responsive to needs of producers; v) stimulate agricultural production by reducing market interventions. Trade liberalization then becomes the sixth phase of the process, eventually followed by the relaxation of controls on external capital flows. It should be noted that the first and second phases require either initial foreign exchange reserves (a happy but unlikely circumstance) or early inflows of foreign aid.

[19] This sequence is a modified version of that found in J. Williamson. 1987. Discussion on Session One: Bank-Fund Papers. *In* V. Corbo, M. Goldstein and M. Khan, eds., *Growth-oriented adjustment programmes*, Washington, IMF and World Bank.

The fallacy of composition

The central element in the liberalization sequence is the reduction of trade restrictions and devaluation. The role of devaluation is to increase the production of tradables with particular emphasis placed on the expansion of exports. However, the proposition that devaluation, even if sustained in real terms, would substantially increase export volumes has been questioned. One of the more important criticisms, called "the fallacy of composition", is based on external demand-side constraints; that is, that markets would be incapable of absorbing the volume of exports of a commodity without a significant fall in its price if a large — or sufficient — number of countries producing the same commodity jointly undertook such an export-led structural adjustment strategy.[20] What is allegedly true for one country may not be true for many taken together (see Box 3.5). Such a concern is apart from that arising from possible constraints on the supply-side; that is, that the agricultural sector may have but a limited capacity to respond to changes in the structure of incentives, particularly in the short term.

The likelihood of a wave of devaluations resulting in small or even negative foreign exchange gains for all the devaluing countries taken together depends on several factors: i) the rate of growth in world demand for each product; ii) the elasticity of world demand for each product with respect to its price;

[20] See, for example, J.N. Bhagwati, Export promoting trade strategy: issues and evidence. In *OECD Research Observer*, 3:1(1988); and Fishlow, 1985, op. cit.

iii) the market shares of the devaluing countries; and iv) the reaction of countries to loss of export shares.

Promoting exports is obviously facilitated if world demand is reasonably buoyant. During the 1980s, this has not been the case for several agricultural commodities important to developing countries. Annual growth rates in world demand during 1980-90 were estimated at 1.6 percent for coffee and cocoa, 0.7 percent for bananas and 1.7 percent for cotton.[21] Indeed, a typical problem facing African countries attempting to undergo structural adjustment has been to arrest the decline in their share of a market which is growing slowly at best. In such a case, stabilizing the market share would be a sign of success in promoting exports. A liberalization of agricultural imports on the part of the major and mainly developed importing countries would assist in expanding market opportunities, even if the underlying demand growth was weak. Removal of tariffs which discriminate against attempts to diversify into export-oriented agro-industries would also stimulate developing country exports as a whole.

[21] *FAO agricultural commodity projections to 1990.* FAO Economic and Social Development Paper 62. (reprint)

It cannot be stressed too much that the export markets faced by developing countries are far from efficient conveyors of allocative signals. On the contrary, they are often severely distorted by protectionist policies, particularly those of developed countries.

Low price elasticity of demand for many agricultural products lies at the heart of the fallacy of composition argument. Price elasticities are rather low for the quantitatively more important tropical products, namely tea, cocoa and coffee. Here again, protectionism plays an important role: for example, it is capable of cancelling the advantages of high price elasticity characteristic of horticultural products. During the 1980s, producers in Central America were encouraged by bilateral trade arrangements to export ornamental flowers, which apparently enjoy both high price and income elasticities, only to suffer subsequent trade restrictions.

The fallacy of composition is likely to apply to those commodities for which a few countries

BOX 3.5

The fallacy of composition

The premise of the fallacy of composition is that the exports of developing countries, particularly but not exclusively in Africa, are concentrated in a narrow range of primary agricultural commodities. For example, coffee, cocoa and tea accounted for 61 percent of the agricultural exports of sub-Saharan Africa during the period 1981-84.[1]

The argument applies in principle to all exports, but is generally thought less relevant to manufactures because these are characterized by greater product differentiation. Over the last three decades, the composition of the exports of middle-income, oil-importing countries changed dramatically: from 1960 to 1981 the share of manufactures rose from 17 percent to 59 percent, leaving non-energy primary products considerably less than half the total. For the low-income countries (apart from China and India), primary products were 91 percent of the total export value in 1981 (down from 95 percent in 1960). The relative decline of non-oil primary products in the exports of developing countries demonstrates a trade diversification which should reduce vulnerability to the fallacy of composition effect. However, one must be cautious because manufactures are not immune to suffering from world market gluts. Textiles, for example, are a highly differentiated commodity by some criteria; however, if all developing country textile exporters devalued simultaneously, it is not certain

[1] G. Koester, H. Schafer and A. Valdés, External demand constraints for agricultural exports: an impediment to structural adjustment policies in sub-Saharan Africa. In *Food Policy*, August 1989, on which much of this discussion is based.

have substantial world market shares and which, at the same time, account for a major proportion of the exporting country's foreign exchange earnings. Table 3.2 shows that among three major groups of exporting developing countries — Africa, South America and Asia — during the mid-1980s only cocoa (Africa), coffee (Africa and South America), tea (Africa and Asia) and cotton (Africa and Asia) satisfied both criteria: that is, the regions had significant shares of these world markets and the products had significant shares of the regions' agricultural exports. Groundnut oil and sisal were marginal in this respect. Some commodities satisfied the first criterion but not the second. For example, Africa had a 68 percent share of the small global market for palm kernels in 1985-87 but such earnings constituted barely 0.1 percent of Africa's total agricultural exports during those years.

Competing countries may react in one of two ways when confronted by falling prices resulting from increased supplies. They may attempt to retain market shares or diversify their exports and move out of this particular market. Several developing countries in recent years have managed successfully to diversify their export portfolio and, in some instances, have moved into manufactures.

From the perspectives of market share and export importance the fallacy of composition argument would seem to apply only to cocoa and, perhaps, to coffee. Possibly the greatest danger to adjusting countries' attempts to expand their agricultural exports is increased competition from countries investing heavily in new crop production and processing capacity. Such investment not only leads to increased supplies, but also to lower production costs. As long as marginal revenue does not become negative, however, it may still be worthwhile for a country to continue to expand its exports of the commodity. However, export subsidies, explicit or implicit, may be required to induce producers to continue producing despite falling prices.

The reliance of SAPs on devaluation raises a broader issue of "composition" that needs to be addressed by the international community; namely the extent to which an export-oriented strategy for all developing countries (as implied by multilateral-financed programmes) fosters a

---

that the *ex post* price elasticity of demand for textiles would be very great in the aggregate, particularly in light of developed country protectionism.

The source of the "fallacy" can be shown by a hypothetical example. Assume that ten countries devalue simultaneously and the effect in each country is to induce producers to increase output by 10 percent. Now, let us consider two cases. In the first, each of the ten countries exports one product which is different from the product exported by the other nine, and each country accounts for one-tenth of the world market for its product. Each country would increase the world supply of its export by only 1 percent (10 percent of 10 percent). In this case, if demand remains unchanged, one would expect a small drop in the world price of each product; and when this small decline returned to confront the producers, they would probably be content with having raised output by 10 percent.

In the other case, assume each country exports the same product. A 10 percent increase in output by each country would result in a 10 percent increase in world supply. At worst (from the exporters' point of view), the world price might fall so much as to reduce total export earnings, so the collective devaluation would have been self-defeating (in such a case, demand is said to be "inelastic"). Less extreme, but still serious, total export revenue might rise while the price decline communicated back to the producers might induce them to cut back on their initial output expansion because the price has fallen relative to costs. In this situation, even though demand is "elastic" (export revenue rises as the price falls), the net increase in export value may be quite small once all producers have adjusted and the world price has reached a new equilibrium. The second case can be cited as a fallacy of composition: while each producer individually sought to increase output by 10 percent, the collective effect was that each discovered it profitable to do so by less than 10 percent because of the world price fall.

TABLE 3.2 **Regional shares of world exports of selected commodities and shares of these commodities in regional agricultural export values, 1985-87**

Commodity	World exports	Regional shares of world exports			Commodity shares of regional agricultural exports		
		Africa	South America	Asia	Africa	South America	Asia
	$ million	(		%			)
Cocoa beans	**3 008**	**62**	**14**	**12**	**16**	**2**	**1**
Coffee	**12 307**	**21**	**38**	**10**	**22**	**25**	**4**
Tea	**2 194**	**15**	**2**	**68**	**3**	**0.2**	**4**
Palm kernels	**16**	**68**	**9**	**15**	**0.1**	—	—
Groundnuts	**564**	**7**	**11**	**41**	**0.4**	**0.3**	**0.7**
Groundnut oil	**242**	**28**	**22**	**23**	**0.6**	**0.3**	**0.1**
Cotton	**5 992**	**18**	**5**	**25**	**9**	**2**	**4**
Tobacco	**3 963**	**11**	**13**	**17**	**4**	**3**	**2**
Sisal	**57**	**38**	**44**	**4**	**0.2**	**0.1**	—

Source: ***FAO Trade Yearbook 1987*****, vol. 41. Rome, FAO, 1988. Adapted from Koester *et al.*, 1989, op. cit., Table 2.**

cooperative solution to the cause of development. According to the latest World Bank figures, the exports of low- and middle-income countries grew in real terms at an annual average of approximately 5 percent and 7 percent respectively during the period 1982-1988.[22] There is clear evidence that these rates of export expansion, achieved during the longest post-war period of sustained growth in the developed countries, are insufficient to service debts and raise per caput incomes, except for in some countries — mainly in Asia. The 1990s are unlikely to be more favourable than the 1980s to developing country exports. Few would expect the growth of the OECD countries to exceed the average attained during the sustained expansion of the 1980s, and this situation would not bring stronger export demand. If the 1990s bring a liberalization of developed country markets with respect to developing country products, the harbinger of this change has yet to show. Yet with each passing year, loan conditionality for trade liberalization affects more developing countries.

Steady growth in developed countries with no change in protectionism, combined with increased export efforts by a growing number of developing countries, could possibly diminish the expansion of foreign exchange revenues for the aggregate of developing country exports. In such a case, the developing countries' liberalization efforts could lead to a continuous struggle to redistribute market shares. Such a situation would be analogous to the competitive devaluations by the developed countries during the 1930s, widely condemned as "beggar thy neighbour" policies. It is precisely to forestall such a destructive policy conflict that the United Nations and its related agencies were established after the Second World War. In keeping with this objective, the multilateral agencies and the development community at large need to verify the contribution of devaluation to the general welfare of developing countries. While the expertise and information exist to do so, little more than hopeful opinion has been offered to support the policy of spreading currency devaluation throughout the developing world. The fallacy of composition raises the serious possibility that arguments valid for one country do not generally apply to many countries taken together.

[22] *World Development Report 1986*, p. 158. Washington, World Bank, 1986; *World Development Report 1989*, p. 150. Washington, World Bank, 1989.

SOCIAL ISSUES IN ADJUSTMENT PROGRAMMES

Food buyers and food sellers

Structural adjustment programmes seek to reallocate resources and, in so doing, redistribute income. Whether the effect of this redistribution is toward more or less inequality and what its impact is on the poorest populations are matters of intense controversy. The focus of this section is the consequence of adjustment for the rural poor, who make up the majority of the poor in the low-income countries. Generalizations in the absence of systematically collected data must be ventured cautiously. As a first step, it is useful to divide the population in a manner relevant to the policy measures discussed previously.

A central goal of adjustment is to raise the return on tradables which, if successful, implies an increase in prices of tradable food items in relation to the general price level, although not necessarily in relation to all other tradables. A relative rise in food prices, explicitly advocated in some adjustment programmes, is sometimes alleged to carry with it a positive income-distribution effect given that most of the poor are rural and food production is their main economic activity.

This argument leads to an important question: other factors remaining equal, does an increase in food prices make most people better or worse off? In terms of direct welfare impact, this may be the most important policy question facing the majority of governments. The initial effect of an increase in food prices is to raise the real incomes of net food sellers and lower the incomes of net food buyers. While subsequent ramifications — "general equilibrium" effects, including the diffusion of higher incomes for food producers, increased employment opportunities and real wage increases — may soften this effect, they cannot counteract it if, indeed, food prices rise in relation to other prices.[23] Thus, in a country where the majority of the population is net food producing, an increase in food prices makes the majority better off, and vice versa.

In a successful adjustment scenario, a conflict between food buyers and sellers may be avoided. Adequate farm incentives and more efficient marketing systems may bring about an increase in productivity and output, causing prices to fall without detriment to producers' incomes, and to the benefit of consumers. However, this discussion focuses only on the first-round effects of food prices, which actually increase faster than other prices, and does not take into account subsequent developments.

Table 3.3 provides estimates of the proportion of net food buyers for 65 countries divided into four regions, for the years 1970, 1980 and 1985. A net food buyer is simplistically defined as anyone not engaged in agriculture. Therefore, the proportions underestimate in that they presume all farm households are at least self-sufficient in food. For Latin America and parts of Asia, this is not necessarily the case. The figures in Table 3.3 demonstrate quite a significant change: in 1970, only 20 of the 65 countries had a non-agricultural majority, but by 1985, the number had risen to 31. Given the underestimation inherent in this measure, it is safe to conclude that in well over one-half of the countries, net food buyers were a majority by the mid-1980s. Only in sub-Saharan Africa was the population still overwhelmingly agricultural by then — Mauritius is the single exception in the region.[24] Even in this region, however, several countries in the table will cross the 50 percent mark in the 1990s with regard to non-agricultural population, with likely candidates being Nigeria, Benin, Côte d'Ivoire, and Cameroon.

The information in Table 3.3 indicates that, for the majority of developing countries outside sub-Saharan Africa, increases in food prices would make most people worse off, at least in the short term, if other all factors remain the same. Furthermore, in these countries increases in food prices almost certainly affect the poor more than the non-poor.

There are a number of mitigating possibilities, however. First, increased food prices may result in higher wages and/or more employment in the food crop sector, thereby transferring income to net food buyers. This effect is probably confined to the farm household

---

[23] P. Hazel. 1987. *Agricultural growth linkages and the alleviation of rural poverty.* New York, Cornell University Press.

[24] A more detailed analysis suggests that the poor in sub-Saharan Africa are not net producers of food to the extent that Table 3.3 would imply. See M.T. Weber *et al.,* Enforcing food security decisions in Africa: empirical analysis and policy dialogue. In *American Journal of Agricultural Economics,* 70:5(1988).

TABLE 3.3 **Non-agricultural population as a share of total population (net food buyers), 1970, 1980, 1985**[1]

Region	Number of countries	Net food buyers[2]			Number of countries with net food buyers above 50% of total		
		1970	1980	1985	1970	1980	1985
		(.............. % ....)					
Near East & North Africa	**9**	**49.1**	**61.3**	**65.2**	**4**	**7**	**8**
Asia	**8**	**40.4**	**47.6**	**51.2**	**1**	**3**	**5**
Latin America & Caribbean	**20**	**58.1**	**65.0**	**67.8**	**13**	**16**	**17**
Sub-Saharan Africa	**28**	**18.3**	**23.9**	**26.4**	**1**	**1**	**1**
	65	**37.2**	**44.5**	**48.7**	**19**	**27**	**31**

[1] The 65 countries in this sample are shown in Part Three, II (Table 3.5).
[2] Simple average of country percentages.
Source: FAO.

population, as measured in the table, given that much of the agricultural wage labour force in Latin America and Asia is made up of net food-buying farm families (microfarms or "minifundistas").

A second mitigating effect is the possibility that increased food prices in the short term will prompt investment and technical change, thereby generating lower food prices in the long term. Empirical evidence suggests that food supply is relatively price-elastic in the long term in most developing countries, but supply is not expanded by raising prices and waiting hopefully. It should be ensured that the adjustment programme that calls forth the higher prices does not cut public expenditure and restrict credit so that the desired supply response does in fact occur in the medium and long term. Even if the eventual supply response is favourable, there still remains the transitional period of high food prices that hurt the poor in most countries outside of sub-Saharan Africa. Therefore, during the transitional period there is the need to adopt measures to aid the poor with the same dedication and vigour used to implement market liberalization and fiscal austerity. A possible measure might be the introduction of programmes that use food aid to target nutritionally-vulnerable households and population groups, both in rural and urban areas. In any event, the high likelihood of a future decline in food prices should not prompt a complaisant dismissal of the certainty of increased suffering by the poor in the short term.

As regards the nutritional impact of higher food prices, the reduction in nutrient consumption may be somewhat mitigated by the substitution of cheap nutrients for expensive ones. Still, the overall negative effect may be significant if other desirable food attributes are foregone, as one would expect to be the case. Further, gender- and birth-order preferences in nutrient allocation may expose the most vulnerable family members, often young girls, to greater risk of malnutrition.[25] In some cases, expenditure-switching policies may reduce domestic food production, namely where the output mix switches in favour of export crops and away from food staples. The effects on food availability at the national level will depend on whether or not increased export earnings enhance the country's ability to increase imports of food.

At the same time, the impact on household food security depends on the level of farm income and the control of its use. In any consideration of impact of price changes on household food security, it is also important to recognize that there are differences in distributional effects within households. Different members of the same household may be subject to different standards of living. While some pooling and sharing of resources takes place within households, this is far from complete and, in periods of stress, more vulnerable household members may suffer disproportionately. This is indicated by recent research showing that in a wide variety of settings in many developing as well as developed countries, men tend to retain a considerable portion of income accruing to them for discretionary spending on personal consumption and leisure activities, whereas income accruing to women is almost exclusively used to meet collective household needs.[26] Men and women frequently have different expenditure priorities and women most often have responsibility for provisioning the

[25] Overall and intrahousehold nutrition are treated in an FAO study: R. Gaiha, *Structural adjustment and rural poverty in developing countries.* 1989. Rome, FAO.
[26] J. Bruce and D. Dwyer, ed. 1988. *A home divided: women and income in the Third World.* Stanford, Stanford University Press.

household while lacking access to the required resources. In addition, these intrahousehold distributional differences suggest that household income by itself is, in many cases, inadequate as an indicator of an individual household member's welfare.

Structural adjustment and the rural poor

The continued and widespread process of urbanization together with falling urban (especially wage) real incomes in many countries has reduced the rural sector's disproportionate share of poverty in developing countries during the 1980s. Nevertheless, in the majority of developing countries most poor households still live in rural areas. Furthermore, the great majority of the rural poor can be presumed to be in farm households, though information on non-farm and farm rural incomes is quite limited. This discussion will refer primarily to the "farming poor", because of both their majority and the presumption that measures to aid them would promote growth of agriculture as a whole, thus spilling over to raise the incomes of the non-farm rural poor.

Key factors in assessing the distributional effects of adjustment programmes are real incomes of the poor, the price of agricultural products, particularly food, and access to government services and transfers.[27] It is also important to recognize the heterogeneity of the rural poor who are differentiated by age and gender and who comprise small-scale farmers, the landless, nomads, pastoralists and fishermen. In addition, many poor farm households derive a significant proportion of their income from non-agricultural activities. Nevertheless, the poor share the common disadvantages of limited assets, poor access to productive resources and public services, and vulnerability to adverse environmental conditions.

The foregoing considerations suggest that the actual welfare effects of policy reforms depend on a number of country-specific features:

- The structure of the agricultural sector in terms of the composition of output (tradables and non-tradables), the role of women in production, and the pattern of land- and other asset-holding.
- The distribution of net food producers and net consumers in the rural population. Evidence indicates that a majority of farm households in many developing countries are net food buyers.[28]
- The functioning of factor markets, particularly the labour market. In many developing countries, especially sub-Saharan Africa, the labour market is likely to be segmented owing to a variety of factors. High transport costs may create geographical segmentation, exacerbated by tribal barriers to integration. There may also be plantations or large-scale farms with a formal labour force of regular wage earners, and a small farm sector which uses mainly family labour but hires or provides seasonal, casual labour. Within each segmented market there may be a greater or lesser degree of specialization by sex and age groups.[29]

While bearing in mind that country-specific characteristics will determine actual outcomes, a generalized account of probable short-term welfare effects on sections of the rural population can be assembled (Table 3.4).[30] Consider agricultural producers with control over land. In the short term, producers of tradables,[31] particularly crops for export and for import substitution, would be expected to benefit from an increase in the relative price of their output, mainly as a result of devaluation (assuming a real depreciation is generated) and the removal of export taxes and price stabilization levies. This outcome presumes that the government and merchants pass the price

---

[27] P. Pinstrup-Andersen. 1989. The impact of macro-economic adjustment: food security and nutrition. *In* Simon Commander, ed., *Structural adjustment and agriculture: theory and practice in Africa and Latin America.* London, ODI.

[28] A. de Janvry, E. Sadoulet and L. Wilcox, *Rural labour in Latin America.* ILO Working Paper 10/6WP79, Geneva, International Labour Organisation, June 1986; M.T. Weber *et al.* 1988. Informing food security decisions in Africa: empirical analysis and policy dialogue. In *American Journal of Agricultural Economics,* 70(5).

[29] L.D. Smith. 1990. *Agricultural labour markets and structural adjustment in sub-Saharan Africa.* Rome, FAO.

[30] This account is based on Commander, 1989, op. cit.; Azam *et al.,* 1989; Norton, 1987; and Pinstrup-Andersen, 1989, op. cit.

[31] The classification of agricultural products into tradables and non-tradables, while convenient, cannot be adhered to rigidly in practice. Agricultural products range widely in their degree of "tradability" i.e. the extent to which world prices determine domestic prices. Moreover, a non-tradable can substitute an imported product and/or become exportable or importable with changes in the relationship between domestic and world market prices.

TABLE 3.4 **Probable short-term effects of major policies on rural incomes**

Category	Percentage of poor	Predominant sex	Major policies	Short-term effetcs	General effect
I. Producers of tradables					
Exports	Low	Male	Real devaluation	Positive	Positive
Domestic (food)	High	Female	Trade liberalization	Positive	
			Removal of subsidies	Negative	
II. Producers of non-tradables	High	Female	Real Devaluation	Negative	Negative
			Removal of subsidies (especially credit)	Negative	
III. Hired labour					
Production of tradables	High	—	(see category I)		Positive
Production of non-tradables	High	—	(see category II)		Negative
IV. Net food buyers	High	—		(see text)	Negative

Source: This table is based on Commander, 1989, op. cit.; Azam *et al.*, 1989; Norton, 1987; and Pinstrup-Andersen, 1989, op. cit.

increase on to farmers, although this is not always the case if parastatals act as market intermediaries. The boost to agricultural incomes may also be tempered by factors such as an increase in input prices through devaluation; trade liberalization and the removal of input subsidies; and an increase in risk if price stabilization schemes have been dismantled — all of which tend to reduce supply response. Overall, real incomes of these producers should rise, with the degree of increase depending on the price responsiveness of supply and on the constraints imposed by non-price factors.

On the other hand, as a result of devaluation and the removal of price distortions, producers of non-tradables suffer a loss, both from the drop in relative price of their output and the rise in production costs. That part of the small farm sector which is market-oriented often engages in the production of both tradables and non-tradables; the net welfare effect depends on whether income gains from the former outweigh losses in the latter. The intrahousehold distribution of gains and losses can also have important implications for the welfare, and especially the nutritional status, of family members. Particular difficulties may arise where men have direct control of cash crop income and how it is to be spent, while women have to manage household consumption with the income from non-tradables.[32] Furthermore, there may be a section of the small farmer population, located in regions with poor infrastructure or in areas unfit for the production of major tradables, which is largely bypassed by the adjustment process.

Turning now to the impact on agricultural labourers, some further generalizations can be made, again bearing in mind the possible complications that poorly functioning markets might introduce. Changes in the labour market would mirror those in production. Workers in the tradables sector would be expected to enjoy an increase in the demand for labour services, with employment and wage rates rising, while the demand for labour in the non-tradables sector would decline. However, expenditure-reducing policies which cut the demand for non-agricultural labour might induce an expansion in the supply of labour in rural areas. This reverse migration would in turn put downward pressure on rural wage rates. The most favourable outcome for labourers is when their employment is required in agriculture's tradable sector and the expansion of the demand for labour overcomes the contraction of employment elsewhere in the economy. Again, aggregate demand is the key.

The rural poor may be affected in other ways by changes in the non-agricultural sector. When contractionary policies reduce employment in the non-agricultural sector, transfers in the form of remittances — an important source of income for the rural poor — will be curtailed. However, an expanding agriculture can stimulate growth in the non-farm sector through the need for marketing, processing and distribution services. Where these goods and services can be provided locally, there will be an injection of income in the rural economy, some of which may be captured by poor, self-employed producers.

This discussion might leave the impression that the effect of adjustment on poverty is indeterminate, implying limited cause for alarm

[32] See footnote 31.

with respect to the probable consequences for the poor. Such is not the case. All scenarios yielding favourable outcomes for the poor require that the overall aggregate demand as well as the overall levels of employment and income do not contract. If they do, the demand for the commodities produced by the poor also contracts, and the labour markets in which the poor participate become burdened with an influx of income-seekers. Without exception, structural adjustment programmes seek to restrict demand through the reduction of government fiscal deficits, and this is frequently associated with reductions in the real value of government expenditure. When a government's contribution to aggregate demand falls in real terms, the total demand for goods and services will fall unless the private sector compensates. In the short term, investment is unlikely to fall because of the rise in real interest rates brought about by a tighter monetary policy. Everything then turns on whether real export demand increases sufficiently to compensate for the decline in government expenditure, plus the falling incomes and unemployment that may be caused by trade liberalization (which exposes domestic producers to greater competition).

In addition to the income and employment effects on cuts in government expenditure, there is the direct impact on the provision of public goods and services. Empirical evidence demonstrates that the reduction of food subsidies hurts the poor much more than the non-poor.[33] In this case, the targeting of specific groups has proved successful in reducing total subsidy outlays while still subsidizing the poor.[34] Cuts in health and education tend to harm the poor disproportionately even when the initial provision of the services is extremely inequitable. This apparent contradiction results from the pattern of political influence in services — meagre though they may be — for the poor who bear a disproportionate share of reductions. In addition, the non-poor are in a better position to turn to the private sector to recover lost access to public health services and education. A recent survey indicates that scaling down public health programmes might have quite a serious effect on infant and child mortality.[35] To the extent that women's unpaid labour is substituted for publicly provided services, either women's crop production capacity is curtailed or their working day is extended even further with adverse consequences for their health a likely result.

Rural poverty alleviation under structural adjustment

Available evidence as well as common sense suggest that structural adjustment programmes are initially likely to worsen the condition of the poor. [36] Therefore, as structural adjustment enters its second decade, the time is perhaps overdue to introduce direct measures to protect the poor along with efforts to close current account deficits and fiscal budgetary overruns.

By their very nature, SALs and SECALs concentrate on the monetized agricultural sector, with considerable emphasis on pricing policy. As such, the supply-side reforms discussed so far are at best irrelevant to the very many farmers who are mainly engaged in the production of non-tradables because of their limited access to resources such as land, water, purchased inputs, transport and appropriate technology. This group also includes the majority of women farmers.

Structural adjustment programmes began as temporary measures and, therefore, equity issues were not given adequate attention. Experience now shows that the macro-adjustment process is often long and difficult. Unless firm action is taken the danger remains that, in their present form, SALs and SECALs will encourage governments and donor agencies to concentrate resources on commercial agriculture and neglect the rest of the rural population. This tendency is accentuated where there is pressure to service external debts.

If concern with poverty and equity are not to be discarded, structural adjustment should be seen as only part of, and not a substitute for, general economic development. To the maximum extent possible, structural adjustment should adopt policies that will encourage

[33] Pinstrup-Andersen found that the removal of food subsidies substantially worsened the distribution of real income in all the surveyed countries: Sri Lanka, Thailand, Egypt and Nigeria.

[34] This conclusion was obtained in a study of Egypt: H. Alderman and J. Von Braun. 1985. *Egypt: implications of alternative food subsidy policies in the 1980s.* Washington, International Food Policy Research Institute.

[35] Gaiha, 1989, op. cit., p. 57.

[36] See, for example, R. Jolly and A. Cornia, eds., The impact of the world recession on children. *World Development*, March 1984; and studies by UNICEF.

general economic development, with particular attention paid to enabling smallholders and the poor to participate in this process.

In particular, emphasis should shift from attempting to compensate for poverty intensification after it occurs toward redesigning the adjustment programmes themselves so that poverty alleviation becomes an integral part of the adjustment process. This in turn would require protecting the poor in the transitional period and also in the long term.[37] This approach is discussed further in the following section.

Groups unable to benefit from the adjustment process may be protected by targeting income or consumption transfers. Where adjustment programmes succeed in accelerating agricultural growth, they can make a significant contribution to poverty alleviation. But poverty groups must be drawn into the restructuring of production and must participate directly in the adjustment process. It cannot be assumed that benefits will automatically spread to the poor.

Conscious targeting of poverty groups is all the more important in light of the growing evidence that poor women are bearing much of the stress of adjustment. In their capacity as household managers, responsible for the daily care and feeding of household members, women have to make fewer resources stretch further. The hidden cost of adjustment is the extra demand it makes on women's time, which is unrecorded through lack of systematic collection of data on women's unpaid labour. Available case-studies show that poor women have no spare time — they already have a working day which is considerably longer than that of men.[38]

Benefiting the poor requires: i) adding to their assets through agrarian reform; ii) improving their productivity by better access to productive resources (fertilizer, irrigation, credit, extension services); iii) improving employment opportunities for those without sufficient land through functioning labour markets, adoption of appropriate technologies, and specific employment schemes; and iv) maintaining health and education services. Agricultural price policies are much too often implemented without detailed information on likely supply responses. Indeed, information from one country is frequently used as a guide in other countries.

A major problem is that SAPs are often implemented in a crisis atmosphere. Governments are struggling to obtain a political consensus to undertake reforms while "vested interest groups" feel unjustly penalized. This is not a situation conducive to carrying out reforms benefiting the poor. It has also been argued that the poor would have been worse off if a SAP had not been implemented, since sooner or later a worse economic crisis would have emerged. But policies aimed at correcting imbalances should be applied in ways which minimize the negative social effects. Rather, the atmosphere of policy reform and adjustment can be exploited to improve the situation of the poor while correcting imbalances. It is a domestic political choice, but the international community can assist by ameliorating the external economic environment.

[37] A useful approach is found in T. Addison and L. Demery, *The economies of rural poverty alleviation. In* Commander, 1989, op. cit.

[38] N. Heyser. 1988. *Economic crisis, household strategies and women's work in Southeast Asia.* Workshop on Economic Crisis, Household Strategies, and Women's Work. Cornell University.

THE IMPLICATIONS OF STRUCTURAL ADJUSTMENT FOR AGRICULTURE AND ECONOMIC DEVELOPMENT AND THE ROLE OF FAO

Lessons emerging from the past decade of SAPs

The preceding chapters describe the process through which national policy-makers and international lending agencies have introduced far-reaching structural reforms since the early 1980s. During the past decade, declining economic productivity, shrinking export earnings, and debt-service costs have dominated discussions on how economic adjustment should proceed. Consequently, the period has witnessed the adoption of structural adjustment and stabilization policies by one country after another in an attempt to attain macro-economic balances by bringing government expenditures in line with national revenues; improve resource allocation through more market-oriented policies; and create a solid foundation for sustainable economic growth.

Developing countries have employed a wide variety of policy instruments to obtain stabilization and adjustment goals. Over the past decade, institutional reforms, trade liberalization measures and price policy shifts have been common. However, while the approaches of adjustment packages have not differed greatly between countries despite their widely varying situations, after ten years of experience, the results have been mixed. Individual country performances have varied according to a number of internal and external factors. Internal factors include: the importance given to specific objectives; the choice of policy instruments, the nature and size of existing economic imbalances and distortions; and political acceptability. External factors have included, at different times: deteriorating terms of trade; rising protectionism; historically high interest rates; declining markets because of recession in the developed countries; as well as other events beyond a government's control such as prolonged droughts. In fact, the 1980s clearly demonstrated that external factors alone are enough to push the adjustment process completely off course.

While economic adjustment often produces results unique to each country's particular circumstances, a number of common lessons have emerged which provide valuable insights for both lending agencies and developing countries. It is now recognized that stabilization and structural adjustment are long-term, complex processes, requiring years before results can be expected. The process is even more problematic when adjustment problems have been ignored for too long and imbalances allowed to develop to a point where the adjustment process becomes long and painful. It is also realized that stabilization policies alone are inadequate to promote renewed economic growth and development, and even macro-economic policy changes coupled with structural adjustment and sectoral reforms have so far proven insufficient in many cases. Policy-makers have drawn two important lessons from these experiences and have attempted to improve implementation procedures as well as adopt new and innovative policy measures which were not initially considered.

First, by the mid-1980s, adjustment programmes gave more attention to issues of implementation and emphasized what the World Bank referred to as "a well defined set of implementable policies rather than broad and rapid programmes of reform".[39] During this period, lending agencies began to provide additional funds as part of adjustment loans for technical assistance to develop implementation management skills, to better define how the public sector can promote private sector activities, and to find ways of creating confidence in the rapidly changing economic structure.

Second, the political difficulties, social problems, and economic stress created by structural adjustment have come to be better appreciated and understood. It is now recognized that, in the past, SAPs have had a negative effect on the "chronically poor" while creating a "new" poor sector, and that the burden of adjustment may fall unequally on already disfavoured categories of the rural population such as women, low-income groups and small farmers. More recent SAPs make social considerations a basic objective of the adjustment programme, not just for equity and humanitarian reasons, but because of a long-overdue recognition of the uneven impact that deteriorating social welfare has on economic growth and the programme's overall success.

---

[39] P. Nicholas. 1989. *The World Bank's lending for adjustment*, Washington, D.C., World Bank.

Special efforts are now being made to minimize the effects of policy reforms on the poor and to address directly the social dimensions of adjustment programmes. For example, the World Bank's social dimensions of adjustment (SDA) project is attempting to find the appropriate mix and sequencing of policy instruments available under adjustment in order to achieve the dual objective of sustainable economic growth and poverty reduction. The UNDP and other UN agencies, including FAO, are also introducing a programme on how to monitor social welfare status during the adjustment process, involving a planned series of four country case-studies. Countries themselves have also been experimenting with different sets of policy prescriptions designed to achieve adjustment while minimizing social hardship. These experiments include individual countries such as Ghana's Programmes of action to mitigate the social costs of adjustment (PAMSCAD) and the ECA's proposed African alternative framework to structural adjustment programmes for socio-economic recovery and transformation (AAF-SAP) launched in 1989.

The experience of the 1980s has shown that, for many countries, the attainment of the desired adjustment objectives — economic recovery, sustainable economic growth and long-term development — is crucially dependent upon a viable solution to the debt issue. Despite many initiatives, including negotiations, debt-relief proposals and, at times, crippling efforts on the part of indebted countries to restore economic balance while repaying debts, to date, little progress has been achieved in reducing the level of indebtedness or even interest charges. Many developing countries have become increasingly net capital exporters to the lending — mainly industrialized — countries.

A growing consensus now suggests that the various debt-relief programmes need to be defined within a broader political dimension that includes an effective partnership between the industrial and the indebted developing countries and presupposes a sharing of the impact of high indebtedness as well as the burden of debt repayment. Debtor country losses have been disproportionately large compared to those incurred by the industrial countries' financial sectors with regard to loan defaults, debt reduction and debt forgiveness.

Similarly, debt repayment efforts through adjustment policies have been heavily one-sided. Many debtor nations have undertaken difficult policy reforms while the major industrial countries have moved rather slowly to reduce their own macro-economic imbalances, open up their markets to debtor country exports or reduce interest charges through some form of debt forgiveness. Many industrialized countries have actually raised their non-tariff barriers over the past decade. Developed countries could also do more to increase general public awareness of the interests shared by creditor and debtor nations. Nevertheless, this does not exonerate debtor countries from the need to pursue reforms and adjustment, while seeking formulas for collective and coordinated action that might enhance their position in intragovernment negotiations.

Another common lesson emerging from the past ten years' experience is that a strong agricultural performance is often fundamental to economic recovery and the adjustment process. The agrarian societies of many developing countries continue to depend on agriculture for employment, income, food, and foreign exchange earnings. While agriculture has been negatively affected by the economic crisis during the 1980s, it has also been the most resilient economic sector in the adjustment period.

In the Latin American region, for example, agricultural exports increased at an annual rate of around 3 percent between 1983 and 1988, compared to only 1.6 percent for total merchandise exports. Similarly, agricultural growth rates for the region largely exceeded those of GDP for the same period. This relative improvement in agricultural economic performance highlights the importance of the sector during a period of economic crisis and structural reform. At the same time, however, rural poverty has increased, spreading to more than 60 percent of rural households. Ironically, one reason for the improved agricultural trade balance in many developing countries has been the fall in real incomes which has reduced the demand for imports.

To understand better how the agricultural sector has performed in individual countries and regions, Part Two of this special chapter (Macro-economic imbalances and agriculture — an empirical review) provides a detailed analysis of changes in a broad range of economic variables for a group of 65 developing countries. The discussion that follows raises some of the issues and questions identified by

researchers attempting to evaluate adjustment programmes.

Evaluating SAPs: methodological problems and issues

The relationship between agricultural performance, policy reforms and adjustment goals is difficult to quantify for a number of reasons. To start with, agreeing on the basis for evaluation is difficult. Comparing the performance of critical variables before and after adjustment measures are implemented is informative and useful, but as a method of comparison it has a strong static bias. For instance, one should be interested in estimating how the economy would have performed had the policies not been implemented and then to compare the resulting outcome with what actually occurred. To date, most studies of this type have utilized simplistic trend extrapolations from data series, thus reducing the usefulness of the comparisons. In addition, the choice of time period greatly affects the results of such comparative studies.

Comparing "adjusting" to "non-adjusting" countries is also problematic because these studies ignore the conditions existing when adjustment policies are initially implemented; the different economic, social and political characteristics of each country; and the different types of policies chosen. Part Two of this chapter argues that adjustment can occur by way of a "healthy" or an "unhealthy" process.

A healthy adjustment pattern is characterized by a reduction in domestic and external imbalances without jeopardizing growth prospects; that is, the domestic gap is reduced by increasing savings rather than reducing investment, while the external gap is reduced by generating export earnings rather than compressing imports. Unhealthy adjustment, on the other hand, closes the external gap through the restriction of imports and investment, thereby threatening the economy's long-term capacity to expand adequately.

Separating the adjustment programmes's effects from those of external factors is the next difficulty facing policy analysts. Relying on simple inspections of data may incorrectly attribute the consequences of non-policy factors, such as weather or international price movements, to adjustment policies. Even when external factors are known and taken into account, there remains the problem of calculating the extent of their effect. Finally, the collection of reliable data remains an obstacle to sound empirical studies.

Despite these limitations, FAO is supporting the development of more practical methodologies so that developing countries may better understand the adjustment process and its potential implications for different social groups in rural areas, as well as for agricultural performance (see Box 3.6).

Evaluations and modelling exercises are needed to assist countries in determining the kinds of policies and resources required to meet development objectives and promote effective economic planning and monitoring. At present, computable general equilibrium models (CGEs) represent the most rigorous approach for accurately assessing the effects of structural adjustment programmes on agriculture. CGEs provide both *ex post* evaluations of adjustment programmes as well as a priori examinations of the short-term effects and long-term implications of policies for growth and development. CGEs are capable of providing information on a wide array of relationships between macro-economic and micro-economic variables and of detailing sectoral interactions. The structure of internationally traded goods, institutional characteristics, production patterns, input-output markets and income distribution may all be incorporated in these models.

Adjustment programmes and development strategies in the future

While more useful and sophisticated models continue to be developed, policy-makers still face the immediate challenges of creating an economic environment favourable to agricultural producers. As outlined earlier, the various types of stabilization and structural adjustment policies are by now well defined. The major task facing governments over the past decade has been finding the appropriate mix of policies and the optimum sequencing for introducing the policy changes.

Today, policy-makers are attempting to improve the usefulness of adjustment programmes by establishing development strategies and policies designed to enhance their effectiveness. One challenge is how to incorporate a broader base of rural society into the adjustment process. Rural women, small farm households, rural labourers, migrant workers and landless families need better defined policies which deal directly with their constraints and their capacity to contribute to economic recovery.

BOX 3.6

Role and contribution of FAO in regard to structural adjustment programmes (SAPs)

Why FAO should be involved in structural adjustment programmes
The key reasons for FAO's involvement in the structural adjustment process include:

i) The mandate of the Organization and the associated scope of its activities: to raise levels of nutrition and standards of living; improve the efficiency of the production and distribution of food and agricultural products; improve the condition of rural populations; and contribute to an expanding world economy.

ii) The major economic weight of the agriculture, forestry and fisheries sector in developing countries, and hence its essential role in the structural adjustment process.

iii) The major impact that the structural adjustment process has on rural people, food security, agricultural production, natural resources and the environment; and the realization that the process can adversely affect standards of living, levels of nutrition and employment opportunities for the poor.

iv) The widespread consequences of structural adjustment programmes across different sectors of the economy. This factor makes consultation and coordination among national agencies, as well as with multilateral and bilateral funding and technical assistance partners, essential. SAPs are truly multisectoral and interdisciplinary, requiring the technical input of all internal and external bodies.

The structural adjustment activities of FAO
Recognizing the need for structural adjustment while being aware of the possible adverse impacts of adjustment programmes, FAO has:

- Conducted studies to improve understanding of the structural adjustment process and its socio-economic impacts and thereby provide a firmer basis for the provision of advice and assistance requested by member countries.
- Provided technical assistance at the request of member countries to help them: assess socio-economic effects of their SAPs; prepare for negotiating a SAP with their financing agencies and donors; conduct sector or subsector policy and strategy reviews; collect and analyse data required for policy formulation; and, through direct assistance and training in policy analysis, strengthen their capacity to monitor and continuously assess the socio-economic effects of policies, including those guiding sector structural adjustment programmes (SSAPs) or SAPs.
- Formulated on behalf of multilateral financing agencies, especially the World Bank, investment projects in the natural resource sector which reflect agreed national structural adjustment programmes and implement specific aspects of sectoral adjustment in agriculture, fisheries and forestry.
- Collaborated with other organizations in structural adjustment studies and reviews. FAO participates in the social dimensions of adjustment (SDA) project in Africa, which seeks the full integration of social dimensions into economic and financial decision-making. It also participates in the Informal Group of Multilateral Agency Representatives on the Impact of Economic Adjustment on Food Security and Nutrition in Developing Countries, a group established at the initiative of the World Food Council (WFC) in collaboration with UNICEF and the International Labour Organisation (ILO) in May 1987, and which also includes the IMF and the World Bank among its members.

FAO can execute and technically support field projects to strengthen national capacity for assessing and monitoring the effects of policies, including those associated with SAPs and SSAPs on food security; nutritional status; employment and welfare of rural people; agricultural production and internal trade; the external agricultural trade account; and the fiscal budget.

Some specific areas where FAO's policy-related technical assistance could be of particular value include:

- reforming or strengthening institutions in the food and agricultural sector and in training local staff;
- undertaking monitoring and evaluation studies, including nutritional surveillance to assess the impact of adjustment;
- establishing appropriate relationships between state marketing agencies and the private sector, as well as defining the role of cooperatives in the production and marketing of agricultural commodities.
- overcoming bottle-necks in the marketing infrastructure, including credit, storage, handling and distribution facilities;
- designing appropriate investment strategies and implementing incentive pricing policies for food and export crops, while taking into account internal physical and institutional bottle-necks, resource limitations, the need to keep consumer prices at reasonable levels, and existing distortions in world markets;
- designing compensatory mechanisms such as special nutritional intervention schemes and employment-creation activities required to mitigate short-term negative effects of adjustment while ensuring that such schemes do not discourage domestic production; and
- assessing the effects of macro-economic factors, both domestic and external, on the food and agricultural sectors and particularly on food security of the poor and vulnerable sections of the population.

Recent FAO activities

The following reports serve as examples of the studies produced by FAO in 1990.

R. Gaiha.
Structural adjustment and household welfare in rural areas — a micro-economic perspective.

A.H. Sarris.
Guidelines for monitoring the impact of structural adjustment programmes on the agricultural sector.

J.M. Boussard.
The impact of structural adjustment on smallholders.

L.D. Smith.
Structural adjustment policy sequencing in sub-Saharan Africa.

A. Thompson.
Institutional changes in agricultural product and input markets and their impact on agricultural performance.

L.D. Smith.
Agricultural labour markets and structural adjustment in sub-Saharan Africa.

C. Kirkpatrick.
The effects of trade and exchange rate policies on production in agriculture.

Salin & Claassen.
The impact of stabilization and structural adjustment policies on the rural sector — case-studies of Côte d'Ivoire, Senegal, Liberia, Zambia and Morocco.

FAO is also involved in numerous training activities aimed at improving policy analysis capacity in developing countries. Training projects are currently under way in various countries in Africa, Asia and Latin America and the Caribbean.

Finally, FAO is participating in a wide variety of sectoral policy analysis and planning assistance projects. The objectives of these projects include reformulating agricultural policies, identifying and evaluating structural adjustment impacts on food and agricultural production and on marketing systems, and implementing new policy and planning processes.

BOX 3.7

Structural adjustment programmes and the environment

In addition to the socio-economic implications of structural adjustment programmes, policy-makers are increasingly concerned about the potentially negative consequences of macro-economic policies for developing countries' natural resources. The debt crisis demonstrated how efforts by some countries to repay loans could lead to unsustainable agricultural practices. For example, encouraging logging or land clearing for agriculture in humid tropical lowlands in order to generate export earnings may only provide a short-term solution for debt repayments. When these areas are not reforested or cultivated properly they rapidly lose their economic potential. Too often, environmentally sensitive areas are never capable of returning to their initial state after being modified by logging or cultivation.

Policy-makers are beginning to identify how structural adjustment policies affect a country's natural resource base and to design programmes and policies to deal more adequately with environmental problems. Both special programmes such as debt-for-nature swaps, and analyses to identify the potential environmental impacts of policy changes are now addressing these important issues. For instance, in 1987, a non-governmental organization purchased $650 000 of Bolivian debt for $100 000 on the secondary debt market in return for a government guarantee to establish three conservation and sustainable-use areas in the Amazon region. All three areas received Congressional Law Status which is the highest legal protection status in Bolivia. At present, five countries (Bolivia, Costa Rica, Ecuador, Madagascar and the Philippines) have participated in debt-for-nature swaps, reducing their external debt by about $100 million.

Similarly, current research is examining how basic market failures in some developing countries limit the usefulness of adjustment policies that would otherwise encourage environmentally sound agricultural practices. These failures include: market prices that do not reflect environmental costs; a divergence of the private and social discount rate leading to a misallocation of resources; and property rights leading to inconsistent economic incentives.

For example, the property rights structure affects not only access to land but also whether that land is farmed in a sustainable manner. Landless agricultural labourers may be forced to farm on fragile hillsides because land market imperfections limit their access to more cultivable farmland. At the same time, insecure land rights influence whether those hillsides are planted with permanent tree crops or annual crops which eventually lead to serious soil erosion.

Each individual policy change associated with structural adjustment programmes can have environmental implications. Fiscal policies, government spending and subsidies influence land use, crop choice, input levels, credit availability and public infrastructure investments. Monetary policies, such as interest rate levels, affect investment behaviour and demand. Likewise, exchange rate and trade policies influence the cost of imported inputs, including machinery and fertilizers as well as the profitability, and hence the scale, of export crop production.

While policy-makers are still unclear about how these individual and combinations of policy changes affect the natural resource base, it is now recognized that the environmental effects of macro-economic policies should be addressed in structural adjustment programmes.

Another challenge is how to accommodate the constant pressure for reduced government expenditures. A more efficient use of limited government resources and sound economic policies that do not discriminate against agriculture are clearly needed. At the same time, agricultural producers still require improved rural infrastructure, research and extension services, and access to credit so that new technologies may be adopted. In addition, the slow improvements in rural education — so painfully gained — must be continued.

One final lesson can be drawn from shifts in the perception of principles and instruments of economic and agricultural development. The past two decades have witnessed radical changes in development strategies.
After the disastrous experience of financial permissiveness and debt accumulation in the 1970s, structural adjustment became the dominant strategy. Whether designed primarily as a development or a loan-repayment strategy, the problems it was intended to address remain unsolved. Indeed, the 1980s have been termed a lost development decade for many countries, and the problem of debt has not subsided. Agriculture has helped buffer the effects of economic shocks, but its contribution to recovery and growth has been uneven. At the same time, beyond the many initiatives and schemes proposed, no comprehensive, alternative strategy has evolved. As we enter the 1990s, the international community, and in particular the developing countries concerned, has yet to find an overall strategy that balances the needs for economic efficiency, debt reduction and the incorporation of large segments of low-income population and rural society into the growth and development process.

PART THREE
STRUCTURAL ADJUSTMENT AND AGRICULTURE

II. Macro-economic imbalances and agriculture — an empirical review

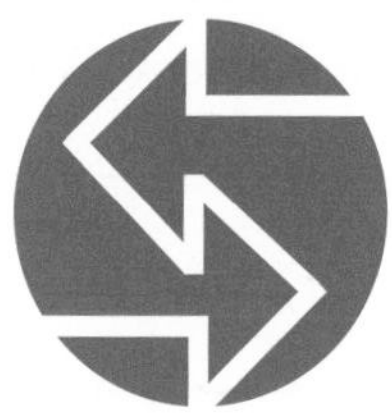

STRUCTURAL ADJUSTMENT AND AGRICULTURE

II. Macro-economic imbalances and agriculture — an empirical review

INTRODUCTION

The purpose of this section is to present an empirical overview of the development experiences of 65 developing countries and to examine the following five major issues:

- How did internal and external macro-economic imbalances of developing countries evolve after the 1981-83 crisis?[1]
- Among countries that succeeded in reducing such imbalances, which succeeded in doing so in an economically "healthy" manner?[2]
- What general features characterized internal and external adjustment in the different country groups?
- What role has agriculture played in the above issues?
- What was the impact of the different adjustment and stabilization patterns on economic and agricultural performance?

The 65 developing countries, selected mainly on the basis of the quality of the data available, have been divided into three analytical groups: "healthy" patterns of internal and external adjustment after the critical period 1981-83 (group A); "unhealthy" adjustment patterns (group B); or, as in the third group of countries, a continuing deterioration in macro-economic imbalances (group C). These country groups are presented in Table 3.5.

As in any grouping of countries, some approximations are inevitable, in particular, the need to use standard reference periods. Some countries have a long history of macro-economic imbalances calling for adjustment while, for others, deficits only became unsustainable and adjustment compelling during different periods in the 1980s. However, it is in the critical period 1981-83 that imbalances and performances had more markedly deteriorated in a large majority of countries, underlining the widespread negative impact of the recessive international economic environment. The years 1981-83 were therefore taken as the most representative "central" time frame for examining macro-economic imbalances at their worst; the years 1970-80 for reviewing previously prevailing conditions; and 1984-87 as "adjustment years", recognizing that the process of adjustment is still unfolding and that this period of time is far too short for a definite assessment.[3]

This exercise does not aim at evaluating SAP performance. The country groupings have been based on simple macro-economic accounting aggregates, devoid of any normative or judgemental assessments of policies behind — and social implications of — adjustment patterns, and regardless of whether or not these policies were introduced in the context of SAPs supported by international financing agencies.

Although the relevance of the groupings varies according to the issue examined, they are used throughout the review as a

[1] Macro-economic imbalances are defined as large and persistent differences between current account (and saving/investment) deficits and long term inflows of capital.

[2] A "healthy" pattern of stabilization and adjustment is defined as one that reduces domestic (saving/investment) and external (export/import) disequilibria without jeopardizing growth prospects. In other words, one that succeeds in closing the "domestic gap" mainly by increasing savings rather than by contracting investment; and closes the "external gap" mainly by generating export earnings rather than by compressing imports. Conversely, economically "unhealthy" stabilization/adjustment is one that fails to switch expenditures in favour of investment and exportables, but resorts to restricting investment and imports, without having the domestic capacity to substitute the latter efficiently.

[3] The year 1987 is the latest for which consistent data series are available. When data series for a specific country were not complete, averages for the respective country groups were estimated either excluding the country or on the basis of extrapolations. All time period and country group averages are simple arithmetic averages, in line with the principle of giving all countries equal weight. When extreme variations in one or a few countries distorted the group averages, these countries were excluded from the average.

TABLE 3.5 **Distribution of countries according to patterns of change in external (export/import) and domestic (saving/investment) balances, 1981-83 to 1985/87**

Region	Number of countries	Improving		Deteriorating
		A (19)	B (29)	C (17)
Latin America and the Caribbean	**20**	**Brazil[1] Chile Colombia Guatemala Guyana Jamaica Paraguay Venezuela[1]**	**Argentina[1] Barbados Costa Rica Haiti Panama Peru Trinidad/Tobago**	**Bolivia Ecuador Honduras Mexico[1] Nicaragua**
Africa	**31**	**Botswana Burkina Faso Gambia Madagascar Mauritania Mauritius Morocco Nigeria**	**Benin Burundi Côte d'Ivoire Congo Kenya Liberia Malawi Rwanda Senegal Sierra Leone Sudan Tunisia**	**Algeria[1] Cameroon[1] Central African Rep. Ethiopia Gabon[1] Guinea-Bissau Mali Niger Tanzania Togo Uganda**
Near East	**6**	**Turkey**	**Cyprus Egypt Jordan Syrian Arab Rep. Yemen Arab Rep.**	
Far East and the Pacific	**8**	**Korea, Rep. of Thailand**	**Bangladesh Fiji Malaysia Philippines Sri Lanka**	**Indonesia[1]**

[1] **Countries with positive balances in most years.**
***Source:* FAO, based on data from World Bank, *World Tables* (1988-89 edition).**

convenient analytical framework. Some consistency is generally found in the direction of the results obtained, however, which confirms the pervasive influence of macro-economic imbalances on overall and agricultural performance.

The main findings of the review are:

- Macro-economic imbalances worsened markedly in the early 1980s, but a large majority of countries reduced them through some form of stabilization and adjustment in more recent years. These are countries in groups A and B.
- Among the 48 countries that reduced such imbalances, only a minority of 19 were able to do so in an economically "healthy" way. These are countries in group A.
- Countries in group A showed a more favourable profile than the others in recent years, by most traditional measures of growth and trade performance, both overall and in agriculture. In particular: i) their growth in GDP was significantly higher, in the aggregate and within regions; ii) their reduction in external imbalances was achieved mainly through a successful export effort. The volume of their exports strongly expanded, more than compensating for a pronounced deterioration in their terms of trade — this feature also applied to agricultural exports. Countries in groups B and C were noticeably less successful on this account; iii) investment was a major casualty of adjustment and stabilization. Nevertheless some countries, particularly in group A, achieved a remarkable domestic saving effort after the period 1981-83, which permitted some recovery in investment. The most outstanding increase in domestic savings was in the Latin America and the Caribbean region, but as much of the savings were used to finance debt-service payments, national savings available for investment were

greatly eroded; iv) fiscal imbalances were reduced through a less traumatic pattern of fiscal management in group A, compared with groups B and C, with a more gradual reduction in expenditure and some recovery in revenues. While a considerable share of fiscal expenditure was being channelled to finance interest payments in many countries, those in group A were able to maintain a stable, if relatively low, share of productive expenditure in recent years; and v) better comparative performance in food production, together with lower reliance on external food supply, enabled countries in group A to achieve greater self-sufficiency in food, yet maintain stable levels of per caput calorie intake.

- Nearly all countries in group A significantly devalued their currencies in real terms, while most of group B did not and currencies in the group C countries generally tended to appreciate.
- Agriculture played a significant role in mitigating the worst effects of the 1981-83 crisis in all groups. The secular decline in the share of agriculture in GDP was arrested in the early 1980s and even reversed subsequently. All groups shared in this trend. Agricultural trade appeared to be more affected than trade in other sectors by the recession of the early 1980s, but its share in total merchandise trade rose more recently.
- Although budgetary difficulties imposed widespread cuts in government productive investment, agriculture was less penalized in relation to other sectors during the 1980s. In most countries central government development expenditure in agriculture rose both as a share of agricultural GDP and as a share of total central government development expenditure.
- Agriculture and external debt: external financing flows to agriculture added little to the debt overhang of developing countries, and the sector maintained positive net transfers throughout the 1980s. Furthermore, agriculture provided a positive contribution to external balances in most cases. Surpluses from agricultural trade enabled a significant share of the aggregate debt-service payments of developing countries to be financed.

BOX 3.8

Two country cases

While entirely different in their socio-economic and political context, the experience of two countries, Chile and Egypt, provide typical examples of opposing patterns of stabilization and adjustment. In the case of Chile, the trade balance turned positive between 1981-83 and 1984-87 thanks to a sharp increase in exports, which exceeded import growth. The trade surplus offset net factor payments, enabling a reduction in the current account deficit as a percentage of GDP. On the domestic balance side, there was a strong expansion in domestic private savings and comparatively lower, but still significant, increases in capital expenditures — despite reduced government expenditure relative to revenues.[1]

In Egypt, by contrast, the external trade deficit declined mainly as a result of a sharp contraction in imports — while the domestic capacity to substitute them efficiently and rapidly was lacking. The improved domestic balance was achieved with much lower levels of domestic savings and investment — the contribution of gross domestic investment to growth of GDP expenditure was actually negative in 1986 and 1987 — and a slightly reduced fiscal deficit. While in Chile private consumption as a percentage of GDP declined from 75 percent to 68 percent, in Egypt it increased from 66 percent to 72 percent — the mirror image of diverging trends in saving rates. While the full impact of these features on GDP growth can only be assessed in a longer-term perspective, it may be noted that such contrasting patterns of stabilization were accompanied by equally contrasting growth trends: in Chile per caput GDP fell 4.8 percent yearly during 1981-83, but rose nearly 3 percent a year in 1985-87, over 5 percent in 1988 and almost 7 percent in 1989; while in Egypt the per caput annual growth rate in GDP declined from 4.2 percent in 1981-83 to 1.4 percent in 1985-87 and turned negative in both 1988 and 1989.

[1]Chile was among the very few countries that moved from a deficit to a surplus fiscal balance position between 1981-83 and 1985-87.

MACRO-ECONOMIC OVERVIEW[4]

External (export minus import of goods and non-factor services) and internal (saving minus investment) balances widely differed among countries, ranging from surpluses representing up to 5 percent of GDP to deficits as large as 40 to 50 percent of GDP (Table 3.6). Behind this diversity it is not necessarily the size or sign of the balance *per se* that determined critical payments situations. Indeed, Table 3.6 shows that several countries, particularly in Latin America and the Caribbean, actually had positive trade and internal balances representing over 3 percent of GDP in the period 1985-87 (and this was the case throughout the 1980s), but they were among those where problems calling for adjustment became more acute. By now it is well known that such problems were rather on the financial side — current account gaps filled through foreign borrowing until the early 1980s and, subsequently, repayment obligations that could not be financed by new capital inflows. Thus, in the following discussion changes in external and internal balances are reviewed in relation to changes in net factor income (i.e. interest payments, foreign investment profits and foreign remittances).

Domestic (saving minus investment) balances
During the 1980s, the overall saving effort (gross domestic savings as a share of GDP) in Africa and Latin America and the Caribbean declined sharply from the levels of the 1970s (Figure 3.1). Savings in these regions were severely constrained by the economic recession and related factors (lower household incomes, higher inflation rates, very low or negative real interest rates which reduced external financing and affected business and public saving). This tendency was less pronounced in the Far East where, although the domestic saving rate during the 1980s fell below the high average levels of the period 1976-80, it remained significantly higher than in the early 1970s. In the group of Near East countries, heavily reliant on external financing flows for investment, the domestic component of saving remained very low, with only a modest increase after 1984.[5]

- The level of gross national savings (gross domestic savings plus net factor income) available to finance investment declined more markedly than domestic savings in all regions because an increasing share of the overall saving effort had to be diverted toward external debt payment obligations.
- While investment rates (gross investment as a share of GDP) during the 1970s were relatively constant at around 23 percent in both Africa and Latin America and the Caribbean, they tended to rise in the Far East and Near East, peaking at 29 percent and 34 percent respectively in 1981. In the following years, however, the volume of investment became a major victim of stabilization and adjustment in all regions. By the period 1985-87 investment/GDP ratios were on average: 18 percent in Latin America and the Caribbean; 23 percent in the Near East; 20 percent in the Far East; and 18 percent in Africa.

In group A, domestic savings and, to a lesser extent, investment and national savings staged a substantial recovery after the period 1981-83 while remaining below the average levels of the 1970s. The most outstanding improvement concerned domestic savings in Latin America and the Caribbean, which enabled the eight countries in the group to close their aggregate saving/investment gap to the smallest margin since 1974. Nevertheless, as much of this remarkable saving effort was used to cover debt-service payments, only a modest and uneven increase was achieved in national savings. In Africa, the improvement observed in the eight countries in group A needs to be qualified in the context of the region as a whole. Indeed, while savings did recover, investment rates merely stabilized at the depressed levels to which they had fallen in 1983.

Among the countries that reduced domestic imbalances in an economically "unhealthy" manner (group B), the seven countries in Latin America and the Caribbean experienced an almost uninterrupted fall in investment and saving rates throughout 1982-87, coupled with a widening net income payment gap. Those in Africa experienced a similar collapse in investment after 1982 despite some recovery in

[4] A set of interdependent accounting identities indicating the fundamental relationships among the internal and external dimensions of adjustment is used as a conceptual framework for this section (see Box 3.10).

[5] The limited country sample for the Near East and Far East and Pacific regions cannot be disaggregated (all but four countries fall in group B). Averages are therefore presented for the region as a whole.

Figure 3.1

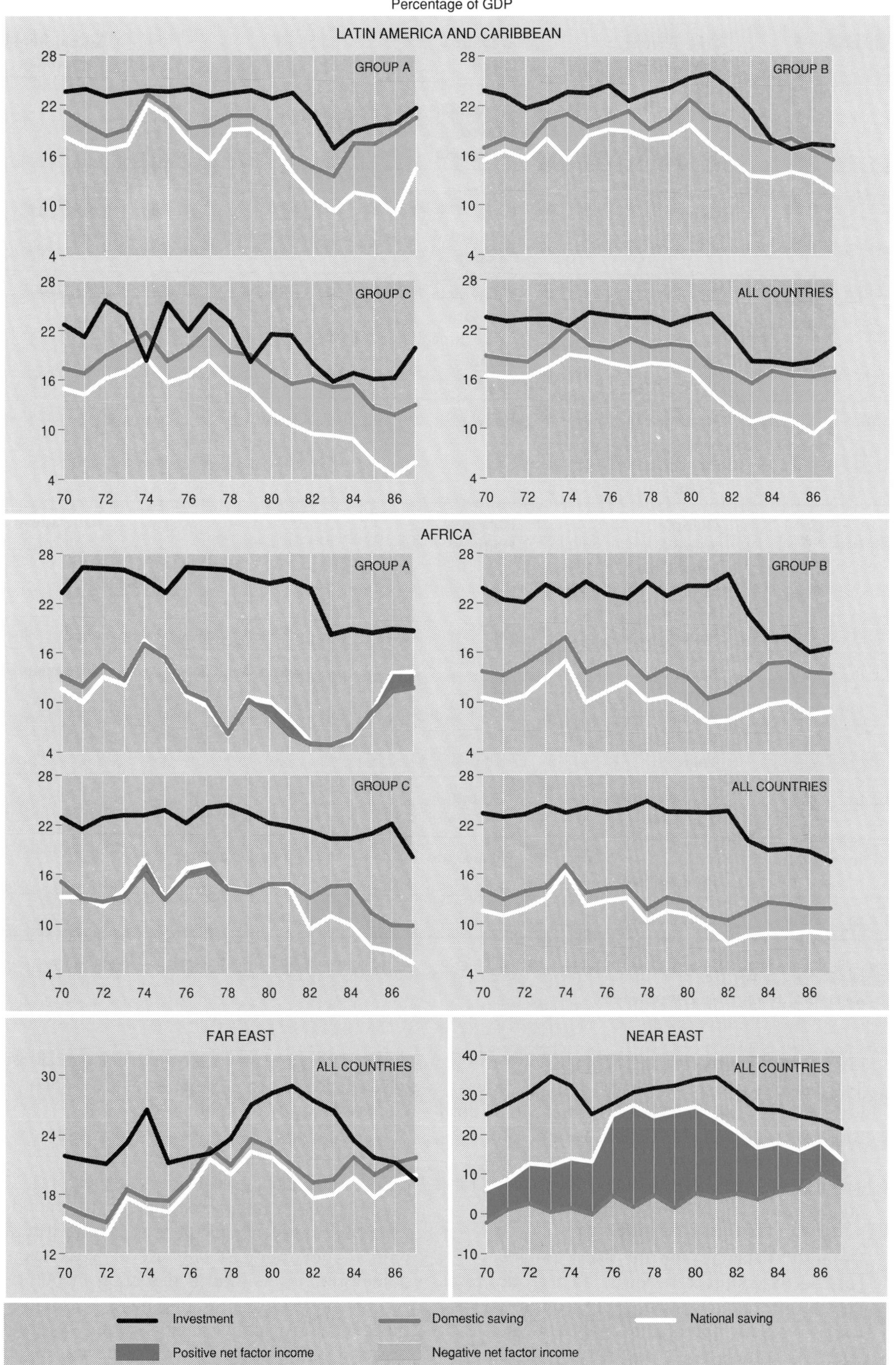

Source: FAO, based on data from World Bank World Tables 1988-89 edition

Note: Net factor income is also broadly seen in Figure 3.6 (external balances) as the gap between net trade and current accounts balances. However, the two NFI figures are not identical as they were derived from independent estimates.

TABLE 3.6 **External and internal balances as a share of GDP, selected developing countries, average 1985-87**[1]

Balance as share of GDP (%)	Region			
	Africa	Latin America & Caribbean	Near East	Asia & Pacific
3.1 and more	**Côte d'Ivoire** **Botswana** **Liberia**	**Argentina** **Brazil** **Chile** **Colombia** **Mexico** **Venezuela**		**Korea, Republic of** **Philippines** **Malaysia**
0.1 to 3.0	**Cameroon** **Gabon** **Nigeria**	**Barbados** **Peru** **Trinidad & Tobago** **Panama**		**Fiji** **Indonesia**
0 to −4.0	**Algeria** **Congo** **Kenya** **Malawi** **Sierra Leone**	**Guatemala** **Honduras** **Costa Rica** **Ecuador** **Jamaica**	**Turkey**	**Thailand**
−4.1 to −7.0	**Madagascar** **Niger** **Tunisia** **Uganda**	**Bolivia**	**Cyprus**	
−7.1 to −10.0	**Ethiopia** **Burundi** **Mauritania** **Morocco** **Senegal** **Sudan**	**Paraguay** **Haiti**		**Bangladesh**
−10.1 to −13.0	**Benin** **Egypt** **Rwanda** **Gambia** **Togo**		**Syrian Arab Rep.**	**Sri Lanka**
−13.1 to −20.0	**Central African Rep.** **Tanzania**	**Guyana**		
−20.1 to −30.0	**Mali** **Burkina Faso**	**Nicaragua**	**Yemen Arab Rep.**	
−30.1 to −50.0	**Guinea-Bissau**		**Jordan**	

[1] **External balances = exports-imports of goods and services; internal balances = saving-investment. The balance of trade and that of domestic saving/investment are *ex post* identical (see Box 3.10).**
Source: **World Bank, *World Tables* (1988-89 edition).**

saving, mainly of domestic origin, during the first half of the decade. Finally, the countries with deteriorating imbalances (group C), in both Latin America and the Caribbean and Africa, combined the worst features of a widening saving/investment gap: reduced investment rates and growing net income payments.

Domestic balances and the fiscal budget

For the 65 countries reviewed as a group, total central government expenditure accounted for a major (about one-quarter in recent years) and, up to the early 1980s, steadily increasing proportion of GDP. Growth in government revenues lagged behind expenditures in virtually all countries and years, the typical overall fiscal deficit being about 5 percent of GDP. (See identity iii) and comment in Box 3.10.)

The economic recession and domestic adjustment measures brought about a pronounced decline in both revenues and expenditures in most countries in the period 1981-83. However, difficulties in raising the tax base — not the least because of weak tax administrations — and a reduction in non-tax income and external grants, more than offset the effects of expenditure restraint. The result was a deterioration in the fiscal deficit, to 7 to 8 percent of GDP in 1982. Subsequently, uneven degrees of improvement were achieved.

The fiscal deficit followed similar patterns in the three country categories. However, while their deficits relative to GDP were comparable during the 1970s, trends and patterns tended to diverge in more recent years (Figure 3.2). Overall, the deficit was reduced in groups A and B, but worsened in group C.

In group A the deterioration in the deficit was less pronounced by the period 1981-83 than in the other two groups, and the subsequent improvement was relatively more significant. Shifts in the revenue/expenditure balances also differed. Group A showed a more gradual reduction in expenditures, coupled with a recovery in revenues during 1985-86. In group B, deficit trimming was mainly achieved through expenditure contraction although revenues also significantly increased in 1986; while in group C government expenditure actually accelerated strongly during 1982-85, outpacing revenue growth and bringing the fiscal deficit to almost 9 percent of GDP in 1985. It is only in 1986, the latest year for which data are available, that government spending was restrained in this group of countries. The broad picture appears to confirm the symmetry between patterns of adjustment — successful or otherwise — in macro-economic balances and those achieved through fiscal management.

The implications of fiscal balances on external accounts may be recalled. Since the fiscal deficit can only be financed internally by a private-sector saving in excess of investment or by an external current account deficit, in a period of depressed domestic saving a marked correlation is to be expected between the fiscal and current account deficits (Figure 3.3).[6]

Changes in government expenditure composition were even more striking than those in fiscal balances. In many countries interest payments on public debt came to represent a disproportionately high item in public budgets. This was particularly the case in Brazil, Mexico, Côte d'Ivoire and the Philippines. This process resulted from a greater reliance on domestic financing of the budget deficit although substituting internal for external debt did little to reduce the servicing burden. The counterpart to soaring interest payments was a reduction in the share of productive and social service expenditures in the public sector budgetary allocation (Figure 3.4).

Inflation is closely related to fiscal deficits. Faced with the inability to generate fiscal resources to finance their budget deficit, many governments resort to issuing currency faster than the increase in money demand at the prevailing prices and interest rates. Inflation may, in turn, worsen the deficit to the extent that expenditures follow the pace of price

[6] In a sample of 35 countries, two-thirds of the fiscal deficit for 21 countries was financed domestically and one-third externally by foreign loans and grants (data for 1983). See Special chapter on Financing agricultural development, in *The State of Food and Agriculture*, 1986.

Figure 3.2

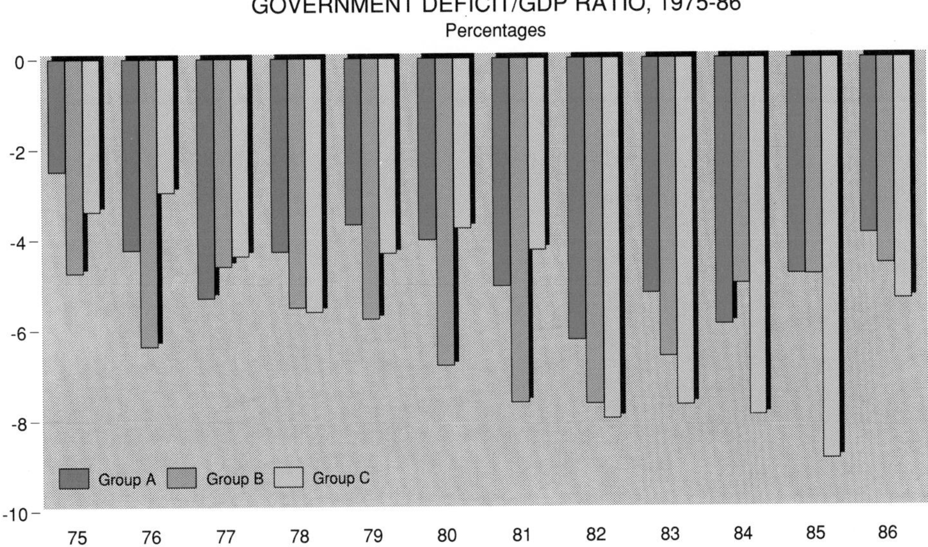

Source: FAO, based on IMF data

increases, but taxpayers tend to delay their payments. This again prompts money creation and closes the vicious circle — fiscal deficit/ price inflation. Such a process has become all too familiar in many countries during the 1980s, mainly in Latin America and the Caribbean but also in Africa and some Asian countries. However, aggregate data do not show any clear relationship between changes in the central government deficit and rates of price increase during the "adjustment" period. Inflation rates steadily accelerated in all regions and country groups, not the least in group A which had shown a relatively more favourable fiscal profile (Figure 3.5). As usual, however, these aggregate observations conceal exceptions. There were several cases of successful anti-inflation programmes, largely based on fiscal reform and austerity, of which Bolivia is the most outstanding recent example.

Figure 3.3

CENTRAL GOVERNMENT AND CURRENT ACCOUNT DEFICITS AS SHARE OF GDP, 1975-86

Percentages

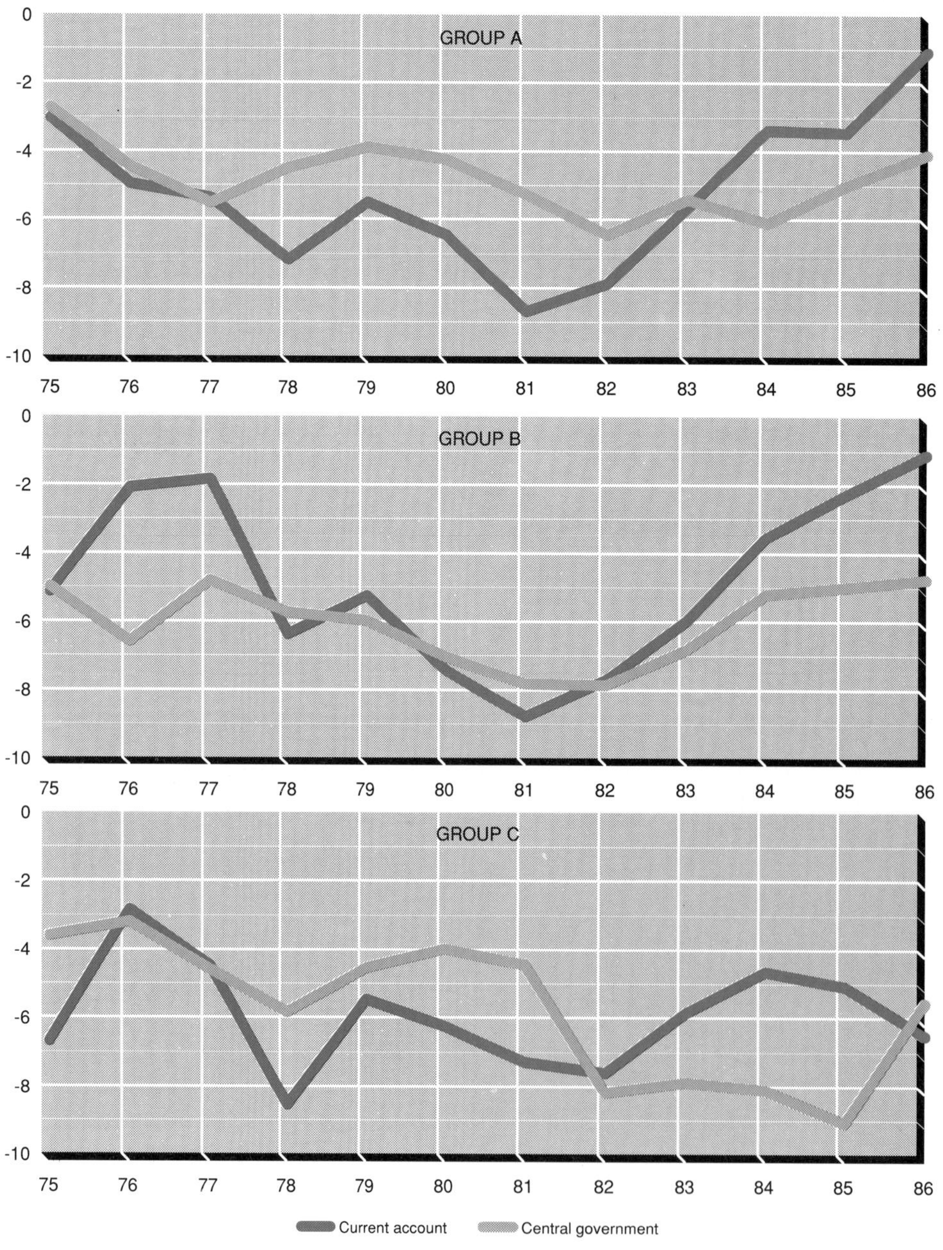

GROUP A

GROUP B

GROUP C

Current account — Central government

Source: FAO, based on IMF data

External (export minus import) balances

There were widely differing patterns in exports, imports and current account balances in the sample of 65 countries reviewed (Figure 3.6). In particular, there were pronounced deficits in trade accounts in the Near East and, to a lesser extent, Africa throughout the 1970s and 1980s, contrasting with comparatively smaller imbalances in their current accounts (implying a surplus in net factor income). This is explained by the presence of several large remittance-recipient countries in the Near East group and low-income economies benefiting from interofficial transfers in Africa. In Latin America and the Caribbean, where deficits in trade balance are consistently smaller than those in current accounts, the gap was closed by factor income payments. In the case of the Far East region, the balances of trade and current accounts have moved close together, underlining the region's relatively small reliance on external sources of financing for investment.

There had been a steady growth in the importance of developing countries' total trade in relation to GDP during the 1970s. This was true for all developing country regions, despite wide variations in the degree of openess of their economies. Aggregate trade balances were generally negative during this period, with the deficit fluctuating around a relatively stable range of 1 to 5 percent of GDP in most years, except in the Near East where it averaged approximately 20 percent.

The economic recession of the early 1980s caused a reversal of these earlier trends. Despite the varying degree of economic openness, a widespread aggravation of external deficits occurred in all regions, together with a

Figure 3.4

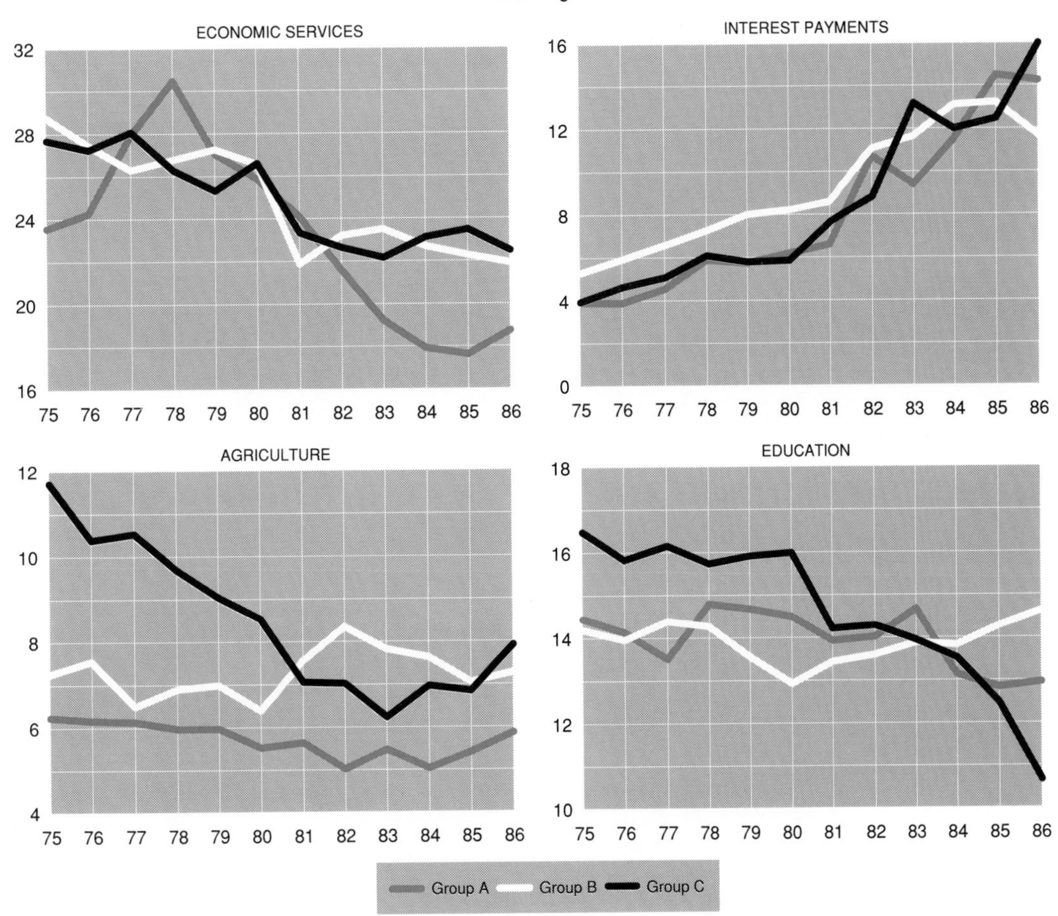

Source: FAO, based on IMF Finance Statistics, 1988

decline in the overall share of total trade in GDP. While widening deficits, mainly the result of an upsurge in imports, had already been noticeable in the late 1970s, the tendency was accentuated during 1981-83, as imports continued to expand but export growth slowed.

At their largest in the period 1981-82, trade deficits as a share of GDP reached 6 percent in Latin America and the Caribbean, 10 percent in Africa, 8 percent in the Far East and as much as 30 percent in the Near East. The ensuing pressure to reduce external imbalances and generate trade surpluses triggered a process of measures to promote exports and contract imports. Between 1981-83 and 1985-87, trade balances improved in three-quarters of all Latin America and the Caribbean countries, two-thirds of those in Africa and all but one — Indonesia — in the Near East and Far East regions. By the years 1985-87 the trade account

Figure 3.5

YEARLY CHANGES IN CONSUMER PRICES, 1976-87

Yearly percentage changes

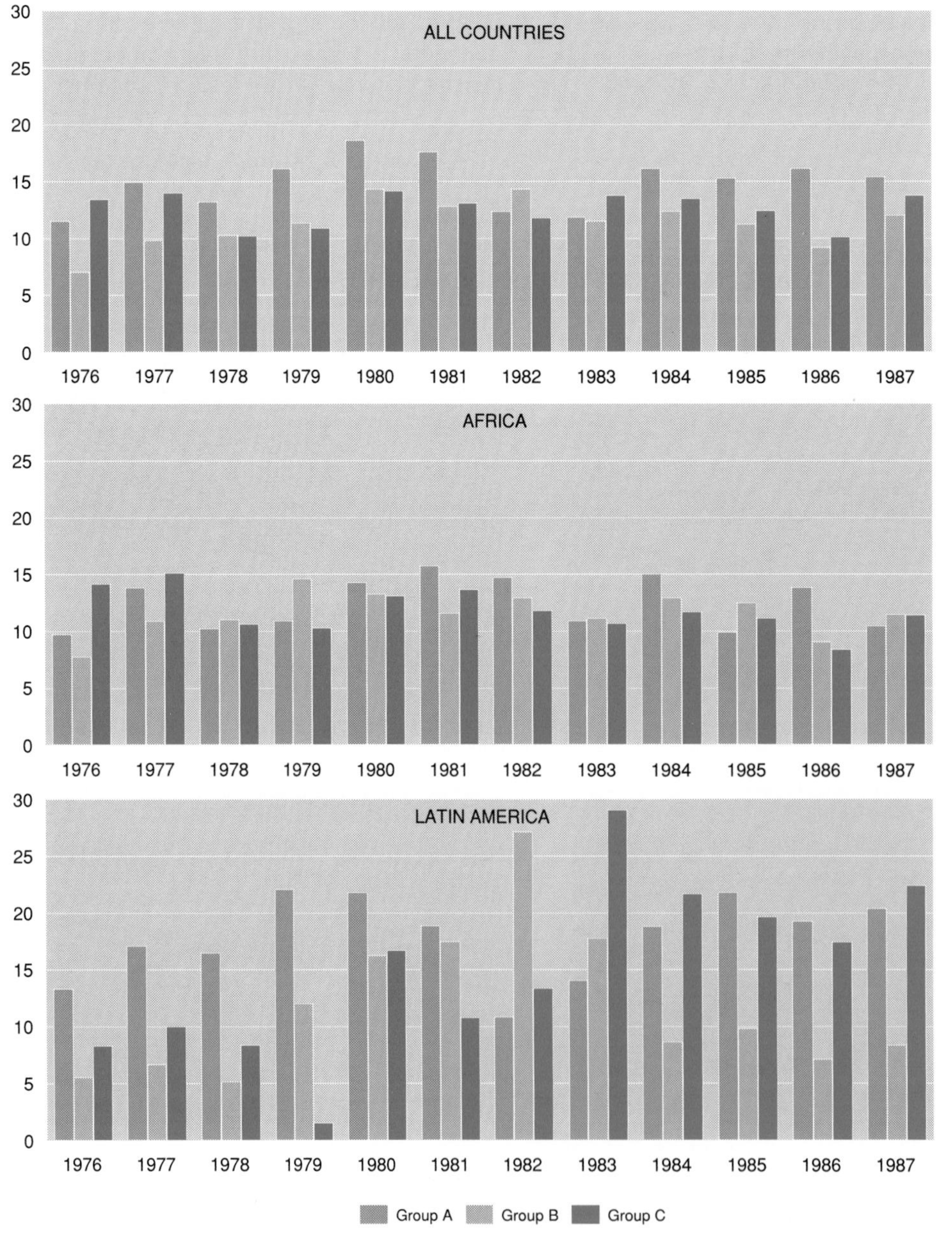

ALL COUNTRIES

AFRICA

LATIN AMERICA

1976 1977 1978 1979 1980 1981 1982 1983 1984 1985 1986 1987

Group A Group B Group C

Source: FAO, based on IMF data

Excluding ten countries for which inflation rates were more than double their respective group average.

Figure 3.6

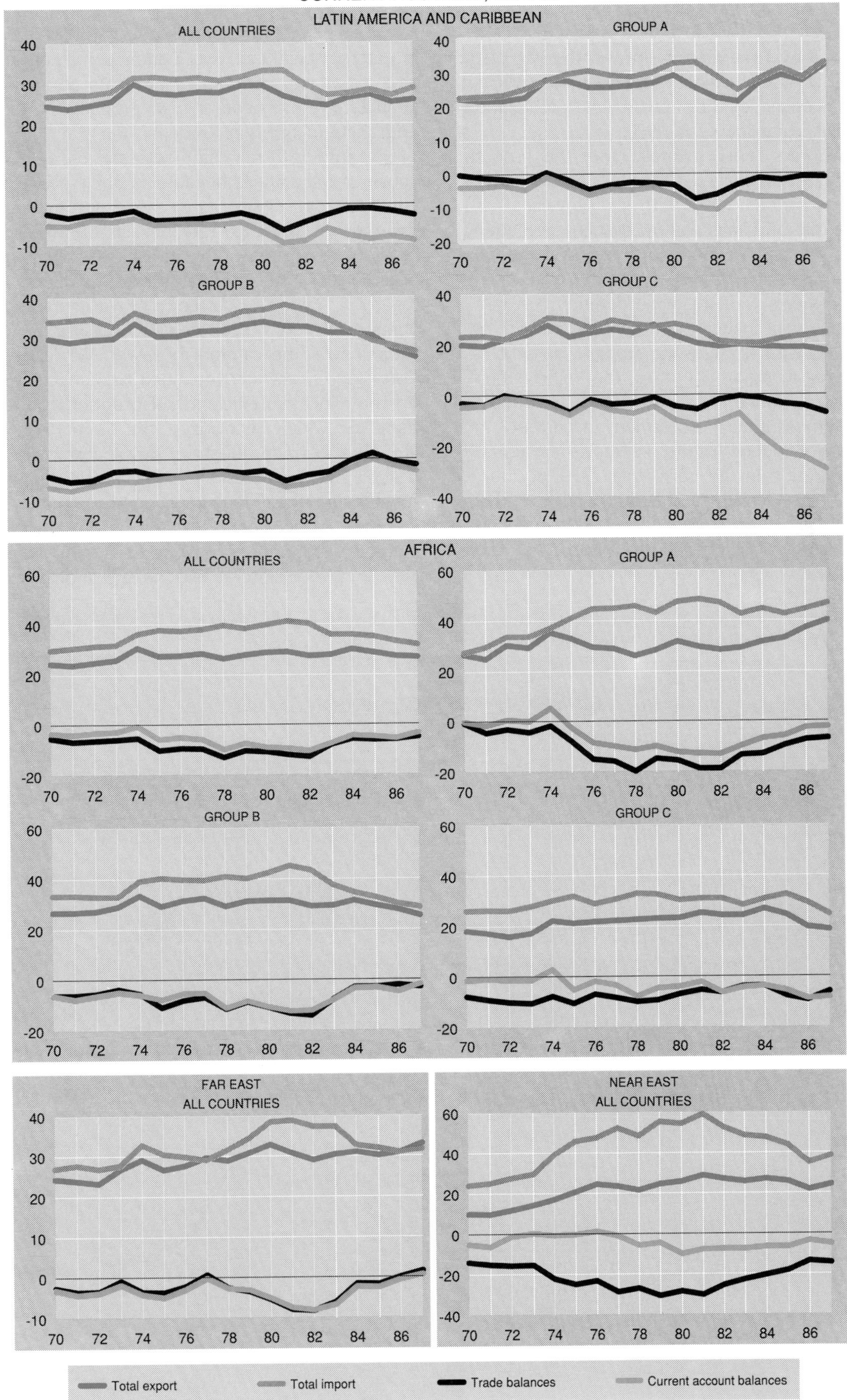

Source: FAO, based on data from World Bank World Tables 1988-89 edition.

had been brought close to a balance in Latin America and the Caribbean and the Far East, while in Africa and the Near East the deficits were reduced to their lowest levels in relation to GDP since the early 1970s.

As noted earlier, in most cases this reduction was initially achieved through a process of import-contraction — with the prevailing economic and market environment preventing a rapid growth in exports, and export-promotion measures requiring some time to bear fruit. Furthermore, with the wide deterioration in international financing and lending conditions coupled with a mounting problem of debt servicing in many countries, the improvement in trade balances was not always rewarded by a commensurate improvement in current accounts. Indeed, Figure 3.6 shows an actual deterioration in the current account position of countries in group A in Latin America and the Caribbean, even though they had succeeded in reducing their trade deficits in an economically "healthy" way.

Among the many factors behind movements in external balances, two influences of common and major importance can be explored in the aggregate: movements in real exchange rates and shifts in terms of trade.

Exchange rate misalignments can be a major cause of imbalances because of their pervasive effects across all sectors of the economy. For this reason, exchange rate policies are at the heart of all stabilization and adjustment programmes. The interpretation of evidence regarding aggregate currency movements and changes in macro-economic disequilibria is subject to some caution, however. While the distorting effects of overvalued exchange rates on the allocative efficiency of production and trade are widely recognized, the ultimate growth and trade effects of devaluation are difficult matters to assess. Some authors have pointed out that, depending on the structure of incomes in the different saving groups, shifts in income distribution following devaluation can lead to a recession, at least in the short term.

Figure 3.7

REAL EXCHANGE RATE, 1970-87

Index (1980 = 100) *

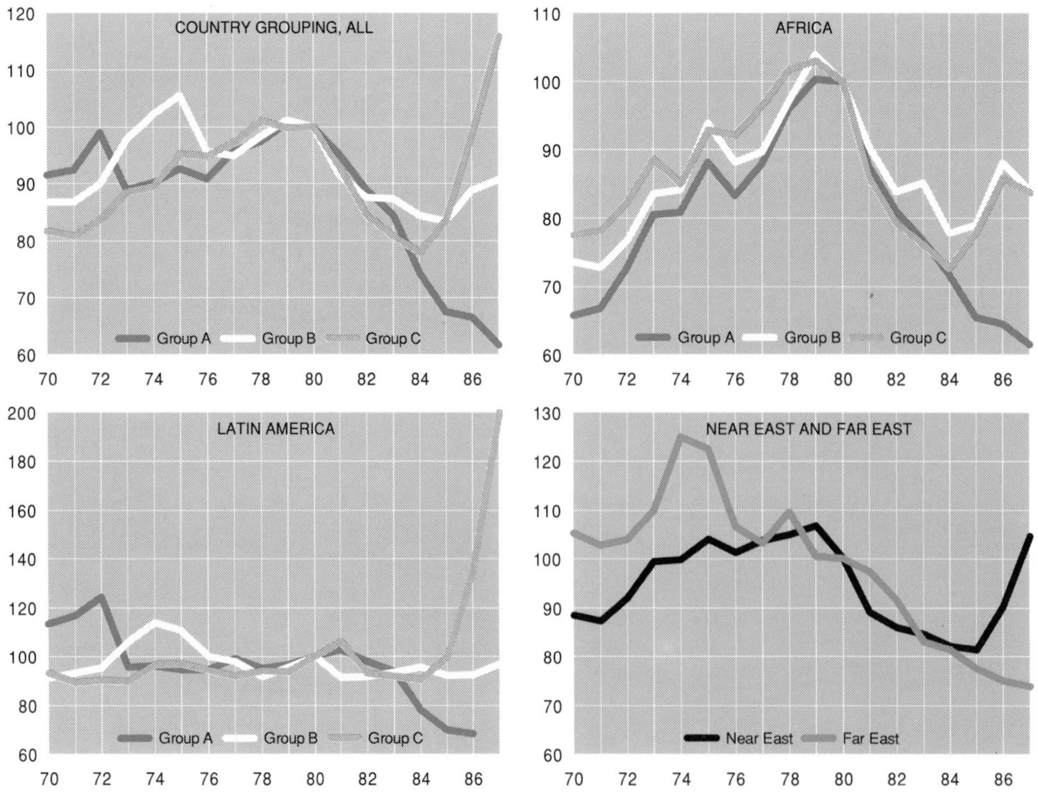

Source: based on IMF data

* Calculated as the ratio of nominal exchange rates *vis-à-vis* the US dollar and local *vis-à-vis* US inflation rates.

A recessionary impact is also to be expected from the restrictive effects on imports of capital/intermediate input goods. It is only under quite restrictive assumptions about intersectoral flexibility, and the ability to check domestic inflation in relation to that of trading partners through appropriate fiscal and monetary policies, that devaluation will yield positive growth and trade results.

Despite these reservations, the experience of the 1980s seems to show a positive relationship between currency devaluation and the ability to reduce external imbalances successfully. In both Africa and Latin America real exchange rates followed similar patterns in the three groups of countries during the 1970s — with a marked tendency for currencies to be overvalued in Africa (Figure 3.7). However, trends widely diverged after 1984. Countries in group A significantly devalued their currencies, but those in group B did not, while currencies in group C significantly appreciated. Real exchange rates in countries in the Far East and Near East regions showed a similar downward slope since the late 1970s but, while in the former the trend continued unabated until 1987, the latter's currencies appreciated in the period 1986-87.[7]

Another factor explaining the diverging experiences of external adjustment was the ability to adjust to adverse movements in international terms of trade. Despite the differences in the structure of their exports, all regions and country groups were hit by a

[7] The presence of extreme cases of appreciation or depreciation sometimes influenced the unweighted country group averages. For instance, the sharp appreciation of currencies in group C in Latin America between 1985 and 1987 was largely caused by the presence of Nicaragua. Similarly, the regional appreciation in the Near East during 1986-87 was largely influenced by the experiences of Egypt and the Syrian Arab Republic. However, the diverging patterns in the three groups remain evident, if somewhat attenuated, even after elimination of these extreme cases. In particular, the strong currency devaluation in group A was common to nearly all countries.

Figure 3.8

TERMS OF TRADE AND PURCHASING POWER OF EXPORTS, 1975-87

(Index 1980 = 100)

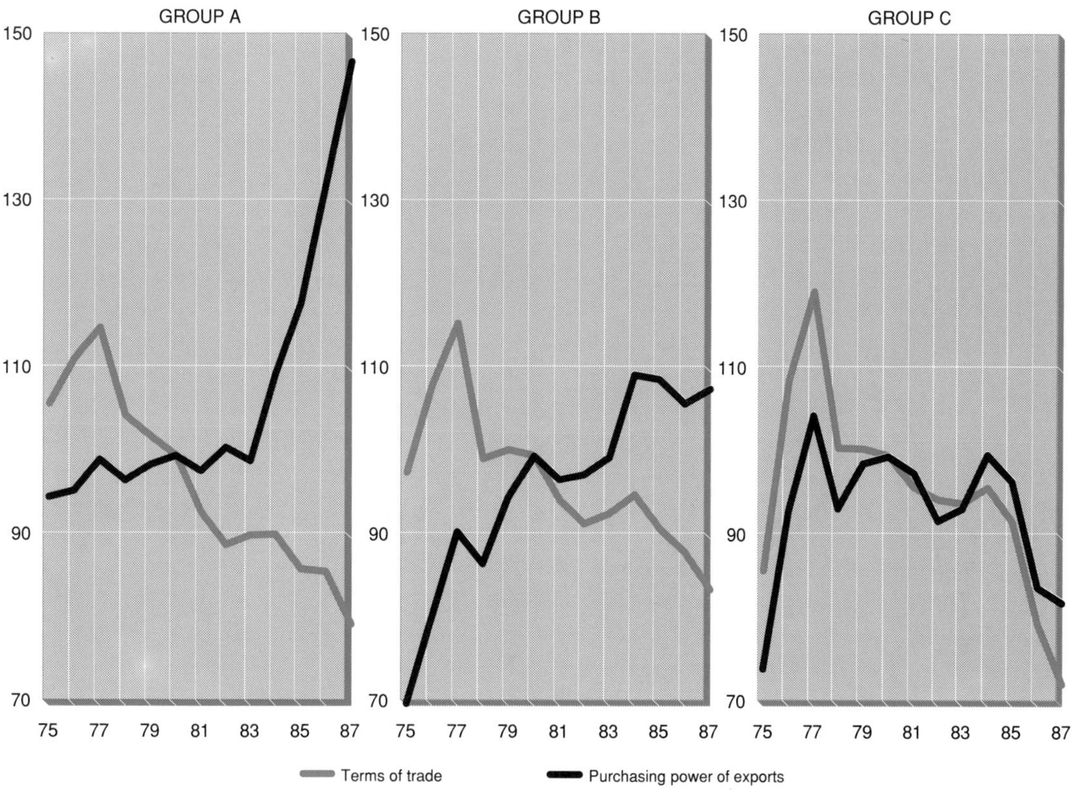

GROUP A
GROUP B
GROUP C
150
130
110
90
70
75 77 79 81 83 85 87
Terms of trade
Purchasing power of exports

Source: FAO, based on IMF data

severe deterioration in the prices of their exports compared to those of imports (net barter terms of trade). From 1980 to 1987 the decline was in the order of 20 percent in group A, 15 percent in group B, and 27 percent in group C. What clearly made the difference between these groups was their capacity to expand the volume of their exports so as to maintain or even expand their purchasing capacity (Figure 3.8). Group A was very successful in this respect with its income terms of trade (or purchasing capacity of exports) improving by nearly 50 percent between 1980 and 1987. By contrast, in groups B and C, net barter fell almost *pari passu* with income terms of trade, implying little compensation through expanded export volumes. It is in Africa that group A achieved the most significant improvement in income terms of trade, particularly during 1985-87, although some improvement also benefited countries in group B. Strong gains were widespread in group A, the notable exception being Nigeria where the fall in oil prices severely reduced the purchasing power of its exports. Improvements in net income terms of trade were found less frequently in group B (such cases were mainly concentrated in Asian countries) and even less so in group C.

Adjustment and growth

To the extent that imports and investment — the most likely twin victims of adjustment — are basic ingredients for growth, their association with changes in GDP can be explored, though bearing in mind that shifts in the allocation or efficiency of investment and the composition of imports are equally important determining factors for growth. In addition, the growth response may lag so that conclusions regarding recent trends can only be tentative.

Despite these reservations, the typical experience of "growth-disequilibrium-adjustment", common to many developing countries, can be clearly traced from Figure 3.9. At the aggregate regional level the overall tendency was an upward trend in imports and investment during the 1970s, supporting variable

Figure 3.9

CHANGES IN GDP AND
IMPORTS AND INVESTMENT AS A SHARE OF GDP, 1970-87

LATIN AMERICA AND CARIBBEAN

(Percentages)

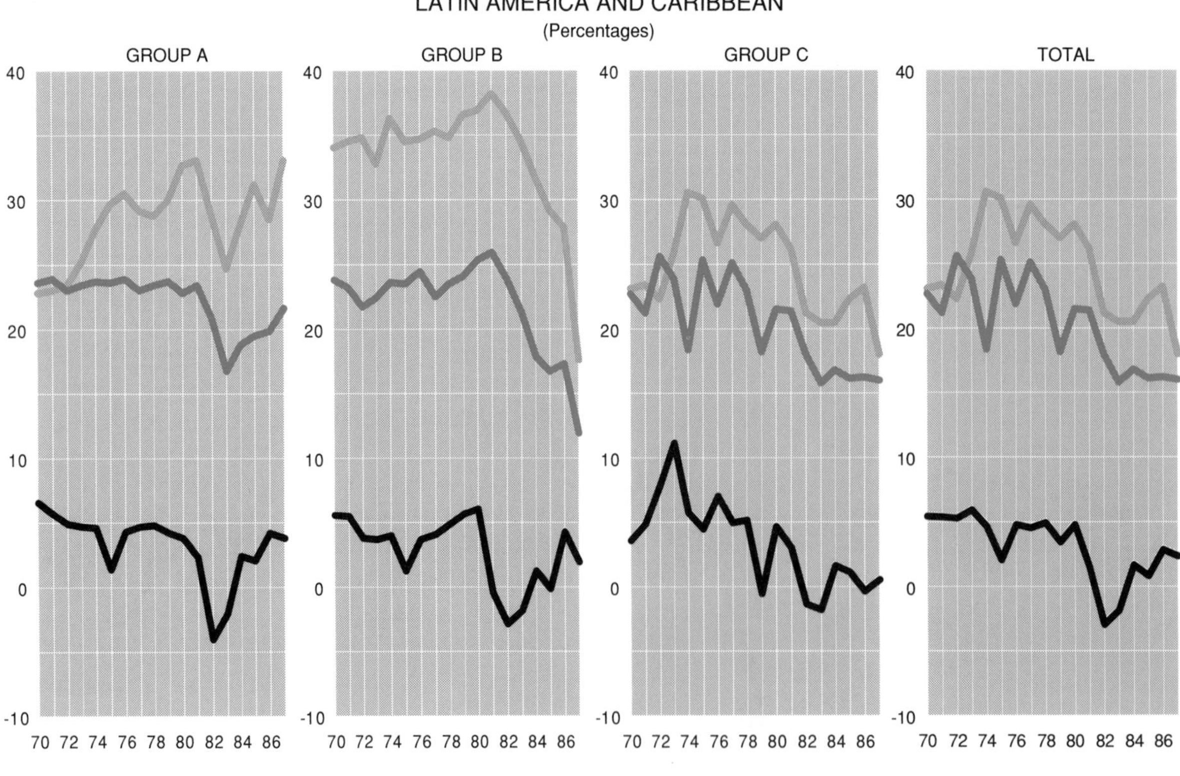

Source: FAO, based on World Bank data

* Yearly changes ** As % of GDP

(cont.)

but by and large (except in Africa) high and sustained rates of GDP growth; a marked downturn in the three variables in the early 1980s; and a very uneven recovery in recent years with all three variables remaining at a lower plateau than during the 1970s. Behind this overall pattern, varying situations are observed within regions. The contrasting growth performances between groups of countries in recent years were particularly significant in the context of adjustment. GDP growth in group A in Latin America and the Caribbean averaged 3.1 percent yearly between 1985 and 1987 compared to 2.2 percent for group B and only 0.9 percent for group C. In Africa the respective growth rates were 4.8 percent, 2.4 percent and 2.2 percent during the same three-year period. Comparisons are less meaningful for the fewer Far East and Near East countries reviewed, most of which fall into category B. However, the Republic of Korea, Thailand and Turkey, the three countries in group A in these regions, were among those achieving the most impressive growth performances in recent years.

IMBALANCES, ADJUSTMENT AND AGRICULTURE

The role of agriculture in the process of adjustment is examined here in terms of the sector's contribution to growth; as well as its capacity to generate trade surpluses; its position with regard to external debt; and its position with regard to government expenditure. The net effects of changes in agricultural production and trade on food availability and self-sufficiency are finally explored.

GDP and agricultural growth

A first indication of the role of agriculture in the process of adjustment can be seen from movements in the share of agricultural GDP in total GDP (Figure 3.10). The secular decline in this share, an expected consequence of development, was arrested during the economic crisis of the early 1980s and even reversed during the following years. This pattern was common to Latin American and African

Figure 3.9 (cont.)

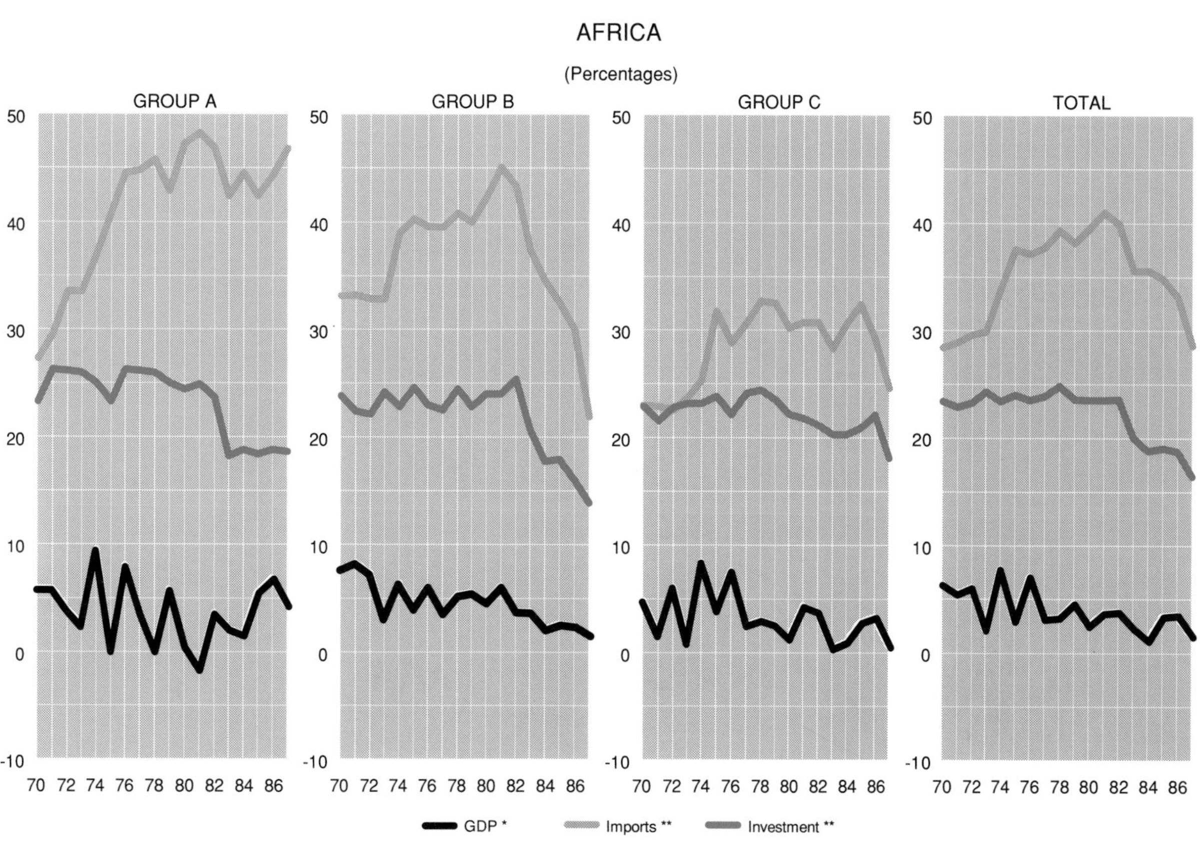

Source: FAO, based on World Bank data

* Yearly changes ** As % of GDP

(cont.)

countries, with all subgroups sharing in it; it was less so in the Near East and Far East regions where the agricultural GDP/total GDP ratio levelled off after the period 1981-83.

This can be further explored by comparing yearly changes in GDP, agricultural GDP and the share of agriculture in total GDP in the four developing country regions (Figure 3.11). Agricultural growth lagged behind that in other sectors in most years, more markedly during the 1970s, but its contribution to growth was positive except in years of exceptionally severe production shortfalls.

These features would generally confirm the traditional role of agriculture as a support rather than an engine of economic growth in periods of expansion, as the 1970s generally were; and, by virtue of the sector's inertial qualities, its capacity to absorb shocks and preserve resources for recovery in periods of depression.

With regard to production performances, a common feature of most country and agricultural product groups was a slow-down in the rate of output growth during the 1980s in relation to the previous decade (Table 3.7). Notable exceptions were African countries in group A which achieved faster growth in total food and agricultural production during the 1980s. It is only in group A that the growth rate in food production during the 1980s came close to that of population in Africa, and still exceeded it by a substantial margin in Latin America and the Caribbean.

Agriculture and external balances

Agricultural exports. Against the background of generally depressed overall trade conditions during the early 1980s, agriculture maintained or gained relative importance in total merchandise trade of developing countries between 1981-83 and 1985-87. In developing countries as a whole there was a slight increase in the share of agricultural exports in total exports from 26 percent to about 27 percent; a rising share of manufactures from 11 percent to

Figure 3.9 (cont.)

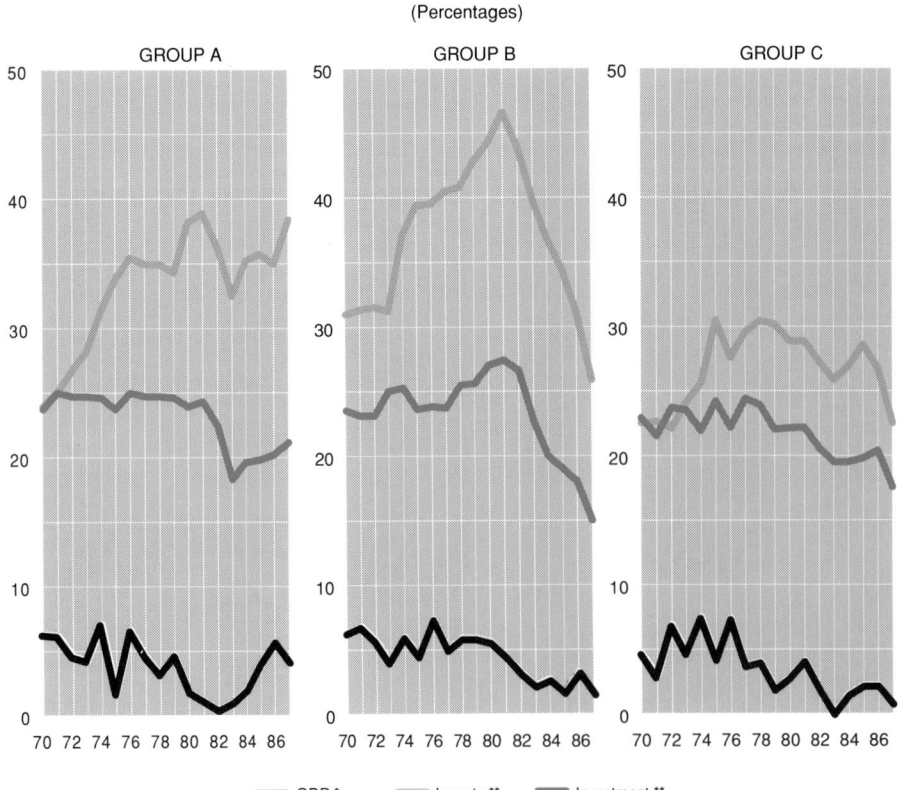

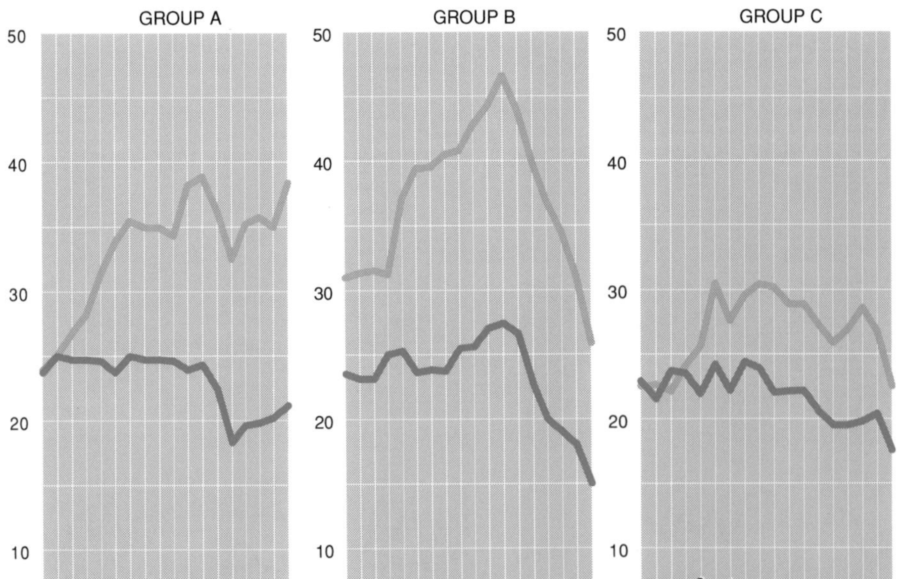

Source: FAO, based on World Bank data

* Yearly changes ** As % of GDP

TABLE 3.7 **Average yearly changes in agricultural production, total and selected agricultural products, 1970-80 and 1981-89**

Item	Africa 1970-80 A	B	C	Africa 1981-89 A	B	C	Latin America & Caribbean 1970-80 A	B	C	Latin America & Caribbean 1981-89 A	B	C	Near East 1970-80 All	Near East 1981-89 All	Far East 1970-80 All	Far East 1981-89 All
Agriculture	1.1	3.1	1.5	2.7	2.5	1.8	3.6	1.2	3.5	3.0	0.9	2.2	3.2	1.9	3.5	3.0
Food	1.1	3.3	1.8	2.7	2.3	1.8	4.1	1.3	3.8	3.1	0.9	2.3	3.4	2.1	3.8	3.2
Crops	0.8	2.8	1.3	2.7	2.3	1.7	3.7	2.1	2.6	3.5	1.7	1.0	3.5	1.8	3.4	2.8
Cereals	0.9	1.5	2.6	5.2	3.4	3.1	4.8	0.8	3.9	3.3	2.4	1.4	3.4	—	2.9	3.4
Stimulants	−0.4	5.6	−0.6	1.5	5.4	1.5	5.2	3.1	3.9	5.7	2.3	4.0	9.8	8.1	3.7	4.3
Sugar	2.9	10.2	1.8	3.8	4.9	2.7	5.7	2.1	1.6	5.8	−1.2	0.2	5.1	5.0	2.9	0.4
Oil crops	−2.9	4.2	−0.7	4.0	0.2	0.6	12.6	10.6	3.1	5.9	6.6	1.0	11.8	3.6	7.1	4.4
Livestock	2.5	4.1	1.9	2.7	3.2	2.1	3.6	0.6	5.1	2.2	0.2	3.7	2.3	2.3	4.7	4.7

Source: FAO.

TABLE 3.8 **Total merchandise and agricultural trade as a share of GDP, developing regions, 1970-80, 1981-83 and 1985-87**

Item	Period	Latin America & Caribbean Total	A	B	C	Africa Total	A	B	C	Far East & Pacific All	Near East All
		(% of GDP)									
Total exports	1970-80	21.9	21.7	22.6	21.3	21.2	21.9	23.4	18.0	22.2	10.2
	1981-83	19.9	18.6	20.9	20.5	20.8	19.6	23.0	19.0	23.9	13.7
	1985-87	18.2	21.0	15.8	17.2	22.0	24.7	26.0	15.7	25.1	11.4
Agricultural exports	1970-80	8.7	8.5	7.4	11.0	10.0	12.2	10.6	7.9	9.3	4.6
	1981-83	6.8	6.6	6.5	7.7	7.4	7.7	7.6	6.8	6.7	4.0
	1985-87	6.6	7.9	5.0	6.8	7.2	7.2	8.3	5.9	6.2	2.7
Total imports	1970-80	26.3	22.4	33.5	22.5	27.3	28.7	31.1	21.8	26.1	34.0
	1981-83	25.9	23.3	31.9	19.7	30.0	33.2	33.9	23.2	31.3	45.2
	1985-87	22.3	23.7	22.6	19.6	26.4	32.0	25.9	23.6	26.7	33.1
Agricultural imports	1970-80	3.8	3.4	3.7	2.7	5.6	7.2	5.7	4.3	5.5	9.3
	1981-83	3.6	3.1	4.7	2.9	6.4	10.0	6.0	4.7	4.5	10.0
	1985-87	2.9	2.6	3.7	2.2	5.7	9.5	4.7	4.5	3.7	7.3

Source: FAO.

13 percent; and a pronounced fall in that of fuels from 64 percent to 51 percent. Agriculture contributed a greater share of total export earnings in all developing country regions except the Far East.

Agricultural trade was particularly affected by the economic recession of the early 1980s. Indeed, the share of agricultural exports in GDP declined proportionally more than that of total merchandise exports during 1981-83 in relation to the average levels of the 1970s (Table 3.8).

By 1985-87 this pattern shifted. While the share of agricultural exports in GDP continued to decline in all regions on the aggregate, the decline was less marked than that in other sectors. There were significant variations to this trend among country groups, however. In Latin America and the Caribbean, agricultural trade appeared to have played a major role in both successful and unsuccessful attempts at adjustment and stabilization. In group A countries in Latin America the agricultural

BOX 3.9

Expenditure in agriculture

Data limitations prevent the comprehensive analysis of the effects of domestic adjustment on agricultural saving and investment, particularly from private sources. Some insight can be gained nevertheless by examining government agricultural investment.

Agricultural development expenditure has a typically low share of total central government expenditure (about 6 to 7 percent of the total) and a correspondingly low share of agricultural GDP.

From a regional perspective, the highest shares were in the Far East and Pacific and Near East countries (about 8.5 percent in each), with the share falling to 7.5 percent in Africa and 6.5 percent in Latin America and the Caribbean.

There were wide variations between countries in these averages, ranging from 20 to 25 percent in Guyana, Botswana, Cyprus and Sri Lanka to less than 2 percent in Costa Rica, Sierra Leone, Bolivia and Uganda during 1984-86. Trends are perhaps more interesting than country differences in levels of development expenditure. Comparing the averages for 1975-80 and 1981-86, the share of government development expenditure in agriculture increased in 26 out of 43 countries (61 percent), declined in only 14 countries and remained stable in three. On the other hand, agricultural current expenditure remained stable, or even tended to decline, suggesting that most governments attempted to protect the investment component of agricultural expenditure. The perceived importance of agricultural investment in most countries is thus indicated by the fact that, despite strong pressure to reduce government expenditure, and a sharp reduction in the part devoted to development activities, agriculture maintained or increased its share, both in real terms (in relation to agricultural GDP) and *vis-à-vis* government expenditures on other sectors.

Government development expenditure in agriculture as share of agricultural GDP, 1975-80, 1981-83 and 1984-86

Country groups (No. of countries)	1975-80	1981-83	1984-86
	(......................	%	......................)
A (11)	7.5	9.1	7.9
B (12)	5.1	7.6	8.2
C (11)	4.9	4.5	5.2

Source: IMF, *Government Finance Statistics*, several issues.

trade/GDP ratio increased, while in groups B and C it declined. In Africa, by contrast, the relatively successful overall trade performance of group A countries was mainly spearheaded by non-agricultural exports, as the ratio of agricultural exports to both GDP and total merchandise exports continued to fall in all countries of this group. In the Far East and Pacific and Near East regions, agricultural exports lagged behind total export growth, with these two regions experiencing the sharpest fall in the contribution of agricultural exports to GDP in relation to the averages of the 1970s. This was the case both for countries where export earnings had an important agricultural component — Sri Lanka, Fiji, the Philippines — and for those where this was relatively less so.

This uneven performance of the sector is to be seen in the context of the international market conditions for agriculture prevailing during this period. It may be recalled that various structural and temporal factors combined to depress commodity prices throughout the decade. They included: excess supply in relation to effective demand; increased market competition following the entry of new exporters; and widespread export subsidies and protectionism. The recession of the early 1980s accentuated the effects of these factors. Export prices for agricultural products fell strongly, both from the levels of the early 1980s and in relation to prices of other tradables.

The general observations made with regard to total terms of trade largely apply to agricultural terms of trade — expectedly, given the significant weight of agricultural trade to the total in most countries reviewed. That is, while a large majority of countries faced a deterioration in net barter agricultural terms of trade, many of them succeeded in expanding the volume of their agricultural exports, thus underlining the responsiveness of the sector on the side of supply, and the pressure to generate export earnings despite adverse conditions. Out of the 65 countries reviewed, only four showed an unambiguous upward trend in agricultural terms of trade between 1980 and 1988 (Burkina Faso, Mauritius, Senegal and Bolivia). All the others faced a deterioration in terms of trade, in some cases partially compensated by an increased export effort, as shown by the diverging trends in agricultural export volume and value. Figure 3.12 reveals a significant difference in the three groups, however. Group A increased the volume of its exports by about 40 percent between 1980 and 1985-87 (in contrast with the stagnation of the previous five years) and, despite depressed prices, export value also rose by 25 percent during the same period. Group B achieved a 20 percent increase

Figure 3.10

AGRICULTURAL SHARE OF TOTAL GDP, 1975-87
ALL COUNTRIES
Percentages

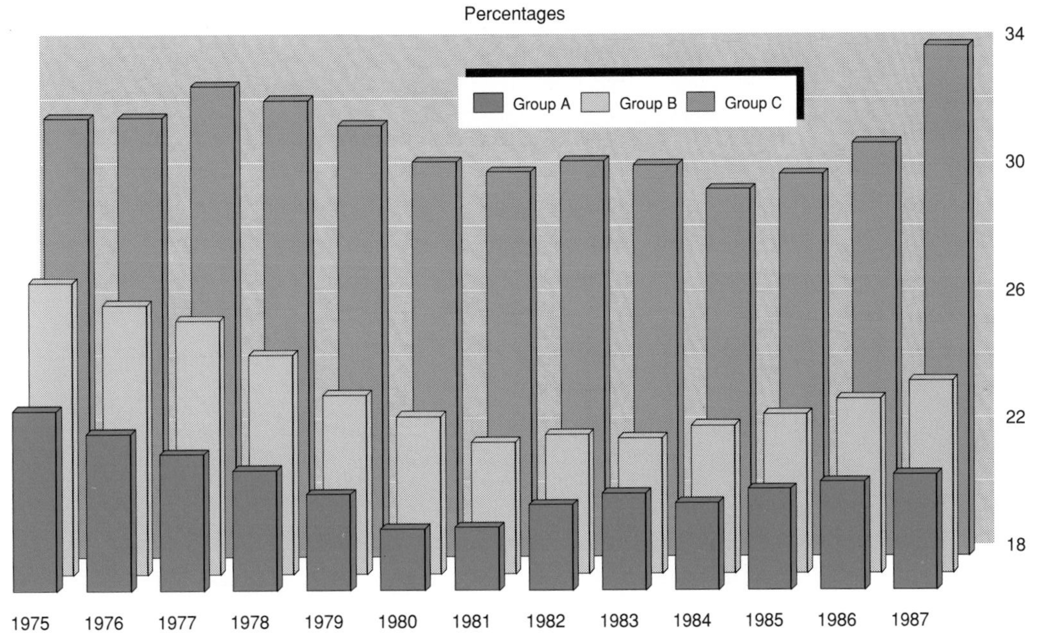

Figure 3.11

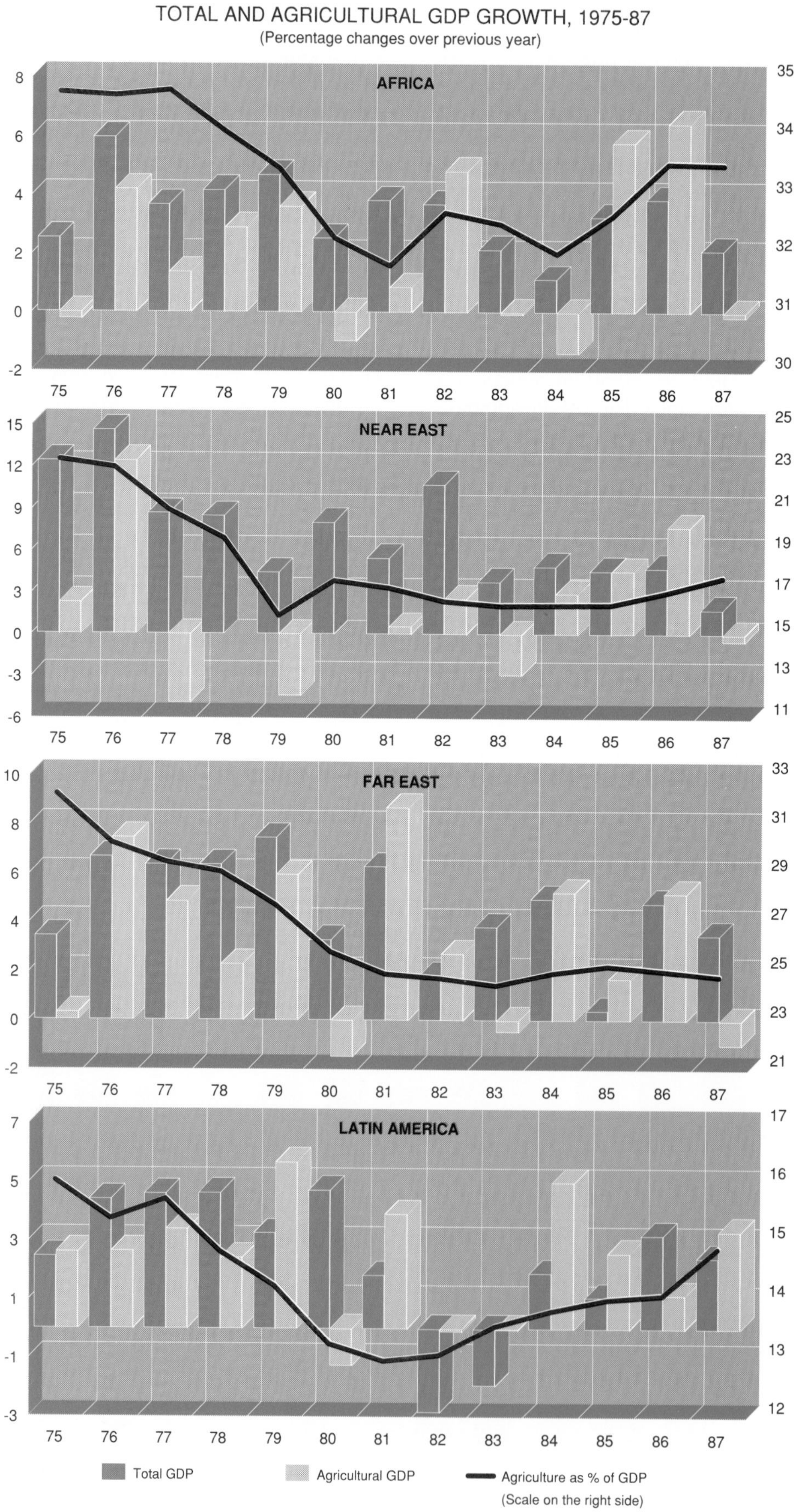

TOTAL AND AGRICULTURAL GDP GROWTH, 1975-87

(Percentage changes over previous year)

AFRICA

75 76 77 78 79 80 81 82 83 84 85 86 87

NEAR EAST

75 76 77 78 79 80 81 82 83 84 85 86 87

FAR EAST

75 76 77 78 79 80 81 82 83 84 85 86 87

LATIN AMERICA

75 76 77 78 79 80 81 82 83 84 85 86 87

Total GDP

Agricultural GDP

Agriculture as % of GDP (Scale on the right side)

in the volume of exports but this achievement was unrewarded in terms of increasing export earnings; while in the third group, volumes only showed a tentative revival after years of decline and export values fell about 15 percent.

A similar picture is provided by net barter and income terms of trade in the three groups of countries (Figure 3.13). It shows that, while the behaviour of net barter agricultural terms of trade did not markedly differ in the three groups, their respective capacity to expand the volume of shipments determined contrasting trends in their purchasing power. With some qualification, this general trend is also observed at the regional level (Table 3.9).

What was the potential value of export earnings from agriculture foregone as a result of the decline in export prices? In other words, what is the difference between what countries actually received in payment for their agricultural exports during the 1980s and what they would have received from the same volume of exports had prices remained at 1979-81 levels?[8] It can be seen that, despite wide variations, losses were considerable in all years and regions (Table 3.10). Countries in the Near East experienced the largest potential losses, equivalent to 30 to 60 percent of their actual export earnings from agriculture. Even in Latin America and the Caribbean, however, where losses in relation to actual agricultural exports were lowest, their amount was substantial in the context of the region's financial problems. As a reference, the total debt-service payments (payments on total long-term, short-term and IMF credit) in 1987 amounted to $46 208 million in the group of 20 Latin American and Caribbean countries. Therefore, the potential losses in agricultural export earnings that year — $4.83 billion — would have covered over 10 percent of those payments.

Agricultural imports. The other dimension of agricultural trade adjustment concerns food and

[8] This theoretical exercise is only intended to illustrate the major impact of price changes on export earnings. Indeed, with generally inelastic demand and expanded volumes of exports, international prices were unlikely to remain at the relatively high levels of 1979-81.

Figure 3.12

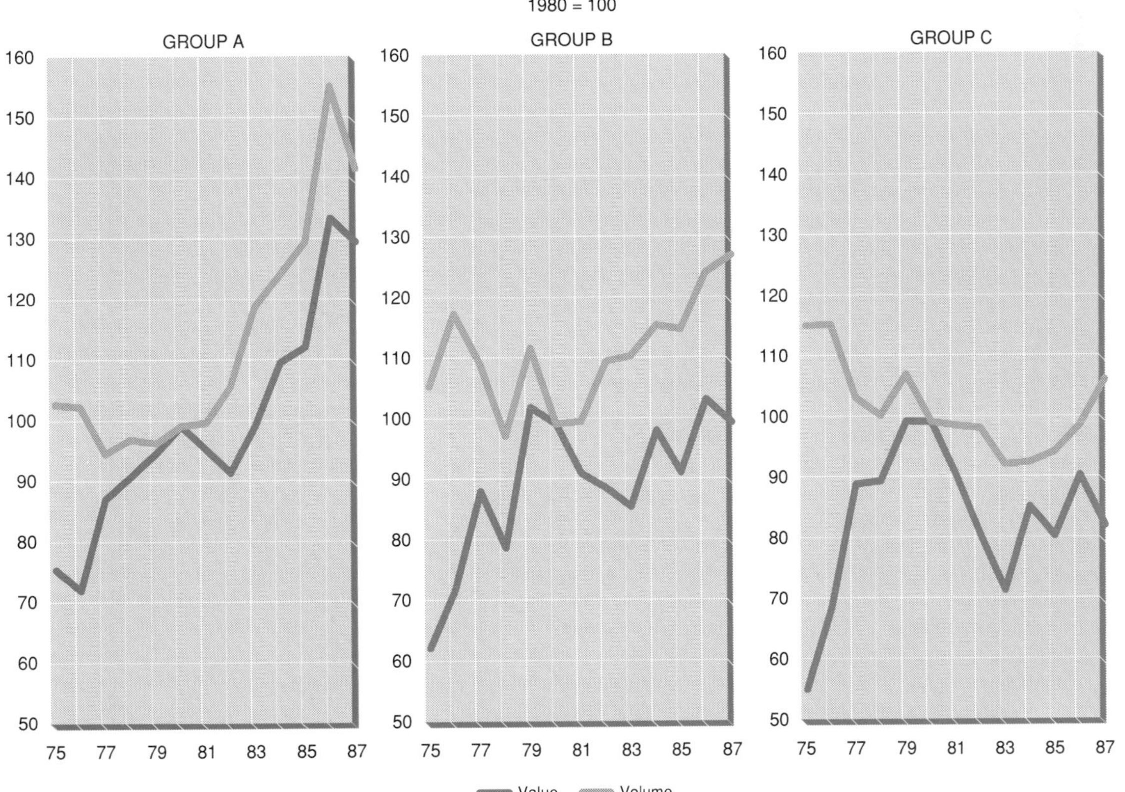

TABLE 3.9 **Net barter (NB) and income (I) terms of trade, 1985-87** (base 1980 = 100)

Country groups	Latin America & Caribbean		Africa		Near East		Far East	
	Net barter	Income	Net barter	Income	Net barter	Income	Net barter	Income
A	**95**	**130**	**122**	**147**				
B	**90**	**83**	**93**	**128**	**79**	**84**	**90**	**97**
C	**92**	**89**	**87**	**82**				

Source: FAO.

agricultural imports. A common feature to all developing regions, and all but a few individual countries, was a sharp slow-down in the growth of such imports during the 1980s compared to the average rates of the 1970s (Table 3.11).

This slow-down occurred despite markedly lower import unit values, underlining the financial incapacity of many countries to take advantage of relatively abundant supplies and low market prices. It also took place during a period of constrained domestic supply conditions, as the increase in food production during 1981-88 barely kept pace with population growth in the Near East and failed to do so in much of Latin America and the Caribbean and Africa. Also, while there is some correlation in individual countries between levels of domestic food production in a given year and food import volumes the following year, this was less so after 1981, suggesting a greater importance of financial constraints influencing import decisions in many countries.[9]

Turning to the 65 selected countries used in this review, the slow-down in agricultural import growth during the 1980s is also seen in nearly all regions and groups (Table 3.12).

Drastic reductions were experienced in group A, particularly in Latin America where, in contrast to the relatively sustained growth in

[9] For instance, Latin America and the Caribbean reduced its volume of food imports by a cumulative 14 percent between 1983 and 1987, but food production only rose by a 1.3 percent yearly average (or 5.2 percent cumulative) during the same period.

Figure 3.13

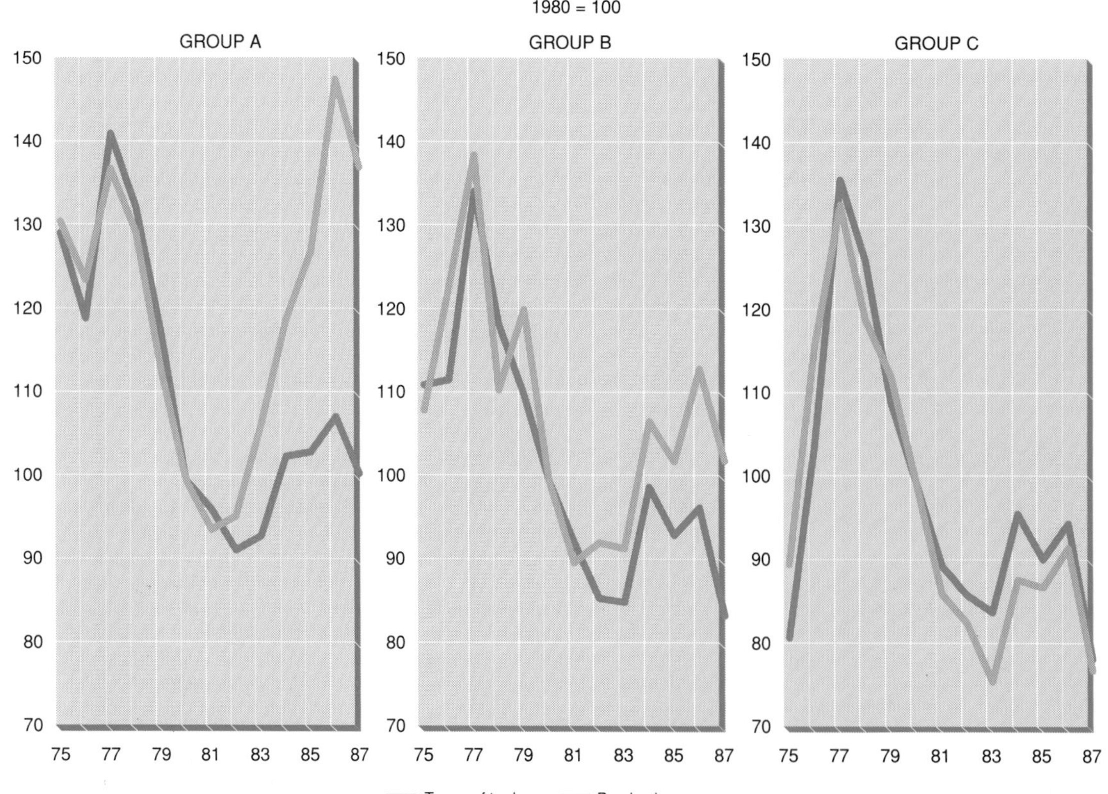

TABLE 3.10 **Potential losses in export earnings from changes in agricultural export prices, millions of dollars and as percentage share of agricultural exports**[1]

Regions and country groups		1982	1983	1984	1985	1986	1987	1988
Latin America & Caribbean								
A	*$ million*	−1 957.7	−2 033.8	294.9	−2 579.5	−328.6	−2 486.8	−890.8
	%	17.6	17.3	2.2	20.3	2.7	21.0	6.8
B	*$ million*	−1 029.1	−2 356.0	−1 403.0	−2 591.3	−1 968.5	−1 266.5	−511.0
	%	17.5	34.5	19.6	38.4	33.9	25.8	7.8
C	*$ million*	−469.9	450.2	12.2	−510.3	11.9	−1 077.6	−1 141.7
	%	16.8	19.8	4.1	16.7	0.3	30.9	31.3
All countries	*$ million*	−3 456.7	−4 840.0	−1 819.0	−5 681.1	−2 285.2	−4 830.9	−2 543.5
	%	17.4	23.2	7.7	25.2	10.3	23.8	10.9
Africa								
A	*$ million*	−185.7	−75.2	−11.0	−6.5	66.0	42.4	−2.8
	%	16.2	5.9	1.0	0.6	5.1	3.2	0.2
B	*$ million*	−835.9	−1 086.4	−193.8	−512.8	−335.9	−1 004.6	−792.0
	%	25.9	34.8	4.7	13.0	7.6	26.6	20.4
C	*$ million*	−408.2	−397.7	−237.8	−392.6	−32.8	−531.2	−563.2
	%	22.1	22.7	11.9	21.7	1.5	30.0	31.1
All countries	*$ million*	−1 429.8	−1 559.3	442.6	−911.9	−302.6	−1 493.4	−1 358.0
	%	23.0	25.3	6.1	13.4	3.8	21.7	19.1
Near East								
All countries	*$ million*	−1 101.6	−1 594.2	−1 515.0	−854.2	−912.9	−991.1	−2 579.8
	%	30.0	46.4	43.2	27.4	28.0	28.5	65.1
Far East								
All countries	*$ million*	−2 396.4	−1 801.8	−303.2	−2 798.7	−4 351.0	−3 766.1	−2 523.0
	%	34.8	22.5	3.1	34.1	57.0	43.5	23.8

[1] Obtained as the difference between total agricultural exports at current prices and agricultural exports at 1979-81 prices. Negative numbers show that exports in a given year would have been larger by that amount, had the prices remained at 1979-81 levels.

Source: FAO.

TABLE 3.11 **Average annual changes in agricultural imports, all developing countries, 1971-80 and 1981-88**

Region	1971-80		1981-88	
	Value	Volume	Value	Volume
	(................................	%	................................)	
Africa	22.3	7.3	−1.6	2.2
Near East	29.7	10.7	2.5	4.6
Far East	18.0	5.5	5.1	5.1
Latin America & Caribbean	21.5	10.4	−2.3	−1.7
Asian CPE	28.9	5.5	2.8	5.6
Total developing countries	22.4	8.7	1.5	2.9

Source: FAO.

TABLE 3.12 **Average annual changes in agricultural imports, sample developing country groups, 1971-80 and 1981-88**

Region	1971-80		1981-88	
	Value	Volume	Value	Volume
	(................................	%	................................)	
Latin America & Caribbean				
A	17.6	7.1	−9.5	−3.9
B	15.4	5.0	−1.9	1.7
C	21.8	13.1	−6.3	−1.6
Africa				
A	22.5	8.4	−6.0	1.8
B	17.4	5.1	−2.7	1.9
C	17.5	6.8	0.2	5.1
Near East, all countries	4.3	4.3	1.9	5.8
Far East, all countries	5.8	5.0	−2.3	3.5

Source: FAO.

imports of other merchandise products, agricultural imports were reduced by over one-fourth in volume during 1981-88. Such reductions could not be justified by vastly improved domestic food supply conditions. As seen above, growth in food production in group A did fare generally better than the others, but still remained below that of the previous decade.

In group B the growth in food imports slowed almost in line with other imports, despite a poor food production performance in the aggregate; while in group C, also characterized by generally faltering production, restraint on growth of food imports was more visible.

Regarding the financial implications, i) the position of food *vis-à-vis* other imports; that is, an indication of shifts in import priorities; and ii) the financial burden represented by food imports in terms of their weight in earnings from total exports were examined (Table 3.13).[10] The Far East and Near East regions fared best in this respect, maintaining a relatively high growth in food import volumes throughout the 1980s, but strongly reducing the weight of these imports during the period 1985-87, in relation to both total imports and (even more so) export earnings. By contrast, in many countries in Latin America and the Caribbean and Africa, the severity of agricultural import restraint appeared to be out of proportion with the financial relief — if any — that this potentially socially harmful course brought about. Groups A and B in Africa did somewhat reduce the agricultural import/total trade ratios during 1985-87 in relation to the 1970s, and this was achieved through sharp reductions in the value of their food imports.

In the "deteriorating" group, however, the collapse in export earnings far outweighed the effect of agricultural import restraint, with agricultural imports in the period 1985-87 rising

[10] These ratios are approximate to the extent that food imports include food aid. In developing countries as a whole, food aid accounted for about 10 percent of total cereal imports in recent years, the share for low-income food-deficit countries being 15 to 20 percent. About three-quarters of these were in the form of grants.

TABLE 3.13 **Agricultural imports as share of total merchandise imports (a) and as share of total exports (b), 65 selected developing countries**

Region and country groups	1969-71 (a)	1969-71 (b)	1973-75 (a)	1973-75 (b)	1979-81 (a)	1979-81 (b)	1985-87 (a)	1985-87 (b)
	(			%				)
Africa								
All countries	**16.1**	**17.8**	**18.1**	**18.4**	**15.6**	**16.4**	**17.2**	**18.2**
A	**16.4**	**16.3**	**18.1**	**12.6**	**15.2**	**13.6**	**15.7**	**14.2**
B	**18.9**	**22.1**	**18.0**	**22.6**	**14.7**	**20.8**	**16.3**	**17.8**
C	**13.1**	**14.9**	**18.2**	**20.0**	**16.8**	**14.7**	**19.5**	**22.7**
Latin America								
All countries	**9.9**	**10.9**	**11.5**	**14.3**	**11.2**	**12.3**	**11.1**	**9.0**
A	**11.6**	**10.1**	**11.2**	**10.8**	**11.5**	**11.4**	**10.2**	**7.4**
B	**10.9**	**11.5**	**10.5**	**11.6**	**9.8**	**10.4**	**11.8**	**11.2**
C	**7.6**	**11.2**	**12.8**	**20.5**	**12.2**	**15.0**	**11.3**	**8.4**
Near East								
All countries	**21.2**	**31.5**	**21.9**	**47.5**	**19.4**	**48.4**	**18.6**	**39.4**
Far East								
All countries	**24.2**	**29.1**	**22.1**	**22.7**	**14.2**	**13.4**	**10.8**	**9.5**

Source: FAO.

to an equivalent of almost one-fifth of total imports and one-fourth of total export earnings. In Latin America and the Caribbean, where the cut in imports of agricultural products was particularly drastic, these still accounted for about 10 percent of the total value of imports and exports in the years 1985-87, although the ratio fell considerably more in group A.

External debt and agriculture

To what extent have external financing flows channelled to agricultural-related projects contributed to the debt overhang of developing countries? This can be discussed by assessing the share of agriculture in total debt and debt-service payments, as well as the interest and principal component of such payments (Table 3.14).[11] In order to assess these indicators against the perspective of total debt, the table also shows the overall debt-service ratio (total debt service as a share of total exports of goods and services).

It can be seen that external agricultural debt accounted for a minor part of the total debt overhang in all country groups. This was so despite the fact that agriculture received a substantial part of total debt-creating external financial flows. In a sample of 35 countries it was found that gross disbursements of external loans to agriculture during 1980-84 accounted for about 7 percent of total gross disbursements; however, given the typically higher concessionality of loans to agriculture — most of which are provided by official rather than private sources — debt-service payments on agricultural loans only accounted for about 4 percent of total debt-service payments.[12]

Table 3.14 confirms those findings. The agricultural sector's share of total debt was in the range of 2 to 7 percent during 1984-88 (lowest in Latin America and the Caribbean, highest in Africa), with an even smaller share of agricultural debt servicing in total debt-service payments. Also, agriculture contributed far less than other sectors to the dramatic aggravation of the debt problem during the 1980s, as its share of total debt and debt servicing tended to decline in relation to the 1970s. These shares are out of proportion with the importance of the agricultural sector in these countries' economies; as seen earlier their share of agriculture in GDP is typically in the order of 25 percent, with agricultural exports contributing 7 percent of GDP.

Furthermore, while developing countries as a whole had become net providers of resources after 1983 (see Box 3.1), the fact that agriculture enjoyed more favourable terms of borrowing permitted the sector to maintain a positive net transfer position throughout the 1980s (Table 3.15).

Finally, the role of agriculture in the context of debt should also be considered in terms of

[11] Data refer to long-term official credits (bilateral and multilateral) and private credits officially guaranteed (supplier credits, nationalization, financial institutions and public loans).

[12] See Special chapter in *The State of Food and Agriculture, 1986*.

TABLE 3.14 **Indicators of agricultural external debt, 65 selected developing countries, averages 1974-80, 1981-83 and 1984-88** (percentage shares)

Region		Agricultural external debt/ total external debt			Agricultural debt service/ total debt service			Agricultural principal repayments/ total principal repayments			Agricultural interest payments/ total interest repayments			Memorandum: Total debt service/ total exports G & S		
		1974-80	1981-83	1984-88	1974-80	1981-83	1984-88	1974-80	1981-83	1984-88	1974-80	1981-83	1984-88	1974-80	1981-83	1984-88
		(%)														
Africa	All	7.9	7.3	7.0	6.3	6.4	4.9	5.7	6.0	4.1	7.1	6.8	6.0	8.9	19.3	29.2
	A	9.4	7.7	7.7	9.2	9.3	6.8	7.4	8.6	5.0	11.2	9.5	8.6	7.8	17.9	23.2
	B	10.7	9.2	7.5	8.9	8.6	5.7	8.0	9.2	5.3	10.2	8.1	6.2	10.4	20.3	27.7
	C	4.8	4.9	5.5	3.2	3.0	3.0	3.5	3.2	3.0	2.7	2.6	2.9	8.0	19.3	35.2
Latin America & Caribbean	All	3.0	2.2	2.5	3.1	1.9	2.3	3.4	2.4	2.9	2.7	1.5	1.9	22.0	32.6	31.9
	A	1.5	1.2	1.7	1.7	0.9	1.9	2.1	1.2	2.8	1.3	0.8	1.5	22.5	33.4	34.0
	B	3.5	1.9	2.0	3.9	2.1	1.1	4.1	3.4	1.3	3.6	1.4	0.9	19.2	24.2	24.7
	C	5.1	3.5	3.9	5.0	3.2	3.4	5.0	4.1	4.0	5.3	2.7	3.0	25.4	43.0	38.5
Far East & Pacific	All	4.8	4.4	6.2	3.6	2.6	3.9	3.3	2.9	3.5	4.0	2.4	4.8	12.2	18.1	24.8
Near East	All	3.1	3.0	3.3	3.2	2.5	3.9	3.4	3.2	4.7	3.2	1.9	3.0	12.2	15.9	20.8

Source: World Bank and FAO, Policy Analysis Division (ESP).

TABLE 3.15 **Net transfers of resources,[1] averages 1974-80, 1981-83, 1984-88**

Region	Agriculture			Total		
	1974-80	1981-83	1984-88	1974-80	1981-83	1984-88
	(..........		*$ billion*			..........)
Africa	**0.7**	**0.6**	**0.6**	**6.1**	**4.2**	**−0.9**
Latin America	**0.4**	**0.2**	**0.1**	**8.2**	**−4.2**	**−25.0**
Asia	**0.6**	**1.0**	**0.4**	**5.8**	**7.2**	**−4.0**
Near East	**0.1**	**0.2**	**0.1**	**2.8**	**1.4**	**0.2**

[1] Net transfers are net flows minus interest payments (or disbursements minus total debt-service payments).
Source: World Bank and FAO.

the sector's contribution to external balances. The fact that developing countries are net agricultural exporters in the aggregate, and in a majority of individual cases, implies that the sector is a net gainer of foreign exchange and a positive contributor to debt financing. Overall, surpluses from agricultural trade for developing countries during 1982-88 represented 10 percent of their aggregate debt-service payments.[13] This was so despite the fact that Near East countries as a group remain major net deficit countries, notwithstanding a pronounced slow-down in food imports during the 1980s; while Africa showed alternating surplus and deficit situations in agriculture during the 1980s, the net result being a rough balance overall. By contrast, agricultural trade surpluses in the other developing regions were major sources of external financing and debt alleviation. In Latin America and the Caribbean the net surplus from the sector during 1982-88 totalled over $147 billion, that is nearly one-half of the aggregate debt-service payments of the region during this period. In the Far East the agricultural trade surplus was about $40 billion, representing 20 percent of debt-service payments by the region during the same period.

Food availability and self-sufficiency

The impact on food availability of changes in domestic food production and trade is briefly examined through changes in per caput calorie consumption or dietary energy supply (DES) (Table 3.16). The overall trends were lower average growth in DES during 1980-86 than during the previous decade in Latin America and the Caribbean and Africa, but higher growth in Near East and Asia and the Pacific. However, while the successful performance of

TABLE 3.16 **Changes in dietary energy supply, 1970-80 and 1980-85 yearly averages, 65 selected developing countries**

Region and country groups	1970-80	1980-85
	(........	% ....)
Latin America and the Caribbean		
A	**0.5**	**0.0**
B	**0.4**	**0.7**
C	**0.5**	**0.5**
All countries	**0.5**	**0.3**
Africa		
A	**0.6**	**0.2**
B	**0.6**	**−0.4**
C	**0.8**	**−0.6**
All countries	**0.6**	**−0.5**
Near East		
All countries	**1.4**	**2.1**
Far East		
All countries	**0.6**	**0.8**

Source: FAO.

the two latter regions mainly reflected the situations in a few countries (in particular Cyprus, Jordan, Bangladesh and Sri Lanka), the experiences in Africa and Latin America were lower growth or a decline in DES in nearly two-thirds of the countries. The African situation was particularly poor, as DES fell during the 1980s in three countries in group A, nine in group B and seven in group C.

While the obvious main influence in changes in DES was food production performance, it should be recalled that DES also reflects a variety of other factors, including the net effect of food trade flows (for instance, group A in Latin America achieved some increase in per caput food production during 1980-85, but sharply reduced the volume of food imports) and changes in the composition and utilization of food supply.

With regard to changes in staple food self-sufficiency in the different groups, the main

[13] The data covering developing country agricultural trade (FAO) and debt (World Bank) are not identical. However, the 102 countries included in the debt-service figure (111 countries in the World Bank's DRS less nine countries in Europe and the Mediterranean) account for over 95 percent of total agricultural trade by developing countries, as defined by FAO.

Figure 3.14

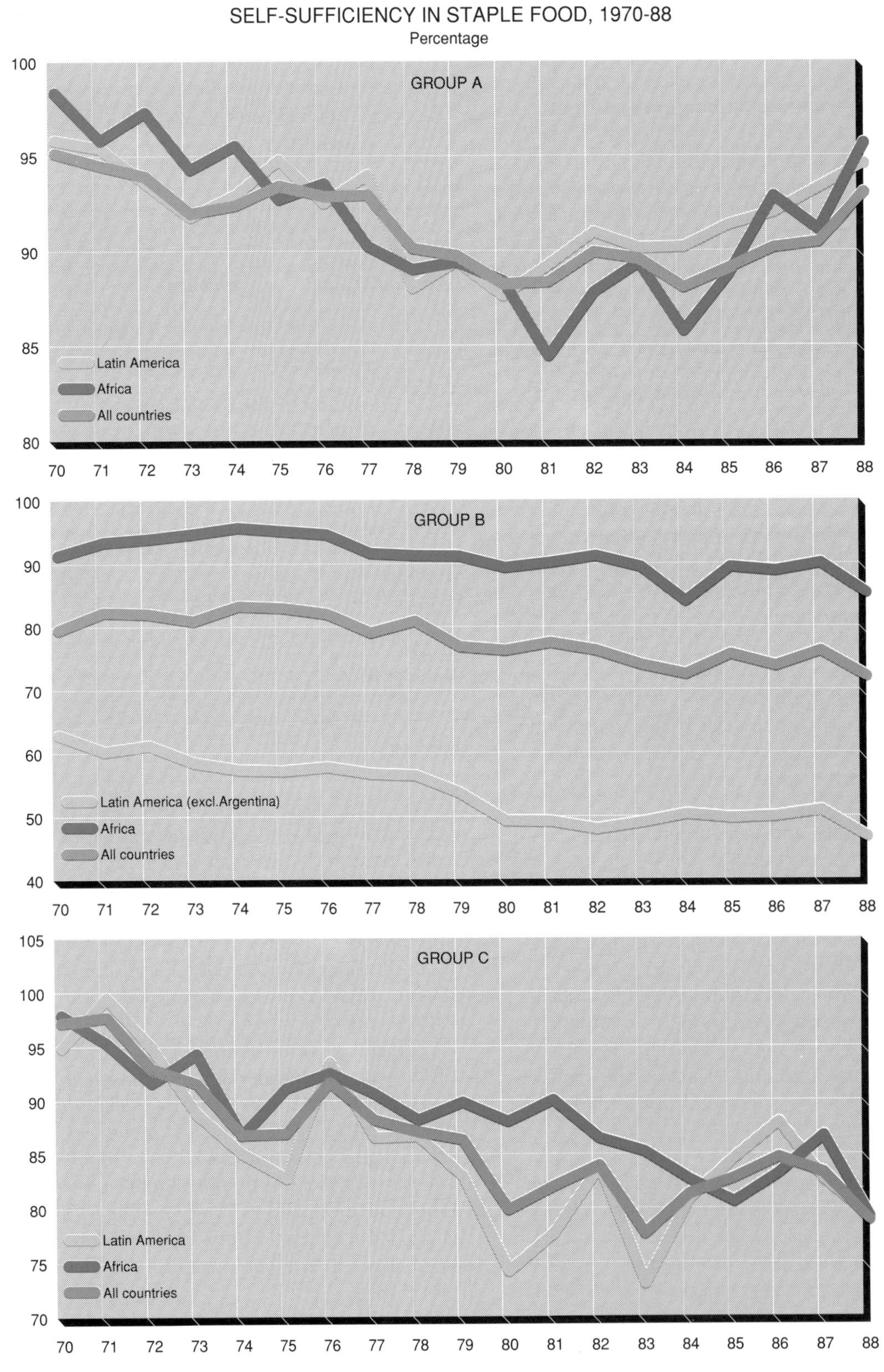

SELF-SUFFICIENCY IN STAPLE FOOD, 1970-88
Percentage

GROUP A

Latin America
Africa
All countries

100 95 90 85 80

70 71 72 73 74 75 76 77 78 79 80 81 82 83 84 85 86 87 88

GROUP B

Latin America (excl.Argentina)
Africa
All countries

100 90 80 70 60 50 40

70 71 72 73 74 75 76 77 78 79 80 81 82 83 84 85 86 87 88

GROUP C

Latin America
Africa
All countries

105 100 95 90 85 80 75 70

70 71 72 73 74 75 76 77 78 79 80 81 82 83 84 85 86 87 88

patterns were:[14] a declining trend in self-sufficiency until the early 1980s in group A but a strong reversal of this trend in more recent years; within wide fluctuations in group B, a pronounced fall after 1985; and a steady decline in self-sufficiency in the third group, although year-to-year fluctuations became more pronounced during the 1980s. These trends are remarkably consistent within the three country groups in Latin America and the Caribbean and Africa (Figure 3.14).

[14] Staple food includes cereals, roots and tubers and pulses converted into a grain equivalent. Self-sufficiency is defined as the ratio between domestic production and total supply (supply = production + imports − exports).

BOX 3.10

Analytical framework

The internal and external components of the macro-economic balance are related by the following *ex post* accounting identity:

i) $(X-M) + R = Sn - I$

where:
- X = exports of goods and non-factor services
- M = imports of goods and non-factor services
- R = net factor income, i.e. interest payments on external debt; profits and dividends on foreign direct investment; and remittances from residents working abroad
- I = total investment, including variations in stocks
- Sn = national savings, i.e. business, and households at home and abroad, and government savings.

This identity is derived from the national accounting identity:

ii) $Y = C+G+I + (X-M)$

where Y is national income, C is consumption expenditure by the private sector and G government current expenditure. Saving may be defined as $S = Y-C-G$. If saving is disaggregated into its private (Sp) and government (Sg) component, the latter being the difference between fiscal revenues (T) and expenditures (G), i.e. $Sg = T-G$, the relation i) can be written:

iii) $(X-M)+R = Sp + (T-G)-I$

This identity implies that, other things remaining unchanged, an increase (or vice versa, a decline) in government current expenditures (relative to net taxes, private saving and total investment) will result in a deterioration (or an improvement) in the trade balance. Conversely, a budget deficit may also reflect external factors: for instance, a fall in export earnings (X) caused by a deterioration in terms of trade may result in lower tax income (T) and a worsening fiscal balance.

Turning to equation i), for a country experiencing excessive macro-economic imbalances, stabilization and adjustment measures will generally tend toward the closure of the external account gap through an increase in exports and/or a reduction in imports (left side of the equation); and closure of the investment-savings gap through an increase in savings and/or a reduction in investment (right side). Reducing such imbalances inevitably entails some degree of economic and social hardship in the short to medium term. In a longer perspective, the difference between a growth-promoting and a recessive path of adjustment will hinge on the ability of the different countries to minimize reductions in imports and investments, and to close the gap through greater exports and domestic savings instead.

Becoming more or less self-sufficient in food may or may not be a positive development for a specific country, depending on its comparative advantage in food production and import financing capacity, and the implications of changes in the components of self-sufficiency on domestic food supply. In other words, an increase in self-sufficiency supports successful adjustment only if it is the consequence of improved domestic food supply conditions allowing large exports, import cuts and increasing — or at least stable at acceptable levels — per caput food supply. Such cases were the exception rather than the rule during the 1980s when both domestic supply and trade conditions combined to depress calorie intake in most countries. Nevertheless, the positive "exceptions" were more concentrated in group A. Indeed, the scenario for the countries in group A, as compared to the other groups, was better performance in food production together with reduced supply from external trade (lower growth, or decline, in the volume of food imports coupled with larger volumes of food exports), leading to greater self-sufficiency but stable levels of DES.

ANNEX TABLES

1. VOLUME OF PRODUCTION OF MAJOR AGRICULTURAL, FISHERY AND FOREST PRODUCTS

	1979	1980	1981	1982	1983	1984	1985	1986	1987	1988	1989	ANNUAL RATE OF CHANGE 1979-89
	....THOUSAND METRIC TONS..											PERCENT
WORLD												
AGRICULTURAL PRODUCTS												
TOTAL CEREALS	1553333	1565693	1650349	1709405	1643288	1802469	1839358	1855581	1789884	1740188	1870161	1.72
WHEAT	428368	445972	455081	481638	492958	516033	504618	535298	509719	505959	540229	2.06
RICE PADDY	376768	399221	412100	423670	451488	469460	472257	471326	464977	490013	508448	2.71
BARLEY	158075	159876	151556	163984	161646	172155	176183	181461	179746	166477	168934	1.18
MAIZE	420396	397898	452580	452259	349926	455098	485778	481091	452869	399892	472326	.96
MILLET AND SORGHUM	91126	82327	100572	95625	94125	94762	105624	95498	91559	94255	88219	.19
ROOT CROPS	588279	537014	556140	559274	562178	594059	579579	583186	586770	571591	591062	.50
POTATOES	299125	241654	268543	266118	265432	292040	283709	291072	284698	269843	277111	.44
CASSAVA	118915	124708	129244	129863	126499	132883	136616	133765	136881	141588	148067	1.75
TOTAL PULSES	41301	40701	41459	45685	47192	49289	50443	52668	54872	55227	57457	3.76
CITRUS FRUIT	53110	58822	59097	58124	62116	57587	60762	63553	66608	68234	71796	2.41
BANANAS	35087	37072	37948	38056	37349	39414	40200	42072	44282	44357	44313	2.41
APPLES	36642	34197	32881	41622	39541	40182	38895	42098	38916	41784	40902	1.71
VEGETABLE OILS,OIL EQUIV	51366	50063	53820	57255	53469	59543	64664	64768	67670	67922	71523	3.70
SOYBEANS	88678	81036	88201	92088	79461	90718	101135	94389	100755	93453	107056	2.02
GROUNDNUTS IN SHELL	18027	16908	20432	17929	19006	20002	20864	21121	21509	24968	23040	3.10
SUNFLOWER SEED	15246	13648	14359	16429	15698	16609	18877	20820	20475	20318	21112	4.60
RAPESEED	10646	10762	12485	15210	14146	16709	19244	19857	22688	22092	22484	8.77
COTTONSEED	26434	26613	28652	28032	27401	34910	32202	28230	30954	33815	31102	2.00
COPRA	4229	4439	4632	4786	4689	3874	4441	5446	5076	4510	4594	.87
PALM KERNELS	1634	1774	1795	2161	1991	2354	2599	2760	2714	2911	3358	7.13
SUGAR (CENTRIFUGAL,RAW)	88364	84222	93279	102817	98361	100090	99291	101225	101800	103686	104495	1.75
COFFEE GREEN	4947	4812	6038	4973	5567	5153	5840	5163	6350	5507	5657	1.36
COCOA BEANS	1676	1666	1740	1607	1602	1761	1953	2080	2045	2457	2411	4.25
TEA	1818	1873	1875	1946	2043	2185	2299	2293	2422	2487	2483	3.62
COTTON LINT	13938	13848	15248	14931	14249	18220	17359	15171	16525	18282	17460	2.49
JUTE AND SIMILAR FIBRES	3702	3557	3610	3209	3429	3569	6328	4405	3502	3330	3704	.82
SISAL	509	533	495	497	408	446	492	469	401	381	426	-2.52
TOBACCO	5416	5305	5965	6894	5944	6490	7022	6076	6169	6817	7175	2.23
NATURAL RUBBER	3838	3797	3785	3807	4110	4184	4338	4560	4732	4981	4984	3.21
TOTAL MEAT	130573	134598	136943	138146	142381	146011	151310	155364	160325	164984	168260	2.63
TOTAL MILK	456086	462465	464562	475418	495205	498605	509376	518795	517913	526717	533652	1.68
TOTAL EGGS	26526	27278	27928	28766	29276	30749	32456	33622	34512	35753	35487	3.32
WOOL GREASY	2705	2774	2836	2867	2883	2867	2958	3003	3079	3152	3192	1.56
FISHERY PRODUCTS 1/												
FRESHWATER + DIADROMOUS	7699	8073	8598	8901	9746	10303	11241	11860	13103	13774	14453	6.83
MARINE FISH	55030	55317	56995	58353	57949	63267	64216	69458	68511	72062	73628	3.25
CRUST+ MOLLUS+ CEPHALOP	8211	8715	8867	9371	9359	9810	10147	10611	11930	12341	12711	4.46
AQUATIC MAMMALS	4	5	3	3	3	3	2	3	3	4	4	-.14
AQUATIC ANIMALS	203	126	220	280	434	264	306	380	386	328	334	7.78
AQUATIC PLANTS	3193	3354	3079	3126	3289	3605	3723	3467	3565	4156	4266	2.84
FOREST PRODUCTS 2/												
SAWLOGS CONIFEROUS	645129	614143	581111	561153	623527	661142	667199	697689	716754	722500	722541	2.16
SAWLOGS NONCONIFEROUS	255853	263120	254221	242405	251516	259584	251903	265602	283337	286311	286311	1.29
PULPWOOD+PARTICLES	357288	370764	372394	361990	369698	381536	377102	397280	410116	416036	416036	1.57
FUELWOOD	1422257	1482809	1524996	1558851	1590104	1638629	1672626	1707020	1735425	1764647	1764859	2.20
SAWNWOOD CONIFEROUS	346927	333769	315612	311543	327361	343331	349007	361553	377312	377842	377842	1.67
SAWNWOOD NONCONIFEROUS	110244	113441	110788	107854	110839	114357	115428	118402	124006	124596	124596	1.42
WOOD-BASED PANELS	106469	101115	100348	96212	105492	108526	111772	117393	121477	125381	125381	2.49
PULP FOR PAPER	125869	128856	128808	123474	132359	140224	141214	146793	152701	157616	157674	2.63
PAPER+PAPERBOARD	169352	170220	170954	167272	177227	189954	192648	201990	215835	231131	231424	3.65

1/ NOMINAL CATCH (LIVE WEIGHT) EXCLUDING WHALES
2/ EXCEPT FOR PULP FOR PAPER AND PAPER AND PAPERBOARD, ALL FOREST PRODUCTS ARE EXPRESSED IN THOUSAND CUBIC METRES

1. (Cont.) VOLUME OF PRODUCTION OF MAJOR AGRICULTURAL, FISHERY AND FOREST PRODUCTS

	1979	1980	1981	1982	1983	1984	1985	1986	1987	1988	1989	ANNUAL RATE OF CHANGE 1979-89
	......................................THOUSAND METRIC TONS......................................											PERCENT
WESTERN EUROPE												
AGRICULTURAL PRODUCTS												
TOTAL CEREALS	164363	177560	167233	181315	173598	211609	195691	191110	186279	195792	196074	1.67
WHEAT	60247	70024	66271	73690	73720	92695	80179	81148	80897	85193	88417	3.31
RICE PADDY	1831	1702	1597	1705	1519	1750	1933	2012	1925	1954	1968	1.88
BARLEY	52830	57235	50636	53714	49748	62856	58834	53697	52899	57069	53022	.28
MAIZE	32384	31280	32622	35505	34533	36438	37681	39890	36540	37629	37706	1.98
MILLET AND SORGHUM	648	617	600	505	460	491	393	384	393	446	548	-3.66
ROOT CROPS	52002	49186	48603	48371	42526	50514	51552	48645	48531	46965	47064	-.44
POTATOES	51857	49040	48465	48240	42403	50406	51437	48542	48434	46871	46970	-.43
TOTAL PULSES	1791	1873	1639	1952	2140	2742	3337	4007	4899	5800	5573	14.99
CITRUS FRUIT	6495	6629	6777	6740	8650	6413	8036	8725	7787	8650	9108	3.34
BANANAS	436	511	522	492	500	489	454	531	512	457	453	-.23
APPLES	10655	10701	7646	12696	9089	10924	9206	10741	9058	11090	9791	-.08
VEGETABLE OILS,OIL EQUIV	2677	3310	2930	3762	3642	4250	4675	4804	6707	5790	5810	8.86
SOYBEANS	102	66	118	233	300	389	523	1130	2044	1868	2244	45.25
GROUNDNUTS IN SHELL	21	19	15	14	17	16	18	19	16	17	16	-1.03
SUNFLOWER SEED	1276	1302	1219	1736	1895	2484	3008	3769	4753	4486	3988	16.44
RAPESEED	1696	2543	2523	3295	3141	4160	4388	4371	6537	5966	5706	12.78
COTTONSEED	284	333	366	285	329	363	419	527	484	660	625	8.51
SUGAR (CENTRIFUGAL,RAW)	15787	15732	19077	18002	14945	16551	16564	16790	15756	16467	17329	.03
COTTON LINT	146	178	196	156	176	196	238	291	254	354	311	8.41
TOBACCO	439	401	435	462	436	481	494	494	475	455	453	1.08
TOTAL MEAT	28488	29513	29686	29740	30200	31033	31169	31574	32398	32591	32155	1.30
TOTAL MILK	133805	135988	136329	139996	144287	142550	141084	142054	136940	133782	135045	
TOTAL EGGS	5395	5443	5536	5692	5562	5479	5562	5503	5425	5513	5314	-.14
WOOL GREASY	157	160	159	159	163	166	173	177	182	190	202	2.44
FISHERY PRODUCTS 1/												
FRESHWATER + DIADROMOUS	210	260	249	267	274	288	322	360	355	416	438	6.94
MARINE FISH	10042	9960	10017	9545	9747	10196	9917	9632	9488	9768	10009	-.20
CRUST+ MOLLUS+ CEPHALOP	1017	1136	1197	1264	1371	1279	1411	1436	1479	1397	1425	3.17
AQUATIC ANIMALS	2	1	1	1	1	1	1	1				-12.52
AQUATIC PLANTS	278	250	208	226	222	242	249	264	272	319	331	2.83
FOREST PRODUCTS 2/												
SAWLOGS CONIFEROUS	96073	97381	90791	89591	94371	96228	95221	95420	94792	99057	99057	.44
SAWLOGS NONCONIFEROUS	23882	24240	23838	22524	21723	22843	22796	23332	23281	24744	24744	.28
PULPWOOD+PARTICLES	83932	83788	86401	84045	82462	86245	87144	91118	99243	102564	102564	2.25
FUELWOOD	35526	37305	38303	38905	39520	39921	40173	40483	39544	39478	39478	.86
SAWNWOOD CONIFEROUS	53613	54877	50554	50134	52307	53470	51566	51750	52395	53799	53799	.09
SAWNWOOD NONCONIFEROUS	12724	12437	11472	11210	10631	11284	11228	11317	11307	11639	11639	-.62
WOOD-BASED PANELS	26607	26602	24960	23577	23901	24225	24457	25236	26418	29556	29556	1.17
PULP FOR PAPER	26736	26647	26489	25045	26880	29161	29299	30044	31403	32530	32603	2.53
PAPER+PAPERBOARD	45174	44736	44707	43738	45571	49971	50106	51867	54610	58748	58867	3.19

1/ NOMINAL CATCH (LIVE WEIGHT) EXCLUDING WHALES
2/ EXCEPT FOR PULP FOR PAPER AND PAPER AND PAPERBOARD, ALL FOREST PRODUCTS ARE EXPRESSED IN THOUSAND CUBIC METRES

1. (Cont.) VOLUME OF PRODUCTION OF MAJOR AGRICULTURAL, FISHERY AND FOREST PRODUCTS

	1979	1980	1981	1982	1983	1984	1985	1986	1987	1988	1989	ANNUAL RATE OF CHANGE 1979-89
	....	....	....	....	....	THOUSAND METRIC TONS	....	....	....	....	....	PERCENT
USSR AND EASTERN EUROPE												
AGRICULTURAL PRODUCTS												
TOTAL CEREALS	251009	264130	233882	269542	268617	260496	269375	297567	291638	275334	295624	1.69
WHEAT	113566	127688	107425	113780	107436	105104	110359	126687	117932	123249	131034	1.00
RICE PADDY	2584	2934	2666	2651	2818	2932	2815	2921	2940	3128	2821	1.08
BARLEY	62927	59219	51413	59740	64483	58199	62261	70162	74216	61013	67035	1.67
MAIZE	32920	30619	31776	40048	35967	37864	35077	39015	36012	32448	33018	.50
MILLET AND SORGHUM	1744	2077	2035	2718	2709	2151	3154	2570	4267	3429	4377	8.40
ROOT CROPS	163116	111251	135403	129664	135629	147334	134596	150729	136554	122498	128089	-.46
POTATOES	163113	111249	135399	129661	135627	147332	134593	150725	136551	122496	128087	-.46
TOTAL PULSES	5052	7130	5290	7803	9872	10220	10808	9411	11543	10613	11282	7.95
CITRUS FRUIT	340	161	313	286	415	367	156	332	194	461	365	2.26
APPLES	11302	8567	10002	13278	13120	11931	11713	13824	8958	10539	10279	-.01
VEGETABLE OILS,OIL EQUIV	4435	4362	4363	4674	4555	4478	4760	5127	5319	5403	6036	2.97
SOYBEANS	1042	1118	907	1007	953	1001	864	1281	1446	1416	1547	4.37
GROUNDNUTS IN SHELL	6	7	9	9	8	8	6	9	10	11	12	5.46
SUNFLOWER SEED	7208	6324	6636	7350	6904	6528	7082	7605	8025	7915	9061	2.49
RAPESEED	574	1226	1097	1064	1312	1718	1932	2330	2329	2550	2949	14.95
COTTONSEED	5615	6095	5896	5685	5645	5276	5361	5046	4979	5319	5247	-1.52
SUGAR (CENTRIFUGAL,RAW)	12229	10842	10943	12450	13563	13434	12923	12706	13404	12204	14043	1.62
TEA	118	130	137	140	146	151	152	146	156	123	143	1.16
COTTON LINT	2514	2813	2905	2800	2597	2354	2792	2660	2475	2770	2666	-.25
JUTE AND SIMILAR FIBRES	48	52	45	45	45	45	45	45	45	47	47	-.47
TOBACCO	627	545	574	637	670	704	698	713	628	537	455	-1.02
TOTAL MEAT	25283	25130	24866	24768	26073	26940	27333	28614	29425	30179	30480	2.31
TOTAL MILK	133738	131285	127674	129251	137219	140471	141472	144836	146133	149231	151874	1.66
TOTAL EGGS	5513	5643	5834	5862	6062	6178	6298	6524	6636	6739	6680	2.12
WOOL GREASY	573	559	574	571	584	595	578	601	583	604	597	.60
FISHERY PRODUCTS 1/												
FRESHWATER + DIADROMOUS	14	15	16	19	17	16	16	17	15	15	15	-.19
MARINE FISH	81	119	99	106	113	99	87	95	102	109	105	.44
CRUST+ MOLLUS+ CEPHALOP	3	1				6	9	9	6	8		
FOREST PRODUCTS 2/												
SAWLOGS CONIFEROUS	154849	155724	155698	153520	156432	158709	157347	165092	168613	168855	164337	.92
SAWLOGS NONCONIFEROUS	33545	33594	33619	33109	33368	34357	33003	32610	33333	33545	32276	-.24
PULPWOOD+PARTICLES	55277	55992	55666	56524	57323	58951	58714	61664	62292	65326	62390	1.61
FUELWOOD	91301	92415	96413	99038	95838	100756	103258	104366	103634	99438	95138	.82
SAWNWOOD CONIFEROUS	102829	101494	100809	100153	100268	100630	101194	103222	103882	103849	100879	.14
SAWNWOOD NONCONIFEROUS	18638	18260	18269	18060	18272	18430	18202	17881	17417	17482	16680	-.81
WOOD-BASED PANELS	17005	17464	17598	17988	18563	19480	19682	20662	20801	21471	21582	2.63
PULP FOR PAPER	11489	11607	11774	12052	12869	13261	13432	13342	13339	14920	14610	2.60

1/ NOMINAL CATCH (LIVE WEIGHT) EXCLUDING WHALES
2/ EXCEPT FOR PULP FOR PAPER AND PAPER AND PAPERBOARD, ALL FOREST PRODUCTS ARE EXPRESSED IN THOUSAND CUBIC METRES

1. (Cont.) VOLUME OF PRODUCTION OF MAJOR AGRICULTURAL, FISHERY AND FOREST PRODUCTS

	1979	1980	1981	1982	1983	1984	1985	1986	1987	1988	1989	ANNUAL RATE OF CHANGE 1979-89
	....	....	....	....	....	THOUSAND METRIC TONS	....	....	....	....	....	PERCENT
NORTH AMERICA DEVELOPED												
AGRICULTURAL PRODUCTS												
TOWREACEREALS	[illegible]	[illegible]	[illegible]	[illegible]	[illegible]	[illegible]	[illegible]	[illegible]	[illegible]	[illegible]	[illegible]	-1.04
RICE PADDY	5985	6629	8289	6969	4523	6296	6122	6049	5879	7253	7007	.12
BARLEY	16821	19257	24033	25198	21289	23324	25263	27926	25311	16526	20456	.82
MAIZE	206659	174400	212895	215702	111972	201705	232415	214854	188157	130562	197597	-.93
MILLET AND SORGHUM	20509	14716	22247	21212	12384	22004	28456	23829	18563	14648	15694	-.76
ROOT CROPS	18895	16762	18680	19565	18253	19833	22137	19737	21222	19445	20077	1.36
POTATOES	18285	16263	18097	18889	17702	19241	21460	19159	20692	18946	19553	1.44
TOTAL PULSES	1299	1676	1954	1720	1161	1373	1492	1699	2374	1649	1811	2.21
CITRUS FRUIT	12092	14954	13703	10938	12344	9790	9549	10012	10874	11571	11959	-1.99
BANANAS	2	2	3	3	2	4	4	4	5	6	5	10.73
APPLES	4121	4553	3933	4162	4283	4213	4073	3953	5378	4628	5034	1.70
VEGETABLE OILS,OIL EQUIV	15756	11883	13250	14340	10893	13015	14190	13232	13393	11426	12732	-.98
SOYBEANS	62183	49612	54742	60459	45253	51592	58140	53840	54007	43306	53664	-1.18
GROUNDNUTS IN SHELL	1800	1045	1806	1560	1495	1998	1870	1677	1640	1806	1811	2.11
SUNFLOWER SEED	3528	1863	2201	2514	1497	1783	1512	1276	1235	861	882	-11.23
RAPESEED	3411	2483	1849	2218	2593	3412	3498	3787	3850	4323	3152	5.02
COTTONSEED	5242	4056	5803	4304	2791	4671	4789	3448	5234	5499	4242	-.05
SUGAR (CENTRIFUGAL,RAW)	5167	5438	5774	5384	5219	5476	5527	6197	6798	6377	6125	2.13
COFFEE GREEN	1	1	1		1	1	1	1	1	1	1	1.66
COTTON LINT	3185	2422	3406	2605	1692	2827	2924	2119	3214	3355	2663	.34
TOBACCO	771	918	1048	975	760	873	782	596	601	691	716	-3.69
TOTAL MEAT	26290	27156	27552	26933	27871	28141	28773	29184	29795	30777	31254	1.64
TOTAL MILK	63626	66099	68182	69691	71166	69490	72760	72845	72717	74069	74300	1.40
TOTAL EGGS	4424	4463	4477	4456	4359	4382	4379	4419	4494	4434	4281	-.17
WOOL GREASY	49	49	51	50	48	45	41	40	40	42	42	-2.52
FISHERY PRODUCTS 1/												
FRESHWATER + DIADROMOUS	433	476	502	485	501	493	577	527	662	665	686	4.43
MARINE FISH	3106	3153	3122	3518	3774	3949	4181	4482	5347	5311	5577	6.78
CRUST+ MOLLUS+ CEPHALOP	1376	1350	1558	1378	1324	1647	1452	1379	1515	1555	1584	1.17
AQUATIC ANIMALS	10	2	2	10	10	9	11	17	24	31	34	25.06
AQUATIC PLANTS	195	191	78	103	29	63	109	82	126	113	114	-2.28
FOREST PRODUCTS 2/												
SAWLOGS CONIFEROUS	298266	260961	238884	220996	276510	304302	310133	333140	348289	348493	348493	3.71
SAWLOGS NONCONIFEROUS	42727	43206	39834	29093	36240	37061	35511	42142	43671	44601	44601	1.22
PULPWOOD+PARTICLES	157282	163894	164429	156026	161024	165399	158513	171356	175472	176865	176865	1.15
FUELWOOD	71933	95976	107410	107595	108119	120638	125203	126047	123102	123102	123102	4.23
SAWNWOOD CONIFEROUS	122060	109483	98688	94908	109365	122153	127361	136114	149275	147711	147711	3.95
SAWNWOOD NONCONIFEROUS	18432	18650	17087	12357	14415	15957	15376	18365	20167	20202	20202	1.96
WOOD-BASED PANELS	36649	31026	32011	28338	34842	36378	38250	40834	40782	40641	40641	2.90
PULP FOR PAPER	63750	65241	65672	61122	65863	69877	68364	72386	75621	77914	77914	2.31
PAPER+PAPERBOARD	70896	70229	71502	67307	72157	76588	75407	79703	83589	86225	86225	2.44

1/ NOMINAL CATCH (LIVE WEIGHT) EXCLUDING WHALES
2/ EXCEPT FOR PULP FOR PAPER AND PAPER AND PAPERBOARD, ALL FOREST PRODUCTS ARE EXPRESSED IN THOUSAND CUBIC METRES

1. (Cont.) VOLUME OF PRODUCTION OF MAJOR AGRICULTURAL, FISHERY AND FOREST PRODUCTS

	1979	1980	1981	1982	1983	1984	1985	1986	1987	1988	1989	ANNUAL RATE OF CHANGE 1979-89
	....	....	....	....	....	THOUSAND METRIC TONS	....	....	....	....	....	PERCENT
OCEANIA DEVELOPED												
AGRICULTURAL PRODUCTS												
TOTAL CEREALS	24143	17159	24472	15066	31940	29719	26361	25941	21107	22976	23163	1.29
WHEAT	16483	11162	16686	9168	22317	18981	16477	17158	12706	14260	14455	.42
RICE PADDY	692	613	728	854	519	634	864	716	613	751	805	1.10
BARLEY	3967	2910	3721	2295	5236	6125	5513	4167	3878	3657	4417	2.60
MAIZE	348	307	325	382	282	392	466	465	383	345	340	1.58
MILLET AND SORGHUM	1162	936	1231	1355	987	1929	1395	1448	1458	1683	1188	3.18
ROOT CROPS	1012	1091	1089	1168	1127	1327	1277	1250	1316	1390	1338	2.95
POTATOES	1001	1071	1075	1157	1117	1314	1264	1239	1301	1376	1322	2.98
TOTAL PULSES	175	209	225	315	322	617	860	921	1602	1563	1659	29.36
CITRUS FRUIT	489	566	509	534	525	587	637	643	614	508	525	.97
BANANAS	125	124	130	140	146	145	134	178	181	165	167	3.67
APPLES	525	510	549	520	534	513	629	632	686	694	714	3.69
VEGETABLE OILS,OIL EQUIV	159	120	126	118	105	157	266	223	186	213	219	6.83
SOYBEANS	99	82	73	77	53	89	110	105	90	75	113	2.07
GROUNDNUTS IN SHELL	62	39	43	58	23	47	42	43	48	39	32	-2.67
SUNFLOWER SEED	186	142	139	115	104	170	293	215	137	179	170	2.52
RAPESEED	41	18	15	7	18	33	88	84	89	66	100	21.66
COTTONSEED	79	136	161	191	164	190	410	382	330	445	449	17.69
SUGAR (CENTRIFUGAL,RAW)	2963	3330	3435	3536	3170	3548	3379	3371	3440	3679	3793	1.47
COTTON LINT	53	83	99	134	101	141	249	258	214	284	286	17.64
TOBACCO	19	18	17	15	15	16	14	12	13	14	12	-4.01
TOTAL MEAT	4102	3799	3812	3855	3923	3583	3777	3816	4063	4233	3975	.37
TOTAL MILK	12202	12248	12079	12203	12593	13711	14089	14440	13625	13948	13944	1.83
TOTAL EGGS	268	265	278	274	275	264	250	249	251	254	254	-.92
WOOL GREASY	1025	1066	1082	1080	1073	1091	1188	1188	1237	1255	1277	2.25
FISHERY PRODUCTS 1/												
FRESHWATER + DIADROMOUS	5	5	4	4	4	6	5	5	5	6	6	3.98
MARINE FISH	171	227	257	253	289	308	308	381	477	563	597	12.17
CRUST+ MOLLUS+ CEPHALOP	93	113	121	150	158	166	152	136	149	135	140	2.87
AQUATIC PLANTS	18	15	16	11	11	18	14	13	16	18	18	1.27
FOREST PRODUCTS 2/												
SAWLOGS CONIFEROUS	7021	8443	8607	8357	7703	7308	8267	8297	8398	9392	9392	1.71
SAWLOGS NONCONIFEROUS	5846	5881	6077	5725	4569	4556	4911	4784	4795	5173	5173	-1.91
PULPWOOD+PARTICLES	8330	9890	10177	9513	9865	10455	11137	11577	11577	11577	11577	2.93
FUELWOOD	1447	1458	1818	2118	2524	2924	2924	2930	2930	2936	2936	8.10
SAWNWOOD CONIFEROUS	2743	3101	3370	3414	3141	3163	3496	3595	2996	3297	3297	.93
SAWNWOOD NONCONIFEROUS	1986	2069	2145	2013	1790	1739	1830	1801	1838	1881	1881	-1.19
WOOD-BASED PANELS	1073	1166	1215	1228	1053	1210	1292	1330	1498	1521	1521	3.51
PULP FOR PAPER	1699	1824	1913	1896	1794	1917	2065	2032	2039	2246	2246	2.48
PAPER+PAPERBOARD	1942	2104	2151	2188	2101	2214	2316	2267	2192	2401	2401	1.66

1/ NOMINAL CATCH (LIVE WEIGHT) EXCLUDING WHALES
2/ EXCEPT FOR PULP FOR PAPER AND PAPER AND PAPERBOARD, ALL FOREST PRODUCTS ARE EXPRESSED IN THOUSAND CUBIC METRES

1. (Cont.) VOLUME OF PRODUCTION OF MAJOR AGRICULTURAL, FISHERY AND FOREST PRODUCTS

	1979	1980	1981	1982	1983	1984	1985	1986	1987	1988	1989	ANNUAL RATE OF CHANGE 1979-89
	....	....	....	....	....	THOUSAND METRIC TONS	....	....	....	....	....	PERCENT
AFRICA DEVELOPING												
AGRICULTURAL PRODUCTS												
TOTAL CEREALS	45585	47492	46693	50495	47398	44028	59653	62559	57370	62961	63011	3.72
WHEAT	4655	5419	4391	5613	4592	4717	6586	6989	6444	6468	7158	4.47
RICE PADDY	5828	6226	6307	6528	6631	6719	7209	7573	7734	7649	7995	3.14
BARLEY	3769	4464	2866	4435	2882	3113	5522	5873	3974	5063	5238	4.03
MAIZE	13621	13295	15192	15204	14495	14572	18441	19321	16498	20535	20874	4.50
MILLET AND SORGHUM	16063	16424	16516	17005	17425	13755	20565	21378	21359	21856	20335	3.44
ROOT CROPS	82173	83715	86912	91100	88436	93664	98950	100531	95335	97621	100050	2.01
POTATOES	3210	3367	3118	3466	3759	3384	4609	4286	4201	4116	4368	3.58
CASSAVA	47178	48128	50860	53363	52287	55264	58156	58707	58366	59864	62258	2.74
TOTAL PULSES	5105	4580	4626	5204	4947	4276	4857	5943	5042	5907	5893	2.06
CITRUS FRUIT	2504	2753	2663	2585	2481	2631	2453	2779	2617	2894	3122	1.26
BANANAS	4164	4478	4597	4605	4581	4567	4797	4991	5185	5420	5509	2.51
APPLES	97	116	134	154	194	233	262	278	333	389	402	15.83
VEGETABLE OILS,OIL EQUIV	3672	3831	3810	3910	3832	3847	4096	4469	4465	4408	4581	2.28
SOYBEANS	178	204	184	202	170	182	202	204	238	260	262	3.58
GROUNDNUTS IN SHELL	3334	3204	3603	3664	3189	3239	3392	3983	4036	3899	4107	2.20
SUNFLOWER SEED	150	140	134	136	139	155	170	184	230	246	253	6.82
RAPESEED	46	52	64	60	74	73	83	93	110	113	118	9.98
COTTONSEED	888	888	842	853	920	1080	1172	1325	1334	1499	1526	6.87
COPRA	181	179	175	189	203	201	213	218	228	237	238	3.32
PALM KERNELS	649	700	682	685	594	654	690	692	642	615	645	-.51
SUGAR (CENTRIFUGAL,RAW)	3498	3534	3761	3900	3984	3967	3967	4192	4224	4410	4321	2.23
COFFEE GREEN	1091	1152	1270	1196	1109	994	1183	1240	1255	1232	1244	.94
COCOA BEANS	1030	1026	1072	882	887	1058	1033	1134	1182	1442	1349	3.38
TEA	197	186	195	208	218	225	260	251	255	273	291	4.52
COTTON LINT	485	504	469	489	551	605	673	733	761	825	874	6.94
JUTE AND SIMILAR FIBRES	8	8	8	9	9	9	9	9	10	10	10	1.74
SISAL	159	168	146	142	124	115	103	106	102	94	105	-5.48
TOBACCO	259	275	214	234	253	296	277	274	301	301	321	2.64
NATURAL RUBBER	198	201	206	202	207	228	238	250	280	294	340	5.32
TOTAL MEAT	4469	4559	4674	4797	4800	4887	5058	5142	5183	5308	5431	1.91
TOTAL MILK	8693	8725	8912	9307	9579	9659	9960	10827	11111	11505	11741	3.34
TOTAL EGGS	611	644	674	731	798	810	869	950	994	1038	1078	6.11
WOOL GREASY	70	73	76	82	95	98	109	112	117	122	124	6.57
FISHERY PRODUCTS 1/												
FRESHWATER + DIADROMOUS	1322	1287	1258	1323	1393	1435	1393	1508	1535	1582	1639	2.54
MARINE FISH	1516	1530	1702	1654	1749	1657	1671	1848	1856	1927	1967	2.45
CRUST+ MOLLUS+ CEPHALOP	71	95	115	133	183	184	185	197	204	197	200	10.12
AQUATIC ANIMALS	1	1	1			1	1	1				-5.61
AQUATIC PLANTS	5	5	5	5	5	5	5	5	5	6	5	.42
FOREST PRODUCTS 2/												
SAWLOGS CONIFEROUS	1032	1279	1266	1305	1129	1250	1261	1353	1492	1502	1543	3.08
SAWLOGS NONCONIFEROUS	16418	17462	17224	16246	15963	16528	16579	16340	16019	16275	16275	-.45
PULPWOOD+PARTICLES	2171	2002	2008	2037	2109	2297	2380	2171	2616	2944	2944	3.80
FUELWOOD	301712	314197	323016	333971	345313	355742	365746	377111	388575	400673	401134	3.00
SAWNWOOD CONIFEROUS	542	592	624	642	591	666	654	681	766	791	791	3.60
SAWNWOOD NONCONIFEROUS	4432	5211	5259	5015	4729	4999	5398	5486	5798	5810	5810	2.20
WOOD-BASED PANELS	937	1119	1122	1179	1213	1197	1283	1374	1396	1428	1428	3.80
PULP FOR PAPER	409	435	471	359	381	387	415	424	465	471	471	1.27
PAPER+PAPERBOARD	344	378	399	400	417	449	505	574	613	642	644	7.01

1/ NOMINAL CATCH (LIVE WEIGHT) EXCLUDING WHALES

2/ EXCEPT FOR PULP FOR PAPER AND PAPER AND PAPERBOARD, ALL FOREST PRODUCTS ARE EXPRESSED IN THOUSAND CUBIC METRES

1. (Cont.) VOLUME OF PRODUCTION OF MAJOR AGRICULTURAL, FISHERY AND FOREST PRODUCTS

	1979	1980	1981	1982	1983	1984	1985	1986	1987	1988	1989	ANNUAL RATE OF CHANGE 1979-89
	......					THOUSAND METRIC TONS					......	PERCENT
LATIN AMERICA												
AGRICULTURAL PRODUCTS												
TOTAL CEREALS	84007	88456	104496	105151	99688	106768	110655	105876	111920	108539	105772	2.11
WHEAT	15103	14874	15202	22727	20133	21917	20201	21675	22321	20160	23080	4.08
RICE PADDY	14348	16401	15676	17475	14754	16940	16990	17609	18187	19989	19396	2.67
BARLEY	1330	1302	1262	1147	1161	1331	1262	1276	1596	1429	1629	2.19
MAIZE	39751	45242	55311	47806	47098	50817	55808	52593	56377	53767	51235	2.18
MILLET AND SORGHUM	12281	9572	16059	14785	15083	14233	15234	11621	12024	11891	8972	-1.84
ROOT CROPS	45631	43830	46239	45677	41863	44034	44945	46954	46243	46589	48231	.59
POTATOES	10988	10355	11862	11801	10100	12161	11602	11213	11796	13674	12809	1.74
CASSAVA	30941	29940	30956	30469	28305	28145	29616	32062	30591	29212	31506	.09
TOTAL PULSES	4569	4310	5329	5448	4335	5113	5061	4857	4573	5220	4762	.40
CITRUS FRUIT	16360	19223	20161	20739	20722	21624	23212	22492	24462	25009	26719	4.04
BANANAS	15678	16184	16340	16596	15884	16960	17038	17370	18556	18284	17901	1.55
APPLES	1704	1702	1769	1816	1801	2239	2209	2064	2635	2666	2895	5.70
VEGETABLE OILS,OIL EQUIV	5868	6525	6350	6207	6634	7426	8652	7967	7919	9142	9709	4.93
SOYBEANS	15464	19814	20499	18655	20331	24445	27170	22254	26472	30089	33768	6.56
GROUNDNUTS IN SHELL	1389	1099	1012	867	799	890	995	828	949	875	631	-4.37
SUNFLOWER SEED	1551	1757	1353	2068	2463	2268	3521	4280	2381	3149	3423	9.32
RAPESEED	75	96	64	32	17	17	46	111	112	139	129	8.89
COTTONSEED	3125	2948	2727	2552	2256	3018	3419	2737	2364	3457	2664	-.03
COPRA	214	235	227	281	282	244	248	263	258	295	271	2.03
PALM KERNELS	324	325	311	306	290	301	317	301	302	316	331	-.02
SUGAR (CENTRIFUGAL,RAW)	26274	26391	27202	28920	29379	29707	28863	28508	27780	28454	27028	.42
COFFEE GREEN	3257	2981	4075	3064	3747	3509	3881	3168	4254	3464	3463	1.04
COCOA BEANS	572	552	561	607	572	535	733	725	568	677	684	2.17
TEA	44	51	39	49	54	56	63	59	63	53	58	3.15
COTTON LINT	1730	1634	1519	1423	1251	1673	1917	1504	1321	1948	1523	.17
JUTE AND SIMILAR FIBRES	108	107	129	89	95	106	95	90	92	87	56	-4.51
SISAL	339	352	339	347	276	307	365	336	275	262	297	-2.03
TOBACCO	797	732	690	758	708	721	703	692	689	751	754	-.33
NATURAL RUBBER	43	46	51	54	57	58	62	58	58	64	63	3.49
TOTAL MEAT	14580	15079	15821	15825	15884	15438	16038	16128	16847	17584	18090	1.77
TOTAL MILK	33749	35445	35857	36600	36401	36767	38152	38782	40693	42876	43347	2.35
TOTAL EGGS	2429	2578	2619	2739	2698	2912	3123	3456	3570	3684	3504	4.46
WOOL GREASY	301	306	314	317	314	297	293	311	317	317	317	.30
FISHERY PRODUCTS 1/												
FRESHWATER + DIADROMOUS	235	297	323	338	444	469	461	455	581	567	581	9.02
MARINE FISH	9129	8605	9274	10338	8007	10756	12359	14697	12586	14925	15575	6.50
CRUST+ MOLLUS+ CEPHALOP	630	539	532	570	602	656	668	670	737	746	764	3.39
AQUATIC ANIMALS	54	50	49	36	30	46	77	57	33	31	33	-3.35
AQUATIC PLANTS	129	124	152	222	213	213	235	181	167	199	207	3.91
FOREST PRODUCTS 2/												
SAWLOGS CONIFEROUS	26802	29294	28493	29038	30038	31453	32405	31529	31810	31868	31868	1.63
SAWLOGS NONCONIFEROUS	27100	30174	29789	29631	30049	31057	31493	33678	34431	34478	34478	2.28
PULPWOOD+PARTICLES	26641	29274	29132	29006	30745	31331	31606	32064	32246	32251	32251	1.72
FUELWOOD	229893	236060	239855	244625	250845	256862	261960	266721	272330	278171	278171	2.03
SAWNWOOD CONIFEROUS	12149	11671	11498	11174	12064	12575	12972	12684	13466	13472	13472	1.73
SAWNWOOD NONCONIFEROUS	12167	13708	14479	14006	14353	15073	15180	15968	16244	16349	16349	2.62
WOOD-BASED PANELS	3737	4283	4421	4362	4472	4596	4548	4755	5192	5316	5316	3.04
PULP FOR PAPER	4439	5408	5261	5566	6106	6192	6516	7055	7152	7577	7562	5.10
PAPER+PAPERBOARD	7026	7730	7451	7723	7962	8764	9093	9951	10514	11290	11106	5.11

1/ NOMINAL CATCH (LIVE WEIGHT) EXCLUDING WHALES
2/ EXCEPT FOR PULP FOR PAPER AND PAPER AND PAPERBOARD, ALL FOREST PRODUCTS ARE EXPRESSED IN THOUSAND CUBIC METRES

1. (Cont.) VOLUME OF PRODUCTION OF MAJOR AGRICULTURAL, FISHERY AND FOREST PRODUCTS

	1979	1980	1981	1982	1983	1984	1985	1986	1987	1988	1989	ANNUAL RATE OF CHANGE 1979-89
	....	....	....	....	....	THOUSAND METRIC TONS	....	....	....	....	....	PERCENT
NEAR EAST DEVELOPING												
AGRICULTURAL PRODUCTS												
TOTAL CEREALS	53834	55760	59162	57852	55061	54887	62229	65377	63600	73814	56986	1.83
WHEAT	30726	30702	31755	31880	30225	30684	33146	36266	36820	39416	32467	1.90
RICE PADDY	4739	4705	4862	5036	4565	4591	4988	4983	5025	4445	4932	.13
BARLEY	8234	9573	10471	10587	10146	10281	11604	12399	11804	15445	9138	2.99
MAIZE	5309	5546	5536	5721	6004	6218	6662	6077	6378	7256	6848	2.76
MILLET AND SORGHUM	3625	4160	5502	3732	3294	2324	5047	4912	2807	6649	3156	.13
ROOT CROPS	6497	7217	7503	7773	7747	8044	9211	9759	10198	9503	8084	3.46
POTATOES	6008	6756	7039	7279	7282	7605	8755	9333	9825	9160	7789	3.89
CASSAVA	127	122	125	125	125	100	90	80	80	65	15	-13.35
TOTAL PULSES	1685	1855	1917	2282	2474	2331	2623	3259	3401	3737	3002	7.72
CITRUS FRUIT	3740	3602	3703	4272	4431	4325	4031	4670	4993	4909	5122	3.51
BANANAS	260	291	315	356	361	391	419	460	514	534	507	7.44
APPLES	2359	2539	2513	2966	3212	3539	3475	3444	3377	3507	3386	4.05
VEGETABLE OILS,OIL EQUIV	1395	1667	1338	1550	1319	1413	1344	1605	1571	1816	1479	1.11
SOYBEANS	195	145	209	319	340	301	357	425	476	371	389	9.80
GROUNDNUTS IN SHELL	977	814	840	610	524	495	398	477	582	726	366	-6.36
SUNFLOWER SEED	634	794	630	652	763	758	867	1008	1161	1217	1325	7.77
RAPESEED	43	12	6	2						1	3	-27.81
COTTONSEED	2332	2284	2226	2335	2460	2520	2479	2339	2274	2435	2286	.21
SUGAR (CENTRIFUGAL,RAW)	2587	2492	3104	3748	3803	3707	3682	3758	4069	3554	3413	3.32
COFFEE GREEN	5	5	5	4	4	5	5	5	5	6	6	2.78
TEA	133	128	76	103	137	160	179	196	187	209	182	7.36
COTTON LINT	1372	1360	1334	1392	1441	1504	1465	1374	1327	1422	1382	.17
JUTE AND SIMILAR FIBRES	13	13	13	10	10	10	10	10	10	10	10	-2.82
TOBACCO	274	295	238	277	305	242	245	230	249	273	309	-.16
TOTAL MEAT	3287	3448	3645	3887	3980	4128	4283	4211	4285	4318	4395	2.83
TOTAL MILK	15239	15668	16545	16654	16809	14839	16711	16377	16461	16613	16377	.49
TOTAL EGGS	708	741	837	904	965	1028	1100	1165	1159	1234	1269	6.15
WOOL GREASY	179	183	189	194	195	176	185	180	187	186	180	-.12
FISHERY PRODUCTS 1/												
FRESHWATER + DIADROMOUS	161	174	176	185	197	207	202	209	294	292	302	6.58
MARINE FISH	699	772	811	880	968	978	996	1024	1105	1143	1061	4.57
CRUST+ MOLLUS+ CEPHALOP	36	47	36	40	42	51	47	49	55	56	50	3.66
AQUATIC MAMMALS	2	3	1									
FOREST PRODUCTS 2/												
SAWLOGS CONIFEROUS	4718	4964	5218	5214	4190	4150	4059	4393	4054	4053	4053	-2.43
SAWLOGS NONCONIFEROUS	1523	1315	1366	1366	1371	1353	1340	1144	1339	1339	1339	-.91
PULPWOOD+PARTICLES	1043	672	714	712	765	513	380	740	726	726	726	-1.87
FUELWOOD	40693	41838	40973	41413	41810	40947	38304	39100	39764	40486	40486	-.41
SAWNWOOD CONIFEROUS	4114	4127	4107	4101	3787	3794	3792	3791	3786	3784	3784	-1.06
SAWNWOOD NONCONIFEROUS	1146	1139	1121	917	1142	1719	1725	1722	1722	1722	1722	6.23
WOOD-BASED PANELS	797	652	629	623	654	888	978	979	980	980	980	4.95
PULP FOR PAPER	463	494	487	487	517	588	588	588	588	588	588	2.73
PAPER+PAPERBOARD	737	774	832	821	674	808	763	762	786	748	748	-.23

1/ NOMINAL CATCH (LIVE WEIGHT) EXCLUDING WHALES
2/ EXCEPT FOR PULP FOR PAPER AND PAPER AND PAPERBOARD, ALL FOREST PRODUCTS ARE EXPRESSED IN THOUSAND CUBIC METRES

1. (Cont.) VOLUME OF PRODUCTION OF MAJOR AGRICULTURAL, FISHERY AND FOREST PRODUCTS

	1979	1980	1981	1982	1983	1984	1985	1986	1987	1988	1989	ANNUAL RATE OF CHANGE 1979-89
	......	......	......	......	......	THOUSAND METRIC TONS	......	......	......	......	......	PERCENT
FAR EAST DEVELOPING												
AGRICULTURAL PRODUCTS												
TOTAL CEREALS	250762	273697	290004	276050	317477	318663	323716	325577	310946	349054	367171	3.17
WHEAT	46459	44140	49540	50449	57213	58446	57997	62820	58334	60801	70401	3.99
RICE PADDY	162613	186944	193463	184162	209545	211011	222091	216890	209735	236207	244529	3.33
BARLEY	3819	2592	3366	2937	2901	2810	2302	2591	2361	2290	2402	-3.89
MAIZE	17060	19227	20325	18141	22257	23852	22313	24476	20476	26816	25910	3.74
MILLET AND SORGHUM	20766	20746	23266	20303	25514	22486	18953	18739	19975	22889	23870	.16
ROOT CROPS	54697	57900	60337	59122	59556	63885	64979	56826	64029	70193	73767	2.25
POTATOES	12458	10923	12430	12818	12997	15353	16046	13876	16057	17661	18579	4.70
CASSAVA	33735	39291	40231	39162	39121	41100	41650	36238	41503	46071	48717	2.28
TOTAL PULSES	14252	11679	13368	13914	15561	15745	15374	16579	15145	14549	16358	2.09
CITRUS FRUIT	3326	3719	4179	4263	4654	4711	4743	5588	5400	5551	5693	5.24
BANANAS	12177	13095	13597	13200	13140	13919	13909	14377	14273	14284	14478	1.45
APPLES	1208	1179	1462	1586	1684	1660	1708	1667	2084	1760	1815	4.47
VEGETABLE OILS,OIL EQUIV	11650	11924	13762	14300	14064	15440	16332	17331	17434	19964	22126	6.10
SOYBEANS	1351	1442	1480	1408	1602	2270	2481	2714	2647	3570	3974	12.05
GROUNDNUTS IN SHELL	7153	6441	8677	6831	8639	8085	6892	7709	7507	11283	10142	3.30
SUNFLOWER SEED	47	75	184	306	377	569	441	653	924	660	583	28.74
RAPESEED	2351	1942	2797	2872	2699	3087	3599	3198	3054	3888	4877	6.76
COTTONSEED	4229	4214	4436	4421	3357	5077	5612	5143	5251	5963	6286	4.36
COPRA	3394	3760	3782	3770	3711	3041	3450	4522	4154	3589	3415	.38
PALM KERNELS	601	691	738	1086	1021	1301	1486	1659	1672	1879	2266	13.89
SUGAR (CENTRIFUGAL,RAW)	12841	9665	12019	17965	16832	13935	14180	14586	16424	17668	19278	4.45
COFFEE GREEN	528	602	616	648	630	578	686	668	730	722	851	3.49
COCOA BEANS	44	54	71	90	111	132	147	184	295	292	378	23.83
TEA	890	911	923	887	919	1020	1048	1015	1082	1117	1102	2.54
COTTON LINT	2114	2007	2201	2211	1680	2540	2808	2574	2589	2981	3141	4.52
JUTE AND SIMILAR FIBRES	2950	2792	2748	2485	2704	2598	4037	3491	2722	2548	2957	.64
TOBACCO	1002	950	990	1082	1135	1074	1100	1083	1010	897	1054	.05
NATURAL RUBBER	3431	3380	3346	3342	3611	3644	3770	3958	4105	4287	4171	2.69
TOTAL MEAT	5581	5818	6017	6149	6485	6908	7226	7415	7863	8304	8658	4.56
TOTAL MILK	41870	43542	46627	48560	52170	55549	58903	61235	62560	66438	70220	5.36
TOTAL EGGS	1937	2090	2152	2296	2432	2495	2640	2802	2914	3094	3212	5.11
WOOL GREASY	75	80	77	80	83	86	89	93	99	102	90	2.76
FISHERY PRODUCTS 1/												
FRESHWATER + DIADROMOUS	698	697	705	724	763	806	817	824	812	843	864	2.38
MARINE FISH	583	608	632	643	649	591	625	620	1211	1231	644	4.81
CRUST+ MOLLUS+ CEPHALOP	99	111	118	123	118	133	149	167	144	141	150	4.19
AQUATIC ANIMALS									2	4		16.66
FOREST PRODUCTS 2/												
SAWLOGS CONIFEROUS	3631	3382	3463	3524	3744	3752	3868	3867	3987	3957	3986	1.59
SAWLOGS NONCONIFEROUS	78791	81434	78401	79735	82966	83444	77696	82821	97759	101753	104206	2.73
PULPWOOD+PARTICLES	3388	3436	3168	3123	3430	3433	2789	2769	3128	3070	2687	-1.89
FUELWOOD	456678	466779	476791	486977	498079	508847	519085	530149	541226	552325	563468	2.13
SAWNWOOD CONIFEROUS	3454	3148	3864	4032	4610	4260	4500	5109	5540	5769	6064	6.35
SAWNWOOD NONCONIFEROUS	23401	26184	25388	28810	30287	30472	31642	31954	35556	35967	37922	4.60
WOOD-BASED PANELS	6047	5654	6286	7200	8331	8549	9230	10198	11269	12624	13682	9.36
PULP FOR PAPER	1842	2065	2652	2790	3116	3300	3478	3533	3584	3640	3621	6.70
PAPER+PAPERBOARD	3764	3831	4254	4253	4772	5250	5464	6210	7075	7781	8191	8.66

1/ NOMINAL CATCH (LIVE WEIGHT) EXCLUDING WHALES
2/ EXCEPT FOR PULP FOR PAPER AND PAPER AND PAPERBOARD, ALL FOREST PRODUCTS ARE EXPRESSED IN THOUSAND CUBIC METRES

1. (Cont.) VOLUME OF PRODUCTION OF MAJOR AGRICULTURAL, FISHERY AND FOREST PRODUCTS

	1979	1980	1981	1982	1983	1984	1985	1986	1987	1988	1989	ANNUAL RATE OF CHANGE 1979-89
	THOUSAND METRIC TONS											PERCENT
ASIAN CENT PLANNED ECON												
AGRICULTURAL PRODUCTS												
TOTAL CEREALS	314278	303442	310397	341926	373276	394775	369530	381802	390483	388679	403389	2.89
WHEAT	63093	55566	60058	69042	82195	88478	86696	90907	88521	86610	91699	5.04
RICE PADDY	163772	160820	165518	185638	193628	203691	194491	197993	199211	198315	209523	2.54
BARLEY	3825	2847	3278	3418	3229	3537	2982	2816	3052	3250	3459	-.59
MAIZE	63980	67244	63934	65363	73171	78791	69194	76247	85235	85723	84733	3.22
MILLET AND SORGHUM	13900	12318	12514	13665	15994	14853	11754	10104	10150	10569	9672	-3.54
ROOT CROPS	156128	158102	143702	148940	159350	155562	143028	139838	154265	147456	148856	-.49
POTATOES	27905	28352	26874	27892	29821	30345	28945	27379	29317	34087	33796	1.73
CASSAVA	6496	6863	6691	6640	6876	7002	6698	6532	6089	6232	5855	-1.13
TOTAL PULSES	7131	7169	6885	6782	6492	6833	6019	5944	5900	6601	5366	-2.30
CITRUS FRUIT	1163	1353	1475	1688	2077	2260	2665	3410	4252	3578	5624	16.29
BANANAS	1128	1235	1281	1429	1385	1587	2002	2602	3489	3383	3528	13.89
APPLES	3331	2843	3501	2941	4083	3515	4208	3953	4906	4987	5152	5.67
VEGETABLE OILS,OIL EQUIV	5066	5691	7066	7710	7779	8829	9566	9247	9923	8912	8493	5.57
SOYBEANS	7843	8339	9747	9480	10217	10182	11022	12151	12997	12162	10795	4.22
GROUNDNUTS IN SHELL	2993	3784	4018	4122	4145	5073	6959	6174	6518	5993	5661	7.35
SUNFLOWER SEED	340	910	1332	1286	1341	1705	1733	1544	1241	1180	980	6.33
RAPESEED	2404	2386	4067	5657	4288	4206	5607	5882	6605	5044	5440	8.41
COTTONSEED	4425	5425	5948	7210	9290	12534	8313	7102	8514	8323	7604	4.99
COPRA	61	64	65	70	98	112	115	119	116	123	132	8.97
PALM KERNELS	43	40	41	45	47	46	48	49	50	52	53	2.65
SUGAR (CENTRIFUGAL,RAW)	3690	3840	4486	5176	4841	5744	6657	6686	5542	6332	6609	5.97
COFFEE GREEN	31	46	37	39	42	40	73	125	137	196	248	23.21
TEA	325	349	389	446	451	466	484	514	564	601	593	6.22
COTTON LINT	2212	2712	2974	3605	4645	6267	4157	3551	4257	4162	3802	4.99
JUTE AND SIMILAR FIBRES	574	583	666	572	564	796	2114	772	645	630	609	2.24
SISAL	8	8	3	3	3	19	18	19	17	17	16	16.81
TOBACCO	1025	995	1591	2279	1485	1908	2553	1832	2070	2749	2953	9.77
NATURAL RUBBER	162	164	177	204	232	249	258	284	315	321	318	8.09
TOTAL MEAT	14032	15537	16345	17476	18309	19965	22412	24203	25193	27862	29342	7.72
TOTAL MILK	3178	3380	3559	3964	4267	4730	5221	5796	6404	6894	6521	8.69
TOTAL EGGS	2988	3151	3258	3472	3756	4771	5815	6036	6419	7460	7742	11.31
WOOL GREASY	174	196	210	223	214	202	197	201	224	238	254	2.36
FISHERY PRODUCTS 1/												
FRESHWATER + DIADROMOUS	1067	1178	1305	1492	1772	2185	2872	3281	3950	4423	4819	18.00
MARINE FISH	2129	2171	2181	2413	2336	2552	2595	3117	3247	3351	3496	5.63
CRUST+ MOLLUS+ CEPHALOP	845	877	874	1004	1087	1171	1291	1576	2090	2553	2713	13.36
AQUATIC ANIMALS	13	9	18	18	19	20	21	26	59	32	32	13.63
AQUATIC PLANTS	1507	1590	1387	1408	1524	1664	1714	1346	1354	1645	1657	.52
FOREST PRODUCTS 2/												
SAWLOGS CONIFEROUS	30973	30984	27923	28442	29419	33965	34591	35011	34977	34520	32041	1.70
SAWLOGS NONCONIFEROUS	20031	19665	18473	18779	19383	22283	22670	22957	22916	22626	23372	2.33
PULPWOOD+PARTICLES	4991	5172	4652	4786	4981	5690	6610	6921	6888	7178	7440	5.13
FUELWOOD	179790	183295	186901	190662	194541	198483	202491	206570	210729	211423	212133	1.80
SAWNWOOD CONIFEROUS	13318	13887	14511	15162	15695	17410	18270	17819	17788	17748	16887	3.00
SAWNWOOD NONCONIFEROUS	8025	8323	8652	9019	9291	9432	9898	9653	9637	9615	9152	1.61
WOOD-BASED PANELS	2328	2320	2475	2523	2709	2524	2599	2881	3443	3877	3696	5.20
PULP FOR PAPER	4697	4930	4967	4940	5412	5974	7229	7511	8149	8218	9676	7.81
PAPER+PAPERBOARD	6392	6942	7017	7576	8461	9624	11270	12019	13135	14280	15472	9.91

1/ NOMINAL CATCH (LIVE WEIGHT) EXCLUDING WHALES

2/ EXCEPT FOR PULP FOR PAPER AND PAPER AND PAPERBOARD, ALL FOREST PRODUCTS ARE EXPRESSED IN THOUSAND CUBIC METRES

2. INDICES OF FOOD PRODUCTION

	TOTAL						PER CAPUT					
	1985	1986	1987	1988	1989	CHANGE 1988 TO 1989	1985	1986	1987	1988	1989	CHANGE 1988 TO 1989
	................1979-81=100................					PERCENT	...............1979-81=100................					PERCENT
WORLD	114	116	116	118	121	3.21	105	105	103	102	104	1.43
DEVELOPED COUNTRIES	108	109	108	106	110	4.15	104	105	103	100	103	3.51
WESTERN EUROPE	107	108	109	108	108	.56	105	107	107	105	105	.25
EUROPEAN ECON COMMUNITY	107	108	110	108	109	.46	106	107	108	106	106	.18
BELGIUM-LUXEMBOURG	107	116	114	116	118	1.85	107	116	113	116	118	1.80
DENMARK	123	120	118	119	123	3.20	123	120	118	119	122	3.16
FRANCE	107	106	109	110	107	-2.79	105	104	106	106	102	-3.15
GERMANY FED.REP. OF	108	116	111	114	112	-1.44	109	117	111	114	112	-1.72
GREECE	111	107	99	103	105	1.23	108	104	95	100	101	1.11
IRELAND	113	114	115	109	104	-4.56	108	109	109	102	97	-5.29
ITALY	101	101	104	101	102	.79	100	99	103	99	100	.64
NETHERLANDS	108	119	115	112	125	12.00	106	116	111	107	119	11.38
UNITED KINGDOM	110	110	108	106	105	-1.59	109	109	107	105	103	-1.85
AUSTRIA	107	108	108	114	112	-1.81	107	108	107	114	111	-2.02
FINLAND	114	113	99	102	115	12.32	111	110	96	99	111	11.98
ICELAND	108	101	98	90	88	-2.22	102	95	91	83	80	-3.08
MALTA	114	118	110	104	106	1.77	120	123	115	108	109	.90
NORWAY	107	100	108	112	112	-.01	106	98	105	108	108	-.44
PORTUGAL	107	105	112	87	113	28.86	103	100	106	83	107	28.63
SPAIN	110	107	121	114	113	-.61	107	104	116	109	108	-.97
SWEDEN	108	106	95	93	101	8.15	108	105	94	92	99	7.40
SWITZERLAND	108	110	106	107	107	-.03	106	107	102	103	103	-.50
YUGOSLAVIA	100	112	106	101	101	.10	96	107	101	96	95	-.48
USSR AND EASTERN EUROPE	110	118	116	117	119	2.41	106	112	110	110	111	1.73
EASTERN EUROPE	109	115	111	113	113	-.34	107	112	108	110	109	-.52
ALBANIA	108	109	110	112	115	1.99	98	97	96	96	96	.20
BULGARIA	95	107	101	101	104	3.76	94	106	100	99	103	3.62
CZECHOSLOVAKIA	117	119	121	125	127	1.60	116	117	119	123	124	1.40
GERMAN DEMOCRATIC REP.	116	115	117	112	114	1.38	117	116	118	113	114	1.32
HUNGARY	107	109	108	115	112	-3.01	107	109	109	116	113	-2.83
POLAND	109	117	111	114	113	-.89	105	111	105	107	107	-.87
ROMANIA	108	118	111	116	112	-3.68	106	115	108	112	107	-4.19
USSR	110	118	119	119	122	2.90	106	112	112	110	113	2.02
NORTH AMERICA DEVELOPED	109	104	101	93	104	11.38	103	98	94	87	96	10.46
CANADA	113	123	116	104	114	9.27	107	117	109	97	104	7.99
UNITED STATES	108	102	100	94	103	9.78	102	96	93	87	94	8.91
OCEANIA DEVELOPED	107	108	106	110	107	-2.88	100	100	97	100	96	-3.88
AUSTRALIA	106	108	104	108	108	-.31	99	100	94	97	96	-1.48
NEW ZEALAND	116	111	112	117	109	-7.36	111	106	107	111	103	-7.61
OTHER DEV.ED COUNTRIES	104	105	106	105	108	3.50	99	99	99	97	99	2.65
ISRAEL	126	115	129	113	118	4.30	115	104	115	99	101	2.65
JAPAN	109	109	104	99	101	2.09	105	104	99	95	96	1.67
SOUTH AFRICA	94	97	103	106	112	5.71	85	85	88	89	92	3.42

2. (Cont.) INDICES OF FOOD PRODUCTION

	TOTAL						PER CAPUT					
	1985	1986	1987	1988	1989	CHANGE 1988 TO 1989	1985	1986	1987	1988	1989	CHANGE 1988 TO 1989
	................1979-81=100				....	PERCENT	...............1979-81=100				....	PERCENT
DEVELOPING COUNTRIES	120	123	125	130	133	2.42	108	108	108	110	111	.29
AFRICA DEVELOPING	114	119	116	120	122	1.66	98	100	94	94	93	-1.46
NORTH WESTERN AFRICA	126	133	127	127	132	3.46	109	112	104	102	102	.67
ALGERIA	127	126	131	122	124	1.70	110	105	106	95	94	-1.43
MOROCCO	126	154	125	161	159	-1.65	110	131	104	130	125	-4.09
TUNISIA	135	114	138	100	118	18.08	119	98	115	82	94	15.43
WESTERN AFRICA	117	124	119	124	126	1.43	100	103	95	96	94	-1.84
BENIN	133	140	127	147	161	9.45	114	117	103	115	122	6.01
BURKINA FASO	128	143	131	147	143	-2.75	113	123	109	120	113	-5.33
COTE D¢IVOIRE	120	122	128	139	132	-4.50	97	95	96	99	91	-8.35
GAMBIA	106	119	111	106	128	20.73	92	100	91	84	99	17.39
GHANA	132	133	138	142	151	6.42	111	107	108	108	111	3.34
GUINEA	105	109	109	108	109	.54	93	95	92	89	88	-1.95
LIBERIA	116	117	122	124	121	-2.46	99	97	97	96	91	-5.50
MALI	110	119	112	125	129	2.95	96	100	92	99	99	-.02
MAURITANIA	100	104	106	110	111	.65	88	89	88	89	87	-2.09
NIGER	99	103	91	121	107	-11.75	86	87	75	96	83	-14.40
NIGERIA	122	130	120	123	126	2.23	104	107	95	94	93	-1.25
SENEGAL	122	127	139	121	137	13.34	108	109	117	99	109	10.22
SIERRA LEONE	102	111	109	105	108	3.31	91	96	93	87	87	.74
TOGO	106	108	109	121	132	9.13	91	90	88	95	100	5.84
CENTRAL AFRICA	109	113	114	116	116	.40	95	95	93	92	90	-2.52
ANGOLA	102	103	104	103	101	-1.74	90	89	87	84	80	-4.48
CAMEROON	112	116	117	120	121	1.58	98	99	97	96	96	-.99
CENTRAL AFRICAN REP	98	106	104	108	110	2.05	87	92	89	90	89	-.46
CHAD	116	116	114	126	120	-4.47	103	101	97	104	97	-6.83
CONGO	113	113	119	122	122	.53	99	97	99	99	97	-2.21
GABON	103	106	108	110	110	.57	84	84	82	81	79	-2.78
ZAIRE	115	118	119	120	121	.62	98	99	96	94	92	-2.57
EASTERN AFRICA	110	113	111	116	118	1.84	95	95	90	91	90	-1.30
BURUNDI	116	121	125	131	118	-10.15	101	102	103	104	91	-12.65
ETHIOPIA	97	106	100	102	105	2.42	89	95	88	88	89	.26
KENYA	120	135	131	142	149	4.95	98	106	98	102	103	.58
MADAGASCAR	112	116	117	117	121	3.03	97	97	94	91	91	-.20
MALAWI	104	105	103	111	115	3.40	89	87	83	86	86	-.01
MAURITIUS	111	119	120	110	110	.19	102	108	108	98	97	-1.01
MOZAMBIQUE	99	101	100	103	104	1.02	87	87	84	84	82	-1.66
RWANDA	112	103	104	102	95	-6.42	95	84	82	78	70	-9.55
SOMALIA	116	125	124	131	135	3.29	97	101	97	98	98	-.05
TANZANIA	113	114	117	119	126	5.65	94	91	90	88	90	1.89
UGANDA	111	105	113	113	115	1.66	94	86	89	86	85	-1.83
ZAMBIA	114	119	119	144	145	.95	93	94	90	105	102	-2.67
ZIMBABWE	130	128	92	129	119	-7.69	112	107	75	102	91	-10.58
SOUTHERN AFRICA	100	104	109	113	112	-.39	86	86	87	88	84	-3.52
BOTSWANA	94	90	83	93	99	6.04	78	73	65	70	72	2.37
LESOTHO	99	92	95	113	93	-18.29	87	78	78	90	72	-20.58
SWAZILAND	111	130	125	128	125	-2.12	94	106	99	98	93	-5.45
LATIN AMERICA	114	113	117	123	125	1.80	102	99	100	104	103	-.28
CENTRAL AMERICA	111	114	113	117	122	4.49	99	98	96	96	98	2.11
COSTA RICA	104	107	109	109	114	4.88	90	90	90	87	89	2.21
EL SALVADOR	99	102	96	106	102	-3.44	94	95	88	95	90	-5.43
GUATEMALA	118	123	122	129	135	5.01	103	104	100	103	105	2.02
HONDURAS	103	104	113	115	120	4.44	86	84	88	87	88	1.24
MEXICO	114	113	112	117	121	3.78	101	98	95	97	99	1.57
NICARAGUA	91	90	90	80	80	.05	77	73	71	61	59	-3.25
PANAMA	110	112	116	105	106	1.59	99	98	100	88	88	-.46
CARIBBEAN	104	106	107	110	113	2.50	97	97	97	98	99	1.00
BARBADOS	79	88	78	81	76	-6.55	78	86	76	79	73	-7.19
CUBA	109	110	107	113	116	3.12	106	107	103	108	110	2.28
DOMINICAN REPUBLIC	107	107	108	109	123	12.76	95	92	92	90	99	10.32
HAITI	110	112	114	111	112	.69	101	100	101	96	95	-1.20
JAMAICA	111	113	115	102	91	-10.46	103	104	104	91	80	-11.84

2. (Cont.) INDICES OF FOOD PRODUCTION

	TOTAL						PER CAPUT					
	1985	1986	1987	1988	1989	CHANGE 1988 TO 1989	1985	1986	1987	1988	1989	CHANGE 1988 TO 1989
	....	....	1979-81=100	....	....	PERCENT	....	....	1979-81=100	....	....	PERCENT
SOUTH AMERICA	116	114	119	126	127	1.09	104	100	102	106	105	-.95
ARGENTINA	97	95	100	108	100	-7.58	91	88	91	97	88	-8.72
BOLIVIA	117	116	123	131	128	-1.81	103	98	102	105	101	-4.49
BRAZIL	124	119	129	139	143	2.65	111	104	110	117	117	.58
CHILE	106	112	115	121	129	6.89	98	101	102	106	111	5.13
COLOMBIA	106	115	117	125	136	8.31	96	101	101	106	112	6.13
ECUADOR	117	119	124	134	142	5.57	101	100	101	107	110	2.69
GUYANA	84	88	82	79	69	-11.84	76	78	72	68	59	-13.31
PARAGUAY	130	117	133	153	160	4.60	111	97	107	119	121	1.62
PERU	110	110	120	128	124	-3.50	97	94	100	104	98	-5.88
URUGUAY	106	106	106	114	124	8.47	102	101	100	108	116	7.64
VENEZUELA	106	118	112	118	106	-9.91	92	99	92	94	83	-12.19
NEAR EAST DEVELOPING	114	118	118	124	115	-6.83	99	100	97	99	90	-9.46
NEAR EAST IN AFRICA	112	113	118	125	124	-1.03	97	95	96	99	96	-3.60
EGYPT	120	126	134	135	138	2.36	105	107	110	109	108	-.14
LIBYA	147	130	148	150	159	6.14	118	101	110	108	110	2.53
SUDAN	109	103	91	121	96	-20.75	93	86	74	95	73	-22.95
NEAR EAST IN ASIA	114	119	119	124	113	-8.80	100	101	98	99	87	-11.46
AFGHANISTAN	92	78	82	82	82	-.66	101	86	90	88	84	-4.50
CYPRUS	96	90	95	111	110	-.58	91	84	88	102	100	-1.58
IRAN	120	125	127	116	113	-3.12	99	99	97	85	80	-6.21
IRAQ	146	141	133	127	136	7.21	122	114	103	96	99	3.58
JORDAN	151	137	158	168	145	-13.35	125	109	121	124	103	-16.63
LEBANON	119	121	140	133	141	6.03	119	120	138	128	132	3.14
SAUDI ARABIA	292	301	313	370	368	-.62	234	232	232	263	252	-4.45
SYRIA	109	123	106	145	84	-41.86	92	100	84	110	62	-43.91
TURKEY	109	115	116	122	114	-6.64	96	99	98	101	93	-8.33
YEMEN ARAB REPUBLIC	98	118	108	125	117	-6.81	85	99	89	99	90	-9.64
YEMEN DEMOCRATIC	101	103	106	102	103	.71	88	87	87	81	79	-2.38
FAR EAST DEVELOPING	121	123	122	132	138	4.53	108	107	104	110	113	2.37
SOUTH ASIA	122	124	123	135	142	4.74	108	107	104	112	115	2.43
BANGLADESH	112	113	112	113	121	6.65	97	96	93	91	95	3.85
INDIA	124	126	124	139	146	5.07	111	110	107	117	120	2.96
NEPAL	118	112	124	138	140	1.68	104	96	104	112	111	-.77
PAKISTAN	121	130	131	139	147	5.33	100	104	101	103	105	1.94
SRI LANKA	111	109	97	101	101	-.06	102	99	86	89	88	-1.27
EAST SOUTH-EAST ASIA	121	122	121	127	133	4.03	109	107	105	108	111	2.15
BURMA	139	140	141	140	142	1.30	126	124	122	119	118	-.80
INDONESIA	129	139	139	146	150	2.33	117	124	122	126	127	.76
KOREA REP	111	110	101	108	105	-2.87	103	101	91	97	93	-3.96
LAO	138	141	137	135	148	9.84	123	123	117	112	120	7.14
MALAYSIA	146	151	156	170	190	11.96	130	131	133	141	154	9.41
PHILIPPINES	100	102	102	105	110	5.03	88	87	85	85	87	2.51
THAILAND	121	111	111	122	127	4.05	109	99	97	106	108	2.56
ASIAN CENT PLANNED ECON	128	133	139	140	144	3.30	120	123	127	126	128	1.74
CHINA	128	134	139	140	144	3.28	121	124	128	127	129	1.78
KAMPUCHEA,DEMOCRATIC	168	169	167	187	182	-2.59	148	146	140	153	145	-4.95
KOREA DPR	114	117	119	122	124	1.89	101	101	101	100	100	-.46
MONGOLIA	109	117	112	114	116	1.89	95	99	92	91	90	-1.31
VIET NAM	123	124	130	132	138	4.30	110	109	111	111	113	1.99
OTHER DEV.ING COUNTRIES	110	113	108	109	115	5.11	97	98	91	89	92	2.54

3. INDICES OF AGRICULTURAL PRODUCTION

	TOTAL						PER CAPUT					
	1985	1986	1987	1988	1989	CHANGE 1988 TO 1989	1985	1986	1987	1988	1989	CHANGE 1988 TO 1989
	................1979-81=100................					PERCENT	................1979-81=100................					PERCENT
WORLD	114	115	116	118	121	2.99	105	104	103	103	104	1.21
DEVELOPED COUNTRIES	108	109	108	106	110	3.70	104	104	103	100	103	3.06
WESTERN EUROPE	107	109	109	108	108	.51	105	107	107	105	106	.20
EUROPEAN ECON COMMUNITY	107	108	110	109	109	.41	106	107	108	106	106	.13
BELGIUM-LUXEMBOURG	107	116	113	116	118	1.91	107	116	113	116	118	1.87
DENMARK	123	120	118	119	123	3.20	123	120	118	119	122	3.16
FRANCE	107	106	109	110	106	-2.84	104	103	106	106	102	-3.20
GERMANY FED.REP. OF	108	116	110	114	112	-1.44	109	117	111	114	112	-1.72
GREECE	113	111	102	107	108	1.14	110	107	98	104	105	1.02
IRELAND	113	114	115	109	104	-4.41	108	109	109	102	97	-5.15
ITALY	101	101	104	102	102	.81	100	99	103	100	100	.66
NETHERLANDS	108	119	115	112	125	11.94	106	116	111	107	120	11.33
UNITED KINGDOM	110	110	109	107	105	-1.50	109	109	107	105	103	-1.76
AUSTRIA	107	108	108	114	112	-1.81	107	108	107	114	111	-2.01
FINLAND	114	113	99	102	115	12.32	111	110	96	99	111	11.98
ICELAND	107	101	98	90	88	-2.15	101	95	91	83	80	-3.02
MALTA	114	117	110	104	106	1.76	119	123	115	108	109	.89
NORWAY	107	100	108	112	112	-.01	106	98	105	108	108	-.44
PORTUGAL	107	105	111	87	112	28.72	102	100	106	83	107	28.49
SPAIN	110	107	121	115	113	-1.32	107	104	116	110	108	-1.68
SWEDEN	108	106	95	93	101	8.15	108	105	94	92	99	7.40
SWITZERLAND	108	110	106	107	107	-.12	106	107	102	103	102	-.58
YUGOSLAVIA	100	112	106	101	101	.18	97	108	101	96	95	-.39
USSR AND EASTERN EUROPE	110	117	115	116	118	2.16	106	112	109	109	110	1.49
EASTERN EUROPE	109	115	111	113	112	-.61	107	112	108	109	108	-.79
ALBANIA	109	111	110	114	114	.11	98	98	96	97	95	-1.65
BULGARIA	95	105	100	98	98	.15	94	104	99	97	97	.01
CZECHOSLOVAKIA	117	119	121	125	127	1.74	116	117	119	122	124	1.54
GERMAN DEMOCRATIC REP.	116	115	118	113	114	1.31	117	116	118	113	115	1.26
HUNGARY	107	108	108	115	111	-3.00	107	109	109	116	113	-2.82
POLAND	110	117	111	114	112	-1.34	105	111	105	107	105	-1.32
ROMANIA	108	118	111	116	112	-3.57	106	115	108	112	107	-4.09
USSR	110	117	117	117	120	2.65	105	111	110	109	111	1.78
NORTH AMERICA DEVELOPED	108	102	101	94	103	10.05	103	97	94	87	95	9.14
CANADA	113	123	116	103	113	9.07	108	116	109	96	103	7.79
UNITED STATES	107	100	99	94	102	8.29	101	94	93	87	93	7.43
OCEANIA DEVELOPED	109	110	109	113	111	-2.06	102	102	100	102	99	-3.06
AUSTRALIA	110	112	109	114	115	.24	103	103	99	103	102	-.94
NEW ZEALAND	113	108	109	112	105	-6.28	109	104	103	106	99	-6.53
OTHER DEV.ED COUNTRIES	103	103	105	103	106	3.09	98	97	98	96	98	2.25
ISRAEL	124	109	119	107	107	.63	114	99	105	93	92	-.97
JAPAN	107	106	102	97	99	1.65	103	102	97	93	94	1.23
SOUTH AFRICA	95	97	102	106	111	5.35	85	85	87	89	91	3.07

3. (Cont.) INDICES OF AGRICULTURAL PRODUCTION

	TOTAL						PER CAPUT					
	1985	1986	1987	1988	1989	CHANGE 1988 TO 1989	1985	1986	1987	1988	1989	CHANGE 1988 TO 1989
	................1979-81=100................					PERCENT	................1979-81=100................					PERCENT
DEVELOPING COUNTRIES	121	122	125	130	133	2.42	109	108	108	110	110	.29
AFRICA DEVELOPING	114	119	117	121	123	1.86	98	100	94	95	93	-1.26
NORTH WESTERN AFRICA	127	134	129	129	134	3.43	111	113	106	103	104	.64
ALGERIA	128	127	132	123	125	1.71	110	106	107	96	95	-1.42
MOROCCO	126	154	126	162	159	-1.57	111	131	104	131	126	-4.01
TUNISIA	134	114	138	100	118	17.78	118	98	115	82	94	15.14
WESTERN AFRICA	117	124	119	123	126	2.16	99	102	95	95	94	-1.14
BENIN	137	146	130	153	166	8.48	118	122	106	120	126	5.08
BURKINA FASO	129	146	133	149	144	-2.95	114	126	111	121	115	-5.53
COTE D¢IVOIRE	117	118	125	131	129	-1.46	95	92	93	94	89	-5.43
GAMBIA	107	119	110	106	127	20.57	92	100	90	84	98	17.24
GHANA	131	131	136	140	149	6.42	109	106	106	106	110	3.34
GUINEA	105	109	109	109	111	1.55	93	95	93	90	90	-.96
LIBERIA	115	117	122	121	122	.20	98	97	98	94	91	-2.93
MALI	111	120	114	126	132	4.45	96	101	94	101	102	1.43
MAURITANIA	100	104	106	110	111	.65	88	89	88	89	87	-2.09
NIGER	99	103	91	121	107	-11.73	86	87	75	96	83	-14.38
NIGERIA	122	130	120	123	126	2.41	103	106	95	94	93	-1.08
SENEGAL	122	127	139	121	136	12.76	108	109	117	99	108	9.66
SIERRA LEONE	101	112	111	104	107	2.72	90	97	94	86	86	.16
TOGO	108	111	113	125	135	8.69	93	93	92	98	103	5.41
CENTRAL AFRICA	109	114	114	117	116	-.87	95	96	93	93	90	-3.76
ANGOLA	99	101	100	100	98	-2.12	87	86	84	81	78	-4.85
CAMEROON	111	117	115	122	119	-2.43	97	100	95	98	93	-4.90
CENTRAL AFRICAN REP	98	107	104	109	112	2.08	88	93	89	91	90	-.44
CHAD	116	116	116	128	123	-4.41	104	101	99	106	99	-6.78
CONGO	113	112	119	121	122	.56	99	96	99	98	96	-2.17
GABON	103	106	108	110	111	.58	84	84	82	81	79	-2.77
ZAIRE	115	119	120	121	122	.62	99	99	97	94	92	-2.57
EASTERN AFRICA	110	113	112	117	120	2.09	95	95	91	92	91	-1.05
BURUNDI	116	119	126	129	116	-9.82	101	101	103	103	90	-12.33
ETHIOPIA	96	104	100	101	104	3.41	88	94	88	87	88	1.23
KENYA	122	134	130	143	146	2.34	100	105	98	103	101	-1.92
MADAGASCAR	112	116	116	116	120	3.44	96	97	93	91	91	.19
MALAWI	110	109	106	115	120	4.08	94	90	85	89	90	.65
MAURITIUS	112	119	120	111	110	-.51	103	109	108	98	97	-1.69
MOZAMBIQUE	96	101	100	102	103	.69	85	86	84	83	81	-1.98
RWANDA	113	106	108	106	101	-5.10	96	87	86	81	75	-8.28
SOMALIA	116	125	124	131	135	3.29	97	101	97	98	98	-.05
TANZANIA	110	114	116	119	125	5.28	91	91	89	88	89	1.53
UGANDA	111	106	114	114	116	1.63	94	86	90	87	86	-1.85
ZAMBIA	115	120	122	146	149	1.59	94	95	93	107	104	-2.05
ZIMBABWE	133	130	108	134	128	-4.37	114	109	87	105	98	-7.36
SOUTHERN AFRICA	100	104	108	111	111	-.45	85	86	86	87	84	-3.58
BOTSWANA	94	90	83	93	99	5.99	78	73	65	70	72	2.32
LESOTHO	101	95	97	114	94	-17.30	88	80	80	91	73	-19.61
SWAZILAND	111	128	124	126	123	-1.94	94	105	98	97	91	-5.27
LATIN AMERICA	114	111	116	121	122	1.10	102	97	99	102	101	-.97
CENTRAL AMERICA	108	110	110	113	116	2.32	95	95	93	93	93	-.01
COSTA RICA	106	110	113	115	119	3.68	92	92	93	92	93	1.04
EL SALVADOR	86	83	83	79	70	-10.34	81	77	76	70	62	-12.19
GUATEMALA	110	111	111	117	126	6.90	95	94	91	94	97	3.86
HONDURAS	102	103	110	116	119	2.84	86	83	86	88	88	-.31
MEXICO	113	112	112	116	118	1.84	100	98	95	96	96	-.33
NICARAGUA	91	85	83	75	73	-3.57	77	69	65	58	54	-6.74
PANAMA	111	113	118	107	109	1.11	100	99	101	90	90	-.93
CARIBBEAN	104	105	106	109	111	1.68	98	97	97	98	98	.19
BARBADOS	79	88	78	81	76	-6.55	78	86	76	79	73	-7.19
CUBA	109	110	107	113	116	3.16	107	107	104	108	111	2.32
DOMINICAN REPUBLIC	109	107	108	109	114	4.50	97	93	92	90	92	2.24
HAITI	108	110	110	108	109	.46	99	99	97	94	92	-1.42
JAMAICA	111	114	115	103	92	-10.61	103	104	104	91	80	-11.99

3. (Cont.) INDICES OF AGRICULTURAL PRODUCTION

	TOTAL						PER CAPUT					
	1985	1986	1987	1988	1989	CHANGE 1988 TO 1989	1985	1986	1987	1988	1989	CHANGE 1988 TO 1989
	...............1979-81=100...............					PERCENT	...............1979-81=100...............					PERCENT
SOUTH AMERICA	116	112	118	124	125	.76	104	98	101	104	103	-1.28
ARGENTINA	97	96	100	108	100	-7.71	91	88	90	97	89	-8.85
BOLIVIA	116	114	121	129	126	-2.16	101	97	100	104	99	-4.84
BRAZIL	125	115	129	135	138	2.24	112	100	111	113	113	.17
CHILE	106	112	115	121	129	6.78	98	101	102	105	111	5.03
COLOMBIA	103	111	113	120	127	6.58	93	98	97	101	105	4.44
ECUADOR	118	120	123	136	142	4.11	102	101	101	108	110	1.27
GUYANA	84	88	82	79	69	-11.82	76	78	72	68	59	-13.30
PARAGUAY	137	120	129	159	169	5.87	117	99	104	124	128	2.86
PERU	109	109	117	125	122	-2.40	96	94	97	102	97	-4.80
URUGUAY	105	108	109	115	123	6.84	102	103	103	109	116	6.03
VENEZUELA	107	118	113	119	107	-9.30	92	99	92	95	84	-11.60
NEAR EAST DEVELOPING	113	116	117	122	115	-6.38	98	98	96	98	89	-9.02
NEAR EAST IN AFRICA	111	110	114	119	119	-.38	96	93	93	95	92	-2.96
EGYPT	116	120	125	125	129	2.90	101	102	103	101	101	.39
LIBYA	146	130	148	150	159	6.11	118	101	110	107	110	2.50
SUDAN	112	104	94	120	97	-19.13	96	87	76	95	74	-21.38
NEAR EAST IN ASIA	114	118	118	123	113	-8.39	99	100	97	99	88	-11.06
AFGHANISTAN	91	78	83	83	81	-3.02	100	86	91	89	83	-6.77
CYPRUS	96	90	95	111	110	-.51	91	84	88	101	100	-1.52
IRAN	120	125	127	116	113	-2.64	98	99	97	85	80	-5.74
IRAQ	146	141	132	126	135	7.40	122	114	103	95	98	3.76
JORDAN	149	134	156	166	145	-12.97	124	107	120	123	103	-16.27
LEBANON	117	119	138	130	138	6.00	118	118	135	126	130	3.11
SAUDI ARABIA	289	299	311	366	364	-.62	232	230	230	261	249	-4.44
SYRIA	111	124	107	144	88	-38.86	94	101	84	110	65	-41.02
TURKEY	109	114	115	122	114	-6.55	96	98	97	101	93	-8.25
YEMEN ARAB REPUBLIC	98	117	109	125	117	-6.57	86	99	89	99	90	-9.40
YEMEN DEMOCRATIC	101	104	107	103	104	.68	88	88	88	82	80	-2.41
FAR EAST DEVELOPING	121	122	121	131	137	5.09	108	106	103	109	112	2.92
SOUTH ASIA	122	123	123	134	142	5.73	108	107	104	111	115	3.40
BANGLADESH	113	115	113	112	119	6.10	99	97	93	90	93	3.32
INDIA	124	124	123	137	146	6.08	111	109	105	116	120	3.95
NEPAL	117	112	122	135	138	1.78	103	96	102	110	109	-.68
PAKISTAN	124	134	136	143	150	4.97	103	107	105	106	108	1.59
SRI LANKA	109	107	96	101	99	-1.97	100	97	86	89	86	-3.15
EAST SOUTH-EAST ASIA	120	121	121	127	132	3.81	108	107	105	108	110	1.93
BURMA	138	139	139	138	140	1.24	125	123	120	117	116	-.86
INDONESIA	128	138	138	145	148	2.41	116	123	121	125	126	.84
KOREA REP	109	109	100	107	104	-2.60	101	100	90	95	92	-3.69
LAO	137	140	136	135	148	9.83	122	122	116	112	120	7.12
MALAYSIA	133	138	142	153	165	7.25	118	120	121	128	134	4.80
PHILIPPINES	100	103	103	106	111	5.30	88	88	86	86	88	2.78
THAILAND	123	114	113	124	129	4.58	111	101	99	107	110	3.08
ASIAN CENT PLANNED ECON	131	134	141	142	147	3.05	123	124	128	128	130	1.49
CHINA	131	134	141	143	147	3.00	124	125	130	129	131	1.51
KAMPUCHEA,DEMOCRATIC	171	173	171	191	187	-2.24	151	149	143	156	149	-4.60
KOREA DPR	114	117	120	122	125	1.99	101	101	101	101	100	-.37
MONGOLIA	107	114	109	111	114	1.86	93	97	89	89	88	-1.34
VIET NAM	123	125	130	133	138	4.30	110	109	111	111	113	1.99
OTHER DEV.ING COUNTRIES	110	111	109	110	116	5.59	97	96	91	90	93	3.00

4. VOLUME OF EXPORTS OF MAJOR AGRICULTURAL, FISHERY AND FOREST PRODUCTS

	1978	1979	1980	1981	1982	1983	1984	1985	1986	1987	1988	ANNUAL RATE OF CHANGE 1978-88
	....	....	....	....	....	THOUSAND METRIC TONS	....	....	....	....	....	PERCENT
WORLD												
AGRICULTURAL PRODUCTS												
WHEAT+FLOUR,WHEAT EQUIV.	84921	81565	99524	105209	104988	111830	116426	105130	96408	110680	119210	2.69
RICE MILLED	9559	11634	12956	13061	12047	11502	12872	11408	13019	12946	12185	1.33
BARLEY	14585	14104	16226	20278	18346	17755	23006	21899	26231	22292	21069	5.12
MAIZE	68794	76097	80303	79442	70049	69087	68679	69725	57640	64450	66681	-1.89
MILLET	315	296	215	242	196	191	165	193	158	177	202	-5.15
SORGHUM	10923	11365	11166	14466	13725	11732	12438	13337	8579	7946	9279	-2.95
POTATOES	4037	4632	4919	4948	5182	4783	4788	5009	5476	6245	6480	3.54
SUGAR,TOTAL (RAW EQUIV.)	26139	26687	27505	29347	30744	29480	28558	28380	27714	28568	29124	.63
PULSES	2116	2350	2810	3148	2960	3195	3363	3706	4824	5369	6042	10.14
SOYBEANS	24062	25489	26877	26219	28928	26592	25790	26152	27637	29245	26122	.84
SOYBEAN OIL	2610	2953	3196	3489	3405	3652	4030	3503	2992	4019	3914	2.99
GROUNDNUTS SHELLED BASIS	746	744	723	831	739	782	740	851	951	895	934	2.52
GROUNDNUT OIL	418	503	477	322	450	529	302	320	367	359	338	-3.21
COPRA	709	440	461	415	438	252	287	388	405	333	291	-5.75
COCONUT OIL	1334	1142	1216	1357	1270	1325	984	1236	1650	1470	1331	1.35
PALM NUTS KERNELS	181	160	201	138	136	120	131	98	110	120	135	-4.58
PALM OIL	2404	2846	3617	3229	3776	4017	4318	5221	6242	5781	5927	9.59
OILSEED CAKE AND MEAL	21875	23221	25689	27792	27625	32126	28496	30547	33969	36650	39062	5.40
BANANAS	7044	6954	6956	6996	7210	6335	6937	6807	7353	7631	7779	.86
ORANGES+TANGER+CLEMEN	5182	4945	5104	4941	4955	4807	5269	4926	5928	5464	5302	.93
LEMONS AND LIMES	970	917	986	923	1000	935	996	1040	1066	1061	1027	1.23
COFFEE GREEN+ROASTED	3440	3792	3738	3732	3959	4031	4229	4425	4086	4469	4250	2.19
COCOA BEANS	1086	930	1065	1336	1252	1208	1354	1386	1504	1566	1636	4.96
TEA	886	903	984	951	927	975	1080	1082	1089	1101	1159	2.63
COTTON LINT	4472	4359	4828	4263	4430	4272	4235	4131	4699	5518	4805	1.02
JUTE AND SIMILAR FIBRES	496	559	519	573	512	508	495	384	524	510	342	-2.72
TOBACCO UNMANUFACTURED	1439	1374	1353	1491	1430	1338	1390	1384	1332	1348	1353	-.55
NATURAL RUBBER	3317	3422	3329	3148	3115	3450	3642	3647	3707	4071	4207	2.44
WOOL GREASY	890	937	907	952	874	893	882	904	945	1011	996	.82
BOVINE CATTLE 1/	7580	7409	7042	7187	7687	7108	6769	6490	7121	7125	7154	-.67
SHEEP AND GOATS 1/	14776	15269	18641	17608	18437	20576	19631	18725	19348	21607	21965	3.39
PIGS 1/	7951	8421	10746	9846	9357	9583	10119	10277	11862	12239	12672	3.97
TOTAL MEAT	7097	7828	8084	8853	8576	8929	8777	9035	9880	10127	10743	3.48
MILK DRY	602	662	872	868	816	743	822	839	872	1044	1117	4.51
TOTAL EGGS IN SHELL	606	656	746	806	825	794	839	766	755	781	764	1.66
FISHERY PRODUCTS												
FISH FRESH FROZEN	3853	4249	4437	4569	4630	5094	5295	6019	7010	7067	7236	6.81
FISH CURED	392	428	441	464	431	408	406	426	442	442	446	.51
SHELLFISH	993	1157	1076	1147	1248	1437	1601	1654	1751	1812	1852	6.91
FISH CANNED AND PREPARED	853	891	1027	1075	943	916	994	1034	1137	1156	1156	2.62
SHELLFISH CANNED+PREPAR	113	115	138	150	162	184	196	205	229	240	257	8.93
FISH BODY AND LIVER OIL	693	724	741	727	686	730	949	994	792	780	771	1.82
FISH MEAL	2051	2313	2332	2124	2639	2302	2639	3206	3236	3087	3140	4.74
FOREST PRODUCTS 2/												
SAWLOGS CONIFEROUS	29768	31748	27904	22480	26310	29382	30884	32586	32744	35793	38822	2.94
SAWLOGS NONCONIFEROUS	48311	45958	42006	32978	33265	32252	29597	29980	28870	33026	32084	-4.27
PULPWOOD+PARTICLE	32616	35824	39944	38596	33427	33657	37527	38919	41056	44881	50198	3.02
FUELWOOD	1894	2243	2780	2248	2392	2784	2653	2097	2068	2196	2270	-.09
SAWNWOOD CONIFEROUS	65879	68743	65938	60656	61439	70576	72754	73472	73656	78824	81354	2.29
SAWNWOOD NONCONIFEROUS	11994	13380	12545	10952	10923	12507	12579	11888	12813	14933	17555	2.50
WOOD-BASED PANELS	16401	16680	16323	16758	15443	17388	18233	19284	20606	23451	25544	4.39
PULP FOR PAPER	17489	18709	19756	18755	17314	19810	20334	20599	22090	23399	24192	2.95
PAPER AND PAPERBOARD	30210	33294	35108	35370	33688	36744	39803	40915	43385	46974	51011	4.74

1/ THOUSAND HEAD
2/ EXCEPT FOR PULP FOR PAPER AND PAPER AND PAPERBOARD, ALL FOREST PRODUCTS ARE EXPRESSED IN THOUSAND CUBIC METRES

4. (Cont.) VOLUME OF EXPORTS OF MAJOR AGRICULTURAL, FISHERY AND FOREST PRODUCTS

	1978	1979	1980	1981	1982	1983	1984	1985	1986	1987	1988	ANNUAL RATE OF CHANGE 1978-88
	THOUSAND METRIC TONS											PERCENT
WESTERN EUROPE												
AGRICULTURAL PRODUCTS												
WHEAT+FLOUR,WHEAT EQUIV.	13773	16091	19923	23693	22408	23811	27408	29646	27688	29590	30525	7.60
RICE MILLED	839	874	943	999	933	941	984	1198	1190	1156	945	2.61
BARLEY	8634	7197	8052	10796	7416	8390	11526	12791	13762	11050	13249	5.84
MAIZE	4869	5050	5474	4808	5743	7705	7809	7025	9310	9529	8248	7.38
MILLET	12	13	16	20	20	26	21	24	18	15	23	4.49
SORGHUM	262	308	206	240	269	159	165	190	124	191	134	-6.80
POTATOES	2798	3016	3455	3543	3666	3517	3526	3778	4174	4773	4522	4.55
SUGAR,TOTAL (RAW EQUIV.)	4448	4632	5628	6147	6466	6078	5631	5261	5561	6569	7152	3.03
PULSES	353	450	458	448	419	606	814	1240	1205	1430	1477	17.12
SOYBEANS	237	353	327	160	207	127	88	95	153	287	304	-3.37
SOYBEAN OIL	1099	1208	1204	1272	1380	1387	1427	1323	1271	1446	1148	1.11
GROUNDNUTS SHELLED BASIS	28	14	18	24	25	17	24	24	33	41	57	8.96
GROUNDNUT OIL	45	64	79	68	74	99	62	61	56	51	63	-.58
COPRA	4	1	2		1							-37.57
COCONUT OIL	119	61	43	58	87	60	57	51	54	67	64	-2.47
PALM NUTS KERNELS	1	2	3	1	2					1	1	-13.62
PALM OIL	97	92	123	114	94	123	131	141	171	156	149	5.64
OILSEED CAKE AND MEAL	3438	3957	4247	4921	5330	6420	6112	6364	5589	6819	5734	5.81
BANANAS	41	43	43	48	46	35	47	35	81	113	49	5.64
ORANGES+TANGER+CLEMEN	1921	1907	1799	1659	1880	1702	2439	1957	3024	2512	2385	4.02
LEMONS AND LIMES	505	483	512	433	574	449	532	542	597	566	518	1.46
COFFEE GREEN+ROASTED	102	130	106	122	126	142	165	202	209	232	265	9.91
COCOA BEANS	34	31	44	48	52	52	66	76	78	74	43	6.98
TEA	50	46	43	44	43	51	56	56	52	55	52	2.12
COTTON LINT	71	60	57	55	75	69	69	98	78	156	107	7.45
JUTE AND SIMILAR FIBRES	19	16	17	17	15	16	14	14	13	11	12	-4.51
TOBACCO UNMANUFACTURED	223	234	197	210	247	249	265	243	254	309	281	3.13
NATURAL RUBBER	21	21	16	14	15	16	23	23	22	28	42	6.61
WOOL GREASY	60	65	69	61	57	69	65	62	63	79	81	1.99
BOVINE CATTLE 1/	3322	3291	3412	3620	3546	3493	3537	3422	3779	3731	3370	.70
SHEEP AND GOATS 1/	1732	1384	1418	927	784	1196	1142	1415	1553	1926	2033	3.34
PIGS 1/	3421	4004	4777	4747	4537	4737	4688	4751	6685	7109	6837	6.38
TOTAL MEAT	2822	3173	3673	3900	3788	4076	4303	4453	5027	4968	5203	5.79
MILK DRY	450	515	660	673	599	531	641	624	616	773	832	4.08
TOTAL EGGS IN SHELL	382	444	506	538	601	596	586	541	548	557	559	2.79
FISHERY PRODUCTS												
FISH FRESH FROZEN	1395	1691	1652	1796	1885	1993	1956	2124	2283	2318	2423	5.00
FISH CURED	253	275	275	302	271	265	269	281	284	289	295	.82
SHELLFISH	263	277	277	325	312	346	406	409	378	366	380	4.28
FISH CANNED AND PREPARED	263	267	261	277	268	272	276	289	293	284	282	.96
SHELLFISH CANNED+PREPAR	36	38	42	47	57	72	75	83	82	84	90	10.62
FISH BODY AND LIVER OIL	270	296	333	335	270	265	270	392	274	267	257	-.85
FISH MEAL	869	948	918	843	822	930	1003	925	852	783	829	-.76
FOREST PRODUCTS 2/												
SAWLOGS CONIFEROUS	1899	2395	2937	2735	2429	2494	2786	3282	2906	3423	3490	4.60
SAWLOGS NONCONIFEROUS	2017	2055	2257	2128	1928	2011	2335	2458	2639	2873	3929	5.25
PULPWOOD+PARTICLE	6846	8321	10313	10737	9666	8772	10596	12087	13595	13671	12955	5.93
FUELWOOD	551	797	965	745	1010	1241	1172	940	910	1004	1086	4.42
SAWNWOOD CONIFEROUS	18051	20349	19783	17142	18334	20620	20377	19637	19183	19395	19498	.44
SAWNWOOD NONCONIFEROUS	2756	2514	2395	2037	1896	2017	2428	2261	2240	2524	2676	.11
WOOD-BASED PANELS	6737	7386	7047	6696	6312	6459	6894	7153	7450	7835	8345	1.55
PULP FOR PAPER	6705	6857	6661	6219	5616	6749	7086	7197	7298	7775	7936	1.97
PAPER AND PAPERBOARD	15659	17385	17423	18108	17770	19661	21939	22707	23381	26048	28815	5.81

1/ THOUSAND HEAD

2/ EXCEPT FOR PULP FOR PAPER AND PAPER AND PAPERBOARD, ALL FOREST PRODUCTS ARE EXPRESSED IN THOUSAND CUBIC METRES

4. (Cont.) VOLUME OF EXPORTS OF MAJOR AGRICULTURAL, FISHERY AND FOREST PRODUCTS

	1978	1979	1980	1981	1982	1983	1984	1985	1986	1987	1988	ANNUAL RATE OF CHANGE 1978-88
	......	......	......	......	......	THOUSAND METRIC TONS	......	......	......	......	......	PERCENT
USSR AND EASTERN EUROPE												
AGRICULTURAL PRODUCTS												
WHEAT+FLOUR,WHEAT EQUIV.	3969	5002	4170	4380	5092	4042	3680	4758	3847	3585	4169	-1.34
RICE MILLED	13	24	33	25	28	38	64	41	89	76	27	12.53
BARLEY	222	232	336	247	276	276	277	276	226	314	289	1.44
MAIZE	1493	554	1325	1770	1326	860	694	977	961	848	628	-4.82
MILLET	3	5	6	3	5	4	3	2	4	5	12	3.92
SORGHUM	7	7	5	9	6	4	4	6	9	23	25	11.18
POTATOES	371	655	322	323	299	185	141	268	302	497	722	.83
SUGAR,TOTAL (RAW EQUIV.)	953	717	738	631	807	762	871	1024	1240	1083	824	3.25
PULSES	135	145	122	122	112	118	193	231	274	361	399	12.86
SOYBEANS	6	30	5	4	5	5	11	6	5	33	15	6.28
SOYBEAN OIL	7	10	17	14	20	15	35	25	12	7	2	-6.63
GROUNDNUTS SHELLED BASIS		1	1				2					
COCONUT OIL		1	1									
OILSEED CAKE AND MEAL	53	20	27	91	115	120	64	205	174	323	226	25.57
ORANGES+TANGER+CLEMEN			1	2	2	1	1	2	2	1		
COCOA BEANS						5	12					
TEA	17	17	20	18	17	26	30	19	6	5	7	-10.68
COTTON LINT	865	807	863	928	970	847	695	720	769	813	776	-1.53
JUTE AND SIMILAR FIBRES										1	5	
TOBACCO UNMANUFACTURED	89	102	103	90	88	85	81	80	89	79	90	-1.49
NATURAL RUBBER											1	
WOOL GREASY	2	3	3	1		1	1	1	1	1	1	-10.45
BOVINE CATTLE 1/	544	676	577	460	607	705	707	658	676	785	925	4.27
SHEEP AND GOATS 1/	3800	4719	4598	3720	3654	4179	4232	3166	3075	3776	4453	-1.34
PIGS 1/	1158	1152	1144	1713	1091	973	857	1120	1151	1177	1218	-.66
TOTAL MEAT	620	744	738	779	715	758	832	889	942	912	933	3.71
MILK DRY											1	
TOTAL EGGS IN SHELL	114	104	90	78	59	55	65	42	38	44	42	-10.42
FISHERY PRODUCTS												
FISH FRESH FROZEN	559	591	612	493	418	543	540	607	823	844	867	4.82
FISH CURED	15	21	17	11	6	18	6				1	-42.58
SHELLFISH	19	26	11	24	51	114	135	113	78	93	85	22.47
FISH CANNED AND PREPARED	37	33	37	36	30	38	39	66	68	69	69	8.76
SHELLFISH CANNED+PREPAR	1	1	2	1	2	2	1	1	2	2	2	3.97
FISH BODY AND LIVER OIL	1	1	1									
FISH MEAL	21	20	22	12	9	12	8	11	12	12	12	-6.29
FOREST PRODUCTS 2/												
SAWLOGS CONIFEROUS	10281	8774	7430	6783	7025	7762	8085	8271	9791	9327	8881	.80
SAWLOGS NONCONIFEROUS	296	404	384	285	289	315	232	193	194	214	236	-5.93
PULPWOOD+PARTICLE	11375	11667	11463	11529	9631	10909	12616	12617	12948	13922	15141	2.72
FUELWOOD	141	143	183	94	70	92	121	132	149	171	201	2.86
SAWNWOOD CONIFEROUS	10782	9956	9513	9363	9630	9697	9476	9701	10238	9978	10571	.17
SAWNWOOD NONCONIFEROUS	752	600	597	539	487	536	564	389	338	488	361	-5.92
WOOD-BASED PANELS	1875	1842	1827	1683	1548	1598	1437	1488	1623	1818	1721	-1.05
PULP FOR PAPER	926	827	895	896	982	1162	1217	1227	1366	1326	1292	5.28
PAPER AND PAPERBOARD	1779	1664	1732	1697	1745	1775	1806	1795	1927	1999	2035	1.72

1/ THOUSAND HEAD
2/ EXCEPT FOR PULP FOR PAPER AND PAPER AND PAPERBOARD, ALL FOREST PRODUCTS ARE EXPRESSED IN THOUSAND CUBIC METRES

4. (Cont.) VOLUME OF EXPORTS OF MAJOR AGRICULTURAL, FISHERY AND FOREST PRODUCTS

	1978	1979	1980	1981	1982	1983	1984	1985	1986	1987	1988	ANNUAL RATE OF CHANGE 1978-88
	THOUSAND METRIC TONS											PERCENT
NORTH AMERICA DEVELOPED												
AGRICULTURAL PRODUCTS												
WHEAT+FLOUR,WHEAT EQUIV.	50841	47174	54495	61342	61264	63319	65263	43528	42833	55214	62600	.30
RICE MILLED	2279	2301	3054	3133	2540	2385	2141	1940	2392	2472	2260	-1.46
BARLEY	4249	4654	4195	6853	7097	7258	5876	2938	7586	8468	4873	2.74
MAIZE	50550	59414	63923	56067	49658	48083	49584	44345	27473	41097	46815	-4.32
MILLET	23	15	60	24	28	41	55	39	74	58	47	10.79
SORGHUM	5184	5950	8050	8032	6051	5325	6828	7239	4149	5009	6532	-1.45
POTATOES	282	289	344	395	461	363	296	321	319	353	442	1.81
SUGAR,TOTAL (RAW EQUIV.)	149	135	654	1187	154	323	397	436	544	673	358	8.73
PULSES	390	470	913	1141	854	679	635	646	851	930	1030	5.54
SOYBEANS	20794	20951	21882	21980	25652	22791	19641	17671	21576	21513	18124	-1.20
SOYBEAN OIL	916	1110	1081	809	911	786	1043	588	540	624	892	-4.46
GROUNDNUTS SHELLED BASIS	381	356	285	146	201	224	266	311	276	221	159	-4.07
GROUNDNUT OIL	40	5	18	20	10	2	7	17	35	3	3	-11.32
COCONUT OIL	9	5	19	14	13	11	21	19	18	39	40	16.18
OILSEED CAKE AND MEAL	6793	6845	8009	7471	6917	7517	5551	5599	7379	8258	8652	.84
BANANAS	201	197	205	217	210	188	202	197	163	188	180	-1.50
ORANGES+TANGER+CLEMEN	356	318	482	443	353	497	374	412	417	403	357	.40
LEMONS AND LIMES	237	173	171	176	135	163	148	144	148	152	146	-3.31
COFFEE GREEN+ROASTED	59	79	79	70	60	46	63	52	77	60	76	-.38
COCOA BEANS	9	9	7	14	14	16	12	11	14	17	14	6.28
TEA	5	5	5	4	4	5	5	13	22	15	3	9.29
COTTON LINT	1347	1527	1823	1269	1392	1126	1367	1001	662	1195	1173	-4.62
JUTE AND SIMILAR FIBRES	1										1	-2.65
TOBACCO UNMANUFACTURED	364	299	293	300	290	264	275	274	247	226	240	-3.53
NATURAL RUBBER	20	21	28	18	16	20	35	41	37	37	56	10.27
WOOL GREASY				1	1	1	1	1	1	1	1	8.47
BOVINE CATTLE 1/	592	436	424	441	563	440	479	506	355	399	868	1.05
SHEEP AND GOATS 1/	153	135	144	225	287	226	332	382	145	67	204	-.16
PIGS 1/	201	145	254	171	342	483	1362	1171	515	435	960	19.48
TOTAL MEAT	721	777	973	1073	987	926	956	1013	1150	1285	1443	5.44
MILK DRY	7	5	36	37	29	37	19	49	30	12	21	8.05
TOTAL EGGS IN SHELL	39	30	61	87	64	31	25	22	19	35	49	-4.80
FISHERY PRODUCTS												
FISH FRESH FROZEN	383	424	480	638	801	918	1167	1465	1913	1895	1922	20.20
FISH CURED	63	64	76	87	89	70	65	70	79	70	77	.67
SHELLFISH	93	133	115	88	80	80	71	83	95	96	104	-1.38
FISH CANNED AND PREPARED	63	64	81	93	68	82	96	85	100	102	88	4.04
SHELLFISH CANNED+PREPAR	11	11	11	11	11	4	3	3	6	6	15	-5.61
FISH BODY AND LIVER OIL	110	101	137	117	98	191	188	133	92	94	95	-1.19
FISH MEAL	58	21	86	50	22	76	25	38	39	43	35	-2.23
FOREST PRODUCTS 2/												
SAWLOGS CONIFEROUS	15565	17865	15135	11676	15269	17395	18441	19320	18316	21212	24104	4.31
SAWLOGS NONCONIFEROUS	522	630	784	751	506	755	761	602	779	879	1163	4.92
PULPWOOD+PARTICLE	8216	9463	9887	8382	6605	6422	5846	5613	5933	6249	8001	-3.79
FUELWOOD	170	98	63	108	85	85	90	89	82	76	104	-2.75
SAWNWOOD CONIFEROUS	34492	35407	33612	31770	31423	38296	40879	42219	42232	47162	49090	4.11
SAWNWOOD NONCONIFEROUS	1341	1025	1190	1209	1083	1340	1373	1172	1513	2174	3663	8.46
WOOD-BASED PANELS	2061	2053	2312	2533	2088	2401	2668	2754	2948	3295	4384	6.38
PULP FOR PAPER	8132	8906	9838	9261	8531	9428	9611	9791	10917	11909	12307	3.49
PAPER AND PAPERBOARD	11062	12336	13742	13149	11941	12846	13288	13378	14558	15455	16105	2.85

1/ THOUSAND HEAD
2/ EXCEPT FOR PULP FOR PAPER AND PAPER AND PAPERBOARD, ALL FOREST PRODUCTS ARE EXPRESSED IN THOUSAND CUBIC METRES

4. (Cont.) VOLUME OF EXPORTS OF MAJOR AGRICULTURAL, FISHERY AND FOREST PRODUCTS

	1978	1979	1980	1981	1982	1983	1984	1985	1986	1987	1988	ANNUAL RATE OF CHANGE 1978-88
	THOUSAND METRIC TONS											PERCENT
OCEANIA DEVELOPED												
AGRICULTURAL PRODUCTS												
WHEAT+FLOUR,WHEAT EQUIV.	11134	6933	14955	10677	10998	8312	10647	15782	16171	14898	12285	4.21
RICE MILLED	277	241	457	281	596	405	246	341	178	186	298	-3.59
BARLEY	1375	1757	3047	1650	1599	852	3231	5482	4399	2345	1530	5.50
MAIZE	32	75	37	52	24	73	30	164	117	103	29	6.31
MILLET	15	18	14	11	25	19	18	16	16	24	46	7.03
SORGHUM	385	516	580	463	1271	445	772	1594	1234	818	415	6.05
POTATOES	20	18	23	21	23	26	21	24	19	19	20	-.12
SUGAR,TOTAL (RAW EQUIV.)	2481	1842	2203	2563	2502	2551	2361	2529	2760	2481	2786	2.17
PULSES	36	45	72	64	71	106	78	100	219	480	441	27.02
SOYBEAN OIL						1						
GROUNDNUTS SHELLED BASIS	2	2	12	4	4	8		5	3	3	3	-3.11
GROUNDNUT OIL					1				1			
PALM OIL									2			
OILSEED CAKE AND MEAL		1	1		1	1	2	1	13	16	1	54.11
ORANGES+TANGER+CLEMEN	22	25	38	32	28	32	25	30	36	48	51	5.96
LEMONS AND LIMES			4	1	2	1	1	1	5	4	2	15.02
COCOA BEANS						1	1	1			1	2.57
TEA	1											-3.80
COTTON LINT	10	24	49	59	79	129	81	140	241	251	176	31.95
TOBACCO UNMANUFACTURED	1		1	1		1						-16.39
NATURAL RUBBER			1					1	2	3	1	20.58
WOOL GREASY	630	705	650	680	642	660	659	709	733	799	784	1.89
BOVINE CATTLE 1/	71	107	74	109	121	120	96	67	181	125	153	5.56
SHEEP AND GOATS 1/	4143	3898	6172	5763	6097	7035	6350	6262	6554	8416	7748	6.18
PIGS 1/	1	1	2	1		1	3				2	-4.16
TOTAL MEAT	1664	1814	1494	1602	1493	1666	1351	1323	1361	1642	1639	-1.12
MILK DRY	125	123	157	137	157	146	148	152	202	227	214	5.63
TOTAL EGGS IN SHELL	1	1	1	1	1	3	6	2	2	1	1	-3.63
FISHERY PRODUCTS												
FISH FRESH FROZEN	32	54	81	95	88	98	94	97	119	122	120	10.63
FISH CURED			1	1	2	1		1	3	3	3	20.26
SHELLFISH	20	32	65	57	70	68	78	70	61	62	66	8.58
FISH CANNED AND PREPARED		1	3	2	4	5	4	4	4	4	4	19.76
SHELLFISH CANNED+PREPAR	2	2	2	2	2	3	3	3	3	3	3	6.60
FISH BODY AND LIVER OIL	5	4					2	2	1	1	1	19.59
FISH MEAL				1			4	1	2	2	2	
FOREST PRODUCTS 2/												
SAWLOGS CONIFEROUS	936	1236	971	529	479	508	452	361	389	429	820	-7.40
SAWLOGS NONCONIFEROUS	2	1	4	4				1	1	22	22	17.10
PULPWOOD+PARTICLE	5074	5357	7064	6647	6240	6105	7345	7376	7188	8069	9381	4.78
SAWNWOOD CONIFEROUS	367	509	617	546	515	401	381	489	401	348	411	-2.49
SAWNWOOD NONCONIFEROUS	30	41	54	35	34	35	41	36	34	16	26	-5.16
WOOD-BASED PANELS	52	104	142	138	99	113	93	79	98	167	169	5.09
PULP FOR PAPER	435	464	475	518	421	471	459	428	504	483	483	.52
PAPER AND PAPERBOARD	332	359	418	447	340	361	342	353	336	330	330	-1.34

1/ THOUSAND HEAD
2/ EXCEPT FOR PULP FOR PAPER AND PAPER AND PAPERBOARD, ALL FOREST PRODUCTS ARE EXPRESSED IN THOUSAND CUBIC METRES

4. (Cont.) VOLUME OF EXPORTS OF MAJOR AGRICULTURAL, FISHERY AND FOREST PRODUCTS

	1978	1979	1980	1981	1982	1983	1984	1985	1986	1987	1988	ANNUAL RATE OF CHANGE 1978-88
	THOUSAND METRIC TONS											PERCENT
AFRICA DEVELOPING												
AGRICULTURAL PRODUCTS												
WHEAT+FLOUR,WHEAT EQUIV.	46	31	17	19	23	9	8	3	6	4	45	-13.87
RICE MILLED	13	12	22	18	14	8	20	13	12	4	15	-4.47
BARLEY		2									263	
MAIZE	652	365	70	245	380	782	262	352	792	785	523	9.14
MILLET	31	78	46	41	6	1	2	2		4		-44.56
SORGHUM		53	12	3	15	25	30	11	5	21	5	11.59
POTATOES	58	50	55	36	30	49	63	60	84	72	56	3.98
SUGAR,TOTAL (RAW EQUIV.)	1296	1659	1586	1491	1683	1683	1569	1628	1806	1779	1677	1.90
PULSES	150	149	220	127	166	191	123	71	123	156	135	-3.20
SOYBEANS	36	1	1	1		1		3	1	1	4	-2.25
SOYBEAN OIL	2	1								1	1	-3.12
GROUNDNUTS SHELLED BASIS	65	82	86	36	56	96	56	45	55	53	74	-1.80
GROUNDNUT OIL	94	160	92	38	162	210	109	48	95	119	149	1.20
COPRA	52	45	32	22	20	14	11	18	18	20	14	-10.70
COCONUT OIL	9	14	15	18	21	21	23	32	34	29	29	12.21
PALM NUTS KERNELS	152	123	140	107	97	87	98	50	87	108	117	-4.27
PALM OIL	96	64	140	85	84	70	75	92	135	157	96	3.28
OILSEED CAKE AND MEAL	457	667	480	362	492	490	336	386	439	442	585	-.84
BANANAS	344	292	243	205	187	193	193	211	193	183	191	-4.80
ORANGES+TANGER+CLEMEN	878	679	855	715	662	594	582	640	642	577	670	-2.88
LEMONS AND LIMES	2	2	1	1	2	7	6	7	3	3	2	11.06
COFFEE GREEN+ROASTED	925	1011	895	965	1053	939	914	973	1076	910	959	.17
COCOA BEANS	778	601	759	976	826	783	894	831	957	930	945	2.93
TEA	182	197	180	168	190	200	195	226	224	231	241	3.04
COTTON LINT	312	324	336	340	316	349	372	377	482	491	515	5.23
JUTE AND SIMILAR FIBRES		1										
TOBACCO UNMANUFACTURED	139	132	172	189	148	144	173	175	173	174	179	2.21
NATURAL RUBBER	145	142	138	146	151	156	186	185	204	215	254	5.93
WOOL GREASY	4	3	4	4	4	4	5	3	2	1	1	-11.72
BOVINE CATTLE 1/	1181	1271	1415	1461	1461	1206	1136	974	812	751	691	-6.60
SHEEP AND GOATS 1/	3066	3049	3646	3412	3574	3001	2488	3397	3087	3007	2649	-1.49
PIGS 1/	1	1	1			1			3	3		.53
TOTAL MEAT	98	96	47	44	44	48	52	45	36	52	47	-5.95
MILK DRY	2	4										-17.65
TOTAL EGGS IN SHELL			1				2	1				-4.33
FISHERY PRODUCTS												
FISH FRESH FROZEN	128	117	113	144	154	146	164	187	171	175	179	4.76
FISH CURED	12	12	13	11	12	9	12	15	14	18	11	1.89
SHELLFISH	48	34	34	74	76	130	135	140	152	157	149	17.90
FISH CANNED AND PREPARED	62	77	79	94	82	101	100	105	105	104	104	4.70
FISH BODY AND LIVER OIL	6	7	4	10	1	8	5		4	4	4	-11.34
FISH MEAL	31	24	20	22	6	15	9	8	3	3	3	-22.69
FOREST PRODUCTS 2/												
SAWLOGS CONIFEROUS	2	2										
SAWLOGS NONCONIFEROUS	6211	6175	5971	4599	4723	4547	5076	4217	3658	3596	3741	-5.54
PULPWOOD+PARTICLE	75	112	84	173	173	173	173	173	173	412	673	18.17
FUELWOOD	51	51	5				28					
SAWNWOOD CONIFEROUS	116	126	108	105	81	79	82	77	79	89	83	-4.08
SAWNWOOD NONCONIFEROUS	706	680	611	522	554	598	681	794	777	818	828	3.05
WOOD-BASED PANELS	261	236	272	283	265	288	300	307	286	283	266	1.15
PULP FOR PAPER	218	240	240	229	192	202	252	244	244	264	284	2.00
PAPER AND PAPERBOARD	16	24	21	11	19	12	14	13	7	14	7	-8.30

1/ THOUSAND HEAD
2/ EXCEPT FOR PULP FOR PAPER AND PAPER AND PAPERBOARD, ALL FOREST PRODUCTS ARE EXPRESSED IN THOUSAND CUBIC METRES

4. (Cont.) VOLUME OF EXPORTS OF MAJOR AGRICULTURAL, FISHERY AND FOREST PRODUCTS

	1978	1979	1980	1981	1982	1983	1984	1985	1986	1987	1988	ANNUAL RATE OF CHANGE 1978-88
	......THOUSAND METRIC TONS......											PERCENT
LATIN AMERICA												
AGRICULTURAL PRODUCTS												
WHEAT+FLOUR,WHEAT EQUIV.	1833	4427	4620	3964	4042	10410	7491	9762	4125	4326	3905	5.38
RICE MILLED	702	563	525	606	510	512	530	559	571	482	468	-2.27
BARLEY	18	58	72	32	24	59	95	86	51	36	150	10.59
MAIZE	5927	5990	3556	9198	5837	7321	5733	7129	7422	4001	4235	-1.46
MILLET	196	139	63	136	101	96	58	93	32	60	67	-10.38
SORGHUM	4625	3899	1545	5075	5369	5332	4278	3332	1960	1005	1511	-9.82
POTATOES	67	77	61	45	44	32	50	55	56	33	30	-6.33
SUGAR,TOTAL (RAW EQUIV.)	12429	12726	12025	12702	13052	12953	12851	12321	11385	11007	11978	-.91
PULSES	464	395	336	286	281	358	412	349	391	306	322	-1.44
SOYBEANS	2845	3814	4493	3909	2877	3270	5170	7171	4469	5577	5999	6.60
SOYBEAN OIL	570	609	840	1355	1024	1369	1413	1510	1103	1734	1700	10.51
GROUNDNUTS SHELLED BASIS	52	97	98	86	61	101	104	138	138	131	145	8.38
GROUNDNUT OIL	155	209	207	80	113	104	57	109	50	104	75	-9.37
COPRA		2			5			1				-20.61
COCONUT OIL	9	8	4	5	6	6	17	4	5	5	7	-1.51
PALM NUTS KERNELS	9	7	5	1	4	4	3	2	1	1	1	-18.34
PALM OIL	4	5	3	7	15	17	27	35	31	33	32	30.75
OILSEED CAKE AND MEAL	7676	7497	8891	10912	10498	12344	12163	13506	12750	13486	15516	7.08
BANANAS	5520	5366	5358	5471	5652	5082	5493	5371	5896	6175	6323	1.34
ORANGES+TANGER+CLEMEN	269	312	306	316	383	418	409	479	536	532	521	7.56
LEMONS AND LIMES	47	74	53	51	34	56	65	154	113	130	141	12.38
COFFEE GREEN+ROASTED	1960	2179	2232	2148	2259	2426	2533	2622	2108	2691	2343	1.91
COCOA BEANS	211	226	183	201	241	229	211	289	239	257	260	2.71
TEA	41	39	44	35	43	54	54	53	49	48	46	2.54
COTTON LINT	903	733	636	600	599	509	481	637	372	435	551	-5.54
JUTE AND SIMILAR FIBRES	1	2	2		1		1	2				
TOBACCO UNMANUFACTURED	274	276	255	271	273	274	290	305	270	262	299	.63
NATURAL RUBBER	6	4	3	2	3	3	2	2	4	2	3	-4.79
WOOL GREASY	107	80	105	125	108	88	79	67	78	75	67	-4.52
BOVINE CATTLE 1/	1551	1277	754	716	962	717	451	556	1022	1057	838	-3.74
SHEEP AND GOATS 1/	125	98	65	312	195	634	462	14	89	20	310	-5.50
PIGS 1/	24	16	1			10	4	6	55	17	13	18.66
TOTAL MEAT	840	815	738	992	1027	983	778	806	805	627	818	-1.46
MILK DRY	10	4	4	11	19	17	2	1	2	3	14	-7.90
TOTAL EGGS IN SHELL	2	4	12	14	6	3	4	6	10	5	2	-.49
FISHERY PRODUCTS												
FISH FRESH FROZEN	346	362	419	375	401	380	368	485	413	397	397	1.31
FISH CURED	4	6	8	5	5	5	5	5	5	5	5	.04
SHELLFISH	144	168	137	125	165	174	178	173	171	175	174	2.31
FISH CANNED AND PREPARED	79	81	143	170	95	55	65	49	72	77	77	-4.69
SHELLFISH CANNED+PREPAR	2	5	4	6	4	6	7	7	8	7	7	9.42
FISH BODY AND LIVER OIL	71	108	103	76	137	25	139	207	187	180	180	10.05
FISH MEAL	830	1020	1052	962	1495	1022	1292	1901	1962	1875	1875	9.12
FOREST PRODUCTS 2/												
SAWLOGS CONIFEROUS	689	968	1029	377	906	1024	902	1271	1162	1271	1400	6.98
SAWLOGS NONCONIFEROUS	60	86	114	65	54	55	68	47	37	43	83	-4.41
PULPWOOD+PARTICLE									323	1347	2737	
FUELWOOD	152	214	167	71	23	57	10	7	6	60	6	-28.05
SAWNWOOD CONIFEROUS	1477	1678	1718	1319	1102	1172	1217	1004	1153	1342	1208	-3.17
SAWNWOOD NONCONIFEROUS	727	1121	1130	994	892	851	911	894	774	861	813	-1.64
WOOD-BASED PANELS	487	488	625	606	608	584	650	651	655	697	869	4.34
PULP FOR PAPER	715	1024	1318	1374	1302	1566	1532	1515	1511	1407	1598	5.66
PAPER AND PAPERBOARD	276	351	398	497	404	651	939	778	920	828	1327	15.16

1/ THOUSAND HEAD
2/ EXCEPT FOR PULP FOR PAPER AND PAPER AND PAPERBOARD, ALL FOREST PRODUCTS ARE EXPRESSED IN THOUSAND CUBIC METRES

4. (Cont.) VOLUME OF EXPORTS OF MAJOR AGRICULTURAL, FISHERY AND FOREST PRODUCTS

	1978	1979	1980	1981	1982	1983	1984	1985	1986	1987	1988	ANNUAL RATE OF CHANGE 1978-88
	THOUSAND METRIC TONS											PERCENT
NEAR EAST DEVELOPING												
AGRICULTURAL PRODUCTS												
WHEAT+FLOUR,WHEAT EQUIV.	2131	877	540	652	717	1145	1046	804	873	2025	4621	8.98
RICE MILLED	223	211	259	159	59	78	158	121	226	214	159	-1.46
BARLEY	50	88	229	424	1026	662	321	100	171	48	678	5.34
MAIZE	43	111	155	40	53	10	6	12	7	13	15	-22.24
MILLET	4	2	2	3	8	2	4		2	5	2	-4.47
SORGHUM	66	197	286	256	423	186	25		31	534	237	
POTATOES	291	316	453	393	463	456	511	355	377	357	495	2.29
SUGAR,TOTAL (RAW EQUIV.)	55	37	45	71	224	341	629	369	78	127	76	12.03
PULSES	256	304	299	500	573	658	609	386	560	800	1278	12.88
SOYBEAN OIL			3	5	16	11	11	5	1		1	22.21
GROUNDNUTS SHELLED BASIS	111	52	51	108	101	24	31	19	6	10	77	-16.21
GROUNDNUT OIL	35	16	33	16	18	2	12	3	2	22	9	-14.35
COCONUT OIL				1								-11.51
PALM OIL					1				5	27	17	
OILSEED CAKE AND MEAL	225	214	261	145	105	104	133	29	73	111	192	-8.87
BANANAS	4	14	19	20	11	10	12	12	13	14	14	4.22
ORANGES+TANGER+CLEMEN	609	592	591	698	637	617	611	558	434	494	441	-3.36
LEMONS AND LIMES	138	144	191	190	191	202	199	137	152	159	174	.24
COFFEE GREEN+ROASTED	3	3	2	6	5	5	8	4	4	2	5	4.29
TEA	10	16	15	17	5	7	5	6	2	2	1	-22.47
COTTON LINT	768	669	608	532	584	623	648	488	611	425	409	-4.45
TOBACCO UNMANUFACTURED	84	77	94	138	110	75	72	105	85	114	81	.17
NATURAL RUBBER									1			85.86
WOOL GREASY	9	8	7	3	6	6	5	5	8	11	14	4.68
BOVINE CATTLE 1/	12	21	13	60	112	77	51	18	13	2	1	-20.09
SHEEP AND GOATS 1/	1209	1421	2026	2858	3505	3710	3866	3353	4039	3713	4217	12.11
PIGS 1/	1	3										
TOTAL MEAT	15	15	22	74	96	78	97	70	58	55	48	13.58
MILK DRY		1	1		1	1						18.46
TOTAL EGGS IN SHELL	7	10	13	17	27	42	70	73	52	51	31	22.10
FISHERY PRODUCTS												
FISH FRESH FROZEN	6	17	15	28	27	28	31	29	30	30	30	11.81
FISH CURED	1	1	1	1	1	1	1	1	1	1	1	1.86
SHELLFISH	8	9	10	5	7	7	8	9	9	9	10	2.00
FISH CANNED AND PREPARED	4	5	8	3	2	1	1	1	1	1	1	-17.60
SHELLFISH CANNED+PREPAR	1	2	3	4	5	7	8	5	7	7	7	16.40
FISH BODY AND LIVER OIL				1	1	2	9	7	5	5	5	77.60
FISH MEAL			1				1					
FOREST PRODUCTS 2/												
SAWLOGS CONIFEROUS	1	1	1	2	7	11	20	15	24	1	5	18.98
SAWLOGS NONCONIFEROUS	5	3	4	36	36	35	100	76	24	23	18	23.21
FUELWOOD	22	20	31	24	16	24	11	11	11	11	11	-9.22
SAWNWOOD CONIFEROUS	60	103	84	96	94	126	107	82	48	50	82	-2.88
SAWNWOOD NONCONIFEROUS		2	3	6	12	7	8	5	8	57	26	38.24
WOOD-BASED PANELS	26	24	19	19	24	27	19	19	19	18	25	-1.26
PAPER AND PAPERBOARD	10	16	21	35	35	41	71	56	65	65	65	20.11

1/ THOUSAND HEAD

2/ EXCEPT FOR PULP FOR PAPER AND PAPER AND PAPERBOARD, ALL FOREST PRODUCTS ARE EXPRESSED IN THOUSAND CUBIC METRES

4. (Cont.) VOLUME OF EXPORTS OF MAJOR AGRICULTURAL, FISHERY AND FOREST PRODUCTS

	1978	1979	1980	1981	1982	1983	1984	1985	1986	1987	1988	ANNUAL RATE OF CHANGE 1978-88
	......THOUSAND METRIC TONS......											PERCENT
FAR EAST DEVELOPING												
AGRICULTURAL PRODUCTS												
WHEAT+FLOUR,WHEAT EQUIV.	967	801	510	295	157	250	359	447	241	495	273	-7,72
RICE MILLED	3031	4965	5331	6033	6050	5525	7018	5888	6913	6717	6913	5,81
BARLEY	13	73	259	275	907	250	1655	130	2		1	-38,08
MAIZE	2198	2146	2342	2721	3030	2861	3476	2933	4101	1779	1467	-,73
MILLET	1	6	2	2	1		2	11	3	1	2	2,49
SORGHUM	166	170	208	288	317	248	327	339	270	149	28	-7,31
POTATOES	55	99	106	72	69	61	60	55	65	60	77	-2,28
SUGAR,TOTAL (RAW EQUIV.)	2822	3269	2722	2930	4093	3580	2970	2968	2663	2797	2661	-1,15
PULSES	245	291	312	338	376	345	337	492	582	456	421	6,58
SOYBEANS	30	27	27	27	27	33	34	43	40	99	126	14,29
SOYBEAN OIL	7	6	27	32	49	78	89	47	62	201	154	35,53
GROUNDNUTS SHELLED BASIS	24	40	55	113	106	89	66	72	84	91	84	9,01
GROUNDNUT OIL	6	16	5	5	6	28	8	10	10	10	11	4,17
COPRA	445	193	234	172	232	75	72	153	193	157	116	-8,27
COCONUT OIL	1112	976	1061	1192	1064	1144	779	1045	1448	1255	1126	1,31
PALM NUTS KERNELS	13	23	45	24	15	14	13	25	5	3	1	-23,56
PALM OIL	2168	2638	3303	2963	3487	3709	3951	4811	5755	5299	5514	9,74
OILSEED CAKE AND MEAL	2582	3291	3054	3091	3220	3553	2876	2932	4090	3497	3231	1,85
BANANAS	832	920	971	922	982	683	841	826	892	816	903	-,63
ORANGES+TANGER+CLEMEN	65	89	78	50	62	75	74	65	64	68	62	-1,17
LEMONS AND LIMES	1	2	1	7	2	2	2	3	2	3	2	7,36
COFFEE GREEN+ROASTED	339	335	370	371	403	405	482	517	544	490	538	5,45
COCOA BEANS	24	32	41	65	88	91	121	143	177	246	330	28,49
TEA	459	445	539	546	488	475	554	544	533	544	583	1,91
COTTON LINT	128	133	396	415	326	410	204	351	841	943	583	16,58
JUTE AND SIMILAR FIBRES	466	520	465	514	451	457	432	290	425	431	278	-4,26
TOBACCO UNMANUFACTURED	224	212	198	259	238	202	193	166	174	136	132	-5,22
NATURAL RUBBER	3080	3179	3101	2924	2886	3205	3340	3339	3376	3721	3785	2,14
WOOL GREASY	1			1	1			1	5	2		11,66
BOVINE CATTLE 1/	78	66	60	36	39	76	40	58	53	50	57	-1,86
SHEEP AND GOATS 1/	70	100	120	60	26	155	241	232	257	196	246	15,86
PIGS 1/	15	19	18	24	130	160	113	222	329	468	611	49,43
TOTAL MEAT	68	95	90	103	127	97	107	128	163	201	255	11,14
MILK DRY	7	10	13	10	10	10	10	13	20	27	33	12,99
TOTAL EGGS IN SHELL	6	5	5	11	8	6	16	17	18	23	25	18,39
FISHERY PRODUCTS												
FISH FRESH FROZEN	579	586	625	581	496	591	519	597	762	790	803	3.25
FISH CURED	30	27	28	27	29	28	36	38	39	42	40	4.69
SHELLFISH	317	362	314	329	379	385	448	474	548	594	625	7.49
FISH CANNED AND PREPARED	49	47	55	80	100	113	144	177	261	281	294	22.87
SHELLFISH CANNED+PREPAR	35	35	50	55	61	68	73	77	91	101	102	11.71
FISH BODY AND LIVER OIL	3	2	2	1	1	1	2	2	2	3	3	2.27
FISH MEAL	141	164	153	151	141	153	155	160	193	195	211	3.33
FOREST PRODUCTS 2/												
SAWLOGS CONIFEROUS	270	396	327	291	127	109	107	38	129	102	102	-14.57
SAWLOGS NONCONIFEROUS	38457	35843	31534	24005	24286	23128	19372	20789	19848	23616	21284	-5.76
PULPWOOD+PARTICLE	860	736	1003	1033	963	1122	793	927	796	1111	1210	2.07
FUELWOOD	731	799	1181	1164	1086	1229	1146	842	832	798	784	-1.18
SAWNWOOD CONIFEROUS	425	481	410	254	197	138	186	214	272	407	362	-2.77
SAWNWOOD NONCONIFEROUS	5463	7236	6415	5511	5838	7003	6469	6239	7043	7914	9080	3.26
WOOD-BASED PANELS	3342	3159	2933	3590	3428	4772	5290	6027	6763	8571	9062	12.50
PULP FOR PAPER	2	6	6	10	8	9	16	16	13	20	55	26.26
PAPER AND PAPERBOARD	154	153	298	309	228	223	249	327	549	879	1068	18.55

1/ THOUSAND HEAD
2/ EXCEPT FOR PULP FOR PAPER AND PAPER AND PAPERBOARD, ALL FOREST PRODUCTS ARE EXPRESSED IN THOUSAND CUBIC METRES

4. (Cont.) VOLUME OF EXPORTS OF MAJOR AGRICULTURAL, FISHERY AND FOREST PRODUCTS

	1978	1979	1980	1981	1982	1983	1984	1985	1986	1987	1988	ANNUAL RATE OF CHANGE 1978-88
	THOUSAND METRIC TONS											PERCENT
ASIAN CENT PLANNED ECON												
AGRICULTURAL PRODUCTS												
WHEAT+FLOUR,WHEAT EQUIV.	8	9	4	9	6	67	75	19	124	32	105	34.32
RICE MILLED	2096	1836	1637	948	994	1279	1603	1305	1447	1640	1099	-2.63
BARLEY	1	2	1			7		58	34	31	30	
MAIZE	230	240	104	141	96	92	1043	6388	5656	3945	3976	53.93
MILLET	30	20	5	1	2	2	2	5	9	5	3	-10.06
SORGHUM		10	1		3	4	4	564	752	186	384	
POTATOES	62	81	77	80	89	78	72	61	52	48	83	-2.29
SUGAR,TOTAL (RAW EQUIV.)	493	514	657	440	463	258	189	405	463	505	284	-4.40
PULSES	76	90	71	111	103	134	159	188	595	438	531	24.31
SOYBEANS	113	306	140	139	160	367	847	1162	1393	1736	1550	34.80
SOYBEAN OIL	6	4	4		1	2	9	1	1	3	10	10.48
GROUNDNUTS SHELLED BASIS	30	49	84	250	136	201	180	214	325	328	309	23.48
GROUNDNUT OIL	13	18	21	57	55	72	39	57	108	41	19	9.21
COPRA					1	2	2	11	9		3	
COCONUT OIL						4	4	7	3	1	2	
PALM NUTS KERNELS					1			1	2			
OILSEED CAKE AND MEAL	31	49	87	208	339	1135	961	1321	3182	3380	4529	68.36
BANANAS	101	117	109	103	112	134	135	143	104	136	109	1.56
ORANGES+TANGER+CLEMEN	81	73	70	54	57	62	52	67	79	91	99	2.39
COFFEE GREEN+ROASTED	5	5	4	1	10	15	13	11	13	17	14	17.58
COCOA BEANS									3	3	3	
TEA	109	126	125	107	126	148	169	157	192	193	217	7.04
COTTON LINT	33	22	2	1	17	131	218	261	558	755	468	69.03
JUTE AND SIMILAR FIBRES	8	20	35	42	44	36	48	77	85	66	47	17.67
TOBACCO UNMANUFACTURED	35	35	32	28	30	35	32	27	24	25	42	-1.28
NATURAL RUBBER	41	50	39	38	41	47	50	50	55	60	59	4.08
WOOL GREASY	22	24	23	21	16	16	14	13	19	14	15	-4.88
BOVINE CATTLE 1/	181	224	272	263	257	252	257	220	225	225	251	.65
SHEEP AND GOATS 1/	443	463	448	330	312	438	515	502	547	487	105	-4.49
PIGS 1/	3129	3079	4548	3189	3256	3217	3091	3007	3123	3030	3031	-1.37
TOTAL MEAT	210	246	251	250	274	271	284	295	327	374	344	4.95
MILK DRY									1	1	1	
TOTAL EGGS IN SHELL	42	51	54	56	57	57	60	56	64	60	50	1.88
FISHERY PRODUCTS												
FISH FRESH FROZEN	129	134	146	169	168	183	189	199	229	229	229	6.37
FISH CURED	5	9	8	6	7	6	8	7	7	7	7	.13
SHELLFISH	57	72	70	76	75	90	102	137	205	205	205	14.95
FISH CANNED AND PREPARED	22	33	43	34	38	42	38	41	39	39	39	3.36
SHELLFISH CANNED+PREPAR	14	10	10	12	9	11	13	13	17	17	17	4.35
FISH BODY AND LIVER OIL							1					-3.02
FISH MEAL	1	1	1	1		1	2	1	2	2	2	9.92
FOREST PRODUCTS 2/												
SAWLOGS CONIFEROUS	32	27	21	33	29	38	35	5	1	1	1	-33.68
SAWLOGS NONCONIFEROUS	42	45	45	33	35	36	59	39	14	12	10	-12.80
SAWNWOOD CONIFEROUS	28	19	10	12	13	13	12	12	12	16	12	-3.99
SAWNWOOD NONCONIFEROUS	103	48	34	26	56	55	53	56	56	55	59	.71
WOOD-BASED PANELS	1244	1096	885	957	834	884	614	565	513	560	515	-8.76
PULP FOR PAPER	44	46	49	86	81	64	30	68	92	81	106	6.72
PAPER AND PAPERBOARD	116	89	149	174	165	139	217	304	386	440	485	17.56

1/ THOUSAND HEAD
2/ EXCEPT FOR PULP FOR PAPER AND PAPER AND PAPERBOARD, ALL FOREST PRODUCTS ARE EXPRESSED IN THOUSAND CUBIC METRES

5. WORLD AVERAGE EXPORT UNIT VALUES OF SELECTED AGRICULTURAL, FISHERY AND FOREST PRODUCTS

	1978	1979	1980	1981	1982	1983	1984	1985	1986	1987	1988	ANNUAL RATE OF CHANGE 1978-88
	......US $ PER METRIC TON......											PERCENT
AGRICULTURAL PRODUCTS												
WHEAT	131	163	186	188	173	162	157	145	135	114	140	-2.42
WHEAT FLOUR	199	225	284	294	245	197	215	208	195	179	198	-2.58
RICE MILLED	346	325	383	445	344	309	296	283	249	240	315	-3.60
BARLEY	137	145	175	175	161	144	147	121	107	103	130	-3.51
MAIZE	117	128	150	154	128	142	149	126	117	104	127	-1.26
POTATOES	157	188	185	178	186	168	209	124	151	172	164	-1.21
SUGAR CENTRIFUGAL RAW	341	355	538	505	403	422	415	389	419	381	383	-.34
SOYBEANS	250	271	264	282	243	256	278	218	200	200	263	-1.96
SOYBEAN OIL	617	675	625	542	483	498	715	644	411	356	466	-3.98
GROUNDNUTS SHELLED	661	679	698	964	668	621	733	601	594	603	566	-2.32
GROUNDNUT OIL	946	965	781	998	647	568	986	919	658	563	612	-4.09
COPRA	374	587	400	312	261	354	583	332	139	225	316	-6.09
COCONUT OIL	627	937	651	536	461	556	1028	591	290	397	539	-4.98
PALM NUTS KERNELS	262	357	267	235	222	263	331	232	111	102	154	-8.61
PALM OIL	554	617	563	528	441	442	660	505	290	327	425	-4.90
PALM KERNEL OIL	617	896	653	540	450	574	908	535	261	391	501	-5.67
OLIVE OIL	1364	1638	1983	1808	1782	1504	1362	1174	1631	1950	1946	.69
CASTOR BEANS	318	341	318	324	284	291	376	275	192	200	304	-3.51
CASTOR BEAN OIL	801	803	970	856	825	908	1119	709	586	703	910	-1.33
COTTONSEED	177	183	179	199	136	130	175	141	106	119	146	-4.16
COTTONSEED OIL	607	682	628	627	529	526	751	639	446	429	498	-3.12
LINSEED	217	281	311	326	285	275	287	266	206	173	276	-2.12
LINSEED OIL	380	543	611	662	533	417	527	625	477	305	395	-2.67
BANANAS	156	168	186	199	204	214	213	220	242	254	266	4.97
ORANGES	266	345	357	347	330	326	299	334	343	397	408	2.21
APPLES	408	399	435	409	432	336	329	323	401	426	440	-.31
RAISINS	1080	1563	1677	1477	1212	1079	941	923	1070	1185	1255	-2.60
DATES	387	390	415	588	654	712	881	809	757	641	748	7.58
COFFEE GREEN	3168	3149	3290	2238	2307	2285	2552	2535	3673	2245	2407	-1.84
COCOA BEANS	3138	3283	2663	1771	1590	1636	2099	2076	2181	2041	1698	-4.42
TEA	2058	1937	2060	1906	1772	1993	2649	2186	1863	1939	1897	-.02
COTTON LINT	1357	1528	1623	1719	1443	1521	1693	1457	1135	1205	1571	-1.32
JUTE	338	383	378	313	284	263	333	503	283	229	315	-1.95
JUTE-LIKE FIBRES	247	248	259	190	235	310	304	259	219	184	263	-.45
SISAL	378	483	601	558	516	433	418	405	425	409	404	-2.01

5. (Cont.) WORLD AVERAGE EXPORT UNIT VALUES OF SELECTED AGRICULTURAL, FISHERY AND FOREST PRODUCTS

	1978	1979	1980	1981	1982	1983	1984	1985	1986	1987	1988	ANNUAL RATE OF CHANGE 1978-88
	....US $ PER METRIC TON											PERCENT
TOBACCO UNMANUFACTURED	2630	2740	2822	2952	3239	3129	2972	2919	2929	2914	3043	,89
NATURAL RUBBER	945	1243	1296	1162	871	1018	1058	851	906	1146	1730	1,10
RUBBER NATURAL DRY	915	1180	1312	1066	797	963	965	755	788	903	1115	-1,90
WOOL GREASY	2221	2463	2825	2956	2919	2517	2626	2514	2445	2668	4271	2,51
CATTLE 1/	355	418	439	423	400	379	365	375	428	463	533	1,85
BEEF AND VEAL	2160	2390	2514	2377	2443	2213	1954	1863	2028	2681	2936	,58
MUTTON AND LAMB	1390	1592	1761	1863	1809	1597	1515	1421	1469	1570	1795	-,04
PIGS 1/	104	111	106	108	113	99	94	88	93	96	96	-1,79
BACON HAM OF SWINE	2247	2630	2894	2744	2640	2345	2226	2236	2741	3057	3256	1,57
MEAT CHIKENS	1295	1361	1430	1338	1162	1031	1072	1032	1206	1258	1259	-1,41
MEAT PREPARATIONS	1607	2105	2529	2414	2150	2106	1983	1845	2010	2354	2144	,53
EVAP COND WHOLE COW MILK	746	849	931	926	939	898	785	775	969	1068	1071	2,13
MILK OF COWS SKIMMED DRY	742	843	1047	1106	1055	864	786	783	1021	1182	1617	3,88
BUTTER OF COWMILK	2246	2281	2468	2631	2704	2395	2006	1683	1844	1746	1983	-3,36
CHEESE OF WHOLE COWMILK	2532	2769	2933	2663	2568	2429	2188	2237	2807	3261	3454	1,43
FISHERY PRODUCTS												
FISH FRESH FROZEN	1135	1242	1254	1288	1208	1090	1057	1046	1205	1345	1364	,52
FISH CURED	1868	2114	2450	2584	2249	2029	1815	1871	2536	3143	3180	3,22
SHELLFISH	3375	3686	3921	3809	3792	3575	3297	3383	4275	4428	4608	1,99
FISH CANNED AND PREPARED	2038	2301	2349	2401	2313	2400	2264	2323	2635	2825	2883	2,59
SHELLFISH CANNED+PREPAR	3797	4490	4694	4354	4201	4350	4065	3963	5110	5512	5695	2,65
FISH BODY AND LIVER OIL	434	426	432	399	343	344	347	301	260	256	280	-5,57
FISH MEAL	427	401	469	473	371	429	391	294	347	359	378	-2,56
FOREST PRODUCTS												
SAWLOGS CONIFEROUS 2/	62	83	89	81	73	63	63	61	65	74	83	-,64
SAWLOGS NONCONIFEROUS 2/	57	93	105	88	87	85	72	70	76	87	91	,38
PULPWOOD+PARTICLE 2/	25	27	36	40	35	30	30	29	32	36	41	2,37
FUELWOOD 2/	21	27	34	34	29	25	26	26	29	33	33	1,66
SAWNWOOD CONIFEROUS 2/	108	131	138	127	114	114	110	105	117	128	137	,16
SAWNWOOD NONCONIF. 2/	164	216	245	223	209	215	201	195	226	233	236	1,45
WOOD-BASED PANELS 2/	228	283	316	294	280	268	248	251	264	309	321	1,00
PULP FOR PAPER	282	361	444	451	411	356	416	353	386	502	582	3,75
PAPER AND PAPERBOARD	454	505	571	567	556	504	521	528	594	688	759	3,44

1/ U.S. DOLLARS PER HEAD
2/ U.S. DOLLARS PER CUBIC METRE

6. VOLUME OF IMPORTS OF MAJOR AGRICULTURAL, FISHERY AND FOREST PRODUCTS

	1978	1979	1980	1981	1982	1983	1984	1985	1986	1987	1988	ANNUAL RATE OF CHANGE 1978-88
	....	....	....	....	....	THOUSAND METRIC TONS	....	....	....	....	....	PERCENT
WORLD												
AGRICULTURAL PRODUCTS												
WHEAT+FLOUR,WHEAT EQUIV.	80117	85545	98025	103071	107888	105897	114863	103281	96536	108017	117523	2.64
RICE MILLED	10155	12146	13059	13838	11489	11902	11456	12432	12745	12274	11408	.30
BARLEY	14749	14767	15086	18682	18655	17746	23004	21788	23503	21538	19695	4.46
MAIZE	68112	75185	79538	80423	69530	69481	67666	68760	58588	64361	66351	-1.81
MILLET	386	366	285	223	283	267	227	281	200	202	216	-5.41
SORGHUM	10433	10195	11037	13691	13569	11017	13152	11762	8421	7660	8705	-2.86
POTATOES	3906	4569	4665	4697	5145	4844	4817	5298	5572	6237	6513	4.18
SUGAR,TOTAL (RAW EQUIV.)	24539	26474	27374	28330	29577	27894	28176	27384	26658	28057	28164	.66
PULSES	2066	2356	2924	3210	3167	3270	3537	3907	4843	5180	6021	10.03
SOYBEANS	23411	26125	27048	26294	28702	26871	25696	25906	27169	29588	26881	.97
SOYBEAN OIL	2404	2873	3239	3255	3792	3677	4028	3449	3017	3903	3907	3.34
GROUNDNUTS SHELLED BASIS	805	777	713	727	814	766	755	814	902	916	950	2.15
GROUNDNUT OIL	475	474	513	359	416	516	323	333	358	372	383	-3.16
COPRA	804	458	465	393	477	251	306	366	407	339	282	-6.52
COCONUT OIL	1255	1198	1125	1400	1291	1294	1052	1136	1493	1434	1373	1.28
PALM NUTS KERNELS	169	161	182	161	123	127	125	99	106	117	93	-6.03
PALM OIL	2318	2701	3408	3220	3686	3917	3901	4873	5973	5762	5914	9.80
OILSEED CAKE AND MEAL	21972	23854	25374	27075	28471	33091	29271	31853	34160	37693	39296	5.59
BANANAS	6882	7044	6742	6786	6796	6190	6643	7145	7303	7561	7840	1.15
ORANGES+TANGER+CLEMEN	4969	5071	5239	5018	5158	5105	5259	4950	5299	5427	5555	.78
LEMONS AND LIMES	961	965	991	970	1049	1003	997	1008	1017	991	1044	.57
COFFEE GREEN+ROASTED	3438	3915	3794	3815	3886	3988	4048	4210	4231	4548	4263	2.06
COCOA BEANS	1096	1026	1063	1242	1270	1259	1325	1464	1400	1470	1542	4.02
TEA	832	891	908	883	888	915	1051	1010	1069	1026	1167	2.94
COTTON LINT	4503	4521	5069	4421	4503	4356	4496	4593	4805	5604	5049	1.23
JUTE AND SIMILAR FIBRES	492	572	574	531	572	518	461	380	523	527	343	-2.95
TOBACCO UNMANUFACTURED	1425	1394	1410	1443	1410	1369	1434	1382	1361	1407	1325	-.46
NATURAL RUBBER	3350	3492	3392	3279	3132	3428	3696	3660	3711	4074	4310	2.33
WOOL GREASY	883	919	852	857	818	823	820	909	923	991	926	.82
BOVINE CATTLE 1/	7212	7208	6672	6919	7287	6786	6658	6620	7034	7003	6789	-.40
PIGS 1/	7749	8084	10498	9715	9020	9357	9987	10218	11821	12230	12576	4.30
TOTAL MEAT	6942	7572	7909	8425	8698	8669	8541	9019	9900	9946	10307	3.57
MILK DRY	538	592	677	695	677	645	724	708	813	932	1064	5.49
TOTAL EGGS IN SHELL	636	674	740	780	825	820	847	785	745	778	767	1.44
FISHERY PRODUCTS												
FISH FRESH FROZEN	3698	4058	4144	4388	4445	4695	4842	5309	5986	6271	6533	5.76
FISH CURED	330	371	400	424	364	434	424	465	432	462	475	3.00
SHELLFISH	1061	1224	1121	1143	1241	1347	1494	1600	1756	1877	1948	6.52
FISH CANNED AND PREPARED	871	904	1020	1078	950	920	942	1011	1090	1138	1147	2.17
SHELLFISH CANNED+PREPAR	162	163	174	184	201	223	238	257	276	292	307	7.31
FISH BODY AND LIVER OIL	653	762	752	732	796	731	951	1098	819	803	863	2.62
FISH MEAL	2089	2471	2263	2052	2599	2324	2520	3104	3243	3221	3307	4.87
FOREST PRODUCTS 2/												
SAWLOGS CONIFEROUS	29858	31516	28054	23842	26400	30362	31057	33089	32712	35992	37152	2.67
SAWLOGS NONCONIFEROUS	47694	48277	42243	34891	32704	32986	30879	29890	31104	31942	33548	-4.18
PULPWOOD+PARTICLE	34187	39249	43086	41400	36495	37710	41454	42094	44786	48926	50209	2.84
FUELWOOD	2769	2908	3112	2533	3158	3545	3924	4096	3878	3844	3494	3.81
SAWNWOOD CONIFEROUS	65298	67388	63311	58325	59443	67704	70499	72550	73756	76203	77620	2.23
SAWNWOOD NONCONIFEROUS	11669	13257	12662	11264	10847	11941	12330	12316	12674	15609	15914	2.31
WOOD-BASED PANELS	15866	16789	15657	16649	15469	16855	17820	18786	20453	23809	25172	4.55
PULP FOR PAPER	17563	18800	19316	18517	17299	19597	20441	20746	22105	23942	24685	3.20
PAPER AND PAPERBOARD	30472	32181	33797	34193	33844	35678	39315	40224	43370	47557	50705	4.97

1/ THOUSAND HEAD
2/ EXCEPT FOR PULP FOR PAPER AND PAPER AND PAPERBOARD, ALL FOREST PRODUCTS ARE EXPRESSED IN THOUSAND CUBIC METRES

6. (Cont.) VOLUME OF IMPORTS OF MAJOR AGRICULTURAL, FISHERY AND FOREST PRODUCTS

	1978	1979	1980	1981	1982	1983	1984	1985	1986	1987	1988	ANNUAL RATE OF CHANGE 1978-88
	.....	.....	.....	.....	.....	THOUSAND METRIC TONS	.....	.....	.....	.....	.....	PERCENT
WESTERN EUROPE												
AGRICULTURAL PRODUCTS												
WHEAT+FLOUR,WHEAT EQUIV.	13384	12981	14122	13336	13943	10586	12635	15531	16641	14965	15487	1.83
RICE MILLED	1460	1299	1290	1490	1687	1559	1702	1901	1796	1665	1597	2.70
BARLEY	6567	5105	5247	5966	6194	6665	5119	4562	5022	5076	5362	-1.71
MAIZE	24755	25117	23448	21740	21102	18873	15992	15035	10940	10468	11443	-9.24
MILLET	234	186	120	130	162	129	145	142	127	128	126	-3.87
SORGHUM	1453	1196	1273	1103	2149	685	1145	244	174	771	619	-13.25
POTATOES	2565	2808	3051	3026	3228	3167	3235	3629	3936	4492	4384	5.31
SUGAR,TOTAL (RAW EQUIV.)	3521	3460	3139	3063	3195	3148	3789	3151	3112	3289	4690	1.31
PULSES	907	1055	1014	924	1067	1306	1429	1867	2094	2887	2946	13.36
SOYBEANS	14201	15311	16249	14414	16454	15009	13575	13843	13802	15420	13387	-.93
SOYBEAN OIL	559	580	675	643	681	743	702	682	604	634	619	.62
GROUNDNUTS SHELLED BASIS	541	528	414	389	431	386	396	424	459	464	480	-.64
GROUNDNUT OIL	325	407	446	297	349	396	255	274	273	299	303	-3.15
COPRA	515	294	253	184	280	113	132	133	148	123	100	-12.47
COCONUT OIL	395	390	414	561	537	512	372	419	591	573	550	3.05
PALM NUTS KERNELS	153	137	147	140	106	96	100	81	97	110	84	-5.49
PALM OIL	781	856	831	723	735	859	717	828	1159	1100	1105	3.68
OILSEED CAKE AND MEAL	15320	16704	17396	18205	19294	21471	19780	22424	23715	23749	23506	4.57
BANANAS	2526	2459	2221	2172	2178	2018	2183	2306	2458	2629	3019	1.45
ORANGES+TANGER+CLEMEN	3141	3228	3228	2969	3186	3117	3299	3010	3415	3618	3642	1.31
LEMONS AND LIMES	428	432	429	416	452	451	431	449	459	479	480	1.18
COFFEE GREEN+ROASTED	1703	1955	1930	1999	1997	2062	1999	2098	2151	2319	2305	2.41
COCOA BEANS	590	569	611	664	721	649	738	793	778	790	860	3.99
TEA	250	278	296	244	287	266	306	277	290	262	281	.56
COTTON LINT	1216	1150	1259	1017	1148	1246	1232	1343	1344	1518	1289	2.04
JUTE AND SIMILAR FIBRES	157	182	132	120	97	85	88	54	74	54	54	-11.66
TOBACCO UNMANUFACTURED	785	743	701	679	670	683	670	678	636	672	635	-1.58
NATURAL RUBBER	861	925	892	838	844	830	865	929	923	958	978	1.01
WOOL GREASY	437	444	399	394	353	316	395	422	409	430	409	-.12
BOVINE CATTLE 1/	3473	3530	3405	3211	3478	3401	3335	3695	3840	3856	3546	.97
PIGS 1/	3875	4382	5202	5496	4680	4889	4877	4973	7253	7221	7000	5.41
TOTAL MEAT	3776	3790	3761	3504	3778	3889	3835	4198	4388	4548	4708	2.46
MILK DRY	116	137	156	133	145	147	146	136	131	197	233	4.18
TOTAL EGGS IN SHELL	366	399	431	431	444	441	467	466	482	526	506	3.01
FISHERY PRODUCTS												
FISH FRESH FROZEN	1335	1474	1599	1604	1711	1564	1616	1803	1898	2206	2404	4.88
FISH CURED	168	193	199	176	174	217	214	249	227	255	260	4.28
SHELLFISH	345	367	413	407	468	509	586	623	648	751	797	9.01
FISH CANNED AND PREPARED	285	311	333	335	315	347	359	380	425	465	477	4.93
SHELLFISH CANNED+PREPAR	73	80	87	86	90	97	97	107	120	135	142	6.44
FISH BODY AND LIVER OIL	584	666	666	637	706	607	813	957	702	691	750	2.31
FISH MEAL	1102	1241	1182	1026	1288	1230	1163	1462	1525	1502	1495	3.38
FOREST PRODUCTS 2/												
SAWLOGS CONIFEROUS	4094	4547	5103	4507	4660	4456	4356	4756	4391	5062	5333	1.23
SAWLOGS NONCONIFEROUS	7715	8044	8424	6889	6139	6174	6337	6034	6112	5986	7461	-2.29
PULPWOOD+PARTICLE	15037	17463	20877	22039	19447	19140	22527	23775	24526	27928	28059	5.39
FUELWOOD	1673	1784	2016	1539	1851	2238	2490	2631	2390	2276	2088	3.67
SAWNWOOD CONIFEROUS	23684	27274	25507	21507	22714	23839	22948	21753	24888	26258	27521	.51
SAWNWOOD NONCONIFEROUS	5620	6724	6088	4933	4891	5386	5322	5516	5643	6588	6696	.80
WOOD-BASED PANELS	8440	9652	8951	8956	8462	8980	9484	9981	11074	11713	12758	3.52
PULP FOR PAPER	9435	10034	10014	9531	8807	9611	10063	10301	11028	11745	11958	2.20
PAPER AND PAPERBOARD	13602	15046	15107	15728	15742	17301	18745	18578	20632	23134	25319	5.87

1/ THOUSAND HEAD

2/ EXCEPT FOR PULP FOR PAPER AND PAPER AND PAPERBOARD, ALL FOREST PRODUCTS ARE EXPRESSED IN THOUSAND CUBIC METRES

6. (Cont.) VOLUME OF IMPORTS OF MAJOR AGRICULTURAL, FISHERY AND FOREST PRODUCTS

	1978	1979	1980	1981	1982	1983	1984	1985	1986	1987	1988	ANNUAL RATE OF CHANGE 1978-88
	....	....	....	....	....	THOUSAND METRIC TONS	....	....	....	....	....	PERCENT
USSR AND EASTERN EUROPE												
AGRICULTURAL PRODUCTS												
WHEAT+FLOUR,WHEAT EQUIV.	13101	16167	21293	24583	27316	26829	31394	24387	18252	21841	25471	3.88
RICE MILLED	710	940	994	1599	1127	601	490	586	734	844	728	-3.62
BARLEY	4137	4559	4311	6019	3258	3531	3326	5907	6573	4326	4360	1.19
MAIZE	17809	20175	18863	22097	14985	7861	13431	18014	9566	10747	13555	-5.69
MILLET	1	1	1	1	1	1		1	7	3		1.27
SORGHUM	830	229	1567	3967	2709	2078	1990	1452	39	155	400	-15.54
POTATOES	301	512	297	330	481	158	131	245	265	482	733	1.81
SUGAR,TOTAL (RAW EQUIV.)	4668	4933	5825	6397	8146	7029	6939	5843	6356	6380	5451	1.58
PULSES	39	41	62	85	60	35	77	39	80	41	59	1.38
SOYBEANS	1409	2360	1707	1653	1906	1938	1205	1113	2533	2229	1730	.66
SOYBEAN OIL	103	126	154	198	313	255	203	401	137	293	143	5.25
GROUNDNUTS SHELLED BASIS	57	46	54	61	67	54	76	74	82	83	91	5.98
GROUNDNUT OIL		2	1		1	1		1		1	1	-.24
COPRA	26	18	20	10	14	14	5	2	5	4	4	-19.38
COCONUT OIL	66	58	89	77	99	79	68	60	67	83	61	-.64
PALM NUTS KERNELS	4	3	4									
PALM OIL	58	113	112	184	384	329	292	250	249	279	190	11.84
OILSEED CAKE AND MEAL	3699	4033	4599	5331	5069	6664	4010	4187	3930	6898	7177	3.96
BANANAS	299	298	269	232	155	167	200	206	119	174	186	-6.14
ORANGES+TANGER+CLEMEN	719	690	748	688	645	599	616	697	690	641	641	-1.02
LEMONS AND LIMES	326	309	333	308	363	289	272	272	275	232	257	-3.08
COFFEE GREEN+ROASTED	181	203	232	211	210	215	247	256	222	246	262	2.80
COCOA BEANS	202	198	201	199	178	243	246	253	252	247	241	3.01
TEA	71	79	102	116	107	110	129	151	146	174	173	8.95
COTTON LINT	681	718	743	638	693	764	841	868	731	695	777	1.18
JUTE AND SIMILAR FIBRES	70	79	93	111	122	93	45	57	87	85	65	-2.32
TOBACCO UNMANUFACTURED	135	133	178	196	201	189	202	185	167	140	122	-.53
NATURAL RUBBER	433	437	441	418	360	446	435	374	341	362	262	-3.62
WOOL GREASY	182	188	182	174	173	219	135	153	157	177	164	-1.54
BOVINE CATTLE 1/	77	169	173	162	160	183	209	167	166	138	209	4.07
PIGS 1/	507	442	479	844	565	637	519	732	536	969	846	5.29
TOTAL MEAT	267	645	956	1226	1091	1132	923	848	896	758	714	4.15
MILK DRY	29	42	71	78	90	47	58	70	85	74	59	5.43
TOTAL EGGS IN SHELL	43	47	43	34	36	31	28	21	24	14	13	-11.80
FISHERY PRODUCTS												
FISH FRESH FROZEN	204	236	280	180	151	433	524	585	626	685	695	16.09
FISH CURED	15	15	20	26	20	40	30	36	33	37	43	10.73
SHELLFISH		2										
FISH CANNED AND PREPARED	39	34	38	39	37	34	33	48	56	67	63	6.08
FISH BODY AND LIVER OIL	6	5	23	13	26	24	32	46	37	31	28	19.13
FISH MEAL	384	476	310	233	291	218	283	341	337	339	399	-.18
FOREST PRODUCTS 2/												
SAWLOGS CONIFEROUS	960	720	1050	960	498	655	629	598	711	610	560	-4.66
SAWLOGS NONCONIFEROUS	442	416	454	487	385	367	375	405	301	300	331	-3.92
PULPWOOD+PARTICLE	1345	1446	1583	1390	1248	1286	1323	1248	1222	1205	1179	-2.09
FUELWOOD	27	25	25	25	20	25	25	12				
SAWNWOOD CONIFEROUS	3228	2644	2665	2884	2544	2685	2983	3671	3165	2856	2491	.15
SAWNWOOD NONCONIFEROUS	326	268	274	331	213	226	222	214	187	152	150	-7.13
WOOD-BASED PANELS	1132	1045	1137	1115	939	832	766	797	897	938	920	-2.74
PULP FOR PAPER	1053	1021	1173	1093	1031	1101	1067	1063	1015	1124	1133	.27
PAPER AND PAPERBOARD	1709	1784	2044	1968	1965	1732	1689	1717	1566	1513	1571	-2.07

1/ THOUSAND HEAD
2/ EXCEPT FOR PULP FOR PAPER AND PAPER AND PAPERBOARD, ALL FOREST PRODUCTS ARE EXPRESSED IN THOUSAND CUBIC METRES

6. (Cont.) VOLUME OF IMPORTS OF MAJOR AGRICULTURAL, FISHERY AND FOREST PRODUCTS

	1978	1979	1980	1981	1982	1983	1984	1985	1986	1987	1988	ANNUAL RATE OF CHANGE 1978-88
	THOUSAND METRIC TONS											PERCENT
NORTH AMERICA DEVELOPED												
AGRICULTURAL PRODUCTS												
WHEAT+FLOUR,WHEAT EQUIV.	1	6	8	11	74	63	110	282	279	392	487	76.89
RICE MILLED	82	91	94	106	126	128	141	174	207	221	264	12.39
BARLEY	108	157	140	127	198	141	146	105	135	201	256	4.22
MAIZE	476	849	1228	1276	807	352	541	567	937	348	527	-5.22
MILLET											2	7.19
SORGHUM	1				2		7				2	-1.46
POTATOES	235	242	212	340	344	280	303	330	305	388	405	5.14
SUGAR,TOTAL (RAW EQUIV.)	4833	5401	4587	5453	3466	3654	4150	3662	3199	2265	2125	-8.09
PULSES	43	39	43	61	47	48	55	51	56	64	72	4.79
SOYBEANS	325	351	483	382	468	315	285	247	166	247	131	-9.12
SOYBEAN OIL	35	22	12	9	4	35	17	42	15	26	297	16.29
GROUNDNUTS SHELLED BASIS	66	63	55	72	61	67	69	69	79	73	61	1.19
GROUNDNUT OIL	6	5	5	4	4	6	5	4	6	9	19	7.66
COPRA										1	1	
COCONUT OIL	503	527	422	476	427	475	400	474	558	534	470	.45
PALM OIL	173	163	137	138	132	168	161	251	288	199	169	4.02
OILSEED CAKE AND MEAL	426	491	431	443	457	536	701	763	797	888	987	9.45
BANANAS	2543	2659	2669	2794	2935	2785	2922	3352	3350	3367	3212	2.91
ORANGES+TANGER+CLEMEN	303	297	320	326	304	325	301	295	325	325	283	-.15
LEMONS AND LIMES	34	36	38	43	38	40	51	66	61	66	80	8.87
COFFEE GREEN+ROASTED	1195	1277	1190	1104	1150	1089	1178	1233	1283	1319	1051	-.04
COCOA BEANS	226	179	162	264	213	233	218	292	224	283	259	3.44
TEA	91	101	107	107	103	97	109	97	110	93	106	.35
COTTON LINT	59	61	65	63	52	61	59	57	54	42	41	-3.44
JUTE AND SIMILAR FIBRES	17	23	10	18	18	17	11	17	16	16	15	-1.61
TOBACCO UNMANUFACTURED	173	188	191	176	167	153	214	177	207	220	168	.93
NATURAL RUBBER	846	862	695	759	713	773	906	927	874	934	960	2.10
WOOL GREASY	15	11	14	20	16	20	23	17	24	32	31	8.83
BOVINE CATTLE 1/	1337	758	731	816	1085	1004	801	894	1407	1295	1401	3.91
PIGS 1/	204	137	248	147	295	448	1322	1227	502	447	839	19.56
TOTAL MEAT	875	912	854	766	866	808	866	1010	1064	1170	1190	3.46
MILK DRY						2	3	4	4	5	6	
TOTAL EGGS IN SHELL	18	21	12	12	11	22	30	19	20	14	12	-.39
FISHERY PRODUCTS												
FISH FRESH FROZEN	800	776	699	735	676	700	686	760	810	823	798	.68
FISH CURED	34	31	26	35	33	32	33	32	33	33	34	.66
SHELLFISH	146	155	146	156	175	213	222	235	261	261	263	7.37
FISH CANNED AND PREPARED	89	95	99	104	112	126	153	187	218	224	229	11.58
SHELLFISH CANNED+PREPAR	40	41	39	47	54	69	73	84	76	77	84	9.19
FISH BODY AND LIVER OIL	9	9	12	10	8	9	8	10	11	11	14	2.38
FISH MEAL	40	82	45	56	79	68	81	234	171	172	181	17.15
FOREST PRODUCTS 2/												
SAWLOGS CONIFEROUS	2043	2458	2146	1674	1772	2683	2887	2837	2826	2710	2705	3.86
SAWLOGS NONCONIFEROUS	409	502	471	415	335	424	585	576	645	757	844	6.98
PULPWOOD+PARTICLE	2516	2504	2249	2348	2000	2409	2173	1976	2805	2142	2527	-.18
FUELWOOD	352	377	268	137	113	113	161	160	154	160	172	-7.04
SAWNWOOD CONIFEROUS	28675	26582	22839	22542	21694	28483	31316	34407	33653	34403	33648	3.90
SAWNWOOD NONCONIFEROUS	1431	1571	1422	1557	912	1246	1407	1432	1496	2072	1630	2.00
WOOD-BASED PANELS	3956	3336	2378	2851	2283	3366	3548	3956	4268	4392	4678	4.45
PULP FOR PAPER	3522	3857	3528	3563	3245	3645	4085	4069	4150	4489	4419	2.51
PAPER AND PAPERBOARD	8394	8220	8314	7773	7538	8434	10381	10973	11622	12494	12786	5.41

1/ THOUSAND HEAD
2/ EXCEPT FOR PULP FOR PAPER AND PAPER AND PAPERBOARD, ALL FOREST PRODUCTS ARE EXPRESSED IN THOUSAND CUBIC METRES

6. (Cont.) VOLUME OF IMPORTS OF MAJOR AGRICULTURAL, FISHERY AND FOREST PRODUCTS

	1978	1979	1980	1981	1982	1983	1984	1985	1986	1987	1988	ANNUAL RATE OF CHANGE 1978-88
	THOUSAND METRIC TONS											PERCENT
OCEANIA DEVELOPED												
AGRICULTURAL PRODUCTS												
WHEAT+FLOUR,WHEAT EQUIV.		32	54	53	51	71	126	67	73	45	137	49.56
RICE MILLED	8	8	8	9	10	12	15	19	21	29	33	16.55
BARLEY											7	51.86
MAIZE	3	3	4	5	11	14	9	11	9	9	6	11.55
MILLET		1	1	1	1	1	1	1	1	1	1	
SORGHUM					4							
POTATOES							1					
SUGAR,TOTAL (RAW EQUIV.)	166	172	151	120	147	157	169	172	149	182	167	.98
PULSES	13	12	14	13	16	16	22	12	11	8	10	-3.31
SOYBEANS	15		13	41	10	23	36	38			31	-3.81
SOYBEAN OIL	29	26	32	29	45	53	48	31	21	36	40	1.60
GROUNDNUTS SHELLED BASIS	12	4	5	9	12	6	13	8	9	9	7	1.60
GROUNDNUT OIL	2	3		1	1	1	1	1	1	1	1	-3.36
COPRA	5	7	4	6	6	4						
COCONUT OIL	18	19	17	16	20	20	22	20	19	19	24	2.10
PALM OIL	23	28	26	24	20	4	7	9	15	48	72	2.73
OILSEED CAKE AND MEAL	30	7	12	19	10	52	11	38	33	60	26	11.74
BANANAS	38	35	37	36	36	40	30	60	37	45	45	2.54
ORANGES+TANGER+CLEMEN	18	14	16	16	17	18	24	21	18	25	21	3.80
LEMONS AND LIMES		1	1	1	1	3	3	4	2	2	2	21.29
COFFEE GREEN+ROASTED	26	35	41	38	42	39	37	37	39	36	43	2.14
COCOA BEANS	17	15	14	15	13	13	10	7	6	2	1	-20.84
TEA	30	30	32	28	30	28	28	27	26	25	25	-2.26
COTTON LINT	4	2	2	2	1	1	1	3	1			-19.56
JUTE AND SIMILAR FIBRES	11	12	9	11	8	8	6	8	9	7	8	-3.88
TOBACCO UNMANUFACTURED	16	13	15	15	14	14	14	14	14	16	16	.23
NATURAL RUBBER	52	53	54	50	47	40	40	44	43	45	47	-2.07
WOOL GREASY	1	1										-10.06
BOVINE CATTLE 1/	1	1	1				1	2			1	-1.20
TOTAL MEAT	1	2	4	4	4	5	8	7	5	6	6	13.38
MILK DRY	1			1		1		1	1		2	28.99
FISHERY PRODUCTS												
FISH FRESH FROZEN	21	22	29	33	33	29	35	41	38	40	42	6.76
FISH CURED	3	5	4	4	4	4	5	5	5	4	5	2.75
SHELLFISH	2	4	4	6	6	8	8	9	10	13	12	17.87
FISH CANNED AND PREPARED	26	22	27	27	28	25	31	30	31	31	27	1.92
SHELLFISH CANNED+PREPAR	7	6	5	7	8	8	8	9	9	9	9	5.26
FISH BODY AND LIVER OIL	1	1			1			1	1	1	1	3.21
FISH MEAL	3	4	14	8	8	11	8	13	8	12	15	11.68
FOREST PRODUCTS 2/												
SAWLOGS CONIFEROUS	2					1						-2.81
SAWLOGS NONCONIFEROUS	17	11	2	1	7	1	1	1	2	4	4	-13.29
FUELWOOD	2	2	1	1	1	1						
SAWNWOOD CONIFEROUS	638	682	697	781	881	642	823	1113	1044	860	1304	5.95
SAWNWOOD NONCONIFEROUS	311	304	317	306	290	210	282	317	265	257	265	-1.76
WOOD-BASED PANELS	89	99	88	104	111	79	102	112	121	106	106	2.01
PULP FOR PAPER	239	280	281	286	262	220	243	208	239	280	280	-.35
PAPER AND PAPERBOARD	584	671	739	736	794	558	670	899	813	817	841	2.88

1/ THOUSAND HEAD
2/ EXCEPT FOR PULP FOR PAPER AND PAPER AND PAPERBOARD, ALL FOREST PRODUCTS ARE EXPRESSED IN THOUSAND CUBIC METRES

6. (Cont.) VOLUME OF IMPORTS OF MAJOR AGRICULTURAL, FISHERY AND FOREST PRODUCTS

	1978	1979	1980	1981	1982	1983	1984	1985	1986	1987	1988	ANNUAL RATE OF CHANGE 1978-88
	THOUSAND METRIC TONS											PERCENT
AFRICA DEVELOPING												
AGRICULTURAL PRODUCTS												
WHEAT+FLOUR,WHEAT EQUIV.	7963	7752	9067	8979	9371	9423	10187	11047	10115	9326	10805	2.85
RICE MILLED	1877	2227	2299	2573	2829	2725	2535	2759	2903	2932	2230	2.48
BARLEY	647	419	302	459	680	397	743	583	129	133	1096	-3.52
MAIZE	1154	1288	2329	2358	2336	1732	2804	2711	2123	2057	2258	5.05
MILLET	83	101	106	35	55	71	31	85	29	29	21	-12.34
SORGHUM	150	133	113	155	161	232	411	382	104	60	213	.95
POTATOES	233	307	237	211	269	448	347	425	312	230	284	2.13
SUGAR,TOTAL (RAW EQUIV.)	2043	2100	2265	2320	2101	2399	2093	2192	2671	2943	2310	2.15
PULSES	118	210	219	161	153	221	232	226	245	186	291	5.07
SOYBEANS	22	31	25	11	35	16	20	22	18	27	13	-3.03
GROUNDNUTS SHELLED BASIS	322	353	320	339	448	408	368	289	288	229	299	[illegible]
GROUNDNUT OIL	10	10	16	16	21	28	11	4	28	7	3	-8.03
COPRA	4	4	3	2	2	3	5	2	2	3	2	-4.77
COCONUT OIL	10	9	7	14	12	10	17	10	9	9	7	-1.54
PALM OIL	106	99	166	241	289	259	187	209	352	232	238	8.53
OILSEED CAKE AND MEAL	122	157	188	241	260	230	298	298	491	498	526	14.98
BANANAS	31	17	18	26	57	26	29	10	13	8	9	-11.04
ORANGES+TANGER+CLEMEN	12	12	10	9	10	9	9	7	10	9	9	-2.70
LEMONS AND LIMES	1	1	1	1	1	1	1	1	1	1	1	.52
COFFEE GREEN+ROASTED	83	76	80	103	67	115	97	103	58	134	74	.97
COCOA BEANS	1	1	1	1	1	1	1	2	1	2	1	6.11
TEA	56	70	57	69	52	60	61	70	78	70	78	2.62
COTTON LINT	42	48	44	64	83	91	98	98	79	101	98	9.50
JUTE AND SIMILAR FIBRES	58	58	64	50	49	59	40	58	52	89	32	-1.68
TOBACCO UNMANUFACTURED	64	63	58	49	49	52	46	56	72	57	60	.21
NATURAL RUBBER	21	20	21	26	23	23	23	24	26	26	31	3.12
WOOL GREASY	4	3	2	2	1	2	2	2	4	1	2	-3.10
BOVINE CATTLE 1/	776	835	824	894	839	920	1007	730	599	480	454	-5.37
PIGS 1/	1	1	1	2	2	3	4	1		1		-11.36
TOTAL MEAT	139	137	141	147	220	186	216	224	251	222	234	6.66
MILK DRY	35	35	51	73	53	69	92	106	112	122	127	14.60
TOTAL EGGS IN SHELL	44	35	50	52	71	78	49	47	20	11	7	-14.20
FISHERY PRODUCTS												
FISH FRESH FROZEN	589	695	764	894	792	832	685	690	758	664	670	-.21
FISH CURED	33	48	56	95	40	50	28	43	41	41	41	-2.09
SHELLFISH	4	3	3	3	3	1	4	3	2	2	2	-3.00
FISH CANNED AND PREPARED	147	127	136	155	114	102	54	59	62	54	53	-11.61
FISH BODY AND LIVER OIL	3	2			1	1						-23.75
FISH MEAL	27	24	24	25	32	36	51	28	38	38	38	5.08
FOREST PRODUCTS 2/												
SAWLOGS CONIFEROUS	32	73	94	84	110	169	139	91	52	59	59	.67
SAWLOGS NONCONIFEROUS	197	204	326	225	241	321	318	325	323	335	343	5.35
FUELWOOD				1	41	43	33	33	33	33	33	
SAWNWOOD CONIFEROUS	763	1019	905	1409	1541	1859	1817	1505	1433	1041	1262	3.96
SAWNWOOD NONCONIFEROUS	202	203	194	232	193	183	187	209	236	243	252	1.99
WOOD-BASED PANELS	263	316	359	332	261	294	202	206	153	131	131	-9.35
PULP FOR PAPER	102	104	120	135	116	149	147	148	132	109	118	1.48
PAPER AND PAPERBOARD	519	529	537	662	577	569	539	553	564	613	619	1.08

1/ THOUSAND HEAD
2/ EXCEPT FOR PULP FOR PAPER AND PAPER AND PAPERBOARD, ALL FOREST PRODUCTS ARE EXPRESSED IN THOUSAND CUBIC METRES

6. (Cont.) VOLUME OF IMPORTS OF MAJOR AGRICULTURAL, FISHERY AND FOREST PRODUCTS

	1978	1979	1980	1981	1982	1983	1984	1985	1986	1987	1988	ANNUAL RATE OF CHANGE 1978-88
	....					THOUSAND METRIC TONS					....	PERCENT
LATIN AMERICA												
AGRICULTURAL PRODUCTS												
WHEAT+FLOUR,WHEAT EQUIV.	10792	10718	12098	12067	11126	11928	12276	11391	9133	10045	8895	-1.88
RICE MILLED	431	1339	1064	794	612	908	662	1062	2114	867	689	3.07
BARLEY	358	323	551	448	339	531	537	466	398	402	436	1.30
MAIZE	4714	3954	8988	7027	3417	8162	5499	4059	5674	6611	5789	.99
MILLET	4	6	3	2	3	4	1	1			24	-9.66
SORGHUM	1442	1876	2927	3578	3226	3830	3168	3383	1562	1655	3019	1.08
POTATOES	205	251	336	198	190	184	197	162	269	185	168	-2.94
SUGAR,TOTAL (RAW EQUIV.)	898	717	1912	1620	1409	1604	1200	392	430	706	788	-7.18
PULSES	291	284	816	878	739	527	526	607	567	454	477	2.01
SOYBEANS	971	952	1205	2235	2198	1385	2858	2122	1432	2089	1809	6.50
SOYBEAN OIL	351	372	432	433	675	551	791	580	586	445	541	4.20
GROUNDNUTS SHELLED BASIS	14	11	13	13	19	10	19	24	7	11	41	4.27
GROUNDNUT OIL	85	9	2	4	1	2	2	1	1	1	4	-25.36
COCONUT OIL	39	15	25	19	21	16	15	10	22	46	47	3.38
PALM NUTS KERNELS		2	1	1	1	3	2					
PALM OIL	8	6	14	10	5	5	8	6	14	14	21	7.09
OILSEED CAKE AND MEAL	647	710	966	964	1131	1142	1214	1225	1245	1528	2031	9.61
BANANAS	287	391	435	446	325	231	227	217	268	265	256	-4.78
ORANGES+TANGER+CLEMEN	22	44	57	33	26	20	18	17	19	22	28	-5.59
LEMONS AND LIMES	6	4	3	5	2	3	5	5	3	3	2	-4.94
COFFEE GREEN+ROASTED	58	93	49	56	64	49	56	44	43	44	49	-4.35
COCOA BEANS	3	2	3	10	13	3	6	9	5	5	5	6.89
TEA	16	19	16	14	15	14	13	15	14	14	13	-2.28
COTTON LINT	71	91	79	94	79	79	122	119	200	221	230	12.67
JUTE AND SIMILAR FIBRES	12	18	36	34	14	14	14	4	47	46	29	4.40
TOBACCO UNMANUFACTURED	16	17	29	24	20	19	16	16	18	17	19	-1.48
NATURAL RUBBER	181	181	187	181	158	165	201	197	218	248	262	3.69
WOOL GREASY	7	9	13	12	13	8	9	12	8	8	6	-2.91
BOVINE CATTLE 1/	583	926	417	463	427	158	224	179	116	113	306	-15.11
PIGS 1/	32	21	10	26	55	16	8	9	4	7	205	-1.74
TOTAL MEAT	374	366	342	415	338	264	298	373	878	599	563	6.09
MILK DRY	139	122	161	161	150	126	153	120	171	183	302	4.74
TOTAL EGGS IN SHELL	11	18	19	18	26	13	10	10	12	14	21	-1.15
FISHERY PRODUCTS												
FISH FRESH FROZEN	109	134	111	97	100	84	65	84	136	135	136	.96
FISH CURED	46	47	56	53	53	51	48	47	49	47	48	-.51
SHELLFISH	9	12	11	14	13	10	12	11	12	12	11	.67
FISH CANNED AND PREPARED	61	75	93	88	75	43	53	65	58	58	58	-3.29
SHELLFISH CANNED+PREPAR	1	2	2	2	1		1	1	1	1	1	-8.81
FISH BODY AND LIVER OIL	36	67	43	64	35	69	62	34	24	24	24	-7.48
FISH MEAL	107	138	163	126	103	61	86	47	58	58	59	-10.14
FOREST PRODUCTS 2/												
SAWLOGS CONIFEROUS	34	54	128	156	162	160	69	48	58	58	58	-2.29
SAWLOGS NONCONIFEROUS	105	65	57	30	29	30	75	82	73	76	49	.42
PULPWOOD+PARTICLE		31	35	24	16	16	8	8	8	8	8	
FUELWOOD	4	4	5	7	5	3	4	6	6	6	6	2.87
SAWNWOOD CONIFEROUS	1715	1524	2184	1874	1477	1666	1944	1479	1548	1492	1730	-1.15
SAWNWOOD NONCONIFEROUS	679	692	917	642	652	597	734	596	700	599	337	-4.38
WOOD-BASED PANELS	304	401	493	499	482	479	399	349	342	381	401	-.75
PULP FOR PAPER	530	653	740	762	735	645	766	774	809	970	926	4.39
PAPER AND PAPERBOARD	1869	1856	2395	2437	2278	1958	1798	1732	1876	1925	1868	-1.36

1/ THOUSAND HEAD
2/ EXCEPT FOR PULP FOR PAPER AND PAPER AND PAPERBOARD, ALL FOREST PRODUCTS ARE EXPRESSED IN THOUSAND CUBIC METRES

6. (Cont.) VOLUME OF IMPORTS OF MAJOR AGRICULTURAL, FISHERY AND FOREST PRODUCTS

	1978	1979	1980	1981	1982	1983	1984	1985	1986	1987	1988	ANNUAL RATE OF CHANGE 1978-88
	....	....	....	....	.........THOUSAND	METRIC TONS.........	....	....	....	....	....	PERCENT
NEAR EAST DEVELOPING												
AGRICULTURAL PRODUCTS												
WHEAT+FLOUR,WHEAT EQUIV.	10319	10698	12860	14011	14156	16474	19465	17514	16007	19273	18166	6.19
RICE MILLED	1548	1887	1821	2025	2006	2251	2362	2313	2302	2756	2085	3.83
BARLEY	852	1493	2364	3292	4988	3871	9213	7762	9001	9004	5887	23.47
MAIZE	1850	2369	2685	3753	3711	4087	4190	4694	4841	5087	3968	8.73
MILLET	4	4	2	2	3	4	4	5	1	1	1	-13.19
SORGHUM	255	121	133	133	340	71	317	34	43	3	36	-24.41
POTATOES	231	282	353	426	462	420	430	334	300	303	359	1.32
SUGAR,TOTAL (RAW EQUIV.)	2401	3465	3263	3405	3946	3429	3925	3280	3254	3563	3016	1.07
PULSES	213	258	257	359	339	304	332	261	328	339	367	3.60
SOYBEANS	138	180	99	116	108	94	79	226	185	267	347	8.61
SOYBEAN OIL	281	381	442	504	529	717	667	630	656	670	456	6.12
GROUNDNUTS SHELLED BASIS	6	8	16	9	8	7	7	7	6	11	8	-.56
GROUNDNUT OIL	1	1	3	1	1	1						-16.39
COPRA	1										1	
COCONUT OIL	7	4	14	12	16	13	13	14	50	22	23	16.35
PALM NUTS KERNELS									1			
PALM OIL	164	187	148	291	378	420	465	511	676	562	660	17.00
OILSEED CAKE AND MEAL	459	442	406	543	674	860	1122	1260	1288	1185	1550	15.33
BANANAS	294	324	306	322	292	276	277	205	183	174	250	-5.13
ORANGES+TANGER+CLEMEN	472	512	545	622	634	631	644	531	421	354	507	-1.98
LEMONS AND LIMES	45	77	79	77	80	88	102	84	76	66	87	2.73
COFFEE GREEN+ROASTED	42	40	46	56	74	75	63	63	49	62	65	3.90
COCOA BEANS	4	1	2	5	5	6	5	5	7	7	5	13.72
TEA	205	188	168	171	168	194	239	216	230	204	302	3.71
COTTON LINT	21	41	22	24	27	27	29	55	71	154	100	18.07
JUTE AND SIMILAR FIBRES	24	41	20	25	37	34	32	34	28	29	30	1.29
TOBACCO UNMANUFACTURED	52	60	47	61	75	77	83	70	69	68	76	3.64
NATURAL RUBBER	46	37	41	52	65	86	82	80	82	89	97	10.03
WOOL GREASY	17	18	18	19	13	18	25	26	22	29	28	5.66
BOVINE CATTLE 1/	393	386	507	739	735	632	650	496	432	574	348	-.38
TOTAL MEAT	582	676	992	1328	1323	1272	1262	1208	1126	1084	1053	4.65
MILK DRY	23	40	39	44	48	43	56	45	53	58	58	6.97
TOTAL EGGS IN SHELL	84	75	107	153	153	153	171	131	91	92	89	.35
FISHERY PRODUCTS												
FISH FRESH FROZEN	69	55	77	107	121	137	151	114	133	134	134	8.36
FISH CURED	3	3	3	6	5	9	5	7	6	6	6	9.57
SHELLFISH	1	2	2	2	2	2	2	4	1	2	2	5.20
FISH CANNED AND PREPARED	55	51	70	64	56	55	49	53	43	44	44	-3.33
SHELLFISH CANNED+PREPAR		1	1	2	3	2	2	1	1	1	1	4.62
FISH BODY AND LIVER OIL	1	1	1	1	1		2					-21.25
FISH MEAL	56	58	77	147	113	106	163	101	131	131	131	8.24
FOREST PRODUCTS 2/												
SAWLOGS CONIFEROUS	176	126	173	205	275	319	316	435	214	1211	1089	20.45
SAWLOGS NONCONIFEROUS	68	42	57	46	5	6	11	41	109	144	173	11.66
PULPWOOD+PARTICLE	36	40	14	38	52	69	57	42	106	47	47	7.91
FUELWOOD	163	119	126	146	183	169	198	156	212	212	212	5.03
SAWNWOOD CONIFEROUS	2441	2689	3242	3498	3938	4179	4563	4139	3144	3099	2747	1.42
SAWNWOOD NONCONIFEROUS	620	469	630	550	630	758	811	838	610	587	618	1.73
WOOD-BASED PANELS	804	931	1072	1425	1588	1324	1450	1510	1260	1253	1224	3.51
PULP FOR PAPER	127	113	121	111	110	178	171	196	169	189	227	7.12
PAPER AND PAPERBOARD	889	905	975	1042	1008	1006	1205	1119	1048	1052	1827	4.40

1/ THOUSAND HEAD
2/ EXCEPT FOR PULP FOR PAPER AND PAPER AND PAPERBOARD, ALL FOREST PRODUCTS ARE EXPRESSED IN THOUSAND CUBIC METRES

6. (Cont.) VOLUME OF IMPORTS OF MAJOR AGRICULTURAL, FISHERY AND FOREST PRODUCTS

	1978	1979	1980	1981	1982	1983	1984	1985	1986	1987	1988	ANNUAL RATE OF CHANGE 1978-88
	....	....	....	....	....	....	THOUSAND METRIC TONS	....	....	....	....	PERCENT
FAR EAST DEVELOPING												
AGRICULTURAL PRODUCTS												
WHEAT+FLOUR,WHEAT EQUIV.	8058	8808	8897	7811	9709	11447	10313	9985	11678	10287	14378	4.54
RICE MILLED	3465	3392	4497	4404	2083	3121	2426	2553	1201	1353	2681	-8.57
BARLEY	107	106	206	270	916	451	1624	97	7	123	41	-13.45
MAIZE	3360	4328	4120	4740	5051	6442	5071	5651	5573	6771	7301	6.51
MILLET	1	2	3	3	6	4	5	4	5	5	7	15.76
SORGHUM	49	144	62	178	445	234	421	391	296	36	69	2.24
POTATOES	117	143	155	145	147	160	149	150	164	138	150	1.24
SUGAR,TOTAL (RAW EQUIV.)	1866	1935	2607	2807	2310	2138	2273	4393	3895	4570	3420	8.08
PULSES	167	207	207	377	380	438	535	559	1125	900	1434	23.07
SOYBEANS	489	728	874	1093	1219	1137	1354	1459	1678	1790	2022	12.89
SOYBEAN OIL	583	841	1004	981	976	856	1155	721	537	1058	1354	2.58
GROUNDNUTS SHELLED BASIS	28	39	67	93	152	144	82	110	165	179	180	17.58
GROUNDNUT OIL	42	36	38	34	36	55	38	44	41	44	44	1.61
COPRA	163	74	115	110	81	46	84	135	146	108	80	-.79
COCONUT OIL	158	91	58	151	83	90	88	71	99	92	104	-1.68
PALM NUTS KERNELS	6	10	15	6	3	12	5	4	1	1	3	-18.84
PALM OIL	847	1058	1757	1436	1561	1679	1865	2560	2791	2824	2775	12.12
OILSEED CAKE AND MEAL	804	965	1005	1026	1339	1550	1755	1252	2041	2203	2417	11.12
BANANAS	57	69	59	49	59	51	71	69	74	80	81	3.56
ORANGES+TANGER+CLEMEN	222	208	238	273	249	287	253	254	275	295	296	2.89
LEMONS AND LIMES	4	6	7	8	8	9	10	13	14	14	16	12.54
COFFEE GREEN+ROASTED	19	27	19	36	51	72	96	91	107	80	86	19.46
COCOA BEANS	12	17	27	45	60	61	50	57	58	78	107	19.27
TEA	77	84	86	97	94	110	127	114	113	125	124	5.00
COTTON LINT	860	827	888	775	791	863	993	972	1131	1380	1308	5.17
JUTE AND SIMILAR FIBRES	64	80	119	109	165	151	180	136	152	160	101	5.94
TOBACCO UNMANUFACTURED	64	69	82	88	69	63	68	66	63	68	84	-.06
NATURAL RUBBER	193	215	182	208	226	199	271	249	295	322	399	6.80
WOOL GREASY	29	30	33	39	34	38	40	46	67	65	63	9.05
BOVINE CATTLE 1/	324	355	356	366	359	353	301	294	287	315	319	-1.63
PIGS 1/	3123	3095	4552	3194	3414	3357	3250	3268	3518	3579	3679	.57
TOTAL MEAT	279	297	228	266	352	360	329	326	320	357	428	3.93
MILK DRY	143	159	161	163	151	171	173	179	197	225	223	4.22
TOTAL EGGS IN SHELL	68	75	76	75	80	79	89	88	94	104	115	4.65
FISHERY PRODUCTS												
FISH FRESH FROZEN	185	229	210	258	280	294	410	443	663	660	720	15.60
FISH CURED	21	21	28	22	26	24	53	40	32	33	32	5.70
SHELLFISH	119	180	123	116	132	140	143	164	191	206	231	5.55
FISH CANNED AND PREPARED	84	79	96	78	92	51	59	47	53	57	59	-5.50
SHELLFISH CANNED+PREPAR	16	14	18	16	21	22	23	18	20	21	21	3.47
FISH BODY AND LIVER OIL	4	5	2	2	3	3	5	22	19	20	21	25.61
FISH MEAL	131	164	148	158	251	171	183	191	210	202	223	4.26
FOREST PRODUCTS 2/												
SAWLOGS CONIFEROUS	2426	2128	1536	1186	1548	2116	2073	2217	2432	2795	3217	5.05
SAWLOGS NONCONIFEROUS	9371	9355	6526	5985	5415	5789	4986	4337	6152	5548	6335	-4.40
PULPWOOD+PARTICLE		2	2	1		3	117	118	83	30	71	
FUELWOOD	489	519	560	588	741	749	727	773	721	698	628	3.44
SAWNWOOD CONIFEROUS	235	80	87	72	45	46	49	37	37	34	60	-12.09
SAWNWOOD NONCONIFEROUS	1829	2345	1850	1762	1910	1840	1778	1361	1492	2176	2724	.42
WOOD-BASED PANELS	575	610	724	821	680	794	652	583	683	1103	1168	4.66
PULP FOR PAPER	696	735	728	815	791	1090	1055	1124	1305	1319	1376	7.97
PAPER AND PAPERBOARD	1830	1995	2072	2247	2349	2313	2537	2504	2612	3133	3388	5.48

1/ THOUSAND HEAD
2/ EXCEPT FOR PULP FOR PAPER AND PAPER AND PAPERBOARD, ALL FOREST PRODUCTS ARE EXPRESSED IN THOUSAND CUBIC METRES

6. (Cont.) VOLUME OF IMPORTS OF MAJOR AGRICULTURAL, FISHERY AND FOREST PRODUCTS

	1978	1979	1980	1981	1982	1983	1984	1985	1986	1987	1988	ANNUAL RATE OF CHANGE 1978-88
	THOUSAND METRIC TONS											PERCENT
ASIAN CENT PLANNED ECON												
AGRICULTURAL PRODUCTS												
WHEAT+FLOUR,WHEAT EQUIV.	10271	11756	13243	15688	15565	12643	11234	6620	7623	15547	17157	-.02
RICE MILLED	250	619	652	441	578	205	535	619	941	1043	595	7.66
BARLEY	336	704	402	354	509	481	430	369	477	536	410	.32
MAIZE	3064	5412	4438	3287	4117	5569	3015	3108	3714	5249	4593	.86
SORGHUM	473	517	417	840	767	534	597	564	810	726	98	-4.93
SUGAR,TOTAL (RAW EQUIV.)	1587	1368	1114	1294	2373	2130	1456	2088	1433	2067	3943	6.99
PULSES	68	58	72	91	124	88	92	89	108	97	95	4.22
SOYBEANS	1172	1696	1529	1682	1516	1420	1361	1503	2098	2314	2332	4.94
SOYBEAN OIL	137	143	136	56	63	36	25	43	192	424	156	4.28
GROUNDNUTS SHELLED BASIS	2	1			6				1			
COPRA		1	3	3	7	1	3	7	17	15	12	36.77
COCONUT OIL	19	27	31	26	31	26	27	33	49	32	61	7.73
PALM NUTS KERNELS				2	1							
PALM OIL	14	48	63	26	24	24	23	70	216	277	427	31.23
OILSEED CAKE AND MEAL	55	1	9	14	15	33	50	48	41	66	65	25.11
BANANAS						20	20	40	36	41	21	
ORANGES+TANGER+CLEMEN	1		2	1	1	5	2	3	4	11	10	34.38
LEMONS AND LIMES									1			
COFFEE GREEN+ROASTED	6	5	6	7	17	30	17	19	6	6	27	10.31
COCOA BEANS	15	17	17	4	23	10	12	6	27	14	17	1.66
TEA	6	5	5	4	4	5	6	6	19	17	21	16.25
COTTON LINT	818	835	1235	1023	824	521	369	362	464	608	458	-8.68
JUTE AND SIMILAR FIBRES	39	36	47	24	43	36	24	3	42	29	1	-21.09
TOBACCO UNMANUFACTURED	19	22	32	54	46	21	28	45	29	34	49	4.89
NATURAL RUBBER	300	333	358	220	232	337	312	256	328	473	572	4.62
WOOL GREASY	28	51	58	78	95	94	74	115	119	120	113	12.60
BOVINE CATTLE 1/			2		1	1	4	10	7	7	5	55.89
PIGS 1/	4	3	3	5	3	3	2	2	2	2	3	-7.13
TOTAL MEAT	11	18	16	23	27	28	32	38	49	62	77	19.33
MILK DRY	44	45	28	30	28	29	32	37	50	56	49	3.34
FISHERY PRODUCTS												
FISH FRESH FROZEN	4	4	2	3	3	3	5	5	3	3	3	1.24
FISH CURED	1	1	1	2	2	2	3					-6.87
SHELLFISH	9	14	20	2	5	5	4	8	9	9	9	-2.46
FISH CANNED AND PREPARED	1	1	2	2	2	2	3	3	3	3	3	10.34
SHELLFISH CANNED+PREPAR		1	1	1	1	1	1	1	2	2	2	18.99
FISH BODY AND LIVER OIL	2	2	1	1	1	1	3	3	3	3	3	11.96
FISH MEAL	142	168	155	162	312	263	370	507	526	526	526	16.97
FOREST PRODUCTS 2/												
SAWLOGS CONIFEROUS	389	422	630	1181	3115	5391	6776	7576	5976	5574	7580	39.25
SAWLOGS NONCONIFEROUS	7170	6810	6509	5286	4762	5431	4634	4630	4310	4703	4732	-4.52
PULPWOOD+PARTICLE	728	1069	843	1957	1192	2005	1563	1912	2752	2719	1946	11.95
SAWNWOOD CONIFEROUS	29	29	31	10	6	11	15	9	8	22	33	-3.20
SAWNWOOD NONCONIFEROUS	56	96	139	197	293	423	519	529	800	1158	1160	34.88
WOOD-BASED PANELS	24	36	51	260	287	314	710	551	592	1280	990	47.28
PULP FOR PAPER	208	210	427	525	440	683	672	737	823	1088	1267	18.52
PAPER AND PAPERBOARD	411	427	650	662	510	678	634	987	1401	1641	1206	13.66

1/ THOUSAND HEAD
2/ EXCEPT FOR PULP FOR PAPER AND PAPER AND PAPERBOARD, ALL FOREST PRODUCTS ARE EXPRESSED IN THOUSAND CUBIC METRES

7. INDICES OF VALUE OF EXPORTS OF AGRICULTURAL, FISHERY AND FOREST PRODUCTS

	1979	1980	1981	1982	1983	1984	1985	1986	1987	1988	1989	ANNUAL RATE OF CHANGE 1979-89
	....	....	....	....	....	1979-81=100	....	....	....	....	....	PERCENT
WORLD												
AGRICULTURAL PRODUCTS	91	105	105	95	93	98	92	99	108	122	128	2.29
FOOD	88	104	107	96	92	97	90	96	105	119	125	2.00
FEED	85	101	114	108	114	100	83	103	116	142	140	3.23
RAW MATERIALS	97	103	100	93	93	100	93	95	114	137	146	3.36
BEVERAGES	105	110	85	88	89	103	106	132	113	118	116	2.39
FOREST PRODUCTS	94	107	98	89	91	97	96	111	140	164	165	5.64
DEVELOPED COUNTRIES												
AGRICULTURAL PRODUCTS	89	105	106	96	92	95	88	96	110	125	133	2.52
FOOD	87	105	107	96	91	93	85	93	106	120	127	2.04
FEED	88	102	110	108	116	96	84	98	110	120	114	1.32
RAW MATERIALS	98	102	100	98	94	102	98	100	122	146	156	4.15
BEVERAGES	101	103	96	96	93	98	107	134	149	162	165	5.97
FOREST PRODUCTS	93	107	100	91	92	99	98	115	144	169	169	6.01
WESTERN EUROPE												
AGRICULTURAL PRODUCTS	92	105	103	97	94	96	97	119	140	150	157	5.08
FOOD	90	106	104	96	92	94	95	117	137	148	156	4.98
FEED	86	96	118	128	133	118	107	105	131	125	133	2.72
RAW MATERIALS	105	100	96	96	97	104	106	124	151	158	160	5.54
BEVERAGES	101	103	96	95	93	98	109	139	156	169	174	6.62
FOREST PRODUCTS	93	109	98	89	88	96	97	120	157	183	183	7.12
USSR AND EASTERN EUROPE												
AGRICULTURAL PRODUCTS	98	102	100	94	86	81	80	87	93	101	99	-.41
FOOD	99	102	99	90	80	79	77	81	84	93	93	-1.30
FEED	113	96	91	116	106	71	95	105	199	154	100	3.06
RAW MATERIALS	93	103	104	104	96	83	86	98	110	121	119	1.64
BEVERAGES	104	103	94	97	103	96	94	106	113	113	95	.53
FOREST PRODUCTS	97	104	99	97	100	100	99	109	113	113	113	1.57
NORTH AMERICA DEVELOPED												
AGRICULTURAL PRODUCTS	86	104	110	95	94	99	76	69	76	98	104	-1.15
FOOD	84	104	112	95	94	98	73	63	68	91	96	-2.21
FEED	88	107	105	93	107	83	69	95	96	119	102	.45
RAW MATERIALS	95	104	101	96	92	109	96	83	108	130	143	2.65
BEVERAGES	94	110	95	91	79	91	88	123	105	127	125	2.74
FOREST PRODUCTS	93	105	102	91	96	104	100	111	140	169	169	5.89
OCEANIA DEVELOPED												
AGRICULTURAL PRODUCTS	82	107	111	103	88	91	95	94	98	123	142	2.58
FOOD	77	109	114	104	88	90	94	89	87	96	113	.32
FEED	127	74	100	103	95	73	60	74	103	144	145	2.16
RAW MATERIALS	95	103	102	99	90	93	99	105	124	184	208	6.61
BEVERAGES	74	94	132	139	157	177	158	182	283	527	607	20.23
FOREST PRODUCTS	83	106	111	94	83	89	85	87	114	128	128	2.60

7. (Cont.) INDICES OF VALUE OF EXPORTS OF AGRICULTURAL, FISHERY AND FOREST PRODUCTS

	1979	1980	1981	1982	1983	1984	1985	1986	1987	1988	1989	ANNUAL RATE OF CHANGE 1979-89
	1979-81=100											PERCENT
DEVELOPING COUNTRIES												
AGRICULTURAL PRODUCTS	94	104	102	92	94	106	100	105	103	117	118	1.79
FOOD	90	102	108	95	95	108	103	102	104	116	119	1.85
FEED	83	100	117	109	112	104	82	109	122	163	164	4.87
RAW MATERIALS	97	104	99	85	92	97	87	88	101	124	129	2.00
BEVERAGES	107	113	80	85	87	106	105	131	94	95	90	-.01
FOREST PRODUCTS	101	110	89	82	88	83	84	92	121	143	143	3.61
AFRICA DEVELOPING												
AGRICULTURAL PRODUCTS	105	105	90	83	78	90	88	105	93	93	89	-.60
FOOD	102	106	92	81	73	84	81	95	98	90	84	-.91
FEED	132	92	76	79	82	52	46	47	51	77	83	-5.17
RAW MATERIALS	92	100	108	92	93	106	92	96	110	131	141	3.08
BEVERAGES	114	106	80	84	81	95	98	126	78	80	76	-2.00
FOREST PRODUCTS	91	124	85	71	68	69	66	73	82	94	94	-.93
LATIN AMERICA												
AGRICULTURAL PRODUCTS	93	104	102	91	97	105	101	102	91	104	102	.32
FOOD	90	101	109	92	99	108	105	95	89	102	102	.19
FEED	79	95	125	104	122	109	85	104	115	171	169	5.12
RAW MATERIALS	95	100	104	93	82	85	81	61	67	87	96	-2.46
BEVERAGES	107	116	77	86	87	102	104	133	95	94	85	-.27
FOREST PRODUCTS	81	111	109	90	96	115	98	107	130	181	181	6.46
NEAR EAST DEVELOPING												
AGRICULTURAL PRODUCTS	88	100	112	110	104	110	96	104	114	127	130	2.54
FOOD	80	99	120	124	116	119	105	122	144	159	158	5.29
FEED	94	121	85	60	49	68	21	38	54	81	67	-5.67
RAW MATERIALS	100	100	100	91	89	98	85	78	69	77	86	-2.90
BEVERAGES	103	84	112	72	73	81	62	52	44	60	118	-3.89
FOREST PRODUCTS	78	86	136	152	164	205	156	134	140	148	148	4.78
FAR EAST DEVELOPING												
AGRICULTURAL PRODUCTS	91	105	104	92	94	110	95	99	107	129	136	2.84
FOOD	87	100	112	100	94	118	102	96	105	129	141	2.94
FEED	87	105	108	114	99	98	78	105	115	126	143	2.76
RAW MATERIALS	95	109	95	75	91	91	79	89	106	136	136	2.93
BEVERAGES	95	115	90	81	95	137	121	132	108	114	108	2.18
FOREST PRODUCTS	109	107	84	81	89	76	83	90	125	143	143	3.56
ASIAN CENT PLANNED ECON												
AGRICULTURAL PRODUCTS	92	105	103	100	106	125	140	168	186	205	210	9.39
FOOD	90	106	104	96	95	112	132	155	162	169	181	7.52
FEED	20	103	177	152	198	161	165	348	452	711	526	29.33
RAW MATERIALS	110	102	87	114	145	172	176	206	263	295	305	13.59
BEVERAGES	106	105	89	107	116	151	141	164	179	195	202	8.39
FOREST PRODUCTS	101	96	103	87	93	85	84	95	118	139	139	3.28

8. INDICES OF VOLUME OF EXPORTS OF AGRICULTURAL, FISHERY AND FOREST PRODUCTS

	1979	1980	1981	1982	1983	1984	1985	1986	1987	1988	1989	ANNUAL RATE OF CHANGE 1979-89
	1979-81=100											PERCENT
WORLD												
AGRICULTURAL PRODUCTS	92	101	106	105	105	109	108	107	114	117	120	1.94
FOOD	92	102	106	104	104	108	106	104	111	114	116	1.61
FEED	89	99	113	121	119	116	124	126	134	147	154	4.65
RAW MATERIALS	100	101	99	97	99	102	102	108	116	111	118	1.75
BEVERAGES	100	98	103	104	103	110	115	105	111	111	117	1.52
FOREST PRODUCTS	101	102	97	93	103	107	109	114	124	135	135	3.45
DEVELOPED COUNTRIES												
AGRICULTURAL PRODUCTS	91	103	106	103	102	106	101	100	111	112	114	1.36
FOOD	91	104	106	102	101	106	100	98	109	111	113	1.29
FEED	91	99	109	119	125	109	114	116	131	125	121	2.50
RAW MATERIALS	100	102	99	99	98	101	101	102	110	108	114	1.20
BEVERAGES	100	96	104	105	103	111	115	105	106	110	114	1.24
FOREST PRODUCTS	100	102	98	95	104	110	112	117	126	137	137	3.68
WESTERN EUROPE												
AGRICULTURAL PRODUCTS	91	99	110	109	116	125	132	136	143	142	150	4.92
FOOD	91	100	109	106	114	124	131	139	145	146	153	5.28
FEED	90	90	119	151	151	141	160	126	149	126	140	3.59
RAW MATERIALS	98	98	104	100	109	121	124	126	137	127	134	3.72
BEVERAGES	101	94	105	106	102	111	117	106	107	110	116	1.38
FOREST PRODUCTS	101	100	99	97	107	116	118	121	132	144	144	4.30
USSR AND EASTERN EUROPE												
AGRICULTURAL PRODUCTS	104	98	98	100	96	95	100	97	103	109	101	.38
FOOD	103	99	99	100	96	98	105	103	107	115	110	1.24
FEED	112	96	92	105	104	71	131	133	238	165	79	3.65
RAW MATERIALS	97	101	102	106	101	88	96	100	104	110	105	.60
BEVERAGES	99	100	101	104	113	113	107	96	102	97	84	-.98
FOREST PRODUCTS	103	100	97	98	102	104	104	112	113	115	115	1.67
NORTH AMERICA DEVELOPED												
AGRICULTURAL PRODUCTS	92	103	104	100	98	99	80	75	91	97	96	-1.16
FOOD	92	103	105	100	98	99	79	72	89	95	94	-1.45
FEED	90	106	104	99	111	92	88	113	123	128	112	2.19
RAW MATERIALS	98	105	97	96	91	95	90	86	93	95	102	-.53
BEVERAGES	87	111	102	98	95	96	104	113	116	134	147	3.84
FOREST PRODUCTS	99	104	98	92	102	105	106	114	126	138	138	3.74
OCEANIA DEVELOPED												
AGRICULTURAL PRODUCTS	78	123	99	105	94	107	140	139	135	119	112	3.31
FOOD	75	125	100	106	93	108	142	141	135	118	110	3.33
FEED	150	67	83	102	98	67	74	96	124	128	115	1.91
RAW MATERIALS	110	98	93	100	106	100	110	122	134	127	129	3.15
BEVERAGES	86	100	115	113	160	165	157	191	274	382	338	15.54
FOREST PRODUCTS	94	105	101	87	88	89	88	89	93	104	104	.27

8. (Cont.) INDICES OF VOLUME OF EXPORTS OF AGRICULTURAL, FISHERY AND FOREST PRODUCTS

	1979	1980	1981	1982	1983	1984	1985	1986	1987	1988	1989	ANNUAL RATE OF CHANGE 1979-89
	1979-81=100											PERCENT
DEVELOPING COUNTRIES												
AGRICULTURAL PRODUCTS	95	97	108	110	114	116	125	124	121	131	135	3.33
FOOD	97	96	107	108	115	116	126	123	117	123	124	2.61
FEED	87	98	115	123	115	120	131	133	136	163	178	6.01
RAW MATERIALS	101	101	99	94	101	102	103	115	123	116	123	2.40
BEVERAGES	99	100	101	103	102	109	115	107	117	113	120	1.91
FOREST PRODUCTS	107	102	91	87	94	94	93	98	113	123	123	2.18
AFRICA DEVELOPING												
AGRICULTURAL PRODUCTS	100	101	99	101	96	86	94	98	96	94	98	-.52
FOOD	98	103	99	102	98	86	94	99	97	93	95	-.67
FEED	129	95	76	89	97	56	73	69	70	89	93	-2.65
RAW MATERIALS	100	98	102	95	96	105	104	116	120	128	147	3.67
BEVERAGES	104	94	102	104	88	88	100	98	86	89	93	-1.15
FOREST PRODUCTS	108	106	86	85	84	93	86	78	78	82	82	-2.56
LATIN AMERICA												
AGRICULTURAL PRODUCTS	98	92	111	104	121	116	127	109	106	114	113	1.45
FOOD	100	89	111	103	120	114	126	105	98	105	99	.32
FEED	85	98	117	113	132	129	142	134	139	157	172	5.91
RAW MATERIALS	103	102	95	85	83	80	89	70	74	80	87	-2.61
BEVERAGES	100	102	98	103	110	117	122	104	135	118	128	2.66
FOREST PRODUCTS	92	105	103	94	113	128	117	123	120	156	156	4.89
NEAR EAST DEVELOPING												
AGRICULTURAL PRODUCTS	88	97	115	136	137	135	112	125	145	193	173	6.08
FOOD	85	96	119	143	143	141	118	131	156	209	187	7.05
FEED	101	122	77	53	54	67	20	44	53	88	79	-4.47
RAW MATERIALS	107	99	94	99	100	105	88	101	82	77	75	-2.96
BEVERAGES	102	85	112	91	89	100	82	65	44	52	104	-4.90
FOREST PRODUCTS	90	85	124	138	163	219	173	166	188	189	189	8.08
FAR EAST DEVELOPING												
AGRICULTURAL PRODUCTS	94	100	106	113	108	119	119	131	130	137	156	4.44
FOOD	92	101	106	118	110	128	122	140	133	136	157	4.58
FEED	88	99	113	132	101	115	123	120	118	143	174	4.66
RAW MATERIALS	98	101	101	97	105	106	107	124	136	128	143	3.94
BEVERAGES	89	105	106	100	100	117	120	121	119	129	131	3.37
FOREST PRODUCTS	111	101	88	84	90	84	87	95	117	122	122	2.11
ASIAN CENT PLANNED ECON												
AGRICULTURAL PRODUCTS	96	110	94	99	109	123	178	214	201	199	191	9.64
FOOD	102	111	87	93	99	113	171	185	165	154	162	7.05
FEED	23	97	180	165	205	190	220	470	512	658	450	28.88
RAW MATERIALS	106	98	96	121	190	244	264	477	601	453	308	19.97
BEVERAGES	105	105	91	112	130	145	136	162	168	189	191	7.55
FOREST PRODUCTS	107	93	101	93	95	84	88	90	100	102	102	-.03

9. INDICES OF VALUE OF IMPORTS OF AGRICULTURAL, FISHERY AND FOREST PRODUCTS

	1979	1980	1981	1982	1983	1984	1985	1986	1987	1988	1989	ANNUAL RATE OF CHANGE 1979-89
	......1979-81=100										......	PERCENT
WORLD												
AGRICULTURAL PRODUCTS	91	104	104	96	92	97	93	99	108	121	126	2.19
FOOD	89	104	107	97	92	96	92	96	105	119	126	2.01
FEED	88	100	113	109	113	100	88	101	120	148	137	3.36
RAW MATERIALS	98	104	98	91	90	100	95	95	115	134	137	3.06
BEVERAGES	104	109	87	88	87	97	99	125	108	111	110	1.70
FOREST PRODUCTS	95	108	97	94	91	97	96	111	140	158	158	5.19
DEVELOPED COUNTRIES												
AGRICULTURAL PRODUCTS	95	104	101	94	90	95	93	103	113	124	126	2.47
FOOD	93	104	103	95	89	94	92	101	115	126	130	2.70
FEED	88	100	112	107	110	93	83	96	116	138	126	2.46
RAW MATERIALS	102	103	96	90	90	99	95	93	110	124	125	2.14
BEVERAGES	104	110	86	87	86	95	98	125	107	109	110	1.61
FOREST PRODUCTS	98	108	94	91	88	94	94	111	142	161	161	5.38
WESTERN EUROPE												
AGRICULTURAL PRODUCTS	99	106	96	93	88	90	91	109	124	134	132	3.26
FOOD	98	106	96	93	87	87	89	108	128	137	138	3.62
FEED	89	101	110	110	105	95	85	100	113	127	111	1.55
RAW MATERIALS	104	104	91	89	88	99	102	101	116	128	128	2.75
BEVERAGES	104	110	86	86	86	92	95	130	117	122	119	2.71
FOREST PRODUCTS	94	110	96	89	84	88	87	112	146	168	168	5.95
USSR AND EASTERN EUROPE												
AGRICULTURAL PRODUCTS	84	102	114	99	94	97	90	82	84	94	101	-.66
FOOD	80	101	119	101	90	97	90	79	78	86	95	-1.38
FEED	82	94	124	100	133	78	64	68	127	170	187	4.70
RAW MATERIALS	98	105	98	92	101	98	94	89	97	107	104	.21
BEVERAGES	96	112	92	90	94	102	102	107	97	96	107	.43
FOREST PRODUCTS	87	106	107	99	90	90	93	88	91	86	86	-1.42
NORTH AMERICA DEVELOPED												
AGRICULTURAL PRODUCTS	98	102	100	92	86	104	103	110	108	112	115	1.80
FOOD	94	100	106	94	90	109	110	111	118	124	129	3.05
FEED	104	93	103	91	114	142	118	128	143	206	200	7.66
RAW MATERIALS	99	95	105	90	79	99	79	81	92	103	112	.24
BEVERAGES	106	109	85	89	80	95	97	119	92	87	86	-.85
FOREST PRODUCTS	103	97	100	102	115	134	137	143	167	189	189	7.67
OCEANIA DEVELOPED												
AGRICULTURAL PRODUCTS	90	106	104	111	99	118	117	115	113	132	161	3.94
FOOD	91	103	106	128	111	136	134	129	128	154	198	5.86
FEED	48	87	165	75	338	107	192	171	309	164	372	15.39
RAW MATERIALS	88	107	105	90	78	90	82	78	85	103	111	.12
BEVERAGES	90	114	96	92	87	103	111	116	101	105	116	1.63
FOREST PRODUCTS	85	104	110	122	83	103	117	115	130	142	142	4.20

9. (Cont.) INDICES OF VALUE OF IMPORTS OF AGRICULTURAL, FISHERY AND FOREST PRODUCTS

	1979	1980	1981	1982	1983	1984	1985	1986	1987	1988	1989	ANNUAL RATE OF CHANGE 1979-89
	......					1979-81=100					......	PERCENT
DEVELOPING COUNTRIES												
AGRICULTURAL PRODUCTS	81	105	114	102	98	103	93	88	93	113	126	1.41
FOOD	79	105	117	104	98	102	90	83	85	102	117	.34
FEED	81	100	119	123	141	163	132	154	164	253	243	10.09
RAW MATERIALS	89	106	106	93	91	102	97	102	131	160	171	5.43
BEVERAGES	104	99	97	89	96	111	108	116	118	125	118	2.57
FOREST PRODUCTS	84	104	112	108	105	106	103	112	133	147	146	4.39
AFRICA DEVELOPING												
AGRICULTURAL PRODUCTS	81	105	114	99	93	92	92	84	80	89	104	-.76
FOOD	79	106	115	101	93	92	91	82	74	85	101	-1.28
FEED	78	98	124	115	92	123	114	152	179	224	255	10.60
RAW MATERIALS	95	100	105	103	101	105	110	104	111	121	121	2.06
BEVERAGES	93	105	102	71	86	77	82	87	120	89	102	.59
FOREST PRODUCTS	82	94	124	108	111	98	94	92	94	102	102	.11
LATIN AMERICA												
AGRICULTURAL PRODUCTS	78	110	112	87	84	86	76	76	71	85	95	-1.58
FOOD	74	111	114	87	85	85	76	73	66	77	91	-2.26
FEED	82	103	115	118	121	123	93	107	127	221	145	5.35
RAW MATERIALS	95	106	99	82	73	101	90	96	113	134	135	3.35
BEVERAGES	130	89	80	75	52	51	52	76	63	67	68	-4.52
FOREST PRODUCTS	75	111	115	114	89	83	79	86	92	95	90	-.95
NEAR EAST DEVELOPING												
AGRICULTURAL PRODUCTS	76	101	123	117	113	127	110	99	101	115	129	2.05
FOOD	73	102	124	118	111	125	106	94	95	108	125	1.43
FEED	83	87	130	135	199	240	250	239	222	377	345	15.65
RAW MATERIALS	90	95	116	106	124	132	122	120	144	139	151	4.67
BEVERAGES	99	102	99	105	115	151	131	122	121	162	132	4.05
FOREST PRODUCTS	78	103	119	116	107	111	108	106	111	109	109	1.37
FAR EAST DEVELOPING												
AGRICULTURAL PRODUCTS	86	103	112	100	104	113	104	99	111	146	158	4.12
FOOD	84	103	113	100	103	107	99	89	96	127	135	2.25
FEED	84	105	111	124	141	166	102	162	170	233	277	10.15
RAW MATERIALS	92	101	107	98	100	121	113	116	157	208	231	8.64
BEVERAGES	95	97	108	112	136	182	195	213	176	169	171	7.82
FOREST PRODUCTS	92	104	104	101	111	118	115	133	174	205	205	8.36
ASIAN CENT PLANNED ECON												
AGRICULTURAL PRODUCTS	86	106	108	102	83	73	61	68	91	120	134	1.03
FOOD	88	101	111	110	89	75	58	63	83	109	130	-.19
FEED	51	94	155	99	121	185	204	94	115	205	258	10.26
RAW MATERIALS	81	117	102	83	70	67	66	78	108	139	140	3.15
BEVERAGES	94	94	112	67	117	93	118	125	207	368	274	13.46
FOREST PRODUCTS	77	109	114	111	147	160	168	204	264	276	276	13.57

10. INDICES OF VOLUME OF IMPORTS OF AGRICULTURAL, FISHERY AND FOREST PRODUCTS

	1979	1980	1981	1982	1983	1984	1985	1986	1987	1988	1989	ANNUAL RATE OF CHANGE 1979-89
	....					1979-81=100					....	PERCENT
WORLD												
AGRICULTURAL PRODUCTS	94	100	105	106	105	109	109	106	114	117	118	1.86
FOOD	94	101	105	105	103	108	106	104	111	114	116	1.54
FEED	93	96	111	125	121	116	131	128	141	153	141	4.45
RAW MATERIALS	99	102	99	96	98	103	104	108	116	112	117	1.81
BEVERAGES	100	98	102	103	103	108	112	107	110	110	117	1.52
FOREST PRODUCTS	102	101	97	95	101	106	108	115	126	132	132	3.33
DEVELOPED COUNTRIES												
AGRICULTURAL PRODUCTS	97	99	104	105	100	104	107	104	109	112	112	1.25
FOOD	97	99	103	102	97	102	104	100	104	107	109	.84
FEED	94	95	111	125	119	111	127	122	135	142	129	3.50
RAW MATERIALS	102	100	98	96	98	104	106	105	108	104	105	.79
BEVERAGES	100	98	102	103	102	107	112	107	110	109	116	1.44
FOREST PRODUCTS	104	101	95	92	99	105	107	115	125	132	132	3.31
WESTERN EUROPE												
AGRICULTURAL PRODUCTS	99	100	101	108	101	100	107	108	113	114	111	1.37
FOOD	100	101	98	102	97	96	101	103	108	109	108	.94
FEED	94	95	111	128	115	113	128	124	132	133	118	2.78
RAW MATERIALS	104	100	96	95	97	103	108	111	112	104	106	1.05
BEVERAGES	102	97	102	103	101	102	109	106	111	113	120	1.67
FOREST PRODUCTS	103	101	96	95	102	106	106	117	127	137	137	3.67
USSR AND EASTERN EUROPE												
AGRICULTURAL PRODUCTS	89	97	114	105	94	103	102	83	91	98	103	-.27
FOOD	88	96	116	107	91	106	105	83	88	95	100	-.55
FEED	86	99	115	114	156	86	97	92	159	186	199	6.39
RAW MATERIALS	98	102	101	96	108	106	103	93	95	92	87	-1.12
BEVERAGES	95	103	102	99	98	105	108	85	85	85	89	-1.66
FOREST PRODUCTS	93	105	103	95	90	90	95	87	85	84	84	-1.84
NORTH AMERICA DEVELOPED												
AGRICULTURAL PRODUCTS	100	96	104	99	101	117	121	123	121	124	134	3.29
FOOD	99	96	105	98	99	115	119	122	118	122	134	3.16
FEED	106	94	100	100	126	164	173	167	188	231	206	9.66
RAW MATERIALS	105	90	106	93	97	111	113	106	112	112	126	2.19
BEVERAGES	99	100	101	106	108	120	126	125	127	115	123	2.60
FOREST PRODUCTS	106	97	96	88	107	122	130	136	144	145	145	5.07
OCEANIA DEVELOPED												
AGRICULTURAL PRODUCTS	94	107	99	114	120	128	128	118	132	148	175	5.02
FOOD	94	109	97	119	122	139	133	121	136	161	191	5.82
FEED	56	94	150	85	393	82	298	259	461	205	375	17.75
RAW MATERIALS	99	103	98	91	84	87	95	86	87	90	93	-1.05
BEVERAGES	94	104	102	111	102	105	112	111	102	118	125	1.87
FOREST PRODUCTS	96	100	104	113	83	100	125	117	115	128	128	2.95

10. (Cont.) INDICES OF VOLUME OF IMPORTS OF AGRICULTURAL, FISHERY AND FOREST PRODUCTS

	1979	1980	1981	1982	1983	1984	1985	1986	1987	1988	1989	ANNUAL RATE OF CHANGE 1979-89
	....	....	....	....	....	1979-81=100	....	....	....	....	....	PERCENT
DEVELOPING COUNTRIES												
AGRICULTURAL PRODUCTS	89	103	108	109	114	118	111	112	124	127	132	3.00
FOOD	89	103	108	109	114	117	110	110	121	124	128	2.64
FEED	86	100	114	132	150	180	188	212	221	292	291	13.11
RAW MATERIALS	92	106	102	98	97	100	98	115	133	130	145	3.94
BEVERAGES	100	96	105	103	108	114	112	109	116	120	123	2.19
FOREST PRODUCTS	93	101	106	104	111	112	111	116	130	134	133	3.49
AFRICA DEVELOPING												
AGRICULTURAL PRODUCTS	90	102	108	110	109	117	123	113	110	115	121	2.02
FOOD	90	103	107	110	109	117	122	112	109	114	119	1.84
FEED	80	96	124	129	115	148	152	246	239	250	273	13.06
RAW MATERIALS	97	102	101	102	104	109	116	117	125	122	128	2.87
BEVERAGES	90	96	114	99	110	109	118	106	118	99	121	1.74
FOREST PRODUCTS	90	93	117	105	116	106	101	95	90	98	98	-.47
LATIN AMERICA												
AGRICULTURAL PRODUCTS	81	111	108	92	106	102	93	90	93	94	94	-.54
FOOD	81	111	108	91	106	101	91	87	89	88	90	-1.05
FEED	82	109	109	126	127	133	136	139	171	227	158	7.40
RAW MATERIALS	96	106	98	91	87	109	109	143	144	146	158	5.66
BEVERAGES	114	93	92	88	68	66	65	71	74	72	77	-3.72
FOREST PRODUCTS	84	110	106	99	88	83	78	84	89	87	83	-1.75
NEAR EAST DEVELOPING												
AGRICULTURAL PRODUCTS	87	99	114	122	128	151	137	135	147	135	151	4.68
FOOD	87	99	114	122	127	150	135	134	146	133	149	4.52
FEED	92	89	118	146	186	260	270	269	255	312	315	14.64
RAW MATERIALS	87	96	117	113	130	129	124	117	144	123	138	3.67
BEVERAGES	100	95	104	117	119	137	119	115	114	147	123	2.73
FOREST PRODUCTS	86	101	113	121	122	134	129	108	112	120	120	2.02
FAR EAST DEVELOPING												
AGRICULTURAL PRODUCTS	92	104	104	107	120	120	119	127	134	158	155	4.98
FOOD	92	105	102	107	120	118	118	121	125	150	142	4.09
FEED	88	101	111	138	159	184	163	240	256	336	365	15.21
RAW MATERIALS	96	100	104	101	106	118	113	132	159	158	177	6.33
BEVERAGES	94	96	109	118	137	165	161	169	169	172	197	7.73
FOREST PRODUCTS	101	98	101	99	116	119	122	142	169	175	175	6.91
ASIAN CENT PLANNED ECON												
AGRICULTURAL PRODUCTS	97	99	104	112	102	83	71	81	125	133	128	1.95
FOOD	98	97	105	113	102	83	68	80	126	133	125	1.73
FEED	59	90	152	102	131	239	530	190	141	367	482	18.40
RAW MATERIALS	87	119	94	88	77	64	65	85	104	97	115	.61
BEVERAGES	96	100	103	141	249	117	146	185	258	380	326	13.75
FOREST PRODUCTS	87	104	110	106	145	155	173	206	257	243	243	12.15

11. THE IMPORTANCE OF AGRICULTURE IN THE ECONOMY

COUNTRY	AGRIC.POPULATION AS % TOTAL POPULATION 1988	AGRIC.EXPORTS AS % TOTAL EXPORTS 1988	AGRIC.IMPORTS AS % TOTAL IMPORTS 1988	SHARE OF TOTAL IMPORTS FINANCED BY AGR.EXPORTS % 1988
ALGERIA	25		31	
ANGOLA	71	1	20	2
BENIN	63	65	21	21
BOTSWANA	65	3	14	4
BURKINA FASO	85	63	17	21
BURUNDI	92	97	12	61
REPUBLIC OF CAMEROON	63	56	16	41
CAPE VERDE	45	41	27	1
CENTRAL AFRICAN REPUBLIC	65	29	8	15
CHAD	77	79	5	27
COMOROS	80	85	32	35
CONGO	60	1	15	2
COTE D¢IVOIRE	58	74	20	86
DJIBOUTI	78	26	32	2
EGYPT	42	12	22	3
EQUATORIAL GUINEA	58	31	15	24
ETHIOPIA	76	82	30	30
GABON	69		17	1
GAMBIA	82	27	48	12
GHANA	51	58	13	43
GUINEA	76	5	18	5
GUINEA-BISSAU	80	36	45	14
KENYA	78	69	7	37
LESOTHO	81	16	20	2
LIBERIA	71	29	25	41
LIBYAN ARAB JAMAHIRIYA	14		23	
MADAGASCAR	77	53	19	40
MALAWI	77	92	6	66
MALI	82	77	17	39
MAURITANIA	65	7	55	13
MAURITIUS	24	36	11	28
MOROCCO	38	16	15	12
MOZAMBIQUE	82	56	24	7
NAMIBIA	37			
NIGER	88	15	22	13
NIGERIA	65	4	9	6
REUNION	12	82	19	10
RWANDA	92	98	11	29
SAO TOME AND PRINCIPE	65	73	28	29
SENEGAL	79	19	25	14
SEYCHELLES	78	2	15	
SIERRA LEONE	64	28	60	20
SOMALIA	72	83	26	16
SOUTH AFRICA	16	7	5	8
SUDAN	63	95	26	45
SWAZILAND	68	42	11	37
UNITED REP. OF TANZANIA	80	68	7	22
TOGO	70	30	29	26
TUNISIA	26	8	18	5
UGANDA	82	98	5	66
ZAIRE	67	15	26	22
ZAMBIA	70	1	4	2
ZIMBABWE	69	35	4	50
BARBADOS	7	28	19	9
BELIZE	34	56	21	38
BERMUDA	3		16	
CANADA	4	8	6	8
COSTA RICA	26	58	10	54
CUBA	20	85	11	58
DOMINICA	29	79	18	58
DOMINICAN REPUBLIC	38	45	13	22
EL SALVADOR	39	64	14	38
GRENADA	29	74	22	26
GUADELOUPE	11	84	21	11
GUATEMALA	52	68	8	51
HAITI	61	29	45	16
HONDURAS	58	69	13	66
JAMAICA	31	25	16	14
MARTINIQUE	9	70	19	12
MEXICO	31	12	15	12
NICARAGUA	40	92	14	25
PANAMA	26	39	15	19
TRINIDAD AND TOBAGO	8	6	19	7
UNITED STATES OF AMERICA	3	13	5	9

1 1. (Cont.) THE IMPORTANCE OF AGRICULTURE IN THE ECONOMY

COUNTRY	AGRIC.POPULATION AS % TOTAL POPULATION 1988	AGRIC.EXPORTS AS % TOTAL EXPORTS 1988	AGRIC.IMPORTS AS % TOTAL IMPORTS 1988	SHARE OF TOTAL IMPORTS FINANCED BY AGR.EXPORTS % 1988
ARGENTINA	11	62	5	104
BOLIVIA	43	10	19	10
BRAZIL	26	28	7	59
CHILE	14	13	6	19
COLOMBIA	29	47	9	48
ECUADOR	32	28	8	36
FRENCH GUIANA	27	4	19	1
GUYANA	23	39	14	41
PARAGUAY	48	91	9	84
PERU	38	10	21	9
SURINAME	17	15	13	14
URUGUAY	14	42	9	51
VENEZ@ELA	11	1	14	1
AFGHANISTAN	56	45	26	31
BANGLADESH	70	12	31	5
BHUTAN	91	8	11	4
BRUNEI DARUSSALAM	54	1	21	1
MYANMAR	48	18	8	12
CHINA (MAINLAND)	69	16	10	14
CYPRUS	22	28	11	11
HONG KONG	1	5	9	5
INDIA	63	17	12	12
INDONESIA	46	17	10	25
IRAN, ISLAMIC REP. OF	28	4	20	4
IRAQ	22	1	21	1
ISRAEL	5	10	8	8
JAPAN	7		14	1
JORDAN	7	9	21	4
CAMBODIA	71	83	23	12
KOREA, DEM. POP. REP.	35	7	14	5
KOREA, REPUBLIC OF	24	2	10	2
KUWAIT	1	1	20	1
LAO POP. DEM. REP.	72	4	15	1
LEBANON	10	24	33	7
MALAYSIA	32	24	11	31
MALDIVES	65		9	
MONGOLIA	32	20	5	13
NEPAL	92	37	17	12
OMAN	42	1	20	2
PAKISTAN	54	28	17	19
PHILIPPINES	47	17	11	14
QATAR	2		20	
SAUDI ARABIA	41	2	17	2
SINGAPORE	1	6	7	5
SRI LANKA	52	43	19	28
SYRIAN ARAB REPUBLIC	25	13	20	8
THAILAND	62	32	5	26
TURKEY	46	26	7	21
UNITED ARAB EMIRATES	3	1	15	2
VIET NAM	62	18	9	8
YEMEN ARAB REPUBLIC	64	32	33	1
YEMEN, DEMOCRATIC	34	3	13	1

11. (Cont.) THE IMPORTANCE OF AGRICULTURE IN THE ECONOMY

COUNTRY	AGRIC.POPULATION AS % TOTAL POPULATION 1988	AGRIC.EXPORTS AS % TOTAL EXPORTS 1988	AGRIC.IMPORTS AS % TOTAL IMPORTS 1988	SHARE OF TOTAL IMPORTS FINANCED BY AGR.EXPORTS % 1988
AUSTRIA	5	4	7	3
BELGIUM-LUXEMBOURG	2	10	11	10
BULGARIA	13	10	7	10
CZECHOSLOVAKIA	10	3	9	3
DENMARK	5	25	11	26
FINLAND	9	3	6	3
FRANCE	5	16	10	15
GERMAN DEMOCRATIC REP.	9	2	8	2
GERMANY, FED. REP. OF	3	5	12	7
GREECE	23	30	19	15
HUNGARY	13	21	9	23
ICELAND	6	2	9	2
IRELAND	14	26	13	31
ITALY	7	7	15	6
MALTA	4	4	12	2
NETHERLANDS	4	23	16	24
NORWAY	6	2	6	2
POLAND	20	9	15	11
PORTUGAL	18	7	14	4
ROMANIA	19	5	4	6
SPAIN	11	17	10	11
SWEDEN	5	2	7	2
SWITZERLAND	4	3	7	3
UNITED KINGDOM	2	7	10	5
UNION OF SOV. SOC. REP	14	3	17	3
YUGOSLAVIA	21	9	10	9
AUSTRALIA	5	35	5	35
FIJI	40	45	15	35
FRENCH POLYNESIA	14	5	18	
KIRIBATI	14	66	38	17
NEW CALEDONIA	47		16	
NEW ZEALAND	10	58	7	62
PAPUA NEW GUINEA	69	20	12	21
SOLOMON ISLANDS	47	23	18	26
TOKELAU	14			
TONGA	14	43	25	6
VANUATU	47	65	16	19

12/A. RESOURCES AND THEIR USE IN AGRICULTURE

COUNTRY	ARABLE LAND AS % OF TOTAL LAND 1987	IRRIGATED LAND AS % OF ARABLE LAND 1987	FOREST LAND AS % OF TOTAL LAND 1987	AGRIC.POPULATION PER HA OF ARABLE LAND 1987	AGRIC.LAB.FORCE AS % OF AGRIC.POPULATION 1987
ALGERIA	3	5	2	.8	23
ANGOLA	3		43	1.8	42
BENIN	17		33	1.5	48
BOTSWANA	2		2	.6	34
BURKINA FASO	11	1	25	2.3	54
BURUNDI	52	5	3	3.4	53
REPUBLIC OF CAMEROON	15		53	1.0	39
CAPE VERDE	10	5		4.0	37
CENTRAL AFRICAN REPUBLIC	3		58	.9	49
CHAD	3		10	1.3	35
COMOROS	44		16	3.9	45
CONGO	2	1	62	1.6	40
COTE D¢IVOIRE	11	2	20	1.8	38
DJIBOUTI					46
EGYPT	3	100		8.2	27
EQUATORIAL GUINEA	8		46	1.0	42
ETHIOPIA	13	1	25	2.4	43
GABON	2		78	1.6	45
GAMBIA	17	7	17	3.8	46
GHANA	12		36	2.5	37
GUINEA	6	4	41	3.1	46
GUINEA-BISSAU	12		38	2.2	47
KENYA	4	2	6	7.2	41
LESOTHO	11			4.2	47
LIBERIA	4	1	22	4.5	38
LIBYAN ARAB JAMAHIRIYA	1	11		.3	25
MADAGASCAR	5	29	25	2.8	44
MALAWI	25	1	46	2.5	44
MALI	2	10	7	3.4	32
MAURITANIA		6	15	6.2	32
MAURITIUS	58	16	31	2.4	39
MOROCCO	19	15	12	1.1	30
MOZAMBIQUE	4	3	19	3.9	53
NAMIBIA	1	1	22	1.0	30
NIGER	3	1	2	1.6	52
NIGERIA	34	3	16	2.1	38
REUNION	22	9	35	1.3	37
RWANDA	45		20	5.3	49
SAO TOME AND PRINCIPE	39			1.8	39
SENEGAL	27	3	31	1.0	44
SEYCHELLES	22		19	8.7	44
SIERRA LEONE	25	2	29	1.4	35
SOMALIA	1	12	14	5.3	40
SOUTH AFRICA	11	9	4	.4	32
SUDAN	5	15	20	1.2	32
SWAZILAND	10	38	6	3.0	41
UNITED REP. OF TANZANIA	6	3	48	3.8	49
TOGO	26		25	1.6	41
TUNISIA	30	6	4	.4	32
UGANDA	34		29	2.0	45
ZAIRE	3		77	3.3	38
ZAMBIA	7		39	1.0	32
ZIMBABWE	7	7	52	2.2	40
BARBADOS	77			.6	51
BELIZE	2	4	44	1.1	33
BERMUDA			20		49
CANADA	5	2	38		50
COSTA RICA	10	22	32	1.4	34
CUBA	30	26	25	.6	41
DOMINICA	23		41	1.4	47
DOMINICAN REPUBLIC	30	14	13	1.8	31
EL SALVADOR	35	16	5	2.6	31
GRENADA	38		9	2.2	47
GUADELOUPE	25	7	42	.9	45
GUATEMALA	17	4	37	2.4	28
HAITI	33	8	2	4.2	48
HONDURAS	16	5	31	1.5	30
JAMAICA	25	13	17	2.8	44
MARTINIQUE	19	30	25	1.5	46
MEXICO	13	21	23	1.1	34
NICARAGUA	11	7	31	1.1	31
PANAMA	8	5	52	1.0	36
TRINIDAD AND TOBAGO	23	18	43	.8	39
UNITED STATES OF AMERICA	21	10	29		44

12/A. (Cont.) RESOURCES AND THEIR USE IN AGRICULTURE

COUNTRY	ARABLE LAND AS % OF TOTAL LAND 1987	IRRIGATED LAND AS % OF ARABLE LAND 1987	FOREST LAND AS % OF TOTAL LAND 1987	AGRIC.POPULATION PER HA OF ARABLE LAND 1987	AGRIC.LAB.FORCE AS % OF AGRIC.POPULATION 1987
ARGENTINA	13	5	22	.1	36
BOLIVIA	3	5	51	.9	31
BRAZIL	9	3	66	.5	37
CHILE	7	23	12	.3	34
COLOMBIA	5	9	49	1.7	32
ECUADOR	10	21	43	1.2	30
FRENCH GUIANA			83	3.9	35
GUYANA	3	26	83	.5	36
PARAGUAY	5	3	39	.9	33
PERU	3	33	54	2.1	30
SURINAME		84	92	1.0	32
URUGUAY	8	7	4	.3	39
VENEZUELA	4	8	35	.6	36
AFGHANISTAN	12	33	3	1.0	30
BANGLADESH	68	24	16	8.2	29
BHUTAN	3	26	55	10.0	44
BRUNEI DARUSSALAM	1	14	48	18.8	42
MYANMAR	15	11	49	1.9	44
CHINA (MAINLAND)	10	46	12	7.7	59
CYPRUS	17	20	13	1.0	46
HONG KONG	8	38	12	10.0	52
INDIA	57	25	23	3.0	40
INDONESIA	12	35	67	3.8	42
IRAN, ISLAMIC REP. OF	9	39	11	1.0	30
IRAQ	12	32	4	.7	27
ISRAEL	22	63	5	.5	38
JAPAN	13	61	67	1.9	52
JORDAN	5	11	1	.5	23
CAMBODIA	17	3	76	1.8	48
KOREA, DEM. POP. REP.	20	49	74	3.2	45
KOREA, REPUBLIC OF	22	59	66	5.0	46
KUWAIT		25		7.1	38
LAO POP. DEM. REP.	4	13	56	3.1	48
LEBANON	29	29	8	.9	30
MALAYSIA	13	8	60	1.2	42
MALDIVES	10		3	42.5	36
MONGOLIA	1	3	10	.5	47
NEPAL	17	28	17	7.0	42
OMAN		85		12.0	29
PAKISTAN	27	77	4	2.9	27
PHILIPPINES	27	19	37	3.5	36
QATAR				1.7	45
SAUDI ARABIA	1	36	1	4.5	29
SINGAPORE	5		5	10.0	48
SRI LANKA	29	28	27	4.6	37
SYRIAN ARAB REPUBLIC	31	12	3	.5	25
THAILAND	39	20	28	1.7	55
TURKEY	36	8	26	.9	47
UNITED ARAB EMIRATES		26		2.4	51
VIET NAM	20	28	40	6.1	47
YEMEN ARAB REPUBLIC	7	18	8	3.5	24
YEMEN DEMOCRATIC		49	5	6.6	26

12/A. (Cont.) RESOURCES AND THEIR USE IN AGRICULTURE

COUNTRY	ARABLE LAND AS % OF TOTAL LAND 1987	IRRIGATED LAND AS % OF ARABLE LAND 1987	FOREST LAND AS % OF TOTAL LAND 1987	AGRIC.POPULATION PER HA OF ARABLE LAND 1987	AGRIC.LAB.FORCE AS % OF AGRIC.POPULATION 1987
AUSTRIA	18		39	.3	55
BELGIUM-LUXEMBOURG	25		21	.3	41
BULGARIA	37	30	35	.3	51
CZECHOSLOVAKIA	41	4	37	.3	53
DENMARK	61	16	12	.1	55
FINLAND	8	3	76	.2	50
FRANCE	35	6	27	.2	48
GERMAN DEMOCRATIC REP.	47	3	28	.3	57
GERMANY, FED. REP. OF	31	4	30	.3	57
GREECE	30	29	20	.6	44
HUNGARY	57	3	18	.3	48
ICELAND			1	2.0	61
IRELAND	14		5	.5	39
ITALY	41	25	23	.3	47
MALTA	41	8		1.1	37
NETHERLANDS	27	58	9	.7	41
NORWAY	3	11	27	.3	47
POLAND	48	1	29	.5	58
PORTUGAL	30	23	40	.7	43
ROMANIA	46	31	28	.4	57
SPAIN	41	16	31	.2	38
SWEDEN	7	4	68	.1	45
SWITZERLAND	10	6	26	.6	59
UNITED KINGDOM	29	2	10	.2	49
UNION OF SOV. SOC. REP	10	9	42	.2	50
YUGOSLAVIA	30	2	37	.7	50
AUSTRALIA	6	4	14		47
FIJI	13		65	1.2	34
FRENCH POLYNESIA	20		31	.3	32
KIRIBATI	52		3	.3	36
NEW CALEDONIA	1		38	3.8	31
NEW ZEALAND	2	51	27	.6	44
PAPUA NEW GUINEA	1		84	6.7	47
SOLOMON ISLANDS	2		91	2.4	33
TOKELAU					32
TONGA	67		11	.4	32
VANUATU	12		1	.5	31

12/B. RESOURCES AND THEIR USE IN AGRICULTURE

COUNTRY	AGRICULTURAL GFCF $ PER HA ARABLE LAND 1987	AGRICULTURAL GFCF $ PER CAPUT OF AGRIC.LAB.FORCE 1987	FERTILIZER USE PER HA ARAB.LAND KG/HA 1987	NOS. OF TRACTORS PER 000 HA ARABLE LAND 1987	OFFICIAL COMMITM. TO AGRICULTURE $ PER CAPUT 1988
ALGERIA			32	11	.6
ANGOLA			3	3	4.1
BENIN			5		20.2
BOTSWANA	3.1	17.2	1	2	25.3
BURKINA FASO	.3	.3	6		12.0
BURUNDI			2		5.6
REPUBLIC OF CAMEROON			7		16.2
CAPE VERDE			3		38.7
CENTRAL AFRICAN REPUBLIC					9.2
CHAD			2		17.1
COMOROS					12.1
CONGO			3	1	14.5
COTE D¢IVOIRE			9	1	12.4
DJIBOUTI					14.2
EGYPT	532.1	239.7	351	18	9.9
EQUATORIAL GUINEA					5.0
ETHIOPIA			4		3.5
GABON			5	3	6.8
GAMBIA			15		24.5
GHANA			4	1	5.8
GUINEA			1		13.3
GUINEA-BISSAU					14.4
KENYA	54.8	18.9	42	4	5.9
LESOTHO			13	5	31.2
LIBERIA			9	1	2.9
LIBYAN ARAB JAMAHIRIYA	400.0	5934.0	42	14	
MADAGASCAR			2	1	7.6
MALAWI			20	1	7.4
MALI			6		16.2
MAURITANIA			6	2	21.7
MAURITIUS	92.5	100.0	307	3	11.5
MOROCCO			38	4	17.1
MOZAMBIQUE			2	2	5.6
NAMIBIA				4	
NIGER			1		16.4
NIGERIA			9		.1
REUNION			261	38	
RWANDA			2		4.3
SAO TOME AND PRINCIPE				3	
SENEGAL			4		23.4
SEYCHELLES				6	1.5
SIERRA LEONE					1.6
SOMALIA			4	2	10.8
SOUTH AFRICA	38.6	282.7	54	14	
SUDAN	8.1	22.1	4	2	14.5
SWAZILAND	61.7	46.8	40	20	4.4
UNITED REP. OF TANZANIA	3.4	1.8	9	4	8.7
TOGO			8		4.1
TUNISIA	72.9	526.1	22	6	10.7
UGANDA				1	6.6
ZAIRE			1		2.4
ZAMBIA			18	1	11.6
ZIMBABWE	22.1	26.7	50	7	9.3
BARBADOS			94	18	2.4
BELIZE			74	19	101.0
CANADA	46.7	4386.9	48	16	
COSTA RICA	175.1	366.9	181	12	11.8
CUBA			197	18	.1
DOMINICA			176	5	41.9
DOMINICAN REPUBLIC			56	2	2.0
EL SALVADOR	12.7	15.1	126	5	6.9
GRENADA				2	38.7
GUADELOUPE			143	37	
GUATEMALA	35.0	51.5	66	2	8.5
HAITI			3	1	2.8
HONDURAS	61.8	162.2	19	2	21.1
JAMAICA			91	11	12.9
MARTINIQUE			1150	40	
MEXICO			75	7	10.9
NICARAGUA			43	2	13.4
PANAMA			66	11	
TRINIDAD AND TOBAGO	75.4	228.2	45	22	22.3
UNITED STATES OF AMERICA	65.9	3977.8	94	25	

12/B. (Cont.) RESOURCES AND THEIR USE IN AGRICULTURE

COUNTRY	AGRICULTURAL GFCF $ PER HA ARABLE LAND 1987	AGRICULTURAL GFCF $ PER CAPUT OF AGRIC.LAB.FORCE 1987	FERTILIZER USE PER HA ARAB.LAND KG/HA 1987	NOS. OF TRACTORS PER 000 HA ARABLE LAND 1987	OFFICIAL COMMITM. TO AGRICULTURE $ PER CAPUT 1988
ARGENTINA			5	6	10.2
BOLIVIA			2	1	13.1
BRAZIL			49	9	2.4
CHILE			54	7	.1
COLOMBIA			94	6	4.5
ECUADOR			23	3	9.2
FRENCH GUIANA			167	30	
GUYANA			27	7	7.7
PARAGUAY			7	5	10.4
PERU			62	4	4.1
SURINAME			161	18	.1
URUGUAY			42	24	22.1
VENEZUELA	118.7	573.1	158	12	.6
AFGHANISTAN			10		
BANGLADESH			77	1	6.3
BHUTAN			1		9.4
BRUNEI DARUSSALAM			57	10	
MYANMAR			12	1	.6
CHINA (MAINLAND)	10.3	2.3	232	9	.9
CYPRUS	718.4	1644.9	135	86	65.4
HONG KONG				1	
INDIA	39.5	32.5	52	4	1.0
INDONESIA			107	1	3.3
IRAN, ISLAMIC REP. OF	120.3	419.5	66	7	
IRAQ	223.4	1171.6	40	8	1.6
ISRAEL	651.4	3562.0	224	56	5.8
JAPAN			433	404	
JORDAN			36	14	3.2
KOREA, DEM. POP. REP.			312	31	.5
KOREA, REPUBLIC OF	994.1	422.2	392	9	.8
KUWAIT			82	28	
LAO POP. DEM. REP.			1	1	12.3
LEBANON			67	10	.3
MALAYSIA	193.1	418.5	160	3	5.3
MALDIVES					45.5
MONGOLIA			18	9	.3
NEPAL			23	1	13.8
OMAN			41	3	
PAKISTAN	30.4	38.4	83	8	9.3
PHILIPPINES			61	3	9.9
QATAR			173	22	
SAUDI ARABIA			368	2	
SINGAPORE			1833	19	
SRI LANKA			109	15	13.9
SYRIAN ARAB REPUBLIC	111.7	870.0	40	9	.1
THAILAND			29	7	3.3
TURKEY			64	23	1.0
UNITED ARAB EMIRATES	1434.2	2369.6	163		
VIET NAM			65	6	.3
YEMEN ARAB REPUBLIC	31.7	38.6	6	2	9.8
YEMEN DEMOCRATIC			12	26	2.9

12/B. (Cont.) RESOURCES AND THEIR USE IN AGRICULTURE

COUNTRY	AGRICULTURAL GFCF $ PER HA ARABLE LAND 1987	AGRICULTURAL GFCF $ PER CAPUT OF AGRIC.LAB.FORCE 1987	FERTILIZER USE PER HA ARAB.LAND KG/HA 1987	NOS. OF TRACTORS PER 000 HA ARABLE LAND 1987	OFFICIAL COMMITM. TO AGRICULTURE $ PER CAPUT 1988
AUSTRIA	850.6	5465.5	221	216	
BELGIUM-LUXEMBOURG	628.8	5570.7	510	147	
BULGARIA			180	13	
CZECHOSLOVAKIA			303	27	
DENMARK	516.5	8853.0	233	63	
FINLAND	620.4	6532.3	216	100	
FRANCE	286.7	3609.9	299	78	
GERMAN DEMOCRATIC REP.			337	33	
GERMANY, FED. REP. OF	810.4	4961.1	421	197	
GREECE	97.3	366.2	154	47	
HUNGARY			260	10	6.6
ICELAND	2332.5	1866.0	2917	1638	
IRELAND	340.9	1655.3	681	165	
ITALY	865.2	5239.0	189	108	
MALTA	346.2	900.0	46	34	
NETHERLANDS	2502.5	9286.3	688	208	
NORWAY	1276.3	8740.0	270	176	
POLAND			222	71	
PORTUGAL	80.0	278.6	103	29	1.7
ROMANIA			130	17	
SPAIN			99	34	
SWEDEN	331.6	5292.4	136	62	
SWITZERLAND			431	262	
UNITED KINGDOM	190.2	2192.7	355	74	
UNION OF SOV. SOC. REP			118	12	
YUGOSLAVIA	108.2		133	131	
AUSTRALIA			29	7	
FIJI	36.6	89.7	90	18	35.1
FRENCH POLYNESIA			12	2	10.2
KIRIBATI					44.6
NEW CALEDONIA			60	64	64.1
NEW ZEALAND	576.0	2060.6	709	153	
PAPUA NEW GUINEA			38	3	18.0
SOLOMON ISLANDS					48.1
TONGA	87.2	683.3		2	56.9
VANUATU					26.7

13. MEASURES OF OUTPUT AND PRODUCTIVITY IN AGRICULTURE

COUNTRY	INDEX OF FOOD PRODUC.PER CAPUT 1979-81=100 1987-89	INDEX OF TOT.AGR. PRODUC.PER CAPUT 1979-81=100 1987-89	PER CAPUT DIETARY ENERGY SUPPLIES 1986-88	INDEX OF VALUE OF AGRIC.EXPORTS 1979-81=100 1986-88
ALGERIA	97	98	2786	98
ANGOLA	84	81	1806	63
BENIN	114	117	2164	84
BOTSWANA	68	68	2296	144
BURKINA FASO	115	117	2036	90
BURUNDI	98	98	2338	75
REPUBLIC OF CAMEROON	96	95	2180	154
CAPE VERDE	126	125	2703	119
CENTRAL AFRICAN REPUBLIC	90	90	2050	111
CHAD	101	103	1840	188
COMOROS	110	110	2067	115
CONGO	98	98	2577	125
COTE D¢IVOIRE	96	92	2449	96
DJIBOUTI				117
EGYPT	109	102	3343	153
EQUATORIAL GUINEA				99
ETHIOPIA	91	90	1715	352
GABON	81	81	2521	115
GAMBIA	92	92	2357	161
GHANA	109	107	2201	99
GUINEA	90	91	2029	130
GUINEA-BISSAU	137	137	2543	123
KENYA	101	101	2117	97
LESOTHO	80	81	2308	102
LIBERIA	95	93	2408	83
LIBYAN ARAB JAMAHIRIYA	109	109	3479	86
MADAGASCAR	93	92	2258	62
MALAWI	85	87	2098	67
MALI	97	99	2147	153
MAURITANIA	88	88	2621	135
MAURITIUS	100	100	2749	81
MOROCCO	120	120	2856	73
MOZAMBIQUE	83	83	1622	162
NAMIBIA	95	93	1890	110
NIGER	86	86	2372	92
NIGERIA	96	96	2106	30
REUNION	72	72	2967	133
RWANDA	77	81	1831	133
SAO TOME AND PRINCIPE	70	70	2352	103
SENEGAL	106	106	2208	92
SEYCHELLES			2337	124
SIERRA LEONE	89	89	1838	100
SOMALIA	97	97	2009	92
SOUTH AFRICA	90	89	3082	130
SUDAN	87	88	2071	84
SWAZILAND	97	95	2587	111
UNITED REP. OF TANZANIA	90	89	2229	78
TOGO	89	93	2149	119
TUNISIA	96	96	2971	97
UGANDA	87	87	2068	68
ZAIRE	94	94	2115	114
ZAMBIA	97	99	2066	35
ZIMBABWE	90	97	2283	97
BARBADOS	76	76	3188	105
BELIZE	92	92	2627	87
BERMUDA			3004	122
CANADA	103	103	3451	116
COSTA RICA	89	93	2781	89
CUBA	106	107	3103	89
DOMINICA	147	147	2884	119
DOMINICAN REPUBLIC	94	91	2359	117
EL SALVADOR	90	69	2396	74
GRENADA	89	89	2959	125
GUADELOUPE	116	116	2713	132
GUATEMALA	103	94	2327	90
HAITI	93	91	1992	116
HONDURAS	88	87	2138	55
JAMAICA	92	92	2579	93
MARTINIQUE	116	116	2844	132
MEXICO	98	97	3123	75
NICARAGUA	63	59	2373	85
PANAMA	92	94	2484	99
TRINIDAD AND TOBAGO	86	84	2983	78
UNITED STATES OF AMERICA	92	91	3644	120

13. (Cont.) MEASURES OF OUTPUT AND PRODUCTIVITY IN AGRICULTURE

COUNTRY	INDEX OF FOOD PRODUC.PER CAPUT 1979-81=100 1987-89	INDEX OF TOT.AGR. PRODUC.PER CAPUT 1979-81=100 1987-89	PER CAPUT DIETARY ENERGY SUPPLIES 1986-88	INDEX OF VALUE OF AGRIC.EXPORTS 1979-81=100 1986-88
ARGENTINA	91	91	3168	56
BOLIVIA	102	101	2096	88
BRAZIL	115	112	2703	69
CHILE	107	106	2581	29
COLOMBIA	102	98	2544	81
ECUADOR	106	106	2302	82
FRENCH GUIANA			2778	152
GUYANA	70	70	2423	44
PARAGUAY	115	118	2784	57
PERU	101	99	2277	121
SURINAME	91	91	2775	103
URUGUAY	106	108	2746	61
VENEZUELA	88	89	2534	68
AFGHANISTAN	88	88	2110	120
BANGLADESH	93	92	1925	135
BHUTAN	121	121		228
BRUNEI DARUSSALAM	115	115	2839	171
MYANMAR	120	119	2545	41
CHINA (MAINLAND)	128	130	2637	75
CYPRUS	99	99		108
HONG KONG	61	61	2883	139
INDIA	113	112	2104	123
INDONESIA	124	122	2645	74
IRAN, ISLAMIC REP. OF	87	87	3124	81
IRAQ	98	98	2950	100
ISRAEL	106	97	3133	95
JAPAN	97	94	2822	117
JORDAN	117	116	2884	105
CAMBODIA	146	150	2162	36
KOREA, DEM. POP. REP.	108	109	3172	90
KOREA, REPUBLIC OF	96	94	2867	116
KUWAIT			3127	113
LAO POP. DEM. REP.	116	116	2614	42
LEBANON	107	106	3275	85
MALAYSIA	142	127	2665	110
MALDIVES	98	98	2140	102
MONGOLIA	91	89	2481	97
NEPAL	107	106	2034	205
OMAN				153
PAKISTAN	103	106	2167	130
PHILIPPINES	86	87	2238	114
QATAR				112
SAUDI ARABIA	249	247	2805	90
SINGAPORE	86	86	2882	128
SRI LANKA	87	87	2297	90
SYRIAN ARAB REPUBLIC	86	87	3142	75
THAILAND	104	105	2288	131
TURKEY	97	97	3084	350
UNITED ARAB EMIRATES			3489	113
VIET NAM	111	112	2217	71
YEMEN ARAB REPUBLIC	98	98	2277	75
YEMEN, DEMOCRATIC	83	83	2314	86

13. (Cont.) MEASURES OF OUTPUT AND PRODUCTIVITY IN AGRICULTURE

COUNTRY	INDEX OF FOOD PRODUC.PER CAPUT 1979-81=100 1987-89	INDEX OF TOT.AGR. PRODUC.PER CAPUT 1979-81=100 1987-89	PER CAPUT DIETARY ENERGY SUPPLIES 1986-88	INDEX OF VALUE OF AGRIC.EXPORTS 1979-81=100 1986-88
AUSTRIA	109	109	3476	126
BELGIUM-LUXEMBOURG	116	116	3901	120
BULGARIA	100	96	3650	160
CZECHOSLOVAKIA	121	121	3540	97
DENMARK	120	120	3605	120
FINLAND	101	102	3120	106
FRANCE	105	105	3312	120
GERMAN DEMOCRATIC REP.	114	115	3855	79
GERMANY, FED. REP. OF	112	112	3528	120
GREECE	100	103	3702	202
HUNGARY	113	112	3635	76
ICELAND	85	85	3361	124
IRELAND	105	106	3688	114
ITALY	100	101	3571	133
MALTA	105	105	3258	82
NETHERLANDS	110	110	3303	135
NORWAY	109	109	3266	104
POLAND	106	105	3434	50
PORTUGAL	100	100	3284	115
ROMANIA	109	109	3327	45
SPAIN	111	112	3494	125
SWEDEN	94	94	3031	116
SWITZERLAND	102	102	3623	121
UNITED KINGDOM	105	105	3218	111
UNION OF SOV. SOC. REP	112	111	3382	93
YUGOSLAVIA	98	98	3570	81
AUSTRALIA	95	101	3347	127
FIJI	80	80	2785	85
FRENCH POLYNESIA	77	76	2856	138
KIRIBATI			2952	84
NEW CALEDONIA	95	92	2919	112
NEW ZEALAND	107	103	3476	122
PAPUA NEW GUINEA	97	97	2227	87
SOLOMON ISLANDS	82	82	2140	118
TONGA	87	87	2964	119
VANUATU	80	79	2533	86

14. CARRYOVER STOCKS OF SELECTED AGRICULTURAL PRODUCTS

	CROP YEAR ENDING IN				
	1986	1987	1988	1989	1990
	MILLION METRIC TONS				
CEREALS					
DEVELOPED COUNTRIES	289.6	319.7	276.0	184.3	167.2
CANADA	14.4	18.5	13.5	9.7	10.9
UNITED STATES	181.2	203.8	169.4	86.1	61.1
AUSTRALIA	6.1	4.1	3.1	3.2	3.2
EEC	36.1	31.6	28.7	28.5	30.6
JAPAN	5.2	5.8	5.6	5.4	5.1
USSR	31.0	38.0	39.0	36.0	39.0
DEVELOPING COUNTRIES	136.8	135.3	121.5	123.0	131.9
FAR EAST	98.8	90.9	81.7	82.5	93.3
BANGLADESH	1.0	0.7	1.5	1.2	1.1
CHINA	52.0	46.3	47.5	43.4	42.3
INDIA	17.0	15.0	5.4	4.4	11.1
PAKISTAN	2.0	3.1	1.6	4.4	11.1
NEAR EAST	16.9	19.7	16.2	17.0	14.6
TURKEY	0.4	0.9	1.0	0.8	0.7
AFRICA	8.8	11.6	8.7	10.3	10.6
LATIN AMERICA	12.0	12.9	14.6	13.0	13.0
ARGENTINA	0.7	0.7	1.3	1.3	0.7
BRAZIL	3.0	4.6	5.7	4.8	5.1
WORLD TOTAL	426.4	455.0	397.5	307.4	299.1
WHEAT	160.2	166.5	140.1	113.2	116.0
RICE (MILLED)	58.1	54.5	45.0	46.8	52.3
COARSE GRAINS	208.2	233.9	212.3	147.4	130.8
SUGAR (RAW VALUE)					
WORLD TOTAL AS OF 1 SEPT.	39.0	36.2	33.4	29.7	29.3
	THOUSAND METRIC TONS				
SKIM MILK POWDER					
UNITED STATES	311.5	80.3	24.1	22.5	...
EEC	824.7	524.0	81.2	101.4	...
TOTAL OF ABOVE	1 136.2	604.3	105.3	123.9	...
CHEESE					
UNITED STATES	357.9	204.5	180.4	149.5	...
EEC	357.5	407.8	307.1	211.3	...
TOTAL OF ABOVE	715.4	612.3	487.5	360.8	...
BUTTER					
UNITED STATES	113.7	66.8	97.5	124.6	...
EEC	1 394.8	1 835.2	163.2	108.2	...
TOTAL OF ABOVE	1 508.5	1 902.0	260.7	232.8	...

15. ANNUAL CHANGES IN CONSUMER PRICES: ALL ITEMS AND FOOD

REGION AND COUNTRY	ALL ITEMS 1970 TO 1975	ALL ITEMS 1975 TO 1980	ALL ITEMS 1980 TO 1985	ALL ITEMS 1988 TO 1989	FOOD 1970 TO 1975	FOOD 1975 TO 1980	FOOD 1980 TO 1985	FOOD 1988 TO 1989
	PERCENT/YEAR							
DEVELOPED COUNTRIES								
WESTERN EUROPE								
AUSTRIA	7.4	3.8	4.8	2.5	6.7	4.4	4.1	1.3
BELGIUM	8.3	6.4	13.3	3.1	7.5	4.6	7.5	3.1
DENMARK	9.5	10.4	7.9	4.8	10.7	...	8.1	3.8
FINLAND	2.0	10.6	8.5	6.6	12.4	10.8	9.3	3.6
FRANCE	8.8	10.4	9.6	3.6	9.6	10.0	9.7	4.3
GERMANY, FED. REP.	6.2	4.0	3.8	2.8	5.6	3.3	3.2	2.3
GREECE	13.1	16.3	20.5	13.7	14.7	17.6	20.6	18.2
ICELAND	24.8	42.0	50.5	21.1	28.3	41.0	53.1	19.2
IRELAND	13.0	12.9	12.1	4.0	14.3	13.7	10.0	4.7
ITALY	11.4	3.0	13.8	6.3	11.6	15.6	12.5	3.9
NETHERLANDS	8.6	6.1	4.0	1.1	6.9	...	3.3	0.8
NORWAY	8.3	8.4	8.9	4.5	8.3	7.4	6.6	2.7
PORTUGAL	15.3	...	23.9	12.6	16.3	21.0	24.2	14.4
SPAIN	12.0	18.6	12.3	6.8	12.1	16.0	12.3	7.7
SWEDEN	7.8	10.5	8.9	6.4	7.9	10.7	11.7	5.7
SWITZERLAND	7.9	2.4	4.1	3.2	7.3	2.9	4.9	1.7
UNITED KINGDOM	12.3	14.4	6.8	7.7	15.1	13.9	5.5	5.7
YUGOSLAVIA	19.3	18.2	45.7	1 251.8	19.1	19.4	47.1	1 207.0
NORTH AMERICA								
CANADA	7.4	8.4	7.3	5.0	11.1	9.9	5.9	3.7
UNITED STATES	6.7	8.9	5.2	4.8	9.5	7.6	3.8	5.7
OCEANIA								
AUSTRALIA	10.2	10.6	8.4	7.6	9.8	12.0	7.8	8.8
NEW ZEALAND	9.8	14.8	11.3	5.7	9.4	16.8	9.6	...
OTHER DEVELOPED COUNTRIES								
ISRAEL	23.9	60.0	193.7	20.2	25.1	65.0	192.9	20.5
JAPAN	12.0	6.5	2.6	2.2	13.0	5.5	2.6	2.2
SOUTH AFRICA	9.3	12.0	13.7	14.7	11.7	13.0	12.9	10.9
DEVELOPING COUNTRIES								
LATIN AMERICA								
ARGENTINA	59.5	100.0	207.9	3 079.5	58.0	...	327.0	3 050.8
BAHAMAS	9.5	6.9	5.5	...	11.8	7.7	5.1	...
BARBADOS	18.6	10.0	6.1	6.2	21.0	9.1	6.1	9.1
BOLIVIA	23.7	17.0	51.6[1]	15.2	27.2	16.4	...	14.3
BRAZIL	23.5[2]	46.0	133.7	1 430.9	25.9[2]	49.0	142.8	1 344.9
CHILE	225.4	70.0	41.0	17.0	245.5	70.0	18.0	20.7
COLOMBIA	19.5	23.0	21.9	...	24.0	25.0	22.5	...
COSTA RICA	13.7	8.1	36.3	16.5	3.7	9.6	38.5	17.9
DOMINICAN REPUBLIC	11.1	8.3	10.6[1]	51.2[3]	13.3	3.4	8.6[1]	60.3[3]
ECUADOR	13.7	11.7	27.2	75.6	18.4	11.2	35.6	87.9
EL SALVADOR	8.4	...	14.0	17.7	8.8	...	14.3	26.9
GUATEMALA	2.9	10.7	...	11.4	3.3	9.4	...	10.9
GUYANA	8.2	12.8	19.6[1]	4.0	12.2	14.1	26.5[1]	4.1
HAITI	13.7	8.0	8.8	7.0	15.5	9.3	6.6	7.5
HONDURAS	6.5	9.2	7.1	10.2	8.0	9.6	4.2	12.0
JAMAICA	14.9	22.0	...	...	17.2	24.0	15.7	20.0
MEXICO	12.4	21.0	18.9	20.0	13.9	19.5	63.7	20.3
PANAMA	7.8	6.9	9.6	-0.1	9.9	6.6	3.6	-1.4
PARAGUAY	12.6	14.7	3.1	...	15.4	14.9	...	...
PERU	12.1	37.0	100.2	3 398.6	13.9	50.0	87.8	2 632.7
PUERTO RICO	8.8	5.6	2.9	3.7	12.6	5.5	2.8	5.1
SURINAME	8.2	11.5	6.4	...	9.5	12.2	4.8	...
TRINIDAD & TOBAGO	13.7	12.9	13.1	11.4	17.1	11.1	14.8	22.5
URUGUAY	73.4	55.0	43.7	80.4	76.0	55.0	43.1	80.5
VENEZUELA	5.5	11.4	10.5	84.5	8.5	15.7	13.6	125.8
FAR EAST								
BANGLADESH	39.0[4]	7.6	10.1	10.0	42.0[4]	5.0	10.9	8.8
BURMA	17.8	3.8	4.5	37.8[5]	21.0	2.6	4.2	45.8[5]
INDIA	13.2	1.3	6.9	7.9	14.2	0.8	6.7	8.1
INDONESIA	21.3	...	10.1	6.4	25.2	...	8.4	8.0
KOREA, REP.	14.3	17.2	6.3	5.7	16.8	17.2	5.4	6.8
MALAYSIA	6.7	4.6	4.5	2.8	10.4	3.7	2.5	3.7
NEPAL	10.3	6.7	11.6	8.8[5]	9.8	6.1	4.1	...
PAKISTAN	15.2	9.0	7.6	7.8	16.6	8.0	7.5	8.7
PHILIPPINES	18.7	12.0	20.6	10.6	20.1	11.0	20.2	12.9
SRI LANKA	8.0	9.9	12.6	11.6	9.1	10.7	12.6	10.3
THAILAND	9.8	10.4	4.6	6.3	11.9	10.6	3.0	10.2

15. (Cont.) ANNUAL CHANGES IN CONSUMER PRICES: ALL ITEMS AND FOOD

REGION AND COUNTRY	ALL ITEMS 1970 TO 1975	ALL ITEMS 1975 TO 1980	ALL ITEMS 1980 TO 1985	ALL ITEMS 1988 TO 1989	FOOD 1970 TO 1975	FOOD 1975 TO 1980	FOOD 1980 TO 1985	FOOD 1988 TO 1989
	PERCENT/YEAR							
AFRICA								
ALGERIA	5.1	12.4	7.1[1]	9.3	7.2	15.7	4.0[1]	9.0
BOTSWANA	...	12.4	8.5	11.7	...	13.8	...	10.2
BURKINA FASO	...	...	5.5[6]	-0.5	...	...	6.5[6]	-4.1
BURUNDI	...	18.3	8.9	...	...	16.2	9.4	...
CAMEROON	10.2	10.7	11.6	...	11.5	11.8	...	...
CENTRAL AFRICAN REPUBLIC	...	...	10.1[7]	0.7	...	...	10.1	1.0
ETHIOPIA	3.7	15.7	6.6	7.5	2.7	19.2	6.8	6.7
GABON	11.4	12.9	10.1	...	2.7	...	...	...
GAMBIA	10.5	10.2	12.0	8.2	12.8	9.7	13.4	6.9
GHANA	17.4	70.0	118.5	24.2	20.3	45.0	51.3	23.7
COTE D'IVOIRE	8.2	16.7	11.7	1.3[8]	9.3	19.3	4.2	1.4[8]
KENYA	13.9[4]	9.8	15.3	10.9	14.7[4]	10.2	12.9	10.4
LESOTHO	14.7[4]	15.1	13.7	...	16.4[4]	18.6	13.1	...
LIBERIA	12.1	8.8	3.4	5.8	13.7	8.1	2.1	10.9
MADAGASCAR	9.7	9.2	20.0	9.0	12.0	9.0	19.8	9.2
MALAWI	8.9	9.2	...	12.4	10.7	9.5	...	15.7
MALI	...	...	...	0.1	12.7	16.1	9.0	-3.1
MAURITIUS	13.1	16.9	...	12.6	14.7	16.3	...	14.0
MOROCCO	5.4	9.7	9.7	3.1	7.2	9.3	10.3	2.0
NIGER	7.9	14.6	6.8	1.6	10.6	14.8	8.4	-6.0
NIGERIA	11.5	14.4	19.9	40.8	13.1	20.0	21.3	39.7
SENEGAL	13.0	6.8	12.3	0.4	16.5	6.4	11.5	0.8
SIERRA LEONE	8.4	13.8	45.0[1]	...	11.0	12.9	43.1[1]	...
SOMALIA	7.8	...	40.0	...	9.0	...	33.0	...
SWAZILAND	9.3	13.2	13.9	20.5	9.8	14.0	13.7	12.8
TANZANIA	13.1	14.5	30.2	25.8	17.7	13.4	30.5	24.5
TOGO	8.9	8.1	6.3	-1.2	9.7	9.9	5.3	-17.6
TUNISIA	4.8	...	10.2[1]	7.3	5.2	...	10.8[1]	8.4
ZAIRE	18.6	...	...	...	21.2	...	...	...
ZAMBIA	7.1	15.2	19.4	124.7	7.4	13.7	19.9	131.2
ZIMBABWE	...	9.8	15.9	12.9	...	8.4	17.8	14.2
NEAR EAST								
CYPRUS	8.0	...	6.4	3.8	10.2	...	7.2	4.6
EGYPT	5.8	12.9	14.9	21.8	8.6	14.4	15.9	26.6
IRAN	9.6	16.1	16.1	...	10.0	18.9	15.4	...
IRAQ	11.3	...	14.5	...	18.1	...	...	...
JORDAN	6.0	11.6	...	25.7	9.2	9.8	...	20.8
KUWAIT	10.1	7.1	4.6	3.3	15.4	6.1	2.6	3.9
SAUDI ARABIA	...	11.3	-0.1	0.9	...	9.5	0.9	2.0
SUDAN	11.6	16.8	27.2[1]	66.1	12.0	14.2	26.6[1]	44.7
SYRIA	16.7	10.9	12.0	11.4	18.2	...	11.2	-11.9
TURKEY	6.2	50.0	30.0[7]	90.1	7.7	47.0	18.7[7]	70.6

[1] 1980-84
[2] 1972-75
[3] JANUARY-JUNE
[4] 1973-75
[5] JANUARY-JULY
[6] 1983-85
[7] 1981-85
[8] JANUARY-AUGUST

SOURCE: ILO, *BULLETIN OF LABOUR STATISTICS*

16. PER CAPUT DIETARY ENERGY SUPPLIES IN SELECTED DEVELOPED AND DEVELOPING COUNTRIES

COUNTRY	1972-74	1975-77	1978-80	1981-83	1986-88
	CALORIES PER CAPUT PER DAY				
ALGERIA	2004	2239	2528	2618	2788
ANGOLA	2001	2008	2162	2012	1806
BENIN	2090	2050	2154	2082	2164
BOTSWANA	2124	2138	2153	2199	2296
BURKINA FASO	1667	1786	1801	1795	2036
BURUNDI	2205	2392	2355	2395	2338
REPUBLIC OF CAMEROON	2245	2245	2244	2145	2180
CAPE VERDE	2044	2283	2552	2610	2703
CENTRAL AFRICAN REPUBLIC	2251	2193	2108	2039	2050
CHAD	1801	1789	1805	1562	1840
COMOROS	2199	2025	2060	2112	2067
CONGO	2256	2310	2405	2475	2577
EGYPT	2566	2727	2941	3184	3343
ETHIOPIA	1600	1575	1750	1754	1715
GABON	2173	2361	2374	2387	2521
GAMBIA	2074	2015	2018	2130	2357
GHANA	2182	2111	1980	1839	2201
GUINEA	2067	2086	2052	2056	2029
GUINEA-BISSAU	1925	1869	1878	2046	2543
COTE D¢IVOIRE	2342	2322	2498	2543	2449
KENYA	2249	2253	2242	2187	2117
LESOTHO	1961	2173	2374	2326	2308
LIBERIA	2226	2312	2396	2372	2408
LIBYAN ARAB JAMAHIRIYA	2885	3343	3445	3510	3479
MADAGASCAR	2458	2502	2464	2452	2258
MALAWI	2425	2375	2282	2251	2098
MALI	1737	1807	1717	1806	2147
MAURITANIA	1877	1947	2090	2168	2621
MAURITIUS	2430	2606	2702	2740	2749
MOROCCO	2534	2611	2729	2736	2856
MOZAMBIQUE	1772	1802	1794	1742	1622
NAMIBIA	1979	1922	1906	1884	1890
NIGER	1978	2046	2277	2317	2372
NIGERIA	2065	2130	2241	2186	2106
REUNION	2590	2658	2821	2908	2967
RWANDA	1822	1995	1989	2026	1831
SAO TOME AND PRINCIPE	2043	2042	2290	2277	2352
SENEGAL	2266	2291	2358	2355	2208
SEYCHELLES	2199	2163	2312	2305	2337
SIERRA LEONE	2025	2010	2077	2005	1838
SOMALIA	1707	1746	1895	2052	2009
SOUTH AFRICA	2867	2929	2934	2997	3082
SUDAN	2098	2224	2340	2256	2071
SWAZILAND	2371	2480	2480	2564	2587
UNITED REP. OF TANZANIA	1857	2194	2270	2238	2229
TOGO	2104	1959	2123	2127	2149
TUNISIA	2510	2616	2718	2782	2971
UGANDA	2273	2228	2141	2213	2068
ZAIRE	2297	2264	2115	2099	2115
ZAMBIA	2255	2340	2220	2157	2066
ZIMBABWE	2179	2146	2198	2214	2283
ANTIGUA AND BARBUDA	2119	2015	2000	2136	2178
BAHAMAS	2440	2225	2246	2539	2680
BARBADOS	2925	2969	3085	3159	3188
BELIZE	2575	2618	2732	2615	2627
BERMUDA	2971	2904	3032	3102	3004
CANADA	3254	3259	3255	3273	3451
COSTA RICA	2500	2591	2604	2640	2781
CUBA	2641	2647	2783	2970	3103
DOMINICA	2163	2289	2404	2599	2884
DOMINICAN REPUBLIC	2140	2191	2249	2273	2359
EL SALVADOR	1910	2141	2246	2330	2396
GRENADA	2445	2389	2552	2739	2959
GUADELOUPE	2332	2375	2423	2564	2713
GUATEMALA	2105	2151	2154	2224	2327
HAITI	1989	2006	2015	2057	1992
HONDURAS	2119	2162	2201	2166	2138
JAMAICA	2623	2682	2613	2557	2579
MARTINIQUE	2422	2544	2622	2751	2844
MEXICO	2680	2780	2948	3099	3123
NETHERLANDS ANTILLES	2550	2735	2882	2983	2794
NICARAGUA	2370	2406	2339	2326	2373
PANAMA	2309	2335	2288	2409	2484
ST. KITTS AND NEVIS	2246	2289	2310	2378	2822
SAINT LUCIA	2149	2149	2310	2458	2760

16. (Cont.) PER CAPUT DIETARY ENERGY SUPPLIES IN SELECTED DEVELOPED AND DEVELOPING COUNTRIES

COUNTRY	1972-74	1975-77	1978-80	1981-83	1986-88
	CALORIES PER CAPUT PER DAY				
SAINT VINCENT/GRENADINES	2336	2284	2416	2517	2764
TRINIDAD AND TOBAGO	2619	2660	2843	2940	2983
UNITED STATES	3392	3423	3489	3499	3644
ARGENTINA	3171	3239	3259	3163	3168
BOLIVIA	1958	2027	2068	2063	2096
BRAZIL	2487	2514	2595	2621	2703
CHILE	2640	2534	2628	2626	2581
COLOMBIA	2270	2383	2455	2544	2544
ECUADOR	2090	2188	2228	2198	2302
FRENCH GUIANA	2478	2420	2506	2611	2778
GUYANA	2333	2336	2427	2445	2423
PARAGUAY	2673	2723	2798	2791	2784
PERU	2272	2239	2184	2195	2277
SURINAME	2327	2380	2527	2616	2775
URUGUAY	2912	2842	2754	2737	2746
VENEZUELA	2336	2529	2656	2584	2534
AFGHANISTAN	2272	2298	2279	2184	2110
BANGLADESH	1907	1910	1911	1933	1925
BRUNEI DARUSSALAM	2439	2603	2742	2846	2839
MYANMAR	2050	2113	2254	2411	2545
CHINA	2029	2087	2275	2460	2637
HONG KONG	2680	2713	2758	2790	2883
INDIA	2003	1999	2100	2113	2104
INDONESIA	2187	2194	2368	2546	2645
IRAN, ISLAMIC REP. OF	2540	2924	2877	3037	3124
IRAQ	2264	2405	2606	2788	2950
ISRAEL	3066	3073	2999	3024	3133
JAPAN	2802	2766	2785	2806	2822
JORDAN	2455	2400	2563	2753	2884
CAMBODIA	2098	1758	1740	1829	2162
KOREA, DEM. POP. REP.	2591	2787	2980	3054	3172
KOREA, REPUBLIC OF	2680	2783	2839	2837	2867
KUWAIT	2673	2788	3009	3004	3127
LAO POP. DEM. REP.	2166	1993	2297	2502	2614
LEBANON	2475	2526	2788	3017	3275
MACAU	2278	2278	2278	2284	2233
MALAYSIA	2516	2574	2655	2619	2665
MALDIVES	1703	1766	1917	2019	2140
MONGOLIA	2379	2405	2457	2453	2481
NEPAL	1957	1964	1953	1987	2034
PAKISTAN	2049	2188	2232	2208	2167
PHILIPPINES	1863	2108	2268	2267	2238
SAUDI ARABIA	1899	2036	2685	2835	2805
SINGAPORE	2730	2709	2710	2717	2882
SRI LANKA	2157	2207	2292	2240	2297
SYRIAN ARAB REPUBLIC	2517	2542	2843	3152	3142
THAILAND	2263	2296	2310	2291	2288
TURKEY	2904	2987	3039	3017	3084
UNITED ARAB EMIRATES	3136	3371	3383	3362	3489
VIET NAM	2094	2003	2037	2161	2217
YEMEN ARAB REPUBLIC	1981	2066	2176	2224	2277
YEMEN, DEMOCRATIC	1969	1908	2075	2220	2314

16. (Cont.) PER CAPUT DIETARY ENERGY SUPPLIES IN SELECTED DEVELOPED AND DEVELOPING COUNTRIES

COUNTRY	1972-74	1975-77	1978-80	1981-83	1986-88
	CALORIES PER CAPUT PER DAY				
ALBANIA	2568	2584	2727	2769	2743
AUSTRIA	3272	3271	3350	3420	3476
BELGIUM-LUXEMBOURG	3507	3511	3577	3765	3901
BULGARIA	3494	3551	3591	3660	3650
CZECHOSLOVAKIA	3408	3396	3406	3491	3540
DENMARK	3372	3339	3486	3504	3605
FINLAND	3178	3117	3086	3080	3120
FRANCE	3124	3162	3249	3216	3312
GERMAN DEMOCRATIC REP.	3370	3486	3609	3711	3855
GERMANY, FED. REP. OF	3210	3208	3345	3327	3528
GREECE	3400	3458	3499	3612	3702
HUNGARY	3375	3438	3478	3524	3635
ICELAND	3041	2987	3077	3192	3361
IRELAND	3603	3544	3607	3675	3688
ITALY	3522	3401	3600	3479	3571
MALTA	3098	3123	3081	3013	3258
NETHERLANDS	3253	3228	3320	3286	3303
NORWAY	3118	3135	3318	3242	3266
POLAND	3473	3570	3584	3345	3434
PORTUGAL	3029	3068	3051	3158	3284
ROMANIA	3199	3377	3397	3292	3327
SPAIN	3080	3248	3323	3338	3494
SWEDEN	2883	2990	3016	3059	3031
SWITZERLAND	3577	3466	3567	3585	3623
UNITED KINGDOM	3264	3255	3227	3170	3218
UNION OF SOV. SOC. REP	3319	3369	3378	3370	3382
YUGOSLAVIA	3370	3508	3529	3597	3570
AUSTRALIA	3109	3280	3286	3257	3347
FIJI	2729	2720	2780	2781	2785
FRENCH POLYNESIA	2746	2737	2776	2812	2856
KIRIBATI	2632	2719	2862	2921	2952
NEW CALEDONIA	2865	2882	2895	2912	2919
NEW ZEALAND	3479	3401	3373	3404	3476
PAPUA NEW GUINEA	2082	2077	2143	2190	2227
SAMOA	2168	2340	2390	2435	2474
SOLOMON ISLANDS	2124	2131	2145	2172	2140
TONGA	2720	2819	2896	2937	2964
VANUATU	2554	2552	2543	2526	2533

17. ANNUAL AGRICULTURAL (BROAD DEFINITION) SHARES OF TOTAL OFFICIAL COMMITMENTS TO ALL SECTORS, BY MULTILATERAL AND BILATERAL SOURCES, 1980-88

	1980	1981	1982	1983	1984	1985	1986	1987	1988[1]
					PERCENT				
TOTAL COMMITMENTS									
MULTILATERAL AGENCIES[2]	38	34	35	35	29	31	33	29	29
WORLD BANK[3]	35	32	32	38	27	28	32	22	24
REGIONAL DEVELOPMENT BANKS[3]	44	38	36	25	26	32	34	40	32
OPEC MULTILATERAL[3]	15	11	17	21	25	28	35	23	19
BILATERAL									
DAC/EEC	10	10	10	11	11	12	12	12	11
OPEC BILATERAL	...	...	...	...	...	...	...	...	...
ALL SOURCES	...	...	...	...	...	...	...	...	...
CONCESSIONAL COMMITMENTS ONLY									
MULTILATERAL AGENCIES[2]	51	54	49	48	47	55	42	55	52
WORLD BANK[3]	46	56	45	52	49	54	29	35	37
REGIONAL DEVELOPMENT BANKS[3]	63	61	56	38	33	51	51	63	65
OPEC MULTILATERAL[3]	30	16	30	26	47	65	45	73	21
BILATERAL	13	14	16	14	15	15	15	(15)	(16)
DAC/EEC	16	18	17	17	17	16	16	16	(16)
OPEC BILATERAL	1	4	12	4	6	5	6	4	6
ALL SOURCES	19	21	22	20	21	22	19	(20)	(21)

[1] PRELIMINARY
[2] INCLUDING, UNDP, CGIAR, FAO (TF/TPC), IFAD
[3] EXCLUDING COMMITMENTS TO CGIAR

18. PERCENTAGE DISTRIBUTION OF OFFICIAL COMMITMENTS TO AGRICULTURE (BROAD DEFINITION), BY MULTILATERAL AND BILATERAL SOURCES, 1980-88

	1980	1981	1982	1983	1984	1985	1986	1987	1988[1]
	PERCENT								
TOTAL COMMITMENTS									
MULTILATERAL AGENCIES	60	60	59	63	55	59	61	53	46
WORLD BANK[2]	35	35	35	44	29	36	38	28	25
REGIONAL DEVELOPMENT BANKS[2]	16	17	15	11	17	15	15	18	15
OPEC MULTILATERAL[2]	1	1	2	2	3	2	3	2	1
OTHER[3]	8	6	7	6	6	6	5	5	5
BILATERAL	40	40	41	37	45	41	39	47	54
DAC/EEC	39	37	35	35	43	39	37	46	53
OPEC BILATERAL	1	3	6	2	2	2	2	1	1
ALL SOURCES (MULTILATERAL + BILATERAL)	100	100	100	100	100	100	100	100	100
CONCESSIONAL COMMITMENTS ONLY (ODA)									
MULTILATERAL AGENCIES	45	42	40	41	37	42	31	37	33
WORLD BANK[2]	21	21	20	18	19	24	11	14	14
REGIONAL DEVELOPMENT BANKS[2]	12	12	7	11	7	8	9	14	11
OPEC MULTILATERAL[2]	1	1	2	2	2	2	3	2	2
OTHER[3]	11	9	11	10	8	8	8	7	6
BILATERAL	55	58	60	59	63	58	69	63	67
DAC/EEC	53	54	51	56	60	56	66	62	66
OPEC BILATERAL	2	4	9	3	3	2	3	1	1
ALL SOURCES (MULTILATERAL + BILATERAL)	100	100	100	100	100	100	100	100	100

[1] PRELIMINARY
[2] EXCLUDING COMMITMENTS TO CGIAR
[3] INCLUDING UNDP, CGIAR, FAO(TF/TCP), IFAD

SOURCE: FAO AND OECD

19. DAC COUNTRIES: BILATERAL ODA COMMITMENTS FROM INDIVIDUAL COUNTRIES AND PROPORTION TO AGRICULTURE (BROAD DEFINITION), 1983-88

	BILATERAL ODA TO ALL SECTORS						PROPORTION OF ODA TO AGRICULTURE					
	1983	1984	1985	1986	1987	1988[1]	1983	1984	1985	1986	1987	1988[1]
	US$ MILLIONS						PERCENT					
AUSTRALIA	536	694	532	532	527	927	5	0	9	13	13	6
AUSTRIA	183	79	60	126	147	(100)	2	1	3	5	3	(5)
BELGIUM	187	180	132	318	404	430	5	1	23	14	18	11
CANADA	1 139	1 575	1 172	1 179	1 644	1 911	25	2	24	24	19	20
DENMARK	260	288	340	480	416	642	22	38	33	29	38	31
FINLAND	96	171	233	276	222	399	24	3	11	19	41	37
FRANCE	4 380	4 403	3 756	4 822	5 493	(6 366)	11	0	10	12	10	(10)
GERMANY, FED. REP.	2 271	2 800	2 427	3 337	4 303	4 841	15	4	15	18	13	18
IRELAND	14	13	17	25	27	22	-	-	...	8	7	14
ITALY	882	903	1 178	2 327	3 135	3 040	20	0	17	16	21	21
JAPAN	3 483	3 968	4 076	4 342	7 343	12 326	17	9	25	18	13	19
NETHERLANDS	901	902	731	1 299	1 709	1 809	23	7	18	29	34	25
NEW ZEALAND	40	41	47	34	51	78	15	5	26	15	10	8
NORWAY	288	350	346	548	514	313	17	35	23	20	20	16
SWEDEN	526	576	566	779	900	1 078	24	6	25	19	21	20
SWITZERLAND	239	218	307	329	462	519	55	22	32	27	26	29
UNITED KINGDOM	927	1 009	731	1 081	1 441	1 691	12	4	14	17	9	9
UNITED STATES	6 989	8 144	9 157	8 746	7 412	7 928	14	4	11	11	13	11
TOTAL/DAC COUNTRIES	23 341	26 314	25 808	30 580	36 150	(44 420)	15	6	17	18	15	(16)

[1] PRELIMINARY

SOURCE: OECD

20. PERCENTAGE DISTRIBUTION OF OFFICIAL COMMITMENTS TO AGRICULTURE, BY PURPOSE, 1981-1988

	1981	1982	1983	1984	1985	1986	1987	1988[1]
	PERCENT							
LAND AND WATER DEVELOPMENT[2]	17	23	20	22	23	19	15	18
AGRICULTURAL SERVICES	7	12	15	16	11	19	16	11
SUPPLY OF INPUTS	5	6	6	7	4	3	7	8
CROP PRODUCTION	6	8	7	7	6	6	4	6
LIVESTOCK	2	1	2	2	3	2	2	4
FISHERY[3]	3	2	2	2	2	2	2	2
RESEARCH, EXTENSION, TRAINING[4]	5	5	6	9	7	7	8	5
FORESTRY	2	3	2	3	5	2	3	5
AGRICULTURAL ADJUSTMENT AND UNALLOCATED	1	1	2	4	7	9	5	10
TOTAL NARROW DEFINITION	48	61	62	72	69	69	62	69
RURAL INFRASTRUCTURE	11	15	12	7	8	8	11	10
MANUFACTURE OF INPUTS[5]	9	4	1	5	2	5	2	7
AGRO-INDUSTRIES	5	3	6	3	4	3	3	4
INTEGRATED RURAL AND REGIONAL DEVELOPMENT	27	17	19	13	17	15	22	10
TOTAL BROAD DEFINITION	100	100	100	100	100	100	100	100

NOTE: THIS TABLE NOW INCLUDES FORESTRY IN THE NARROW DEFINITION

[1] PRELIMINARY
[2] INCLUDING RIVER DEVELOPMENT
[3] INCLUDING INPUTS SUCH AS FISHING TRAWLERS, FISHING GEAR
[4] INCLUDING COMMITMENTS TO CGIAR
[5] MOSTLY FERTILIZERS

SOURCE: FAO COMPUTERIZED DATA BANK ON EXTERNAL ASSISTANCE TO AGRICULTURE

21. DISTRIBUTION OF OFFICIAL COMMITMENTS TO AGRICULTURE (BROAD DEFINITION) FROM ALL SOURCES, BY REGION AND ECONOMIC GROUPS, 1981-88

	1981	1982	1983	1984	1985	1986	1987	1988[2]
	PERCENT							
TOTAL COMMITMENTS								
FAR EAST AND PACIFIC	42	48	42	46	46	40	40	45
AFRICA	28	29	26	28	26	27	34	27
LATIN AMERICA	23	18	24	18	19	24	22	20
NEAR EAST	7	5	9	7	10	9	4	8
TOTAL DEVELOPING REGIONS	100	100	100	100	100	100	100	100
LOW-INCOME FOOD-DEFICIT COUNTRIES[2]	68	71	68	75	72	63	78	77
CONCESSIONAL COMMITMENTS								
FAR EAST AND PACIFIC	49	46	48	51	49	48	46	49
AFRICA	32	40	31	34	30	37	40	31
LATIN AMERICA	12	9	12	8	12	7	8	10
NEAR EAST	8	6	9	8	8	8	6	10
TOTAL DEVELOPING REGIONS	100	100	100	100	100	100	100	100
LOW-INCOME FOOD-DEFICIT COUNTRIES[2]	82	83	83	85	82	80	89	88
NON-CONCESSIONAL COMMITMENTS								
FAR EAST AND PACIFIC	28	52	33	39	40	32	29	36
AFRICA	21	11	19	18	18	18	22	14
LATIN AMERICA	46	34	40	36	30	41	49	46
NEAR EAST	5	3	8	7	12	9	-	4
TOTAL DEVELOPING REGIONS	100	100	100	100	100	100	100	100
LOW-INCOME FOOD-DEFICIT COUNTRIES[2]	40	50	47	56	53	48	55	48

[1] PRELIMINARY
[2] 74 COUNTRIES DEFINED BY THE WORLD BANK ACCORDING TO GNP PER HEAD OF $1 135 IN 1989

SOURCE: FAO COMPUTERIZED DATA BANK ON EXTERNAL ASSISTANCE TO AGRICULTURE

4649